B J Gillespie
Archer Pest Control
653-6023

URBAN ENTOMOLOGY

WALTER EBELING
PROFESSOR OF ENTOMOLOGY AND ENTOMOLOGIST
IN THE EXPERIMENT STATION,
UNIVERSITY OF CALIFORNIA,
LOS ANGELES

Slightly Revised 1978

DIVISION OF AGRICULTURAL SCIENCES
UNIVERSITY OF CALIFORNIA
BERKELEY

This book is published by the University of California, Division of Agricultural Sciences, and is sold at a price based on returning a portion of the production cost. By this means, it is possible to make available publications which, because of relatively high production cost or limited audience, would otherwise be beyond the scope of the Division's publishing program.

To simplify this information, it is sometimes necessary to use trade names of products or equipment. No endorsement of named products is intended, nor is criticism implied of similar products not mentioned.

Originally published 1975. Edition of 1978 slightly revised.

Library of Congress Catalog Card Number: 78-72961
Printed in the United States of America

International Standard Book Number: 0-931876-19-2

To the Pest Control Operator,
in Recognition of His Contributions
to Health, Comfort, and Quality of Life
in the Urban Community

Preface

THE UNPRECEDENTED RATE of urbanization in the United States and many other nations calls for readjustment of entomological priorities to appropriately reflect the current needs and legitimate demands of our burgeoning metropolitan populations.

The term "urban entomology," designated as the title for this volume, compares well with such time-honored terms as "agricultural entomology," "forest entomology," "medical entomology," or "veterinary entomology" to define a specific area of entomological research, extension, and professional practice. This fact is slowly being recognized in the reassignment of research and extension priorities in some of our universities and governmental agencies. It is of interest in this connection to note that a joint task force of the United States Department of Agriculture, state universities, and land grant colleges proposed a revision of the traditional definition of the insect problem areas, and suggested the following 5 distinct categories of research which, although not mutually exclusive, deserve different research approaches (Anonymous, 1969*b*).

1. Problems associated with farm and other outdoor labor.
2. Problems resulting from modification of the environment by man.
3. Insect pests of man in and around his buildings.
4. Problems associated with urbanization.
5. Insect pest problems in recreational areas.

The last 3 of the above problem areas form, in varying degrees, the principal theme of the present volume, although the second category pertains to urban as much as to rural pest problems. Traditionally, pest problems of the urban community have been too narrowly defined as "insect pests of man in and around his buildings" (the third category).

However, the more recent encroachment of cities upon agricultural lands and the resulting proximity of large human populations to cattle and poultry operations have gradually renewed interest in fly problems that may have been all but forgotten for several decades, just as increased impoundment of water in flood control projects has created mosquito and midge problems in some areas.

Besides all this, people who have built their homes in forested or chaparral-covered suburban areas have encountered new pest problems not found in the central city. These are merely a few examples of how "urbanization" and "modification of the environment" have extended the scope of the pest problems of the urban community.

Stanley Rachesky, of Chicago, one of the very few state extension entomologists in the United States dealing exclusively with pest problems of a nonagricultural and noncommercial nature, informs us that over a 4-year period he received 22,419 telephone calls, 1,715 office visitors, and sent out over 100,000 brochures, even though his extension responsibilities were confined only to Chicago and its environs.

Along with their increasing awareness of and interest in pest problems, as indicated by these statistics, urban populations have become concurrently aware of the necessity for pest control materials and methods that are reasonably safe to man and that result in minimal contamination of the environment, increasing thereby the scope and the challenge in the field of urban entomology. Increasing environmental awareness results in change and innovation in pest control materials and procedures that may in some respects be similar in urban and agricultural pest control, but will be distinct in other respects.

Increasing affluence and mobility have resulted in recreational areas becoming a part of the average family's yearly environment. For most of our population, the time spent in leisure will probably eventually approximate or exceed the time spent at work. Current trends already show that much of this leisure time will be spent in outdoor recreational areas. The participants will be overwhelmingly urban populations. It is difficult to convince people whose automobile windshields are splattered with gnats and midges as they drive through great swarms of these insects, or whose picnic foods, as well as their own bodies, are attacked by wasps, bees, and ants, or who pick up chiggers and ticks while walking on park trails, or who are annoyed and even tortured by mosquitoes, gnats, and biting flies, or whose summer cabins are infested with termites, carpenter ants, or carpenter bees, or invaded by rats, mice, ants, and crickets, or whose lakefront homes are defaced by midge excrement, that their problems are less important than the pest problems of the city or of agriculture.

Thousands of sportsmen who may never see a cockroach or termite may have a genuine interest in the habits and habitats of the hellgrammites they use as fish bait. One normally thinks of mealworms and house crickets as pests, but to thousands of anglers they are known only as live fish bait, reared in enormous numbers by firms specializing in bait production.

The chapter titles indicate food and habitat preferences of pests and the way in which the pests affect man and his pets. For each pest, such subjects as distribution, description, life cycle, and control are discussed in the chapter into which the species fits most appropriately. However, the same species may be of some importance in several different ways, and is discussed in more than one chapter.

The title *Urban Entomology* seems to imply a consideration of the pests of ornamental plants, but this vast and important subject has been deliberately excluded, except for a brief discussion of pests of house plants in chapter 11. The entomology of ornamental plants had best be left to other workers with more appropriate backgrounds of interest and experience.

To aid in identification, field keys are presented only for orders or groups for which practical field keys have not been included in earlier books. The key to yellowjackets and hornets of North America in chapter 9 was kindly prepared by the author's co-workers, R. E. Wagner and D. A. Reierson, and the key to reduviid bugs (also in chapter 9) was prepared by S. F. Wood. The woodwasp key in chapter 5 was developed by combining 2 previously published keys dealing with different genera. Keys for other insect groups are available elsewhere, for example: *Recognition of Structural Pests and Their Damage* (Sweetman, 1965); and *Pictorial Keys: Arthropods, Reptiles, Birds, and Mammals of Public Health Significance* (CDC, 1966). Keys to termites may be found in *The Termites of the United States* (Weesner, 1965); and those to some arthropods commonly associated with disease transmission (cockroaches, fleas, ticks, mites, and certain groups of flies) in *Herms's Medical Entomology* (James and Harwood, 1969). Other keys may also be found in the laboratory manuals for many university-level courses dealing with the various taxonomic groups of arthropods treated in this book. A simplified key to the principal orders of insects in the adult stage is given in chapter 4.

The entomologist's traditional prerogative has been utilized—that of including under the term "entomology" many arthropods that are not insects, notably ticks, mites, spiders, and scorpions. Even some vertebrates (rodents, bats, pest birds, and venomous snakes) are discussed because they are a part of the urban pest control problem and are usually dealt with by the same persons who control insect pests. Snakes are included principally because the great quantity of venom they inject justifies certain emergency treatments that involve some risks to the victim of snake bite that are not justified by spider bites or scorpion and insect stings. It is advisable to assemble the information on the treatments for bites or stings of all venomous animals under one cover.

The thousands of questions we have received by telephone or correspondence from the inhabitants of a vast metropolitan area during the past 30 years have been reliable indications of the entomological interests of the urban community. It has been encouraging to find that many of these questions have supported the thesis that "man does not live by bread alone." Although the majority of questions have concerned the principal pests: termites, cockroaches, ants, powderpost beetles, pantry pests, flies, mosquitoes, gnats, fleas, bed bugs, lice, conenoses, mites, ticks, spiders, rats, mice, pest birds, etc., many are prompted by an innate interest in the curious and unusual things that one is likely to observe in and around his home or his place of business or recreation. To provide answers to such questions, the text digresses at times from the more prosaic pest problems of the urban community by discussing such subjects as Jerusalem crickets, vinegarroons, solpugids, primitive wood-eating cockroaches, the song of the male cricket, the bee dance, how ants find their way home, the details of a bumble bee nest, the bizarre mating rituals of scorpions and spiders, how to care for pet tarantulas, the aquatic skills of rats, and the sonar system of bats—to name just a few.

In some groups of indigenous insects, such as mosquitoes, midges, ants, and solitary wasps and bees, space does not permit an adequate discussion of all species throughout the United States. An attempt to do so would result in this book becoming encyclopedic and would thereby defeat its purpose. In such cases, typical examples are chosen from the fauna of California, or California and other western states. Usually the same or related species, with similar biologies and control measures, occur in other parts of the country. Some of the more important pest species that do not now occur in California but may eventually be accidentally introduced, such as the brown cockroach, the Formosan termite, the imported fire ant (red and black), and the brown recluse spider, are discussed in considerable detail. Most of the important urban pests are nation-wide or even world-wide in distribution.

Of the information that can be found on pest control in books such as this one, a substantial portion is of lasting value, including, of course, information on the description, distribution, biology, habits, and ecology of the various pest species. Likewise, certain principles of pest control remain indefinitely valid, even though specific pesticides may be periodically replaced. For example, in subterranean termite control, construction measures designed to minimize the possibility of infestation, as well as structural alterations to control existing infestations, will probably always be of prime importance. The control of house flies and mosquitoes depends largely on measures taken to eliminate food and breeding sources, both on an individual family and on a community-wide basis. Whenever appropriate, emphasis is placed on the use of pesticides in such a way as to minimize their adverse effects on natural enemies and to integrate their use with sanitation and the pestproofing of buildings at the time of construction. Such basic considerations do not change with changing chemicals. Discussion of specific pesticide compounds currently recommended is held to a minimum in this volume, because continual changes result in such information rapidly becoming obsolete. Current recommendations in publications of federal, state, university, and industry agencies or organizations should be consulted. The user of pesticides is encouraged to consult the appropriate state authority regarding the current registration status and recommendations regarding the proper utilization of any pesticide concerning which he may have any doubts.

On the other hand, information on pesticide-application procedures has lasting value. These procedures are based on the biologies and habits of pests, and vary with the species against which they are directed. One may use potentially effective pesticides but fail because of improper methods of application.

Information regarding pesticide formulation, such as the advantages and disadvantages of solutions, emulsifiable concentrates, emulsions, wettable powders, dusts, mists, aerosols, and vapors, or combinations of 2 or more of these formulations, is also of lasting value. The same can be said regarding the use and care of the different types of pesticide-application equipment.

In a volume with as broad a scope as the present one, appreciation is obviously owed to the many specialists who kindly consented to review certain chapters or sections. A list of these specialists follows. (For those who are not University of California staff members, the institution with which they are affiliated is given.) N. B. Akesson; J. R. Anderson; J. N. Belkin; W. S. Bowers, United States Department of Agriculture; J. E. Brooks, New York Department of Health; L. R. Brown; M. W. Cummings; A. S. Deal; F. A. Gunther; R. M. Hawthorne, California Department of Agriculture; B. J. Kaston, San Diego State College; E. F. Legner; W. W. Middlekauff; M. S. Mulla; K. F. Murray, California Department of Public Health; G. T. Okumura, California Department of Agriculture; R. J. Pence; D. A. Reierson; F. E. Russell, M.D., University of Southern California; R. E. Ryckman, Loma Linda University; R. R. Snelling, Los Angeles County Museum of Natural History; W. P. Stephen, Oregon State University; R. W. Thorp; R. E. Wagner; W. G. Waldron, Los Angeles County Health Department; D. L. Wood; and S. F. Wood, Los Angeles City College. Taxonomists of the USDA Systematic Entomology Laboratory, United States National Museum, checked all scientific names.

Space does not permit individual acknowledgment of the hundreds of entomologists and other scientists whose published papers and, in many cases, generous personal communications, provided the bulk of the information presented in this volume. Thanks are also due to the many pest control operators and their official associations, through which university research personnel are brought into contact with the practical problems of a vast and important industry.

Special appreciation is expressed to Robert E. Wagner and Donald A. Reierson for painstakingly reviewing the entire manuscript, correcting errors, and making many valuable suggestions for improvement. These workers, together with Roy J. Pence, were also responsible for nearly all the color photographs used in the 8 plates as well as the original black-and-white photographs that are reproduced throughout the text, except where otherwise acknowledged in the figure legends. The plate or figure numbers of the photographs each of these entomologists contributed are given at the end of chapter 14, preceding the Literature Cited section.

I wish to express my deep appreciation of the kindness of William W. Paul, manager of the University of California's Agricultural Publications office at Berkeley. He made available excellent facilities for the publication of this extensive manuscript, including a number of color plates as well as 391 black-and-white figures. Such a large amount of illustrative material, particularly when it included the use of color, would not be admissible with most publication media. While Mr. Paul's well-earned retirement occurred in the summer of 1974, he has made it possible to produce this book in complete form.

My thanks are also due to Grant V. Wallace, of Mr. Paul's office, for his kind and helpful cooperation. He has given me the benefit of his many years in the printing industry and as an editor of scientific publications by carefully scrutinizing the manuscript as well as by developing the format. His background as an amateur entomologist has enabled him to check all the scientific terminology to advantage.

WALTER EBELING
January, 1975

Contents

Chapter 1 · *Entomological Organization and Legislation*

We are living in an age and in a part of the world where an entire younger generation has come into being without having seen a "wormy" apple. The presence of a single house fly in a home is cause for alarm and outrage. The louse and the bed bug, once passively accepted symbols of human wretchedness, are virtually unknown. There are large areas now practically free of mosquitoes where these insects once made life unbearable and where malaria and other mosquito-borne diseases were a constant threat.

Freedom from insect pests is so much a part of modern life that the technology, organization, and legislation responsible for this modern blessing are generally little understood and appreciated by most people. Those who have given the subject any thought may be aware that the life history, behavior, and ecology of insect pests had to be studied and the marvels of chemistry had to be harnessed to combat them, for science popularizers have skillfully and effectively brought this aspect of the problem to the attention of many readers. However, there is another aspect of man's triumph against insect pests that is equally important, but apparently not so amenable to popularization, and that is entomological organization and legislation.

Organization and legislation are absolutely essential to bring science and technology to fruition in the service of mankind. They must be effective on international, intranational, state, county, and local levels, and they require harmonious and efficient relationships between governmental and private agencies. The subject of this chapter will be the history and present status of many of these agencies and the part they play in the total pest control effort. There has long been a need to bring the widely scattered information on this subject together into one place.

INTERNATIONAL AGENCIES

Probably no field of human activity has been so successful as science in the long and often disappointing effort to bring the nations of the world together to work harmoniously toward objectives that promote their common welfare. Prominent among the many examples of such successful enterprises has been the work of certain agencies of the United Nations in a concerted effort to suppress, and in many instances eradicate in large areas, arthropod vectors of diseases such as malaria, yellow fever, filariasis, dengue, and encephalitis. Several organizations are now involved in the continually increasing exchange of information and technical assistance between nations.

It is important that international cooperation be continued and extended. The need for it can probably be best illustrated with the malaria-eradication program. Reservoirs of parasites in the human population can be eliminated before the mosquito vectors become insecticide-resistant. Without international cooperation, the results achieved by one country could be largely nullified by the continued transfer of parasite carriers and vectors from other countries in which control programs were not being practiced. Also, the transfer of resistant vector strains from one country to another can be avoided by internationally coordinated control programs.

World Health Organization

In 1948, a specialized agency of the United Nations called the World Health Organization (WHO) was established. Its headquarters are in the Palace of Nations, Geneva, Switzerland. Through WHO, the public health and medical professions of more than 120 countries exchange their knowledge and experience, and collaborate in an effort to achieve the highest possible level of health throughout the world.

This agency does not deal with problems that individual countries or territories can solve by utilizing their own resources, but rather with those problems that can be solved satisfactorily only through cooperation of all or several countries, such as eradication or control of malaria, bilharziasis, smallpox, and other communicable diseases.

Advice on technical and scientific matters is given by expert committees and other international groups of experts who serve without remuneration in their personal capacity and not as representatives of government or other bodies. However, due regard is given to adequate geographical distribution of these experts. The collective reports of such groups (e.g., the Expert Committee on Insecticides and the Expert Committee on Malaria) are published in the *WHO Technical Report Series,* the annual subscription price for which was $15.00 in 1971. The reports do not express the views of WHO, but are taken into consideration in developing research and control programs. Among other WHO publications are the *Bulletin of the World Health Organization, Chronicle of the World Health Organization, WHO Working Papers,* and *WHO Newsletter.* In the United States, publications of WHO may be obtained from the Columbia University Press, International Documents Services, 1960 Broadway, New York. The same agency can provide a 146-page 1947–1967 catalogue of all WHO publications.

The United Nations International Children's Emergency Fund (UNICEF) was established to improve the health and well-being of children. In this organization, a high priority has been given to malaria control, and UNICEF has provided large quantities of insecticides and other supplies for malaria programs. Technical direction of such programs is provided by WHO. The latter made its greatest contribution through its role in the control of malaria throughout the world. Malaria had been responsible for more illness and poverty than any other disease. In 1948, WHO established a Malaria Section at Geneva, and within 5 years 27 malaria control demonstration teams were, or had been, in the field in 24 countries. The teams conducted demonstrations, studied particular problems, and trained local personnel in malariology and modern methods of control, so that after 2 years the host countries could take over the work. The Expert Committee on Insecticides sponsored by WHO established specifications for chemicals and application equipment to be used for malaria control programs (Brown, 1953).

The well-known success of WHO's operations was, of course, not only dependent on the efficacy of its excellent antimalarial program, but also on the spectacular efficacy of DDT in mosquito control. The dosage of DDT recommended by WHO was 200 mg per sq ft (929 sq cm), applied twice a year. In some areas, less was applied at more frequent intervals. The application was principally made to the insides of houses. The work could be done by large forces of unskilled and inexpensive labor. The cost of control measures (20 to 50 cents per capita) was less than the earlier outlay for antimalarial drugs, which were thus rendered superfluous.

Other important bonus dividends from DDT spraying included: control of species of mosquitoes that transmitted other diseases, such as yellow fever and filariasis; the elimination of plague as a consequence of the control of rat fleas; the eradication of sand fly fever by eliminating *Phlebotomus* flies; and a striking decrease in the abundance of house flies and flyborne dysenteric diseases (Brown, 1953).

In recent years, WHO has become increasingly preoccupied with the problem of insect resistance. By 1962, there was clear evidence of developed resistance to insecticides in 81 arthropod species of medical or veterinary importance, and by 1968 the number had increased to 102. Resistant species were found among such groups as mosquitoes, house flies, stable flies, blow flies, black flies (*Simulium*), lice, fleas, bed bugs, the German cockroach, and ticks (WHO, 1970). In 1970, the Seventeenth Report of the World Health Organization's Expert Committee on Insecticides, No. 443 in the Technical Report Series, was devoted to resistance and the changing problems it had brought into current focus.

Pan-American Health Organization

The Pan-American Sanitary Bureau was established in 1902 by all 21 of the American republics. In 1949, it became the WHO Regional Office for the Americas. In 1959, its name was changed to the Pan-American Health Organization (PAHO). Its purpose is to provide for member nations epidemiologic information, technical assistance, and fellowships, and to stimulate international cooperation in research. Prominent among current objectives is the eradication of malaria and of the yellowfever mosquito (*Aedes aegypti*) from the Americas. This organization publishes a monthly *Boletín de la Oficina Sanitaria Panamericana.* Headquarters are at 525 Twenty-Third Street N.W., Washington, D.C. 20037.

International Cooperation Administration and Agency for International Development

In 1955, the International Cooperation Administration (ICA) was formed, with headquarters in Washington, D.C., to administer technical and economic foreign aid programs of the United

States government. It provided technical skills, administrative services, pesticides, spraying equipment, vehicles, and other necessities to facilitate bilateral programs with underdeveloped countries, the host country providing local currency, resources, and personnel. The ICA worked in close cooperation with WHO, and these agencies worked together toward the goal of world-wide eradication of malaria. It was hoped that as this objective became realized, greater attention could be directed toward the control of other mosquito-borne diseases such as filariasis, yellow fever, encephalitis, and dengue, as well as many serious diseases carried by other arthropods.

The ICA was abolished by the Foreign Assistance Act of 1961, and its functions were transferred to the Agency for International Development (AID) of the United States Department of State. Its policies and programs were included among the fields of activity of that agency, the purpose of which is to carry out United States overseas programs of economic and technical assistance to less-developed countries designed to bring these countries to a level of self-sufficiency. The great extent of the program in which AID has been involved is indicated by the fact that of the 1.7 billion persons in 1968 living in areas once considered malarious, 1.35 billion were receiving some degree of protection, ranging from maintenance programs where malaria had been completely eliminated to DDT house spraying and other antimosquito and/or antiparasite work in areas still infested (Johnson, 1969).

FEDERAL AGENCIES

Before the year 1854, the federal government had become officially involved in entomology only to the extent of sponsoring a few exploratory expeditions and natural history surveys, in the course of which insects were collected and classified. These insects were harbored in the collection of the Smithsonian Institution, which was later designated as the United States National Museum. American entomologists rely heavily on the insect taxonomists of the USNM for identification of insect species, although in some states they may choose to send their specimens to state or university specialists in the various insect groups. In the USNM, the majority of the entomologists are personnel of the United States Department of Agriculture (USDA) Systematic Entomology Laboratory.

In 1854, a Bureau of Agriculture was established in the United States Patent Office, and a prominent and productive amateur naturalist, Townsend Glover, was appointed Entomologist. The designation "bureau" was changed to "department" in 1862, the same year the first Morrill Act was passed by Congress, providing for the donation of public lands to the states and territories for the endowment, support, and maintenance of colleges of agriculture and mechanic arts. It was the passage of the Hatch Act in 1887 which provided for the establishment of an agricultural experiment station in each state. The experiment stations soon became places where a large proportion of the entomological investigations in the United States were made. Many of these investigations dealt with the pest problems of the household and of public health. State and county departments of agriculture cooperated with the federal government in survey, quarantine, and regulatory activities. State and county departments of public health also contributed with survey and regulatory activities, direct health services, and control and eradication activities through such agencies as the bureaus of vector control.

United States Department of Agriculture

Some of the research dealing directly with household and other urban pests is done in the Insects Affecting Man and Animals Branch of the Entomology Research Division of the USDA. Research on such insects as house flies, mosquitoes, cockroaches, bed bugs, lice, fleas, silverfish, ants, and ticks is done in Florida, where a laboratory was originally established in Orlando in 1942, but was moved to the new ultra-modern facilities on a 12-acre site on a campus of the University of Florida in Gainesville in 1963. The laboratory has the responsibility for developing measures for the control of and protection against pests of medical importance to the armed services, but most developments are also applicable to civilian problems.

It was this laboratory that did the research on the radiosterilization and release of laboratory-reared male insects. This research eventually led to the eradication of the screwworm fly from the southeastern states and to other notable successes. The laboratory is also doing much research on chemosterilization techniques and on new conventional pesticides and repellents.

A USDA laboratory in Kerrville, Texas, directs its major efforts toward livestock pests. Among the pests investigated that are of interest to many homeowners is the brown dog tick. In Fresno, California, is a laboratory that has conducted re-

Fig. 1. Field test of the resistance of electrical cable insulation against subterranean termites. A cooperative experiment of the United States Forest Service and the United States Department of Defense. (Courtesy Southern Forest Experiment Station, Gulfport, Mississippi.)

search with yellowjackets, begun when the laboratory was situated in Corvallis, Oregon, in which the first synthetic attractants for these insects were developed while research was being done on fly and gnat control.

The Stored-Product Insects Research Branch conducts a research program to develop preventive and control measures for insects that infest, damage, and contaminate harvested agricultural commodities and products milled, processed, or manufactured from them; on preventing infestations in the facilities in which food, feed, seed, and fiber are handled, stored, processed, packaged, and transported; on developing and improving chemical analytical methods; and on the fate of pesticide residues and their effects on the quality and composition of agricultural commodities, derived products, and the environment. Much research is done in this branch that relates to the subject matter in various chapters of this book, particularly chapters 7 and 8.

Headquarters of the Stored-Product Insects Research Branch are located in Hyattsville, Maryland. The location and research activities of each of the field units are as follows: Basic Biology Investigations, Gainesville, Florida; Dried Fruit and Tree Nut Insects, Fresno, California; Dairy-Product Insects, Madison, Wisconsin; Midwest Grain Insects, Manhattan, Kansas; Peanut and Southern Corn Insects, Tifton, Georgia; Rice Insects Investigations, Beaumont, Texas; Stored-Product Insects Research and Development Laboratory, Savannah, Georgia; and Tobacco Insects, Richmond, Virginia.

The USDA research on structural woodboring insects is done by entomologists of the United States Forest Service in various locations. Probably of greatest interest to the homeowner is the research done at the Southern Forest Experiment Station at Gulfport, Mississippi, discussed in greater detail in chapter 5. The field tests at Gulfport are made according to standard procedures that determine the suitability of a new compound for federal registration as a soil insecticide for the protection of buildings from attack by subterranean termites. Figure 1 shows a field plot in which electrical cable insulation is being tested for termite resistance. There has been limited damage to underground cables and plastic pipes in the United States, but most of this kind of damage occurs in the tropics. In the Wood Products Insects Work Unit, the greatest emphasis has always been on research on the biology, prevention, and control of subterranean termites, but much work has also been done on powderpost beetles, and currently a program of research concerns anobiid beetles, with emphasis on life histories, habits, nutritional requirements, attractants, and diseases.

In the Forest Products Laboratory at Madison, Wisconsin, research is directed toward preventing insect and fungal biodegradation of wood after it has been put to use, and the study of beneficial and detrimental effects of microorganisms.

The Forest Service also operates 2 Forest Insect and Disease Laboratories. Research on the gypsy moth, *Porthetria dispar* (L.), is done in the laboratory at Hamden, Connecticut, and research on shade-tree insects is carried on in one at Delaware, Ohio.

In the Pesticide Chemicals Research Laboratories of the Entomology Research Division, fundamental work is done on the chemistry of pesticides and adjuvants as well as on the physiology of arthropod pests, particularly with reference to the mode of action of pesticides and the physiological causes of pest resistance. In cooperation with other USDA branches and divisions, chemical and biological analyses are made to determine the extent to which pesticidal residues occur in food crops, soils, and milk or other animal products. Field stations are located in various parts of the country.

The USDA has published an Agriculture Handbook, No. 331, *Suggested Guide for the Use of Insecticides to Control Insects Affecting Crops, Livestock, Households, Stored Products, and Forest Products.* As the title implies, this publication contains information related to pest problems of the urban community as well as to agriculture. The information is presented in tabular form. In the section on household insects, for example, for any common household pest, information is found under the headings "Insecticides," "Strength" (spray or dust), "Where to Apply," and "Remarks and Precautions." This highly useful publication is sold by the Superintendent of Documents, Government Printing Office, Washington, D.C., 20402 ($1.25 per copy in 1971).

The USDA Plant Quarantine Division enforces the provisions of the National Plant Quarantine Act to prevent the introduction into the United States and the spread therein of injurious pests. Federal officials work in close cooperation with state and county quarantine officials throughout the nation. Although agricultural pests gave the initial impetus to the establishment of a Plant Quarantine Division, the division provides a barrier against accidental introduction or spread of pests of interest to the urban community as well as to agriculturists.

Environmental Protection Agency

The provisions of the Federal Insecticide, Fungicide, and Rodenticide Act of 1947, and the Pesticide-Residue Amendment (Public Law 518) to the Food, Drug, and Cosmetic Act, formerly administered by the Pesticide Regulation Division of the Agricultural Research Service of the USDA, began on December 2, 1970, to be administered by the newly established federal Environmental Protection Agency (EPA). This agency also issues licenses and registers pesticide labels according to the procedures discussed in chapter 2.

After a product has been registered, a field inspector of the Pesticide Regulation Division of the EPA may at any time draw a sample for testing. If the product does not conform with the analysis submitted by the licensee, the license may be revoked.

The *Summary of Registered Agricultural Pesticide Chemical Uses* is frequently revised, and provides current information on official pesticide tolerances, maximum dosages, waiting periods following last application and harvest, and other restrictions important to the safe and effective use of pesticides. The *Pesticides Monitoring Journal* is published quarterly as a source of information on pesticide levels relative to man and his environment. The *Health Aspects of Pesticides Abstract Bulletin* is a monthly publication of the Technical Services Division of the EPA that reviews major literature, from more than 1,150 domestic and foreign journals, pertaining to the effects of pesticides on humans. These publications can be obtained from the Superintendent of Documents, U. S. Government Printing Office, Washington, D.C. 20402.

The State Services Branch, Division of Pesticide Community Studies, of the EPA is offering *Training Courses in Pesticide Technology*, designed to provide training assistance and services in the environmental and health aspects of pesticides to personnel from state, federal, and local governmental agencies and other public and private agencies concerned with pesticide usage. The courses are presented in Chamblee, Georgia, as scheduled, or in locations throughout the United States upon request by cosponsoring state, federal, local, or other requesting agencies. Tuition fees are charged and enrollees can qualify for a certificate of attendance by attending all sessions of a particular course.

The courses in Atlanta are titled *Safety and Pesticide Usage, Principles of Chemical Epidemiology*, and an introductory and advanced course in *Pesticides and Public Health*. Courses are presented by EPA personnel, and in addition, consultants from other governmental agencies and from educational institutions and industry will lecture on those subjects in which they have special competence. The same courses may be taken elsewhere, or the content of the "field courses" can be planned in cooperation with the cosponsoring organization, and qualified local speakers may be invited to lecture.

The Division of Pesticide Community Studies also offers consultative services in the various aspects of pesticide technology. Requests for such services or for information concerning the courses in Atlanta, or the "field courses" that may be scheduled elsewhere, should be directed to the Chief of the State Services Branch of EPA, 4770 Buford Highway, Chamblee, Georgia 30341, attention Training Unit.

United States Public Health Service

The United States Public Health Service (USPHS) in the Department of Health, Education, and Welfare (HEW) has contributed greatly to our knowledge of the biology and control of a wide variety of arthropods that can be considered to be pests of public health importance—a large proportion of the pests of the urban community. The principal activities of the USPHS in this field are centered in an agency founded in 1946, known

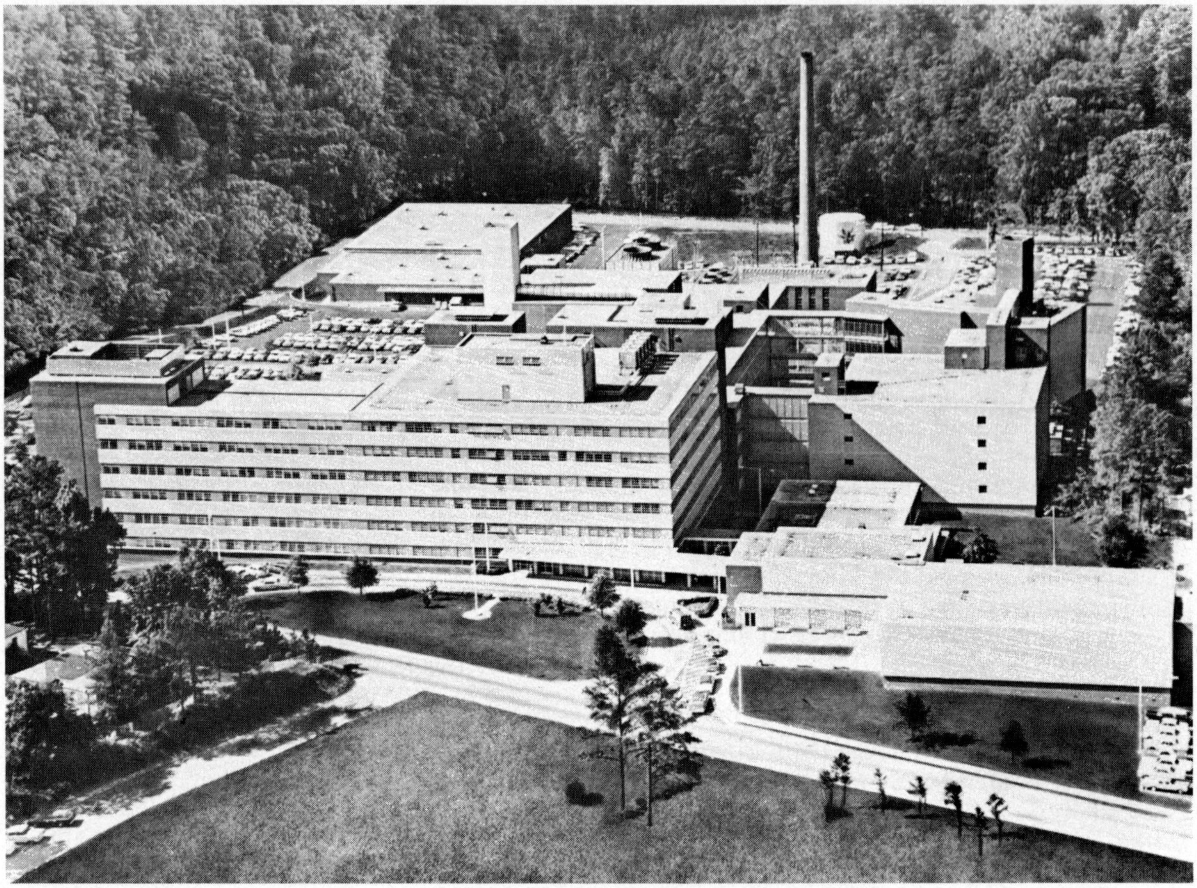

Fig. 2. Buildings of the United States Public Health Service, Center for Disease Control, Atlanta, Georgia. (Courtesy Center for Disease Control.)

for many years as the National Communicable Disease Center (NCDC) in Atlanta and Savannah, Georgia. In recent years its name was changed to Center for Disease Control (CDC). The headquarters address is 1600 Clifton Road N.E., Atlanta, Georgia 30322.

In Atlanta, the CDC (figure 2) conducts basic investigations, provides operational guidance, maintains surveillance to detect potential outbreaks of communicable diseases, provides assistance to combat epidemics and natural disasters, and assists in the training of public health personnel. In the Toxicology Section of the Technology Branch, research is done on the hazards of pesticides to humans and on ways of alleviating them.

The center has prepared many training guides, each pertaining to a particular group of arthropods, such as flies, mosquitoes, household and stored-food insects, ectoparasites of humans and domestic animals, bees, wasps, ants, scorpions, spiders, and other venomous arthropods, as well as snakes, rats, mice, pest birds, and bats, in which identification of the pest species is facilitated by means of pictorial keys and in which useful biological information and control recommendations are given. The pictorial keys in particular, besides being published in the training guides, often appear elsewhere, as in books on pest control or in the magazine *Pest Control*. In *Pictorial Keys: Arthropods, Reptiles, Birds, and Mammals of Public Health Importance,* all keys for 29 groups of arthropods and vertebrates are included in a single 192-page publication.

The Technical Development Laboratories (TDL) of the Technology Branch of the CDC were until recently located in Savannah. Here, attention was given to "studies on bionomics of arthropod vectors of diseases, to the development and evaluation of insecticides and equipment for insecticide dispersal, to investigations to determine basic reasons for insect resistance to insecticides, to the development of measures for controlling resistant insect species, and to investigations in the chemistry and physics of pesticides" (Anonymous, 1961*a*). Although TDL dealt primarily with research and development relating to mosquitoes, some work was done with house flies, conenose bugs, fleas, and other vectors of disease. Some investigation was also done on the control of domestic rats. The Savannah laboratory of the TDL has been closed. Some of the activities were transferred to the Vector Biology and Control Branch at Chamblee, Georgia. Correspondence should be addressed to Vector Biology and Control Branch, Tropical Disease Program, Center for Disease Control, Building 31/Chamblee, Atlanta, Georgia 30333.

An annual report of the findings of the CDC and various cooperating agencies relating to the control of household pests and pests of public health importance is titled "Public Health Pesticides," and is published every year as an exclusive March or April feature in *Pest Control* magazine, a technical trade journal published monthly at 9800 Detroit Avenue, Cleveland, Ohio 44102. In this report one may find the latest recommendations on the control of such pests as mosquitoes, house flies, other pest flies, fleas, bed bugs, ticks, chiggers, lice, cockroaches, scorpions, spiders, wasps, rats, and mice.

The Arbovirus Ecology Laboratory of the CDC Arbovirology Unit was active in the 1971 campaign against Venezuelan equine encephalitis in Texas. Currently, CDC support of research on vectorborne diseases is primarily centered in the Fort Collins, Colorado, laboratory of the Arboviral Disease Section of the Ecological Investigations Program.

Headquartered at the CDC Technology Branch in Atlanta is the State Aids Section. Here, specialized technical consultation and operational services on vector control are given to state health departments, Public Health Service regional offices, and to various other agencies.

The United States Public Health Service maintains a Division of International Health that promotes and provides technical assistance in disease control programs in foreign countries. The Director of this agency is in the Office of International Health, ASH, Rooms 18–57, 5600 Fishers Lane, Rockville, Maryland 20852. The programs of the DIH are financed by the Agency for International Development. An example is mosquito-eradication work directed against malaria, yellow fever, and filariasis.

Occupational Safety and Health Act of 1970

Every employer in the United States must comply with occupational safety and health standards promulgated under the Williams-Steiger Occupational Safety and Health Act (OSHA) of 1970. He must provide a place of employment free from recognized hazards. Inspectors for the Secretary of Labor are authorized to enter any work areas, and any employee or representative of employees may request an inspection if he believes health hazards exist. Regional offices of the Occupational Safety and Health Administration are located in Boston, New York, Philadelphia, Atlanta, Chicago, Dallas, Kansas City, Denver, San Francisco, and Seattle.

Among commonly recognized safety measures are elimination of fire hazards and provision for adequate firefighting equipment, proper instruction to workmen in safe use of electrical equipment and machinery, proper ventilation, provision for change of clothing if there is likelihood of contamination of clothing with hazardous chemicals, provision of first-aid kits, and proper instruction for obtaining medical attention for more serious injuries. In the case of pest control operators, employees who apply pesticides should be thoroughly familiar with the data and the instructions on the registered label for each pesticide and be guided by this information in the handling and application of the pesticide. Members of the National Pest Control Association (see later in this chapter) have an enviable safety record. The association has published 53 articles in its *Technical Releases* to members, most of them within the past decade, in which guidance is provided for the safe use of equipment and pesticide chemicals, including specific instructions for each of the commonly used pesticides and pesticide formulations.

The "Medical Services and First-Aid" section of the OSHA reads as follows:

> (*a*) Employer shall insure the ready availability of medical personnel for advice and consultation on matters of plant health.
>
> (*b*) In the absence of an infirmary, clinic, or hospital in near proximity to the work place which is used for the treatment of all injured employees, a person or persons shall be adequately trained to render first aid. First-aid supplies approved by the consulting physician shall be readily available.

It follows from the quoted section that every pest control firm should have contact with a local physician in general practice or occupational health. The physician should know what pesticides and solvents the firm uses. It is also helpful to supply him with pesticide labels and, if possible, published material pertaining to pesticides that is designed for physicians. In some areas it is possible to make arrangements with a physician with specific training in occupational health.

Regulations of the OSHA require that an employer provide for first-aid personnel at the work site if no infirmary, clinic, or hospital is in close proximity. The American Red Cross presents an approved instruction program. If a pest control firm sends 1 or 2 employees to attend a 3-day seminar, they can chart their progress through a programmed workbook and are then tested in the subject matter. They in turn are then able to present a multimedia 1-day program to other employees.

The national headquarters of the American Red Cross are in Washington, D.C. 20006, and there are regional offices in San Francisco, St. Louis, Atlanta, and in Alexandria, Virginia.

Poison Control Centers

The United States Public Health Service has assisted in developing a nation-wide network of Poison Control Centers or Poison Information Centers, usually established in local hospitals. Their staff members are specially trained in the treatment of poison cases. In California, no information on antidotes or treatment is given to the public, but only to physicians. However, anyone who calls the center will be given the address of the nearest emergency hospital equipped for handling poison cases at any hour, day or night. In some other states, this information is available to all.

A list of all Poison Control Centers, with addresses and telephone numbers, is every year brought up to date for all the states in *Pesticide Handbook—Entoma,* published by College Science Publishers at State College, Pennsylvania 16801. There were 643 centers in the United States in 1973.

Institute of Allergy and Infectious Diseases

Some highly specialized research on the factors involved in disease transmission by vectors and on the nature of infection chains is done by scientists in the Institute of Allergy and Infectious Diseases, National Institute of Health, Bethesda, Maryland. This work is done both at the institute's headquarters and its Rocky Mountain Laboratory in Hamilton, Montana.

Federal Bureau of Reclamation

The Federal Bureau of Reclamation cooperates with state and local mosquito-abatement agencies in their efforts to eliminate sources of mosquito production. The bureau submits plans of proposed impoundments, canals, drains, and other facilities to state health departments. Recommendations are also made for eliminating or minimizing potential mosquito-breeding sources. When the bureau's structures result in the creation of mosquito problems, it will reimburse local agencies for the cost of resulting mosquito control programs.

Tennessee Valley Authority

The Tennessee Valley Authority (TVA) is a federal agency created in 1933 for the development of the Tennessee river system. The investigations and operations of its Division of Health and Environmental Science (formerly the Division of Health and Safety) have not only minimized the hazards of insectborne diseases, particularly malaria, in the Tennessee Valley, but have provided the basis for improved water-management principles and procedures for water use wherever water is impounded in potentially malarious areas. Biological and control studies of aquatic vegetation as it relates to mosquito control have been particularly noteworthy in this connection. Scientists employed by the division have published hundreds of papers, principally on mosquitoes, but also on a wide variety of other aquatic fauna. Important basic investigations have also been conducted on insecticides and insecticide-application equipment. Aside from the field of vector control, the Division of Health and Environmental Science has also done much research on employee health and safety, industrial and air hygiene, occupational health, and water quality.

One of the important contributions of TVA, relatively early in the life of that agency, was to demonstrate that mosquito control and wildlife conservation programs, traditionally considered to be antagonistic, could be developed together harmoniously. In impounded water projects like those in the Tennessee Valley, most of the interests of malaria control and wildlife conservation have been shown to be mutual rather than antagonistic (Hinman and Hess, 1949).

National Research Council

The National Academy of Sciences, under its congressional charter, established the National

Research Council in 1916 as a cooperative organization of American scientists. Its purpose has been to integrate the work of individual scientists and to coordinate work on scientific problems.

Chemical-Biological Coordination Center

Affiliated directly with the executive board of the National Research Council is the Chemical-Biological Coordination Center. Its headquarters are at 2101 Constitution Ave. N.W., Washington, D.C. 20418. This organization was established in 1946 with the following objectives: (1) to assemble and organize information which relates the structure of chemicals to their biological activity, (2) to sponsor the preliminary testing ("screening") of chemicals on a variety of animals and plants in order to determine the biological effects of the compounds, and (3) to prepare reviews of literature on (*a*) the effect of chemical structure upon various biological actions, and (*b*) test methods used in the study of such actions. The center is sponsored by the Army, Navy, United States Public Health Service, and the American Cancer Society. The data obtained by the center are available to representatives of its sponsoring agencies and to authorized scientists upon request.

The center has a system of recording coded information on cards that are filed in such a way that a specific chemical problem can be segregated from other considerations and the collected information assembled quickly by a sorting machine. The investigator, by using this information retrieval system, may save himself months or even years of literature research.

Agricultural Board

Another agency of the National Research Council is the Agricultural Board, which has a Committee on Plant and Animal Pests comprising 11 scientists from various federal organizations, universities, and private companies. This committee in turn has a Subcommittee on Insect Pests comprising 13 scientists, likewise from federal organizations, universities, and private companies. The subcommittee has enlisted the cooperation of 71 scientists to collaborate with them in the preparation of a series of extensive reports on the principles of controlling pests and diseases of plants and animals.

The published reports of the Subcommittee on Insect Pests are in 6 volumes, titled: (1) *Plant-Disease Development and Control;* (2) *Weed Control;* (3) *Insect-Pest Management and Control;* (4) *Control of Plant-Parasitic Nematodes;* (5) *The Vertebrates That Are Pests: Problems and Control;* and (6) *Effects of Pesticides on Fruit and Vegetable Physiology.* The reports were sponsored by the following organizations: Agricultural Research Service of the United States Department of Agriculture, Agency for International Development of the United States Department of State, National Agricultural Chemicals Association, Rockefeller Foundation, and the Bureau of Sport Fisheries and Wildlife of the United States Department of the Interior.

More recently, a conference was held under the aegis of the Agricultural Board, and was devoted to the capabilities and limitations of the process of biological degradation as a means of disposing of natural, pesticidal, and various other man-made compounds. The conference papers and discussions were published in 1972 in a 352-page volume titled *Degradation of Synthetic Organic Molecules in the Biosphere.*

The publications of the Agricultural Board are available from the Printing and Publishing Office of the National Academy of Sciences, 2101 Constitution Ave. N.W., Washington, D.C. 20418.

Building Research Advisory Board

In the National Research Council's Division of Engineering and Industrial Research, there is a Building Research Advisory Board (BRAB). In 1956, the BRAB conducted for the Federal Housing Administration (FHA) an extensive review and correlation of the existing data and knowledge pertaining to the technical and practical problems in combating industrial decay and termites. A special advisory committee of 15 members was chosen from such federal, university, and private organizations or agencies as might have research or economic interests in the subjects investigated. The report of the "BRAB Committee" was published under the title, *Protection Against Decay and Termites in Residential Construction.* The FHA could utilize the advice of the BRAB Committee for the revision of its Minimum Property Requirements (MPR) that pertain to the prevention or control of industrial decay or termites if it wished to do so, but the report was not prepared as a compilation of MPR's. The final determination of MPR's is the responsibility of the various existing technical bodies within the building industry or government.

Department of Defense

Military entomology as a discipline involves not only disease-vector and pest arthropods. but also economic pests of vegetation, clothing, and subsistence products, as well as pests of other kinds, such as rodents, reptiles, pigeons, and other feral animals, snails, and weeds. Many entomologists are working on research, development, and control problems relating to arthropods as they affect the armed services. Of 122 entomologists in the Army, there are about 22 civilians; in the Navy, there are 22 in uniform and 23 civilians; and in the Air Force, 20 are in uniform and 6 are civilians (A. W. Morrill, Jr., correspondence).

The organization for research, development, and pest control in the armed services is constituted as follows: The Armed Services Pest Control Board, located at Walter Reed Army Medical Center, Forest Glen, Washington, D.C. 20012, advises the Office of Naval Research (ONR) and the Army Medical Research and Development group in the office of the Surgeon General, and has a billet at the USDA laboratory in Gainesville, Florida, which is filled in rotation by the 3 services. The board also operates the Military Entomological Information Service, an information retrieval system at Forest Glen, with tri-service billets and one civilian entomologist. The board encourages and supports research on pest problems, and arranges for extensive tests with new materials and procedures.

There are Army medical laboratories in each Army area, as well as one in Europe and another in Japan. Their research is mostly limited to vector survey and bionomics, but there is some field evaluation of control techniques.

The Army Medical Equipment Research and Development Laboratory, Fort Totten, New York, took over the functions of the Army Engineer Research and Development Laboratories, which were located at Fort Belvoir, Virginia. Much work is currently being done in the field of aerial spraying.

The Entomology Branch of the Army Medical Corps has on its staff entomologists, physiologists, toxicologists, and chemists who are conducting valuable basic research on arthropod pests. Their efforts are nearly all directed toward insect physiology and toxicology as related to the modes of action of pesticides. Excellent papers have been published by this group.

The Entomological Branch, Office of the Chief of Engineers, is primarily concerned with the control of insect-infested products and buildings on military bases, but also assists in the control of flies, mosquitoes, and other pests that affect man directly.

In the Office of the Surgeon General of the Army, there is a Preventive Medicine Division that investigates insecticides, repellents, and other means of controlling arthropods. as they affect the armed services. Approximately 100 entomologists in this division, stationed with Army units in various parts of the United States and abroad, are concerned primarily with problems of special military significance, but often assist civilians who desire information about pests and their control.

The Department of the Navy, Division of Preventive Medicine, Bureau of Medicine and Surgery, Washington, D.C. 20390, conducts intensive research and extensive control operations relating to pest abatement. Any organization involved in pest control or abatement can obtain valuable assistance and advice from this group of approximately 30 entomologists.

The United States Navy Disease Vector Ecology and Control centers are located at Jacksonville, Florida, and Alameda, California. At these centers the emphasis is primarily on the training of technicians, but there is also considerable evaluation of equipment and testing of materials for field pest control. There are also medical research units and civilian entomologists in each field division of the Naval Facilities Engineering Command. These do varying amounts of field-control evaluation of materials and techniques. There are some 22 entomologists, in all, involved in these operations throughout the continental United States, the Pacific areas, and Europe.

Navy Environmental Preventive Medicine units are located in Norfolk, Virginia; San Diego, California; Hawaii; and Naples, Italy. They do research on pesticides and control methods under their differing local conditions, but also in combat areas. There are also Naval Medical Research units at Bethesda, Maryland; Taipei, Taiwan; and Cairo, Egypt, which do research on speciation, distribution, pathology, taxonomy, and some on pest control.

Air Force Regulation 91–21 states the policies, responsibilities, and procedures for pest control at Air Force installations. Among the responsibilities for pest control are the following:

(1) Insure that effective preventive and corrective pest control measures are established and accomplished at all installations under its jurisdiction.

(2) Insure that the environmental impacts of proposed pest control measures are assessed, not only for the short-term benefits to accrue from the program but also for any cumulative and long-term effects anticipated from the program.

The Air Force has a much smaller staff than the Army or Navy in the field of pest control. Research and development in pest control are done mainly at Eglin Air Force Base in Florida and Kirtland AFB in New Mexico. Birds are serious flight hazards, and much work is being done at Kirtland on such control devices as sound, lights, and microwaves. Some medical entomologists are doing research at Environmental Health Laboratories at Kelly AFB in Texas and McClellan AFB in California.

As with all branches of the armed services, the Air Force depends heavily on research and development accomplished by the USDA and USPHS under grants or other funding arrangements. Some work for the armed services is also done by the Department of the Interior in the field of rodent control for the most part (Walter Barrett, personal correspondence).

STATE AGENCIES

The states have an important role in the investigation and control of arthropods that annoy or transmit disease to man and his animals, and this includes the majority of the pests of the household, or that can be considered to be pests of the urban community. Their function may be that of technical supervision and assistance or research or sometimes financial assistance, depending on the state agency involved. States may perform this function through their departments of public health or agriculture, various departments of the state university and agricultural experiment station, or through all of these agencies. The states may also furnish technical assistance and advice in the forming of new pest-abatement districts.

Some states, such as California and Florida, do considerable work on operational investigations and even basic research. In California, the Bureau of Vector Control and Solid Waste Management in the Department of Public Health does some research directly and some through the University of California. It may cooperate in various operational investigations with certain districts of the state which have facilities, staff, and funds. Florida has a special laboratory and staff for both basic and operational investigations.

California Bureau of Vector Control and Solid Waste Management

This bureau of the State Department of Public Health was organized in 1947 as a consolidation of the mosquito control and plague surveillance functions of the Division of Environmental Sanitation, and was originally known simply as the Bureau of Vector Control (BVC). Other states have agencies with similar functions. There are 24 states with at least 1 organized mosquito-abatement (or control) district, and there are 229 agencies organized for mosquito abatement in the United States (Reeves, 1971). In 1946, the California Legislature had established a state subsidy for local mosquito control agencies to cope with the state-wide mosquito problem, and the BVC was assigned the responsibility for administering the program.

By furnishing technical guidance, the BVC had a prominent role in facilitating the remarkable growth of local mosquito-abatement districts (MAD) throughout the state. In 1946, there were about 30 local districts, covering approximately 5,000 sq mi (12,950 sq km) of mosquito-infested areas and spending a total of about $500,000 annually. In 1971, there were 64 such agencies, covering over 40,000 sq mi (103,600 sq km) and spending a total of over $9,000,000 annually.

In 1951, the bureau acquired responsibility for solid waste management with the creation of 1 state and 9 regional water-pollution control boards. Fly and rodent problems associated with solid wastes were the primary bases for the bureau's receiving this assignment. It was not until 1968, however, that the bureau's title was expanded to identify this important added function. The present Bureau of Vector Control and Solid Waste Management has a professional staff of 42, the majority being entomologists, augmented by several zoologists and engineers.

The bureau's pest-suppression program has 5 elements: zoönoses (diseases communicable from animals to man under natural conditions); water-related vectors; solid waste-related vectors; and field-wilderness vectors. The staff engages in surveys, consultations, investigations, technical development, demonstrations, education, pest-identification services, surveillance, and emergency control activities where required. About 40 vector-borne diseases and over 1,100 animal species are included in the state-wide program prospectus.

The relation of the Bureau of Vector Control and Solid Waste Management to mosquito-abatement districts is in the field of surveillance, advice, and consultation. The role of the abatement districts in preserving a reasonably pest-free environment for Californians is discussed in considerable detail in the section on mosquito control. (See chapter 9, under "Mosquito-Abatement Districts.")

In the course of their official activities, the bureau's personnel have made many substantial scientific contributions within the broad scope of vector control. The bureau prepared a list of about 1,000 publications, principally contributed by its own staff, indicating the extent of their scientific contributions in basic and applied aspects of vector control. Although the majority of their papers were published in journals such as *California Vector Views, Mosquito News, Proceedings of the California Mosquito Control Association* and, more recently, in *National Pest Control Operators News,* many of their papers can be found in a wide variety of other publication media, depending on whether the papers are of a technical, semitechnical, or popular nature.

In 1969, the bureau prepared a syllabus titled *Community Pest and Related Vector Control* in cooperation with individuals within other state and university departments, local public health agencies, and the Pest Control Operators of California, Inc. The syllabus can be purchased from the latter at 3444 West First Street, Los Angeles, California 90004.

State Department of Agriculture

As the name implies, a state department of agriculture is organized, staffed, and financed with service to the state's agricultural interests as the principal objective. Nevertheless, there is considerable "spinoff" from its activities that is of value in the field of household and public health insect control. Likewise, in the activities of state entomologists in quarantine, survey, rodent control, and eradication of accidentally introduced pests, household pests and those of direct interest to the urban community can be the objects of their attention as well as agricultural pests, although the latter happen to be much greater in numbers of species.

Anyone entering the country from a foreign area is checked by both customs and plant quarantine officials. If the customs official finds agricultural products or insects, he asks the quarantine official to check them. Not only any suspicious or questionable plant material, but any insect would be intercepted and identified, and state public health officials would be notified if it were advisable to do so. Likewise, if an exotic venomous spider, insect vector, or other harmful organism were detected in the regular agricultural surveys, it would be identified and the Department of Public Health would be notified. The State Department of Public Health is empowered to take eradication measures against an insect vector of disease if, in its judgment, a hazard to the public exists and local agencies appear to be taking no action or are unable to cope with the problem.

State Universities and Agricultural Experiment Stations

Little entomological work had been done by the states before the passage of the federal Hatch Act on March 2, 1887, which provided for the establishment in each state of an agricultural experiment station. With the organization of these stations in 1888, basic and applied research in entomology in this country obtained one of the sound bases for its subsequent unparalleled development. Many of the experiment stations operated in close connection with the state agricultural colleges organized under the Morrill Act of 1862.

The entomological research programs in the agricultural experiment stations have from the beginning included projects pertaining to the biology and control of pests of the household and the urban community. Publications pertaining to such pests can be found in technical and semitechnical bulletins and manuals or Extension Service circulars and leaflets of the state universities and agricultural experiment stations. Also, departments of zoology, public health, and the medical schools in many universities provide for research and teaching related directly or indirectly to pest problems of urban communities.

Professor J. J. Davis, as Chairman of the Department of Entomology at Purdue University, was the first university entomologist to recognize the importance of the rapidly expanding structural pest control industry. Structural pest control operators were invited to Purdue for their first technical conference in 1937, just 4 years after the founding of the National Pest Control Association (see later in this chapter), and thereby Professor Davis became the founder of formal pest control education. Since then, the well-known 3-day Purdue Conference has convened annually at the university, and has continued to be devoted to the original objective of education in the many technical and business aspects of structural pest control. Purdue University sponsors the conference in cooperation with the National Pest Control Association and the Department of the Interior, and there are usually between 450 and 500 people in attendance.

In 1946, Purdue initiated a 4-year structural pest control curriculum in entomology. Subsequently, 2-year college programs in pest control, correspondence courses, and extension programs were developed. Certain other universities eventually developed courses in structural pest control in their curricula for university credit, as well as short courses and regional conferences for pest control operators (Osmun and Butts, 1966). However, only Purdue has established a major in pest control. On the other hand, most state universities provide training in economic entomology and zoology which can be combined with courses in business management to provide training leading to the Bachelor's degree for those seeking managerial responsibilities in pest control.

Students who are interested in enrolling in universities or institutes offering courses designed specifically for training in the scientific and technical aspects of structural pest control may wish to obtain additional information concerning the courses that are offered and the requirements for enrollment. They may contact the Department of Entomology in any of the following universities: University of Florida, Gainesville, Florida 32601; University of Georgia, Athens, Georgia 30601; North Carolina State University, Raleigh, North Carolina 27607; Pennsylvania State University, University Park, Pennsylvania 16802; Purdue University, West Lafayette, Indiana 47907; also, the Department of Biology at Bowling Green State University, Bowling Green, Ohio 43402; and the Department of Inspection and Services, California State Polytechnic University, Pomona, California 91766.

There are also universities and colleges which offer an Associate in Applied Agriculture degree, but with majors with such titles as "Urban and Industrial Pest Control," or they may offer institute-level courses usually incorporating some on-the-job training. Those with well-established programs of this kind are North Carolina State University; Pennsylvania State University; State University Agriculture and Technical College, Farmingdale, New York 11735; Ferris State College, Big Rapids, Michigan 49307; and Broward Junior College, Fort Lauderdale, Florida 33314.

Adult Extension Courses involving 12 to 15 sessions of 2 to 3 hours are offered at the following: Rutgers University, New Brunswick, New Jersey 08903 (Short Course in Entomology); New York City Community College, 300 Jay Street, Brooklyn, New York 11201 (Pest Control); State University Agriculture and Technical College, Farmingdale, New York 11735 (Structural Pest Control); and Exterminating Industry Institute, Inc., 80 Fifth Ave., New York 10011 (Pest Control). Correspondence courses in various aspects of structural pest control are offered by the University of California at Berkeley, Pennsylvania State University, and Purdue University.

Pi Chi Omega, a national professional fraternity of pest control operators, was founded and incorporated in 1950 by a group of university entomologists and prominent pest control operators. Membership is by invitation, and is restricted primarily to college graduates who majored in entomology and are actively engaged in pest control work, or who are in a supervisory, executive, or professional capacity related to pest control (Osmun and Butts, 1966).

The California Structural Pest Control Board

In California, the business of controlling pests that invade or destroy structures is regulated by the Structural Pest Control Act, enacted by the State Legislature in 1935. Amended on several occasions, it remains the law under which licensed individuals or companies engaged in structural pest control operate. The act is administered by the Structural Pest Control Board, under the jurisdiction of the Department of Professional and Vocational Standards.

The 6 members of the board are appointed by the Governor for 4-year terms. Two members must have been licensed pest control operators for a period of not less than 5 years preceding the date of their appointment. The act requires that the board shall elect a president to serve for 1 year, and that it shall meet annually during the month of October. However, special meetings may be called by the president or by any 3 members.

With the approval of the Director of the Department of Professional and Vocational Standards, the board is required to appoint a registrar, fix his compensation, and prescribe his duties. He is the executive officer and secretary of the board. With the approval of the board and director, and subject to the provisions of the State Civil Service Act, the registrar may appoint and fix the compensation of such other assistants as may be necessary. These include the deputy registrars, located in Sacramento and Los Angeles. Complaints against pest control operators based on alleged unsatisfactory work can be brought to the attention of the deputy registrars in the same way as complaints against other trade and professional services regulated by the Department of Professional and Vocational Standards.

The Structural Pest Control Board, based on the results of written examinations, issues licenses to pest control operators. The board also has the authority to suspend or revoke a license. After receiving a verified complaint in writing from any person, the board must investigate the actions of the pest control operator against whom the complaint is directed. Disregard for the building laws of the state or any of its political subdivisions, or of the safety laws, labor laws, health laws, or compensation insurance laws of the state, or the negligent handling or use of any pesticide without due respect for public safety are grounds for disciplinary action by the board. False advertising, misrepresentation, or performing or soliciting of structural pest control work for structural or household pests other than those for which the pest control operator or his field representatives are licensed are also grounds for disciplinary action.

The 3 licensed branches of structural pest control are (1) Fumigation, (2) General Pest Control, and (3) Termite Control. A pest control operator may obtain his license for work in any 1, 2, or all 3 of the branches, provided he qualifies from the standpoint of required written examinations and years of experience for the branch or branches in question.

Termite operators (branch 3) are required to fill out a standard structural pest control inspection report, including the findings of the inspector pertaining to subterranean, drywood, or dampwood termites, woodboring beetles, fungus infections, faulty grade levels, earth/wood contacts, shower-pan leaks, cellulose debris in the crawl space, excessive moisture conditions, and any inaccessible areas. The inspection form includes a space for a diagram of the inspected premises, with standard symbols to indicate the presence of wood-destroying organisms or conditions that would be conducive to them. The termite operator is also required to fill out a standard notice of work completed (and not completed) upon completion of any job. (The property owner may request that certain recommended work not be done, and in such jobs this must be reported by the termite operator.)

Copies of the inspection and completion forms must be sent to the homeowner and also to the Structural Pest Control Board. If within 2 years after the work was performed, the homeowner becomes aware of an infestation not reported by the inspector, or if the infestation was reported and treated but not controlled, the termite operator is required to do the necessary remedial work. An investigator from the Structural Pest Control Board might make his own investigation of the case if authorized to do so by the registrar or any of his deputies. Investigators are located in Sacramento, Los Angeles, San Francisco, Fresno, Riverside, Panorama City, and Anaheim. If a termite operator has been implicated in too many cases of fraudulent or substandard work, he may lose his license to do termite work.

Pest control operators are generally quite willing to rectify any valid customer claims of faulty or inadequate work done by their employees. Only after giving the operator such a chance to correct mistakes and not receiving satisfaction does the Structural Pest Control Board take action on citizen complaints.

The headquarters of the Structural Pest Control Board are at 1021"O" Street, Sacramento, and 107 South Broadway, Los Angeles.

COUNTY AGENCIES

County Health Departments

The Los Angeles County Health Department (LACHD) will be discussed as an example of the approximate organization and function of county health departments throughout the United States. The first full-time county health department was established in Jefferson County, Kentucky, in 1908. Los Angeles County had the second full-time health department in the nation, a full-time health officer having been appointed by the Los Angeles County Board of Supervisors in 1910. In 1913, the LACHD was placed under civil service regulations by a special county charter.

Among the 7 bureaus of the LACHD, the activities of the Bureau of Environmental Sanitation pertain most closely to pests and pest control. This bureau comprises several divisions. The General Services Division maintains a staff of sanitarians in all 24 "health districts" in the county. The staff investigates sanitation complaints and enforces public health laws related to environmental health. Problems involving flies, mosquitoes, fleas, cockroaches, bed bugs, and other insects are investigated, and steps are taken to eliminate them.

The Specialized Services Division contains 10 sections. The Central Housing and Institutions Section is responsible for the inspection of hotels, motels, jails, mental institutions, private colleges, boarding homes for the aged, private and parochial schools, day nurseries, and children's homes to assure that such places conform to applicable public health laws and ordinances. This section also makes housing surveys and investigates carbon monoxide poisonings.

The Entomology Section provides consultation for the Bureau of Environmental Sanitation as well as other bureaus of the LACHD on entomological and and other biological matters. Entomologists in this section identify arthropods, birds, and other animals of public health importance and advise on control methods. They maintain a surveillance program on potential vectors of disease, and have established a liaison with physicians and hospitals on vector problems. They also maintain a reference and training collection of specimens, literature, exhibits, and photographs for public information and in-service education.

Other helpful activities of the Entomology Section include applied research on projects pertinent to the needs of the bureau and assistance to other county agencies and departments, universities, and school districts. Surveys and entomological investigations are conducted as the need arises (Sutherland, 1965).

The Bureau of Environmental Sanitation also includes a Vector Control Section which conducts rodent surveys and exterminates rodents in the unincorporated territory of Los Angeles County and in the incorporated cities that contract for these services. The section also enforces ordinances pertaining to rodent control, primarily for the control of communicable diseases potentially spread by rodents.

County Agricultural Commissioner

The County Agricultural Commissioner's office, also known as the County Department of Agriculture, is unique to California counties and is comparable, insofar as pest quarantine functions are concerned, to the office of state entomologist in many states. The California Agricultural Code is now the major source of statutory authority for the work of the commissioners' offices, although authority may in some cases be derived from the county charters or ordinances.

As in the case of the State Department of Agriculture, the activities of the counties are of interest principally to agriculture (Ebeling, 1959). Such activities as quarantine, rodent control, and eradication of new pests can also pertain to pests of household or public health importance or be of general interest to the residents of the urban community.

MOSQUITO-CONTROL ASSOCIATIONS

Membership in the American Mosquito-Control Association, Inc., with its headquarters address at P.O. Box 278, Selma, California 93662, is open to anyone interested in mosquito control and related work who wishes to participate in its promotion and improvement, whose application has been endorsed or sponsored by 2 members in good standing, and who has been elected by a majority of the Board of Directors. Members pay annual dues of $10.00 (1971), receive *Mosquito News*, and are entitled to hold office and participate in annual meetings.

There are 12 state or regional mosquito-control associations in the United States. The California Mosquito-Control Association, Inc., is headquartered at 1737 West Houston Avenue, Visalia, California 93277.

The proposed bylaws of the association were adopted at the 34th Annual Conference in 1966. Article II states:

> The objects of this Association are to promote cooperation among those directly and indirectly concerned with and interested in mosquito control and related subjects, to stimulate the development of improved methods and techniques, to disseminate information in relation thereto, and to aid in the advancement of this field in California and elsewhere.

There are 3 types of membership. *Corporate members* include mosquito-abatement districts (MAD), pest-abatement districts concerned principally with mosquito control, and other local governmental agencies directly engaged in mosquito abatement in the state, such as county health departments, flood-control districts, pest-control districts, etc. Each member has 1 vote, cast by its designated representative. *Associate members* are individuals, agencies, or other organizations interested in or concerned with mosquito abatement. They receive copies of the *Proceedings of the Annual Conference*, notices and minutes of association meetings, and may participate in meetings and discussions and serve on committees, but they have no vote. Their dues are $5.00 per year (1971). *Sustaining members* are individuals and/or organizations who desire to contribute to the furtherance of mosquito control through the association. They have the same rights and privileges as associate members, and likewise have no vote. The dues of sustaining members are $100.00 per year, and funds from this source are used to support the annual conference and the publication of its proceedings. This is a definitely constructive feature.

Special Committees

Coordinated with the activities of the mosquito-control associations are those of various special committees working in the interest of vector control. In California, for example, the general fields of interest of the committees include Research (a joint committee of the State Department of Public Health and the University of California, dealing with arthropods of importance to public health); Conservation (California Mosquito Suppression—Fish and Wildlife Management Coordination Committee); Bureau of Vector Control and Solid Waste Management of the State Department of Public Health; State Department of Agriculture; State Department of Fish and Game; and University of California (staff members on various campuses concerned with arthropods of medical and veterinary importance and related vector problems).

THE PESTICIDE INDUSTRY

In 1969, the entire pesticide market in the United States, comprising some 390 different chemicals, was valued at 1 billion dollars at the manufacturers' price level. It was expected to increase to 2 billion dollars by 1975. These chemicals are being used to combat crop losses, estimated at 11.2 billion dollars per year, which result from insects, weeds, fungi, nematodes, and other pests, and an additional loss of 2 billion dollars that occurs during crop storage (von Rümker *et al.*, 1970). Estimates of the total loss to the United States economy resulting from household and public health pests are not available, but the annual loss (damage plus cost of corrective and preventive work) from termites alone is probably well over the estimated half-billion dollars reported by Ebeling (1968).

Industrial Research and Development

As might be expected, the pesticide industry itself supports a massive research and development effort. An industry questionnaire developed by a committee formed by the National Research Council revealed that the cost of bringing a single compound to the market as a pesticide (insecticide, fungicide, nematocide, herbicide, rodenticide, etc.) averaged $4,060,436 in 1969 compared with $1,196,000 estimated in an earlier survey for 1956, an increase of 240%. On the other hand, the chance for success for an experimental material is decreasing (1 in 3,600 in 1964; 1 in 5,040 in 1969). These figures underscore the fact that the search for new pesticides is meeting with diminishing returns. This has resulted principally from the increasingly stringent biological criteria by which compounds are selected: great selectivity, short persistence, and safety to man and other nontarget organisms—fish, wildlife, and domestic animals—and to the environment in general. The committee concluded that by the time a new pesticide is registered and reaches the market, much more is known about its properties and effects than is known about most other chemicals that are used in our homes and factories (von Rümker *et al.*, 1970).

Of the thousands of new compounds that are developed by the chemical industry, and among countless industrial by-products, a few materials will be found each year to be of some value in pest control, either as pesticides or as adjuvants. The majority of the large chemical companies have accordingly organized entomological research departments where compounds of special interest to the company may be "screened" in order to segregate those which show some promise as pesticides. Many other compounds, of course, are synthesized directly as pesticides, based on what is already known concerning promising molecular combinations and arrangements.

A survey was made of 33 pesticide manufacturers, representing firms accounting for 81% of the total pesticide sales for 1969. These companies reported that they had spent $69,900,000 in pesticide research and development in 1970—an increase of 33% over 1967 expenditures. The various types of expenditure and the amount spent in each, in millions of dollars, were as follows: synthesis and screening, 22; field testing and development, 22.3; toxicology and metabolism, 9.1; formulation and chemical development, 12.3; and registration, etc., 4.2. The companies participating in the survey synthesized 28,000 compounds in 1970 for the purpose of identifying pesticidal activity, an increase of 19% over 1967. (The total number of compounds screened was 62,800.) Of the synthesized compounds, 561 were selected for development, 38 were submitted for regulatory clearance, and 11 were registered for marketing. In the 33 companies, the personnel engaged in research and development of pesticides in the United States in 1970 included 539 with a Doctoral degree, 235 with a Master's degree, 584 with a Bachelor's degree, and 913 others. When compared with 1967, the increase in number of employees ranged downward from 23% in the first to 6% in the last of the above 4 categories (NACA, 1971). From 70 to 80% of the current expenditure of $5,000,000 or more, over a period of 6 to 8 years, to develop a new pesticide to the point of commercial usage is for the salaries of the large staff required for research and development (Anonymous, 1972*c*).

As the cost of bringing a pesticide to the market increases, the necessity for industry-sponsored pesticide research and development to be confined to the giants of the pesticide chemical business increases accordingly. The delays occasioned by increasing difficulties of registration make it desirable or even imperative for industry to have direct control over the personnel involved (Gunther, 1971). An increasingly larger percentage of men and women trained in entomology and its supporting disciplines (chemistry, biology, ecology, toxicology, physiology, statistics, etc.) are finding employment with commercial companies. The research laboratories (figure 3) and even field experiment stations of these companies are often large and well equipped; in equipment, personnel, and quality of work, they compare favorably with corresponding features found in governmental agencies and the universities.

As is to be expected, the number of new pesticides being developed by the research staffs of the commercial companies is now far greater than the number being developed by federal agencies or universities. The latter have done considerable research with compounds that the various exploratory techniques of the commercial companies have indicated to be promising as insecticides. This research may be along the line of molecular rearrangement, new combinations, or related compounds that may be selected or synthesized in an effort to obtain greater pesticidal efficacy. Governmental and university researchers also augment the work done by the pesticide industry in laboratory and field experiments on insecticidal efficacy, compatibility with commonly used insecticides, effects on man and domestic animals and wildlife, phytocidal effect, effect on beneficial insects, general environmental impact, special equipment for application, and various practical problems that must be worked out separately for each pest species in agricultural and structural pest control. The increasing complexity of the pesticide problem and increasing public awareness of environmental considerations result in continual changes in the relative research and extension responsibilities of industry, government, and universities. An attempt to delineate these responsibilities in greater detail would be beyond the scope of this chapter. A good discussion of the subject from the viewpoint of the universities was provided by Gunther (1971). A special and unchanging responsibility of the universities, of course, is the training of scientists and technicians.

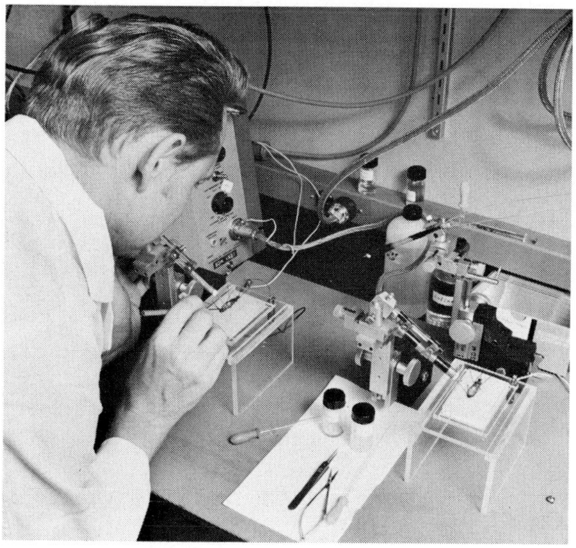

Fig. 3. Toxicological investigation in a research laboratory of private industry. (Courtesy Shell Development Company, Modesto, California.)

The frequent relationship of the work of commercial companies and that of universities in conducting entomological research has led to the establishment of grants and fellowships in which the investigations of entomologists, often graduate students, are financially supported by the companies. Much important work has been accomplished in this manner in fields of research of mutual interest to university personnel and the companies concerned in these cooperative projects.

National Agricultural Chemicals Association

A large number of American chemical companies, either directly or indirectly interested in insecticides and fungicides, formed an association in 1933 to further the common interests of the group in keeping abreast of developments pertaining to entomology, plant pathology, and chemicals used in pest and disease control and, in addition, to keep the interested public informed in regard to these matters. Through meetings, and more especially through its bimonthly publication, *NACA News*, the association has successfully attained this purpose, and has also provided a source of timely news of nation-wide interest to economic entomologists. Frequently, entomological news items concerning control equipment, the insecticide supply outlook, research developments, pest problems of various parts of the country, meetings, insecticide legislation, and matters of general concern to the chemical industry can be found in *NACA News* before they appear in any other publication.

The headquarters of the NACA are at 1155 Fifteenth Street N.W., Washington, D.C. 20005. Its membership includes the following 8 groups: basic producers, formulators, custom applicators, equipment manufacturers, suppliers, regional associations, individuals (scientists, dealers, and others), and allied industries. The membership includes nearly all the important pesticide companies, accounting for practically all the pesticide production of the United States (Anonymous, 1953).

Among its many functions, the NACA offers means for the exchange and coordination of information on technical matters, such as toxicological data, compatibility of insecticides and fungicides, identification of trade-name products, packaging and labeling, nomenclature of new and complex materials, analysis methods, residue determination and removal, ground and air application methods, and adaptability of equipment to new chemicals.

Chemical Specialties Manufacturers Association

This organization, formerly known as the National Association of Insecticide and Disinfectant Manufacturers, was organized in 1914 (H. W. Hamilton, 1947*a*). Its present address is 50 East 41st Street, New York 10017. Together with federal and state officials, who are welcome at its meetings, members of the CSMA Insecticide Scientific Committee and various special committees discuss and help to solve mutual problems. The interests of the group tend toward the field of household insecticides. The CSMA established the Peet-Grady method for evaluating liquid household insecticides, and established standard specifications for these products which were also adopted by the National Bureau of Standards and the United States Department of Commerce. It has always strived to promote the development of new products and means for standardizing and testing these products. The organization has compiled all economic poison laws and regulations in a single book which it keeps up to date by annual supplements. Copies of this compilation can be obtained at a nominal cost.

American Council of Independent Laboratories

There are commercial testing laboratories qualified to perform chemical analyses on bulk or packaged pesticides; pesticide residues in or on raw or processed agricultural products; food, water, and other materials; and tests required in cases of suspected poisonings. Some of these laboratories can make the field trials and other tests required for registration and label approval of new pesticides.

The American Council of Independent Laboratories, 1026 Seventeenth Street N.W., Washington, D.C. 20036, can supply information on firms which are members of their organization. However, a list of firms, their addresses, and the type of analysis or testing they do is given in the *Pesticide Handbook—Entoma,* noted earlier.

LEGISLATION, STANDARDIZATION, AND RESEARCH

Association of American Pesticide Control Officials

This association, under the name "Association of Economic Poisons Control Officials," was organized during a meeting of the Western Cooperative Spray Conference in Portland, Oregon, in 1947. A constitution and set of bylaws were tentatively drafted, with the object of promoting uniformity and effective legislation, definitions, regulations, and enforcement of laws relating to the control, sale, and distribution of insecticides, fungicides, and other economic poisons. An Executive Committee was selected to give representation from all sections of the country. Forty-three states endorsed the plan. The membership consists of research workers employed by state, territory, dominion, or federal agencies who are engaged in the investigation of "economic poisons" and of officials and their deputies charged with the active execution of laws regulating the sale of such compounds, now more commonly referred to as "pesticides." In recent years, the association has been most active in drafting model legislation relating to pesticide use and updating model bills on the registration of pesticides. A statement of the current association policies and functions, rewritten in 1967, is quoted here.

> The purpose of the corporation shall be to establish and maintain an association through which officials of any state, territory, dominion, federal, or other governmental agency on the North American continent, Hawaii, and Puerto Rico, and employees thereof charged with a responsibility in enforcing the laws regulating the production, labeling, distribution, sale,

or use of pesticides may unite to promote uniform and effective legislation, definitions, rulings, and enforcement of laws relating to the control of the sale, distribution, and use of pesticides; to encourage and sponsor the adoption, by all member agencies, of the most effective and adequate methods of analysis of pesticides; to develop high standards of pesticide inspection techniques and procedures; to promote adequate labeling and safe use of pesticides; to provide facilities and opportunities for free exchange of information, discussion, and cooperative study of problems confronting members of the association; and to cooperate with members of industry in order to promote the usefulness and effectiveness of pesticide products.

Each year, the association meets with the Association of American Feed Control Officials and the Association of American Plant Food Control Officials in different parts of the nation during the first week in August. It issues an annual report called the *Official Publication* of the association. This includes information of general interest to control officials. During the last few years, a list of the commonly used pesticides and their properties and characteristics was added. In 1966, the association published the *Pesticide Chemicals Official Compendium,* selling (in 1973) for $15.00 a copy. This compendium comprises 1297 pages in loose-leaf form, without binder, and includes most of the currently used pesticides.

The office of the current Secretary-Treasurer of the association is located in the Virginia Department of Agriculture and Commerce at Richmond.

Association of Official Analytical Chemists

This association is a professional organization of local and federal government regulatory scientists that is devoted to developing, testing, and publishing methods for the analysis of fertilizers, foods, feeds, pesticides, drugs, cosmetics, hazardous substances, and other materials related to agricultural and public health pursuits. Its name was changed from Association of Official Agricultural Chemists in 1965, but without change in policy or scope of activities. The address of the association is P.O. Box 540, Benjamin Franklin Station, Washington, D.C. 20044.

The AOAC provides regulatory scientists with accurate and reproducible methods of analysis that are required for the enforcement of laws and regulations. Hundreds of scientists, designated as "associate referees," are studying methods of analysis in over 50 categories, ranging from "Agricultural Liming Materials" to "Waters." Many of the referees are industry scientists whose specialized knowledge and experience are useful in the testing of methods and materials.

The association maintains formal and informal cooperative arrangements with other scientific societies, such as the American Public Health Association, American Society for Testing Materials, Collaborative International Organization for Standardization, and Pesticides Analysis Committee of the Ministry of Agriculture in the United Kingdom.

All state and municipal chemists, including those of universities and experiment stations, are considered active members of the association. Federal organizations represented in the association are the Food and Drug Administration, Public Health Service, Environmental Protection Agency, Department of Agriculture, Department of Defense, Internal Revenue Service, Department of the Interior, and National Bureau of Standards. Scientists of corresponding agencies in other North American countries are also members. No formal membership application or acceptance is required, no formal membership list is maintained, and no membership dues are assessed.

The *Handbook of the AOAC* is a basic reference intended primarily for the use of referees and associate referees throughout the world who have been appointed to develop and validate analytical methods. The *Journal of the AOAC* contains much information of interest to entomologists, including many scientific papers on standardized procedures for detecting insect parts and other contaminants resulting from insect and rodent infestation of grain and other stored food products. A leaflet listing books and other publications of the AOAC may be obtained from the association's headquarters.

American Conference of Governmental Industrial Hygienists

The ACGIH, organized in 1938, deals with administrative and technical aspects of worker health protection. Its members are professional personnel in governmental agencies engaged in occupational health activities. Foreign countries are represented, giving the organization an international scope. This organization has contributed substantially to the development and improvement of official industrial health services to industry and labor. It holds annual and such other meetings as may be necessary to effectuate its purposes. The papers presented at the formal meetings and the proceedings of the 2 business sessions are published in the *Transactions of the American Conference of Governmental Industrial Hygienists.*

The work of standing committees has been published in special publications, such as *Industrial Ventilation—A Manual of Recommended Practice; The Air Sampling Instruments Manual; Guide to Records for Health Services in Small Industries; A Guide to Records and Reports for Evaluating Environmental Conditions in Industry;* and a loose-leaf *Manual of Analytical Methods.*

Each year, the Committee on Threshold Limits issues a list of recommended threshold limit values (TLV's) as guides for the control of toxic dusts, fumes, gases, vapors, and mists. These values are under continuous review, and are accepted and utilized throughout the world. The third edition, titled *Documentation of the Threshold Limit Values for Substances in Workroom Air,* was published in 1971. It provides documentation for almost 500 threshold limit values.

The ACGIH is represented on committees of such organizations as the American Industrial Hygiene Association, the U.S.A. Standards Institute, and the American Society for Testing Materials. It also maintains liaison with other professional groups with which it has common interests. In 1960, under the joint sponsorship of the ACGIH and the American Industrial Hygiene Association, the American Board of Industrial Hygiene was established to certify qualified workers in the field of industrial hygiene.

The ACGIH issues a list of its official publications and their prices. Most of these may be purchased from the Secretary-Treasurer, ACGIH, P.O. Box 1937, Cincinnati, Ohio 45201. The *Respiratory Devices Manual* and *Industrial Ventilation—A Manual of Recommended Practice* may be purchased from the Committee on Industrial Ventilation, Post Office Box 453, Lansing, Michigan 48902.

Crop Protection Institute Biological Research Center

This institute, generally referred to as the "CPI" or "CPI Biological Research Center," was organized in 1921 under the auspices of the National Research Council and in affiliation with the American Association of Economic Entomologists (as it was known at that time). Formerly a report of the institute was published in the *Journal of Economic Entomology,* but the CPI now sends its reports, covering a wide field of agricultural research, directly to its sponsoring individuals or organizations.

The CPI has been instrumental in securing cooperation among commercial companies, experiment stations, and other biological research institutions for solution of problems connected with the development and use of pesticides and the equipment for their application. CPI's original Board of Governors consisted of 3 entomologists, 3 plant pathologists, and 3 chemists, these being selected from various state universities and experiment stations and other research institutions. The present Board of Directors consists of representatives of a much wider range of agricultural sciences. The address of the institute is P.O. Drawer S, Durham, New Hampshire 03824.

Boyce Thompson Institute for Plant Research, Inc.

This institute, located at 1086 North Broadway, Yonkers, New York 10706, was organized as a nonprofit membership corporation in 1924. The corporation consists of 11 members who appoint a Board of Directors that determines the general policies and administers the operations of the institute through a managing director. Financial support is derived primarily from an endowment established by the founder, William Boyce Thompson. Other funds are obtained in the forms of fellowships and grants to serve the interests of private donors, the federal government, industry, and various associations.

Scientific findings of the research staff of the institute are released to the public in technical journals, primarily a quarterly periodical, *Contributions from Boyce Thompson Institute.* Many important entomological papers may be found in these journals. Prominent among them are papers on the physiological effects of insecticides, investigated by histological and bioassay techniques. Noteworthy among the current activities of the institute are its extensive investigations on pest management in forests, involving the utilization of synthetic pheromones and host terpenes to lure bark beetles to trap trees (McNew, 1970).

THE PEST CONTROL INDUSTRY

Pest control in the urban community is coming increasingly within the sphere of private industry as the years go by. Pest control operators have acquired high degrees of professionalism and competence, and the majority of the American people have found from experience that commercial pest control is not only a great convenience but in many cases the only truly satisfactory remedy for urban pest problems. Much of the credit must be given to organization, legislation, standardization, and education as sponsored by groups of pest control operators.

National Pest Control Association

The National Pest Control Association (NPCA) originated at a meeting in Washington, D.C., in 1933, where pest control operators (PCO's) representing 200 firms adopted a constitution and bylaws and drafted a National Code of Fair Competition under the terms of the National Recovery Act (NRA), and was thus a product of the Great Depression. First known as the National Association of Exterminators and Fumigators, it adopted its present name at the 1937 national meeting. The NPCA had no sooner begun to establish itself on a sound basis in the postdepression period than it was hit by the material shortages brought about by World War II. However, it thereby obtained its first official recognition as an "essential" industry, assuring to PCO's a supply of what were then the tools of the trade: gasoline, tires, repair parts for autos and equipment, baits, arsenic, pyrethrum, sodium fluoride, strychnine, paradichlorobenzene, and thallium sulfate. Many PCO's of the postwar period were men who had obtained pest control experience while with the armed services during the war. The advent of DDT, followed by other chlorinated hydrocarbons and other classes of synthetic pesticides, was a great boon to the industry.

The association now includes about 1,500 members in the United States and 38 foreign countries, and does about 400 million dollars in business annually. Its new headquarters are located at 8150 Leesburg Pike, Suite 501, Tison's Corners, Virginia. It is governed by a Board of Directors of 80 members which meets twice a year and acts on broad policy determinations recommended by the Executive Board of 7 members which meets 4 times a year or, in urgent situations, at the call of the president.

The principal responsibility for directing this complex trade organization rests with the Executive Secretary and his supporting staff. The NPCA's Table of Organization provides for Technical Management and Production and Training departments. A Research Council of 9 members provides guidance from the association membership for the staff. There are also committees (Bird Management, Financial Management, Fumigation, Insect Control, Marketing Management, Operations, Personnel Management, Rodent Control, Sanitation, and Wood-Destroying Organisms) made up of PCO's working in the field. These play important roles in developing industry standards and making the latest technical and management information available to members. Among the functions of these committees is the development of *Policy* and *Good Practice* statements. The *Policy Statement* describes what a responsible PCO will do in a specific situation, and the *Good Practice Statement* describes how the work should be done. Safety and efficiency are emphasized, and the statements are regularly reviewed to take into consideration the latest changes in technology.

A fund for research is budgeted annually and supplemented through the contributions of members for the purpose of supporting research, usually at universities, on the biology, ecology, behavior, and control of the major household pests as well as on management studies. Such projects are currently being sponsored in 11 universities throughout the country.

The Buettner Pest Control Foundation, named in honor of the first executive secretary of the association, is a legal entity wholly separate from the NPCA. The foundation receives and administrates funds exclusively for scientific, literary, and educational activities pertaining to structural pest control. It also is authorized to provide scholarships and to loan funds to educational institutions in the United States and Canada.

The NPCA holds a series of workshops throughout the year in various regions of the country. The workshops cover such subjects as the identification, biology, and ecology of pests; pesticides and their formulations; pesticide application equipment; and control procedures. The workshops enable PCO's and the association staff members to meet on a personal basis to discuss practical and scientific problems. The association's technical staff also attempts to answer by correspondence any special and individual questions from members, or even from the general public.

One of the outstanding contributions of the association is the preparation by its technical staff of the *Technical Releases*. These loose-leaf publications contain up-to-date technical information regarding pests and the newest developments in pest control, and are sent to association members at the rate of 2 or 3 per month. Probably, there is no other aspect of pest control in the United States in which pest control operators are being briefed in such a thorough and timely manner on the latest scientific and technological developments. At the end of the year, an index is prepared and is sent to the members. The assembled volumes of *Technical Releases* are particularly useful as timely references in the field of pest control technology.

Along with the *Technical Releases,* there are management reports and service letters which pertain to nontechnical phases of pest control, such as actions by federal agencies, safety, training aids, business practice, announcements of new publications, and the association's Policy and Good Practice statements.

The NPCA has also developed reference standards and approved methods for pest control work which have been published in such manuals as *Approved Reference Procedures* (for subterranean termite control), *Serviceman's Manual, Insurance Checklist,* and *Uniform Accounting Manual.*

Pest Control Operators of California, Inc.

In several states, pest control operators have formed their own associations, although members may also belong to the national association (NPCA). As with the NPCA, the aim of the state associations is to increase the professional competence and enhance the public image of the members, to agree on work standards, and to develop and support legislation and legal agencies, such as the Structural Pest Control Board (in California), by means of which members doing substandard work can have their licenses rescinded. The largest of the state associations is in California, and a discussion of the PCOC should suffice as an example of the general nature of this type of organization.

The PCOC was organized in 1942. Its present headquarters are located at 3444 West First Street, Los Angeles, California 90004. The organization is divided into 15 "districts," representing the various geographical areas of the state. There are currently 382 licensed pest control operators in California, and approximately half of them belong to the association. All members have the right to use and display the seal of the association.

The PCOC sponsors an annual 2-day "Cal Poly" Conference in December, in cooperation with the Agricultural Biology Department, California State Polytechnic University, on the college campus in Pomona. The conference consists of talks on the biology and control of pests, on pesticides and pesticide-application equipment, on business practices, and there are workshops and demonstrations. An average of about 130 PCO's and a variable number of "trainees" attend the conference. During the 2-day session, there are training courses attended by people who wish to take the examination required for the pest control operator's license, or even by licensed operators who wish to review certain subject matter. In 1969, the PCOC published a 224-page book titled *Community Pest and Related Vector Control,* with 19 chapters by experts from governmental agencies, educational institutions, and the pest control industry.

An Educational Program for Certification of Pest Control Operators

In past years, there have been 3 training courses of 10 weeks given each year at the PCOC headquarters in Los Angeles. These were originally taught as University of California extension courses, but have later been taught by personnel of a large California pesticide distributor for a fee of $40.00 (in 1971) per course. The money obtained from these fees was given to the PCOC. The same company sponsored seminars on pests and pest control in various parts of the state. The objective was to impart to the trainee the information he or she would be required to have in order to pass the examination given by the Structural Pest Control Board that is required for a license. As already stated, examinations may be taken in any one of the 3 branches of pest control: Fumigation, General Pest Control, or Termite Control, or in any combination of them. Then in 1972, PCOC expanded its educational program to include 5 technical courses: Wood-Destroying Organisms, General Pest Control, Fumigation, Principles of Construction, and Safe Chemical Handling. Its Education Committee has as its goal the publication of the courses as textbooks.

In April, 1973, the PCOC cooperated with the California State Polytechnic University (Cal Poly) at Pomona in offering the first courses in an accreditation program for pest control operators and students. On the basis of 10 classroom hours per 1 credit, successive phases of certification can be attained, with the highest, Phase IV (Master Certification), requiring 15 to 20 3-unit courses. The program is expected to develop in a series of steps so that all the courses for Phase IV will be offered by early 1975. The courses will be offered in any area of the state where a minimum of 15 people can be enrolled. An Accreditation Advisory Board, consisting of 5 persons from the

PCOC, 5 academicians from relevant disciplines in Cal Poly, and 2 from Cal Poly's Extension Administration, is expected to have considerable influence on the development of the accreditation program, but the ultimate academic and administrative decisions will be made by Cal Poly's Extension personnel (Young, 1973).

ENTOMOLOGICAL SOCIETIES AND CONFERENCES

As the relative importance of the pest problems of the urban community has increased, the number of people engaged in urban entomology, and particularly the extent of their scientific training and professional competence, have increased proportionately. Therefore, there has been a greater desire among such people not only to participate in the societies or conferences sponsored by their trade or professional associations, but also to gain a greater insight into the broad field of entomology as an important branch of biological science. To expedite this objective, many are joining entomological societies and conferences that are available on a local, state, or national level.

Entomological Society of America

In 1953, the American Association of Economic Entomologists, founded in 1889, joined with the Entomological Society of America, founded in 1906, to pool and consolidate the interests of the 2 organizations. The office of the Executive Secretary of the society is at 4603 Calvert Road, College Park, Maryland 20740. Within the national society, which holds its annual meeting in December, there are 5 geographic branches (Pacific, Southeastern, Eastern, North Central, and Southwestern) that meet separately, generally in June.

Application for membership in the society may be made by anyone who has an interest and a background in either the economic or the purely academic phases of entomology. The candidate must be endorsed by 2 members, each of whom must furnish a letter stating his background. The election of new members takes place at the annual meeting. The annual dues of $15 (1972) must accompany the application. The dues also pay for the *Bulletin of the Entomological Society of America*. Other official publications of the society are the bimonthly *Journal of Economic Entomology,* which is devoted to applied subjects; the quarterly *Annals of the Entomological Society of America,* which is more academic; and the new *Environmental Entomology* (1972). These journals contain many of the papers read at the annual meetings of the society and its various branches. The *Journal of Economic Entomology* constitutes the most important single source of information on the progress of applied entomology in the United States.

American Registry of Certified Entomologists

In 1971, the Entomological Society of America established an American Registry of Certified Entomologists designed to identify specialists with the training and technical ability to advise the public on entomological matters pertaining to man and his environment. Certification was considered desirable in view of increasingly stringent laws and regulations governing pest management, restrictions as to choice and use of pesticides, prescription programs for pest control, and increasing pressure to permit only qualified individuals to engage in these programs. People who sell pesticides, pest control advisers, pest control operators, or those who render special services, including insect identification, consultation, marketing of biotic agents (growth stimulants, hormones, etc.), and others who have acquired special training and competence, can be registered in their professional specialties.

The society outlined the education and experience required to qualify a person for certification, including various combinations of major studies in entomology. The number of years of experience required as a practicing entomologist were specified. The 16 specialty classifications in which registration can be obtained include agriculture, insect behavior, biological control, ecology and/or population dynamics, forest entomology, general entomology, medical and/or veterinary entomology, pest management, pesticide development and analysis (includes pesticide residue work), physiology and/or morphology, regulatory entomology, survey, taxonomy and/or systematics, toxicology, and urban and industrial entomology.

Other Entomological Societies

Some of the other entomological societies in this country and Canada and their publications are: American Entomological Society (*Transactions*); Entomological Society of Ontario (*Canadian Entomologist*); Brooklyn Entomological Society (*Bulletin*); Cambridge Entomological Club (*Psyche*); Washington Entomological Society, (*Proceedings*); New York Entomological Society (*Journal*); Entomological Society of Hawaii (*Proceedings*); Florida Entomological Society (*Florida*

Entomologist); Society of Natural History of San Diego (*Transactions*); Pacific Coast Entomological Society *(Pan-Pacific Entomologist)*; and California Academy of Sciences (*Proceedings of the California Academy of Sciences*). The Lorquin Entomological Society, connected with the Los Angeles County Museum, publishes occasionally in various journals.

California Entomological Clubs

The Entomological Club of Southern California was organized in 1926, and at present (1975) the membership is over 200. Meetings are held on the first Fridays in March, June, September, and December at the Los Angeles County Arboretum lecture hall in Arcadia. Papers and informal talks of a widely varied nature, but usually of economic significance, are presented. A group of similar purpose and organization, the California Entomological Club, was organized in 1930 for the northern part of the state. It has 220 members, and meets on the first Fridays in February, May, and November, at Davis, Sacramento, and Berkeley, respectively.

Chapter 2 · *Pesticides and Their Uses*

ALTHOUGH THE WORDS "pesticide" and "insecticide" are commonly used interchangeably, the latter is used correctly only to mean "insect killer." During the last 2 decades, the word "pesticide" has come to be used with increasing frequency to refer to all products formerly called "economic poisons"—not only insecticides, but also acaricides (miticides), nematocides, molluscicides, rodenticides, fungicides, algaecides, herbicides (weed and brush killers, defoliants, and desiccants), repellents, and usually growth regulators and attractants.

Pesticides that are used to control pests only by being consumed are traditionally called "stomach poisons" because the poisons are absorbed through the alimentary tract. They are used against insects with chewing mouthparts. They may be applied to plants as sprays or dusts to control insects such as beetles and caterpillars, or in baits to control insects such as grasshoppers, crickets, cockroaches, flies, ants, and yellowjackets. Examples of "stomach poisons" include inorganic compounds such as the arsenicals and fluorine compounds, but also some organic compounds such as Kepone®, mirex, and carbaryl. (See table 1 in this chapter.)

Arthropods that have piercing-sucking mouthparts obtain their food by imbibing plant juices, and therefore do not ingest surface deposits of "stomach poisons." Both chewing and piercing-sucking pests may be controlled by spraying or dusting them, or the surfaces on which they are likely to rest or crawl, with a "contact poison" or "contact pesticide." Motile pests that are not hit by the spray or dust at the time of treatment will contact the deposit or residue of toxicant with their tarsi and other body parts.[1] The toxicant must penetrate the cuticle and body wall of the pest to be effective. Fortunately, most organic insecticides and acaricides penetrate the cuticles of insects and mites much more readily than they penetrate the skin of man and other vertebrates. They also generally exert their toxic effect against coldblooded animals such as insects and mites much more rapidly than against man or other warmblooded animals.

A third type of pesticide which also can control either type of feeder is the "systemic pesticide" or "systemic." Systemics may be applied to the plant surface, to be absorbed by foliage, then transported to a limited extent in the plant via the plant sap. Unfortunately, the new growth that develops after such a treatment is not usually protected by the pesticide. However, systemics applied to the soil, to be absorbed by the roots, will protect all parts of the plant including growth developing subsequent to treatment. The same may be said of systemics applied to trunks or canes of plants, e.g., dimethoate emulsifiable solution (Cygon® 2E) applied to rose canes for long-term protection against rose aphids.

The Need for Pesticides

Man shares with all living animals and plants a potentially hostile environment. The number of people now inhabiting the planet, to say nothing of the vastly increased numbers predicted for the near future, could not be fed, clothed, and protected from pestilence and disease without a multitude of technologies, including the use of pesticides. The astounding success of agricultural technology in the United States is attested by the fact that in this country consumers spend only about 17% of their income on food—less than any other people anywhere. Freedom from dreadful insectborne diseases, such as malaria, yellow fever, dengue, elephantiasis, typhus, bubonic plague, cholera, and dysentery, is taken for granted. Unfortunately, many people are unaware that we enjoy this blessing not only because of important advances in our knowledge of the biology and habits of insect vectors such as mosquitoes, lice, fleas, and flies, but also because of the advanced state of pest control technology, including the development and use of pesticides.

Fortunately, insects and other arthropods generally have a greater innate sensitivity for most

[1] The word "deposit" refers to the coating of pesticide initially laid down on a surface by the treatment. The word "residue" refers to the pesticide, regardless of its location on or within the treated substrate. On plants, for example, some of the pesticide may penetrate below the surface and be translocated by the sap stream. It may also undergo chemical change or degradation. Deposits become residues as soon as they are affected by weathering, metabolic degradation, or other processes (Gunther and Blinn, 1955).

of the currently used pesticides than do warm-blooded animals. They can be destroyed without harm to humans when pesticides are applied according to the directions on the label of the container. This does not mean that people can completely avoid taking at least minute quantities of pesticide into their bodies. Some pesticide will be consumed in food, but in this respect the public appears to be well protected by legislative controls. Pesticides are not tolerated by state and federal agencies at concentrations greater than 1% of the maximum levels found to be safe in extensive experiments with rats and other laboratory animals. If the same criterion were used for naturally occurring substances, many common vegetables, fruits, and nuts would have to be banned as foods, for they contain many naturally occurring poisons. This is not surprising, in view of the fact that plants cannot flee from their predators, and natural selection may be expected to result in species that are protected by chemicals of no value in plant metabolism, but that are poisonous to animals that feed on the plants (Whittaker and Feeny, 1971; Leopold and Ardrey, 1972). Toxic substances in plants have been classified according to their types of physiological action: enzyme inhibitors, physiological irritants, allergens, compounds that alter hormone systems, and vitamin and amino acid antagonists (Leopold and Ardrey, 1972). Poisons in common plant products regularly consumed by man and other mammals, such as solanine, cyanogens, glycosides, oxalates, saponins, antitrypsin, phytic acid and phytates, goiterogens, pressor amines, protease inhibitors, favism substances (from horsebeans), toxic alkaloids, hemagglutinins, lathyrogens, irritant oils, estrogens, protoanemonin, and mimosine, sometimes occur in as much as 10-fold greater quantities than would be tolerated if they were pesticide residues (Kingsbury, 1964; Liener, 1969; Anonymous, 1970*i;* Whittaker and Feeny, 1971; Leopold and Ardrey, 1972). The same may be said of certain beverages and condiments. Likewise, seafood would be considered to contain too much poison (zinc and arsenic). Smoked foods contain small amounts of the polynuclear aromatic hydrocarbons (carcinogenic). Common foods such as eggs, butter, and milk contain vitamins A, D, and K and several essential minerals that could not be tolerated at levels of a hundred times normal consumption. [Vitamin D is considered to be therapeutic at a dose of 400 units, but has been found to be toxic at a 5-fold greater dose (Mayer, 1972).]

All the above foods are wholesome, and pose no threat to the consumer. Minute quantities of absorbed or consumed poisons are excreted with no harm whatever. The possibility of misuse of pesticides, with greater than normal residues remaining on edible produce and reaching the consumer, is ever present, but is not likely to happen often, if only for the reason that it is to everyone's interest that official recommendations be followed. If, despite official regulations, a shipment of produce with excessive residue reaches the market, it will be widely dispersed, and it is unlikely that any 1 person will consume a significant amount of it.

Nearly all illnesses and deaths to people who handle pesticides are the results of carelessness or accidents, but there has been some degree of hazard to field workers who have come into contract with sprayed foliage even as long as 2 or 3 weeks after treatment. Ironically, this difficulty has been experienced with the less persistent, but much more toxic, pesticides that have been substituted for the "persistent" ones which are now gradually being phased out of agricultural usage. The result has been longer periods of waiting after pesticide treatments before workers are allowed to enter a treated orchard or field. According to the federal Food and Drug Administration, pesticides account for 4.9% of all cases of accidental poisoning (fatal and nonfatal), the same percentage as is attributed to cosmetics. The highest percentages of accidental deaths are caused by pharmaceuticals (57.9%) and cleaning and polishing agents (12%) (Anonymous, 1972*c*).

A more serious indictment against pesticides is that within a few years after its initial use, a pesticide may begin to lose its effectiveness as not only the "target" or primary pest begins to develop resistance, but quite possibly also certain "nontarget" or secondary species. (See under "Resistance" later in this chapter.) The latter are greater in number of species, but are normally kept at population density levels below "economic zero" by their natural enemies. To complicate the matter further, the natural enemies (predators and parasites) of both primary and secondary pests are reduced in numbers by the pesticide. The usual result is that after a few years of effective control there is a "resurgence" of the primary pest and "secondary outbreaks" of nontarget species that were originally of little or no economic importance. This is not meant to imply that all pesticides are failing—many continue to be successful. Others, declining in effectiveness, are being replaced by new compounds that are at least initially effective.

Another adverse effect of pesticides is environmental pollution, resulting in such adverse effects as "dieoffs" of birds and fish. Urban areas appear to contribute more than rural areas to pesticide contamination of our fresh-water and sea-water resources (Burnett, 1971). Probably the disposal of discarded pesticide in the sewer system, and the drainage of pesticide from treated areas into storm drains, thence into streams, rivers, lakes, and oceans, are the principal causes of pesticide pollution originating in urban areas. Those pesticides with a long residual life—the so-called "persistent pesticides" (generally organochlorine compounds), even when present in plants, soil, or water in extremely small concentrations—appear to be "biologically magnified," that is, increasingly concentrated in organisms higher up in the food chain. In some polluted aquatic environments, for example, hazardous concentrations might be reached in fish-eating birds. However, the extent and seriousness of this particular problem are controversial, and at least some reports of environmental disaster have appeared to be exaggerated, to judge from later and possibly more objective investigations.

Integrated Control

Despite all the difficulties arising from the use of chemical pesticides, they retain an important advantage not possessed by other methods of combating pests: they can be used to eliminate pests rapidly when a few days' delay might result in prohibitive economic crop loss and no other control method (cultural control, biological control, juvenile hormone analogs, etc.) would act quickly enough to prevent this loss. Therefore, the current tendency is to combine the best qualities of chemical pesticides with those of other control methods in a program of "integrated control" or "pest management." An example of the way in which integrated control can be effectively utilized in the solution of an urban pest problem (flies in poultry ranches on the outskirts of cities) is presented in chapter 6, under "Control of Flies on Poultry Ranches."

In integrated control, all suitable methods of pest control are utilized, but attention is focused not only on the immediate problem but also on possible subsequent developments and on the effect of the treatment on the total environment. In integrated control, one relies as much as possible on such natural pest-regulating forces as natural enemies and weather, using pesticides only when necessary and preferably only "selective" pesticides that kill the target species, with a minimum of disruption of the natural balance of the pest and its natural enemies. Such a control program might involve the rearing of certain species of natural enemies and their release when needed. Cultural control measures, such as crop rotation, water management, destruction of alternate host plants, trap crops, earlier or later planting dates, and early or rapid harvesting, are utilized to whatever extent these are appropriate and applicable in a particular pest problem. The development of acceptable new plant varieties that are more or less resistant to pests has met with outstanding success in some instances in the past, and should be an important aspect of an integrated control program. In integrated control, the goal may not necessarily be the complete elimination of pest damage, but merely to keep it from exceeding acceptable levels. Pesticides desired in integrated control are those least disruptive to the natural balance, i.e., those that break down rapidly, are reasonably selective in their action, and have a minimal tendency to induce resistance. An important obstacle is that it costs millions of dollars to develop and test a new pesticide, and therefore there is little or no incentive to develop one that does not have a "wide spectrum" of pesticidal activity. Therefore, financial help by governmental agencies to encourage research and development of selective insecticides may be essential. The environment we wish to preserve is shared by us all. In one way or another, the public is going to pay for environmental protection.

Integrated control is a more complex and sophisticated program than one which depends on chemical control alone. An idea of the mass of information that has been developed and the degree of expertise that has been attained can be obtained by reading Publication 1695 of the National Academy of Sciences, titled *Insect-Pest Management and Control.* Many scientists collaborated over a 4-year period to prepare this 508-page volume. Some of the modern trends in "pest management" are indicated by such chapter titles as "Control by Parasites, Predators, and Competitors," "Microbial Control of Insects," "Management by Genetic Principles," "Cultural Control," "Physical and Mechanical Control," "Antimetabolites, Feeding Deterrents, and Hormones," "Chemical Attractants and Insect Behavior," "Chemical Repellents," "Sterilization,"

"Integrated Systems of Pest Management," and "Organization and Education" (Subcommittee on Insect Pests, 1969).

In the last-named chapter of the volume, recommendations are made concerning the type of training required for effective teaching, research, extension, and the supervision of practical control programs in this new field of insect-pest management. In recognition of the special interdisciplinary background required of an entomologist who is called upon to diagnose a pest problem and recommend treatment, the California State Legislature passed Senate Bill 1021, which became effective July 1, 1972. The law now requires the licensing of pest control advisers, exempting only federal, state, and county officials and University of California personnel engaged in official duties relating to the agricultural use of pesticides. However, any person making a pest control recommendation must make it in writing, and a copy must be given to the grower. The written recommendations must be kept on file for a year.

The role of pesticide chemicals is changing along with continual advances in pesticide technology and in our understanding of the complex biotic interrelationships involved in pest outbreaks, but apparently for the foreseeable future the need for pesticide chemicals will remain. R. F. Smith (1972), one of the leading proponents of integrated control, stated that he strongly advocated "a greatly increased interest in and support for biological control and other neglected controls appropriate to the better management of pest populations, but this does not even infer the elimination of chemical controls. That, of course, would be inane because it does not take into account the facts of the real world."

A concise yet well-documented discussion of "Pesticides in the Modern World" is contained in a 59-page symposium prepared by members of a cooperative program of agro-allied industries with the Food and Agriculture Organization and other United Nations organizations (Anonymous, 1972*c*). This symposium was published in 5 languages: English, French, German, Russian, and Spanish. It is of particular value because of its timely and objective discussion of the "risk-benefit ratio."

As with many other technologies upon which mankind at the present population density and cultural heritage depends, obvious benefits must be balanced against possible risks, with every effort made to minimize the latter.

Pesticide Nomenclature

Unfortunately, a pesticide label may contain no indication of the toxic ingredients in a labeled container other than their chemical names. These names may mean little to anyone except a pesticide chemist or someone involved professionally or commercially with the preparation or use of pesticide formulations. However, pesticides can often be identified, at least to their major groupings, by certain combinations of letters in their chemical names that may act as clues (Pence, 1967). Thus, the organochlorine compounds (chlorinated hydrocarbons) are identified by "chloro-" in the chemical name; for example, *gamma*-1,2,3,4,5,6-hexachlorocyclohexane, which is lindane. The organophosphorus compounds are identified by the terms "phosphate," "phosphono-," "phosphoro-," or "phosphoric," and include: (1) aliphatic derivatives of phosphorus compounds, such as dichlorvos, malathion, trichlorfon, dimethoate, and many others; (2) aryl (phenyl) derivatives, such as parathion, dicapthon, ronnel, and fenthion; and (3) heterocyclic derivatives, such as diazinon and chlorpyrifos (Dursban®).

The organocarbamates, or simply "carbamates," are usually easy to identify, for they nearly always end with the designation "carbamate," like *o*-isopropoxyphenyl methyl carbamate for the familiar insecticide propoxur (Baygon®). Zectran® and carbaryl (Sevin®) are other common organocarbamates. An empirical formula (the simplest ratio of elements in a compound) also gives clues as to the class of a pesticide, particularly among synthetic organic compounds. One need only know the symbols for a few common chemical elements. Carbon (C) or hydrogen (H), and nearly always both elements, are common to all organic compounds. The clue to the class of pesticide will be found in the symbols for other elements: chlorine (Cl) in the organochlorine compounds; phosphorus (P), or both P and sulfur (S), in the organophosphorus compounds; nitrogen (N) in the organocarbamates and in the alkaloids such as nicotine; and both Cl and S (but no P) in most of the sulfonates, sulfites, etc. Many pesticides also include oxygen (O) and sometimes bromine (Br). Empirical formulas are less useful to identify botanicals such as pyrethrum. They are generally characterized by the large number of C and H atoms which, particularly in polycyclic compounds, indicates a very complex molecular structure. Their complexity often accounts for their generally rapid decomposition and thus for their characteristically brief residual activity.

Table 1, adapted from data developed by Kenaga and Allison (1969), and with additional data obtained from the Environmental Protection Agency (EPA) of the federal government, lists the generic name (such as "chlordane," "malathion," etc.), chemical name, empirical formula, group, use, acute oral toxicity expressed as LD_{50}, and the "toxicity category" for insecticides and acaricides that are either likely to be of interest to the user of household insecticides and acaricides or are so well known that they frequently appear in pest control literature. (See table 2. Also, under "Acute Toxicity" in this chapter.)

The chemical names of 738 pesticides are tabulated by Gunther *et al.* (1968) in accordance with the principles of chemical nomenclature employed by the journal *Chemical Abstracts.* This table also gives water-solubility data, which are important in view of the claim that runoff waters and soil waters are transferring large amounts of pesticides to canals, streams, lakes, and oceans. The same chemical name can sometimes be found written in different ways in the world literature, but its empirical formula is always the same. Trade names are indicated for pesticides for which the generic name has not yet been designated or is not yet well known. The trade name begins with a capital letter and is followed by the superscript symbol ®, indicating a registered trademark.

SOME CHARACTERISTICS OF PESTICIDES

The list of pesticides in table 1, if it were to be compared with a similar list prepared immediately before World War II, would be a reminder of the pesticide development that has taken place in the relatively brief intervening period. Not only have the organic compounds increased in relative importance, but the synthetic organic compounds are now much more abundant and important than the natural ones. The synthetic product is usually as effective as the natural product it replaces, and often more so. This trend parallels the changes in the fields of dyes and drugs, for example, for synthetic compounds can be produced when and where needed, without regard to season or locality, and can be manufactured with consistent quality standardization. Eventually, if not immediately, economic considerations will favor the synthetic product. In the case of pesticides, increased efficacy has been principally the result of longer residual life of the compound but, as will be discussed later, longer residual life has not been an unmixed blessing.

Following is a brief discussion of the organic pesticide groups listed in table 1, as well as the few inorganic pesticides which are still used to a limited extent in pest control programs.

Chlorinated Aryl Hydrocarbons and DDT Relatives

It was the search for better mothproofing compounds in the laboratories of J. R. Geigy in Switzerland which led to the discovery of the insecticidal properties of DDT (the "initials" or "acronym" for dichloro diphenyl trichloroethane). The discovery of DDT ushered in a new era in insecticide chemistry. Its importance in the control of insect vectors of the causal organisms of many important human diseases, particularly malaria, led to the Nobel prize in medicine for Dr. Paul Müller, who discovered its outstanding insecticidal properties in 1939.

As might be expected, the insecticidal efficacy (low dosages, long effect) of DDT led to intensive investigations of other configurations of organochlorine compounds for insecticidal activity. This was soon followed by benzene hexachloride, the most insecticidally active isomer of which is lindane, and by chlordane and toxaphene. A number of DDT-like diphenyl aliphatics, including a few with acaricidal properties, and the cyclodienes (aldrin, dieldrin, endrin, heptachlor, and Thiodan®) were then developed.

The prolonged residual effectiveness of these chlorinated hydrocarbons, although valuable in a pesticide, has resulted in certain public health problems. The organochlorine pesticides are similar to other halogenated compounds, such as chloroform, in that large doses may affect the heart or central nervous system of warmblooded animals, and smaller doses, if administered over a prolonged period, can injure filtering organs such as the liver and kidneys. The more stable compounds, such as DDT and dieldrin, can reach relatively high concentrations in some animals near the top of a food chain. However, adverse effects on human beings are unlikely to occur except in occasional instances of accidental or intentional swallowing of large quantities of these pesticides.

The organochlorine compounds include some of the safest pesticides. The oral LD_{50} for methoxychlor, which is similar to DDT in its chemical

(Turn to page 36)

TABLE 1
Organic Insecticides and Acaricides*

Common or trade name†	Chemical name	Empirical formula	Group‡	Use§	LD50 (mg/kg)	Toxicity category‖
Abate®	*O,O*-dimethyl phosphorothioate *O,O*-diester with 4,4′-thiodiphenol	$C_{16}H_{20}O_6P_2S_3$	OP	Ins.	1,000–3,000	III
Acaralate®	isopropyl 4,4′-dichlorobenzilate	$C_{17}H_{16}Cl_2O_3$	C	Acar.	5,000–34,600	IV
Acarol®	isopropyl 4,4′-dibromobenzilate	$C_{17}H_{16}Br_2O_3$	C	Acar.	5,000	IV
aldicarb	2-methyl-2-(methylthio) propionaldehyde (*O*-methylcarbamoyl) oxime	$C_7H_{14}N_2O_2S$	OC	Acar. Ins. Nem. Sys.	*ca.* 1	I
aldrin	1,2,3,4,10,10-hexachloro-1,4,4*a*,5,8,8*a*-hexahydro-1,4-*endo-exo*-5,8-dimethanonaphthalene	$C_{12}H_8Cl_6$	C	Ins.	39–60	I
allethrin	2-allyl-4-hydroxy-3-methyl-2-cyclopenten-1-one ester of 2,2-dimethyl-3-(2-methyl propenyl)-cyclopropane carboxylic acid	$C_{19}H_{26}O_3$	B	Ins.	680–1,000	III
Animert V-101®	(See *tetrasul*)					
Aramite®	2-(*p-tert*-butylphenoxy)-1-methylethyl 2-chloroethyl sulfite	$C_{15}H_{23}ClO_4S$	S	Acar.	3,900–6,300	III
azinphosethyl	*O,O*-diethyl *S*-4-oxo-1,2,3-benzotriazin-3(4*H*)-ylmethyl phosphorodithioate	$C_{12}H_{16}N_3O_3PS_2$	OP	Acar. Ins.	7–18	I
azinphosmethyl	*O,O*-dimethyl *S*-4-oxo-1,2,3-benzotriazin-3(4*H*)-ylmethyl phosphorodithioate	$C_{10}H_{12}N_3O_3PS_2$	OP	Acar. Ins.	7–18	I
barthrin	6-chloropiperonyl 2,2-dimethyl-3-(2-methylpropenyl) cyclopropane carboxylate	$C_{18}H_{21}ClO_4$	B	Ins.	23.6 ml	I
Baygon®	(See *propoxur*)					
benzene hexachloride	1,2,3,4,5,6-hexachlorocyclohexane, mixed isomers, and a specified percentage of *gamma*	$C_6H_6Cl_6$	C	Ins.	600–1,250	III
BHC	(See benzene hexachloride)					
Bidrin®	3-hydroxy-*N,N*-dimethyl-*cis*-crotonamide, dimethyl phosphate	$C_8H_{16}NO_5P$	OP	Ins. Sys.	22–75	I
binapacryl	2-*sec*-butyl-4,6-dinitrophenyl 3-methyl-2-butenoate	$C_{15}H_{18}N_2O_6$	M	Acar.	136–225	II
Black Leaf 40®	(See nicotine sulfate)					
bromophos	*O*-(4-bromo-2,5-dichlorophenyl) *O,O*-dimethyl phosphorothioate	$C_8H_8BrCl_2O_3PS$	OP	Ins.	3,750–7,700	III

*Adapted from Kenaga and Allison (1969).

†Common names are given in *italic*. A compound may also be known by one or more trade names.

‡The major groups to which a pesticide belongs are designated thus: B, botanical; C, chlorinated hydrocarbon (organochlorine); M, miscellaneous; MP, mothproofing agent; OC, organocarbamate; OP, organophosphorus; and S, sulfonates, sulfides, sulfones, sulfonamides, and sulfites.

§The uses for pesticides are abbreviated thus: Acar., acaricide; Ins., insecticide; Nem., nematocide; Sys., systemic for plants; and Sys. *a*, systemic for animals.

‖The roman numerals I, II, III, and IV represent four categories of toxicity to humans. They are based on the LD50's of the pesticides to animals (usually rats), as shown in table 2. The chemicals in each successive category of toxicity are, on an average, about one-tenth as toxic as those in the preceding category. See also pages 63–64 of this chapter.

TABLE 1 (Continued)

Common or trade name	Chemical name	Empirical formula	Group	Use	LD_{50} (mg/kg)	Toxicity category
bromophosethyl	O-(4-bromo-2,5-dichlorophenyl) O,O-diethyl phosphorothioate	$C_{10}H_{12}BrCl_2O_3PS$	OP	Acar. Ins.	52–270	II
BUX®	m-(1-ethylpropyl) phenyl methylcarbamate mixture (1-4) with m-(1-methylbutyl) phenyl methylcarbamate	$C_{13}H_{19}NO_2$	OC	Ins.	87–170	II
carbaryl	1-naphthyl methyl carbamate	$C_{12}H_{11}NO_2$	OC	Ins.	307–986	II
carbofuran	2,3-dihydro-2,2-dimethyl-7-benzofuranyl methylcarbamate	$C_{12}H_{15}NO_3$	OC	Ins. Sys.	5	I
carbophenothion	S-[(p-chlorophenylthio)methyl] O,O-diethyl phosphorodithioate	$C_{11}H_{16}ClO_2PS_3$	OP	Acar. Ins.	6–100	I
chlorbenside	p-chlorobenzyl p-chlorophenyl sulfide	$C_{13}H_{10}Cl_2S$	S	Acar.	2,000–10,000	IV
chlordane	1,2,4,5,6,7,8,8-octachloro-3a,4,7,7a,-tetrahydro-4,7-methanoindane	$C_{10}H_6Cl_8$	C	Ins.	283–590	II
chlorfenvinphos	2-chloro-1-(2,4-dichlorophenyl) vinyl diethyl phosphate	$C_{12}H_{14}Cl_3O_4P$	OP	Ins.	12–56	I
chlorobenzilate	ethyl 4,4′-dichlorobenzilate	$C_{16}H_{14}Cl_2O_3$	C	Acar.	700–3,200	III
chlorpyrifos	O,O-diethyl O-(3,5,6-trichloro-2-pyridyl) phosphorothioate	$C_9H_{11}Cl_3NO_3PS$	OP	Acar. Ins.	97–276	II
Ciodrin®	*alpha*-methylbenzyl 3-hydroxycrotonate dimethyl phosphate	$C_{14}H_{19}O_6P$	OP	Ins.	125	II
CO-RAL®	(See *coumaphos*)					
coumaphos	O-(3-chloro-4-methyl-2-oxo-2H-1-benzopyran-7-yl) O,O-diethyl phosphorothioate	$C_{14}H_{16}ClO_5PS$	OP	Sys. *a*	16–41	I
cubé	(See rotenone)					
Cygon®	(See *dimethoate)*					
DDD	(See TDE)					
DDT	1,1,1-trichloro-2,2-bis(p-chlorophenyl) ethane	$C_{14}H_9Cl_5$	C	Ins.	87–500	II
DDVP	(See *dichlorvos*)					
Delnav®	(See *dioxathion*)					
demeton	mixture of O,O-diethyl S-(and O-)-2-[(ethylthio)ethyl] phosphorothioates	$C_8H_{19}O_3PS_2$	OP	Acar. Ins. Sys.	2–12	I
derris	(See rotenone)					
diazinon	O,O-diethyl O-(2-isopropyl-4-methyl-6-pyrimidyl) phosphorothioate	$C_{12}H_{21}N_2O_3PS$	OP	Acar. Ins. Sys. *a*	66–600*	II
Dibrom®	(See *naled*)					
dicapthon	O-(2-chloro-4-nitrophenyl) O,O-dimethyl phosphorothioate	$C_8H_9ClNO_5PS$	OP	Acar. Ins.	284–650	II

***The LD_{50} to rats of the "stabilized" diazinon, now used in all formulations, is between 300 and 400 mg/kg.**

TABLE 1 (Continued)

Common or trade name	Chemical name	Empirical formula	Group	Use	LD50 (mg/kg)	Toxicity category
dichlorvos	2,2-dichlorovinyl dimethyl phosphate	$C_4H_7Cl_2O_4P$	OP	Fum. Ins.	25–170	I
dicofol	4,4′-dichloro-*a*-(trichloromethyl) benzhydrol	$C_{14}H_9Cl_5O$	C	Acar.	575–1,331	III
dieldrin	1,2,3,4,10,10-hexachloro-6,7-epoxy-1,4,4*a*,5,6,7,8*a*-octahydro-1,4-*endo-exo*-5,8-dimethanonaphthalene	$C_{12}H_8Cl_6O$	C	Ins.	46	I
dimethoate	*O,O*-dimethyl *S*-(*N*-methylcarbamoyl-methyl) phosphorodithioate	$C_5H_{12}NO_3PS_2$	OP	Acar. Ins. Sys. Sys. *a*	155–500	II
dimethrin	2,4-dimethylbenzyl 2,2-dimethyl-3-(2-methylpropenyl) cyclopropane carboxylate	$C_{19}H_{26}O_2$	B	Ins.	40 ml	I
dimetilan	1-(dimethylcarbamoyl)-5-methyl-3-pyrazolyl dimethyl carbamate	$C_{10}H_{16}N_4O_3$	OC	Ins.	25–64	I
dinocap	2-(1-methylheptyl)-4,6-dinitrophenyl crotonate	$C_{18}H_{24}N_2O_6$	M	Acar.	980–1,190	III
dioxathion	*S,S′-p*-dioxane-2,3-diyl *O,O*-diethyl phosphorodithioate (*cis* and *trans* isomers)	$C_{12}H_{26}O_6P_2S_4$	OP	Acar. Ins.	19–176	I
Dipterex®	(See *trichlorfon*)					
disulfoton	*O,O*-diethyl *S*-2-[(ethylthio) ethyl] phosphorodithioate	$C_8H_{19}O_2PS_3$	OP	Ins. Sys.	2–12	I
Di-Syston®	(See *disulfoton*)					
Dursban®	(See *chlorpyrifos*)					
endosulfan	6,7,8,9,10,10-hexachloro-1,5,5*a*,6,9,9*a*-hexahydro-6,9-methano-2,4,3-benzodioxathiepin 3-oxide	$C_9H_6Cl_6O_3S$	C	Acar. Ins.	30–110	II
endrin	1,2,3,4,10,10-hexachloro-6,7-epoxy-1,4,4*a*,5,6,7,8,8*a*-octahydro-1,4-*endo-endo*-5,8-dimethanonaphthalene	$C_{12}H_8Cl_6O$	C	Ins.	3–45	I
EPN	*O*-ethyl *O-p*-nitrophenyl phenylphosphonothioate	$C_{14}H_{14}NO_4PS$	OP	Acar. Ins.	7–65	I
ethion	*O,O,O′,O′*-tetraethyl *S,S′*-methylene bisphosphorodithioate	$C_9H_{22}O_4P_2S_4$	OP	Acar. Ins.	13–208	I
Ethyl Guthion®	(See *azinphosethyl*)					
famphur	*O*-[*p*-(dimethylsulfamoyl)phenyl] *O,O*-dimethyl phosphorothioate	$C_{10}H_{16}NO_5PS_2$	OP	Sys. *a*	35–62	I
fenson	*p*-chlorophenyl benzene sulfonate	$C_{12}H_9ClO_3S$	S	Acar.	1,560–1,740	III
fenthion	*O,O*-dimethyl *O*-[4-(methylthio)-*m*-tolyl] phosphorothioate	$C_{10}H_{15}O_3PS_2$	OP	Ins. Sys. *a*	178–310	II
formetanate (hydrochloride)	*m*-[(dimethylamino)methylene]amino phenyl methylcarbamate, hydrochloride	$C_{11}H_{16}ClN_3O_2$	OC	Ins. Acar.	19–22	I

TABLE 1 (Continued)

Common or trade name	Chemical name	Empirical formula	Group	Use	LD50 (mg/kg)	Toxicity category
formothion	*S*-[(formylmethylcarbamoyl)methyl] *O,O*-dimethyl phosphorodithioate	$C_6H_{12}NO_4PS_2$	OP	Ins. Acar. Sys.	375–535	II
Gardona®	(See *tetrachlorvinphos*)					
Genite®	2,4-dichlorophenyl benzene sulfonate	$C_{12}H_8Cl_2O_3S$	S	Acar.	980–2,140	III
Guthion®	(See *azinphosmethyl*)					
heptachlor	1,4,5,6,7,8,8-heptachloro-3*a*,4,7,7*a*-tetrahydro-4,7-methanoindene	$C_{10}H_5Cl_7$	C	Ins.	100–162	II
Imidan®	*O,O*-dimethyl *S*-phthalimidomethyl phosphorodithioate	$C_{11}H_{12}NO_4PS_2$	OP	Acar. Ins.	147–299	II
Karathane®	(See *dinocap*)					
Kelthane®	(See *dicofol*)					
Kepone®	decachlorooctahydro-1,3,4-metheno-2*H*-cyclobuta[*cd*]pentalen-2-one	$C_{10}Cl_{10}O$	C	Ins. Acar.	95–140	II
Korlan®	(See *ronnel*)					
Lethane 60®	2-thiocyanatoethyl laurate	$C_{15}H_{27}NO_2S$	M	Ins.	500	III
Lethane 384®	2-(2-butoxyethoxy) ethyl thiocyanate	$C_9H_{17}NO_2S$	M	Ins.	90–300	II
lindane	*gamma*-1,2,3,4,5,6-hexachlorocyclohexane	$C_6H_6Cl_6$	C	Ins.	76–200	II
malathion	diethyl mercaptosuccinate, *S*-ester with *O,O*-dimethyl phosphorodithioate	$C_{10}H_{19}O_6PS_2$	OP	Ins.	885–2,800	III
menazon	*S*-[(4,6-diamino-*s*-triazin-2-yl)methyl] *O,O*-dimethyl phosphorodithioate	$C_6H_{12}N_5O_2PS_2$	OP	Sys. *a*	1,200–1,950	III
Mesurol®	(See *methiocarb*)					
methidathion	*O,O*-dimethyl phosphorodithioate, *S*-ester with 4-(mercaptomethyl)-2-methoxy-Δ^2-1,3,4-thiadiazolin-5-one	$C_6H_{11}N_2O_4PS_3$	OP	Ins. Acar.	65	II
methiocarb	3,5-dimethyl-4-(methylthio)phenol methyl carbamate	$C_{11}H_{15}NO_2S$	OC	Acar. Ins.	46–47	I
methomyl	methyl *N*-[(methylcarbamoyl)oxy] thioacetamidate	$C_5H_{10}N_2O_2S$	OC	Acar. Ins. Sys.	17–24	I
methoxychlor	1,1,1-trichloro-2,2-bis(*p*-methoxyphenyl) ethane	$C_{16}H_{15}Cl_3O_2$	C	Ins.	5,000–7,000	IV
methyl parathion	*O,O*-dimethyl *O*-*p*-nitrophenyl phosphorothioate	$C_8H_{10}NO_5PS$	OP	Ins.	9–42	I
mevinphos	methyl 3-hydroxy-*alpha*-crotonate, dimethyl phosphate	$C_7H_{13}O_6P$	OP	Ins. Sys.	3–7	I
mirex	dodecachlorooctahydro-1,3,4-metheno-1*H*-cyclobuta[*cd*]pentalen	$C_{10}Cl_{12}$	C	Ins.	235–702	II
Mitin FF®	sodium 5-chloro-2-(4-chloro-2-(3-(3,4-dichlorophenyl)-ureido)phenoxy) benzene sulfonate	$C_{19}H_{11}Cl_4N_2NaO_5$	S	MP	750–1,380	III

TABLE 1 (Continued)

Common or trade name	Chemical name	Empirical formula	Group	Use	LD_{50} (mg/kg)	Toxicity category
Mitox®	(See *chlorbenside*)					
Morestan®	6-methyl-2,3-quinoxalinedithiol cyclic *S,S*-dithiocarbonate	$C_{10}H_6N_2OS_2$	M	Acar.	3,000	III
Morocide®	(See *binapacryl*)					
naled	1,2-dibromo-2,2-dichloroethyl dimethyl phosphate	$C_4H_7Br_2Cl_2O_4P$	OP	Acar. Ins.	430	II
Nemafos®	*O,O*-diethyl *O*-2-pyrazinyl phosphorothioate	$C_8H_{13}N_2O_3PS$	OP	Ins. Nem.	9–16	I
Neopynamin®	2,2-dimethyl-3-(2-methylpropenyl) cyclopropane carboxylic acid esters with *N*-(hydroxymethyl)-1-cyclohexene-1,2-dicarboximide	$C_{19}H_{25}NO_4$	B	Ins.	20,000	IV
nicotine sulfate	salt of nicotine *l*-1-methyl-2-(3-pyridyl)-pyrrolidine	$(C_{10}H_{14}N_2)H_2SO_4$	B	Ins.	50–91	II
Omite®	2-(*p-tert*-butylphenoxy) cyclohexyl 2-propynyl sulfite	$C_{19}H_{26}O_4S$	S	Acar.	2,500	III
ovex	*p*-chlorophenyl *p*-chlorobenzene sulfonate	$C_{12}H_8Cl_2O_3S$	S	Acar.	2,000–2,050	III
Ovotran®	(See *ovex*)					
oxydemeton-methyl	*S*-[2-(ethylsulfinyl)ethyl] *O,O*-dimethyl phosphorothioate	$C_6H_{15}O_4PS_2$	OP	Acar. Ins. Sys.	65–75	II
parathion	*O,O*-diethyl *O-p*-nitrophenyl phosphorothioate	$C_{10}H_{14}NO_5PS$	OP	Acar. Ins.	3–30	I
Penta®	pentachlorophenol	C_6HCl_5O	M	Ins.	125–210	II
Perthane®	1,1-dichloro-2,2-bis(*p*-ethylphenyl) ethane	$C_{18}H_{20}Cl_2$	C	Ins.	6,600–8,170	IV
phorate	*O,O*-diethyl-*S*-[(ethylthio)methyl] phosphorodithioate	$C_7H_{17}O_2PS_3$	OP	Acar. Ins. Sys.	1–5	I
phosalone	*S*-[(6-chloro-2-oxo-3-benzoxazolinyl)-methyl] *O,O*-diethyl phosphorodithioate	$C_{12}H_{15}ClNO_4PS_2$	OP	Ins. Acar. Sys.	100–180	II
Phosdrin®	(See *mevinphos*)					
phoxim	phenylglyoxylonitrile oxime *O,O*-diethyl phosphorothioate	$C_{12}H_{15}N_2O_3PS$	OP	Ins.	8,500–8,800	IV
Prolate®	(See Imidan®)					
propoxur	*o*-isopropoxyphenyl methyl carbamate	$C_{11}H_{15}NO_3$	OC	Ins.	95–175	II
propyl thiopyrophosphate	*O,O,O,O*-tetrapropyl dithiopyrophosphate	$C_{12}H_{28}O_5P_2S_2$	OP	Acar. Ins.	1,224–2,730	II
pyrethrins	Four esters: pyrethrins I and II; cinerins I and II	$C_{21}H_{28}O_3$ $C_{22}H_{28}O_5$ $C_{20}H_{28}O_3$ $C_{21}H_{28}O_5$	B	Ins.	820–1,870	III
Rabon®	(See *tetrachlorvinphos*)					
resmethrin	(5-benzyl-3-furyl)methyl 2,2-dimethyl-3-(2-methylpropenyl)cyclopropane carboxylate	$C_{22}H_{26}O_3$	B	Ins.	2,500	III

TABLE 1 (Continued)

Common or trade name	Chemical name	Empirical formula	Group	Use	LD_{50} (mg/kg)	Toxicity category
Rogor®	(See *dimethoate*)					
ronnel	*O,O*-dimethyl *O*-2,4,5-trichlorophenyl phosphorothioate	$C_8H_8Cl_3O_3PS$	OP	Ins. Sys. *a*	906–3,025	III
rotenone	1,2,12,12*a*, tetrahydro-2-isopropenyl-8,9-dimethoxy-[1] benzopyrano-[3,4-*b*]furo-[2,3-*b*] [1]benzopyran-6 (6*aH*)-one	$C_{23}H_{22}O_6$	B	Ins.	100–132	II
Ruelene®	4-*tert*-butyl-2-chlorophenyl methyl methylphosphoramidate	$C_{12}H_{19}ClNO_3P$	OP	Ins. Sys. *a*	660–1,000	III
ryania	Ground *Ryania speciosa* stemwood (alkaloid ryanodine)	$C_{25}H_{35}NO_9$	B	Ins.	750–1,200	III
sabadilla	Seeds of *Schoenocaulon officinale* (veratrine alkaloids)	$C_{32}H_{49}NO_9$ (cevadine) $C_{36}H_{51}NO_{11}$ (veratridine)	B	Ins.	4,000	III
SBP-1382®	(See *resmethrin*)					
Sevin®	(See *carbaryl*)					
Spectracide®	(See *diazinon*)					
Supracide®	(See *methidathion*)					
Systox®	(See *demeton*)					
TDE	1,1-dichloro-2,2-bis(*p*-chlorophenyl)ethane	$C_{14}H_{10}Cl_4$	C	Ins.	4,000	III
Tedion®	(See *tetradifon*)					
Temik®	(See *aldicarb*)					
tepp	tetraethyl pyrophosphate	$C_8H_{20}O_7P_2$	OP	Ins.	0.5–2	I
tetrachlorvinphos	2-chloro-1-(2,4,5-trichlorophenyl) vinyl dimethyl phosphate	$C_{10}H_9Cl_4O_4P$	OP	Ins.	4,000–5,000	III
tetradifon	*p*-chlorophenyl 2,4,5-trichlorophenyl sulfone	$C_{12}H_6Cl_4O_2S$	S	Acar.	5,000–14,700	IV
tetrasul	*p*-chlorophenyl 2,4,5-trichlorophenyl sulfide	$C_{12}H_6Cl_4S$	S	Acar.	6,810–8,250	IV
Thanite®	isobornyl thiocyanatoacetate	$C_{13}H_{19}NO_2S$	M	Ins.	1,000–1,603	III
Thiodan®	(See *endosulfan*)					
toxaphene	chlorinated camphene containing 67–69% chlorine		C	Ins.	80–90	II
trichlorfon	dimethyl (2,2,2-trichloro-1-hydroxyethyl) phosphonate	$C_4H_8Cl_3O_4P$	OP	Ins.	560–630	III
Vapona®	(See *dichlorvos*)					
Zectran®	4-dimethylamino-3,5-xylyl methyl carbamate	$C_{12}H_{18}N_2O_2$	OC	Ins.	15–63	I
Zinophos®	(See Nemafos®)					
Zytron®	*O*-2,4-dichlorophenyl *O*-methyl isopropyl phosphoramidothioate	$C_{10}H_{14}Cl_2NO_2PS$	OP	Ins. Sys.	270	II

structure, is 5,000 to 7,000 (table 1). Methoxychlor at the top of the food chain in a model ecosystem was found to be at a level 1,500 times that in the water, compared to 90,000 times for DDT. The very pronounced selective insecticidal action of methoxychlor is believed to be the result of the much greater efficiency of the *O*-dealkylation of the compound in the mammalian liver than in insects (Kapoor *et al.*, 1970).

The relatively long residual activity of DDT, when compared with most other synthetic insecticides and with the botanicals, such as pyrethrum, rotenone, and nicotine, is indispensable in insect control programs such as mosquito control in antimalarial programs. For other uses, particularly in plant protection, its prolonged residual activity resulted in the long-term elimination of the insect enemies of certain pests such as mites, which were not killed by DDT. This led to an intensified search for new and better acaricides, and it happened that most of these were also organochlorine compounds such as dicofol, (Kelthane®), chlorobenzilate, Dimite, and Neotran. The addition of sulfur (S) resulted in many new acaricides, listed in table 1 as sulfonates, sulfides, sulfones, sulfonamides, and sulfites. These were excellent acaricides, and were all of low mammalian toxicity.

The organochlorine acaricides often give excellent results against mites because of either high ovicidal activity, long residual efficacy, a minimal interference with the activity of natural enemies, or a combination of these characteristics. They are often added simultaneously to insecticides to provide mite control, or they may be applied after the insecticide treatment.

Organophosphorus Compounds (Contact)

The insecticidal developments in the field of organochlorine compounds, one quickly following the other, would in themselves have kept insecticide chemists and entomologists well occupied for years in synthesizing and testing new compounds. Never before had there been such feverish interest in the development of new pesticides; thousands of technical, semitechnical, and popular articles testify to the enormous amount of work that has been done. The seizure of German chemical patents after World War II revealed the fact that another group of synthetic organic compounds, the "phosphates," gave evidence of being equally spectacular as insecticides. In addition, they were often good acaricides and were effective against aphids, whereas the organochlorine compounds were generally ineffective against the majority of the mites and aphids.

The organic phosphates and other organophosphorus insecticides are primarily contact and stomach poisons, although their vapors may act as fumigants. Their toxic action is believed to be due usually to their metabolic conversion *in vivo* to active enzyme inhibitors called anticholinesterases (Metcalf and March, 1952). Cholinesterase or acetylcholinesterase (AChE) is an enzyme that is necessary for the sequence of events involved in transmission of nerve stimuli. Normally, it breaks down a chemical messenger, acetylcholine (ACh), which helps transmit impulses from one nerve cell to another and from nerve cells to muscle glands and other structures. The inhibition of AChE by an anti-cholinesterase, such as an organophosphorus insecticide, results in an increase in the blood level of ACh. An

TABLE 2

Federal Insecticide, Fungicide, and Rodenticide Act Categories of Toxicity*

Category	Signal words	Route of administration		
		LD_{50}‡		LC_{50}§
		Oral (mg/kg)	Dermal (mg/kg) (24-hour exposure)	Inhalation (μg/1)
I	Danger—Poison	0–50	0–200	0–2,000
II	Warning	50–500	200–2,000	2,000–20,000
III	Caution	500–5,000	2,000–20,000	....
IV	None†	5,000	20,000	....

*Adapted from Environmental Protection Agency, Pesticide Regulation Division.
†None required, based on acute toxicity. However, the nature of the product and its use pattern may require appropriate precautionary statements.
‡LD_{50} is the dose level that will kill 50% of test animals. Data were compiled from a minimum of 14 days' observation. The animals were forced to fast for the oral studies.
§LC_{50} is the air concentration that will kill 50% of test animals exposed for 1 hour. Data were compiled from a minimum of 14 days' observation. Vapor or gas may be expressed in ppm.

excessive level of ACh causes continuous stimulation of nerve cells and a disruption of all body functions controlled by the nervous system. The change in ACh level in the blood is reversible. Adverse effects to warmblooded animals result only from an excessive level of this compound, not from prolonged very low-level exposure such as might be encountered as pesticide residues in treated foodstuffs at normal harvest times. Some compounds, such as dichlorvos, may themselves bring about this enzyme inhibition; they are "direct" rather than "indirect" inhibitors. When man is exposed to excessive amounts of the indirect inhibitors, the first symptoms may not be apparent for several hours after the individual has left work, but may not always be associated with occupational poisoning.

The physiological action of acetylcholine, the physiological role of cholinesterase, the mechanism of toxic action of phosphate ester pesticides, the manifestation of phosphate ester poisoning in man, and the mechanism of protective action of atropine, are simply and effectively explained with the aid of diagrams in the University of California's *Pesticide Information and Safety Manual* (Bailey and Swift, 1968).

Most of the organophosphorus pesticides are far more toxic to man and warmblooded animals than are the organochlorine pesticides. On the other hand, malathion and phoxim are among the safest of all the synthetic pesticides. Certain other organophosphorus compounds are also relatively safe (see table 1).

The risk to humans in the use of a pesticide depends on the quantity of material required and the means of application for insect control, as well as the mammalian toxicity. The majority of organophosphorus pesticides are often used at extremely low concentrations as compared with other materials, and this lessens the danger involved in applying them. Public health hazards, such as persisting residues in foods, are minimized by the comparatively rapid decomposition of many of the organophosphorus pesticides now in general use in pest control.

Organophosphorus Compounds (Systemic)

The term "systemic insecticides" was proposed in 1947 to describe those insecticides that are absorbed by plant roots or leaves and translocated to other plant areas in sufficient quantity to be insecticidal (Martin and Shaw, 1947).

The potential significance of systemic insecticides did not become widely recognized until Gerhard Schrader, who had developed parathion and other organic phosphate insecticides, developed some toxicants that he originally referred to as "endotherapeutic" because they permeated treated plants. Ripper *et al.* (1951) emphasized the significance of the use of such systemic insecticides in relation to biological control. They pointed out that certain systemics, by not leaving a surface residue, preserved the beneficial insects and thus resulted in more prolonged control. In addition, resistant mutants of the pest population that survived treatment were destroyed by the beneficial insects, thus decreasing the possibility of the development of resistant strains.

There is another important advantage to systemics when they can move into new plant growth as the plant develops, as when they are applied to the soil and taken into the plant via the root system. They may then be important in the control of insectborne virus diseases, for only a very few insects are required to infect the plants with virus; the temporary harborage provided by pesticide-free new growth obviously cannot be tolerated.

The absorption and distribution of a systemic pesticide by a plant decreases the need for uniform application and thus permits the use of low-volume mechanical equipment under circumstances in which high-volume equipment, and possibly manual application, would otherwise be required.

Organocarbamates

Organocarbamates, often referred to simply as "carbamates," are salts or esters of carbamic acid, NH_2COOH .The first hint of the efficacy of carbamates as pesticides may have been the establishment of the chemical structure of the carbamate physostigmine or eserine, the toxic ingredient of the beans of the poisonous West African Calabar bean, *Physostigma venenosum*. It was found to be an ester of a derivative of carbamic acid (Stedman and Barger, 1925). It was not until the discovery of the role of acetylcholine in the transmission of nerve impulses and the role of cholinesterase in degrading acetylcholine that the pharmacological properties of physostigmine were understood, for like the organophosphorus pesticides, it acts by blocking cholinesterase. A number of medicinal carbamates were developed, but they were nonpesticidal, apparently because of their low lipoid solubility (Kolbezen *et al.*, 1954) and because they were ionized or ionizable and were therefore not able to penetrate the ion-impermeable sheath of the insect nerve (O'Brien and Fisher, 1958). Some pesticidal activity was shown by *N*-dimethyl carbamates, and later even greater activity by the *N*-methyl carbamates, of which carbaryl (Sevin®) was the first important example. Zectran® and propoxur are among the other important carbamates, the latter being widely used in cockroach control.

Since the principal carbamate pesticides are "direct inhibitors" of cholinesterase, the unmetabolized compounds inhibit the enzyme. Thus, symptoms of poisoning in man may appear rapidly, and generally a person applying the pesticide is forced to stop work long before a dangerous dose can be absorbed. Recovery is rapid and complete, chronic or cumulative effects are extremely improbable, and in the extensive experience noted in WHO (1967), there have been no cases of severe poisoning after occupational exposure to carbamates.

Insecticides of Plant Origin (Botanicals)

A considerable number of organic contact insecticides, such as nicotine, pyrethrum, and rotenone, are derived from plants. They have long been important materials for controlling insect pests, and are still widely used. However, since World War II they have gradually declined in importance in comparison with the synthetic organic compounds. The botanicals are almost exclusively of value because of their contact action, and are rarely used as stomach poisons. With the exception of the fumigants, they are exceptional among insecticides in having practically no "residual effect" in insect control, since they decompose quickly upon exposure to air and light. Obviously, direct contact with the insect is nearly always essential for obtaining control with botanicals.

Nicotine is an alkaloid obtained from the tobacco plant, and the insecticidal preparation "nicotine sulfate" is is a solution containing 40% of the alkaloid. Pyrethrins are extracted from flower heads of *Chrysanthemum cinerariaefolium* and *C. coccineum*, and rotenone comes from the roots of *Derris and Lonchocarpus* (Leguminosae). Certain liliaceous plants containing toxic alkaloids—principally sabadilla and hellebore root (*Schoenocaulon officinale* and *Veratrum* spp.)—are used in the preparation of a number of other botanical insecticides.

A *synergist* is a chemical that may have little or no toxicity in itself, but when combined with a toxicant may increase its effectiveness. Most of the work with synergists has been directed toward synergizing pyrethrins and carbamates. Certain compounds, when added to pyrethrins, increase toxicity to such an extent as to reduce by 10- or 15-fold the quantity required for "knockdown." Pyrethrins are rapidly broken down in an insect's body as the result of an oxidative enzymatic process, forming a nontoxic derivative. If the synergist piperonyl butoxide is added to pyrethrins, it also is oxidized in the insect, and the rate of oxidation of pyrethrins is reduced (Casida, 1969). Although usually associated with pyrethrins, synergism may occur with other insecticides, such as among mixtures of organochlorine compounds (Dicke and Paul, 1951) and among mixtures of organophosphorus (OP) insecticides, such as between tetrachlorvinphos (Gardona®) and other OP compounds.

The following 3 common botanical insecticides interact at different sites of action in insects: (1) pyrethrins act on the nerve axon; (2) nicotine, on the synapses of the central nervous system; and (3) rotenone, on the respiratory chain of tissues, including nerves and muscles. They not only differ from one another in their toxic action, but also from the organochlorine compounds and organophosphorus insecticides. They persist as active residues for a much shorter period than most synthetic insecticides, and this may account for the fact that insect resistance to them is rare. There is currently much intensive investigation of the mode of action of these botanical insecticides with the hope of synthesizing other insecticides with similar properties. Some progress has already been made in this direction, particularly with pyrethroid compounds (Yamamoto, 1970).

An advantage gained from rapid decomposition of botanical insecticides is the correspondingly reduced hazard to humans from the residues, even in the case of a material high in mammalian toxicity, such as nicotine. Rapid decomposition also allows for a more rapid reappearance of beneficial insects in a sprayed area.

The safety of a compound to mammals also depends on the rate of breakdown or detoxification in the body which, in the case of pyrethrins, is very rapid. Pyrethrins are considered to be the safest insecticides to use, and have even been used without ill effects by oral administration as anthelmintics. They are also poorly absorbed through the skin, and are of low inhalation toxicity. Nevertheless, some people do not like to apply pyrethrins. There appears to be an allergenic factor in pyrethrins causing contact dermatitis, hay fever, or an asthma-type reaction in a few sensitive persons, particularly those with a broad allergic background. Persons allergic to ragweed (*Ambrosia*) are likely to be allergic to pyrethrins.

The rapid "knockdown" characteristic of pyrethrum has resulted in the addition of its toxic ingredients to some of the formulations of synthetic insecticides of great toxicity but less rapid "knockdown" ability. In household pest control, pyrethrins are often used to "flush out" insects, particularly cockroaches, to reveal the locations in which they are hiding and thereby increase the efficacy of a treatment. Flushing out cockroaches or other insects can also be employed to obtain an estimate of relative population densities before and after treatment as a means of determining the efficacy of a control measure.

Compounds such as allethrin and resmethrin (SBP-1382®) are synthetics closely related to the natural pyrethrins, and have similar uses. Resmethrin is a synthetic pyrethroid that is highly toxic to many household and agricultural pests, but has a very low order of toxicity to mammals. SBP-1390, an isomer of resmethrin, has been shown in laboratory tests to be the most potent pyrethroid so far developed, and has a favorable degree of stability in water (Anonymous, 1972*a*). Bioallethrin and bioresmethrin are new synthetic pyrethroids that can be formulated as concentrates for dilution with oil and water, and have a long shelf life. Bioallethrin has the more rapid knockdown property, but bioresmethrin is approximately 10 times more insecticidal than bioallethrin and 17 times more so than pyrethrins. The 2 pyrethroids are compatible with each other, with pyrethrins, and with lindane, DDT, and other insecticides.

Effect of the Substrate on the Efficacy of a Pesticide

Many investigators have shown that the physical and chemical natures of the substrate—the surface on which an insecticide is deposited—can have striking effects on the pesticidal activity and residual life of a pesticide. The results of these investigations were summarized by Slominski and Gojmerac (1972). In their own investigations, these authors tested 21 different substrates, and found the residual toxicity of malathion to black carpet beetles, *Attagenus megatoma* (F.), to vary from 0 to 27 weeks, depending on the substrate. None of 4 insecticides tested were toxic longer than 1 week on concrete. Their efficiencies could be improved, however, by painting the concrete with a wall-whitening agent (talc or calcium carbonate).

Formulations

Prominent among the factors affecting the performance of insecticides is the manner in which they are formulated for commercial use. The principal proprietary insecticide formulations are the wettable powders, emulsifiable concentrates, emulsions, solutions, aerosols, dusts, granules, baits, and fertilizer–pesticide mixtures. It is beyond the scope of this chapter to discuss the highly specialized formulations, such as insecticidal paints and polishes, insecticide-impregnated fabrics and papers, seed dressings, and others. The common and widely accepted insecticides are usually prepared in all, or nearly all, of the formulations mentioned above.

Dusts

A pesticidal dust may consist of the toxic ingredient alone, without dilution, as in the case of calcium arsenate, sodium fluoride, sulfur, silica aerogels, boric acid, ground pyrethrum flowers, or tobacco leaves. A diluent (carrier, extender, filler, or dispersant) is usually added to facilitate the passage of finely divided pesticide through dusting machinery and onto the dusted surfaces, and to make possible an even distribution of the minimum effective quantity of pesticide per unit of treated area to reduce the hazard to plants or to improve the chemical or physical properties of the pesticide. A pesticidal dust may contain from 0.25 to 10% of toxicant. The diluent itself may have insecticidal or fungicidal properties, like sulfur, but generally it is inert (pyrophyllite, talc, various clays, fuller's earth, frianite, gypsum, etc.). Porous diluents tend to absorb and greatly prolong the residual life of unstable toxicants such as pyrethrins. The usual criteria considered in the selection of a diluent are its sorptivity, compatibility with the toxicant, possible catalytic properties causing decomposition, particle size, abrasiveness, specific gravity, flowability, wettability, availability, and cost.

From the standpoint of its pesticidal efficacy per unit of toxicant deposited in plant-pest control, a dust is ordinarily not so efficient as a spray formulation. However, for control of household pests, dusts have some important advantages in certain inaccessible and confined areas, such as attics, wall voids, voids under cabinets or kitchen appliances, and cracks and crevices. In such places, dusts can be more easily and thoroughly applied than liquid formulations. Dusts do not penetrate into the treated substrate as do the residues left by sprays and other liquid formulations. In inaccessible locations, inorganic dusts such as silica aerogels, boric acid, or sodium fluoride leave deposits that remain pesticidal for many years, sometimes for the life of the treated building. Liquid formulations are generally preferable in visible areas of homes and most office or commercial buildings because dusts are usually removed soon after they are applied in such areas, for the dust deposit is conspicuous and unsightly. Outdoor dusting, particularly by aircraft, has virtually ceased in many places because of excessive drift of insecticide to nontarget areas.

Wettable Powders

A wettable powder could be prepared simply by making an insecticide dust wettable. It is a dry mixture of a toxicant, an inert diluent, and a small quantity (usually 2 to 5%) of wetting agent. As with other concentrates, wettable powders are prepared to contain the maximum amounts of toxicant for purposes of economy in shipping, and concentrations of 25 or 50% of toxicant are the most common. Some toxicants, such as technical grade DDT, are waxy materials that cannot be ground by themselves but must be mixed with inert diluents in order to be ground into small particles. The usual purpose of the dust diluents and other additives in the formulation is to improve the dispersion and suspensibility of the toxicant in the spray tank, the distribution and insecticidal efficacy of its deposit on the sprayed surface, and its adhesiveness.

Wettable powders have the advantage of being relatively inexpensive. Although they may not result in so high an initial kill of insects or mites as may be obtained by the use of emulsifiable concentrates or emulsions, the long-term effectiveness of the deposits remaining on sprayed surfaces is ordinarily considerably greater. This is believed to be owing, at least in part, to the fact that the discrete particles forming the residue left by a suspension are more readily picked up by the tarsi and other parts of the bodies of insects (LaPlante, 1949). The formulation most readily picked up by insects is a dust.

Emulsifiable Concentrates

This type of concentrated formulation, designed to be mixed with water, is a single-phase liquid system containing a solvent, a pesticide (commonly at a concentration of 25%), and 1 or more surface-active agents (emulsifier and wetting agents). The latter are usually of the non-ionic class, and are consequently little affected by the hardness of the water. The solvents used usually include either a material such as toluene or xylene that evaporates rapidly after application, leaving a deposit of the pesticide only, or of a type such as petroleum oil or alkylated naphthalene that evaporates slowly, leaving an oily residue. The solution emulsifies spontaneously when added to water, and little or no agitation is required to keep it continuously emulsified after it is once uniformly dispersed throughout the spray tank.

Emulsions

If 2 liquids which differ considerably in density and are almost entirely insoluble in each other are mixed, they will separate into distinct layers or "phases." If the liquids are agitated, one will momentarily become uniformly dispersed throughout the other in the form of small globules. The liquid which forms the globules is called the *disperse* or *dispersed* phase, and the liquid through which the globules are distributed is called the *continuous* or *dispersing* phase.

The lack of stability of the mixture is believed to result from the high interfacial tension between the 2 phases. Emulsifiers lower the tension by coating each globule of the dispersed liquid with a colloidal layer. For example, if oil and water are shaken together, the oil and water globules will at first be uniformly mixed, the water being the *continuous* and the oil the *disperse* phase. However, the oil and water will soon separate ("break out") into distinct layers. If a surface-active agent such as soap is added to the mixture, the interfacial tension between water and oil will be reduced, the 2 liquids will act more as if they were the same liquid, and the rate of "layering out" will be greatly reduced; an emulsion will be formed.

Solutions

An insecticide may be dissolved in water, as in the case of nicotine sulfate or trichlorfon, or it may be dissolved in an organic solvent, as in the case of the fly sprays. Aqueous concentrate solutions are further diluted with water before use, while the solutions in organic solvent, usually deodorized kerosene (base oil) or other petroleum fractions, may be atomized without further dilution, or may be mixed with water with the aid of a suitable emulsifier. The presence of a high proportion of organic solvent usually increases the insecticidal efficacy of the formulation because penetration of the body is enhanced.

Aerosols

The word "aerosol" is a convenient term to denote a system of solid or liquid particles of ultramicroscopic size dispersed in a gas (Roark, 1942). A great variety of pesticides may be used in aerosols. If solid particles of colloidal dimension are dispersed in air, a smoke is formed, whereas a liquid dispersed in colloidal form is called a fog. Aerosols can therefore be classified either as smokes or fogs. Such formulations are considered to be "vapors" rather than "aerosols" when their particle sizes are less than 0.1 micron in diameter (Yeomans, 1948).

Granulated or Pelleted Pesticides

Granulated or pelleted formulations were developed to avoid the loss of pesticide through drift that accompanied the application of sprays and dusts. In addition, granulated materials will filter through a canopy of foliage that would prevent a large proportion of a spray or dust from reaching the ground. Little or no toxic residue is left on crops. Hazard to bees and other beneficial insects is minimized (Anderson and Atkins, 1962). Granules can be applied more rapidly and economically than liquid formulations and with less hazard to applicators. The regulated release of toxicant from granules is advantageous in some pest control situations (Mulla, 1963*a*).

Common clays, highly absorptive clays, bentonite, diatomaceous earth, tobacco by-products, corncob grit, and many other materials have been used as carriers of pesticides used in granulated formulations. The pesticide in the liquid state or in solution is added to the granular carrier. The carriers have contained 2.5 to 25% of toxicant.

The particle-size range of granulated pesticides has been stated to be 8- to 60-mesh United States sieve size (2.38 to 0.25 mm diameter) (Sferra, 1962), but there has been a tendency to raise the lower limit to 40-mesh (0.42 mm) (USDA, 1962). A typical 1-lb sample of 30/60-mesh attaclay contains about 13.5 million granules. If 1 lb of the material were evenly distributed over 1 acre, each sq ft would contain 300 granules, and 1 sq in. would have more than 2 granules. When distributed at 30 per sq in. (about 15 lb per acre), the granules would be about 0.2 in. apart (Farrar, 1953). (Metric equivalents: 1 lb, 0.45 kg; 15 lb, 7 kg; 1 sq in., 6.45 sq cm; 1 sq ft, 929 sq cm; 1 acre, 0.405 ha; 0.2 in., 5 mm.)

There is a controlled release of toxicant from the granule in its intact form and as it disintegrates. Granules tend to result in less hazard from toxic residues than do other pesticide formulations. They can be effective against insects feeding on parts of the plant aboveground or below the surface of the soil (Farrar, 1953; Walker, 1971). Granulated pesticides have been found to be highly effective in the control of mosquitoes, sand flies, wireworms, cutworms, rootworms, turf or pasture insects, and many others.

When granulated insecticides are used against larvae of mosquitoes and other aquatic insects, a quick kill of the larvae is necessary so that they will not have time to transform to pupae and adults. Aside from such obvious factors as the particle size and sorptivity of the granule, the rate and magnitude of the release of toxicant can also be determined by the type of petroleum solvent used to incorporate the toxicant into the granules. With nearly all the petroleum solvents investigated, the higher the flash point, minimum and maximum boiling points, distillation range, and evaporation time of the petroleum solvent, the slower the release of toxicant (Mulla and Axelrod, 1960).

A recent development is the coating of dichlorvos-impregnated granules of plastic with a food substance which masks the presence of the insecticide and makes the granules acceptable to rodents. Rodents carry such granules to their burrows where the dichlorvos fumes kill rodent-infesting fleas. Dichlorvos has not yet been registered for this use, pending the acquisition of the extensive data required for federal registration of any pesticide. However, preliminary tests have raised hopes that the granules will eventually provide a means by which fleas capable of transmitting sylvatic plague can be controlled in recreational areas without resort to the time-consuming dusting of individual rodent burrows and runways, which are often difficult to find (Anonymous, 1971*b*).

Pesticide Baits

Poison baits are among the oldest pesticide formulations. They obviate the necessity for wide, and sometimes indiscriminate, distribution of pesticides. The current effort to minimize the impact of pesticides on the environment is leading to renewed interest in baits.

One of the older types of pesticide bait consists of poison added to wheat bran moistened with water, and usually mixed with an attractant, such as crude molasses. The poison may be a highly toxic and soluble form of arsenic, such as white arsenic, sodium arsenite, paris green (or even the practically insoluble calcium arsenate), but may also be one of the modern synthetic organic pesticides, usually an organochlorine compound. Half of the bran may be replaced by other materials, such as hardwood sawdust or corncob grits. This type of bait is used in the control of grasshoppers, crickets, earwigs, cutworms, armyworms, the imported fire ant, certain beetles, and snails and slugs.

Another type of pesticide bait with a long history of effective usage is ant bait. For ants that eat sweets, such as the Argentine ant, a syrup weakly poisoned with sodium arsenite is commonly used. It is important that the syrup be only weakly poisoned, so that the worker ants can carry some of it back to the nests to feed to the queens and larvae before they themselves succumb to the poison. For ants that prefer fatty foods, the poison must be added to such fatty substances as ground meat or grease.

Bait sprays consisting of sugar and some poison, such as an arsenical, sodium fluosilicate, or tartar emetic, have long been used in the control of fruit flies (Trypetidae), being sparingly sprayed onto the host tree. More recently, a bait spray of malathion and yeast protein has been developed for fruit flies.

Although attractants and toxicants in water can provide effective control of house flies when sprinkled on infested surfaces, house flies consume dry granulated sugar even more readily than aqueous sugar solutions. This led to the development of dry baits consisting of granulated sugar with malathion, diazinon, or trichlorfon (Gahan, Wilson, and McDuffie, 1954). The toxicants were mixed with sugar and ground in a mortar .The resulting baits were sifted onto floors and other locations where flies congregated. Only 1 gram of insecticide was required for each application of bait in a typical dairy barn with 6,000 sq ft (667 sq yd or 558 sq m) of ceiling and wall surface, an enormous area to treat if liquid sprays were used. The inexpensiveness of the dry-sugar bait, ease of preparation, low cost of equipment, and rapidity of application resulted in a highly practical treatment for house flies.

A bait consisting of corncob grits (the carrier), soybean oil (the attractant), and mirex (the toxicant) was developed for aerial application in a formulation that would not drift away from target areas. Aerial application at the rate of 1.25 lb of bait per acre (0.568 kg per 0.405 ha)—1.67 g of actual toxicant—has been successful in the control of the imported fire ant, *Solenopsis invicta*.

Microencapsulated baits are among the more recent developments. Their ability to withstand weathering makes them superior to the baits described earlier when used outdoors during periods of heavy rainfall. The absence of a bulky carrier results in their being particularly suitable for application by aircraft. A bait consisting of microdroplets of vegetable oil with 2% mirex, encapsulated within thin gelatin or plastic coats, has been used successfully in the control of the imported fire ant (Markin and Hill, 1971).

Baits have recently come into use for the control of yellowjackets. Fish-flavored pet food has been found to be a preferred bait base. The attractancy of the bait is approximately tripled by the presence of an artificial attractant (heptyl butyrate or heptyl crotonate) in an attractant generator in the bait station. Among the 12 insecticides tested, mirex, an organochlorine compound (table 1), was the only one not exhibiting at least some degree of repellency; it actually somewhat increased the attractancy of the bait. Excellent control of the western yellowjacket, *Vespula pensylvanica,* was obtained in large test areas by using a single bait station, containing about 1 lb of bait per 5 acres (0.454 kg per 2 ha). The bait was acceptable to the yellowjackets as long as it remained. They fed it to their larvae in the underground nests, resulting in complete destruction of the colonies' occupants (Wagner and Reierson, 1969).

Much research is currently being directed toward chemical attractants to lure insects to conventional insecticides, thereby obviating the necessity of distributing insecticides over wide areas and the possibility of hazard to nontarget organisms and of environmental pollution (Knipling, 1968). Examples have been: (1) the use of protein hydrolyzate (Hagen, 1953; Steiner *et al.,* 1961), and later the male lure, methyl eugenol (Steiner *et al.,* 1965), combined with insecticides in baits for trypetid fruit flies; (2) mirex–soybean oil bait for the control of fire ants (Lofgren *et al.,* 1962; Markin and Hill, 1971); and (3) the use just noted of heptyl butyrate along with mirex–pet food bait (in bait stations) for the control of yellowjackets (Wagner and Reierson, 1969).

Dichlorvos Strips and Flea Collars

Dichlorvos can be formulated in a 20% concentration in a plastic resin (polyvinyl chloride) that delays volatilization and provides long-lasting control. It is particularly useful in reaching otherwise inaccessible areas. The strips come in 2 × 10-in. and 2 × 2-in. sizes (5 × 25-cm and 5 × 5-cm). The larger strip can be suspended in a room to provide control of such insects as flies, gnats, and mosquitoes for as long as 3 months. One strip suffices for a room of 1,000 cu ft (28 cu m). The strips should not be used where infants or ill or aged persons are confined, or in areas where food is being prepared or served (CDC, 1972). The 2-in. strip can be placed in closets, cabinets, drawers, garbage cans, and other small, enclosed places.

A room-size (1,000 cu ft or 28 cu m) test chamber was developed in which practical home conditions could be simulated, including comparable absorbing surfaces and air-renewal rates. The temperature in the chamber was maintained at 26.6 ± 1.1° C (80° F), and relative humidity at 50 ± 5%. With a dichlorvos resin strip hung from the ceiling in the usual way, a nearly 100% kill of house flies, exposed in cages, could be obtained in about 70 minutes. When the flies were exposed as free-flying insects, comparable mortality occurred in 45 minutes. Dichlorvos concentration in the air was 0.1 μg per l. A dichlorvos vapor concentration of 0.5 μg per l for 3 weeks did not cause an adverse effect on plasma cholinesterase activity in rats exposed in the chamber for that period (Batth *et al.*, 1973).

The 10-in. (25-cm) resin strip in suitable containers has controlled adult mosquitoes (*Culex pipiens quinquefasciatus*) for 10 to 15 weeks in sewer catch basins when suspended approximately 10 in. above the water line in each basin. The 10-in. strips placed in the water controlled the larvae for periods as long as 17 weeks (Maddock *et al.*, 1963; Brooks and Schoof, 1964*a, b;* Silveira *et al.*, 1964).

Dichlorvos is commonly used in a polyvinyl chloride plastic band to serve as a flea collar for dogs and cats (see chapter 9, in the section on fleas). The dog collar has a concentration of 8.37% of the toxicant and the cat collar has 4.37% (Muller, 1970).

Dust Desiccants

Dust desiccants kill arthropods by removing a portion of the very thin, superficial wax layer that normally protects them from a lethal rate of water loss. The wax can be removed by abrasion if the arthropod crawls over dust particles having a hardness number above 6 in Moh's scale, but nonabrasive dusts can be effective if they are highly sorptive. Sorptive dusts generally possess a higher degree of pesticidal efficacy than abrasive dusts against a given arthropod under the usual practical conditions of usage in pest control (Ebeling and Wagner, 1959*a*; Wagner and Ebeling, 1959; Majumder *et al.*, 1959; Majumder and Krishnamurthy, 1960; Ebeling, 1961, 1971; Melichar and Willomitzer, 1967). The requirements for an effective sorptive dust desiccant are: (1) high specific surface, (2) pores sufficiently large to admit the large wax molecules (an excellent oil absorbent, e.g., Molecular Sieves, is not necessarily a good absorbent for wax), and (3) low sorptivity for atmospheric moisture (Ebeling, 1961, 1971).

Certain silica aerogels and acid- and heat-activated clays have shown the most promise to date. Silica aerogels have long been used commercially when blown into attics and other enclosed spaces for the prevention or control of drywood termites and certain other cryptobiotic insects. Silica aerogels and acid- and heat-activated clays have shown promise when mixed with stored grain for protection against grain-infesting weevils and certain other insects (LaHue, 1970; Ebeling, 1971). Sorptive dusts have been found to be effective against more than 30 species of animal ectoparasites (Tarshis, 1961; Melichar and Willomitzer, 1967).

Control by Sterilization

The discovery by Runner (1916) and Muller (1927) that insects could be rendered sterile by X-ray treatments set the stage for subsequent developments in insect control by the release of sterile males. X rays and gamma rays eject electrons from atoms through which they pass and break chemical bonds that join atoms in molecules. In genes and chromosomes, this causes physical and chemical aberrations that affect their function. Knipling (1960*a*, 1962) discussed the theoretical basis for the superiority of inducing sterility in a large proportion of an insect population when compared with insecticidal action of similar intensity. The first demonstration that this phenomenon could be utilized in a practical way was made by entomologists of the United States Department of Agriculture, using as a test insect the screwworm *Cochliomyia hominivorax* (Coquerel), a blow fly that oviposits in wounds on warmblooded animals, particularly cattle, sheep, goats and hogs. Male and female pupae of this insect, preferably after 5 days of development, were sterilized with 5,000 r (röntgens) of X rays or gamma rays. Males required only 2,500 r. They competed for mates about equally with normal males (Bushland and Hopkins, 1951, 1953). The mated females did not mate again, and laid infertile eggs. This suggested that the release of overwhelming numbers of sterile males in a wild population of screwworms to mate with normal females might be a successful control measure. Theoretically, repeated releases of large numbers of sterile males should result in eradication of the species in the areas of the releases. This was first demonstrated in actual practice on the island of Curaçao (Baumhover *et al.,* 1955). This successful experiment was followed in 1958–59 by the release of more than 2 billion sterile flies not separated by sexes over a period of about 18 months in an area of some 70,000 sq mi (181,300 sq km) in the southeastern United States.

Incredibly, millions of pounds of horse meat and whale meat were used to rear the screwworms to the pupal stage. The irradiation was accomplished with a cobalt-60 gamma-ray source loaned by the Oak Ridge National Laboratory. The sterile flies were distributed by crop-duster airplanes. The screwworm appeared to be eradicated, with the only recurrences caused by imported infested animals (Knipling, 1960*b*).

The success of sterile-male release programs led to interest in chemosterilants—chemical compounds that cause sexual sterility in insects. Currently, the compounds of greatest interest are the radiomimetic materials—those that imitate radiation in their effects on reproductive cells (e.g., aphamide, apholate, metepa, and tepa). One way of utilizing these compounds is to apply them to laboratory-reared insects in their food or by tarsal contact. The insects can then be released as just discussed for radiosterilized insects. However, chemosterilants have an advantage over radiosterilization in that they can be used in baits to sterilize species of insects that occur in numbers that are so enormous that eradication by laboratory rearing and releasing techniques is not practicable. Another advantage is that they sterilize both males and females, increasing the frequency of matings between sterilized and unsterilized insects (Knipling, 1962; Smith *et al.,* 1964).

Pheromones, Attractants, and Repellents

A pheromone is a substance secreted by an animal that influences the behavior of other individuals of the same species. It may be produced and secreted in certain glands in the same way as a hormone, but it is released to the outside, and not into the blood. It is secreted not for its effect inside the organism, but for its effect on other individuals of the species. The word "ectohormone" was used for a time by some investigators to designate this substance, but the word is obviously incongruous and was replaced by "pheromone," from the Greek *pherein* (to carry) and *hormon* (excited, stimulated) (Karlson and Butenandt, 1959). The literature on behavioral responses to insect pheromones was reviewed by Shorey (1973).

Among insects, pheromones have been termed *olfactorily active,* such as sex attractants, repellents, aggregation scents, marking scents, and alarm or defensive secretions; or as *orally active,* such as the queen substance of bees and the active principles of sexual termites.

Sex attractants, produced and released by either sex to attract or excite the opposite sex for mating, are among the most potent physiologically active substances known. They can be detected by insects over great distances and in such minute amounts as to be almost unbelievable. For example, the antennae of the male silkworm moth, *Bombyx mori,* have 40,000 receptor cells to catch the molecules of bombykol (hexadeca-10-*trans*-12-*cis*-dien-1-ol), the female sex attractant. Electrophysiological experimentation has led to the conclusion that one of these cells is able to inform the central nervous system that a single bombykol molecule has contacted its receptive site. Males could react if only 40 of these 40,000 specialized cells received 1 strike of a bombykol molecule per second. The threshold of typical male response is reached in an airstream containing 200 molecules of bombykol per cc. The antennal pore plates of the honey bee drone contain some olfactory cells specialized as receptors for queen substance (9-oxo-dec-*trans*-2-enoic acid). The electrophysiologically determined threshold of these cells is similar to that of the male silkworm moth for bombykol (Boeckh *et al.,* 1965; Schneider, 1966).

Sex attractants, in conjunction with pesticides if necessary, have long been considered as possibly providing more selective and even more effective means of controlling insects than have been provided by pesticides alone. Another important use for pheromones is as survey tools: sensitive means of detecting accidentally introduced pests from other regions within the country or from foreign countries, even when the pests are present in very small numbers, thereby enhancing the possibility of their early eradication. For example, disparlure, the synthesized sex attractant of the female gypsy moth, has been used for detecting the spread of this insect. The gypsy moth is a serious defoliator of forest, shade, and orchard trees in the northeastern United States. Since the banning of DDT for its control, it has spread rapidly to the south and west and threatens to become a pest throughout the nation. The sex attractant has been isolated, identified, and synthesized (Bierl *et al.*, 1970). It is *cis*-7,8-epoxy-2-methyloctadecane, known by the common name just given (Bierl *et al.*, 1972). Survey traps were baited with 100 mg of disparlure, to which was added a diluent (trioctanoin) that increased its attraction and persistence (Beroza *et al.*, 1971 *a, b*). After 3 months of aging in the field, 1 mg of disparlure, formulated with trioctanoin, continued to attract male gypsy moths (Bierl *et al.*, 1972). Males responding to the disparlure got stuck on a gummy material within small cardboard traps, placed at 1.4-km intervals in lines about 1.6 km apart (Beroza and Knipling, 1972).

Pheromones may also be used as monitoring devices to determine the status of a pest in a field or orchard—such factors as the emergence of adults, the peak period of their flight, the time to expect the first egg hatch, and fluctuation in infestation levels. Such information not only improves the timing of pesticide treatments but may also improve the chances for combining pesticidal methods with alternative methods of pest control in an "integrated" control program. The use of sex attractants, such as muscalure for flies, mosquitoes, and gnats, to lure the insects to poisons or to areas of concentration where they might be killed by applying small amounts of unstable insecticide is an obvious means of utilizing these compounds in ways in which they might eliminate the environmental impact of long-lasting insecticidal residues.

Kits containing traps and pheromone caps are already available for large-acreage trials by qualified researchers and technical personnel for such insects as the redbanded leafroller moth, codling moth, oriental fruit moth, eastern and western spruce budworm moths, and gypsy moth.

As might be expected on the basis of the worldwide desire to reduce the environmental impact of pest control procedures, the literature on sex attractants has been voluminous; 425 references on the subject appeared from 1837 to 1965, and more than 400 in the following 4 years. Up to 1965, 159 insect species were listed in which sex pheromones were produced by the female and 53 species in which they were produced by the male. In his recently revised book, *Insect Sex Pheromones,* Jacobson cited approximately 1,400 references (Jacobson, 1972).

Of the 212 pheromone-producing insects listed by Jacobson *et al.* (1970), 70% were species of Lepidoptera. Shorey (1970) pointed out that the sex pheromones of 7 species of Lepidoptera, in 4 superfamilies, were all unsaturated primary alcohols or their derivatives, with significant similarity in chemical structure. This type of structure may have been derived from some primitive type. In Lepidoptera, the sex pheromone glands are usually located in the intersegmental membrane between the eighth and ninth abdominal segments, but sometimes between the seventh and eighth.

In another large insect order, Coleoptera, by the year 1970, attractant pheromones had been identified in only 8 species in 4 families (Silverstein, 1970). Nevertheless, research on attractants for bark beetles (Scolytidae) has been intensive during the past decade, and the potential for behavioral control of this very destructive group of forest insects is considered to be very good (Wood, 1970).

One of the most interesting discoveries in pheromone research is that plant substances are necessary for sex pheromone release in some insects. For example, an aldehyde (*trans*-2-hexanal) in red oak leaves stimulates the female of the polyphemus moth, *Antheraea polyphemus,* to release her sex pheromone. In the laboratory, the female moth will not mate unless this compound is present. She receives the volatile substance via her antennae and releases a sex pheromone that attracts the male. Leaves of other food plants of the polyphemus moth contain *trans*-2-hexanal, but do not elicit mating because they also release masking odors that block the activity of the aldehyde (Riddiford, 1967).

Sex pheromones *per se* have not yet been found in plants. However, a sex pheromone mimic has been isolated and identified from needle and cone distillates of certain species of spruce and fir, and has been identified as D-bornyl acetate. This compound in hexane solution, absorbed in 1 cm^2 of filter paper and held in a 2-l rearing jar with forceps, elicited the typical sexual display in male American cockroaches, *Periplaneta americana.* Also, merely by soaking various parts of flowering plants in hexane or methanol and exposing the extract on filter paper, behavior characteristic of exposure to the natural sex pheromone could be provoked among isolated groups of cockroaches. Eighteen species of plants, out of 100 species investigated, have been found to contain the sex pheromone mimics, but none of these hydrocarbons is believed to be identical with the natural cockroach pheromone, which is far more active than the mimics. However, it is hoped that the structural characterization of the active plant hydrocarbons, along with what is known about such pheromone mimics as D-bornyl acetate, may provide leads in the elucidation of the structure of the cockroach sex pheromone (Bowers and Bodenstein, 1971).

Synthetic pheromones can be used effectively only if they are competitive with natural sources of attraction. Insect pheromones or their synthetic mimics may have to be supplemented by plant extractives or their synthetic mimics. For example, for combating bark beetles in forests, synthetic pheromones must be combined with host terpenes, as in douglure, a mixture of frontalin,[2] camphene, and alpha pinene, and they are most effective when applied to a host tree. The infested lumber is then harvested, and the beetles are destroyed mechanically in the debarking process in the lumbermill. When timely harvesting of the trap tree is not possible, the tree may still be rendered unsuitable for brood survival or development by mechanical or chemical means. Thus, cacodylic acid can be injected into baited trees and adjacent trees within 15 to 25 ft (5 to 8 m), provided they are 6 in. (15 cm) or more in diameter at breast height (DBH). Cacodylic acid checks brood development without harming natural enemies of the pest species. The simplicity and efficacy of the trap-tree harvest method indicate that further investigation is justified toward adoption of the technique as a routine practice in Douglas-fir management (Vité, 1970; Knopf and Pitman, 1972).

The previously discussed high potency of disparlure in survey traps suggested its use in various ways to deplete a population of gypsy moths. Male gypsy moths have difficulty in locating females when competing synthetic attractants are present in the environment; unmated females remain alive for less than a week. Mating of wild gypsy moths was significantly suppressed by broadcasting microencapsulated disparlure, by aircraft, at rates of 2.5 to 15 g per hectare (Cameron *et al.*, 1974).

It appears that the most immediate possibility for the practical utilization of a pheromone is in connection with the control of the pink bollworm, *Pectinophora gossypiella* (Saunders), a serious pest of cotton in many parts of the world. A chemical sex attractant for the male moth is produced by the female (Flaschenträger *et al.*, 1957). Tests with the attractant, extracted with methylene chloride, demonstrated that it competed successfully with live females in luring males and that the extract in the moth trap was still highly attractive after 32 days (Ouye and Butt, 1962). Before the identification of the pheromone, a biologically active (though less potent) pheromone mimic, hexalure, was used in a substantial cotton acreage in the summer of 1973 to confuse male moths and prevent matings. This caused a reduction in larval damage to cotton bolls in comparison with the most effective commercial insecticide treatments (Shorey *et al.*, 1974). Subsequently, the natural pheromone was found to be a mixture of geometrical isomers of 7,11-hexadecadienyl acetate (Hummel *et al.*, 1973). Plans have been made to use the synthetic compound (gossyplure) for the treatment of an extensive acreage of cotton in the Coachella Valley of California during the summer of 1974. Applied at 2-week intervals in special volatilizers, only 4 grams per acre per season are now believed to be sufficient for satisfactory control (H. H. Shorey, personal communication). If this large-scale experiment proves to be as successful as the preliminary tests indicated, it will have the distinction and the considerable historic significance of marking the first successful use of a pheromone to actually control a major pest.

Many insect pheromones have already been isolated, identified, and synthesized, and an even more imposing array of compounds, particularly among attractants and repellents, has been discovered merely by accumulating or synthesizing many chemicals and testing their attractancy or repellency for insects. The greater the number and variety of chemicals tested, the better the chances of finding a compound that shows some degree of promise. Related compounds may then be synthesized. This approach, called "volume screening," has resulted in such attractants as trimedlure for the Mediterranean fruit fly, cuelure for the melon fly, methyl eugenol for the

[2] Frontalin is a powerful aggregating pheromone generated by bark beetles. It has been identified in the hindguts of Douglas-fir beetles. *Dendroctonus pseudotsugae* Hopkins (Pitman and Vité, 1970).

oriental fruit fly, heptyl butyrate for yellowjackets, amlure for the European chafer, methyl cyclohexane propionate and eugenol for the Japanese beetle, and ethyl dihydrochrysanthemumate for the rhinoceros beetle. A malathion–protein hydrolyzate spray was used for the eradication of the Mediterranean fruit fly in Florida in 1957 (Beroza, 1970).

Since a good attractant is usually highly specific for a given species, one might expect a great diversity in the chemical structures of the many compounds that have been synthesized. This diversity is indeed exemplified in the range of chemical structures developed to date, including esters, ethers, ketones, phenols, unsaturates, aromatics, carbocyclics, and heterocyclics. Although it is not known why insects respond to these synthetic compounds, it is presumed that they represent to the insect a source of food or a natural sex attractant (Beroza, 1970).

More than 20,000 chemicals have been tested as repellents by USDA scientists alone, but only about 0.2% of these were safe enough to use on skin. Currently, the most widely used repellent for application to skin is deet (*N,N*-diethyl-*m*-toluamide). It is effective against fleas, mosquitoes, chiggers, ticks, deer flies, sand flies, biting gnats, and land leeches, protecting against most species for 3 to 8 hours. As with attractants, attempts to relate chemical structure to efficacy have not been particularly fruitful. With attractants and repellents, there is currently a strong tendency to develop the potentials of natural products (Beroza, 1970; Wood *et al.*, 1970).

Possibly even more important than the relatively modest contributions toward pest control that have been made to date through utilization of attractants and repellents is the fact that the way has been opened for the utilization of the survival mechanisms of insects in order to destroy or control them.

Insect Juvenile Hormone Analogs

The approximately 3,000 species of insect pests amount to only about 0.1% of the existing insect species. Biologists have long been troubled by the tendency of currently used insecticides to kill nontarget species that may be harmless or actually beneficial to man, e.g., bees and insect predators and parasites. The tendency of some insecticides to persist in the environment and increase in concentration as they pass along the food chain, thus presenting a hazard to other organisms, including man, is another indictment against them. Also, the ability of insects to become resistant to the currently used insecticides, resulting in the need for increasingly higher doses to effect adequate control, is a very serious problem in the control of public health and agricultural pests. The utilization of hormones or their analogs or mimics appears to provide a reasonable chance of developing insecticides with which the above difficulties might be avoided.

It is characteristic of much of the current research on insecticides that one of the developments of greatest promise derives from basic investigations of hormones regulating the biological process of metamorphosis—one of the most amazing and puzzling phenomena of insect life. The hormones involved are *brain hormone* (BH), *molting hormone* (MH), and *juvenile hormone* (JH). These are the 3 internal secretions used by insects to regulate growth and metamorphosis from the immature to the adult stages.

Molting in insects is initiated by the secretion of BH by certain neurosecretory cells of the brain. The hormone enters the hemolymph by way of the *corpora cardiaca,* one of the pairs of head glands innervated by the brain. The BH stimulates the ecdysial glands to produce and release MH, which in its pure crystalline form is referred to as *ecdysone.* The structures of several ecdysones have been elucidated and the compounds have been synthesized. The MH initiates all molting processes and induces metamorphosis. It controls adult development, possibly in synergy with BH (Röller and Dahm, 1968; Berkoff, 1969).

Some MH-active compounds have been isolated and characterized which to date have all been steroids structurally related to the ecdysone molecule. They have been isolated from a wide variety of animals and plants. The wood of an Australian timber tree, *Podocarpus elatus,* was reported to be a rich source of crustecdysone, an MH-like compound of which 2 mg had been isolated from a ton of crayfish, *Jasus lalandei,* by Horn *et al.* (1966), who believed that ecdysone might be its precursor. This tree is particularly resistant to insect attack, and it is believed that the hormone was developed by natural selection to interfere with the growth processes of phytophagous insects (Galbraith and Horn, 1966).

Natural ecdysones are complex compounds, but simpler synthetic analogs have been prepared which are much more potent and will inhibit immature development, reproduction, and diapause of many insects (Robbins *et al.*, 1970).

The JH (figure 4, *A*) is synthesized by the *corpora allata,* 2 small glands near the insect brain (Wigglesworth, 1934, 1936, 1940; Engelmann, 1968). Carroll Williams, of Harvard University, who located the source of "brain hormone" in 26 neurosecretory cells in the brain of the giant silkworm moth *Platysamia cecropia* (L.) in 1946, also made an ether extract of macerated abdomens of the adult males of this species in the early 1950's, obtaining a "golden oil" containing juvenile hormones (Williams, 1956).

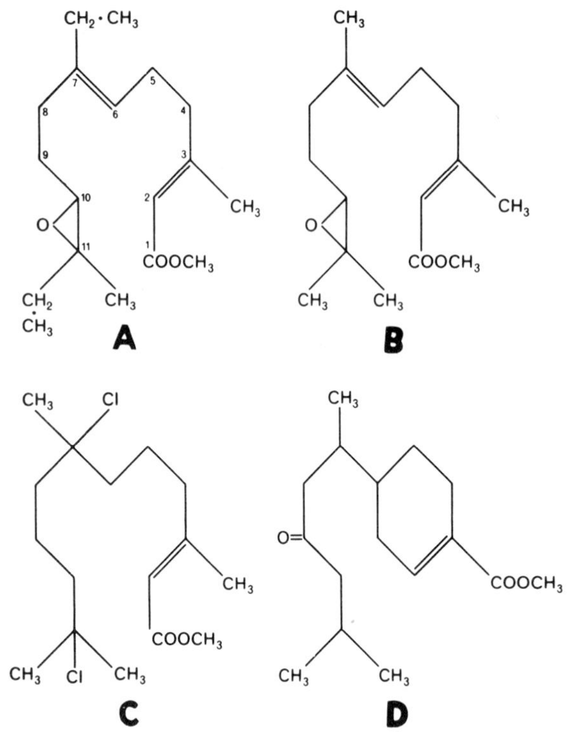

Fig. 4. Chemical structures of juvenile hormones, analogs, and mimics. *A*, juvenile hormone from the giant silkworm moth, *Platysamia cecropia*; *B*, synthetic analog of *A*; *C*, a highly potent synthetic terpenoid compound; *D*, the "paper factor" found in the balsam fir. Note similarity of the 4 compounds. (From Williams, 1967.)

The JH modifies the expression of a molt, in that it suppresses adult differentiation and favors formation of larval structures, acting in conjunction with ecdysone. Excessive JH during the ultimate nymphal or larval stage causes supernumerary molts. The pupal molt probably takes place entirely in the absence of any JH, and absence of the hormone results in a pupa-to-adult molt controlled only by ecdysone. A decline in JH in hemimetabolous insects is followed by a nymph-to-adult molt.

Excessive JH can also affect caste differentiation. The successful development and vitality of termite colonies depend on the proper ratio of the different castes and developmental stages to one another. The ratio was drastically altered by adding 0.2 ml of an acetone solution of a synthetic juvenile hormone and 2 juvenile hormone analogs to filter paper on which termites (*Reticulitermes lucifugus santonensis* Feytaud) were feeding in a petri dish. A large proportion of larvae and pseudergates developed into presoldiers and/or pseudergate–soldier intercastes, even exceeding other castes in numbers. Normally, the ratio of soldiers to other individuals is 1:100–300 (Hrdý and Křeček, 1972).

The action of JH also expresses itself in such processes as lipid metabolism, protein synthesis, and general respiratory metabolism. Some investigators believe the JH of all insects may be identical. JH activity has been found in extracts from tissues of many other invertebrates as well as from vertebrates, plants, and microorganisms (Röller and Dahm, 1968).

A compound of moderate JH activity (figure 4, *B*) was synthesized by a team of researchers headed by W. S. Bowers of the USDA in 1965 (Bowers *et al.,* 1965). Another team, headed by Herbert Röller of the University of Wisconsin, isolated and identified the cecropia hormone. They found that it had the empirical formula $C_{18}H_{36}O_2$, corresponding to a molecular weight of 284 and a structural formula that appears at first sight to be relatively simple (figure 4, *A*) but which can exist in 16 different molecular configurations, only 1 of which is the authentic hormone (Röller *et al.,* 1967). This hormone differed from the one synthesized 2 years earlier by Bowers by only 2 carbon atoms. A gram of pure hormone would be enough to result in the death of about a billion mealworms.

Bowers synthesized the compound *trans,trans*-10,11-epoxy methyl farnesenate, and speculated that this compound would be nearly identical to the juvenile hormone when it was isolated and identified. In 1972, Ken Judy and John Siddall of Zoecon Corporation detailed their isolation and identification of this very same compound from the blood of the tobacco hornworm, *Manduca sexta* (L.). This compound was also identified as the major juvenile hormone from organ cultures of adult females. It is possible that the true juvenile hormone will turn out to be 10,11-epoxy methyl farnesenate (W. S. Bowers, correspondence).

Farnesol Derivatives and Other Sesquiterpenoids

Schmialek (1961) isolated the acyclic sesquiterpenoid alcohol, farnesol, and its aldehyde, farnesal, as the active principles from certain juvenilizing extracts from the feces of *Tenebrio molitor*. These 2 terpenes did not appear to have sufficient activity to suggest that they were insect JH. However, certain derivatives of farnesol proved to be the most active among a large number of substances with JH activity that had been found or synthesized (Schmialek, 1963).

Many natural and synthetic farnesol derivatives, and sesquiterpenoids in general, were evaluated for JH activity following Schmialek's discoveries. One mixture of isomers was 1,600 times more active than farnesol. (Bowers *et al.*, 1965; Krishnakumaran and Schneiderman, 1965; Levinson, 1966). A mixture of terpenoids 1,000 times more active than cecropia extract was synthesized by Law *et al.* (1966). At 10 ppm, the mixture was able to prevent the emergence of yellowfever mosquitoes, *Aedes aegypti,* and at 10 mg per 3 × 3-cm (1.25 × 1.25-in.) woolen pad, it was lethal within 2 hours to human body lice, *Pediculus humanus humanus,* placed on the pads (Vinson and Williams, 1967). One of the 6 components (figure 4, *C*) of the mixture was synthesized by M. Romaňuk and his co-workers at the Czechoslovak Academy of Sciences (Romaňuk *et al.*, 1967).

Following the discovery that an epoxide derivative of farnesate esters showed high JH activity, related compounds were synthesized in an effort to find the most potent in this class of compounds (Bowers *et al.*, 1965). Among a group of 19 terpenoid amide derivatives were 2 which when added to starch at 20 ppm and fed to German cockroaches, *Blattella germanica,* caused most of the adults to die within a few days. Gravid females dropped their oöthecae prematurely, and few, if any, nymphs hatched from the eggs and in any case they did not survive to adulthood. Baits exposed in a test cage retained their efficacy for more than 6 months. The same 2 compounds arrested the development of larvae of *Tenebrio molitor* and *Tribolium confusum* and the eggs laid by treated adult females, but other species, such as *Plodia interpunctella,* seemed to be unaffected by either compound (Cruickshank and Palmere, 1971).

Compounds with Selective JH Activity

All the above compounds are selective in that they kill only insects, but are not selective to the extent of killing only a single species or taxonomic group. This degree of selectivity appears possible, however, from investigations stemming from an observation made by the Czechoslovakian entomologist, Karel Sláma, while working in C. M. Williams' laboratory. He found that the nymphs of the linden bug, *Pyrrhocoris apterus,* a European species, failed to mature when their cages were lined with paper toweling or any other paper made in North America. The insects continued to grow as nymphs or to molt into adult-like forms that retained many nymphal characteristics. Sláma and Williams (1965, 1966) discovered that the "paper factor" was contained in the balsam fir tree (*Abies balsamea*), and accompanied the pulp all the way to the printed page. Paper manufactured in Europe or Japan had no effect on the bugs. Bowers and his associates isolated and characterized the active material of the paper factor, which proved to be the methyl ester of an unsaturated fatty-acid derivative (figure 4, *D*), and was called *juvabione* (Bowers *et al.*, 1966).

Some investigators have hypothesized that pyrrhocorid bugs may have at one time been pests of balsam firs, and that these trees were able to develop chemical mimics of the insects' JH. However, Bowers (correspondence) informs us that juvabione occurs in the heartwood of the balsam fir, and one would not expect that bugs could obtain it in the tree sap.

Morphogenetic Effects from Synergists and Other Compounds

As explained under "Insecticides of Plant Origin" in this chapter, synergists are nontoxic compounds which, when added to such insecticides as pyrethroids and carbamates, increase their toxicity. Bowers (1968) combined several of these synergists with 10,11-epoxyfarnesenic acid methyl ester, to see if he could get synergism of the hormone. Milkweed bugs (*Oncopeltus fasciatus*) and yellow mealworms (*Tenebrio molitor*) were used as test insects. He did not find any synergism, but discovered that the synergists alone caused morphogenetic effects. Figure 5 shows the typical effects obtained in Bowers' experiments. All active compounds were derivatives of methylene dioxyphenyl, and differed principally with respect to the side chain substituents. Those lacking a side chain (safrole, piperonyl alcohol, piperonal) were ineffective. The activity of the synergists was owing to the compounds themselves, and not to a stimulation of the production of the insect's own JH. Even greater morphogenetic ac-

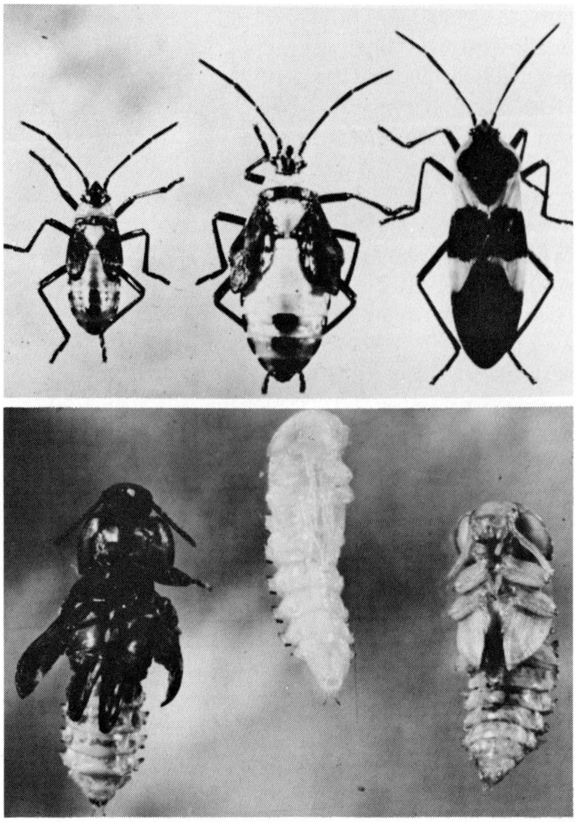

Fig. 5. Typical examples of juvenilization of insects resulting from treatment with juvenilizing compounds formed by combining synergists with 10,11-epoxyfarnesenic acid methyl ester. *Top,* last-instar nymph (*left*) of the milkweed bug transforms into a supernumerary giant nymph (*center*) instead of a normal adult (*right*); *bottom,* a normal pupa (*center*) of the yellow mealworm molts to a pupal-adult intermediate (*left*), or undergoes a second pupal molt to form a second pupa (*right*). (From Bowers, 1968.)

tivity was shown by compounds that combined chemical features of the synergists with terpenoid structures known to possess JH activity (Bowers, 1969). Seven of the new synthetic JH compounds were used as synergists for pyrethrins against house flies. None was as effective as the well-known pyrethrin synergist piperonyl butoxide, suggesting that JH compounds of this structural type should be evaluated as synergists for insecticides (Fales *et al.,* 1970). Piperonyl butoxide at 0.0001% concentration in water substantially increased the efficacy of a synthesized mixture of hormonelike substances against fourth-instar mosquito larvae (Wheeler and Thebault, 1971).

Saturated and unsaturated alcohols with 8 to 15 carbon atoms in straight or branched chains, and also the methyl ethers of several simple saturated alcohols, especially dodecyl methyl ether, caused morphogenetic effects against *Tenebrio molitor* (Bowers and Thompson, 1963). Similar effects were observed when dermestids, termites, and cockroaches were treated with diluted dodecyl alcohol and fatty-acid derivatives (Pence and Viray, 1969). Morphogenetic effects on *Tribolium* pupae have been caused by application of tricalcium phosphate to the larvae. These effects were markedly increased by the addition of most sugars, but were inhibited by trehalose (Bano and Majumder, 1965, 1968; Majumder and Bano, 1964, 1966). These results indicate the diversity of chemical structures that can duplicate the morphogenetic effects of the insect's own hormones.

Juvenile hormone controls many aspects of insect biology besides metamorphosis. Diapause (see that heading in chapter 4), which is vital to many species, can be terminated by JH (Bowers and Blickenstaff, 1966; Connin *et al.,* 1967). The JH compounds also regulate lipid metabolism, the production of sex pheromones, and the development of ovaries. The usefulness of such compounds should be increased by a greater understanding of their biosynthesis and metabolism (Bowers, 1971).

Bowers (1971) also suggested that JH chemicals might best be applied during the most sensitive periods of an insect's development or in situations in which the insect would be in continuous contact with the chemical, such as in the treatment of stored grain, of breeding sites of mosquitoes and house flies, and of soil against soil-inhabiting insects. However, the elimination of the first generation of a multiple-brooded pest might be accomplished by field applications.

Sláma (1971) suggested the following advantages in the use of juvenile hormone analogs: (1) they are not generally toxic to insects, and hopefully are also not toxic to higher animals, and they have nontoxic degradation products; (2) they have selective action against the target insect; (3) they can be disseminated into the pest population by sexual contact and cause permanent sterility of females; (4) some are enormously active, requiring only a few milligrams per metric ton (2,205 lb) of insect weight, and would thus minimize environmental contamination; and (5) compounds with short or long persistence can be prepared. Some disadvantages of juvenile hormone analogs when compared with conventional pesticides are: (1) treatment will not be followed by immediate destruction of all harmful stages of the target insect, but will only prevent its reproduction; and (2) the hormone must persist in active form until the sensitive critical periods of all developing stages and instars have passed.

Bowers (1969) has suggested the possibility of using JH analogs as fumigants against sensitive pest species. *Tenebrio* pupae confined to the vapors of certain terpenoid ethers with an exceptionally high order of morphogenetic activity molted to pupal-adult intermediates and second pupae.

The versatility of JH analogs should be increased as a result of the discovery of the great systemic action of the peptidic analog of the juvenile hormone ethyl pivaloyl-*L*-alanyl-*p*-aminobenzoate against the red cotton bug, *Dysdercus cingulatus,* feeding on treated sunflower plants. This JH analog appears to be translocated in active form throughout the plant, and is imbibed by any sucking insect. Juvenile hormone analogs of other types also appear to be active as systemic insecticides (Babu and Sláma, 1972).

So much progress has been made on hundreds of synthetic analogs and variants of the natural hormone molecule that new compounds must now be active at around one-billionth of a gram per insect, in the laboratory, in order to be serious candidates for further investigation. The compounds have been found to be harmless in toxicity tests on rats, mice, rabbits, fish, and poultry. Nevertheless, practical effectiveness, stability, and ease of synthesis of the compounds are factors that require much investigation, and thus prolong the period that an impatient public must wait for substitutes for the toxic compounds on which reliance is now placed (Grosser, 1970).

Fumigants

Fumigants may be applied in enclosed spaces, such as fumigation chambers, with or without vacuum; in buildings and greenhouses; under gastight tarpaulins in the fumigation of trees or other plants, grain, fruits, and food products; in burrows of animal pests; or for soil fumigation, especially with soil-injection equipment.

Fumigants were among the first insecticides used by man, the fumes of burning sulfur, for example, having been used since periods of early historical record. Many substances may be used as fumigants, and they may be applied in the form of a solid, liquid, gas, or reactive solid (Phostoxin), but they must have the ability to volatilize readily and to form a toxic concentration of vapor in an enclosed space. The lower the toxicity of a compound, the more volatile it must be to have practical value as a fumigant. Among organic compounds, the principal compounds used primarily as fumigants have been those of simple structure and therefore with the required degree of volatility. In a homologous series of organic compounds, toxicity usually increases up to a point as the molecular weight increases, but volatility decreases and eventually becomes so low as to render the compound ineffective as a fumigant.

By far the greatest use for fumigants is in the treatment of enclosed spaces, such as homes, mills, packing plants, warehouses, ships, freight cars, and greenhouses. Tarpaulins or canvas tents may be pulled over the material or area to be fumigated, and the gas may be liberated in the resulting enclosure. The fumigants most commonly used in the fumigation of large enclosed or confined spaces are HCN, methyl bromide, sulfuryl fluoride, chloropicrin, ethylene oxide, carbon disulfide, sulfur dioxide, and Phostoxin. Naphthalene and paradichlorobenzene (PDB) are commonly used fumigants for the protection of fabrics from insect damage in small, confined spaces, or for protection of insect collections from dermestid beetles, but they are being replaced by more effective volatile insecticides, such as dichlorvos and lindane.

For anyone wishing to read about fumigation more extensively than space permits in this chapter, a manual on fumigation for the control of insects aboveground, i.e., excluding soil fumigation, prepared by Monro (1969), is suggested. Basic principles, some of the more general applications, and precautions and protective devices are discussed in this very useful manual.

Advantages and Disadvantages of Fumigants

Fumigants are "wide-spectrum" pesticides, killing all species of arthropods and rodents that are likely to be found in a building. Fumigants penetrate to many areas of a building not reached by sprays and dusts, even penetrating to the burrows of wood-infesting insects, and pests are rapidly eliminated. A gas does not leave unsightly, odorous, or hazardous residues. Fumigation also has certain disadvantages. For the fumigation of buildings, or even for the proper and safe use of fumatoriums, highly skilled, experienced, and licensed fumigators are required. Fumigated buildings must be sealed and their occupants must leave, sometimes for 24 hours or longer. The cost of fumigation is generally higher than for other pest control procedures. These disadvantages do not pertain to the use of fumigants in small containers, such as the use of the vapors of cleaning solvent or dichlorvos resin strips in a closed garbage can to fumigate small, infested materials. Fumigants leave no protective residue, and reinfestation is possible immediately after treatment.

General Precautions

All fumigants that have been used successfully against insects have also been found to be poisonous to other coldblooded animals, to warm-

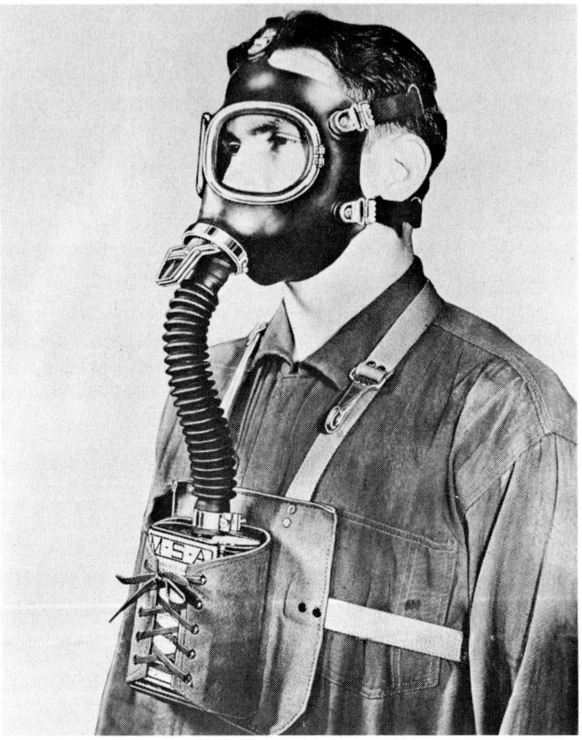

Fig. 6. Full-face gas mask with canister. (Courtesy Mine Safety Appliances Company.)

blooded animals, including man, and to plants. A concentration of 0.03 vol./% (300 ppm) of hydrocyanic acid gas (hydrogen cyanide, HCN) is sufficient to kill human beings. In the fumigation of buildings, gas masks (figure 6) with the correct filter cartridge for the particular gas being used are usually employed in applying the fumigant and when entering the building afterward to open doors and windows, if these cannot be opened from the outside.

Toxic compounds, such as hydrocyanic acid, methyl bromide, and ethylene oxide, that are gaseous at or immediately above ordinary fumigation temperatures, must be confined in strong metal cylinders. The dangers involved in storing, transporting, and releasing such fumigants make fumigation operations involving their use especially hazardous to all but specially trained and experienced persons.

Gases may be absorbed through the human skin. For example, it is not safe to work in an atmosphere containing 2 vol./% of hydrocyanic acid gas, even though an effective gas mask is used, for sufficient gas can be absorbed through the skin at this concentration to cause poisoning. Carbon disulfide, hydrogen sulfide, carbon tetrachloride, ethylene dichloride, formaldehyde, and nicotine vapor are among other gases that may also be absorbed through the skin.

Fire hazard is another factor that calls for the utmost care in the use of fumigants. Hydrocyanic acid gas, carbon disulfide, and ethylene oxide are especially hazardous because of their flammability. Carbon dioxide, carbon tetrachloride, or methyl bromide may be mixed in specified proportions with fumigants to eliminate the danger of explosion. Different mixture ratios by volume of these gases are required for the various fumigants.

Fumigants may also be injurious to plants, and great care must be exercised in order to apply a fumigant so that it may be used in concentrations and under conditions of temperature, humidity, and period of exposure such as to be effective against insects without at the same time being injurious to plants. As in the case of some of the sprays and dusts, the "margin of safety" between insecticidal and phytocidal effect may be rather small, and requires on the part of the fumigator a thorough knowledge of his pest control operation and of the risks and responsibilities that are involved. In the fumigation of living plants with fumigants that dissolve in water, such as HCN, the danger of injury is increased by the accumulation of water on the plants. For this reason, it is important to avoid a falling temperature during the exposure, for moisture may condense on the foliage, absorb some of the gas, and result in injury to the plant. Fortunately, dry food substances are not significantly contaminated by the common fumigants, but there is danger of poisoning foods that contain considerable quantities of water.

Hydrocyanic Acid Gas

Hydrocyanic acid or hydrogen cyanide (HCN) is a colorless, very volatile liquid, with a specific gravity of 0.697 g/cc at 18° C (65° F), which boils at 26° C (78.8° F) to form a gas that is slightly lighter than air (sp. gr. 0.9483 g/l). It has a slight odor of peach pits or bitter almonds, but not everyone can detect it. HCN is not considered to be dangerously flammable or explosive unless used at about 4 lb per 1,000 cu ft (1.82 kg per 28 cu m), a volume concentration of 5.6%. Nevertheless, all open flames should be extinguished in buildings that are being treated. Much of the type of fumigation done with HCN in former years is now done with safer gases; in fact, its use for fumigation is not permitted in some cities.

HCN, being highly poisonous to man and other warmblooded animals, should be applied only by those who are thoroughly trained in the procedures of safe application and in first aid.

Fifty mg of HCN is considered to be a lethal dose for human beings. No more than 50 to 60 ppm can be inhaled for an hour without serious consequences. A concentration of 300 ppm rapidly results in death.

Hydrocyanic acid gas may be generated "on the job" in commercial fumigation of dwellings and greenhouses by using the "pot method," in which 1-oz (28-g) sodium cyanide "eggs" are dropped into a sulfuric acid solution in an earthenware crock. The following proportions of ingredients are used:

Commercial (60° Baumé) H_2SO_4	1.5 pt (0.71 l)
Water	3.0 pt (1.42 l)
Sodium cyanide	1.0 lb (0.454 kg)

For large spaces, such as mills and warehouses, pressurized liquid HCN may be piped to various parts of the building.

Portable cyanide generators in which hot HCN gas can be produced are available. The gas can be pumped into a building or enclosure, and the operator does not have to enter the building until it is ready be to be ventilated. The generators are made in sizes that make possible the fumigation of as much as 10,000 or 25,000 cu ft (283 or 708 cu m) of space with a single charge. Other models can be conveniently used to fumigate rodents in their burrows.

When calcium cyanide dust is used, as in greenhouse fumigation, HCN is generated by the interaction of atmospheric moisture and the calcium cyanide. When the dust is spread in thin layers, the gas is evolved even if the relative humidity is below 25%, and the rate of evolution increases with increasing relative humidity.

A convenient method of fumigating with HCN is with waferlike discoids, consisting of liquid HCN absorbed into some inert material. The impregnated discoids are sealed in cans. Each discoid contains about 0.5 fl oz (1 tbsp or 15 cc) of liquid HCN, which evaporates on exposure to air. The cans may be precooled with dry ice (solidified carbon dioxide) before opening, so as to retard the generation of gas, thus providing greater safety. The discoids should be placed on several thicknesses of paper or a shallow pan to avoid injury to the floor.

Methyl Bromide

Methyl bromide (CH_3Br) is a fumigant that boils at 4.5° C (40° F) and is a gas at ordinary room temperatures. The colorless liquid has a specific gravity of 1.731 g/cc at zero C (32° F) (water = 1.000 at 4° C or 39° F), and the gas has a specific gravity of 3.27 g/l (air = 1). The gas is nonflammable and has only a slight chloroform-like odor, even when in toxic concentrations.

Methyl bromide is considered to be the most universal fumigant yet developed. Its use as a fumigant apparently began in France in 1932, when it was added to ethylene oxide to eliminate the fire hazard of the latter, but it was found to be even more effective than the ethylene oxide. It was later discovered to be more toxic to the granary weevil than HCN. Where it is sufficiently toxic to insects, it has certain conspicuous advantages over HCN, such as having greater penetrating ability, being generally less harmful to plants, and being nonflammable.

It should be borne in mind that the greater penetrativity of methyl bromide results in its escaping readily from imperfect enclosures. This is a disadvantage from the safety standpoint, for unlike HCN and some other gases, methyl bromide is not readily detected by its odor. It is less toxic than HCN, but it is a cumulative poison, whereas recovery from a sublethal dose of HCN is complete.

A device known as the halide leak detector or halide lamp is available for approximate determination of methyl bromide concentrations in air. These lamps are readily available, usually from refrigeration supply dealers, for they are also used for detecting leaks of freon-type refrigerants. In either case, a clean piece of copper burns with a green to blue flame if organic halide vapor is present. The color changes from green to greenish blue or blue as the concentration of halide gas increases (Monro, 1969).

Methyl bromide is one of the least injurious to plant tissues of all the important fumigants, and does not affect the germination of seeds. Green vegetables and other foliaged plants are fumigated regularly with methyl bromide as a part of standardized quarantine practice. With a sufficiently high dosage, of course, any fumigant will kill plant life. Methyl bromide is sometimes used to kill lawn grass, seeds, nematodes, soil insects, etc., when preparing to plant a new lawn. One lb per 100 sq ft (0.45 kg per 9 sq m) of lawn is applied under a tarpaulin.

One of the characteristics of methyl bromide is that it often kills insects very slowly. For example, in one recorded instance, potato tuber moth larvae were moribund 24 hours after being fumigated, but some did not die until 5 or 6 days had elapsed (Mackie and Carter, 1937). Methyl bromide is toxic to insects in all stages of development, and thus has an obvious advantage over some other fumigants, such as sulfuryl fluoride, that are characteristically ineffective against the eggs and/or pupae.

Methyl bromide, generally at 2 lb per 1,000 cu ft (0.91 kg per 28 cu m), is extensively used in the fumigation of homes and other buildings for the control of drywood termites. For this and many other fumigation procedures, methyl bromide is heated to increase the speed of vaporization and penetration. Special equipment is available for this purpose (Steinweden, 1945).

The most serious criticism against this material for home fumigation is that it can cause persistent mercaptan odors in sulfur-containing materials, such as rubber, leather, and animal hair, and in iodized salt. Despite efforts to remove these materials before fumigation, some are occasionally overlooked; in any case, the removal of all such substances, e.g., foam-rubber padding under rugs and carpets, may be difficult and time-consuming.

Gastight tarpaulins are sometimes placed over piles of grain, boxes, fruit, nursery stock, or lumber and wooden articles infested with termites or beetles, and the gas is released under the tarpaulins. Much greater care must be exercised to prevent leakage from under the tarpaulin than in the case of fumigation with HCN gas, which is not only much lighter but less penetrating than methyl bromide. "Sand snakes," made of canvas tubing 5 or 6 ft long and 3 to 4 in. in diameter (1.5 or 2 m; 7.5 to 10 cm), are filled with sand, and are carefully arranged to keep the canvas tightly pressed to the floor around the pile of fumigated material. Since there is no way of controlling temperature, fumigation of fruits and plants under a tarpaulin should not be done in sunlight, for the temperature may rise too high.

Temperatures of 80 to 85° F (27 to 29° C) are recommended for fumigation with methyl bromide, although by increasing the dosage, fumigation may often be successfully done at lower temperatures. Humidity has little effect on its insecticidal effectiveness, but enough humidity should be maintained to prevent desiccation of plants. Exposures are seldom for less than 90 minutes, since methyl bromide acts slowly, and its toxicity is increased with an increased exposure period. The usual timing for the fumigation of foliaged nursery stock, greenhouse plants, fruits, and vegetables is 2 hours. Dormant deciduous nursery stock and dormant bulbs are treated for 3 or 4 hours. Low dosages usually 1 lb per 1,000 cu ft (0.45 kg per 28 cu m) for 24 hours, are usually used for grains, beans, peas, cereals, raisins, and dried fruits in general. Air should be circulated during methyl bromide fumigation to prevent stratification of the gas. An initial period of 10 to 15 minutes of circulation in a chamber is generally sufficient. For exposures of 24 hours or longer, such as in warehouses, fans are not used as a rule, since the exceptionally long period of exposure offsets the lack of circulation (Steinweden, 1945).

Sulfuryl Fluoride (Vikane®)

Sulfuryl fluoride (SO_2F_2) is a nonflammable, colorless, odorless compound that boils at a little over −55° C (−67° F) at 760 mm Hg pressure. The specific gravity of the liquid is 1.36 g/cc at 21° C (70° F), and the specific gravity of the gas is 3.72 g/l at zero C (32° F). It is highly stable, and is a more penetrating gas than methyl bromide. Even though it is more expensive than methyl bromide, under certain circumstances sulfuryl fluoride is preferred for house fumigation for drywood termites. Sulfuryl fluoride gives a good kill of drywood termites without causing disagreeable odors. It is generally used at a dosage of 1 lb per 1,000 cu ft of space (0.45 kg per 28 cu m).

Bess and Ota (1960), using the powderpost (drywood) termite (*Cryptotermes brevis*) as the test insect, found the penetration of sulfuryl fluoride into wood to be so superior to that of methyl bromide that they suggested the possibility of using higher than usual dosages of the gas to reduce the time required for fumigation, which would be a great advantage. Occupants are now usually required to stay out of a fumigated building for 24 to 48 hours.

Phostoxin®

Phostoxin is a fumigant that can be applied as tablets or pellets. In either case, the active ingredient is aluminum phosphide 55%, and the inert ingredients are ammonium carbamate 41% and edible paraffin 4%. When the tablets are taken out of their container, the ammonium carbamate reacts with moisture in the atmosphere to release ammonia and carbon dioxide; these gases prevent the self-ignition of the highly flammable phosphine gas (PH_3) released when the atmospheric moisture contacts the active ingredient, aluminum phosphide. The only remaining product is a harmless dust, aluminum hydroxide. Each tablet weighs about 3 grams and releases 1 gram of phosphine; pellets weigh 0.6 gram and release 0.2 gram of phosphine. Phostoxin can be applied directly to raw agricultural products, but when treating finished food products, the fumigant must be applied in such a way that the tablets, pellets, or dust remaining after the phosphine is generated do not come into direct contact with the food. For example, the tablets or pellets can be placed in a paper envelope (Rutledge, 1967).

Phosphine is a colorless gas with a strong odor resembling decaying fish. It is toxic to all forms of animal life, but the strong odor can be detected long before a toxic concentration of gas evolves. The pellets do not release appreciable phosphine for 1 to 2 hours, and the tablets for 3 to 4 hours, after they have been taken from their container. This delay in the generation of phosphine has generally made possible the application of the product without the use of a gas mask. The Auer Tester can be used to detect phosphine and determine whether or not an area is safe to enter after a fumigation (Rutledge, 1967).

Phostoxin can be used for the fumigation of enclosed spaces up to the size of a large warehouse (Rutledge, 1967). It can be used to fumigate processed foods in transit in boxcars or covered hopper cars. The gas can penetrate dense material, such as flour or wheat, and has no adverse effect on seed germination or on the baking quality of wheat. It was found to be effective as a fumigant against 6 species of granary beetles (Lindgren *et al.,* 1958), and for the fumigation of large stocks of natural raisins in sweat boxes (Nelson, 1970).

In another investigation, 2 large bins of wheat containing the sawtoothed grain beetle, *Oryzaephilus surinamensis,* and 9 species of mites were fumigated with Phostoxin. Six days after fumigation, the only species surviving was one of the mites, *Aeroglyphus robustus* (Banks). Three other mites, *Acarus siro, Glycyphagus destructor,* and *Spinibdella bifurcata,* reappeared within 1 to 3 months in both bins of wheat, possibly because of the elimination of a natural enemy, the predaceous mite *Cheyletus eruditus* (Sinha *et al.,* 1967).

When tested against 2 important species of mites attacking stored food products, ethylene dibromide gas was found to be 17 times more toxic to *Tyroglyphus putrescentiae,* and 6 times more toxic to *Caloglyphus krameri,* than phosphine (Muthu *et al.,* 1970).

All metals, especially copper or copper-containing alloys, are severely corroded by phosphine. High humidity, high salt content of the air, as in areas near the ocean, and the ammonia given off during the decomposition of the proprietary product are all factors that increase the chemical reaction, that is manifested by discoloration and formation of acids. Any copper-containing equipment, particularly electrical apparatus, may be severely damaged. A heavy coating of paraffin wax will protect copper connections and wiring from corrosion.

The respirator suggested for protection against phosphine vapors requires the type of canister containing specially activated charcoal for protection against acid gases. The standard-sized canister of this type will give protection against as much as 0.5% phosphine by volume in air, in contrast to the 2% often specified for other gases (Monro, 1969).

Vacuum Fumigation

The rate of penetration of gas into the material to be fumigated is often an important factor, and is of course related to the sorptive capacity of the treated material. The rate of penetration of the fumigant is greatly increased if fumigation is done in an airtight chamber (figure 7) from which the air has been exhausted. This type of treatment is called *vacuum fumigation.* The fumigant may be removed quite completely from the fumigated material by drawing a second vacuum after the proper period of exposure to the gas and then breaking this vacuum with air. This process is known as "air-washing."

Vacuum fumigation can be used to advantage when speed is an important factor, for usually the period required for effective kill of insects protected by plant materials (e.g., in baled cot-

Fig. 7. A vacuum fumigator containing gardenia plants. (Courtesy California State Department of Agriculture.)

ton) is reduced to a small fraction of that required under natural atmospheric conditions. However, the amount of fumigant needed is greatly increased when much plant material is present, because of greater sorption of the gas by the fumigated material. Another advantage of vacuum fumigation is that workmen are not exposed to the gas.

It appeared at one time that vacuum fumigation would be far more important than we find it to be today. Its de-emphasis has been brought about largely by methyl bromide. The great penetrating ability of this fumigant has obviated the necessity for the use of vacuum in the majority of cases. However, vacuum fumigation with HCN is sometimes employed for the treatment of nursery stock in observance of quarantine regulations. Some pest control operators have fumigating chambers for atmospheric fumigation of furniture and other household articles infested by drywood termites or powderpost and other species of beetles. Methyl bromide is generally used as the fumigant.

A list of approved fumigation chambers in the various counties of California for vacuum or atmospheric fumigation can be obtained from the State Department of Agriculture.

RESISTANCE

The World Health Organization's Expert Committee on Insecticides (WHO, 1957) defined resistance as follows:

> Resistance to insecticides is the development of the ability in a strain of insects to tolerate doses of toxicants which would prove lethal to the majority of individuals in a normal population of the same species. The term "behavioristic resistance" describes the development of the ability to avoid a dose which would prove lethal.

Resistance may appear either through genetic mutations which may have occurred in the past or because of a mixture of resistant and susceptible individuals already present in the original population. The susceptible individuals are eliminated by repeated insecticide treatments. The resistant individuals may be extremely rare in the original wild population, but they are more likely to survive insecticide treatments, and the survivors tend to have progeny that have the genetic characteristics for resistance that were possessed by their parents. The biochemical, physiological, and genetic aspects of resistance were reviewed by Crow (1957), Perry (1964), Oppenoorth (1965), Terriere (1968), Tsukamoto (1969), Plapp (1970), Brown (1971), Georghiou (1969, 1971*b*), and Brown and Pal (1971).

Insects possessing *physiological,* as contrasted with *behavioral,* resistance survive because of one or more physiological or biochemical mechanisms in which they are more efficient than susceptible insects of the same species. The resistant insect may be able to more effectively slow down the penetration of an insecticide into its body or more effectively detoxify it, by one or more pathways, or eliminate it, or it may possess all three mechanisms and possibly others not yet identified. Any one of these mechanisms may by itself be responsible for only a relatively low level of resistance, and one that could be tolerated. However, the interactions between the mechanisms usually result in resistance that is far beyond any level that would be expected from their apparent potential as isolated defense mechanisms (Georghiou, 1971*a*, *b*).

The tendency of insect species to become resistant in varying degrees to an insecticide after repeated treatments has posed one of the most difficult of all the problems of economic entomology. Only a few years after the introduction of DDT and subsequently developed synthetic insecticides into public health programs, the main problem was the development of resistance to them by many of the insects they formerly controlled, including such important species as mosquitoes, house flies, lice, and fleas. The only genera of insects of public health importance for which resistance had not been reported were certain Diptera (*Simulium, Phlebotomus,* and *Glossina*) (Brown, 1958). However, by 1971 3 species of *Simulium* had become resistant to DDT. There are now more than 225 species of insects with resistance to at least some insecticides. Some of the more important pests have already been treated with representatives of all the major groups of pesticides, and are resistant to almost all the available control chemicals.

In insects with a brief life cycle (many generations per year), such as house flies, resistance was noticed within a few years after the beginning of treatments with DDT. It required somewhat longer for the German cockroach, *Blattella germanica,* to become resistant to chlordane; it has 3 or 4 generations per year, the number of generations house flies could have in 2 weeks in summer. The American cockroach, *Periplaneta americana,* which has only about 1 generation per year, has not yet become resistant to chlordane and other organochlorine insecticides. Unfortunately, only 4 species of natural enemies are recorded in the literature to have developed some degree of resistance to pesticides—all predatory mites in the family Phytoseiidae (Georghiou, 1972).

The characteristically prolonged residual effectiveness of the organochlorine insecticides is believed to be a factor that has caused the resistance problem to become more acute since this group of insecticides first came into use with the advent of DDT following World War II. Even though an insecticide when originally deposited may result in the death of every insect coming in contact with it, the gradual reduction of its effectiveness because of weathering or decomposition results in a progressively larger percentage of insects receiving a nonlethal dose. The survivors may be insects that migrated into the area after it was treated or individuals that developed from life stages that were immune or inaccessible at the time of treatment. On the other hand, insecticides with little or no residual effectiveness, used at concentrations that result in practically 100% kill of the exposed insects, provide little chance for survival of even the more resistant species. When using insecticides with residues of gradually decreasing effectiveness, survival of exposed insects will occur, and some natural selection of resistant individuals is inevitable.

Resistance of a pest to a particular pesticide in a group (e.g., chlordane in the cyclodiene compounds) generally results in resistance to other pesticides in that group, a phenomenon known as "cross-resistance." Also, when the first insecticide is replaced by another, the insect becomes resistant to the second while its resistance to the first often continues to increase. If a third insecticide is used, the insect becomes resistant to the latter while its resistance to the first 2 insecticides may also continue to increase; this process continues apparently *ad infinitum.* This phenomenon has been called "co-selection" (Georghiou, 1971*a, b*). Co-selection prevents a successful return to an insecticide used earlier and for which resistance might otherwise have declined during the years it was not used. Continued selection of resistant individuals or strains with one class of chemicals may result in resistance to compounds in other chemical classes. For example, a normal strain of German cockroaches, when selected for diazinon resistance, developed resistance to topical applications of diazinon (26 ×) and malathion (27 ×), and became cross-resistant to 3 other major classes of organic insecticides represented by DDT (> 40 ×), propoxur (18 ×), and pyrethrins (23 ×). When the selected strains were injected with insecticide, they manifested far less resistance in the case of diazinon (6 ×) and DDT (2 ×) than when the insecticide was topically applied, indicating that reduced penetration may have been in large measure a resistance mechanism. However, injection of insecticide resulted in no decreased resistance, when compared with topically applied insecticide, in the case of propoxur (14 ×) or pyrethrins (34 ×) (Collins, 1973). Newly synthesized compounds are not necessarily the solutions to resistance problems unless tests have demonstrated that "cross-resistance" to the anticipated category of these new compounds is not present.

Just because a pest is resistant to a given insecticide in one area does not mean that it is necessarily resistant in other areas. Therefore, when there is evidence of incomplete control, such factors as the pest's life history or habits and the control techniques used by the pest control operator should be checked as possible factors accounting for poor control before genetic resistance is seriously suspected. Likewise, the nature of the surface to which a pesticide is applied can have a great effect on its pesticidal activity and period of residual efficacy (Slominski and Gojmerac, 1972).

There may be a point beyond which selection for increased resistance is no longer possible, but the dose of insecticide required at that point may be hundreds of times greater than that required for normal (susceptible) strains. The upper limits beyond which resistance can no longer be augmented by further selection are usually much higher for the organochlorine than for the organophosphorus compounds. However, even a 10-fold increase in resistance of 1 or 2 important pests might result in a pesticide no longer being useful because the increased dosage required might result in excessive cost, injury to plants, and destruction of natural enemies and bees, or because the pesticide residue remaining after treatment might exceed the legal tolerance.

PRECAUTIONS IN USING PESTICIDES

Many pesticides used in the home or garden can injure or kill people or pets if not handled and applied properly. Years of experience have shown that children under 5 years of age are the most likely to be victims of pesticide poisoning. Pesticides should be kept in the original or a properly labeled container out of reach of children, pets, or people who cannot read the language in which the labels are written. Records show that most poisonings, whether to adults, children, or pets, are caused by carelessness and can be prevented.

Some people are allergic to many common substances, including pesticides. Even DDT, an insecticide with a remarkable record of safety to humans, has occasionally been reported to be severely allergenic (Vanat and Vanat, 1971). An example of allergy to organophosphorus insecticides is the dermatitis and even systemic illness experienced by some dogs and cats wearing flea collars made of polyvinyl chloride plastic impregnated with dichlorvos. Some people handling these collars (or dichlorvos resin strips) may also develop dermatitis (Anonymous, 1970*h*; Muller, 1970). (See chapter 9, in the section on fleas.)

Although pesticide pollution of rivers, lakes, and oceans is commonly associated with agricultural usage, there is much reason to believe that careless use or disposal of pesticides in urban areas is a major source of water pollution (Burnett, 1971). Much of this probably results from disposal by homeowners of unused or unwanted pesticides via the city sewer system. The draining of unused spray water from a spray tank and allowing it to run into the street water-drainage system is an example of this type of pollution. It is therefore encouraging to see on pesticide labels such admonitions as "Do not contaminate water by cleaning of equipment or disposal of wastes."

Bailey and Swift (1968) prepared a *Pesticide Information and Safety Manual* containing much timely information on pesticides and how to use them safely and effectively. The oral and dermal toxicities of practically all pesticides are given. Antidotes for external and internal use by nonmedical people as well as information for physicians are contained in this valuable manual, which is available from the University of California Agricultural Extension Service, Berkeley, California 94720. Another useful manual, prepared by the California State Department of Public Health in cooperation with the State Department of Agriculture and the University of California Agricultural Extension Service, is titled *Tentative Guidelines for the Safe Handling and Disposal of Used Pesticide Containers in California* (Rogers and Cornelius, 1970). This manual deals principally with the problems of the commercial pest control operator. Some advice to the general public concerning the safe handling and disposal of pesticide containers is also given.

The USDA Handbook No. 321, titled *Safe Use of Agricultural and Household Pesticides* (USDA, 1967), is quoted at some length in the paragraphs that follow. Among the suggested precautions are some general recommendations and others pertaining to the home and garden use of pesticides. These are given verbatim.

General Precautions

1. *Read the label.* This is the first rule of safety in using any pesticide—*read the label and follow the directions and precautions printed on it.*

2. Store pesticides in closed, well-labeled containers where children or pets cannot reach them. Do not store them under the sink, in the pantry, or in the medicine cabinet. Do not store them near food of any kind.

3. Store application equipment as you do pesticides —out of the reach of children or pets. Label each duster or sprayer with the name of its pesticide contents.

4. Do not save or re-use empty pesticide containers. Dispose of containers promptly, as follows: Do not puncture pressurized containers. Do not burn them; they may explode and cause bodily injury or start fires. Wrap the empty containers, including bags and cartons, in heavy layers of newspaper or other paper. If you have trash-collection service, deposit them with the trash for removal. If you do not have this service, crush metal containers and break glass containers; take them to a public dump, or bury them at least 18

inches [about one-half meter] deep at a place where they will not contaminate water sources.[3]

5. Do not apply more pesticide than the label recommends. Overdosage is wasteful, and may be dangerous.

6. If you use poisoned bait to control rats, mice, or other pests, either indoors or outdoors, place it where children or pets cannot find it.

7. When opening a liquid pesticide container, keep your face away from, and to one side of, the cap or lid.

8. Mix or prepare dusts or sprays outdoors or in a well-ventilated room.

9. In handling *any* pesticide, avoid contact with the skin. Do not get pesticide near your mouth, eyes, or nose.

10. If pesticide gets in your eyes, flush the eyes with water for 5 minutes; get medical attention.

11. Never smoke, eat, or drink while handling a pesticide. After finishing the work, wash exposed skin surfaces with soap and water.

12. If you spill pesticide on your clothing, launder the clothing before wearing it again.

13. If you become ill during or shortly after using a pesticide, call a physician immediately.

14. If you accidentally swallow some pesticide, call a physician at once. From the container label, read to him the names of the active chemical ingredients; follow his instructions for first-aid treatment.

Home Precautions

In addition to the preceding *general precautions*, housekeepers should observe the following precautions to insure the proper and safe use of pesticides in the home.

1. Apply each form of pesticide for its proper purpose.

Surface Sprays. These sprays have coarse particles that dampen or wet the surfaces to which they are applied. When the spray dries, a thin deposit of pesticide remains. The deposit usually is invisible. Buy surface sprays in pressurized containers, or buy a liquid insecticide and apply it with a household hand sprayer that produces a continuous, coarse spray.

Space Sprays and Aerosols. These 2 forms of sprays are designed for application into the air. They are effective against flying insects, and also may be used to penetrate the hiding places of other insects and drive them into the open where they may be killed with a surface spray or dust. The particles or droplets of a space spray are finer than those of a surface spray, and float in the air for a time. The particles of an aerosol are finer than those of a space spray, and float in the air for a longer time. Do not apply a coarse surface spray into the air of a room; it may be hazardous if so applied. Use a space spray or aerosol that is designed and labeled for this purpose.

Dusts. Pesticidal dusts usually contain the same active ingredients as sprays. They are used for surface applications, and may be blown by a household hand duster into cracks, corners, and other places hard to reach with sprays.

Poisoned Baits. As the name implies, poisoned baits are materials on which a pest will feed, to which poison has been added. They are used to control rodents such as rats and mice, and other pests. Poisoned baits are applied by placing them where they will be encountered by the pest to be controlled.

2. Some sprays sold in pressurized containers may be labeled for both surface and space applications. If you use one of these products for spraying in the kitchen or pantry, first place cooking and eating utensils and food where they will not be contaminated by falling particles of spray.

3. Some people are allergic to the materials in space sprays and aerosols. If you find that you are allergic to a particular pesticide formulation, select another formulation suitable for the purpose.

4. Do not use inside the house any pesticide that requires a respirator or gas mask for its application.

5. Keep children away when you are applying pesticide.

6. Do not spray entire floors, walls, or ceilings with pesticide. This would be wasteful, and would leave residues where they would come in contact with human beings or pets.

7. Keep persons or pets out of treated rooms until the rooms have been aired.

8. Remove aquariums, and birds, cats, dogs, and other pets and their food and water bowls before spraying or dusting.

9. When spraying, dusting, or dipping pets, use only a pesticide labeled for the purpose.

10. Unless the container label states otherwise, assume that all prepared sprays are flammable. Do not spray near an open flame. Turn off gas pilot lights before spraying. Do not smoke while spraying.

11. Sometimes fumigation is needed to rid a house of rodents, termites, bed bugs, or other pests. Fumigation should be done *only by a professional exterminator or other experienced person.*

12. Occupants must leave a house before it may be fumigated, and they should not return until the fumigated house has been thoroughly aired. Airing usually requires 24 to 48 hours, the time depending on the type of fumigant used. The entire operation—fumigating and airing—requires from 2 to 4 days.

Garden Precautions

The information that follows applies to both flower and vegetable gardening. It is fundamentally the same advice given for commercial farming. The difference between home gardening and commercial farming is basically in *scope*. The householder usually pursues his

[3] This is deep enough to protect the pesticide from being dug or plowed up, but there are no data to indicate whether this is the best depth from the standpoint of moisture migration. The answer to the home pesticide disposal problem is being sought through incineration (R. E. Enlow, correspondence).

gardening as a hobby. He cultivates a small piece of ground—a backyard or a lot—for pleasure rather than profit. He may handle pesticide concentrates that are very toxic before being diluted for application, but he does not (or should not) handle, mix, apply, or store on his premises the extremely toxic chemicals that are often used in commercial farming. In addition to the preceding general precautions, the home gardener should observe those that follow.

1. When mixing or handling concentrates, wear rubber gloves. Rinse them well with water before removing.

2. Use special containers for mixing pesticides. Never use food or beverage containers.

3. Apply a pesticide only on the plants for which it is recommended. Apply it only against the specific pests it is designed to control. The container label will tell.

4. Do not apply dosages greater than those recommended on the container label.

5. Do not apply dust or spray on a windy day. When there is a breeze, apply the dust or spray from the windward side of the area being treated, so the breeze will blow the pesticide away from you, not onto you.

6. Do not allow pesticides to drift onto adjoining property. Do not apply pesticides near wells or cisterns or any other water sources into which they may run or be washed by rain. Do not clean application equipment, empty unwanted pesticide, or dispose of empty containers near these places.

7. To prevent harmful residues from remaining on edible plants, it may be necessary to allow a waiting period between the time of the last pesticide application and the time edible parts of the plant are picked. The length of the waiting period varies for different pesticides on different plants. Read the container label for this information.

8. Do not apply pesticide to a lawn when people or pets are on it. After treating a lawn to control soil pests, sprinkle it with water to wash the pesticide from the grass into the soil. This will reduce the hazard to children and pets; it will not reduce the effectiveness of the treatment. Do not allow children or pets on the lawn until the grass has dried.

Special Precautions for Using Pesticides That Exert Cholinesterase Inhibition

As already explained, the public health hazard from certain highly toxic organophosphorus residues is minimized because of the relatively rapid breakdown of most of these toxicants. It should not be inferred that these compounds have a low chronic toxicity, for just the opposite has been found to be true. The person applying the pesticide is continually exposed to fresh deposits of a compound of high acute and high chronic toxicity (see under those headings, later in this chapter). A high degree of precaution must be taken when using these compounds, but their great value against certain pests often justifies the added expense and inconvenience of the extreme safety measures that must be followed.

A prime requisite for the safe use of compounds such as the organophosphorus pesticides, which exert their toxic effect by cholinesterase inhibition, is that the sprayman should understand that the immediate effect of his first exposure to these compounds is not necessarily the most important factor. The absence of symptoms during the first few times he uses the materials may lead him to underestimate the importance of continuing all the precautions that are essential for the safe use of such pesticides. The greatest likelihood of poisoning is from the cumulative effects of many low-level exposures over a prolonged period. Thus, it is important for the pest control supervisor to explain to the operator why all the recommended precautions should be continued without modification so long as cholinesterase-inhibiting materials are used. The fact that the poisonous compounds can be taken into the body through the skin and by inhalation as well as by ingestion should continually be emphasized. Instructions for use of the more poisonous pesticides can be obtained from appropriate governmental agencies and pesticide manufacturers.

Full-face gas masks (figure 6), equipped with tested filters and cartridges, are recommended for the protection of those formulating or mixing pesticides in closed spaces, and those applying pesticides, including pilots of pesticide-application aircraft. Respirators are not advised for these particular uses (Lemmon, 1951). A respirator (figure 8) is effective only for filtering out particulate material, such as dust particles and liquid droplets, but gives no protection against gases. It should be used only for those chemicals for which it is designed, as stated on the instruction sheet that comes with the equipment.

The importance of washing thoroughly after exposure to pesticides should not be overlooked. Soap and water destroy phosphate chemicals rapidly. After applying pesticides, or immediately following accidental contamination, an operator should take a thorough shower, including the hair, and change to clean clothing. The contaminated protective clothing and respirators should be washed immediately.

Where work is regularly done with toxic organophosphorus pesticides, arrangement should be made for the services of a physician with a good understanding of these chemicals who is located reasonably close to where the work with the pesticides is performed. The physician can arrange for and interpret blood tests that can detect the accumulating effects of organophosphorus pesticides, and he can assist in educating personnel regarding the health hazards involved in their work. He can also administer prompt and adequate treatment if accidental poisoning should occur. If and when cholinesterase activity is found to be too low, personnel may be removed from contact with organophosphorus chemicals until the level is restored (Irma West, 1959).

Symptoms and Antidotes

If the upper and lower portions of the respiratory tract are exposed, common symptoms are salivation, nasal discharge, excessive bronchial secretion, and bronchospasm. If the poison is ingested, common early symptoms are gastrointestinal spasm, with colic, hypersecretion, vomiting, and diarrhea (WHO, 1967).

The antidote for poisoning by organophosphorus or organocarbamate compounds is atropine in massive doses, when compared with the quantity used for other medical purposes. It should be injected immediately after the appearance of any serious signs of anticholinesterase poisoning. Atropine is a complex alkaloid derived from *Atropa belladonna* (Solanaceae), and is itself poisonous, especially so if the patient has not first been poisoned by an anticholinesterase (Grob, 1963).

Certain oximes, when used in combination with atropine, provide more effective therapy than atropine alone when the poisoning results from some organophosphorus anticholinesterases. Oximes reactivate phosphorylated cholinesterase, and thus tend to correct the biochemical defect rather than merely to relieve symptoms. They may reduce the quantity of atropine required. As far as is presently known, oximes should never be administered to persons poisoned by carbamates (WHO, 1967).

Bailey and Swift (1968) discussed antidotes that could be safely used by nonmedical people and those suggested for use by physicians or under their supervision. Neither atropine nor protopam chloride (2-PAM) should be carried by workers for first aid, nor should they be used to *prevent* poisoning. They should be administered only by a physician or under his supervision. Employers should bear in mind that they cannot lawfully assume responsibility as dispensers of prescription drugs. The manual prepared by Bailey and Swift not only contains much information of value to physicians, but also lists published references suggested by medical authorities that advise physicians regarding the diagnosis and treatment of poison victims. If a Poison Information Center is unable to provide information regarding the ingredients and properties of a pesticide, the manufacturer can do so. Therefore, it is important to know the name of the pesticide causing the poison symptoms, the name of the manufacturer, and the telephone number of the office from which the desired information can be obtained.

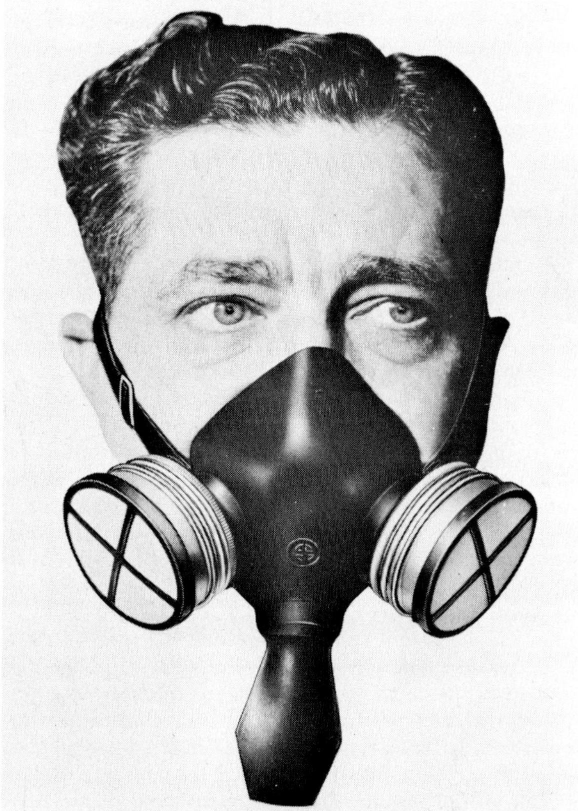

Fig. 8. Type of respirator giving protection against dusts or mists of most pesticides. (Courtesy Mine Safety Appliances Company.)

Automatic injectors are available for the administration of atropine, and should be on hand during spraying operations, but the World Health Organization advises that they should not be issued to the individual spraymen. Information on these devices may be obtained from Vector Biology and Control, World Health Organization, Avenue Appia 1211, Geneva.

Following are the recommendations of the World Health Organization pertaining to the administration of atropine when mild, moderately severe, and severe signs or symptoms of an organophosphorus compound are observed.

When signs or symptoms produced by an organophosphorus compound are *mild,* 1–2 mg of atropine (adult dose) should be injected intramuscularly and this dose repeated, if necessary, at 30-minute intervals until symptoms are relieved. Any patient sick enough to receive even one dose of atropine should be under medical observation for at least 24 hours, because the symptoms may reappear, especially in poisoning by organophosphorus compounds. Mild symptoms due to carbamate compounds have been known to respond to the administration of atropine by mouth or to remit spontaneously.

When symptoms of poisoning are *moderately severe,* 2–4 mg of atropine should be injected intravenously or, if this is not feasible, intramuscularly. Repeated doses of 2 mg at 10-minute intervals should be given until symptoms are relieved.

When *severe* signs and symptoms are present (particularly respiratory difficulties, marked slowing of the heart, or convulsions), 4–6 mg of atropine should be injected intravenously or, if this is not feasible, intramuscularly, and repeated doses of 2 mg should be given at intervals of 5 to 10 minutes until the severe symptoms decrease markedly or cease. In severe anticholinesterase poisoning, the effect of each injection of atropine may be transient, lasting only 10–30 minutes. The patient must, therefore, be observed as closely as possible for recurrence of signs of poisoning, and atropine must be repeated at appropriate intervals for at least 48 hours if his clinical condition requires it. In severe poisoning, as much as 50 mg or more of atropine may be required the first day. Respiratory failure requires immediate artificial respiration.

It should be remembered that treatment with atropine may reduce heat loss, owing to inhibition of sweating. Under hot environmental conditions this may be dangerous, particularly in young children. It is therefore advisable that the hospitalized patient be kept cool and quiet. In no circumstances must a person who has received even one dose of atropine be allowed to perform muscular work, and he must be kept under observation for at least 24 hours or until he is fully recovered.

Protection of the Consuming Public

It is estimated that about 1.25 billion pounds of pesticides are manufactured in the United States annually, yet there appear to be no authentic records of ill effects to the consumers of pesticide-treated food products. Among those who handle and apply pesticides, the remarkably few examples of death or illness are attributable to accidents or misuse of the poisonous materials. More deaths and illnesses occur every year from many other chemicals and various kinds of industrial equipment. The remarkable safety record for pesticides indicates the great care and foresight exercised by the state and federal authorities who legislate and enforce the pesticide tolerances, the chemical industry that produces and supplies the materials, and the entomologists who recommend the treatments.

No responsible person questions the fact that pesticides are generally more or less poisonous to man. However, all authentic scientific evidence supports the view that hazards are negligible when pesticides are used so as to leave residues that are below the official tolerances determined by the federal Food and Drug Administration. General environmental contamination is another subject that is too involved to be adequately discussed here. However, it is important to bear in mind that under some socioeconomic conditions, such as in malaria control in developing countries, no economically feasible alternative to an inexpensive, long-lasting residual insecticide such as DDT has been developed to this date. When such an insecticide can be safely used without significant environmental contamination, as in the spraying of the interiors of homes for protection against *Anopheles* mosquitoes, the banning of the insecticide for such a purpose without a satisfactory substitute control measure that has been proved to be effective is unwarranted and irresponsible.

All evidence indicates that the American consuming public is protected not only by legislation and analytical surveillance, but also by the far-sighted concern of the pesticide industry itself, to an extent that is probably unparalleled elsewhere in the world today.

REGISTRATION AND LABELING

Federal and state registration provides the user of pesticides and the consumer of pesticide-treated products a high degree of protection. The federal Insecticide, Fungicide, and Rodenticide Act, administered by the Environmental Protection Agency (EPA), requires that the manufacturer of a pesticide provide scientific evidence that the product will be effective against the pest or pests listed on the label, and will not injure human beings, crops, livestock, or wildlife when used as directed. If, after much additional investigation by the pharmacology and toxicology staff of the EPA and extensive laboratory and field tests by specialists in appropriate areas of competence (entomologists, plant pathologists, bacteriologists, nematologists, physiologists, weed control specialists, animal biologists, etc.), the EPA review staff decides the product would be useful and can be used safely when directions are carefully followed, the product may be registered. The magnitude of the task is indicated by the fact that in 1971 there were about 35,000 registered pesticide products in the United States, and about 450 to 600 new applications for registration, amendments to registration, and registration renewals were received each week by EPA (Ruckelshaus, 1972*a*).

If the pesticide is to be used on food or feed crops, the Food and Drug Administration (FDA) of the Department of Health, Education, and Welfare is initially required to determine how much residue, if any, can safely be permitted to remain on the treated crops. The FDA publishes a regulation that specifies the amount of residue, known as the "tolerance," that may legally remain on the crops for which the pesticide is intended. The "tolerance" is one-hundredth of the largest amount that causes no effect on test animals. This amount of residue, as a maximum, is considered to be harmless.

Pesticides used in a way not involving food or feed, or which are very low in toxicity, may be exempted from tolerance requirements. The EPA will not register the product for use on a particular crop until it has sufficient evidence that the pesticide residue on that crop from applications required for satisfactory pest control will not exceed the levels defined by the FDA. The opinions of other federal agencies, such as the Department of Agriculture, the Public Health Service, and the Fish and Wildlife Service of the Department of the Interior, are also solicited before EPA officials decide on a registration application.

If there is any doubt in the mind of the user of pesticides as to the current registration status of the compounds he wishes to use and for the purpose he wishes to use them, he should consult the appropriate state authority, for the registration of pesticides is under constant review by both state and federal agencies.

Acute Toxicity

The legal tolerance depends in part on the toxicity of the pesticide to human beings, but is determined largely by analogy. Acute toxicity is determined by administering the poison (usually orally) to rats, rabbits, or other animals. A determination is made of the quantity of toxicant in a single dose for each animal, in milligrams per kilogram of body weight of the animal (mg/kg), that would kill 50% of the animals tested. This quantity is often referred to as the median lethal dose (LD_{50} or MLD). Thus, the single-dose or acute oral toxicity of DDT to white rats is about 250 mg/kg.

It is actually a convenient coincidence that toxicants, on a milligram per kilogram of body-weight basis, have approximately the same effects on white rats and human beings. Fatalities from a single dose of DDT are not likely, except from accidental ingestion of a rather large quantity of the material; even then, most of the DDT would escape the body in the feces unless much oil or fat were also present. Examples of pesticides having what are considered to be low acute toxicities are pyrethrins and synthetic pyrethroids, methoxychlor, Perthane, TDE, malathion, ronnel, phoxim, Gardona, and the chlorinated acaricides. (See table 1, earlier in this chapter.)

By comparing the $LD_{50's}$ of various pesticides with that of a well-known material such as DDT, an idea can be gained of their relative toxicities. Those with LD_{50} values higher than that of DDT are safer, at least as far as their acute oral toxicity is concerned, and those with lower values are more toxic. Since nearly all the LD_{50} values shown in table 1 pertain to rats, a fairly reliable direct evaluation of the relative toxicities of the materials can be made by comparing their LD_{50} values. The LD_{50} value gives little or no information about the possible cumulative effects of a pesticide. The human body has greater ability to rid itself of some compounds than it does of others.

Occupational poisoning caused by nonfumigant pesticides is said to show a much closer correlation with acute dermal LD_{50} values than with oral toxicity. Nevertheless, oral LD_{50} values are considered to be useful for comparing objectively the inherent toxicity of different compounds. The following tabulation was suggested by Hayes (1963) as a very general guide for estimating the probable lethal oral dose for a grown person for compounds within certain LD_{50} ranges.

Acute oral LD_{50} for any animal (mg/kg)	*Probable lethal oral dose for a human adult*
Less than 5	A few drops
5 to 50	"A pinch" to 1 tsp (5 cc)
50 to 500	1 tsp to 2 tbsp (5 cc to 30 cc)
500 to 5,000	1 oz to 1 pt (1 lb) (30 cc to 473 cc)
5,000 to 15,000	1 pt to 1 qt (2 lb) (473 cc to 946 cc)

(1 lb equals 0.454 kg.)

Chronic Toxicity

After the single dose or acute toxicity has been determined, there remains another consideration from the public health standpoint, and that is the long-term effects or chronic toxicity. To determine the chronic toxicity of a pesticide, laboratory animals are fed food to which the pesticide has been added in small amounts, far below the LD_{50}, to determine any chronic effects, i.e., those that appear only after a period of weeks or months. Chronic toxicity may be exhibited as injury to the liver, gall bladder, kidneys, or other vital organs. It may also be shown as a reduction in weight, appetite, vitality, or other established criteria. The chronic toxicity is usually expressed as the level of toxicant in parts per million (ppm) of food that produces a given response over a certain period. Usually, a determination is also made of the maximum concentration in ppm that fails to induce undesirable symptoms or death.

The Label

The container labels of registered pesticides provide certain information in accordance with the federal Insecticide, Fungicide, and Rodenticide Act. A label must contain the brand name or product name, ingredient statement, warning or caution statement, directions for use, net contents, the name and address of the manufacturer, and the USDA registration number. Inspectors for the EPA throughout the country collect samples of pesticides marketed interstate, from their containers, and send them to a laboratory for testing to make sure they contain all the ingredients in the amounts listed on the label. The products are also checked to see that they are not adulterated with chemicals or other materials *not* listed on the label. A registration may be suspended or canceled at any time.

Registration and approval of labels must also be obtained from each state, and state officials also may suspend or cancel registration of a product if necessary to protect the public interest.

About half of the registered pesticides are designed for use in or around homes, apartment buildings, and industrial plants. The investigations required to supply the information on a label for a new pesticide require 2 to 5 years of research and testing, and cost the manufacturer from $500,000 to $8,000,000, not to mention the cost of government and state laboratory and field research. It behooves the user of the product to read the label and follow instructions for safe and effective use.

Labels must bear warnings and precautionary statements appropriate to the degree of danger. Pesticide products are considered to fall into 4 categories with respect to toxicity, category I being the most toxic. (See table 2, earlier in this chapter.) Labels for pesticides in this category must bear a skull and crossbones, the word "POISON" printed in red, an antidote statement, emergency first-aid instructions that can be followed by a layman, and the instruction, "If swallowed, or if symptoms of poisoning occur, call a physician immediately." According to this system of categorizing toxic hazard, the chemicals in each successive category of toxicity are about one-tenth as toxic as chemicals in the preceding one, and the caution statement on the label is appropriate for the degree of hazard. Products in category IV have little or no toxicity and may not require a caution or antidote statement on the label, but if they are irritating to the skin, nose, throat, or eyes, the label should so indicate (USDA, 1967).

In a publication titled *Pesticides: The Issues, The Alternatives,* the University of California, Division of Agricultural Sciences, suggests that the user of pesticides look for certain symbols or "signal words" on the label of a pesticide package, such as those outlined here.

> A skull and crossbones shows the product is highly toxic. Anywhere from a pinch to a teaspoonful probably will kill an average person.
>
> The word WARNING shows it is moderately toxic. The probable lethal dose is from a teaspoonful to a couple of tablespoonfuls.
>
> The word CAUTION shows it is somewhat less toxic, but still quite dangerous. From 1 ounce to a pint [28 g to 0.473 l] probably will kill.
>
> If the label contains none of these words or symbols, but only the statement, KEEP OUT OF THE REACH OF CHILDREN, the product is in the least-toxic category. It might take as much as a quart [0.946 l] to be lethal.

It is important to use the correct dosage of pesticide, as indicated on the label. A pesticide that is highly toxic to human beings but may be used at a low dosage may present no greater hazard to health than a less toxic preparation that must be used at a higher dosage.

Certain pesticides may require special caution statements, e.g., if they are flammable or if the contents are under pressure, as in the aerosol bomb, in which case they should not be placed near heat or open flame.

WEIGHTS AND MEASUREMENTS

Even in the United States, temperatures, weights, and measurements given in scientific publications are usually expressed in the metric system. Since the layman is ordinarily more familiar with the United States system, the following tabulations are included to enable him to convert from one system to the other. The conversions given are sufficiently accurate for all practical purposes. Many involve the use of long decimals, and are "rounded out" here. For further convenience, a legend follows, giving the meanings of the various abbreviations employed in the tables.

United States Terms: a, acre; cu, cubic; dr, dram; fl, fluid; ft, foot; gal, gallon; in., inch; lb, pound; mi, mile; oz, ounce; pt, pint; qt, quart; sq, square; tbsp, tablespoonful; tsp, teaspoonful. (A period follows "in." to distinguish it from the preposition.)

Metric Terms: cc, cubic centimeter; cm, centimeter; dm, decimeter; g, gram; ha, hectare; kg, kilogram; km, kilometer; l, liter; m, meter; ml, milliliter; mm, millimeter.

Temperature

To convert from Centigrade to Fahrenheit,
°F = (°C × 9/5) + 32.
To convert from Fahrenheit to Centigrade,
°C = (°F − 32) × 5/9.

Weight

1 oz	28.35 g
1 lb	0.454 kg
1 g	0.03527 oz
1 kg	2.205 lb

1 kg is 1,000 g; 1 g is 1,000 mg.

Area Measure

1 sq in.	6.45 sq cm
1 sq ft	929 sq cm
1 sq yd	0.836 sq m
1 acre (a)	4,047 sq m or 0.405 ha
1 sq mi	2.59 sq km
1 sq cm	0.155 sq in.
1 sq m	10.76 sq ft
1 ha (10,000 sq m)	2.471 a
1 sq km	0.3861 sq mi

Linear Measure

1 in.	25.4 mm
1 in.	2.54 cm
1 ft	30.48 cm
1 yd	0.91 m
1 mi	1.61 km
1 mm	0.03937 in.
1 cm	0.3937 in.
1 m	3.281 ft
1 km	3,281 ft
1 km	0.6214 mi

1 km is 1,000 m; 1 m is 100 cm; 1 cm is 10 mm; 1 mm is 1,000 microns (μ).

Cubic Content

1 cu in.	16.387 cc
1 cu ft (1,728 cu in.)	28.3 cu dm
1 cu yd (27 cu ft)	0.765 cu m
1 cc	0.061 cu in.
1 cu m (stère)	1.308 cu yd

Liquid Measure

1 fl oz	29.57 ml
1 pt	0.473 l
1 qt	0.946 l
1 gal	3.785 l
1 gal	3,785 ml
1 ml (1 cc)	0.0338 fl oz
1 l (1,000 ml)	1.057 qt
1 l	0.2642 gal

For practical purposes, a milliliter (ml) and a cubic centimeter (cc) are the same (1.00000 ml = 1.00028 cc), but when correctly used, "ml" is employed only in the measurement of liquids and "cc" only in the measurement of solids or gases.

Volumetric Measure for Powders

1 tsp or 1.33 fl dr	5 cc
1 tbsp or 3 tsp	15 cc
1 fl oz or 2 tbsp	30 cc
1 cup or 16 tbsp	240 cc
1 pt or 2 cups or 16 fl oz	480 cc

The "teaspoonful," "tablespoonful," and "cup" are standard measuring devices that can be purchased at hardware stores or ten-cent stores. The metric equivalents are rounded.

Chapter 3 · *Equipment Used for Applying Pesticides**

A PESTICIDE'S APPLICATION is as important as its toxicity to the pest against which the treatment is directed. There is usually a greater choice of pesticides than of the equipment and procedures by which they can be applied. The equipment used may include a sprayer, mist blower, fogger (for cold or thermal fogs), aerosol generator, vaporizer, or duster. The pest problem may be such that 2 or more of these types of equipment should be used on the same job. It is important to know the appropriate means of pesticide application for each pest species and each chemical or formulation, and how to use the apparatus safely, efficiently, and effectively. Intelligent procedures in the application of pesticides are based on a thorough knowledge of the biology and habits of the pests against which they are to be applied; such matters are discussed in other chapters. Precautions to be observed in the use of various types of pesticides are discussed in chapter 2.

HAND SPRAYERS

Homeowners, commercial pest control operators, and public health service personnel frequently use small, easily transportable equipment, such as hand sprayers. These, as well as the power sprayers discussed later, are designed for applying 3 types of pesticide sprays: space spray, residual (or surface) spray, and dual-purpose spray. A space spray is directed against such insects as flies and mosquitoes that are all present in a room and can be reached at the time of treatment. The sprayer should have a nozzle producing fine droplets that remain floating in the air a long time. The spray should be directed upward to fill the room or enclosure with an aerosol or mist.

Some insects, such as cockroaches or silverfish, may be hidden in out-of-sight areas during the day when the spray is applied. The object is to deposit enough insecticide to kill them when they crawl over the residue at night. For some places inside a building, such as on baseboards, pipes, screens, and similar surfaces, an inexpensive paintbrush can be used to apply insecticide solutions or emulsions. A light film is usually sufficient. For spraying, a type of sprayer or a type of nozzle or nozzle adjustment should be chosen that will produce a coarse-droplet spray, and the surfaces over which the insect is expected to crawl should be wet almost to the point of runoff.

For a dual-purpose spray, the object is to produce droplets of intermediate size that will float about sufficiently to produce a space spray but will also leave a reasonably effective residue.

Atomizer or "Flit Gun"

This is the smallest and simplest type of hand sprayer. It may be effective for household or even garden spraying, for often the pests on garden plants are confined to limited areas, as for example when aphids are grouped about the stems of rosebuds. The atomizer consists of a hand-operated, air-compressing piston pump fastened to a small tank (figure 9). In one type of atomizer, the compressed air moving rapidly over the end of the liquid tube causes a reduced pressure to occur at this point, and the atmospheric pressure on the tank contents forces liquid out of the tube. The liquid is then propelled by the air-stream to the target. Such devices operate by

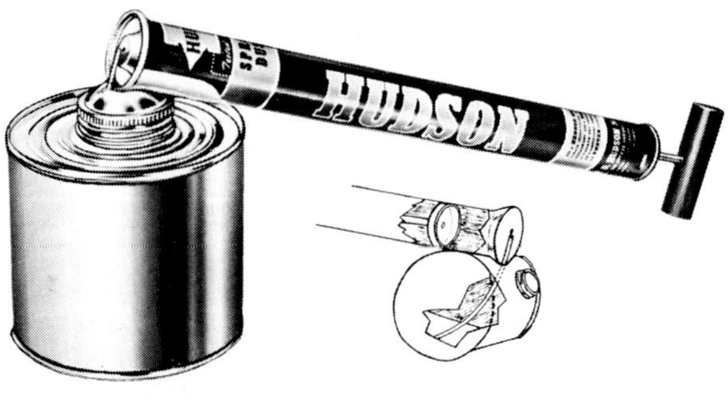

Fig. 9. Atomizer or "flit gun," illustrating the utilization of the venturi effect for the atomization of liquids. (Courtesy Hudson Manufacturing Company.)

* This chapter is expected to be principally of service to readers in the United States, for which reason all weights and measurements are stated primarily in units of the English system—inches, feet, gallons, etc. However, these are always followed by their metric equivalents (usually rounded) for the benefit of those who work with the latter system.

what is known as the "venturi principle." The spray stream is intermittent, corresponding to the strokes of the piston. More commonly used is the compression-type atomizer, in which a continuous action is provided by a buildup of air pressure in the reservoir above the liquid. Instead

Fig. 10. A portable atomizer operated by compressed air on the same principle as a paint gun. (Courtesy Spraying Systems Company.)

of the liquid being aspirated, a positive pressure is put on the liquid, and the tube carries it into the airstream.

Inside the house, the atomizer is satisfactory for applying an insecticide residue to a limited area, as in "spot treatment" for ants or for partial control of cockroaches or silverfish, for applying a residue to a screen door where flies or mosquitoes often congregate, or for treating pantries, cabinets, closets, and furniture. Agitation of the spray mixture in the sprayer reservoir is accomplished by the motion incidental to the spraying or, in the case of weakly emulsified or suspended materials, by continual, deliberate shaking.

The venturi principle is also utilized in the paint-gun type of equipment shown in figure 10, in which the handle inlet connects to an air hose. A steady stream of finely atomized spray is delivered.

Water-Hose Attachment

The venturi effect just described can also be obtained from a stream of water from a garden hose. This fact has led to the development of a variety of hose sprayers or other hose attachments (figure 11). The reservoir jar of the hose sprayer is usually filled with a prediluted water-miscible mixture of the insecticide. However, the fluid in the container does not necessarily have to be mixed with water; it can be a technical material in almost any solvent, including water or, if the material is normally a liquid, no dilution is required. Even wettable powders can be used, suspended in water and shaken while in use. Instructions on the reservoir jar inform the user how much actual pesticide is required to result in the desired percentage in the final diluted spray mixture.

The hose sprayer provides a convenient way of spraying lawns for fleas or for other lawn-infesting insects, for it is not necessary to prepare and carry about large quantities of the dilute spray mixture. These sprayers can also be used to spray plants, shrubs, and trees, or for the application of fertilizers or micronutrients.

Pistol Sprayer

In a type of hand-operated *hydraulic* sprayer (figure 12), the operator squeezes a trigger which activates a plunger. The spray liquid inside the pump is thereby displaced and forced out of a discharge nozzle by hydraulic pressure, much as it is in the "gun" used for oiling machinery. When the trigger is released, the pump is again filled with liquid. This device has a capacity of 1 pt or 1 qt—about 0.5 or 1 l. It is satisfactory for spraying a limited area, as just described under "Atomizer." The pistol sprayer pumps directly from the container, and the contents are not diluted by either air or water, so it is ordinarily used only where a high dosage can be applied—generally not for treating plants or animals.

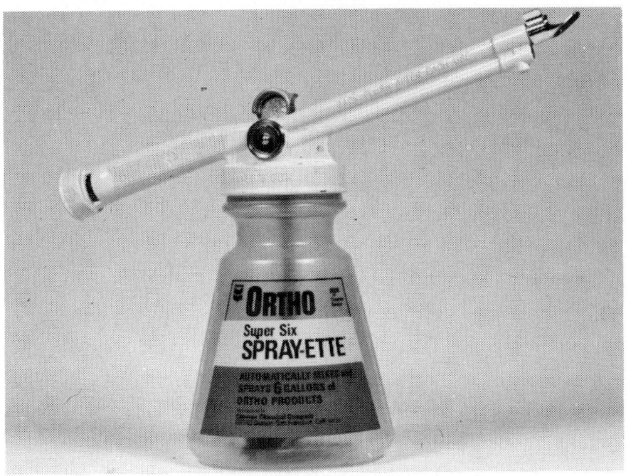

Fig. 11. "Hose-attachment" sprayer. (Courtesy Chevron Chemical Company.)

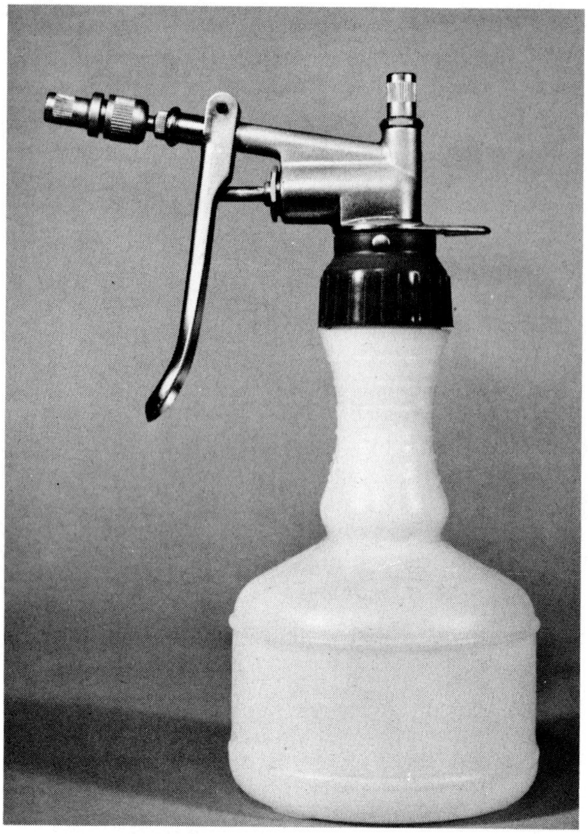

Fig. 12. Hand-operated hydraulic "pistol sprayer." (Courtesy McCarl Company.)

Bucket or Stirrup Pump

(Not illustrated)

This simple and inexpensive hydraulic sprayer consists of a plunger pump equipped with a spray hose, wand, and nozzle. The pump may be clamped into a bucket containing the spray material. Agitation is usually furnished by a rod connected with the handle that operates the pump, or by a jet of liquid forced into the bucket through an orifice at the bottom of the pump.

The bucket pump is quite commonly used in some foreign insect control programs, but is not widely used in the United States.

Trombone Sprayer

An interesting modification of the bucket or stirrup pump is the *trombone* sprayer (figure 13), in which the spray liquid is drawn from the bucket and discharged through the nozzle of the sprayer by a plunger and cylinder which move on each other with a trombonelike action. The spray-liquid container may be attached to the sprayer, as shown in figure 13 (bottom).

Knapsack Sprayer

The term "knapsack" refers to the mounting rather than the mechanical type of the sprayer. What is ordinarily meant is a sprayer with a tank holding 3 to 5 gal (about 11 to 19 l) of dilute spray liquid, equipped with straps, and carried on the back like a knapsack. It is operated by a constant-pressure pump of either the plunger or the diaphragm type, both of which are positive-displacement mechanisms. With these types, the tank does not have to withstand any pressure; it merely holds the liquid. The pump is mounted inside the tank, and is operated by a long handle that extends over the shoulder or under the arm of the operator, who is able to pump with one hand and direct the spray with the other. Agitation of the mixture is maintained by an attachment to the handle. Knapsack sprayers of this

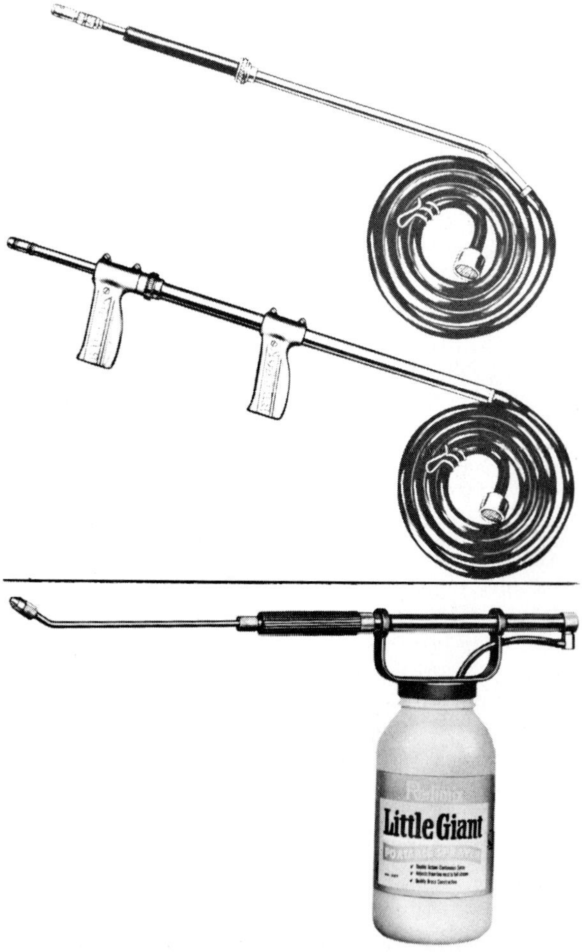

Fig. 13. Trombone sprayers. (*Top,* courtesy Hudson Manufacturing Company; *bottom,* courtesy Burgess Vibrocrafters, Inc.)

kind are used for garden or limited agricultural spraying, but not for structural or household pest control.

The compressed-air sprayer described to follow can be mounted on a knapsack frame, and can then be referred to as a "knapsack sprayer," but it is not used that way in structural pest control in the United States. However, for outdoor spraying in public health or agricultural pest control work, particularly in developing countries, it is commonly carried on a knapsack frame.

Compressed-Air Sprayers

The hand-operated compressed-air sprayer (figure 14) is the basic spray equipment of most commercial pest control operators doing structural pest control and of most public health insect control projects. The tank, preferably of stainless steel, has a capacity that usually ranges from 0.5 to 2 gal (about 2 to 7.5 l). The tank is filled with the desired concentration of spray liquid to about 75% of its capacity. When the plunger of the pump is operated, air is forced through a check valve at the bottom of the pump cylinder (figure 14, *right*), and rises to the top of the tank, exerting pressure on the liquid. When the discharge valve is opened, liquid is forced through a delivery tube, usually about 4 ft (1.25 m) of hose, and out through a spray wand or gun. Pressure should be maintained so as not to drop too far below 40 psi, otherwise there will not be enough pressure to break the liquid into very small droplets and to provide for the desired spray patterns. It may occasionally be necessary to shake the spray tank vigorously to insure a uniform mixture of insecticide, particularly if it has been set aside for an extended period before all the liquid has been sprayed out. Emulsifiable concentrates that form very stable emulsions are the formulations most commonly used in compressed-air sprayers.

The Spray Pattern

The kind of spray pattern obtained depends on the type of nozzle used, and should be appropriate for the kind of spraying to be done. For example, a pinstream nozzle may be selected to direct the spray deeply into cracks and small openings in a straight, narrow stream. To provide an even coating of insecticide over a flat surface, a *fan-* or *flat-spray, solid-cone,* or *hollow-cone* nozzle would be appropriate. There are also adjustable nozzles that can produce various types of spray patterns. It appears that the fan- or flat-

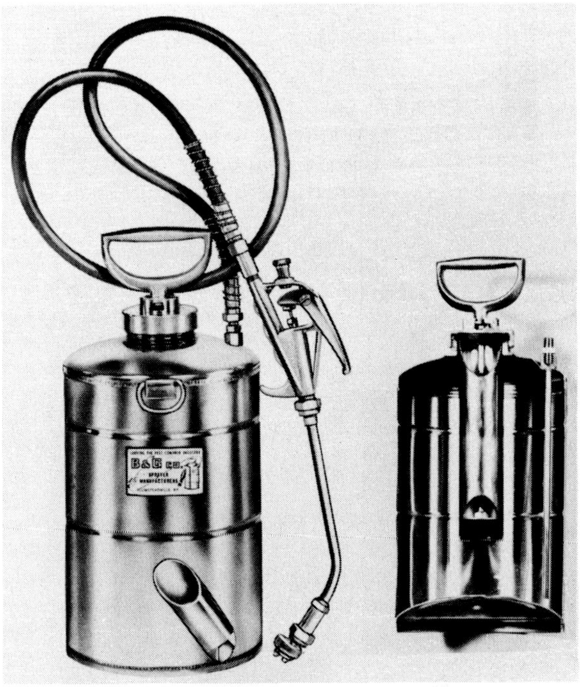

Fig. 14. Compression sprayer and inside view. (Courtesy B & G Company.)

spray nozzle is the most commonly used by structural pest control operators. By orienting a fan-type spray properly, one can effectively spray into cracks and crevices, but not as deeply as with a *pinstream.*

Care of Compressed-Air Sprayers

After compressed-air sprayers are used, they should be flushed out with water and detergent. Water should be pumped through the hose until it appears clear. The hose should be regularly inspected for weak spots, particularly near each end. Weak hoses should be replaced or repaired to avoid accidental bursting. The leather plunger in the piston should be occasionally lubricated with neatsfoot oil so that it will retain its shape. Also, check the strainer screen occasionally to prevent an accumulation of debris that might interfere with the normal discharge of spray.

Pest control operators who use a compressed-air sprayer should not only rinse it daily, but should also clean it thoroughly once a week, using a brush and an ammonia-containing household detergent. Some insecticides, particularly organophosphorus compounds, form acids when left in a tank too long, causing "pinhole" leaks to develop, so it is important to remove insecticide deposits. Leftover spray mixtures should be deeply buried or inactivated with lime. Above all, they should not be poured where they can become a part of the city sewage.

Nozzle tips on spray guns sometimes become clogged with extraneous materials. If the tip is soaked in a good solvent and cleaned with a small brush (nothing more rigid than toothbrush bristles), the accumulated sediment is usually removed. No metal object, such as a pin or wire, should be inserted into the orifice of the tip, for it is precision-machined and can be easily damaged, with a resulting distortion of the spray pattern (Brehm, 1968). The World Health Organization (WHO) suggests the use of stainless-steel tips, which are much more durable than brass or plastic.

POWER SPRAYERS

Power sprayers are used primarily by commercial pest control operators. They may be carried in the service truck or be permanently truck-mounted, and are suitable for the treatment of large industrial plants, "pretreatment" of the soil for termite prevention during the construction of a building (figure 92, chapter 5), spraying the crawl space under buildings at any time after construction, and for the treatment of lawns or ornamental shrubs and trees.

Power sprayers make possible the use of large spray tanks, ranging up to hundreds of gallons; an even, continuous discharge; and more rapid, less tiring work than with hand sprayers. Usually, 200 to 300 gal (755 to 1,135 l) is the maximum that can be carried on ordinary automotive tires. A spray rig with a tank holding more than 300 gal requires special mounting and industrial tires.

Power sprayers are commonly operated by hydraulic pressure, generated by pumps that are driven by gasoline or electric motors. If a pest control operator intends to spray trees as well as to do structural control, he uses a reciprocating pump, which can deliver 30 to 40 gal (114 to

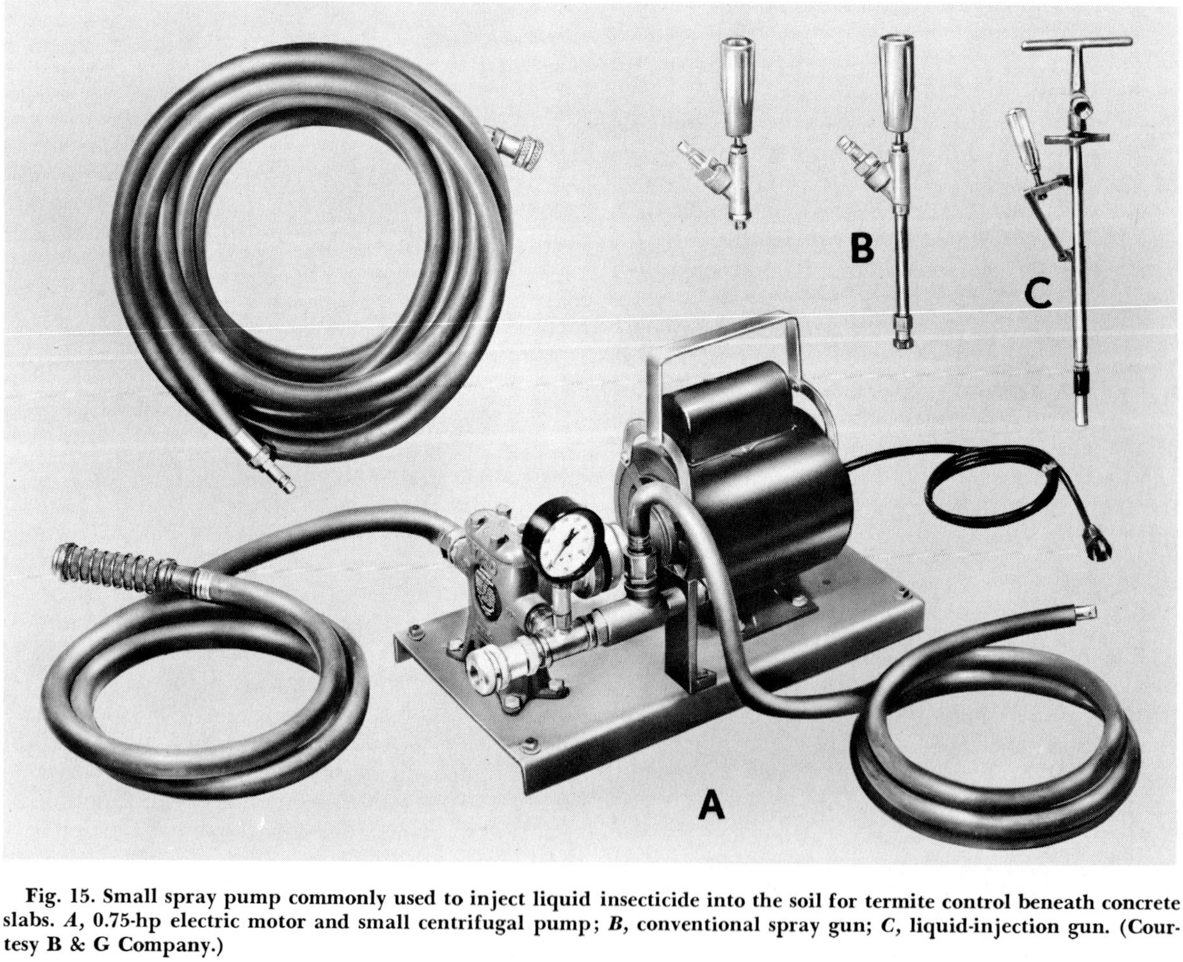

Fig. 15. Small spray pump commonly used to inject liquid insecticide into the soil for termite control beneath concrete slabs. *A*, 0.75-hp electric motor and small centrifugal pump; *B*, conventional spray gun; *C*, liquid-injection gun. (Courtesy B & G Company.)

150 l) per minute at pressures up to 800 psi. For other jobs, he is more likely to use gear pumps delivering 2 to 8 gal (7.5 to 30 l) per minute at 75 to 80 psi, or roller pumps and diaphragm pumps delivering 2 to 25 gal (7.5 to 95 l) per minute at pressures up to 350 psi. Another type of power sprayer employs a pressurized spray tank, with the motive power supplied by a compressor driven by gasoline or electricity. This sort of equipment is suitable only if a single toxicant is going to be used in quantity, since the tank must be emptied of air, cleaned, recharged, and repressurized for each change of chemical.

All pressurized equipment requires either state inspection by the Industrial Safety Commission and approval at the manufacturing plant or, if homemade, inspection and approval if used by anyone other than the owner. PRESSURIZED TANKS ARE DANGEROUS!

Termite control sometimes requires a specialized type of equipment for injecting liquid insecticides into soil or beneath concrete slabs. Figure 15 shows a small centrifugal pump, adjustable up to 300 psi of pressure, and delivering up to 3 gal (11 l) per minute. This model is operated by a 0.75-hp electric motor, but models operated by a gas engine are also available. A specially designed liquid-injection gun and rod (figure 15, *C*), with a nonclogging, stainless-steel, soil-rodding tip that sends the insecticide solution forward or laterally, can be attached to the spray hose and pushed into the soil beneath slabs through which holes have been drilled vertically, or the rod may be pushed beneath the slab horizontally via holes drilled in the foundation wall (figure 96, chapter 5). Four-ft (1.25-m) lengths of pipe, threaded at both ends, can be attached until the total length of the rodding device may be as much as 40 ft (12 m). The same equipment can be used, of course, for termite pretreatment spraying, lawn and shrubbery spraying, mosquito and fly control, weed control, etc., using conventional spray guns (figure 15, *B*). The liquid spray can be drawn from any size container that is convenient.

Care of Power Sprayers

Great advances have been made in power sprayers, principally in the improvement of lubrication by enclosing working parts. This keeps out dust, and makes possible a more nearly self-oiling system. New designing for easy accessibility of valves and plungers has greatly simplified maintenance, and the use of high-grade metals has prolonged the life of working parts. However, this kind of equipment is expensive, and it behooves the owner to understand the operation of each type of sprayer and the kind of maintenance that is required to keep it in good working order. The manufacturer or supplier of pesticide-application equipment can provide information on its construction, maintenance, and operation. Instruction manuals and diagrams that are supplied should not be misplaced or discarded.

AEROSOL OR MIST APPLICATORS

In order to understand what is meant by "aerosol generators," "foggers," or "mist machines," it is necessary to know what is meant by "aerosol," "fog," "mist," and other types of liquid emanations from insecticide-application equipment, particularly from the standpoint of particle sizes.

Definition of Aerosol

An aerosol is a suspension of very small liquid or solid particles in air or gas. The particles are so small that their sizes are usually expressed in units known as *microns.* A micron (μ) is 0.001 millimeter (mm). A hundred microns (0.1 mm) is 0.004 inch, or about the thickness of the paper on which this is printed.

The term "mass median diameter" (MMD), expressed in microns, is commonly used to indicate droplet size. This is the number that divides the aerosol or mist into 2 equal parts by mass, one half containing droplets smaller than this diameter and the other half containing larger droplets. This happens to be the droplet-size designation most frequently found in the literature. A more meaningful designation, and the one that is recommended by WHO (1971) is "volume median diameter" (VMD). It is the volume of the drop, rather than its mass, in which we are interested. The definition is the same, except that the word "volume" is substituted for "mass."

Technically, aerosols or fogs are composed of particles ranging in size from 0.1 to 50 μ VMD. Systems with larger particles are mists (50 to 100 μ VMD), fine sprays (100 to 400 μ VMD), and coarse sprays (400 + μ VMD). Some confusion may result from the fact that some application machines can be adjusted to produce either an aerosol or a mist. Fumes or smokes are solid systems with even smaller particles (ranging in size from 0.001 to 0.1 μ) than aerosols, and a vapor, which is a gaseous form, is made up of particles below 0.001 μ in size.

To determine the size of droplets produced by a given aerosol- or mist-generating equipment,

No. 2 diesel oil is used as the standard test liquid. Droplets are made to impinge upon some suitable surface, such as an oil matrix for water drops, or a cellulose–soap mix for oil, and are then measured. These are the best, but the most practical system is to use heavy, clay-impregnated papers, such as Kromcote or other hard-surface paper. A colored dye in the spray is necessary. For the measurement of drops larger than 100 μ, as in sprays, plates coated with magnesium oxide are used. Drops impinging on these plates produce small craters, and the craters are measured by high-power microscopy (N. B. Akesson, correspondence). Teflon® has been introduced recently as a surface for deposition of aerosol droplets (Anderson and Schulte, 1971).

For measuring aerosol-droplet size, the World Health Organization (WHO) recommends a clean glass slide (1×3 in.; 2.5×7.5 cm), coated with a layer of magnesium oxide, which may be prepared by burning magnesium ribbons a short distance below the slide. The lower limit of droplet size that can be detected with a slide coated with magnesium oxide is about 10 μ. A magnification of 100× is recommended for aerosol droplets. Measurement is facilitated by the use of a special eyepiece graticule. All the craters, from one edge of the slide to the other, are measured as the slide is moved across the field by the mechanical stage of the microscope. At least 200 craters should be measured (WHO, 1971).

Coverage and Insecticide-Depositing Properties of Aerosols

Aerosols are particularly suitable for the treatment of enclosed spaces, for in these they cannot drift away from the target area. Despite their small size and consequent low deposition rate, the particles of an aerosol remain suspended long enough to result eventually in adequate insecticide deposit, particularly on small, flying insects. A small quantity of insecticide solution applied as an aerosol can quickly fill a room, and no insect that is exposed (not hidden in wall voids, voids under cabinets, in cracks, etc.) can escape the treatment. The amount of liquid involved is so small and the quantity of insecticide deposited is so minute that staining of the room and its contents is not a problem. Aerosols provide a rapid and clean way to use a minimum quantity of insecticide liquid for maximum coverage.

The significance of the droplet size generated by aerosol canisters and other aerosol-generating equipment, in relation to coverage, is probably most readily visualized by considering the increase in the number of droplets per milliliter as droplet diameters are reduced. A milliliter is approximately the volume of a small grape.

Diameter of droplets (microns)	*Number of droplets per ml*
1,000	1,920
100	1,910,000
50	15,278,000
10	1,910,000,000
6	8,849,557,500

The number of droplets varies inversely as the cube of the diameter (d^3), so a reduction of droplet size from 50 to 10 μ increases the number of droplets 125-fold: $[(50/10)^3 = 125]$. A reduction of 10 to 6 μ increases the number of droplets 4.57-fold to about 8.85 billion droplets per ml of liquid. It is obvious from these figures that droplet size is an important factor affecting the dispersal or "coverage" obtained per unit of insecticide liquid, as well as the opportunity afforded for convection currents to disperse the insecticide in a building thoroughly. However, an equally important factor is how efficiently the aerosol droplets are deposited on the bodies of insects.

Particle Impaction

With the small quantities deposited by aerosol equipment, it is important that the insecticide be deposited on the bodies of the insects at the time of treatment, for the small amount of residue deposited by this type of equipment on the surfaces on which insects crawl about is of little or no value in their control. Persons with some knowledge of the physics of droplet impaction may have doubts as to the ability of droplets of the size ranges discussed in this section to deposit on insect bodies. Objects that obstruct the path of a horizontally moving aerosol are surrounded by a boundary of moving air called the "layer of resistance." Larger droplets can penetrate this layer because they have enough linear momentum or inertia to resist a change in direction, but smaller droplets ($< 20\ \mu$) float off in the direction of the moving boundary air current (Sell, 1931; Yeomans *et al.*, 1949).

The impact efficiency of a liquid drop increases as the increase in velocity (V) of the drop and the square of the drop diameter (d) and inversely as the diameter (D) of the object upon which it is impacting: Efficiency $= Vd^2/D$. The formula explains why flying insects collect more aerosol droplets than do insects that are not moving, and why impact efficiency so greatly increases with increase in drop size. The value *D* in the denominator of the equation explains why it is necessary to have greater velocity and drop size for efficient impaction against large objects. It also explains why small drops impact well only on small insects and on insect setae.

Despite the limitations implied in the formula, in a confined space the prolonged period of contact of even a resting or slowly crawling insect with the slowly settling aerosol may result in its accumulating a toxic dose of insecticide. Even an insect as large as a cockroach can be killed before the aerosol finally settles out of the air. (With pyrethrin aerosols, some may recover after becoming paralyzed.) However, aerosols are not very effective against cryptobiotic insects such as cockroaches. Many cockroaches are flushed from their hiding places in wall voids, voids under cabinets, the insulated hollow walls of refrigerators, cracks, crevices, etc., but evidently many are also able to avoid lethal contact with the aerosol. Also, an aerosol leaves very little residual deposit of insecticide on surfaces upon which cockroaches not contacted by the aerosol would subsequently crawl. In any case, even if many insects are initially killed, the long-term effect of an aerosol is very poor against cockroaches, although it may be excellent against small, flying insects such as house flies, mosquitoes, and gnats.

Fortunately, dense body hairs or setae, which can protect an insect from dust particles, do not protect it from liquid aerosol droplets. An investigation made with the aid of laser holography, using exposures as low as 20 billionths of a second, showed that droplets of pyrethrins in base oil that were 10 μ in diameter, or even much smaller, impinged on ventral tactile or sensory setae of a German cockroach held in such a way that the setae were in the path of a horizontally moving aerosol. The droplets migrated toward the bases of the setae, increasing in size as they collided with other droplets (figure 16) and finally became deposited in the setal pits. Figure 16, bottom, is the setal pit of a western spruce budworm, *Choristoneura occidentalis* Freeman.

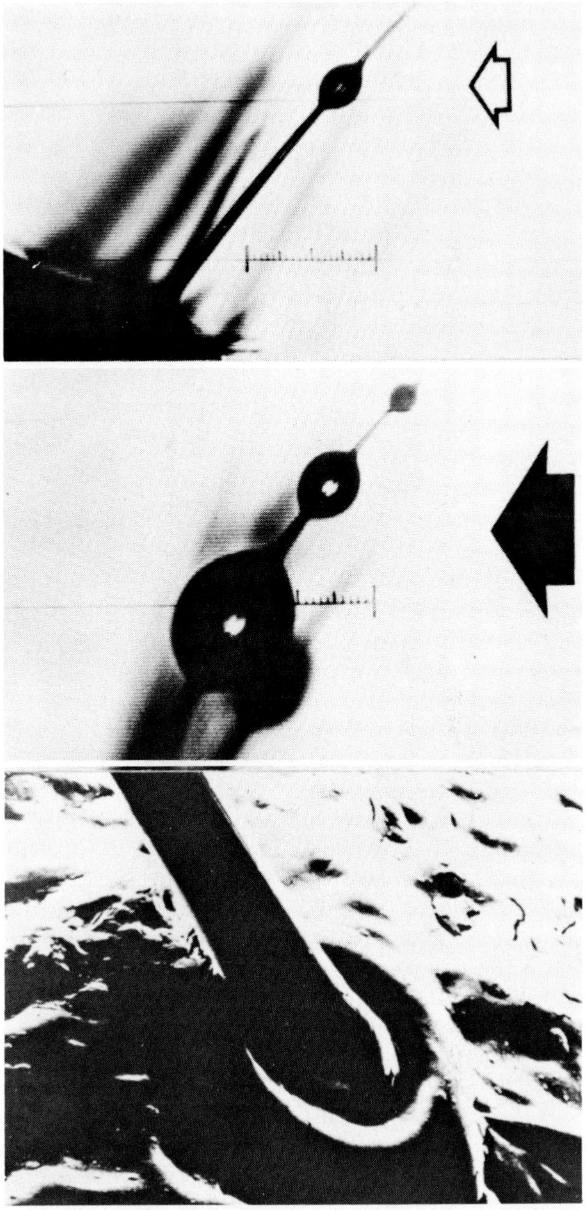

Fig. 16. Impaction of aerosol droplets on a seta of a German cockroach. Arrows show direction of movement of aerosol cloud. Forty-micron reticles are shown (*top* and *center*). The 10-micron droplet (*top*) moves toward the base of the seta, enlarges as it is struck by other droplets or collides with others along the seta (*center*), and finally reaches the setal pit. (The setal pit in the figure is that of the western spruce budworm.) Top and center photos courtesy Laser Holography, Inc.; bottom photo, courtesy R. B. Roberts.

The layer of cuticle covering the seta is very thin. Also, in the setal pit the cuticle is much thinner than in surrounding areas, and a cytoplasm continuum (figure 40, chapter 4) forms a pathway for the rapid movement of insecticide from beneath the setal pit to epidermal cells below the cuticle. Penetration of toxicant to sites of toxic action within the insect is greatly facilitated. Quraishi and Poonawalla (1969), using radioautograph techniques, showed that insecticide deposited on the insect cuticle ordinarily migrates laterally to portals of entry within thin membranes, such as setal pits and intersegmental conjunctivae, in order to be able to enter an insect's body. The impaction of aerosol-size droplets directly on the setae evidently provides still another avenue of approach to a portal of entry.

CAUTION!

Aerosols are a hazard to the lungs, especially in the droplet size range generated by the aerosol canister, in which the bulk of the liquid is released in droplets under 10 μ in diameter. Other types of aerosol generators discussed in this chapter also release a significant part of the liquid in this size range. Droplets 0.5 to 5 μ in diameter tend to be retained by the lungs. Aerosols of the smallest droplet sizes are also the ones in which the liquid contains the highest concentrations of insecticide to compensate for the greatly reduced volume applied in a given area. This is another factor that increases the health hazard.

Pyrethrins are commonly used in aerosols, and although they are relatively nontoxic and nonpersistent, they are irritating. This is also the case with the freon carrier used in aerosol canisters. Some people suffer considerable throat irritation when inhaling pyrethrin aerosol or dust. Hay fever and asthma symptoms tend to be aggravated. The possibility of psychogenic distress is another factor that should not be overlooked.

Inhalation of aerosols of any kind should be avoided as much as possible, and anyone who is likely to have more than momentary exposure should be provided with a suitable respirator (figure 8, chapter 2). Rooms that are fogged should not be occupied until the fog has settled or the rooms have been ventilated.

Aerosol Generators

Aerosols may be used whenever the pest is sufficiently susceptible to extremely small quantities of toxicant. They make possible the rapid treatment of large areas with relatively small quantities of material, thus providing a means of eliminating certain types of pests that would otherwise be economically impractical to control. They are also used to kill insects in enclosed spaces by direct contact or by depositing a small quantity of insecticide on exposed surfaces. The residue left by an aerosol is generally ineffective for outdoor treatments to control insects. Nevertheless, the ever-increasing use of the outdoors for leisure activities around the swimming pool and in patio parties, cook-outs, etc., has resulted in increased interest in aerosol generators or "foggers" as a convenient and rapid method of ridding the area of flying pests for a few hours. The absence of appreciable residue from aerosols is in one way an advantage in certain areas, for there is less spotting of cars and windows than when mists are applied.

Aerosols are generally affected by wind and ground thermals, and usually can be effectively used outdoors only on cool days, or in the morning or late in the evening. The wind velocity must not exceed 5 miles (8 km) per hour. For some uses, aerosol generators should be adjusted to generate droplets 10 μ or larger in diameter; otherwise, the aerosol will rise too rapidly or insects may fly through it without picking up a lethal dose of insecticide.

Because so much aerosol application, both indoors and outdoors, is directed against mosquitoes, the optimum droplet sizes found by various investigators may be of interest. The most suitable droplet size for killing adult mosquitoes in a wind-tunnel was shown to be from 11.9 to 20.4 μ with wind velocities of 2 to 8 miles (3.2 to 13 km) per hour (Latta *et al.,* 1947). In a confined (still) atmosphere, the best droplet size for adult mosquitoes was found in 2 separate investigations to be 11 μ (LaMer *et al.,* 1947), and 15.8 μ (Yeomans *et al.,* 1949). Insects that fly through an aerosol with droplets under 10 μ in diameter accumulate many more droplets than those that remain motionless. For a good kill of mosquitoes in aerosols with such small droplets, it is necessary for the insects to fly about actively. Therefore, the addition to an aerosol liquid of a compound that exerts a strong irritating action, such as a pyrethrin, results in an increase in the amount of insecticide accumulated by the mosquitoes (David and Bracey, 1946). Mount (1970) concluded from a review of the literature that for ground aerosols, a droplet-size range of 5 to 10 μ MMD (mass median diameter) and for aerial aerosols a range of 10 to 25 μ MMD would be optimum, but added that these values needed to be confirmed by additional research.

Aerosol-generating equipment may be categorized into various types, depending on the method of generating and propelling the liquid.

Thermal-Aerosol Generators

There are 2 basic types of aerosol generators, one producing thermal aerosols and the other cold aerosols. In the thermal-aerosol generator, an oil-base insecticide formulation is introduced into a chamber that is heated electrically or by a flame to a temperature sufficient to cause immediate vaporization of the oil. Heat can also be produced by steam-velocity injection or by using the exhaust gases of a jet or internal-combustion engine. The droplets produced range from 0.2 to 9 μ in diameter. Only a few examples of models commonly used in and around the home will be described.

Portable Thermal-Aerosol Generators. Even small, light, portable generators (figure 17) can deliver a large cloud of dense aerosol that spreads through the air and rolls over the ground, killing flies, mosquitoes, gnats, and other insects on contact. The generator shown in the figure has a built-in electric pump that forces the solution of insecticide in oil through an atomizing nozzle that sprays the liquid into a heated aerosol-producing chamber. The insecticide solution is instantly vaporized, and a dense aerosol surges from the end of the "barrel," where a built-in thermostat maintains proper vaporizing heat. Operating temperature is reached in 6 minutes. The barrel of this generator is detachable, converting the unit into a sprayer for applying water-base insecticides and leaving enough residue to control crawling pests. This type of equipment is convenient and effective for "fogging" in attics, basements, garages, sheds, buildings, patios, flower beds, and around pools.

Fig. 17. Portable electric thermo-fogger. (Courtesy Burgess Vibrocrafters, Inc.)

Fig. 18. Wheel-mounted thermo-fogger. (Courtesy Burgess Vibrocrafters, Inc.)

For operations beyond the reach of an electric cord, portable models of pulse-jet aerosol generators (Swingfog and Dynafog) are available. The aerosol is generated by a pulse-jet heater, using gasoline and operating at 80 pulsations per second. An oil solution of insecticide is pushed from a tank to an outlet nozzle and then to an aerosol-mixing tube under air pressure tapped from the jet unit (Brown and Morrison, 1955).

Mounted Thermal-Aerosol Generators. There are many types and brands of wheel-mounted thermal generators for more extensive pest control operations. A 3-hp, 4-cycle gasoline engine supplies both the heat and power for the one shown in figure 18. This generator can produce up to 22,000 cu ft (623 cu m) per minute. The wheels and handle are easily removed, so that the aerosol-producing unit can be mounted on a jeep, pickup truck, or tractor. By adjusting the rate of flow to as much as 45 gal (170 l) per minute, droplets can be produced with VMD as high as 75 μ.

Because of the high temperatures required for the operation of thermal generators, carbon is formed in the heat chambers. The carbon should be removed after each use of the machine and before it has had time to harden and to accumulate in a thick layer. This will avoid interference of carbon deposits with the normal flow of insecticide through the machine. The carbon should be removed with care so as to avoid damage to the heat chambers or to small orifices which may be present.

Steam Aerosol Generators

The ordinary steam jenny used for steam-cleaning heavy equipment has been modified to perform as an insecticide "fogger." Superheated steam generated in a stainless-steel or copper coil is passed out of the machine through 1 or more nozzles. The insecticide, which may be in oil solution, emulsion, or aqueous suspension, is injected into the steam blast. The high velocity of the steam results in a shearing of the liquid to produce an insecticidal aerosol. Adjustments can be made to produce droplet sizes of any desired VMD, from 10 to 100 μ, depending on the insecticide formulation and on steam temperature (Glasgow and Collins, 1946; Gentner *et al.*, 1948; Scott and Littig, 1959).

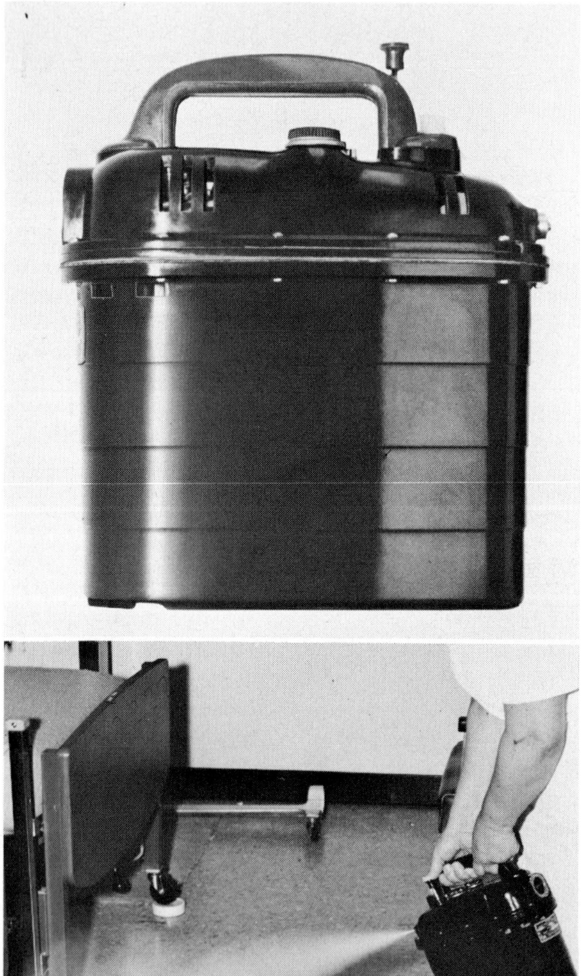

Fig. 19. An "electro-jet" aerosol generator utilizing superheated dry steam as a carrier for insecticides or germicides. (Courtesy Bishop-Wisecarver Corporation.)

Reasons for using steam are: (1) the temperature does not rise as high (superheated steam under pressure may rise in temperature to 400–500° F or 204–260° C), hence the damage even to organochlorine insecticides is not so high as with straight oil-burning machines; and (2) the steam evidently hastens condensation, making the drops more visible and presumably larger.

Figure 19 shows a type of aerosol generator that utilizes superheated dry steam as a carrier for insecticides or germicides. It has become a popular aerosol generator among pest control operators in California. Softened or distilled water and insecticide solution are each poured into a tank, as indicated on the label next to the filler cap. The machine weighs about 15 lb empty and 17 lb when filled with water and insecticide liquid (about 7 and 8 kg, respectively).

The equipment is connected to a 115-volt outlet and allowed to heat until the indicator light goes out. The first heating requires about 8 minutes, and reheating requires 4 or 5 minutes. This results in a 125-psi pressure, and enables one to make 200 half-in. (12-mm) strokes of the plunger. A longer stroke may be used to generate greater pressure and volume of liquid to force the aerosol into cracks, crevices, and hard-to-get-at places. Professional pest control operators generally use 2 machines on a job, so that one is being reheated while the other is in use. To operate, the machine is held in the right hand and the plunger is pushed down with the thumb, as shown in figure 19. Continuous pumping will produce a steady, high-pressure steam jet. When treating for cryptobiotic insects such as cockroaches and silverfish, it should be directed toward their hiding places.

Cold-Aerosol Generators

Because of the adverse effect of high temperatures required in thermal-aerosol generators on certain insecticides such as pyrethrins, the use of mechanical generators, cold-aerosol generators, or cold foggers can be particularly advantageous. However, they are also used for other types of insecticides, and are gradually replacing the thermal machines. Another indictment against thermal-aerosol generators is the highly visible and nondepositing very fine smoke which impairs the visibility of motorists and causes accidents when allowed to drift across highways. Thermal-aerosol generators give a distinct bimodal droplet size distribution, separated by a wide range. Thus, a thermal fogger may give a peak of 10 to 15 μ and another in the smoke range of 0.1 to

0.5 μ. Droplets in the 0.1- to 1-μ range in an aerosol cloud greatly affect visibility (N. B. Akesson, correspondence).

The cold-aerosol generators do not produce as much visible aerosol, and therefore are preferred in present-day vector control as well as for general home and garden use. For the treatment of enclosed spaces, of course, the thermal and cold types of equipment are equally suitable. Both portable and vehicle-mounted equipment is available for the generation of cold aerosols. It is often stated that one needs only to adjust certain types of "mist blowers" to produce aerosols, but in actual fact most mist blowers will not produce drops below 100 VMD. To do this requires more energy in the form of heat, or mechanical atomizers (spinners), or gas pressure (paint sprayers).

Chemical Flashing or Freon-Type Aerosol Generators. Insecticide aerosols released from pressurized canisters have shared the great popularity of aerosol products in general. They have advantages greatly prized by the modern public —they are convenient and easy to use and are not "messy." Against mosquitoes, gnats, house flies, and most flying insects, they are also remarkably effective. The aerosol droplets strike the insect directly. Aerosols are generally ineffective against those insects that are expected to crawl over the residue deposited by the aerosol. Only by holding the valve of an aerosol canister close to the treated surface when applying an insecticide can a reasonably good deposit be obtained. An aerosol canister is available with a plastic extension tube 5.75 in. (14.5 cm) long and with an outer diameter of 0.75 in. (2 cm). With this canister one can concentrate heavier deposits of liquid more closely to the desired surfaces and can also apply the liquid into cracks and crevices more accurately and with less waste of material. This type of treatment is being recommended for cockroach control as the "Whitmire Prescription Treatment."

The optimum range of droplet size for an insecticidal aerosol is 5 to 20 μ. The smallest and most convenient unit for generating a cold aerosol is the aerosol canister, originally called the "aerosol bomb." In these canisters, aerosols are produced by the rapid volatilization of a compressed and liquefied gas (usually freon), which is called the *propellant* and is contained in the canister. When liquefied, the gas is a solvent. To this solvent has been added 10 to 20% of a slowly volatilizing dilute insecticide oil solution. When pressure is released, the solution is discharged from the pressurized canister through a valve, the propellant is instantly converted from a liquid to a gas (a process called "flashing"), and an aerosol is formed as the toxicant is dispersed in very finely divided droplets of oil (petroleum distillate or other solvent).

The aerosol canister is a convenient device for producing small amounts of aerosols, and is especially suitable for indoor use. The container was originally a heavy steel or brass canister, strong enough to withstand a gauge pressure of about 70 psi at 70° F and 120 psi at 100° F (21 and 38° C). To reduce the cost of the containers, the thick-walled canisters were replaced by thin-walled, cylindrical canisters similar to beer cans, and gauge pressures were reduced to 25–38 psi at 70° F with little or no decrease in insecticidal efficiency. However, propellants of higher boiling points, such as an 11:6 mixture of Freons 11 and 12,[4] or methylene chloride, were required. Auxiliary solvents, such as polymethyl naphthalenes or cyclohexanone, are usually required to dissolve the insecticide completely.

Only 2 or 3 g of aerosol per 1,000 cu ft (28 cu m) are required to knock down the usual household flying insects. Directions on the canister state how long the valve should be released to provide this amount of aerosol. The canister should not be placed near flame, for the excessive heat may cause it to explode.

Insecticides used in aerosol canisters generally include a fast "knockdown" agent, such as the pyrethroids (pyrethrins and allethrins) or dichlorvos (DDVP); and synergists, such as piperonyl butoxide, sulfoxide or sesame oil extract, etc., may be added to increase their potency. The fast-knockdown agents mentioned have a very brief residual life, and the pyrethroids have the further disadvantage that insects may recover after their initial paralysis. Therefore, low-toxicity insecticides are frequently added that have longer residual activity and from which insects are less likely to recover once they have received a sufficient dose for knockdown. For example, an aerosol formulation may contain 0.2% pyrethrins, 1% synergist, and a chlorinated hydrocarbon, organophosphorus, or carbamate insecticide. An aerosol is sometimes desired that contains only pyrethrins and synergist. The concentration of pyrethrins is then usually 0.5%.

[4] These propellant gases were originally referred to as "Freon-11" (trichloromonofluoromethane) and "Freon-12" (dichlorodifluoromethane), but are now more frequently called Propellant-11 and Propellant-12.

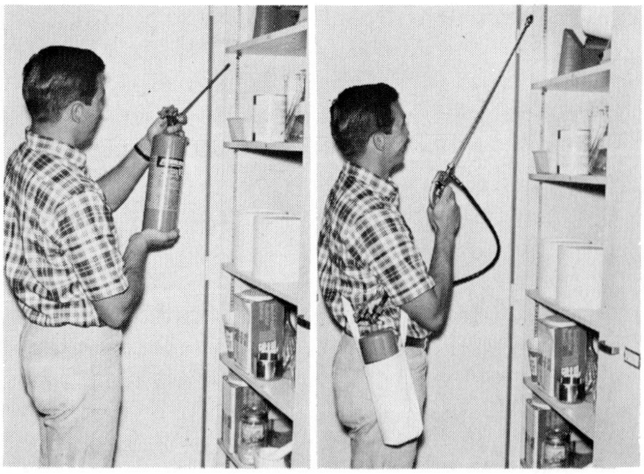

Fig. 20. Five-lb aerosol canister with a 13-in. applicator tube (*left*), and a 25-in. applicator tube on a 38-in. flexible hose, with the aerosol canister in a carrying sack (*right*). (Courtesy Virginia Chemicals, Inc.)

Pest control operators often find it advisable to use a 5-lb (2.25-kg) aerosol canister rather than the more common 1-lb (0.45-kg) can. The larger canister is more economical, and allows application for extended periods with the same container. For one type, either a 13-in. (33-cm) fixed applicator tube or a 25-in. (63-cm) applicator tube on a 38-in. (1-m) flexible hose can be attached to the canister. When using the flexible hose, the canister can be placed in a carrying sack and suspended from the shoulder (figure 20, *right*). The hose and long applicator tube allow one to spray into the highest shelves, spaces between shelves, spaces between cabinets and ceilings, spaces above high molding, etc., without excessive reaching or kneeling, and without having to lay the canister on its side. However, it must be borne in mind, when using the illustrated equipment, that the canister with the hose attachment is constructed to release aerosol twice as fast as the one with the stationary applicator tube. With 5 lb (2.25 kg) of liquid, only 28 to 40 minutes of intermittent use is possible with the canister with the hose attachment, compared to 75 minutes with the one having the 13-in. tube.

A 10-lb (4.5-kg) canister has been developed recently that contains only pure technical pyrethroid dissolved in the propellant fluorocarbon gas, thus greatly increasing the area that can be treated with a single filling. Particle sizes of the aerosol range from 0.5 to 32 μ in diameter, with 83% by weight below 15 μ. When treating for cockroaches, the canister may be placed on the floor, and a 5- or 8-ft (1.5- or 2.4-m) hose allows the operator to reach their usual hiding places.

An advantage of aerosol canisters or any type of easily portable aerosol-generating equipment in household insect control is that the aerosol can be directed into the usual areas of concentration of insects, such as in pantries, cabinets, closets, under sinks, and under furniture or appliances to concentrate the insecticide in areas of major infestation. In cockroach control, when pyrethrins, pyrethroids, or dichlorvos are used, the insects are flushed out, usually lie paralyzed on the floor, and can be swept up and destroyed in order to avoid possible recovery. This procedure is particularly effective when combined with a conventional treatment with an insecticide of long-lasting residual efficacy. In cockroach control, the types of aerosol-generating equipment that merely project the aerosol into the general area, and cannot be manipulated to direct the aerosol into the usual hiding places of the insects, have not been found to contribute any measurable degree of added effectiveness to residual treatments when evaluated a month after treatment (Reierson, 1973).

Hi-Fog Machine. This aerosol generator (figure 21) is used by many structural pest control operators. Any liquids can be used effectively except for those containing wettable powders. High pressure generated by compressed nitrogen gas is used to force an aqueous or base-oil solution of an insecticide through a "turbulence chamber," thus producing the atomization. The nitrogen gas is permanently retained. The energizing force is separated by a piston from the liquid being dispersed; nothing is released but the atomized solution, without a propellant gas.

A high-pressure, hydraulic, electrically operated pump is an independent unit that is used to fill the applicator. The liquid is lifted from any container and injected into the pressure chamber, thus depressing the separating piston against the nitrogen gas. Without liquid in the chamber, the nitrogen gas exerts a pressure of 200 psi. By pumping slightly more than 3 qt (3 l) of liquid into the chamber, the pressure may be

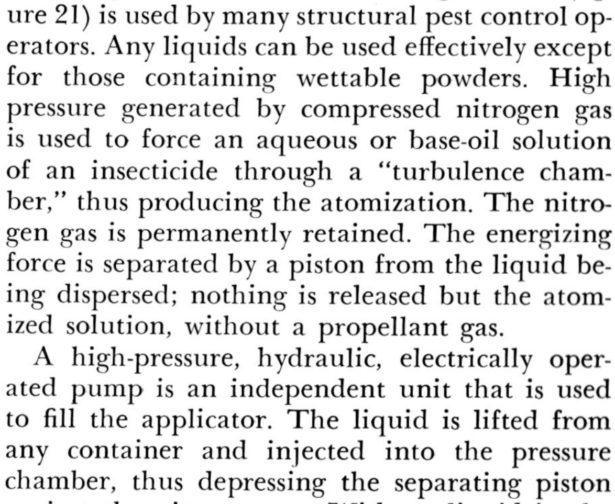

Fig. 21. The Hi-Fog machine, a high-pressure aerosol generator built on a backpack. (Courtesy Hi-Fog Company.)

increased to a recommended 600 psi. The apparatus is so designed that the rate of discharge is uniform at all pressures.

The discharge applicator is provided with flexible pressure tubing at the end of which is a rod and nozzle. The applicator is 18 in. (46 cm) long, weighs 17 lb (7.72 kg), and the cylinder is made of stainless steel. The discharge applicator can be either carried in a backpack, using a 3.5-ft (1-m) flexible hose, or it may be placed on the floor and a 10-ft (3-m) hose can be used to reach a considerable distance on either side of the equipment. In this way, the equipment can be used like the 10-lb (4.5-kg) aerosol canister described earlier, but with the added advantage that it is refillable, greatly reducing the cost of treatment.

Important advances have been made in the Hi-Fog machine during the past decade. The earlier machine produced droplets with a VMD of between 50 and 75 μ with base oil. Despite the name "Hi-Fog," it was actually a mist generator. The currently available Hi-Fog machine can be used with either water or base oil. Two nozzles are supplied. The one with the smaller orifice delivers water emulsions with a droplet size between 10 and 15 μ VMD, and the one with the larger orifice delivers droplets of an average of around 30 μ VMD. With base oil, of course, the droplet sizes are much less. The spray pattern can be adjusted at the nozzle for a fan or cone spray, or a "pinstream." A motorized unit of greater capacity can be carried about in a pickup truck and used for outdoor work. Even with water, either Hi-Fog unit can be correctly referred to as ULV (ultralow-volume) equipment. Using base oil, the equipment can be used to apply space sprays.

There are certain advantages in using water as the insecticide carrier when residual deposits are desired, as in cockroach control. Fire hazard is eliminated. With the Hi-Fog machine, the water droplets do not float about excessively. They settle out of the air rapidly, for they are not mixed with a propellant gas. Since a long-floating fog is eliminated, the inhalation of irritating and toxic droplets is avoided. With the nozzle of the Hi-Fog machine delivering droplet sizes of 10 to 15 μ, we have deposited only 30 to 50% as much insecticide as with compressed-air sprayers and have obtained significantly less effective control of cockroaches, based on evaluations made 3 months after treatment. A slower movement of the rod, larger droplet sizes, or increased concentration of insecticide will, of course, result in a heavier deposit. We have found that, with reasonably uniform distribution of spray, the degree of cockroach control is dependent on the amount of insecticide applied. With fogging equipment one must apply approximately as much insecticide on a treated area as with compressed-air sprayers in order to obtain the same degree of long-term control.

Nearly all insecticides are repellent to cockroaches. The repelled insects tend to seek hiding places in areas not treated in the usual "spot treatments" applied for cockroaches, silverfish, and other cryptobiotic insects. Therefore, the operator should attempt to cover all possible hiding places with the spray. It is also important to be able to supply sufficient insecticide rapidly on all areas over which cockroaches are likely to crawl without wasting the material or staining the sprayed surfaces. Equipment with which water-droplet size can be accurately controlled is an important contribution toward this goal.

Electrically Powered Mechanical Cold-Aerosol Generators. There are various types of electrically operated equipment that produce cold aerosols with a VMD under 25 μ and are generally designated as "cold foggers" by their manufacturers. The liquid is mechanically broken into aerosol-size droplets. Figure 22 shows 2 examples of such equipment. These generators are used for the dispersion of pesticides, germicides, and deodorants in enclosed spaces, such as warehouses, factories, schools, farm buildings, and nurseries. With this equipment, fogging with disinfectants has been found to be an effective supplementary technique in contamination control in hospital and institution maintenance. The cold aerosols of smaller droplet sizes work well indoors. Cold aerosols are also widely used outdoors for adult mosquito, gnat, and fly control. These are the "coarse aerosols," with a VMD of 20 to 75 μ.

The generator shown in figure 22 (*top*) is 13 in. (33 cm) high, and weighs 13 lb (6 kg). It holds 1 gal of liquid, and is adjustable to 3 rates of liquid output: 1, 2, and 4 gal (3.8, 7.5, and 15 l) per hour. Only 0.1% of the droplets are over 100 μ in diameter, 25.6% are between 50 and 100 μ, and 74.3% are below 50 μ. The manufacturers claim that the smallest droplets are 5.5 μ in diameter. (*See next page.*)

There is currently much interest in developing equipment with a flexible hose by means of which aerosol can be directed into areas of high insect concentration or deeply into otherwise

Fig. 22. Electrically operated mechanical aerosol generators. (*Top,* courtesy AFA Corporation; *bottom,* courtesy Root-Lowell Corporation.)

Fig. 23. *Above,* a mechanical or cold-aerosol generator, courtesy Buffalo Turbine Agricultural Equipment Company; *below,* a diagram of working parts, courtesy N. B. Akesson.

inaccessible insect harborages, as in cockroach control. Figure 22 (*bottom*) shows an electrically operated machine that weighs 11 lb (5 kg) and has a tank that holds 1.5 gal (5.7 l). It can be conveniently carried by means of a strap 2 in. (5 cm) wide. The flexible hose has an over-all length of 2 ft (60 cm), including the nozzle. A flow-control needle valve gives precise discharge volume control, ranging from 0.025 to 13.2 oz (0.074 to 390 cc) of liquid per min. The machine discharges droplets of 9.3 μ VMD, and no droplets are more than 30 μ VMD. The air-blast speed is 74 mi (119 km) per hour, and the "effective discharge carry" in still air is 100 ft (30 m).

Mechanical Cold-Aerosol Generators Driven by Gasoline Engines. In one type of cold-aerosol generator of "cold fogger," the insecticide liquid is broken up mechanically in a spinning perforated metal cage (figure 23) into aerosol-size droplets. It weighs only 14.5 lb (6.6 kg), the frame, shroud, and fan being of aluminum, and is driven by a gasoline engine. The fan has a capacity of 750 cu ft (21 cu m) per minute, a horizontal range of 60 ft (20 m), a vertical range of 30 ft (10 m), and an exit velocity of 90 mi (145 km) per hour. The pressurized tank has a capacity of 40 oz (1.2 l) of insecticide, enough to cover an area of 5 acres (over 2 ha). The liquid is broken up mechanically into very fine droplets, said to be nearly all under 33 μ in diameter. According to figures supplied by the manufacturer, the highest percentage (22) of droplets is in the 12.5-μ range, and the VMD is about 10 μ. Adjustments can be made to produce other droplet-size ranges.

Another type of machine that may be used in the "fogging" of residences and other buildings is the Micro-Gen cold-aerosol generator. The smallest model (figure 24) weighs 25 lb (11.3 kg) and has 1 nozzle; the medium model weighs 125 lb (57 kg) and has 2 nozzles; and the largest weighs 250 lb (113 kg) and has 4 nozzles. These machines operate on the principle of the Belvoir-type[5] aerosol generator (figure 25). Air and liquid are forced through 2 static air baffles that twirl them and break the liquid into very fine droplets. The aerosol is discharged at only moderate velocity. Droplets of 15 μ VMD were produced in tests with Belvoir-type equipment (Akesson, 1971).

Fig. 24. Model S1W-5 Micro-Gen cold-aerosol generator. (Courtesy Micro-Gen Corporation.)

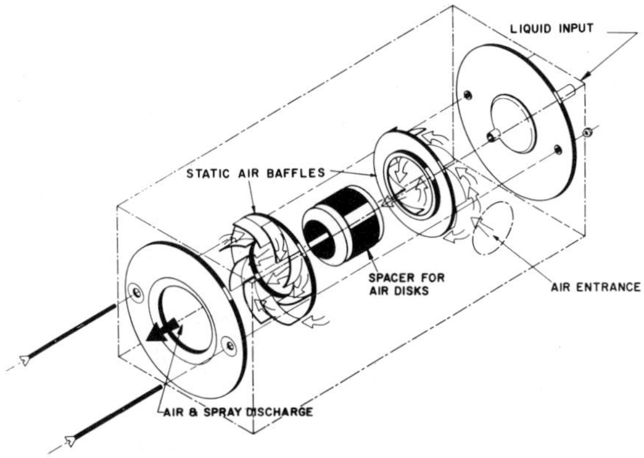

Fig. 25. Exploded view of the Belvoir atomizer, illustrating the principle of the Micro-Gen aerosol generator. (Courtesy N. B. Akesson.)

The smallest wheel-mounted Micro-Gen unit (Model S1W-5, shown in figure 24) is probably the most suitable one for treating apartments, houses, and most commercial buildings because of its small size and better maneuverability. This model is 12 in. (30 cm) wide, 16 in. (40 cm) long, 12 in. high, has a 2.5-hp gasoline engine, an emission rate of 0.5 to 5 oz (15 to 150 cc) per minute, and a capacity of 3 pt (1.4 l) of liquid insecticide. The larger units would appear to be more useful for purposes such as treatment for flies in barns, mosquito abatement, golf-course maintenance, and treatment of recreation areas. Micro-Gen equipment also includes a small, portable unit (HCS 1-2) that weighs 20 lb (9 kg), has a 2.5-hp, 2-cycle engine, a tank capacity of 1 qt (1 l), and discharges 0.5 to 2 oz (15 to 60 cc) per minute. It has a 4-ft (120-cm) hose, which is a useful feature when treating cryptobiotic insects such as cockroaches and silverfish.

[5] The mechanical principle involved was utilized in the construction of a military nonthermal aerosol generator by the Army Engineer Research and Development laboratories at Fort Belvoir, Virginia (Morrill *et al.*, 1955), with a nozzle credited to A. H. Yeomans. Mount *et al.* (1969) used this type of equipment mounted on a truck or as an ultralow-volume (ULV) aerial nonthermal aerosol generator for mosquito control in Alaska.

PRECAUTIONS

Petroleum products are the usual diluents for insecticides used in generating aerosols. Only the insecticides that are registered by the appropriate federal or state agencies for indoor "fogging" and only the diluents approved by the manufacturer of the fogging machine should be used. A petroleum diluent will result in an oil-enriched atmosphere. With a thermal-aerosol generator, for example, 1 gal (3.8 l) of kerosene can render 2,000 cu ft (57 cu m) of air explosive. Even at a concentration of 1 gal in 50,000 cu ft (1,416 cu m) of space, well below the lower explosive limit, an open flame can ignite the oil, and therefore this concentration should not be exceeded. The appearance of the usual aerosol of this density is indicated by the fact that a concentration of 0.67 gal (2.5 l) per 50,000 cu ft reduces visibility to 2 ft (60 cm) (NPCA, 1970).

Before fogging, all pilot lights and other open flames should be extinguished. Gas and electric current should be turned off at the main valve and switch box, respectively. Windows should be closed, and all other openings to the outside sealed off. The treated building should be vacated, and all domestic animals should be removed. A paper covering will prevent oil droplets from settling on house plants. All food should be removed from the building, unless it is in containers that exclude aerosol droplets or insecticide vapors. During the fogging operation, the treated building should be locked, and warning signs, stating that the building has been fogged, should be placed at all entrances. The building should not be entered for at least 4 hours following treatment, and should then be ventilated until considered by an experienced person to be safe for occupancy. The period required for ventilation will, of course, depend on the type of building, the outside air velocity, etc. After the building has been thoroughly ventilated, pilot lights should be relighted (NPCA, 1970).

A small building may be fogged entirely from outside entrances. If it is necessary to work under conditions in which the aerosol vapors may be inhaled, an organic vapor respirator (figure 6, chapter 2) approved by the United States Bureau of Mines should be worn. An approved type of fire extinguisher should be readily accessible and ready for use.

The electric cords of any type of electrically powered equipment should always be heavy-duty and checked for breaks or cracks in the insulation or plugs. One should not attempt to patch electric cords with tape; the cord should be replaced or the broken portion removed and reattached properly to the plug. When the operator has a choice, he should use a larger electric motor than is required for the job, so as to have adequate power without overloading.

Electrically operated equipment should be wired with a 3-wire circuit, with the third wire always grounded when in use. It is convenient to plug 3-wire extension cords and tools into a 3-prong (Durotronic) adapter. For use on a 2-hole receptacle, press a latch on the adapter, and the third prong swings out of the way. Another handy safety device is the "ground monitor" or "receptacle circuit tester." When this is plugged into a 3-hole receptacle, 3 lights indicate whether the ground wire, hot wire, or neutral wire is operating, if the hot and ground wires are reversed, if polarity is reversed, or if the wiring is correct.

Mist Blowers

As with the aerosol generators, the principal advantage of using a mist blower is that a small amount of concentrated pesticide can be rapidly and economically applied to a large area. For example, a large city or large portion thereof may be rapidly treated during disasters or in periods of high insect incidence. Mist blowers do not produce sufficient atomization energy to produce droplets below about 75 μ VMD. They generate a mist that may result in spotting of cars and windows and, particularly when oil solutions are used, shrubbery and trees may be damaged if the machines are not operated properly. On the other hand, the larger droplets produced by mist blowers allow their use during the day, and in fact a light breeze provides for improved distribution of insecticide. Also, mist blowers deposit more insecticide than do the foggers, and therefore provide longer-lasting residues.

Portable Mist Blowers

Portable mist blowers may consist merely of containers for the spray liquid, which can be pressurized sufficiently to cause the liquid to be finely atomized as it passes through a small orifice. The paint-gun or DeVilbiss-type atomizer lends itself well in experimental work for producing small quantities of fine mists, such as those used in Peet-Grady chambers or their more modern equivalents, and in turntable tests. The twin-fluid DeVilbiss atomizer can break up a liquid down to aerosol-size droplets, and is in fact one way to produce an aerosol with or without blowers. On the other hand, the most commonly used mist blowers have hydraulic nozzles only, and an air blower for propelling the spray. In these mist blowers, the blower stream aids in atomization, but in not quite the same way as does a paint sprayer, in which the air is introduced before the liquid is discharged as an internal mix, or directly after discharge as an external mix. But liquid and air are brought together at practically the point of discharge, in contrast with the mist blower, which may be thought of as a nozzle-in-the-airstream machine.

Another type of portable mist machine draws the liquid formulation into a stream of air by venturi action, in the manner of the "flit gun" already described. A number of small, electrically powered units (figure 26) operate on this principle. Electric mist blowers such as those shown in the figure convert either water- or oil-based liquids into fine, penetrating mists, with an effective range of over 75 ft (23 m), and they can be adjusted to generate droplets ranging from 15 to 60 μ VMD. It appears, from the range of possible droplet sizes, that the designations "mechanical fog generators" or "foggers" would be more appropriate than "mist blowers," but the latter term has been adopted in the trade for some reason.

The output of electric mist blowers may be adjusted from zero to 7.5 gal (28 l) per hour, depending on the viscosity of the liquid. Metering valves permit a wide range of output rates and mist patterns, from a light, airborne mist to a moist, residual spray. With an extension tube accessory, mist blowers can draw liquids directly from a bulk container, and can lift them over 6 ft (2 m).

Motorized Mist Blowers. Most mist blowers are motorized, and are similar in principle to the high-concentration–low-volume spray blowers used in orchard spraying. A large fan produces a column of air into which liquid insecticide is injected. The liquid is atomized and carried to the target area by the airstream. However, with a mist blower the droplets are small enough at the outset, and are discharged into an airstream with velocity high enough to break them up further into the sizes generally regarded as "mists." This gives the equipment a high degree of versatility, for larger droplets may be generated for spraying shade or orchard trees (figure 27), or smaller droplets may be selected for area control, as for mosquitoes.

With one type of mist blower, atomization of the liquid may be augmented by a system of disks rotating at high speed. It can be adjusted to produce droplets of different sizes. A blower with a capacity of about 4,500 cu ft (128 cu m) per minute forces the liquid from the disks. The machine, which can deliver up to about 250 gal (946 l) per hour, is mounted on a base which

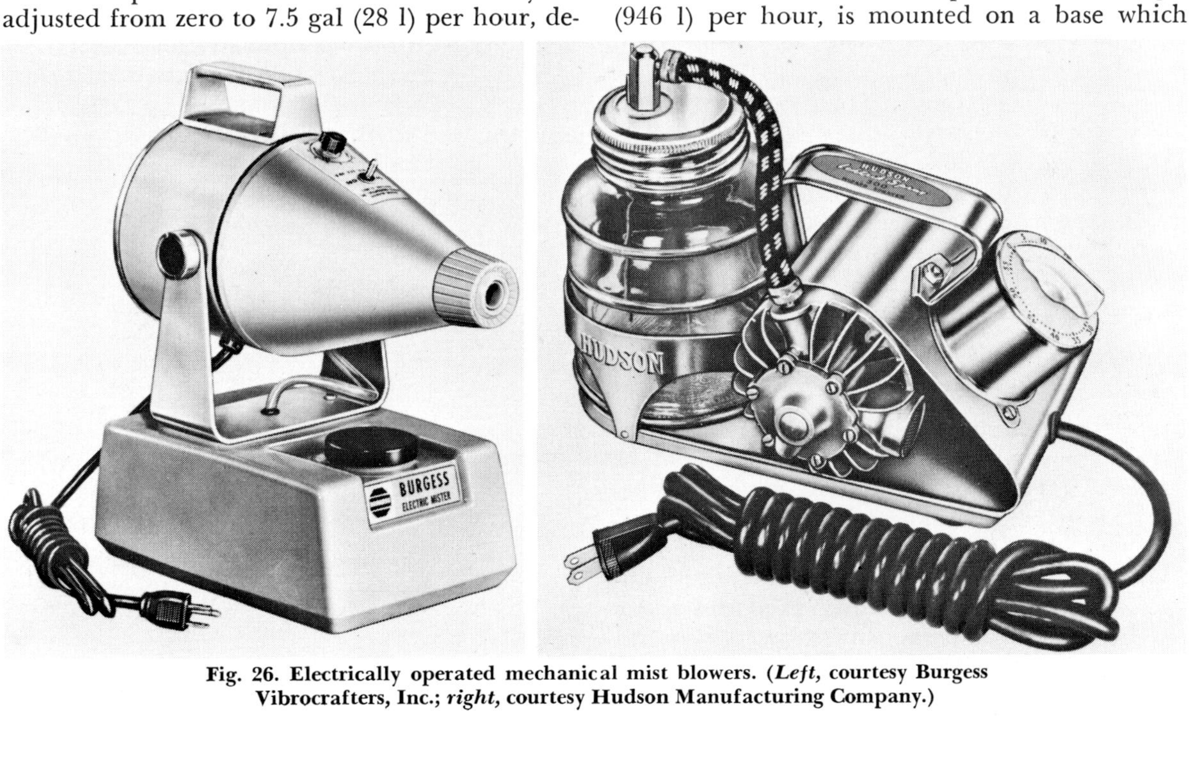

Fig. 26. Electrically operated mechanical mist blowers. (*Left*, courtesy Burgess Vibrocrafters, Inc.; *right*, courtesy Hudson Manufacturing Company.)

Fig. 27. Mist blower in operation. (Courtesy John Bean Division of Food Machinery and Chemical Corporation.)

allows a 20° tipback and a rotation of 360°. Another type is equipped with an airplane engine driving a pusher-type propeller. This creates an air blast that receives the insecticide from an independently operated supply system linked to a 550-gal (2,080-l) tank. A circulating pump and feed pump are driven by a universal power shaft. The entire nacelle can be swung in a full circle horizontally, and can be directed by remote control from the operator's station above the truck-driver's cab.

Electrically Operated Thermal Vaporizers, Generators, or Projectors

For the application of insecticides to confined spaces, "insecticide vaporizers," also known as "thermal generators" or "electric vaporizing devices," have been developed (Spear, 1952; Hensill and Leland, 1952). In this equipment, chlorinated hydrocarbon insecticides, particularly lindane, are held in electrically heated containers at a temperature permitting slow, continuous volatilization. Continuous lindane vaporizers are permitted in commercial establishments that do not serve, handle, or process food, but not in homes or in food establishments (Anonymous, 1971*d*).

Another type of equipment that may be designated as "lindane-vapor generator" or "lindane-vapor projector" (figure 28) can in 5 minutes convert an ounce (28 g) of pure, dry lindane crystals into an estimated 300 billion particles, enough to kill insects in 10,000 cu ft (283 cu m) of space. In this equipment, the lindane is converted into vapor rapidly and at a high temperature. In both types of equipment, the lindane acts as a fumigant, moving into cracks and crevices and giving complete coverage. The model shown in the figure weighs 28 lb (12.7 kg), and is 14 in. high, 13.5 in. long, and 9.5 in. wide (36, 34, and 24 cm). The hopper holds 1 lb (0.45 kg) of lindane, so that frequent filling is not necessary. The machine is started by flipping a switch, and a time clock turns it off automatically.

HAND DUSTERS

Very simple and inexpensive equipment may be perfectly satisfactory for the application of insecticide dust for occasional, limited use in the home or garden. In the absence of any manufactured equipment, dust can be applied to low-growing plants by shaking a dust-laden cloth over them. Better still, one may place the dust inside a burlap, cheesecloth, or muslin sack and shake the dust out over the plants. Dust may also be sprinkled from a perforated can.

Much use has been made of small, inexpensive, plastic squeeze-bottles (figure 29, *D*) designed as mustard or ketchup dispensers, for application

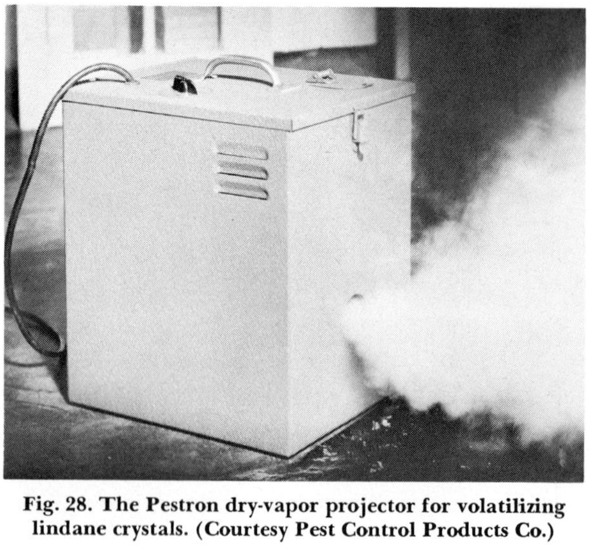

Fig. 28. The Pestron dry-vapor projector for volatilizing lindane crystals. (Courtesy Pest Control Products Co.)

Fig. 29. Hand dusters. *A* and *B*, bulb dusters; *C*, bellows-type powder blower (Getz gun); *D*, plastic "squeeze bottle."

of dusts in the home. When purchasing these squeeze-bottles, one should select the type with the widest possible opening, after the screw-top with spout is removed, so as to facilitate filling the bottle with dust. Brooks and Scott (1960) listed the following uses for plastic dusters: dusting personnel for body, head, or crab lice; dusting laboratory or pet animals for ectoparasites; dusting burrows for diseasebearing ectoparasites before killing rodents; applying finely powdered silica aerogel for control of fleas and cockroaches; dusting electrical apparatus (such as fuse boxes) and other equipment subject to spray damage when treating for cockroaches and silverfish; dusting anthills; applying granular larvicides for control of mosquitoes in small bodies of water; laying tracking patches during rodent surveys; filling bait stations with liquid rodenticides; and fumigating rodent burrows.

Plastic squeeze-bottles can also be used to apply dusts into cracks and crevices, wall voids, voids under cabinets, etc. However, persons who have to do considerable dusting may choose to purchase one of two types of small, hand-operated dusters commonly used by professional pest control operators: *bulb-type* (figure 29, *A* and *B*) and *bellows-type* (figure 29, *C*). With them, one can apply a greater volume of dust with greater force and effectiveness than with the squeeze-bottle dusters and, for most purposes, they are more convenient and efficient.

The *telescope-type* hand duster usually comprises the package in which the dust is sold, and consists of one cardboard cylinder within another. With the *plunger-type* hand duster (figure 30), the forward action of a plunger forces a current of air down the tube, carrying the dust with it. With this equipment, there is less control of the quantity of dust discharged and the placement of the dust than with the types of dusters shown in figure 29. It is more suitable for dusting in the garden than for use indoors.

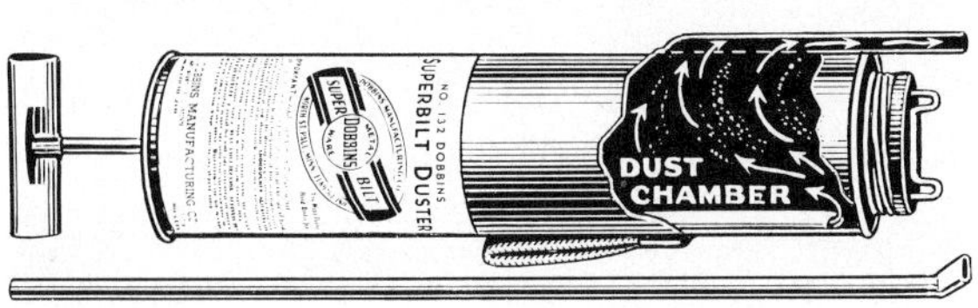

Fig. 30. Plunger-type hand duster.

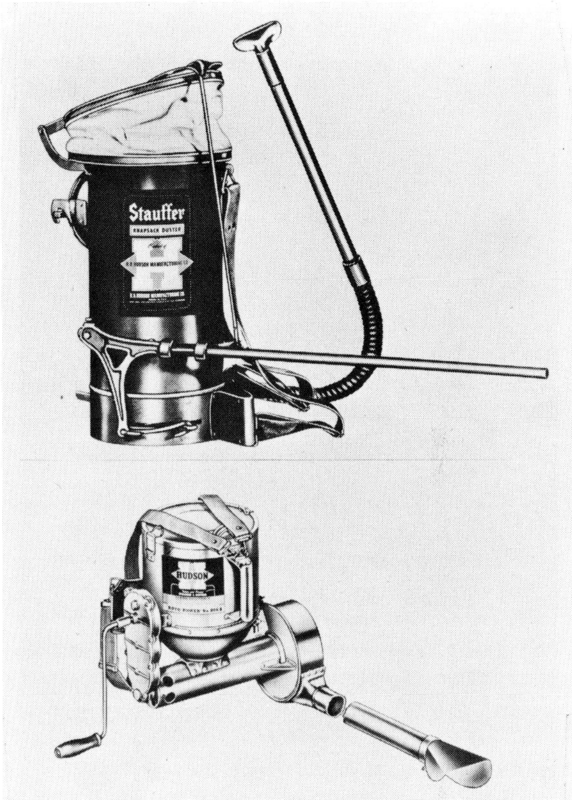

Fig. 31. Knapsack dusters. Bellows type (*upper*) and blower-fan type (*lower*). (Courtesy Hudson Manufacturing Company.)

Manually Operated Knapsack Dusters

In household pest control, the various kinds of knapsack dusters are suitable mainly for blowing dusts into large, enclosed spaces, such as attics, and the crawl space under buildings. They may also be used outdoors for the control of certain types of pests. A large area can be treated more rapidly with dusts than with liquid sprays. How-

Fig. 32. An electrically operated blower-fan type of duster. (Courtesy Neil A. Maclean Company, Inc.)

ever, dusts have a tendency to float away to nontarget areas, and this must be taken into consideration when deciding between dusts and sprays in outdoor pest control.

Devices such as bellows and blower dusters (figure 31) are used for propelling the air necessary for knapsack dusters. The bellows duster is operated by a lever that extends to the front and side of the operator. Air is forced through a discharge chamber by the expansion and compression of a bellows attached to the top of the duster. The blower duster has a crank that operates a blower fan and agitator. The fan forces a draft of air through a chamber connected to a dust hopper. The agitator in the hopper feeds the dust into the air chamber, after which it passes down through the delivery tube.

An electrically operated blower-fan duster (figure 32) has been used extensively for dusting large areas, such as attics and the crawl spaces under buildings.

(See also, figure 100, chapter 5.)

Modified Water-Type Fire-Extinguishers

The same areas in which hand-operated or electric knapsack dusters are used, as well as areas away from sources of electric current, can be just as effectively dusted with a water-type fire-extinguisher (figure 33). The tank is filled to about 80% of capacity with the dust, and is then pressurized to 100 psi from an air-compressor source. One drawback of this equipment is that dust eventually accumulates under the valve seat and may result in a small leakage of dust after the valve is closed. The valve seat must then be taken out and cleaned. One should check to see that the valve is clean before starting a dust job. A type of fire-extinguisher with a valve that gives a minimum of trouble in this respect is currently being sold in large numbers to pest control operators. When equipped with a spray nozzle, this "duster" can also be used for liquid spraying.

An advantage of the slightly modified water-type fire-extinguisher over the electric dust blower is that it has a narrow (7-mm) spout at the end of the delivery hose that makes it possible to blow dust into narrow cracks and crevices and into small holes drilled into plaster lath and kick panels to provide access to wall voids, voids under cabinets, etc., when "insectproofing" buildings with inorganic powders, such as silica aerogel or boric acid, during the construction of a building (figure 101, chapter 5).

(See under the heading "Power Sprayers" for precautions to be taken when operating pressurized equipment.)

At present, with the greater emphasis that is being placed on avoidance of environmental contamination, dusting with power dusters is no longer practiced in vector control, and has been almost completely abandoned in agricultural pest control. Dusts are much more likely to drift away from the target area than are sprays.

Fig. 33. Water-type fire-extinguisher used for blowing dust.

AIRCRAFT APPLICATION

The use of aircraft by the United States Bureau of Entomology and Plant Quarantine for the application of larvicides for the control of mosquitoes in Louisiana in 1923 was among the first uses of aircraft in pest control. Subsequent development of aircraft application always received a large part of its stimulus from mosquito-abatement work, and in World War II, airplane application of dusts and liquid concentrates in the mosquito-infested Pacific Islands proved to be the only practicable method of large-scale mosquito eradication in combat zones. The opportune discovery of the amazing insecticidal properties of DDT played an important role, of course, in the development of aircraft application. Gradually, the number of pesticides suitable for aircraft application increased, and there are now thousands of spray aircraft in operation throughout the country.

The main advantage of aircraft application is that it is rapid and saves labor; it also makes possible the application of pesticides when the soil is too wet to use ground equipment, a particularly important consideration in areas of frequent rain during the pest control season. Other obstacles to ground equipment, such as ditches and fences, are not hindrances in the case of aircraft application. In agriculture, aircraft application is advantageous for the treatment of closely growing and tall crops which may be damaged by even the high-clearance type of ground equipment. The ability of airplanes to apply insecticides with only a fraction of the fuel that would be used by ground rigs to treat a similar acreage is an important factor in the fuel-shortage crisis.

On the other hand, aircraft application has certain definite disadvantages. The drift of sprays into areas where they are objectionable has posed some difficult problems to pest control operators, even with ground equipment. With airborne equipment, this problem is greatly accentuated. The degree of thoroughness of application that can be obtained with ground equipment is usually not attainable with airborne equipment, even with the use of greater amounts of material for a given area. Applications from low flight elevations must be based on the deposit required in areas of the swath obtaining minimal deposit rates, even though other portions of the swath will then obtain deposits in excess of requirements (Chamberlain *et al.,* 1955).

Aircraft application continues to be important in mosquito control. When applying adulticides, applications should be made when the mosquitoes are most active, but these should not also be the hours when local bee populations are active (Burgoyne and Akesson, 1969).

Apparatus for Spraying

Concentrated sprays of pesticides have been applied by aircraft for about 50 years, and equipment has become highly specialized. In recent years, many safety and convenience features have been developed. A modern airplane may have a gross weight of 3,800 lb (1,725 kg), a 230-hp engine, and a hopper capacity of 200 gal (757 l), but heavier models with much larger hoppers are available (Joseph, 1973).

The spray pumps on aircraft may be of the rotary-gear, centrifugal, or turbine type. The rotary-gear pumps use a combination of meshed gears to move the liquid. They are satisfactory with solutions and emulsions, but wettable powder suspensions cause excessive wear of the teeth and sidewalls, and may even cause the pump to "seize." Since these pumps may create a positive pressure of as high as 500 psi, a pressure-relief valve must be installed between the pressure side of the pump and the shutoff valve.

Centrifugal pumps can be used with all kinds of spray chemicals with minimum wear, and since they are not of the positive-displacement type, they require no relief valve. Their top pressure is seldom above 70 psi. They may have either small impellers with wide vanes to move a high volume at low pressure, or large-diameter impellers with narrow vanes to move a lower volume at higher pressure. The best pump for aerial spraying is believed to be somewhere between these extremes (USDA, 1954).

The turbine-type spray unit can also handle all kinds of spray materials without undue wear, can produce somewhat higher pressures than the centrifugal pump and, unlike the latter, will pump in either direction. Like the centrifugal pump, however, it is not self-priming, and so must be mounted lower than the spray tank or be primed by hand.

Some pesticides that are most effective when applied as coarse sprays may be applied without pumps by using a gravity-feed system. In fact, any type of spray may be applied by gravity feed if certain kinds of atomizing devices are used instead of standard nozzles. When using a gravity-feed system, pressure and rate of flow will be decreased as the tank is emptied, but one can compensate for this by installing a float chamber or variable orifice between the tank and the nozzles. The rate of flow can also be kept constant by the use of a "negative-pressure vent system."

Power sources for spray-plane pumps include wind-driven propellers, hydraulic motors, electric motors, or the accessory-drive pad of the aircraft engine (USDA, 1954). The spray may be delivered by means of booms and nozzles similar to those used on ground equipment, rotating brushes or disks, or by venturi tubes. If abrasive mixtures are to be used, it is desirable to use orifice disks made of stainless steel or coated with ceramic material.

Fixed-Wing Planes

With fixed-wing planes, the principal factors affecting width of the spray swath are (1) fineness of the spray, since a finer spray spreads farther, and (2) length of spray boom and length of wings (wing length and boom length are usually about equal). The height of the plane aboveground is a factor affecting width of swath, particularly if the droplets in the spray are below 250 μ VMD.

Droplet size is influenced by the relative angle between air flow and nozzle-output flow. When the plane is in motion, smaller droplets are produced when the nozzle orifice faces forward than when it faces rearward. However, when facing forward, the nozzles should be directed downward far enough to keep the spray from being blown back into the nozzle body or supporting structure.

A rotating assembly is preferred by some operators because it can handle a larger variety of liquids. This consists of a horizontal shaft mounted parallel to the plane's longitudinal axis, with a fan on the front end and disks or a brush on the other. Spray liquid flows by gravity to the center of the brush or disk unit. When the plane is in flight, the unit is spun by a fan, and the liquid is thrown out by centrifugal force. The boom type of dispersing device (figure 34) is usually favored because it delivers a more uniform spray pattern than the rotating unit. Also, rotating devices are more expensive, and tend to break down.

Fig. 34. Fixed-wing aircraft with boom type of spray-dispersing device. (Courtesy Piper Aircraft Corporation.)

Spraying Over Urban Areas. Some aerial spraying for mosquitoes involves low flights over urban areas to apply "low-volume" sprays, a term generally considered to mean less than 0.5 gal (2 l) per acre (0.405 ha). Government regulations require the use of multiengine aircraft for such purposes.

A compact, electrically driven spraying unit for light, twin-engine aircraft was discussed by Burgoyne and Akesson (1969). Thirty gal (114 l) of spray liquid were carried by this aircraft in a fiberglass belly-tank. Eight nozzles, vertically oriented, were spaced at equal distances along a boom. The flow rate was 128 fl oz (3.8 l) per minute at 40 psi of pressure. At a flight altitude of 100 to 150 ft (30 to 45 m) and air speed of 160 mi (258 km) per hour, and spraying a 50-ft (15-m) swath, 0.75 fl oz (22.5 cc) per acre was applied.

Helicopters

Helicopters lend themselves well to insecticide application, especially in relatively small areas, areas bounded by tall windbreak trees, etc. Their commercial use has been limited by their relatively high initial cost as compared with fixed-wing planes. In addition, the maintenance costs of helicopters are rather high. This has limited their purchase and use, even in areas where, if other things were equal, they would be vastly superior to other means of pesticide application.

Helicopters provide for better control of dosage, uniformity of insecticide coverage, and for prevention of insecticide drift to nontarget areas. The current demand for pest control with minimal environmental disturbance may lead to greater interest in helicopters for this purpose. To quote from the report of the Commission on Pesticides and Their Relationship to Environmental Health (Mrak Commission) to the Secretary of the Department of Health, Education and Welfare: "Increased engineering development effort is needed for the design of equipment for, and the adaptation of helicopters to, the aerial spraying of pesticides" (Mrak, 1969).

When spraying from a helicopter, the balancing of the equipment is an important factor. Balance is usually maintained by installing 2 spray tanks, 1 on each side of the fuselage, directly under the rotor. The spray is pumped simultaneously from each tank. Rotating devices for the dispersion of the spray are not practical because of the slow flight speed. Boom-and-nozzle assemblies are mounted forward of the nose, directly in line with the rotor axis, or a short distance aft. The air-blast outlets of the engine-cooling blower system can be used to atomize the liquid. The spray is dispersed by the rotor's air blast.

Agricultural Aviation Literature

The International Agricultural Aviation Centre, The Hague, Netherlands, publishes the journal *Agricultural Aviation,* which contains articles on recent developments and has a good review section that abstracts all agricultural aviation literature.

The Centre has also published the proceedings of 4 International Agricultural Aviation meetings (1959, 1962, 1966, 1969), as well as a *Handbook for Agricultural Pilots* which lists specifications for all fixed-wing and rotary-wing aircraft.

Chapter 4 · *Classes of Arthropod Pests of the Urban Community*

MORE THAN a million species of animals have been discovered and named. They have been divided into the most inclusive of the taxonomic groups, called *phyla* (singular *phylum*) from the Greek *phylon,* meaning tribe or race. For examples, animals with a backbone, insects, snails, starfish, earthworms, roundworms, flatworms, jellyfishes, sponges, and amoebae are members of different phyla.

The members of each phylum have certain common characteristics. The largest phylum, Arthropoda, includes animals having a segmented body, bilateral symmetry, paired, jointed appendages, usually terminating in claws, a chitinized exoskeleton secreted by the epidermis and molted periodically, striated muscles, a tubular alimentary canal, a ventral nervous system, and a dorsal heart.

Within a phylum there are smaller groups of animals called *classes,* which also have common characteristics that distinguish them from other groups. This is evident by merely recalling the common names of representative animals in the different arthropod classes, such as crayfish, crabs, lobsters, and sowbugs (Crustacea); centipedes (Chilopoda); millipedes (Diplopoda); spiders, mites, ticks, scorpions, and harvestmen (Arachnida); insects (Hexapoda); and several others; each group brings to mind a distinct image.

Some of the arthropod classes contain economically important pests, and will therefore be briefly characterized in the present section merely to present in one place an over-all view of the phylum. They will be discussed further, in detail depending on their importance, in appropriate chapters and sections of this book. Within a class there are groups of animals called *orders,* with common characteristics that distinguish them from other groups. Again, the common names of representative groups of animals give a good idea of what is meant by the term. For example, in the class Arachnida are the spiders (Order Araneida) and related orders: scorpions, whip scorpions, pseudoscorpions, sun spiders, harvestmen, and the ticks and mites (figure 35). In the class Hexapoda are such important orders as silverfish, cockroaches, crickets and their allies, termites, thrips, lice, bugs, butterflies and moths, beetles, flies, fleas, and the bees, wasps, and ants. Each order has clearly evident similarities among its *families,* and each is different in important ways from the other orders.

Directional Orientation

Certain terms are used in biology to designate the various regions of an organism. The following terms will be found to be useful: *anterior,* at or toward the front; *posterior,* at or toward the hindmost part; *dorsal,* of or pertaining to the upper surface; *ventral,* of or pertaining to the undersurface; *proximal,* the part of an appendage nearest the body; *distal,* the part of an appendage farthest from the body; *caudal,* pertaining to the posterior or anal extremity; also, many words, such as lateral, marginal, etc., which are self-explanatory.

In description, one may employ adjectives formed from such words as thorax, abdomen, and pleuron—thoracic ganglia, abdominal appendages, and pleural regions.

CLASS CHILOPODA (CENTIPEDES)

The centipedes (figure 342, chapter 10) are the closest relatives of the insects for, like the insects, they have a single pair of antennae, breathe by a series of tracheae, and the reproductive organs open at the posterior end of the body. The body is flattened, and consists of 15 to 173 segments, each with a single pair of legs except the last two and the one just behind the head. The segment in back of the head bears a pair of ventral venom-secreting claws called *maxillipeds,* with which centipedes kill their prey. The antennae are moderately long, and consist of at least 12 segments.

Centipedes in the United States cause a wound that may be painful, but is seldom serious (see under "Centipedes" in chapter 9). The larger species of *Scolopendra* in tropical countries may reach 30 cm in length, and the amount of venom issuing from their claws is very painful and even dangerous to man.

CLASS DIPLOPODA (MILLIPEDES)

Millipedes (figures 342 and 343, chapter 10) differ from centipedes in that almost every segment bears 2 pairs of legs, the body is cylindrical instead of flattened, there are no venom-bearing claws, the antennae are short, and the reproductive organs open far forward on the body. Millipedes crawl slowly, whereas centipedes crawl rapidly. On the seventh segment of the male, one or both pairs of legs are modified as copulatory organs. A pair of scent glands on each segment, opening laterally, discharge a repugnant secretion. In some species, it can be discharged to a distance of nearly a meter and, in the case of some tropical species, has caused blindness in children (Hegner and Engemann, 1968).

CLASS ARACHNIDA (Spiders, Mites, Ticks, Scorpions, Whip Scorpions, Pseudoscorpions, Solpugids, and Harvestmen)

Most of these arthropods are terrestrial forms, although a few mites are aquatic. They have tracheae or book lungs for breathing, simple eyes (ocelli), no antennae or wings, and have 4 pairs of thoracic legs. They have a chitinous exoskeleton, and the body is divided into 2 portions—an unsegmented cephalothorax and an abdomen. The cephalothorax may be broadly joined to the abdomen or joined by a narrow pedicel.

Arachnids (figure 207, chapter 9) have 6 pairs of cephalothoracic appendages. The first pair, the *chelicerae,* are either 2- or 3-segmented, and may bear pincers for holding and crushing prey. However, they may also serve as venom-bearing fangs, intromittent organs for transferring sperm, or spinnerets. The second pair of appendages are 6-segmented, and are called *pedipalps.* These may also serve as intromittent organs, as pincers, for crushing food, as sensory organs, or for other functions.

Arachnids have 4 pairs of walking legs, usually 7-segmented. Each leg is composed of a *coxa, trochanter, femur, patella, tibia, metatarsus,* and *tarsus* (figure 207, chapter 9). Most arachnids are carnivorous, but because they have no mandibles, they commonly predigest their food in the mouth and then suck the juices into the stomach. Some species crush their food with their chelicerae. The black widow spider cannot crush its prey, but pierces it and injects a digestive fluid, then sucks out the digested food. Spiders breathe by means of "book lungs" (from 1 to 4 pairs) or tracheae, or sometimes both. Arachnids are oviparous, except for the scorpions and certain mites. There are 11 orders of living arachnids, of which 7 will be considered of more or less importance as household pests, although some are harmless and are of interest only because they may be confused with bona fide pests.

Order Araneida (Spiders)

Whereas the body of an insect is divided into 3 parts—head, thorax, and abdomen—as has just been noted, the body of the spider is divided into 2 parts, the head and thorax being intimately fused into a cephalothorax and the abdomen being usually soft, rounded, and unsegmented (figure 35). The cephalothorax and abdomen are separated by a narrow pedicel, as is also true of some

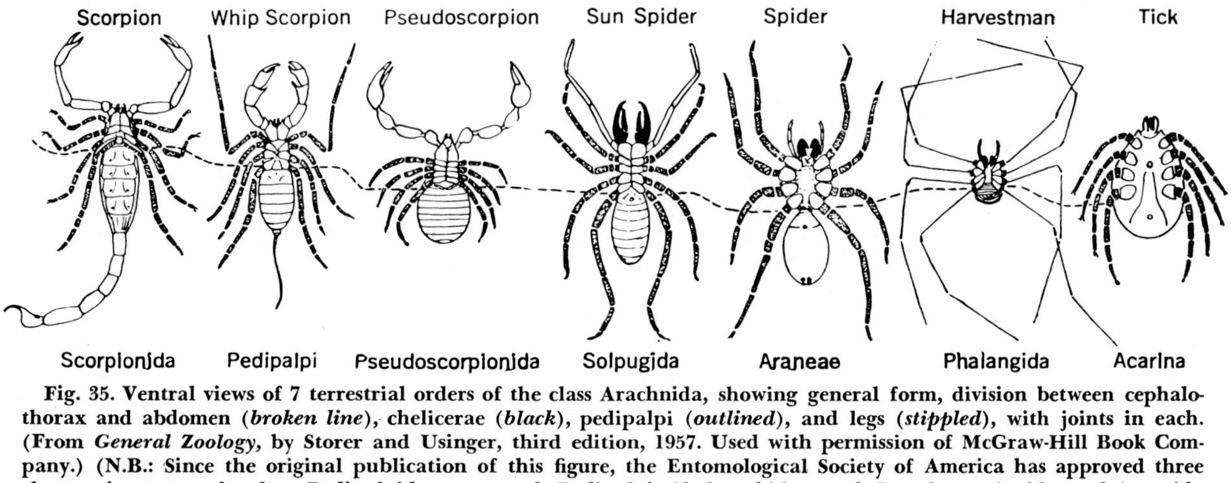

Fig. 35. Ventral views of 7 terrestrial orders of the class Arachnida, showing general form, division between cephalothorax and abdomen (*broken line*), chelicerae (*black*), pedipalpi (*outlined*), and legs (*stippled*), with joints in each. (From *General Zoology,* by Storer and Usinger, third edition, 1957. Used with permission of McGraw-Hill Book Company.) (N.B.: Since the original publication of this figure, the Entomological Society of America has approved three changes in names of orders. Pedipalpida now equals Pedipalpi; Chelonethida equals Pseudoscorpionida; and Araneida equals Araneae.)

other arachnids, including the solpugids and whip scorpions. The cephalothorax of spiders is provided with a dorsal, hardened shield or carapace, which bears the simple eyes (ocelli) at the front end. The eyes are usually arranged in 2 rows of 4 each, but there may be fewer eyes or none at all.

In the spiders, the female pedipalps (figure 207 [palpus], chapter 9) are short and leglike, but in the males they have become modified to form copulatory organs. The last segment is enlarged and knoblike, and is used for transfer of sperm. The spiders are the only arachnids in which the chelicerae are provided with venom glands that open at the tip of the fang. In some spiders, these glands are located entirely within the chelicerae, and in others they extend backward into the first section of the cephalothorax, as diagrammatically illustrated in figure 208, chapter 9.

The organs of respiration consist of leaflike external book lungs, or book lungs and tracheae. Most spiders have 1 pair of book lungs, but tarantulas have 2 pairs. They open on the ventral side of the abdomen as slitlike clefts. If a tracheal system is present, the spiracle is usually located just in front of the spinnerets.

Most spiders spin silken webs. The webs are spun from as many as 6 spinnerets, which are transformed vestigial legs and are located ventrally near the tip of the abdomen (figure 207, chapter 9). Webs may serve as nests or as traps for prey. A spider's silk must serve the dual purpose of transporting her to any part of the web, and also of trapping insect prey. She may have to secrete 2 kinds of web, dry and sticky, supplied by different glands in her abdomen. (Other glands supply the strong webbing for her egg sac.) For example, in the orbweaver's web, the central "lookout," where she spends most of her time, and also the "spokes" that radiate to the outer rim of the web, consist of dry webbing over which she can run rapidly, upside-down, aided by hooks, toothed claws, and bristles at the ends of her legs. She avoids the sticky threads that join the spokes and are for the purpose of catching the insect prey. She can perceive the slightest disturbance of any part of the web, her highly developed tactile sense compensating for her poor eyesight. After reaching her prey, the spider quickly envelops it in fresh threads, and bites it with her sharp upper jaws, injecting a venom that quickly kills it (von Frisch, 1960).

After the prey dies, the threads of the web entangling it are severed, and the prey is transported to the central "lookout" and hung up by a short thread. To consume her victim, the spider bites it, meanwhile secreting a digestive fluid, then sucks up the dissolved contents together with the digestive fluid. Dissolved muscles and intestines of the prey are imbibed within a few hours, and the indigestible remants are dropped out of the web (von Frisch, 1960).

Webbing may also be used for transportation when it is borne aloft by air currents. Gertsch (1949) made the interesting observation that although spiders were wingless, a young spider (spiderling) could float its threads on the breeze, often reaching tremendous heights and sailing for long distances, thus colonizing new areas at a rate not possible even for winged insects. He stated that the rigging of ships 320 km from the nearest land had been "showered with tiny aeronauts riding on silken streamers."

Some spiders do not spin true webs, but live in burrows (trapdoor spiders) or crevices (wolf spiders).

The courtship and mating of spiders is so strange and complicated as to be almost beyond belief. They are described in great detail in W. J. Gertsch's fascinating book, *American Spiders* (Gertsch, 1949). Soon after he becomes mature, the male spider transmits some seminal fluid to a delicate web which he has spun. He then collects the sperm and stores it in bulbs of his pedipalps (palpi) until the opportunity for mating arises. Herms *et al.* (1935) observed the sperm-induction procedure in the mating of black widow spiders and the courtship and copulatory behavior of that species, lasting over 2 hours. This was described in detail by McCrone and Levi (1964). Most bizarre is the sperm-induction sequence practiced by tarantulas. It is so complicated and replete with ritual that it sometimes requires 3 or 4 hours (Gertsch, 1949).

Spiders are predators, mostly on insects ensnared in their webs. A few species are very venomous. In the United States, the bite of the black widow spider and the brown recluse spider or related species may be very painful, and can lead to serious general systemic disturbances, or even death. The aid of a physician should be obtained as soon as possible.

Order Acarina (Mites and Ticks)

The body divisions of the Acarina do not correspond to those of other arachnids, for the head, thorax, and abdomen are fused together, forming an unsegmented body (figure 35). This fact has resulted in a special terminology. The *gnathosoma* or *capitulum* represents the first 2 primitive segments, and bears the chelicerae and pedipalps. The *idiosoma* includes the rest of the body. Many mites have transverse, fine lines, but these have no relationship to segmentation. The appendages of most Acarina are somewhat similar to those of spiders, except that the newly hatched young have only 3 pairs of legs. (Mites of the family Eriophyidae have only 2 pairs.) Most mites have a tracheal system opening through 1 to 4 pairs of spiracles, located near the bases of the appendages in the anterior half of the body. Some species have no tracheae, and obtain oxygen by diffusion through the cuticle. The majority of species are oviparous (egg-laying), but some are ovoviviparous, the eggs hatching within the body of the female. There are 4 stages in the development of mites: the egg, the hexapod larva, the octapod nymph, and the adult. In each postembryonic juvenile stage, there may be 1 or more instars.

Most mites are very small, and many species are pests of plants or stored food products. Other species are parasites of man and animals, and some are disease vectors or are responsible for allergies and asthma. Horsfall (1962) listed 36 genera of mites in 19 families in which there are 1 or more species of medical importance to man or domestic animals.

Ticks are the largest members of the Acarina. They are all bloodsuckers and, when fully engorged, some species attain a length of 1 cm and a few are considerably larger. They are serious pests of domestic animals, and are the only vectors of some severe diseases of man and animals.

Order Scorpionida (Scorpions)

Like the spiders, scorpions have 2 body regions, the *cephalothorax* or *prosoma*, and the *abdomen*. However, the abdomen is segmented, and consists of a large, anterior preabdomen or *mesosoma* of 7 segments and a long, narrow, tail-like posterior postabdomen or *metasoma* of 6 segments, terminating in a vesicle or *telson* bearing a venomous stinger. The anterior part of the cephalothorax bears 3 groups of eyes: a median group of 2 eyes and a lateral group of 2 or 3 eyes at each margin. The cephalothorax also bears small, chelate chelicerae, large, conspicuous, pincerlike pedipalps, and 4 pairs of legs. The general body form and thornlike rigid stinger of the scorpion identify it for most people. Figure 36 shows a scorpion of the family Vaejovidae. It ranges throughout the southwestern United States and northern Mexico, and is 11 to 12 cm long—the largest scorpion in the United States.

Fig. 36. Giant hairy scorpion, *Hadrurus hirsutus* (Wood), in the defense or attack position.

Scorpions breathe by means of 4 pairs of book lungs in the ventral blood sinus. Paired coxal glands, opening in the coxae of the third pair of legs, provide for excretion. They do not lay eggs, but give birth to active young. These climb upon the back of the mother and remain fastened by their pincers until after the first molt, nourished by yolk material stored in their bodies. All scorpions are venomous, but most species are not dangerous. In Arizona, however, there is a species that has caused more deaths over a 36-year period than all other venomous animals of that state, including rattlesnakes (see chapter 9, opening paragraphs).

Order Pedipalpida (Whip Scorpions)

In this order, the cephalothorax bears 8 eyes; the chelicerae are small, 2-jointed, and chelate; and the pedipalps are strong, and have 6 joints (figure 35). In many species, the last 2 segments of the pedipalps are modified to form pincers used in seizing prey, tearing it apart, then passing it to the chelicerae to complete the maceration. The first pair of legs are not for locomotion. These legs are very long, and the long, many-segmented tarsi are used as feelers. These harmless arachnids are nocturnal, hiding during the day beneath leaves, rocks, or logs, usually in damp places, although there are a few desert species (Storer and Usinger, 1957; Barnes, 1963).

There are 2 suborders. In the Uropygi, there are 2 glands that open one on each side of the anus and discharge a stream of fluid when the animal is irritated. This fluid has the odor of vinegar, which suggested to the Spanish-speaking settlers of the Southwest the name *vinagrón* (from the Spanish *vinagre*), later modified to "vinegarroon" in English. The last abdominal segment bears a *flagellum,* which is very long with some species. The body length may range from only 2 mm to as much as 6.5 cm, as in *Mastigoproctus giganteus* (Lucas) (figure 37), which occurs in

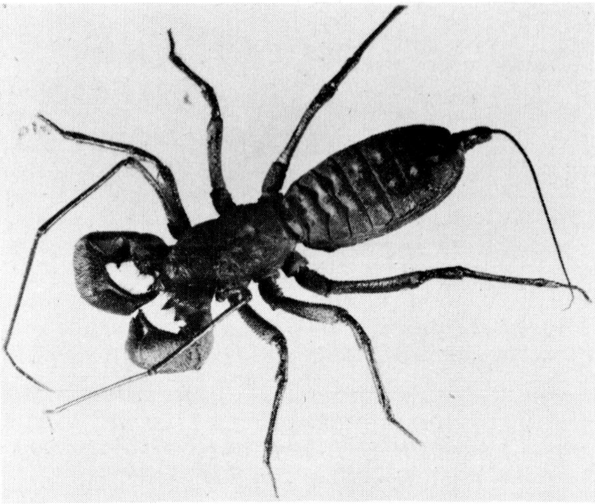

Fig. 37. Whip scorpion (vinegarroon), *Mastigoproctus giganteus.*

the southern United States from coast to coast. (The whip in this species can be twice as long as the body proper.) *Mastigoproctus giganteus* gets water from its prey and by drinking from a moist substrate which is located by the sensitive, antenniform forelegs. The whip scorpion probably does not have a protective epicuticular wax layer, such as that possessed by insects and many other arthropods, to protect it from desiccation. At 26° C (79° F), it readily loses water to the air at relative humidities up to 95% (Crawford and Cloudsley-Thompson, 1971).

In the suborder Amblypygi, the abdomen bears neither odoriferous glands nor a whip. In the United States the principal genus is *Tarantula,* although it is not a tarantula, which is a large spider. *Tarantula* is about 2 cm long, and has been reported from California, Texas, and Florida (Storer and Usinger, 1957).

Order Chelonethida (Pseudoscorpions)

The pseudoscorpions (figure 35) are relatively small, ranging from 4 to 10 mm in length. They resemble scorpions, but have no postabdomen or stinger, and are harmless. The abdomen is 11- or 12-segmented, and is rounded posteriorly. Their conspicuously large pedipalps bear lateral claws called chelae, with a poison gland for disabling their prey. Unlike scorpions, the pseudoscorpions do not possess book lungs. They have 2 pairs of ventral tracheal spiracles on the third and fourth abdominal segments. Like scorpions, they have a pair of glands that open on the coxae of the third pair of walking legs and provide for excretion.

Order Phalangida (Harvestmen)

The harvestmen or daddylonglegs (figures 35 and 38) are distinguished most readily by their long, slender legs. When they crawl, the small body appears to be suspended from the long, arched legs. The cephalothorax is broadly joined with the short, segmented abdomen, has 1 pair of simple eyes, and a pair of odoriferous glands. The pedipalpi are leglike. A pair of tracheal tubes opens on the second abdominal sternite. A pair of coxal glands, opening on each side between the third and fourth coxae, provide for excretion.

The harvestmen easily lose their legs or parts of their legs, and cannot regenerate them. They drink frequently, and must have water readily available. They are harmless, and should not be confused with true spiders.

Order Solpugida (Solpugids)

The solpugids, also called "sun spiders," are harmless to humans. They are predaceous arachnids that occur mostly in dry tropical and in warm temperate regions. Of the 200 known species, about 60 are found in the southwestern United States (Hickman, 1967).

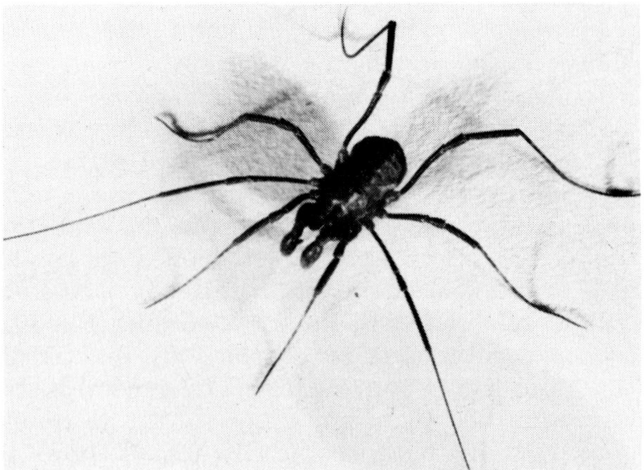

Fig. 38. A harvestman or daddylonglegs.

Solpugids (figures 35, 39) are most strikingly characterized by their enormous, 2-jointed chelicerae, with the pincers articulating vertically and a rounded abdomen without a tail-like postabdomen, telson, or stinger. They do not have venom glands. The pedipalps are simple, nonchelate, leglike, and terminate in an adhesive organ. Solpugids seize their prey with their pedipalps and pass it to the chelicerae, where it is

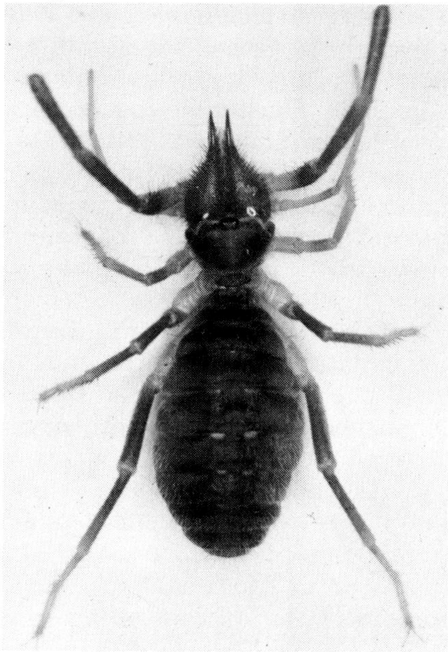

Fig. 39. A common solpugid, *Eremobates pallipes* (Say).

crushed. The macerated and predigested tissues of the prey are then sucked into the mouth and esophagus. Only the last 3 pairs of legs are ambulatory; the first pair is tactile. The abdomen is 10- or 11-segmented. Paired spiracles are located between the second and third coxae and in the third and fourth abdominal segments, leading to a well-developed tracheal system. A pair of coxal glands and a pair of Malpighian tubules provide for excretion.

CLASS HEXAPODA (INSECTS)

The hexapods comprise the largest and most highly developed group in the phylum Arthropoda. Some authorities estimate that over a million species have already been described, more than all other animal species combined. Insects also occur in incredible numbers. Williams (1964) estimated a total of 10^{18} live individual insects on the earth at any given moment. Adult hexapods may be distinguished from other arthropods by a number of structural characteristics, including a segmented body divided into 3 regions, the head, thorax, and abdomen. Adult hexapods possess 1 pair of antennae, 3 pairs of legs, and most have 2 pairs of wings, although some have 1 pair and others none. No other invertebrates have wings. Furthermore, insects have a front lip (*labrum*), 1 pair of mandibles, 1 pair of maxillae, and a hind lip (*labium*). Their gonopore is posterior, and they usually undergo a change of form (*metamorphosis*).

Respiration in hexapods usually takes place by means of a system of tubes called *tracheae,* that divide and subdivide internally until they terminate in the delicate network of *tracheoles* that reaches every organ, tissue, and cell of the body. Hexapods have a simple, open circulatory system.

The modern tendency is to divide the class Hexapoda into 3 subclasses. Subclass Myrientomata contains 1 order (Protura), subclass Oligoentomata contains 1 order (Collembola), and the subclass Insecta (Euentomata) contains the remaining 23 orders. This modern classification is based on the evidence that the 3 subclasses represent distinct evolutionary lines (Belkin, 1972). However, since the present work has primarily economic rather than taxonomic objectives, for the sake of convenience all orders of hexapods will henceforth be referred to as "insects." Likewise, for the sake of brevity and convenience, the older concept of all Hexapoda being divided into 2 subclasses, Apterygota (primitively wingless insects without metamorphosis) and Pterygota (winged insects, with some secondarily wingless) will be retained.

The Exoskeleton

The skeleton of an insect is external. It forms a shell-like support, and provides protection for the internal organs. This outer shell is light, but very strong. This is demonstrated by the tremendous impact that an insect can withstand without injury—as, for example, when colliding with a solid, stationary object during flight. The hard material on the exterior comprises a thin covering of the entire insect known as the *cuticle* (figure 40). It is secreted by a single, continuous layer of living cells, the *epidermis* (sometimes called "hypodermis"), which is supported on a noncellular *basement membrane.* It was once assumed that the polysaccharide *chitin* was the main constituent of the cuticle of insects and other arthropods, and such terms as "chitinous exoskeleton" are often found in fairly recent literature. However, although chitin may comprise as much as 60% of the cuticle in a few cases, it may also be present in amounts as small as 1 or 2%. Arthropod cuticle may be thought of as a continuous

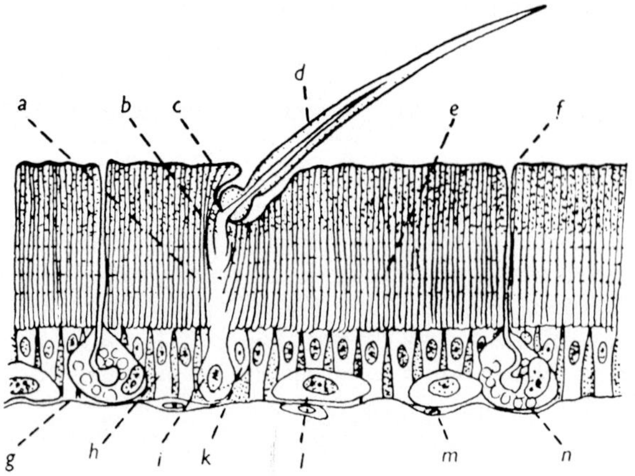

Fig. 40. Section of insect integument: *a*, laminated endocuticle; *b*, exocuticle; *c*, epicuticle; *d*, sensillum; *e*, pore canals; *f*, duct of dermal gland; *g*, basement membrane; *h*, epidermal cell; *i*, trichogen cell; *k*, tormogen cell; *l*, oenocyte; *m*, hemocyte adherent to the basement membrane; *n*, dermal cell. (From Wigglesworth, 1948.)

protein layer that usually contains chitin within and waxes on the outer surface (Richards, 1947). A review of the literature on the structure of insect cuticle was prepared by Locke (1974), and a study on its chemistry was written by Hackman (1974).

The *procuticle* comprises the bulk of the cuticle, and is immediately above the epidermis. It is essentially a hydrophilic chitin–protein complex containing a considerable quantity of water. The outer layer of the cuticle is called the *epicuticle.* This consists of a proteinaceous *cuticulin* about 1 μ in thickness, containing no chitin, and covered with a layer of lipid, usually a hard wax, averaging about 0.25 μ in thickness. This wax, and particularly 1 or more packed and oriented monolayers of it at the cuticulin–wax interface, is the principal barrier to water loss. After some insects molt, dermal glands secrete a fluid that spreads over the surface of the wax layer to form a "cement" that appears to be similar to shellac (Beament, 1955). Until somewhat recently, it was not known how the wax was brought to the surface of the cuticulin. Arising from the epidermal cells and extending outward through the procuticle of most insects, and sometimes into (but not completely penetrating) the overlying cuticulin, are minute ducts called *pore canals.* By means of the electron microscope, it was shown that these canals terminated in *wax canals* that branched out and reached the outer surface of the cuticulin, depositing a liquid wax similar in appearance to the liquid crystalline phases of lipid–water systems (Locke 1965, 1974; Gluud, 1968). By means of sorptive dusts, a sufficient amount of wax can be removed from the solidified epicuticular wax layer to cause a lethal rate of water loss, providing means of controlling some insects by nonchemical methods. (See chapter 2, under "Dust Desiccants.")

Setae (bristles or hairs) are sensilla that arise from pits in the cuticle surface in a sort of ball-and-socket arrangement (figure 40, *d*). Setae are

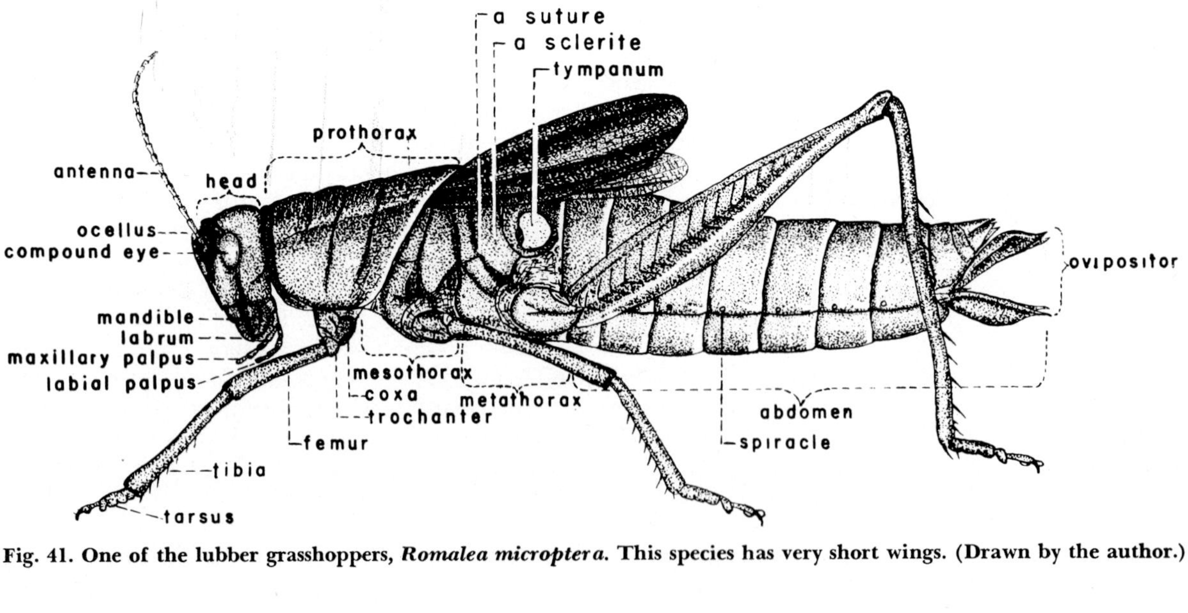

Fig. 41. One of the lubber grasshoppers, *Romalea microptera*. This species has very short wings. (Drawn by the author.)

formed by 2 cells: a sensillum-forming cell (*trichogen*) and a socket-forming cell (*tormogen*) (figure 40, *a, i, k*) and, of course, whatever cells that give rise to the sensory nerve(s). In the pit, the cuticle (articular membrane) is relatively thin, and so is the cuticle of the seta, etc., particularly in the case of some of the chemoreceptors. Cuticles of olfactory sense hairs are perforated by many minute pores (Slifer *et al.*, 1959; Steinbrecht and Müller, 1971). Cytoplasmic filaments, arising in the trichogen and tormogen cells, pass up through the cuticle to enter the seta, forming a cytoplasm continuum in close contact with chemicals that may be deposited on the surface of the cuticle, and thereby forming one of the pathways for entry of insecticides through the integument.

The insect cuticle does not readily decompose, but retains its original form after the insect dies, which makes possible the preservation of insects for hundreds of years with their original lifelike appearance. Thus, the mounting and preservation of insects are comparatively easy.

An insect's exoskeleton cannot be readily dissolved away by strong acids, alkalis, or solvents, but fortunately for man's attempts to control insects by means of contact poisons, it can be penetrated by solvents, surface-active materials, vapors, and gases. In fact, it can be penetrated by organic compounds of all kinds, even when they are solids and are not dissolved in a liquid solvent. Fortunately, the rate of penetration of organic insecticides into insect cuticle is much greater than into mammalian skin. Even inorganic insecticides can penetrate the cuticles of some insects, such as cockroaches, at a lethal rate. The *spiracles,* which are openings for the tracheae, form a channel of entry for liquids of low surface tension, and for vapors and gases.

Figures 41 and 42 show that insects are divided by constrictions into ringlike plates called *segments.* Thin, flexible portions of the cuticula called *conjunctivae* connect the segments. These conjunctivae may be folded into the body of the insect, allowing for expansion between segments much in the manner of an expanding accordion. Other constrictions or infoldings of the cuticle called *sutures* run in various directions on the surface of the exoskeleton. The internal infoldings often provide points for muscle attachments. The areas bounded by sutures are called *sclerites.* The shape and position of the sclerites are often used in the description of species in some insect groups.

In addition, the hairs, spines, setae, ridges, and other exoskeleton structures are used in taxonomy. External structures may be merely projections of the cuticle, or they may be associated with the underlying body structures, such as the sense organs.

The dorsal or upper face of each segment is called the *tergum,* the ventral or lower face is called the *sternum,* and each lateral face is called a *pleuron.* Each of these faces is made up of 1 or more *sclerites.* The dorsal part of a segment may be called a *tergite* and the ventral part a *sternite*—terms usually restricted to the abdomen (figure 42).

If reference is made to the thorax, the term *notum* is generally employed instead of tergum. Thus, one refers to the dorsal part of the first segment of the thorax as the *pronotum,* but the corresponding part on the ventral area as the *prosternum.*

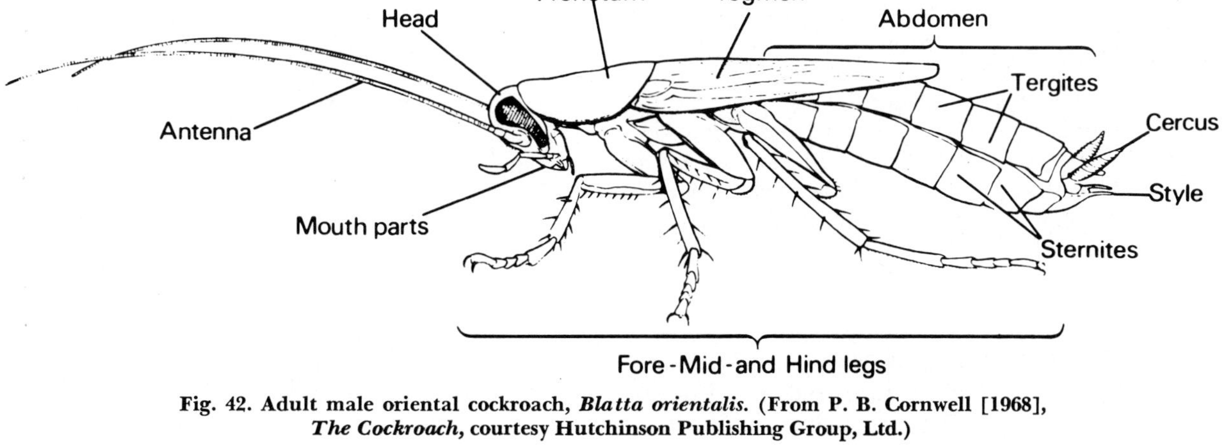

Fig. 42. Adult male oriental cockroach, *Blatta orientalis.* (From P. B. Cornwell [1968], *The Cockroach,* courtesy Hutchinson Publishing Group, Ltd.)

Molting

An insect can increase in size during the molting process until the hardening cuticle restricts further expansion. At molting, the old cuticle is largely dissolved by special molting fluid secreted by glands or by the epidermal cells in

Fig. 43. A cockroach (*Blatta orientalis*) molting. The black portion is the shed skin.

general. New, soft cuticle is deposited beneath the remnant of the old cuticle. The old, chemically weakened cuticle is split by internal blood and air pressure and muscular action, the split occurring along some specific line of least resistance which differs in its location among the various insect species, but frequently is along the top of the thorax and head. In figure 43, a cockroach (*Blatta orientalis*) is shedding its old skin (black). The insect is white immediately after molting, but becomes black within a few days. Molting appears to be a difficult and laborious process, and is a critical period in the life of an insect. The old cuticle cast off by an insect is called the "cast skin" or *exuvia*.

The Head

The head appears to comprise a single segment, but embryological observations indicate that it is a fusion of 5 or 6 segments possessed by primitive insect progenitors. The arrangement of the skeletal features of a cockroach's head is fairly well defined and will, therefore, be used as a generalized model. As seen from figure 44, the principal external regions of the head are designated as *vertex, front* or *frons, gena, clypeus,* and *labrum*. The eyes, antennae, and mouthparts are borne on the head.

Eyes

An adult insect usually possesses both compound and simple eyes. The compound eyes are among the most characteristic features of adult insects. They are composed of many small, hexagonal lenses or facets fitted closely together. Each facet is the exposed face of an independent lens and an independent eye unit called an *ommatidium*. Depending upon the species, there may be from 50 to 30,000 ommatidia in a single compound eye.

Besides the 2 compound eyes, an adult insect usually has 3 simple eyes or ocelli, but may have 2, 1, or none, depending upon the species. Simple eyes have only a single facet. The larvae of some orders of insects, such as the flies, beetles, butterflies, and ants, never have compound eyes. They may have from 1 to 6 or more simple eyes on each side of the head. Sometimes, several simple eyes form a cluster. Nymphs of bugs, grasshoppers, and dragonflies may have compound eyes.

Insects cannot change the focus of their eyes (accommodation), and they have no protection equivalent to the eyelids and eyelashes of vertebrates. Protection is afforded by the cuticle that covers the entire insect body, and which continues in an unbroken sheet over the eyes. Over the eye surface, however, the cuticle is thin and transparent, admits and refracts light, and is curved to form a lens.

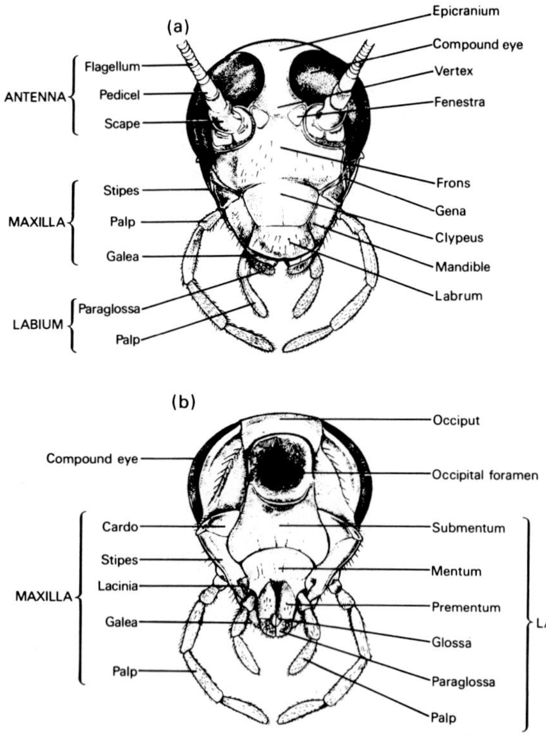

Fig. 44. Head and mouthparts of the American cockroach (*Periplaneta americana*) as seen from the front (*a*) and the back (*b*). (From P. B. Cornwell [1968], *The Cockroach*, courtesy Hutchinson Publishing Group, Ltd.)

Antennae

The antennae of insects are extremely variable in form. Various names, for the most part self-explanatory, have been more or less generally accepted to designate the different types of antennae: setaceous (bristlelike), filiform (threadlike), moniliform (beadlike), serrate (sawlike), pectinate (comblike), clavate (clubshaped), capitate (knobbed), lamellate (bearing many plates), and plumose (featherlike). Many forms of the above types, as well as miscellaneous unclassified forms, may be found among the insects.

Antennae may be used as tactile organs, or may possess organs for smelling or hearing. In general, they are used as sensory aids for locating food and finding mates, and it appears that the ants use their antennae to communicate with others of their species.

Mouthparts

Insects are divided into groups of those having chewing (mandibulate) mouthparts and those having various modifications of the primitive chewing mouthparts, including many that are used only for sucking or siphoning.

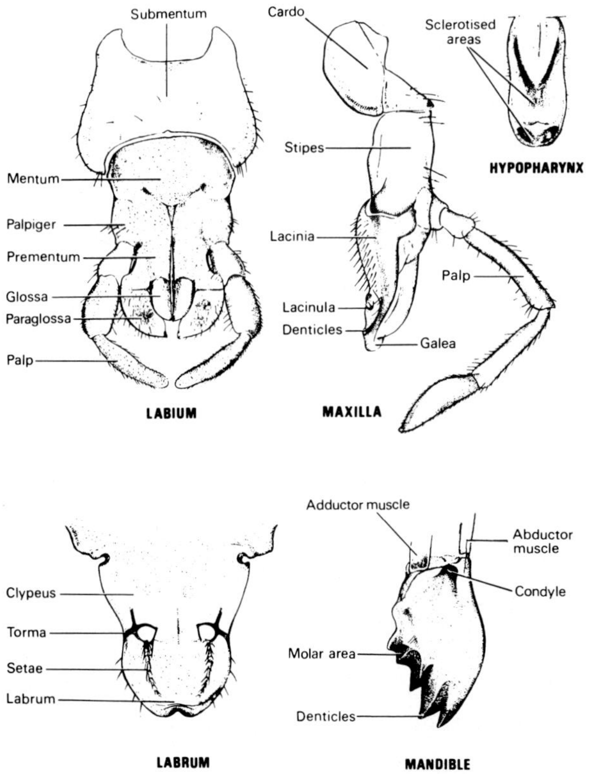

Fig. 45. Detailed structure of the mouthparts of the American cockroach (*Periplaneta americana*). (From P. B. Cornwell [1968], *The Cockroach*, courtesy Hutchinson Publishing Group, Ltd.)

Chewing Mouthparts. Examples of insects with chewing mouthparts are silverfish, cockroaches, grasshoppers, crickets, termites, earwigs, psocids, biting lice (Mallophaga), dragonflies, scorpionflies, beetles, ants, sawflies, woodwasps, and larvae of moths, butterflies, and primitive flies.

The mouthparts of the cockroach are of a fairly generalized type, and may be used as an example of the chewing mouthparts (figures 44 and 45). The upper lip (*labrum*) and lower lip (*labium*) oppose each other in a vertical plane. The labrum is a sclerite of the head, but the labium is a true mouthpart appendage. The actual biting and chewing appendages are the *mandibles* and *maxillae* (figure 45). They operate in a horizontal plane, i.e., from side to side. The mandibles are the first pair of "jaws" that can be seen when the labrum and labium are removed. They are strong, heavily sclerotized, often darkened, and possess teeth and grinding surfaces. They are operated by 2 sets of muscles, one set that closes them against each other and another set that pulls them apart.

The second pair of "jaws" are the maxillae. As shown in the figure, a maxilla is made up of a number of parts. The part modified for tearing food apart is called the *lacinia*. The maxillae bear segmented sensory palps provided with tactile hairs and also organs of smell and taste. In cockroaches, chemoreceptors involved in feeding (taste) are confined to the mouth. The Hemiptera, Lepidoptera, and Diptera also have chemoreceptors on the tarsi, while the Hymenoptera have similar receptors on the antennae as well as on the mouthparts (Frings and Frings, 1949).

The lower lip or *labium* performs a function similar to the lower lip of vertebrates. In addition, it bears a pair of sensory palps whose function is similar to those of the maxillary palps. As might be surmised from the paired palps, the labium was formed by the fusion of primitively separate appendages, the "second maxillae."

The *epipharynx* is a sensory organ attached to the inner surface of the labrum that is believed to contain end-organs of taste and is somewhat analogous to the palate of higher animals. It is of interest principally because it becomes modified into highly specialized structures in some of the sucking insects.

The *hypopharynx* is a tonguelike organ that forms the floor of the mouth cavity. The salivary duct opens in a pocket between the hypopharynx and the labium. The mouth opens between the hypopharynx and labrum.

Various Types of Sucking Mouthparts. Chewing mouthparts are the most primitive type, from which the various sucking types have evolved to form various interesting structural modifications. Sucking mouthparts are thus thought to be homologous with chewing mouthparts, even though they usually differ greatly both in appearance and function. The following types of sucking mouthparts among adult insects are distinguished by Metcalf *et al.* (1962).

1. The *rasping–sucking* type of mouthparts is possessed by the thrips. This type is considered by some authorities to be intermediate in structure between the chewing and piercing–sucking types. The mouthparts of thrips are somewhat asymmetrical, the right mandible being vestigial. The styletlike left mandible and the maxillae can be protruded through a circular opening at the conical apex of the head. To extract plant juices, a thrips thrusts its head downward and backward in the manner of a pickax, the stylets piercing the leaf or fruit cuticle. The plant sap is then withdrawn through the mouth cone which, when closely pressed to the plant surface, can probably provide considerable suction.

2. The *piercing–sucking* type of mouthparts is characterized by a tubular, usually jointed, beak, which encloses the stylets (figure 46, *A*). The stylets are modifications of the mandibles and maxillae of the chewing insects, sometimes supplemented or replaced by the labrum–epipharynx and hypopharynx. There are several subtypes of the piercing–sucking mouthparts, namely, the "bug or hemipterous subtype" (figure 46, *A, B*); the "louse or anoplurous subtype" (figure 46, *D*); the "common biting-fly or dipterous subtype [including in the latter the mosquitoes (figure 46, *H*) and horse flies (figure 290, chapter 9)]; the "special biting-fly or muscid subtype" [including the horn fly and stable fly (figure 46, *F, G*) and tsetse fly (figure 46, *E*)]; and the "flea or siphonapterous subtype" (figure 46, *C*). Figure 47 shows the successive stages in the insertion of the feeding bristles (mandibular and maxillary bristles) of the mouthparts of a bug (Hemiptera), according to Snodgrass (1935).

3. The *sponging* type of mouthparts is illustrated by the house fly (figure 46, *I*), blow flies (figure 46, *J*), and others. In such flies, the end of the fleshy, elbowed, and retractile labium is specialized into a pair of large, spongelike organs, the *labella*. The labella contain many furrows or channels called *pseudotracheae,* through which liquid material is drawn, possibly by capillarity. Liquid food is drawn up from the pseudotracheae through the food channel into the esophagus. It is usually not possible for insects with mouthparts of this type (nonbloodsucking Muscidae, Syrphidae, and many other Diptera) to pierce the skin, but some have labellar teeth with which they can abrade the skin, e.g., the eye gnats (*Hippelates*).

4. In the *siphoning* type, found in the adult Lepidoptera (moths and butterflies), the proboscis is derived almost entirely from the galeae of the maxillae of the more primitive mandibulate insect. The 2 elongated maxillae fit together to form a tube which, in the resting position, is coiled and held close to the underside of the head (figure 48). The labial palpi are often large and conspicuous. Except for a few species, notably the fruit-piercing moths of some tropical countries, moths and butterflies cannot pierce the epidermis of animals or plants. They usually feed by uncoiling the proboscis and sucking exposed liquids, such as nectar, through it. Müller (1873) illustrated the proboscis of a sphingid moth from Brazil found to be 11 in. (28 cm) long, and called attention to the fact that Darwin had predicted that a moth with such a long proboscis must exist in order to pollinate certain orchids with flowers of this length.

5. In the *chewing–lapping* type of mouthparts (figure 49), illustrated by the bees and wasps, the mandibulate origin of the labrum and mandibles is read-

Fig. 46. Mouthparts of some sucking insects. *A*, head of seventeen-year cicada side view (*left*), and view of newly emerged specimen with lower parts of head separated; *B*, mouthparts of female bed bug, from left: mandibular and maxillary stylets of one side, labium, distal part of a mandibular stylet, and position of head and beak with stylets deeply inserted while feeding; *C*, head and mouthparts of a flea, from left side (*left*), and anterior view; *D*, median vertical section of louse head (*above*) and stylets in relative position; *E*, head and proboscis of tsetse fly; *F*, proboscis of horn fly (*Sitona*); *G*, proboscis of stable fly (*Stomoxys*); *H*, head and mouthparts of female mosquito (*left*), and with stylets separated from labium; *I*, head of a female house fly with proboscis extended; *J*, undersurface of the labellar disk of a female blow fly; *K*, cross section of proboscis of a mosquito (*left*), and distal parts of the stylets. *Detailed legend: a*, subapical aperture or labial groove; *Aclp*, anteclypeus; *Ant*, antenna; *Bk*, beak; *Br*, brain; *CbP*, ciborial pump; *clp*, clypeus; *cnl*, canaliculi of labellum; *dStl*, dorsal stylet; *E*, eye; *fc*, food canal; *fm*, food meatus; *Fr*, frons; *FrG*, frontal ganglion; *g*, labial groove; *Ga*, galea; *hf*, hypostomal fold; *HL*, free lobe of hypopharynx; *Hphy*, hypopharynx; *Hst*, hypostoma; *Hstl*, haustellum; *iStl*, intermediate stylet; *Lb*, labium; *Lbl*, labellum; *LbPlp*, labial palpus; *Lc*, lacinia; *Lm*, labrum; *Lor*, lorum; *Md*, mandible; *Mds*, mandibular stylet; *mcl*, labellar muscles; *Mth*, true mouth; *mth*, mouth of sucking pump; *Mx*, maxilla; *Mxae*, maxillae; *MxL*, maxillary lobe (stipes); *MxPlp*, maxillary palpus; *MxS*, maxillary stylet; *Nv*, nerve; *O*, ocellus; *Oe*, oesophagus; *Phy*, pharynx; *Prb*, proboscis; *Prstm*, prestomum; *Rst*, rostrum; *Sac*, sac containing the stylets; *sc*, salivary canal of hypopharynx; *SlDct*, salivary duct; *SoeG*, suboesophageal ganglion; *sof*, suboral fold of hypopharynx; *Stl*, fascicle of stylets; *t*, prestomal teeth; *Tra*, trachea; *VNC*, ventral nerve cord; *vStl*, ventral stylet. (From Snodgrass, 1944.)

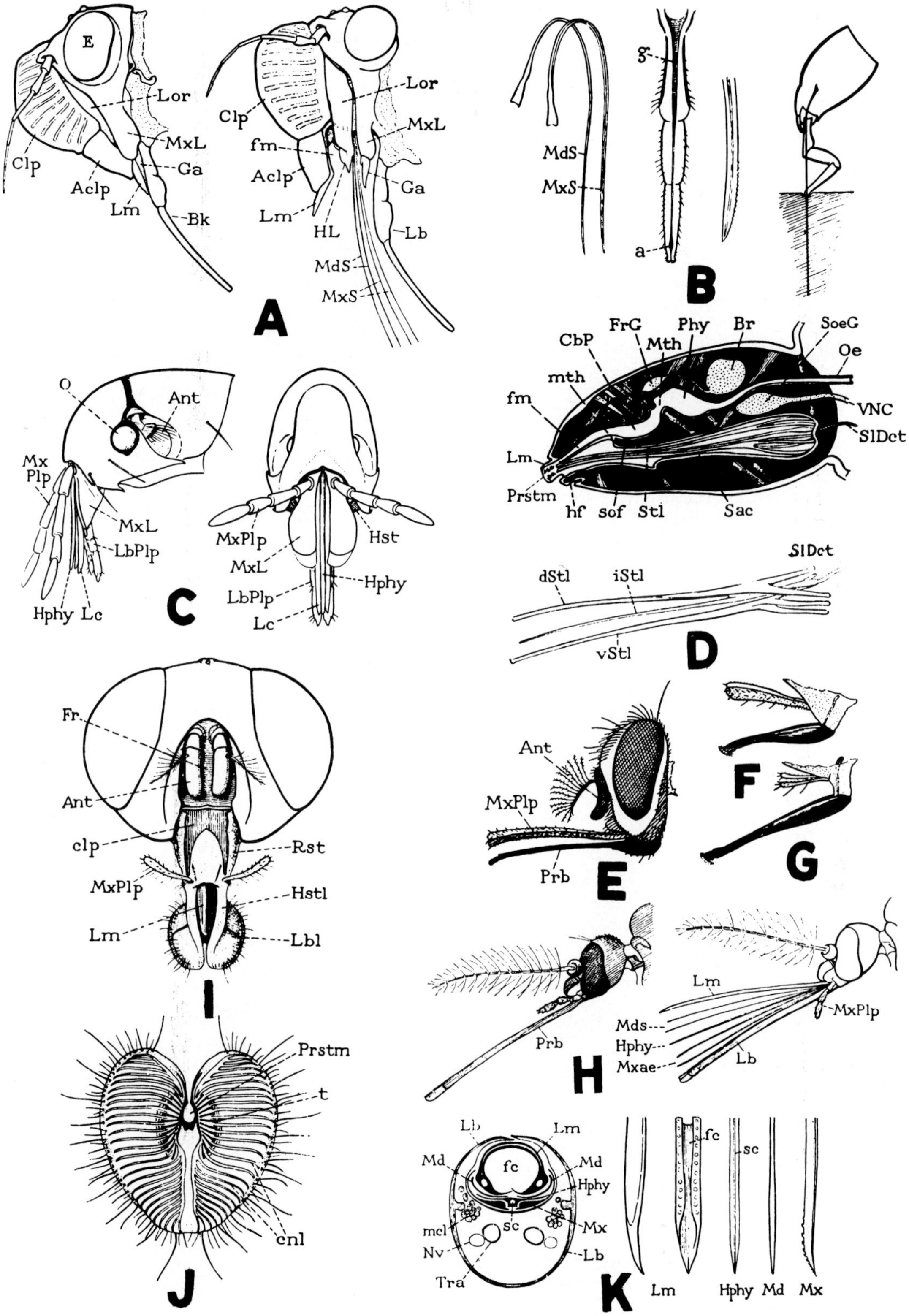
E
Lor
Clp
MxL
Ga
Aclp
Lm
Bk
fm
HL
Lb
MdS
MxS
A
g
a
B
O
Ant
Mx
Plp
MxL
LbPlp
Hphy
Lc
C
MxPlp
Hst
CbP
FrG
Mth
Phy
Br
SoeG
Oe
mth
VNC
SlDct
Prstm
hf
sof
Stl
Sac
dStl
iStl
vStl
D
Fr
clp
Rst
Hstl
Lbl
I
E
F
G
Prb
H
Mds
Mxae
t
cnl
J
fc
Md
sc
mcl
Nv
Tra
K

ily recognizable. The mandibles may be modified to serve as tools for a variety of tasks performed by the various species possessing such mouthparts. The maxillae and labium are elongated, forming a long tongue which is of great importance in feeding.

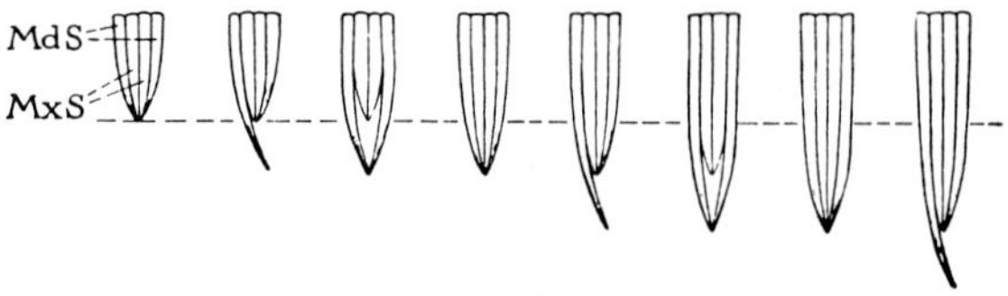

Fig. 47. Successive stages in the insertion of the mandibular and maxillary bristles of the mouthparts of a bug (Hemiptera). (From Snodgrass, 1935.)

The type of mouthparts possessed by the adult insect does not necessarily indicate the type possessed by the young of the same species. Because of their gradual or incomplete metamorphosis, in the majority of the more primitive orders of insects the adults have mouthparts of much the same appearance and function as those of the immature forms. The principal orders of this type are the Protura, Thysanura, Collembola, Orthoptera, Isoptera, Dermaptera, Psocoptera, Thysanoptera, Phthiraptera, and Hemiptera. (However, in the Coleoptera, with complete metamorphosis, larvae and adults have chewing mouthparts.) In the Odonata, Trichoptera, Hymenoptera, and Diptera, the mouthparts of the immature insects usually differ markedly from those of the adults. In the Lepidoptera, the mouthparts of the larvae are always different from those of the adults, the larvae having mandibulate and the adults having haustellate (sucking) mouthparts.

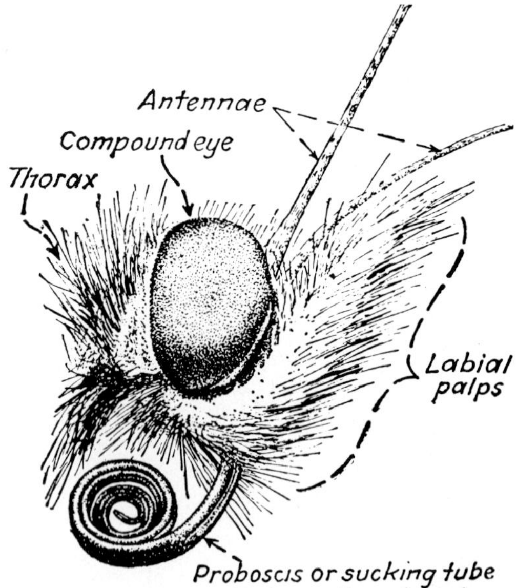

Fig. 48. Siphoning type of mouthparts possessed by butterflies and moths. (From *Destructive and Useful Insects*, by Metcalf *et al.* [1962]). Used with permission of McGraw-Hill Book Co.

The cocoon or attachments for the pupa of Lepidoptera are formed of silk threads spun by silk-spinning organs in the mouth of the larva.

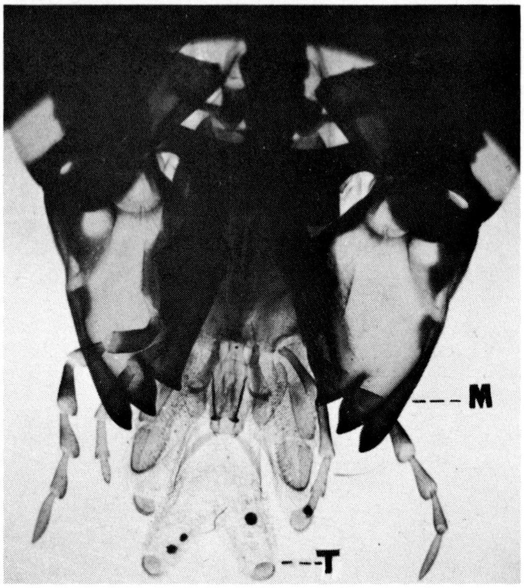

Fig. 49. The "chewing–lapping" mouthparts of a yellowjacket (*Vespula pensylvanica*), showing the mandibles (*M*) and lapping tongue (*T*).

Adult Neuroptera and Coleoptera have biting mouthparts, and the immature life stages have mouthparts similar to those of the adults except that in some species (antlions, lacewing larvae, and diving beetles) the mandibles and maxillae are grooved to facilitate the sucking up of the liquids of the insects' prey after the body has been pierced.

The Thorax

The second of the 3 body regions, called the *thorax,* is composed of 3 segments:[6] the *prothorax,* the *mesothorax,* and the *metathorax* (figure 41). In the grasshopper (figure 41) and the cockroach (figure 42), only the pronotum is visible from above; the wings cover the mesonotum and metanotum. Each thoracic segment bears a pair of legs, and in the case of winged insects, the mesothorax and metathorax bear the wings.

[6] In the Hymenoptera, the first abdominal segment (propodeum) is incorporated in the thorax, so the thorax is functionally (though not morphologically) 4-segmented.

Also, in the case of adult insects, the latter 2 segments usually each bear a pair of spiracles. The locomotory appendages of adult insects are thus confined to the thorax, but among the immature insects, locomotion may be aided by 2 to 8 pairs of fleshy, unjointed *prolegs* (figure 185, chapter 7) on the abdominal segments (Lepidoptera and Hymenoptera), or locomotory appendages may be entirely absent (as in some Coleoptera, most Hymenoptera, and most Diptera).

Legs

Three pairs of jointed legs are practically always present in adult insects, and are generally present in the immature stages. Nearly all adult animals having 6 legs are insects,[7] and the name of the class, Hexapoda, is based on this fact. The 5 parts of the legs are held together by intersegmental membranes at the joints, and are operated by internal muscles. The 5 parts, beginning next to the thorax, are called *coxa, trochanter, femur, tibia,* and *tarsus* (figure 41). These terms should be remembered, because they are often encountered in entomological literature. The trochanter in some hymenopterous species has 2 segments, and the tarsus may have 1 to 5 segments.

There are usually 2 claws on the tarsus, and there are frequently pads (*pulvilli*) beneath the claws that offer greater purchase against smooth objects. The ability of the pulvilli to cling to smooth surfaces is usually enhanced by many small hairs which sometimes, as in the case of the house fly, exude a sticky substance. Between the claws, there may be another type of pad or lobe called the *arolium* which has considerable suction force, and enables the insect to cling to objects with little or no aid from the claws. When the arolium is "hairlike," as in flies, it is called an *empodium*. Among the thrips (Thysanoptera), the claws are very small or may be entirely absent, and the tarsus terminates in a large bladder-like arolium. Among the scale insects, sucking lice, and some biting lice, there is only a single claw on the tarsus. The claws and the associated structures at the tip of the tarsus are presently considered to be a separate segment, the *pretarsus*.

Depending on the mode of life to which an insect has become adapted, great variation can be seen in the morphology of the legs of different insect species. A rapidly running insect like the tiger beetle (Cicindelidae) will have long, slender legs, while an insect that leaps, such as a grasshopper, flea, or flea beetle (*Haltica*) will have enlarged femora or coxae, at least on the hind legs. The legs of insects that dig into soil or wood are usually short and stout, and may be armed with spurs. The forelegs may be specialized for the grasping of prey, as in the case of the praying mantis and ambush bugs. Aquatic insects often have legs adapted for swimming. Arboreal insects, such as the cerambycid beetles, may have greatly expanded and hairy tarsi and strongly developed claws and pulvilli to afford the maximum grip.

Wings

Among invertebrates, only insects have wings, and insects are the only animals with 2 pairs of wings. Wings, like the 3 pairs of legs, are features which can be used to separate most insects readily from the many small invertebrate species with which they could be confused. Only adult insects have functional wings.[8] Some groups (proturans, springtails, and silverfish) appear never to have had wings in the course of their evolutionary history, while others (lice, bed bugs, fleas, some aphids, and ants; females of the mealybugs and scales; and certain lepidopterous and coleopterous families) have no wings, but had winged ancestral forms. The absence of wings in this latter group of insects does not indicate a primitive evolutionary status, but on the contrary, suggests a high degree of morphological specialization for their particular modes of existence.

Most insects can flex their wings against the top of the body while they are resting, and also fold their wings if they are broad. It is usually the broader hindwings that are folded, and if they are too long, the tips may be folded under so as to be protected by the forewings. Dragonflies and damselflies (order Odonata) and mayflies (order Ephemerida) lack the mechanism for flexing their wings, which are therefore held outstretched laterally or held vertically over the body. This is considered to be a primitive condition, but some of the higher insects, such as butterflies, have secondarily lost the ability to flex their wings.

Typically, there are 2 pairs of wings: one pair on the mesothorax and another pair on the meta-

[7] Some adult mites of the family Tenuipalpidae (*Larvacarus* and *Phytoptipalpus*) and the males of some Podapolipodidae also have only 3 pairs of legs. (The females are frequently legless.)

[8] The mayflies, before they become adults, molt their wings once.

thorax. The posterior pair, in the case of the true flies (Diptera), is replaced by vestigial appendages called *halteres,* which are believed by some entomologists to function as balancing organs. In the male scale insects, the hind pair of wings is replaced by 2 spines, 1 on each side, which hook into the weakly developed pair of wings of the mesothorax, giving the forewings added rigidity. Among the Coleoptera, the mesothoracic wings (*elytra,* singular *elytron*) are horny and shieldlike. When in flight, most beetles hold their elytra to each side like the wings of a glider, depending on the membranous second pair of wings for propulsion. The second pair of wings may be longer than the first pair, but are folded away under the elytra when the beetles are not flying.

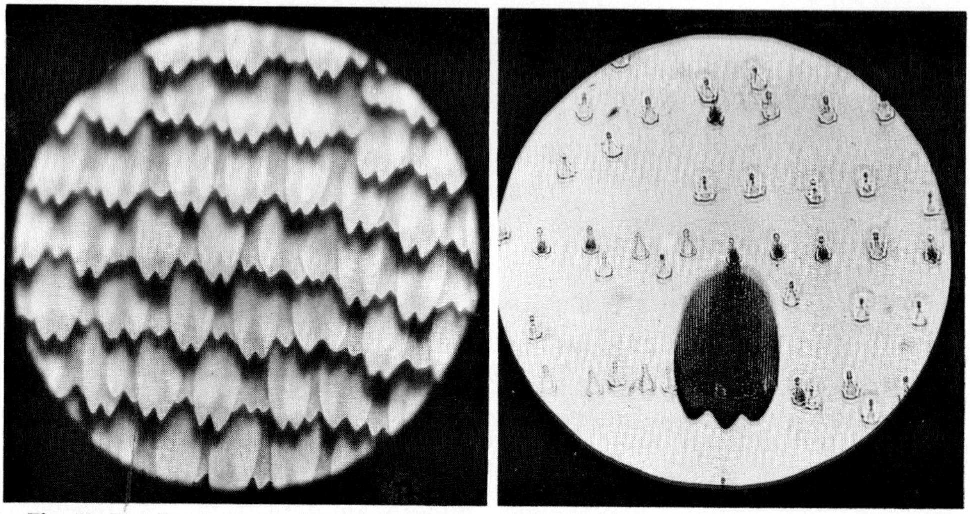

Fig. 50. Details of a butterfly's wing structure. *Left,* shinglelike "scales"; *right,* a single scale, with its pedicel fitting into a "socket" in the wing membrane. Other sockets have had the scales removed.

The exoskeleton and musculature of the thorax are highly developed to suit the needs of the great muscular activity involved in flight. This is most conspicuous among Odonata, Diptera, and aculeate Hymenoptera, and these insects are among the most active of the insect fliers. The wings of flies and bees move so rapidly that they become invisible.

The wingbeat frequency of one species of midge (*Forcypomyia*) has been recorded at 133,080 beats per minute, or 0.00045 second per wingbeat.

An examination of the wing pads of a nymph of the lower winged insects or of the chrysalis of Lepidoptera will reveal that an insect's wing is simply a double-layered hollow sac folded out from the body wall, in the lumen of which are nerves, tracheae, and body fluid. The double-layered nature of the wings may also be easily demonstrated by placing small, adult beetles in a damp location until the wings have absorbed enough water to become bloated, thus revealing their saclike construction.

The color of the wings of Lepidoptera depends on the color and arrangement of myriads of shinglelike "scales" attached to their wings (figure 50). The shingled arrangement of the scales strengthens an otherwise membranous structure that would be inadequate for rapid flight. Each scale is often strengthened by minute corrugations (figure 50, *right*) that diffract light rays and result in the brilliant color and iridescence usually noted in butterflies. Each scale has a pedicel at its base which serves as an anchor point. The pedicel is enlarged in the middle, and forms the "ball" that fits, in a fixed position, into the "socket" cavity in the cuticle of the wing membrane (figure 50, *right*).

The Abdomen

Most of the segments of the third region of the body, called the *abdomen,* are somewhat similar in appearance (figures 41, 42). The tergites and sternites are joined by thin, membranous conjunctivae in the same way that successive abdominal segments are joined. The conjunctivae permit expansion, and the abdomen of a gravid female may be distended to fantastic proportions, as in the case of the termites.

Adult insects usually do not have any abdominal appendages, except for those called *genitalia,* used in reproduction (figure 41, ovipositor). The genitalia are discussed in this chapter, under the heading, "Reproductive System." *Cerci* and *caudal filaments* are borne on the eleventh segment of some insect species (figure 42, and figure 324, chapter 9). If 11 abdominal segments are present, the respiratory spiracles normally occur on the first to eighth segments. In many species of insects, a fusion of posterior abdominal segments takes place, and it may be impossible to recognize more than 5 or 6 segments.

The larvae of some insects have fleshy, unjointed projections of the abdomen, known as *prolegs* (figure 185, *C, D,* chapter 7), which are used as legs and are considerable aids to locomotion for some of the more elongate forms. The larvae of Lepidoptera never have more than 5 pairs of prolegs, but those of certain Hymenoptera, the sawflies (Tenthredinidae), may have 6 to 8 pairs.

Orthopterans, such as grasshoppers, locusts, katydids, and crickets, that produce stridulatory sounds, have well-developed, thin, membranous tympanic organs (figure 41, *tympanum*) for detecting these sounds. Receptive tympanic sensilla are located on the front tibiae or the first tergite of the abdomen. Insects that lack such complex tympanic organs, such as cockroaches, can detect air movements by means of sensilla that may cover the cerci. The cerci of the American cockroach have a partly auditory function (Pumphrey and Rawdon-Smith, 1936*a, b*).

Internal Anatomy

The cockroach will again serve as a generalized form of insect, this time to illustrate the internal anatomy of the subclass Insecta (figure 51).

Digestive System

As already stated, the exoskeleton may be likened to a shell serving as an outer skeleton. The alimentary tract can be thought of as a tube within this outer shell, extending toward the rear from the mouth to the anal opening. This analogy is more striking in the case of the larvae of the higher insects. Between the exoskeleton and the alimentary tract is the body cavity, or *hemocoele.*

The alimentary canal is divided into 3 regions: the fore-intestine, the mid-intestine, and the hind-intestine. The fore- and mid-intestines are connected by the *cardiac valve,* and the mid- and hind-intestines are connected by the *pyloric valve.* The fore- and hind-intestines are divided into several more or less distinguishable parts, which are well illustrated by the dissected cockroach shown in figure 51. The parts of the cockroach's fore-intestine are the *pharynx* (which begins at the mouth), the *esophagus,* the *crop,* and the *pro-*

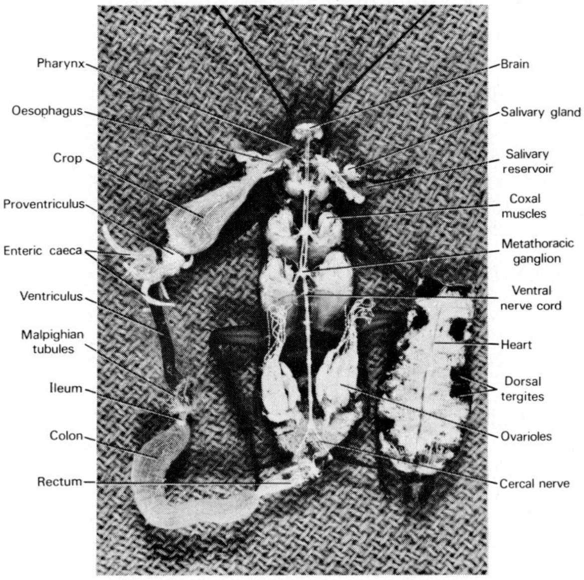

Fig. 51. Alimentary canal, central nervous system, and ovarioles of an adult female American cockroach (*Periplaneta americana*). Also, the tergites of the abdomen have been turned back to show the dorsal vessel or "heart." (From P. B. Cornwell [1968], *The Cockroach,* courtesy Hutchinson Publishing Group, Ltd.)

ventriculus, or *gizzard.* The gizzard occurs in insects that eat hard substances, and it appears to have a grinding and straining function like the gizzard of birds. The crop serves as a food reservoir, and is absent in some insects, or may be represented as a side sac, or *diverticulum.* The pharynx is a well-muscled pump in insects having sucking mouthparts.

The mid-intestine or *ventriculus* is not differentiated into regions, but a variable number of evaginations of variable size and shape and in various locations arise from it. These are called *enteric* or *gastric caeca;* and digestive enzymes may be produced in them, or they may represent additional digestive surface.

In the hind-intestine of the cockroach, the *ileum,* the *colon,* and the *rectum* may be distinguished. *Malpighian tubules* discharge into the anterior end of the hind-intestine at the pyloric region. The tubules vary in number and length, but the point at which they are attached to the alimentary canal can always be used as a means of establishing the point of juncture between the mid-intestine and hind-intestine. The functions of the Malpighian tubules are primarily to remove the waste products of metabolism (water and uric acid) from the blood stream, empty them into the hind-intestine, and adjust the ionic concentration of the blood (homeostasis).

The length and complexity of the alimentary tract of an insect species are usually related to the kind of food the insect eats. Long and complex systems are found among the herbivorous insects, while those of the entomophagous species and the insects feeding on concentrated food are relatively short and simple.

Circulatory System

The "heart" of an insect, stated in simplest terms, is a flexible dorsal tube (figure 51), closed at the posterior end and open at the anterior end, with small, valvelike openings or *incurrent ostia* (sing., *ostium*) located along the sides of successive chambers. Blood enters the heart through the ostia as the alary muscles dilate the heart chamber, but the ostia close when circular muscles of the chamber contract, and their valvelike arrangement prevents the blood from returning to the *hemocoeles* or posterior chambers. Because the contractions start at the posterior chamber of the heart and pass forward, the blood is pushed anteriorly, passing through the *aorta,* a constricted portion of the anterior part of the heart. The blood is pumped from the aorta into a series of cavities or sinuses and veins throughout the insect. The blood bathes the "brain," and from there it passes posteriorly through the hemocoeles, bathing the internal organs and passing into the appendages and wings. The circulation of the blood is aided by the contraction of a *ventral diaphragm,* and often by *accessory hearts* at the bases of legs and wings. One function of the blood is to carry digested food materials from the alimentary canal to all tissues of the body. In contact with the alimentary canal, the blood picks up the products of digestion by osmosis. While the blood is bathing the network of Malpighian tubules, waste products are removed. Insect blood has no red blood corpuscles and plays no part in oxygen transport, although the plasma of chironomid larvae does contain hemoglobin which may aid in respiration.

In the greater number of cases, both oxygen and carbon dioxide are exchanged directly with the external environment through the tracheal system.

Nervous System

In common with all arthropods, the nervous system in the body of insects is located ventrally (figure 51). Primitively paired *ganglia* in each segment are usually fused together, as are also the double cords connecting the ganglia. In the head, above the esophagus, lies the supraesophageal ganglion or *brain,* linked to the smaller subesophageal ganglion by paired connectives which pass around the esophagus.

Since so much of the nervous activity of insects is on a simple receptor–effector basis, the brain does not serve as an important coordinating center as it does among the vertebrates. If the brain

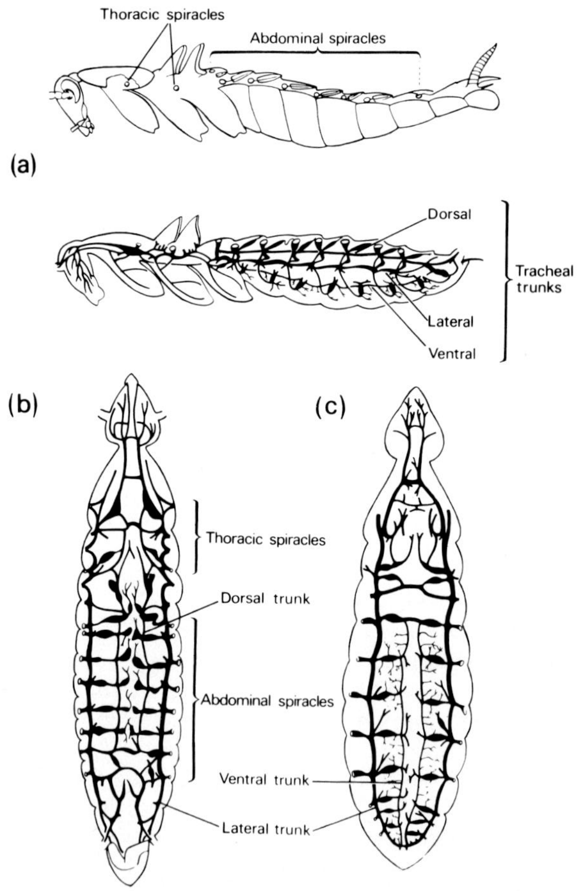

Fig. 52. Respiratory system of the German cockroach (*Blattella germanica*). *a*, side view; *b*, dorsal arrangement (male); *c*, ventral arrangement (female). (From Cornwell, 1968, after Haber, 1926.)

of an insect is removed, it may continue to live for a considerable period, and can carry on many reflex activities. If a bee is captured and confined in a transparent container (e.g., a Mason jar) when returning to the hive with a load of nectar or pollen, it will die from stress in 2 or 3 hours (Pence *et al.,* 1963). However, the excised abdomen will live for 2 or 3 days, and has been found to be an excellent subject to monitor physiologic response to injected drugs or insecticides, free of psychogenic influences, by means of a recently developed biosonic analyzer (Pence, 1969; Pence and Lomax, 1971).

Respiratory System

The *spiracles* (figure 52) are the external openings of the air tubes, known as *tracheae,* which divide and subdivide as they penetrate to all portions of the body. The tracheae eventually divide into microscopic tubes, usually less than 1 μ in diameter, called *tracheoles.* These reach every cell in the body. The cells obtain oxygen and give off the waste gases of metabolism by diffusion through the walls of the tracheoles.

By microscopic examination of insect tissue, the presence of the tracheae can always be determined by the fact that they are lined with cuticula in fine, spirally arranged thickenings called *taenidia,* which prevent the collapse of the delicate tracheal tubes.

Many insects, for example, parasitic insects spending their larval lives in the bodies of their hosts, respire through their cuticles, even though they may have a tracheal system. Some insects may live for long periods without oxygen. Wigglesworth (1939) reported an oestrid larva remaining alive 17 days immersed in oil. Specimens of California red scale, *Aonidiella aurantii* (Maskell), have lived as long as 40 days after being placed on glass slides, covered with refined spray oil, and then sealed with glass cover slips. The majority survived under these conditions for 10 days or more. If the cover slips were not placed on the scale insects after they were bathed in oil, they became desiccated and died much more rapidly; none of them survived for a period longer than 10 days (Ebeling, 1945).

Reproductive System

The *genitalia* are the external organs of reproduction, with all their appendages. In the females they may be modified into a highly specialized exserted or retractile egg-laying organ or *ovipositor.* It is not present in all orders of insects, but when present it may be used to deposit eggs into the ground, into plant tissues, or into the bodies of animals. The most remarkable modifications of the ovipositor may be found in the Hymenoptera. In the hymenopterous parasites, it is greatly elongated and slender. The eggs are then compressed and stretched to an extreme degree in

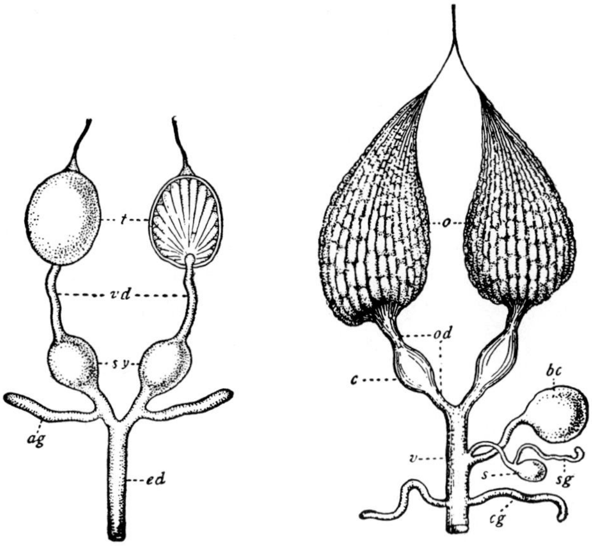

Fig. 53. Reproductive organs of an insect—*left,* male; *right,* female: *ag,* accessory glands; *bc,* bursa copulatrix; *c,* egg calyx; *cg,* colleterial gland; *ed,* ejaculatory duct; *o,* ovary; *od,* oviduct; *s,* spermatheca; *sg,* spermathecal gland; *sv,* seminal vesicles; *t,* testes; *v,* vagina; *vd,* vasa deferentia. (Reprinted from John Henry Comstock: *An Introduction to Entomology.* Copyright 1933, 1936, 1940, 1949 by Comstock Publishing Company, Inc. Used by permission of Cornell University Press.)

their passage through the very narrow channel of the bristlelike ovipositor (Fulton, 1933). In the stinging Hymenoptera (ants, wasps, and bees), the ovipositor is modified into a stinger. The venom is ejected at the tip of the stinger, and the eggs are ejected from the opening of the genital chamber which lies at the base of the ovipositor (Snodgrass, 1935).

The female internal reproductive organs (figure 51, and figure 53, *right)* consist primarily of a pair of *ovaries.* Each ovary normally consists of several *ovarioles* within which the eggs are developed. From the ovaries issue a pair of *oviducts,* which are usually united and lead to the *vagina.* There is almost always a saclike pouch called the *seminal receptacle* or *spermatheca* (figure 53), which receives and stores the sperm. Since some insects mate only once, the spermatheca performs an important function. In some insects (e.g., Lepidoptera), there is also a *bursa copulatrix* which receives the sperm first. The *colleterial glands* secrete a viscid material used to "cement" eggs together, waterproof them, attach them to objects, or to form envelopes that may encapsulate a number of eggs.

The number of eggs laid by an insect may vary from a few to many thousand. The queen honey bee may lay 2,000 to 3,000 eggs per day for weeks at a time, and the termite queen can lay millions of eggs. Not all insects lay eggs. Insects which lay eggs are said to be *oviparous.* Their eggs may be fertilized, as they pass through the vagina, by sperm stored in the spermatheca. If the eggs produced by the female bear viable offspring although the eggs are not fertilized, reproduction is said to be *parthenogenetic.* The developing embryo within the egg is nourished by means of the yolk, as in the case of birds. Unlike the birds, however, insects do not incubate the eggs and, in fact, rarely care for the eggs in any way after they are laid. They usually lay eggs in, on, or near food that will be consumed by the hatched young.

Many species of insects give birth to active, living young. Such insects are said to be *viviparous.* Among such species, the production of young may also be by parthenogenesis as, for example, among the aphids. The term *ovoviviparous* refers to the type of reproduction in which eggs with well-developed shells hatch within the body of the female and active young are then produced. The most recent tendency is to consider that the term "ovoviviparous" is superfluous, because all insects lay eggs, whether these hatch within the body or, as is more often the case, they are laid by the female and hatch outside.

The males of ants, bees, and social wasps are produced from unfertilized eggs and the females and workers from fertilized eggs. The males of some parasitic Hymenoptera and aleyrodids are also produced from unfertilized eggs.

In a diagram of the male and female reproductive organs, the structural analogy can be seen (figure 53). Comparable to the ovaries are the male *testes,* in which sperm is produced, leading to the *vasa deferentia* which unite to form the *ejaculatory duct. Accessory glands* secrete fluid with which the sperms are mixed. The external genitalia consist functionally of the *genital claspers* and the *aedeagus,* which includes the *penis.*

The external genitalia, especially of the male, are used extensively in the description of species because they often show more constant and more distinct differences than any other structural characters to be found on the insect.

Musculature

Insects have a large number of muscles. The external skeleton allows for a large area of attachment for muscles compared to the bony internal skeleton of vertebrates. In man, there are said to be 696 muscles, but Lyonet found over 4,000 in the caterpillar of the goat moth, *Cossus ligniperda.*

The muscles of insects are soft, but quite strong, and they are cross-striated, alternate light and dark bands crossing the fibers. There are *skeletal* and *visceral* muscles. In the latter, the transverse striations are not so distinct as in the skeletal muscles.

Insect muscles are very strong in relation to the insect's body weight. Insects are able to lift from 15 to 25 times their body weight. Their muscles contract and relax with phenomenal rapidity, as illustrated by the great speed of wing vibrations. Insect wing muscle is the most active tissue known, and consumes the most fuel and oxygen per unit of protein. The insect tracheal system facilitates the diffusion of oxygen and carbon dioxide through muscle tissue several thousand times faster than diffusion in the liquid phase (Weis-Fogh, 1964).

Metamorphosis

Metamorphosis means "change in form," and with respect to the development of animals, it refers to change in form during postembryonic growth.

In some of the lower forms of insects, there is no change in form; in others, there are varying degrees of change, up to such striking transfor-

Fig. 54. Various instars and stages of a thysanuran (*Thermobia domestica*), an example of ametabolous development.

mations as may be observed between a grub and an adult beetle, a caterpillar and a butterfly, or a maggot and an adult fly.

Development Without Metamorphosis

The insects of the 3 orders lowest in the scale of evolution, the Protura, Thysanura, and Collembola, do not go through any change in form after hatching, any more than do the vertebrates. Their development is without metamorphosis (ametabolous development) (figure 54). A young silverfish or springtail, for example, appears so much like the adult that one would immediately infer that the newly hatched young and the adult are the same species. The young insect increases in size, but its shape changes very little. The 3 orders in which no metamorphosis occurs comprise the subclass Apterygota, the insects of which have no wings and are descended from wingless ancestry.

Gradual Metamorphosis

Since no insect has visible wings immediately after birth, any insect that develops wings must undergo some degree of metamorphosis during its life history. Therefore, metamorphosis may be found among all the insects of the subclass Pterygota. These are insects that have wings in the adult stage or, if the wings are absent, they were lost during the course of evolution from winged ancestry. Cockroaches, grasshoppers, earwigs, termites, lice, thrips, bugs, aphids, and scales are representative of the insects that undergo *gradual* metamorphosis (paurometabolous development) (figure 55). Since the immature stages, commonly called *nymphs* (see the next paragraph), require about the same kind of habitat and food as the adults, internal anatomical changes during the life history of these insects are not on the same scale as for insects for which food and habitat change drastically. The most conspicuous change other than an increase in size during the development of insects with gradual metamorphosis is the difference in the appearance of the externally developed wing buds. Such insects are placed in the Exopterygota (Division A), of the subclass Pterygota later in this chapter.

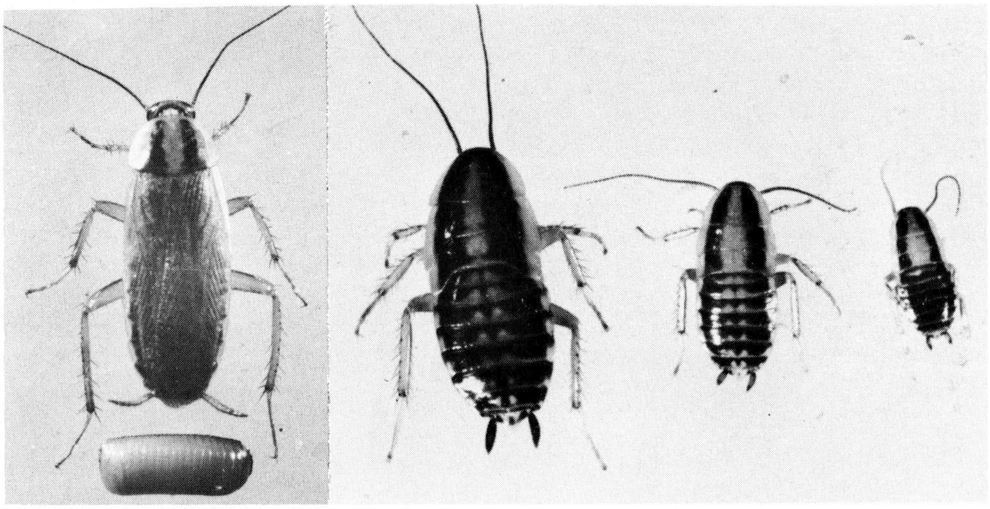

Fig. 55. Various life stages of the German cockroach (*Blattella germanica*), an example of gradual metamorphosis.

European entomologists have not been so adamant as those in the United States in using only the term "nymph" for the immature stages of the paurometabolous insects. Some refer to them as "larvae," or may use the term "nymph" for some insects (e.g., cockroaches and earwigs) and the term "larva" for others (e.g., thrips and scale insects). Some use the term "larva" for early instars and "nymph" for later instars of the immature forms of the same paurometabolous insects, such as the termites. European entomologists did the pioneer work on termite caste development, and used the term "larva" to refer to termite nymphs before they acquire wing buds and the term "nymph" for individuals succeeding the apterous nymphs and possessing external wing buds. For the sake of standardization, American entomologists are using the same terminology for termites.

Incomplete Metamorphosis

Three orders of aquatic insects, the Ephemerida (mayflies), Plecoptera (stoneflies), and Odonata (dragonflies), are placed by some authors in a group called the Hemimetabola, a group having *incomplete* metamorphosis (hemimetabolous development). These insects undergo radical changes in habitat during their life histories (figure 56).

The nymphs (naiads) and adults of insects with incomplete metamorphosis live in entirely different habitats. The aquatic nymphs have tracheal anal or rectal gills; their legs are modified for clinging, climbing, or burrowing; their bodies are modified for swimming; and their mouthparts are modified for taking food in the water. Special adaptations are required to enable the adults to escape from the last nymphal skin. The adults, being aerial, require different methods for capturing food (dragonflies and damselflies) or they may take little or no food, but inflate the alimentary canal with air to aid in flight (mayflies and most stoneflies) (Matheson, 1951).

Fig. 56. Incomplete metamorphosis of a dragonfly: *a*, nymph; *b, c, d*, adult in progressive stages of emergence from the nymphal skin; *e*, adult free of the nymphal skin and resting while its wings become fully expanded and hardened. (Reprinted from Robert Matheson: *Entomology for Introductory Courses*, second edition. Copyright 1951 by Comstock Publishing Company, Inc. Used by permission of Cornell University Press.)

Obviously, the morphological changes dependent upon a change from aquatic to terrestrial existence must of necessity be greater than those which accompany the changes occurring among insects with gradual metamorphosis, which undergo no change in habitat or food habits. The 3 orders involved are placed in a separate division of the Exopterygota later in this chapter, but this is not meant to indicate their phylogenetic relationship to other insects. An arrangement of orders according to modern concepts of phylogenetic relationships may be found in *Fundamentals of Entomology,* Part 1 (Belkin, 1972).

Complete Metamorphosis

The higher orders of insects, including the 4 that have the greatest number of described species, the Coleoptera (beetles), Lepidoptera (butterflies, moths), Hymenoptera (ants, wasps, bees), and Diptera (flies, mosquitoes), undergo what is known as *complete, complex,* or *indirect* metamorphosis (figure 186, chapter 7), and are placed in a group called the Holometabola (having holometabolous development). The young of insects with complete metamorphosis are known as *larvae,* and are so different from the adult that no one unfamiliar with their life history would suspect them of being merely a different stage in the development of the same individual. For example, what is there about the habitat, food habits, and appearance of a codling moth larva, the "worm" of the wormy apple, that would lead an uninformed person to suspect that it is an immature stage of the codling moth? For such radical transformations, the insects with complete metamorphosis have a life stage called the *pupa,* that is not possessed by any of the groups already discussed. The pupal stage is a quiescent, transient one, in which there is usually little locomotory movement and no feeding. The larva generally pupates in a place where the pupa will be hidden or inconspicuous—the pupa, being motionless, is also defenseless. A protection for the pupa is often spun, or prepared in some other way, by the larva. On the other hand, some dipterous pupae in the Culicidae and Chironomidae are very active, as evidenced by the familiar mosquito "wigglers."

All activity during the pupal stage is concentrated on the complex chemical changes associated with simultaneously breaking down (histolysis) and building up (histogenesis) tissues before the insect can emerge from the pupal skin, transformed for its radically new form and life habits. For example, in the pupa of a moth, sexual organs develop, chewing mouthparts are transformed into coiled, siphoning mouthparts, wings develop, compound eyes and antennae are formed, and in general the insect whose simple, cylindrical larval form was specialized for feeding and storing energy is transformed into a highly active and sensitive organism, specialized for flying about, securing food, finding a mate, and locating a place for oviposition.

One of the advantages of complete metamorphosis is that the larvae are specialized in habits and form for feeding, growing, and storing energy, and the adults are specialized for mating, reproducing, and effecting a dispersal of the species.

The means of locomotion, sense organs, and reproductive organs that the larva did not require are now essential to the adult. In addition, the pupae may provide a means of passing through seasons of adverse weather conditions or periods of food shortage.

Insects with complete metamorphosis develop their wing buds and the buds for all other adult structures internally in the larval instars. For this reason, they are placed in the division Endopterygota of the subclass Pterygota in the synopsis of orders at the end of this chapter. The buds often become external in the pupal stage, but are hidden under the pupal sheath in the case of obtect (smooth) pupae, when they are not seen until the adult stage.

Neither the pupa nor the adult ever grows. The layman often supposes that a small house fly is a young fly that has not yet completed its growth. However, differences in the sizes of adults merely indicate variations in the amount of growth attained by the larvae. This, in turn, was largely dependent upon the food supply.

The change of form from larva to adult never fails to excite the wonder and imagination of all people; the development of mammals, including man, appears so simple and uneventful by comparison. In fairness to the mammals, however, it should be pointed out that their greatest changes in form occur during embryonic development—changes which, in fact, recapitulate the greater part of their long and eventful phylogenetic development. The morphologic changes of the higher insects are more spectacular because they occur during the postembryonic development of the animal.

The Terms "Instar" and "Stage"

Now that some understanding has been gained of the meaning of "metamorphosis," the distinction between the terms *instar* and *stage* can be more readily understood. The cuticle of an insect allows for only limited growth; for further growth, the insect must split and shed its cuticle, and this process is called *molting* or *ecdysis.* Before the old cuticle is shed, a new cuticle is already forming, and will allow for another limited increase in body size. This process usually continues until the insect becomes an adult, although in a few insects such as the proturans (Protura), springtails (Collembola), and silverfish (Thysanura), molting continues even after the adult stage is reached.

An *instar* is the form in which an insect appears after each molt. The period between any 2 molts is called a *stadium* (plural, *stadia).* The life *stages* of an insect are the periods of its life which are radically different from each other in appearance and usually also in behavior and activity, particularly so in the case of holometabolous insects. For example, a cockroach has 3 stages: egg, nymph, and adult; a moth has 4 stages: egg, larva, pupa, and adult.

Differences Between Nymphs and Larvae

Since the naiads of aquatic insects with incomplete metamorphosis are of comparatively little interest to the majority of entomologists, the accompanying tabulation gives only the factors which differentiate nymphs and larvae (*p. 112*).

Some Adult Morphological Specializations

In many of the holometabolous insects, the adults do little feeding; the energy supply for their relatively brief adult life is carried over from the larval stage in the fat bodies. Some females (e.g., mosquitoes and fleas) must feed in order to develop eggs. However, adult specializations are chiefly for reproduction and distribution of the species. In the higher insects, one form of specialization is sexual dimorphism. In some insects, the 2 sexes are so different that they could easily be mistaken for separate species. In others, the male may be partially or completely eliminated, and reproduction is by *parthenogenesis* (e.g., in aphids). In some insects, the normal adult morphological features are absent; characters of the immature stage are carried over to the sexually mature and reproducing individuals. This phenomenon, called *neoteny,* is frequently associated with parthenogenetic reproduction and vivipar-

Nymph	Larva
1. Appearance similar to that of adult.	1. Appearance radically different from that of adult; typically cylindrical.
2. Habitat and food same as for adult.	2. Habitat and food often greatly different from those of adult.
3. Successive nymphal instars become progressively more like adult in appearance.	3. Last larval instar appears no more like the adult than the first instar.
4. Has no inactive stage between nymph and adult.	4. Has an inactive pupal stage for transformation from larva to adult.
5. Wing buds develop externally.	5. Wing buds develop internally.
6. Has few, if any, structures different from those of adult.	6. Has temporary structures, such as prolegs, not retained by adult.
7. Mouthparts same type as those of adult.	7. Mouthparts often different from those of adult (e.g., caterpillars, chewing; moths, sucking).
8. Has compound eyes if compound eyes are possessed by adult.	8. Never has compound eyes.

ity (rearing active young instead of eggs), and is restricted to the female.

In social insects, different *castes* perform particular functions in the colony. Information on castes and their functions can be found elsewhere in this book in discussions of the biologies of termites, ants, wasps, and bees.

Diapause

Common to all forms of life, and apparently as necessary as any other life function, is the phenomenon of rest or *diapause,* involving various forms of cessation of activity. It may take the form of a temporary cessation of development, or a temporary cessation of activity, or both.

In insects, diapause most often occurs in the egg or pupal stages, although it may occur in other stages of development. It may be caused by changes in temperature, moisture, food, water, or oxygen. Diapause may be triggered by a lengthening (vernal) photoperiod or by a shortening (autumnal) photoperiod, or may be imposed by such internal factors as heredity, enzymes, or hormones. Some believe that diapause may be brought about by autointoxication or by the accumulation of some chemical, somewhat like muscular fatigue in higher animals. Diapause that enables an insect to resist cold is called *hibernation;* if it enables an insect to resist heat or drought, it is called *aestivation.*

CLASSIFICATION: THE ORDERS OF INSECTS

Binomial Nomenclature

It has been shown that the members of the class Hexapoda have in common a number of characteristics which definitely separate them from other classes of the phylum Arthropoda. Likewise, the class Hexapoda can be divided into a number of *orders,* each of which includes insects with certain common characteristics that indicate more or less similar morphological and/or phylogenetic relationships. For example, there are striking points of difference noted in grasshoppers (order Orthoptera), which have 4 wings (the front pair generally thickened), chewing mouthparts, and gradual metamorphosis, as compared with the flies (order Diptera), which have a single pair of wings, sponging or piercing mouthparts, and complete metamorphosis. The various orders can usually be readily distinguished, even by relatively inexperienced students of entomology.

The orders are subdivided into families, the families into genera (sing. genus), the genera into

	Man	*Head Louse*
Kingdom	Animal	Animal
Phylum	Chordata	Arthropoda
Class	Mammalia	Hexapoda
Order	Primates	Phthiraptera
Family	Hominidae	Pediculidae
Genus	*Homo*	*Pediculus*
Species	*sapiens*	*humanus*
Subspecies	Caucasian	*capitis*

species, and the species sometimes are subdivided into subspecies or varieties. The accompanying classification of man in comparison to that of one of his external parasites, the head louse, *Pediculus humanus capitis* De Geer, will illustrate the significance and the relative magnitude of the various categories universally accepted in both plant and animal classification. However, only a relatively few species of insects have been divided into taxonomically distinct subspecies.

In addition to the categories shown in the box, additional subdivisions have been established to facilitate the classification of certain groups whose large size or structural peculiarities make further division desirable. For example, orders may be divided into suborders, a number of families may be grouped into a superfamily, or a family may be divided into a number of subfamilies. A subfamily may in turn be divided into tribes, each consisting of a number of genera. The suffix for the name of the superfamily is "-oidea." This should be distinguished from "-odea," which is used as a suffix for the names of some of the orders and suborders. The suffix for the name of the family is "-idae," for the subfamily, "-inae," and for the tribe, "-ini." Thus, one of the familiar tarantula hawks of the southwestern United States, *Pepsis formosa* (Say), is in the superfamily Vespoidea, family Pompilidae, subfamily Pepsinae, and tribe Pepsini (see figure 246, chapter 9.)

Binomial nomenclature gives to each species of plant or animal a name of 2 terms, of which the first is the *genus,* and should always begin with a capital letter, and the second is the *species,* and should always begin with a small letter. Both terms are italicized in scientific literature, and are accordingly underscored (underlined) in manuscripts. It is a system of nomenclature first standardized by the great Swedish naturalist Linnaeus about the middle of the eighteenth century. In binomial nomenclature, the name for the human species is *Homo sapiens,* and the name for the head louse is *Pediculus humanus.* There are two subspecies or varieties of *Pediculus humanus: P. h. capitis,* the head louse; and *P. h. humanus,* the body louse.

Some people are inclined to think of scientific names as pedantic gestures, or of having perhaps some vague academic significance of small concern to anyone besides the taxonomist. The common names of most animals, especially insects, are highly variable, even within a relatively limited area, and may mean nothing to people in other areas within the same country, to say nothing of those in other parts of the world. Likewise, common names may give little or no indication of the taxonomic relationship of the insects to one another. Therefore, only the scientific names have a universal significance—referring to the same plant or animal the world over.

In the Entomological Society of America, there is a Committee on Common Names of Insects. This committee has developed a list of approved common names of insects and a few other arthropods, and also their scientific names, according to the latest concepts of taxonomists in each insect group. This list is periodically revised, expanded, and sent to members of the society. By following the recommendations of the committee, American and Canadian entomologists are standardizing nomenclature in their published papers.

The dictionary is not always a reliable authority on common names of insects. One finds such words as "bedbug," "housefly," and "honeybee." These should be written as 2 words—"bed bug," "house fly," and "honey bee." The bed bug is a true bug (Hemiptera), the house fly is a true fly (Diptera), and the honey bee is a true bee (a superfamily in the order Hymenoptera). On the other hand, the words "doodlebug," "dragonfly," and "silverfish" are properly written as 1 word because the doodlebug is not a bug, the dragonfly is not a fly, and the silverfish is not a fish. The words "bedbug," "housefly," and "honeybee" would be equivalent to the word "Johnsmith."

Rules of Binomial Nomenclature

Before the time of Linnaeus, the name of the insect was a short Latin description. It can be imagined how cluttered with names entomological literature would be if such a system of nomenclature had been retained. Linnaeus suggested that the name of an animal (or plant) be composed of the compounded names proposed for the genus and for the species of the animal in question. These were always Latin names, for Latin was the universal language among scholars of that time. Latin has an additional value for use in "scientific names," in that its form and usage are now stereotyped, and can be used with unchanged meaning by people of all languages and in all parts of the world. We hear much nowadays about the desirability of a universal language. A universal language has been used for centuries by biologists in the naming of plants and animals, and if this had not been the case, the chaotic condition of biological literature would be difficult to imagine. Certainly, natural science would have been greatly retarded. Chemists are also in world-wide agreement as to chemical nomenclature, having adopted a universal system at a conference held at Geneva, Switzerland, in 1892. (While Linnaeus used only Latin, latinized forms of Greek and other languages were later often used in binomial nomenclature. Many generic names are more recently of Greek derivation.)

Linnaeus grouped the insects into 7 orders: Coleoptera, Hemiptera, Lepidoptera, Neuroptera, Hymenoptera, Diptera, and Aptera. He described and named a surprisingly large number of species, as a perusal of general entomological literature will show. In 834 pages, Linnaeus described 4,379 animals, including 1,937 insects. By now, about a million species of insects have been named and described, and about 6 or 7 thousand new species are being added in the world's taxonomic literature each year. There are probably at least as many species of insects still unnamed and undescribed.

Linnaeus perfected the binomial system of classification, and used it in his famous general treatise on animals, called *Systema Naturae.* The tenth edition of this publication, which appeared in 1758, is universally accepted as the starting point in binomial nomenclature. Any specific name proposed before the date of publication of the tenth edition of *Systema Naturae* is not considered valid. In all the naming of species subsequent to 1758, however, the rules of priority are strictly followed. If a species is described and named, and subsequent research reveals that someone else had already described and named the same insect, the later name becomes invalid. Taxonomists are no more infallible than any other specialists, and the constant revision of initially presumed taxonomic relationships of species, genera, and families of insects to one another is a task that continues apace with the description of new species.

For complete information on a scientific name, the name of the "author," the researcher who first described and named the species, follows the name of the species, but his name is not underscored or written in italics. In most entomological literature, the author's name is abbreviated. However, the modern tendency is to write out the author's name completely, since there are now too many names to abbreviate without duplication or confusion. Likewise, beginning students and those not familiar with a given insect group are not likely to know the meaning of the majority of the abbreviations. However, an abbreviation for Linnaeus or Fabricius is acceptable.

The house fly may be taken as an example to illustrate the correct form of binomial nomenclature. It is one of the many species which was named by Linnaeus. The scientific name of the house fly is *Musca domestica* Linnaeus (or *Musca domestica* Linn., or L.). One knows from the scientific name that Linnaeus described the house fly as the species *domestica* and that he placed it in the genus *Musca.* Let us take another example, the codling moth of apples, pears, and walnuts, another well-known cosmopolitan insect pest. The scientific name of this insect is *Laspeyresia pomonella* (Linnaeus). The parentheses around "Linnaeus" show that he named the codling moth *pomonella,* but that he placed it in some other genus (*Tinea*). Later, a specialist in this particular group of the Lepidoptera found that the codling moth showed greater taxonomic relationships to insects which belong in the genus *Laspeyresia,* a genus that had not yet been recognized in Linnaeus' time. Since then, the species *pomonella* has actually been placed in 3 successive genera, *Cydia, Carpocapsa,* and *Laspeyresia,* the latter being currently recognized by entomologists and approved by the Entomological Society of America.

Relation of Scientific Names to Entomological Investigation

The first step in investigating an insect problem is to establish the identity (scientific name) of the insect in question. If the name of the insect is not known to the investigator, or if he is not absolutely certain of its identity, he may send specimens to the United States National Museum for identification, or if he knows to what group (family or genus) the insect belongs, he may send specimens directly to a taxonomist who specializes in that particular group of insects. When he is certain of the scientific name of the insect, the investigator proceeds to examine the entomological literature for information on it.

The literature may reveal to an investigator much that is already known about an insect, such as its economic importance, geographic distribution, hosts, life history, seasonal history, habits, tropisms and ecological relationships, climatic or other physical factors favorable or unfavorable to its abundance, natural enemies, and previous investigations on control measures. The obtaining of all this wealth of information is predicated on the correct identification of the insect species and the ability to use the world's entomological literature on the basis of a standardized system of nomenclature. No matter in what country the literature originated, the insect should always have the same scientific name, for insects are named according to universally accepted standards of nomenclature. Without a universally accepted system of classification and naming of insects, no other phase of entomological investigation could be systematically pursued.

Some Characteristics of the Phylum Arthropoda

As shown in the various categories of classification, the phylum is the most inclusive of the subdivisions of the animal kingdom. In the Arthropoda, the various classes (subdivisions of a phylum) share these characteristics in common: bilateral symmetry; segmented bodies; paired, jointed appendages; protective cuticle; a dorsal blood vessel; and a ventral nervous system. Some, but not all, of these characteristics are possessed by some groups of animals farther down in the scale of evolution. For example, the members of the phylum Annelida, to which the earthworms belong, possess bilateral symmetry; segmented bodies; protective cuticle; a dorsal blood vessel; and a double ventral nerve cord, segmented into a series of ganglia. However, they do not have jointed appendages. No other phylum of animals possesses *all* of the characteristics of Arthropoda. Within the phylum Arthropoda, however, there are some points of difference which relegate different groups of animals to different classes. The class Hexapoda, for example, is the only group of arthropods that has 3 body regions, 3 pairs of legs, and wings (as adults).

Use of Keys for Identification of Insects

When an insect taxonomist or systematist is given a large group of insects to separate into their respective orders, families, genera, and species, he first attempts to sort out the species he recognizes from earlier experience. For aid in determining the relationships of the insects of whose identification he is not certain, the taxonomist will often use "keys," by means of which the insects can be traced through a successive series of morphological characters. The keys are so arranged as to facilitate the determination of scientific names and systematic relationships, and have been prepared for all the categories of insect classification. They are based on the work of a large number of specialists in the various groups of insects, and are, of course, subject to revision as new information is gained regarding the taxonomic relationships of the various groups. The use of a key depends on some knowledge of the structural characteristics and terminology of the group of animals (or plants) for which it has been worked out.

The accompanying simple and admittedly incomplete key was designed to enable students to "key out" the principal orders of adult insects for a course in economic entomology. The orders Protura and Collembola are included among the insects, although modern taxonomists consider them to belong to different subclasses. In this simplified key the diplurans (Diplura) are considered to be in the order Thysanura and the cockroaches are considered to be in the order Orthoptera, although the cockroaches are placed in Dictyoptera in chapter 6. Belkin (1972) has combined keys for adult and immature insects.

SIMPLIFIED KEY TO THE PRINCIPAL ORDERS OF INSECTS IN THE ADULT STAGE

1. Wings usually present 2
   Wings absent 17
2. Mouthparts formed for biting 3
   Mouthparts formed for sucking 13
3. Tip of abdomen with paired, forcepslike appendages (earwigs) .... Derm ap'tera
   Tip of abdomen without paired, forcepslike appendages 4
4. Forewings membranous or leathery, not horny in texture 5
   Forewings horny, shieldlike, forming cover for hindwings (beetles) .... Cole op'tera
5. Forewings leathery, hindwings folded fanlike (grasshoppers, crickets, cockroaches) Orth op'tera
   Forewings membranous, hindwings not folded fanlike 6
6. Abdomen usually constricted at base, often with a stinger or specialized ovipositor (ants, bees, wasps) .... Hymen op'tera
   Abdomen not constricted at base, without a stinger or specialized ovipositor 7
7. Antennae short, inconspicuous 8
   Antennae long or conspicuous 9
8. Hindwings smaller than forewings; abdomen with 2 or 3 threadlike filaments (mayflies) .... Ephem er'ida
   Hindwings not smaller than forewings, abdomen without long, threadlike filaments (dragonflies, damselflies) .... Odo na'ta
9. Tarsi with less than 5 segments 10
   Tarsi 5-segmented 11

(Continued)

10. Hindwings broader than forewings, folded in repose (stoneflies) .... Plec op'tera
    Hindwings same width as forewings, not folded (termites) .... Is op'tera
11. Wings naked or very slightly hairy 12
    Wings densely clothed with hair (caddisflies) .... Trich op'tera
12. Head prolonged into beak, with mouthparts at tip (scorpionflies) .... Mec op'tera
    Head not prolonged or beaklike (lacewings, antlions) .... Neur op'tera
13. Wings densely clothed with scales; mouthparts coiled below head (butterflies, moths) .... Lepid op'tera
    Wings not clothed with scales; mouthparts not coiled below head 14
14. Mouthparts sunken into head capsule, asymmetrical; usually very tiny insects; wings with hair fringe (thrips) .... Thysan op'tera
    Mouthparts external in repose, symmetrical .. 15
15. Wings usually 4; mouthparts forming a jointed beak 16
    Wings 2; mouthparts styletlike or spongelike, not forming a jointed beak (flies) .... Dip'tera
16. Basal portion of forewings thickened, leathery, or horny; distal portion membranous (true bugs) .... Hem ip'tera
    Forewings with same consistency throughout (aphids, scale insects, leafhoppers, cicadas, psyllids) .... Hom op'tera
17. Abdomen without conspicuous appendages .. 18
    Abdomen with conspicuous "tails" or "spring" 19
18. Body flattened dorsoventrally; prothorax small, ringlike (lice) .... Ano plur'a
    Body flattened laterally; prothorax large and hoodlike (fleas) .... Siphon ap'tera
19. Abdomen 10- or 11-segmented, terminating in forceps (pincers) or long filaments (silverfish, bristletails) .... Thysa nu'ra
    Abdomen with not more than 6 segments, often terminating in a springing apparatus (springtails) .... Collem'bo la

Most arthropod orders in the key may be keyed out to important urban pest species by illustrated keys in manuals for students in university courses in medical entomology, such as the one by Furman and Catts (1970).

SYNOPSIS OF THE INSECT ORDERS

For the sake of brevity and convenience, in this synopsis all the primitively wingless hexapods are grouped in the subclass Apterygota. However, as already stated, a good argument could be made for placing the Protura and Collembola in separate subclasses, Myrientomata and Oligoentomata, respectively, with the remaining orders in the subclass Insecta (Euentomata) (Belkin, 1972).

Subclass I. Apterygota. Primitively wingless insects without metamorphosis.

Protura. Proturans

Wings. None.

Mouthparts. Piercing–sucking, retracted into a pouch.

Miscellaneous. Very small, slender, white insects. Abdomen 12-segmented or more, with a pair of appendages on the first 3 segments; tracheae may be present or absent; no visible antennae; blind. They are principally found in leaf mold. Proturans crawl only on the second and third pair of legs, holding the first pair in front of and above the head as tactile organs (figure 57).

Collembola. Springtails, Snowfleas

Wings. None.

Mouthparts. Chewing type, sunken into the head.

Miscellaneous. Very small to minute; compound eyes; eyes are grouped, ocelli-like. Antennae short, rarely have more than 5 or 6 segments; 6 abdominal segments, the first usually provided with a collophore (suckerlike ventral tube), the fourth usually bearing a pair of appendages that constitute the springing organ or *furcula,* and the third bearing a pair of appendages that operate to act as a catch to hold the furcula when it is at rest; tracheal system usually absent; Malpighian tubules absent. Adults molt many times after reaching sexual maturity.

The Collembola (figures 336 and 337, chapter 10) are sometimes very abundant, and some species are injurious to plants. They may be found in mushroom beds, maple sap buckets, on the surface and in the soil of fields and woodlands wherever there is shade, and moving about on the surface of water in swimming pools, but they can be found in greatest abundance under boards or logs, where they find both darkness and moisture. They are nocturnal. The characteristic of their habits for which they are best known is their ability to spring into the air by means of their furcula, which is doubled under the abdomen and held in place by the "catch," then released like a spring.

Diplura. Japygids, Two-Pronged Bristletails

Wings. None.

Mouthparts. Chewing, partially retracted.

Miscellaneous. Small, narrow, rarely over 10 mm in length, unpigmented because of their secluded and subterranean habits. Antennae long, filiform, with beadlike segments; eyes and ocelli absent; abdomen 11-segmented. Diplurans may have long cerci with beadlike segments as in the antennae, and without the median caudal filament that characterizes the Thysanura, or they may have a pair of forceps or pincers (*Japyx*). Molting continues after sexual maturity is reached.

Thysanura. Silverfish, Bristletails, Firebrats

Wings. None.

Mouthparts. Chewing.

Miscellaneous. Small insects, with elongated, flattened, naked, or scaly bodies; abdomen with 10 or 11 segments; 3 long, styliform, many-segmented abdominal appendages consisting of a pair of cerci and a median caudal filament, and bearing 2 to 8 pairs of ventral appendages (styli); antennae long and filiform; compound eyes present in some species, degenerate in others, absent in some. The Thysanura have an unusually large number of molts, both before and after maturity; they may pass through 45 to 60 molts before death. (See chapter 8.)

This order of universally distributed insects is represented by the silverfish (figure 206, chapter 8), which are common household pests and are injurious to bookbindings, wallpaper, and similar articles containing starch or glue size.

Subclass II. Pterygota. Winged insects, with some secondarily wingless.

Division A. Exopterygota. Wing pads develop externally in nymph. Paurometabolous insects with gradual metamorphosis.

Orthoptera. Grasshoppers, Crickets, and Locusts and, according to some authorities, the Cockroaches, Walkingsticks, and Mantids

Wings. Four, sometimes greatly reduced or lacking. When present, the forewings are thickened and modified into somewhat hardened *tegmina,* though with distinct venation; hindwings are broad, fanlike, membranous, and are folded longitudinally when at rest.

Mouthparts. Chewing.

Miscellaneous. For this brief and elementary treatment of orders, the older concept of the Orthoptera is retained, although there are good grounds for the splitting of the order into a number of other orders; in subsequent chapters, the cockroaches and mantids are considered as a distinct order (Dictyoptera) (see chapter 6, under "Cockroaches"), as are also the walkingsticks (Phasmida) and grylloblattids (Grylloblattodea).

Among the Orthoptera are found some of the largest of all insects. A grasshopper from Venezuela is as long as 17 cm, and walkingsticks attain a length of 25 cm in Africa and South America. While the forewings (tegmina) of most Orthoptera are very drab, the hindwings are often brightly colored. The katydids (figure 58) and crickets are "singing" insects, the males being able to make sounds of more or less musical quality by means of stridulating structures. The locusts (family Acrididae), since times of earliest historic record, have caused widespread agricultural losses and even occasional famine. The migratory locust, *Locusta migratoria* L., occurs over vast areas of the Eastern Hemisphere. It is normally confined to its breeding grounds, but occasionally, according to Uvarov (1928), crowding results in the production of a migratory phase. During this phase, the locusts sweep over vast areas in incredible numbers, often destroying practically all the vegetation. The Rocky Mountain grasshopper, *Melanoplus spretus* (Walsh), has done great damage in the Great Plains region of the United States and Canada. A general lessening of the severity of its attacks is believed to have resulted from cultivation in its usual breeding grounds, which results in the destruction of the egg cases (oöthecae) in the soil.

Cockroaches (plate III) are predominantly tropical and subtropical insects, but some species have been disseminated into temperate regions, being able to survive in homes, bakeries, restaurants, and similar protected places where food is available. They are very objectionable because they frequent dark, filthy places, have a repugnant odor, and may be carriers of disease.

Other Orthoptera are noteworthy because of their interesting forms and habits. The praying mantis, so called because of the manner in which it holds its

Fig. 58. Two citrus-infesting katydids in California. *Top,* forktailed bush katydid, *Scudderia furcata* Brunner; *bottom,* broadwinged katydid, *Microcentrum rhombifolium* (Saussure).

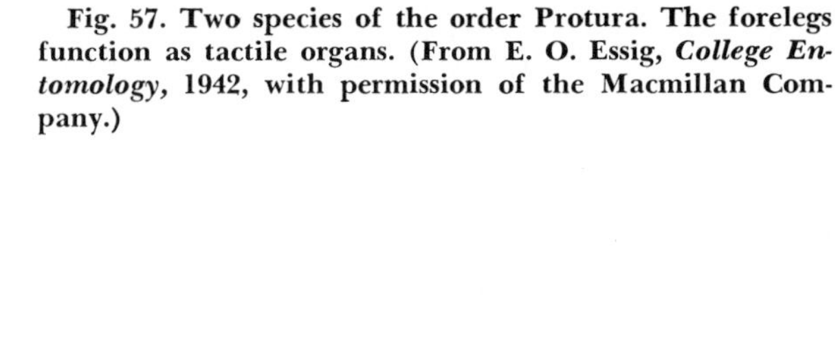

Fig. 57. Two species of the order Protura. The forelegs function as tactile organs. (From E. O. Essig, *College Entomology,* 1942, with permission of the Macmillan Company.)

Fig. 59. The walkingstick, an orthopteran, presents a good example of protective mimicry, as shown by close inspection of this photograph.

enlarged forelegs, uses these forelegs for capturing and holding its prey. The walkingsticks (figure 59) closely resemble the twigs or leaves of their environment, and afford some of the most striking examples of protective mimicry.

Isoptera. Termites

Wings. Four, when present, with both pairs of similar size and shape, laid flat on back when at rest, extending far beyond the tip of abdomen, and capable of being shed by means of basal fractures.

Mouthparts. Chewing.

Miscellaneous. Social and polymorphic species, living in colonies composed of winged and wingless reproductive forms together with numerous wingless, sterile soldiers and workers. Abdomen wide where it joins the thorax, and sometimes ending in a pair of very short cerci.

Termites, sometimes called "white ants," can sometimes be seen in large numbers in logs in forests, in wood lying in contact with the ground, or in the timbers of buildings. The subterranean species of the family Rhinotermitidae are very destructive, either to wooden structures that are in contact with the ground or that can be reached by means of shelter tubes built by the termites from soil and body secretions. Building timbers are often so badly infested with termites that, while the exterior may appear to be sound, the interior may be a mass of tunnels and excrement. The insects will frequently not be seen until spring, when great swarms of winged reproductives appear. The latter are often mistaken for "flying ants."

A termite colony is an example of regimentation *par excellence.* The colony consists of a number of *castes.* Only a few of the eggs develop into individuals that can reproduce; the remainder devote themselves to the care of the "kings" and "queens" and their myriads of offspring (figure 60). The wings of the reproductive individuals are used for only a single flight, after which they break off near their bases. Mating then takes place. Some of the eggs develop into wingless castes known as *soldiers* and *workers.* The soldiers have massive heads with greatly enlarged mandibles, or have other modifications for the defense of the colony.

The digestive tracts of some termites contain protozoa possessing enzymes that break down the cellulose into products that termites can assimilate. If these protozoa are removed, the termites will eventually die of starvation. Certain species that have no intestinal fauna to aid in digestion feed on fungi cultivated in their nests.

Dermaptera. Earwigs

Wings. Usually 4; forewings modified into very short, leathery tegmina without veins; hindwings semicircular, membranous, with veins highly modified and disposed radially; when at rest, the hindwings are folded both lengthwise and crosswise; wingless forms are common.

Mouthparts. Chewing.

The earwigs are easily recognized by their short, leathery forewings and the large, forcepslike cerci at the end of the abdomen. The European earwig, *Forficula auricularia* L. (figure 356, chapter 12), is a widely distributed species that breeds in garbage dumps, manure piles, lawn cuttings, leaf piles, and other kinds of debris, and has but 1 generation per year. Earwigs feed on dead or living animal matter, but also on many kinds of plants. They often become household pests. The female may brood (stay with and care for) her eggs, a rare phenomenon in nonsocial insects.

Embioptera. Embiids

Metamorphosis. Gradual in male; absent in female.

Wings. Males usually winged, but some are wingless or have vestigial wings. When winged, they have 2 pairs that are membranous, with reduced venation. The females are wingless.

Mouthparts. Chewing.

Miscellaneous. The embiids are solitary or gregarious insects, living in silken tunnels; cerci 2-segmented, usually asymmetrical in the male. These insects (figure 61) comprise a small order of fewer than 150 species. They live under stones and other objects, the silk of their tunnels being spun by glands located on the swollen first segments of the tarsi of the forelegs. Their webbed galleries are the most conspicuous indicators of their presence. Embiids generally shun the light of day, and leave their tunnels only at night; however, the males are commonly collected at lights (Ross, 1965). Their food is thought to be chiefly dead and decayed vegetable matter. The males may eat animal food.

Psocoptera. Psocids, Booklice, Barklice

Wings. When present, 2 pairs, membranous; forepair the larger and with extensive pterostigmata; when at rest, held rooflike over abdomen.

Mouthparts. Chewing.

Miscellaneous. Small or minute in size; antennae relatively long; prothorax small; cerci absent; tarsi 2- or 3-segmented.

One group of psocids, all formerly included in the family Atropidae, includes the booklice, which are found in stored books, especially in dark, damp areas. The booklice have no wings or ocelli, and have more than 13 segments in the antennae (figures 334, 335, chapter 10). Other psocids live on the bark of trees and on foliage. Molds and fungi are eaten by these species, and none is known to be of economic importance. Psocids sometimes become a household nuisance by breeding in cereals and starchy materials. They sometimes occur in great numbers in a house.

Thysanoptera. Thrips

Metamorphosis. Gradual, but deviates from the usual type of gradual metamorphosis in that the last 2 nymphal stages (prepupa and pupa) are quiescent.

Wings. Four narrow, membranous wings, nearly veinless, fringed with hairs; some species are wingless.

Mouthparts. Rasping–sucking; asymmetrical, there being only a single mandible.

Miscellaneous. Tarsi 1- or 2-segmented, terminating in a protrusible bladder.

The suborder Terebrantia includes those species of the order in which the abdomen of the female terminates in a sawlike ovipositor. This kind of ovipositor is used for inserting her kidneyshaped (reniform) eggs into the epidermis of leaves or fruit. These species are generally quite active, the greenhouse thrips (figure 346, chapter 11) being a conspicuous exception, and include the most important pests. Males may be absent, but even when present, reproduction may occur by parthenogenesis.

The suborder Tubulifera includes large, inactive species, in which the female has no ovipositor, the eggs being laid on the plant surface. The last segment of the abdomen of both sexes is tubular, as contrasted with the coneshaped terminal abdominal segment of the female Terebrantia or the bluntly rounded terminal segments of male Terebrantia. The Tubulifera may often be found in the flower heads of the Compositae, under bark, in galls, in moss and turf, or under leaves, where they may prey on small insects and mites or feed on plant exudates or on dead and decaying vegetable material. A few species are injurious to cultivated plants.

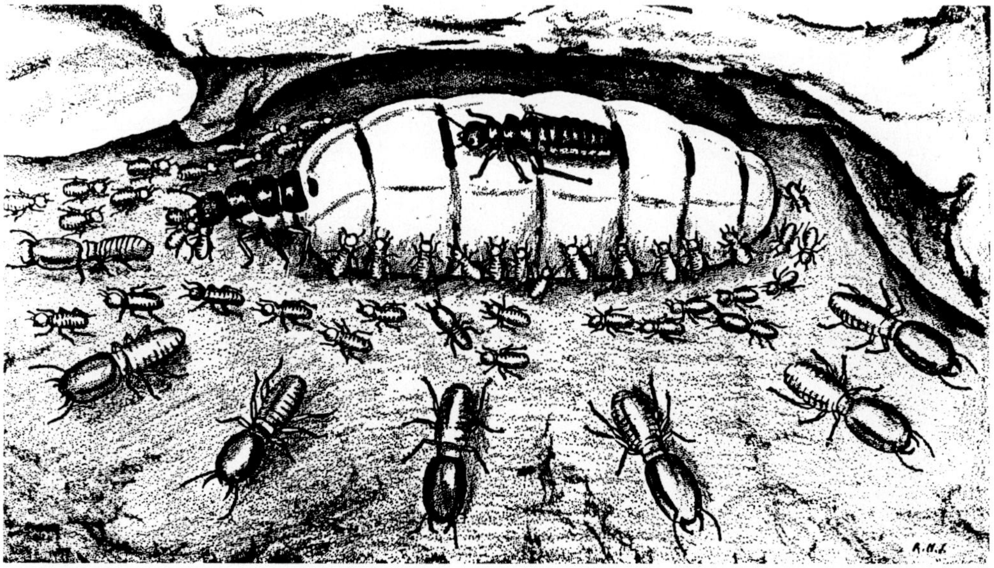

Fig. 60. The queen in a community of tropical termites, with the king on her abdomen, surrounded by many workers grooming and feeding her and by soldiers on guard. (Reprinted from Robert Matheson: *Entomology for Introductory Courses,* second edition. Copyright 1951 by Comstock Publishing Co., Inc. Used by permission of Cornell University Press.)

Phthiraptera. Biting Lice, Sucking Lice

Wings. None.

Mouthparts. Greatly modified; adapted for chewing in suborder Mallophaga, for piercing–sucking in suborder Anoplura.

Miscellaneous. These are small, flattened, wingless insects, living as external parasites of birds and mammals; legs short, with tarsi adapted for clinging to the host; antennae short, 3- to 5-segmented; eyes reduced or absent.

The order Phthiraptera may be divided into 2 suborders, the Mallophaga (biting lice) and the Anoplura (sucking lice) (figure 294, chapter 9). Some authors, however, regard the Mallophaga as a distinct order. Both orders exist continuously on the bodies of warm-blooded animals. They glue their eggs to the hairs or feathers of their hosts. The biting lice subsist on bits of hairs or feathers, skin scales, or the dried blood from scabs, and some species can obtain blood by puncturing the bases of young feathers. The great majority of biting lice occur on birds, but 3 families are restricted to mammals, attacking, among other hosts, cattle, horses, sheep, goats, dogs, and cats. Nearly all species infesting mammals have only 1 tarsal claw which, with modification of the end of the tibia, forms a grasping organ for holding on to the hairs of its host. The prothorax is free and well developed. The species on birds have 2 tarsal claws. Compared with sucking lice, biting lice are of little importance.

The members of the suborder Anoplura are all parasitic on mammals, and their mouthparts are modified for sucking, but are completely withdrawn into the head when not in use. Their tarsi have only 1 segment that terminates in a single claw, much in the manner of the biting lice occurring on mammals. The sucking lice can be distinguished from biting lice not only by the difference in mouthparts, but also because the head of the sucking lice is narrower and more pointed, and the thoracic segments are fused. The head and body lice of man, *Pediculus humanus* var. *capitis* and *P. humanus* var. *humanus* (plate VIII, 2, and figure 294, chapter 9), the "cooties" of World War I fame, are carriers of typhus and trench fever, and have caused great losses of human life, especially in times of war and pestilence. In World War II, the timely development of DDT, a remarkably effective insecticide against lice, reduced casualties attributable to louse-borne diseases to a very low number. The crab louse (figure 295, chapter 9) is the only other louse parasite of humans, but different species of lice attack cattle, horses, dogs, hogs, and goats.

Fig. 61. Apterous female of *Embia californica.*

Hemiptera. Bugs, Leafhoppers, Aphids, Mealybugs, Scales

Metamorphosis. Gradual, except in some highly specialized forms.

Wings. Four, or secondarily wingless (2 wings in male Coccoidea).

Mouthparts. Piercing–sucking, with palpi lacking. The labium is modified into a dorsally grooved sheath which is usually jointed; within it lie needlelike stylets, the modified mandibles and maxillae.

Miscellaneous. Antennae, with rare exceptions, are 2- to 10-segmented; tarsi are 1- to 3-segmented, with 1 (Coccoidea) or 2 claws, with or without *arolia* (cushionlike pads); abdomen with few to 10 segments; cerci absent.

Suborder Heteroptera

This suborder is often given order status (Hemiptera), and comprises a reasonably well-defined group of insects, characterized primarily by the forewings (hemelytra) being thickened in the basal half, while the apical half is thin and membranous (figures 363 and 364, chapter 12). The wings lie flat on the back when at rest, and the membranous apical halves of the forewings overlap. The tarsi are usually 3-segmented. The beak is attached well forward.

The aquatic bugs of this suborder have short antennae that can be concealed in grooves on the underside or back of the head. The ones most likely to attract attention are the giant water bugs or electric-light bugs (Belostomatidae). These are very large forms that feed on insects, tadpoles, and small fish. The females of 2 genera lay their eggs on the backs of the males. The females secrete a waterproof glue with which the eggs are attached. The males carry the eggs about until they hatch.

The terrestrial Heteroptera have conspicuous antennae that are longer than the head and extend in front of it. The predaceous species include the semi-aquatic water striders, the assassin bugs, bed bugs, damsel bugs, ambush bugs, some stink bugs, and some lygaeids and mirids.

The plant-infesting bugs include the stink bugs, leaf bugs, lace bugs, chinch bugs and squash bugs. Among the stink bugs is the harlequin bug, *Murgantia histrionica* (Hahn), a very destructive pest of cruciferous plants in the southern states. The tarnished plant bug, *Lygus pratensis* (L.), a representative of the family Miridae, is a serious pest of fruits and field crops. Best known of the family Lygaeidae is the very destructive chinch bug, *Blissus leucopterus* (Say), which annually causes millions of dollars of damage to corn and wheat crops. Among the Coreidae, the well-known squash bug, *Anasa tristis* (De Geer), is a serious pest of cucurbitaceous plants, and the boxelder bug, *Leptocoris trivittatus* (Say) (plate VIII, 8), infests box elder, ash, maple, and many fruit trees, and also becomes a nuisance by hibernating in houses or invading them in large numbers on warm winter days (see chapter 12, under "Bugs (Hemiptera)"). In the Cimicidae is the bed bug, *Cimex lectularius* L. (plate VIII, 4).

Suborder Homoptera

The suborder Homoptera is characterized by the wings, when present, being membranous, or sometimes with a somewhat leathery texture, but never with the basally thickened forewings (hemelytra) which characterize the Heteroptera. The hindwings are shorter and wider than the forewings. When these insects are at rest, they usually hold their wings rooflike over the abdomen. The beak of the Homoptera is attached ventrally near the back of the head, and often appears to arise from between the front coxae. All Homoptera are plant feeders, and the excrement of many species becomes practically pure sugar. It is known as "honeydew," and can support the growth of the familiar sooty-mold fungus which blackens the foliage and fruit of plants upon which the honeydew-producing species appear.

The aphids or "plantlice" are Homoptera that are economically important on the majority of agricultural plant crops. Their bodies may be naked, or may be covered with a cottony or mealy wax, as with the woolly apple aphis and mealy plum aphis. The mealybugs (figure 347, chapter 11) nearly always have this type of body covering.

The armored scale insects are so named from the separate "scale" that the members of the family Diaspididae secrete to form a protective "armor" that covers the body completely on its upper surface (figure 349, chapter 11). An analogous term applies to the unarmored scales (figure 348) of the family Coccidae, even though they do not secrete a separate scale or armor. However, the adult females have a leathery or hard exoskeleton that is either smooth, roughened, naked, or covered with a thin layer of wax. In both families, metamorphosis is incomplete in females and complete in males, which have 2 wings. The mature females are unable to move.

In the majority of the citrus-growing regions of the world, the Homoptera are more important pests than all other insect orders combined, the scale insects usually being the worst offenders, often attacking other fruit trees as well. The aphids, leafhoppers, and whiteflies (Aleyrodidae) (figure 350) are also important plant pests the world over.

Division B. Exopterygota. Wing pads develop externally in nymph or naiad. Hemimetabolous insects with incomplete metamorphosis.

Ephemerida. Mayflies

Wings. Generally 4, membranous. Hindwings sometimes absent; when present, smaller than the forewings. Wings triangular in outline, generally gauzy, with many longitudinal cross veins, and held vertically when at rest. The Ephemerida (and Odonata) have an archaic type of wing attachment to the thorax which does not permit them to flex the wings against the body.

Mouthparts. Those of adults vestigial or absent; those of the nymphal aquatic forms (naiads) chewing.

Miscellaneous. Antennae tiny and setaceous; adults with 2 or 3 very long, slender, many-segmented "tails." Forelegs longer. Naiads aquatic, with paired tracheal gills on sides and back of abdomen; 2 or 3 slender "tails."

Mayflies are the only pterygotes with a *subimago.* When the last nymphal instar is ready to transform, it contains within its skin a fully developed adult, almost sexually mature. It rises to the surface of the water and out flies a mayfly, the subimago. It alights, rests for a day, then molts, and thereafter exists as a true sexually mature adult.

The delicate, gauzy-winged, long-tailed adults of the mayflies (figure 324, chapter 9) may swarm in large numbers around lights or near their breeding grounds. At the right season, one may observe these swarms, which are actually mating flights or "dances," near the breeding grounds. The swarm, consisting only of males, rises and falls until a receptive female approaches. She is then seized by a male and the pair depart. The "dance" of the males continues, with females being seized as soon as they enter the swarm. They lay their eggs soon after mating.

The scientific name of the order is derived from the fact that the mayfly adult lives one of the briefest (or most *ephemeral*) existences known in the winged state among insects, a period of 4 days or less, depending upon the species. Because the adults rarely if ever consume food, their intestinal tract is modified so as to contain large amounts of air for buoyancy. It no longer functions as a digestive tract.

Odonata. Dragonflies, Damselflies

Wings. Four membranous, elongate, finely netveined wings of about equal size.

Mouthparts. Chewing. In naiads, the labium of the mouthparts is modified into a prehensile organ used in capturing prey, and folding like a mask over the face when not in use.

Miscellaneous. Antennae very small, setaceous; compound eyes large; abdomen long and slender; pleural sclerites slanted backwards so the wing attachments are behind the leg attachments; naiads aquatic, respiring by means of rectal or caudal gills. Adults and naiads predaceous. The Odonata are divided into the Anisoptera (dragonflies) and the Zygoptera (damselflies).

These 2 well-defined suborders differ in a number of morphological and behavioral characteristics. The dragonflies (figure 56, this chapter) are strong fliers. Their hindwings are broader, at the base, than are the forewings. The wings are not folded when at rest, but are held in a horizontal position at right angles to the sides of the body. The damselflies are feeble fliers. The 2 pairs of wings are of the same size and shape, and both are narrow at the base. When at rest, the wings are held parallel to the abdomen, or are slightly uptilted. The eyes of the dragonflies do not project from the side of the head, whereas the eyes of the damselflies do project, and are constricted at the base.

Dragonflies lay their eggs on water or on aquatic plants, whereas damselflies deposit them in the stems of aquatic plants. Dragonfly naiads respire through tracheal gills inside the rectum, and are propelled by the forcible ejection of water from the anus. Damselfly naiads respire by means of 3 leaflike tracheal gills that project from the end of the abdomen.

Both the nymphs (naiads) and adults of the Odonata are predaceous. The Odonata are probably beneficial as predators on mosquitoes, horse flies, and gnats, and certainly serve as food for fish.

However, there is a harmful stage. The naiads are said to prey upon fry of fish, and certain species are intermediate hosts for flukes which are parasitic in poultry and waterfowl.

Plecoptera. Stoneflies

Wings. Generally 4 membranous wings, held flat over the back in repose; in the majority of genera, the hindwings are much larger than the forewings, and folded in plaits (longitudinal folds) on the abdomen when at rest; venation variously modified.

Mouthparts. Chewing, though weakly developed and frequently vestigial in adult.

Miscellaneous. Antennae long, filiform, with 25 to 100 segments; abdomen of adult bears cerci that are usually long and many-segmented; immature forms (naiads) aquatic, with tufted tracheal gills commonly present, though variably located.

The drab-colored stoneflies may be seen resting on stones, trees, and bushes, or flying for short distances over the water near swift streams with stony bottoms. The naiads may be found under stones or trash, in streams or along their margins. Their cast skins are a common sight on stones and shrubs projecting over a stream or lake shore. They are of value to man as a native food for fish.

Division C. Endopterygota. Wings develop internally as imaginal buds before being everted in the pupa. Holometabolous insects with complete metamorphosis.

Neuroptera. Lacewings, Antlions, Dobsonflies

Wings. Four membranous wings of nearly equal size, usually finely netveined, held rooflike over the abdomen when at rest.

Mouthparts. Chewing, though secondarily adapted in many larvae for sucking the blood of other insects.

Miscellaneous. Tarsi 5-segmented; larvae predaceous; the aquatic forms usually possess abdominal gills; some pupate in earthen cells, but most species spin silken cocoons in which the pupa lies freely.

The family Corydalidae includes the dobsonflies, the largest of the Neuroptera. They have a large, square prothorax, long antennae, and the male has very long mandibles. The larvae, sometimes called hellgrammites, are aquatic and are well known to fishermen, especially those fishing for bass or trout. They live under water, usually under stones in swift currents, and are predatory on other insect life. *Corydalus cornutus* (L.) (figure 62) occurs in eastern North America and is the largest species. It has a wingspread of 10 or 12.5 cm, and the male has very long mandibles. The larva has paired lateral filaments on the abdomen and short tracheal tufts at the bases of the first 7 of these

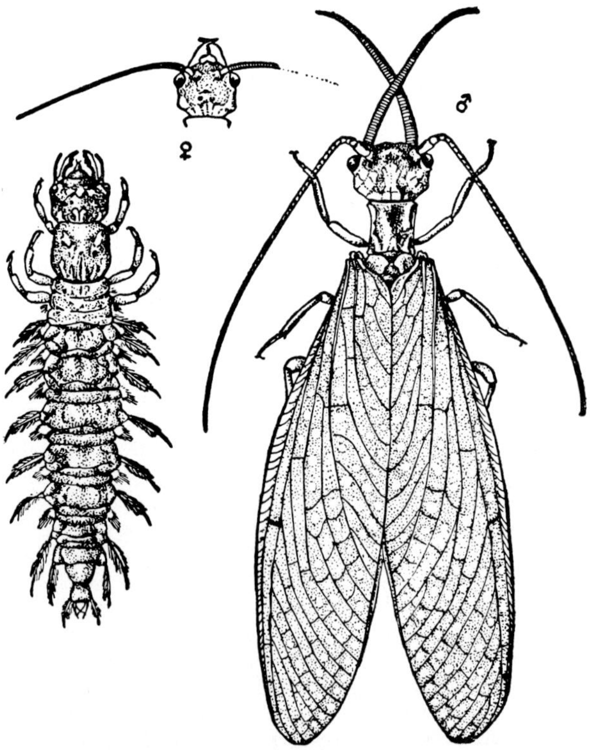

Fig. 62. A dobsonfly, *Corydalus cornutus*. *Left,* larva (commonly called a "hellgrammite"); *right,* adult. (From Ross, 1944.)

Fig. 63. Green lacewing, *Chrysopa californica*.

filaments. The family Corydalidae is represented in California by only a single rare species in the genus *Corydalis,* but there are 6 species in *Protochauloides* and 2 in *Dysmichohermes* (Chandler, 1956).

The family Chrysopidae contains the lacewings. The green lacewing, *Chrysopa californica* Coquillett, (figure 63), is a beautiful insect, with gauzy green wings folded rooflike over the body and with golden eyes. The female places her eggs on the ends of hairlike stalks about 12 mm in length. This may be an adaptation to keep them out of reach of the cannibalistic larvae. The larvae are thysanuriform (alligatorlike), hairy, and with large, sicklelike mandibles. They are often important as predators of aphids.

The larvae of the Myrmeleontidae (antlions or doodlebugs) dig conical pitfalls in sand to trap insects, mainly ants, which they devour as food. Only their mouthparts protrude at the bottom of these pits. After their victims are drained of their body juices, they are cast far beyond the borders of the pit.

Mecoptera. Scorpionflies

Wings. Four or none. When winged, have 2 pairs of similar netveined (with generalized wing venation) and membranous wings, held longitudinally and horizontally in repose.

Mouthparts. Chewing, situated at the end of a beak or snout that is prolonged ventrally into a conspicuous structure that characterizes the adult Mecoptera.

Miscellaneous. Predaceous insects; small to medium-sized and slender; antennae long, filiform, many-segmented; legs long, slender; compound eyes large, widely separated.

The common name of these insects, "scorpionfly," is derived from the shape of the upturned terminal segments of the males of the family Panorpidae, which resemble the corresponding part of a scorpion. The larvae and adults of the panorpids are scavengers on animal food, and are found in moist locations, usually near streams. The larvae resemble caterpillars. The members of the family Boreidae are wingless, and the abdomen of the male terminates in large claspers.

Trichoptera. Caddisflies

Wings. Four similar membranous wings with generalized venation, usually densely hairy, held rooflike over the back in repose.

Mouthparts. Modified by reduction for chewing in the adult, the mandibles being absent, with palpi well developed.

Miscellaneous. Antennae long, filiform; legs long, tarsi 5-segmented. Larvae aquatic, usually living in cases and breathing by means of abdominal gills.

The caddisflies are medium-sized, mothlike insects, with hairy, brownish, or dull-colored wings. They are found by streams and lakes, often in large numbers. They attract popular interest primarily because of the cases which the larvae build around their bodies for protection. What appears to be a mass of small sticks and pebbles at the bottom of a quiet pool may, upon further investigation, be found to be in motion, with the head, thorax, and legs of the tiny larva protruding from this protective case (figure 325, *D,* chapter 9). The cases of certain species may be attached to stones. The larvae of one trichopterous family (Limnephilidae) build the familiar "log-cabin" type of structure out of tiny sticks.

Lepidoptera. Moths, Skippers, Butterflies

Wings. Usually well developed, rarely vestigial, covered with overlapping scales or hairs; venation of fore- and hindwings usually dissimilar.

Mouthparts. Siphoning in adults; chewing in larvae.

Miscellaneous. Body of adult covered with scales and hairs; compound eyes large; antennae variable, often serrate, hooked, or knobbed. Larvae (caterpillars) cylindrical in shape, with 3 pairs of thoracic legs and 2 to 5 pairs of prolegs on the abdomen; prolegs terminate in series of tiny hooklets (crochets); pupae with appendages usually fastened to body, often in silken cocoons, although butterflies in general, and some moths, do not construct cocoons.

This order is among the largest in numbers of species, and is also one of the most destructive of all insect orders. Among the destructive species are the codling moth, oriental fruit moth, cutworms, corn earworm, pink bollworm, European corn borer, tomato hornworm, webworms, leaf rollers, tussock moths, gypsy and browntail moths, flour and meal moths, and clothes moths.

Nearly all of the larvae feed on plants or plant products, the clothes moths (chapter 8) and the greater wax moth (chapter 12) being exceptions. The silkworm moth is a species of considerable value to man.

The general characteristics of the larvae and adults (figure 185, chapter 7) of the Lepidoptera are well known to everyone, and further description for the sake of identification is unnecessary. However, certain features connected with the structure and habits of these insects are not so well understood. The patterns and colors that constitute the chief attraction of the Lepidoptera to the majority of people are understood by only a few. The arrangement of the hairs or scales (figure 50) that usually cover the wings and all other parts of the body accounts for their colors and color patterns. The scales vary in structure from simple hairs to broad, striated plates of various sizes and shapes. The Lepidoptera are also highly specialized in other ways. The coiled proboscis is probably the most highly specialized of all types of insect mouthparts.

The Lepidoptera are generally divided into 2 suborders, the Jugate and the Frenate. In the Jugate, the 2 wings on each side are united by the *jugum,* a small lobe at the base of the forewing. The fore- and hindwings are similar in venation. In the Frenate, the hindwing is smaller than the forewing, and has a reduced venation. The 2 wings on each side are united by the *frenulum,* a spine or group of spines arising at the humeral angle of the hindwing, or simply by the expanded humeral angle of the hindwing. The suborder Jugate contains only 3 rare families. The Frenate can be divided into a series of superfamilies on the basis of differences in wing venation, the presence or absence of a frenulum, and other taxonomic features. These superfamilies are divided into the Macrolepidoptera and the Microlepidoptera, the latter made up of many families of small moths, and including nearly all the species discussed as pests in this book: the limacodids (slug caterpillars) and megalopygids (flannel moths) in chapter 9; the pyralids (snout moths), olethreutids (codling moth), tortricids (leaf rollers), cosmopterygids (the pink scavenger caterpillar), gelechiids (Angoumois grain moth), oecophorids (the white-shouldered and brown house moths) in chapter 7; and the tineids (clothes moths) in chapter 8.

The Macrolepidoptera include the large moths, butterflies, and skippers. The moths have large bodies, and the wings, when at rest, are held horizontally, rooflike, or wrapped about the body. Their antennae are varied in structure; they may be filamentous or feathery, but are rarely clubshaped at the tip. They are mostly night-fliers. The pupae are often protected by cocoons, which the larvae spin of silk produced by modified salivary glands.

The butterflies have slender bodies, and hold their wings vertically when at rest. They have no special structure to hold the fore- and hindwings together, this usually being accomplished by an expansion of the hindwing near its base. They have threadlike antennae that are clubshaped at the tip. Butterflies (and skippers) have no ocelli, while the moths often have two. They are primarily day-fliers. Their larvae are usually not so hairy as those of some moths, but they may possess spines. Most butterfly pupae are without cocoons, and some are attached to twigs, leaves, and other objects by the *cremaster,* a stout spine or hooked area at the posterior end of the pupal abdomen.

The skippers (family Hesperiidae) are so called because of their manner of flying. They dart suddenly from place to place. In contrast to the butterflies, the wing area is small when compared with body size. The larvae are usually naked, and have a constriction just behind their large heads.

An advantage found by the amateur collector of Lepidoptera is that the majority of species are so distinct that they can be identified by means of illustrations in a number of books that may be purchased in bookstores, and are obtainable in most libraries.

Coleoptera. Beetles, Weevils

Wings. Four; rarely 2 or none; when winged, the forewings are modified into horny or leathery "wing covers" or *elytra* that meet in a straight line along the middle of the back (figure 115, chapter 5). The membranous hindwings are folded when at rest.

Mouthparts. Chewing.

Miscellaneous. The adults are heavily sclerotized or rarely larviform; larvae wormlike, usually with 3 pairs of thoracic legs and not more than 1 pair of prolegs; pupae with appendages free, rarely in cocoons.

At least as far as the number of described species is concerned, the order Coleoptera is the largest of the insect orders. The order contains about 350,000 described species, more than a fourth of all the described species of animals (Arnett, 1968). In common with Hemiptera and Lepidoptera, the order Coleoptera contains a large number of species that are injurious to agricultural crops, but among the beetles are also found many beneficial predators, such as the lady beetles and ground beetles. Among the injurious species may be mentioned the June beetles and their larvae (white grubs), Colorado potato beetle, Japanese beetle, alfalfa weevil, cucumber beetles, flea beetles, Mexican bean beetle, bean and pea weevils, plum curculio, granary weevils, rice weevils, and mealworms. Among the forest pests are the bark beetles and flatheaded and roundheaded borers; and among the pests of cured lumber are the powderpost beetles.

The Coleoptera commonly have been divided into 2 suborders, the Adephaga and the Polyphaga. All the commonly found Adephaga are predaceous, including 2 terrestrial families: tiger beetles (Cicindelidae) and ground beetles (Carabidae); and 2 families of aquatic beetles: diving beetles (Dytiscidae) and whirligig beetles (Gyrinidae). The latter are so called because of their strange gyrations when they circle round and round one another as they swim or skate on the water. The Adephaga are distinguished from the other Coleoptera by the separation of the pronotum from the propleuron (lateral portion of the prothorax) by a distinct suture (notopleural suture), and by the first visible abdominal segment being deeply cleft by the hind coxal cavities. Sclerites and sutures are often used in the taxonomy of Coleoptera because the sclerites are fitted together with remarkable exactness and precision. The larvae of the Adephaga have 6 segments in each leg and 2 claws at the end of the leg.

The Polyphaga are divided into a number of superfamilies. All have the first ventral segment of the abdomen in a single piece, not cleft by the hind coxal cavities, and lack the notopleural suture. The legs of the larvae have 5 or fewer segments, and always end in a single claw. The division of the suborder into its superfamilies is based mainly on the nature of the antennae and on tarsal segmentation. The following list of superfamilies is based on a system of classification of the Polyphaga adopted by Metcalf *et al.* (1962). It is adequate only in a general way, or for beginning students or amateurs. Whereas only 7 superfamilies are recognized here, some authorities may recognize as many as 18, including 112 families (Borror and DeLong, 1971).

1. Shortwinged beetles (Brachyelytra). These beetles have short elytra, exposing much of the abdomen, and have 5-segmented tarsi. They include the rove beetles (Staphylinidae) and the carrion beetles (Silphidae), which for the most part are scavengers.

2. Clubhorned beetles (Clavicornia). The antennae are clavate and the tarsi are 5- or 3-segmented. Includes the water scavenger beetles (Hydrophylidae), the flat bark beetles (Cucujidae), lady beetles (Coccinellidae), the tarsi of which are 3-segmented, and the dermestids (Dermestidae).

3. Sawhorned beetles (Serricornia). With serrate antennae and 5-segmented tarsi. Includes the clerids (Cleridae), the fireflies or lightning beetles (Lampyridae), the cantharids (Cantharidae), the flatheaded borers (Buprestidae), and the click beetles or wireworms (Elateridae).

4. Beetles with different-jointed tarsi (Heteromera). Five segments in tarsi of the first and second pair of legs; 4 segments in the third pair. Darkling beetles (Tenebrionidae) and blister beetles (Meloidae).

5. Leafhorned beetles (Lamellicornia). These beetles have 5-segmented tarsi throughout. The lamellate antennae consist of a cylindrical basal part and a number of flattened, leaflike segments at the extremities. Includes the stag beetles (Lucanidae) and the scarabs or June beetles (Scarabaeidae).

6. Plant-eating species (Phytophaga). The tarsi in this group are apparently 4-segmented. Includes longhorned beetles or roundheaded borers (Cerambycidae), the leaf beetles (Chrysomelidae), and the pea and bean weevils (Bruchidae).

7. The snout beetles (Rhynchophora). The head often bears a long snout. In this group, the tarsi are again apparently 4-segmented. The antennae are clubbed and elbowed. Included are the weevils (Curculionidae) and bark beetles (Scolytidae).

Strepsiptera. Stylopids or Twisted-Winged Insects

Wings. Only the males are winged; forewings are reduced to clubshaped appendages resembling halteres; the hindwings are relatively large, fanshaped, with radiating veins, and are folded longitudinally when at rest.

Mouthparts. Vestigial or lacking.

Miscellaneous. The adult female is larviform, and lacks legs, wings, eyes, antennae, and mouthparts.

The Strepsiptera comprise a very small order of 5 families, and are primarily of interest because of the peculiarities of their biology. They were formerly considered as a family (Stylopidae) of the order Coleoptera. Some taxonomists believe they should be returned to that order. The larvae are endoparasitic on bees, wasps, and a few Hemiptera and Orthoptera. The female develops and remains within the host, and the males fly about, seeking a mate. Both males and females extrude the anterior portions of their bodies between the abdominal segments of their hosts, and can be seen if an infested host is carefully examined. A few free-living female Strepsiptera have been found, but their life histories are unknown.

Hymenoptera. Ants, Wasps, Wasplike Parasites, and Bees

Wings. Four wings (rarely 2) or none, dissimilar in size, membranous, usually with few veins.

Mouthparts. Chewing or chewing–lapping.

Miscellaneous. Ovipositor is always present in females, and is modified for sawing, piercing, or stinging; larvae usually legless (except Symphyta); pupae with appendages free, commonly in cocoons; many species are social.

Although currently the order Hymenoptera is third in size from the standpoint of the number of species which have been described and named, probably great numbers of inconspicuous species, especially among parasites, have escaped the attention of taxonomists, so it is likely that there are more species of this order in existence than of any other. Some authors consider the Hymenoptera to be the most highly evolved of the insect orders because of the high degree of development and specialization of social organization found in this group, including the care of the young. The striking ability of some Hymenoptera to learn through experience has been taken by some scientists to indicate that the rudiments of intelligence may be found among these insects. On the other hand, some authorities consider flies and fleas as having reached a higher evolutionary status than the Hymenoptera, basing their opinions on the greater degree of morphological specialization in these orders.

The social organization of the Hymenoptera has long excited wonder and admiration, and has been the subject of much investigation. In colonies or nests of Hymenoptera, as with the termites, the reproductive function has been limited to a few specialized individuals, the kings and queens, but unlike the termites, the workers of the Hymenoptera are exclusively females. The males are therefore called drones, their only function being the fertilization of the queen. The caste formation and development of various Hymenoptera, particularly of ants, wasps, and bees, are discussed elsewhere in this book in considerable detail.

Among the insects, the Hymenoptera are distinguished by many species having a highly specialized modification of the ovipositor into an organ known as the *stinger*. Only the females can sting, and the pain of the sting is caused mostly by the reaction that the victim has to the venom injected into the wound made by the stinger.

The order Hymenoptera has been divided into 2 suborders, the Symphyta (Chalastogastra) and Apocrita (Clistogastra).

Suborder Symphyta

This suborder contains at least 12 families, including the Tenthredinidae (sawflies) and Siricidae (woodwasps) discussed below. (See figure 138, chapter 5.) The anterior segments of the abdomen are as broad as the following segments, and are joined to the thorax throughout their entire width. The trochanters are 2-segmented. The ovipositors are sawlike (sawflies) or awl-like (woodwasps) for cutting plant tissues in oviposition. The larvae are caterpillarlike, but can be differentiated from lepidopterous larvae in that they have 6 to 8 pairs of prolegs, while the lepidopterous larvae never have more than 5. Moreover, the prolegs of the sawfly larvae never possess hooklets (crochets) like those of lepidopterous larvae. The sawfly larvae have a single pair of ocelli, while caterpillars have several on each side of the head.

Suborder Apocrita

In this suborder, the second abdominal segment is constricted into a slender petiole connecting the thorax to the abdomen. The ovipositor may be specialized for boring into plant tissue, placing eggs within other insects as done by the parasites (figure 64), or it may

Fig. 64. A hymenopterous parasite (*Macrocentrus*) ovipositing on a lepidopterous larva.

be modified into a stinger associated with venom glands. The larvae are always legless and grublike, having a reduced head and mouthparts, antennae and palpi with at most 1 segment, and usually lack ocelli.

The Apocrita may be divided into 3 groups: (1) the ants, with the petiole or pedicel (a slender stalk) of the abdomen with 1 or 2 swellings or nodes; (2) the wasps or wasplike species, with the hind tarsus usually slender and cylindrical and no nodes on the abdomen; and (3) the bees, with branched or plumose hairs on the thorax, and with broader and more hairy bodies than the ants and wasps. Nearly all are winged.

Among the Hymenoptera in general, there are few agricultural pests, and the majority of these are Symphyta (sawflies). In the Apocrita nearly all species are either beneficial to man or harmless. The honey bee, of course, is of great value as a producer of honey and beeswax, as well as being an important pollinator of fruit crops, many of which would have no commercial value without insectborne pollination. Ants, wasps, and bees can be pests because of their venomous stings. Ants may be injurious because of their ability to keep predators and parasites away from injurious insects, or to care for them in other ways. In return, they feed on the "honeydew" of certain injurious species that they protect, such as soft scales, aphids, mealybugs, and whiteflies. Ants are occasionally phytophagous, and damage plants. The Apocrita, however, are predominantly beneficial. Hymenopterous parasites are the principal entomophagous members of the insect world. They are among the "wasplike" species of the suborder, and include the families Ichneumonidae, Braconidae, Evaniidae, Chalcididae, and Proctotrupidae, as well as others that are not so well known.

Diptera. True Flies, Mosquitoes, Gnats, Midges

Wings. Two membranous wings, borne on the mesothorax; hindwings are modified into clubshaped organs (halteres).

Mouthparts. Sponging or piercing–sucking.

Miscellaneous. Head, thorax, and abdomen are very distinct (i.e., neck and petiole very narrow); larvae legless, usually maggotlike, with head greatly reduced and mouthparts composed of highly modified mandibles and maxillae which form the "mouthhooks" in higher forms; pupae with appendages commonly enclosed in a puparium or case in higher forms.

The Diptera differ from the other orders of insects in that the species with wings have only a single pair. The vestigial second pair (halteres) are believed by some entomologists to be balancing organs. A haltere may be seen on the asilid shown in figure 65. The few species in other orders that have only a single pair of wings do not possess halteres. In the Diptera, all but a few species have halteres, even if they have lost the front pair of wings. Since only the forewings are functional and are attached to the mesothorax, this segment makes up the larger part of the thorax. The thorax is highly developed, and among the insects, certain flies are the most rapid fliers. Many species also show a marked specialization of the abdomen, with a great reduction in the number of segments.

Flies are not conspicuous agricultural pests. Among the more economically important Diptera are the fruit flies, seed midges, and the Hessian fly. The Diptera rank first, however, as vectors (carriers) of diseases affecting man and animals. Mosquitoes are vectors of the causative organisms of malaria, yellow fever, dengue, filariasis, and encephalitis. Horse flies can transmit filariasis and anthrax. The tsetse fly is the vector of sleeping sickness. Various diseases are transmitted by Diptera of the family Psychodidae (*Phlebotomus* spp.), others by the black flies (*Simulium* spp.), and the ceratopogonids. The house fly is a mechanical carrier of the causative organisms of typhoid, dysentery, cholera, poliomyelitis, anthrax, yaws, conjunctivitis, and the eggs of several cestodes and nematodes. House flies, flesh flies, and others may cause myiasis, the invasion by fly maggots of the organs, tissues, and cavities of man and animals.

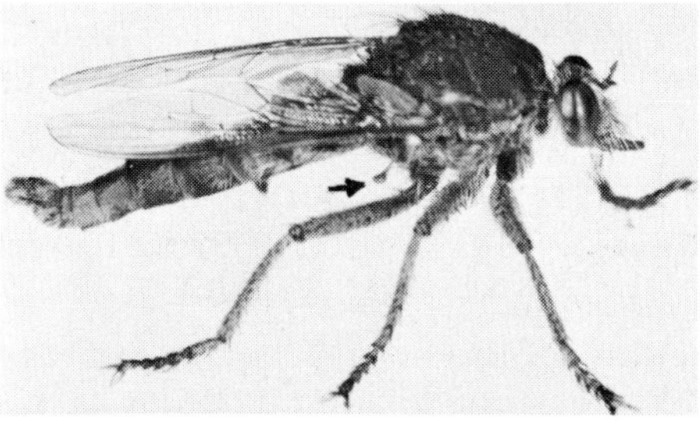

Fig. 65. A species of robber fly, a dipterous predator (Asilidae). Arrow shows a haltere.

The Diptera are divided into 3 suborders: Nematocera, Brachycera, and Cyclorrhapha (Furman and Catts, 1970).

Suborder Nematocera

These are mostly small, slender insects, including mosquitoes, midges, black flies, and moth flies (Psychodidae). They have slender, many-segmented antennae (6 to 39 segments) that are usually longer than the thorax, and with the segments beyond number 3 generally very small and often closely joined. The palpi have 4 or 5 segments, and are pendulous (figure 46, *H,* this chapter, and figure 287, chapter 9). The larvae have a distinct head, eyes, and true mandibles (Snodgrass, 1944).

Suborder Brachycera

The members of this suborder are mostly medium to large, robust, and have large wings, such as the deer flies, horse flies, and snipe flies. The antennae are usually shorter than the thorax, and mostly 3-segmented, with the last segment elongate. When the arista is present, it is generally terminal. The palpi have 1 or 2 segments (figure 290, chapter 9). The larvae often have the head invaginated into the thorax, and have mouthhooks that work in a vertical direction (Snodgrass, 1944).

Suborder Cyclorrhapha

These insects are variable in size, but are usually robust and have medium wings. The house fly, stable fly, flesh flies, blow flies, bot flies, vinegar flies, and eye gnats (Chloropidae) are well-known examples. Flies in this suborder have very short, 3-segmented antennae, with the last segment greatly swollen and the arista dorsal (figure 46, *E,* this chapter). The head usually has a frontal lunule (crescent) and a ptilinum, an inflatable organ that can be thrust out through a frontal suture just above the base of the antenna. The cap of the puparium is forced off by the inflated ptilinum. The bladderlike organ is then withdrawn into the head.

In the Nematocera and Brachycera, formerly considered to be superfamilies of the suborder Orthorrhapha, the adult flies usually emerge from the pupal skin through a T-shaped or straight split down the back. The pupae are generally naked. In the suborder Cyclorrhapha (circular-seamed flies), the larva pupates in the last larval skin, and the adult emerges from the anterior end, pushing off a circular lid.

The Cyclorrhapha do not have mouthparts with a biting mechanism. Most species collect their liquid food via the "pseudotracheae" of the broad labellar disk (figure 46, *J,* this chapter), but in some species (horn flies, stable flies, tsetse flies), the labium forms a piercing organ (figure 46, *F, G*) (Snodgrass, 1944).

Siphonaptera. Fleas

Wings. None.

Mouthparts. Piercing–sucking, with 2 pairs of palpi.

Miscellaneous. Bodies small and laterally compressed; hind legs enlarged for jumping; coxae elongate; tarsi 5-segmented; adults are external parasites on warmblooded animals; larvae are slender and cylindrical, without legs or eyes, but with well-developed heads; the pupae have their appendages free, and are enclosed in cocoons.

When examined with a hand lens, fleas are easily distinguished from other insects (figure 303, *A,* chapter 9). They have a laterally compressed body that is, in itself, almost infallible as a taxonomic character. In addition, the hard, polished body contains many hairs and short spines that are regularly arranged and directed backward. Fleas have no compound eyes, and often lack even simple ones. They lay their eggs upon their hosts, as do the lice, but the eggs are not fastened and will drop off, hatching on the ground or floor. The larvae feed on dead animal or vegetable matter and adult flea excrement. They pupate in a silken cocoon, to which particles of dirt usually adhere.

There are only 5 families in this small but important order, and 2 of them are of special importance. The Pulicidae (common fleas) includes the human flea, *Pulex irritans* L., and the fleas of cats and dogs, *Ctenocephalides felis* (Bouché) and *C. canis* (Curtis). Fleas are most likely to infest houses where pets are kept.

These insects are vectors of bubonic plague. The chief vector is the oriental rat flea, *Xenopsylla cheopis* (Rothschild), which feeds mainly on rats, but also attacks man. The sticktight flea, *Echidnophaga gallinacea* (Westwood), is an important pest of poultry that also attacks many other animals. The chigoe flea, *Tunga penetrans* (L.) (see chapter 9, under "Fleas"), is considered not to have been permanently established in the United States, but is found in many tropical and subtropical regions throughout the world.

Chapter 5 · *Wood-Destroying Insects and Fungi*

Among the wood-destroying insects there are a few species that attack healthy, living trees, but many more attack trees that are weakened by disease, fire, or old age, or are blown down or felled for lumber. Included among these insects are the bark beetles and ambrosia beetles (Scolytidae and Platypodidae), some weevils and snout beetles (Curculionidae), the flatheaded borers (Buprestidae), most roundheaded borers (Cerambycidae), some caterpillars (Cossidae), and the woodwasps (horntails) of the family Siricidae. The larvae of some of these pests require many years to develop, and the adults may emerge from seasoned wood in buildings or furniture long after construction. However, none of the above insects, except the cerambycid *Hylotrupes bajulus* (L.) (the oldhouse borer), reinfests seasoned lumber from which the adults have emerged.

Other groups of insects can infest dry and seasoned wood. These include the termites (order Isoptera), powderpost beetles (Lyctidae), false powderpost beetles (Bostrichidae), deathwatch beetles (Anobiidae), carpenter ants (Formicidae: Formicinae), and the oldhouse borer just mentioned. These insects may be very damaging because they can reinfest timbers until the wood is completely hollowed out, leaving only a deceptive outer shell. When wood is infested by the beetles in this group, this outer shell is perforated with the exit holes of the successive generations of insects.

There are other insect species that do superficial or only occasionally serious damage to wood, but which are included in this book under more appropriate headings. For example, the spider beetles (Ptinidae) and hide beetles (Dermestidae) can bore into wood when seeking a place to pupate, and they sometimes severely damage structures used for storage of their principal food materials—grain, hides, and offal (the inedible parts of butchered animals).

Fungi probably cause as much damage as termites to wood structures. They require about the same environmental conditions for their growth and development as the subterranean termites (*Reticulitermes*) and dampwood termites (*Zootermopsis*) do for theirs. The elimination of these conditions results in control for both termites and fungi.

Fungi also are important in supplying important nutrients or even the entire food supply of certain wood-destroying insects.

ASSOCIATION OF WOOD-DESTROYING INSECTS WITH BACTERIA, FUNGI, AND PROTOZOA

Structural timber would appear to be, from a physical and nutritional standpoint, a particularly adverse habitat for insects. But, as indicated by the foregoing discussion, many species in several orders have successfully adapted themselves to it. Therefore, it is not surprising that many of these insects are dependent upon, or have taken advantage of, various types of symbiotic relationships with bacteria, fungi, and protozoa. Graham (1967) made a list of the families and genera of arboricolous insects known to have mutualistic associations with fungi. They occur in the orders Isoptera, Homoptera, Coleoptera, Hymenoptera, and Diptera. He also listed the principal kinds of fungi associated mutualistically with insects in trees. They occur among the Phycomycetes, Ascomycetes, Basidiomycetes, and Deuteromycetes (or Fungi Imperfecti). Becker (1971) reviewed the literature up to 1971 pertaining to the physiological influences on wood-destroying insects, not only of substances produced by microorganisms but also of many other compounds occurring naturally in wood.

Becker (1968) found that some strains of *Aspergillus flavus, Penicillium funiculosum,* and *Trichoderma viride* were highly toxic to the larvae of *Hylotrupes bajulus* in small, sapwood samples of *Pinus sylvestris* L. On the other hand, of the soft-rot fungi tested, *Fusarium aquaeductuum* and *Phialophora aurantiaca* accelerated the larval growth rate 10- to 30-fold compared with the control insects.

Young larvae of the deathwatch beetle, *Xestobium rufovillosum* (De Geer), an anobiid, are absolutely dependent on fungi that develop in their tunnels in dry wood, becoming less dependent as they mature (Campbell and Bryant, 1940). On the other hand, larvae of the cerambycid *Ergates faber* (L.) become increasingly dependent on fungi with each molt (Becker, 1942, 1943). Some anobiids are so dependent on fungus-infected wood that they discontinue feeding at the point where fungal infection stops (Becker, 1950, 1951). The ambrosia beetles of the Scolytidae and Platypodidae feed on fungi grown in their brood galleries; these galleries are the extent of their damage to the wood. The ambrosia beetle *Xyleborus ferrugineus* (F.) can go through 1 generation on a diet lacking either fungus or ergosterol, the major sterol of fungi. Apparently, there is an adequate carry-over of sterol from the preceding generation that fed on diets with fungus. However, the second generation fails to pupate if it is not supplied with either fungi or ergosterol in its diet (Norris *et al.,* 1969).

Dominik (1966) found that, among the species of beetles attacking building timbers, some attacked wood heavily infected with fungi, some attacked sound, uninfected wood, and others attacked either infected or uninfected wood. He observed that a single genus (*Anobium*) contained species that required and species that shunned fungus-infected wood. *Anobium punctatum* (De Geer) occurs outdoors in Great Britain where wood has resisted fungal attack, possibly because of sheltered location or good drainage. Old branch scars on trunks of trees or areas of exposed sapwood are almost always infested, and the freedom from fungal decay in such situations led Hickin (1963*a*) to suspect that some fungicidal material had been secreted into the wood by the developing larvae.

Digestion of cellulose by coleopterous larvae was once believed to be dependent on symbiosis with cellulose-secreting microorganisms. This has since been shown to be untrue, although such microorganisms may be important in connection with the vitamin requirements of the insects. There is no longer any doubt that cellulase is present in the digestive enzymes of many, but not all, woodboring beetle larvae. The nutrition of woodboring Coleoptera appears to depend upon more factors than the enzyme complex in the larval gut, and much further research is still needed (Parkin, 1940).

A remarkably close relationship between an insect and a fungus is seen in the woodwasp *Sirex* and the Basidiomycete *Oidia*. The spores of the fungus are carried by the adult siricid in 2 small, invaginated, intersegmental sacs (figure 140) at the base of the ovipositor, described by Büchner (1928, 1930). During oviposition, the spores are deposited in the egg tunnel and develop to produce a mycelium that penetrates the surrounding wood. The woodwasp larva is believed to be benefited by the action of the fungal enzyme, which continues to digest wood particles in the alimentary tract of the insect (Francke-Grosmann, 1939, 1957; Parkin, 1942; Middlekauff, 1960; Stillwell, 1966; Morgan, 1968).

The ambrosia beetle *Platypus wilsoni* (Swaine) (figure 133, *right*) possesses punctures concentrated on the caudal half of the pronotum that were shown to be mycangia (Farris and Funk, 1965). The male of the scolytid beetle *Gnathotrichus sulcatus* (LeConte) (figure 133, *left*) carries 3 species of symbiotic fungi in special enlargements (mycangia) of the forecoxal cavities. This is the only ambrosia beetle that has been recorded as carrying more than 2 symbiotic fungi (Funk, 1970). Graham (1967) discussed the results of an investigation by W. Schedl, who reported 8 different types of mycangia in 26 species of ambrosia beetles examined:

> In the head, there may be mandibular, pharyngeal, or external ventral pockets; in the thorax, there may be tubular "glands," dorsal pits, pockets between the pro- and mesonotum, coxal cavity spaces, or pockets in the elytra. Most occur only in females of a species, but some occur only in males. At present, the process by which spores are transferred into and out of the mycangia is not understood.

The relationship of termites and fungi has been the subject of many investigations. Fungi in decaying wood may furnish nitrogen, vitamins, and other substances beneficial to termites, and may also destroy toxic volatile material or extractives in the wood. Fungi are also known to produce attractants or feeding stimulants. On the other hand, they can remove certain nutrients or produce toxic metabolites. Whether fungi are beneficial or not depends on the species of termite and fungus and other variables (Becker and Kerner-Gang, 1964; Smythe *et al.,* 1971).

Representatives of all the classes of fungi are associated with termites. Slight deterioration of wood by certain Basidiomycetes of the brown-rot type increases its nutritional value to termites (Light and Weesner, 1947; Becker, 1948, 1965), and the same may be said of certain Ascomycetes and Fungi Imperfecti, the soft-rot fungi (Becker, 1948). Other species or even other strains of the same species may have no effect, or may even be toxic to termites (Lund and Engelhardt, 1962;

Becker and Kerner-Gang, 1964; Kovoor, 1964; Becker, 1965; Smythe and Coppel, 1966*a*). For rearing termites in the laboratory, Becker (1969*b*) suggested the use of wood blocks with about 3 to 10% weight loss owing to attacks by brown-rot Basidiomycetes, such as *Coniophora puteana*, *Lenzites* spp., *Polyporus* spp., or *Merulius lacrymans*. The mycelium of certain nontoxic fungi can be added with additional benefit.

Workers of the eastern subterranean termite, *Reticulitermes flavipes*, when exposed to both decayed and sound wood of 6 wood species, ate 1.4 to 2 times more decayed than nondecayed wood. During an 8-week test, survival of termites was highest on wood decayed by *Daedalea quercina* and *Poria oleracea*. Termite survival was better on ovendried decayed wood than on less dry sound wood for all wood species except a walnut (*Juglans nigra*), on which there was no survival (Smythe *et al.*, 1971).

The question is sometimes asked whether subterranean termites will feed on sound wood that is not infected by fungi. To obtain information on this subject, Pence (1957) collected *Reticulitermes hesperus* from sound, white Douglas-fir lumber and placed the insects on sterilized and ovendried strips of the same wood, dyed with India ink. Within a few hours, their intestinal tracts were filled with the dyed wood. When given a choice, they fed on black wood exclusively in the presence of light, but fed on either black or white (undyed) wood indiscriminately when placed in darkness. There is no doubt that termites attack and destroy perfectly sound timber, but the galleries they form in the wood are soon infected with fungi, which probably are a useful amendment to their diet.

The eastern subterranean termite was attracted by aqueous extracts of 6 species of wood decayed by the fungus *Lenzites trabea*, to 3 species decayed by *Poria cocos*, and to 1 each decayed by *Lentinus lepideus* and *Daedalea quercina* (Smythe *et al.*, 1971). Attractancy of wood decayed by *L. trabea* should not be construed to imply usefulness of the fungus to termites under natural conditions; in fact, Smythe *et al.* (1971) found it to be associated with generally decreased termite survival. However, there has been much recent investigation on the subject of attractant fungi because of the possibility that their extracts, combined with suitable insecticides, may provide specific control, or that fungal parasites and pathogens may be used in biological control (Esenther and Gray, 1968; Sands, 1969).

Unique genera and species of oxymonad, trichomonad, and hypermastigote flagellates (protozoa) that occur in the intestinal tracts of termites are found only in the 4 most primitive termite families: Mastotermitidae, Hodotermitidae, Kalotermitidae, and Rhinotermitidae. The great majority of these flagellates ingest wood particles, and appear to be indispensable for the survival of the termites. Although there appears to be no doubt that the flagellates are responsible for cellulase activity in the hindgut of the lower termites, it seems that they do not contribute substantially to the nitrogen (for protein) requirements of the insects. Possibly, the bulk of the nitrogen is supplied by fungi present in the wood (Honigberg, 1970). The higher termites (Termitidae) possess protozoa such as found in other insects and in other invertebrates, and even in the large intestines of some vertebrates, but they do not depend on them for the digestion of wood. In fact, as a rule they do not feed solely on wood or cellulosic material, and what wood they do eat is usually much more decayed than the wood eaten by the lower termites.

The function of the fungus in the fungus combs of the Termitidae "appears to be mainly the breakdown of lignin, but it probably also supplies nitrogenous materials and possibly other factors, such as vitamins" (Sands, 1969).

TERMITES

Termites are social insects of the order Isoptera. They live in colonies comprising winged and wingless reproductive forms and numerous wingless sterile workers, nymphs, and soldiers. Despite the fact that social insects account for only a small percentage of the structural pests, 2 groups of social insects, termites and ants, are among the first 3 in importance. The other group, the cockroaches, comprises ancient insects from which the termites are believed to have descended.

Termites feed on wood, and throughout a large area of the world they are the most destructive insects to wood structures. This is a measure of the importance of their natural role in nature—breaking down and returning to the soil and atmosphere the enormous tonnage of dead and fallen trees and other cellulosic material that is continuously accumulating on the earth's surface. They are important pests of agricultural crops, forest nursery seedlings, and range grasses, and also damage an enormous amount of stored food and household furniture and commodities, including even most plastics (Snyder, 1935, 1955*b;* Harris, 1961; Ebeling, 1968). Damage from termites plus the cost of controlling them probably amounts to approximately a half billion dollars per year in the United States alone (Ebeling, 1968). Our knowledge of the history of South America would probably have been much more complete if it had not been for termites, for they are said to have eaten most of the books more than a century old (Howse, 1970).

Termites are found in tropical, subtropical, and in most temperate climatic zones. They are increasing their range and density northward, being favored by the accelerated urbanization incident to the "population explosion," particularly in areas where central heating of buildings affords them a particularly favorable environment for the establishment of colonies.

Much has been written about termites, yet until recently the largest and most comprehensive work, particularly on termite biology, was a book titled *Termites and Termite Control.* It was edited by C. A. Kofoid, and published by the University of California Press in 1934 (partially revised in 1965), and includes the contributions of 35 experts on termite biology, taxonomy, and control. In 1969 and 1970, a treatise of 2 volumes, titled *Biology of Termites* and edited by Kumar Krishna and Frances M. Weesner, was published. Twenty-five scientists contributed to this exhaustive treatise, which covered many aspects of termite biology, systematics, distribution, behavior, social organization, and control. *Termites—Their Recognition and Control,* by Harris, published in 1961; and *Termites—A World Problem,* by Hickin, published in 1971, are books on the distribution, classification, economic importance, and control of termites, as well as wood-preservation procedures, written from a world standpoint. An excellent popular book on termites was written by Howse (1970).

The lower termites (e.g., drywood and dampwood termites) can be easily collected and then cultured and observed under glass. The higher termites are more difficult to culture, but successful techniques for culturing have been developed for certain species in all families, and have been described by Becker (1969*a*). An observable termite colony should be a fascinating subject for classroom demonstration.

Climatic Limitations

Although termites were able to extend their range to approximately the 50° F (10° C) annual mean isotherm north and south of the equator, only time will tell how much farther they may be able to go because of the improved microclimate afforded them by heated buildings.

In Wisconsin, the approximate northern limit for the eastern subterranean termite, *Reticulitermes flavipes,* coincides with an annual minimum isotherm of –30° C (–22° F). Upward movements of termites from the soil cease near zero C, and they overwinter primarily at a soil depth of from 3 to over 4 ft (1 to 1.5 m). In this way, the insects can escape adverse weather conditions, such as dryness or low temperature. It appears that the most adverse effect of winter is to confine them to a zone below an adequate source of food, which is usually near the ground surface. The survival of termites in cold climates depends on their ability to repopulate during the warm season (Esenther, 1969).

The situation in Wisconsin is probably typical of all areas in which subterranean termites are extending their range into colder regions. Esenther (1969) points out that only "man-oriented" colonies have been found in Wisconsin. He believes that they were introduced originally on infested lumber and spread mainly through new colony formation facilitated by their subterranean tunneling. Unlike the situation in the warmer southern parts of the range of the species, dissemination by flight must be relatively insignificant in the north. The rate of development of an incipient colony in the north is too slow, and therefore the termites are not likely to survive the rigors of the first winter to form alates. Thus, reproduction is by neotenic individuals, those which attain sexual maturity without attaining the alate (winged) form.

Becker (1970*a*, *b*) has observed differences between *Reticulitermes flavipes* in Wisconsin and Hamburg, Germany, when compared with what he considers to be bioecological races of that species in South Carolina and Hallein, Austria. The northern forms were the most active gallery builders, and demonstrated daily rhythms of activity that were lacking in the southern forms.

Biology and Colony Formation

Termites are relatively primitive from an evolutionary standpoint, but their social organization is the most complex among the insects. Along with the ants and the more highly organized bees and wasps, they belong to the truly social or *eusocial* insects. The common traits of eusocial insects are: (1) cooperative caring for the young, (2) a division of labor in which more or less sterile individuals work on behalf of fecund individuals, and (3) an overlap of at least 2 generations in life stages so that during some period of their life, the offspring can assist their parents. Species of insects lacking all these traits are called *solitary*. There are also many species among the wasps and bees that are in various stages of sociality between solitary and eusocial (Wilson, 1971).

A society can be developed only if its members are long-lived. This in turn depends on an adequate and continuous food supply. Termites solved this problem by acquiring the ability to use wood-cellulose as food. The principal termite pests in the United States are in the families Rhinotermitidae and Kalotermitidae, the members of which depend on protozoa (mastigophoran flagellates) in their hindguts to break down cellulose. Termites do not possess these protozoa when they are born; they must obtain them by proctodeal feeding, that is, feeding upon liquid intestinal content taken from the anal aperture of an older termite. Every time the termite molts, the lining of the hindgut is shed, along with the entire body cuticle, and the protozoa are lost. Refaunation takes place by proctodeal feeding (Andrew, 1930; Honigberg, 1970).

In some areas of the world, termites belong primarily to the family Termitidae. The termitids are responsible for most of the earthen termite mounds, some as much as 10 meters in height, which form a characteristic feature of many landscapes of the African and Asian tropics. These termites have no intestinal fauna of the types that can aid in digestion. They may consume grass, leaves, humus, the manure of herbivorous animals, and decaying wood. The Macrotermitinae have spongelike fungus combs in their nests. They are constructed of chewed wood and feces, and are built up to fit the chambers of the nest. These termites feed on fungus (*Termitomyces*) in the combs. The function of the fungus in digestion appears to be mainly the breakdown of lignin, but probably other factors are supplied, such as nitrogenous materials and vitamins (Sands, 1969). Thus, termites have an abundant, continuous food supply, and this, coupled with their longevity, the potential immortality of the colony, and their ability to care for their eggs and young and protect the colony against natural enemies and the elements, allows for the development of enormous numbers of individuals. For example, in a large colony of the moundbuilding termite *Nasutitermes exitiosus* (Hill) in Australia, 11.05 kg of termites were removed, representing 2.5 million insects, of which about 87% were of the wood-destroying worker caste (Gay and Wetherly, 1970). Such enormous numbers of termites result in a great capacity for destruction of wood structures, with no seasonal letup in tropical areas. Most of the foregoing factors also favor another social insect, the ant, as a successful and persistent pest of man. Ants may not have as constant a food supply, at least not in nature, but they can store foods in their well-protected nests.

Termites live what is known as a "cryptobiotic" mode of life. They live in enclosed passageways, either entirely in the wood in which they feed or partly within the wood and partly within soil. At certain times of the year, depending on the species, a certain percentage of the colony develops wings and changes from the whitish color of the nymphs to the distinctive dark or black color of the winged reproductives (alates). The latter fly off to form new colonies.

The alates are the members of the colony most likely to be seen by the homeowner. They are the potential kings and queens. The homeowner often confuses alates with "winged ants." The abdomen of the termite is broadly joined to the thorax, while the thorax and abdomen of the ant are joined by a narrow petiole or "waist" (figure 66). The termite has straight, beadlike antennae, while those of the ant are elbowed. Unlike the castes they left behind, the termite alates are heavily pigmented. The fore- and hindwings of the alate termite are approximately equal in

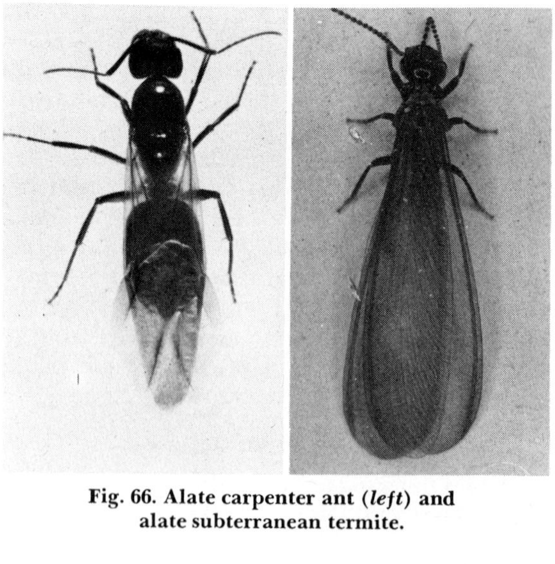

Fig. 66. Alate carpenter ant (*left*) and alate subterranean termite.

Fig. 67. Wings of a carpenter ant (*top*), and of a dampwood termite.

length, and usually extend from 25 to 33% of their length beyond the end of the abdomen when folded. The hindwings of the alate ant (figure 67) are much shorter than the forewings, and when folded they rarely extend beyond the end of the abdomen.

After the flight of the alates, their wings break off near the base. Males and females pair off and begin a small excavation for a new nest. Subterranean termites (Rhinotermitidae), for example, may excavate their nest in wood found after digging into the ground, or between a piece of wood and damp ground, or in a crevice in wood on damp ground. (Galleries eventually extend deeply into the ground.) The pair then mate, and the first eggs are laid. The egg-laying capacity of the queen increases as she grows older. Queens of some moundbuilding tropical termites can lay as many as a thousand eggs per day for as long as 25 years.

Several years may pass before all castes are present in a new colony. The complete colony consists of the primary pair of reproductives (royal pair) and three castes: (1) the workers, which feed on wood or fungi and, by regurgitation and excretion, provide food for the young and the other castes; (2) the soldiers, which in the United States are usually large-headed individuals with massive jaws that guard the nest entrances and the royal pair; and (3) usually 2 kinds of supplementary or substitute reproductives known as *neotenics.* These may be either lightly pigmented and with short wing pads (brachypterous) or very lightly pigmented and without wing pads (Krishna, 1969).

Chemicals that are secreted to the outside of the bodies of insects for caste regulation, attraction, communication, trail-marking, etc., are called *pheromones.* A colony of social insects (termites, ants, wasps, or bees) maintains its social cohesiveness primarily through the utilization of such chemicals. They are produced in specialized tissues known as *exocrine glands.* In response to specific stimuli, these glands evacuate their contents into the environment (Blum, 1970).

Termites continually groom one another by means of their mouthparts to obtain desired secretions or exudates containing pheromones. Among the pheromones they obtain in this manner are some that are believed to inhibit the formation of additional members of the sex or caste from which the hormones are obtained, thus serving as a regulatory mechanism to prevent a disproportionate ratio of males, females, and soldiers in a colony (Lüscher, 1956*a, b,* 1961; Weesner, 1956). Lüscher (1961) stated that the queen can inhibit sexual development of other potential reproductives, even if her abdomen is covered with varnish, thus covering all integumental glands and the genital opening, but that inhibition is no longer possible if the anus is blocked. He concluded that the inhibitor substance must be given off with the excrement.

There is considerable evidence that the "royal pheromone" that prevents *Kalotermes flavicollis* (F.) from undergoing the final molt is produced by the mandibular glands of the sexual forms. In 20 trials, when one of these glands from a primary sexual female was implanted in the abdominal cavity of a nymph shortly before the final molt, adult differentiation was blocked or inhibited, depending on how close to the final molt the nymph was at the time of implant (Lebrun, 1972).

Reproductives can also stimulate the development of a caste. For example, when a group of nymphs of *Kalotermes flavicollis* is separated from soldiers, some will differentiate into soldiers. The number of soldiers produced is much greater when reproductives are present. The effect of reproductives on soldier production was graded as follows: pair of reproductives (king and queen) > 2 queens = 1 queen = 2 kings > 1 king (Springhetti, 1970). Miller (1969) pointed out that, among the lower termites (all families but Termitidae), there is no evidence that the various castes are genetically different; their caste destinies are the expression of social and environmental factors. Even "workers" can become sexuals, and in laboratory colonies have been able to reconstitute all castes of the colony when they were sufficiently numerous.

The role of pheromones in the structure and formation of termite colonies, particularly when the order Isoptera as a whole is considered, is extremely varied and complex. The amazing extent to which an understanding of the role of pheromones has already developed from the pooled results of world-wide investigations is concisely discussed by Howse (1970). No doubt an even broader understanding will result from current investigations.

Supplementary reproductives (neotenics) are required for rapid increase in numbers of termites in a colony. When groups of workers and nymphs of the western subterranean termite, *Reticulitermes hesperus,* were separated from the mother colony, they formed a new colony in 6 to 8 weeks, utilizing supplementary queens developed from some of the short-winged nymphs found in every large colony (in addition to the nymphs that develop into the alates that leave the colony). A supplementary queen can produce more eggs (60 to 80) in a day at the height of egg-laying than the primary queen in the first 2 years of the colony's development (Pickens, 1934*a*).

Trail-Marking Substances

It has long been known that termites follow "odor trails." The odor trails may serve varied purposes. It has been observed, for example, that breaks in the nest structure of *Zootermopsis* stimulate the laying of trails to the breaks so as to recruit workers for repair work. The intensity of the stimulus the workers receive from the odor trails determines the number recruited (Stuart, 1967). This primitive trail-laying mechanism was apparently adapted secondarily for foraging purposes. The trail-marking pheromone may at the same time be a food attractant (Smythe *et al.,* 1967*a, b;* Ritter and Coenen-Saraber, 1969), and in this capacity it offers some potential as a possible means of termite control. The pheromone is secreted by the sternal glands of the workers and soldiers of all termite families (Lüscher and Müller, 1960; Stuart, 1961, 1963, 1964, 1969; Noirot and Noirot-Timothée, 1965; Mosconi-Bernardini and Vecchi, 1966; Smythe and Coppel, 1966*b;* Stuart and Satir, 1968; Moore, 1969; Noirot, 1969; Howse, 1970; Mertins *et al.,* 1971). One such gland is situated on each of the third, fourth, and fifth sternites in the primitive *Mastotermes* of Australia's Northern Territory, on the fourth sternite of *Stolotermes, Porotermes,* and *Hodotermes,* and at the base of the fifth sternite (figure 68) on all other termites. In the sternal gland, there appears to be no duct associated with the glandular cells. The pheromone is probably secreted by the cells, and passes through the fine pores of the cuticle, collecting in a reservoir formed by the overlapping sternal plate of the preceding segment (figure 68). Possibly, the extent to which the aperture at the posterior end of this reservoir is opened is regulated by the pressure exerted by the abdomen as it is pressed against the ground (Stuart, 1969).

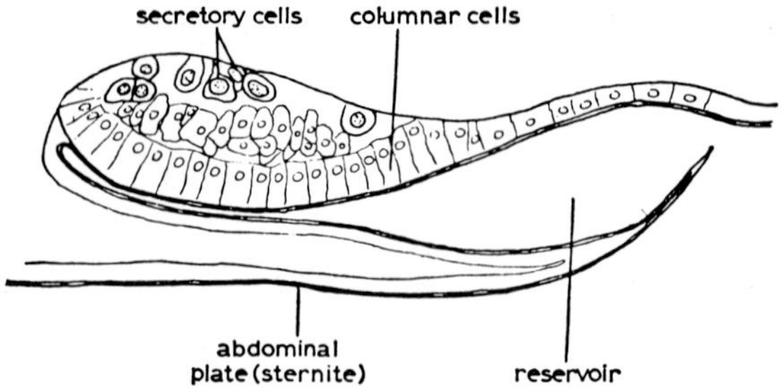

Fig. 68. Longitudinal section through the sternal gland of *Zootermopsis nevadensis.* [From Howse (1970), after Stuart (1964), with permission of Hutchinson and Company, London.]

Extracts of the trail-marking pheromone from either *Reticulitermes flavipes* or *R. virginicus* were attractive to both these species and to *R. hesperus,* but not to the dampwood termite, *Zootermopsis angusticollis* (Smythe and Coppel, 1966*b*). Several synthetic analogs of the trail-marking pheromone of *R. virginicus* have been prepared, and the molecular structures to which these compounds owe their pheromone-mimicking characteristic have been identified (Tai *et al.,* 1971).

Insect pheromones are generally rather species-specific. Therefore, it is of special interest that 6 subterranean termite species (*Reticulitermes flavipes, R. virginicus, R. hesperus, R. tibialis, Coptotermes formosanus,* and *Leucotermes speratus* [the latter from Japan]) all responded to 4 trail-marking pheromone analogs (Matsumura *et al.,* 1972). The nonspecificity of these compounds would be advantageous in any attempt to use them as lures for trapping purposes in a control program. Two of the pheromone analogs have been found to be easily synthesized (Tai *et al.,* 1971).

There are many other nonpheromone substances, some found in nature and some artificially produced, that have effects similar to those caused by pheromones. For example, wood rotted by the fungus *Lenzites trabea* produces an attractant for *Reticulitermes flavipes* as well as other species of *Reticulitermes* and *Coptotermes*. The fungus induces trail-following by termites similar to that induced by the trail-marking pheromone secreted by the sternal glands of these insects (Esenther *et al.*, 1961; Esenther and Coppel, 1964; Allen *et al.*, 1964*b*; Smythe *et al.*, 1965, 1967*a*, 1967*b*; Esenther, 1969). Column chromatography of the unsaponifiable lipids from pine wood on which *L. trabea* was cultured yielded 2 well-separated fractions that were highly active in choice tests and in a trail-following test with *Reticulitermes lucifugus* (Rossi). The unsaponifiable lipids of the workers yielded only a single active fraction, but it corresponded to one of the fractions obtained from the wood (Ritter and Coenen-Saraber, 1969).

The attractance of 8 compounds formed in wood by wood-rotting fungi (Basidiomycetes) was tested, using 5 termite species as test insects. Five compounds were generally attractive, and 3 deterred termites or had no attraction. Acids were usually attractive, but aldehydes were attractive in only a few cases (Becker, 1964). Many organic compounds have been found to be attractive to *R. flavipes* (Watanabe and Casida, 1963).

A trail-marking scent for *Nasutitermes exitiosus* (Hill) was isolated and identified by B. P. Moore in Australia (Anonymous, 1967*d*), and was found to be an unsaturated diterpenoid hydrocarbon; only 10^{-8} gram was required to lay 10 m of trail. Other species of *Nasutitermes* followed the trail, but distantly related species did not. Because diterpenoid compounds occur in certain essential oils, Moore searched for possible attractants in vegetation. He found a substance in the oil from the Western Australian sandalwood, *Eucarya spicata*, which attracted *Nasutitermes*.

At the United States Forest Service's Wood Products Insect Laboratory in Gulfport, Mississippi, it was observed that *Reticulitermes flavipes* followed marks made by a certain ballpoint pen with blue ink, but other available ballpoint pens did not have this effect. This phenomenon was thoroughly investigated by Becker and Mannesmann (1968). In their investigation, they used 55 termite species from 21 genera of 4 families. They found 3 ballpoint inks that served as trail-markers for many termite species, and 3 others that attracted fewer species. Figure 69 shows workers of *R. lucifugus* following a spiral line made with a ballpoint pen with ink containing a glycol compound. The termites tend to follow a tangent to the curve, and must make repeated corrections in their direction of movement. The inks were less effective for *Mastotermes* and species of Kalotermitidae than for species of Rhinotermitidae and Termitidae. Nine glycol compounds, including some used in ink for ballpoint pens, proved to be trail-marking substances of varying degrees of efficacy. Diethylene glycol monoethylether and diethylene glycol monobutylether were very effective for almost all termite species, including Kalotermitidae. Also, some of the decomposition products (aldehydes and acids) produced when wood was attacked by Basidiomycetes were found to be attractants and trail-markers. *Reticulitermes lucifugus* is unable to react to scent trails if its antennae are partially amputated unless at least 8 segments are retained.

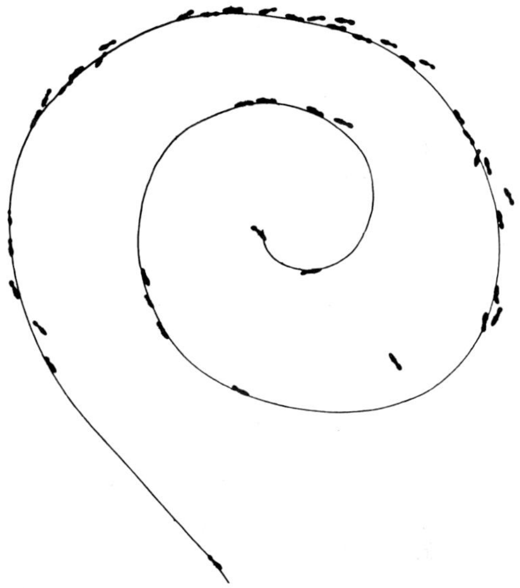

Fig 69. Worker termites (*Reticulitermes lucifugus*) follow a spiral line made with a ballpoint pen, the ink containing an attractant glycol compound. (From Becker and Mannesmann, 1968.)

The possibility of practical utilization of termite attractants in wood baits is discussed under "Termite Baits" later in this chapter.

Western Subterranean Termite, *Reticulitermes hesperus* Banks (Rhinotermitidae)

This is the principal subterranean termite in California. It ranges from British Columbia south to western Mexico and east to Idaho and Nevada. Winged reproductives (alates) are dark brown to brownish black, and have brownish-gray wings (plate I; figure 70). Including the wings, they are 8 to 9 mm long, and the body, without the wings, is about 5 mm long. If a piece of termite-infested wood is opened in the winter or spring in southern California, the alates may be seen, already darkly pigmented, waiting for a warm, sunny day after a rain to emerge from the nest and fly away. The percentage of alates increases with increasing age of the colony. All other members of the colony except the queen are whitish in color. Densely packed in the termite nests are hundreds of workers and a smaller number of soldiers and supplementary reproductives (neotenics) (figure 70). The queen is always the center of considerable activity by other castes that are feeding, grooming, and guarding her and carrying away and caring for her eggs (plate I; figure 71). The best way to see this is to break open a piece of infested wood that has long lain on or beneath damp ground. Nests in soil may be as deep as 6 or 7 meters below the surface.[9]

[9] The term "workers" is retained because of common usage, although there is reason to believe that termites in the family Rhinotermitidae, in which *Reticulitermes* is one of the genera, have no definitive workers. Weesner (1965) stated that in *Reticulitermes* those forms that had generally been considered to be workers might instead be "fairly size-stable individuals functioning as workers, but still capable of molting and differentiating into slightly larger individuals, soldiers, or alates." True workers exist in the family Termitidae.

Fig. 70. Western subterranean termite, *Reticulitermes hesperus*. *A*, supplementary reproductive nymph; *B*, winged reproductive (alate); *C*, gravid queen; *D*, gravid supplementary queen; *E*, soldier; *F*, worker. (From Ebeling and Pence, 1965.)

Colony Development

The growth of a colony from the primary pair of reproductives is slow. Only a few eggs are laid the first year, and they require an average of over 50 days to hatch. The first 2 instars require only 14 to 18 days each. The second and third instars require 1 and 2 months, respectively, and the fifth instar may be the dominant one, and last for as long as 2 years. In sufficiently large colonies that have a large amount of fraternal feeding (trophallaxis), large, well-matured workers and reproductive nymphs of the sixth instar may develop. There may be a seventh instar of still larger workers, and in this instar the perfect reproductive stage is attained in the reproductive caste. Even under the most highly favorable conditions, flights of alates cannot be expected before the third or fourth year. Later, the supplementary reproductives can be expected to greatly accentuate the growth of the colony, and tens of thousands of individuals can develop. (Pickens, 1934*a*).

The question is often asked as to how many years are required for a newly constructed building to show the first signs of subterranean termite infestation. If the infestation started from a primary pair of reproductives, at least 3 or 4 years would pass before even a few swarmers could be seen, even if the building became infested at the time of construction. Evidence of structural damage should not appear for an even longer period. On the other hand, if the building were built over or close to a strong existing colony, perhaps located in a buried stump or roots, substantial damage from termites could be noticeable within a year.

Shelter Tubes

When conditions of food, moisture, and temperature are satisfactory, subterranean termites can live without any contact with the ground, although this is rare. In most areas, the ground serves as a protection against extremes of surface temperatures, as well as being a moisture reservoir. Subterranean termites are known to change the depth of their nests in the soil to accommodate to changing temperature or moisture requirements.

Subterranean termites in the United States generally span the distance between the ground and the first wood members of the substructure of a building by working their way up through intervening wood or by building "shelter tubes" over concrete foundation walls. Intermittent sections

Fig. 71. Western subterranean termite, *Reticulitermes hesperus*. Queen surrounded by nymphs and a soldier.

of shelter tubes can also be seen on structural wood members. Periodically, the workers must return via the shelter tubes to their galleries in moist soil to regain moisture through their cuticles to replenish what was lost in the relatively dry wood where they have been feeding.

The tubes do not conduct moist air to the areas of termite activity, as is often recorded in the literature. A humidity sensor, inserted into termite galleries in joists only 18 in. (45 cm) aboveground and directly connected to the ground by shelter tubes, showed that the humidity was identical with that of the atmosphere surrounding the joists (Ebeling, 1968). However, the protection that the shelter tubes provide against air movement probably helps to preserve the termites' moisture to some extent. Likewise, shelter tubes are not built to keep out light, for termites have no aversion to light if they are protected, as in termitaria under glass. The principal function of shelter tubes appears to be protection against natural enemies, especially ants.

The worker termites build shelter tubes from particles of sand or earth, or minute particles of wood, or both, coating them with a gluelike substance that they secrete. Fecal material is also used as a cement. (For *Zootermopsis,* excrement is likely to be the principal construction material.) The particles may be tightly packed, or they may adhere so loosely as to form a coarse filigree. The western subterranean termite uses at least 4 types of shelter tubes. Usually, all types can be seen in the crawl space under an infested house on a "raised foundation."

Utility or Working Tubes. These form runways through which termites feeding on the wood above the foundations can return periodically to the moist atmosphere of their subterranean galleries in order to replenish lost body moisture. These tubes are wide, flattened, and usually extend from the soil to the wooden construction above it (figure 72, *A*).

If the tubes are broken as rapidly as they are repaired by termites, and if the insects have no other sheltered passageways through which to return to their nests, and if there is no source of moisture other than the damp soil below, those termites above the broken portions of the tubes will perish.

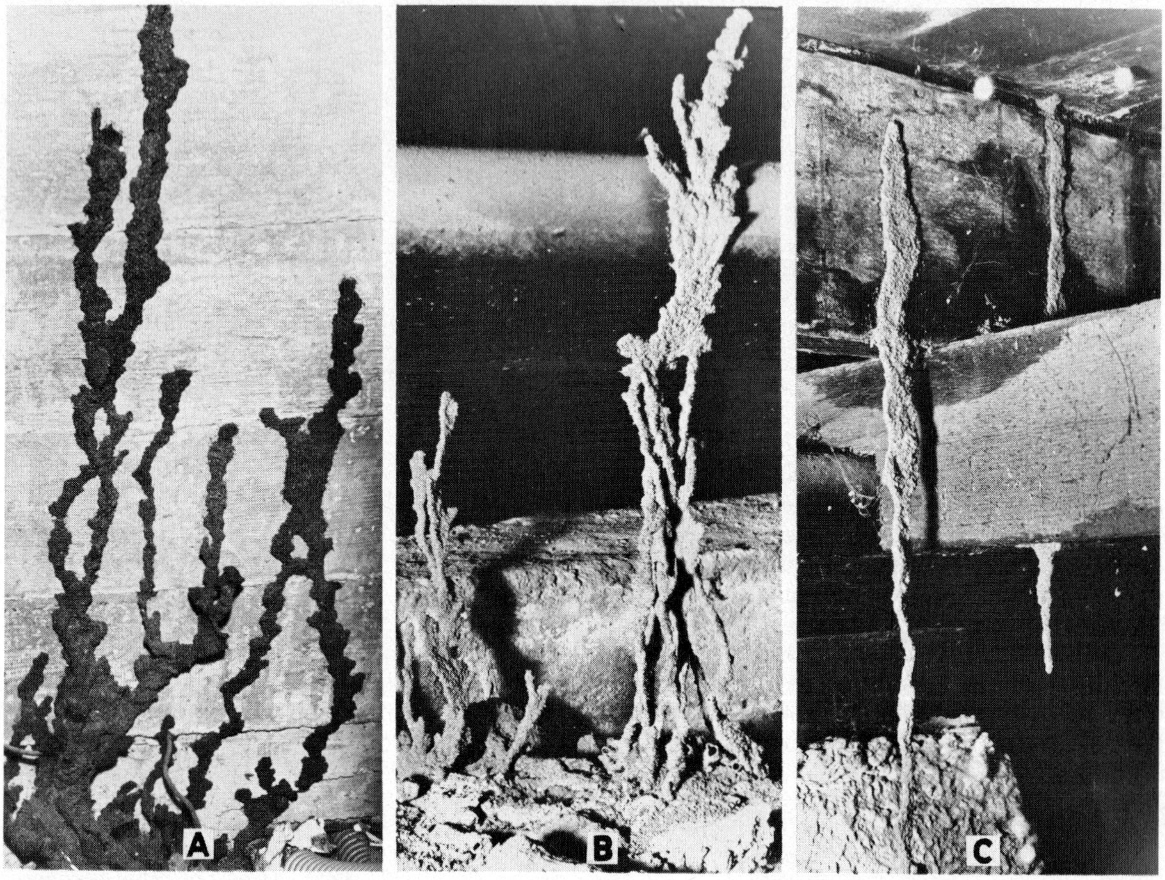

Fig. 72. Shelter tubes of the western subterranean termite. *A*, utility or working tubes; *B*, exploratory or migratory tubes; *C*, suspended or drop tubes. Note 2 egg sacs of black widow spider at top. (From Ebeling and Pence, 1965.)

Exploratory or Migratory Tubes. These (figure 72, *B*) are similar to utility tubes, but are not so strong, usually do not reach the wood above, and have small exit holes. Figure 73 shows one of these tubes being constructed by a worker termite.

Suspended or Drop Tubes. These are utility tubes, built downward from a wood member to the ground. They are lighter colored than the other types of tubes because they contain more wood fiber (figure 72, *C*). At the top of the figure, note the 2 egg sacs of the western black widow spider attached beneath the subflooring.

Three types of shelter tubes are shown in figure 74: *A*, a utility tube against the concrete foundation; *B*, a utility tube reaching from soil to subfloor, but not attached to the foundation; and *C*, a suspended or drop tube. Note the much lighter color of the latter. Also, note the block of wood, *D*, which was left lying on the ground and became infested with subterranean termites.

Fig. 73. A subterranean termite worker building an exploratory or migratory shelter tube. (From Ebeling and Pence, 1965.)

Swarming Tubes. These are constructed at swarming time to provide for the exit of alates. They usually extend aboveground about 4 to 8 in. (10 to 20 cm), but under particularly favorable circumstances may extend considerably farther, and may reach wood members of the substructure. Swarming tubes are often found around or beneath floor furnaces or in other warm places.

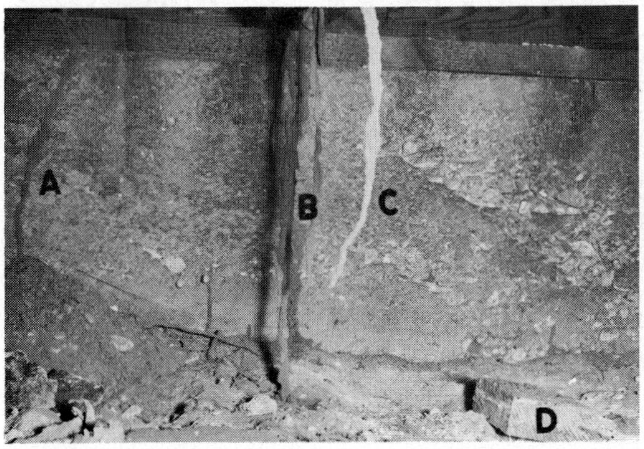

Fig. 74. Termite signs in a crawl space. *A,* shelter tube against concrete foundation; *B,* shelter tube free of foundation; *C,* drop tube suspended from subfloor (note light color); *D,* a block of wood infested by termites.

Formation and Defense of Galleries in Wood

Subterranean termites primarily attack the soft spring growth of the infested wood, in contrast to drywood termites and other nonsubterranean species which burrow indiscriminately across and with the grain of the wood, excavating broad

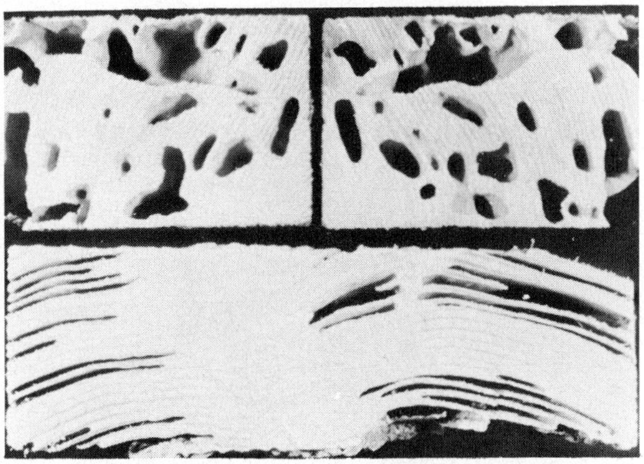

Fig. 75. Cross section of Douglas-fir lumber, showing characteristic pattern of chambers and tunnels of drywood termites (*top*) and subterranean termites (*bottom*).

pockets or chambers connected by tunnels (figure 75). Termites tend to feed in a structural wood member until only the harder wood portions and a fragile outer shell remain (figure 76).

The workers of subterranean termites have strong mandibles (figure 77), very large in relation to the size of the head. They are able to tear up and consume wood, but do not eat all the wood chewed out of a gallery. They transport some of it to the rear, and may pile it into a natural cavity in wood or soil, or into a large, unused gallery; much of it is mixed with fecal plaster and packed along the walls of the gallery. Workers also sustain the other castes by oral or anal feeding, care for the eggs, feed the very young nymphs (larvae), queens, kings, and sol-

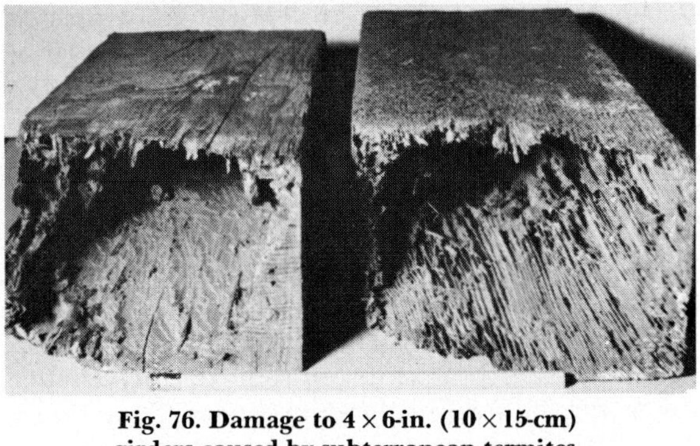

Fig. 76. Damage to 4 × 6-in. (10 × 15-cm) girders caused by subterranean termites.

diers, and in fact, they perform all the duties of the colony except for reproduction and defense. The soldiers (plate I, 3; fig. 70, *E*) are responsible for defense. Thrusting their large heads and powerful jaws into breaks in the gallery system, they can defend it against the ant, the termite's worst enemy. Only in case of major destruction of the gallery system is the termite in danger from insect predators. After swarming, the alates are, of course, easy prey for predatory insects and birds.

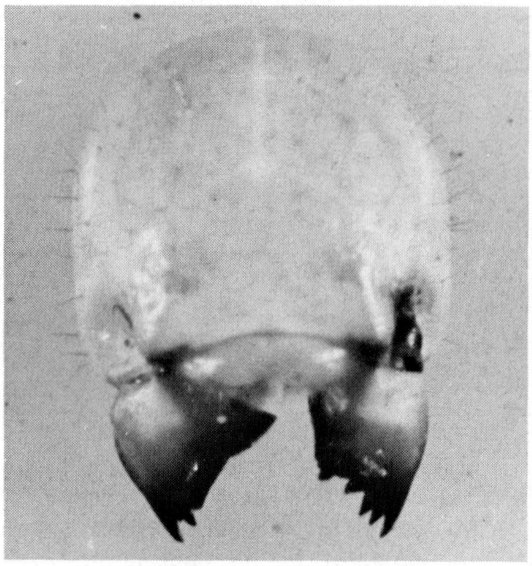

Fig. 77. Head of western subterranean termite worker, showing relative size of mandibles. Note absence of eyes; the workers are blind.

The First Signs of Infestation

In California, the rainy season is in late fall, winter, and early spring. During this period, flights of subterranean termites take place on warm, sunny days after a rain. (Flights during this period can also be initiated by heavy irrigation or sprinkling in the vicinity of a house). The alates may be the first indication of infestation observed by the homeowner, and it is important that he be able to distinguish them from winged ants (figures 66 and 67) and from the alates of the drywood termite, *Incisitermes minor.* The total length of the alates of the drywood termite is 11 to 12 mm (about a fourth longer than that of the subterranean termites), the body is dark brown, and the head and thorax are reddish.

Another common indication of the presence of subterranean termites is the occurrence of dark areas or blisters in flooring. One can easily crush these areas with a knife or screwdriver. Shelter tubes may be seen in the crawl space. Sometimes, the distinctive sound of the soldiers may be heard. If they are disturbed, as by a knock on an infested wall, floor, or subfloor wood member, the soldiers jerk their bodies and heads violently up and down and produce an audible tapping as the head capsule strikes the roof or floor of their gallery.

Possible Extent of Damage

Subterranean termites may infest the "mudsill," usually the first structural wood member encountered on their way up from their subterranean galleries. If the sill is of termite-resistant wood or has been chemically treated, the termites generally build their shelter tubes over this wood member, and attack support structures that rest on the mudsill (girders, joists, and cripples), as well as adjacent wood structures, such as flooring and studding (figure 78, *left*). These wood members may be hollowed out by the termite feeding. Shelter tubes, constructed of earth, bits of wood, and excrement, extend up into the infested timbers (figure 78, *right*), and are among the definite indications of subterranean termite infestation. The infested timber becomes hollowed out to the extent that it is vulnerable to ants and other predators, and the termites must construct shelter tubes to protect themselves from their attacks.

Other Indigenous Species of Subterranean Termites

Another species of *Reticulitermes, R. tibialis* Banks, may be found in some inland areas, such as the San Joaquin and Sacramento valleys and certain high deserts in California, and in general

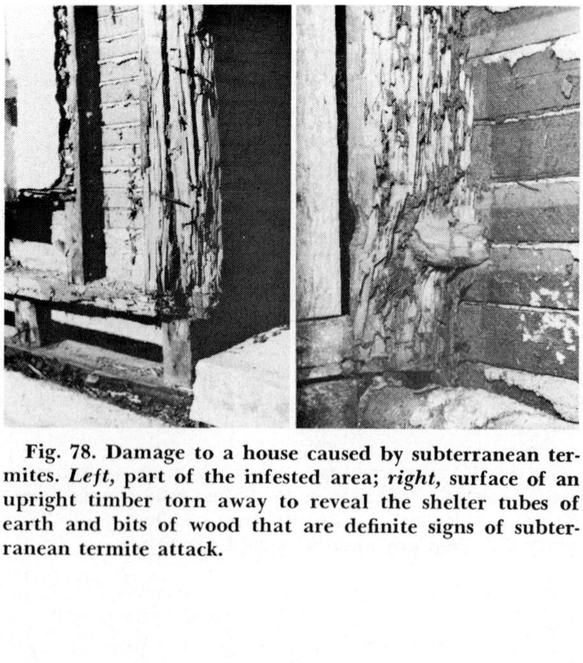

Fig. 78. Damage to a house caused by subterranean termites. *Left,* part of the infested area; *right,* surface of an upright timber torn away to reveal the shelter tubes of earth and bits of wood that are definite signs of subterranean termite attack.

the inland areas of the Pacific Coast eastward to the Rocky Mountain region and Texas. The alates can be distinguished from those of *R. hesperus* by their pale, almost whitish, wings, with brownish veins in the forearea, as compared with the dark wings of *R. hesperus* (Weesner, 1965). The head of the soldier of *R. tibialis* is short, broad, and dark, compared with the long, narrow, and pale head of the soldier of *R. hesperus* (Pickens, 1934*b*). Where these species overlap in distribution, *R. hesperus* prefers cool, shady, moist places, while *R. tibialis* requires open, sunny, drier locations. The 2 species are equally able to damage wood structures.

In the deserts of Arizona, southeastern California, and Mexico another subterranean termite, *Heterotermes aureus* (Snyder), occurs which can be very destructive, although much of its damage is probably attributed to species of *Reticulitermes*. The alates, which are nocturnal fliers, can be distinguished from those of *Reticulitermes* by their pale color. The soldiers of the 2 species are difficult to distinguish.

In the East, *Reticulitermes flavipes* (Kollar) is the most important termite pest. It extends along the Atlantic Coast north to Maine, and occurs to some extent in Canada, generally up to the line where the average annual minimum temperature does not fall below –20° F. It extends southward throughout the eastern and midwestern states to the Gulf and into Mexico and Guatemala. It is the most common and widely distributed termite in North America, but does not occur in the western United States. It is similar in appearance to *R. hesperus*, but is somewhat smaller, and it has similar habits (Smith and Johnston, 1962; Weesner, 1965).

At its western limits, *R. flavipes* overlaps the range of *R. tibialis;* in the Great Lakes region, its range meets that of *R. arenicola* Goellner; and to the south it overlaps the range of *R. virginicus* (Banks) and *R. hageni* Banks. It can be distinguished from these species on the basis of size or color of body and wings, color of tibia, location of ocelli, or a combination of these features (Weesner, 1965).

Formosan Termite, *Coptotermes formosanus* Shiraki

Species of *Coptotermes,* present throughout the tropics, have an even greater capacity for destruction than *Reticulitermes.* The Formosan termite, a subterranean species, long very destructive in Hawaii, was reported from Houston, Texas, in 1965, and the known infestations were apparently exterminated. The following year, it was again found in Houston, as well as in Galveston and in several areas in Louisiana (Krishna, 1966), and it appeared to have become firmly established in the United States. It has since been found in incipient infestations in other states, including California.

All infestations noted have been in harbor cities, initially in boats, ships, dredges, piers, and floating drydocks, indicating that this species came to the United States aboard ship. It will probably eventually have the same range of distribution in latitude in this country as it has in other areas of the world.

The Formosan termite is a more vigorous and aggressive species than the indigenous North American subterranean termites, indicated by more extensive tube and tunnel building, the rapidity with which a new food source is located and attacked, and the greater variety of materials attacked. *Coptotermes formosanus* was compared with 2 indigenous termite species of the eastern United States. It ate more and survived better than *Reticulitermes flavipes,* which in turn ate more and survived better than *R. virginicus* (Smythe and Carter, 1970). *Coptotermes formosanus* was found to be much more aggressive in tunnel building than *R. flavipes* and *R. virginicus.* This characteristic, coupled with its greater tolerance to the currently used soil insecticides (aldrin, chlordane, dieldrin, and heptachlor) results in its being more difficult to control. *Coptotermes formosanus* could penetrate 5 cm of soil containing 500 ppm of insecticide, whereas *Reticulitermes* could not. This is near the recommended concentration for chlordane, and double the concentration recommended for aldrin, dieldrin, and heptachlor. (See table 1, chapter 2.) Pending the completion of field investigations recently in progress, the dosages currently prescribed for soil treatment directed against our native termite species should be increased when treating for *C. formosanus* (Beal and Smith, 1971).

The most obvious characteristics that distinguish the Formosan termite from native species of subterranean termites are its larger size and pale yellow body color, the oval shape of the head of the soldier (figures 79, 80) compared with the rectangular head of the *Reticulitermes* soldier (plate I, 3; figure 70, *E*), and the hairy wings compared with the absence of hairs on the wings of most of the native species.

The habits of the Formosan termite are similar to those of native subterranean termites. They make nests in wood in or on the ground, in hollows they have excavated from tree stumps or posts, or in the hollow spaces in the walls, floors, or attics of buildings. Earthen shelter tubes are

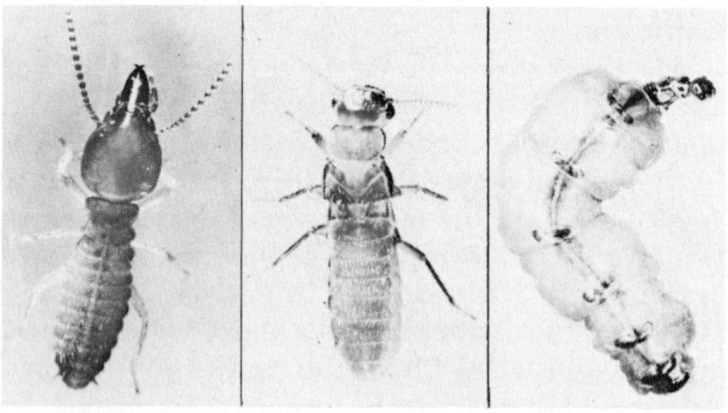

Fig. 79. Formosan termite, *Coptotermes formosanus. Left,* soldier; *center,* reproductive after shedding wings; *right,* egg-laying queen. (Courtesy A. A. LaPlante.)

built over objects the termites cannot penetrate, such as concrete or pressure-treated wood, although they can penetrate through cracks in the surface of the treated wood to reach interior portions that have not been treated chemically.

The enormous area covered by some Formosan termite gallery systems was revealed by an excavation of such a system in Louisiana in an earthfill known to be only 10 years old (King and Spink, 1969). The galleries of a single colony totaled about 1,900 ft (580 m) in length, and covered about 1.4 acre (0.57 ha). The galleries ranged from 5 to 117 cm in depth, but in some areas they are known to extend down to a depth of 3 m. The primary nest (figure 81) of the gallery system was found 48 cm below the soil surface. The nest was 53 cm wide and 48 cm high, and was built of carton, a mixture of soil and masticated wood that had been cemented together by the saliva and excrement of the termites. The same kind of material lined the galleries.

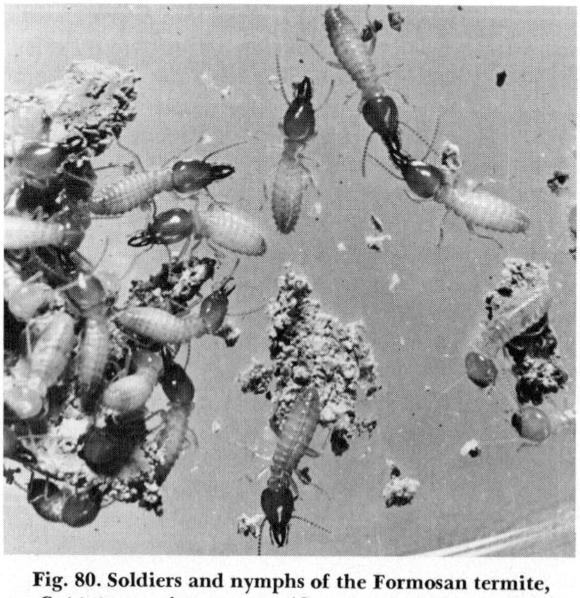

Fig. 80. Soldiers and nymphs of the Formosan termite, *Coptotermes formosanus.* (Courtesy A. A. LaPlante.)

The nest was encased in a 7-mm layer of cemented sand, and rested on a layer of cemented carton and sand. Small cavities separated by thin

Fig. 81. Nest of Formosan termites *in situ.* (From King and Spink, 1969.)

walls of the carton were found throughout the nest. In the center of the nest were 6 physogastric supplementary queens, about 14 mm long, and in near-by cavities were several batches of eggs. In the outer layers of the nest, soldiers were the dominant form, but workers became more numerous toward the center.

The primary king and queen, along with workers and soldiers, were found in a small piece of cypress wood. The gravid queen was 17.5 mm long and 4.5 mm wide at the broadest part of the abdomen. From the nest, galleries extended to adjacent small pieces of wood.

As with other subterranean termites, if only a portion of a Formosan termite colony is destroyed, even if that portion includes the queen (figure 79), supplementary reproductives will develop from individuals in the isolated group.

Warm, sultry evenings, especially following rain, are favorable for extensive flights. The flights usually begin before sundown and end before midnight. Alates of *Coptotermes formosanus* are strongly attracted to lights. Indigenous subterranean termites (*Reticulitermes* spp.) fly only during the day.

Western Drywood Termite, *Incisitermes minor* (Hagen) (Kalotermitidae)

Drywood termites establish their colony and continue to live in nondecayed wood having little moisture and, unlike subterranean termites, they never require contact with the ground. In some regions, such as southern California and the Caribbean area, they are the most important termite pests.

In California, the western drywood termite is found under natural conditions as far north as Mendocino County and the Sacramento Valley. It is abundant in coastal regions, and extends eastward into Arizona in peripheral desert regions, in mountain canyons, and along streambeds. Colonies of this species have been transported to, and at least temporarily established in, various parts of the United States. Drywood termites are very amenable to accidental distribution because they may infest commonly transported articles, such as boxes, crates, and furniture, can tolerate low moisture conditions for long periods, and the colony is often small, infesting only a small volume of wood, and can therefore be readily transported for long distances.

Unlike the powderpost termite, *Cryptotermes brevis,* the western drywood termite is seldom seen in furniture and other small wooden products. However, an interesting observation on such an infestation was made by R. E. Wagner in October, 1973 (personal communication). In a bookcase having no contact with wooden members of the house structure, he found many predrilled 3-mm holes, 5 of which had been plugged with the typical drywood termite sealoff. A recently established royal pair of drywood termites was found about 10 mm deep in each hole, evidently individuals from the annual flight of alates. The house had been fumigated about 10 years earlier, but had become reinfested.

In newly developed residential tracts in California, some of the orchard trees are retained in the dooryards. The dead limbs on such trees often harbor drywood termites, particularly if the trees are walnut, and these may serve as sources of infestation for the newer houses. In southern California, homes in new residential tracts tend to become infested by drywood termites sooner and in greater numbers than by subterranean termites (Ebeling, 1968). Most often, drywood termites infest these homes as alates originating in older buildings in near-by areas. However, infested telephone poles, posts, and piles of lumber or firewood can also be sources of further infestation.

The lumberman, builder, and homeowner and, in lawsuits, the lawyer, often inquire as to where, in the sequence of events from the planing mill to the finished home, infestations of such insects as drywood termites, woodboring beetles, and woodwasps originate. It is important to know the life histories and habits of such insects.

Description of Drywood Termite Alate. The alate (plate I, 4; figure 82) is dark brown, and has smoky-black wings with black veins. It may be distinguished from the alate of the western subterranean termite by its larger size (about 11 to 12 mm long) and by the fact that it has a reddish-brown head and thorax, while the subterranean termite is brown to brownish black throughout.

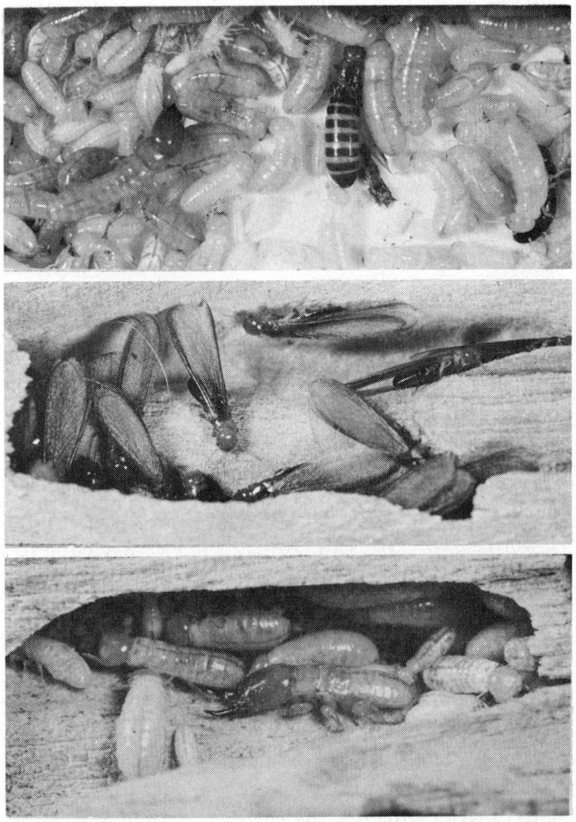

Fig. 82. Western drywood termite, *Incisitermes minor*. *Top*, queen, soldier, and nymphs; *center*, alates; *bottom*, soldier and nymphs.

Habits. The first evidences of drywood termite infestation are usually piles of brownish fecal pellets (figure 83) below "kickout" holes or chinks and cracks in the infested wood, particularly where outer walls of the wood member have become excessively thin from prolonged infestation.

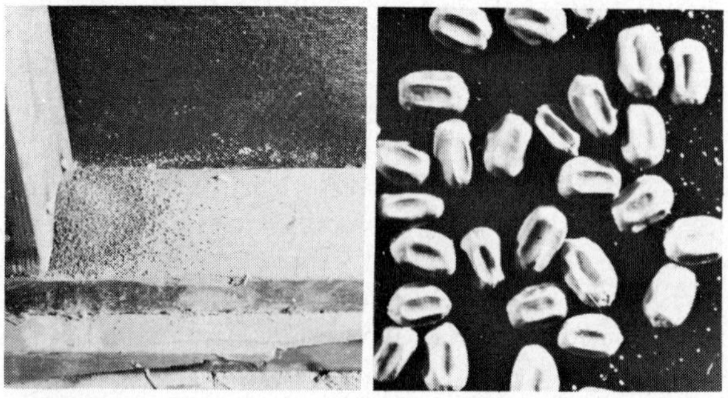

Fig. 83. Fecal pellets of the western drywood termite, *Incisitermes minor*. *Left*, typical pile of pellets on a ceiling plate; *right*, enlarged view of pellets. (From Ebeling, 1968.)

The pellets are elongate, averaging about 0.85 mm in length, with rounded ends, and with 6 flattened or roundly depressed surfaces. Longitudinal ridges occur at the angles between the 6 surfaces. The shape of the fecal pellet is the result of pressure exerted by 6 plates of rectal epithelium and 6 rectal grooves (Child, 1934). Another indication of the insects' presence may be the flight of alates during warm, sunny days in the fall months.

Besides infesting the dead branches of common native trees and shade and orchard trees, drywood termites also infest utility poles, posts, and piles of lumber (particularly sapwood of redwood) in lumberyards. They generally enter houses through attic vents or shingled roofs but, particularly in hot, dry localities, they are often found in the substructure, where they may have entered via the foundation vents. They attack rafters, ridgepoles, and sheathing in the attic, windowframes and sills, door and window jambs, doorsills and, in the substructure, mainly the floor joists and adjoining structural timbers. They may also infest wooden furniture or other wooden materials within the home.

Pence (1956*b*) found that under extremely dry conditions, individuals of *Incisitermes minor* in a wood cavity sealed themselves in thoroughly with carton and huddled together to conserve moisture. One individual in such a group survived in kiln-dried wood placed in a silica gel desiccator for 245 days. Its abdomen was then completely flat from loss of water. When given access to water, the termite drank until it became turgid and then continued its life normally. Because most kalotermitids can live in dry wood and do not require shelter tubes leading to the ground, they can damage wooden furniture, even if it is moved about frequently. This, plus the ability of a colony to exist in a very small piece of wood, results in kalotermitids and another nonsubterranean termite species being easily dispersed from one area to another. They often appear in areas far removed from regions in which they are indigenous, and it is then urgent that such localized infestations be eradicated.

The Royal Pair

After a flight of winged reproductives and subsequent breaking off of their wings, a mated pair of drywood termites will select a place to enter wood. Pairs work together to make a hole (figure 84), and then seal themselves in (Harvey, 1934). The hole serves as the entrance to the "royal cells," and is about 10 mm deep. The "royal pair" enlarge the hole, and the queen lays her first eggs. Generally, 2 to 5 nymphs hatch from the first eggs and begin an enlargement of the burrow. The nymphs (plate I, 5; figure 82) perform the duties of the worker caste of the higher termites. The queen then lays more eggs in the advanced part of the main passage of the burrow.

Colony Composition

Toward the end of the second year, after the colonizing pair has entered the wood, the colony generally consists of the primary king and queen, one soldier, and a dozen or more nymphs

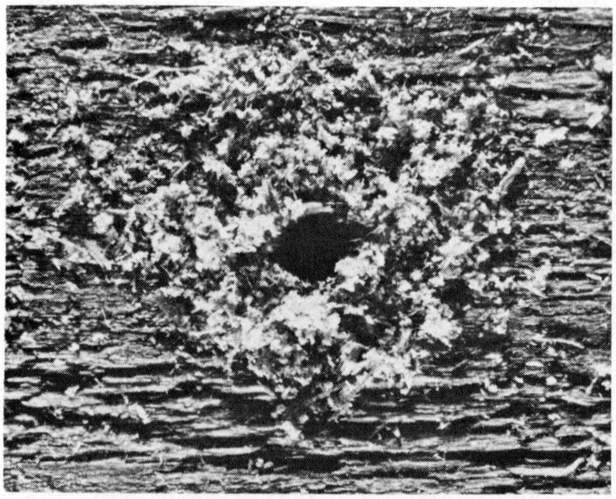

Fig. 84. Entrance hole of the "royal pair" of the western drywood termite.

(Harvey, 1934). By this time, the abdomen of the queen has become broader and longer, and she can lay a greater number of eggs (figure 85). From late spring to late fall, the primary queen lays from 1 to 12 eggs each day for 7 to 10 days, ceases egg-laying for a month or more, and then resumes at the same rate as before. Harvey (1934) found that at a temperature of 80° F (27° C) and a relative humidity of 83%, the eggs in 1- and 2-year-old colonies were hatched in an average of 77 days. He believed that maximum egg-laying capacity is reached when the queen is 10 to 12 years old, and then declines rapidly. A secondary queen then takes over her functions.

Fig. 85. Primary queen of the western drywood termite and a clutch of eggs.

The time required for *Incisitermes minor* to develop from egg to alate or soldier is believed to be more than 1 year. It has been estimated that a colony 15 years old might contain a primary king and queen, 1 or more supplementary reproductives, 120 soldiers, and 2,600 nymphs. Thus, the colonies are small compared with those of the subterranean termites, which may contain millions of individuals.

The relatively small colonies result in correspondingly less rapid and less severe damage to buildings than that which is caused by most subterranean termites, but proliferation of colonies in a building can result in extensive infestation.

Southeastern Drywood Termite, *Incisitermes snyderi* (Light)

This species ranges from South Carolina to Florida and west to eastern Texas. It is lighter in color than *I. minor,* and is not so severe a pest, but does similar damage. Its flights occur in May or June at nightfall.

Desert Drywood Termite, *Marginitermes hubbardi* (Banks)

These termites replace *Incisitermes minor* at lower elevations and in more arid regions of the deserts of southeastern California, in Arizona, and in northwestern Mexico. The alates are yellow to light brown and, including wings, are about 13 mm long. They may be distinguished from the alates of most other drywood termites of the areas by their very pale coloration, round ocelli, and by the third antennal segment, which is not longer than the second or fourth. In the soldier, the third antennal segment is clublike and is almost as long as all the succeeding segments combined, making identification easy. The alates emerge at night, usually just after a rain, and tend to collect at lights by the thousands. This species readily invades buildings, and is sometimes very abundant in favorable locations. Its damaging potential may increase as cities and suburbs extend into desert regions where it can exist in large cacti and in trees (Weesner, 1965). It is very destructive to poles, posts, floors, ceilings, window and door frames and sills, furniture, boxcars, and other wooden objects in coastal or low-lying towns of the west coast of Mexico. It is much more common in wooden structures than in wood occurring in nature (Light, 1933).

Powderpost Termite, *Cryptotermes brevis* (Walker)

This species (plate I, 6) was accidentally introduced into the United States, and is well established in Hawaii, Florida, and parts of Louisiana. It has not been recorded in any natural habitat, but only in man-made structures. It is widely distributed throughout tropical and semitropical areas of the world. Like other drywood termites, powderpost termites can thrive in very dry wood, and need no connections with the ground. They can attack buildings, but are particularly noted for their severe damage to floors, woodwork, furniture, and small wooden objects.

During a 4-year period, quarantine officials intercepted *Cryptotermes brevis* 47 times in cargo entering the continental United States from the Bahamas, Cuba, Hawaii, Honduras, Philippines, and South Africa, and they finally intercepted an extensively infested ship. Piles of fecal pellets were found in wooden bunks and along paneled corridors. To avoid the damage to electronic equipment that would result from the use of methyl bromide, sulfuryl fluoride was used as a fumigant. A total of 282 lb (128 kg) of gas was used to fumigate 70,480 cu ft (1,997 cu m) of space (Mahaney, 1972).

Powderpost termites have been said to have smaller fecal pellets than the drywood termites just discussed (Light, 1934*b*). However, we were unable to distinguish pellets taken from a live colony of powderpost termites in 5-ply Philippine mahogany veneer, imported from Puerto Rico 5 years earlier, from the pellets of *Incisitermes minor;* both appeared to be identical in size and shape.

The alates of *Cryptotermes brevis* are 10 to 11 mm in total length, and their wing membranes are colorless, with brownish veins. The caste that most conspicuously distinguishes this species is the soldier. The head is mostly black, almost as broad as long, and high in front. The strongly truncated head is distinctly concave, and is rough in front (figure 86).

Pacific Dampwood Termite, *Zootermopsis angusticollis* (Hagen) (Hodotermitidae)

Dampwood termites do not require contact with damp ground, but do require wood with a high moisture content. They may be found most easily under the bark of fallen coniferous trees—those that have not lain on the ground so long as to have very loose bark. Infestations in forest cabins are common. They are also common in beach areas where water tables are high, and where water used for irrigating lawns and shrubs is likely to keep the earth under buildings, and therefore the wood in contact with earth, damp for longer periods than in locations farther inland. The moist soil and high humidity near beaches combine to make beach houses particularly susceptible to dampwood termite attack.

Description. *Zootermopsis angusticollis* occurs in coastal forests from British Columbia to Baja California. The alates (plate I, 7) may be as long as 25 mm, including wings—about twice the length of the drywood termite. Their color ranges from yellowish brown to cinnamon-brown to chestnut. The soldiers (figure 87) may be 15 to 20 mm long. The head is somewhat narrowed in front, unlike that of a related species, *Z. nevadensis,* or of *Incisitermes minor.* Figure 88 shows a soldier severing the head of a carpenter ant with its powerful mandibles. The nymphs are white or somewhat cream-colored, but the abdomens of the older nymphs are colored by the intestinal contents, which can be seen through the integument (plate I, 8; figure 87).

As with drywood termites, fecal pellets, colored according to the kind of food eaten by the termite, may be found throughout the colony. The pellets are hard, usually elongate, rounded at both ends, but are only slightly hexagonal and lack the longitudinal ridges that give drywood termite pellets their strongly sculptured appearance. The pellets of *Z. angusticollis* are about 1 mm long, compared with an average of 0.85 mm for drywood termites. If the wood is extremely damp, the pellets are often spherical or irregular, and usually stick to the sides of the galleries. In drier wood, the pellets collect in the bottom of various chambers, or the termites may expel them the way drywood termites do.

Swarming. Most swarming takes place during August, September, and October, on warm, sultry evenings just before sunset, but alates may occasionally be found flying in every month of the year. They often appear after the early rains, and are attracted to lights.

Fig. 86. Powderpost termite, *Cryptotermes brevis* (soldier).

Fig. 87. Pacific dampwood termite, *Zootermopsis angusticollis*. *Top*, soldier; *bottom*, nymphs. (Courtesy Stennett S. Heaton.)

Damage. Damage from dampwood termites can be greater than from subterranean termites, for they have more of a tendency to work their way upward from the foundation to the roof rafters. In one instance in San Francisco, they were found on the fifth floor of a concrete fireproof building (Mallis, 1969). Infestations of dampwood termites may be found wherever there are wood–earth contacts. Poles, piling, bridge timbers, and other structures built over or near water are subject to attack. Down timber or, in the city, old, infested buildings serve as reservoirs for swarmers and invite termite attack.

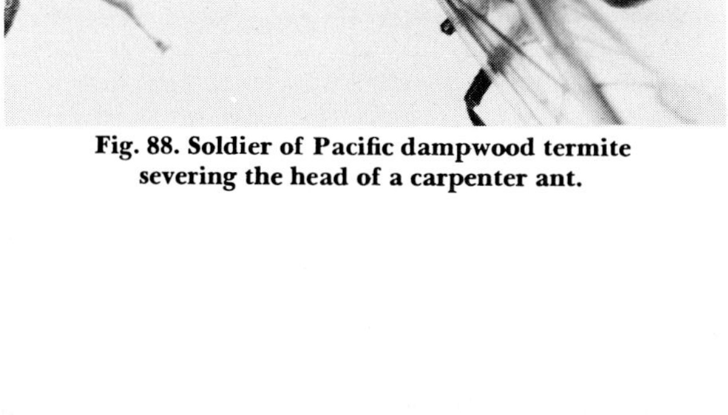

Fig. 88. Soldier of Pacific dampwood termite severing the head of a carpenter ant.

Although *Zootermopsis* species have been called "rottenwood termites," Weesner (1965) pointed out that colonizing pairs are usually situated in sound wood, including felled logs, standing dead trees, and dead portions of live trees, and that only mature colonies invade rotten wood.

Nevada Dampwood Termite, *Zootermopsis nevadensis* (Hagen)

This termite is smaller and darker than *Z. angusticollis*, but its over-all length may still be over 20 mm. It occurs along the Pacific Coast from British Columbia to within Monterey County, California, and eastward into the Sierra Nevada and the mountainous regions of Idaho, Montana, and northern Nevada. Thus, only *Z. angusticollis* occurs in southern California, but in that region it can be found up to 1,800 m.

In the coastal areas, *Z. nevadensis* swarms mostly in summer and early fall, but in inland areas at higher elevations, the peak of the swarming appears to occur in the spring. This species seems to have about the same capacity for destruction of wood as *Z. angusticollis*, but it does not occur in the more densely populated areas and therefore is not so often recognized as a pest.

Zootermopsis laticeps (Banks)

This species occurs in southeast Arizona and southwest New Mexico, inhabiting dead wood, usually stumps or large logs, along streambeds or water-storage ponds. This is a spectacularly large termite, the alate being 25 to 30 mm long, with wings 6 mm wide, and the soldier, 16 to 22 mm long. The soldier's head is very large, and bears tremendous mandibles. The antennae of both alate and soldier have 20 or more segments. These termites swarm during July and August, and are nocturnal in habits (Light, 1934*c;* Nutting, 1965; Weesner, 1965).

Desert Dampwood Termite, *Paraneotermes simplicicornis* (Banks) (Kalotermitidae)

This species is in the same family as the drywood termites, but infests wood at or below grade level in the southwestern United States. It sometimes girdles young citrus trees and grapevines below the soil line in desert areas.

Florida Dampwood Termite, *Prorhinotermes simplex* (Hagen) (Rhinotermitidae)

This termite inhabits the southern tip of Florida and the Caribbean area, and occasionally does some damage to buildings. The soldiers are difficult to distinguish, by superficial examination, from those of *Coptotermes formosanus,* described earlier in this chapter (Miller, 1967).

Family Termitidae

In the United States, there are few visible indications of the presence of termites apart from the damage they cause. However, in many other parts of the world, some species of this family build earthen mounds above ground level that vary from a few centimeters to as much as 10 m in height. There are 4 genera of Termitidae (*Gnathamitermes, Amitermes, Anoplotermes,* and *Tenuirostritermes*) in the southwestern deserts, but they do not build mounds and are of little or no economic importance.

Gnathamitermes perplexus (Banks) is a large, brown species, reaching a length of 16 mm, including wings. It is the common desert termite in California and Arizona, and extends into almost all the southwestern desert regions (Weesner, 1965). It is primarily a soil-dweller, but builds extensive earthen workings aboveground to cover vegetation, stumps, posts, and wooden structures. This behavior, plus the superficial damage the termite can do to wood, results in it being an occasional pest. A similar species, *G. tubiformans* (Buckley), occurs in Texas, New Mexico, and Mexico.

Amitermes spp. are somewhat smaller than *Gnathamitermes,* rarely exceeding 12 to 13 mm in length, including wings. The alates are dark, and even black. The soldiers have mandibles usually less than 75% as long as the width of the head, and the outer margins are recurved from the base to the tip. Unlike *Gnathamitermes, Amitermes* spp. do not build extensive earthen structures aboveground. They attack wood directly, entering at the point where it contacts the ground, and are greater economic hazards than other desert-dwelling members of the family Termitidae (Weesner, 1965). *Amitermes* spp. have been known to girdle and kill young citrus trees.

Control of Subterranean Termites

Termite control is difficult and relatively expensive. A good knowledge of termite biology and habits is required for proper inspection and diagnosis, and specialized tools, equipment, and experience in their use are often required for structural alterations and control procedures. It is usually desirable to hire a termite operator for the initial control work. The same operator can usually be retained with a "control contract" that guarantees periodic inspection and treatment if they should be required.

Subterranean and drywood termites require entirely different control measures. Subterranean termites have their nests in the ground, and control consists of structurally or chemically isolating the wood structures of a building from these nests.

Such measures have no value against drywood termites. Drywood termites remain in infested wood, and must be controlled there. Fumigation of a building for drywood termites would kill the subterranean termites that happened to be in the wood structure at the time, but would not destroy the entire colony, and in fact would not kill the reproductive or potentially reproductive individuals. Therefore, the control of subterranean and drywood termites will be discussed in separate sections.

Inspection

Successful treatment generally depends on a thorough and accurate diagnosis. Although swarming termites (flights of winged reproductives called alates) are often a sign that wood members of a building are infested, this is not true in all cases. For example, termites may inhabit an earth-filled porch, feeding in buried wood that was carelessly shoveled in when the original fill was made (figure 89). If the porch was properly sealed away from the foundation, the termites may not have had access to wood members of the building, and the alates may have emerged from cracks in the porch cap or other openings reaching the earth fill.

Important signs of subterranean termite infestation are shelter tubes (figure 72, this chapter). Shelter tubes are proof of termite infestation, but their absence does not necessarily mean that a structure is free of termites. The insects may reach sills and other wood members through cracks or voids in the foundation wall, under the outside stucco, or from earth-filled porches, steps, terraces, or patios. Among California homes with no wood contact with the ground that are infested with subterranean termites, more than half of the infestations originate in earth-filled extensions of the main foundation.

Moisture caused by condensation or leakage invites termite attack or dry rot. This is particularly true in subframing adjacent to the first-floor showers, bathrooms, toilets, sinks, laundry facilities, and other plumbed areas. All moist areas of a building should be examined with particular care and the source of moisture eliminated. Blisters or darkened areas on the flooring sometimes indicate the presence of termites. Such areas are easily crushed, sometimes revealing the light-colored termite workers and a few soldiers.

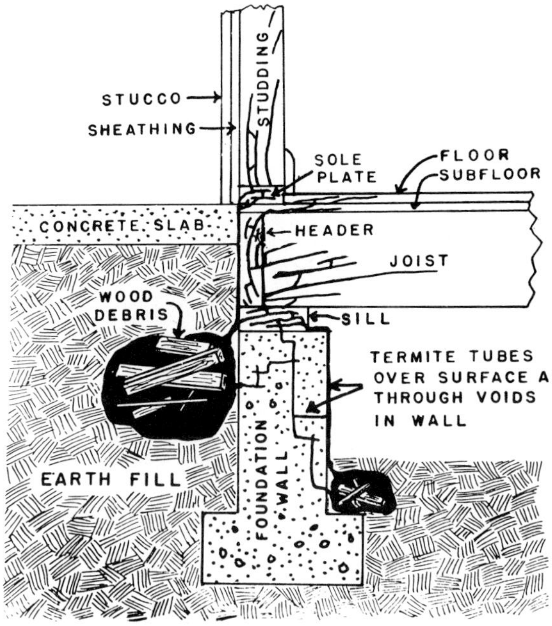

Fig. 89. Wood debris in soil near a foundation—a common source of subterranean termite infestation. (From Ebeling and Pence, 1965.)

A Job for Professionals. Locating termite colonies in a building and determining the best way to combat the insects is very difficult and strenuous work. The homeowner can seldom inspect so thoroughly as the trained and experienced inspector that a professional termite operator would send out to do the work. The inspector must crawl over dust, mud, and debris, often in a space barely high enough to admit his body. He must crawl the entire distance around the inside of the foundation, inspecting and probing the wood structure, and must investigate areas under showers, sinks, and in the vicinity of floor furnaces.

Locating the source of a termite infestation is particularly difficult when the building has been constructed on a concrete slab, for the inspector cannot crawl beneath the floor as he can when inspecting joist-type construction. Generally, the termite operator must guess where to drill through the slab to inject his soil insecticide.

A reliable operator will provide the homeowner with a diagram of the inspected premises, indicating the points of infestation and hazardous conditions, and an estimate of control and repair cost. In areas where drywood termites are a potential problem, inspection for these insects should be included, and usually requires access to the attic area. On the diagram, the inspector will indicate with the letter "S" the approximate locations where subterranean termites or their damage are found. The letter "K" (for the old genus *Kalotermes*) is indicated where drywood termites are found or suspected.

It takes a long time for termites to cause appreciable damage in addition to that which is revealed by the inspection. There is rarely cause for an operator to impress the homeowner with the urgency of immediate action. If a great deal of work is recommended, it may be wise to investigate the reliability of the operator and to compare his diagnosis and competitive bid for the job with those of other termite operators.

Along with his diagram of the locations of termites and termite-conducive conditions, the inspector will make an estimate of the cost of control and if necessary, for repair or replacement of damaged wood. Some termite operators offer free inspections, others charge an inspection fee if no further work is done on the inspected premises, and some operators include an inspection fee in their estimate of the work to be done. The homeowner may arrange for a periodic inspection service done by experienced personnel, usually once a year.

The Standard Report Form. In California, all termite control companies, in accordance with state law, must file a standard report form with the Structural Pest Control Board for every piece of property their agents inspect. This is mandatory, regardless of whether an infestation is found or an inspection fee is charged. Copies of all reports filed within the preceding 3 years are available from the board for a nominal fee. The condition of a property with regard to structural pests is thus fully documented and disclosed.

The termite reports are usually requested by the escrow company during real estate transactions. They are also generally required by any lending institutions. However, a termite inspection is not legally required when a house changes

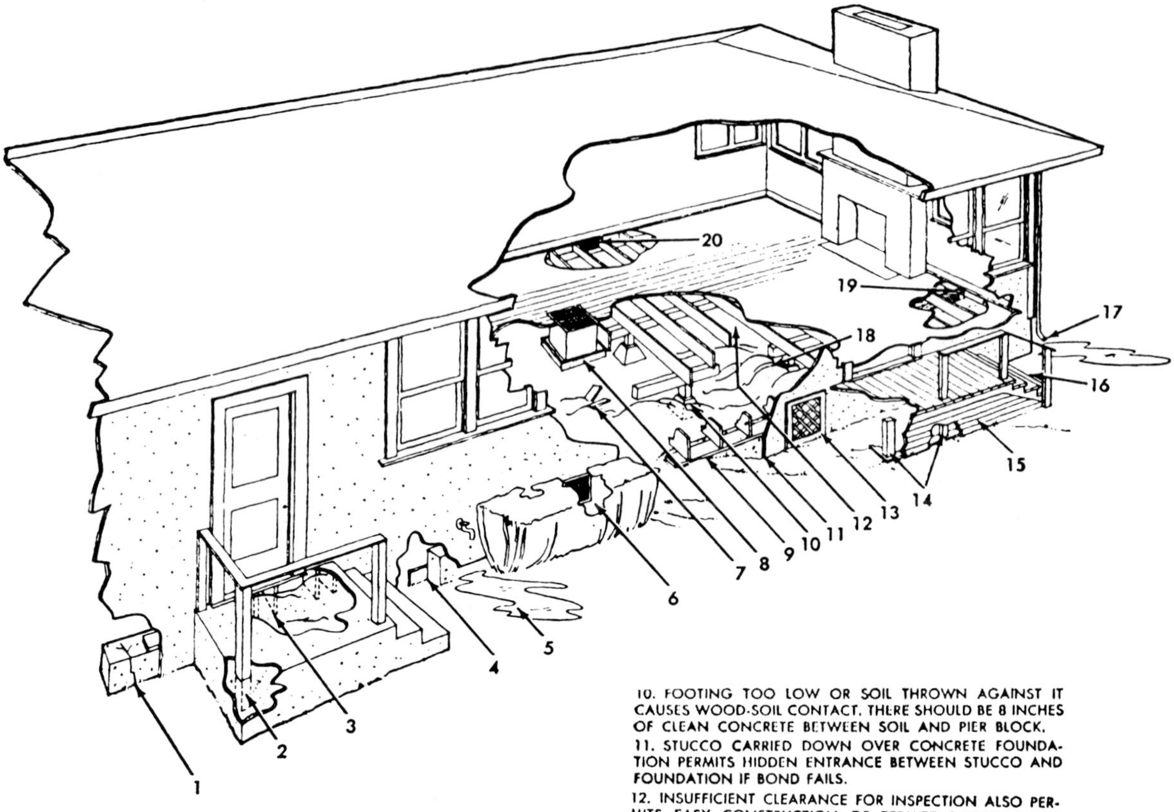

10. FOOTING TOO LOW OR SOIL THROWN AGAINST IT CAUSES WOOD-SOIL CONTACT. THERE SHOULD BE 8 INCHES OF CLEAN CONCRETE BETWEEN SOIL AND PIER BLOCK.
11. STUCCO CARRIED DOWN OVER CONCRETE FOUNDATION PERMITS HIDDEN ENTRANCE BETWEEN STUCCO AND FOUNDATION IF BOND FAILS.
12. INSUFFICIENT CLEARANCE FOR INSPECTION ALSO PERMITS EASY CONSTRUCTION OF TERMITE SHELTER TUBES FROM SOIL TO WOOD.
13. WOOD FRAMING OF CRAWL HOLE FORMING WOOD-SOIL CONTACT.
14. MUD SILL AND/OR POSTS IN CONTACT WITH SOIL.
15. WOOD SIDING AND SKIRTING FORM SOIL CONTACT. SHOULD BE A MINIMUM OF 3 INCHES CLEARANCE BETWEEN SKIRTING AND SOIL.
16. PORCH STEPS IN CONTACT WITH SOIL. ALSO WATCH FOR LADDERS AND OTHER WOODEN APPURTENANCES.
17. DOWNSPOUTS SHOULD CARRY WATER AWAY FROM BUILDING.
18. IMPROPER MAINTENANCE OF SOIL PILED AGAINST PIER FOOTING. ALSO MAKES CAREFUL INSPECTION IMPOSSIBLE.
19. WOOD GIRDER ENTERING RECESS AND FOUNDATION WALL. SHOULD HAVE 1 INCH FREE AIR SPACE ON BOTH SIDES AND END AND BE PROTECTED WITH A MOISTURE IMPERVIOUS SEAL.
20. VENTS PLACED BETWEEN JOISTS TUNNEL AIR THROUGH SPACE WITHOUT PROVIDING GOOD SUBSTRUCTURAL AERATION. VENTS PLACED IN FOUNDATION WALL GIVE BETTER AIR CIRCULATION.

1. CRACKS IN FOUNDATION PERMIT HIDDEN POINTS OF ENTRY FROM SOIL TO SILL.
2. POSTS THROUGH CONCRETE IN CONTACT WITH SUBSTRUCTURAL SOIL. WATCH DOOR FRAMES AND INTERMEDIATE SUPPORTING POSTS.
3. WOOD FRAMING MEMBERS IN CONTACT WITH EARTHFILL UNDER CONCRETE SLAB.
4. FORM BOARDS LEFT IN PLACE CONTRIBUTE TO TERMITE FOOD SUPPLY.
5. LEAKING PIPES AND DRIPPING FAUCETS SUSTAIN SOIL MOISTURE. EXCESS IRRIGATION HAS SAME EFFECT.
6. SHRUBBERY BLOCKING AIR FLOW THROUGH VENTS.
7. DEBRIS SUPPORTS TERMITE COLONY UNTIL LARGE POPULATION ATTACKS SUPERSTRUCTURE.
8. HEATING UNIT ACCELERATES TERMITE DEVELOPMENT BY MAINTAINING WARMTH OF COLONY ON A YEAR-ROUND BASIS.
9. FOUNDATION WALL TOO LOW PERMITS WOOD TO CONTACT SOIL. ADDING TOP SOIL OFTEN BUILDS EXTERIOR GRADE UP TO SILL LEVEL.

Fig. 90. Twenty points of faulty home construction and maintenance that may lead to infestation by subterranean termites. (From Ebeling, 1959, after Eleventh Naval District, San Diego, California.)

ownership. The buyer or his lending agent must ask for it. It is a good idea to have a termite inspection before buying a house, even if the buyer has to pay for it.

The termite reports made by inspectors also serve as a basis on which state investigators may examine the competency of termite operators against whom complaints of substandard work or fraudulent practices have been lodged. Should comparison of the reports of several pest control firms indicate that the complaints against one of these firms are valid, further direct investigation is conducted, and can ultimately result in revocation of licenses held by the firm or its representative or both.

Faulty Home Construction and Maintenance

No method of construction or treatment has yet been devised that is an absolute guarantee against infestation by subterranean termites. The builder and homeowner can take certain measures to decrease the possibility of infestation. Figure 90 is a sketch of a residence, showing where such preventive measures can be taken.

Wood-to-Ground Contacts. Wood members should never be allowed to rest on the ground, but should rest on concrete foundations or piers. These are ordinarily not susceptible to termite penetration, although the insects may gain access to wood through cracks that develop in the concrete (figure 90 [1]). Termites can build shelter tubes over the surface of concrete, but the tubes are then exposed and may be destroyed. The ground below the tubes may then be chemically treated.

Specifications of the Uniform Building Code should be followed, particularly those pertaining to the minimum clearance between wood members of the substructure and the ground, the size, location, and screening of vents, and the installation of termite shields where required. The rules for foundation clearance and ventilation also pertain to wood-frame porches and steps. Nowhere should wood be allowed to come in contact with the ground (figure 90 [2], [3], [14], [15], and [16]). Attention to building code specifications will be of little avail if earth is piled against the foundation or pier posts after construction (figure 90 [18]). Planter boxes built on the ground should be separated from the main structure by at least 4 in. (10 cm); if possible, they should be kept below the level of the first wood in contact with the foundation wall.

To eliminate breeding places for termites in or on the soil, buried stumps, logs, and roots should be removed from the building site. Scrap lumber should be removed from the ground under the building after its completion (figure 90 [7]).

Care should be taken not to include wood scraps with the earth used to fill extensions of the foundation, such as porches, terraces, patios, and steps. This is very important, for termite colonies in these fills appear to be the sources of over half the subterranean termite infestations of structures. Even if these fills are properly sealed away from the foundation at the time of construction, earthquakes, sonic booms, or subsidence of the earth may break the seal, allowing termites within them to reach the mudsill and other wooden members.

Foundation forms and stakes should be completely removed (figure 90 [4]). If parts of forms and broken stakes have been left in the ground and cannot be removed, they may be soaked with an oilborne toxicant.

Avoidance of Excess Moisture. Because subterranean termites are favored by moist soil, every practicable site-grading and construction measure should be employed to prevent accumulation of moisture around and under the house. In addition, vents (figure 90 [20]) should be sufficiently large and numerous to remove moist air from the area beneath the house, and should be placed so as to provide cross-ventilation wherever practicable. Shrubbery should not be allowed to obstruct the vents (figure 90 [6]). Sprinkling against the sides of stucco houses should be avoided or minimized because the resulting dampness favors termite infestation. Termites have access to the mudsill from outside the foundation if stucco has become separated from it. Stucco will adhere to the foundation more firmly if the latter is "dashed" with a rich cement mixture before the stucco is applied.

All leaking faucets or other plumbing leaks should be corrected, and downspouts should carry water away from the building (figure 90 [5], [17]). There is often no practicable repair measure for shower-stall leaks. The shower-stall pan may be replaced with copper, lead, fiberglass, or masonry pans. A watertight wall should overlap the pan interior. The wood used in framing the pan should be pressure-treated with appropriate chemical preservatives to avoid termite infestation or fungal infection.

If drainage and ventilation do not reduce soil moisture sufficiently, it is helpful to cover exposed soil with heavy roofing paper. In the case of slab-on-ground construction, a "vapor barrier," usually consisting of a sheet of some synthetic material impervious to water, may be placed on the ground before the concrete for the slab is poured.

Besides reducing the possibility of termite infestation, measures that decrease under-area moisture also decrease the likelihood of decay from fungal infection. They may also prevent warping of floors and doors, result in greater comfort in summer by reducing humidity in the living space of the home, and save fuel in winter by preventing dampness.

Chemical Treatment of Lumber

Lumber "pressure-treated" with an approved wood preservative such as pentachlorophenol (C_6Cl_5OH) or Wolman salts resists attack by termites and fungi. It is considered too expensive to treat all lumber, even in the substructure, so treated lumber is ordinarily used only where it is most subject to attack by termites or decay.

Since most of this lumber is structurally vital and difficult to replace, the cost of treatment is well justified.

Lumber thus treated offers the additional advantage of forcing the termites to tube around it and thus expose their activities to view. In an experiment in which pressure-treated wood, with untreated wood on top of it, was in contact with the ground for 4 years, termites tubed over the treated wood in 5 out of 20 samples, and damaged the untreated wood on top. Where more than 0.0625% of aldrin, chlordane, dieldrin, or heptachlor was added to the wood preservative, termites were not able to tube over the treated wood during the same period (Beal, 1969).

When wood is subjected to leaching, it is best protected by oils or oilborne preservatives. When the treated wood will not be in contact with the ground or water, or where it may require painting, waterborne preservatives are generally used. Chemicals and their uses are given in (1) *Federal Use Specification TT-W-571d* (current revision), (2) *Standards of the American Wood Preservers' Association,* and (3) *Standards of the National Woodwork Manufacturers' Association.* A brush coat of a termite- and fungus-resistant solution is helpful for treating sills and other wood members subject to termite attack if they have not been pressure-treated, but it is not so effective in all cases as pressure treatment.

Sills also have been constructed of naturally durable or naturally resistant wood, having extractives making them resistant and/or repellent to termites.

As used in the United States, grade-marked "foundation-grade" California redwood or 100% heartwood of southern tidewater red cypress, or very pitchy southern pine (lightwood), or heartwood of eastern red cedar (less resistant than the above) are suggested when durable wood is desired (St. George *et al.*, 1963). It is generally recognized, however, that pressure-treated wood is more reliable.

Wood–Cement Composite

A structural building board composed of wood elements coated with Portland cement, frequently referred to as "century board," was found to be resistant to 2 termite species, *Coptotermes formosanus* and *Zootermopsis angusticollis.* The resistant property of wood coated with cement was found to be the result of the creation of a physical barrier into which the termites were unable to penetrate (Allen and Dolan, 1970).

Termiteproof Concrete

The addition of long-lasting toxicants to concrete foundations can deter subterranean termites, and is often useful even where soil treatment is made. In Australia, Gay and Wetherly (1959) substituted a 0.5% emulsion of dieldrin for the normal mixing water to prepare a termiteproof concrete. They found that workers of *Nasutitermes exitiosus* and *Coptotermes lacteus,* confined in cylinders of treated concrete, were affected after 2 hours of contact, and that complete mortality occurred in 24 hours; exposure of 1 hour was sufficient for complete mortality. They also found that while the termites could make their way through a "no-fines" concrete containing only coarse aggregates, this type of concrete, when treated with dieldrin, was not penetrated. F. J. Gay (correspondence) stated that laboratory tests had shown aldrin to be as effective as dieldrin when used in the same way and at the same concentration. The strength of the concrete was not adversely affected by the treatment, and accelerated weathering procedures did not destroy its insecticidal effectiveness.

In the United States, a 75% dieldrin wettable powder was added to water used in concrete, so that concentrations of the toxicant in cement blocks were approximately 0.1 and 1.6% by weight. A week later, the insecticide in these blocks resulted in a 100% mortality for *Reticulitermes flavipes* workers having contact with them for 1 minute, and for *Nasutitermes columbicus,* nasutes exposed 10 minutes (Allen *et al.*, 1961). After 16 months, the newly cracked surfaces of laboratory-aged mixtures were equivalent in toxicity to the original surfaces of new mixtures (Allen *et al.*, 1964*a*).

Information has recently become available on tests with insecticide-treated blocks placed on soil in the termite-infested pine forest in Gulfport, Mississippi, where so many tests had been made by the USDA Southern Forest Experiment Station with insecticide-treated boards and stakes (see "Insecticides and Concentrations," following). Concrete blocks of the standard mix for slab-on-ground construction, containing chlordane and dieldrin wettable powders (WP) and emulsifiable concentrate (EC) at concentrations from 0.04 to 1.33%, were placed on ground where subterranean termites (*Reticulitermes* spp.) were known to be abundant. Wooden bait boards 1 × 6 × 6 in. (2.5 × 15 × 15 cm) were placed under the concrete blocks to encourage termite activity, and on top of the blocks to better determine the presence of termites. Measures were employed to provide humid, shaded conditions for 1 year, and then wooden covers were placed on the units to simulate conditions under a house. After 5 years of exposure under field conditions, termites had not tubed over any of the dieldrin-treated blocks, and on only 1 of 10 blocks treated with 0.125 and 0.25% chlordane WP. Five of the 10 untreated blocks were completely tubed over in 2 years (Beal, 1971).

Under actual foundation construction practice, treatments would probably be even more effective than the field tests indicated, for about a third of the water in concrete runs off the top of the slab during construction, and would be present in the soil for long periods, as soil treatments have indicated (Johnston *et al.,* 1971). However, at the time of publication of the report, the use of insecticides for termiteproofing of concrete had not yet been registered by appropriate state and federal agencies (Beal, 1971). It appears that dieldrin is banned for any type of termite control other than soil treatment, so it is unlikely that anyone will seek registration of the insecticide for use in concrete.

Termite Shields

Metal shields have usually been of little or no practical value against termites, principally because they are seldom installed properly. Even when shields are of proper design and materials and are properly installed, termites may build tubes over their edges. Some termite operators have found the sharp edges of the shields to be hazardous when they are inspecting or doing repair work under a building. Shields appear to be favored nesting places for Argentine ants. It is not uncommon for disintegration of the metal to take place within a few months when such nests are present, presumably because of the formic acid given off by the ants (Scott, 1960). The undersides of shields are often filled with carton by termites, thus presenting no barriers to them (Isherwood, 1957).

Correction of Structural Deficiencies

After a building is constructed, termite control may involve structural alterations or repairs that correct certain adverse environmental conditions.

Grade. Older buildings sometimes have only "crawling ditches" to allow inspection and chemical treatment. If a building does not have at least 18 in. (45 cm) of clearance between the soil and the lowest horizontal wooden members of the substructure, the inside grade should be lowered by excavation. In areas with high water tables, lowering the inside grade and digging ditches are undesirable if excessive dampness and/or standing water develop.

Where frame or stucco construction is used, additional foundation material is sometimes added to bring the sill at least 6 in. (15 cm) above grade. As a rule, a few linear feet of the mudsill and cripples are removed at a time, the new foundation is built up to the desired height, and a new substructure is built upon it. In certain cases, particularly where low foundations and resultant inadequate clearance beneath the building exist, the entire structure may be elevated and the whole foundation is raised to meet it.

Gutters. When wood sills on foundation walls are at or below grade, a gutter may be required to lower the soil or grade level to below the sill. The bottom of the gutter should be at least 4 in. (10 cm) below the sill. Gutters are usually made of preformed concrete slabs or poured concrete. Preformed slabs are used as retaining walls and as the bottoms of gutters. Joints should be properly filled with hot pitch, and the gutter should be at least 8 in. (20 cm) wide to allow removal of debris.

Flash Wall. The flash wall (figure 91) is a solid, poured-concrete facing, firmly bonded to the

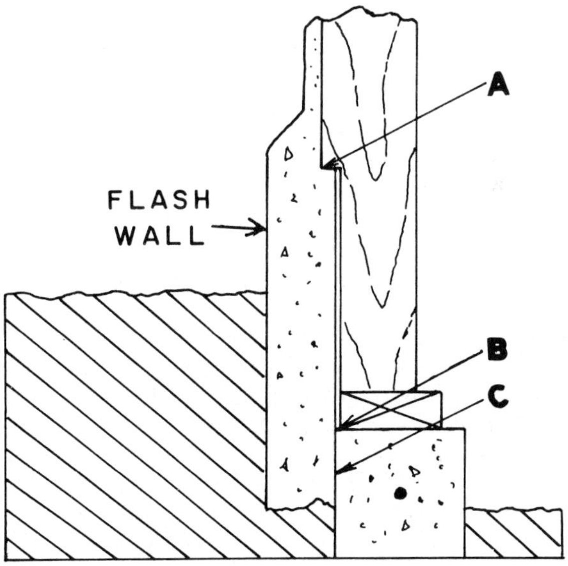

Fig. 91. Conventional flash wall, showing points of weakness. Water may leak down the wall or enter through stucco at *A* and collect at *B*, where it can be responsible for termite or fungus problems. At *C* there is a "cold joint" that can provide access for moisture or for termites. (From Ebeling, 1968.)

foundation, and extending at least 4 in. (10 cm) below the top of the foundation and a minimum of 3 in. (8 cm) above the natural grade level. It should be a minimum of 3 in. thick, and the top must be sloped outward to provide adequate drainage away from the wall of the house.

The flash wall is used as a relatively inexpensive alternative to raising the foundation when the latter procedure would normally be indicated —as when the sill is below the top of an adjacent concrete slab or below the earth grade. It is most justified when raising the foundation would be excessively expensive, as when the floor joists rest on the sill. Flash walls have often proved unsatisfactory because water may enter from above, and the flash wall is likely to separate from the foundation, thus providing a hidden access for termites.

Porch "Cut-and-Seal"

The earth fill that is contained in concrete porches, terraces, patios, and steps has already been discussed as a potential source of subterranean termites. Although the earth is originally sealed away from the mudsill, this seal may be broken by earthquakes, sonic booms, gradual subsidence of the earth, or by temperature extremes. Termites then have access to the wood. To remedy this, termite operators usually cut out a strip of the slab of the porch, terrace, etc., adjacent to the wall of the building, and excavate the earth fill until at least 4 in. (10 cm) of the foundation are exposed. The exposed soil is always saturated with a suitable liquid insecticide, and reinfestation in the treated area is unlikely. The exposed portion of the foundation and the cut edges of the slab caps are then cleaned, and cracks (if any) in the foundation are flushed with toxicant and sealed. Concrete is then poured into the excavated area (figure 92). This procedure is particularly common in California and is called the "cut-and-seal" or "sealoff."

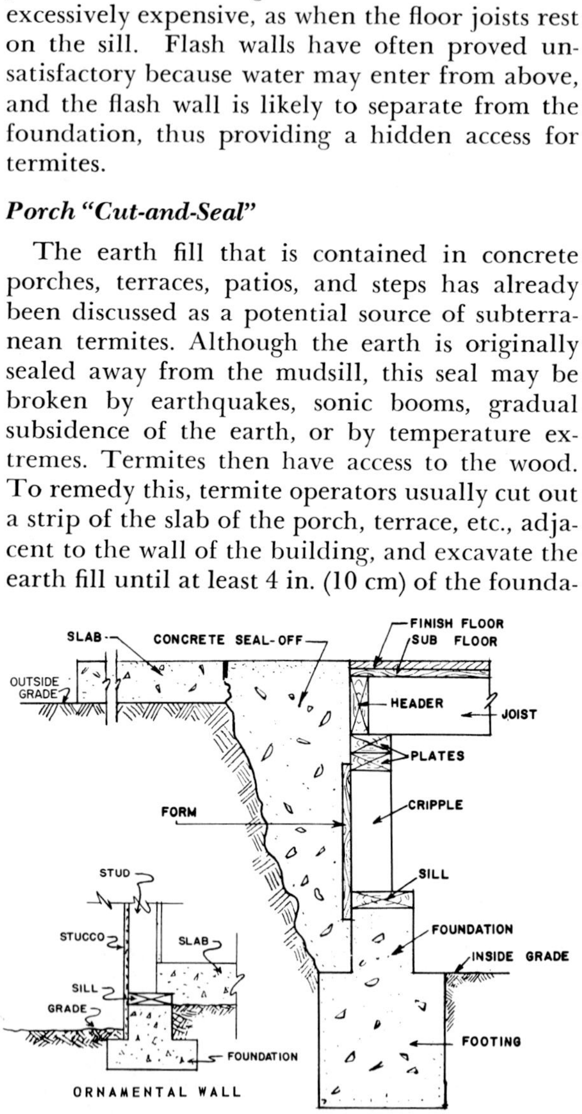

Fig. 92. Concrete cut-and-seal "sealoff" used to close foundation away from earth-filled porch. *Lower left,* ornamental wall sometimes built around a slab porch; this may require a sealoff. (From Ebeling, 1968, after Heal, 1951.)

An ornamental wall adjacent to a porch slab (figure 92, *lower left*) was once commonly constructed in California. The sill in this wall is vulnerable to termite infestation. Mechanical alteration consists of removing a section of slab along the wall and sealing the earth away from the foundation with concrete, as just described. The soil is first treated with liquid insecticide and cracks, if any, in the foundation wall are flushed with toxicant and sealed.

Treatment of Wood Substructure with Pentachlorophenol Emulsion

Pentachlorophenol, C_6Cl_5OH, is one of the principal toxicants used for pretreatment of timber destined for residential construction, telephone poles, railway ties, fenceposts, etc. "Mayonnaise-type" oil emulsions containing 10% pentachlorophenol are known as "Woodtreat-TC," and have long been used for applying a maximum quantity of termiticide and fungicide on subarea wood members not pretreated before construction. Currently Woodtreat-TC also contains 0.5% technical heptachlor (0.36% actual heptachlor).

Using a paintbrush or a specially designed "caulking gun," mayonnaise-type emulsions can be applied in a thick layer containing about 20 times more toxicant than can be applied with a single brush coat of an unemulsified oil solution. For brush application, the emulsion should be stirred before use until it has a creamy consistency, for it is then easier to apply.

As the emulsion breaks, the toxic solution is gradually released to penetrate into the wood. Termites have a tendency to burrow as closely as possible to the wood surface without causing the infested wood member to collapse. Therefore, penetration of toxicant into the galleries is quite likely. Woodtreat-TC can be applied to such wood members as mudsills or floor plates, either for prevention of infestation by termites or to get rid of an existing infestation. It is particularly effective when applied generously where 2 wood members come together, as at the intersection of a girder, joist, or cripple on a mudsill, or where a stud meets a floor plate.

Aside from preventing conditions favorable to subterranean termites, such as a damp subarea or a dangerously high grade level, either outside or inside the foundation, the application of Woodtreat-TC to termite-infested or termite-susceptible wood members is probably the most feasible thing the homeowner can do for himself. The author controlled 2 subterranean termite infestations in his own home, one of them originating from an earth-filled patio, with Woodtreat-TC which was applied liberally on the sill and at the intersection of the sill and cripples in the entire infested area beneath the house. In his garage, he applied the material at the intersections of all studs and the mudsill. The infestation of subterranean termites had been so severe that the tapping noises[10] of termite soldiers could be heard easily every time the garage door was opened. The treatment appears to have been completely successful. Although a portion of the sill is below grade level and becomes wet during every heavy rain, the infestation has not reappeared in the 12-year period since the treatment was applied.

When applying Woodtreat-TC, inhalation of vapor should be avoided as much as possible, and if skin is accidentally contaminated, the material should be removed with soap and water. If much work with it is contemplated, chemical-resistant gloves and a respirator designed to remove organic vapors should be worn.

Pretreatment of Soil with Insecticides

One of the most important means of isolating a building from subterranean termites is the insecticide-treated soil barrier. Soil treatment is most effectively done before and during construction of the foundation, and this is called "pretreatment." It is of particular importance when concrete slab-on-ground construction is employed, for treatment of the soil is difficult after the concrete has been poured. Concrete slab was once stated to be a termite barrier, but experience has proved that this is not the case.

There are 3 types of slab-on-ground construction: monolithic, suspended slab, and floating slab. Figure 93, *A,* represents a cross section of the monolithic type, in which the floor (cap) and footing are poured in one continuous operation so that there are no "cold joints" that might eventually form cracks and allow access for subterranean termites. At *B* is shown a "suspended slab" extending completely across the top of the foundation. The foundation and the floor are constructed as independent units. Even though a vertical crack may develop in the foundation, this type of construction prevents enlargement of the crack. With both these types of construction, chances for entry of termites around the foundation are minimal, but other avenues of termite entry (such as cracks in the cap slab, or apertures around utility pipes) are as common as with other types of slab-on-ground construction.

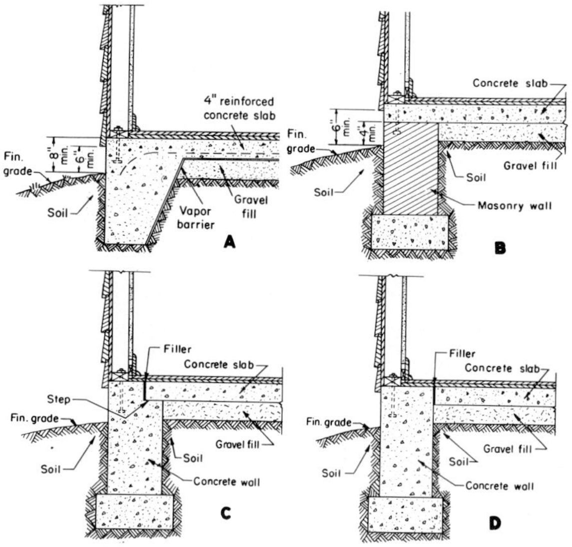

Fig. 93. Slab-on-ground construction. *A*, monolithic; *B*, suspended slab; *C* and *D*, floating slab type of construction in which the slab rests on a ledge of the foundation (*C*) or is independent of it (*D*). The "filler" in the expansion joints at *C* and *D* usually consists of a roofing-grade coal-tar pitch or rubberoid bituminous sealer. (From St. George *et al.*, 1963.)

In figure 93, *C* and *D* show "floating slab" construction, in which the slab or floor either rests on a ledge of the foundation or is independent of it. Entry for termites is made possible by expansion joints. The difference in the coefficient of expansion between the concrete and the filler in the expansion joints results in a pulling away of these materials from each other as soon as there is sufficient temperature change. This is also true of the supposedly termiteproof fillers or caulking substances placed around utility pipes extending through a slab.

[10] The tapping noise is made by the termite soldier by banging its massive head up and down against the ceiling and floor of the gallery. This is considered to be an alarm signal (Stuart, 1969). The tapping noises are detected through the subgenual organs in the termites' legs (Howse, 1970).

Fig. 94. Standard field tests of the efficacy of insecticides, applied to soil, for protecting wood against termite attack. *Top*, untreated sapwood pine boards measuring 1 × 6 × 6 in. (2.5 × 15 × 15 cm) laid on top of insecticide-treated soil, each held in place by a brick; *bottom*, 2 × 4 × 18-in. (5 × 10 × 45-cm) untreated sapwood pine stakes driven to a depth of 12 in. (30 cm) into treated soil; *insets*, individual test units. (Courtesy Southern Forest Experiment Station, Gulfport, Mississippi.)

Termites also penetrate through cracks that can form anywhere in the slab. The incidence of such cracks is greatly minimized by wire-mesh reinforcement of the concrete cap.

The principal difficulty with slab-on-ground, as far as termites are concerned, is that this type of construction makes treatment more difficult and expensive than where raised-foundation (joist-type) construction is employed.

Insecticides and Concentrations for Soil Treatment. The recommended insecticides for either pretreatment or for treatment of known infestations at any period after construction are the same: aldrin at 0.5%, chlordane at 1%, dieldrin at 0.5%, and heptachlor at 0.5%. These insecticides may be used either as water emulsions or in solution in an oil not heavier than No. 2 fuel oil. Oil solutions should not be used under concrete slabs, for they may penetrate to the surface of the slab, causing unsightly stains or weakening the adhesives used for parquet flooring or tiles.

The insecticides are purchased in the form of emulsifiable concentrates that differ in the percentage of toxicant each contains. The concentrate must be diluted with the correct volume of water to obtain a solution for application that contains the recommended amount of toxicant for termite control. Emulsifiable concentrates blend readily with water. The globules of insecticide are largely sieved out by the first 0.75 in. (2 cm) of soil through which they pass (Ebeling and Pence, 1958; Beal and Carter, 1968). The insecticide penetrates to a greater depth in moist soil than it does in dry soil (Beal and Carter, 1968). Despite the shallow insecticide barrier, good protection is obtained against termites. The remarkable efficiency of this barrier is indicated by experiments of the United States Forest Service's Southern Forest Experiment Station at Gulfport, Mississippi (Johnston, 1960). Boards were laid on treated soil, or wood stakes were driven into it, in forested land heavily infested with the eastern subterranean termite, *Reticulitermes flavipes* (figure 94). Aldrin, chlordane, dieldrin, and heptachlor have given complete protection for periods of 23 to 27 years to this date (1975), depending on when the material was applied (Johnston *et al.*, 1971; Smith *et al.*, 1972). Test boards and stakes have always been destroyed by termites within a year in untreated plots in these tests. Tests with granular formulations were begun much later, but the granules have provided a 100% control for a 12-year period where 0.031% dieldrin or 0.062% aldrin, chlordane, or heptachlor were used. Granules with a lower percentage of insecticide have already failed (Smith *et al.*, 1972). [Labels for aldrin and dieldrin granules to be used for termite soil treatment have been voluntarily withdrawn by the sole registrant (Ruckelshaus, 1972*b*).]

In Hawaii, experiments were begun in 1958 to investigate the relative persistence of 7 insecticides which were mixed with 3 different types of soil and poured into prepared holes in randomized plots. Dosages of insecticide were as recommended by the manufacturer, and at 25% of those dosages. In 6 months, and at yearly intervals for 9 years, samples were used for laboratory bioassay tests, with the subterranean termite *Cop-*

totermes formosanus used as the test insect. The relative persistence of the insecticides, based on the percentage of the quantity applied that remained after 7 years, was in the order DDT > dieldrin > chlordane > aldrin and converted dieldrin > heptachlor and heptachlor epoxide > lindane > sodium arsenite. The relative toxicity of the treated soil samples to termites was in the order dieldrin > aldrin > heptachlor > chlordane > DDT > lindane > sodium arsenite (Bess *et al.*, 1966). Later, bioassays showed that also after 9 years, "dieldrin and aldrin were clearly superior to the other termiticides in the test" (Bess and Hylin, 1970).

Fig. 95. Soil treatment for subterranean termites before pouring of concrete slab.

When the emulsifiable concentrate is diluted with a fuel-oil carrier, the insecticide stays in solution regardless of how deeply the solution penetrates. Therefore, fuel-oil solutions are useful when it is desirable for the insecticide to penetrate to a greater depth in a limited area, for instance, around a stake that cannot be removed. A limited degree of wood penetration, particularly in the direction of the grain, is also possible with fuel-oil solutions.

In the United States Forest Service tests in Mississippi, benzene hexachloride (BHC), at a concentration of 0.8% of the gamma isomer (lindane), was 100% effective for only 8 years, but heavy rainfall (about 2 m annually) in the test area may have adversely affected lindane, which is soluble in water to the extent of 10 ppm. The initial toxicity of lindane to termites is extremely high (Ebeling and Pence, 1958), and this insecticide is often used in southern California in attempts to exterminate termites under slabs. Lindane is relatively volatile, but when the objective is to create an effective barrier throughout the area beneath a concrete slab, volatility could be beneficial, even though the more volatile insecticides are generally the least persistent under other conditions (Bess *et al*, 1966).

Under the slab-on-ground foundations of 32 houses, Fahey and Osmun (1971) recovered 49% of the lindane applied 10 years earlier. This was a greater percentage than was recovered from the other organochlorine insecticides, although the latter are ordinarily more persistent. The percentages of aldrin, heptachlor, chlordane, and dieldrin recovered were 35.1, 37.8, 41.2, and 46.2, respectively.

Application Procedure and Dosage. Termite operators generally apply insecticide emulsions with power sprayers (figure 95), but high pressure is not required, and it is possible to apply the emulsions with a sprinkling can. To treat the ground where a concrete slab is to be poured, including attached porches, apply 1 gal (about 4 l) of the diluted insecticide emulsion to 10 sq ft (about 1 sq m) as an over-all treatment. Over-all treatments applied to washed and ungraded gravel fills, or fills of an absorbent material (e.g., cinders), should be increased by 0.5 gal per 10 sq ft (2 l per 1 sq m). Apply 2 gal per 5 linear ft (7.5 l per 1.5 m) to critical areas, such as along the inside of foundation walls and around utility pipes, entrances, and interior partition foundation walls; also, use 2 gal per 5 linear ft along the outside of the foundation.

Ideally, a narrow trench should be dug around the foundation to within about 1 ft (30 cm) above the foundation footing. The soil should be treated as it is replaced at the rate of 1 gal (4 l) per 5 linear ft for each foot of depth from grade level to footing, but allowing 2 gal per 5 linear ft for the first foot in depth from ground level. The treated backfill from the trench should then be covered with a layer of soil. According to the Environmental Protection Agency's instructions, the trench should be no wider than 6 in. (15 cm) (Ruckelshaus, 1972*b*).

Although the dosage for the over-all treatment under a building should average 1 gal per 10 sq ft (about 4 l to 1 sq m), in actual practice, termite operators tend to apply more insecticide emulsion where the chances for termite penetration are greatest, as around utility pipes, and less, or even no emulsion, in other areas. According to EPA instructions, if concrete foundation slabs are to be poured on treated soil, a polyethylene sheeting or other waterproof material shall be placed over the treated soil, unless the concrete is to be poured on the day of the treatment. This is to avoid washing away of insecticide by rain (Ruckelshaus, 1972*b*).

Unless treated ground or fill is to be promptly covered with a vapor barrier or by the slab, precautions must be taken to prevent disturbance of the treated soil that might break the continuity of the insecticide barrier. To avoid surface flow of the spray liquid away from the application site, treatments should not be made when the soil or fill is very wet, or after heavy rains.

In the treatment of joist-type houses at the time of construction, generally not all the crawl space is sprayed. Two gal (7.5 l) of the dilute emulsion is applied per 5 linear ft (1.5 m) to the inside of foundation walls and around piers and utility entrances.

Along the outside of foundation walls, including the part opposite entrance platforms, porches, etc., apply 2 gal per 5 linear ft where the foundation is deep, and apply 1 gal per 10 sq ft (4 l per 1 sq m) of soil surface as an over-all treatment only when the attached porches, entrance platforms, utility entrances, etc., have covering slabs on fill or ground. Treat all voids of masonry block walls and piers, using 1 gal per 5 linear ft applied from grade to footing.

Is Soil Treatment for Subterranean Termites an Environmental Hazard? To be effective, soil treatment for subterranean termites should result in a very long-lasting barrier. As stated earlier, tests made by the United States Forest Service have already demonstrated that certain chlorinated hydrocarbons have the ability to create such a barrier for at least 21 to 25 years, and no one knows how much longer. Environmentalists may inquire as to the possibility of pollution from such long-lasting residues. Experiments have shown that there is neither downward nor lateral migration of insecticide that is of any practical consequence.

In tests with emulsions of 0.5% dieldrin and heptachlor and 1% chlordane made with 7 soils from various parts of the United States and at a dosage of 1 pt per sq ft (about 0.5 l per 930 sq cm) of soil surface, in most tests more than half of the insecticide was found in the upper 0.75 in. (2 cm) of soil. The emulsion penetrated least in arid soils from Arizona and eastern Oregon and most in soil from Missouri. An increase in soil moisture at the time of application increased penetration, but in any case only extremely small percentages of insecticide reached to a depth of 5 in. (13 cm) within 24 hours or within 4 years (Beal and Carter, 1968; Carter *et al.,* 1970). Considering the inconsequential migration of the insecticides from the application site and their insolubility in water, soil treatment for termite control does not appear to offer any significant threat of environmental pollution.

Soil residues can be physically removed, as when soil is eroded by water in floods, etc., but this is rarely a problem in treated soil under buildings.

Rate of Infestation of Newly Constructed Residences

The decision of the prospctive buyer of a new home as to whether the cost of a pretreatment is a justifiable insurance for his investment depends on how soon it is likely to become infested with termites. Extensive field experiments were made in large areas in southern California to determine the rate of infestation by termites in newly constructed residences built in tracts that had just been devoted to citrus or walnut growing. Soil was treated in lots on which houses with adjoining garages were to be built, using at least 8 insecticides in each of 12 tracts where slab-on-ground foundations were constructed, and in each of 10 tracts where joist-type (raised) foundations were used. In each tract, untreated lots equal in number to treated ones were left as checks (Ebeling and Wagner, 1962).

In cooperation with the California Structural Pest Control Board, a study was made of the incidence of termite infestation for 11 years after the lots were treated. On 90 treated and 86 untreated lots inspected, on which slab foundations were used, no infestation of subterranean termites was found. Of the 77 inspected treated lots in which the joist-type foundations were used for the houses, 1 infestation of structural timber was found in a garage (slab foundation) where the ground had been treated with 2% toxaphene. The infestation was obviously the extension of a vigorous colony over which the building was constructed.

No termites were discovered in any of the houses. However, they were found in wood debris on the ground under 5 of the houses. Of the 72 joist-type houses on untreated soil, infested wood debris was found under 7, and termites were also found in a form board that had not been removed after the concrete was poured. Thus, in 11 years, subterranean termites were found in the wood structure of only 1 building (a garage) out of a total of 167 treated and 158 untreated properties (Ebeling, 1968). Subsequent inspection reports of termite operators, obtained from the Structural Pest Control Board, showed the dates of transfer of ownership of the above properties and indicated that they had changed ownership averaging about once every 3.5 years. The above data show that a very long period is required for subterranean termites to become established in southern California in new tracts when construction is in accord with modern building codes. On the other hand, in many areas where houses are 35 to 50 years old, nearly every house in the block has had one or more infestations of both subterranean and drywood termites. It must be remembered that most of these older structures would fall far short of current building regulations, particularly relating to earth–wood contacts, grade levels, and clearance. These factors, along with the greater period of exposure to termite attacks, are probably responsible for the record of heavy infestation by subterranean termites in the older buildings.

In some areas, as in the forested regions of the Gulf states, much of the ground is continuously infested with subterranean termites. The termite colony may be very large when it first attacks a new building, having developed to that size on subterranean and surface food that was available before the land was cleared for construction. In such areas, a large percentage of new houses can be expected to be infested in less than 2 years, and in some cases damage from termites is evident before construction is complete (Redd, 1957).

Drywood termites infest a greater percentage of residential properties in southern California, and reach them sooner than do subterranean termites. In 11 years following construction, drywood termites were found in 35 out of 325, or 10.8%, of inspected properties (Ebeling, 1968). They tended to spread in the direction of the prevailing winds from infested buildings, poles, dead tree branches, etc. In garages that usually had their doors left open during the day, the incidence of infestation was much greater than in those in which the doors were usually kept closed.

Treatment After Construction

Treatment after construction is not greatly different from pretreatment for buildings on raised foundations, except for the earth-filled extensions of the foundation, such as porches, patios, etc. For these, it is necessary to drill through the concrete cap to apply insecticide. Sometimes termites eventually build shelter tubes from the ground to the nearest timber, or there may be many swarming tubes, particularly around a floor furnace. Sprays can then be applied to the entire subfloor area, as well as to the inside of foundation walls. The entire subfloor area is also generally sprayed if termites have been found in scraps of wood on the ground, even though the wood is removed.

A Special Problem with Slab Foundations. Postconstruction treatment of buildings with slab-on-ground construction is much more difficult than pretreatment. Holes must be drilled either downward through the concrete or horizontally through the foundation in order to inject an insecticide. For vertical injection, the proximity of the holes to one another should depend on the nature of the earth or fill below the slab. If the slab is underlain by heavy soil, 3 ft (1 m) between the holes may be appropriate;

Fig. 96. Horizontal rodding of slab-on-ground foundation. *A*, 22-ft (6.6-m) iron pipe at the moment of entry into a 1.12-in. (2.8-cm) hole in the foundation (*arrow*); *B*, pipe after being pushed to nearly its entire length under the slab floor, and 2 more holes (*arrows*) drilled 6 ft (2 m) apart; *C*, 2 holes in the slab foundation, one (*left*) filled with a neoprene cone that fits over the pipe and plugs the hole to prevent backflow. The hole at right, with liquid flowing out, shows that the liquid can spread laterally for about 6 ft. (From Spitz, 1958.)

if underlain by sand fill or gravel, 6-ft (2-m) spacing may be adequate. A sound procedure is to drill holes about 6 ft apart and observe whether the liquid is forced out of adjacent holes. If not, more holes should be drilled until there are enough to provide reasonably complete coverage of the subslab area.

Because of the hazard of drilling through radiant-heating pipes and plumbing, or the possibility of destroying a vapor barrier, as well as the expense and hazard incurred in removing tile, parquet, etc., many termite operators treat under slabs through the foundation wall.

Horizontal Rodding. Horizontal subslab rodding involves drilling holes through the foundation at a height that allows entry immediately below the usual 4-in. (10-cm) cap of the slab. Figure 96, *A*, shows a 22-ft (6.6-m) iron pipe being inserted into one of these holes. The pipe was pushed in to a distance of 20 ft (about 6 m) with a back-and-forth motion while insecticide was flowing at 4 to 5 gal (15 to 19 l) per minute. A neoprene cone or stopper over the pipe prevented premature backflow of the liquid (figure 96, *C*). The figure also shows that the liquid had spread laterally for a distance of 6 ft (2 m). The lateral spread is believed to be uniformly greater than 3 ft (1 m) as the pipe is pushed in, and it is also believed that the grid pattern formed by pushing the pipe in from all sides of the foundation results in soil saturation (Spitz, 1958).

Ethylene Dibromide for Subslab Treatment. Ethylene dibromide gas moves readily under slabs in coarse gravel, but is greatly restricted in clay soils with fine particles. A high content of organic matter also restricts the movement of the gas, as in peat or muck soils. The moisture content of the soil is another important factor in retarding gas penetration. No prediction of results can be made unless much is known about these factors at the time of treatment, and in any case fumigation is not a panacea for the control of termites under slabs. The hazard to building occupants from the subslab ethylene dibromide injection must also be considered. A fumigant may find its way into the heating-system ducts beneath a slab and from there into the house (Osmun, 1958).

Termite Baits

The attraction of subterranean termites to wood decayed by a brown-rot fungus, *Lenzites trabea,* was discussed earlier in this chapter. Esenther and Gray (1968), using a bait of wooden blocks infected with *Lenzites trabea* and immersed for 10 seconds in a 1% solution of mirex in toluene, were able to attract subterranean termites (*Reticulitermes flavipes*) to the wood, and markedly suppress termite foraging activity and damage in treated areas, compared with nontreated ones. Trials were first made in southern Ontario, Canada, near the northern border of the range of subterranean termites. Later research revealed that even in southern Mississippi, in an area where termites were abundant and severely destructive, mirex-attractant blocks effectively suppressed termites for 3 years. This control method appears to show considerable promise (Esenther and Beal, 1974).

Control of Drywood Termites

Although the following control recommendations are based on experience with *Incisitermes minor* in California, they should also apply for the most part to the control of *Cryptotermes brevis,* for the behavior of the 2 species is similar. In either case, control differs greatly from that used for subterranean termites, because procedures employed for preventing access of subterranean termites to the wooden members of the foundation are not applicable. Conversely, fumigation of the building as practiced in drywood termite control is of little value against subterranean termites, for although subterranean termites that happen to be aboveground are killed, a large part of the colony, including the reproductive forms, is in the ground and is not affected.

Inspection

In most of California and in many southern states, a complete termite inspection includes, besides the inspection of the subarea for either subterranean or drywood termites, the inspection of the attic for drywood termites. Infestations of drywood termites may be revealed by fecal pellets that have dropped from "kickout" holes in infested wood members. The pellets may be in conical piles if they have dropped only a short distance (figure 83), or they may be scattered if they have fallen from a greater distance, as from a ridgepole, rafter, or sheathing, to the floor of the attic. If infested wood is broken open, pellets may tumble out of galleries and pockets in the wood in large numbers. Fecal pellets that are seen in an attic may have been pushed up from below to the tops of ceiling joists and plates, but usually they have fallen from wood members above the place in which they are found.

Sometimes, holes ("royal cells") made by the primary pairs of reproductives can be seen. They are about 2 mm in diameter, and proceed straight into the wood at right angles to the surface. Such holes are usually sealed with wood fragments, and a small festoon of frass can often be seen hanging from them.

Once a pair of reproductives enters an attic, the insects crawl about extensively, and may bore directly into the wood or crevices between sheathing and rafters, between rafter ends and a plate or ridgepole, between plates and studding ends, between studding and siding, between building paper and studding or rim joists or, in the substructure, between mudsills and foundations, to mention only a few of the possibilities. About 90% of the infestations begin in some crack in the wood or in wood joints. The termites may also be in lumber used for construction. In southern California in the fall months, particularly in October, piles of lumber in lumberyards are often attacked by large swarms of drywood termite alates. The entrance holes of the primary pairs may be numerous, and are easily seen.

The attics of flatroofed houses may occasionally be too narrow for a man to crawl into, or they may not be accessible. In such cases, a termite operator indicates in his report what sections of the attic were inaccessible, and may recommend that an access be cut. He is not held accountable for control of termites in inaccessible areas of the attic. If the house is to be fumigated, the termites in such areas will be killed anyway.

In coastal areas, drywood termites are not found in subfloor areas as often as in attics. In areas of very high summer temperatures, as in the San Joaquin Valley of California, they are usually found only in subfloor areas, and not in attics.

Drywood termites are frequently found in windowframes, sills, and internal trim. Pellets are usually seen there by the homeowner if an infestation is present. The windowframes and sills are often so completely eaten out that they collapse and have to be replaced.

The "Drill-and-Treat" Method

Control methods for drywood termites depend primarily on the extent of infestation. If the infestation is sufficiently localized and does not appear to extend into the walls, inaccessible parts of the attic, etc., treatment may be different from that used if it is widespread and partly inaccessible. In the latter case, fumigation is the only practicable procedure. Otherwise, the "drill-and-treat" method of control may be employed.

Termite galleries are located by probing suspected timbers with a screwdriver, ice pick, or other sharp instrument. Quarter-inch (7-mm) holes are then drilled into the infested wood members at about 1-ft (30-cm) intervals to provide access to the galleries. An insecticide dust such as "Kali-dust" (50% calcium arsenate) may be blown into the holes by means of a "Kaligun" (figure 97), or a liquid fumigant may be injected. An ounce (28 cc) of dust is enough for 15 to 30 holes; too much dust may plug the galleries and decrease the effectiveness of the treatment. The holes are plugged after treatment.

When a liquid fumigant is used, about one-third tsp (2 ml) of the liquid, usually a mixture of ethylene dibromide and DDT in a petroleum solvent of high flash point, is used per hole. An entire gallery system usually requires 0.5 to 2 tbsp (7 to 30 ml) of liquid. A 1-qt (1-l) fumigant injector pressurized with a small CO_2 bomb is commercially available, but liquid fumigants can also be satisfactorily applied with a small gun oiler or other low-pressure oilcan. When treating wood into which it is not desirable to drill, 1 or 2 coats of ethylene dibromide solution, applied with a paintbrush, will often be effective. Ethylene dibromide should be employed only by experienced pest control operators because of the hazards involved in its use.

Insecticide dust blown into drywood termite galleries travels a great distance in them and, in addition, is widely distributed by termites carrying the dust on their tarsi. Some termite operators believe that the gas released by liquid fumigants is more widely distributed than dust throughout the system of galleries contacted by the injected liquid.

Examination of the records of 4 termite operators in southern California revealed that in many houses treated for drywood termites, live termites were again found in a few years (figure 98), particularly if the drill-and-treat method had been used (Ebeling and Wagner, 1964). Even though some termites survive a treatment, it is not a total failure, for infestation and future damage may be greatly diminished by expert operation.

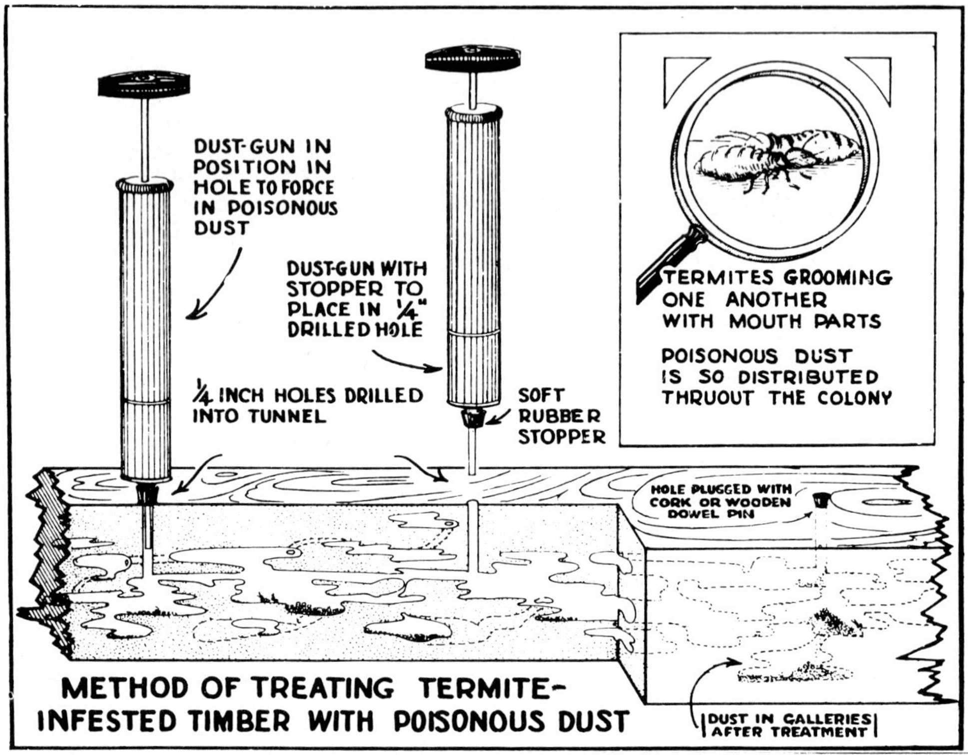

Fig. 97. Injection of poisonous dusts into drywood termite galleries according to the "drill-and-treat" method of control. (From Light *et al.*, 1930.)

Pentachlorophenol Emulsion

The use of a proprietary pentachlorophenol emulsion (Woodtreat-TC) for control of subterranean termites was just discussed under "Treatment of Wood Substructure." This emulsion can also be applied with a paintbrush or caulking gun for the control of drywood termites. In attics, heavy wrapping paper or several layers of newspaper should first be laid down to catch any dropped emulsion that otherwise might soak through the ceilings of the rooms below. The old fecal pellets can be vacuumed, or other newspapers can be spread under the treated areas to catch any pellets that fall from the treated wood. Such pellets would indicate an incomplete treatment, and the areas should be retreated.

An application of Woodtreat-TC is more rapid and foolproof than the drill-and-treat method, because in the latter, galleries are sometimes hard to detect. They may be stopped up, and dust or fumigants may not fully penetrate the gallery system. The Woodtreat can be brushed on beyond the limits of the suspected area, because the time and expense of doing so are negligible.

Treatment for drywood termites sometimes involves application of Woodtreat-TC to painted wood. Some paints are not penetrated effectively, and thus painted surfaces should be roughened or scratched until there are breaks showing bare wood before applying the emulsion. Very few paints, with the exception of certain marine products, will adhere effectively to areas treated with Woodtreat-TC until at least 12 to 18 months after treatment. Treated areas should then be cleaned and sanded lightly before painting.

Fumigation

Fumigation is universally recognized as the most effective treatment for drywood termites. Frame houses must be covered with a gastight tarpaulin, usually of 7-oz nylon coated with rubber, neoprene, or plastic (figure 99). Stucco structures with flat roofs are generally sealed with a special gastight paper over outer doors, windows, vents, and wherever the fumigator believes a satisfactory seal can be obtained without the labor of tarping. (Fumigating should be done only by licensed pest control operators.)

The procedures for house fumigation for drywood termites were first developed with HCN or methyl bromide (Gunn *et al.*, 1947), and the treat-

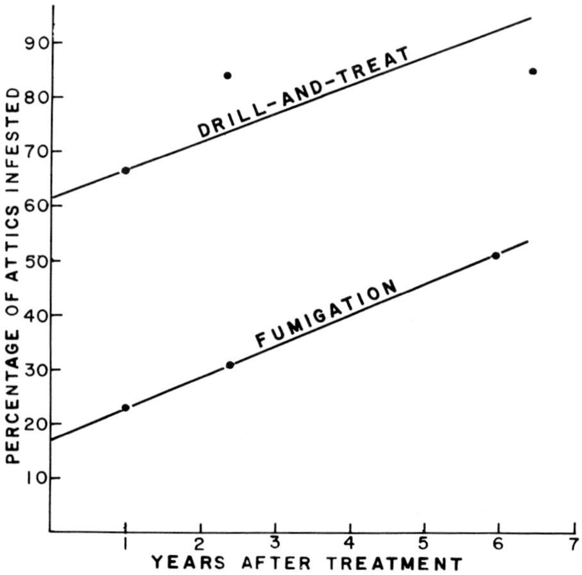

Fig. 98. Percentage of attics infested with drywood termites at various periods after fumigation or "drill-and-treat." (From Ebeling and Wagner, 1964.)

Fig. 99. House covered with a gastight tarpaulin during fumigation for drywood termites. (Courtesy J. B. Eyer.)

ment was first shown to be effective against *Incisitermes minor* (Pencille, 1947; Hodel, 1949). HCN was soon discontinued for house fumigation because of the fire hazard. Sulfuryl fluoride was found to have certain advantages, and was recommended for both *Incisitermes minor* and *Cryptotermes brevis* in Hawaii (Stewart, 1957, 1966; Bess and Ota, 1960). Minnick *et al.* (1972) demonstrated that the findings of Bess and Ota with reference to the success of sulfuryl fluoride as a fumigant for *C. brevis* in Hawaii were applicable for this species under Florida conditions.

The fumigating gases currently most commonly used in the United States are methyl bromide, sulfuryl fluoride, and acrylonitrile formulations. Methyl bromide is generally used at 2 lb per 1,000 cu ft (0.91 kg per 28 cu m) of building space, and sulfuryl fluoride at 1 lb per 1,000 cu ft (0.45 kg per 28 cu m). Acrylonitrile with 30% carbon tetrachloride and 30% chloroform is used at 2 lb (0.91 kg) per 1,000 cu ft for drywood termites and 3 lb (1.36 kg) per 1,000 cu ft for powderpost beetles. The gas is generally allowed to remain for 24 hours, but there are no rigid rules as to gas dosages or the period required for fumigation. Both are influenced by such factors as temperature, air movement, porosity of the soil under the building, etc. Much depends on the discretion and judgment of the fumigator.

The gas most often used is methyl bromide. In recent years, sulfuryl fluoride, said to be even more penetrating and effective against both *Incisitermes minor* and *Cryptotermes brevis* than methyl bromide (Stewart, 1957, 1966; Bess and Ota, 1960; Bess, 1971), and having the added advantage that it is not necessary to remove any furnishings containing foam rubber from the house, has been used increasingly for fumigating drywood termites. Currently, methyl bromide costs much less, but rubber products containing more than a minimal concentration of sulfur must be removed, otherwise they will give rise to mercaptan odors that may persist for years. Foam rubber is the worst offender, and is present in some of the pads under carpets. Some kinds of leather are susceptible, and shoes, for example, must be removed from the building. The fumigator generally uses sulfuryl fluoride rather than having to remove foam-rubber carpet padding, etc., because of the labor involved in such removal.

On the basis of experiments with sulfuryl fluoride in over 100 buildings, Bess (1971) concluded that a fumigator can guarantee a fumigation against *Cryptotermes brevis,* provided gas readings are obtained so that the ounce-hour-exposure (OHE) factor can be determined and an adequate exposure assured. In about 90% of the buildings, a dosage of 1 lb per 1,000 cu ft (0.45 kg per 28 cu m) was sufficient to provide an adequate OHE, but in several buildings it was inadequate. Fumigations could be done with confidence only when gas-concentration readings were made. Under the climatic conditions of Honolulu, Hawaii, it was found that the OHE should be higher during the winter and spring than during the summer and fall, and should be increased under unusual circumstances, such as when extensive plywood paneling, cabinets, or moist timber were present. In some cases, additional fumigant may be introduced into parts of a building where it is especially needed.

One of the great inconveniences of fumigation is that the occupants of a building, including pets and plants, must stay out for a day or two. The gas must remain in the building for an extended period to be effective, and the building must be thoroughly aerated afterward.

If the entire building is not to be fumigated, and some furniture or other movable wooden objects are infested with drywood termites or powderpost beetles, the articles may be taken to a fumatorium for treatment (see figure 119, later in this chapter). Some pest control operators and, in California, the County Agricultural Commissioners' offices (County Departments of Agriculture) have fumatoria.

Residual Protection

The drill-and-treat or Woodtreat-brushing methods already discussed provide residual protection only for the limited areas treated, and fumigation provides none whatever. The average rates of reinfestation following treatment are shown in figure 98. One of the suggested solutions to this problem has been the pressure treatment of all framing lumber, instead of only the sills, the latter being the present practice. The increased cost of construction has been estimated to be only 2% (Coaton, 1948; Hunt, 1949). The necessity for treating cut ends, notches, and bored holes at the time of construction to prevent points for termite entry makes the pressure treatment of all framing lumber an impractical procedure, in the opinions of many people. Some investigators have advocated brushcoating the lumber (Wolcott, 1955) or spraying the rough framing of houses (Hunt, 1959) or the accessible wood of finished buildings with wood preservatives. Glue-line treatment of plywood panels has been found to be effective (Ebeling, 1968). None of these approaches to the problem have been followed in commercial practice, apparently.

In the late 1950's, highly sorptive clays, diatomites, and silica gels or aerogels,[11] dusted onto wooden blocks, were found to protect them indefinitely against drywood termites. Insects are protected from a lethal rate of desiccation by a very thin film of lipid, usually a hard wax, that covers their entire epicuticle. Certain sorptive powders with high specific surface and sufficiently large pores and with not too great a sorptivity for water can absorb sufficient epicuticular lipid to cause a lethal rate of water loss (Ebeling and Wagner, 1959*a, b;* Wagner and Ebeling, 1959; Ebeling, 1961, 1971).

Certain silica aerogels possessing a monomolecular layer of ammonium or magnesium fluosilicate have unique physical properties, including an electrostatic charge that results in efficient deposit on dusted surfaces and efficient "pickup" by insects crawling over them. To be insecticidal, enough silica aerogel must be deposited on a surface so that an insect crawling over it will pick up a lethal quantity on its lower body surfaces; for termites, a barely visible film suffices.

[11] Silica gels of lowest bulk density and of greatest porosity are called aerogels. The silica aerogel referred to here, known as Dri-die 67®, has a density of 4.5 lb per cu ft (2 kg per 28 cu dm) as packed. Dri-die 67 has a very low sorptivity for water, even at high humidities. It has advantageous physical properties for dusting confined spaces and voids, and possesses an insecticidal efficacy far superior to that of hundreds of other powders tested, including many silica gels and aerogels. In this book, wherever the term "silica aerogel" is used, it refers only to powders or dusts having physical characteristics and insecticidal efficacies approximately equivalent to those of Dri-die 67.

Application of Silica Aerogel During Construction. If silica aerogel is dusted onto wooden blocks, even in such small quantities that termites placed on them can survive for 2 weeks, the insects do not feed on the wood, although nontreated wood is vigorously attacked. Silica aerogel is generally used at 1 lb per 1,000 sq ft (0.45 kg per 93 sq m) in dusting attics for prevention of *Incisitermes minor*. The alates generally gain access to a house via the attic, and crawl about extensively before attempting to bore into the wood. In crawling about, they soon pick up a lethal dose of desiccating dust, and may die in as brief a period as 2 hours (Ebeling and Wagner, 1959*a, b*).

The dust is usually applied from the crawl hole that provides access to the attic with an electric blower that generates an airstream of high velocity or with a water-type fire-extinguisher, commercially available for the application of insecticide dusts (figures 100, 101; figures 32, 33, chapter 3). Fire-extinguishers are filled with dust to within a few inches of the top, and then pressurized to 100 psi. Because they are equipped with a Schraeder valve, they may be pressurized at a service station or by other sources of air or gas, such as small electric air compressors or tanks of nitrogen, provided the latter are equipped with the appropriate type of outlet. They can discharge a high volume of dust at high velocity and, like the electric blowers, can dust the average attic from the crawl hole. The unusually light weight of silica aerogel results in an even distribution of dust throughout the attic and into its extremities.

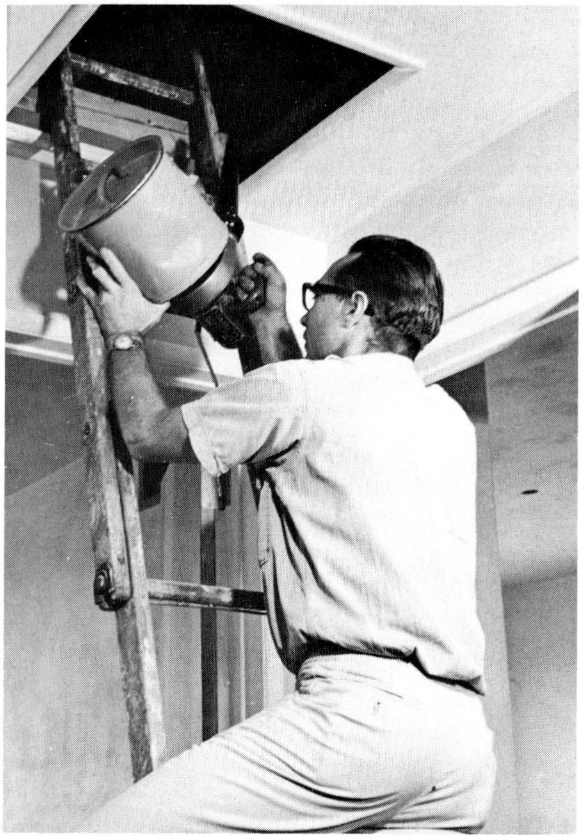

Fig. 100. Electric blower used to propel silica aerogel dust into an attic from the crawl hole.

Fig. 101. Applying silica aerogel dust into a wall void by means of a water-type fire-extinguisher.

The success of attic dusting for prevention of infestation by drywood termites during recent years, as confirmed by many termite operators, suggests the advisability of blowing the dust into attics at the time of construction. As stated before, records supplied by the California Structural Pest Control Board showed that out of 325 houses built in new tracts in California in 1956–57, 10.8% were infested with drywood termites 11 years later (Ebeling, 1968). With these colonies already established in the tracts, the rate of infestation can be expected to increase with age. Practically all houses in the older residential areas of southern California are, or have been, infested with drywood termites.

In addition to attic dusting, the protective film should be extended to include wall voids and other enclosed spaces. This is best done during construction, when holes can be made in the plaster lath about 4 ft (1.2 m) above the floor and between every 2 studs (Ebeling and Wagner, 1964). The holes may be made with a sharp pick or a half-in. (13-mm) drill. Three grams of silica aerogel are blown into each hole (figure 101). If the plaster lath is already perforated, there is no need to drill holes. Even with perforated lath, which is currently rarely used, the bulk of the dust settles on the interior of the void or falls to the floor plate. If insulation is present, we recommend that the dust be discharged downward, between the plaster lath and the paper that encloses the insulation, to be deposited principally on the floor plate. When "dry-wall" construction is used over conventional wood-stud framing, the sheets of "sheetrock" must be installed horizontally, which provides a horizontal juncture of sheets 4 ft (1.2 m) above the floor. Holes through which treatments are made are drilled at this juncture and between all studs.

In "dry-wall" construction with metal studs, such as used in concrete and steel construction for partition walls, the dust is blown into a hole drilled at the vertical juncture of 2 sheets of "sheetrock." Because there are no solid studs to prevent lateral movement of the dust, a single hole is enough for a horizontal distance of about 10 ft (3 m). Tape is routinely applied to cover the junctures of the sheets of dry wall; the holes are thus covered before the wall is painted.

For dusting houses, 3 grams of silica aerogel have been applied per interstud void. For dry wall on metal studs, 9 g were blown into every hole drilled (i.e., between the 2 sheets of sheetrock). This usually amounted to about 1.25 lb per 1,000 sq ft (0.57 kg per 93 sq m) of floor space. Combined with attic dusting (1 lb [0.45 kg] per 1,000 sq ft), the treatment should provide much protection against drywood termites for the life of the building. This treatment should also be effective for complete or partial prevention of other cryptobiotic pests, such as cockroaches, silverfish, fungus beetles (Lathridiidae), psocids, odorous house ants, Pharaoh ants, carpenter ants, bees, woodwasps, cluster flies, face flies, tropical rat mites, bird mites, and spiders.

Silica aerogel repels cockroaches, and although it serves the purpose of eliminating the dusted areas as harborages and breeding places for these insects, it is not so effective as boric acid dust for eliminating cockroaches that have entered a building without contacting dusted areas. Cockroaches will repeatedly enter protected areas dusted with boric acid, and these act as traps (see chapter 6). Boric acid is recommended in wall voids, etc., at least in kitchens, bathrooms, and food-storage areas where cockroaches can be expected to be the principal pests. A film of boric acid will also prevent infestation of wood by drywood termites (Reierson, 1966).

Application of Silica Aerogel After Construction. Most of the silica aerogel that has been used to date has been applied long after construction was complete, usually following a fumigation for drywood termites and for the purpose of preventing reinfestation. Silica aerogel has no effect on termites already in the wood—they must be controlled by fumigation or one of the localized treatments that have been described. Following treatment, most termite operators recommend attic dusting with silica aerogel to prevent reinfestation. If some of the drywood termites survive treatment or are undetected, which is quite likely if one of the localized treatments has been used, attic dusting is justified to limit the proliferation of surviving colonies. In the fall, a certain percentage of the termites (the alates) in existing colonies can fly to other parts of a building and start new colonies; this would be prevented on any wood surface with a film of silica aerogel. Also, termites that gain access to the attic after treatment, via vents, under shingles, or through cracks, would likewise be prevented from entering the dusted wood.

Some pest control operators are still using conventional insecticide dusts as they did before fluorinated silica aerogel became available. Tests have demonstrated that the toxicant soon loses its insecticidal efficacy, particularly in the high temperature of the attic, and after a few weeks the only remaining deterrent to termites is the diluent, which is relatively inefficient. The great superiority of the silica aerogel Dri-die 67 to 5% chlordane dust in its ability to knock down and to kill drywood termites (*Incisitermes minor*) in attic dusting, not only immediately after application, but particularly a month later, has been repeatedly demonstrated (Ebeling and Wagner, 1959*b*). Likewise, Minnick *et al.* (1972) found in laboratory tests that a Dri-die 67 deposit killed all powderpost termites (*Cryptotermes brevis*) in 1.5 hour, regardless of the exposure time, whereas a 20-minute contact with a 1% dieldrin spray residue resulted in complete mortality of the termites in 14 hours. The particular advantage of the silica aerogel is that it is inorganic and not subject to decomposition, and should protect the dusted wood against termite attack for the life of the building.

Control of Dampwood Termites

Prevention and control of dampwood termites consists of the same measures taken against subterranean termites and dry rot: construction designed to avoid contact of wood and soil, elimination of excessive moisture conditions, and treatment of the soil with an effective insecticide. (See under "Control of Subterranean Termites.")

DECAY FUNGI

Decay fungi grow on and in wood, and destroy the wood substances as they grow. Generally, conditions that are favorable for subterranean termites are also favorable for decay fungi, and vice versa. Temperature affects the activity of both termites and decay fungi and thereby affects the rapidity of their destructive action. Subterranean termites and decay fungi both require an abundance of moisture, but termites can obtain their moisture in galleries situated in moist soil and can then feed on dry wood, whereas most decay fungi cannot colonize and decay wood which is below the fiber saturation point, a moisture content of about 30% (Amburgey, 1972). However, some fungi can transport their own water from damp soil.

Dry Rot, *Poria incrassata*

This is the most destructive fungus. It becomes established in wood on or under damp soil, such as in foundation forms left alongside a foundation or under a porch slab. It can transport its own water from the moist soil to normally dry wood. Both coniferous and broadleaf species of wood are attacked, but the fungus is generally confined to structural timbers in buildings or to stored lumber. Poles, posts, ties, or slash in the forest are rarely attacked. The fungus occurs throughout most of the United States, but particularly in the southern and Pacific Coast states (Humphrey, 1923; Boyce, 1961).

Description. Water is conducted from damp ground through strands of mycelium comprising the rhizomorph, which is rootlike in appearance (figure 102, *bottom*). The rhizomorph varies in color from white to brown to brownish black, depending on its age, and varies from 3 mm to as much as 5 or 8 cm in thickness. The water-conducting strands can extend for 30 ft (9 m) or more across the surface of concrete, brick, or other impenetrable material and upward to the second or third floor of a building, spreading out into flattened veins embedded in sheets of mycelium. Over this entire area, the wood is moistened by the fungus and made susceptible to decay, even though it would otherwise be too dry.

The wood is moist while the fungus is alive, and in a cross section of infested timber, progress of the fungus is indicated by a dark stain (figure 102, *top*). The common name, "dry rot," presumably derives from the appearance of the wood after the fungus is dead. The severely decayed wood is brown, and breaks up into irregular chunks resulting from shrinkage cracks both across and in the direction of the wood fiber (figure 102, *bottom*). Sheets of white mycelium can often be seen in the cracks. Owing to the extraction of cellulose by the fungal hyphae, the wood becomes very light in weight and loses its

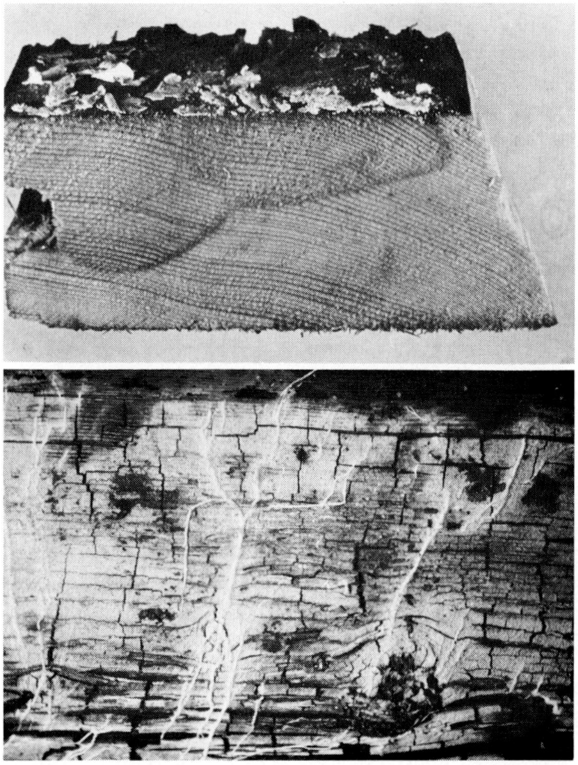

Fig. 102. Dry-rot fungus, *Poria incrassata*. *Top,* cross section of an infected timber, with dark stain showing limits of fungal infection; *bottom,* fungal rhizomorphs and typical checkered appearance of infected wood after it has become dry.

structural strength. The dry material has a distinct odor where the more succulent fruiting bodies occur, reminding one investigator of "drying slippery-elm bark" or of certain of the fleshy *Hydnums* (fungi) when they are dried out (Humphrey, 1923). The fruiting bodies can reactivate the infection if the wood becomes damp again.

Paperlike, fanshaped sheets of mycelium that appear on wood in damp places are often the first indications of decay. Masses of mycelium may protrude along the bottoms of baseboards or where timbers are joined. Leathery, crustlike sporophores are produced annually. They vary in color, from orange through olivaceous to deep purplish black, depending on their age and location, and all these colors may be represented in order from the margin to the center of a single fructification (Boyce, 1961).

Type of Wood Attacked and Rate of Destruction

It happens that the rough frame lumber most commonly attacked by dry-rot fungus is coniferous, but Humphrey (1923) pointed out that the fungus is an omnivorous saprophyte that can attack and destroy almost any of the commercial woods of the United States, even such reputedly highly durable ones as cedar, cypress, juniper, sequoia (redwood), catalpa, greenheart, black locust, sassafras, white oak, black walnut, and black cherry, all of which are severely attacked and sometimes completely destroyed.

Rapidity of Destruction

Another important characteristic of *Poria incrassata* is the rapidity with which it can destroy structures. A frame house in Florida suffered severe loss in 3 years, and in 5 years about 25% of the house was visibly affected. A 3-year-old house in Kentucky suffered severely from decay of subfloor timbers, flooring, and casings. In another house in Louisiana, much of the front wall and 750 sq ft (69 sq m) of flooring had to be replaced within 3 years after construction. In a store in that state, heart cypress paneling laid over a brick wall lasted only 6 months (Humphrey, 1923).

Other Decay Fungi

The same kind of decay as produced by *Poria incrassata* is produced by the teardrop fungus, *Merulius lacrymans,* which occasionally causes much damage to structural timbers in the northern part of the United States and in Canada. The 2 rots cannot be distinguished unless sporophores are present. In Europe, where *P. incrassata* does not occur, *M. lacrymans* is the "dry-rot fungus," and is very important.

A considerable amount of what has been said about *P. incrassata* pertains also to *M. lacrymans,* including the rhizomorph and its water-conducting capacity. The specific name *lacrymans,* which means "weeping," refers to the characteristic the fungus shows in active growth under damp conditions, when innumerable globules of water are seen, "which sparkle in the light of a torch like a large number of teardrops" (Hickin, 1963*c*).

The mycelium of the teardrop fungus grows rapidly, and is snowy white, but where the edge of an expanse of mycelium comes into contact with drier air or is exposed to light, it becomes bright yellow (Hickin, 1963*c*). When the mycelium of the closely related *Poria incrassata* grows through the cracks in the floor into a lighted room, the mycelium turns orange in color, and this color change has been likewise observed in partially lighted basements (Humphrey, 1923).

The prodigality of nature is indicated by the estimate that 1 sq m of a sporophore of *M. lacrymans* will produce over 50 million spores per minute over a period of many days, and that other wood-decay fungi have been found to be even more fruitful.

Many species of decay fungi attack buildings. In New York state, *Poria incrassata* is not known to occur, but in an extensive survey, 11 other species of *Poria* were found attacking wooden buildings, resulting in 22.8% of the incidence of decay. In the same family of fungi (Polyporaceae) *Lenzites saepiaria* made up 12.4% of the total and, in the family Hydnaceae, *Odontia spathulata* was found with about the same frequency (12.8%) (Silverborg, 1953). The latter species produces a white, stringy rot, unlike the brown, cubical type associated with the other common building-decay fungi. In the foregoing investigation, *Lenzites trabea* was found in 4.6% of the instances of decay. This fungus came prominently to the attention of entomologists in the early 1960's, when an extract from *L. trabea*-rotted wood was found to be an attractant to termites (Esenther *et al.,* 1961). It was later discovered that this extract induced trail-following by the eastern subterranean termite, *Reticulitermes flavipes,* similar to that induced by a trail-marking pheromone secreted by the sternal gland of this insect (Smythe and Coppel, 1966*a*). Trail-marking substances isolated and purified from *R. virginicus* and from wood infested with *L. trabea* were found to be identical (Matsumura *et al.,* 1969).

Molds and Stains

Wood may be attacked by molds, which are superficial fungi, generally green, yellowish green, red, or black, depending on the species. Molds do not damage wood and can be easily scraped off, but they indicate that conditions are right for the growth of decay-producing fungi and that corrective measures should be taken.

Wood is also affected by sap-stain or blue-stain fungi. They penetrate more deeply than molds, but do not significantly weaken the wood. The fungi are usually bluish, and cannot be scraped off wood like molds. They do no damage unless they occur where they may be considered to be unsightly, but like the molds, they indicate that conditions are favorable for the growth of the more destructive decay fungi (Anonymous, 1963).

Control of Decay Fungi

The same measures and precautions during and after construction that decrease the susceptibility of subarea structural timbers to subterranean termites also pertain to the control of decay fungi. They include not only the measures practiced to eliminate excessive moisture, but also the use of pressure-treated lumber and the removal of all wood scraps or other cellulose debris that might provide food for fungi or termites.

All that is required for the control of a water-conducting decay fungus such as *Poria incrassata* is to break the connection between the source of water and the infected structural wood member. The source of water should be eliminated, if possible, and the water-conducting rhizomorphs should be removed. The wood will then dry out and the fungus will die. Replacement of the infected wood member is required only if the wood is too weak to support its load. Verrall (1968) advised against dependence on surface applications of wood preservatives, or on such supposedly decay-resistant woods as redwood, as controls against fungi that are "water-conductors."

Wood used in construction should not have a moisture content of more than 20%. Depending on the atmospheric conditions in the area in which it is dried, air-dried wood may have a moisture content between 6 and 24%, whereas kiln-dried wood has a content of from 6 to 12%. Wood that is not pressure-treated with a wood preservative should not be allowed to contact the soil. The caulking around windows, doors, bathtubs, and sinks should be maintained to prevent accumulation of water in walls. Paint decreases penetration of water into wood except at joints between the wood members (Amburgey, 1972).

In many homes, condensation of moisture within insulated walls and floors favors wood decay. Warm, moist air passing into walls comes in contact with wood members with temperatures below the dew point of the air, resulting in deposit of excess water. Measures that may be taken to reduce or prevent condensation vary in different geographical regions, and are discussed in chapter 10, under "Sources of Moisture in Buildings."

POWDERPOST BEETLES, FALSE POWDERPOST BEETLES, AND DEATHWATCH BEETLES

The insects discussed in the preceding part of this chapter do not make their presence known by exit or entrance holes at the surface of infested wood. One must usually break into the wood and uncover their galleries in order to locate the infested areas. The insects discussed in the following pages make their presence known by exit holes, sometimes so numerous that the surface of the wood looks as if it had been riddled by a charge of birdshot. The most important group of insects that attack wood in this way includes 3 closely related families of beetles (Lyctidae, Bostrichidae, and Anobiidae), the larvae of which reduce wood to a mass of powdery or pelleted frass. They attack dead wood almost exclusively, especially dried and cured lumber, and several generations can reinfest the same piece of wood until only the outer shell remains, riddled with exit holes. Beneath the surface of the infested wood are frass-filled galleries or tunnels, usually coursing along the grain of the wood.

The damage caused by these 3 families of beetles can be identified as follows: The true powderpost beetles (Lyctidae) loosely fill their galleries with very fine powder, similar in appearance to face powder; the false powderpost beetles (Bostrichidae) tightly pack their galleries with a coarser boring dust, often containing small wood fragments; and the deathwatch beetles (Anobiidae) fill their galleries with small pellets. The frass in the galleries of the bostrichids and anobiids is not only coarser than that of the lyctids, but tends to stick together. Although all these beetles have sometimes been collectively referred to as "powderpost beetles," in a more definitive sense the term should apply only to the lyctids, for their larvae are the only ones that produce a fine, powderlike dust in their galleries.

There is a tendency for lyctids to attack new hardwood, anobiids old hardwood and softwood, and bostrichids new hardwood and softwood, but frequent exceptions to this generalization have been noted. However, often the first step in determining the family of beetles involved is to determine the type of wood being attacked.

Probing or tapping the infested area of the wood will usually reveal one of the types of frass described above. In the United States, the fine powder of the lyctid beetles is most often encountered. In the wood structure of the building, beetle infestations occur most often in hardwood cabinets and flooring. On the other hand, the presence of subterranean termites usually first comes to the attention of the homeowner when he sees blisters or dark stains in flooring or other wood or where the paper-thin surface covering the infested wood is crushed, as when furniture is moved about.

When the structural damage is caused by termites, there are no exit holes in the vicinity, and their irregular tunnels are not filled with powder. Drywood termite tunnels may hold many pellets, but the pellets are larger than those produced by deathwatch beetles (anobiids), are characteristically sculptured, and are blunt at each end (figure 83, this chapter). The presence of drywood termites usually comes to the attention of the occupants of a building when the pellets are pushed out of occasional "kickout" holes in wood members, such as the frames of doors or windows or exposed timbers.

There can be no confusion between the typical appearance of termite workers and the wormlike larvae of beetles. The winged reproductives of termites are often seen during their brief, fluttering flights. After they have landed and are crawling about, they are often mistaken for "winged ants."

On the other hand, all of the adult lyctid and anobiid beetles are strong fliers and are attracted to light. They may, therefore, often be discovered at or near windows when an infestation occurs in a room. Even experienced entomologists often mistake the furniture beetle, *Anobium punctatum,* for a fly, since in flight it closely resembles the latter, even to the extent of flying around light bulbs (Hickin, 1963*a*).

Powderpost Beetles (Lyctidae)

Lyctids are small beetles ranging from 2 to 7.5 mm in length. They are reddish, various shades of brown, or black; have a prominent head not covered by the prothorax; short, 11-segmented antennae, each with a 2-segmented terminal club; and tibiae with distinct spurs (figure 103).

To distinguish between species of Lyctidae, a description of certain features of the elytra is generally utilized. The *interspace* (figure 103) is the surface between striae; a *stria* is a longitudinally depressed line or furrow, frequently punctured, extending from the base to the apex of the elytra; and a *carinula* (diminutive of *carina*) is a keel-like ridge.

Lyctids are cosmopolitan, each geographical region having indigenous as well as established introduced species. Of the 66 known species, 10 occur in the continental United States and 6 of these are common and important as pests. In the United States, species of *Lyctus* and *Trogoxylon* are widely distributed. Species of *Minthea* and *Lyctoxylon* are often intercepted in wood products in maritime ports of entry. Representative species of these 4 genera are shown in figure 104. In this country, lyctids are second only to termites in their destructiveness to wood and wood products, but confine their attacks to large-pored hardwoods such as oak, ash, hickory, mahogany, and bamboo. Powderpost beetles may be found

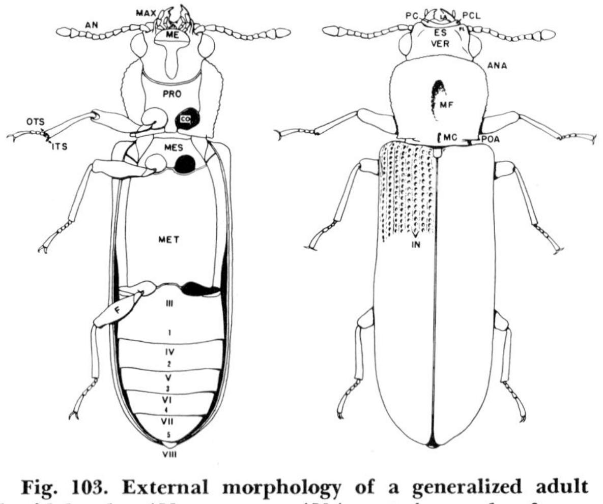

Fig. 103. External morphology of a generalized adult lyctid beetle. *AN*, antenna; *ANA*, anterior angle of pronotum; *CO*, coxal cavity; *ES*, epicranial suture; *F*, femur; *FL*, frontal lobe; *IN*, interspace; *ITS*, inner tibial spur; *LA*, labrum; *MAX*, maxilla; *MC*, median canaliculation; *ME*, mentum; *MES*, mesosternum; *MET*, metasternum; *MF*, median fovea; *OTS*, outer tibial spur; *PC*, postclypeus; *PCL*, postclypeal lobe; *POA*, posterior angle of pronotum; *PRO*, prosternum; *VER*, vertex. (From Gerberg, 1957*a*.)

in hardwood flooring, hardwood timbers, plywood, and wood articles such as crating, furniture, antiques, tool handles, spokes of wheels, gunstocks, etc., causing millions of dollars' worth of damage every year.

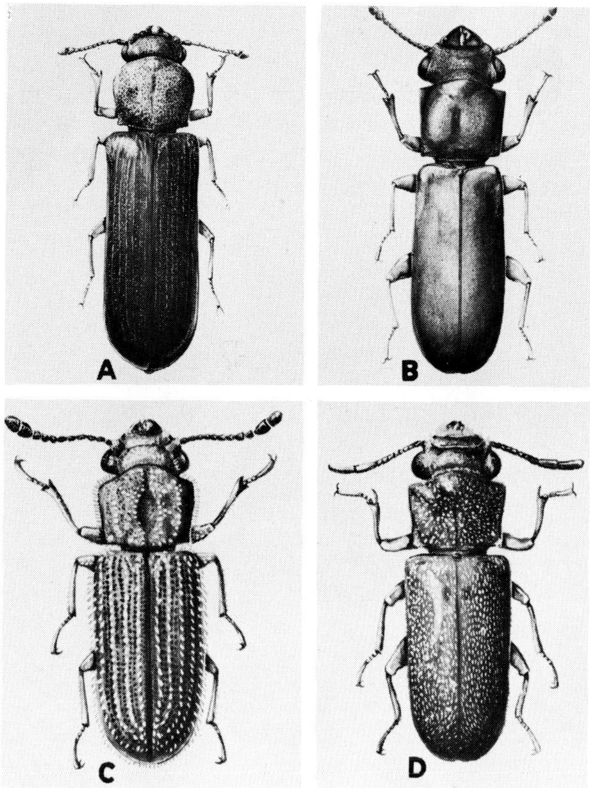

Fig. 104. Powderpost beetles. *A, Lyctus planicollis; B, Trogoxylon parallelopipedum; C, Minthea rugicollis; D, Lyctoxylon japonum.* (From Gerberg, 1957*a*.)

The damage the beetles cause consists of reducing sound wood to fine powder, and is usually not evident until the emergence holes of the adults are seen (Gerberg, 1957*b*).

Lyctids are generally brought into buildings in wood that contains their eggs or larvae. Lumber should be examined for signs of infestation. Once in a home, hardwood items are usually finished with paint, varnish, shellac, sealer, or wax, which treatments prevent oviposition by adult beetles.

If the adults' emergence holes are found in such finished items, it may usually be assumed that the infestation began before the finish was applied. From an economic standpoint, the 2 most important lyctid species in the United States are *Lyctus planicollis* and *Trogoxylon parallelopipedum.*

Southern Lyctus Beetle, *Lyctus planicollis* LeConte

This is a predominantly Nearctic species (plate II, 1; figures 104, 105, 106), found throughout the United States and in parts of Mexico. It is about 5 mm long, and is black when fully chitinized. The males are much smaller than the females. The prothorax usually has a broad, shallow, median depression. The interspace is composed of 2 regular series of elongate punctations, and the striae are separated by rows of fine, long hairs. The tenth antennal segment of *L. planicollis* is wider than long, whereas in the closely related *L. cavicollis* it is not wider (Gerberg, 1957*a*).

Western Lyctus Beetle, *Lyctus cavicollis* LeConte

This species is widespread throughout the United States. It varies from 2.5 to 5 mm, but usually is not over 3 mm in length. It is slender,

Fig. 105. Southern lyctus beetle, *Lyctus planicollis.* Adult, and pupa in chamber.

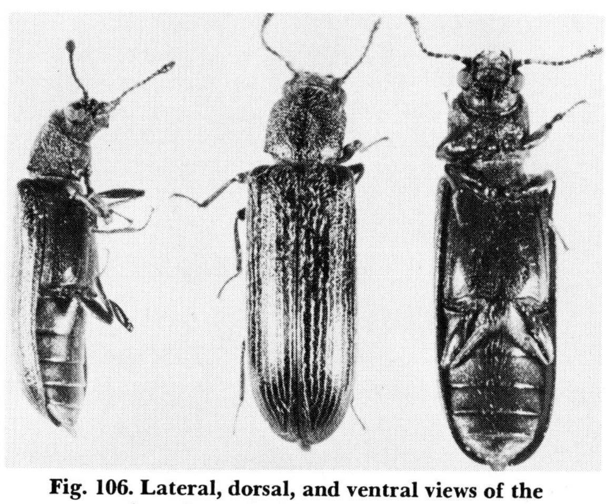

Fig. 106. Lateral, dorsal, and ventral views of the southern lyctus beetle, *Lyctus planicollis.*

and rusty brown. The anterior portion of the prothorax is narrower than the combined bases of the elytra. As in *L. planicollis,* the striae punctures are in a double row. In California, *L. cavicollis* mines the seasoned wood of eucalyptus, hickory, oak, and orange, and is also destructive to any products made of these woods. It often infests oak firewood, and can spread to hardwood flooring, paneling, and furniture.

Brown Lyctus Beetle, *Lyctus brunneus* (Stephens)

This is a cosmopolitan species that may have been originally Neotropical, but has shown the ability to adapt itself readily to other zoogeographical regions. Its length can be 2.2 to 7 mm, but is usually 4 to 5 mm. It is slender, brown, with black head and prothorax. The interspaces of the elytra have a single row of punctations, and not all interspaces are separated by longitudinal carinulae.

European Lyctus Beetle, *Lyctus linearis* (Goeze)

This is another cosmopolitan species that resembles *L. cavicollis* in size and color, but can be distinguished by having single rather than double rows of punctations of the elytral striae, and by its prominent frontal lobes. It is commonly found in commercial products of ash, wild cherry, hickory, oak, locust, poplar, and walnut, and in California it has been reared from dead grape canes and limbs of oak, orange, and sycamore (planetree) (Essig, 1926). It is common in the eastern United States, but not so much so in the western states (Linsley, 1943*b*).

Velvety Powderpost Beetle, *Trogoxylon parallelopipedum* (Melsheimer)

The velvety powderpost beetle (figure 104, *B*) is Nearctic, occurring throughout the United States. It is 2.5 to 4.3 mm long, and rusty red-brown to black. The elytra are about twice as long as wide, and have a fine, golden pubescence. It attacks seasoned hardwoods, but has also been found boring into the trunks of living honeylocust (*Gleditsia triacanthos*) in Indiana (Gerberg, 1957*a*).

Minthea rugicollis (Walker)

The genus *Minthea* is characterized by the erect, flattened hairs along the denticulate lateral margins of the pronotum (figure 104, *C*). The elytra have 6 or 7 rows of erect, flattened, whitish hairs, usually separated by a row of fine, reclining hairs. The adults are 2 to 3 mm long. The terminal segments of the antennae are slightly longer than the preceding ones, and the lateral margins of the pronotum bear about 19 flattened hairs. This species is found throughout tropical regions. It has been intercepted in quarantine in the United States in wooden articles, and even in such products as ivory nuts (*Phytelephas macrocarpa;* Palmaceae) and dried roots of *Derris elliptica* (Leguminosae) (Gerberg, 1957*a, b*).

Lyctoxylon japonum Reitter

This genus can be readily distinguished by the elongate antennal club: both segments are longer than broad, and the terminal segment is narrower than the preceding one (figure 104, *D*). The beetles are 1.5 to 2 mm long. The species is apparently distributed throughout the Orient, but may have become established in Panama. It is often intercepted in quarantine in the United States, usually in bamboo (Gerberg, 1957*a*).

The Life Cycle of Lyctid Beetles

Adult lyctids mate soon after they emerge from infested wood. Oviposition begins 2 or 3 days later, and continues for a week or two, most of the eggs being laid within the first 7 or 8 days. The female has a long, flexible ovipositor that she inserts deeply into the pores of hardwoods, laying 1 or several eggs in each pore. Investigators generally report that the female places her eggs only in the spring in wood vessels or pores. These are exposed when the wood is cut, or may be opened by the beetle herself. Cymorek (1965) observed that *Lyctus brunneus* and *Lyctus africanus* Lesne did not restrict their oviposition to the vessels of the wood, but also oviposited in all kinds of cracks and crevices. He also found that both species could oviposit and develop in corn and wheat grains in the laboratory.

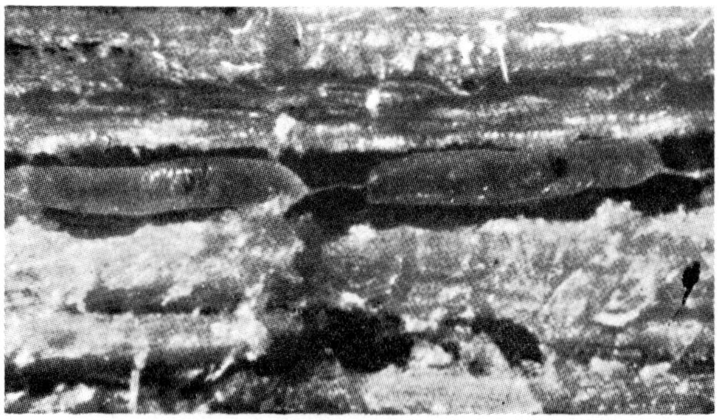

Fig. 107. Two eggs of *Lyctus planicollis* in place. (Photo by J. G. Pratt, from Christian, 1940*b*.)

From 15 to 20 eggs per female have been reported for *L. linearis* and 17 to 21 for *L. brunneus*. The lyctid egg is about 1 mm long, translucent white, and cylindrical, with rounded ends, and bears a threadlike process at the anterior end

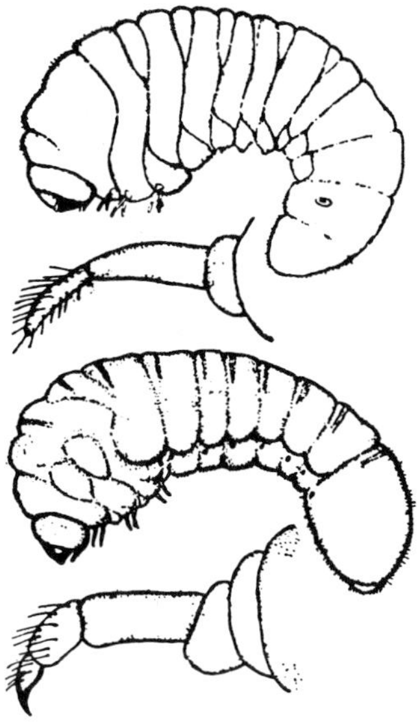

Fig. 108. Differences between lyctid and anobiid larvae. *Top*, larva and leg of *Lyctus* sp.; *bottom*, larva and leg of *Anobium punctatum*. (Credit to Hickin, 1963*a*.)

(figure 107). Eggs are never deposited in polished, waxed, varnished, or painted surfaces (Gerberg, 1957*a*).

The incubation period can range from 1 to 3 weeks, depending on temperature. The emerging larva is white, straight-bodied, and bears a pair of small spines at the rear. After the first molt, the larva assumes a curved form. The young larva usually tunnels with the grain of the wood, but later takes an irregular course, sometimes intersecting its earlier track or the tunnels of other larvae. The tunnels are packed with a fine, powderlike dust. The larvae do not penetrate to the surface but, like termites, leave a thin, unbroken surface layer. Outdoors, the larvae grow chiefly during spring and summer, but in heated rooms they can develop continuously.

A mature larva is usually less than 5 mm long, curved, and enlarged at the thorax. Its antennae are 3-segmented and have an accessory appendage. The abdomen bears 8 spiracles, and the most posterior pair is 6 times larger than the others (figure 108). The legs are distinct, 3-segmented, and unclawed. In contrast to lyctid larvae, the larvae of anobiids do not have an enlarged posterior spiracle and they have 5-segmented legs, with the terminal segment clawed (figure 108). The mature lyctid larva bores to a point near the surface of the wood and builds a pupal chamber (figure 105). The pupal period lasts from 12 days to a month. After metamorphosis, the adult chews its way to the surface, forms an exit hole 2 or 3 mm in diameter, and generally pushes some of the fine dust out of the hole when emerging (figure 109). Small piles of the dust are common near new emergence holes. Although the larva confines its burrowing to sapwood, if the adult has no other way of emerging, it can bore its way through heartwood, softwood, asbestos, lead sheeting, or plaster (Tooke, 1949).

Adult lyctids conceal themselves in cracks and holes in the wood during the day, but they become active at night, fly readily, and are positively phototactic (Gerberg, 1957*b*). Indoors, they may be seen crawling on floors, windowsills, furniture, and other surfaces.

The life cycle, from egg-laying to emergence of adults, ordinarily requires 9 to 12 months, but under exceptionally favorable conditions of high temperature and high starch content of wood, that period may be reduced to only 6 or 7 months.

Fig. 109. Southern lyctus beetle, *Lyctus planicollis*. *Top*, adult emerging from wood; *bottom*, an emergence hole and damage to ash molding.

Two generations a year are not uncommon in the southern United States, but under adverse conditions of temperature and nutrients, the life cycle may be prolonged to 2.5 to 4 years, or longer (Gerberg, 1957*a, b*). Development of *Lyctus brunneus* was greatly prolonged in compressed oak sapwood when compared with the same wood before compression (Cymorek, 1967). There can be a great difference in the length of the life cycle of different species; the average for *Trogoxylon parallelopipedum,* for example, being 3 to 4 months compared with 9 to 12 months for *Lyctus planicollis* (Gerberg, 1957*a*).

Wright (1960) investigated the life cycle of *L. planicollis* in a laboratory under uniformly high temperature (31° ± 3° C av. 88° F) and at 85 ± 5% relative humidity. He reared the beetles on red-oak wood chips or wood flour. The mean incubation period for eggs was 8.2 days. The mean length of the larval stage was not determined, but some larvae remained alive as long as 42 days. A prepupal stage, not recorded earlier, lasted from 2 to 5 days. The mean pupal period was 8.1 days. The mean length of adult males was 3.5 mm and of adult females, 3.7 mm. The 2 sexes occurred in approximately equal numbers. The mean adult life of males was 33.1 days and of females, 35.1 days. Adult beetles lived about 2 weeks after emergence from the wood. The number of eggs deposited per female ranged from 18 to 122, with a mean of 51.4.

Factors Affecting Lyctid Infestation

Starch Content of Wood. Lyctid beetles primarily infest the sapwood of hardwoods. Ash, hickory, maple, and oak are preferred, but many other hardwoods are sometimes damaged. Bamboo, which is a grass, not a hardwood, is readily attacked in the Orient, frequently by *Lyctus brunneus.* Their chief source of food is starch, and whatever sugar and protein may be present; they do not digest cell walls, for the larvae cannot digest cellulose, hemicellulose, lignin, or pentosans—these pass through the intestine undigested. Lyctids will not oviposit in sapwood with a starch content of less than 3%, and the greater the concentration of starch, the better they thrive.

The starch content of lumber depends on the tree species, the season the tree is cut, and the method by which the lumber is dried. Lyctids prefer recently dried rather than old wood because the starch content of lumber decreases as it is converted to lignin in the long process of natural seasoning.

Water Content of Wood. A definite range of water content is also necessary. Lyctids can live in wood with a water content of between 8 and 32%. Since the water content of green wood is commonly about 50% and can be even higher, it is obvious why attack by powderpost beetles is generally confined to partially or wholly seasoned wood (Christian, 1940*a, b;* 1941; Tooke, 1949). The moisture content of seasoned wood in the United States is said to average between 12 and 15%. The greatest lyctid beetle activity is found in wood with a moisture content of from 10 to 20% (NPCA, 1961).

Pore Diameter in Wood. The diameters of the pores or vessels in the wood can be a limiting factor, for they must be large enough to allow insertion of the ovipositor, otherwise oviposition cannot take place. The average size of the ovipositor of *Lyctus brunneus* is 0.078 mm; of *L. linearis,* 0.083 mm; and of *L. planicollis,* 0.076 mm (Parkin, 1934). Hardwoods with pores or vessels with diameters greater than these dimensions are subject to infestation. Softwoods or conifers do not have pores, usually have a low starch content, and generally are immune to infestation.

Natural Enemies. Natural enemies of lyctus beetles are common in infested lumber in mill-yards (Snyder, 1926; Christian, 1941). Among 6 species taken from an infested piece of wood, Christian found 2 predaceous clerids, *Tarsostenus univittatus* (Rossi) and *Monophylla terminata* (Say), as well as 4 parasitic Hymenoptera, *Monolexis fuscicornis* Foerster [=*M. lycti* (Cresson)], *Sclerodermus macrogaster* Ashmead, *Heterospilus* n. sp., and *Ecphylus* n. sp.

False Powderpost Beetles (Bostrichidae)

The bostrichids are small to large beetles, usually elongate and cylindrical, and are brown, reddish brown, or black. Unlike the lyctids, in most bostrichid species the head is deflexed and not visible from above, being hidden from view beneath a large thorax that gives the beetle a humpbacked appearance. The thorax is noticeably roughened in most species. The antennae are short in both families, but the bostrichids have 3 or 4 enlarged, sawtoothed, terminal segments, compared with 2 for the lyctids. The first abdominal segment of bostrichids is equal in length to each of the others, whereas in the lyctids it is longer than the others (figure 103). The tibiae have distinct spurs.

The Lyctidae and Anobiidae do not contain large-sized species, whereas the bostrichids have a considerable number of them. The most striking example is the palm borer, *Dinapate wrightii* Horn, that is 4 to 5.5 cm long and infests native fan palms in California deserts and Baja California.

The bostrichids differ from the other 2 families mentioned in that the *adults* bore into the wood, preparing "egg tunnels" instead of ovipositing in surface cracks or pores. The female oviposits in pores leading from the tunnels. The larvae also bore tunnels. Thus, the bostrichids' tunnels vary greatly in size and shape. Like the lyctids, bostrichids can continue to breed in a piece of wood as long as it affords adequate sustenance. However, some species attack and breed in both hardwoods and softwoods.

The bostrichids are most abundant in the tropics, and are not so important as the lyctids and anobiids in temperate regions. Several species mine dead, seasoned, or cured hardwoods or hardwood furniture. Some species of *Amphicerus* and *Polycaon* burrow into the small twigs of living fruit trees, usually at the crotch or bud axil, but are usually not abundant enough to cause serious damage.

Black Polycaon, *Polycaon stoutii* (LeConte)

This is a cylindrical, coal-black beetle, 11 to 22 mm long and rather atypical of the bostrichids (figure 110). Unlike most of this family, the head

Fig. 110. Black polycaon, *Polycaon stoutii*.

is prominent and exserted. The pronotum is oval, not hoodlike, and has no tubercles. The punctures on the head and pronotum are coarse, and on the elytra they are fine. This species occurs in the Pacific Coast states and in Arizona, where it is the largest member of the genus. It attacks almost any of the softwoods and several of the hardwoods. As a pest, it is most common in 3-ply panel used for making desks and similar articles. The beetles fly at night, and can attack stacked plywood. The larvae burrow in the inner portion of the panel in a softwood such as basswood, where they require 1 or more years to complete their development. Their presence is not discovered until the adults emerge, burrowing through the outer layers of hardwood such as oak or mahogany, leaving emergence holes about 7 mm in diameter. The beetle does not infest finished products after manufacture, and its emergence holes will indicate that a portion of the wood was infested before manufacture (Doane *et al.*, 1936; Linsley, 1943*b*).

Although *Polycaon stoutii* is generally reported in the literature as occurring in furniture, particularly veneer furniture, Mallis (1969) stated that it would also attack cured hardwoods in the lumberyard and buildings in mountainous areas. Pest control operators have captured and obtained identification of adults emerging from hardwood flooring over a subfloor of Douglas fir, the beetles having originated in the fir.

Leadcable Borer, *Scobicia declivis* (LeConte)

Leadcable borers are common along the Pacific Coast, particularly in California and southern Oregon, occurring at all elevations up to 5,300 ft (1,600 m). Outdoors, they normally infest dead and seasoned oak (figure 111), but they will also infest acacia, eucalyptus, maple, California laurel, elm, chestnut, and other hardwoods. They can reinfest the wood from which they emerge. They occasionally attack newly painted houses, apparently being attracted to them, and have been known to infest hardwood paneling and floors. They are attracted to corks in bottles containing alcoholic beverages, or to hard liquors or wine in oak casks, and can damage the casks severely.

Description. The leadcable borer (figures 111, 112) is 5 to 6 mm long, cylindrical, brown or reddish brown to black (usually black), with reddish mouthparts, legs, and antennae. The antennae are 8-segmented, with the 3 terminal segments much enlarged, and most of the head is concealed from above by the hoodlike pronotum, as is generally the case with the bostrichids. The prothorax has many tubercles on the anterior half, and the elytra have deep, linear punctures.

Life Cycle. The life cycle of the leadcable borer was investigated in great detail by Burke *et al.* (1922). The female bores an egg gallery, begin-

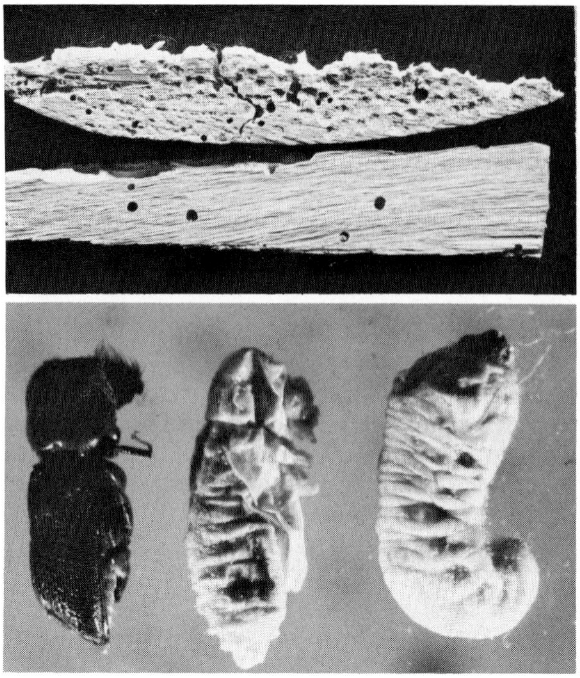

Fig. 111. Leadcable borer, *Scobicia declivis. Bottom,* adult, pupa, and larva; *top,* damage to seasoned oak.

ning at any place where leverage can be obtained, such as a knot, scar, bark scale, crevice, or between 2 pieces of wood. After the gallery reaches a depth of some 8 mm, it turns abruptly and runs parallel to the surface and across the grain of the wood for approximately 6 cm. Mating usually takes place when the tunnel has reached about the length of the female. The eggs are deposited in pores at a distance of 1 to 2.5 mm from the gallery—a single egg in each pore. The female apparently stops at any suitable pore to oviposit as she goes back and forth to throw out the borings. Some eggs are found so close to the gallery entrance that it appears the female can also back down into the gallery to oviposit.

The dull-white or cream-colored egg is very elongate, averaging 2.1 mm in length, including the stem, and 0.14 mm in width. It hatches in about 3 weeks. The larva mines parallel with the grain of the wood, packing its borings behind it, and becoming full grown (figure 112) after 6 molts and in a period of about 9 months. By the time it is ready to pupate, it has mined out a gallery 50 to 60 cm long. No distinct pupal cell is made. The prepupal larval skin gradually becomes transparent, and the pupa (figure 111) can be seen within for about 6 days before the skin is shed. The pupal stage lasts about 14 days, and the adult remains in the wood for a month to 6 weeks, hardening and maturing and becoming darker brown and finally black. The adult then eats its way out of the outer wood and bark. After emerging, the males live from 10 to 18 days and the females 30 to 44 days, provided they have access to wood; otherwise, both sexes die in 5 to 7 days. There is 1 generation a year. Adult beetles may be seen from the middle of April to the last of September, but they are most common in July and August.

Types of Damage. The usual damage to wood is a complete riddling caused by the egg galleries of the adults and the larval tunnels packed with a meal-like frass, leaving nothing but the bark and enough solid wood to hold the tunnels together (Burke *et al.,* 1922).

The leadcable borer derived its name from its ability to bore through the lead sheathing of aerial telephone cables, making holes 2.5 mm wide. Much damage and great losses occur when moisture enters the cables, causing short circuits. According to Linsley (1943*b*), the adults appear to be greatly stimulated by heat, and sometimes bore into lead or asphalt roofing materials because of high temperatures. Such attacks, as well

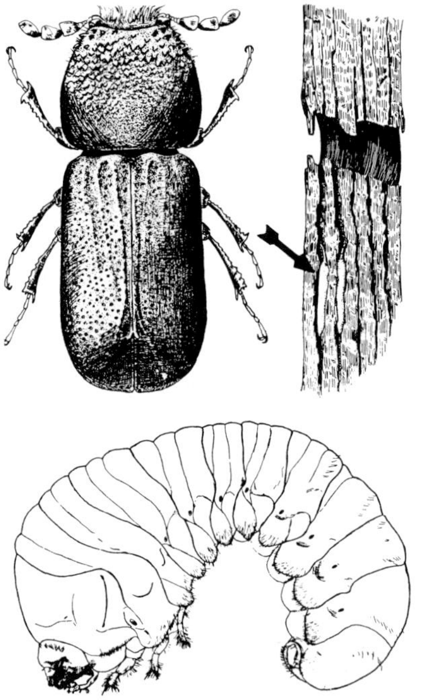

Fig. 112. Leadcable borer, *Scobicia declivis.* Adult; eggs (*arrow*) in pores, about 2.5 mm from the egg gallery; and full-grown larva. (From Burke *et al.,* 1922.)

as attacks on lead cables, can be precipitated by forest fires, particularly in oak, where the beetles may occur in large numbers.

Various species of beetles that bore into lead and other metals include other bostrichids, as well as certain anobiids, lyctids, ptinids, buprestids, cerambycids, anthribids, bruchids, curculionids, dermestids, and tenebrionids. Insects of other orders that can do the same are termites; woodwasps (Siricidae); certain carpenter moths (Cossidae) (Burke *et al.*, 1922; Chamberlin, 1949); and some species of carpenter bees (*Xylocopa*) (H. Scott, 1932; Middlekauff, 1960).

In May and June, 1972, many houses in 2 new residential tracts in Fresno, California, were infested with leadcable borers. Adult beetles emerged from the acoustic layers of the ceilings and from the wall plaster—as many as 50 beetles in 1 house—leaving holes about 2 mm in diameter. The probable sources of the beetles were piles of heavily infested fig prunings in an orchard situated between the 2 tracts. *Scobicia declivis* was reared from portions of dried fig branches taken from these piles. Also reared was another bostrichid, *Amphicerus cornutus* (Pallas), similar to *S. declivis* in superficial appearance but much larger (11 to 13 mm long). The leadcable borer had not been recorded before from the fig. The beetles probably entered the attics via the attic vents or by crawling beneath shingles. With none of their favored wood species to attack, they tried to escape from the attics by crawling down through holes made in the ceiling plates for electrical conduits. The conduits were in the corners of the rooms, and the beetles' emergence holes in the plaster were likewise always found in these corners. The infestations ceased after the piles of fig prunings were burned (Ebeling and Reierson, 1973*b*).

Other Bostrichids

Probably the most common bostrichid in the eastern United States is the redshouldered shothole borer, *Xylobiops basilaris* (Say), which can attack practically any freshly cut and partially seasoned hardwood, and has been found attacking partially seasoned pecan lumber. It principally attacks the wood of hickory and persimmon, although it has been found in elm and other species. It has been found in lumber shipped to England (Craighead, 1950). St. George (1970) considers *X. basilaris* to be the only powderpost beetle in the eastern states that seriously damages wood products such as log cabins, rustic bridges, or wood used in the manufacture of rustic furniture, rustic fences, arbors, and wooden tool handles. Sometimes this beetle causes a 50% loss in these wood products. In the usual manner of bostrichid beetles, the adult bores in from the outside and down through the bark of freshly cut wood.

Fig. 113. Bamboo borer, *Dinoderus minutus*.

It mines a tunnel in the wood for a short distance, following the annual rings, and this serves as an egg tunnel. Most of the larval boring is in the sapwood, which is ultimately destroyed (Chamberlin, 1949; St. George, 1970).

In the Midwest, *Scobicia bidentata* (Horn) is commonly found in the freshly cut wood and lumber of many hardwood species.

The bamboo borer, *Dinoderus minutus* (F.) (figure 113), is a cylindrical brown beetle, 3 to 4 mm long, which is indigenous to Asia but is now cosmopolitan in distribution. It attacks bamboo furniture, poles, and ornaments, as well as drugs, spices, stored grain, and flour. Powder sifting from cracks in bamboo and adult beetles found in buildings after emerging from bamboo articles are usually the first indications of infestation. It is particularly destructive in the West Indies (Plank, 1948).

Heterobostrychus aequalis (Waterhouse) is occasionally found in wood products imported from the Orient. The specimen shown in figure 114 was taken from imported mahogany furniture. This appears to be the most common of the larger bostrichids in India and parts of southeast Asia, where it bores into packing cases, boxes, plywood, furniture, and lumber, reducing the wood to a powder to a depth of 2 or 3 in. (5 or 7.5 cm) in severe infestations. In January, 1967, *H. aequalis* was found in oak lumber in Fort Lauderdale, Florida, and infestations were later found in Florida in lumberyards in the Fort Lauderdale and

Fig. 114. A bostrichid (*Heterobostrychus aequalis*) occasionally found in wood products from the Orient.

Miami areas and in an oak hammock near Hallandale (Woodruff, 1967). The species is now believed to be well established in Florida in scattered and localized areas (R. E. Woodruff, correspondence).

Heterobostrychus aequalis is 6 to 13 mm long, and reddish brown to brownish black. The anterior half of the strongly convex pronotum has broad, toothlike, marginal projections. The elytra are nearly tubular in shape back to the somewhat excavated apical declivity, which in the males has 2 incurved, hooklike projections. The surface of the elytra has deep punctures arranged in fairly distinct rows (Woodruff, 1967).

Deathwatch Beetles (Anobiidae)

There are about 260 species of anobiids in the United States, the majority of which feed on dead wood. However, 2 species, the cigarette beetle (*Lasioderma serricorne*) and the drugstore beetle (*Stegobium paniceum*) are important pests of stored food products. Anobiids are widely distributed, and in some parts of the world, for example in Europe and New Zealand, they are much more important than in the United States. They attack mostly sapwood, but can extend their attacks into heartwood, and they attack both hardwoods and softwoods.

Anobiids range from 1.1 to 8 mm in length, and are highly variable in body form, ranging from narrow and elongate to nearly circular from the dorsal aspect. When viewed from above, the hoodlike or "bellshaped" pronotum usually conceals the head, as in the case of the Bostrichidae and Ptinidae, but is unlike that of the true powder-post beetles (Lyctidae), in which the head is not hidden when viewed from above. The last 3 segments of the antennae are usually lengthened and broadened, or simply lengthened, but in some species the antennae are either serrate (saw-toothed) or pectinate (combshaped). (The antennal clubs of lyctids have only 2 segments.) The tibiae lack the spurs found on the tibiae of the lyctids and bostrichids. The tarsi are 5-segmented, with divaricate (widely separated) claws.

Illustrations of some of the anobiids that have been recorded as pests in the United States are shown in figure 115.

Furniture Beetle, *Anobium punctatum* (De Geer)

The potential for damage from the furniture beetle in the United States is indicated by the estimate made by Hickin (1963*a*) that at least 50% of the buildings of Great Britain contained an active infestation by this species, and possibly 20% of those built since World War II were infested. In that region, the furniture beetle is the principal species involved in "woodworm" damage—a large portion of the business of pest control operators. The beetle attacks structural timbers, paneling, flooring, and furniture In Germany, it is said to infest "nearly every building," and in New Zealand, it is believed to occur in every house that has been in existence more than 15 years. As the common name implies, the insect commonly infests furniture and other wooden articles.

In the United States, the furniture beetle is widely distributed, but has nowhere attained the importance it has in Europe. It is fairly common in the East, but not along the Pacific Coast, where infestations are usually isolated in imported furniture. The species is confined to the temperate regions of the world, and is not likely to become a pest in tropical regions except at high altitudes where the climate may be suitable for its development.

Even in temperate regions where it is well established, the incidence of infestation and the parts of a building infested can be greatly influenced by differences in climate within a small area. Thus, in Norway *A. punctatum* was found as frequently in attics as in cellars near the coast where summer temperatures were low, whereas relative frequency of infestation in attics decreased with increasing distance from the coast because of increasing summer temperatures, affecting attics more than cellars (Knudsen, 1967).

The furniture beetle attacks only well-seasoned lumber, provided the surface offers ovipositional sites, and it attacks wood only if the bark is removed. Exit holes of this species are never seen in the bark (Hickin, 1963*a, b*). Oviposition is greatly reduced on normally favored wood if it is finely sandpapered, painted, or varnished, all of which eliminate oviposition sites (Kelsey *et al.*, 1945). According to Hickin (1949), it is rare for softwood to be attacked until about 20 years, and the sapwood of oak about 60 years, after it has been cut. However, exit holes have been found in birch plywood as little as 5 or 6 years after being cut.

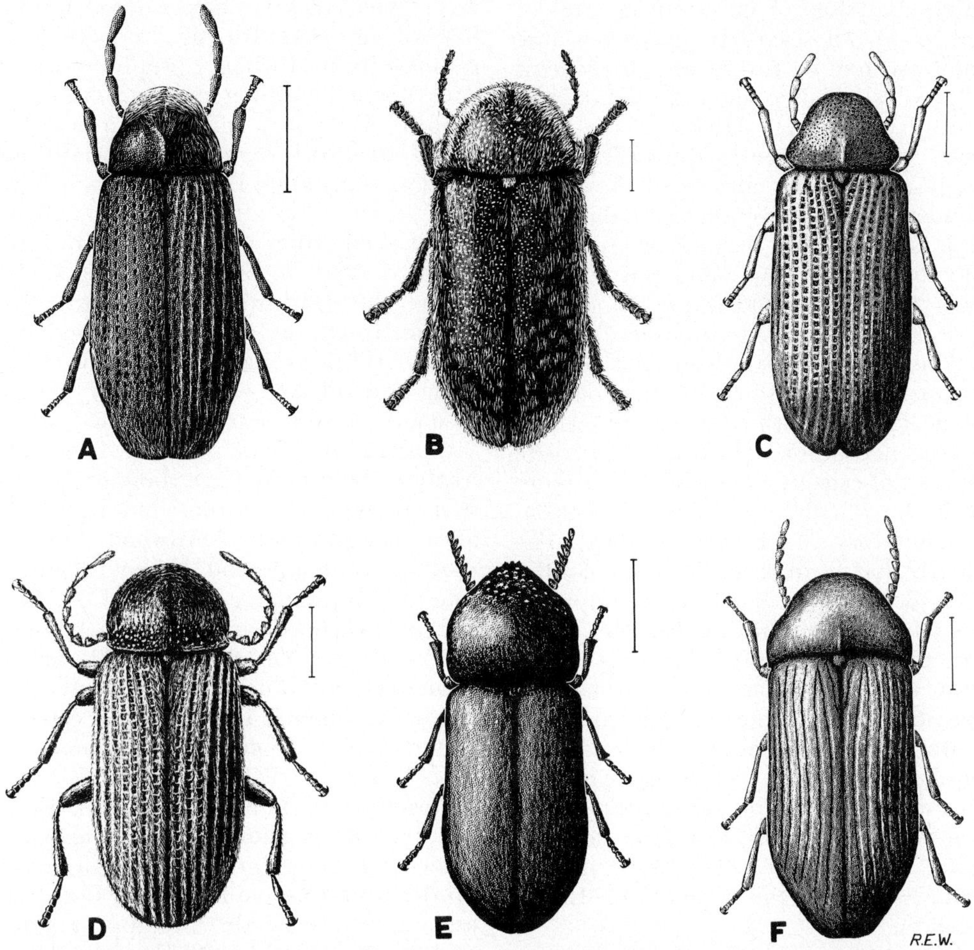

Fig. 115. Representative species of each of 6 genera of Anobiidae. *A, Anobium punctatum; B, Xestobium rufovillosum; C, Hemicoelus carinatus; D, Priobium sericeum; E, Ptilinus ruficornis; F, Xyletinus peltatus.* The line at the side of each specimen represents 1 mm. (After White, 1962.)

Description. The furniture beetle (figure 115, *A*) is 4 to 6 mm long, cylindrical, reddish brown to dark brown, with longitudinal rows of pits on the elytra. These punctures are clearly visible through the fine, yellowish hairs that cover the beetle. As in all anobiids, the head is not visible from above; this distinguishes adult anobiids from the lyctids. The head has a prominent frontal protuberance, and the antennae are 11-segmented, with the last 3 segments enlarged so that the 3 together are longer than the first 8 segments combined (White, 1962).

The grayish-white larva is about 6 mm long. Spinules occur on the first 7 abdominal segments, and are absent on the last 3 segments. On the first 6 segments, the bands (transverse markings) have 2 rows of spinules, and on the seventh segment the band has only a single row. A tubular projection or air tube on each spiracle is about equal to the length of the spiracle itself, and is longer than the projections on the spiracles of other common anobiid larvae. As stated earlier, anobiid larvae differ from lyctid larvae in not having a greatly enlarged posterior spiracle and in having 5-segmented legs, the terminal segment with a claw (figure 108). The tunnels formed by the larvae, mostly along with the grain of the wood, are loosely filled with frass. The frass consists of cigar-shaped pellets composed of minute fragments of the chewed wood. In severe infestations, large amounts of frass may be forced out of old exit holes, but "powdering" does not occur to the same extent as with lyctid beetles (Hickin, 1963*a*).

Life Cycle. Adults emerge in the spring from pupal cells just below the surface of the wood. They mate soon after emergence, and a day later the female lays her eggs in a crack or crevice in the wood (figure 116), in rough end-grain, rough-sawed timber, unplaned wood where the grain is open, or just inside old emergence holes (Hickin, 1963*a*, *b*). The eggs are easily seen with the unaided eye, appearing like groups of "oval pearls," but are often slightly deformed when forced into crevices during deposition (Hickin, 1963*a*). The average number of eggs laid has been reported to be 18 by Kelsey *et al.* (1945), 28 by Hickin (1963*a*) and, in the laboratory, 54.8 by Spiller (1964). The eggs ordinarily hatch in 6 to 10 days. Spiller (1949) observed either no hatching or impaired hatching at relative humidities below 60%.

Outdoors, the larvae require about a year for development, and will pupate for 2 to 3 weeks in a cell just below the surface of the wood. Indoors, the life cycle may require 2 to 3 years or more; this may result in a staggering of generations, so that all stages may be found in the wood at any one time. The longer indoor developmental period is presumed to be the result of a lower moisture content of the wood (N. E. Hickin, correspondence).

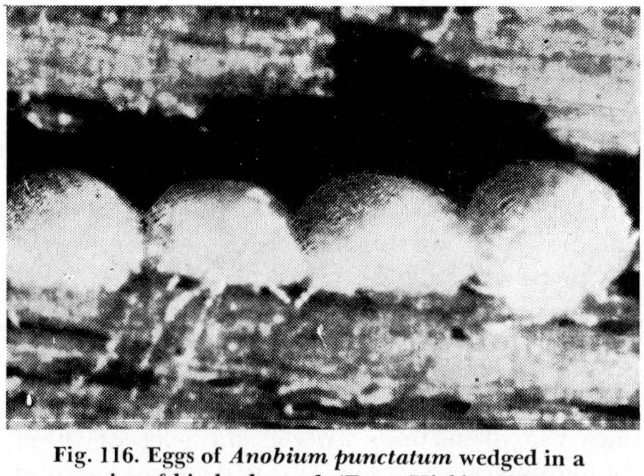

Fig. 116. Eggs of *Anobium punctatum* wedged in a crevice of birch plywood. (From Hickin, 1963*a*.)

Deathwatch Beetle, *Xestobium rufovillosum* (De Geer)

The deathwatch beetle is so named because of a mating signal produced by each sex by striking its head against the surface on which it stands. A female captured immediately after emergence was placed in a small box, where she continued to live for 10 weeks. During this period, she would respond in kind when a tapping noise was made with a pencil within a few meters of the box. This tapping was once considered to be a foreboding sound or death signal by superstitious people (Gahan and Laing, 1932). Deathwatch beetles seldom if ever fly, although they will sometimes flutter their wings when thrown into the air.

Wood must be partially predigested by fungi before the deathwatch beetle can utilize it. In nature, it attacks several hardwood species. Indoors, it infests hardwood timber that was at some time infected with fungi. In England, where so many buildings are constructed of oak, this is the principal timber infested. The timbers most likely to be attacked are those soaked by rain or that are in poorly ventilated locations. Infestation always originates in hardwood, but can extend into adjacent softwood. Heavy furniture is often infested, particularly if made of oak or chestnut. Deathwatch beetles can also attack books (Fisher, 1937; Hickin, 1963*a*). In the United States, *X. rufovillosum* is much less important as a pest than it is in Europe. It is common in certain parts of eastern North America, but not along the Pacific Coast (Linsley, 1943*b*).

Description. The deathwatch beetle adult (figure 115, *B*) is about 7 mm long, dark grayish brown, with a pattern of yellowish, scalelike hairs on the pronotum and elytra; if the elytra are old and worn, the beetle may appear more reddish and shining (Hickin, 1963*b*). *Xestobium rufovillosum* does not have longitudinal rows of small pits on the elytra, such as are present in the furniture beetle, *Anobium punctatum*. The antennae are 11-segmented, but much shorter than those of *A. punctatum*, and the last 3 segments are somewhat enlarged and as long as the preceding 5 combined (White, 1962). Circular emergence holes of the deathwatch beetle are about 3 mm in diameter—about twice as large as those of the furniture beetle. The frass consists of relatively large lenticular or "bunshaped" pellets, characteristic of the species (Parkin, 1933).

The creamy-white larva of the deathwatch beetle is the largest of the anobiid larvae, being up to 11 mm long. The head is yellowish, and the mandibles are nearly black. The eyes consist of 2 black spots on each side of the head, compared with 1 black spot in *A. punctatum* larvae. Spinules occur dorsally on the first 8 abdominal segments, and laterally and ventrally on the ninth and tenth segments. Unlike the spiracles of *A. punctatum,* those of *X. rufovillosum* have a very short, tubular projection or air tube (Parkin, 1933).

Life Cycle. In England, where the deathwatch beetle has been investigated extensively, the adults begin to appear in April and May, and their tapping noise is then commonly heard. Unlike the furniture beetle, the adults may remain in their pupal cells until the following spring. They mate soon after emergence, and then die in a few weeks.

The female lays her eggs in small clusters, usually of 3 or 4, on rough surfaces, in cracks, or just inside the emergence holes. When first laid, the eggs are sticky and adhere together. They are whitish, oval, smooth throughout, and about twice as large as those of the furniture beetle. Usually, 40 to 60 eggs are laid, but up to 201 have been recorded. They hatch in 2 or 3 weeks. Under favorable conditions of food and temperature, the deathwatch beetle may go through its life cycle in a year, but this period may be prolonged to as much as 10 years. Outdoors, in oak infected with fungi, the life cycle has varied in duration from 3 to 7 years, and indoors it is believed to average about 4.5 years (Hickin, 1963*a*).

Other Anobiids

The eastern deathwatch beetle, *Hemicoelus carinatus* (Say) (=*Hadrobregmus*) (figure 115, *C*), is reddish brown to blackish brown, and is 3.4 to 6.8 mm long. The antennae are 10-segmented, with the last 3 segments (club) as long as or longer than the pronotum in the male, but shorter than the pronotum in the female. It is widely distributed in eastern North America, and has been recorded from ash, basswood, maple, beech, and elm, and in sills, joists, beams, and flooring. The larvae often cause much damage (Payne, 1936; Linsley, 1943*b;* Craighead, 1950; Wright, 1959). Payne called attention to the fact that the damage caused by *H. carinatus* resembled that caused by lyctid beetles in hardwoods, except that the emergence holes were a little larger and the frass was somewhat coarser.

The California deathwatch beetle, *Hemicoelus gibbicollis* (LeConte) (=*Hadrobregmus*) (figure 117), is similar in appearance to the eastern species, but has 11-segmented antennae, and the pronotum is distinctly narrower than the base of the wing covers, whereas in *Hemicoelus carinatus* they are about the same width. Linsley (1943*b*) stated that the California deathwatch beetle bred in old, well-seasoned Douglas-fir studs, joists, and

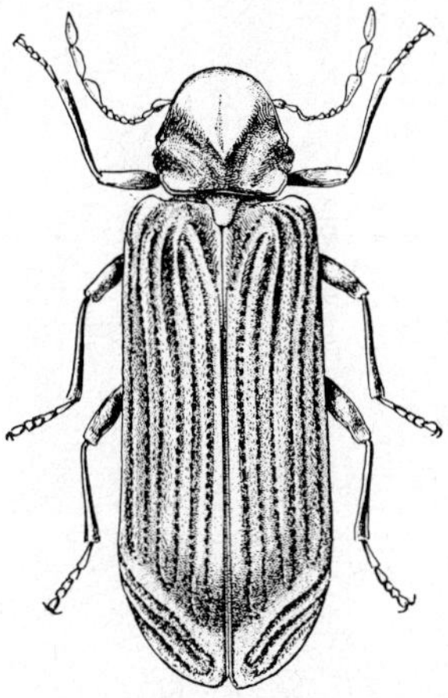

Fig. 117. California deathwatch beetle, *Hemicoelus gibbicollis.* (From Doane *et al.,* 1936. Used with permission of McGraw-Hill Book Company.)

supporting timbers along the Pacific Coast, and was especially common in basement timbers of buildings 20 years old or more (figure 118). Hatch (1946) found serious damage from this species to the weathered boards of spruce in the flooring of a porch. In the areas where it occurred, Linsley considered *H. gibbicollis* to be the most serious pest among the powderpost beetles in the broad sense of the term.

Ernobius mollis (L.) appears somewhat like a large *Anobium punctatum.* The short, golden hairs of the newly emerged adult produce a light shade of golden brown, but when some of the hairs are lost, the beetle becomes more uniformly brown. The elytra are less horny, and lack the longitudinal rows of pits characteristic of *A. punctatum. Ernobius mollis* oviposits only in softwoods, and then only when the bark is attached. Hickin (1963*a*) found that when the bark was stripped off, the infestation rapidly died out. In the United States, however, *E. mollis* occasionally does much damage to pine and spruce flooring (Craighead, 1950).

White (1962) listed 7 species of *Ernobius* occurring in the eastern United States, including *E. granulatus* LeConte. This species has unusually large eyes. It is 2.3 to 4.3 mm long, reddish brown, with elytra yellowish apically, appendages and abdomen yellowish brown, and with a very fine,

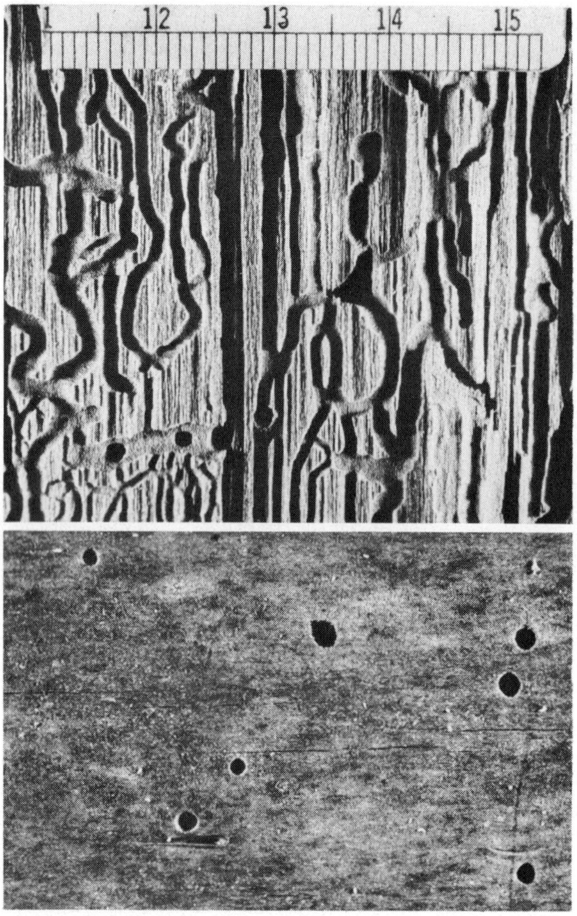

Fig. 118. Damage by California deathwatch beetle, *Hemicoelus gibbicollis*. *Top*, galleries, exit holes, and metric scale; *bottom*, exit holes at surface of Douglas-fir lumber.

short, yellow pubescence. The last 3 antennal segments are very long. *Ernobius granulatus* has been recorded from many eastern states.

Despite the usual propensity of anobiids to attack old, seasoned wood, some species can destroy weakened shoots of young pine trees, and are scavengers in the cones of conifers. In California, the species most commonly involved is *Ernobius punctulatus* (LeConte). The beetles are most abundant where dead pine branches have been allowed to accumulate (Doane *et al.*, 1936).

Priobium sericeum (Say) (= *Trypopitys*) (figure 115, *D*) is 4.9 to 6.2 mm long, with the head deeply retracted within the prothorax. It is reddish brown to brown, covered with a short, yellowish pubescence. Antennal segments 3 to 7 are distinctly serrate, 8 to 10 are somewhat serrate, and segment 11 is rather narrow and elongate. The combined length of the last 3 segments of the antennae equals the combined length of the 5 or 6 preceding ones (White, 1962). This species is occasionally found in flooring, sills, and in buildings in the eastern states.

In southern California, *Priobium punctatum* (LeConte) (= *Trypopitys*), a species closely resembling the furniture beetle, *Anobium punctatum*, has been found on oak door casings and flooring and in maple wainscoting. This species has also been found in Monterey cypress and pine (Doane *et al.*, 1936).

Ptilinus ruficornis Say (figure 115, *E*) is 2.8 to 4.8 mm long. The male is completely black, or black with light-brown elytra. The female is bright reddish brown to dark reddish black, with the pronotum bright reddish brown. Both sexes have reddish-brown to reddish-yellow appendages. The antennae of the female are distinctly serrate, whereas those of the male are actually branched, with each branch, on segments 5 to 10, nearly or quite as long as the combined length of segments 1 to 10. The pronotum of both sexes is roughened anteriorly, but particularly so in the female. This species infests beech, maple, oak, sycamore, mesquite, and other hardwoods, and is a rather common and injurious pest of woodwork and of stored wood products (Craighead, 1950). In California, another species, *Ptilinus basalis* LeConte, along with the bostrichid *Scobicia declivis*, attacks California laurel (*Umbellularia californica*) so readily that laurel has had to be nearly abandoned for interior finishing, for which this very beautiful wood is well suited (Doane *et al.*, 1936).

Xyletinus peltatus (Harris) (figure 115, *F*) is 3.4 to 6.3 mm long, reddish brown to brown, with a fine, short, yellowish pubescence. Its eyes are large, those of the male separated by about 2 times, and those of the female about 3.3 times, the width of the eye as seen from the front. The 11-segmented antennae are moderately serrate, and the last 3 segments are slightly enlarged. The elytra are finely, not deeply, striate (furrowed), with the interval between striae feebly convex (White, 1962).

Xyletinus peltatus is widely distributed throughout the eastern states, and is the most frequently collected species of the genus (Wright, 1959; White, 1962). Williams (1973) considered it to be the most common anobiid in the southeastern states. It often does great damage to cellar joists and flooring in damp buildings, particularly when they are unoccupied or closed up. This species appears to require moist conditions and associated fungi. When conditions are optimum, it can attack both softwoods and hardwoods so severely that complete structural failure will occur. However, it has been known to manifest distinct preferences for certain wood species. Williams (1973) gave an example of such a preference in lumber stored in an old earthen-floor barn infested with *X. peltatus*. The wood species were yellow poplar, cypress, and western pine. The cypress and western pine were not infested, but radiographed sample boards of yellow poplar contained an average of 72 larvae per sq ft (929 sq cm), and yellow-poplar molding contained about twice that number.

Prevention and Control of Powderpost Beetles

The following discussion pertains to infestation by the true powderpost beetles (Lyctidae), false powderpost beetles (Bostrichidae), and deathwatch beetles (Anobiidae). Prevention of losses from these beetles should begin at the lumbermill, and be continued in lumberyards, builders' lots, and other areas in which lumber may be stored before use.

An extremely important phase of prevention is sanitation. In nature, beetles breed in old and dried wood, such as dead branches and limbs of trees, and material of this kind should be burned if it exists near piles of lumber. The same may be said of old lumber and other materials that are subject to infestation. Storage areas should be inspected periodically. Infested lumber should be burned, and pallets, stacking sticks, stakes, platforms, and shelves should be protected against attack. Paraffin wax, varnish, shellac, and paint will fill the pores in which lyctid beetles lay their eggs, and thereby prevent infestation. Infestation by lyctid beetles can be avoided merely by using only softwoods for the pallets, stacking sticks, etc. However, oak is preferred, despite its susceptibility to attack, because it is stronger, more durable, and does not shatter or splinter so easily as coniferous woods.

Firewood stacked outdoors is frequently a source of beetles that can infest structural wood after the firewood has been carried into the house. Other woodborers that may not be able to infest seasoned wood, but are nevertheless nuisances and sources of anxiety when found crawling or flying about in the home, can also emerge from firewood.

Green lumber can be protected against powderpost beetles by submerging it in an insecticidal cold-water emulsion. Similar protection for seasoned wood can be obtained with a solution of insecticide and deodorized kerosene. Insecticides that have proved to be effective in this connection are 5% toxaphene, 2% chlordane, 0.5% dieldrin, and 0.5% lindane. Immersion for 3 minutes gives protection for at least 3 years. A 5-minute dip in 5% pentachlorophenol in a light fuel-oil base is also said to be effective, and this same compound, added to the insecticide emulsions or solutions just mentioned, acts as a fungicide. Protection for 1 season can be obtained with only a 10-second immersion in these insecticides or a 15-second dip in a 5% borax solution heated to 180° F (82° C). Following any of these treatments, the treated lumber or wood products can be painted or varished after a reasonable drying period (NPCA, 1961).

Some homeowners would certainly not want to have insecticide-treated lumber used for such articles as kitchen cabinets and baby cribs. A concentration of 0.3% dieldrin or aldrin added as an emulsifiable concentrate to glue prepared for plywood manufacture has been found to provide good protection of plywood against *Anobium punctatum* in New Zealand (Harrow *et al.*, 1970).

All stages of lyctid beetles can be killed by heat in kilns. In table 3, the periods of exposure of wood of various thicknesses that are required to check damage by powderpost beetles at different temperatures and relative humidities are given. The required periods of exposure increase with increasing thickness of the wood, decreasing temperature, and decreasing relative humidity (St. George, 1970).

The type of treatment selected for control of powderpost beetles will depend on a number of circumstances, among them the size of the infested area. If the infestation is localized, the application of a toxic solution to the infested wood surface or a liquid gas volatilized under a tarpaulin may be considered. Infested furniture can be fumigated in a gastight fumatorium. For the treatment of a large infested area, such as an entire house, tent fumigation, such as practiced for drywood termite control, may be the only practical procedure. Even if one finds only a few beetles, believed to be a localized infestation, a complete fumigation of the entire structure is the most reliable way to control the infestation. It is often difficult to locate incipient infestations. However, even with complete fumigation of a building, incomplete control of lyctid beetles occurs in about 10% of the treatments.

TABLE 3
Schedule for Treating Wood to Check Damage by Powderpost Beetles*

Relative humidity	Lethal temperature required	Thickness of timber†	Time required to overcome lag after kiln has attained lethal temperature	Additional margin of safety	Time then held at lethal temperature	Total period of exposure after kiln has attained required conditions
Per cent	*°F and C*	*Inches*	*Hours*	*Hours*	*Hours*	*Hours*
100	130–54	1.00	0.50	0.50	1.50	2.50
		2.00	2.00	0.50	1.50	4.00
		2.50	3.25	0.50	1.50	5.25
		3.00	4.50	0.50	1.50	6.50
80	125–52	1.00	0.50	0.50	2.00	3.00
		2.00	2.00	0.50	2.00	4.50
		2.50	3.25	0.50	2.00	5.75
		3.00	4.50	0.50	2.00	7.00
	120–49	1.00	0.50	1.50	6.00	8.00
		2.00	2.00	1.50	6.00	9.50
		2.50	3.25	1.50	6.00	10.75
		3.00	4.50	1.50	6.00	12.00
	115–46	1.00	0.50	7.50	30.00	38.00
		2.00	2.00	7.50	30.00	39.50
		2.50	3.25	7.50	30.00	40.75
		3.00	4.50	7.50	30.00	42.50
60	125–52	1.00	0.50	1.00	4.00	5.50
		2.00	2.00	1.00	4.00	7.00
		2.50	3.25	1.00	4.00	8.25
		3.00	4.50	1.00	4.00	9.50
	120–49	1.00	0.50	2.00	7.00	9.50
		2.00	2.00	2.00	7.00	11.00
		2.50	3.25	2.00	7.00	12.25
		3.00	4.50	2.00	7.00	13.50
	115–46	1.00	0.50	9.00	36.00	45.50
		2.00	2.00	9.00	36.00	47.00
		2.50	3.25	9.00	36.00	48.25
		3.00	4.50	9.00	36.00	49.50

*Adapted from St. George (1970).
† For metric conversions, consider 1 in. to be 25 mm.

Toxic Solutions

For treatment of raw wood, such as unfinished hardwood flooring, unfinished paneling, or structural timbers already infested with powderpost beetles, 2% chlordane or 0.5% lindane, dissolved in a moderately volatile light oil such as deodorized kerosene, have been successful when sprayed or painted on the wood at 1 gal per 100 sq ft (4 l per 9 sq m) of surface or until the wood can no longer absorb the liquid. (Solutions of 5% DDT and 0.5% dieldrin are no longer registered for this purpose.) On finished wood, the entire surface should be treated evenly with a coarse, fan-type spray, but not to the point of runoff or puddling. Repeated applications may be necessary to obtain absorption of the recommended quantity. Each successive application should be made after the preceding one has been absorbed, but before the surface is completely dry. Unless the treatment is applied to the ground floor, care should be exercised to avoid staining the ceiling and walls below the treated floor. Treated floors should not be walked on, and no objects should be placed on any treated surface for at least 24 hours. Since the treatment can destroy wood finishes, it is advisable to treat a small, inconspicuous area for preliminary observation before treating infested finished wood. Damaged surfaces must be resanded, refinished, and rewaxed. Sanding, of course, can also be done before treatment, particularly if the floor happens to need it. A few beetles may occasionally emerge after treatment, but the toxic residue will prevent reinfestation. When parquet or block flooring is bound to concrete slab by adhesive, the oil carriers for insecticides can deteriorate the adhesives. In such cases, the entire building should be fumigated.

Smith (1968) was able to prevent infestation of ash lumber by *Lyctus* powderpost beetles during air-drying, which commonly requires 3 to 18 months. Ash is highly susceptible to infestation. Boards were dipped in lindane emulsions of various concentrations for 10-second periods and then stood on end in a rack to dry for 1 hour. Twenty boards, 1.2 m long and 10 to 20 cm wide, were used for each lindane concentration. The boards were then placed in stacks, with 1.5-cm pine spacers between them to keep them apart. About 12 months after treatment, the boards were exposed to infestation by introducing infested boards from laboratory-reared colonies. All surfaces were examined annually. With 0.06% lindane emulsion, the number of *Lyctus* beetle holes varied from none the first year to 6.5 the third year, compared with 22 and 295, respectively, in the untreated controls. At 0.25% concentration of lindane, no holes were found during the 3-year period. Since the normal storage period was not longer than 18 months, Smith concluded that a 10-second dip in 0.06% lindane emulsion was adequate for protecting lumber from *Lyctus* beetles during air-drying. A 25% increase in lindane concentration sufficed for protection against ambrosia beetles, which are discussed later.

For the treatment of timbers such as those used in log cabins, rustic bridges, rustic furniture, and the like, a 0.5% lindane solution in petroleum oil, sprayed onto the surface of the bark or wood, is effective for the control of bark beetles or borers that have only recently attacked and have not penetrated deeper than the outer surface of the wood. Wood that has been attacked for a longer period should be soaked for at least 5 minutes.

Toxic Solutions for Anobiid Beetles

Anobiid infestations usually begin in crawl spaces, particularly if the soil is wet, for the wood then contains more moisture, favoring infestation. Fortunately, the infestation spreads slowly. Control is least difficult and least expensive when the beetles are still confined to the substructure and before they have spread upward into the walls (Williams, 1973). Extensive research was done by Spink *et al.* (1966) at Louisiana State University on the control of the anobiid *Xyletinus peltatus*. The external signs of infestation by this beetle were entrance holes less than 1.6 mm in diameter and exit holes about 3 mm in diameter. The number of holes might exceed 200 per sq ft (929 sq cm). Dieldrin at 1% and chlordane at 2% concentrations, in either water or kerosene, gave complete control when applied with a 1-gal (4-l) compression sprayer to the inner sides of 2 joists and the subflooring between them. The liquids were sprayed back and forth over the infested wood until no more liquid was absorbed. There were 4 replications of each treatment. Pretreatment and post-treatment evaluations were based on the number of pellets falling from the infested areas. The tests were stopped after a year.

Another test was made in 6 houses. Dieldrin at 0.15% and 1% and chlordane at 1% and 2% concentrations in water or kerosene were applied to subfloor wood members infested with *Xyletinus peltatus*. The liquids were applied with a motorized 15-gal (57-l) sprayer, with pressure held to a minimum. The spray was applied in a solid, cone-type pattern, using a No. 5 nozzle disk with a 2-mm aperture. The sprays were applied back and forth over the wood until the timbers were saturated. Approximately 4 gal (15 l) of kerosene solution or 4.5 gal (17 l) of water solution were required for treating 100 sq ft (9 sq m) of infested wood surface. No spreaders or stickers were added to the spray liquids because the tests were made to determine whether sprays as used for termite control and applied with similar equipment, but with reduced pressure, could control anobiid beetles. Evaluations of efficacy were made as in earlier tests, and no further evidence of infestation was seen after any of the treatments.

When the beetle larvae bore into wood, they fill their entrance holes with frass. Tests with water-soluble dye have shown that the frass acts as a wick to carry the insecticide deeply into the wood.

Fumigation for Powderpost Beetles

Methyl bromide fumigation for drywood termites was discussed under the heading "Fumigation." Fumigation for powderpost beetles can be done in the same way, but instead of a concentration of 2 lb per 1,000 cu ft (0.91 kg per 28 cu m) of building space, 3 lb (1.36 kg) is generally recommended for a 24-hour period, provided the temperature is over 60° F (16° C), the space being fumigated is properly sealed, and a minimum of 0.5 lb per 1,000 cu ft (0.23 kg per 28 cu m) of gas remains at the end of the 24-hour period of fumigation. Under some circumstances, a higher initial

Fig. 119. A pest control operator's fumatorium with door open. (Courtesy Columbia Pest Control Company.)

dosage or additional gas sometime during the fumigation period may be required. Methyl bromide fumigation should be done only by a licensed and experienced operator.

Movable items such as furniture can be fumigated in vaults or fumatoriums designed for this purpose. Some pest control firms have such facilities (figure 119). Because of tighter sealing and controllable temperature, either the dosage or exposure can be decreased below that which is required for effective building fumigation. For vault fumigation with methyl bromide, the usual relationships of dosage and period of exposure are, per 1,000 cu ft (28 cu m): 1 lb (0.45 kg) for 24 hours, 3 lb (1.36 kg) for 8 hours, or 9 lb (4.08 kg) for 3 hours. When it is possible to provide partial vacuum, dosage or time can be even further decreased (NPCA, 1961).

Beetle-infested timbers may also be placed in a pile, covered with a tarpaulin, and fumigated with methyl bromide at the rate of 4 lb per 1,000 cu ft (1.82 kg per 28 cu m) of air space. The gas should be confined for about 72 hours. Methyl bromide is an odorless gas and deadly poisonous if inhaled (St. George, 1970).

Sulfuryl fluoride (Vikane), often used for fumigation of drywood termites, is not recommended for control of lyctid beetles because of its relatively low efficacy against the eggs. About 10 times the dosage used in drywood termite fumigation is required to kill lyctid eggs.

An aluminized, neoprene, portable, gastight fumigation tent, with walls and floor continuous, and supported by a collapsible tubular metal pipe frame (figure 120) was developed in England for on-the-spot fumigation of small commodities. It is 7 ft wide, 7 ft high, and 9 ft deep (about 2 m and 2.75 m) and has a 300-cu ft (8.4-cu m) capacity. It has 2 types of gastight seals for the front opening: an industrial metal zipper or a pressuretight channel that completely seals all edges of the flexible flap opening. A 10-in. (25-cm) fan is operated in the tent for a period of 3 minutes, immediately after the fumigant is introduced, to thoroughly distribute the gas (Burns Brown and Heseltine, 1964). A metering device is supplied with the tent that delivers the entire contents of a 1-lb (0.45-kg) can of methyl bromide, supplying a concentration that is the equivalent of a little over 3 lb per 1,000 cu ft (1.36 kg per 28 cu m). As stated above, in vault fumigation the period of exposure for this concentration of methyl bromide is normally 8 hours.

If a lower rate of dosage is required for the commodity fumigated, a shorter exposure period should be utilized, rather than using less gas.

Fumigation is very expensive, and when an infestation of powderpost beetles is known to be confined to a section of hardwood flooring, a localized treatment may be appropriate. S.T.D.,[12] a proprietary 15.17% solution of the liquid fumigant ethylene dibromide and 4.03% chlordane, in deodorized kerosene, may be mopped evenly over the infested area, using just enough to wet it without leaving puddles. On upper floors, care must be taken to avoid using so much liquid that some of it will stain the ceiling below. The treated area is covered with a polyethylene tarpaulin, the edges of which are taped down to the dry floor surrounding the wet area, and the tarp is left in place for 24 hours. It is then removed, and the windows are opened to allow for complete ventilation.

Occupants are kept out of the room(s) being fumigated during this 24-hour period and the additional period required to completely remove the gas (Ehmann, 1961; NPCA, 1961). Repeated

[12] Product of Neil A. Maclean Company, 765 Landess Avenue, Milpitas, California 95035.

treatments may be required, but pest control operators in the Los Angeles area advise us that the surprisingly few "callbacks" after localized treatment for powderpost beetles indicate to them that their treatments are generally successful. Such pest control operators usually offer a 1-year guarantee on their work.

In another form of localized treatment for powderpost beetles, some pest control operators place a tarpaulin over infested areas, such as hardwood wall cabinets, and release methyl bromide under the tarp.

The localized treatment with ethylene dibromide, when properly performed, usually will not harm a varnished floor surface but will dissolve the wax coating.

In cases where parquet or block flooring is laid directly on a slab to which it is bound by adhesive, the liquid fumigant should be applied in a pan under the tarp so that the ethylene dibromide will contact the floor only as a gas and not as a liquid; otherwise, the adhesive under the floor may be dissolved. Plywood is more difficult to penetrate with a gas because of the glue layers, and increased dosages and extremely careful application techniques are required to ensure good results.

The homeowner may decide to consider removal and destruction of infested wood as an alternative to fumigation (NPCA, 1961).

ROUNDHEADED BORERS (CERAMBYCIDAE)

The family Cerambycidae includes many species of large, slender, woodboring beetles. The adults usually have very long antennae that may be much longer than the body, giving them the common name, "longhorned beetles." The 2 most important cerambycid pests of buildings, the newhouse borer and oldhouse borer, do not have particularly long antennae. All adult cerambycids have large, strong, toothed mandibles.

The larvae are somewhat cylindrical, and have a large and fairly cylindrical thorax, resulting in burrows of a similar shape in the infested wood, hence the name, "roundheaded borers." The larvae have small heads, but powerful mandibles. Larval legs are lacking or vestigial.

Newhouse Borer, *Arhopalus productus* (LeConte)

The newhouse borer is a cerambycid beetle found throughout most of the western United States and western Canada. It occurs in pines and firs, but particularly in Douglas fir ("Oregon pine" of lumbermen). It breeds in dead or dying trees, and is especially abundant in fire-swept areas, probably being attracted to the trees mainly during the first summer after a fire. When lumber from such trees is used for building houses, the emergence holes of the adults may appear in hard-

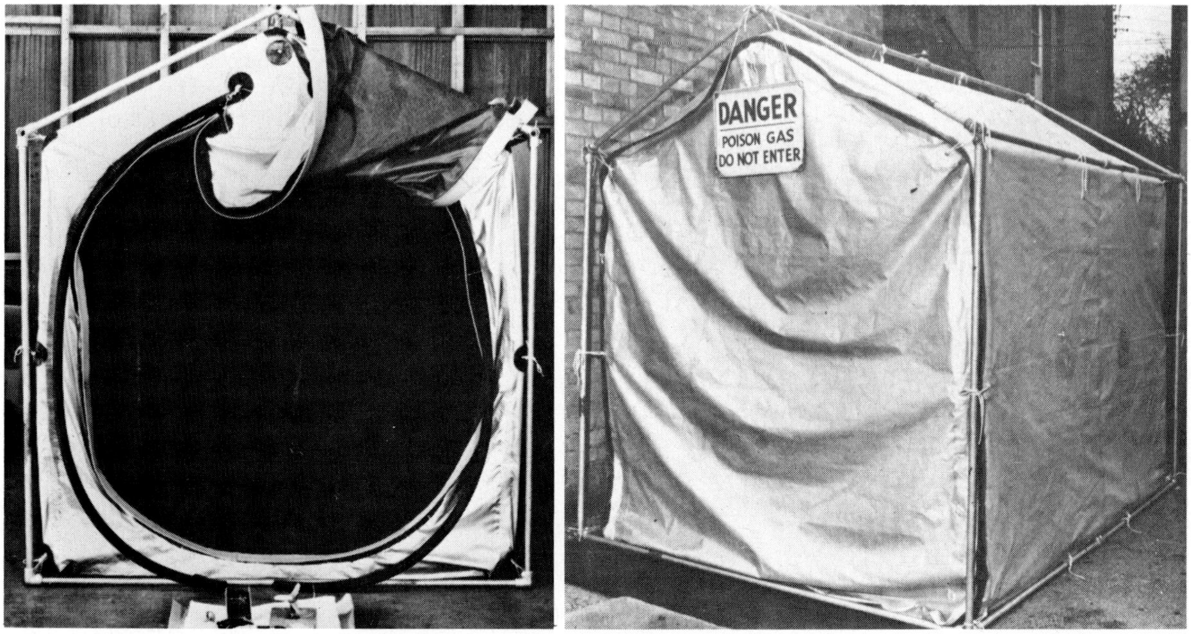

Fig. 120. Portable gastight fumigation tent. *Left,* tent opened for entry; *right,* tent after closure. (Courtesy RFD-GA, Ltd., Godalming, Surrey, England.)

wood flooring, linoleum, plaster, plasterboard, or composition roofing.

Biology. The adult (figure 121) is a narrow, black beetle, 2 to 3 cm long. The larva is yellowish white, and is about 4 cm long when full grown. The female lays eggs in deep crevices of the bark.

Fig. 121. Adults and an immature larva of the newhouse borer, *Arhopalus productus*.

Because the bark has been removed, the female does not oviposit in wood that has been converted into lumber. Therefore, there is no danger of reinfestation from beetles that have emerged (Wood, 1964). In trees, the larvae initially tunnel at the juncture of sapwood and bark. Later, they tunnel entirely in the sapwood, and finally move into the heartwood. They always pack their tunnels tightly with fibrous frass. If 2 pieces of recently cut lumber are in contact with each other, the larvae sometimes tunnel from one piece into the other. In isolated pieces of lumber, the larvae occasionally bore to the surface, plug the hole, and continue tunneling in the interior. The full-grown larvae construct large pupal cells, and when the adults bore their way out of the wood, they make clean-cut oval holes at the surface about 6 mm in diameter. These are easily distinguished from the circular holes made by emerging woodwasps. Also, surrounding each emergence hole of the newhouse borer there is usually an accumulation of long drillings that looks like a pile of cigarette tobacco.

Duration of Infestation. Although it takes at least 2 years for the newhouse borer to complete its life cycle, the appearance of adults in a house may occur within a few months after construction is completed, and rarely after 1 year. The reason appears to be that as the wood dries out, it becomes unsuitable for larval development. Only those larvae that are nearly full grown are able to survive and reach maturity under these dry conditions (Eaton, 1959).

Types of Lumber Infested and Extent of Infestation. Apparently, the most commonly infested lumber is Douglas-fir subflooring. Any material, such as hardwood flooring or other floor covering on the infested lumber, may be perforated in order to provide an escape route for the emerging adult. Eaton and Lyon (1955) once found 1 to 8 (average 4) visible emergence holes per house in a tract in which there was some evidence that the Douglas-fir subflooring had come from a fire-killed stand of timber. The beetles sometimes emerged from infested wood framing in the walls or ceilings of buildings, or from sheathing. In one building, 34 exit holes were counted in the plaster. The infestations are never extensive enough to cause serious mechanical weaknesses in a structure. Nevertheless, the holes, just as those caused by woodwasps, may be causes of consternation and worry for the homeowner.

The incidence of infestation by such insects as the newhouse borer and woodwasps has increased in recent years. This is because of the increased demand for wood, which has made it profitable to cut timber that in past times would have been rejected, particularly since this kind of timber can be obtained at little or no cost. It is perfectly sound for structural purposes. Air-drying of the lumber for over 1 year, or kiln-drying, would solve the problem, and the complete use of such timber would result in a greater utilization of our natural resources.

Western lumber destined for eastern markets is generally kiln-dried because removal of water decreases its weight. Beetles and other insects are killed by the process. Lumber harvested in the West which is to be used on the West Coast is often air-seasoned for just a few weeks. To allow a long period for air-seasoning and storage would likely cause the death of most wood-infesting insects, but increase the cost of production.

Control of the Newhouse Borer

Wood (1964) pointed out that radiographic techniques were available, so that the presence and extent of woodborer infestations in finished lumber could be determined. It may be desirable to use this technique before investing in an expensive control measure, such as fumigation. Fumigation with methyl bromide gas has been successful in the United States and Europe for the control of the oldhouse borer, *Hylotrupes bajulus*, which has habits similar to those of the newhouse borer and does similar damage. Therefore, it is reasonable to suppose that methyl bromide can be successfully used against the latter species.

Infestation of a building by newhouse borers is seldom, if ever, sufficiently severe to justify the great expense of fumigation. Filling of exit holes or repair of damaged areas after the adults have emerged is the most economical solution. In any case, by the time the infestation has been detected it is often too late for any measures aimed at the destruction of the insects themselves.

Oldhouse Borer, *Hylotrupes bajulus* (L.)

The oldhouse borer apparently originated in northern Europe, where it has been found in 40 to 50% of the houses in some surveys. It has spread from there to many parts of the world, including most of the states of the Atlantic seaboard and in various other areas as far west as Minnesota and Texas. An August mean temperature of about 73° F (23° C) is an important factor favoring the development of the oldhouse borer, and might be a useful criterion in predicting the future distribution of the beetle in North America (Anonymous, 1967*c*). There are well-documented records of the oldhouse borer having been transported to new areas in infested wooden boxes and packing cases. From France, the insect was transported to England in pine flooring, and to Australia, in the pine framing of prefabricated houses (NPCA, 1965). Thus, while this serious pest has not yet been established in the western United States, there is no reason to believe it could not be accidentally introduced.

Unlike the newhouse borer, the oldhouse borer can be present in old houses as well as new ones and, most importantly, can reinfest timbers from which it emerges. It can thus cause great structural damage.

Korting (1962) stated that lumber in houses 60 years and more of age was less subject to infestation than in new houses. Moreover, adult female beetles in the older houses are smaller and have fewer progeny. Patton (1931) reported that, in Denmark, roofs of timber covered with metals more than 20 years were most subject to attack. Houghton (1939) stated that in the United States, *H. bajulus* was most likely to attack attic and roof timbers, but infested framing and flooring also. In Norway, *H. bajulus* was found mainly in attics, and only in districts of the highest mean temperatures and the longest periods of relatively high temperatures during the summer (Knudsen, 1967). In Massachusetts, houses that are centuries old, some of historic interest, are being infested. Sometimes the infestations are so severe that the occupants of infested houses hear "gnawing" or "clicking" sounds that are caused by the beetles (Becker, 1954). Like the newhouse borer, the oldhouse borer attacks only softwoods, feeding mainly on the sapwood, but the adults may bore through other types of wood or plaster in order to emerge.

Whereas both *Arhopalus productus* and *Hylotrupes bajulus* can be introduced into a building in infested lumber at the time of construction, of the 2 beetles only *H. bajulus* can infest the building after construction. At least in the United States, it is strictly a structural pest, and has not been found in logs and stumps (Snyder, 1955*a*), presumably because it is an introduced species that has not yet become established in nature. [Another species, *H. ligneus* (F.), has been found

Fig. 122. Oldhouse borer, *Hylotrupes bajulus*. *Top*, female (*left*) and male; *bottom*, larva *in situ*. (Courtesy United States Forest Service.)

damaging rustic furniture in cabins (Thompson, 1932).] Although infestation by *H. bajulus* can occur wherever susceptible wood is present, in the United States the majority of infestations are said to be in attic framing in the northeastern area and in the substructure along the mid-Atlantic Coast (NPCA, 1965).

Description. The adult females of *Hylotrupes bajulus* may reach a length of 2.5 cm, but the males are only about half as long. The adults are slightly flattened, grayish black to very dark brownish black, with many gray or yellowish-gray hairs on the head and anterior part of the body. However, these may be rubbed off on older specimens. Two elevated black, shiny knobs on the prothorax give the dorsum an appearance like a face with a pair of eyes (figure 122). About a third of the way posterior on the elytra, and centrally located, there are 2 grayish, transverse marks.

The full-grown larva (figure 122) is grayish white, from 2 to 4 cm long, has a broad thorax, and tapers posteriorly. There are deep folds between the abdominal segments, and there is a deep groove running lengthwise in the center of the prothorax. There are 3 black ocelli in a row on each side of the very small head, but a hand lens is required to see them. The stout, very dark jaws are relatively prominent. The pupa is about the size of the adult, and is at first creamy-white and then becomes light brown.

About 150 to 200 white to grayish-white, spindleshaped eggs about 2 mm long are laid in checks, cracks, crevices, or irregularities of the wood. Stacks of lumber are said to be excellent oviposition sites. It takes 2 or 3 weeks for the eggs to hatch. The larvae feed in the dry sapwood from 2 to 10 years (usually 3 to 5) until the sapwood is completely destroyed. They fill their mines loosely with frass composed of tiny pellets and fine, powdery material. The frass occupies a greater volume than the wood from which it was produced, and this causes the surface of the infested wood to have a blistered or rippled appearance (Hickin, 1963*a*). A rhythmic rasping or chewing sound made by the larvae may be the first indication of their presence, for their tunnels seldom break through to the surface, even though the interior may be severely mined.

Life Cycle. The adults may remain in the tunnels prepared by the larvae for 7 to 10 months before emerging, but then live for only a brief period. They appear in the summer, and oviposition takes place at that time. In the United States, the period required for the life cycle of the oldhouse borer may be from 3 to 5 years in the southern states, whereas northward from the latitude of Washington, D.C., an additional 2 to 3 years may be required. Most of the period is spent in the larval stage, for the egg and pupal stages each last only about 2 weeks, and the adults live 8 to 16 days. The majority of adults fly in June and July. Adult beetles often emerge from attic timbers a year or two earlier than from basement wood in the same building, because of the higher attic temperatures (Patton, 1931; Craighead, 1950; St. George *et al.*, 1957; Hickin, 1963*a;* McIntyre and St. George, 1961).

Signs of Infestation

There are a number of indications when oldhouse borers are active in a building. In the early stages of an infestation, the rasping or ticking sounds made by the larvae while boring may be heard, or a blistering of the wood when the larva works close to the surface may be seen. If the wood surface is probed where tunneling is suspected, the larvae or their powdery borings may be located. Only after the buildings are at least 3 to 5 years old in the southern states, or 5 or more years old northward from the latitude of Washington, D.C., will the emerging adults, or their 7-mm, broadly oval emergence holes, be evident (McIntyre and St. George, 1961).

Relation of Protein Content to Susceptibility of Wood to Attack

Becker (1963) determined that a minimum of 0.2% of protein was required, in the softwoods investigated, to support an oldhouse borer population. From that point, larval development increased in direct proportion to an increase in protein content. The suitability of the wood for the development of larvae decreased as it aged, probably because of a change in the nutritive value of proteins during storage, a decrease in vitamin content, and other chemical or structural changes in the wood. The suitability of the wood for larval development could be fully restored by the addition of protein and vitamin B. Nevertheless, as just stated, some very old buildings are being severely attacked. The older the building, the greater the chance that it may have an infestation, despite the decreasing nutritional value of the wood.

During the first stages of decay, wood infected by wood-destroying fungi, such as *Lenzites,* shows an increase in protein content and becomes more favorable for larval development. Likewise, pine wood contains more protein than spruce, and in areas of Europe where pine is widely used in construction, oldhouse borer infestations are more frequent and severe than in areas where spruce predominates. A reduction in quantity and quality of protein and a loss of vitamin B can be quickly accomplished by artificially applied gamma radiation. This suggests a new approach to the protection of wood from attack by the oldhouse borer (Becker, 1963).

Larvae of the oldhouse borer were unable to develop in 4 commonly used types of fiberboard, whereas termites (*Reticulitermes, Heterotermes, Coptotermes,* and *Kalotermes*) were able to attack and destroy them, and chemical preservatives were required to protect them from termite attack (Becker, 1969*b*).

Control of the Oldhouse Borer

In both Europe and the United States, fumigation of infested buildings is considered to be an effective treatment for *Hylotrupes bajulus* infestations (Rasmussen, 1967; McIntyre and St. George, 1961), but of course fumigation leaves no protective residue. Methyl bromide at 3 lb, or HCN gas at 2 lb per 1,000 cu ft (1.36 or 0.91 kg per 28 cu m) has been used satisfactorily (St. George *et al.,* 1957). Fumigation must be done only by qualified pest control operators. Borer larvae can be killed by applying 2% chlordane in deodorized kerosene to the infested wood with a brush or, for larger areas, a sprayer, applying the liquid at the rate of 1.5 pt per cu ft (0.71 l per 28 cu dm) of wood treated. For example, one would apply 1.5 pt to 12 linear ft of a 2 × 6-in. roof rafter, or 18 linear ft of a 2 × 4-in. stud, or 12 sq ft of subflooring.[13] Wood absorbs kerosene solutions when it is dry and the temperature is at least 80° F (27° C). For spot treatments, 0.5% lindane may be used instead of the chlordane solution. If the infested wood does not at first absorb the required quantity of toxic solution, the treatment must be reapplied until the necessary quantity has been absorbed. The solution will not penetrate properly if the wood has been painted, varnished, or waxed. Unlike fumigation, treatment with toxic solutions is not only a control but also a preventive (McIntyre and St. George, 1961).

In South Africa, similar solutions have been applied on infested timbers, but lumber used to replace timbers structurally damaged by *Hylotrupes bajulus* was impregnated with a preservative such as copper naphthenate or zinc naphthenate, preferably applied by vacuum pressure (Durr, 1956).

The pentachlorophenol emulsion Woodtreat-TC, which has been used successfully against subterranean and drywood termites in structural timbers, may also be effective against newhouse and oldhouse borers (Ebeling, 1968).

Other Cerambycids

Any insect that infests old, weakened, or fire- or insect-killed trees is likely to be incorporated into buildings constructed from lumber obtained from such trees if it is not kiln-dried. Among such insects are many species of cerambycids besides those already discussed.

One of the most conspicuous and best-known species found in lumber is *Ergates spiculatus* (LeConte) (figure 123, *A*), a large, slender beetle, 4.5 to 6 cm long, reddish brown, with head and thorax darker than the elytra. The lateral margin of the prothorax is armed with a few large and many small, sharp spines. The adults are often attracted to lights. The creamy-white larvae are 5 to 7 cm long. According to Linsley (1962), *E.*

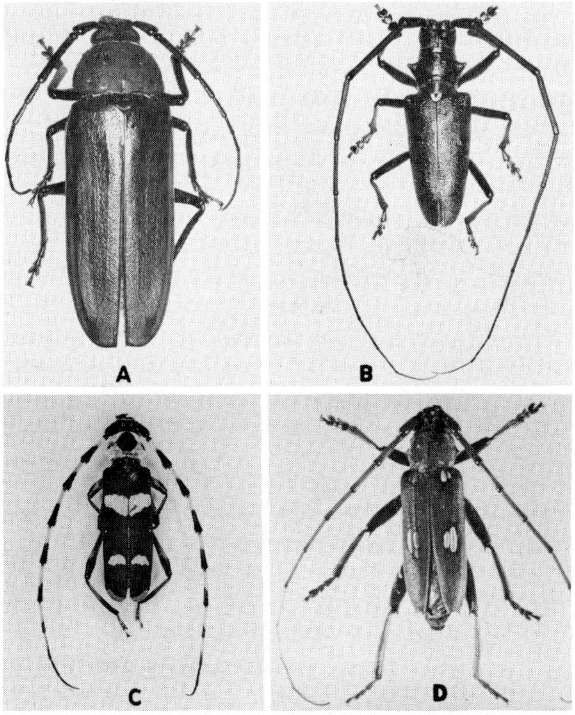

Fig. 123. Some common cerambycids. *A, Ergates spiculatus; B, Monochamus scutellatus; C, Rosalia funebris; D, Eburia quadrigeminata.*

[13] Metric equivalents: 0.71 l per 28 cu dm; 0.71 l to 3.65 linear m of a 5 × 15-cm roof rafter; 5.5 linear m of a 5 × 10-cm stud; about 1.10 sq m of subflooring.

spiculatus "not only attacks the wood of recently killed or felled trees, but also fallen logs, stumps, and power and telephone poles," and "has often ruined fire-killed Douglas fir before salvage operations could begin." This species occurs throughout western North America.

Another cerambycid, the black pine sawyer, *Monochamus scutellatus* (Say) (figure 123, *B*), is 1.6 to 3 cm long and shining bronzy black, with a white scutellum, coarsely punctured, a large median "tooth" on each side of the prothorax, and very long antennae, often twice the length of the body. It is widely distributed in the coniferous forests of North America, and is especially abundant in timber following burns (Chamberlin, 1949). The spotted pine sawyer, *Monochamus maculosus* Haldeman, may reach a length of 3 cm or more, with antennae 2.5 times the length of the body. These beetles are brown, with bluish-gray dorsal spots. The prothorax has a prominent median tooth on each side. The larvae are very destructive to both heartwood and sapwood of dying, fire-scorched, and recently felled pines in the western states. The black timberman, *Monochamus oregonensis* (LeConte), ranges from British Columbia to southern California and east into the Rocky Mountain states, sometimes causing losses to fir logs exceeding 10 to 20% if they are left in the woods a few months. This species is 2.5 to 4 cm long, and has a cylindrical, deeply and coarsely punctured body. The color is uniformly black, sometimes with a bronzy reflection, and the scutellum is white. Silvery-white spots are sometimes scattered over the elytra and on the thorax and antennae (Chamberlin, 1949).

The banded alder borer, *Rosalia funebris* Motschulsky, is one of the best-known and most attractive of the Pacific Coast cerambycids. It is 2.5 to 4 cm long, marked with black and white as shown in figure 123, *C*, and has very long antennae that are annulated with black and white. It breeds in ash, alder, and California laurel and, east of the Sierra Nevada, generally in willow. According to Doane *et al.* (1936), this species is responsible for the hollowing-out of the trunks of many of the larger laurel trees in California. It is mentioned here mainly because of its conspicuous and striking appearance, resulting in many requests for identification. However, during periods of very high temperatures, the beetles are attracted to recently painted dwellings after the paint has become dry, apparently attracted to certain volatilized ketones from the paint. They are present on some buildings in such large numbers as to be a definite nuisance (Chemsak and Linsley, 1971; Essig, 1943).

A common eastern species on hickory, locust, and ash, often attracting attention in resort areas, is the fourspotted longhorn, *Eburia quadrigeminata* (Say) (figure 123, *D*). It is 1.5 to 2.5 cm long,

Fig. 124. Nautical borer, *Xylotrechus nauticus*; adult and larva.

light brown to tan, and easily recognized because of the 2 pairs of elevated, ivory-colored swellings on, and the 2 spines at the tip of, each elytron. The life cycle usually requires 2 years, but this species has also been found emerging from flooring, sills, and other wood members 10 to 15 years after the materials were installed (Chamberlin, 1949).

The nautical borer, *Xylotrechus nauticus* (Mannerheim) (figure 124), is a western species, 8 to 15 mm long, and varies in color from grayish brown to nearly black. It has 3 transverse, zigzag white lines on the elytra, and rather short antennae for a cerambycid. The larva is robust in form, and may attain a length of 18 mm. In California, the larvae may be found in live oak, black oak, and sometimes in eucalyptus and madrone, or in walnut or stone-fruit trees in orchards. It is most frequently encountered in the home emerging from firewood cut from such trees.

Various species of *Neoclytus* likewise commonly emerge from firewood and are found in homes. This genus is characterized by the many transverse carinae (elevated ridges) on the disk of the pronotum. The antennae are relatively short, as in *Xylotrechus,* but are more or less clubbed, and the legs are long. Unlike the larvae of *Xylotrechus,* which have no legs, those of *Neoclytus* have small legs composed of at least 2 segments (Linsley, 1964). Probably the most common species of *Neoclytus* on the Pacific Coast is *N. conjunctus* (LeConte). It has either white or yellow markings on a black background. The pronotum is broadly rounded, the elytral apices are rounded, and the head, thorax, and abdomen are hairy (figure 125).

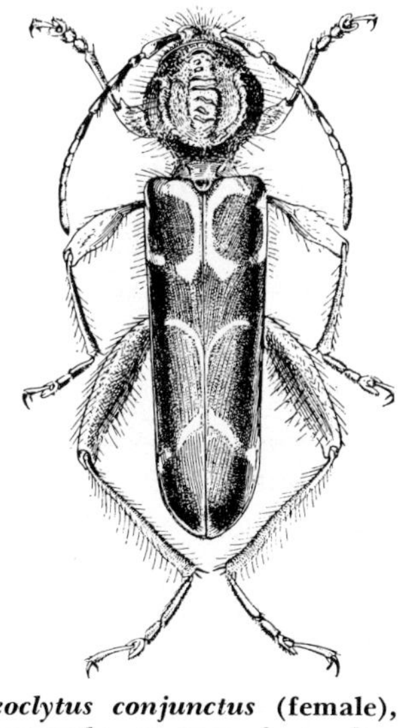

Fig. 125. *Neoclytus conjunctus* (female), a longhorn beetle that commonly emerges from firewood. (From Linsley, 1964.)

It breeds in Oregon ash, various white oaks, madrone, manzanita, eucalyptus, pear, and apple (Doane *et al.,* 1936; Linsley, 1964). Any of these trees cut for lumber should be utilized promptly, otherwise the larvae of *N. conjunctus* will render the wood useless for anything but firewood. Even the value of the infested timber as firewood is reduced because of the honeycombing of the wood by these beetles. Cordwood should be piled and covered with brush, for shaded wood is not attractive to the beetles (Chamberlin, 1949).

In the eastern United States, 2 species of *Neoclytus* that emerge from firewood and are sometimes found in houses are *N. acuminatus* (F.) and *N. caprea* (Say). *Neoclytus acuminatus* is 6.5 to 18 mm long, reddish brown, with 4 yellow bands across the elytra, and long, reddish legs. It attacks unseasoned wood with bark remaining, and frequently causes serious losses of ash, hickory, and oak logs left exposed during the flights of these insects (Craighead, 1950). *Neoclytus caprea* is 12 to 15 mm long, deep purple, with 3 yellow bands across the elytra and 3 narrow, yellow bands across the prothorax (Sweetman, 1965). This species sometimes destroys an entire winter cut of ash logs, honeycombing the sapwood with tunnels that are tightly packed with granular frass (Craighead, 1950).

Wharf Borer, *Nacerdes melanura* (L.) (Oedemeridae)

This insect, which resembles a cerambycid, is found throughout much of the United States and Canada, particularly in coastal areas, along lake shores, and along major inland waterways. The larvae may be found in moist timbers, such as wharf timbers, foundation pilings under buildings, in damp cellars or crawl spaces, in wood around leaky faucets, in buried wood, or in any wood that is damp or soggy, and particularly if it is decayed. In England, it has been found in timbers of many barges and wooden-hulled ships sailing in the region of the Thames estuary. In London, the adults have been seen by the "tens of thousands, running over the pavements of the Strand" without their origin being discovered (Hickin, 1963*a*).

It is not known if wharf borer larvae damage sound wood. They are pests primarily because the adults may suddenly emerge in a building in large numbers and gather around windows, doors, or other sources of light, and thereby become nuisances.

Description. The slender adult is about 10 mm long, and is brown to reddish yellow above, with the tips of the elytra, the eyes, sides of the thorax, legs, and ventral parts generally blackish. There are 3 raised, longitudinal lines on each elytron. The body is covered with a yellow pubescence. The antennae are about half as long as the body

(Balch, 1937; Hickin, 1963*a*). The full-grown larvae are about 3 times longer than the adults, cream-colored, with brown mouthparts, and mandibles that are almost black at the tips. Besides 3 pairs of true legs on the thorax, they have prolegs on the third and fourth abdominal segments. Their legs are larger than those of other wood-boring larvae.

FLATHEADED BORERS (BUPRESTIDAE)

The family Buprestidae is well represented among the timber beetles. The adults are medium to large, usually somewhat flattened, and some are characterized by brilliant, metallic colors with various iridescences. The elytra are usually ridged or roughened. Because of their attractiveness, they are favored by both professional and amateur insect collectors. The larvae lack eyes and legs, are distinctly segmented, and have very small heads and greatly enlarged and usually flattened thoraxes, which latter feature has given rise to the name, "flatheaded borers."

The winding tunnels made by the larvae tend to have approximately the same shape as their thoraxes—about 3 times broader than high. In this respect, the tunnels frequently can be distinguished from the more nearly rounded tunnels of most of the roundheaded borers of the family Cerambycidae. The life cycle of most buprestids usually requires only 1 or 2 years, and is therefore ordinarily completed before the lumber is used, but with several species found in structural lumber, development can continue for a much longer period. Their tunnels, packed with frass, are often seen in softwood lumber, especially in that from trees salvaged after forest fires, windstorms, or bark beetle infestations.

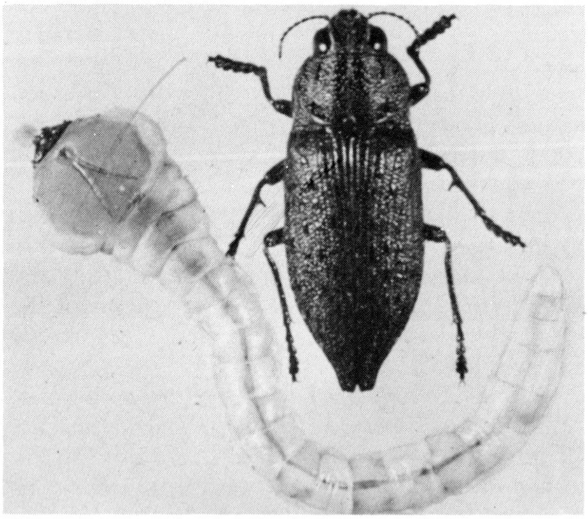

Fig. 126. Adult and larva of a common Pacific Coast buprestid, *Dicerca horni*.

In recent years, buprestid beetle damage to "grapestake" fences has come to our attention. Damaged fences can sometimes be found throughout a newly developed residential tract. If the wood becomes wet, the frass within the burrows expands and extrudes through the oval burrows. This may lead observers to believe that active burrowing is taking place. The wood for such fences was cut from trees that had been infested, but it cannot become reinfested. Buprestid beetles are long-lived, but probably very few of the insects remain in the wood to cause further damage.

Dicerca horni Crotch

This is a common flatheaded borer of the Pacific Coast states. It belongs in a genus of medium-sized buprestids that are characterized by their dull-bronze color and the prolonged tips of the elytra (plate II, 1; figure 126). *Dicerca horni* is a dark, grayish bronze, 13 to 25 mm long, and has small, black, narrow, broken ridges on the dorsum. The larvae are approximately 2.33 times longer than the adults. This species occurs on many species of deciduous trees (including fruit trees) and shrubs, inhabiting dead or dying trees or dead wood on living trees. Adults may be seen from April to September. This is not a pest, but we receive many requests for its identification.

Golden Buprestid, *Buprestis aurulenta* L.

The entomologist may receive many more inquiries regarding the golden buprestid than one would expect merely on the basis of its relative importance as a pest. This is because of the striking beauty of the adult beetle, which attracts much attention and curiosity. The adult beetle (figure 127) is about 2 cm long, and golden green or blue-green in color, with the median suture and margins entirely bordered with copper. The larva is white, and about 3.5 cm long.

Life Cycle and Habits. The golden buprestid prefers Douglas fir, but will also attack several species of pine, spruce, and fir, and is occasionally found in western red cedar. The adults generally oviposit on trees, with the bark still on them, that are dying or have been lightning-struck, fire-scorched, or recently felled. They are particularly attracted to resinous wood. They usually, but not necessarily, oviposit on the bark. They can also oviposit on fire scars, in cracks of freshly sawed lumber, and less frequently an older, dry, or partially dried structural lumber. It is believed possible, but probably rare, for beetles emerging from a building to reinfest it. Upon hatching, the larvae bore into the tree, their winding tunnels continuously increasing in size as the larvae grow during the 2 to 4 years required for their

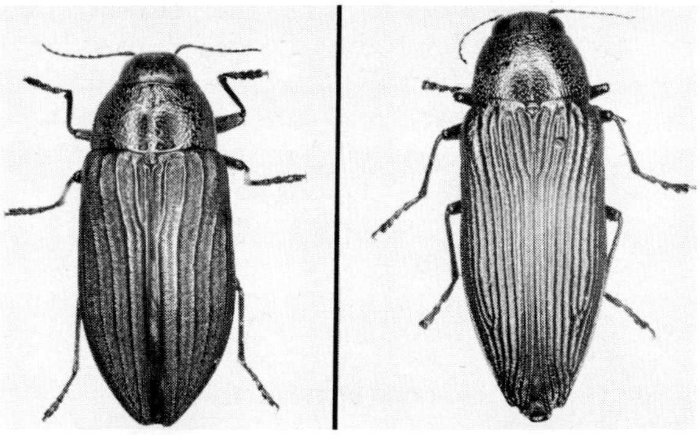

Fig. 127. Golden buprestid, *Buprestis aurulenta (left)*; green buprestid, *B. langii (right)*.

development. They readily mine heartwood as well as sapwood. The tunnels are broadly oval, reach a width of about 10 mm, and vary in length from 1 to 5 m. The tunnels are packed with powdery borings, which aids in distinguishing them from the tunnels of cerambycid (roundheaded borer) larvae that are packed with a loose mixture of fibrous and powdery material (Every and Rudinsky, 1962).

The longevity of the larvae of *B. aurulenta* in furniture and in lumber used in construction is amazing, some having been recorded as living 15 to 20 years (Every and Rudinsky, 1962) and even up to 40 to 50 years or more (NPCA, 1964*b*). Linsley (1943*c*) believed that in the many records of such longevity, the chance that the beetles oviposited long after the structures were built was very remote. For one thing, this species, like other buprestids, mates and oviposits in the bright sunshine of midday, and is not likely to be ovipositing within a building. Also, in some of the instances of prolonged larval life, the beetles were found in linoleum-covered flooring or boring in kitchen cabinets. As further evidence for infestations originating during construction, an adult emerged from a dresser in Germany, where the species does not occur, 13 years after the dresser was manufactured, partially from wood of American origin.

Pupation takes place in oval pupal chambers near the surface of the wood. The adults eat their way out of the wood, leaving the typical oval buprestid exit hole. In nature, adults emerge most frequently in spring and early summer, but inside buildings they emerge most frequently between late fall and late spring. The adults live 3 to 5 months (Every and Rudinsky, 1962).

Control of the Golden Buprestid

Infestations of the golden buprestid are not likely to occur in lumber cut from logs removed from the forest soon after they were felled, and which were healthy when cut. The beetles cannot survive in kiln-dried lumber. The insect can survive in air-seasoned 1-in. (2.5-cm) boards, but it is most often reported from 2-in. (5-cm) tongued and grooved Douglas-fir subflooring. Lumber to which bark is still attached is more likely to be infested than bark-free lumber. Wood preservatives will discourage oviposition to some extent (Every and Rudinsky, 1962). Exit holes can be plugged and damaged surfaces can be refinished. Infestations are only occasionally severe enough to cause serious structural weakness.

Green Buprestid, *Buprestis langii* Mannerheim

This is another attractive beetle (figure 127) that is less common than the golden buprestid, is about the same size, and has similar habits. It is usually bright green, but may be brilliant blue or coppery bronze, the female often with 2, 4, or 6 light spots on the elytra, the tips of which are bidentate. This species breeds in Douglas fir, and does damage similar to that of *B. aurulenta* in structural timbers, piling, poles, siding, and flooring (Chamberlin, 1949).

BARK BEETLES (SCOLYTIDAE)

Bark beetles are small, cylindrical, robust or slender beetles, usually brown, reddish brown, or black; the head partially or completely concealed from above; the antennae short, with a variable number of segments, but usually 11 or 12, with the first one enlarged and the terminal 3 or 4 greatly enlarged and fused into a compact knob or club. They are often about 3 mm long, rarely as long as 8 to 10 mm, and are robust in appearance. These beetles are our most serious forest pests, being attracted to weak, injured, dying, or felled trees. Some species attack fruit, shade, and ornamental trees and shrubs. The small larval tunnels, about 1.5 to 3 mm in diameter, can be seen in lumber or in buildings after construction. Scolytids cannot live in seasoned wood.

Among scolytids, only 1 sex selects the individual host tree in which mating and oviposition take place. In *Dendroctonus,* a monogamous genus, the female excavates the entrance tunnel, and in *Ips,* a polygamous genus, this is done by the male (Wood, 1970).

Bark beetles tunnel through the bark and then between the bark and the sapwood, often scoring the surface of the latter. The adults excavate brood galleries and lay their eggs in tiny niches at intervals along the sides of these galleries. In some species, the brood galleries may be constructed by a single mated pair of beetles; in others, the male may have a harem of 2 or more females, each of which excavates a brood gallery. Fine, brownish-white, sawdustlike frass is pushed out of the entrances of the burrows by the beetles. The larvae bore out on either side of the egg tunnel, but remain between the wood and the bark, making patterns and engravings that are characteristic for each species (figure 128). The larval tunnels are filled with frass, and become larger as the larvae increase in size. Pupation takes place at the end of the tunnel, and the adults bore out through the bark. The exit holes may be so numerous that the tree appears as if it had been riddled with birdshot.

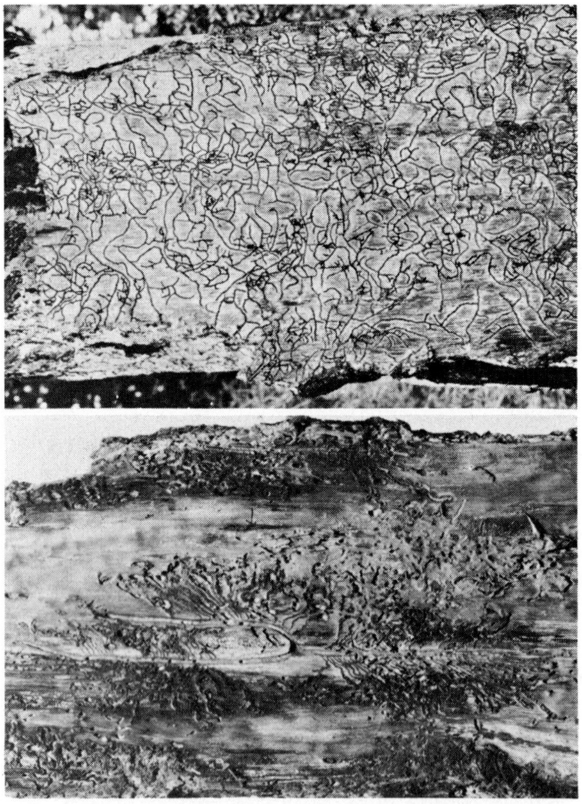

Fig. 128. Galleries of 2 bark beetles, *Dendroctonus brevicomis* (*top*) and *Ips confusus* (*bottom*), when bark has been removed from infested pine trees. (Courtesy D. L. Wood.)

When the bark is pulled away, the engravings are revealed, but the galleries do not penetrate deeply into the wood. In log cabins, rustic wood furniture, or rustic bridges, the engravings are revealed when loose bark falls off. However, before the bark falls off, the frass (the same color as the underbark), which is made up of small, granular, gritty particles, may fall through cracks in the bark. Adult beetles may emerge from exit holes in the bark for as long as a year after the tree is killed. In structures built of infested logs, they may be found at windows where they are attracted by light. They cannot reinfest the dry bark from which they have emerged (St. George, 1970).

The Genus *Dendroctonus*

The principal bark beetles in pine forests are species of *Dendroctonus* and *Ips,* the first-named being the more important. Twenty-four of the 25 species of *Dendroctonus* are in North America. The adults are stout, cylindrical, reddish brown to black, and the species vary from 2.2 to 9.5 mm in length. The largest species is the red turpentine beetle, *D. valens* (LeConte), which is from 5.7 to 9.5 mm long.

Some species of *Dendroctonus* attack weak, dead, or dying trees, but others seem to prefer normal, healthy trees. Infested trees can be detected because of their fading foliage, the presence of pitch tubes with reddish boring dust on the trunk, or loose dust in the crevices of the bark. Egg tunnels and larval tunnels can be seen between the bark and wood (Hopkins, 1909; Chamberlin, 1918, 1925, 1939; Doane *et al.,* 1936; Keen, 1938; St. George, 1970).

Western Pine Beetle, *Dendroctonus brevicomis* LeConte

The most important species of *Dendroctonus* in western North America is *D. brevicomis* (figure 129). The adults of this species are dark brown to black and 3 to 5 mm long. Adults, eggs, and larvae hibernate in or beneath the bark. Activity begins in the spring when the adults attack trees to start new galleries. The adult galleries of *D. brevicomis* are easily recognized because they wind about between the bark and wood and crisscross one another. The beetles mine principally in the phloem (figure 128, *top*), and only slightly engrave the outer sapwood. The larvae feed in the outer bark after completing the first instar. Upon maturing, the adults leave the trees, each through

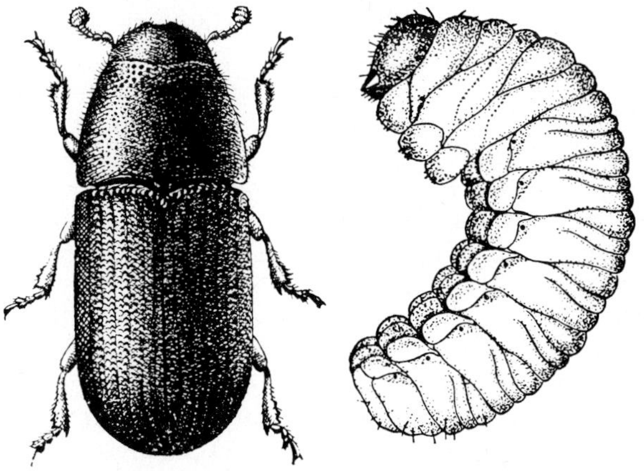

Fig. 129. Western pine beetle, *Dendroctonus brevicomis*, adult and larva. (From Hopkins, 1909.)

a separate exit hole. Ponderosa (yellow) pine and Coulter pine are the only trees attacked. From 1 to 3 generations are produced each year, depending on latitude and altitude.

Mountain Pine Beetle, *Dendroctonus ponderosae* Hopkins

This is a very destructive bark beetle. It is 4 to 7.5 mm long, robust, and dark brown to black. It occurs throughout the West, and as far east as western South Dakota, combining the ranges formerly ascribed to both *D. ponderosae* and *D. monticolae* Hopkins, which are now considered to be one species.

This beetle attacks principally ponderosa pines, and occasionally, in very destructive epidemics, any trees with trunks over 6 in. (15 cm) in diameter.

Southern Pine Beetle, *Dendroctonus frontalis* Zimmerman

This is one of the smaller scolytids, being only 2.2 to 4 mm long. It periodically appears in great numbers in the southern states, and attacks all species of pines, including healthy, vigorous trees of all ages, killing the trees and causing enormous losses.

The Genus *Ips*

The genus *Ips* is next to *Dendroctonus* in importance. These beetles attack mostly pine and spruce, and occasionally fir. They attack primarily young trees and the tops of older ones, while *Dendroctonus* is more destructive to medium-aged or over-mature trees. The adults are medium-sized, shining reddish brown to black, and cylindrical, easily recognized because of the pronounced concavity on the posterior end, with 3 to 6 pairs of stout spines along its margin.

The principal species of *Ips* are the California fivespined ips, *I. confusus* (LeConte), in the West, and the pine engraver, *I. pini* (Say), in the eastern United States and Canada.

California Fivespined Ips, *Ips confusus* (LeConte)

This species is very destructive to the tops of mature trees and to young trees, and attacks ponderosa pine, sugar pine, and most other pine species from southern California north to Oregon, west of the Sierra Nevada and Cascades.

The adults are reddish brown to pitch-black, 4 to 4.5 mm long, and have 5 pairs of spines on the margin of the elytral cavity (figure 130). The male

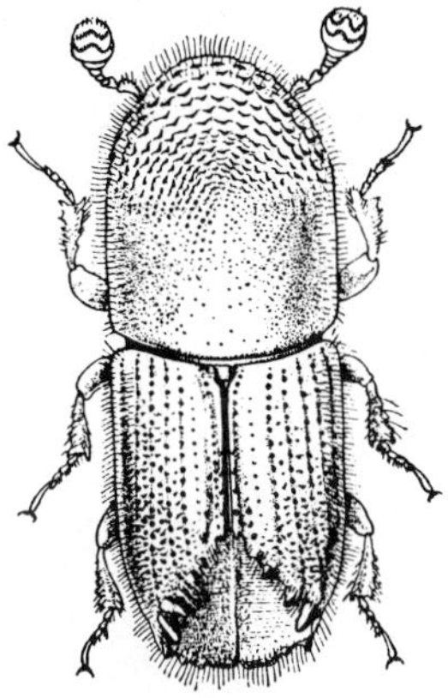

Fig. 130. California fivespined ips, *Ips confusus*. (From Doane *et al.*, 1936. Used with permission of McGraw-Hill Book Company.)

makes the "nuptial chamber," and then is joined by 1 to 5 (usually 3) females. Each constructs an egg gallery, and the galleries radiate longitudinally from the central nuptial chamber, with 1 to 5 from each chamber. A gallery typically has 3 branches in the form of a "Y" (figure 128, *bottom*), mostly in the inner bark and only slightly scoring the outer sapwood. Twenty-five to 50 eggs are laid in each egg gallery. The larvae mine directly away from the main galleries, leaving the mines behind them filled with frass and excrement. Pupation takes place at the ends of the larval mines in cells formed in the inner bark. There are 3 to 5 generations produced during a year.

AMBROSIA BEETLES

In the family Scolytidae, besides the bark beetles, there are beetles that bore directly into the wood for several inches before constructing lateral brood chambers. They do not feed on wood, but on fungi ("ambrosia") that grow upon the walls of the tunnels. Several generations may continue to extend the galleries as long as the tree retains sufficient moisture for the growth of fungi. The wood along the galleries becomes stained. Although these beetles feed on fungi and may be called "ambrosia beetles," they should be distinguished from the ambrosia-feeding beetles of the family Platypodidae. In either family, the species can also be referred to as "wood stainers." They bore directly into the wood and stain it, the stained areas being black, blue, or brown. The tunnels are made by the adults, and are therefore uniform in size, whereas the tunnels of bark beetles increase in width as the larvae grow. Adult ambrosia beetles feed and care for their young, whereas adult bark beetles do not. Emerging ambrosia beetles leave through the entrance holes that have been made by the mother, whereas each emerging bark beetle excavates its own exit hole.

Scolytid Ambrosia Beetles

Although the scolytid bark beetles are more destructive to trees, one seldom sees their damage to finished lumber, except occasionally in rustic timbers with or without attached bark, as used for building cabins, rustic bridges, and the like. On the other hand, the small holes (about 1.5 mm in diameter) and the stains caused by the ambrosia-feeding scolytid beetles or wood stainers are sometimes seen in finished lumber.

Large California Oak Ambrosia Beetle, *Monarthrum scutellare* (LeConte) (= *Pterocyclon*)

An investigation made by Doane and Gilliland (1929) on *M. scutellare* may be used to illustrate the interesting biology of the ambrosia-feeding scolytids. The adults of this species are 3.5 to 4.1 mm long, elongate, cylindrical, and dark brown. The female, but not the male, has a tuft of long hairs on the margin of the antennal club. It is the male that bores through the outer bark in 6 to 10 hours, producing a tunnel about 1.5 mm in diameter. It takes the male 3 or 4 weeks more to excavate the main gallery and the nuptial chamber in the sapwood. The female then bores out the secondary galleries leading from the nuptial chamber, but the male may assist in the task. There are 2 to 4 of these galleries (figure 131), called "compound ambrosial tunnels," and these vary from 5 to 15 cm in length.

Copulation takes place after the galleries are completed. The female excavates niches along the sides of the secondary galleries into which she deposits her eggs. The eggs are creamy-white and nearly 1 mm long. She may lay as many as 80 eggs, but 51 has been found to be the average number. There are 2 periods of egg-laying, one beginning in March and the other in October. The eggs hatch in 10 to 12 days. The larvae excavate "larval cells," pushing small particles of sawdust and excrement out of the mouth of the cell. The female transfers this accumulation of debris from the secondary galleries to the entrance tunnel, and the male pushes it through the entrance tunnel to the outside. The larvae become full grown in 6 to 8 weeks. Pupation takes place in the larval cells, and requires 8 to 12 days. The pupae are creamy-white, and about 3.7 mm long. There are 2 generations per year.

There is a much smaller species of *Monarthrum*, *M. dentiger* (LeConte), only 2 mm long, often found with *M. scutellare* in live oak (*Quercus agrifolia*) and white oak (*Q. lobata*) (Doane *et al.*, 1936).

Striped Ambrosia Beetle, *Trypodendron lineatum* (Olivier)

This ambrosia beetle is widely distributed in the United States and Canada, attacking nearly all the coniferous genera. The adults (figure 132) are about 3 mm long. The prothorax is subglo-

Fig. 131. Typical "compound ambrosial tunnels" of a scolytid ambrosia beetle, *Monarthrum scutellare*, in wood of live oak, *Quercus agrifolia*. (Courtesy D. L. Wood.)

bose, reddish, with numerous transverse rugosities, the latter being most prominent in the female. The antennae are yellowish brown, with a very large knob, much lighter in color than the other segments. The elytra have several rows of punctures, and are very variable in color, ranging from almost completely black to almost completely yellowish brown, but generally black, with broad, brownish-yellow stripes. They can also be predominantly brownish yellow, with broad, black stripes. In the female, the head is convex and in the male it is concave.

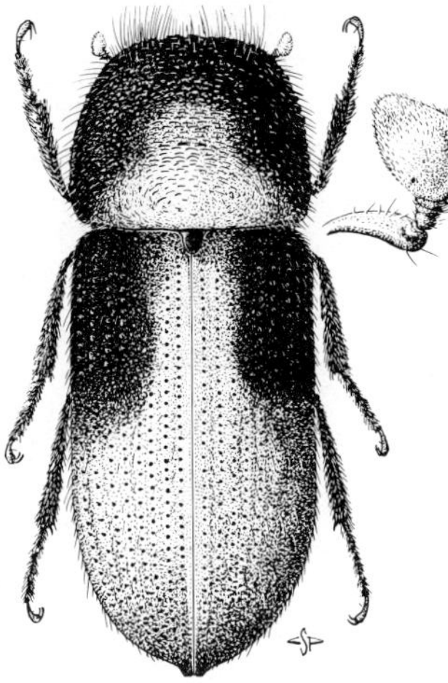

Fig. 132. Striped ambrosia beetle, *Trypodendron lineatum*. (Drawing by C. S. Papp.)

The galleries of *T. lineatum* penetrate the sapwood and heartwood and often branch considerably. The beetle has been found to be a rather serious pest in recently felled timber. To prevent damage, it is essential that the logs be either moved to the mill as soon as possible or rolled into water, if that is more convenient (Chamberlin, 1939).

Western Hemlock Wood Stainer, *Gnathotrichus sulcatus* (LeConte)

This species (figure 133) is 3 mm long and blackish brown, with yellow appendages. It is common in the forests of the Pacific Coast and Rocky Mountains, attacking spruce, fir, Douglas fir, hemlock, pine, redwood, and cedar. Characteristics of *G. sulcatus* and other species of the genus are the long prothorax and the similarity in appearance of the males and females, although the tubercles or teeth at the ends of the elytra of the male are more strongly developed than those of the female, and there are differences in the antennal clubs of the 2 sexes.

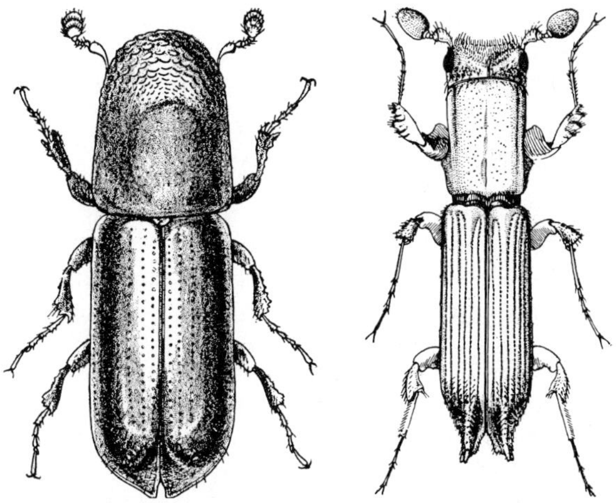

Fig. 133. Two ambrosia beetles. Western hemlock wood stainer, *Gnathotrichus sulcatus* (*left*); Wilson's wide-headed ambrosia beetle, *Platypus wilsoni* (*right*). (From Doane *et al.*, 1936. Used with permission of McGraw-Hill Book Company.)

The adults make a gallery about 1 mm in diameter to a depth of 5 to 8 cm, and then the gallery branches in a horizontal plane, following annual rings in several directions (figure 134). The fine, white boring dust expelled from the gallery collects in crevices of the bark at the base of the tree. The female lays elongate eggs in niches she has cut both above and below the main gallery—1 egg in each niche. The larvae extend the niches as they grow (figure 134, *right*), then pupate in these "cells" or "cradles." There are usually 2 or more broods each year, but they are not regular broods, for all stages of the insect may be found within the galleries at almost any season (Doane *et al.*, 1936). A blue stain (*Ceratostomella*) devel-

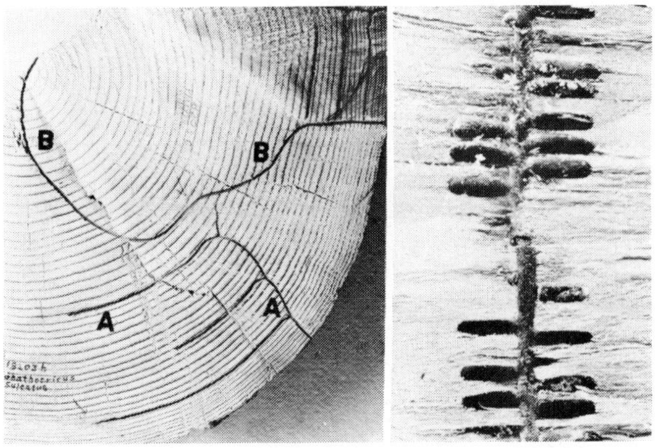

Fig. 134. Gallery systems of 2 ambrosia beetles. *Left*, galleries of *Gnathotrichus sulcatus* (*A*) and *Platypus wilsoni* (*B*); *right*, larval cells dug out along the main gallery by larvae of *Gnathotrichus sulcatus*. (Courtesy D. L. Wood.)

ops rapidly in the mines of *Gnathotrichus sulcatus*, sometimes completely filling the galleries and smothering the beetles (Chamberlin, 1939).

Lesser Shothole Borer, *Xyleborus saxeseni* (Ratzeburg)

This small, cylindrical scolytid is probably the most common species in a genus containing many species of ambrosia beetles. It is common in the forests of Europe and North America, attacking nearly all types of conifers and hardwoods. The females are 2.3 to 3.5 mm long and the males are slightly smaller. The adults are yellow-brown to black, with the elytra obtuse at the apex (figure 135), and with an acute marginal granule on each side. The small, round entrance tunnels lead horizontally to the sapwood and end in a leafshaped cavity in which fungus is grown as food for the larvae (Doane *et al.*, 1936).

Platypodid Ambrosia Beetles

Ambrosia beetles in the family Platypodidae, sometimes called pinhole borers or flatfooted ambrosia beetles, are more slender and elongate than scolytids. They always exceed 3 mm in length, whereas the scolytid ambrosia beetles seldom exceed 3 mm. The platypodids can also be distinguished morphologically from the scolytids because they lack a large spine or projection at the apex of the front tibia that is present in the scolytids. Also, the first tarsal segment is longer than the combined remaining segments, whereas in the scolytids it is shorter (Chamberlin, 1939, 1949). The platypodids are principally tropical beetles, and there is only 1 genus (*Platypus*) in the United States. The typical differences in the appearances of the 2 groups of ambrosia beetles can be seen in figure 133; scolytid (*left*) and platypodid.

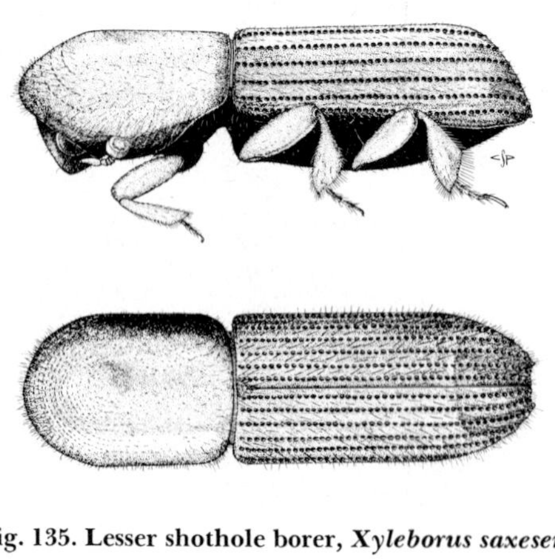

Fig. 135. Lesser shothole borer, *Xyleborus saxeseni*.

Wilson's Wideheaded Ambrosia Beetle, *Platypus wilsoni* (Swaine)

This ambrosia beetle may be used as an example of the few species of *Platypus* occurring in the United States. It attacks dying, weakened, recently dead, and sometimes healthy trees of all Pacific Coast conifers except cypress. It is a slender, brown, shining beetle, 5.5 mm long, with long, yellow hairs. The female has a more or less blunt posterior end, while the male's elytra diverge and are spined (figure 133). Figure 134 shows the type of gallery mined by *Platypus wilsoni*, compared with a gallery system mined by the scolytid ambrosia beetle, *Gnathotrichus sulcatus*, just discussed.

Detection of Ambrosia Beetle Damage in Lumber

Both the scolytid and platypodid ambrosia beetles penetrate deeply into the wood, and their tunnels are free of frass. The larvae develop in small cells adjoining the main galleries, and in most species they are fed fungi by the females, which also carry away the larval feces in order to keep the galleries clean. After pupation, the new adults emerge, not by burrowing through the bark like the bark beetles, but by crawling out of the galleries made by the mother.

Each species of ambrosia beetle usually cultivates only 1 species of fungus, and when the females emerge to fly to other trees to excavate new galleries, they carry the conidia (reproductive parts) of this fungus with them. As the fungus develops, it stains the wood blue, black, or dark brown. The depth to which the tunnels penetrate into the wood depends on the beetle species. Four genera have been ranked in order of increasing depth of tunnels, as follows: *Xyleborus, Trypodendron, Gnathotrichus,* and *Platypus* (Graham, 1963).

The galleries of ambrosia beetles must be moist to enable the fungus to grow, so neither the fungus nor the beetles can survive in dry or seasoned wood. This is unlike the powderpost beetles, which can not only develop in dry, seasoned wood, but may reinfest the wood members from which the adults emerge, and may infest other hardwood in the building. Ambrosia beetles can sometimes survive in wood that is constantly wet, such as in water tanks or wine casks.

In buildings, ambrosia beetle damage can be observed in either softwoods or hardwoods. In the hardwoods, which can also be infested by powderpost (lyctid) beetles (see under that heading), careful examination of the damage will disclose which insect is involved. The galleries of ambrosia beetles are empty, darkly stained, and intercept milled surfaces at various angles, for the wood was cut *after* the galleries were excavated by the beetles. Figure 136 shows the typical appearance of framing lumber milled from coniferous trees infested by ambrosia beetles. In this figure, the wires passed through tunnels in the corner of a stud, extending from one surface to the other, give evidence that the stud was milled from an infested tree rather than being infested after it was milled. The wires show that the tunnels are empty and follow a straight course, whereas the tunnels of powderpost (lyctid) beetles are packed with powder and meander about. The wires also show that some of the tunnels were cut at an angle to the surface of the stud. The galleries of powderpost beetles proceed directly into the wood at a right angle to the milled surface, and in any case they would never be found in softwoods. All these are typical signs of earlier ambrosia-beetle infestation that can be seen in some of the framing lumber in almost any residential tract during construction. The author has seen them in framing lumber milled from white fir, Douglas fir, incense cedar, and hemlock, but never to the extent that the tunnels could be said to be sources of significant structural weakness. Live ambrosia beetles are never present. In fact, ambrosia beetles will never be seen in a house unless they emerge from recently cut firewood (Ebeling and Reierson, 1973*b*).

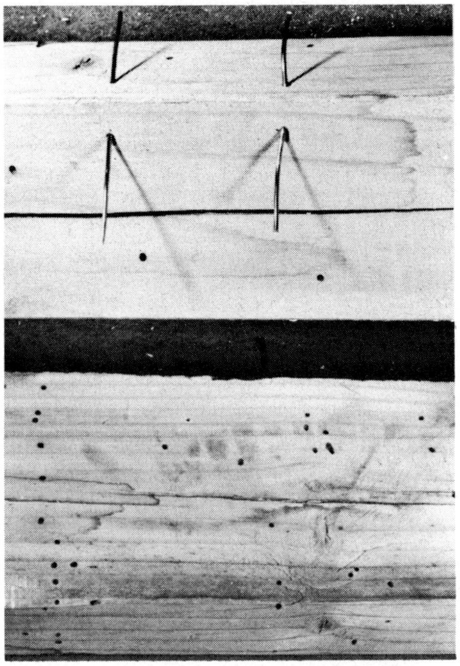

Fig. 136. Ambrosia beetle tunnels in studs that were to be used in house construction.

The discoloration shown in figure 137 was caused by the scolytid *Trypodendron lineatum*. The figure shows the typical staining caused by ambrosia beetles. In this case, the stain from the fungus was some distance from the tunnels.

Control of Bark Beetles and Ambrosia Beetles on Wood with Bark Attached

Much can be done to prevent infestation of logs, poles, or slabs with bark remaining, as used in the construction of rustic cabins, summer houses, arbors, bridges, and fences. Most bark beetles and ambrosia beetles are not likely to infest such wood when their active season begins if the trees are cut in early fall and dry out or season sufficiently during the winter months. However, some scolytids, such as the redshouldered shothole borer, *Xylobiops basilaris* (Say), will attack such wood, and further protection may be desired. A solution of 0.5% lindane in a carrier such as deodorized kerosene may be sprayed onto the wood, after which it should be placed in a sunny place to dry. The most effective treatment is pressure impregnation of the wood, which can be done in commercial wood-preserving plants. It is a good idea to check with your state or county agricultural agent or university extension specialist to be sure that the insecticide intended to be used is registered (St. George, 1970).

Note: Nothing can be done about ambrosia beetles after the damaged lumber is in place, for the damage occurs before the tree is cut.

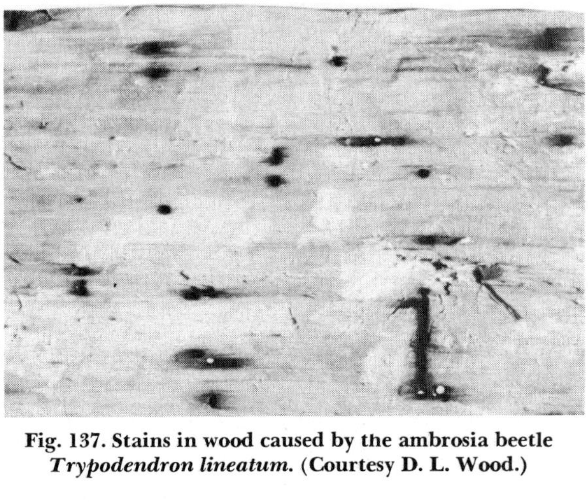

Fig. 137. Stains in wood caused by the ambrosia beetle *Trypodendron lineatum*. (Courtesy D. L. Wood.)

WOODWASPS (SIRICIDAE)

Woodwasps (also known as "horntails") cause damage to new structures by defacing finished surfaces as the adults emerge from infested structural lumber. Instances of woodwasp damage are not frequent, but they are very serious matters when they do occur, sometimes resulting in lawsuits involving up to a half million dollars in damage claims in a single building project.

Woodwasps usually come to the attention of the occupants of infested buildings when they see the large holes left by the emerging adults. The cylindrical emergence holes, about 7 mm in diameter, are found in hardwood floors and walls or in their coverings, such as linoleum, carpeting, nonceramic floor tiles, wallboard, or plaster. Woodwasp emergence can occur in any part of the country where the infested lumber is shipped and used. For example, western species of woodwasps emerged in homes in a new housing tract in Wisconsin in such large numbers as to cause "almost an en masse exodus from the housing project" (Friis, 1961).

Description. Woodwasps are large hymenopterous insects, usually 2.5 cm or more in length. The females are much larger than the males, and both sexes vary greatly in size within a single species. Size can be greatly affected by unknown factors in the host-wood species. While siricids are superficially wasplike in appearance, more careful scrutiny will reveal that the thorax and abdomen are uniform in width and closely joined together, while in the true wasps the abdomen is joined to the thorax by means of a narrow petiole. Woodwasps do not bite or sting, and are harmless in this sense. The adults are usually black or metallic dark blue, or with combinations of black, red, and yellow. Tibiae and tarsi are flattened, and the last abdominal sternite ends in a sharp point.

The common name "horntail" is derived from a dorsal hornlike projection called the *cornus* on the last segment of the abdomen of the female. Originating under the cornus and extending backward is the ovipositor, which in some species may exceed the length of the body (figure 141).

Biology. The biology of the American species of woodwasps has received very little study, but the life history of at least a single Holarctic species, the blue horntail, *Sirex cyaneus,* was carefully investigated by British workers (Chrystal, 1928; Chrystal and Myers, 1928) and was reported by Middlekauff (1960). There is good reason to believe it is typical of woodwasps in general, at least in essential detail.

Adults fly about in bright sunshine, and make a characteristic noisy buzz. Not many males are seen in collections, but this is probably owing to their habit of remaining in treetops, where mating takes place, whereas the females descend to the lower levels to oviposit (Middlekauff, 1960). Woodwasps are attracted to coniferous trees that are weakened or are dying as a result of fire, disease, insect attack, or other debilitating factors. The species *Sirex juvencus californicus* was reared in large numbers from ponderosa pine trees in southern California that showed advanced decline caused by photochemical atmospheric pollution (Cameron, 1968). [Bark-beetle infestations have also been associated with atmospheric pollution (smog) injury to pine trees (Stark *et al.,* 1968).]

The female uses her ovipositor to deposit eggs as deeply at 2 cm into the wood of weakened or dying coniferous trees, or she may oviposit in felled trees. Her body may contain 300 to 400 eggs, but she lays only 1 to 7 fusiform, translucent eggs in each oviposition site, spaced at short intervals while withdrawing her ovipositor. The eggs hatch in 3 or 4 weeks, and the larvae then begin to burrow into the wood at right angles to the oviposition channel.

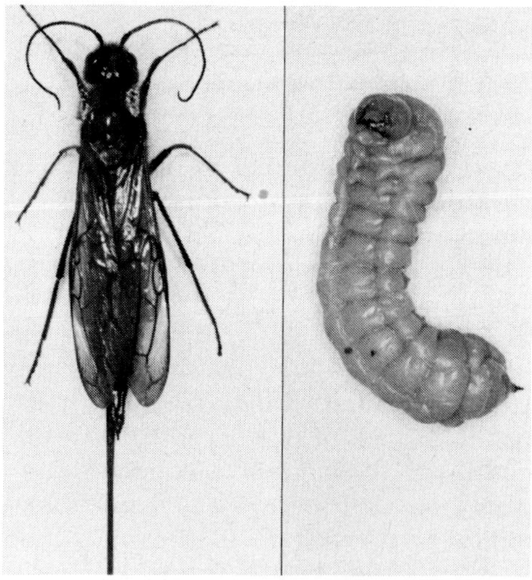

Fig. 138. A woodwasp, *Sirex longicauda;* adult female and larva.

The larva is cylindrical, whitish cream or yellow, slightly S-shaped, and has a small, horny, supra-anal spine on its rear end (figure 138) that is retained through all instars. It has 3 pairs of vestigial thoracic legs, but does not have abdominal prolegs. The larva uses its terminal spine to pack the frass or boring dust in its cylindrical

Fig. 139. A woodwasp, *Sirex juvencus*. *A*. female commencing to oviposit in a balsam fir tree; *B*, oviposition hole containing 2 eggs (*arrow*); *C*, second- and third-instar larvae boring at right angles to oviposition hole; *D*, pupa. (From Stillwell, 1966.)

tunnel, and also uses it as a terminal support by driving it into the walls (Chrystal, 1928).

After first tunneling in the outer sapwood, the developing larva continues toward the heartwood, then returns toward the surface, burrowing in general 25 or 30 cm during its larval life. At intervals, the cast skins from its 3 or 4 molts are incorporated with the frass. Pupation takes place at the end of the tunnel, and requires 5 to 6 weeks. The emerging adult must chew its way through about 2 cm of remaining wood to escape to the outside, leaving a perfectly round exit hole that cannot be distinguished from those of certain cerambycid beetles, e.g., *Monochamus oregonensis* (LeConte). The minimum period for the complete life cycle is 2 years, but it may be 3 or 4 years or even 5 years in some situations.

Stillwell (1966) stated that *Sirex juvencus* in New Brunswick usually oviposited (figure 139, *A*) in 8 to 10 minutes, making oviposition holes 2 to 10 mm long at right angles to the surface, and depositing 1, 2, and occasionally 3 eggs at intervals in the holes (figure 139, *B*) while withdrawing her ovipositor. The eggs, which are about 1 mm long, usually overwinter, although some hatch in late summer and the larvae overwinter in the first or second instar (figure 139, *C*). The larvae molt 5 to 11 times, and pack their cast skins in the frass in the galleries. These galleries vary from 5 to 15 cm long, but are never more than 4 cm from the surface, and pupal chambers (figure 139, *D*) are normally about 2 cm from the surface.

An interesting aspect of the biology of woodwasps is the symbiotic relationship of the females of the genera *Sirex, Urocerus,* and *Tremex* with certain wood-destroying fungi (Francke-Grosmann, 1939; Stillwell, 1966; Morgan, 1968). The adult female becomes infected in the pupal chamber after shedding the pupal skin before emergence (Parkin, 1942). A pair of small, invaginated intersegmental fungal sacs (figure 140, *left*) project into the body cavity at the basal end of the ovipositor. According to Morgan (1968), woodwasp eggs are always deposited with oïdia of a symbiotic fungus and glandular secretions. Stillwell (1966) found fungal oïdia (figure 140, *right*) in the intersegmental sacs of 605 out of 607 *Sirex* and *Urocerus* specimens examined, but no oïdia could be found in the vestigial intersegmental sacs of 18 specimens of *Xeris spectrum*. In his experiments, *Sirex juvencus* did not develop in balsam fir (*Abies balsamea*) logs that were not infected with the fungus, *Stereum chailletii*. Many eggs hatched, but all the larvae died in the first instar. *Stereum chailletii* is associated with much of the sapwood decay in weakened and newly killed balsam fir in New Brunswick and Nova Scotia, but fruiting bodies have not been found in that area. This suggests that the fungus may be propagated solely by woodwasps in the New Brunswick–Nova Scotia area.

As the woodwasp egg is laid, it becomes infected, and the fungus subsequently penetrates the wood surrounding the larva as it feeds. The larva can live for at least 3 months in a pure culture of the fungus (Middlekauff, 1960). According to Morgan (1968), the larvae do not ingest wood. They secrete a saliva that destroys the fungal mycelium, from which they can then extract the necessary nutrients. The fragments of wood are afterward passed along the outside of their bodies to join the accumulation of frass behind them.

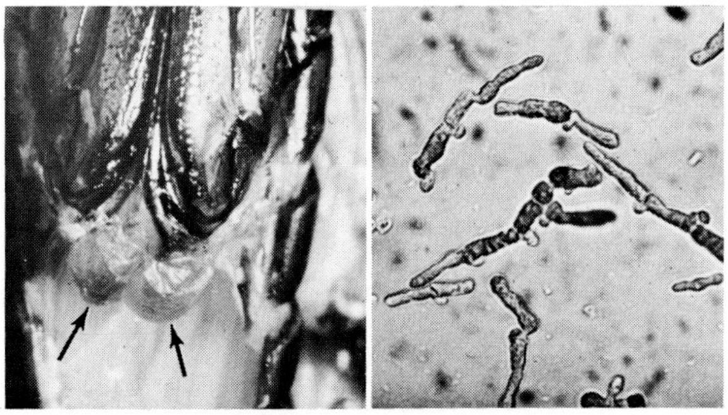

Fig. 140. *Left*, paired intersegmental sacs (*arrows*) of a female *Sirex juvencus* that are located anteriorly to the base of the ovipositor; *right*, fungal oïdia taken from the intersegmental sacs. (From Stillwell, 1966.)

California Species of Woodwasps

There appear to be 12 species of the family Siricidae in California. The subfamily Siricinae contains the genera *Sirex, Urocerus,* and *Xeris.* All species have long, filiform antennae that are much longer than the head and thorax combined. The larvae bore in conifers. The subfamily Tremicinae is represented in California by only a single genus, *Tremex,* in which the antennae are short, slightly swollen in the middle, and at most are as long as the head and thorax combined. The forewings (figure 141, *C, top*) of *Tremex* lack vein 2 r-m and possess an appendicular vein (3*A*); the hindwings (figure 141, *C, bottom*) lack an anal cell; and the hind tibiae have only a single apical spur. The larvae bore in hardwood trees.

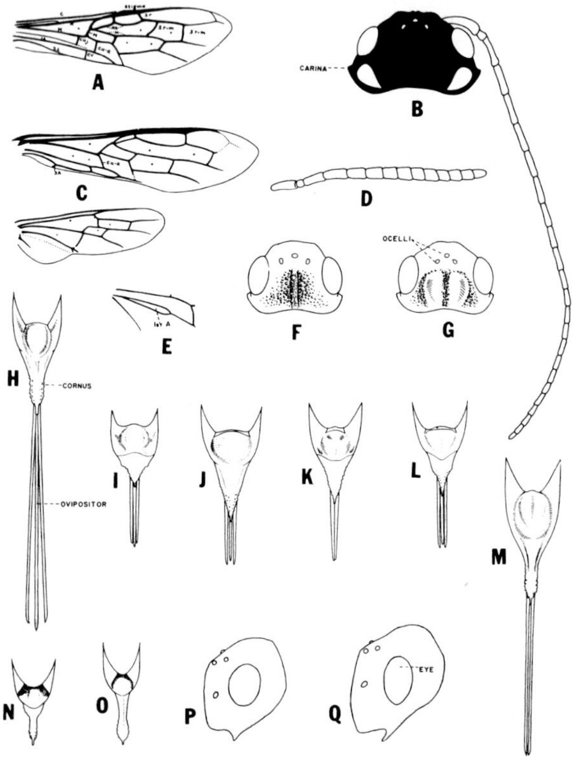

Fig. 141. Taxonomic characters of woodwasps. *A,* forewing of *Sirex; B,* dorsal view of head and antenna of *Xeris; C,* fore- and hindwing of *Tremex; D,* antenna of *Tremex; E,* anal region of hindwing of *Urocerus; F,* dorsal view of head of male *Xeris tarsalis; G,* dorsal view of head of male *Xeris morrisoni; H,* cornus and ovipositor of *Sirex longicauda; I,* cornus and ovipositor of *Sirex obesus; J,* cornus and ovipositor of *Sirex cyaneus; K,* cornus and ovipositor of *Sirex juvencus californicus; L,* cornus and ovipositor of *Sirex behrensii; M,* cornus and ovipositor of *Sirex areolatus; N,* cornus and ovipositor of *Xeris tarsalis; O,* cornus and ovipositor of *Xeris morrisoni; P,* lateral view of head of female *Xeris morrisoni; Q,* lateral view of head of female *Urocerus californicus.* (From Middlekauff, 1960.)

In the adjacent key to the California species of woodwasps of the subfamily Siricinae, keys developed for each of the 3 genera by Middlekauff (1960) and Cameron (1967) have been combined into a single key. The morphological characters utilized in the key are illustrated in figure 141.

Key to the California Species of Woodwasps of the Subfamily Siricinae

1. Females: ovipositor present 2
   Males: ovipositor absent (some species not separable with any degree of certainty) 12

Females

2. Head without pale areas above and behind eyes; cornus of female usually not constricted in middle (except slightly in *Sirex longicauda*) (figure 141, *H*); forewing with at least basal half of vein Cu_1 present (figure 141, *A*) 3
   Head with dorsal surface entirely pale or with at least a pale spot behind each eye; cornus of female usually constricted in middle; forewing with vein Cu_1 usually completely atrophied (figure 141, *C*) 7
3. Ovipositor as long as the forewing, usually longer (figure 141, *H* and *M*) 4
   Ovipositor shorter than forewing (figure 141, *I–L*) 5
4. Legs with tibiae and tarsi reddish brown; 25–26 antennal segments; wings slightly infuscated (smoky gray-brown, with blackish tinge [Roman sepia]) along veins and around apical margin . . . . *Sirex longicauda*
   Legs entirely black or blue-black; 20–23 antennal segments; wings completely infuscated . . . . *Sirex areolatus*
5. Abdomen, except for basal segments, reddish brown; forewing with a transverse fuscous band beneath the stigma and a fuscous apical margin; 2 basal abdominal segments blue-black . . . . *Sirex behrensii*
   Abdomen entirely bluish black 6
6. Legs, except coxae and trochanters, reddish brown; wings usually nearly hyaline except for infuscated apical margin . . . . *Sirex cyaneus*
   Legs metallic blue-black, usually with a reddish-brown area in the dorsal portion of each femur; forewing infuscated on apical margin and across middle to uniformly dark violaceous . . . . *Sirex juvencus californicus*
7. Head lacks a lateral carina behind the eye; ovipositor usually shorter than forewing; hindwing with a closed anal cell; eye is oval (figure 141, *Q*) instead of nearly round; hind tibia with 2 apical spurs 8
   Head with distinctive carina (figure 141, *B*); hindwing without closed anal cell; hind tibia with 1 apical spur; eye nearly round; all species with a white spot behind the eye 10

8. Abdomen black, with segments VII and VIII, and sometimes II, yellow .... *Urocerus gigas flavicornis*
   Abdomen usually entirely black or blue-black . 9
9. Wings golden yellow; flagellum of antennae and basal half of tibiae and metatarsi golden yellow; a large light-colored spot behind the eye .... *Urocerus californicus*
   Wings infuscated, brown; most of flagellum, a spot behind eyes, sometimes lateral spots on the abdomen, and the bases of the tibiae and basitarsi white .... *Urocerus albicornis*
10. Cornus not constricted in middle or only slightly so (figure 141, *N*); tibiae and tarsi usually considerably lighter than femora; antennae basally darker than apical part of flagellum .... *Xeris tarsalis*
    Cornus distinctly constricted in middle (figure 141, *O*); legs mostly uniformly reddish 11
11. Abdomen entirely black .... *Xeris spectrum*
    Abdomen, except for first segment, reddish .... *Xeris morrisoni*

MALES

12. Forewing with at least basal half of vein Cu_1 present 13
    Forewing with Cu_1 usually completely atrophied 17
13. Head, thorax, antennae, basal, and apical abdominal segments metallic blue-black; abdominal segments III–VII yellowish orange or reddish brown 14
    Head, thorax, antennae, and basal and apical abdominal segments black, blue-black, or metallic greenish black; antennae blue-black, or basally reddish brown or yellowish orange; abdomen reddish brown posteriad from segment III or IV. 15
14. Femora, tibiae, and tarsi of fore- and midlegs and the last 2 tarsal segments of posterior legs reddish brown; apex of fore- and hindwings with distinct infuscated band . *Sirex longicauda*
    Legs metallic blue-black; abdominal segments III–VII yellowish orange; only slight infuscation in apices of wings .... *Sirex areolatus*
15. Fore- and midlegs beyond the trochanters reddish brown; posterior legs bluish black except distal segment of the trochanter; basal segments of antennae reddish brown or black; thorax metallic blue or greenish; wings slightly yellow, with faint, smoky band around the margins ....*Sirex cyaneus*
    Legs not as above; basal segments of antennae reddish brown or yellowish orange; wings hyaline 16
16. An indistinct, postocular area reddish brown; coxae dark reddish brown; basal segments of abdomen almost black, remainder of abdomen orangish brown; mid-dorsal suture on first abdominal tergite not distinct; hairs on first and second abdominal tergites quite as dense as on dorsum of the thorax; first submarginal cell of forewing not noticeably yellow . *Sirex behrensii*
    Head unicolorous; coxae and basal segments of the abdomen metallic greenish black; remainder of legs and abdomen reddish brown, rather shiny; mid-dorsal suture on first abdominal tergite usually quite distinct, often slightly opened for part or all its length; density of hairs on first and second abdominal tergites, especially medially, not as great as on dorsum of thorax; costal and first submarginal cell of forewing noticeably golden yellow .... *Sirex juvencus californicus*
17. Head lacks a lateral carina; hindwing with a closed anal cell; eye is oval instead of nearly round; hind tibia with 2 apical spurs 18
    Head with distinctive carina behind the eye; hindwing without closed anal cell; hind tibia with 1 apical spur; eye nearly round; all species with a white spot behind the eye. (Male siricids are normally smaller than females, and the last abdominal sternite ends in a sharp point.) 20
18. Abdomen reddish brown; wings golden yellow; legs and antennae reddish brown .... *Urocerus californicus*
    Abdomen with segments III to VI yellow, the remainder black; wings nearly hyaline 19
19. Hind basitarsus 4.1 to 5.6 times longer than broad .... *Urocerus albicornis*
    Hind basitarsus 6.4 to 8.1 times longer than broad .... *Urocerus gigas flavicornis*
20. Abdomen black .... *Xeris spectrum*
    Abdomen almost reddish 21
21. Postocellar area very rugose; median furrow shallow, broadly rugose (figure 141, *F*) .... *Xeris tarsalis*
    Postocellar area sparsely pitted, not rugose, polished, a distinct, narrowly rugose median furrow (figure 141, G) .... *Xeris morrisoni*

The Genus *Sirex*

Six of the 9 species of *Sirex* that occur in North America can be found in California. The females oviposit in the trunks of many species of native conifers in the genera *Abies, Cupressus, Larix, Libocedrus, Picea, Pinus, Pseudotsuga,* and *Sequoia*. Members of this genus (except for *S. behrensii*) do not have the pale area behind the eyes such as that possessed by *Urocerus* and *Xeris*. The cornus is not normally constricted (figure 141, *H–M*), and the forewing (figure 141, *A*) has at least the basal half of Cu_1 present.

Sirex longicauda Middlekauff (figure 138) appears to be confined to California. It is one of the species commonly found infesting houses (Ebeling and Wagner, 1963). These insects are metallic dark blue, except that the males, first recognized and described by Middlekauff (1962), are reddish brown on abdominal segments III-VII and the tibiae and tarsi of the first 2 pairs of legs. The females are usually about 4 cm long, and the males are slightly over 2.5 cm.

Fig. 142. A woodwasp, *Sirex behrensii*.

Sirex areolatus (Cresson) ranges throughout the Pacific Coast states, as well as Arizona, New Mexico, and Colorado. It is predominantly coastal in distribution in California. Like *S. longicauda,* the female of this species has a steel-blue body, but her legs are bluish black throughout, whereas the legs of *S. longicauda* have red tibiae and tarsi. The 2 species are about equal in size. The male of *S. areolatus* can be separated from *S. longicauda* by the entirely blue-black tibiae and tarsi. Females of *S. longicauda* and *S. areolatus* have a much longer ovipositor than other species of *Sirex* (figure 141, *H* and *M*).

Sirex behrensii (Cresson) occurs in the Pacific Coast states and Nevada. The females can be distinguished by the reddish-brown abdomen beyond the second segment (figure 142) and by the shape of the cornus (figure 141, *L*). The males have a reddish abdomen except for the basal segments; the first 2 pairs of legs are reddish brown except for their coxae and trochanters; and the basal segments of the antennae are red. The legs of *S. behrensii* are colored similarly to those of *S. longicauda,* but the bases of the antennae, an indefinite area behind the eyes, and the apex of the abdomen are reddish brown.

Sirex juvencus californicus (Ashmead) ranges throughout the Pacific Coast states, and has also been recorded from New Mexico. The body and appendages of the female are dark blue, and the ovipositor is short, extending only 2 to 3 mm beyond the tip of the cornus (figure 141, *K*). The wings have a smoky band around the apical margin, and also one across the wings beneath the stigma. The male of this species was not described until 1967 (Cameron, 1967). The distal segments of the antennae, head, thorax, abdominal segments I and II, a triangular area on the anterior dorsum of segment III, and the coxae are metallic greenish black. The basal 5 or 6 segments of the antennae, the remainder of the abdomen, and the legs are reddish brown.

The blue horntail, *Sirex cyaneus* F. (plate II, 2), is generally believed to have originated in North America, where it may be found in most of the continent north of Mexico. This species is among those with an ovipositor shorter than the forewing. Its abdomen is entirely bluish black, and its legs, except for coxae and trochanters, are reddish brown. The wings are nearly transparent except for a smoky gray-brown apical margin. The cornus is not shouldered (figure 141, *J*). The wings of the males differ from those of the females in being slightly yellow, with a faint smoky band around the margins. The last pair of legs of the male are bluish black, in contrast to the reddish-brown legs of the females. The bases of the antennae are reddish brown or black.

The Genus *Urocerus*

Three of the 5 species of *Urocerus* that occur in North America may be found in California. *Urocerus* contains our largest species, attaining in *U. californicus* a length of nearly 4 cm from head to tip of cornus. The females oviposit in the trunks of conifers in the genera *Abies, Larix, Libocedrus, Picea, Pinus, Pseudotsuga, Thuja,* and *Tsuga*.

In common with *Xeris,* species of *Urocerus* possess a distinctive pale spot behind the eye, a constricted cornus, and usually vein Cu_1 in the forewing is completely absent. Unlike *Xeris,* the head lacks a lateral carina, the ovipositor is usually shorter than the forewing, the hindwing has a closed anal cell, the eye is oval (figure 141, *Q*) instead of round; hind tbiae have 2 apical spurs.

Urocerus californicus Norton is the largest and possibly the most common western woodwasp (Middlekauff, 1960). The female may reach 4 cm in length. It is the only California siricid with the combination of a black abdomen and golden yellow wings and antennae. It attacks principally the firs (white and red firs and balsam), species not commonly used in construction, and is therefore not likely to be found as a pest in buildings. The abdomen of the male is reddish brown.

Urocerus albicornis (F.) is relatively uncommon in California. In this species and in the female of *U. californicus,* the abdomen is almost entirely black. However, the wings are smoky brown, in contrast to the golden yellow of the wings of *U. californicus* females. In the male, abdominal segments III to VI are yellow and the remainder are black. The wings are nearly transparent. Belyea (1952) (cf. Middlekauff, 1960) stated that the life cycle from egg to adult required 2 years, but other authors state that it may be completed in 1 year. The larval burrows are said to run in all directions, and to be closely packed with a very fine, dustlike frass.

Urocerus gigas flavicornis (F.) ranges throughout the coniferous belt of North America, in the Pacific Coast states, and south along the Rocky Mountains to Arizona and New Mexico. It is somewhat smaller than *U. californicus,* and abdominal segments VII, VIII, and sometimes II are yellow.

The Genus *Xeris*

There are only 4 known species of *Xeris,* 3 in North America and 1 confined to the Himalayas. Like *Urocerus,* the species of *Xeris* have a pale spot behind the eye, but the ovipositor is longer than the forewing. The hind tibia has 1 apical spur. The eyes are more nearly round (figure 141, *P*) instead of oval, as in *Urocerus.*

The female of *Xeris tarsalis* (Cresson) (= *macgillivrayi* Bradley) can be distinguished from the other 2 species of American *Xeris* by the unconstricted cornus, the tibiae and tarsi being considerably lighter in color than the femora, and the basal part of the antennae being darker than the apical part of the flagellum.

Xeris morrisoni (Cresson) is widely distributed in the mountains of California, but attacks mostly fir and lodgepole pine, and is not likely to be found in commercial lumber. It can be distinguished from *X. spectrum* by the reddish abdomen.

Xeris spectrum (L.) is Holarctic in coniferous forests, extending farther south in mountainous regions. Lodgepole pine is believed to be its favorite host in California. This species may be distinguished from *X. morrisoni* by its entirely black abdomen, in contrast to the reddish abdomen of the latter.

The Genus *Tremex*

The species of *Tremex* can be separated from other siricids by the short antennae. Only a single species with several color phases has been described from North America. Unlike other siricids, species of *Tremex* feed on weakened or dying deciduous trees, not on conifers.

Tremex columba (L.) (the pigeon tremex) is widely distributed over the United States, occurs in southern California, and has been reared from many species and varieties of deciduous trees. In the typical *T. columba,* most commonly found in the northeastern United States and southeastern Canada, the abdomen is black, with ocher-yellow bands and spots along the sides. The common form in Colorado, New Mexico, Arizona, and California is the race *aureus,* in which the ground color of the abdomen is yellow and the markings are black. The wings are golden yellow. The race *sericeus* may be found in the southeastern United States, as far north as Pennsylvania and west into Colorado, New Mexico, Arizona, and California. The entire body is fulvous (tawny), the legs distal to the femora are yellow, and the wings are dark reddish brown. More study will be required to determine whether these 2 color forms should be given subspecies status (Middlekauff, 1960).

In New Brunswick, the female of *T. columba* deposits her eggs most commonly in weakened or injured beech trees from about the middle of August to the middle of October, laying 2 to 5 eggs in each oviposition hole. The blackish eggs may either hatch in 2 to 4 weeks or the next May or early June. The larvae may be 4 cm long, and are believed to be able to complete their growth and transform to pupae within a year. Usually, only the upper trunks and injured branches in the crown of the tree are attacked, and this species is therefore of little economic importance (Stillwell, 1967).

Woodwasp Damage

One of the most important factors contributing to woodwasp damage is forest fire. In the western states, where there is characteristically a long dry season in summer, severe infestations of woodwasps frequently occur in the wakes of fires that annually kill or damage thousands of acres of timber. Fire-killed timber is salvaged and utilized for certain construction purposes, as for example the studs and subflooring. Although kiln-drying will kill woodwasp larvae that have survived milling operations, the relatively low value of the lumber makes this treatment economically impractical.

There is danger of increasing the range of distribution of the various species through commerce. Pacific Coast species have been reported emerging from a new door in Massachusetts and from framing of a new house in Georgia. Australia requires that raw lumber imported for use in crates or prefabricated housing be heated by a steaming treatment at 165° F (74° C) for 10 hours to kill such insect pests (NPCA, 1964*a*).

Woodwasps, as well as certain wood-infesting beetles, are "built in" to the house at the time of construction, and whatever damage they do will be completed within the first 2 or 3 years after the house is built. The exit holes, although conspicuous because of their large size, are not likely to be numerous, although there have been occasional instances of dozens of holes in one building. However, the damage from woodwasps is not the structural weakening of timbers, but the defacement of walls, floors, and other interior surfaces from which the adults emerge.

The exit holes of adult woodwasps are not necessarily found in the lumber originally infested, but in whatever material the adult insects must penetrate in their attempts to leave infested wood and fly away. The adults have powerful jaws and are said to be able to cut through lead sheathing (H. Scott, 1932; Middlekauff, 1960). In one industrial building, a woodwasp had bored through 2.5 cm of plaster.

Some homeowners have reported a small amount of "sawdust" around or under the exit holes of the emerging adults, but most have not observed this. R. W. Stark (correspondence) found no frass around or under holes of woodwasps emerging from fir logs. This was in contrast to the exit holes of the newhouse borer, *Arhopalus productus,* a cerambycid beetle that pushes out long drillings that look like cigarette tobacco from its burrow when emerging.

Neither woodwasps nor newhouse borers can reinfest buildings. This is in contrast to the oldhouse borer, *Hylotrupes bajulus,* another cerambycid that may attack very old buildings and can reinfest the same structural timber until the wood is reduced to a mass of powder. The oldhouse borer also makes holes about 7 mm in diameter, but it appears to be confined to the east coast and Gulf states.

Control of Woodwasps

Woodwasp adults can emerge from wood that has been sprayed or brushed with the commonly available insecticides. However, the emergence holes can be filled with any appropriate filler material, and the filled area can be spot-painted. A painted wall can be redone, of course, but it is impossible to be certain that the last adults have emerged until about 3 years after the house is constructed.

Kiln-drying or vacuum fumigation of the lumber before use would be effective, but merely fumigating it in boxcars or under a tarpaulin has not proved to be of much value (Middlekauff, 1962). Fumigation has likewise proved to be ineffective in buildings (Ebeling, 1968). After some failures with methyl bromide, in one building Vikane (sulfuryl fluoride) gas was used in order to obtain a maximum degree of penetration. It was used at a dosage of 2 lb per 1,000 cu ft (0.91 kg per 28 cu m) with a 48-hour exposure. For the first 30 days after fumigation, emergence of woodwasp adults had ceased, whereas it continued during this period in near-by nonfumigated buildings in the residential tract. However, after the approximately 30-day period, emergence of adult woodwasps was again observed. The author concluded that the gas penetrated to and killed woodwasps that were near the surface, either because they were in the act of emerging or because the lumber had been cut to within a short distance from their burrows. Evidently, many of the insects were not reached by the gas and later emerged (Ebeling, 1968).

As already stated, woodwasp larvae burrow deeply. Even the pupal chambers are 2 cm or more from the surface. Some of the larvae or pupae are likely to be left at considerable distances from the surface in such wooden members of residential construction as studs and floor and ceiling plates. Although the tunnels may be exposed when such lumber is cut in the sawmill, they are tightly blocked with frass and borings by the larvae, and this may prevent the passage of lethal amounts of even the most penetrating fumigants. In Britain, however, timbers up to 10 cm thick are reported to have been successfully fumigated to control woodwasps by using 3 lb per 1,000 cu ft (1.36 kg per 28 cu m) of methyl bromide, applied for 24 hours at not less than 90° F (32° C) (NPCA, 1964*a*).

One pest control operator observed that of the 54 holes made in the plaster walls of 4 houses by emerging woodwasps, none were made in the areas at which the plaster was contiguous to studs. He was able to push a wire probe through each hole and move it about freely within the interstud wall voids. The woodwasps had emerged from the sides of the studs, entered the interstud wall voids, and then bored their way through the plaster.

The silica aerogel Dri-die 67® was blown into the walls at the rate of 6 grams per interstud space (3 g below and 3 g above each fire block) through holes drilled into the plaster walls and subsequently filled with matching plaster.

The number of emergence holes in the treated houses had steadily increased up to the date of treatment, but no new holes were found thereafter. Since the treated houses were only 6 months or less old, and emergence could continue for 2 or 3 years after construction, it was unlikely that the cessation of emergence immediately after treatment was a coincidence. Dusting of this kind can be done most efficiently with a slightly modified water-type fire-extinguisher, large numbers of which are currently being sold to pest control operators for applying insecticide dusts into wall voids, attics, and other enclosed spaces. (See under "Residual Protection" earlier in this chapter.) The 6 grams of Dri-die used per interstud space amounted to 2 lb per 100 linear ft (0.91 kg per 30 m) of wall, or 2.5 lb per 1,000 sq ft (1.14 kg per 93 sq m) of floor space.

The reason that a dust desiccant was recommended was that, being inorganic, it would remain effective against woodwasps as long as needed, and also serve as a preventive against drywood termites, cockroaches, silverfish, carpenter ants, odorous house ants, fungus beetles (Lathridiidae), tropical rat mites, and other cryptobiotic pests.

The pest control operator responsible for the successful treatment just described had subsequent successes with the same method. However, the author was informed by R. E. Wagner that he had observed woodwasp emergence holes in a house at the point of contact of plaster and stud. In that event, the emerging insects would probably not have had contact with dust if the wall voids had been treated. Even in such a situation, the dust desiccant in interstud spaces might be expected to kill most of the woodwasps.

CARPENTER ANTS

General information on the biology of ants may be found in chapter 6. In the present chapter, only those species that damage wooden structures will be discussed. They are called "carpenter ants," and are all included in the genus *Camponotus*. Besides causing considerable destruction to structural timbers, carpenter ants are also nuisances in the same way as any other ants, that is, by crawling about in the house in search of food. They are distributed throughout the United States, and occur at elevations up to 9,000 ft (2,700 m). They are abundant in forests, and can be easily found under the loose bark of dead trees, stumps, or fallen logs.

In the Pacific Northwest and some other areas, carpenter ants are considered to be equal to termites in importance as pests. They often infest cabins in forests. In cities, they usually infest houses in foothill areas containing native chaparral or trees or in estates containing large shade trees, but are sometimes found in homes in crowded residential districts as well.

These ants are predominantly black, and are among the largest species in the United States. The petiole has a single node. They are polymorphic; there are workers of different sizes, called "major," "minor," and "intermediate." These workers can emit a strong formic acid odor. They do not sting, but their bite can be painful, particularly if they inject formic acid into the wound.

Among American carpenter ants, the species for which the biology has been most thoroughly investigated is the black carpenter ant, *Camponotus pennsylvanicus* (De Geer). It is especially common in the eastern and central United States up to 5,500 ft (1,700 m) elevation. In a study of colonies made on the campus of the University of Illinois, Pricer (1908) concluded that sexually perfect individuals are not produced in a colony until it contains approximately 2,000 workers, and that a period of 3 to 6 years or longer is required for a colony to reach this size. The distributions of castes in the 2 largest nests found by Pricer were as follows: In an oak log, he found 3,018 workers, 1 wingless queen which started the colony, 196 winged queens, 177 males, and 842 larvae. In an ash tree, he found 3,212 workers, 1 wingless queen, 233 winged queens, 176 males, and 724 larvae. In his studies, he found 14 colonies with 1,000 or more workers. The smallest colony among the 24 Pricer found consisted of 1 wingless queen, 106 workers, and 93 larvae. Colonies did not produce winged forms until they were more than 2 years old. The winged forms that were produced in summer overwintered in the parental nest, and left for the nuptial flight in May to late July.

Among ants reared in 2 artificial nests in the laboratory at temperatures ranging from 70 to 90° F (21 to 32° C), Pricer found that *C. pennsylvanicus* spent the following numbers of days in the various stages: egg, 24; larva, 21; and pupa, 21—a total of 66 days from egg to adult.

Carpenter ants burrow into wood not to feed on it, but to make nests. In nature, their nests are found in dead trunks of standing trees, stumps, or logs, or under fallen logs and stones, sometimes in galleries extending far underground. They also burrow into structural timbers, and if their galleries are extensive, the ants can cause some damage to buildings. Even species able to burrow into sound timbers prefer to burrow into unsound wood, such as that which has been invaded by dry rot. More commonly, at least in California, they use existing hollow spaces, including wall voids, hollow doors, cracks, crevices, furniture, and termite galleries, in which to make their nests. In forested areas, they often establish colonies in masses of fir needles and other refuse that they accumulate within walls, under floors, in attics, and in other undisturbed areas.

The nest may also be behind the rear end of a drawer in a dresser or cabinet, or behind books in a library. In an area that is seldom disturbed, such as an attic, the nest may be exposed to view, as on the plaster laths between 2 ceiling joists.

Carpenter ants sometimes enter a house at a point at which tree branches contact the roof or other parts of the house. They may also enter through such access points as windows, holes in the foundation, foundation vents, heating ducts and air-conditioners, or along power or telephone lines. They can also be brought into the house in firewood.

Signs of carpenter ant infestation are the large, black ants running about the house or swarms of the winged reproductives in or on the house, usually in the spring; piles of fibrous, sawdustlike borings which the ants have expelled through slitlike openings in the infested wood; the slitlike openings themselves in woodwork, especially in window or door casings; or a faint rustling in walls, floors, and woodwork (Furniss and Every, 1958).

The borings of carpenter ants are similar to those of carpenter bees. The borings of these 2 hymenopterous wood invaders and the fecal pellets of the drywood termite, *Incisitermes minor*, are shown in figure 143. The sizes of the particles can be visualized by the fact that the fecal pellets of drywood termites average about 0.85 mm in length.

A carpenter ant colony in a home may consist of the progeny of a single queen that has invaded the house. It is more likely to consist of individuals that immigrated from an existing colony, especially when the colony is seriously disturbed, as in the clearing of building sites. Or, the ants may have merely wandered in from the outside in search of dead insects, other animal matter, or sweets.

In a survey, 30 American and Canadian pest control operators were asked where they found carpenter ants nesting in houses. Their answers indicated the relative importance of the different locations: inside walls (15), in ceilings (10), under outdoor siding (7), in wood contacting soil near foundations (7), by downspouts near the roof (6), around roof gutter braces (5), under roofing panels (4), in hollow doors (4), in floors (2), in windowsills (2), under a bathroom floor (1), and in padded ceiling insulation (1) (Anonymous, 1967*a*).

All kinds of houses are invaded by carpenter ants, regardless of age or type of construction. When they infest wood, some preference is shown for moist, rotting timbers about the foundations, but sound wood can be mined by some species, anywhere in a house. Carpenter ants excavate galleries that somewhat resemble the work of drywood termites, but they contain no debris and have an almost polished appearance. The galleries extend in all directions, like those of the drywood termites and unlike those of the subterranean termite, which follow the spring growth.

The workers may sometimes be seen in the house, foraging about for food, or swarms of the large winged reproductives may emerge from the colony and try to escape through the windows.

Fig. 143. ***Top left,*** **pellets of western drywood termite,** ***Incisitermes minor; top right,*** **borings of valley carpenter bee,** ***Xylocopa brasilianorum varipuncta; bottom,*** **borings of a carpenter ant,** ***Camponotus clarithorax.***

With either the wingless or winged forms, it is important to know how to distinguish them from termites (figure 66), for control methods differ. The winged forms comprise only a part of the colony—the rest stay behind and continue to develop and extend the infestation.

Some California Carpenter Ants

There are so many species of carpenter ants throughout the United States that either occasionally invade homes foraging for food or, in the case of some species, build their nests within houses, that space would not permit a discussion of all species, even if adequate information were available, which in many areas is not the case. Therefore, the following discussion is confined to the house-infesting carpenter ants of California, although some of the species may be found in other western states and even in the East.

Camponotus nearcticus Emery

This species (figure 144, *A*) has a discontinuous distribution, occurring in the West from British Columbia to California, thence eastward to Idaho. Farther east, it is found from North Dakota to Ontario, and ranges from Colorado to Florida. These ants occur mostly at lower elevations, but have been observed at altitudes as high as 6,000 ft (1,800 m). The workers are 4.5 to 7.5 mm long, smaller than those of most carpenter ants. The color is generally black throughout, but on some specimens, some segments or areas of the body may be largely reddish or brownish. The body is shiny, particularly the abdomen. This species forms small colonies of less than a hundred to several hundred, "in dead twigs and branches of trees, in or beneath the bark of dead or live trees, in insect galls, pine cones, beneath the bark of logs and stumps, in hollow stems of plants, and in wooden posts." Day and night, the workers crawl over shrubs and trees in search of honeydew and dead insects. They often form their nests in houses, but seldom infest household foods (M. R. Smith, 1965).

Camponotus hyatti Henry

This species (figure 144, *B*) is similar to *C. nearcticus* in size and appearance, but is restricted to southern California, where it normally nests in dead limbs of oak. In the chaparral belt, where so many suburban homes are being built, it is often found in them, searching for sweet foods. These ants may sometimes infest termite galleries in buildings (R. R. Snelling, correspondence).

Camponotus clarithorax Emery

This is the species (figure 144, *C*) most commonly found in houses in southern California, and it ranges northward to Oregon. The ants nest in crevices, in interstud wall voids, or under floors between joists, in hollow door panels and window casements, and in termite galleries. If "sawdust" (figure 143) is found that has been pushed out of the structural timbers by this species, it is most likely to be from wood that has been softened by dry-rot fungus infection. These ants feed largely on carbohydrates. *Camponotus clarithorax* is a small species; major workers are 7 to 8 mm and minor workers are 5 to 6 mm long. The head and gaster are dark brownish to nearly black, and the thorax is yellowish to reddish. The anterior margin of the clypeus has a notch in the middle (as in *C. hyatti*), and the cheeks have coarse punctures, from each of which arises a short, stiff hair (R. R. Snelling, correspondence).

Camponotus vicinus Mayr

These carpenter ants (plate II, 3; figure 144, *D*) are widely distributed throughout California in both lowlands and mountains, and may be found under stones in rather dry, sunny places. The ants are large, the workers being 6 to 10 mm, winged males 10 mm, and the queens 13 to 15 mm long. They have a red thorax and black abdomen. The life cycle requires about 2 months, and approximately 2 years will pass before a new colony will begin to produce alates. This species may have large colonies, with over 3,000 individuals. They do not nest in houses, but may sometimes be seen in them, foraging for food. Their nests are usually in the soil at some distance from any house. They occasionally damage lawns by building their mounds in them.

Camponotus semitestaceus Emery

This species (plate II, 4; figure 144, *E*) differs from *C. vicinus* in being paler, with the thorax and first and second segments of the abdomen near the petiole yellowish. Its habits are similar to those of *C. vicinus.*

Camponotus laevigatus (F. Smith)

This species (figure 144, *F*) ranges from Colorado westward, most often in the Transition Zone. This diurnal forager is common throughout California at high altitudes, and its large colonies may be found in dry stumps or logs. It seldom invades houses. The entire insect is jet black and very shiny, with abundant white hairs, and often with strong, bluish reflections. It can burrow into sound wood, and has some potential as a structural pest, but generally utilizes galleries of dampwood termites (R. R. Snelling, correspondence).

Camponotus modoc Wheeler

The workers of *C. modoc* are 5 to 11 mm or more in length, dull black with reddish legs, with a lighter color on the margins of the segments of the gaster. The abdomen is covered with closely appressed, long, golden hairs (R. R. Snelling, correspondence). This species is most common in the northern part of California at altitudes beginning at 3,000 ft (900 m). It is occasionally found at lower elevations, even in the valleys. It can mine sound wood. Large colonies occur in stumps and rotten centers of fir-tree butts, doing little damage to living tissue. These ants sometimes work from near-by stumps into the timbers of cabins or larger houses. This species is pugnacious, and bites viciously when disturbed (Mallis, 1941*a*).

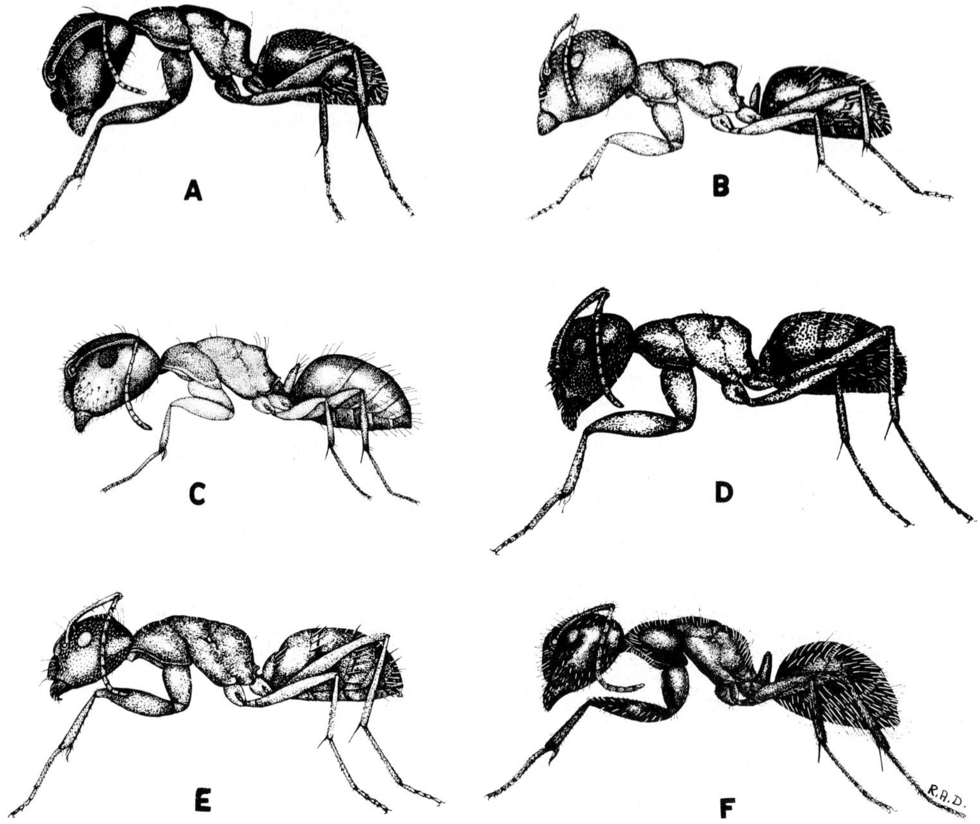

Fig. 144. Some California carpenter ants. *A, Camponotus nearcticus; B, Camponotus hyatti; C, Camponotus clarithorax; D, Camponotus vicinus; E, Camponotus semitestaceus; F, Camponotus laevigatus.* (Courtesy R. R. Snelling.)

Some Eastern Species

The black carpenter ant, *Camponotus pennsylvanicus* (De Geer), is the most common species in the eastern states. The red and Florida carpenter ants, *C. ferrugineus* (F.) and *C. abdominalis floridanus* (Buckley), are also common.

Control of Carpenter Ants

The control of these ants may involve the destruction of colonies within 90 m or so of the house, around the foundation where ants are observed, or even under the house, using 2% chlordane spray or 5% dust, or lindane at lower concentrations. Use enough spray or dust to cover all surfaces on which the ants may crawl. Provided the location of the nest can be determined, 6% chlordane dust in wall voids or other hard-to-reach places inaccessible to liquids has been recommended (Rachesky, 1970*a*).

Inside the house, carpenter ants may be found in secluded but not completely covered places. These can be easily treated. It may be necessary to remove siding, window casings, doorsills, or similar members to reach the ant runways. If the ants are within timbers, try to determine where their galleries are by locating the small, slitlike openings or the piles of "sawdust," then bore small holes into the timbers to reach the galleries, and blow 5% chlordane dust into them (Furniss and Every, 1958). Drione®, a formulation of Dri-die 67® and pyrethrins, is also effective for treating carpenter ant nests, and results in rapid knockdown and kill. The most difficult part of carpenter ant control is often the location of the nest. A doctor's stethoscope is helpful in locating nests in particularly difficult cases.

A pest control operator in southern California has successfully controlled carpenter ants in wall voids by blowing the fluorinated silica aerogel, Dri-die 67. The dust retains its insecticidal properties indefinitely if it does not become wet, and therefore acts as a preventive against possible future infestations of carpenter ants as well as other species of cryptobiotic insects.

CARPENTER BEES (ANTHOPHORIDAE)

There are 2 groups of Hymenoptera whose activities in dead wood and wooden structures have suggested the term "carpenter" in their common names: carpenter ants and carpenter bees. Carpenter ants burrow into wood, expelling typical borings (figure 143), or use ready-made enclosed spaces, not only to make nests in which to lay eggs and care for their young, but also as harborages for themselves. They leave the gallery system, which is sometimes very extensive, to forage for food, but always return to their "home." Carpenter bees likewise burrow into wood, forming a series of chambers, but only to rear their offspring. The chambers, beginning with that farthest from the entrance, are each provisioned with a pollen mass on which a single egg is laid, and are then capped with disks of wood particles. Each pollen mass in the series of burrows is enough to bring a larva to maturity. After sealing the cell nearest the burrow entrance, the adult carpenter bee departs, not to return. After the burrows are abandoned, other carpenter bees may find them and store pollen in them to feed on before hibernation. (They use old carpenter bee tunnels for this purpose.)

Identification. In North America, the wood-boring or carpenter bees are found in 2 genera, *Ceratina* and *Xylocopa*. The former includes the "little carpenter bees," which average about 6 mm in length, are rather slender, and metallic green, blue, purple, or rarely black. They occasionally nest in hollow stems and reeds, but most often in the hollowed-out stems and canes of pithy plants (H. S. Smith, 1907). Their habits are similar to those of *Xylocopa,* but they do not bore into structural timbers, and are usually of little importance to the homeowner.

The genus *Xylocopa* includes the large, robust carpenter bees that somewhat resemble bumble bees, except that the dorsal surface of the abdomen is largely naked, whereas that of the bumble bee is hairy. Also, the females have a dense brush of hairs on the hind legs, as contrasted with the pollen baskets of the bumble bees. Carpenter bees are usually metallic blue-black, with green or purplish reflections, but in some species the males are entirely or partially buff or pale yellow. Other morphological differences between the sexes are: males have 13-jointed antennae, females have 12 joints; males have 1 spine at the tip of the hind tibia, females have 2; males have 7 exposed abdominal segments, females have 6; males have only 2 teeth at the apex of the mandible, females have 2 at the apex and also on the upper margin. The triangular row of spines on the dorsum of the last abdominal segment of the females is not present on the males (Essig, 1926). Keys and detailed descriptions of the genera, subgenera, and species of the New World and Old World large carpenter bees (tribe Xylocopini) were prepared by Hurd and Moure (1963). Keys to California *Xylocopa* were prepared by Hurd (1955).

Biology. The biology of *Xylocopa virginica virginica* (L.) has been described in considerable detail (Rau, 1933; Osmun, 1954; Chandler, 1958). This species ranges from Maine to Florida and westward to Nebraska and Texas. It is black in both sexes, marked with areas of yellow hair. The adults of both sexes hibernate in abandoned nest tunnels, as just mentioned. In areas of low winter temperatures, many may die. In the spring, the survivors emerge from the tunnels to search for nectar for their own food. Mating then takes place, and extends into the nest-construction period. For the purpose of preparing her nest, the female may use an old gallery without further burrowing, or she may lengthen it, or may bore an entirely new gallery or extend one from an entrance used by several bees.

Newly excavated galleries average 10 to 15 cm in length, but a gallery system developed by several bees may be 2 to 3 m in length. A mass of pollen and regurgitated nectar is packed into the bottom or end of the brood gallery, and the female lays an egg on it. She then seals off this portion of the gallery with a disk of chewed wood pulp. This procedure is repeated until a linear series of an average of six cells is formed. The developmental period for *X. v. virginica* is 36 days. Following emergence, the adult bee spends a day or two within the cell in drying and cleansing itself, and perhaps feeding on pollen remains.

Rau (1933) found that with *X. v. virginica,* the first bee to become an adult developed in the first cell to be provisioned. It then cut through all the partitions, by-passing the other developing brood.

The newly emerged carpenter bees have different habits from their parents, that have died by the time the young emerge. They store caches of pollen in abandoned carpenter bee tunnels. The pollen is consumed during inclement weather by bees that are about to hibernate. During this period, there is no mating and no boring in wood, except that an occasional bee may enlarge its hibernating quarters.

Nininger (1916) studied the biology of *X. tabaniformis orpifex* nesting in a small, deserted cabin in the San Gabriel Mountains in southern California. The insects preferred sound redwood, but a few bored into sound Douglas fir. The bee and its typical cells are shown in plate II, 7, and figure 145. Slanting or horizontal surfaces were generally chosen by the bees studied by Nininger, and entrances were made from the undersides of the timbers. The burrow continued a short distance at (or nearly at) right angles to the surface, and then turned sharply to follow the grain of the wood. The burrows varied in length from 2.5 to 30 cm (average 10 to 15). Burrows of greater length were believed to be more than one season's work. A female could lengthen her burrow by 2.5 cm in six days. In their tunneling, the females nearly always remained equidistant from the 2 outer surfaces of a timber, possibly guided by the vibration caused by the burrowing activity.

Fig. 145. The mountain carpenter bee, *Xylocopa tabaniformis orpifex,* and its galleries and cells in a cedar shake. In some cells are immature bees in various stages of development.

When a new site was chosen by a single female, others soon found it, and every available timber became honeycombed with tunnels. The bees did not store food for themselves in the tunnel, but came forth in search of food when temperatures were 20 to 21° C (68 to 70° F) or over. The provisioning of the cell and oviposition were approximately as described for *X. v. virginica.* There were usually 5 or 6 cells completed—about 1 per day.

Developmental periods for *X. t. orpifex* were as follows: The eggs hatched in 7 days. The larvae developed in 22 to 28 days, but remained the next 15 to 19 days with little change. About 40 to 45 days were spent in the pupal stage. Thus, the period from egg to adult ranged from 84 to 99 days. If remnants of larval food were present, the bee might remain for some time within its cell, but otherwise it destroyed the entire series of cell partitions separating it from the tunnel entrance in order to get to the outside. Sometimes the emerging bee waited for the bees in the cells ahead of it to emerge, but this did not appear to be the rule (Nininger, 1916). Davidson (1893) stated that the first adult to emerge came from the last cell to be completed. He stated that males developed in the last cells to be completed, and that they developed more rapidly than the females. Thus, Davidson's observation did not agree with Rau's, mentioned earlier, regarding the order in which the cells were evacuated.

California Species of *Xylocopa*

In America north of Mexico, there are 7 species of *Xylocopa.* In the eastern United States is found *X. virginica virginica* (L)., considered to be the most destructive carpenter bee in the country (Chandler, 1958). Of the 3 species in California, *X. tabaniformis orpifex* Smith (figure 145) causes most of the structural damage, but damage is occasionally reported from *X. tabaniformis androleuca* Michener and *X. californica arizonensis* Cresson (Hurd, 1955, 1958). Nininger (1916) stated that *X. brasilianorum varipuncta* Patton preferred to burrow in decaying wood, but Mallis (1969) had records of its damage to structural timbers. Likewise, in 1970, Roy J. Pence (personal communication) observed the Hawaiian subspecies around structurally sound picnic tables that had been severely mined by the insect.

Mountain Carpenter Bee, *Xylocopa tabaniformis orpifex* Smith

This species is 12 to 17 mm long, and both sexes are black (plate II, 7). The head of the male has white and yellow hairs mixed with the black. Mature larvae are yellowish white, robust, and 7 mm long. The details of the life cycle of this species have already been discussed under "Biology" to exemplify the usual life cycles of carpenter bees.

These bees are found mainly in the foothills and mountains of California, Oregon, Nevada, Arizona, and Baja California. Hurd (1955) observed that the creation of suitable nesting sites by man, in the form of structural timbers, had resulted in the extension of the range of this species to lower altitudes.

Valley Carpenter Bee, *Xylocopa brasilianorum varipuncta* Patton

This bee is 18 to 20 mm long, somewhat larger than *X. t. orpifex*. The metallic black of the female, with brilliant violaceous, brassy, or bronzy reflections, is in sharp contrast to the golden-brown or buff color of the male (plate II, 8; figure 146). This is the species that is most likely to be found in the valleys and lower foothills of California, Arizona, and Baja California. It is said to prefer partially decayed live oak, deciduous oak, pepper (*Schinus molle*), eucalyptus, and other hardwood trees. Several of its burrows may have a common entrance (Nininger, 1916). Figure 147 shows a male bee emerging from a tunnel that had been cut into the exposed transverse section of the sawed-off flowering stalk of a dead Spanish bayonet, *Yucca whipplei*.

California Carpenter Bee, *Xylocopa californica californica* Cresson

This species is 20 to 24 mm long, and both sexes can be predominantly either metallic blue or metallic green, with dusky wings. The pronotum of the male has white, yellow, or orange pile, and the entire first abdominal segment has whitish hairs. These bees nest in redwood and incense cedar in the north Coast Ranges and the Sierra Nevada of California, and in the Cascade Mountains of Oregon.

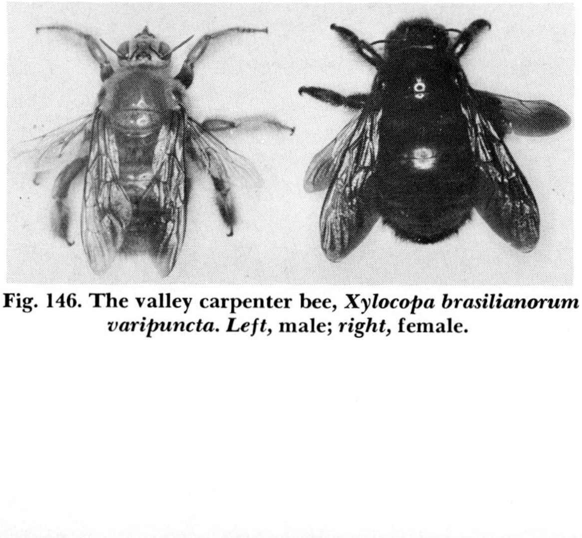

Fig. 146. The valley carpenter bee, *Xylocopa brasilianorum varipuncta*. *Left*, male; *right*, female.

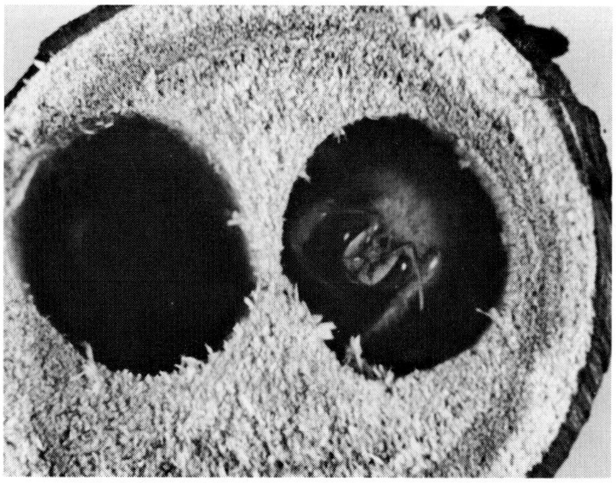

Fig. 147. A male valley carpenter bee emerging from a tunnel in the trunk of a yucca.

The subspecies *X. c. arizonensis* Cresson is 15 to 20 mm long. It is always dark blue, and its wings are blackish, with a strong violaceous tinge; they are more nearly opaque than those of the type. Also, the male has white hairs only on the *sides* of the first abdominal segment, unlike the type, which has them on the entire segment (Essig, 1926; Hurd, 1955). This subspecies occurs in the deserts of Arizona, California, Nevada, New Mexico, Utah, Texas, and Mexico, nesting in various agaves and yuccas.

Still another subspecies, *X. c. diamesa* Hurd, resembles *X. c. arizonensis*, but the wings are paler and less strongly violaceous, and the blue abdomen has traces of green. It occurs in the mountains of southern California and northward to Monterey County, and nests principally in *Yucca whipplei* (Hurd, 1955).

Economic Importance

Chandler (1958) listed the types of damage done by carpenter bees as follows: "(1) the weakening of structural timbers through the excavation of galleries; (2) excavation of galleries in wooden water tanks (especially in arid western areas), resulting in costly leaks; (3) unsightly defecation streaks on houses or other painted structures; and (4) annoyance."

Although severe structural damage by carpenter bees is rare, some old, uninhabited, or little-used buildings in mountain or foothill areas may eventually become so honeycombed with the burrows of these insects that they will collapse. People are sometimes alarmed by the large size of the bees, their resemblance to bumble bees, and their noisiness. The males are inclined to make rapid flights around the head of a person intruding near nesting sites, sometimes hovering a short distance in front of his face, and they are not discouraged by attempts to swat them. However, the males have no stinger, and are harmless. The females do have stingers, but do not use them unless they are handled or accidentally come into contact with a person (Chandler, 1958). During the period when females are searching for nesting sites, they will circle around persons standing in open areas as if inspecting them to determine whether they are posts suitable for colonization.

Control of Carpenter Bees

An insecticide such as 5% chlordane emulsion in an oil can may be squirted into each hole. A few days later, the holes may be plugged with dowel pins or plastic wood. Merely spraying the infested wood with 2% chlordane spray will keep carpenter bees away for a season, but they will probably come back the following year to re-use old holes and make new ones (Anonymous, 1970*f*).

Chapter 6 · *Pests On or Near Food*

THIS CHAPTER deals with three of the most important groups of household pests—the domestic species of the cockroaches, flies, and ants. These are insects whose habits are such that they commonly come in contact with food or crawl about on surfaces on which it is placed, such as shelves, sinks, and breadboards. Cockroaches and flies are known to be mechanical carriers of many kinds of disease germs.

COCKROACHES

Cockroaches are flattened, fast-running, nocturnal insects that seek warm, moist, secluded areas. They are of ancient origin, and have not changed much during the 400 million years since the Devonian period of the Paleozoic era (Jeannel, 1960). They are gregarious and, because they develop from egg to adult by gradual metamorphosis, all stages and instars are found together in their preferred harborages. Cockroaches are now placed in a separate order: Dictyoptera.

These are among the most important of the household pests. In a survey of 8 large housing projects in North Carolina, 2 in each of 4 cities, cockroaches were found in 76.7% of the apartments in which control work was done by project maintenance men and in 12.7% of those that were regularly treated by licensed pest control operators (Wright, 1965*a*). Cockroaches are also known to cause severe injury to plants and their fruits, particularly in greenhouses (Roth and Willis, 1960).

This group of insects is well known to biology students because, like grasshoppers, they are commonly chosen for laboratory demonstrations of a somewhat generalized type of external and internal insect morphology. They are also commonly reared for research purposes, e.g., as test insects in the investigation of new insecticides. Nearly every entomological research group continuously maintains a culture of 1 or more species of cockroaches.

It is fortunate that during the last decade, insects of such great importance as the cockroaches, not only as pests, but also as experimental animals in teaching and research, have become the subjects of some excellent books containing a wealth of information of both academic and practical importance. In *The Biology of the Cockroach* (Guthrie and Tindall, 1968), one chapter deals with insecticides and control. Likewise, *The Cockroach* (Cornwell, 1968) has thus far appeared as Volume I, which contains "An account of the biology of the more common species, including details of their structure, physiology, behavior, and ecology." Volume II is expected to be published in 1975, and will be devoted to cockroach control methods. It should be available by the time our present book has been completed.

Cockroaches as the Ancestors of Termites

Certain morphological, biological, and behavioral characteristics of cockroaches indicate that they were the ancestors of termites, which appeared much later in the scale of geological time. These characteristics include the possession of intestinal protozoa (flagellates) by the primitive wood-eating cockroaches, *Cryptocercus punctulatus* and *Panesthia javanica,* the former being indigenous to the United States. These protozoa convert indigestible cellulose into soluble sugar, as they do in the more primitive families of termites (Trager, 1932; Cleveland *et al.,* 1934; Rau, 1941). Of interest in this connection is the observation by Snyder (1935) that *C. punctulatus* and the termite *Zootermopsis* were found in California inhabiting the same log. Cockroaches and primitive termites also have intracellular symbionts (Brooks and Richards, 1955). This has led to the idea that the common ancestors of cockroaches and termites had both intracellular microorganisms and xylophagous flagellates. Cockroaches, except for the genus *Cryptocercus,* have retained the intracellular symbionts, and *Cryptocercus* has retained the intestinal flagellates. The primitive living termites (all the families except Termitidae), except for *Mastotermes darwiniensis* Froggatt, the most primitive of them all, have kept only the intestinal flagellates (*M. darwiniensis* has both). The Termitidae have lost the xylophagous intestinal flagellates, but have acquired a rich and complex bacterial fauna (Grassé and Noirot, 1959).

Mastotermes darwiniensis has an egg mass similar in appearance to the egg capsule (oötheca) of cockroaches. Its eggs are firmly cemented together by a light-brown, gelatinous secretion that fills the interstices (Hill, 1925; Snyder, 1935).

The ability many cockroaches have to mix bits of debris with mouth secretions in order to hide or disguise their oöthecae reminds one of a similar process used by termites to build nests. There are other interesting similarities (Rau, 1941; McKittrick, 1964, 1965; Cornwell, 1968).

Cryptocercus punctulatus leads a subsocial life in decayed logs, which serve as both food and shelter (Snyder, 1935). An Australian cockroach, *Panesthia,* lives in distinct family groups in burrows in the soil. Each group consists of an adult male and a viviparous female, and from 10 to 20 nymphs in various instars. Soon after reaching maturity, the adults bite off their own tegmina and hindwings; wings are inconvenient for insects that inhabit burrows. This indicates the probable origin of termites' social life in subterranean galleries and their discarding of wings (Tillyard, 1926). The subsequent evolution of termite social organization to such an astounding degree of complexity, superficially so strikingly similar to that of the social Hymenoptera, depended on a concomitant chemical evolution of pheromones and a highly specialized chemical communication among these insects. This is a good example of the important role of pheromones in insect evolution.

The ancient cockroaches folded only a small area (anal lobe) of their hindwings, as can be seen in the fossil genus *Pycnoblattina.* The living *Mastotermes darwiniensis* also folds the anal lobe of its hindwings. The cockroaches gradually evolved more complex folding of the hindwings, while the termites, except for *M. darwiniensis,* completely lost this feature of their wing structure (Tillyard, 1936, 1937; McKittrick, 1964).

McKittrick (1965) believed that cockroaches, walkingsticks, and termites belonged in the same insect order. She pointed out that *Cryptocercus* had xylophagous flagellates in common with primitive termites but not with other cockroaches. On the other hand, the primitive *Mastotermes darwiniensis* has intracellular bacteriocytes in common with cockroaches, but not with other termites. Also, in certain morphological features, *Cryptocercus* differs far less from *Mastotermes* than it does from many of the cockroaches.

The Place of Cockroaches in the Class Hexapoda

The place of cockroaches in the class Hexapoda has been the subject of considerable disagreement among experts during the last half-century. All the common cockroaches were formerly considered to be members of the family Blattidae of the order Orthoptera. This order included, besides the cockroaches, the grasshoppers, crickets, katydids, mantids, walkingsticks, and several other related families. In recent years, several new classifications have been proposed. The author has adopted the classification used by Cornwell (1968), based on the one suggested by Imms (1957) in the ninth edition of his *A General Textbook of Entomology.* According to this classification, the cockroaches should be placed in the order Dictyoptera, which is divided into the suborders Blattaria (cockroaches) and Mantodea (mantids). Blattaria in turn is divided into the superfamilies Blattoidea and Blaberoidea. Blattoidea contains 2 families, one of which (Blattidae) includes the well-known genera *Blatta* and *Periplaneta.* Blaberoidea contains 3 families, one of which (Blattellidae) contains the genera *Blattella* and *Supella.*

One way in which cockroaches differ from the closely related grasshoppers, locusts, and crickets is that they do not have a visible ovipositor. Also, they do not lay their eggs singly or in pods, but in a hardened, purseshaped egg case called the oötheca (plural, oöthecae) (plate III). Most domestic cockroaches drop their oötheca as soon as it is formed, but the German cockroach (*Blattella germanica*), generally the species most frequently found in the household, carries hers until a day or two before the eggs are ready to hatch. She is often seen with an attached oötheca (figure 148).

Fig. 148. Female German cockroach, *Blattella germanica,* with attached oötheca shortly before it was dropped.

Among domiciliary cockroaches, *B. germanica* is the only species with an oötheca subject to desiccation. The eggs are dependent on moisture supplied by the mother. The oöthecae of other species are remarkably resistant to desiccation. Pryor (1940) believed the oötheca of *Blatta orientalis* to be covered with an oily secretion, probably derived from the cuticle of the mother. It was so effective in retarding water loss that eggs hatched in about 46 days from oöthecae that had been kept in an atmosphere of 0% relative humidity from the age of less than 1 day (Roth and Willis, 1955).

Oöthecae also effectively protect the enclosed eggs from insecticides. Not until the eggs are hatched and the nymphs leave the protection of the oötheca and crawl over the insecticide residue are they subjected to the effects of an insecticide treatment, provided the residue retains its efficacy that long.

World-Wide Distribution of Cockroaches

Cockroaches are tropical or subtropical in origin, and can maintain themselves outdoors in such areas. Species that became adapted to living in human habitations—"domestic" or "domiciliary" species—all originated in north or tropical Africa. Species names such as *americana, australasiae, germanica,* or *orientalis,* and the corresponding common names, are therefore misleading.

The distribution of domiciliary cockroaches throughout the world was made possible by their ability to infest the various means of intra- and intercontinental transport, particularly ships. Originally, and probably to this day, their principal means of transportation between continents has been by ships. (Movie fans and readers of classics may be interested in the fact that in 1792, the much-maligned Captain Bligh combated cockroaches on HMS *Bounty* with boiling water.) However, transport of cockroaches in the baggage compartments and kitchens of aircraft has been abundantly documented (Roth and Willis, 1960).

For many years, the principal species infesting ships was the American cockroach, but in more recent times the German cockroach has become the dominant ship-infesting species (Cornwell, 1968).

To judge by the enormous numbers of cockroaches infesting ships before the advent of effective insecticides, these insects must have found abundant food and water. Even without food and water, however, cockroaches can live for considerable periods. Newly emerged adult females of the German cockroach, *Blattella germanica,* have been found to live at 27° C (81° F) and 36 to 40% relative humidity for an average of 20.1 days without food or water, 14.7 days with dry food (dog biscuits) and no water, 35.1 days with water but no food, and 82 days with both food and water. The corresponding figures for the American cockroach, *Periplaneta americana,* were 41.7, 40.1, 89.6, and 190 (Willis and Lewis, 1957). The apparent ability of the 2 species to survive slightly longer without food and water than with dry food alone is interesting, but in any case, cockroaches are well adapted to dispersal over great distances by man, even when food and water are lacking.

Domestic species of cockroaches never or rarely fly, but are readily carried about in and on such items as sacks, cartons, or packages (particularly corrugated cartons) of food, in laundry, or in kitchen appliances and furniture. Beverage cartons are particularly important means of distributing cockroaches. The cartons are often contaminated with spilled syrups or malts, which attract cockroaches. Empty and unrinsed soft-drink or beer bottles in night clubs, restaurants, markets, and homes form a part of the infestation chain. More than 200 cockroach nymphs have been found in a single soft-drink bottle brought to a market for exchange (DeLong, 1962). Cockroaches may be found by the thousands in insulation in the walls of refrigerators and gas or electric ranges, surviving for months if necessary without access to their usual foods and feeding only on cast skins and dead insects. Cockroaches may become established in basements or crawl spaces, particularly if these places are dark and damp, and they may then enter the building around utility pipes, air ducts or ventilators, or under doors.

Economic Importance of Cockroaches

Cockroaches rank with termites as the most important insects that must be dealt with by the owners or occupants of homes or other buildings or by pest control operators hired to combat insects. Flies and mosquitoes may be more important as household pests in some areas, but they breed elsewhere and then invade dwellings. Therefore, in the United States and many other countries, fly and mosquito control is generally considered to be primarily a community or regional responsibility, and the homeowner is generally left with the job of controlling only minor or seasonal infestation.

The Cockroach as a Domestic Nuisance

Cockroaches breed within the home, sharing with man his food, water, shelter, and warmth, and are thus active throughout the year. They consume any human or animal food or beverage, as well as dead animals and plant materials, leather, glue, hair, wallpaper, fabrics, and the starch in bookbindings. They contaminate food, imparting an unpleasant odor and taste, and may transmit disease organisms.

Although cockroaches are nocturnal, if they are abundant a few of them may be seen during daylight hours, particularly if articles in pantries, cabinets, and closets are moved about to disturb their hiding places. Other evidences of a cockroach infestation may be dead cockroaches, cast skins of the various nymphal instars, empty egg cases, and fecal droppings. The droppings are variable in size, and range from as small as "flyspecks" to as large as mouse droppings. Intersections and corners of shelves and the hinges of cabinet, pantry, and even closet doors show fecal stains when there are heavy infestations. The musty odor left on objects cockroaches contact may also be evident in heavy infestations.

Cockroaches as Disease Carriers

Many of the organisms causing diseases of man have been found on the tarsi and other body parts, or in the fecal pellets, of cockroaches, but these insects are most commonly implicated in the transmission of *Salmonella,* the causal agent of food poisoning (Roth and Willis, 1957, 1960; Rueger and Olson, 1969). When the feces of American cockroaches infected with *Salmonella oranienburg* were spread on human foods and on glass, the bacteria survived for the following periods: on corn flakes, 3.25 years; on crackers, over 4.25 years; and on glass slides, over 3.67 years. Mice placed in jars containing minute quantities of infected cockroach feces became themselves infected in as brief a period as 1 day (Rueger and Olson, 1969).

Parasitic toxoplasmosis is a disease that is believed to infect more than a third of all adult Americans at some time during their lives. It is usually a mild disease, but can be extremely serious in pregnant women, causing congenital defects in an unborn child while remaining mild or symptomless in the mother. The protozoan parasite, *Toxoplasma,* can be acquired by eating infected raw or undercooked meat. If the blood serum of an infected person is added to cultures of the parasites, the differing colors of fluorescence of the latter will indicate either a positive or negative reaction and thereby serve as a test for the disease (Anonymous, 1973*a*).

Cats that consume birds or rodents carrying the parasite can become carriers. Filth flies have been found capable of transmitting infectious *Toxoplasma* oöcysts to human food 1 or 2 days after consuming infected cat feces. In laboratory tests, *Toxoplasma* parasites were isolated from the digestive tracts and feces of cockroaches as long as 7 and 10 days, respectively, after the cockroaches had last been in contact with infected cat feces. Filth flies and cockroaches might consume infectious cat feces and contaminate human food. They might also serve as food for birds and small rodents that could in turn be eaten by cats (Anonymous, 1973*a*).

Goodwin (1973) recorded the successful passage of hepatitis B antigen (AB-Ag) through the alimentary tracts of American cockroaches, and identified the antigen in the feces of the insects. The hepatitis antigen was found in the feces of the cockroaches for up to 9 days after their initial exposure to the antigen-positive test meal.

Cockroach Bites

Roth and Willis (1957) discussed 20 reports of cockroaches biting man. However, such instances are rare and not serious; they are discussed here only because questions are often asked concerning the biting ability of cockroaches. There have been a number of reports in the literature of cockroaches being so numerous in ships that they gnawed the skin and nails of the men on board; in one case, the sailors wore gloves while sleeping to prevent hordes of the insects from gnawing off their fingernails (Roth and Willis, 1960). Cockroaches are most likely to gnaw fingernails, eyelashes (of sleeping children), and the callused portions of hands and feet, but can also cause small wounds on softer skin. *Periplaneta americana* and *P. australasiae* are the species most often implicated.

The Cockroach Allergen

Some people are allergic to cockroaches. In one investigation among 253 normal persons, 7.5% showed positive skin tests with extracts of cockroaches (*Periplaneta americana* and *Blatta orientalis*) compared with 28% of an unselected group of 114 allergic patients. Skin-sensitizing antibodies were present in the blood sera of positive reactors. The allergen is thermostabile. There is evidence that in some instances symptoms attributable to food allergy may be caused by cockroach allergen with which the food is contaminated (Bernton and Brown, 1964, 1970*a*, *b*).

In an investigation to test the age of onset of skin reactivity, 38 out of 102 allergic children, ranging from infants to 12 years old, gave positive cutaneous reactions to body extracts of the German cockroach, *Blattella germanica,* compared with only 5 out of 100 nonallergic children. A 4-year-old asthmatic child was the youngest to give a positive reaction (Bernton and Brown, 1970*a*). In a later investigation, it was found that an extract of *B. germanica* caused an attack of asthma in 10 asthmatic persons with skin hypersensitivity to the extract and other allergens, but not in asthmatics without such skin hypersensitivity (Bernton *et al.*, 1972).

An allergen in the feces of *B. germanica* acts as an ingestant when it contaminates food and as an inhalant when dried fecal particles become incorporated with house dust (Bernton and Brown, 1970*b*).

Natural Enemies of Cockroaches

Natural enemies appear generally to be of little importance, although according to Roth and Willis (1960), E. C. Zimmerman stated that the hymenopterous egg parasite *Comperia merceti* (Compere) (Encyrtidae), when accidentally imported, practically wiped out the brownbanded cockroach in Hawaii. In some parts of Honolulu, almost 100% of the oöthecae of this species were parasitized. Muesebeck *et al.* (1951) recorded this parasite from the German as well as the brownbanded cockroach in Florida and Texas. Flock (1941) stated that another related parasite, *Anastatus blattidarum* Ferrière (Eupelmidae), seemed to be effective in controlling the brownbanded cockroach in certain areas in Arizona, and Hull and Davidson (1958) found some oöthecae of this species to be parasitized by *A. blattidarum* among the cockroaches they reared in Ohio. T. Maki considered the hymenopteron *Tetrastichus hagenowii* (Ratzeburg) (Eulophidae) to be an important egg parasite of cockroaches in Taiwan (Roth and Willis, 1960), and Muesebeck *et al.* (1951) recorded *Blatta orientalis, Blattella germanica, Periplaneta americana,* and *P. australasiae* as hosts in Florida and Louisiana.

Roth and Willis (1960) also stated that species from at least 6 families of Hymenoptera had been recorded as developing on cockroach eggs. They quoted Edmunds (1953) as stating that evaniid egg parasites were so abundant in a home in Ohio that the occupants considered them a nuisance, although the oriental cockroaches in the basement did not annoy them. Roth and Willis also listed many predators, including scorpions, spiders, dragonflies, mantids, bugs (e.g., reduviids), beetles (carabids, rhipiphorids [similar to mordellids], dermestids, and others), wasps, ants, toads, frogs, lizards, birds, poultry, and various mammals. Cannibalism among cockroaches, including the devouring of oöthecae, has been noted by many investigators even when food was adequate.

When humidity is too high in cockroach culture colonies, mites may become numerous, and have been known to cause German cockroaches to drop their oöthecae prematurely, resulting in a low percentage of eggs hatching. Control of mites with acaricides has been known to increase the vigor of cockroach colonies (Roth and Willis, 1960). A pterygosomid mite *Pimeliaphilus podapolipophagus* Trägårdh, must actually feed on live cockroaches to survive; it cannot subsist on feces, cast skins, or dead insects (Cunliffe, 1952). It has been accused of biting people, and its presence in homes is an indication of cockroach infestation (Baker *et al.*, 1956).

Roth and Willis (1960) reported 2 species of spiders in the family Theridiidae, the South African *Latrodectus indistinctus* and the North American *L. mactans,* the latter the venomous "black widow," as predators of cockroaches. The author observed another theridiid spider, *Steatoda grossa* (C. L. Koch) (figure 212, chapter 9), which resembles the black widow but is one of its natural enemies, in great abundance in an infestation of German cockroaches confined to experimental mockup closets at the University of California (Los Angeles). Remains of cockroach nymphs were found in their webs. It is also of interest to note in this connection that the related *S. fulva* (Keyserling) blocks the nest entrances of the harvester ant *Pogonomyrmex badius* (Latreille) during the early afternoon, when temperatures are high and the ants are inactive, and feeds on the ensnared ants (Hölldobler, 1970). *Steatoda albomaculata* (De Geer) also feeds on ants (Levi, 1957).

From their extensive review of the literature, Roth and Willis (1960) concluded that with the exception of a few instances of egg parasitism such as just cited with respect to *Comperia merceti* and *Tetrastichus hagenowii,* there had been too little information to enable them to evaluate the effectiveness of biological control in reducing the numbers of pest cockroaches, and that further investigation would be justified.

A large percentage of any cockroach population generally contains internal parasites. Two species of nematodes and 7 of protozoan parasites were found in 105 *Blattella germanica* in New York city.

One of these nematode species, *Blatticola blattae* (Graeffe), was present in 96.2% of the cockroaches collected. Among the protozoa, *Nephridiophaga blattellae* was found in 82.8% of the cockroaches, and 3 other species were abundant (Tsai and Cahill, 1970).

Rearing Cockroaches Experimentally

A continuous supply of several species of cockroaches has been reared for experimental purposes in our laboratory at UCLA. These species are, in the order of what we consider to be their importance in California, the German, oriental, brownbanded, and American cockroach. Later the Australian and the brown cockroach were added. The insects were originally reared in 15-gal (57-l) garbage cans with an electric barrier (Wagner *et al.*, 1964), later replaced by 20-gal (76-l) plastic trash cans (Ebeling *et al.*, 1966). A double layer of strips of aluminum foil with a strip of insulation tape between them, and with the terminals from a 22.5-volt battery connected to the aluminum strips, serves as an electrified barrier to the escape of the cockroaches. In each trash can there is a poultry-watering device, Purrina Dog Chow, and 3 mazes of concentrically rolled and closely approximate layers of corrugated paper to serve as the harborage areas that cockroaches require for optimum development. The trash cans are covered with their usual plastic lids, provided with screens for ventilation.

It is important to prevent contamination of a colony by stray individuals of other species. It is especially important to keep German cockroaches out of colonies of other species because of the ability of this insect to thrive and rapidly increase its numbers at the expense of the others.

In the cockroach rearing room, boric acid powder applied under benches, radiators, appliances, cabinets, etc., will serve to permanently prevent a buildup of the cockroach population, in the rearing room and adjacent rooms, from those insects which escape when they are being removed for experimental work.

Twenty cans of German cockroaches and a can or more of each of the other 5 species are continuously maintained in our laboratory. The temperature is kept near 80° F (27° C) in the rearing room. Provided that a water supply is constantly maintained, the author and his colleagues have had no indication that the maintenance of relative humidity above ambient is of any advantage in the rearing of cockroaches, at least in the rearing cans just described. With ambient humidity and monthly removal of the debris at the bottoms of the cans (with a vacuum-cleaner device), infestation of German cockroaches by parasitic mites has been avoided. American cockroaches drop their egg cases to the bottoms of their rearing cans; therefore, the cans are vacuumed only about twice a year, but their colonies do not produce so much debris and do not seem to be so susceptible to mite infestation as those of the German cockroaches.

German Cockroach, *Blattella germanica* (L.) (Blattellidae)

The German cockroach is generally the most important cockroach species in the United States and many other areas of the world in homes, apartment buildings, restaurants, and in the food-preparation and food-storage rooms of hotels, hospitals, and other private and public buildings in which food is prepared, stored, or served.

Description. Adult German cockroaches are 1.3 to 1.6 cm long, pale brown or tan, and have 2 parallel dark streaks on the pronotum (plate III, 1). They have chewing mouthparts. (See figures 44 and 45, chapter 4, showing the similar American cockroach.) Their movements are very rapid when they are disturbed. They are nocturnal. If a few are seen crawling about in open spaces during daylight hours, this indicates that the infestation is already severe. The female is darker and has a broader abdomen than the male, and is rounded posteriorly. If German cockroaches are placed on their backs, the males are easily distinguished by the yellowish, slender abdomen, tapering gradually to the posterior end. The abdominal cerci of the males have 11 segments and those of the females have 12.

In their investigation of the German cockroach, Willis *et al.* (1958) found the ratio of females to males to be 1.12:1.

In similar work, Ross (1929) observed that the sexes occurred in approximately equal numbers.

Mating Behavior. Roth and Willis (1952) determined that the male of the German cockroach could not detect the female from a distance, even when in close proximity, but had to make physical contact, ordinarily via the antennae. The antennae of the female, as well as other body regions, contain a chloroform-soluble, nonvolatile substance that will stimulate the male sexually. Thus, sex discrimination by males is mainly owing to "contact chemoreception." It follows that the male German cockroach probably cannot be attracted from a distance by means of synthetic sex attractants (provided they are available) as some other insect species can be.

When male and female meet, their antennae touch and vibrate against each other. The male then turns around, raises his wings to expose the orifices of a pair of dorsal glands located on the seventh and eighth tergites, and extends his abdominal segments to expose the openings of the 2 pairs of glands. The glands are not normally visible, being covered by the wings and by the margins of the preceding abdominal sclerites. Both wing-raising and extension of the abdomen are required to uncover the glands. The female eats a secretion from these glands. After she has fed for a few seconds, the male pushes his abdomen farther back, and connection of the genitalia is made. The male then moves out from under the female, and the pair remain attached in a linear position for an average of about 86 minutes (Roth and Willis, 1952). These authors also pointed out that the fact that German cockroaches do not need to be attracted from great distances, as do many other insects for the meeting of sexes, is understandable in view of the habits of cockroaches in general. The chance meeting of opposite sexes is enhanced by the fact that they are negatively phototactic and positively thigmotactic (principally guided by contact), and that they are gregarious, with large numbers seeking the same environment. They seek secluded and particularly very narrow hiding places, such as cracks, crevices, and voids, especially those having optimum temperature and moisture conditions. Attraction to the odor of the species is another factor that favors aggregation. Once the sexes have been brought together by these different stimuli, more refined stimuli lead to sexual discrimination.

Life Cycle. The German cockroach lays more eggs and has more generations per year than the other common domiciliary species, and therefore tends to increase in numbers much more rapidly. There may be 3 or 4 generations per year. In one investigation at a temperature of 76° F (24.4° C), the developmental period varied from 54 to 215 days, with an average of 103 (Gould and Deay, 1940). It has been found to be as little as 60 days for males and 65 days for females at 95° F (35° C) and 90 to 95% relative humidity (Ross, 1929). An individual German cockroach, however, lives an average of 200 days at room temperature (Gould and Deay, 1940), and may live as long as 10 months (Truman, 1961*a*). A comparison of the developmental periods of the 4 principal domiciliary cockroaches at 82° F (28° C) and at ordinary room temperature is shown in figure 149.

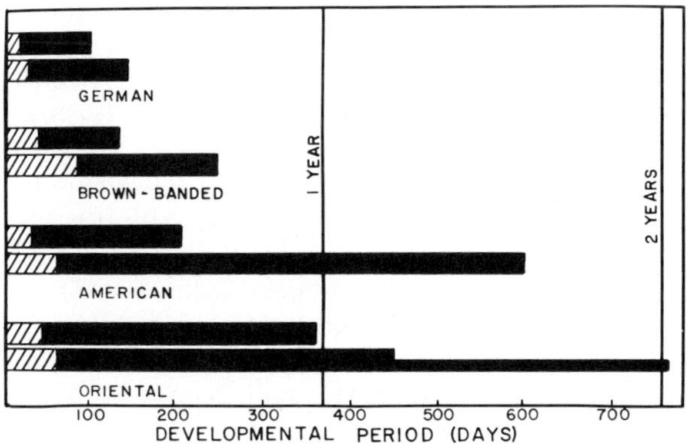

Fig. 149. Developmental periods for 4 species of cockroaches at a constant temperature of 82° F (28° C) and relative humidity of 70% (shorter bar in each pair) and at ordinary room conditions. Reduction in size of bar for the oriental cockroach indicates a possible 2-year cycle for some individuals. (From Gould and Deay, 1940.)

The Oötheca (Egg Capsule). The German cockroach female produces her first oötheca 11 or 12 days after becoming an adult (Ross, 1929). Oöthecae can be produced parthenogenetically (without fertilization), but although embryos develop, the eggs do not hatch (Roth and Willis, 1956). Unlike other domiciliary species, the German cockroach female carries her oötheca for as long as a month, until a day or two before the eggs are ready to hatch, then drops it anywhere. The eggs occasionally hatch while the oötheca is still attached. The female generally deposits 4 or 5 oöthecae, but there can be as many as 8. The females of other cockroach species drop the oötheca as soon as it is fully formed. Thus, the abundance of females with their attached, brownish, purseshaped oöthecae (figure 148) is a characteristic of a German cockroach infestation.

The oötheca is about 8 × 3 mm—rather large in relation to the size of the female. As with all common cockroach species, the eggs are located in 2 parallel rows. The locations of the eggs are indicated by corresponding divisions on the outer wall of the brownish oötheca. When the embryos have developed to the point where they exert enough pressure, the oötheca splits open at the top and the nymphs wriggle out. The female will eat the young nymphs if no water is available to her (Ross, 1929).

There are usually 30 to 40 eggs per oötheca, but there can be as many as 48. In one investigation, an average of 29.9 nymphs were hatched per oötheca in an average of about 28 days at ordinary room temperature (Gould and Deay, 1940), while in another investigation at 95° F (35° C), eggs hatched in 14 days (Ross, 1929). After the fourth oötheca is produced, the number of eggs per oötheca gradually decreases to about 75% of the original number in the seventh and eighth oöthecae (Willis *et al.,* 1958).

The eggs become desiccated if the oötheca is removed from the female any time before its normal time to drop, for they must obtain moisture from the body of the mother. The end of the oötheca that is attached to the female is relatively soft and permeable to water, and is not so heavily sclerotized as the posterior end. Roth and Willis (1955) found that German cockroach oöthecae placed with their anterior ends on wet filter paper gained weight, whereas those with their posterior ends on the same wet paper lost weight, even though the humidity was high in the covered petri dish in which the oöthecae were kept. Probably, the wall of the oötheca in contact with the female's genital pouch is permeable to water (Ross, 1929; Parker and Campbell, 1940).

Even before the oötheca begins to emerge, the abdomen of the female is greatly distended. Once the translucent tip of the oötheca becomes visible, the entire oötheca will be fully developed and entirely visible by the following day, changing from white to pink within a few hours. Within a day or two, it becomes light brown and finally chestnut. It is turned with the keel to the left or right (Haber, 1919; Gould and Deay, 1937, 1940).

Nymphal Instars. Wright (1968) found that among 749 German cockroach nymphs, 5.7% emerged while the oöthecae were still attached to the body of the female, 92.9% emerged within 24 hours after the oöthecae were dropped, and only 1.4% emerged thereafter. Among field-collected German cockroaches, 27.9 nymphs per oötheca emerged from those dropped by females collected in summer, compared with only 9.7 from those collected in winter.

German cockroach nymphs have 6 or 7 instars. As in the case of all insects with gradual or incomplete metamorphosis, there is no abrupt change in appearance between immature and adult forms, except that the adults have wings. Nevertheless, the layman will not necessarily associate the tiny, newly hatched German cockroaches or even the more mature nymphs with the adult insects unless he sees all instars and stages together in considerable numbers. The smaller nymphal instars are sometimes isolated from the remainder of the colony, particularly when they have gained access to a crack or crevice too narrow for the older nymphs and adults.

The first-instar nymph is only 3 mm long. The body is dark gray to almost black, except for the second and third thoracic segments, which are pale brown. The pale-brown band conspicuously characterizes the first-instar nymph. In succeeding nymphal instars, the light band becomes narrower and extends in both directions to become a median longitudinal stripe (plate III, 7). In its anterior extension, it eventually becomes the median pale-brown stripe dividing the "two parallel dark streaks" (already mentioned) that characterize the pronotum of the adult. The remainder of this stripe is covered by the wings.

Completely white nymphs or adults may be seen in a cockroach colony. These are newly molted individuals that have not yet had time to harden their cuticles and to acquire the normal color for the species.

A high degree of mortality occurs during the molting periods. Ross (1929) found mortality to be about 5% during each molt except for the last, when it was 40%. He found that about half the insects died of natural causes before reaching maturity. Molted skins were quickly eaten by the nymphs that emerged from them or by other cockroaches that happened to be near-by. On the other hand, Willis *et al.* (1958) found in their cultures that 85% of the hatched insects reached maturity.

Aggregation Pheromone

As already stated, because cockroaches have a gradual metamorphosis, all immature instars may be seen together with the adults. They tend to congregate in a single area, leaving it to search for food and water, then returning. An aggregation pheromone, present in the feces and on the bodies of the insects (it spreads easily on the greasy epicuticular lipid), is responsible for this tendency to aggregate. The gregarious behavior of nymphs favors their growth and development; they do not develop so well when isolated (Ishii and Kuwahara, 1968). Isolated nymphs that could only see or smell the others, or received contact only with the washings of the body surfaces of other nymphs, or with filter paper contaminated with feces, did not grow at the group rate. This indicated that a "psychological factor" affecting the endocrine system by means of tactile stimulation, chiefly via the antennae, was the principal factor involved in growth acceleration (Izutsu *et al.,* 1970).

Effect of Excessive Crowding. The author investigated the effect of crowding in a "choice box," 30 cm square and 10 cm high, with a 12-mm hole at the top of the partition wall between the 2 halves (Ebeling *et al.*, 1966). When there are only 20 to 40 cockroaches in the choice box, most of them will spend nearly all their time in the dark half, for they are negatively phototropic. The insects tend to congregate at the intersection of 2 plane surfaces or in corners. When such areas are filled, the insects will gather in small, scattered groups on plane surfaces. However, despite the aggregation pheromone, German cockroaches have some aversion to being crowded too closely together. They prefer to leave some space between themselves and their neighbors. They sometimes initially display aggressive behavior toward their neighbors when actual physical contact is made. However, if large numbers are forced to occupy limited space, they soon appear to get used to this situation.

When 200 adult male German cockroaches were placed in a choice box, only 11% were in the light half 3 days later, compared with an average of 16.2% in the light halves of 5 choice boxes, each containing 20 adult male cockroaches. On the other hand, when 959 adult German cockroaches of mixed sexes and 530 nymphs were placed in the dark half of a choice box, they were so crowded that they made physical contact with one another, and in some cases were one on top of the other. They were left that way for 2 hours, and then a cork occluding the 12-mm hole in the partition wall was removed. Large numbers of cockroaches left the dark compartment. On the third day of the experiment, the insects were anesthetized and counted. Of the live insects, 36.1% of the adults and 41.7% of the nymphs were in the light compartment of the choice box, preferring the light to an excessively crowded condition in the dark. If German cockroaches are found in parts of an untreated building far removed from food and water or from preferred habitats, it is an indication of excessive crowding in their customary harborages.

Crowding in Narrow Spaces. Adult German cockroaches can move about in space only 1.6 mm in width or depth (Wille, 1920). When given a choice among 8 spaces between masonite plaques placed one above the other and ranging in distances from 1.6 to 12.7 mm apart, in 1.6-mm increments, 67% of adult German cockroaches gathered in the 4.8-mm space. When the plaques were vertically arranged, this percentage increased to 85. Only a few adults gathered in the 1.6-mm space (Berthold and Wilson, 1967). In our laboratory we found that only the first 3 instars of a population of all ages of German cockroaches crawled from a light to a dark compartment of a 5-cm-wide wooden box for a distance of 2 cm under a strip of masonite that was only 1 mm above the floor of the box. Only 8% of the adult population (6 males and 2 females without oöthecae) crawled through a 2-mm aperture in a 24-hour period, even though the insects were driven by hunger and negative phototropism to seek food and darkness. No adults were able to crawl from the light to the dark compartment when the 2-mm aperture was between the wooden partition wall (1 cm thick) and a masonite ceiling.

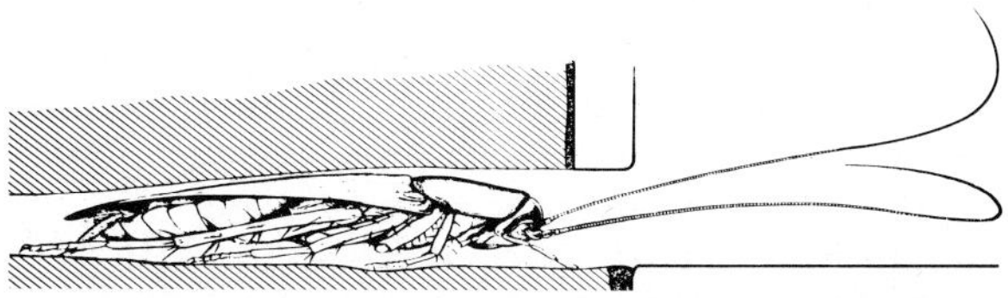

Fig. 150. Characteristic resting position of a German cockroach in a narrow crevice. (From P. B. Cornwell [1968], *The Cockroach*, courtesy Hutchinson Publishing Group, Ltd.)

The author once observed a severe infestation of German cockroaches in the food-storage room of a large "rest home." Leaking or spilled packages of flour, cereal, or similar foods on the shelves provided plentiful food for the insects. Cockroaches were crowded into a space of 2 or 3 mm between a shelf and the wall. They were completely hidden from view, except for a continuous band of waving antennae that extended from the aperture, and could be seen by looking beneath the shelf.

This sort of thing is a common occurrence, well known to pest control operators. However, it appears that the first mention of this phenomenon in the literature was in P. B. Cornwell's *The Cockroach*, and figure 150 is taken from his book.

Habitat

In the living space of a house, German cockroaches are most abundant in the kitchen, and this is also true of other domiciliary species. During the day they may be found under stoves or ranges and refrigerators, and in the insulation in the walls of appliances; under the sink and especially in the "dead space" between the sink and the wall; under and behind the water heater, particularly if it is enclosed in a cabinet; in, under, or behind cabinets, pantries, and closets; behind baseboards and moldings; and in other cracks, crevices, and dark, protected areas. They venture forth at night in search of food and water.

German cockroaches are also likely to be found in the bathroom, particularly if the kitchen and bathroom share a common wall, for then they can infest a common wall void and pass from one room to the other via areas surrounding utility pipes or the louvers and razor-blade receptacle of the built-in medicine cabinet. In moderate to severe infestations, they can be found throughout the house in protected areas, such as closets, dressers (usually above the top dresser drawer), sofas, electrical appliances such as radios and television sets, under stairwells, behind moldings, picture frames, etc. All such places must be thoroughly treated with insecticide for effective control.

Television sets in particular are becoming increasingly important harborages for both German and brownbanded cockroaches. Not only the warmth, darkness, and the edible glue used in construction, but often the bits of human food commonly found in the vicinity of television sets are attractive to cockroaches and combine to provide an ideal harborage.

In severe infestations, enormous numbers of German cockroaches may sometimes be found in the attic and in wall voids. They use the wall voids to move either laterally or vertically from one room to another or from one apartment to another. They may also be found in the crawl space under the house, at the bases of foundation walls, in cracks in the sidewalk, in the lawn and under shrubbery, and in outdoor incinerators or garbage-disposal bins.

German Cockroaches Finding New Habitats

Before German cockroaches became almost universally resistant to chlordane in the 1950's, they were known to inhabit kitchen areas almost exclusively. Since that time, they have extended their habitats until they are now often distributed widely throughout the house, at least in the cases of more severe infestations (Shuyler, 1956). This is a characteristic formerly attributed only to the brownbanded cockroach, *Supella longipalpa* (F.). It may be that because of the switch from chlordane to organophosphorus and carbamate insecticides, German cockroaches were not only more strongly repelled by the latter insecticides, which was easily demonstrated (Ebeling *et al.,* 1966, 1967), but that they were more successful in surviving initial contacts with the toxicants, and found new harborages free of insecticides in a wide variety of locations throughout the infested premises (Ebeling *et al.,* 1968). This new behavior might be inherited because of its survival value.

There are other reasons for the increasing dispersal of German cockroaches throughout a house that arise from changes in furnishings and in man's living habits. Central heating has become common, resulting in temperatures in all areas that are conducive to cockroach breeding, even in winter. Even if some rooms are cooler at night, the insects can find warmth in such appliances as electric clocks, radios, or television sets. Some people eat their meals, or at least light snacks, near the television, which could be in the livingroom, bedroom, or den. A few crumbs, pieces of cookies, etc., can sustain many cockroaches. Many modern homes also provide more sources of water than were formerly found in homes. There may be 2 or 3 bathrooms, as well as air-conditioners with evaporating pans. Beer cans and soft-drink bottles not completely emptied may be left in more locations, now that homes are more extensively utilized than they formerly were, and these also provide sources of both food and water.

Physical Features of the Habitat That Influence Aggregation

The earliest insects had sheltered and restricted habitats that were at least reasonably damp, such as beneath stones and litter or under loose bark, as one might expect on the basis of the relatively poor waterproofing mechanism of their cuticles. Ametabolous or paurometabolous development made a continuously cryptobiotic life style possible. Food and mates were found by chance encounters. It was improved cuticular waterproofing and the development of wings that freed insects from a restricted environment. An improved olfactory sense became advantageous. Eventually, the pheromones and sophisticated chemical communication systems of some of the modern insects were evolved (Moore, 1967).

Among household pests, the silverfish and cockroaches are among the cryptobiotic insects. Although most cockroaches have wings, they use them sparingly or not at all. Accordingly, one might expect that neither silverfish nor cockroaches would be effectively attracted to foods or chemical lures, at least not over distances that might lead to effective chemical trap devices, and this has indeed been found to be true for both these relatively primitive insect orders.

In the case of *Blattella germanica* and *Blatta orientalis,* the species aggregation odor was shown to be largely responsible for their gregarious behavior, but the stimulus was effective for only a short distance (Ledoux, 1945). As stated earlier, sex stimulus, although of chemical origin, is limited in cockroaches and requires actual physical contact. Even food odors are only feebly attractive, and will attract German cockroaches only if they occur in areas normally frequented by the insects or in the routes in which they would normally travel. For example, in a room, cockroaches normally travel mainly along wall intersections, the intersections of the floor and the walls, the intersections of the walls and the ceiling, the intersections of shelves and pantry or cabinet walls, and infrequently in other areas.

When experimenting with German cockroaches, Ebeling *et al.* (1966) found that in a kitchen area in which food was available to the insects, an average of 2.45 per baited trap-jar was trapped in jars placed along the baseboards (intersections of floor and wall), but none in jars placed in the center of the kitchen floor. When all other food was removed from the kitchen, an average of 13.5 insects was trapped in the baited jars placed along the baseboard and 1.15 in the center of the room.

Hunger is one of the factors influencing exploratory activity of cockroaches, but the physical features of a habitat (structural features, temperature, humidity, light) are the principal factors influencing the areas in which the insects will travel.

Factors such as pheromones and food odors can become operative only after these physical features have already attracted the insects. For insecticides or even poison baits to be effective, they must be applied in areas in which cockroaches would normally travel or congregate (Ebeling and Reierson, 1970).

Mass Migrations of German Cockroaches

These mass migrations have been known to occur without the stimulus of insecticide treatment. For example, Howard (1895) reported a "vast army" of German cockroaches leaving a restaurant on a "dark, drizzly day" and crossing a muddy street to other buildings in a migration that lasted for 3 hours. The migrating insects were nearly all females with oöthecae. No unusual activity, such as cleaning or insecticide treatment, had taken place in the restaurant. Dr. Howard concluded that this mass movement of cockroaches was a true migration occasioned by overcrowding in the restaurant. Another instance of active migration of German cockroaches from a restaurant to another building was reported by Mielke (1970).

Field Cockroach, *Blattella vaga* Hebard (Blattellidae)

This species, discovered in Arizona in 1933, is probably of Asian origin, for its nearest relative occurs in India (Hebard, 1935). It may now be found, sometimes abundantly, in irrigated fields and in yards in southern California, Arizona, and Texas, but also sometimes appears in the desert, far from inhabited or cultivated areas. A severe infestation was found at Oroville, in northern California, in 1967, as recorded by Buxton and Freeman, 1968.

Biology. The field cockroach resembles the German cockroach, and is only slightly smaller. It can be distinguished by the blackish-brown area on the face, from the mouthparts to between the eyes, that is absent from the German cockroach. The 2 parallel, longitudinal stripes on the pronotum of *B. vaga* are similar to those of *B. germanica,* but are very dark and more sharply defined. With the aid of a hand lens, the 2 species can be told apart by differences in the subgenital plate (Buxton and Freeman, 1968). The life cycle of *B. vaga* is similar to that of the German species, but Willis *et al.* (1958) found an average of only 28 eggs per oötheca, of which 19 hatched, compared with 37 for *B. germanica,* of which 28 hatched. In *B. vaga,* after the the third oötheca is produced, the number of eggs per oötheca decreases until in the eighth oötheca it is only 10% of the number found in the first 3. (In the case of the German cockroach, there is a reduction of only about 25%.) Of the hatched insects, 73% reached maturity under laboratory conditions, compared with 85% for the German cockroach.

Habits. The field cockroach is so named because it is ordinarily found outdoors, where it feeds on decaying vegetation, but during the drier seasons, it may invade houses in large numbers. Its feeding habits are similar to those of the German cockroach. Of the 2 species, *B. vaga* is more likely to be seen during the day, for it is not repelled by light, and in fact is commonly found in the evening around street lights and in display windows. Infestations have been eliminated by removing decomposing plant material from around the foundations of houses (Flock, 1941).

In Texas, a pest control operator who believed he was treating a house for German cockroaches failed to control the infestation. When the insects were found to be field cockroaches, he sprayed the yard, which was the source of infestation, and obtained control (Walter, 1968).

Brownbanded Cockroach
Supella longipalpa (F.)
[= *supellectilium* (Serville)] (Blattellidae)

This species, like most of the principal domiciliary cockroaches in the United States, is apparently of African origin. It was first recorded in the United States at Key West and Miami, Florida, in 1903, presumably having been introduced from Cuba, where it was known to be abundant even as early as 1862 (Cornwell, 1968). It now probably occurs in all of the United States, with the possible exception of Alaska, particularly in buildings providing moderate to high temperatures. However, it is far less important than the German cockroach as a pest. Even in Egypt, where the species was probably indigenous and where it was called the "furniture cockroach," it was less abundant than either *Blattella germanica* or *Periplaneta americana* (Hafez and Afifi, 1956).

Description. The adult brownbanded cockroach (plate III, 2) is approximately the same size as the German cockroach. The male is 13 to 14.5 mm long, and the female slightly shorter (11 to 12 mm). The abdomen of the female is much broader and more rounded posteriorly than that of the male. The wings of the male cover the abdomen completely, while the wings of the female are short, and never cover the entire abdomen. Males may often be seen flying when a colony is disturbed, but the females cannot fly.

From the dorsal aspect, the adults are dark brown, particularly the head and thorax and the exposed segments of the female, which may in fact range to black. The wings of the female are reddish brown to very dark brown throughout, while those of the male are dark brown at the base, becoming increasingly lighter posteriorly. Both sexes have a band of pale brown at the base of the wings and another band of pale brown a third of the distance from the base, which gives the adults a "banded" appearance. The wings and body of the female are much darker than those of the male. The ventral body surface and legs of the male and the legs of the female are light straw-colored. The lateral margins of the pronotum of both sexes are translucent, and this is also true of a narrower lateral margin of the anterior portion of the wing. The ratio of females to males was found in one investigation to be 1.32:1 (Willis *et al.*, 1958).

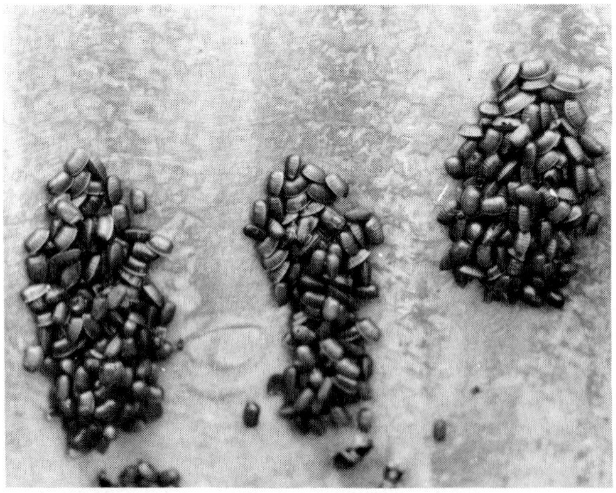

Fig. 151. Oöthecae of the brownbanded cockroach, *Supella longipalpa*, as they are characteristically deposited in clusters on vertical surfaces.

Life Cycle. The developmental period of the brownbanded cockroach is considerably longer than that of the German cockroach, as shown in figure 149. The temperature for optimum development is above 80° F (27° C) (Gould and Deay, 1940), and homes are seldom kept at such high temperatures. Therefore, the biotic potential for the brownbanded cockroach is considerably less than for the German cockroach. In California, it is found in only about 5% as many homes as the German cockroach, but where food is abundant and temperatures are high, the species can become extremely plentiful.

The Oötheca. The females carry their capsules for 24 to 36 hours before attaching them to some object, possibly the kitchen sink, furniture, walls, shelves, bedding, draperies, behind pictures on walls, or other handy places, usually in clusters. Corners under chairs and tables are among the favored places. Individual clusters (figure 151) on the walls of rearing cans in the author's laboratory have at times contained 50 or more egg capsules. One housewife, in attempting to describe them, said they appeared to be "clusters of seeds." This habit of attaching the oöthecae to furniture and other household objects results in the species being readily disseminated over great distances. The oötheca of the brownbanded cockroach measures about 4 × 2.5 mm—the smallest of the common domiciliary cockroach oöthecae. It varies in color from yellowish to reddish brown. A few days after it is formed, the fertile eggs show greenish through the walls of the oötheca, and shortly before hatching, the eyes of the young are visible.

The incubation period for the eggs of the brownbanded cockroach was found to be the longest among 6 cockroach species investigated by Gould and Deay (1940). It averaged 49 days at

82° F (28° C) and 95.6 days at 72.5° F (22.5° C). About 14 fertile oöthecae were formed per female. The maximum number of eggs normally produced per capsule was 18, but the average number hatching was 13.2.

Nymphal Instars. From the dorsal aspect, the nymphs (plate III, 7) are also basically dark brown in color, but with a narrow, pale margin on the thorax. The meso- and metathorax are pale brown, and the abdomen has a broad, pale area. The dark and pale areas of the nymphs are in striking contrast, and easily identify this species. (The nymphs could more appropriately be called "brownbanded" cockroaches than the adults.) There are 6 to 8 nymphal instars. At room temperature, Gould and Deay (1940) found the average period for nymphal development to be 161 days for males and 162 days for females, but at about 85° F (29° C), the average periods were 90 and 95 days, respectively. At room temperature, extremes of developmental periods were 95 and 276 days. The length of adult life varied from 131 to 315 days, with an average of 206 days.

Habits. The brownbanded cockroach was once believed to be unique among domiciliary cockroaches in its tendency to distribute itself throughout a house, even in bedrooms, furniture, and closets, particularly on high shelves, behind pictures on walls, and behind moldings. Gould and Deay (1937) stated, regarding this species: "Except when in search of food, it seldom visits the kitchen, and confines its activities to other parts of the house." About the German cockroach, they stated: "In an infested home, it is confined to the kitchen and lavatories, where it hides behind baseboards, in cupboards, iceboxes, and dark corners, and around water pipes." Apparently, the habits of the 2 species have changed, for now there appears to be no great difference in the way they are distributed in a house or apartment. As already stated, Shuyler (1956) believed the change in the habits of the German cockroach coincided with widespread resistance to chlordane, which in the years following World War II was almost universally used for control of this species. Both species are now most abundant in the kitchen areas, and either can be widely distributed throughout a house or apartment. Sometimes, mixed populations of the 2 species are seen. However, as stated earlier, infestations of brownbanded cockroaches generally occur only in houses or apartments in which temperatures are higher than most people would consider to be comfortable.

Oriental Cockroach, *Blatta orientalis* L. (Blattidae)

In common with the other important domiciliary cockroaches, the oriental cockroach, despite its name, is believed to have originated in north Africa in areas which combine summer heat and moderate winters (Rehn, 1945). Although in some parts of the world, as in Great Britain (Cornwell, 1968), the oriental cockroach is by far the most important cockroach pest, the German cockroach is much more important in the United States.

In California, the oriental cockroach is found in homes more frequently than the brownbanded cockroach, but does not reach such large numbers or become so widely distributed in the house. Nevertheless, pest control operators report more calls for control of the oriental than of the brownbanded cockroach. Its large size and black color cause it to be regarded as a particularly repulsive pest. The housewife prefers to believe that the insect is a "water bug" or "black beetle" rather than a cockroach, but usually loses no time in calling for the exterminator.

Description. The adult females of the oriental cockroach are about 3 cm long. The male is about 2.5 cm long, and is much narrower than the female. Both sexes vary from dark reddish brown to black, usually the latter (plate III, 3). The female has small, rudimentary, functionless wings, and the male, although it has well-developed wings covering about 75% of its abdomen, does not fly. The tarsi of both nymphs and adults lack an arolium (cushionlike pad between the claws), and therefore they cannot climb smooth, vertical surfaces. This species is relatively sluggish in its movements. The ratio of females to males has been found to be 1:1.15 (Willis *et al.*, 1958).

Life Cycle. As shown in figure 149, the developmental period of the oriental cockroach is more than a year, even under indoor conditions, and for some individuals it may be prolonged to about 2 years. The insect may also live for long periods after reaching maturity. Gould and Deay (1940) found that at room temperature, the length of adult lives of 5 females varied from 34 to 181 days. Among 4 males, the length of adult lives varied from 112 to 160 days. Whereas it may require a long time for an appreciable infestation of oriental cockroaches to develop, enormous populations are eventually possible. Ordinarily, the oriental cockroach does not become so abundant as the German cockroach in buildings, but can become very numerous in secluded places, such as basements or crawl spaces, if they are damp.

The species thrives better outdoors than other domiciliary cockroaches in the United States, even under quite severe climatic conditions.

The Oötheca. The female carries her oötheca for about 30 hours after it is formed and filled with eggs. It is dropped, or attached to debris or food material, in some protected place. It is large (10 × 5 mm), and when first deposited it is reddish brown and soft, but becomes black, hard, and brittle. Although the number of oöthecae formed varies greatly, in an observation of 8 females, the average number was 8. A perfect capsule contained 16 eggs (Gould and Deay, 1940). Another investigator observed that females produced from 1 to 4 oöthecae, with the usual number being 2 (Rau, 1924). The incubation period was noted to vary from 42 days at 29.5° to 81 days at 21° C (85 and 70° F), and was approximately 2 months at the usual room temperatures.

Nymphal Instars. The newly hatched nymph (plate III, 7) is only about 6 mm long, and pale brown. It becomes more reddish with succeeding instars, and finally, in the last instars, dark reddish brown to black, like the adults. At ordinary room temperatures, the period for nymphal development averaged 515 days for males and 542 days for females, whereas at 82° F (28° C) it averaged 288 days for males and 310 for females. Willis *et al.* (1958) found the period for nymphal development at 30° C (86° F) to be 164 days for males that had the maximum of 7 instars, and 282 days for females that had the maximum of 10 instars. The oriental cockroach seems to have a seasonal cycle, with the adults appearing in May and June and dying during July (Gould and Deay, 1940; Rau, 1924).

Habits. In the southern areas of the United States, this species has long been known as the one best able to survive outdoors. Its temperature preferences are somewhat lower than those of the German cockroach, and it can thrive in both drier and cooler areas (Gunn, 1935). Like some other cockroach species, *Blatta orientalis* appears to be changing its habits. It can now survive outdoors farther north than it could originally. Shuyler (1956) observed that oriental cockroaches survived outdoors during 13 weeks of almost continuous freezing weather with a light snow cover. They could be found by moving stones, leaf debris, and clods of soil near a house. About 2 weeks after the arrival of the first warm weather of spring, these insects invaded the house.

Although once known to occur mainly under a house, in the crawl space, basement, or cellar, or on the first floor, particularly around sources of water, the oriental cockroach is now reported with increasing frequency inside at higher levels —as far up as the fifth floor (Shuyler, 1956). In southern California, it is often abundant in the dense vegetation of estates in areas in which some of the cleanest and most elegant homes are located. Its entry into such homes may be purely random movement, generally under sliding glass doors or around utility pipes, air ducts, and ventilators leading from a damp crawl space.

In some localities, oriental cockroaches can be collected at any time of the year from the dark, damp, water-meter vaults that are typically located in the ground in alleys and parkways.

Anything done to increase light, improve air circulation, and decrease dampness in dark, damp basements and crawl spaces will reduce the oriental cockroach infestations, for they can otherwise become established in enormous numbers in such areas.

The "Cockroach Odor." The oriental cockroach possesses in abundance the much-despised "roachy" odor that is associated with cockroaches. Those who believe that everything in nature must serve some useful purpose for mankind will be encouraged by the following account of the alleged usefulness of the "essence" of oriental cockroaches, taken from the 1907 edition of the highly respected *Merck's Index:*

> *Constituents:* Blattaric acid; antihydropin; fetid, fatty oil. *Uses:* Internal, in dropsy, Bright's disease, whooping cough, etc. External, as an oily decoction for warts, ulcers, boils, etc. *Doses:* 10–15 grains in dropsy, as powder or pills, or 4 fluid drams' decoction. (Illingworth, 1915.)

American Cockroach, *Periplaneta americana* (L.) (Blattidae)

In most areas of the United States, the American cockroach is the least important of the domiciliary species, but in many commercial and certain other large buildings, such as restaurants, grocery stores, bakeries, packing houses, hotels, office buildings, hospitals, and prisons, it is second only to the German cockroach in abundance. In such buildings, these pests commonly infest the basements, food-storage, and food-preparation rooms. They are frequently found in the holds and messrooms of ships. In the United States, this is the predominant species in city sewer systems. In some parts of the world, as in India, *Periplaneta americana* is the predominant cockroach species rather than *Blattella germanica,* even in households (Nigam, 1933). The American cockroach occasionally causes injury to greenhouse plants, such as cutting through the stems of seedlings and removing the bark of larger plants (CEIR, 1973).

Description. The adult American cockroach (plate III, 4) is about 4 cm long. The males appear to be considerably longer than the females because their wings extend 4 to 8 mm beyond the tip of the abdomen, while those of the females do not. When 25 males were measured from the tip of the head to the tips of the wings, they were found to average 39.2 mm in length. Measured from the tip of the head to the tip of the abdomen, 25 male American cockroaches averaged 34.3 mm and 25 females averaged 34.4 mm. There was no significant difference in the average width of males (12.2 mm) and females (12.0 mm). (See table 4, under "Brown Cockroach.")

The American cockroach is reddish brown throughout, except for a pale-brown or yellow band around the edge of the pronotum. The band is widest at the posterior margin. Both sexes have a pair of slender, jointed cerci at the tip of the abdomen. The cerci of the females have 13 or 14 segments and those of the males have 18 or 19 (Bugnion, 1922). The males have a pair of styli between the cerci. The presence of 2 pairs of appendages at the tip of the male's abdomen, compared with 1 pair in the female, is one of the morphological features separating the sexes. Among laboratory-reared American cockroaches, there have been recorded ratios of females to males of 1:1.07 (Gould and Deay, 1940) and 1.08:1 (Willis *et al.*, 1958). It therefore appears that the sexes may be about equal in numbers.

Life Cycle. The developmental period of the American cockroach is greatly dependent on temperature (figure 149), but it averages about 600 days under ordinary room conditions. After reaching sexual maturity, the average life of the females may be another 400 days (Gould and Deay, 1940).

The preferred temperature for adults and nymphs was 28° C (82.4° F), but they remained active at 21° C (70°F). At 29° C (84° F), the life expectancy averaged over 630 days, but was found to be as long as 1,293 days. The adult life span at 29° C ranged from 90 to 706 days (average 225 days) for females and from 90 to 362 days (average 200 days) for males (Griffiths and Tauber, 1942).

Under suitable conditions, the longevity and high reproductive potential of the American cockroach may result in enormous populations. This is evident to anyone who has reared the species over a period of years. It is also evident from the great numbers that can be found in sewer systems. As many as 5,000 American cockroaches have been vacuumed from a single manhole in the Los Angeles city sewer system (Wagner *et al.*, 1966).

The Oötheca. The oötheca (plate III, 4) measures 8×5 mm, is brown when deposited, and turns black in a day or two. The lateral indentations marking the locations of the eggs are only weakly indicated. Along the upper ridge of the purselike egg case there is a series of "teeth," marking the location of corresponding eggs, and each tooth has a minute opening at its apex.

The female deposits each oötheca near a source of food, usually within a day after it is formed, either dropping it or gluing it to a suitable surface with a secretion from her mouth. If the female has the opportunity to do so, she hides the oötheca with great care in a crevice, or buries it in soft wood or available debris. The oöthecae of the American cockroach require a high relative humidity for successful hatching (Tsuji and Mizuno, 1971). That may be the reason that in semiarid regions such as southern California, this species is found in large numbers only in damp locations, such as sewer systems, damp basements, the dishwashing rooms in restaurants, and warehouses with inadequate ventilation. Often during the latter part of their lives, American cockroaches deposit their eggs unprotected by an egg case (Gould and Deay, 1940), thus anticipating the egg-laying habit of the descendants of cockroaches, the termites (Rau, 1941). From 15 to 90 oöthecae (average 57.6) were produced at the rate of about 1 per week, and each contained 14 to 16 eggs. This constituted an enormous production of oöthecae when compared with most other cockroach species. The average hatch from 534 oöthecae at room temperature was 13.6 nymphs (Gould and Deay, 1940).

In another similar investigation, an average of only 21 oöthecae per female was observed (Griffiths and Tauber, 1942). Nigam (1933) observed that oöthecae were generally produced in summer, beginning in April and May, but very few during winter. At a temperature of 24.4° C (76° F) the average incubation period was 57.4 days, but at 30° C (86° F) it was 31.8 days.

Nymphal Instars. According to Gould and Deay (1940), it was difficult to determine the number of nymphal instars. Molting required only a few minutes, and the cast skin was usually eaten.

From an examination of a large series of nymphs, they concluded that there were probably 13 instars.

The first-instar cockroach, immediately after hatching, consumes its castoff embryonic skin. It is white, then becomes grayish brown. It attains a length of 3.5 mm. In the early instars, the females have a median notch on the posterior margin of the ninth sternite, whereas in the males, this margin is either smooth or only slightly indented (Gould and Deay, 1940).

Fig. 152. American cockroach, *Periplaneta americana* (nymph).

After the first instar, the succeeding instars are almost uniformly reddish brown, although the posterior margins of both thoracic and abdominal segments are of a darker color, giving the insects a transversely striped appearance (plate III, 8; figure 152). Wing pads are first noticeable in the third or fourth instar. These gradually become larger, and in the last nymphal instar they are about 7 mm long, distinct, and show venation. Prominent cerci and much smaller styli are present on nymphs of both sexes. In the next to the last instar, they may be hidden by the seventh sternite in females. As already stated, the styli are not present in adult females, and the presence or absence of styli is a convenient criterion for distinguishing males from females. In the final molt, the completely formed wings appear, requiring 25 to 30 minutes to expand completely. The cast skin is not eaten, although it has usually been eaten in all the preceding molts. Complete coloration requires 4 or 5 hours (Gould and Deay, 1940).

Habits. In southern California, the American cockroach is almost the only species found in the manholes and laterals of the sewer system, but it is seldom seen elsewhere—rarely in houses and apartments. Mallis (1969) estimated that in 177 apartments in 3 Texas cities, 99.7% of the cockroaches were *Blattella germanica,* 0.2% were *Periplaneta* spp., and 0.1% were *Supella longipalpa.* No *Blatta orientalis* were found in apartments. In North Carolina, 4 pest control servicemen made a survey in 3 cities to determine the percentage of apartments infested with the 4 principal domiciliary cockroach species among the infested buildings in housing projects in their regular monthly routes. They found *B. germanica* in 90% of the infested buildings, *S. longipalpa* in 9%, and *P. americana* in 1% (Wright, 1965*a*). As in the Texas survey, no *B. orientalis* were found in multiple-unit apartment buildings. However, when all types of buildings were surveyed, *B. germanica* accounted for 54% of the infestations and *B. orientalis* was second with 34%. Infestations of *Periplaneta americana* were found in 8% and of *Supella longipalpa* in 4% of the infested structures (Wright, 1965*b*). On a military base in North Carolina, percentages of the 4 species were: German, 84.9; American, 14.4; brownbanded, 0.5; and oriental, 0.2 (Wright and McDaniel, 1969).

The author has not been able to provoke American cockroaches to fly, merely by shaking them out of a maze. However, on a sunny day in July, at a temperature of 70° F (21° C) and relative humidity of 62%, some male and female American cockroaches were taken from rearing cans and thrown into the air to observe flight performance. Some fluttered downward to ground positions, but others flew gradually upward, some so far that they were lost from sight. Along the Gulf Coast of Texas, American cockroaches have been observed to be common in palm trees and flying around street lights (Gould and Deay, 1940). In Needles, California, an extremely hot desert area, they are commonly seen flying outdoors, both day and night.

Mass Migrations. In common with *Blattella germanica, Periplaneta americana* sometimes engages in mass migrations. The former migrates by crawling (Howard, 1895; Marlatt, 1908; Guldin, 1967), but the American cockroach can migrate either by crawling or flying. Gould and Deay

(1940) observed that during the summer, American cockroaches had migrated from store buildings for a distance of 4 blocks to some decaying maple trees in which they were infesting cavities. Another instance has been reported of these insects being found near an infested hospital, under the loose bark of large shade trees where they were feeding on exuding sap (Anonymous, 1967*e*). Migrations from restaurants and city dumps have also been observed (Gould and Deay, 1940). Both American and oriental cockroaches are known to be able to enter houses and apartments from sewers via the plumbing; control of cockroaches in sewers has decreased the infestations in near-by homes (Roth and Willis, 1957).

In experiments on the control of cockroaches with 10% DDT dust in houses, Stenburg (1947) found that the treatment was more effective against American than German cockroaches. He attributed this to the fact that American cockroaches traveled greater distances and thereby increased their chances of encountering the dust. He observed that German cockroaches appeared to move only short distances from their resting places, and apparently always returned to these harborages after searching for food and water.

Australian Cockroach, *Periplaneta australasiae* (F.) (Blattidae)

The Australian cockroach closely resembles *P. americana* in size and appearance and frequents similar habitats, but it cannot tolerate as much cold. This is indicated by its abundance in tropical areas and scarcity where the climate is as relatively mild as that of Southern California. The author has only rarely encountered the species in the hundreds of apartments and houses in which experiments have been made in the Los Angeles and San Francisco metropolitan areas. In the glasshouses of a large Los Angeles orchid nursery, it was abundant and was injuring the plants; this has likewise been the case in various other parts of the world where plants were grown in greenhouses. R. E. Wagner (personal communication) found *P. australasiae* in great numbers in the Florida Keys, in the damp areas between the trunks and old stubs of fronds of coconut palm trees.

Description. Measurements were made of widths (distance across wings at widest point when at rest) and lengths (distance between tip of head to tip of abdomen) for 25 males and 25 females of Australian and American cockroaches. These figures are shown in table 4, on the next page. Although there is but little difference between the 2 species in width, the male Australian cockroach is only 83% as long as the male American cockroach, and the female Australian species is 92% as long as her American counterpart. The basic reddish-brown color, and even the pattern of yellowish markings on the pronotum, are similar to those of the American cockroach. The markings on the pronotum are more distinct, and in addition there is characteristically a light-yellow streak on the outer edge at the base of each forewing (plate III, 5). The ratio of females to males is about 1:1.

The early nymphal instars have a variable color pattern of predominantly brown, but with a narrow margin of black, as well as some light-yellow spots on the abdominal tergites. Later instars are basically reddish brown, but strikingly marked with light-yellow marginal spots on both thoracic and abdominal tergites (plate III, 8).

Life Cycle. In experiments made at 30° C (86° F), Willis *et al.* (1958) found the preoviposition period of the Australian cockroach to be 24 days. Between 20 and 30 oöthecae were produced. The interval between successive oöthecae was 9.8 days, the incubation period was 40.3 days, the number of eggs per oötheca was 23.8, and the number hatched per oötheca was 16.4, stated as averages. The percentage of hatched insects that matured was 55, compared with 84% for the American cockroach. Among nymphs reared alone, females had 11 or 12 instars, and the periods for nymphal development were 253 and 410 days, respectively. Males had 10 or 11 instars, and the developmental periods were 306 and 365 days, respectively. For females reared in groups, the period for nymphal development was 213 days, and for males it was 198 days. Thus, nymphs reared in groups matured more rapidly than those reared alone.

Habits. In infested premises, *P. australasiae* occupies habitats similar to those occupied by *P. americana* (Cornwell, 1968). It is a pest of buildings in many areas of the tropics, but as just stated, in southern California the author has found it abundantly only in orchid glasshouses. The most common habitat in the glasshouses was in the space between small pots placed within the size larger pot to provide stability to small plants during watering. There is about 2 cm of space between the bottoms of the 2 pots, and this space is very damp. Australian cockroaches were found principally in these spaces, in large numbers.

The Australian cockroaches were feeding on orchid flowers and tender growth terminals. Although German cockroaches were feeding on these as well, the Australian cockroach is particularly known for its propensity for feeding on tender greenhouse plants. It has been reported to cause damage to plants in the tropical house of the Royal Botanic Gardens at Kew, England (Cornwell, 1968). In Denmark, *P. australasiae* occurred in large numbers in an indoor garden room of a Copenhagen restaurant, primarily in the plant beds, but sometimes it was seen by guests at tables. It had not yet spread to the kitchen or other parts of the restaurant, where the German cockroach was abundant. It was believed to have been introduced with plants from a greenhouse in the Netherlands (Anonymous, 1967*e*).

Brown Cockroach, *Periplaneta brunnea* Burmeister (Blattidae)

This species closely resembles the American cockroach in habits and appearance, but is more confined to the tropics in its geographic distribution. It is common in the southern and southeastern United States from the Carolinas to Florida and westward to eastern Texas (Edmunds, 1957). It has been collected indoors at points farther north, and in 1969 it was found in an apartment near Los Angeles, California, the first record of the insect in California (Reierson and Ebeling, 1970). The apartment was occupied by tenants who had arrived 4 months earlier from San Antonio, Texas, where the predominant species of *Periplaneta* is *P. brunnea,* according to Mallis (1969). It is also known that the brown cockroach is more common than the American cockroach in certain other areas of the South. Another accidental introduction of brown cockroaches from Texas was noted by Waldron and Hall (1972).

Despite the close resemblance of *P. brunnea* and *P. americana,* hybridization does not occur. A mixed colony of these 2 species plus *P. fuliginosa* was maintained for more than 3 years without the production of any recognizable hybrids (Eddleman and Simon, 1969).

Description of Adults. The adults are dark reddish brown (plate III, 6), resembling the American cockroach, but are somewhat darker in color. The 2 species also resemble each other in the yellowish margin of the pronotum. When measuring 25 individuals of each sex among those reared in our laboratory, the lengths and widths of *Periplaneta americana, P. brunnea,* and *P. australasiae* were recorded, and are shown in table 4. The data show that the American and brown cockroach are similar in length, but that the latter is the wider of the 2 species. The wings of the male brown cockroach do not extend beyond the tip of the abdomen as far as those of the American cockroach. Both the brown and the American cockroaches can be distinguished from the Australian cockroach by their slightly larger size and the absence of the yellowish streak on the outer edge of the base of each forewing that so distinctly characterizes *P. australasiae.*

The cercus of the American cockroach is stout basally, and tapers markedly toward the tip (Pratt, 1955). The sides of the last segment are more or less parallel, and 2 or more times as long as the basal width. The cercus of the brown cockroach is also stout, but the appendage is more evenly spindleshaped, with the last segment somewhat triangular and less than twice as long as its basal width. *Periplaneta brunnea* can fly, but its flight is usually of the gliding type.

Life Cycle. Judging only from the enormous production of *P. brunnea* in our rearing room at normal room temperature when compared with other species of *Periplaneta,* it appears to have an alarming reproductive potential. Possibly, there

TABLE 4

Body Dimensions of 3 Species of *Periplaneta*

Species	Sex	Length (mm)*	Width (mm)	Width/length
P. americana	Male	34.3 ± 0.31	12.2 ± 0.12	0.355
P. americana	Female	34.4 ± 0.38	12.0 ± 0.10	0.349
P. brunnea	Male	33.4 ± 0.38	13.1 ± 0.08	0.392
P. brunnea	Female	33.9 ± 0.39	13.5 ± 0.09	0.398
P. australasiae	Male	28.5 ± 0.27	11.5 ± 0.07	0.403
P. australasiae	Female	31.7 ± 0.29	12.4 ± 0.11	0.391

*From tip of head to tip of abdomen.

are ecological factors under actual field conditions that have thus far resulted in this species being generally less abundant than the American cockroach, even in the South.

Copulation occurred within a few hours after a female completed development, and egg deposition began 16 to 20 days later. It continued throughout her life. The periods between oötheca depositions varied greatly, but could be as little as 5 to 6 days. Pope (1953) found that the maximum number of ovipositions was 30, but usually it was less. Edmunds (1957) kept some *P. brunnea* for 20 months, and they were still living and reproducing.

The Oötheca. While the oöthecae of the American and Australian cockroaches are similar in appearance, those of the brown cockroach are strikingly different (plate III). Reierson and Ebeling (1970) found that the oöthecae from a colony of brown cockroaches reared on Purina Dog Chow at 80° F (27° C) were 13.5 mm long (range 12 to 16 mm) and 5 mm wide at the widest point (range 4.5 to 5.5 mm). Those produced by Australian cockroaches reared under the same conditions averaged 10.9 mm long (range 10 to 11.5 mm), except for 1 oötheca that was 13 mm long. Their average width was 4.4 mm (range 4 to 5 mm). The oöthecae of American cockroaches averaged 8.3 mm in length (range 7.5 to 9.5 mm), and were about as wide as those of the brown cockroach. Thus, the oöthecae of the brown cockroach averaged 2.6 mm longer than those of the Australian species, and 5.2 mm longer than those of the American cockroach.

Edmunds (1957), conducting all his experiments at 75° F (24° C), found that *P. brunnea* deposited the oötheca from 20 to 24 hours after it was extruded. The number of eggs per oötheca averaged 24 (range 21 to 28). Reierson and Ebeling (1970) obtained an average of 23.8 eggs per oötheca at 80° F (27° C) and 65 to 75% relative humidity. The incubation period at this temperature was 35 days, compared with 61 to 63 days at 75° in the Edmunds experiments.

The oöthecae are at first brownish, but become black with increasing age, a characteristic of cockroach capsules. In large plastic rearing cans, the brown cockroach glues many of its eggs to the vertical interior walls of the cans or to the corrugated-paper mazes (Ebeling *et al.*, 1966). In glass jars, the female frequently glues her oöthecae to the jar lids. Under similar rearing conditions, the American cockroach glues her oöthecae to the floor of the container. Both species at least partially cover their oöthecae with bits of paper, cotton, or other available debris in an apparent instinct to hide them.

An interesting aspect of the biology of *P. brunnea* was brought to light by Willis *et al.* (1958), who found that eggs hatched from 9% of parthenogenetically produced oöthecae and that 31.7% of the 208 hatched nymphs matured. All were females.

Maternal Care for Eggs. Cockroaches share with the somewhat related termites and earwigs the instinct of maternal care for their eggs. Edmunds (1957) described the phenomenon with respect to *P. brunnea* as follows:

> In ovipositing, the female roach secreted from her mouthparts a frothy white substance which she smeared over the spot in which she was going to deposit the egg capsule. Some females spent from 30 to 40 minutes preparing this frothy bed. The egg capsule was then deposited in the froth and covered with additional froth secreted by the female. Some cockroaches were observed spending as much as 2 hours coating the egg capsule after it was deposited. The substance hardened to become a very strong cementing material. It was so strong that it was difficult to pry the capsule loose without causing it to rupture. For several hours after a capsule was deposited, the female rested with her body over the capsule and drove away any other roaches which approached.

It may be that if the oöthecae of the brown cockroach were not guarded by the female, they would be in danger of being consumed by other cockroaches during the period when she customarily remains with them, since many are later found and eaten. Cockroaches that are injured or weakened in some way are also eaten by others. This may also be observed among other cockroach species.

Nymphal Instars. In the first-instar nymph, whereas the antennal segments of the American cockroach are uniformly brown, the first 8 and last 4 segments of the brown cockroach are white, and the intermediate ones are brown. The first-instar nymph of the Australian cockroach resembles the brown cockroach in this respect. A median translucent area in the mesothorax of the brown cockroach nymph is absent in *Perplaneta americana.* The first 2 segments of the abdomen of the first-instar nymph of the American cockroach are entirely gray, brown, or reddish, while those of the brown cockroach have faint, cream-colored spots on the dorsolateral margins. In later instars, these dorsolateral spots are extended to and include the sixth segment of the abdomen. The intermediate nymphal instars (plate III, 8) of the Australian cockroach are similar to those of the brown cockroach, except that the cream-colored markings are more distinct, and extend throughout the length of the abdomen (Edmunds, 1957; Reierson and Ebeling, 1970).

Edmunds also observed that at 75° F (24° C), out of 25 newly emerged nymphs, the first male completed development in 263 days and the first female in 268 days, but that all nymphs had completed development in 277 days. The entire life cycle of the brown cockroach, from egg to egg, varied from 339 to 351 days—relatively brief when compared with the American cockroach.

Other Cockroach Species

Many people familiar with the common domiciliary species of the United States are not aware of the large number of other species, indigenous or accidentally introduced, that occur in this country. Some are at times abundant in certain regions. Excellent photographs of the species to be described here were presented by Roth and Willis (1960).

Pennsylvania Wood Cockroach, *Parcoblatta pennsylvanica* (De Geer) (Blattellidae)

This cockroach is indigenous to the United States, ranging through the midwestern, southern, and eastern states and north into Canada. Rau (1940) described it as "a trim, pretty creature, with its chestnut-brown color, and thorax and wing pads edged in white," but Hebard (1917) emphasized the variability of the species in different areas, and did not consider a general description of color to be of any value. The males are fully winged, and can occasionally be seen flying about outdoors. The wings of the females are short and nonfunctional. The males are considerably larger than the females, being 2.2 to 3 cm long compared with 1.3 to 2 cm for the females. Houses in rural districts and near wooded areas are commonly invaded, particularly by the males, which fly and are attracted to lights. This species is also frequently seen by motorists when driving through the woods at dusk (Gould and Deay, 1937, 1940).

Parcoblatta pennsylvanica happens to have the widest distribution, but there are at least a dozen other species of the genus that occasionally find their way into houses in the United States, particularly *P. lata* (Brunner) and *P. virginica* (Brunner). Species of *Parcoblatta* are found throughout the country. They are generally seen in wooded areas, but one, *P. desertae* (Rehn and Hebard), is found only in desert and semidesert mountains of the Southwest (Rehn and Hebard, 1909; Hebard, 1917, 1942). *Parcoblatta americana* (Scudder) is the only species of the genus found in California. It is small, the males being 12 to 14 mm long and the females 9 to 13 mm. The females are wingless. The color is generally shining buff or brown, ranging to black in some females in the region of Mount Shasta, California. The smaller and paler specimens are found in the more arid and barren regions (Hebard, 1917).

Tawny Cockroach, *Ectobius pallidus* (Olivier) [= *livens* (Turton)] (Blattellidae)

This species may be found actively running about outdoors in temperate regions, such as southern England and parts of the European continent. It was first discovered in the United States in 1948 on Cape Cod, and has become established in southeastern Massachusetts. It is common in some areas (Gurney, 1953; Smith and Chao, 1956). It does not attack household foods or goods, but could become a nuisance when attracted to dwellings by lights. This is a pale-brown cockroach, smaller than *Blattella germanica* and lacking the dark lines on the pronotum that characterize the latter. The wings have many tiny, dark spots that can be seen only with the aid of a hand lens. The nymphs are also pale brown (Smith and Chao, 1956).

Smoky Brown Cockroach, *Periplaneta fuliginosa* (Serville) (Blattidae)

This species is uniformly shining brownish black or mahogany in color. Both sexes have wings that extend beyond the tip of the abdomen. It is common in the southern states from central Texas eastward, and has been found in heated buildings in isolated locations as far north as Iowa, Indiana, and Illinois. This is the only uniformly dark-colored species of *Periplaneta* in North America, and is also the only species not known to occur in other parts of the world (Hebard, 1917). It flies to lights at night, and is sometimes attracted to lighted porches. It has been recorded as second only to the German cockroach as a pest in houses in southwest Georgia (Haines and Palmer, 1955), but is more commonly found in garages, outbuildings, and woodpiles. It is sometimes a pest in greenhouses (Hebard, 1917; Gould and Deay, 1940; Cornwell, 1968).

Woods Cockroach, *Eurycotis floridana* (Walker) (Blattidae)

This species occurs in the West Indies and southern Florida, where it may be found in almost any outdoor sheltered location—under the bark of dead pine stumps and logs, in tree cavities, under palmetto leaves on the ground, under debris in leaf mold or decaying wood, in stacked lumber and firewood, and often in cavities in limestone rock (Hebard, 1917; Roth and Willis, 1960). It may occasionally be found in houses (Creighton, 1943, 1954). *Eurycotis floridana* is 4 to 4.5 cm long. Adults and nymphs are rich red-brown to almost black, but the nota (dorsal thoracic segments) of the late nymphal instars are often conspicuously marked with broad bands of pale yellow. The tegmina (thickened forewings) of both sexes are very short, extending barely beyond the posterior margin of the mesonotum, and the hindwings are lacking. When alarmed, the adults of both sexes emit a greasy liquid with a most repellent odor (Hebard, 1917; Eisener *et al.*, 1959). A single gland on the underside of the abdomen ejects a spray of it toward the rear.

Neostylopyga rhombifolia (Stoll) (Blattidae)

This cockroach, believed to have originated in Indo-Malaysia (Roth and Willis, 1960), has become established in various parts of the Western Hemisphere, including Mexico. From Mexico, it migrated northward to Nogales, Arizona (just across the border), and it also occurs in Baja California (Hebard, 1917). In California, a single adult was found in a warehouse in Los Angeles in 1958, and another in a box of tomatoes in a market in Santa Rosa in 1965. An adult was found on a tomato fruit in a field near Watsonville in 1966. This species was occasionally found on Travis Air Force Base, California, where cargo planes were arriving from Vietnam during the war (George Buxton, personal communication). The length varies from 2 to 2.7 cm. The tegmina (forewings) of both sexes are small lobes, only about 4 mm long, and hindwings are entirely lacking. The basic color is shining brownish black, marbled with yellowish buff in a distinctive pattern that makes identification easy.

Surinam Cockroach, *Pycnoscelus surinamensis* (L.) (Blaberidae)

This species is distributed throughout the world in the humid tropics and subtropics. In the United States, it is abundant in the Florida peninsula and in the Brownsville area of Texas, but it is also established in New Orleans, San Antonio, and Hawaii. The Surinam cockroach is 2 to 2.5 cm long, and is shining brown to black. In both sexes, the wings extend beyond the tip of the abdomen. The rear margin of the adult's prothorax is sinuate (curving in and out), and the front margin is marked by a pale band in some specimens.

Where the climate is favorable, the Surinam cockroach is generally found outdoors under rocks and boards, under litter, or in damp soils. In Hawaii, it has been found in soil around the roots of pineapple, under mulching paper, or even feeding on pineapple roots (Illingworth, 1927, 1929). In colder regions, this species can sometimes be found in greenhouses, where it can cause damage to plants. In some Philadelphia greenhouses, Surinam cockroaches were present "by the millions." From 30 to 50 thousand rose plants out of 200 thousand were so severely injured by the gnawing off of the bark, young buds, and shoots of the main stems that they had to be discarded (Doucette and Smith, 1926). In a greenhouse in Germany, many of these cockroaches penetrated soil to a depth of 8 to 10 cm, and a few even more deeply. Some insects spent several days at a time in chambers at the bottoms of the holes. Nymphs molted and females produced their young in such chambers (Roeser, 1940). *Pycnoscelus surinamensis* is parthenogenetic in North America and Europe, and the offspring are all females, but it is bisexual in Indo-Malaysia (Roth and Willis, 1956).

Madeira Cockroach, *Leucophaea maderae* (F.) (Blaberidae)

This species is probably native to West Africa, but was first described from the island of Madeira. It occurs in the western Mediterranean area, in many parts of South America, throughout the West Indies, the Bahama Islands, Hawaii, and in a Puerto Rican section of New York City. Nymphs were reported by quarantine officials to be inhabiting a residence on Cheli Air Force Base, Maywood, California, in November, 1972. The adults are pale brown to tawny olive in color, and 4 to 5 cm long—about 7 mm longer than our largest common species, *Periplaneta americana*. The wings of both sexes reach the tip of the abdomen. The forewings are marked with 2 dark-brown lines in the basal area, and the posterior areas are mottled. The nymphs have very small, short spines that are particularly noticeable along the dorsal posterior margin of each segment. The adults crawl about slowly but fly actively, and like *Eurycotis floridana,* give off an offensive odor when disturbed (Cornwell, 1968). The Madeira cockroach has a very long life span, reported to be 2.5 years by Scharrer (1951), who recommended this large insect as a convenient and useful laboratory animal. He stated that experimental operations on it might be significant with respect to mammals, and the species has, in fact, become popular in many research laboratories for this reason.

The Madeira cockroach is frequently intercepted in quarantine at seaports, which is understandable in view of its predilection for fruit, particularly bananas (Sein, 1923). Despite the high temperature requirements of this species, it can adapt itself to sheltered life in temperate climates, as shown by its establishment in New York City in buildings having sufficiently high temperatures. There appears to be no reason why it could not become more widely distributed. The brownbanded cockroach also requires relatively high temperatures, but has nevertheless become widely distributed throughout the United States (Gurney, 1953).

Lobster Cockroach, *Nauphaeta cinerea* (Olivier) (Blaberidae)

This insect has world-wide distribution in tropical areas, where it is known as a domiciliary species. It was first recorded in the southern United States in 1952, and also occurs in Hawaii, where it is known to kill and eat the Pacific beetle cockroach, *Diploptera punctata* (Eschscholtz) (Illingworth, 1942). The lobster cockroach, so named because of the lobsterlike design on its pronotum, is ashy gray in color and 2.5 to 3 cm long. The wings do not cover the abdomen, although they are slightly longer in the male than in the female.

Blaberus craniifer Burmeister (Blaberidae)

This large, broad cockroach is known from Mexico and Central America, but in the United States it has become established only in Key West, Florida (Hebard, 1917). However, cultures may sometimes be seen in zoological laboratories. It is said to be abundant in houses in Cuba. The normal habitat is under rotting logs and vegetation in rain forests. Hebard found the length of the male's body to be (rounded figures) about 4.3 to 5 cm; the width of the pronotum, 2 to 2.2 cm; and the length of the tegmina (forewings), 5 to 5.5 cm. The latter extend beyond the tip of the abdomen, about as shown by the foregoing figures. The corresponding figures for the female were about 5.5 to 6 cm, 2.1 to 2.3 cm, and 5.4 to 5.8 cm, respectively. The general body color is blackish brown, with an even darker area on the pronotum of the adult that bears pale markings that suggest the human eyes, nose, and mouth. The same area on the nymph bears 4 pale spots, and the abdomen has a median double row and lateral double rows of pale spots. In the related *Blaberus atropos* (Stoll), indigenous to South America, the fancied "death's-head" marking is on the mesonotum and metanotum (Hebard, 1917; Roth and Willis, 1960). The adults of *B. craniifer* are long-lived, the males averaging 14 months and the females, 16 months.

Cryptocercus punctulatus Scudder (Cryptocercidae)

This primitive, wingless, wood-eating cockroach is discussed here only because of its great scientific interest. The males and females are of the same size and appearance; 2.3 to 3 cm long; wings are completely lacking; the general color is almost black, tinged brown; the eyes are very small; the pronotum is very convex, except for a wide and deep mediolongitudinal depression, and has a broad margin that is raised above the head, thought to give the head protection while burrowing. This species occurs in the United States in heavily forested areas of the Pacific Coast, and in the Appalachian Mountains from Pennsylvania to northern Georgia. Hebard (1917) described a typical habitat in fir logs in Oregon. The insects were found only where the wood was soggy and the bark could be easily removed. The cockroaches had eaten their way through the sapwood, preferring the places where it was soft, soggy, and decayed. This species is found in hardwood as well as softwood material. The infested wood is generally "soft, punky, and wet," but Cleveland *et al.* (1934) also found *C. punctulatus* in sound logs that had been down for only a few years.

This cockroach is of great interest because of the presence in its alimentary canal of cellulose-digesting, flagellated protozoa, transferred from older insects to the young as in the termites, which are believed to be descended from the cockroaches. The first fecal pellets passed by the older nymphs after a molt contain large numbers of encysted protozoa. When eaten by the newly hatched nymphs, these cysts ensure a heavy population of flagellates within a few days. *Cryptocercus* depends on these flagellates for the digestion of cellulose.

One genus of flagellates (*Barbanympha*) contained the largest and most abundant species found by Cleveland *et al.* (1934) in every one of the thousands of *Cryptocercus* they examined. They contained the enzymes cellulase and cellobiase, which the cockroach itself was unable to produce, and these enzymes converted cellulose into soluble sugar (dextrose) (Trager, 1932). *Cryptocercus* possesses 2 genera of protozoan symbionts (*Trichonympha* and *Leptospironympha*) that are also present in termites (Honigberg, 1970). Transfaunations can be achieved between *Cryptocercus* and *Zootermopsis* by feeding one the intestinal contents of the other (Cleveland *et al.*, 1934; Nutting, 1956).

Control of Cockroaches

Good Housekeeping. Cockroaches cannot live in human habitations without the food and water they provide. Obviously, good housekeeping is an important factor in control. Food particles should not be allowed to remain on shelves or floors. Dishes should not be left unwashed after a meal, particularly overnight. Areas under refrigerators, stoves, sinks, and furniture should be checked to see whether food has accidentally fallen or has been swept into them or dropped there by children or pets. If pets are fed indoors, food particles should not be left in their dishes after feeding, particularly not overnight. Keep cereals, crackers, meats, vegetables, flour, pet foods, and all other foods covered or in a refrigerator at all times between uses. Keep garbage and waste containers covered, or empty them into a covered container every night.

Water is often available as condensation or spillage under the refrigerator. Patches of water often remain for hours in certain locations after a floor mopping or following the use of a washing machine. Other common sources of water for cockroaches are the traps of sinks, washbasins, tubs, and toilet bowls; flush tanks; condensation on cold pipes and windows; leaking pipes and faucets; miscellaneous water-filled containers such as pets' drinking dishes, aquaria, vases, empty beverage bottles; and the juices from soft fruits and vegetables (Roth and Willis, 1960).

Cockroach infestation is favored by a cluttered household. Stored magazines and newspapers, corrugated-paper boxes and piles of paper bags should be eliminated. They not only provide harborage and breeding areas for cockroaches, but also provide areas into which the cockroaches can escape from insecticide residues when control is attempted. Insecticides are repellent to cockroaches in varying degrees.

Application of Insecticides. If liquid sprays are to be applied, and ideally for dusts also, all shelves and drawers should be emptied and cleaned. After treatment, shelf paper may be placed on sprayed areas and on dust deposits. If a relatively safe dust such as boric acid is to be used, it is recommended that the operator reach carefully behind shelf contents and apply the dust with a bellows powder blower ("Getz Gun"), bulb duster, or plastic squeeze bottle (figure 29, chapter 3) in a narrow band along intersections and into corners and crevices. Shelf paper lifted while the powder is applied may then be allowed to fall back into place.

When applying insecticides for cockroach control, pest control operators ordinarily ask the occupants of the treated premises to remove cooking and eating utensils and foodstuffs when there is danger of their becoming contaminated with insecticide. For example, they may be placed on a table in the center of a room and covered with a sheet. Occupants are asked to leave the building and remove their pets during treatment.

Treatment for cockroaches, like so much of household pest control, is "spot treatment." The insecticide is applied to relatively limited areas where cockroaches are likely to hide during the day or where they are likely to travel at night. Most cockroaches congregate in dark, warm, damp places, such as under and behind the refrigerator and stove and under the sink, particularly the "dead space" between the sink and the wall. The most thorough treatment can be made under the refrigerator and stove if the kick panels are first removed, allowing the person applying the spray or dust to see what he is doing. Many cockroaches seek harborage and breed in the insulation in the walls of a refrigerator, stove, or range. These insects should come in contact with insecticide residues when they come out to forage for food and water. Insecticide sprays should also be applied behind window and door frames, under and behind furniture standing against the wall, on the undersides of tables and chairs, on closet and bookcase shelves, and in cracks and apertures around cabinets. Cockroaches are attracted to warmth, and may hide under and behind radiators and water heaters or in television sets and radios. Apply enough spray to wet surfaces thoroughly, but not enough to drip or run. If both sprays and dust are to be used, spray first and apply the dust after the spray dries, forcing the dust into cracks and voids.

When cockroaches are abundant, they are likely to be located in wall voids. In apartment houses, they may migrate from heavily infested to lightly infested or uninfested apartments via these voids. After blowing an insecticide dust into openings around utility pipes or any other openings into walls, particularly in kitchen and bathroom areas, such openings should be sealed to prevent further ingress or egress of the insects. Holes through which to dust wall voids can be drilled through plaster if necessary, and the holes can then be filled with matching plaster. Holes can also be drilled at the top of the kick panel under a kitchen cabinet, or in the floor of the cabinet, in order to dust the subcabinet voids, unless the latter can be reached by removal of drawers or through existing cracks or apertures. After treatment, seal all cracks and crevices where cockroaches may hide. Even very narrow cracks may harbor the small nymphs. If the backsplash behind the sink and on the sink cabinet has parted from the wall, leaving a narrow aperture, this is a particularly favored harborage for cockroaches and should be sealed.

Insecticides

Because the German cockroach is almost universally resistant to chlorinated hydrocarbons, the organophosphorus and carbamate insecticides are now used for control of this species. Suggested insecticides and dosages are: diazinon, 0.5% spray or 1% dust; propoxur (Baygon®), 1% spray or 2% bait; chlorpyrifos (Dursban®), 0.5% spray or 1% dust; dichlorvos, 0.5% spray or 1% bait; malathion, 5% spray or dust; and Kepone®, 0.125% bait. It has been our experience that the reduction in German cockroach infestation from sprays with base-oil solutions or emulsions of the insecticides currently being used against German cockroaches in multistoried apartment buildings can be expected to average around 85%, based on the number of cockroaches trapped before and a month after treatment. This is also approximately the average degree of control accomplished by pest control operators. On the other hand, Burden *et al.* (1972), applying 0.5% chlorpyrifos emulsion spray in ground-level homes constructed of concrete blocks, were able to obtain German cockroach reductions that were "95% within 1 day, 97% within 2 weeks post-treatment, 99% from the third through the eleventh week, and averaged 98% at 12 weeks post-treatment." Wright and Hillmann (1973) applied 4 treatments: 1% chlorpyrifos dust, 0.5% chlorpyrifos spray, 0.5% chlorpyrifos spray plus 1% chlorpyrifos dust, and boric acid powder, then spot-treated at the 2-week and 1-month recheck periods wherever necessary, mostly in harborages missed during initial inspection and treatment. They used approximately 22% as much material in the retreatments as in the initial treatment, and did not use disproportionate amounts in the different treatments. They obtained controls ranging from 97 to 100% for as long as 6 months following the initial treatment.

Pyrethrin aerosols are commonly used to flush cockroaches from their daytime hiding places for pre- or post-treatment estimates of infestation levels in buildings, or to drive the insects from their hiding places so as to facilitate treatment with residual insecticides. Pyrethrins and their synergists are also added to insecticide solutions or emulsions for more rapid knockdown or presumed increase in the efficacy of the sprays. In field tests in which pyrethrins plus synergists were added to 3 insecticide solutions, 1 emulsion, and to boric acid powder, the pyrethrins resulted in more rapid knockdown, but evaluations made a month after treatment did not reveal, on the average, any increased insecticidal efficacy (Ebeling and Reierson, 1973*a*). In addition, when pyrethrins were used in apartment buildings they had the disadvantage of driving cockroaches from the treated apartment to adjacent untreated apartments via the wall voids.

Pyrethrin formulations, particularly dusts, frequently cause allergic reactions in people with such allergic tendencies as hay fever and asthma. Some persons applying certain organophosphorus insecticides suffer respiratory disorders, but the carbamate propoxur has not been reported to have this disadvantage (Lisella, 1973).

Resmethrin is a synthetic pyrethroid that has been suggested as being much less irritating to the throat than pyrethrins. Resmethrin does not seem to flush cockroaches out of their hiding places as quickly as pyrethrins, and the insects appear to be less agitated. However, within 15 to 20 minutes, a greater proportion of the cockroach population is said to be flushed out, providing a more accurate estimate of the infestation level (Grothaus *et al.*, 1972).

If base oil (deodorized kerosene) is used as a carrier in the spray, it should not be applied near a flame, as under a gas-operated water heater; a water-base spray (emulsion) should be substituted in areas of fire hazard. Ideally, base-oil solution should not be applied to plaster, asphalt tile, or linoleum, for it leaves an oily stain on plaster, destroys the waxy finish on tile or linoleum, and leaves a slippery surface that may be hazardous to occupants of the sprayed building. Base oil sometimes penetrates between old floor tiles and weakens their bonding to the floor. In cockroach control, it is necessary to spray beneath and behind the refrigerator and stove and along baseboards, and thereby get base oil on the floor at least in these out-of-sight areas. An experienced serviceman sometimes sprays with base-oil solution along baseboards in other areas when the floor is covered with tile or linoleum, but he sprays very lightly. Most disadvantages of base oil can be avoided by using water as the carrier. However, water emulsions have their disadvantages also. They tend to leave visible residue ("spots") when the emulsion dries. Either base oil or water will leave unsightly residues if applied over greasy or dirty surfaces to the point of runoff, for this may cause streaks to form around the treated area.

The residual efficacy of insecticides applied in liquid sprays is greatly affected by the nature of the treated surface, as shown in table 5. Dust deposits are less affected by the substrate, and this is one of the advantages of dust formulations.

Dusts can be applied most effectively in cracks, crevices, and enclosed spaces such as wall voids, voids under cabinets, and voids under built-in appliances. In visible places, such as along baseboards and moldings and in bathrooms, dust deposits are unsightly and liquid sprays are more appropriate. In the bathroom, there are few places where dusts are not unsightly, but if they are blown into the louvers or razor-blade slots of medicine cabinets, they can be widely distributed in a wall void that is likely to harbor cockroaches. The best results are obtained with both spray and dust, each in appropriate locations. A spray or dust treatment, or a combination of the two, can be effectively supplemented with cockroach bait that may be placed in out-of-sight areas similar to those in which a dust would be applied. Dichlorvos bait initially kills cockroaches more rapidly and completely than Baygon or Kepone, but the latter provide longer residual control. Glycerol has been found to be effective as a feeding stimulant in poison baits, and is much more so than maltose, the most effective of the sugars (Tsuji and Ono, 1970*b*).

Vapors and Gases in Enclosed Spaces

A piece of dichlorvos resin strip 1 sq in. (2.5 cm) in area, attached to the underside of the lid of a 24-gal (91-l) garbage can, results in complete mortality of all active stages of any of the 4 principal species of cockroaches in 24 hours under room-temperature conditions (Ebeling, 1964*b*). A "ministrip" ($2 \times 2.5 \times 0.25$ in.; $5 \times 6.5 \times 0.7$ cm) is now available that has an adhesive substance on one side so that it can be pressed against the bottom of a garbage-can lid. The ministrip, placed in closed kitchen sink cabinets in apartments infested with German cockroaches, resulted in 75 to 90% mortality during the first 7 to 12 weeks, decreasing to 50% mortality at the end of 12 weeks. The results with the ministrip were not significantly different from those with a spray of 1.5% malathion plus 0.5% diazinon (Wright, 1971).

The efficacy of resin strips depends on the concentration of dichlorvos vapor they can generate

TABLE 5
Relative Longevity of the Residues of 5 Insecticide Sprays on 4 Types of Surfaces Against the German Cockroach*
(Number of Days with 70% Control)

Insecticide	Painted metal	Unpainted metal	Masonite	Asphalt tile
Fenthion	1	49	8	1
Diazinon	11	8	11	8
Malathion	14	84	14	14
Dichlorvos	8	1	8	15
Propoxur	42	42	14	21

*From NPCA (1966).

in the area to be treated. In a normally ventilated room, the concentration does not reach a sufficiently high level for cockroach control, but the strips are effective against them in enclosed spaces with very poor ventilation or none.

As indicated by the experiments of Barnhart (1963) and Cantwell *et al.* (1973), dry ice can be used as a safe and inexpensive means of generating a gas for exterminating cockroaches in food service carts, as used, for example, in hospitals. These carts have many areas where cockroaches can hide and be inaccessible to sprays or dusts. Tompkins and Cantwell (1973) developed a dosage and time schedule for killing not only active stages, but also the eggs. The cart was centered on a 12 × 20-ft (3.7 × 6.1-m) sheet of black 6-mil plastic. Twenty pounds (9 kg) of dry ice was broken into pieces of appropriate size for placing at different levels in the cart. The plastic sheet was then drawn up and tied at the top. The volume of the enclosed cart was 71 cu ft (2 cu m). The level of carbon dioxide was monitored with a Gow-Mac Gas Analyzer at 5 and 24 hours. At 78° F (26° C), the minimum concentration of carbon dioxide required for the 24-hour period of the test was 21% in order to kill not only all the female German cockroaches with attached oöthecae that were used as the test insects, but also to prevent hatching of their eggs over the 30-day period of observation. This method of fumigation has the advantage of being safe and not requiring special training of personnel.

Boric Acid for Cockroach Control

Experiments made in our entomology laboratory at UCLA showed that the principal factor affecting the effectiveness of an insecticide against cockroaches was its repellency (Ebeling *et al.*, 1966). The greater the repellency of the insecticide, the greater the chances that a certain percentage of cockroaches would learn to avoid treated areas before picking up lethal doses. Cockroach control is ordinarily accomplished by "spot treatment" of the dark and hidden places normally frequented by cockroaches during daylight hours. Some of the surviving insects in these areas may learn to avoid the treated areas and seek harborage elsewhere in the building. In apartment buildings, some cockroaches may leave treated apartments, and may invade untreated apartments via the wall voids or hallways.

Boric acid powder is the least repellent of the insecticides tested. In buildings treated with boric acid, cockroaches will crawl over the deposits repeatedly. Although they do not die as rapidly as when treated with more toxic insecticides, few if any should be seen after a week or 10 days. Boric acid is inorganic, and in an area in which it can remain dry and is not likely to be removed, it will continue to kill cockroaches indefinitely

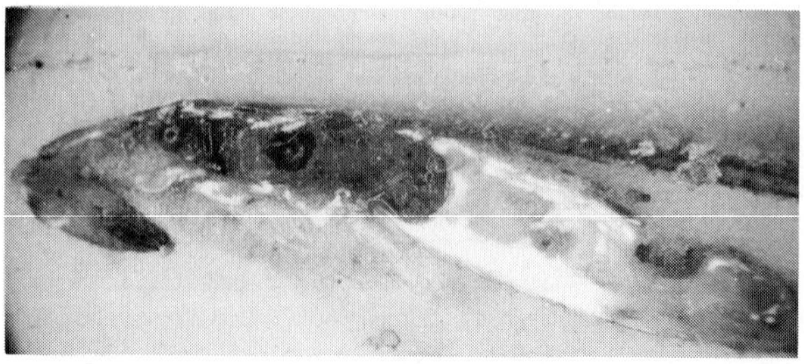

Fig. 153. Median longitudinal section of a German cockroach that had been left on a deposit of stained boric acid until "knockdown." The dark intestinal contents show the point reached by the boric acid within the intestinal tract.

(Ebeling *et al.*, 1966, 1967, 1968; Moore, 1971, 1972). Before using, it must be passed through a sieve or screen to break up lumps and facilitate its passage through the usual dust applicators.

Sodium fluoride is likewise an inorganic powder, and is much more toxic to insects and to humans than boric acid, but it is highly repellent, and has been far less effective in control of cockroaches, both in laboratory devices utilized to simulate typical harborage areas in buildings (choice boxes, mockup wall voids, mockup closets) (Ebeling *et al.*, 1967) and in experiments made in apartment buildings (Ebeling *et al.*, 1968). As applied for cockroaches, boric acid is also effective for control of silverfish.

Boric acid acts in part as a contact insecticide, penetrating through the insect's cuticle. In addition, cockroaches remove the dust from their antennae and legs while passing these appendages through their mouthparts. In this way, much dust is consumed, and boric acid, like sodium fluoride, acts partly as a stomach poison. Figure 153 shows the point to which boric acid, stained with Janus Green dye, generally penetrates into the alimentary canal of an adult German cockroach within 3 hours when the insect is placed on a deposit of 2 cc of boric acid in a petri dish. Repeated trials showed that the powder always filled the crop of a cockroach to capacity within this period, and although the insect might continue to preen for another 2 hours, its esophagus, pharynx, and mouth were packed with powder and no more could be ingested. The powder in the crop is prevented from moving farther into the alimentary canal, except in very small quantities, by the denticles of the armarium of the proventriculus.

In areas in which a dust would not ordinarily be applied, the liquid insecticides discussed earlier can be employed advantageously for a liquid–dust combination treatment. Gupta *et al.* (1973), experimenting with 0.5% chlorpyrifos spray alone or in combination with 2% propoxur bait, 2% diazinon dust, boric acid, or Drione® (silica aerogel plus pyrethrins), found that, in apartments with bad sanitation, only where dusts were used was control of German cockroaches satisfactory 2 months after treatment. They recommended the use of "a combination of inorganic dusts and insecticidal sprays (with knockdown qualities) that can be applied in appropriate and somewhat mutually exclusive areas, so that the efficacy of both the formulations is fully realized."

When boric acid alone is used as the residual insecticide, pyrethrin aerosols can be applied in areas of heaviest cockroach concentration to hasten control of the insects. Aerosol canisters, such as shown in figure 20, chapter 3, are suitable for this purpose, but have the disadvantage of not being refillable. A refillable aerosol generator with a flexible hose, such as the Hi-Fog (figure 21, chapter 3), can be used just as effectively and also more economically in extensive treatments. The cockroaches should then be swept up and destroyed as soon as possible after treatment, otherwise many may recover. Boric acid dust should then be applied in the usual way. The efficacy of the pyrethrin–boric acid treatment, as determined in extensive field experiments in California (Ebeling *et al.*, 1968), was later confirmed in other field experiments (Moore, 1971, 1972).

Japanese entomologists found that 20% boric acid, in starch-bait tablets or granules, was not repellent to German cockroaches. The addition of 0.5% lindane or fenitrothion resulted in an increased initial rate of mortality, but because of their repellency, these insecticides also increased the period required for a 100% kill. Methyl myristate at 0.02 to 0.2% or maltose at 10 to 15% concentration increased the amount of bait consumed. With a given amount of bait, small granules scattered widely were more effective than the larger tablets not so widely distributed (Tsuji and Ono, 1970*a*). Another advantage would be that granules are less likely to be found and consumed by small children and pets.

Application of Boric Acid During Construction. As an inorganic dust, boric acid lends itself well to "insectproofing" at the time of construction. The dust is applied in enclosed spaces, such as attics, dropped ceilings, wall voids, voids under cabinets, and built-in appliances such as ranges and refrigerators. Half-inch (12-mm) holes can then be drilled into wall voids and subcabinet voids to facilitate the application of the dust. The treated areas are not only eliminated as harborage and breeding places for cockroaches but, with the relatively nonrepellent boric acid dust, they act as traps to eliminate cockroaches that may develop in untreated areas of the building. A water-type fire-extinguisher (figure 33, chapter 3) is particularly effective for this kind of dust application. Although cockroaches can still gain entry and develop in the "living space" of a building, they will never become so abundant as they would have if boric acid had not been applied in enclosed and inaccessible areas of the building at the time of construction (Ebeling, 1969).

Toxicity of Boric Acid to Humans. Boric acid has long been used in medications, and although in recent years such use has been greatly diminished, there is much confusion as to its potential hazard when used as an insecticide. Its LD_{50} is relatively high [3,200 and 3,500 mg/kg in male rats and 4,100 mg/kg in female rats (Weir and Fisher, 1972)], but it should be borne in mind that boric acid is used undiluted or only slightly diluted as an insecticide, whereas most insecticides are used only in greatly diluted form. Nevertheless, there appears to be a considerable margin of safety in the use of boric acid. Absorption of the material in unbroken skin is negligible (Pfeiffer, 1951). A standard teaspoonful of boric acid, screened to break up lumps, weighs 2.2 grams. Sublethal doses (e.g., 3 g) are rapidly excreted in the urine (Chittenden and Gies, 1898; Wiley, 1907). When 2 normal women were given approximately 2 grams of boric acid during the third, fourth, and fifth days of a 28-d. metabolism experiment, 93 to 94% of the boron was recovered during the first week, demonstrating its rapid absorption and metabolism (Kent and McCance, 1941). Chronic intoxication has occurred in man after daily ingestion of 4 to 5 grams of boric acid for 3 to 4 weeks and 6 to 20 grams of borax for as long as several months. On the other hand, death has occurred in adults within 46 hours after oral ingestion of as little as 7.5 grams and within 36 hours after the oral ingestion of 15 grams. Although infants have survived doses of 4.5 to 9 grams of boric acid in several hospital accidents, death to newborn babies has resulted from accidental administration of 3 to 13 grams (Pfeiffer, 1951). These figures show that death from boric acid is likely to occur only from accidental ingestion of considerable quantities. As in the case of any insecticide, as well as many common household powders and liquids, containers should be conspicuously labeled and kept out of reach of small children.

Control in Sewer Systems

In the warmer areas of the United States, cockroaches inhabit the sewer systems of many cities, sometimes in enormous numbers (Eads *et al.*, 1954; Schoof and Siverly, 1954; Roth and Willis, 1957; Wagner *et al.*, 1966; Anonymous, 1968*a;* Mackie, 1969). There is some danger of cockroaches moving from sewers to buildings, particularly if sewage backs up into manholes, forcing the insects to flee (Mackie, 1969). The species most commonly involved is the American cockroach, although the oriental cockroach has occasionally been found in large numbers. On the other hand, the oriental cockroach is commonly found in the water-meter vaults in alleys and parkways, while the American cockroach is never found in them.

Good control of the American cockroach in sewer manholes was obtained by blowing Dri-one®, a silica aerogel–pyrethrin formulation, into the manholes at the rate of 0.3 lb (135 g) per manhole. The dust was applied with gasoline-powered centrifugal blower (figure 154) of 500 to 1,500 cfm (14 to 42 cu m per minute) capacity. Even with this small quantity, dust was seen rising from plumbing vents protruding from the roofs of houses on both sides of the street and from manholes up to a block away, indicating good penetration of all sewer-system laterals (Wagner *et al.,* 1966). The same insecticide and procedure were used to control cockroaches in Yakima, Washington (Anonymous, 1969*a*). As in the treatments noted by Wagner, no live ccokroaches were found immediately after treatment, but 2 miles away the sewage-disposal system was flooded with the insects that had been washed from the treated sewer manholes and laterals. Periodic inspections showed little reinfestation over a 2-year period, but additional infested areas found during routine sewer-maintenance work were treated in the same manner, and this method continues to be used. Other cities in the Pacific Northwest have adopted the same treatment.

Fig. 154. Truck-mounted, gasoline-powered centrifugal blower for applying insecticide dust into the sewer system.

In Phoenix, Arizona, good results were obtained when carbaryl (Sevin®) insecticide dust was blown into manholes, using a portable crop duster that gave the dust a 36,000-volt electrical charge as it left the blower spout (Anonymous, 1968*a*).

Application of a 5% chlordane emulsion in sewer-system manholes for American cockroaches and in meter boxes for oriental cockroaches was also found to result in good control in San Diego, California. Even the spraying of only the upper 3 ft (1 m) of the manhole was successful in controlling American cockroaches in one experiment. The control campaign involved the treatment of 14,500 manholes at a cost of 22 cents each and 150,000 water-meter vaults at 4.5 cents each, including the cost of materials, labor, and equipment (Mackie, 1969).

FLIES

Flies belong to the large and important order Diptera. Stone *et al.* (1965) listed 1,971 genera and 16,130 species of flies occurring in America north of Mexico. Among the flies are some of man's greatest tormentors and some of the most important species from the public health standpoint.

All winged flies have only 1 pair of wings, the hind pair being replaced by a pair of clubshaped organs (halteres) used for balancing (figure 65, chapter 4). There are about 384 species of flies that are wingless or have only rudimentary wings (Cole, 1969*b*). Further information on the morphology of adult as well as immature stages of flies, and a discussion of the 3 major taxonomic groups (suborders), may be found in chapter 4. In the present chapter, only synanthropic species, also known as "domestic" or "filth" flies, will be discussed—species that either require man's environment or are greatly benefited by it. There are about 200 such species (Greenberg, 1965), so only the more important are included here. Other Diptera (mosquitoes, gnats, midges, etc.) are discussed in chapter 9.

Except for the psychodids, the flies discussed in this chapter are in the suborder Cyclorrhapha (see near the end of chapter 4), in which all species lack mandibles and most species have no maxillary blades, although the maxillary palpi are retained. Among the nonbiting species are the house fly, the blow flies, flesh flies, eye gnats, and *Drosophila* flies. The *labial labella* (figure 46, chapter 4) of these flies are large, soft, oval lobes that can be spread out to form a broad disk by which food can be collected and conveyed to the food canal of the *haustellum*. The Cyclorrhapha, despite having no specialized piercing mechanism, includes some species, the horn flies and stable flies, that can bite and pierce. The labium forms the piercing organ (figure 46, *F* and *G*), and the labella, instead of being soft, spreading lobes, are a pair of small, hard plates at the tip of the *theca*, and bear eversible teeth internally (Snodgrass, 1944).

In the United States, the house fly is still generally the most abundant species in houses, especially in the southern states, but other flies that resemble it, such as the little house fly (*Fannia canicularis*), can also be troublesome, mostly as "doorway pests." The face fly (*Musca autumnalis*) and the cluster fly (*Pollenia rudis*) are often more abundant than the house fly in houses in some areas, especially the midwestern states (Osmun and Butts, 1966). Face flies are usually house pests only in the overwintering form. Resistance of house flies to various organochlorine and even to some organophosphorus compounds should result in their retaining considerable importance as household pests. One or two treatments per year are not likely to result in control, as they did in former years.

The Influence of Modern Livestock Production Methods on Fly Production and Fly Control

According to Anderson (1966*a*), the "filth fly" problem is continuously increasing in magnitude, because:

(1) There is an increased number of poultry and other livestock; (2) the economics of shipping and marketing influence the location of livestock concentrations near metropolitan consumer centers; (3) man's rapidly expanding suburban populations are bringing more people into closer proximity and daily contact with previously relatively isolated rural communities; and (4) recent changes in agricultural technology economically favor large concentrations of animals.

The result is that in the fringes of large urban areas, we are developing high-density, low-area monocultures[14] of beef and milk cattle, poultry, etc. These, like monocultures of plant crops, tend to create difficult pest problems.

Before the current practice of high concentration of livestock in limited areas, the animals scattered their feces in small piles over large ranges or pastures. This created a favorable ecosystem for minimal fly development. Poultry and hogs could consume various pest arthropods that developed in their feces, but this is no longer possible. In cattle feedlots, all important predators associated with pasture feces have been eliminated, along with almost all coprophagous arthropods that efficiently convert pasture feces into arthropod life (Anderson, 1966*a*). The importance of the latter factor is indicated by the estimates quoted by Anderson of the annual production of manure in the United States—nearly 2 billion tons per year by cattle alone. Sound management in manure disposal is essential for effective fly control programs. A complete utilization of manures as fertilizers would appear to be a worthy goal toward which to work, and consistent with the current desire for "recycling" and efficient utilization of our resources.

Pictorial Key for Recognition of Species

There are many species of flies that are similar to the house fly in size and general appearance. The *Pictorial Key to Common Domestic Flies in the United States* (figure 155), prepared by the Center for Disease Control of the United States Department of Public Health, is useful in helping to distinguish the adults of these flies, as well as to distinguish species in other groups that are easily separated from the species resembling house flies by conspicuous differences in size and/or color, but which should be distinguished from one another. A paper written by Dodge (1954) and his supplemental pictorial key (figure 156) offer helpful advice and additional information in the use of the CDC key. The California Mosquito Control Association published an illustrated field guide, edited by Ecke (1963), with keys to the adults and larvae of the genera of domestic flies of California. The guide is suitable for persons interested in acquiring an introductory working knowledge of the domestic flies. A 10-power hand lens provides sufficient magnification

[14] Monoculture is the cultivation of a single product over an extensive area to the exclusion of other possible uses of the land. This practice decreases the complexity of the ecosystem, and this is generally disadvantageous to the natural enemies of livestock or plant-crop pests.

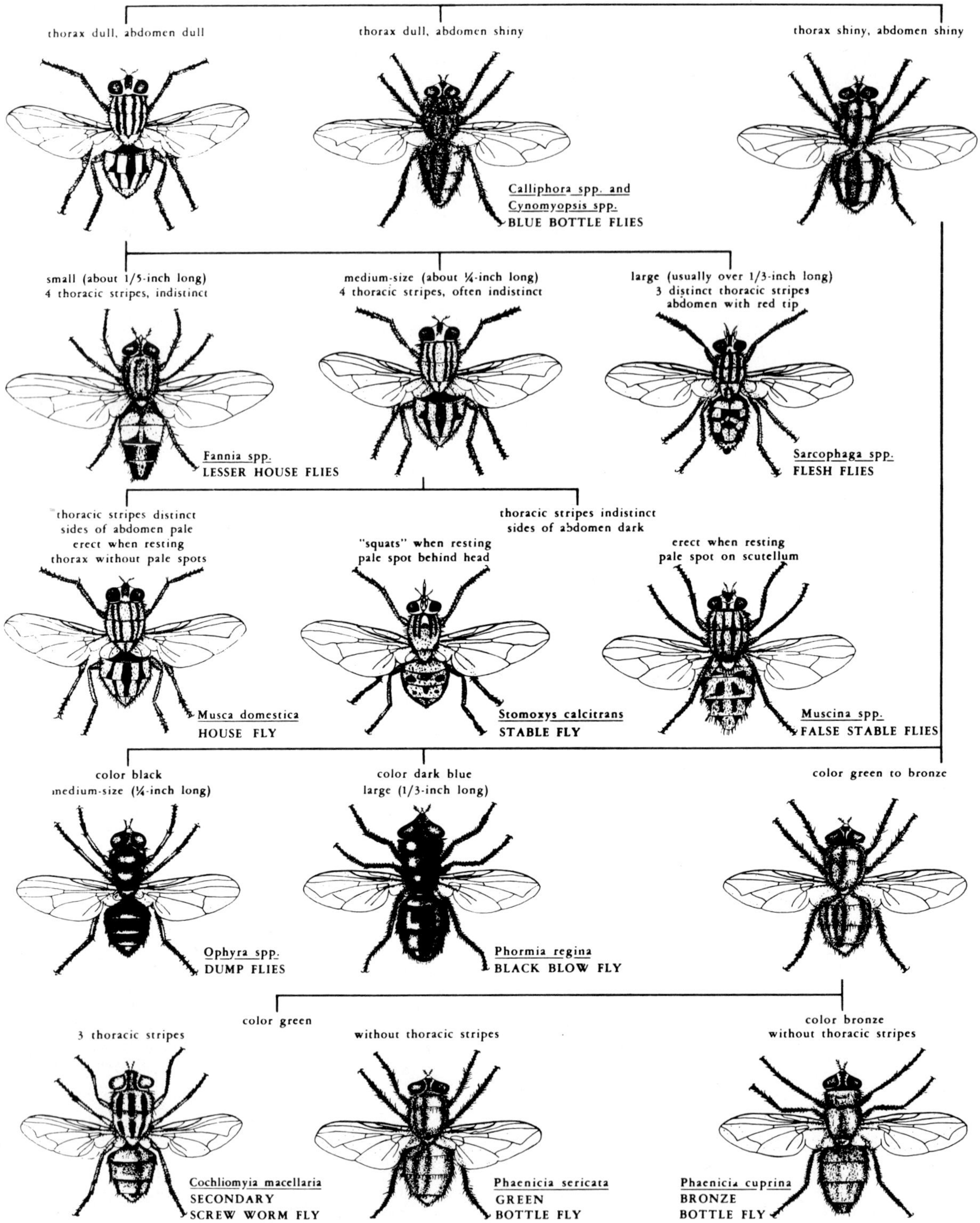

Fig. 155. Pictorial key to common domestic flies in the United States. (From Scott and Littig, 1962.)

for observation of all the morphological characteristics utilized in the keys. Information on life cycle and behavior of the most important species is also provided in this very useful publication.

Simple and practical keys such as those just mentioned contain much useful information, and are especially helpful in identifying species in the

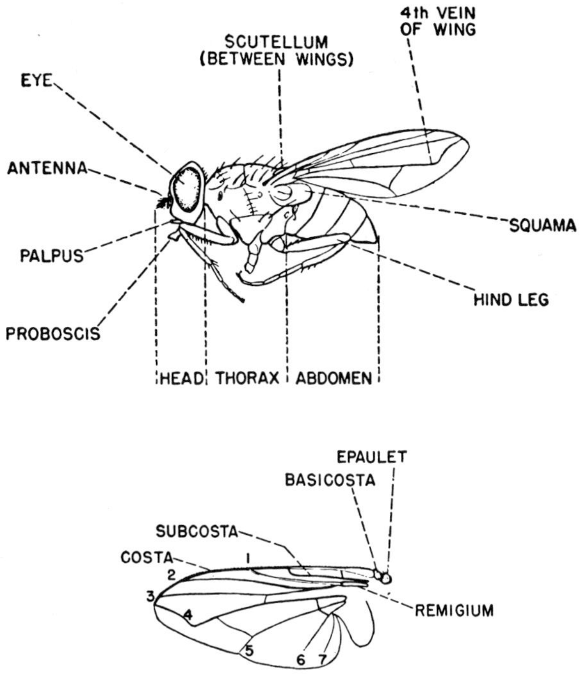

Fig. 156. Morphological characters of the generalized fly (*top*) and numbered wing veins (*bottom*) used for identification of species. (From Dodge, 1954.)

field. However, any instructor will testify that classroom experience shows that they are far from being infallible aids in keying out species.

Some flies can be recognized by superficial characters. Thus, the bluebottle flies of the genus *Calliphora* can be recognized by having a metallic blue abdomen but a dull thorax. The flies with dull-gray or brown to black bodies can be separated from those with the entire body shining metallic or black. Within these 2 major groups of flies, wing venation alone separates smaller groups, after which other structural characters are utilized for separating the species of each group.

Flies with the fourth vein angled (figure 156) are either house flies, *Musca domestica,* which have 4 black stripes on the thorax, or flesh flies, *Sarcophaga* spp., which have 3 stripes on the thorax. (The cluster fly has the fourth vein angled also, but has no stripes on the thorax.) Flies with the fourth vein gently curved include the stable fly, *Stomoxys calcitrans* (piercing mouthparts) and the false stable fly, *Muscina stabulans* (sponging mouthparts). Flies with the fourth vein straight include the little house fly, *Fannia canicularis,* and the latrine fly, *F. scalaris.* Flies with the fourth vein sharply angled include the common greenbottle flies (*Phaenicia* and *Lucilia*) as well as the black blow fly, *Phormia regina.* The former have blackish anterior spiracles and relatively large dorsal thoracic bristles, while the latter has reddish anterior spiracles and smaller dorsal thoracic bristles. Within the genera *Phaenicia* and *Lucilia,* further separations can be made, and the CDC pictorial key includes the structural characters required (see *Phaenicia,* at bottom).

House Fly, *Musca domestica* L. (Muscidae)

The house fly (plate IV, 1) is cosmopolitan within certain latitudes, and is important not only because it annoys man and domestic animals, but also because its habit of crawling on garbage and excrement, as well as on humans and their food, makes it a potential carrier of various disease-provoking pathogens. Pathogens isolated from the house fly include those capable of causing diarrhea, dysentery, typhoid, cholera, poliomyelitis, intestinal worms, eye infections, yaws, anthrax, and tularemia. As with cockroaches, flies have been implicated as vectors of *Salmonella* bacteria, responsible for food poisoning and gastric infections. The house fly appears to be even more important as a potential vector of *Shigella* than of *Salmonella* (James and Harwood, 1969; Greenberg, 1971).

It is not possible to incriminate the house fly as a transmitter of disease organisms with the same certainty with which, for example, *Anopheles* mosquitoes are identified beyond doubt in the transmission of malaria, or lice in the transmission of typhus. Flies are accused of spreading infections that can also be spread with the fingers or in contaminated food or water. They have been found to harbor over 100 different species of pathogenic organisms, and are charged with transmitting over 65 human and animal diseases, but only on circumstantial evidence. For example, outbreaks of diarrheal diseases are closely correlated with seasonal increases in fly abundance. Also, fly control has been closely correlated with a decline in the incidence of disease. Greenberg (1965) reported, in regard to an experiment made by the United States Public Health Service in

Texas, that in towns sprayed with DDT, the rate of dysentery caused by *Shigella* bacteria declined, whereas in towns left unsprayed as controls, the disease rate remained unchanged. After 18 months, the sprayed towns were left unsprayed and the former control towns were sprayed. The relative incidence of dysentery in the 2 groups of towns was then reversed.

The scarcity of soundly documented cases of field transmission of disease by flies results in inadequate knowledge of how important they are in the natural transmission of the many pathogens commonly associated with them. Therefore, it is still principally the fact that they are annoying to have around domestic premises that motivates us to control them (Anderson, 1966*a*).

Description. The general size and appearance of the house fly are well known. It is 4 to 7.5 mm long, and may be distinguished from other house-frequenting flies by wing venation, the 4 dark stripes dorsally and lengthwise on the thorax, and the 2 velvety stripes, silver above and gold below, on its face. The eyes of the female are more widely separated than those of the male (figure 157). With its sponging–sucking mouthparts (figure 46, *I*, in chapter 4), the house fly can ingest only liquid material, but some solid foods can be liquefied by means of regurgitated saliva, and thus ingested. All the aspects of house fly morphology, as well as biology and bionomics, were discussed in great detail by West (1951).

Life Cycle. House flies lay their eggs in warm, moist materials that are suitable for larval growth, such as animal excrement, garbage, piles of lawn clippings, decaying vegetables and fruits, or soil

Fig. 157. Head and thorax of the male (*left*) and female house fly, showing difference in width of space between eyes. (Photo by Stennett S. Heaton, courtesy Shell Chemical Company.)

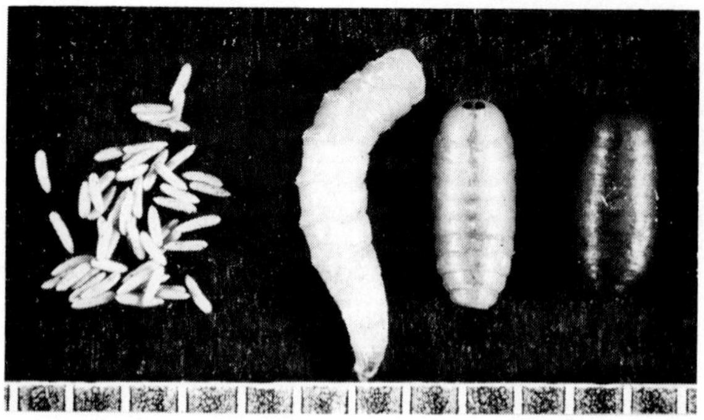

Fig. 158. House fly, *Musca domestica*. Eggs, larva, newly developed pupa, and older pupa. (Each division of the scale below represents one-sixteenth in., or about 1.5 mm.)

contaminated with such material. Immature stages are shown in plate IV, 1, and figure 158. During warm weather, the eggs, which are laid singly but piled together in small masses, can hatch within a day. The cream-colored larvae (maggots), distinguished by their being narrow at the head end, burrow into the food material in which they hatched, and may complete their 3 instars and pupate in a week or less. As the pupation period approaches, the larvae seek the drier outer regions of their manure pile or other breeding medium, or may leave it to pupate a considerable distance away in the ground, in loose material, or under boards or stones. Under warm conditions, the pupal stage may last 4 to 6 days. Under the warmest natural conditions, a house fly may complete its life cycle in about 8 days, but this period is considerably longer at lower temperatures and when the larvae have had a less than optimum food supply. There may be as many as 10 or 12 generations in one summer. The adult flies generally live from 15 to 25 days, depending on temperature (Roy and Brown, 1954). In colder areas, the adults may overwinter. The potential increase in fly population in a single season is obviously enormous.

Sources of Infestation. In rural areas, piles of manure may be the principal sources of house flies, and in urban areas, piles of fermenting lawn clippings may be important sources of infestation. In garbage cans, blow flies generally breed more abundantly than house flies. The proximity of beef and dairy feedlots and poultry houses to residential areas has become an increasingly frequent cause of complaint and litigation.

People bothered with house flies are interested in knowing whether they live within the flight range of these insects from known sources of infestation, such as those just mentioned. An investigation was made of the flight range of flies that had been fed radioactive materials. Although some house flies were recovered as far as 20 miles from the release site, the bulk of the fly population did not move more than 1 or 2 miles (32, 1.6, and 3.2 km) (Schoof, 1959).

Face Fly, *Musca autumnalis* De Geer (Muscidae)

The face fly is native to Europe and Asia and, since the early 1950's, it has spread to the eastern, midwestern, and southern United States and to parts of Canada. It finally moved westward, until now it occurs throughout most of the country, including the Pacific Coast states, having been found in California near the Oregon border in May, 1968 (Poorbaugh and Smith, 1968). Since then, it has moved steadily southward in California. The face fly received its common name from the habit of the adult of feeding on mucus and watery secretions around the eyes, nostrils, and mouths of cattle, causing great annoyance, reduced production of milk, and additional time necessary to finish beef animals for slaughter.

Description. The face fly (figure 159) is only slightly larger than the house fly, and it is difficult to tell them apart. The males are most easily

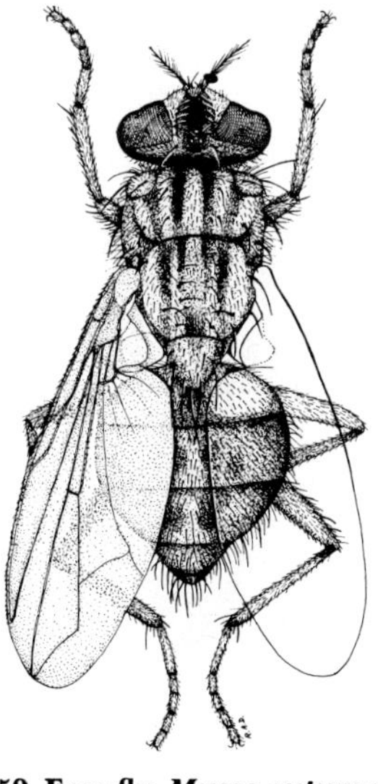

Fig. 159. Face fly, *Musca autumnalis*. (Drawing by Ruth DeNicola.)

distinguished, for the eyes of the male face fly almost touch at the top of the head, while those of the male house fly do not. The males are commonly found on certain flowers.

Biology. The face fly breeds only in undisturbed cow droppings, and does not breed in cow manure that has been heaped into piles, mixed with hay, straw, or urine, or that has been tramped on by cattle in crowded feedlots. The opposite is true of the house fly and stable fly—their larvae feed in such disturbed manure but never develop in undisturbed cow droppings in pastures (Anderson, 1966*b;* Anderson and Poorbaugh, 1968). Mature face fly larvae leave the manure and crawl as far as 30 ft (9 m) before pupating in the soil (Jones, 1969). They are sometimes parasitized by nematodes (Nickle, 1967).

As household pests, face flies resemble the cluster fly (described later in this chapter), for they seek such places as attics, wall voids, or basements as winter harborages, sometimes in enormous numbers. On warm days, many face flies become active and move to the living space of the house.

Stable Fly, *Stomoxys calcitrans* (L.) (Muscidae)

Another fly that superficially resembles the house fly but rarely occurs in houses is the stable fly. Unlike the house fly, it has a stiff proboscis (figure 46, *G,* chapter 4) projecting forward from the lower part of the head, enabling both males and females to inflict painful bites and suck out blood. Stable flies normally are serious pests of domestic animals, but on occasion they also attack humans. They are most likely to bite in the region of the ankle. Socks and stockings offer no protection.

Description. The stable fly (figure 160, *B*) is more robust than the house fly and has a broader abdomen. It can also be differentiated by the difference in wing venation (*A* and *B* of the figure), by a prominent, light-colored area between the longitudinal bands in the middle of the thorax, and by the presence of a number of nearly round, dark spots on the upper surface of the abdomen (figure 160, *B*). When the stable fly is at rest, its wings are held widely apart instead of projecting backward on top of the abdomen, as with the house fly. The stable fly rests with the head cocked upward and abdomen held against the resting surface, whereas the house fly often rests in a crouching position, with head down and body parallel to the surface.

Life Cycle. Stable flies breed in decaying hay, straw, or grain; piles of lawn clippings or rotting fruit; and various manures mixed with grain or bedding. They also breed in the manure of dairy cattle and in poultry droppings when they are "stacked." They are often found in large numbers on beaches, where they lay their eggs in piles of seaweed, and are sometimes called "beach flies." Unlike *Fannia* species, the immature stages of the stable fly somewhat resemble those of the house fly. In warm weather, adults may emerge in 20 to 25 days after the eggs are laid, but this period can be extended to several months at lower temperatures. (See chapter 9 for a related account.)

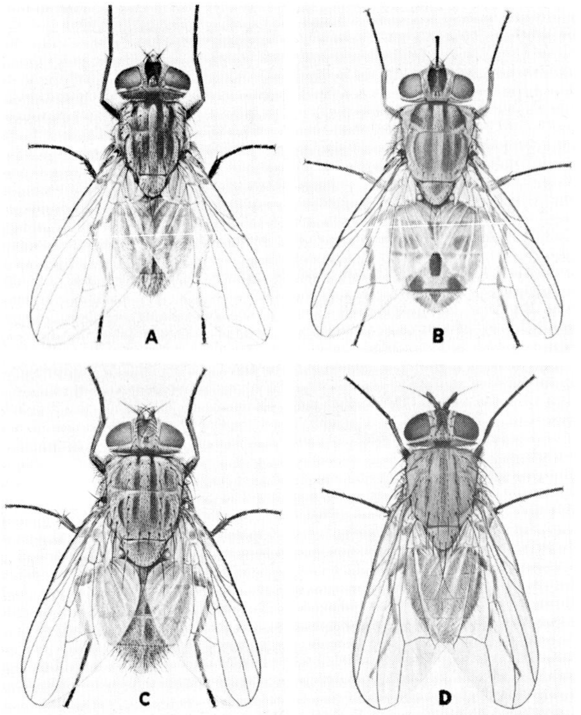

Fig. 160. Domestic flies (females). *A,* house fly, *Musca domestica; B,* stable fly, *Stomoxys calcitrans; C.* false stable fly, *Muscina stabulans; D,* little house fly, *Fannia canicularis.* (From B. Greenberg [1971], *Flies and Disease,* courtesy Princeton University Press.)

False Stable Fly, *Muscina stabulans* (Fallén) (Muscidae)

The false stable fly (figure 160, *C*) superficially resembles the house fly, but it is larger. Note also the difference in wing venation. It is closely related to the house fly, and has similar sponging mouthparts. The female lays her eggs in animal manure, human excreta, and decomposed vegetable matter, including lawn clippings. Twenty-five adult false stable flies were reared from the caked grass taken from beneath the blade guard of a rotary power mower (Ware, 1966). Several generations are produced during the summer months, and hibernation normally takes place in the pupal stage. The females often enter houses to oviposit, scattering from 140 to 200 eggs over various foods, particularly those that are slightly spoiled. Probably it is in this way that man occasionally consumes them and suffers intestinal myiasis (James, 1947).

Muscina assimilis (Fallén) is darker than *M. stabulans,* and its legs are entirely black, while those of the latter species are in part reddish brown. In California, *M. assimilis* is a common species. Adults have been reared from partly decomposed watermelon and other vegetable matter. It is less likely to be found in homes than the closely related *Muscina stabulans.*

Black Garbage Fly, *Ophyra leucostoma* (Wiedemann) (Muscidae)

The genus *Ophyra* contains species that may be found near refuse dumps and poultry ranches and occasionally around cow and hog manures. The adults are not found in or around houses and other buildings, but congregate in trees and shrubs. The larvae are scavengers in decaying vegetation, animal matter, and excrement, but are also known to feed on the maggots of other flies. The black garbage fly (figure 161) is a small, black fly that originally invaded this country from Europe and Asia. One of its larvae may kill from 2 to 20 house fly larvae per day. Fortunately, the *Ophyra* adults rest at different sites from those frequented by adult house flies. This should make it possible to apply insecticides to sites upon which house flies and other nuisance flies rest, and to avoid those upon which *Ophyra* adults rest (Anderson and Poorbaugh, 1964).

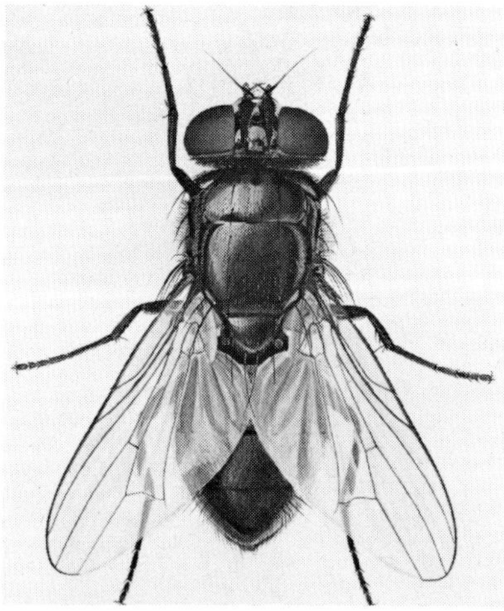

Fig. 161. Black garbage fly, *Ophyra leucostoma* (female). (From B. Greenberg [1971], *Flies and Disease,* courtesy Princeton University Press.)

Little House Fly, *Fannia canicularis* (L.) (Anthomyidae)

Males of this cosmopolitan fly are often found in houses; the females are rarely found indoors. They are more uniformly distributed than the house flies throughout the average home. In one investigation, 65.6% of house flies, but only 40.9% of little house flies, were found in the kitchen (Peters, 1965). Males of the little house fly tend to fly about aimlessly in the middle of the room or some shaded area outside, often hovering or making jerky, zigzag flights. The little house fly appears earlier in the spring than the house fly, and in regions of high temperature it tends to disappear during the summer. In many areas, this difference is possibly explained by the fact that the little house fly becomes active at a lower temperature (40.1° F compared with 44° for the house fly), and also suffers heat paralysis at a lower temperature (105.5° F compared with 115.7° for the house fly) (Nieschulz, 1935). (Centigrade: 4.5, 6.7, 40.6, and 46.5°.) This difference in temperature requirements may account for the fact that in subarctic regions, *Fannia canicularis* often replaces *Musca domestica,* and is, for example, the common house fly of Iceland (Chillcott, 1960). This and other species of *Fannia* are attracted to honeydew, and are often found swarming under aphis-infested vines or trees (Ecke, 1963).

Description. The little house fly (figures 160, *D,* and 162) resembles the house fly, *Musca domestica,* but is smaller (3 to 5 mm long) and more slender. The wings in repose are parallel and partly folded over each other, whereas those of the house fly are held in a diverging position. The abdomen of the female *F. canicularis* is ovoid and gray, while the male's is tapering, with pale-yellow, semitransparent patches.

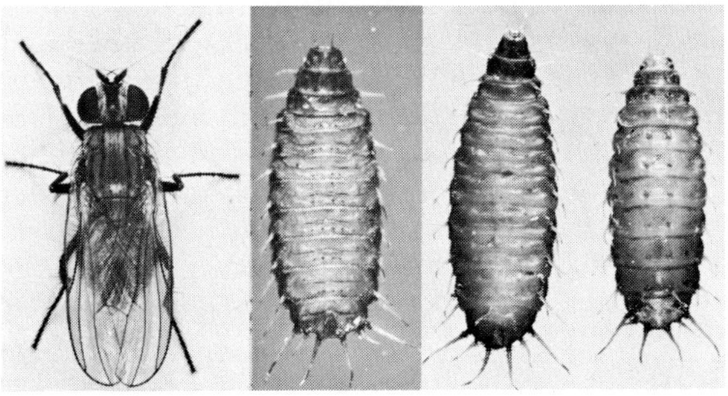

Fig. 162. Little house fly, *Fannia canicularis. Left to right:* adult female; pupa; dorsal (*left*) and ventral views of larva. (Photo by Stennett S. Heaton, courtesy Shell Chemical Company.)

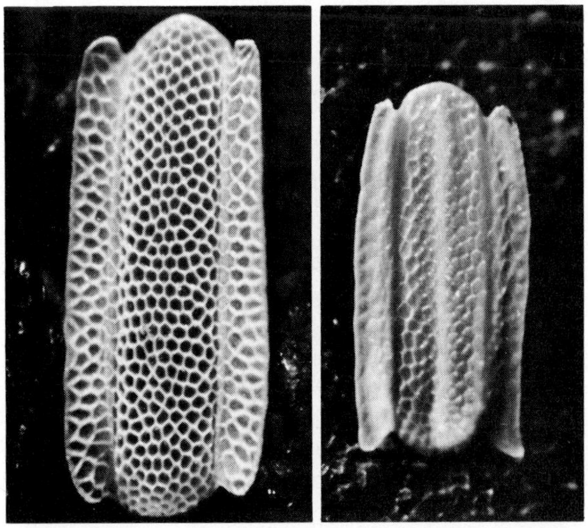

Fig. 163. Eggs of little house fly, *Fannia canicularis* (*left*) and coastal fly, *Fannia femoralis.* (Photo by Stennett S. Heaton, courtesy Shell Chemical Company.)

Life Cycle. The little house fly breeds in animal excrement or decaying vegetable material. Float-like appendages on the eggs (figure 163) enable them to float on top of semiliquid material. The larva has spiny processes along the sides and top of the body. These have fine, whiplike hairs that are said to propel it through semiliquid media (Hickin, 1964). The larvae change from translucent white to light brown as they mature. Like those of the house fly, *F. canicularis* larvae move to a drier environment near the surface of their food medium before pupating. The pupae resemble the third-instar larvae, but are darker and somewhat shorter. At 80° F (27° C) and 65% relative humidity, the approximate durations of the immature stages were recorded as follows: egg, 1.5 to 2 days; larva, 8 to 10 days; pupa, 9 to 10 days. The period from egg to adult was 18.5 to 22 days, and from egg to egg, 22 to 27 days. About 50% of the male fly population died within 14 days, and 50% of the females died within 24 days (Steve, 1960). In another experiment, the life cycle took 24 to 29 days (Lewallen, 1954).

Latrine Fly, *Fannia scalaris* (F.)

The latrine fly resembles the little house fly in appearance and life history, but its thorax is bluish black, contrasted to the thorax of *Fannia canicularis,* which is blackish gray. Also, a rounded protuberance of the median point of the middle pair of legs can be used to distinguish *F. scalaris.* It prefers human to animal excrement, and has been known to cause myiasis in man.

Coastal Fly, *Fannia femoralis* (Stein)

This species, of South American origin, is widely distributed throughout the United States. It is a little over half the size of *Fannia canicularis,* and is shiny black in color (figure 164). It sometimes occurs in large numbers, but is much less important than the little house fly as a pest.

Fannia benjamini Malloch

This species is known from various areas of the West, particularly California. The adult is 3.5 to 4.5 mm long, with a brown-pollinose, unstriped thorax and a dark abdomen without a trace of the yellow translucence found in some species, such as the male of *Fannia canicularis.* The antennae and palpi are yellow, as are also the legs, with the exception of the black tarsi. The female is much lighter-colored than the male.

Fig. 164. Coastal fly, *Fannia femoralis. Top left,* adult female; *top right,* pupa; *bottom,* dorsal, lateral, and ventral views of larva. (Photo by Stennett S. Heaton, courtesy Shell Chemical Company.)

The expansion of city suburbs into the foothills, as well as the increasing use of such areas for recreational purposes, is bringing *F. benjamini* into sharper focus as a pest. It is common in California along streams, particularly in groves of oak trees. It is a very annoying pest, and persists in its attempts to enter the ears, nostrils, and eyes, apparently attracted by perspiration and mucus. This species and *F. canicularis* have been implicated as vectors of a nematode eye worm, *Thelazia californiensis* Price. The known hosts of this nematode, besides man, are dogs, cats, horses, sheep, deer, coyotes, and bears (Burnett *et al.,* 1957).

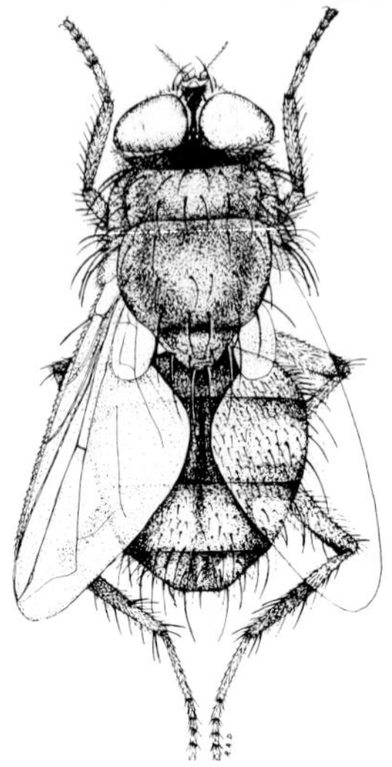

Fig. 165. Cluster fly, *Pollenia rudis.* (Drawing by Ruth DeNicola.)

Cluster Fly, *Pollenia rudis* (F.) (Calliphoridae)

Cluster flies derive their common name from their habit of entering a house in the fall and gathering together in a cluster that may emit a sickly sweetish odor. They tend to gather and hibernate in secluded areas, often in wall voids, and hibernate in them. Cluster flies and face flies (*Musca autumnalis*) often become trapped in large numbers in wall voids, attracting larder beetles (*Dermestes lardarius*). The beetles feed on dead flies and then migrate to other parts of the house, causing the occupants to believe they have an infestation of carpet beetles (W. L. Gojmerac, correspondence).

In the spring, cluster flies again become pests as they migrate from wall voids to the living space, via window pulleys, around baseboards, and through other small openings, and then crawl sluggishly over the walls. They are usually found in windows and other out-of-the-way places, and do not fly about noisily like other house-entering calliphorids. In this respect, they are not such nuisances.

Description. Cluster flies (figure 165) closely resemble house flies, but are larger, more robust, and slower in their movements. They are about 8 mm long, dark gray, with checkered black and silvery abdomens, and with many golden hairs on the thoraxes of newly emerged specimens which may be lost in older ones. The stripes on the thorax are not so prominent as those of the house fly. The wings overlap when at rest. *Pollenia rudis* can be distinguished from all other synanthropic North American calliphorids by its abdomen being nonmetallic in appearance and being covered by a pollenlike dust. A striking characteristic of the biology of cluster flies is that the larvae are parasitic on earthworms. Eggs are deposited in cracks in the soil, hatch in 3 days, and the larvae enter earthworms at almost any point on the body wall. The total developmental period of the fly varies from 27 to 39 days (Webb and Hutchison, 1916). Its breeding habits result in the cluster fly being less of a hazard than other calliphorids as a mechanical carrier of disease organisms (James, 1955).

Control of Cluster Flies

Lawns have been treated with 2.5% heptachlor granules at 3 lb per 1,000 sq ft (1.36 kg per 93 sq m), or with a spray of heptachlor 25% emulsifiable concentrate at 0.25 pt in 3 gal (0.12 l in 11 l) of water per 1,000 sq ft to kill earthworms, and thus prevent the development of cluster flies. The lawn should then be sprinkled to force the toxicant into the soil.

Blow Flies (Calliphoridae)

The blow flies (greenbottle and bluebottle flies) (figure 166) are large scavenger flies, with a metallic blue or green sheen. They may become abundant in and around homes, particularly if they are located near such places as slaughterhouses, meat-processing plants, and garbage dumps. However, it should be borne in mind that some of these insects are stronger fliers than the house fly, and at least a small percentage may travel great distances from the point of dispersal. For example, in one experiment, 13% of the black blow flies, *Phormia regina,* were trapped between 4 and 28 miles (6.4 and 46 km) from the point of origin, and more than 8% at or beyond 11 miles (18 km) compared with only 3% of house flies trapped at or beyond 4 miles (Schoof, 1959).

Blow flies are most active during warm, sunny weather. When it is cool and cloudy, they tend to retreat to resting sites, usually on vegetation adjacent to areas where they oviposit. *Phormia regina* tends to disperse more widely than the other species, and in warm weather it often rests on sunlit walls (Busvine, 1966).

Fig. 166. A blow fly (Calliphoridae).

Some blow flies, such as *Phaenicia sericata* and *Eucalliphora lilaea* (Walker), freely enter houses, and because of their breeding habits and the fact that they feed and oviposit on food, mechanical transmission of disease becomes a distinct sanitary consideration. Meat or fish must be protected from oviposition by blow flies, not only for aesthetic and sanitary reasons, but also because the eggs and larvae of blow flies and some other species, when swallowed, may remain alive for some time, causing acute gastric or enteric disorders (James, 1955).

Blow flies have been known to breed in the carcasses of dead birds or rodents in attics or wall voids of houses. These may be sources, not only of adult flies, but even of the larvae, which are sometimes found crawling about in the living space of the house. Because of their large size, buzzing sound, and tireless flying about throughout the house, blow flies are generally more annoying than house flies. They tend to fly to windows, for like all flies, except perhaps the males of *Fannia,* they are attracted to bright sunlight.

Although blow flies usually lay their eggs on meat or dead animals, when meat is not present they may oviposit on animal excrement or decaying vegetable matter. One female may lay as many as 600 eggs. Larvae first feed on the surface and then burrow down to less-decayed material. The full-grown larvae burrow into the ground to

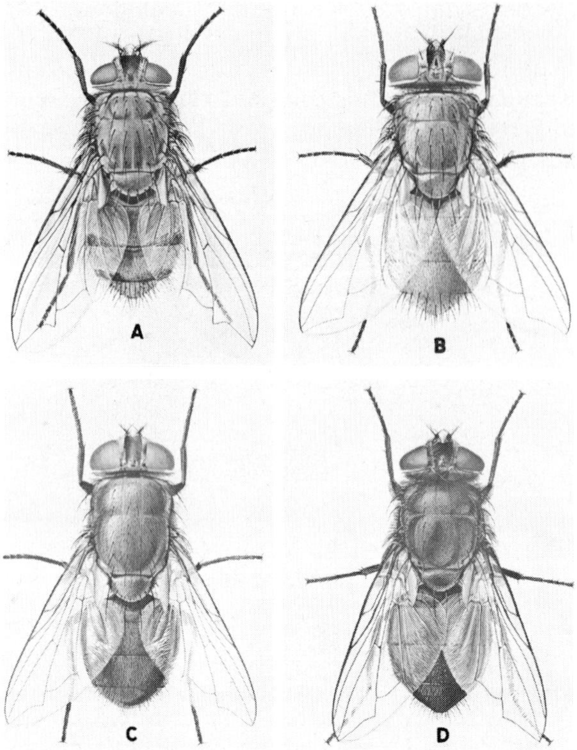

Fig. 167. Blow flies (females). *A, Calliphora vicina; B, Phaenicia sericata; C, Phormia regina; D, Protophormia terraenovae.* (From B. Greenberg [1971], *Flies and Disease*, courtesy Princeton University Press.)

pupate, usually in the top 5 cm of soil. There are several generations a year. The winter is normally spent in soil by full-grown hibernating larvae.

Some Common Species

James (1955) prepared keys to California species of blow flies, and gave geographical distributions. Some common examples of bluebottle flies are *Calliphora vicina* (Robineau-Desvoidy) [= *erythrocephala* (Meigen)] (figure 167, *A*), *C. terraenovae* Macquart (plate IV, 2), and *C. vomitoria* (L.). They are much larger and more robust than the house fly, about 11 mm long, rather bristly, and have metallic-blue abdomens. *Eucalliphora lilaea* not only enters houses, but sometimes breeds there in the bodies of poisoned or trapped mice, and then occurs in large numbers. A striking variability in the size of this species results sometimes from larval overcrowding (James, 1955). Some bluebottle flies (e.g., *C. vicina* and *C. vomitoria*) have been observed to deposit active, newly hatched larvae instead of eggs. Two of the common greenbottle flies, *Lucilia illustris* (Meigen) [= *caesar* (L.)] and *Phaenicia sericata* Meigen (= *Lucilia*) (fig. 167, *B*), are about 10 mm long, less bristly than *Calliphora*, and metallic blue-green. These and many other species occur throughout the United States. Large numbers of maggots of *P. sericata* dropped onto the bed of a resident of North Hollywood, California, falling from a light fixture in the ceiling over a period of several days. These had bred in the carcass of a dead rat.

The black blow fly, *Phormia regina* (Meigen) (figure 167, *C*), is slightly more robust than the greenbottle flies and darker in color. The thorax is almost black, with a metallic bluish-green luster, and the abdomen is darker, being metallic green. *Protophormia terraenovae* (Robineau-Desvoidy) (figure 167, *D*) is even darker, and has a purplish cast. The colors of some common blow or bot flies, as well as certain other synanthropic flies, are well displayed in 15 color plates in *Flies and Disease*, Vol. I (Greenberg, 1971).

Flesh Flies (Sarcophagidae)

The flesh flies are not so likely to enter houses as the related blow flies, but many species, including some carrion and excrement feeders, are considered to be synanthropes (domestic flies). The sarcophagids are medium-sized to relatively large flies. They generally range from about 10 to 14 mm in length, and are robust and dull grayish black, with 3 dark stripes on the thorax. The more familiar species are characterized by a checkerboard pattern of spots that change from black to gray and back with changes in the incidence of light. Sarcophagids are larviparous, i.e., the eggs hatch within the female's body and she deposits the larvae on the food they will be eating. Both the common and family names are derived

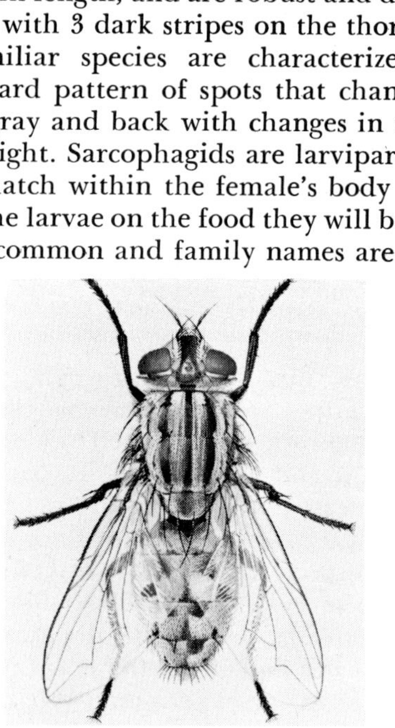

Fig. 168. A flesh fly, *Sarcophaga haemorrhoidalis* (female). (From B. Greenberg [1971], *Flies and Disease*, courtesy Princeton University Press.)

from the fact that the larvae of most species develop in decayed flesh and spoiling meat, although some kinds can also develop in manure or other animal droppings, and in food residue at the bottoms of garbage cans. The life cycle ranges from 8 to 21 days, and adults may be present from spring through late fall (Loomis *et al.*, 1967).

A representative flesh fly is *Sarcophaga haemorrhoidalis* (Fallén) (figure 168). It is nearly cosmopolitan, and in North America it occurs from Oregon to Quebec and south to California and Florida. The adults are 10 to 14 mm long, gray, with the terminalia of the male red. The eyes are reddish brown. The larvae feed on dead insects, carrion, and excrement. Another common and widely distributed flesh fly is *Sarcophaga bullata* Parker (plate IV, 2). It ranges from British Columbia to Quebec and south to California and Florida (Stone *et al.*, 1965).

Drone Flies (Syrphidae)

The drone flies (*Eristalis* spp.) are robust, yellowish flies that mimic bees in appearance. They do not cause serious annoyance to man or animals, and rarely enter homes or alight on food. They are often seen around flower beds, the males hovering in the air, almost motionless, but darting swiftly aside when disturbed. Most syrphids are important predators of aphids, but there are also saprophagous genera that are scavengers in semiliquid, rotting vegetable and animal material or, in the case of *Eristalis,* in putrid or stagnant water or in moist excrement. The larvae are called "rat-tailed maggots" because of the elongate breathing tube that allows them to obtain air even when the remainder of the body is buried in liquid material.

Control of House Flies and Similar Species

The major contribution to fly control is garbage pickup and disposal. In Los Angeles, for example, a city ordinance requires that garbage be wrapped or placed in bags and placed in covered containers to await pickup by the City Sanitation Division once a week, along with prunings, litter, lawn clippings, or other organic materials that are potential sources of fly infestation. The trash and garbage are dumped in areas which are continually being covered with earth by means of giant bulldozers. This is a method of trash and garbage disposal that may be particularly suitable when an earth fill is desired by the community in a particular location, as for filling a ravine or canyon. Incineration is of course a suitable alternative, except where it may contribute substantially to air pollution. City sanitation divisions are relieved of much of the burden of garbage removal by the wide use of automatic garbage-disposal units in the home, for much of the organic waste material thus becomes part of the city sewage.

Not all city ordinances are conducive to maximum fly control. Ordinances pertaining to manure management and disposal or to pest control are helpful only if they are based on sound principles of integrated control, as revealed by the research of qualified entomologists.

Supplementing the community garbage-removal service, decaying organic material and animal excrement on residential properties should be deeply buried or otherwise disposed of by the homeowner, for house flies, as well as stable flies, flesh flies, bottle flies, blow flies, little house flies, and false stable flies, all breed in this type of material.

When urban areas are surrounded by areas of high concentration of poultry and livestock, freedom from flies in the city may depend heavily on manure management, water systems,[15] general farm sanitation, and the wise use of insecticides in an "integrated control" program in such areas (Anderson, 1965; Burton *et al.*, 1965; Smith, 1969; UCAE, 1971-72).

Exclusion at Doorways

The usual type of window or fly screen (18-mesh) on windows, screen doors, and other openings to buildings is a great help in keeping ingress of flies to a minimum. However, screen doors that must be opened many times during the day cannot be depended upon to completely exclude insects, and may be unsatisfactory where human traffic is heavy or where little, if any, insect entry can be tolerated. In such situations, it is possible to exclude flying insects by means of a downward air current at the doorway. An air speed of 1,500 ft (450 m) per minute, approximately 17 mi (27 km) per hour, when the ratio of the distance through the doorway to its height was 0.37, was found to offer a satisfactory combination of efficiency with economy and acceptability. The installation in the doorway of a removable floor grid or grating that can be readily treated with a long-term residual insecticide or a low-friction finish has been suggested for the exclusion of crawling insects (Hocking, 1960).

[15] Water dripping from faulty watering systems results in wetter manure and increased fly problems.

Sprays

Two types of spray are generally used for fly control within buildings: space sprays and residual sprays.

Space Sprays. These are applied in order to produce a fine mist that permeates all the air space within a building, so that the insecticide contacts the insects as they are flying about or at rest. The insecticide may or may not break down rapidly; the objective is to contact all the insects in a room and kill them rapidly. A mist can be produced by applying the insecticide in a petroleum base, such as deodorized kerosene (base oil), with an ordinary hand atomizer (figure 9, chapter 3) or any other kind of atomizer. The same objective is accomplished by using an aerosol canister or fogging machine, which generates even smaller droplets, requiring longer for the droplets to settle out of the air. It is desirable to remove pets and ornamental plants and to keep doors and windows closed after treatment for as long as it is convenient. Cover foods and cooking utensils.

Dichlorvos (DDVP) is commonly used for fogging such buildings as food-processing plants, industrial plants, warehouses, and theaters, using 1 gal of a solution containing 1% of toxicant to 64,000 cu ft of space (about 4 l to 1,813 cu m). Manufacturing buildings should not be in operation, and should be vacated and not reoccupied until they have been well ventilated. Food should be removed from the buildings, and food-handling equipment should be covered with draping paper before fogging—or it should be washed *after* fogging. A suitable mask or respirator should be used by persons applying the fog. Dichlorvos is also applied as a 0.5% spray with a low-pressure sprayer to localized areas that may be infested by flies and other insects that are attracted to food, such as cockroaches and ants. In another procedure, it may be added to a longer-lasting residual spray to obtain rapid knockdown and fuming action in cracks, crevices, and other confined areas, combined with prolonged residual efficacy.

Synergized pyrethrins or allethrin at 0.1% concentration have been used effectively as space sprays, but the newly developed synthetic pyrethroid, resmethrin (SBP-1382®), is as much as 20 times more toxic when used at the same concentration, and no synergist is required. The knockdown rate is slow, but whereas with pyrethrins the flies that are knocked down may recover, with resmethrin they do not. Resmethrin also differs from pyrethrins in that its residues may be effective for several weeks on some types of substrate.

Residual Sprays. These sprays are applied to surfaces upon which flies generally rest. The insecticide is expected to kill flies crawling about on the deposit long after the spray is applied. Great impetus was given to this method of fly control by the original spectacular success of DDT. This insecticide, which came into general use after World War II, was not only highly toxic to house flies and most other household pests, but also had a long residual life. Against house flies, the effectiveness of a spray residue ranged from 8 months on unsized cement surfaces to 18 months on smooth surfaces, such as glass, wallpaper, or finished wood. Within the home, fly control could be obtained for long periods by spraying surfaces on which flies commonly rested, such as window screens. Residual wall and ceiling sprays in barns resulted in about 95% reduction in the fly population. Flies entering during the day were killed overnight.

Within a few years, house flies developed resistance to DDT and other persistent organochlorine insecticides. Later, the little house fly (*Fannia canicularis*) was shown to be resistant also. Both species, taken from 3 poultry ranches in southern California, were also shown to be resistant to a number of organophosphorus compounds, but to a smaller degree than to organochlorine compounds. Still, *Musca domestica* was found to be over 909 times resistant to coumaphos, 182 times to Ciodrin®, 102 times to malathion, and 66.5 times to diazinon. *Fannia canicularis* was found to be over 174 times resistant to coumaphos, 81.9 times to Ciodrin, 59.1 times to malathion, and 6.8 times to diazinon (Georghiou, 1967). The organophosphorus compounds and carbamates do not have the long residual life inherent in most organochlorine compounds, and resistance is compounding the problem. Diazinon, fenthion, dichlorvos, dimethoate, malathion, naled, and resmethrin are recommended as outdoor space sprays (CDC, 1973). Instructions on labels should be carefully followed when using any of these insecticides.

In California, resistance has not yet developed with Gardona, except on a low level in one area. In Denmark, however, where its use has been more extensive, high resistance is reported. The period required for resistance to appear against a new insecticide is in proportion to the frequency of its use. This can be delayed by using methods of suppressing the fly population that do not require insecticides, thus reducing the frequency of their application (Georghiou *et al.*, 1972*a*).

For the control of cluster flies, spraying the outside of a house in the fall with 2% malathion or 1% naled or ronnel has been recommended. Particular attention must be paid to the upper third of the wall, the overhang, and the eaves, especially on the south and west sides (Rachesky, 1969). Residual sprays, such as 2% chlordane, have been applied to the surfaces of attics, basements, closets, storerooms, and other areas where cluster flies and face flies are seen, usually with a 2- or 3-gal (7.5- or 11-l) compressed-air sprayer. To avoid staining wallpaper and furnishings, such sprays should not be applied in bedrooms or in papered or well-furnished rooms. Poisonous sprays should also be avoided in food-preparation areas.

A 10% chlordane dust is highly recommended for treatment of places that are difficult to reach with sprays. A 0.5% solution of dichlorvos may be sprayed into window-pulley openings or wall voids (Anonymous, 1970*e*).

Community Spraying. A community-wide effort in fly control can include spraying. The little house fly, *Fannia canicularis,* was effectively controlled in one town near a large poultry farm by blowing a mist consisting of 8 gal of DDT 25E and 0.5 gal of malathion 57E per 100 gal of water from roads and driveways over lawns, shrubbery, and onto buildings, at the rate of 1 to 10 gal per acre, depending on vegetative cover (30 l; 2 l; 378 l; 3.8 to 38 l). One application was made in the morning on July 28 and another on September 15 (Hansens, 1963). Although such results can also be obtained with currently registered insecticides if the target species is not resistant, it must be borne in mind that beneficial insect species are also eliminated. In the foregoing example, only a nonvector species was involved, for which most complaints resulted from the activities of the hovering males. In the current climate of ecological awareness, a similar control program would probably not be undertaken. Such community-wide spraying programs should be considered only under emergency disaster conditions. Spraying should not be considered as a substitute for a basic sanitary solution at the sources of infestation.

Insecticide recommendations change as flies become resistant to insecticides. Generally, pyrethrum, with a suitable synergist such as piperonyl butoxide to increase its potency, is added to the currently recommended insecticides, and the combination provides good fly knockdown and kill. With the continually changing statuses of pesticides, an appropriate state official should be consulted for advice as to which compounds are currently registered for the projected treatment.

Other Treatments

A property owner with a fly problem may not be responsible for or have any control over the sources of infestation, but may wish to know what he can do to eliminate the flies on his own property. No great detail on methods and materials will be given here because these are continually changing. A few examples are given, mainly to indicate the types of treatment available rather than specific insecticides. Specific recommendations are revised almost yearly, and are available from university extension personnel or federal and state agricultural extension literature.

The current tendency is to resume the use of space sprays for fly control, or to use baits and "vapor strips." However, according to Rachesky (1970*b*), Cygon® (dimethoate 25% emulsifiable concentrate) gives 4- to 6-week residual protection during fly season. Treatments should begin in May or June at the rate of 2 gal per 1,000 sq ft (7.5 l per 93 sq m) or to the point of runoff on ceilings, walls, support posts, and outside around closed doors and windows. Rachesky also recommended the use of dimetilan fly bands in garages, under breezeways, in covered porches, and on loading docks. The bands should be wiped occasionally with a damp cloth to remove dust. They are effective for more than a year. No vapors are given off by dimetilan bands; the fly must contact the band.

Dry baits of dichlorvos or dipterex are often very effective in substantially decreasing the fly population, particularly when scattered around garbage cans and outside of doorways. Inside the home, they may be scattered on floors and near windows, according to directions on the labels. The baits should be widely dispersed, to increase the probability that flies will find them and to decrease the hazard to children and pets. Stable flies and most blow flies are not controlled by these baits, but cockroaches, snails, and slugs are susceptible if they happen to be in treated areas (Howell, 1961). Liquid baits, in which molasses, syrup, or sugar are usually the attractants, may be applied as sprays, using 1 gal per 1,000 sq ft (about 4 l per 93 sq m), when added to any organophosphorus or carbamate insecticide currently recommended.

Baits tend to have a spectacular but brief effect. A permanent liquid-bait station has been suggested that is merely a chicken-watering unit with a cellulose sponge in the trough to prevent clogging by dead flies. Dichlorvos or trichlorfon at 0.1% concentration in a sugar/water solution has been used successfully. A constant source of toxicant can be provided by inserting a dichlorvos resin strip in the reservoir jar (CDC, 1972).

In San Bernardino, California, a county sanitarian observed that little house flies (*Fannia canicularis*) seemed to prefer beer cans to regular garbage. This led to the suggestion that insecticide be placed in about 2 in. (5 cm) of beer in open containers for fly control. This poisoned-bait method was found to be effective (Times, 1972). Obviously, the poisoned beer should be placed in locations where it cannot be reached by children.

Dichlorvos resin strips suspended from suitable room fixtures are effective against small, flying insects and are convenient to use. The strip is a solid polyvinyl chloride resin formulation, impregnated with 20% dichlorvos (DDVP) and related products. It is 2.5 in. wide, 0.25 in. thick, and 10 in. long (6 cm; 7 mm; 25 cm). In the period of its activity, the vapor issuing from the strip is highly toxic to insects, but it is rapidly hydrolyzed by moisture. The strips should not be used in kitchens, restaurants, or other places where food is prepared or served, where infants or aged people are confined, or near fish tanks. One strip is sufficient for 1,000 cu ft (28 cu m) of space, and remains effective for weeks. A smaller strip of the same material has been made available. This "ministrip" is only 2 in. (5 cm) long, and is sufficient for a volume of 100 cu ft (2.8 cu m) for the control of such insects as cockroaches, silverfish, and clothes moths in enclosed spaces. The strips may also be placed in a closed garbage can, attached underneath the lid, for fly control.

An additional procedure is indicated when treating for flies, such as cluster flies, face flies, and blow flies, that spend the greater part of the winter in attics, wall voids, and similar enclosed spaces. Insecticide formulations, such as aerosols, silica aerogels, or the currently available combinations of silica aerogel and pyrethrum that can be very thoroughly dispersed, may be used in such areas.

Dichlorvos resin strips may be suspended in attics, closets, and storerooms, provided they have very little air circulation. If the flies work their way into occupied rooms, a space spray or aerosol containing pyrethrum is suggested, repeating as necessary. Sometimes flies in a house will escape if windows are open and the screens removed (Gojmerac, 1969).

Exposure of house flies for 3 days to baited chemosterilants, with light corn syrup as the attractant, resulted in a high degree of sterility with hempa, metepa, and tepa within acceptable limits of adult mortality. With 1% concentration, 100% sterility of males was induced after 72 hours of exposure. The chemosterilants remained active for at least 20 weeks. The tests were sufficiently encouraging to justify further tests in local house fly control with baited chemosterilants alone, and with an integrated program of baited chemosterilants with an insecticide (Pausch, 1971).

Control of Flies on Poultry Ranches

Poultry production is a major nation-wide industry. Flies breeding in poultry manure can be a serious problem when the poultry ranch is situated rather close to urban areas. The most important house pests among these flies are *Musca domestica* and *Fannia canicularis,* that develop in manure, and *Phaenicia* spp., that thrive on fowl carcasses and broken-egg wastes. All can fly considerable distances—up to 20 miles (32 km)—and may be attracted to food and shelter in homes and backyards (Peters, 1963).

As already stated, because modern poultry production is a monoculture, natural enemies of poultry pests are at a disadvantage, just as are the natural enemies of plant pests in agricultural monocultures. In plant monoculture, changes in cultural practices and modification of the pest control program in such a way as to avoid the usual need for continually increased dosages of pesticides seem to have been first advocated in 1940 by the Swiss entomologist Schneider (1940), in an extensive monoculture of gambier (*Uncaria gambir,* family Rubiaceae) in Malaya, and are now generally referred to as "integrated control." This term appears to have been first used by Smith and Allen (1954), who stated: "If we are to escape the ever-tightening spiral of more complex problems and ever-increasing costs of control, then we must integrate chemical control with the natural factors influencing populations." While admitting that many of the strands of the ecosystem of which man is a part remain to be unraveled, they believed: "We are now in a position to take more intelligent steps toward an integrated control program which will utilize all the resources of ecology and give us the most permanent, satisfactory, and economical insect control that is possible."

As in the control of plant pests, the current trend in combating pests of poultry and livestock is "integrated control." The number and variety of predators and parasites of filth flies is impressive. The predators include many species of beetles, bugs, flies, ants, earwigs, and mites, and there are some effective hymenopterous parasites (Legner and Brydon, 1966; Legner and Olton, 1968, 1970, 1971). The species that are natural enemies of the flies in accumulated excrement or manure piles are quite different from the ones that are natural enemies of the flies in the undisturbed feces of livestock in pastures. The 2 groups of natural enemies are discussed separately in an extensive review of the subject by Legner and Poorbaugh (1972).

The importance of developing manure removal and pest control practices that result in minimal reduction of the influence of these natural enemies of the pest fly population is obvious. Sound integrated control programs must be preceded by studies on the ecology and behavior of the pest species and the complexity of the ecosystem of which they are a part. An important element of fly behavior studies is the investigation of resting habits during the day and night, indoors and outdoors, in urban and in rural areas, and when buildings are screened or not screened (Kilpatrick and Quarterman, 1952; Maier *et al.*, 1952; Keiding, 1963, 1964). A study in poultry ranches in northern California, in which a backpack vacuum-sampling machine and sticky fly tapes were used, resulted in much information on which species of "domestic" or "filth" flies were involved, which species of natural enemies, and also much useful information on their habits, particularly their resting habits (Anderson, 1964, 1966*b;* Anderson and Poorbaugh, 1964). In striking contrast to the widespread dispersion of the pest flies during the day, at night all fly species except *Phaenicia sericata* that remained outdoors rested predominantly on branches of trees and shrubs, and those inside the poultry houses rested predominantly in the general ceiling area. The flies remained at their overnight resting places for 12 to 16 hours, depending on temperature. Sticky tape catches revealed that 85% of resting *Fannia canicularis* and *Musca domestica* were caught in houses and 91% of the beneficial predator *Ophyra leucostoma* and 95% of the ichneumonid parasites were caught in trees and shrubs. The difference in the nocturnal aggregation sites of pest flies and certain of their natural enemies "seems to offer one potential approach to an integrated fly control program on poultry ranches" (Anderson and Poorbaugh, 1964).

The University of California Agricultural Extension Service has published a circular in which integrated control procedures on poultry ranches are recommended (UCAE, 1971–72). A concrete base beneath caged laying hens to catch droppings prevents house flies from pupating in the soil. Natural enemies can then destroy a greater percentage of immature flies. During the season when flies are a problem, only the top portion of manure should be removed, and a 6- to 8-in. (15- to 20-cm) pad should remain on the concrete base, or all manure may be removed on an alternate row basis. In either case, natural enemies from the manure that remains can quickly invade adjacent fresh droppings. If all manure is removed at one time, a period of 6 to 12 months is required to re-establish the natural enemies.

If insecticides are required, they should be applied only to overhead and peripheral portions of the poultry houses. If droppings are sprayed, nearly all the natural enemies that normally infest droppings will be killed. These natural enemies can destroy more than 95% of the fly population. Residual insecticides directed against adult flies and applied to the ceilings and peripheral parts of the poultry houses do not interfere with key natural enemies (Legner and Olton, 1968). Great care should be taken to avoid contamination of feed and water. Diazinon, dieldrin, lindane, and trichlorfon are no longer labeled for use in poultry houses (CDC, 1973).

In an attempt to control flies on poultry ranches without long-range residue buildup and adverse ecological effects, the relatively new organophosphorus insecticide Rabon® (table 1, chapter 2), in an encapsulated formulation mixed with the feed, has been fed to poultry, and this passes through the bird without being assimilated to any appreciable extent. It causes complete mortality of fly larvae in the manure, while leaving no residues in body tissues of the poultry or in the eggs (yolks) (Wasti and Shaw, 1971). Another organophosphorus insecticide, coumaphos, mixed with cow feed, showed promise in the control of house fly larvae in the manure, and the experiments have been continued (Miller *et al.*, 1970).

In one investigation, a dichlorvos formulation (20% Shell Vapona Resin Strip Regrind) successfully controlled fly larvae and adults for about 7 weeks with 3 applications of 2 g of actual toxicant per square meter of breeding area in Florida poultry houses containing caged layers. The investigators emphasized that concurrent treatment of 2 life stages of the flies might speed up the development of resistance to the insecticide, and they believed that an alternative chemical that would be equally effective against both stages was needed (Bailey *et al.*, 1971). However, advocates of integrated control do not believe that insecticides should be used in this way on poultry ranches, not only because of the possibility of rapid development of resistance, but because the treatment is nonselective and kills all beneficial species. They suggest that its use should possibly be restricted to the wet "hot spots" where the greatest concentrations of larvae are found (J. R. Anderson, correspondence).

On the other hand, some investigators have felt that certain insecticides could be used advantageously on poultry droppings. Rodriguez *et al.* (1970) found that diazinon, ronnel, Bayer 38156, and malathion, when applied to poultry manure, were effective against fly larvae and relatively nontoxic to predatory mites [*Glyptholaspis confusa* (Foa), *Macrocheles muscaedomesticae* (Scopali), and *Fuscuropoda vegetans* (De Geer)]. Control of the house fly in poultry manure by these mites, under semifield conditions, ranged from 86 to 99%, depending on the species of mite. The use of sugar-based baits against adult flies was another approach to the integrated control program (Rodriguez *et al.*, 1970). Also, cockerels were allowed to run freely under caged hens, and the only food they obtained was what they could find, primarily fly larvae (Rodriguez and Riehl, 1959*a*, 1962).

Vinegar Flies (Drosophilidae)

These very small flies are cosmopolitan, and are usually known as vinegar flies. They are so small that they can pass through ordinary "fly screen," and are therefore common pests in homes, restaurants, fruit markets, canneries, cellars, or basements containing fermenting liquids (vinegar, cider, wine, beer), and in other places that contain attractive food materials.

Description. All the economically important species are superficially similar in appearance (figure 169). They are dull brownish yellow or brownish black in their over-all aspect. A feathery bristle on the antenna helps to distinguish *Drosophila* from other small flies. The adult of the best-known species, *D. melanogaster* Meigen, is 3 mm long, with red eyes, tan-colored head and thorax, and the abdomen is blackish on top and grayish underneath. Other species that commonly become pests are *D. funebris* (F.), *D. repleta* Wollaston, *D. busckii* Coquillett, *D. affinis* Sturtevant, *D. falleni* Wheeler, *D. tripunctata* Loew, and *D. hydei* Sturtevant.

Life Cycle. Vinegar flies lay their eggs near the surface of fermenting or rotting fruit and vegetable material or in unclean garbage cans. The pearly-white, cylindrical eggs, which can be observed with a hand lens, have 2 to 4 threadlike filaments at one end, and these filaments protrude above the liquid material in which the eggs are laid. The tiny maggots crawl out of the liquid food to some drier substance, preferably loose soil, to pupate. The brown, seedlike pupae have 2 hornlike stalks at the anterior end. About 500 eggs are deposited by a single female, on an average, and since the life cycle may require only 8 to 10 days in warm weather, the reproductive potential is enormous.

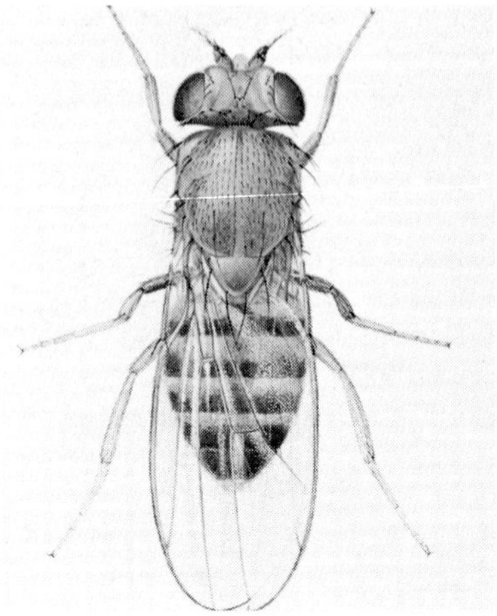

Fig. 169. Vinegar fly, *Drosophila melanogaster* (female). (From B. Greenberg [1971], *Flies and Disease*, courtesy Princeton University Press.)

Control of Vinegar Flies

The most important aspect of *Drosophila* control is the elimination of food and breeding sources. This refers not only to the most obvious materials, such as piles of rotting and fermenting fruits and vegetables, but also to such sources as peelings accidentally swept under appliances or counters, discarded ketchup or milk bottles that have retained a portion of their contents, dishwater or water from floor moppings that has seeped into crevices, garbage-laden water from sinks, or a foul, soured mop. Vinegar flies constitute a serious problem in rural communities that are close to extensive sources of the insects, such as tomato fields or canning plants. Cannery sanitation is extremely important. Egg deposition in boxes of tomatoes in the field or on pallets in the receiving stations of canneries can be prevented for a 24-hour period by the application of pyrethrum dust, principally because of its repellent effect (Stombler *et al.,* 1957).

Piles of culled fruit are important sources of vinegar flies and driedfruit beetles (*Carpophilus* spp.). These pests can be controlled by suitable insecticides, such as liquid sprays or granular formulations. Insecticide granules have to be mixed into the mass of fruits if the pile is more than 24 in. (60 cm) deep. Larvae can be prevented from developing long enough for the fruit wastes to become too dry to attract insects. Sprays of azinphosmethyl, fenthion, or trichlorfon, or granular formulations of malathion, endrin, or heptachlor have proved to be effective (Yerington, 1969). For control of vinegar flies in wine cellars, a thermal aerosol of 0.25% pyrethrins, combined with either piperonyl butoxide or Tropital® as synergists, was found to be effective, but treatments once or twice daily were required to keep the insects under control (Yerington, 1971).

Substantial reduction in vinegar fly populations in tomato fields was obtained by releasing large numbers of the adult flies of both sexes that had been sterilized with 1% aqueous apholate, a well-known insect chemosterilant. The investigators stated that release of sexually sterilized flies in a program such as they described "shows promise of controlling *D. melanogaster* in commercially grown tomatoes, especially if the program is conducted on a community-wide basis" (Mason *et al.,* 1968).

In the home, space sprays, aerosols, or dichlorvos resin strips, as they would be used for house fly control, have been effective against vinegar flies.

ANTS

Ants are among the most common pests of the home. Their nests are found almost everywhere—in lawns, under the house, in the walls of the house, in open ground, under stones or wood, in wood of decaying trees, or in other locations that afford protection. Ants are social insects of the order Hymenoptera, and live in colonies. They cannot live alone, but unlike other social insects (termites, wasps, and bees), they are not dependent on a definite habitation, but can quickly transport their eggs, larvae, and pupae to a more favorable location when necessary. With the exception of the males, ants share with termites the advantage of a long life span and comparative freedom from natural enemies.

Ants enter buildings when seeking sweet or fatty substances in the kitchen, pantry, storeroom, or warehouse. They may also be pests of the lawn or garden because of their nesting habits or because they may destroy plants, seeds, fruits, and nuts. Ants may feed on the sugary excretion (honeydew) of certain plantsucking insects, such as aphids, psyllids, mealybugs, unarmored scales, cottonycushion scales, whiteflies, and treehoppers. In crawling back and forth over the infested plant, the ants interfere with activities, such as oviposition, of the natural enemies of these pests. In addition, ants actually destroy adult parasites and the larvae of predators. They may also carry aphids and other honeydew-excreting insects to plants in order to establish colonies of them. Control of ants is a great aid in garden pest control.

How to Recognize Ants

The principal body regions of ants are distinctly defined by narrow constrictions (fig. 170). The elbowed antennae are composed of a *scape* (long basal segment) and a *funicle* or *funiculus* (terminal segmented portion). At the end of the antenna, the last 1 to 4 segments may be enlarged to form the "antennal club." The abdomen is attached to the thorax via 1 or 2 slender segments that are parts of the abdomen, and form the *petiole* or *pedicel.* Each segment of the petiole has a bump or *node* on its dorsal surface, as can be seen in the drawings in this section. Some genera have 2 nodes (e.g., *Pogonomyrmex, Veromessor, Pheidole, Crematogaster, Monomorium, Solenopsis,* and *Tetramorium*), and other genera, higher in the evolutionary scale, have 1 node (*Liometopum, Iridomyrmex, Dorymyrmex, Tapinoma, Camponotus, Paratrechina, Prenolepis, Lasius,* and *Formica*), so the number of nodes is one of the morphological features by which an ant can be identified. The remainder, and major portion, of the abdomen is called the *gaster.* It may consist of 7 or 8 segments, the male having 1 more than the female.

Keys to the genera and species of the ants of North America were prepared by Creighton (1950).

The Ant Castes

Ants have 3 distinct *castes* (queen, male, worker), and there may be different forms in each caste. The soldier is a large worker. The head and mandibles of the soldier and the thorax of the queen are very large in comparison with those of other forms and castes. Ants are generally wingless, but at mating time swarms of winged males and females (figure 176) usually emerge from the nests, although the reproductives of some species, such as the Argentine and Pharaoh ant, mate and remain in the nest. Those emerging from the nests mate while in flight, and the male dies a day or two later. The females (queens) lose their wings and attempt to establish new colonies after digging a nest in the ground or seeking a cavity under some object, such as a stone or piece of bark. The workers or neuters are sterile females, and may vary greatly in size. They are very different from the females in appearance. They do all the work of the colony, such as building the nest, caring for the queen and the young, and providing food.

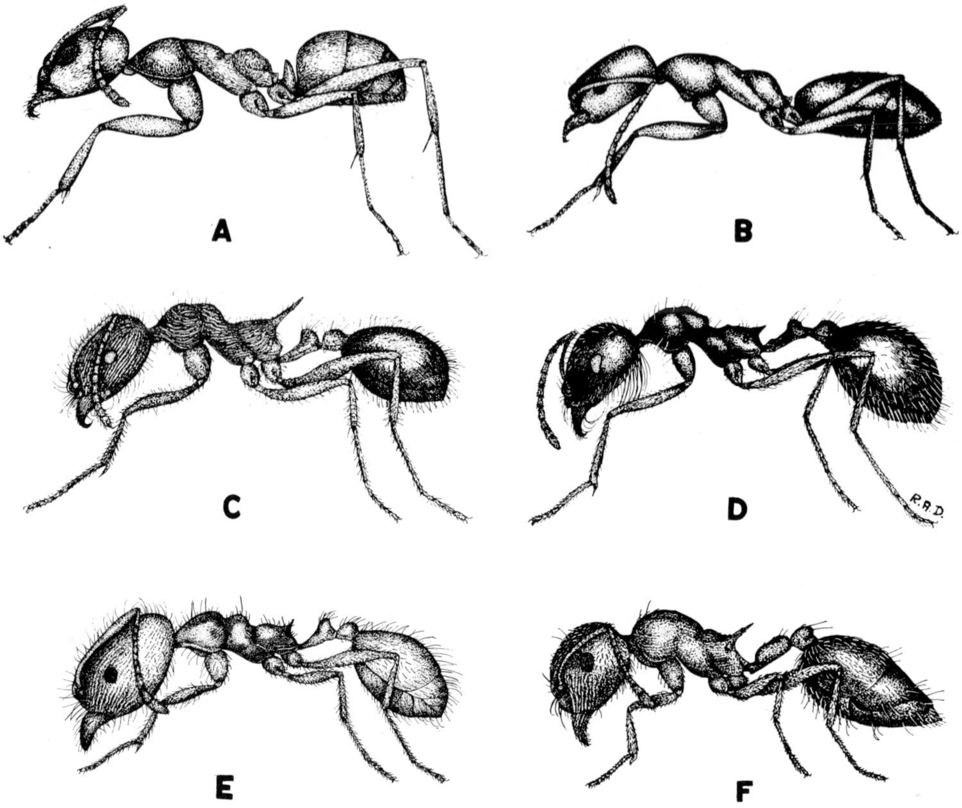

Fig. 170. House-infesting ants. *A,* Argentine ant, *Iridomyrmex humilis; B, Iridomyrmex pruinosus; C,* longspined harvester ant, *Veromessor andrei; D,* jetblack harvester ant, *Veromessor pergandei; E,* western bigheaded ant, *Pheidole hyatti; F,* California acrobat ant, *Crematogaster californica.* (Courtesy R. R. Snelling.)

Trophallaxis is common among the higher ants. The workers gather food, chew and prepare it, then feed it to the larvae. The larvae digest the food and regurgitate the digested food for the adults to consume. Such ants are often seen moving their larvae about from place to place; they are then "carrying their stomachs around." Adult ants, in common with adult bees and wasps, can feed on liquid foods such as nectar and honeydew. They have powerful mandibles that can pierce and chew fruits, seeds, and other insects. However, adult ants have a very slender esophagus and a very small proventricular opening that do not admit the chewed-up particles. The larvae must digest such food and feed the adults (Went *et al.,* 1972).

Ant Trails

In the intensive investigation of insect pheromones in the last 2 decades, odor-trail pheromones have received their share of attention. Among the ants, these chemicals originate in several different glandular tissues. In 3 subfamilies, odor-trail pheromones originate in the hindgut. In the dolichoderine ants, these pheromones originate in Pavan's organ, where they are probably synthesized. In some myrmicine ants, trail pheromones arise in glands associated with the stinger, e.g.,

Dufour's gland of the imported fire ant, *Solenopsis invicta*. In some other myrmicine genera, the venom-gland secretion itself is utilized to lay odor trails. In the myrmicine genus *Crematogaster* (acrobat ants), the trail pheromone is apparently produced in a tibial gland which presumably evacuates through the pretarsus. Some ants emit both volatile and nonvolatile trail pheromones, and are able to lay a trail that will be long-lasting but will also recruit workers quickly (Blum, 1970). Some species, such as the Florida harvester ant, *Pogonomyrmex badius,* utilize odor trails as well as sun orientation, whereas landmarks are of minor importance. Some other species, such as *Cataglyphus bicolor,* orient mainly to visual landmarks, utilizing sun orientation only when the landmarks become ineffective (Hölldobler, 1971). The foraging column of *Veromessor* needs sunlight to orient itself. Many of the foragers break out of the column and wander about aimlessly if they encounter a patch of shade more than a half-meter long (Creighton, 1953).

Hölldobler and Wilson (1970) observed that a scouting ant of the genus *Pogonomyrmex* held its abdomen upward in ordinary locomotion, but returned from a successful scouting trip with its abdomen completely lowered and dragging the extruded stinger over the surface of the ground so as to leave a scented trail for others to follow. They proved that *P. badius* used recruitment pheromones released from the venom gland and orientation or homing pheromones released at least in part from Dufour's gland.

The source of the trail pheromones of the ponerine ants *Leptogenys attenuata* Smith and *L. nitida* Smith is the venom sac of the stinging apparatus. When given a choice, the ants of these 2 species preferred to follow the trail pheromone of their own species, but were also able to follow the trails of the other species (Fletcher, 1971).

M. S. Blum (Smith, 1965) stated that the workers of *Iridomyrmex pruinosus* deposit on their foraging trails a methyl-*n*-amyl ketone substance, and that workers can be induced to follow artificial trails containing the scent of this compound. The Argentine ant, *Iridomyrmex humilis,* relies heavily, if not completely, on scent when following trails. Olfactory stimuli are received in sites located in the funicle, possibly primarily in the first 2 segments (Dechene, 1970).

Ant Bites and Stings

Some species, such as the harvester ants (*Pogonomyrmex)* and fire ants (*Solenopsis*), inflict severe stings. Others merely bite, but they may curve their abdomens forward to inject a secretion into the wound and are, therefore, thought to be stinging (e.g., *Formica* and other Formicinae). The Formicinae are the only ants that secrete formic acid, emptying it into the wound. The venom of ants that sting does not contain formic acid, neither does the venom of wasps and bees; therefore, local application of alkalies results in no relief from pain when one is stung by such insects (Keele, 1963).

Ants that have stingers normally use their venom offensively only to kill prey, but if the nest is attacked, they may use the venom defensively to repel the intruders. The species known principally for their bites or stings are discussed in chapter 9, and the carpenter ants are discussed in chapter 5. In the present chapter, only the species are discussed that are pests mostly because they nest in the house or invade it from the outside, and became nuisances by getting into household foods.

Argentine Ant, *Iridomyrmex humilis* (Mayr)

Argentine ants were first found in the United States in New Orleans in 1891, and now range throughout the southern states as well as California, where they first appeared in 1905. In California, they may be found at the lower elevations in moist locations in urban areas and orchards. There is no record of their presence in Arizona (J. N. Roney, correspondence). These ants invade the home in columns, seeking sweets, and may be traced back to their nests, that are usually in the ground, although they can also nest in treeholes, refuse piles, bird nests, or beehives, to name a few possibilities. Argentine ants bite, but do not sting. This species is not only the most common pest in the home, but is also an important agricultural pest because of its interference with the activities of natural enemies of aphids and similar plant pests as it crawls over a plant in search of honeydew.

Description. The worker is light to dark brown (plate II, 5), 2.2 to 2.6 mm long, and its petiole has 1 node (figures 170, *A,* and 171). The queen is brownish, with a silky pubescence, up to 6 mm long, and is usually wingless. The male is shiny brownish black, winged, less robust than the female, and is up to 5 mm long. There is no perceptible odor if an individual ant is crushed, but if large numbers are crushed, the Argentine ant has a "greasy" or musty odor. In this respect it differs from another common house-infesting ant, the odorous house ant, *Tapinoma sessile,* which also seeks sweets and travels in trails. One can detect the pungent odor of this species by crushing only a single individual (Eckert and Mallis, 1941).

Life Cycle. There are few winged females and consequently few mating flights. Mating generally takes place in the nest, after which the male is eliminated from the nest. There are many reproductive females in the colony, and each lays a large number of eggs. That is the reason for the enormous numbers of individuals produced. The period required for the completion of the life cycle varies greatly with temperature, moisture, nutrition, and other factors, but has been observed to vary from 33 to 141 days, with an average of 74 days (Eckert and Mallis, 1941).

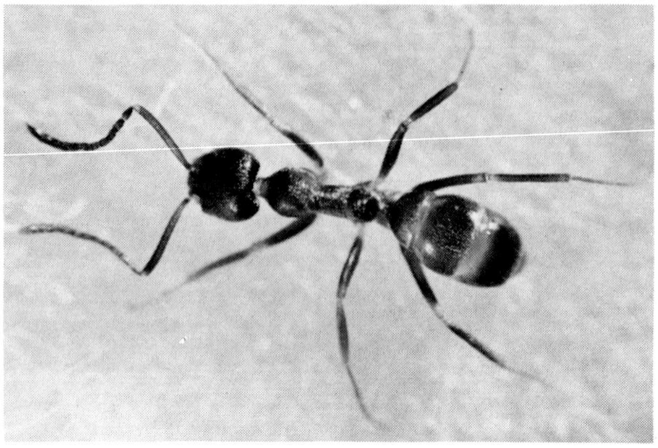

Fig. 171. Argentine ant, *Iridomyrmex humilis* (worker).

During the winter, Argentine ants tend to congregate at greater depths in the soil, down to about 6 in. (15 cm), or in such locations as manholes or basements kept warm by steam pipes. They begin to increase in numbers toward the end of February, and reach their maximum number during July, August, and September. During this period, they have gradually spread out over larger areas, particularly in well-moistened locations such as under shrubbery and along the edges of walks. The nests are then usually from 0.5 to 1.5 in. (12 mm to 4 cm) below the surface. By the middle of October, they have again noticeably decreased in number (Eckert and Mallis, 1941). Argentine ant nests are sometimes under buildings, and soil treatment for subterranean termites is also valuable in ant control.

Markin (1970*a, b*) made a 3-year study of the seasonal life history of the Argentine ant in 6 California citrus orchards, but primarily in a 5-acre (2-ha) block of Valencia orange trees with a slightly southern exposure. The orchard had received no insecticide treatments in more than 10 years. There was an ant nest for each tree in the orchard, but the nests were not permanent structures. Adverse conditions such as flooding or drying resulted in the ants moving to other nesting sites or joining adjacent colonies. Nests were connected with trails, and workers could sometimes be seen carrying the brood or queen from one nest to another.

A great many eggs are produced in late February to early March, and mostly sexual forms develop from them. The sexual forms mature in May, and mate within the nest soon after the females emerge from the pupae. The females then shed their wings and oviposit. The number of queens in the nest remains fairly constant until January or February of the following year, when about 75% of them are killed by workers. Workers are first produced about mid-March, and increase in numbers until a seasonal high is reached in October, when production of workers decreases sharply. The number of workers then decreases steadily, reaching its lowest level the following March or April (Markin, 1968, 1970*a, b*).

Habits. The Argentine ant tends to drive out other ant species wherever it becomes established. When in conflict with species of greater fighting ability, several Argentine ants will attack a single victim, and in that way will make up by their superior numbers what they may lack in fighting prowess.

In the "flash floods" of southern California, these ants have been seen to mass into balls and float down the stream, thus disseminating the species along the course of the streambed.

The workers forage at all times of the day and night, in either light or darkness. They become sluggish at 7° C (about 45° F), and cease activity entirely at 6° C (43° F), but immediately become fully active when again exposed to warmer temperatures (Dechene, 1970). Argentine ants will leave their nests to forage at temperatures from 10 to 30° C (50 to 86° F). They prefer sweet foods, and are remarkably adept at finding them and rapidly establishing their trails. They will also feed on both live and dead insects, meat, injured fruits, seeds, and cereals, particularly cornmeal. These ants have been observed killing adults and nymphs of conenose bugs (*Triatoma*) in small jars, cutting them into bits small enough to pass through the 1.5- to 3-mm holes of the perforated jar lids, and carrying the insect bits to their nests (S. F. Wood, personal communication). They have also been observed killing all active stages of German cockroaches (*Blattella germanica*) in rearing jars and carrying away the comminuted fragments (D. A. Reierson, personal communication).

In addition to invading the home, Argentine ants are common garden pests, for they are among the species that are attracted to honeydew, a sugary liquid excreted by aphids and related plant pests. One investigation in a southern California citrus orchard revealed that honeydew and nectar comprised more than 99% of the material carried into the nests by the workers (Markin, 1970*a*). Argentine ants are among the species, already referred to, that can interfere with the activities of hymenopterous parasites, kill the larvae of predators, and thereby increase insect infestation of plants. The ants can also carry aphids and other sapsucking, honeydew-producing insects to plants in order to infest them. Control of Argentine ants may substantially diminish or even eliminate pests that excrete honeydew as well as others that may be benefited by the decrease in their natural enemies when ants are present.

Iridomyrmex pruinosus (Roger)

This is a native species that ranges from Wisconsin to New York, south to New Mexico and Florida, and also occurs in Colorado, Arizona, Utah, Nevada, California, Oregon, and Idaho. It is common and widely distributed in the southern United States. It infests houses, particularly in the Gulf Coast states, generally invading them from outside, but is also believed to be able to nest within a house (Smith, 1965).

Description. As shown in figure 170, *B,* this species can be distinguished from the Argentine ant by the difference in the shape of its head, the presence of erect hairs on the thorax, and by having a more densely pubescent body. The body color (plate II, 5) is variable. In the eastern states, this species is commonly uniform dark brown or black, whereas the body of the Argentine ant is usually light to dark brown (Smith, 1965). In the western states, the body color is even more variable, some specimens being even lighter in color than *Iridomyrmex humilis* (R. R. Snelling, personal communication). The workers are 1.8 to 2.5 mm long. The odor of crushed specimens of *I. pruinosus* is like "rotten coconut," similar to the odor of the odorous house ant (*Tapinoma sessile*), whereas a mass of crushed specimens of *I. humilis* has a stale and musty or greasy odor.

Habits. *Iridomyrmex pruinosus* prefers open spaces, such as fields, meadows, pastures, or entirely bare areas, but will also nest in open woods away from dense and prolonged shade. The nests may be constructed in exposed soil or under cover of stones or the bark of logs and stumps, and contain small to moderate-sized colonies. Like the Argentine ant, this species feeds on live or dead insects, is particularly fond of honeydew, and forms foraging trails.

Harvester Ants (*Veromessor* spp.)

These ants are predominantly black, and have conspicuous thoracic spines. The petiole has 2 nodes. Their habits are somewhat similar to those of *Pogonomyrmex* (chapter 9), except that they do not sting, their stinger being vestigial and nonfunctional. Homeowners object to their nests in yards and lawns. They are very active, and harvest large quantities of seeds.

Longspined Harvester Ant, *Veromessor andrei* (Mayr)

This species (figures 170, *C,* and 172) is 4.5 to 7 mm long, with a reddish-brown head and thorax, and in some specimens the posterior part of the abdomen is black. The remainder of the body is lighter in color—in southern California, generally reddish brown. It is limited almost entirely to California in its distribution. Wheeler and Wheeler (1973) point out that while *V. andrei* does not sting, its bite is quite tenacious and annoying.

Jetblack Harvester Ant, *Veromessor pergandei* (Mayr)

This species (figure 170, *D*) is 2.5 to 6 mm long, and as the common name indicates, is a beautiful "jet-black, patent-leather-colored" ant (Mallis, 1941*a*). It inhabits the deserts of southern California, northwestern Arizona, and Baja California. The nests of *V. pergandei* are usually surrounded by a crater of excavated soil, whereas those of *V. andrei* usually are not. *Veromessor pergandei*

Fig. 172. Longspined harvester ant, *Veromessor andrei.*

makes a neat pile of chaff with seed husks in the form of a semicircle or ring around the nest, whereas *V. andrei* makes a more ragged pile or scatters the chaff about. *Veromessor pergandei* can tolerate the most extreme heat of the deserts, but *V. andrei* cannot.

Bigheaded Ants (*Pheidole* spp.)

These ants are so named because of the exceptionally large heads of the soldiers in relation to their bodies. They live in situations somewhat similar to those inhabited by fire ants, have similar habits, and are distributed throughout the warmer areas of the United States. As in the case of the fire ant, the petiole has 2 nodes. They resemble fire ants superficially, but can be distinguished from them by their 12-segmented antenna, 3-segmented antennal club, and the large head, particularly of the soldier. The natural foods of *Pheidole* spp. are generally live and dead insects, seeds, and honeydew, but in houses they will feed on meats, grease, peanut butter, fruit juice, and other foods.

Western Bigheaded Ant, *Pheidole hyatti* Emery

This species (figure 170, *E*) may be found throughout the San Joaquin and Sacramento valleys of California, as well as in most of the state below 3,000 ft (900 m) elevation, and east to Texas. These ants are 3 to 6 mm long, uniformly light to dark reddish, with the soldiers sometimes darker and more brownish than the workers. The soldier's head is deeply furrowed in the frontal view. The body has abundant long hairs. *Pheidole hyatti* is common in arid and semiarid areas, and makes nests in the soil, often beneath stones and boards. It is usually nocturnal, and is omnivorous, with a decided tendency to be predaceous (R. R. Snelling, correspondence). The ants are sometimes found in and around houses; in kitchens, they are usually found around the sink. They are especially fond of greasy meats.

Acrobat Ants (*Crematogaster* spp.)

Acrobat ants are characterized by the heart-shaped abdomen, the upper surface of which is flattened and the lower surface curved. They may be black, brown, or yellow, and the petiole has 2 nodes. Their common name is derived from their habit of raising the abdomen over the head and thorax (figure 173, *bottom*). They are slow-moving ants that tend aphids, and are known to construct "tents" or "cowsheds" of earth and

Fig. 173. California acrobat ant, *Crematogaster californica*.

vegetable debris over the aphids from which they obtain the sugary excrement known as "honeydew" (Wheeler, 1906; Eckert and Mallis, 1941; Matheson, 1951). They are generally found under stones or in old stumps, and have sometimes been found to tunnel and nest in wood (Mallis, 1969). They occasionally become pests by invading houses.

Lined Acrobat Ant, *Crematogaster lineolata* (Say)

This ant ranges from southern Canada southward along the Rocky Mountains and eastward to Florida, and has been found at altitudes up to 5,500 ft (1,700 m). It varies in color from light brown through dark brown to black. The scape (first segment) of the antenna extends beyond the posterior border of the head. The workers are 2.5 to 3.5 mm long. Medium-sized to large colonies may nest in exposed soil, under stones or other objects, or in logs, stumps, and dead trees. The ants may also nest in the woodwork of houses and infest household foods, where they show a slight preference for sweets and meats or other high-protein foods. In nature, they feed on honeydew and dead or live insects, including winged termites. The workers bite aggressively, and give off a repulsive odor when alarmed (Smith, 1965).

California Acrobat Ant, *Crematogaster californica* Emery

This species occurs in southern California and Baja California, nesting in soil or rotten wood at elevations below 3,500 ft (1,060 m) (figures 170, *F*, and 173). The workers are 2.5 to 3.5 mm long. The head and thorax may be reddish and the gaster black, but the head and thorax may be variably infuscated. One or two erect hairs occur on the pronotal shoulder. These ants are both diurnal and nocturnal. They tend aphids and mealybugs on both foliage and roots. They are attracted to sweets (R. R. Snelling, correspondence).

Pharaoh Ant, *Monomorium pharaonis* (L.)

This species is native to Africa, but has been spread by commerce to nearly every part of the world. The specific name, from which the common name was derived, resulted from Linnaeus' mistaken belief that the ant was one of the plagues of Egypt in the time of the Pharaohs, mentioned in Scripture, whereas the only plagues mentioned were lice, flies, and locusts (Riley, 1889).

Although the Pharaoh ant occurs throughout the United States and Canada, it is rarely found in California, but appears to be established in at least 2 locations in the state (M. S. Wasbauer, correspondence). It has never been identified as a household pest in California, but is important in some other parts of the country. It is especially common in hotels, large apartment houses, grocery stores, or other places where food is commercially handled. In residential areas, nests may be found outdoors in the lawn or garden and indoors in wall voids, subfloor areas, attics, cracks, crevices, and behind wainscoting, baseboards, plaster, or mantles, under hearthstones, between flooring, or in furniture. The ants tend to seek warm and inaccessible locations. They breed throughout the year, and produce enormous numbers of individuals. They feed on dead and live insects, and often damage the dried specimens in collections. In the home, they appear to prefer grease, fats, and meats, but are practically omnivorous. In the early years of American agriculture, they caused great damage to crops. For example, they injured young corn plants by gnawing the blades and drinking the sap (Forbes, 1920).

Monomorium pharaonis has become particularly well adapted to existence in close association with man by abandoning the nuptial flight and territorial boundaries. Colonies multiply by "budding." If groups of these ants in a large building are joined by odor trails, they may constitute a widely dispersed colony that can contain up to millions of workers and thousands of queens (Bellevoye, 1889; Wilson *et al.*, 1971).

Description. The Pharaoh ant (figure 174, *A*) is one of the smallest ants, 1.5 to 2 mm long, and is light yellow to reddish in color. The petiole has 2 nodes. It is often confused with the thief ant, *Solenopsis molesta,* but averages slightly larger, and can be distinguished by 3 segments in the antennal club, compared with 2 in the thief ant.

Life Cycle. Although the usual sexual forms are winged, nuptial flights have never been seen. Mating takes place in the nest, and the formation of a new colony is not normally accomplished by a single female after a nuptial flight, as with many other ant species. Instead, a large part of a colony will migrate to a new location, the workers carrying with them eggs and young stages. In this way, new nests can be developed throughout a building, and such nests have been found to contain from 400 to 4,800 ants (Smith, 1965; Busvine, 1966).

Peacock and Baxter (1949) developed procedures and equipment for rearing colonies of *Monomorium pharaonis* in the laboratory. In subsequent investigations (Peacock and Baxter, 1950), they found that the various life stages of the workers required the following numbers of days: egg, 5 to 6; larvae, 22 to 24; prepupal, 2 to 3; and pupal, 9 to 12; thus, the life cycle required 38 to 45 days at room temperature. Sexual forms required about 4 days longer. The maximum period of adult life of the workers was 9 to 10 weeks, females lived as long as 39 weeks, and males generally not more than 3 weeks. The females were able to lay from 350 to 450 eggs.

Habits. Pharaoh ants crawl great distances from their nests, and do not always follow definite paths. Therefore, the nests may be difficult to find, increasing the difficulty of control. A shipowner spent $4,000 in trying to eradicate Pharaoh ants from his vessel. They have been stated to be the most persistent and difficult of all our house-infesting ants to control (Smith, 1965).

An interesting aspect of Pharaoh ant infestation is their tendency to nest in odd places, as between sheets of stationery, layers of linen, between clothing in trunks, in an electric iron, in a can of modeling clay, or in piles of trash.

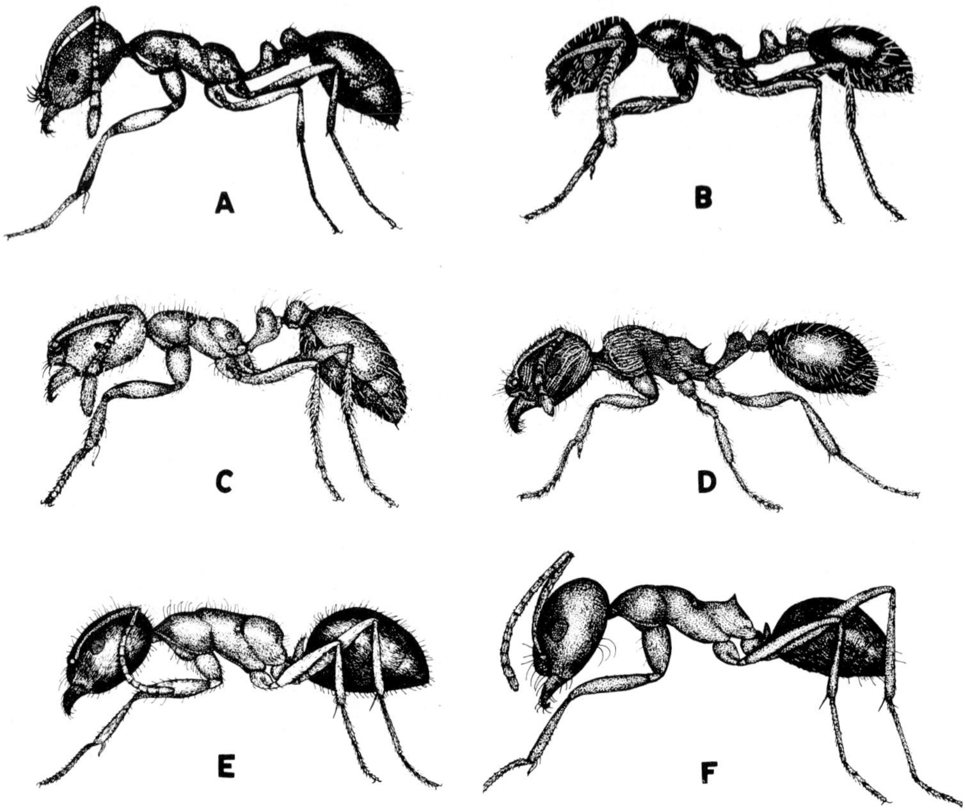

Fig. 174. House-infesting ants. *A,* Pharaoh ant, *Monomorium pharaonis; B,* little black ant, *Monomorium minimum; C,* thief ant, *Solenopsis molesta; D,* pavement ant, *Tetramorium caespitum; E,* velvety tree ant, *Liometopum occidentale; F,* pyramid ant, *Dorymyrmex pyramicus.* (Courtesy R. R. Snelling.)

Many investigators have noted the tendency of the Pharaoh ant to be attracted to water, such as from dripping faucets, and one observer noted that their trails to the earth beneath the house did not necessarily indicate the presence of a nest, for the ants might be seeking moisture in the soil (Mallis, 1969).

Little Black Ant, *Monomorium minimum* (Buckley)

This small, jet-black species (figure 174, *B*) is 1.5 mm long. Its petiole has 2 nodes. It is found throughout the United States. In California, it occurs in the southern part of the state and in the San Francisco Bay area. These ants may nest beneath rocks, in rotting wood, in lawns, or in open areas, where the nests are indicated by very small craters of fine soil. Their colonies are usually moderate to large, and contain many fertile females. Males and winged females may be seen from June to August. The workers are predaceous on other insects, tend honeydew-producing insects, and feed on honeydew as well as nectar and pollen. There have been reports of recently hatched bluebirds and kingfishers being killed by these ants. They may either invade a house from the outside, or establish their nests in the woodwork or masonry of the building. In the house, they feed on both sweets and greasy foods (Smith, 1965).

Thief Ant, *Solenopsis molesta* (Say)

Thief ants are found over most of the United States, and have been found at altitudes up to 5,500 ft (1,700 m). In nature, the nests of *S. molesta* and other tiny thief ants are usually located so that extremely narrow passages from the thief ant nests communicate with the nest passages of a larger ant species. The common name of the thief ant was derived from its habit of steadily pilfering the brood and other food from the nests of larger species (Creighton, 1950). The stinger of the thief ant is rarely used. Although they are in the genus *Solenopsis,* thief ants are not considered to be "fire ants." The latter are discussed in chapter 9.

Description. The workers (figure 174, *C*) are very small, 1.3 to 1.8 mm long, among the smallest ants in the United States. The workers are monomorphic (1 form only). The petiole has 2 nodes. The body is largely smooth and shiny, and ranges from yellowish to light or dark brown. Thief ants superficially resemble Pharaoh ants, but are smaller, lighter in color, and the antenna has a 2-segmented rather than a 3-segmented club. The antennal club is unusually large and elongate, about 1.33 times the length of the remainder of the funiculus. The scape of the antenna extends farther than between the eye and the posterior border of the head (Smith, 1965).

Biology. The outdoor nesting habits of the thief ant are similar to those of other *Solenopsis* species. There may be several hundred to a few thousand ants in a colony. In nature, they nest in exposed soil or under stones and other objects, and in rotting wood. Copulation takes place during nuptial flights, which have been observed from late July to early fall. A single fertile female is able to establish a nest (Eckert and Mallis, 1941; Smith, 1965).

Habits. Workers are predaceous, and feed on both live and dead insects. They also feed on planted or germinating grains, often causing considerable damage. They seek honeydew and tend honeydew-excreting insects. Because they feed on dead rats and mice, they may be carriers of disease-producing organisms to food. They are known to attack young chicks in poultry houses (Eckert and Mallis, 1941). Workers have been seen carrying gravid segments of the poultry tapeworm, *Raillietina tetragona* (Molin), into their nest, suggesting that the ants may be intermediate hosts for the tapeworm (Smith, 1965).

When they are found in the home, thief ants usually have their nests in wall voids or elsewhere in the house rather than entering from outside. Their extremely small size enables them to enter containers not accessible to larger ants. They often infest pantries, cupboards, cabinets, and shelves, even though these have been kept closed and scrupulously clean. They are often found around sinks. However, they are so small that they are sometimes not seen, even when crawling on food.

Thief ants will eat practically anything, but appear to prefer foods with a high protein content. They have more of a tendency than the Pharaoh ant to feed on greases, but will eat sweets also (Eckert and Mallis, 1941; Smith, 1965).

Pavement Ant, *Tetramorium caespitum* (L.)

This species occurs in most of the eastern half of the United States and in California, where it occurs in both the San Joaquin and Sacramento valleys. Nests occur under stones, along edges of curbing, and in cracks in the pavement. The nests may or may not be surrounded by craters. When nesting in houses, these ants are most likely to be found around or between the lower masonry walls of the foundation. In some areas of the eastern states, the pavement ant is considered to be the most important ant pest in houses, where it may be found throughout the year, but in greater numbers during the summer (Mallis, 1969). It feeds on both greasy and sweet foods.

Description. This species is 3 to 4 mm long, and its body varies from light to very dark brown, with lighter legs and antennae. The head and thorax are furrowed with parallel lines (figure 174, *D*), and the posterior dorsal portion of the thorax bears a pair of small spines. The petiole has 2 nodes. The entire body is covered with stiff hairs.

Velvety Tree Ant, *Liometopum occidentale* Emery

The principal pest in the genus *Liometopum* is the California velvety tree ant, *L. occidentale.* This species has been observed in northern California east to Colorado and south to central Mexico (R. R. Snelling, personal communication). In foothill and mountain areas, the trails of this species may extend 200 ft (60 m) or more from the base of a tree or log to a house, where the ants search for sweets and insects. Sometimes velvety tree ants enter a house where a tree limb is in contact with it; the limb should be removed. Although the ants may be seen prowling about during daylight hours, they make their forays for food at dusk or during the night (Eckert and Mallis, 1941).

There is only 1 other species in the genus, *L. apiculatum* Mayr, which ranges throughout approximately the same areas as *L. occidentale,* except that it does not occur in California.

Description. This species (figure 174, *E*) is 2.5 to 6 mm long, with a glistening, velvety-black abdomen, red thorax, and brownish-black head. Except for the 2 species of *Iridomyrmex,* which because of their economic importance and well-known biologies were used to introduce the ant section of this chapter, the preceding ant species all have petioles with 2 nodes. Species of *Iridomyrmex* and *Liometopum,* and all remaining species to be discussed in this chapter, have petioles with 1 node.

Habits. Throughout California, this ant occurs in large colonies in the crotches or hollows of trees, in stumps, and in the ground beneath stones. The insects can be found crawling up trails, up and down oak, alder, elm, sycamore, cottonwood, and other trees, particularly along streams, foraging for insects and honeydew. Areas along streams and under trees also happen to be popular picnic grounds, and the velvety tree ant is therefore the bane of picnickers—biting people, crawling over their food, and producing a disagreeable odor when crushed. The removal of decayed portions of ornamental or shade trees in which the ants nest helps to eliminate these pests.

Pyramid Ant, *Dorymyrmex pyramicus* (Roger)

This small ant is occasionally encountered at lower altitudes throughout California, where it nests in dry and vegetation-free areas. The earth excavated from its nest and gallery system is in a circular crater, usually about 2 to 4 in. (5 to 10 cm) or larger in diameter. It invades the home from the outside, feeds on a wide variety of household foods, but seems to prefer sweets (Eckert and Mallis, 1941). This species is widely distributed over most of the United States, but is more common in the southern half.

Description and Habits. The pyramid ant (figure 174, *F*) is 1.5 to 2.0 mm long, and has a reddish-black head and thorax and black abdomen.

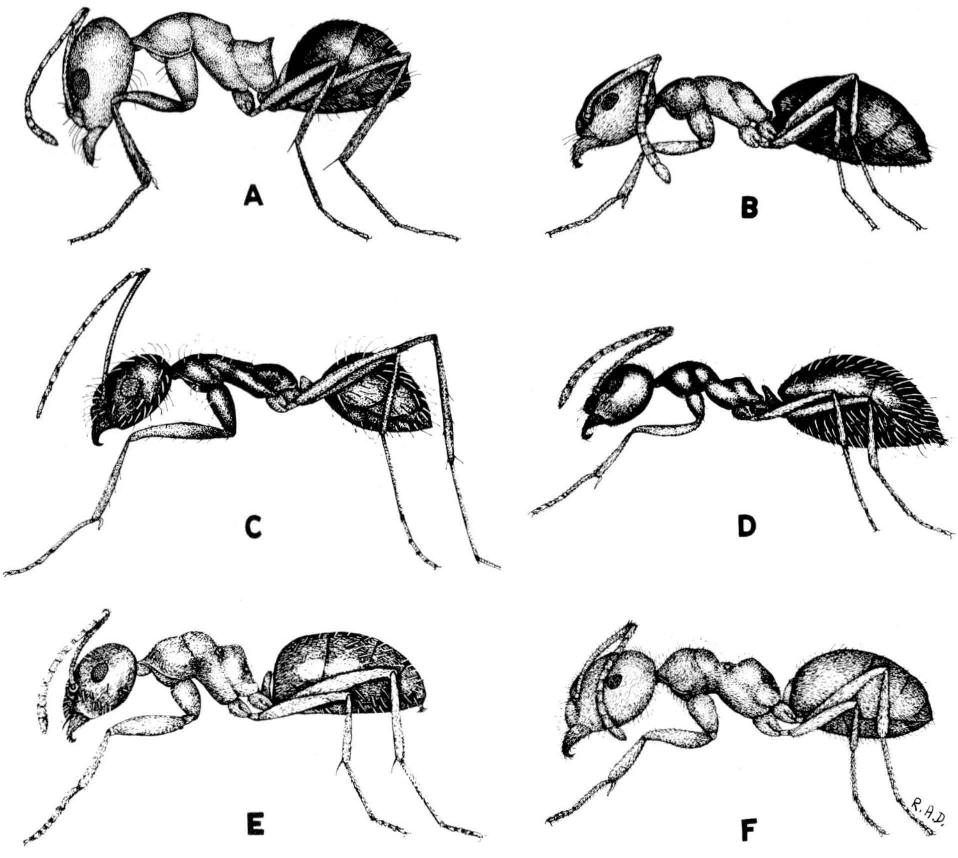

Fig. 175. ***A,*** **bicolored pyramid ant,** ***Dorymyrmex bicolor; B,*** **odorous house ant,** ***Tapinoma sessile; C,*** **crazy ant,** ***Paratrechina longicornis; D,*** **small honey ant,** ***Prenolepis imparis; E, Lasius niger; F, Lasius pallitarsis.*** **(Courtesy R. R. Snelling.)**

Its common name was suggested by the pyramid-like projection on the thorax. The petiole has 1 node. The workers travel in trails, and are fast-moving. They seek honeydew on garden plants, but are also carnivorous and predaceous.

Bicolored Pyramid Ant, *Dorymyrmex bicolor* Wheeler

This species (figure 175, *A*) is similar to *D. pyramicus,* of which it was once considered to be a variety, but it is slightly larger and has a red head and thorax. It is found in the same geographical areas as the pyramid ant and in similar locations, but is of very little importance as a pest (Eckert and Mallis, 1941).

Odorous House Ant, *Tapinoma sessile* (Say)

The odorous house ant, so named because of its unpleasant odor when crushed, occurs throughout the United States, from sea level to 10,000 ft (about 3,000 m) elevation. It is second only to the Argentine ant as a household pest. However, in areas where the Argentine ant is also found, the odorous house ant tends to become less common or to disappear entirely. Nevertheless, it still frequently invades homes or apartments in southern California. Tilden (1969) pointed out that the odorous house ant had a tendency to invade houses late in the year when the honeydew upon which it generally fed became less abundant. Also, rainfall and leaf fall, which reduce the numbers of aphids and other honeydew-excreting insects, may induce the ants to enter houses in search of food. It is most likely to invade houses during rainy weather. This ant does not possess a stinger.

Description. The odorous house ant workers (figure 175, *B*) are 2.4 to 3.25 mm long—slightly larger than the Argentine ant. The petiole is a single segment, with the node vestigial, flattened or inclined, and concealed from above by the base of the gaster. Like the Argentine ant, the odorous house ant travels in trails, and it also seeks the same kinds of food, preferring sweets. It can be distinguished from the Argentine ant by its darker color (uniform brown to black), broader abdomen, by the fact that the petiole and its 1 vestigial node are hidden by the abdomen, and by its odor when crushed. Its pungent, "rotten-coconutlike" odor is detectable even when only a single individual is crushed.

Biology. The odorous house ant nests in many types of habitats—sandy beaches, pastures, woodlands, and bogs. It can also locate in buildings, and often nests in wall voids, particularly around hot-water pipes and heaters (Tilden, 1969). Its nests may be under stones or logs, but also under loose bark, in plant cavities, insect galls, refuse piles, and bird or mammal nests. In the soil, the nests are shallow and indefinite in form (Smith, 1965). The number of ants in a colony may range

Fig. 176. Winged females of the odorous house ant, *Tapinoma sessile*.

from a few hundred to many thousands, and there may be many reproductive females. Mating usually takes place in the nest, but nuptial flights have been observed. Lone females (figure 176) have been observed to establish a colony, although Smith (1965) believed that colonies might also be formed when 1 or more fertile females left the parental colony accompanied by a number of workers. An average colony was found to have 345 eggs, 829 larvae, 533 pupae, 2,319 workers, and 1 to 6 dealated females (Wang and Brook, 1970).

Habits. As just mentioned, the workers travel in trails. They are very active, and have rapid movements. When alarmed, they dash about in an erratic manner, and the entire gaster is elevated. Odorous house ants are avid seekers of honeydew, and have been observed to transport honeydew-excreting insects such as aphids and mealybugs. They also feed on both live and dead insects. Although they will eat almost any type of household food, they prefer sweets.

Crazy Ant, *Paratrechina longicornis* (Latreille)

This introduced species, possibly of Oriental or African origin, is now almost cosmopolitan, and is frequently intercepted in plant quarantine. It is common in the Gulf Coast region, especially in Florida, but is also well established in many towns and cities throughout the United States, including California. In the colder regions of the country, it occurs mainly in apartment buildings, hotels, and greenhouses, and can be a pest all year. Its common name arises from the worker's habit of running in a haphazard manner, as if lacking a sense of direction (Smith, 1965).

Description. The workers are 2.2 to 3.0 mm long. They have slender bodies, with very long antennae and legs (figure 175, *C*). The antennae are 12-segmented, and do not have a club. The body is dark brown to black, with a "peculiar gray to violaceous luster or sheen," and has long, coarse hairs. The petiole has 1 node. The insect has no stinger. (Smith, 1965.)

Habits. Outdoors, the crazy ant nests in trash, cavities in trees, rotten wood, and in the soil under various objects in colonies of moderate to large size, but it can also nest in buildings. The workers run very rapidly, and can even jump, according to some observers. Their keen sense of smell enables them to find food quickly. They are almost omnivorous, feeding on live and dead insects, seeds, honeydew, fruits, plant exudates, and many household foods. In Puerto Rico, workers of *P. longicornis* have been reported to be predaceous on the larvae and adults of the house fly, flesh flies, and other Diptera (Pimentel, 1955). They have been observed to attack the larvae of the oriental rat flea (*Xenopsylla cheopis*) under laboratory conditions, indicating that they may be important in reducing flea infestations under natural conditions (Fox and García-Moll, 1961). On the debit side, the workers are also known to gather small seeds of such crops as lettuce and tobacco from seedbeds, and their fondness for sweets forced the proprietor of a soda fountain in Florida to discontinue business (Smith, 1965).

False Honey Ants (*Prenolepis* spp.)

False honey ants occur throughout the United States. The workers are 3 to 4 mm long, light to dark brown, very shiny in appearance, and have a somewhat triangular abdomen. They seldom invade the home. Their name probably comes from their habit of feeding on honey, honeydew, and nectar until their gasters are greatly extended, almost as much as those of the honeypot ants (*Melophorus*) of Australia and of *Myrmecocystus* of the United States Southwest. (The "repletes" of these species store honey in their gasters, and are sources of sweets for primitive people.) However, honey ants also feed on the exudates of oak galls, live and dead insects, the fluids of dead earthworms, and the young, tender tissues of plants.

Small Honey Ant, *Prenolepis imparis* (Say)

This species ranges throughout much of the United States, is common in California, and often replaces the Argentine ant where the latter has been exterminated. When these ants are found in houses, they have usually come from outdoors, but it is likely that they often nest inside, for on many occasions winged females and males have been collected inside houses from January to March (Smith, 1965). This ant lacks a stinger and seldom bites, for it is not aggressive. A striking characteristic of *Prenolepis imparis* is that it is most active outdoors at temperatures between 35 and 55° F (2 and 13° C), and has been observed to emerge from its subterranean galleries at 6° F (–14.4° C) (Wheeler, 1930).

According to Smith (1965), worker ants feed on such household foods as "sweet corn, corned beef, meats, cakes, breads, sugar, honey, syrup, watermelon, and fruits, but are especially fond of sweets." Outdoors, the ants can damage flower buds. H. M. Armitage observed *Prenolepis imparis* feeding extensively on tender, young growth and the calyxes and unopened petals of orange buds in southern California (Wheeler, 1930).

Description. *Prenolepis imparis* varies in color from pale to dark brown or black, with the head and thorax often lighter in color than the abdomen. The body is smooth and shiny. The thorax is small, slender, and almost cylindrical, and is divided into 2 parts by a strong constriction (figure 175, *D*). The abdominal petiole is composed of a single segment (1 node). The antennae are 12-segmented, do not have a club, and the scape extends beyond the posterior border of the head by half the length of the scape.

The Nest. This ant normally constructs its nest in soil, preferably moist clay or loamy soil in shady woodlands, but seldom under stones or other objects. The nest has a central opening, surrounded by a crater of characteristic earthen pellets. A single gallery leads to the nest, which may be from 18 to 51 in. (45 to 130 cm) deep and have 6 to 40 lateral chambers leading from it. There are seldom more than a few thousand ants in a colony (Smith, 1965).

Life Cycle. In winter, an old colony contains a single reproductive female, winged males and females, and many workers. Usually in March to April, the overwintering males and females make their nuptial flight, although most of the mating appears to take place on the ground. The female then discards her wings and seeks a crevice in the ground in which to oviposit and start a new colony (Smith, 1965).

Species of *Lasius*

The *Lasius* species are relatively unimportant in California. The workers are monomorphic, active, emit a strong formic acid odor, but do not have a stinger. The antennae are 12-segmented and lack a club. The eyes are well developed, and are located closer to the posterior border of the head than to the base of the mandible. If ocelli (simple eyes) are present, they are small and indistinct. The abdominal petiole has only 1 segment, and the petiolar node is narrow in profile and either vertical or not strongly inclined (figure 175, *E* and *F*).

Lasius niger (L.) and *L. pallitarsis* Provancher (= *sitkaensis*)

In California, these 2 species are usually found in evergreen forests. They have similar habits, nesting in soil or rotten logs. They forage either diurnally or nocturnally. They are strongly attracted to sweets, and tend mealybugs and aphids, especially on roots. *Lasius niger* (figure 175, *E*) is 2.6 to 4.1 mm long, and uniformly medium to blackish brown, the scapes of the antennae tending to be yellowish brown. The body is moderately hairy. This species occurs throughout most of the West. In California, it ranges from the northeastern area southward along the Sierra Nevada to the San Gabriel Mountains. *Lasius pallitarsis* (figure 175, *F*) is 2.5 to 4.5 mm long, and is similar to *L. niger*, but is hairier and its color is more variable, ranging from yellowish brown to dark brown. Whereas in *L. niger* the basal tooth of the mandible is on the same plane as the rest of the teeth, in *L. pallitarsis* the basal tooth of one or both mandibles is offset, i.e., not in line with the other teeth. *Lasius pallitarsis* ranges throughout the mountainous areas of the western United States, and also transcontinentally across the northern states and southern Canada (R. R. Snelling, correspondence).

Control of Ants

Recognition of species is important in ant control. The odorous house ant and thief ant may nest in houses, and their trails will lead to the nesting areas, which can then be treated. When they nest in large, inaccessible areas, such as attics, wall voids, and voids under cabinets, dusting with a chlordane or silica aerogel dust may be the most convenient and also the most effective way to obtain complete coverage and good control. Although the Argentine ant nests outdoors, it infests houses more frequently than any other ant species. The housewife may wish to treat for Argentine ants indoors in order to get rid of them quickly, but for effective and long-term control of this and other outdoor-nesting species that may invade the house, the main reliance should be placed on outdoor treatment.

The Argentine, odorous house, thief, jetblack harvester, field, and velvety tree ants travel in trails, and the trails help to trace them to their nests. Repeated spraying or dusting of trails can result in good ant control. Some ants are attracted to sweets only (Argentine, field, and velvety tree ants), and the thief ant is usually attracted to fats only. For a discussion of field ants, see chapter 9, under "Ants (Formicidae)."

The other species discussed in this section are attracted to both sweets and fats, except for the harvester ants (*Veromessor*) which eat mainly seeds. A knowledge of the food habits of ants is necessary in order to decide on what type of bait to use when control is to be accomplished by the use of poison baits.

For control of ants within a house, sprays of 2 to 3% chlordane applied with a sprayer or paintbrush to baseboards and cracks where ants are seen, or a 5% chlordane dust applied behind baseboards or in out-of-sight areas, are effective. It may be most convenient to apply an aerosol containing chlordane, holding the canister close to the sprayed surface so as to wet it visibly. If ant trails are found outside the house, a spray or dust may be applied next to the foundation in the general area of infestation. Aphid-tending species of ants, like the Argentine ant, may be controlled in yards and gardens by spraying or dusting the ground under the infested shrubs or trees. The shrubs or trees themselves should not be sprayed, for they may be injured and natural enemies of the pests may be destroyed. This should not only help to control the honeydew-producing insect pests that are favored by the presence of ants, but should also decrease the ant population in the general area of the home and thereby decrease the chance of infestation in the house.

A "hose-attachment" sprayer (figure 11, chapter 3) is a great convenience, and is a time-saver if large areas need to be sprayed. Best results can be obtained with power sprayers, such as those used by commercial pest control operators, to spray the entire area around a house where nests may be located as well as tree trunks and subfloor areas. Some operators control ants by periodically spraying the entire foundation and the adjacent soil.

The homeowner can effectively use baits, such as 0.125% Kepone® or 0.075% mirex, placing them in ant-nest openings, around foundations, and in other areas frequented by ants. It is important to use the proper percentage of toxic ingredient when placing ant baits. For example, dieldrin at 0.05% concentration in syrup gave complete control of the odorous house ant, but at 0.5 and 1%, the baits were repellent and the ants would not feed on them. Dieldrin happens to have amazing toxicity against the odorous house ant—it required a 67 times greater concentration of chlordane to obtain an equal mortality level (Wang and Brook, 1970). However, dieldrin is no longer registered for such household uses.

Poisoned syrups are still commercially available for ant control, and are effective against species attracted to sweets, such as the Argentine ant. The receptacles containing the syrup should be placed at intervals of 10 to 20 ft (3 to 6 m) around the foundation of the house. The syrups contain sodium arsenite, and it is not advisable to use them if children or pets are in the vicinity.

For the control of Pharaoh ants in hospitals, where they are considered to be potential carriers of disease germs, Eichler and Kleinsorge (1973) found baiting with arsenicals to be the only fully effective procedure. One successful bait had the following ingredients: ground beef, 1,000 g; arsenic, 200 g; and honey, 80 g; mixed by hand, using rubber gloves. Ground liver was the most attractive bait, and did not require the addition of honey. However, it dried out too rapidly in warm rooms, and was therefore more effective when mixed with beef. The bait was placed in locked wooden bait stations. Ground meat was left 3 days before poison bait was used, to accustom the ants to the new food station. Baits were renewed as often as required to avoid a dry bait surface. If an ant trail was disturbed during the baiting period, the insects fled to their nest, and it took several days for them to re-establish a trail to the bait station.

Chapter 7 · *Pests of Stored Food Products*

THESE PESTS cause losses to man by destroying large quantities of grain, corn, legumes (beans, lentils, peas), milled cereal products, flour, bran, macaroni and other paste products, pet foods, dried fruits, dried vegetables, cheese, nuts, candy, and other food materials. They contaminate, and thus eliminate for human consumption, far more food than they eat, and either cause directly or transmit diseases to man and his domestic animals (Scott, 1962–63). Although these marauders attain their greatest importance as pests of stored grain, damaging about 10% of the world's production, and as high as 30 to 40% in some tropical and subtropical countries (Munro, 1966), they are also important in the household as "pantry pests" that find their way into packaged cereals and other foodstuffs. Certain pantry pests may originate in crops when they are still growing in the field, while others may infest food products at any time until they are ready for use.

Pantry pests may fly into buildings from the outside, come from secluded harborages indoors, or migrate into uninfested items from infested ones or from refuse that has collected in cracks and crevices. They can also be carried into a building on materials other than food products. For example, drugstore and cigarette beetles, carpet beetles, spider beetles, psocids, and mites may gain entry via old furniture, rugs, drapes, bedding, or in or on almost any other product of plant or animal origin (Linsley and Michelbacher, 1943).

An idea of the species that can become pantry pests is obtained in surveys of grain elevators and feed mills. In one such survey made in Ohio, 34 species of beetles and moths were found. The granary weevil and the cadelle were the most important of the insects attacking sound, whole grain. The most important secondary invaders were the Mediterranean flour moth, flat grain beetle, sawtoothed grain beetle, mealworms, and 2 species of flour beetles (Triplehorn, 1965). In the southern states, the rice weevil is more important than the granary weevil or the cadelle, and world-wide it is considered to be "undoubtedly the most destructive insect pest of stored grain" (Cotton, 1963). However, as "pantry pests," the insects attacking sound, whole grain are relatively secondary in destructiveness. More harmful in this category are the species that attack broken grain or flour. Many kinds of insects have become adapted to feeding on stored flour, bran, and other processed grain products. In ancient times, only whole grain was stored, and there are many instances recorded of its having remained for remarkably long periods free of insect infestation (Hartnack, 1943).

Another factor affecting the infestation of stored grain or grain products is humidity, as it stimulates the growth of fungi. Sinha (1971) stated that insects that normally infested stored grain rarely occurred in the absence of fungi, and that several species were able to feed and survive on fungal diets.

Beginning in 1953, a summary was made of the results of extensive surveys by county, state, and federal inspectors of insects and mites associated with foods and seeds in storage throughout California. The specimens were identified by taxonomists of the State Bureau of Entomology (Strong and Okumura, 1958). Listed were 102 species of beetles, 27 moths, 1 psocid, 1 fly (cheese skipper), 3 Hymenoptera (the clover seed chalcid and 2 ants), 2 silverfish, 3 mites, and 9 species of natural enemies (parasites and predators) of stored-food pests. The beetles and moths comprised 87% of the arthropods found in the surveys. Beetles feed on stored products as both adults and larvae, but moths feed on them only as larvae. The great majority of pests of stored food products are beetles. A key has been prepared to help identify the species that are commonly encountered (Pratt and Scott, 1962).

Strong (1970) used a survey method that probably gave as good an idea of the relative abundance of pests that could invade the home pantry as it would be practicable to obtain on a large scale. In 30 localities, well distributed throughout California, he placed sheltered food packets in such places as carports, garages, equipment sheds, inactive dairy barns and poultry houses, old livestock barns, and other structures not being used to store dry food products, seeds, or feeds. The food packets consisted of 0.5 lb (0.23 kg) of a mixture of equal parts by weight of poultry laying mash, rolled barley, wheat, and corn, tied up in cheesecloth and covered with hardware cloth to keep out rodents. This is a mixture that would be expected to be a good food medium for insects that feed and develop on cracked grains or flour as well as those that feed and develop on whole grains.

Table 6 lists the species, according to family, of the insects found infesting the food mixtures after they had been exposed in the 30 localities for 4 to 5 months during spring and summer, and also the number of areas in which each species was found. The list provides a good idea of the species that are potentially pantry pests in California, but not necessarily their relative importance, for it includes species that attack whole grains, which are seldom stored in home pantries. Nevertheless, the species one generally associates with the home pantry (cigarette beetle, drugstore beetle, sawtoothed grain beetle, confused flour beetle, red flour beetle, dermestids, ptinids, and the Mediterranean flour, Indian meal, almond, and tobacco moths) were well represented. The list includes only beetles and moths, but with the exception of the psocids (not included because they tended to escape in large numbers from the food mixtures), other orders of insects were of no importance. The relative importance of some of the species might have been changed if dried

TABLE 6
Stored-Product Beetles and Moths Found in Sample Populations from Various Areas of California*
(Numerals indicate the number of areas)

Species	Areas
COLEOPTERA	
Anobiidae	
Lasioderma serricorne (F.), cigarette beetle	15
Stegobium paniceum (L.), drugstore beetle	10
Bostrichidae	
Rhyzopertha dominica (F.), lesser grain borer	21
Cucujidae	
Ahasverus advena (Waltl), foreign grain beetle	1
Cryptolestes spp. = *C. pusillus* (Schönherr), flat grain beetle and/or *C. ferrugineus* (Stephens), rusty grain beetle	22
Oryzaephilus mercator (Fauvel), merchant grain beetle	1
O. surinamensis (L.), sawtoothed grain beetle	30
Curculionidae	
Sitophilus granarius (L.), granary weevil	12
Sitophilus spp. = *S. oryzae* (L.), rice weevil and/or *zeamays* Motschulsky, maize weevil	12
Dermestidae	
Anthrenus spp.	29
Attagenus spp.	29
Megatoma spp. probably *prolixa* Beal	1
Novelsis spp. probably *varicolor* Jayne	5
Perimegatoma spp.	23
Trogoderma glabrum (Herbst)	11
T. grassmani Beal	11
T. inclusum LeConte	10
T. ornatum (Say)	12
T. parabile Beal	17
T. simplex Jayne	29
T. sternale Jayne	27
Lyctidae	
Trogoxylon aequale (Wollston)	3
Ostomidae	
Tenebroides mauritanicus (L.), cadelle	2
Ptinidae	
Ptinus californicus Pic	10
P. clavipes Panzer, brown spider beetle	20
P. cognatus Fall	12
P. extendus Smith	2
P. ocellus Brown, Australian spider beetle	2
P. verticalis LeConte	10
Niptus ventriculus LeConte	1
Trigonogenius globulum Solier, globular spider beetle	5
Tenebrionidae	
Alphitobius diaperinus (Panzer), lesser mealworm	6
A. laevigatus (F.), black fungus beetle	2
Gnathocerus cornutus (F.), broadhorned flour beetle	1
Latheticus oryzae Waterhouse, longheaded flour beetle	15
Tenebrio obscurus F., dark mealworm	8
Tribolium audax Halstead, American black flour beetle	2
T. brevicornis (LeConte)	9
T. castaneum (Herbst), red flour beetle	27
T. confusum Jacquelin duVal, confused flour beetle	16
T. destructor Uyttenboogaart	2
LEPIDOPTERA	
Galleriidae	
Aphomia gularis (Zeller)	2
Gelechiidae	
Sitotroga cerealella (Olivier), Angoumois grain moth	5
Geometridae	
Pyralis farinalis L., meal moth	1
Oecophoridae	
Endrosis sarcitrella (L.), whiteshouldered house moth	7
Hofmannophila pseudospretella (Stainton), brown house moth	5
Phycitidae	
Anagasta kuehniella (Zeller), Mediterranean flour moth	16
Cadra cautella (Walker), almond moth	6
C. figulilella (Gregson), raisin moth	2
Ephestia elutella (Hübner), tobacco moth	12
Paramyelois transitella (Walker), navel orangeworm	8
Plodia interpunctella (Hübner), Indian meal moth	13
Sosipatra rileyella (Ragonot)	4
Vitula edmandsae serratilineella Ragonot, driedfruit moth	5

*Adapted from Strong (1970).

fruits and nuts had been present along with the grains. Linsley (1942) pointed out that the majority of stored-product pests appeared to have evolved from insects infesting 1 or more of the following natural sources:

(1) Seed-infesting species which infest similar material in nature.

(2) Scavengers feeding generally on dead plant or animal products.

(3) Scavengers or semipredators living primarily under bark.

(4) Scavengers or predators inhabiting the nests of bees, wasps, or other aculeate Hymenoptera.

In the present chapter, the pests are divided into groups of similar habits with regard to the ways they affect stored food products; that is, whether they infest whole grain, broken grain, legumes, meats and cheeses, stored foods in general, or dried fruits and nuts. (Acarid mites are discussed separately.) This is followed by a section on the problems connected with insect contamination in milled cereal products. At the end of the chapter, certain control measures are discussed that may be employed while the food is still in the field, in storage, in the food-processing plant, and finally, in the home, where the offending insects and mites are generally called "pantry pests."

INSECTS INFESTING WHOLE GRAIN

Certain beetles and moths that primarily infest grain (wheat, oats, barley, rice, corn [maize], millet, and sorghum) have been placed in 2 groups, based on whether they feed on the whole grain or can feed on the grain only after it has been broken or ground into flour. Pests of whole grains will be discussed first. However, it should be borne in mind that stored whole grain ordinarily has enough broken kernels and grain dust to enable the second group to become established.

Granary Weevil, *Sitophilus granarius* (L.) (Curculionidae)

This is one of the most important and oldest known pests of stored grain. It is believed to have spread from its original home in the eastern Mediterranean area to the cooler regions of the entire world. In the United States, it is more frequently found in the northern states than in the South. It sometimes causes almost complete destruction of grain in grain elevators, farmers' bins, or ships. This species has become domesticated, and has lost its functional wings. It has forsaken both wild and cultivated field grains to infest only those in storage, and depends on man for dissemination. As a "pantry pest" in the home, it is not so important as some other beetles and moth larvae that are primarily pests of grains after they have been broken or milled. The adults have a tendency to wander about, and may first be seen by the housewife somewhat far from the infested materials.

Howe (1965) found that the granary weevil could infest acorns with cracked or removed shells, and preferred them to many common cereals. He suggested that the acorn might have been a host before the advent of primitive agriculture, especially when gathered by squirrels or man and kept in storage at moderate humidity. Another species of *Sitophilus* of larger size is a serious pest of acorns in India (Mathur, 1954). *Sitophilus granarius* may also have been a larger insect before adapting itself to the infestation of

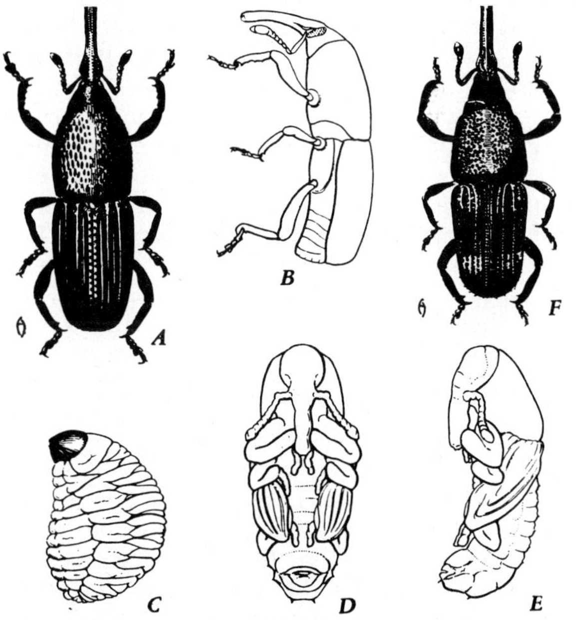

Fig. 177. Granary weevil, *Sitophilus granarius*. *A*, adult; *B*, lateral view of adult; *C*, larva; *D*, ventral view of pupa; *E*, lateral view of pupa; *F*, rice weevil, *Sitophilus oryzae* (adult). (From Linsley and Michelbacher, 1943.)

stored grains. Likewise, the small strains of *S. oryzae* (described next) are believed to be adaptations toward a life in stored products (Kiritani, 1965).

Description. This polished chestnut-brown or blackish beetle (figure 177) is 3 to 4 mm long, and is similar in appearance to the closely related rice weevil, but lacks the pale markings of its elytra, and its vestigial wings are useless. The pits on the thorax are elongate rather than round or irregularly shaped, as they are in the rice weevil. The small, legless larvae of the granary and rice weevils can be distinguished from those of other beetles infesting grain and other stored products by the ventral surface being straight (figure 177), while the dorsal surface is rounded. The larvae of the other beetles are more "wormlike" (figures 178 and 179). The larvae of any beetles can be distinguished from those of moths by not having the short, fleshy, leglike abdominal appendages known as "prolegs" (figures 180 and 185), that aid in locomotion and in clinging to surfaces.

Life Cycle. The female bores a hole in a grain kernel with the stout mandibles at the end of her long, slender snout, deposits an egg in the hole, and seals it with a gelatinous material. She may lay more than 200 eggs under favorable conditions. The eggs hatch in a few days, and the larvae feed on the interiors of the kernels, hollowing them out. There are 4 larval instars, and the developmental period is from 3 to 5 weeks. Pupation within the kernel requires 5 to 16 days. The entire life cycle may vary from as little as a month to as long as 5 months, depending on temperature. There are commonly 4 generations per year. The granary weevil is long-lived, surviving for 7 to 8 months as an adult. The female lays very few eggs at temperatures below 60° F (16° C), but can survive for 2 months or more at 35° F (about 2° C). A temperature of 120° F (49° C) for 1 hour or 130° F (54° C) for 0.5 hour is fatal.

Rice Weevil, *Sitophilus oryzae* (L.)

This insect is primarily a pest in warm countries, and in the United States it is most important in the South. The larva feeds inside a kernel of grain in which both larva and pupa must complete their development. Both larvae and adults eat similar food, but the latter can crawl or fly about and feed on various products. They have been reported to occur on beans, nuts, cereals, wheat products, and grapes, and have been observed sucking the juice from apples and pears, gradually forming cavities in which they concealed themselves (Hinds and Turner, 1911). Unlike the granary weevil, the adults have well-developed wings, and fly from stored grain to fields of corn, wheat, or rice, and start infestations that continue in storage. In the northern states, infestation begins in new grain when it is stored in farm bins, warehouses, or grain elevators.

Description. The adult (plate IV, 5; figure 177, *F*) is 2.5 to 3.5 mm long, somewhat smaller than the granary weevil. It is reddish brown, with 4 faint reddish or yellowish spots on the elytra, round or irregularly shaped punctures on the pronotum, and with the head prolonged into a long, slender snout, bearing the mandibles at its tip.

Life Cycle. The female lives 4 to 5 months, and lays a total of 300 to 400 eggs. Like the granary weevil, she bores a small hole in the kernel with her mandibles, lays an egg in it, and covers the hole with a gelatinous fluid. The small, white, legless larva, with a brownish-black head, has 4 instars. In warm weather, the combined stages from egg to pupa may take as few as 26 days. After emergence, the adult stays in the kernel for 3 or 4 days, hardening and maturing, then bores its way out. This weevil is basically a grain pest, but has been reared from solidified farinaceous products such as macaroni or caked flour (Cotton, 1963). The most favorable temperatures for development are 80 to 86° F (27 to 30° C). Below 45° F (about 7° C), the beetle is dormant. It usually dies after 2 weeks of freezing temperatures, or after an hour at 120° F (49° C).

Lesser Grain Borer, *Rhyzopertha dominica* (F.) (Bostrichidae)

This is one of the smallest of the grain-infesting beetles, but one of the most important. Originally native to the tropics, it has spread through commerce to all parts of the world. It is most prominent in the United States, southern Canada, Argentina, India, and Australia. In the United States, it is particularly widespread in the Gulf states. Both adults and larvae feed within the interiors of nearly all grains, including rice, and the kernels are reduced to mere shells.

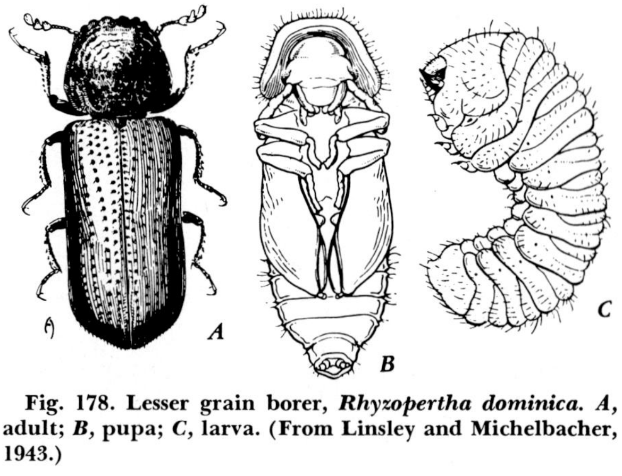

Fig. 178. Lesser grain borer, *Rhyzopertha dominica*. *A*, adult; *B*, pupa; *C*, larva. (From Linsley and Michelbacher, 1943.)

Description. The adult (figures 178 and 179) is about 3 mm long, polished dark brown or black,

Fig. 179. Adult of lesser grain borer, *Rhyzopertha dominica* (*top*), and adult and larva of drugstore beetle, *Stegobium paniceum*. (See the latter under "General Feeders.")

and has a somewhat roughened surface. In common with other bostrichids, which are principally woodborers, *Rhyzopertha dominica* is almost cylindrical, and the head is vertically deflexed under the thorax so that it cannot be seen from above. Likewise, this insect, though small, has powerful jaws with which it can bore directly into wood (USDA, 1968). Wood may have been its original food. It can eat its way into wooden and paper boxes, and may destroy book bindings (Hoffman, 1933; Potter, 1935).

Life Cycle. The female lays 300 to 500 eggs, singly or in clusters, in the loose grain. They hatch in a few days. The larvae molt 2 to 4 times. They may feed on the flour produced by the boring of the adults, or may bore directly into kernels that have been slightly damaged. They complete their growth within the grain, transform to white pupae, and the adults cut their way out (USDA, 1968). The life cycle takes only a month or two, depending on the temperature.

Other Bostrichids

Another bostrichid beetle that is believed to have acquired its grain-feeding habits relatively recently is the larger grain borer, *Prostephanus truncatus* (Horn). It is similar in appearance and habits to the lesser grain borer, but is a little larger—about 4 mm long—and has a smooth, polished surface. It is a tropical insect, not yet widely distributed in the United States, but can occasionally be found infesting corn in the southern states. It has also been reported from California, Texas, and the District of Columbia (USDA, 1968).

The bamboo borer, *Dinoderus minutus* (F.) (figure 113, chapter 5) is another bostrichid, and resembles *Prostephanus truncatus* in appearance, being cylindrical, brown, and 3 to 4 mm long. It infests bamboo, reducing the inner core to a fine powder. It has been recorded from many other hosts, including stored grain, but this is believed to be probably the result of misidentification (G. T. Okumura, personal communication).

Cadelle, *Tenebroides mauritanicus* (L.) (Ostomidae)

The cadelle is believed by some entomologists to be a native of America, for all species of the family Ostomidae are found in the New World. It is one of the best known of the stored-grain pests, in part because it is also a common "pantry pest" in the house, where adults and larvae feed on cereals, breakfast foods, potatoes, shelled and unshelled nuts, or fruit, and prefer to lay their eggs under the flaps of cartons. They can gnaw through sacks and paper packages. Unlike the granary and rice weevils, they feed not only on whole grains, but also on flour and a variety of other products, and are therefore much more important as pantry pests.

The cadelle may have been originally a predator under loose bark, as other species of the family are today. It does attack the larvae of other grain-infesting insects. Another peculiarity of this insect is its habit of burrowing into the woodwork of grain bins or other wooden structures, sometimes causing them to collapse. It may remain there for long periods in large numbers, only to emerge and infest the next load of grain. Other grain pests may also hide in its burrows.

Description. The shiny black adult (plate IV, 7) is elongate, oblong, and flattened, and is one of the largest of the grain-infesting beetles, being about 8 mm long. The prothorax is distinctly separated from the rest of the body by a loose, prominent joint.

Life Cycle. The females lay about 1,000 eggs loosely in flour, grain, or other foodstuff. The fleshy, white to grayish-white larva has a black head and a black plate with two horny black projections at the tip of the abdomen. The life cycle may be as short as 70 days, but it may be much longer under conditions unfavorable to the insect's development. The females generally live for a year, and have been kept alive for more than 3 years.

Angoumois Grain Moth, *Sitotroga cerealella* (Olivier) (Gelechiidae)

The Angoumois grain moth is second only to the rice and granary weevils in its importance as a pest of stored grain in the United States. It attacks grains maturing in the field as well as in storage. Infested grain in storage has a sickening taste and smell that make it unpalatable. Only whole grain is attacked, so other grain products are safe (Linsley and Michelbacher, 1943).

Description. The adults (plate V, 2 and figure 180) are small, buff to grayish or yellowish-brown moths, with a wingspread of 13 to 17 mm. The size of the adult depends on how much food the larva has consumed. The hindwings narrow down to a point apically, and are heavily margined with long hairs.

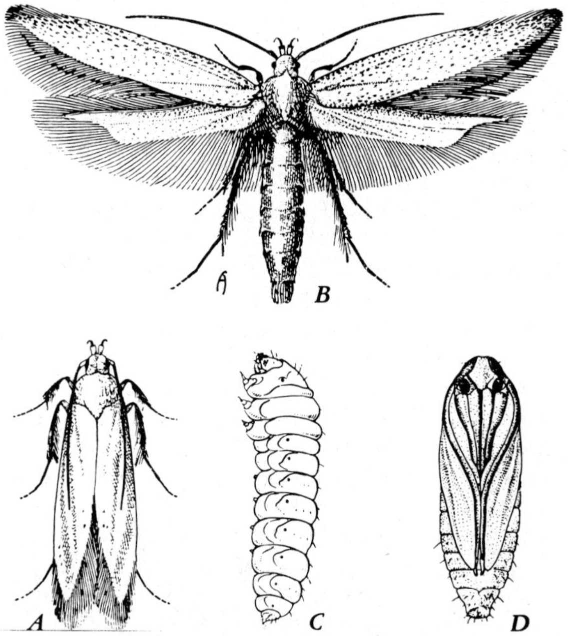

Fig. 180. Angoumois grain moth, *Sitotroga cerealella. A*, adult in normal resting position; *B*, adult with wings spread; *C*, larva; *D*, pupa. (From Linsley and Michelbacher, 1943.)

Life Cycle. Each female lays an average of about 40 tiny white eggs on or near the grain, and in 4 to 8 days these hatch into minute, white larvae that bore into the kernels. The larvae are pale yellowish when mature, with a yellowish-brown head, and are about 5 mm long when full grown. They have poorly developed abdominal prolegs (figure 180, *C*). The larva passes through 3 instars in about 3 weeks, but may hibernate before changing to a reddish-brown pupa in the spring. Pupation takes place in a silken cocoon in the feeding cavity in the kernel. There are 2 to 4 generations a year.

European Grain Moth, *Nemapogon granella* (L.) (Tineidae)

This moth is creamy white, thickly mottled with brown. It is about the size of the Angoumois grain moth, but the mottled appearance distinguishes it from that species. The forewings are mottled with brown and white and the hindwings are gray, narrowed down at the apex like those of the Angoumois grain moth, and with long hairs along the edges. The larvae feed on all kinds of grain, in the field and in storage, and web the kernels together. This species is distributed throughout the northern states, but not abundantly, and is not so destructive as the Angoumois grain moth (USDA, 1968).

Pink Scavenger Caterpillar, *Sathrobrota rileyi* (Walsingham) (= *Pyroderces*) (Cosmopterygidae)

This highly adaptable insect had for many years been known as a scavenger commonly found in cotton in North and South America, the West Indies, and Hawaii. It was necessary to be able to distinguish the small, reddish larvae from those of the pink bollworm, *Pectinophora gossypiella* (Saunders), and this led to the first complete description (Busck, 1917).

Description. The moth (figure 181) has a wingspread of 9 to 12 mm. The forewings are chestnut brown, with whitish or straw-colored streaks edged by irregular black scales. The hindwings are pale gray, and are edged with long fringes. The head is light chestnut brown, the abdomen reddish brown, and the legs light reddish. The full-grown larva (figure 181) is 7 to 8 mm long, the abdomen is deep wine red, the thoracic shield is broad and dark brown, and the head is light brown. The pupa is light yellowish brown, and has a smooth surface.

Economic Importance. This insect has been causing considerable injury to corn in the southern states, beginning in the field and continuing after the corn is placed in storage. The larvae feed on the seeds, husks, and cobs, producing a large amount of frass that is loosely webbed together and fills the interstices between kernels or the cavities of kernels that have been eaten out. Although capable of causing serious damage to corn

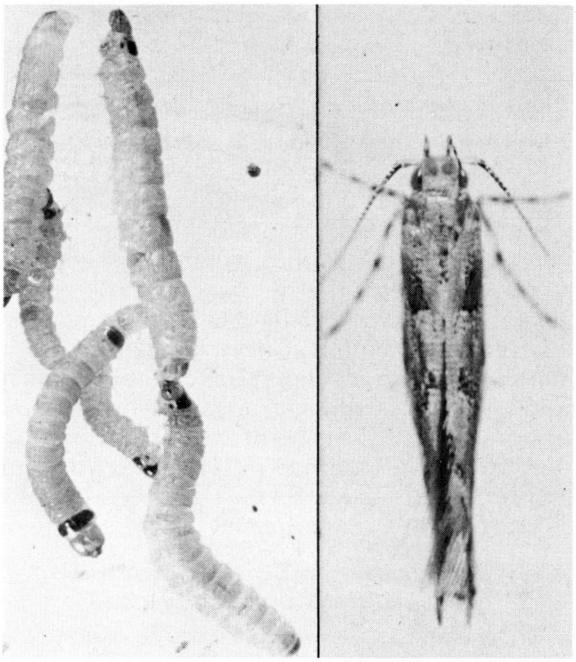

Fig. 181. Pink scavenger caterpillar, *Sathrobrota rileyi*.

in the field and in cribs on southern farms, the insect is seldom an important pest of commercial shipments. On orange trees, it is primarily a scavenger, but when it is abundant, some larvae may feed on oranges, and have sometimes caused severe damage in California and Florida (Ebeling, 1959).

Psocids (Psocoptera)

The psocids are discussed more fully in chapter 10 as one of the groups of insects that are attracted to the household by excessively damp conditions, and particularly by the development of molds. The greatest economic losses from psocid infestations in stored products result from the contamination of processed foods, especially when they are stored for long periods under domestic conditions (Champ and Smithers, 1965). Moderately high humidity is required for the actual survival of psocids, but the development of molds is not necessary for it. Considerable damage can be done to stored grain, for example, by feeding injury to the kernels. A psocid can chew a small hole through the seed coat and completely devour the germ. Heavy populations can develop in stored wheat, particularly if it has 12.5% or more of moisture content and the ambient relative humidity is 65% or more (Watt, 1965).

Finlayson (1933) stored wheat in trays to propagate eggs of the Angoumois grain moth, *Sitotroga cerealella*. Temperature was between 75 and 85° F (24 and 29° C), and the relative humidity varied from 75 to 90%. He observed that within 5 months from the day the first psocids (*Liposcelis divinatorius*) were seen, the 32 boxes in which the trays were stored were completely covered with the psocids; he estimated that there were 200 million of these insects present. After about a month of infestation, the *Sitotroga* population began to decline. The psocids seemed to show no preference between feeding on crushed wheat or *Sitotroga* eggs.

The humidity of the air in a storage facility and the water content of the stored product that are required for heavy infestations of psocids are about the same as for heavy infestations of tyroglyphid (acarid) mites (Solomon, 1946, 1962; Knülle, 1963*a, b*; Cunnington, 1965).

INSECTS INFESTING BROKEN GRAIN

This group is made up of insects that are general feeders on stored food, attacking not only broken grains, flour, and cereals, but such products as dried meat, vegetables, fruit, milk, sugar, candy, tobacco, or drugs. Certain insects will be briefly mentioned that are minor pests of stored foods but are also pests in other ways, and are considered more fully in other chapters. The acarid mites can be important pests of some of these stored-food products, but they are considered separately near the end of this chapter.

Confused Flour Beetle, *Tribolium confusum* Jacquelin duVal, and Red Flour Beetle, *T. castaneum* (Herbst) (Tenebrionidae)

These 2 flour beetles, along with the saw-toothed grain beetles and Indian meal moths, are considered to be the most important pests of stored foods in grocery stores and in the home. *Tribolium* spp. are believed to have been semi-predators originally, feeding on living and dead materials under bark. *Tribolium indicum* Blair continues to live only under bark, and *T. audax* Halstead is usually found under bark, but is occasionally taken in seeds and meal. *Tribolium destructor* (Uytenboogaart) has been recorded only from seeds and other stored products (Good,

1936). *Tribolium confusum* and *T. castaneum* are now pests of stored grains and grain products, peas, beans, shelled nuts, dried fruits, spices, chocolate, drugs, snuff, cayenne pepper, herbarium and museum specimens, and many other items, but they are sometimes found under bark also. They are unable to feed on undamaged grain, but are serious pests in flour mills. In the United States, the confused flour beetle is the more abundant of the 2 species in the northern states, whereas in the South, the red flour beetle is more abundant. They are about equal in numbers in the intermediate regions.

Description. *Tribolium confusum* and *T. castaneum* (plate IV, 6) are small, reddish-brown beetles, about 3.5 mm long. They are similar in size and appearance as well as in life histories and habits. However, they can be distinguished with the aid of a hand lens. The antenna of *T. confusum* is gradually enlarged toward the tip, ending in a "club" of 4 segments, whereas that of *T. castaneum* is abruptly clublike, with the club consisting of 3 segments. *Tribolium confusum* has the sides of the thorax almost straight, while those of *T. castaneum* are curved.

Life Cycle. Each female deposits 400 to 500 clear-white, sticky eggs on or among food particles, in cracks, or through the meshes of sacks containing food, such as cereal products. She lays only 2 or 3 per day, but she is long-lived, and may survive as long as 2 years. The tiny eggs can be removed from flour by sifting it through bolting cloth. They hatch in 5 to 12 days into small, brownish-white larvae, which go through 5 to 12 instars, and reach maturity in as few as 30 days under optimum conditions, but may require up to 4 months. The full-grown larvae are 4 to 5 mm long, slender, cylindrical, wiry in appearance, and white, tinged yellowish. They can be distinguished from the larvae of other species that are somewhat similar in appearance by the prominent, 2-pointed termination of the last body segment.

As the larvae mature, they come to the surface of the food medium to transform into naked white pupae. In heated storehouses and mills, there are 4 or 5 generations annually. Good (1936) stated that the life cycle of *T. confusum* was "somewhat longer" than that of *T. castaneum,* but that the optimum temperature appeared to be the same—30° C (86° F) for both species. The author has observed for years, when rearing the 2 species at ordinary room temperatures in a food medium of 47.5% white flour, 47.5% cornmeal, and 5% brewer's yeast, that the yield of *T. castaneum* was always several times greater than that of *T. confusum* in a given period. In an experiment that started with 100 adults of *T. castaneum* in a 1-gal (4-l) rearing jar and 100 adults of *T. confusum* in another jar, after 32 days there were 2,135 live *T. castaneum* (adults and larvae) compared with 112 live *T. confusum* (adults only).

Flour Beetles as Laboratory Test Insects

Neither of the foregoing beetles can climb up the vertical surfaces of glass containers. Besides this, *Tribolium confusum* adults cannot fly, their wings being nonfunctional, while the *T. castaneum* adults can fly, but seldom do. These features, plus the fact that they can be easily reared in large numbers, make the 2 flour beetles popular test insects in entomological laboratories. If a blotter is placed in the rearing jar, particularly after the food medium is momentarily agitated, large numbers of the beetles will crawl onto it, and can then be brushed into a beaker to await transfer to insecticide residues and other media used in laboratory experiments.

Yellow Mealworm, *Tenebrio molitor* L., and Dark Mealworm, *T. obscurus* F.

These 2 species are similar in appearance, habits, and life histories. Their common names are based on the colors of the larvae infesting stored products. They are nocturnal, and seek moist, dark, undisturbed places. They may be found "in accumulations of grain in neglected corners of mills, under bags of feed in warehouses and feed stores, in grain bins where some of the grain has become damp, or in the litter of chickenhouses and birdhouses where feathers and refuse grain are mixed with excrement" (Cotton, 1963).

Mealworms are of only moderate importance as pantry pests. The larvae are probably best known as fish bait or as food for fish, amphibians, reptiles, birds, and small mammals kept as household pets or in zoos. They are reared commercially in enormous quantities for these purposes. They are also commonly reared by entomologists as test insects for evaluation of insecticides or for other experimental purposes. As a food medium, Cotton (1963) suggested wheat bran, to which would be added a little graham flour and commercial meal scrap, with an occasional addition of raw, fresh vegetables, such as carrots, potatoes, or lettuce. Commercial producers of mealworms utilize joints of the prickly-pear cactus instead of fresh vegetables.

Biology. The adult *Tenebrio molitor* is shiny dark brown to black, while *T. obscurus* is pitch-black (plate IV, 6). Both are more than 12 mm long. The female lays an average of 400 to 500 beanshaped, sticky, white eggs, to which particles of the food medium adhere. The eggs are laid singly or in clusters in the food material. In about 2 weeks, they hatch into slender, white larvae, which soon acquire the characteristic color of their species: bright yellow for the yellow mealworm and dark brown for the dark mealworm. The larvae have a hard, shiny cuticle that gives them a resemblance to the larvae of elaterid beetles (wireworms). After 14 or 15 molts, the larvae pupate near the surface of their food medium. The adults normally emerge in the spring and early summer, and live for 2 to 3 months, and there is 1 full generation per year (Cotton and St. George, 1929; Cotton, 1963).

Mealworms and Canthariasis. In cases of gastrointestinal canthariasis (infestation by beetle larvae), mealworms are the ones most often implicated. Eggs or larvae are ingested in cereals and breakfast foods, and in either case live larvae may eventually be passed in the feces (Palmer, 1946).

Related Species

The lesser mealworm, *Alphitobius diaperinus* (Panzer), is a black or very dark reddish-brown beetle, 5 to 6 mm long (figure 182). The larva is yellowish brown, and resembles the young larvae of the true mealworms. This species may become a pantry pest in damp situations. It feeds on such materials as damp and moldy grain, milled products, and spoiled foods (Cotton and Good, 1937).

Fig. 182. Lesser mealworm, *Alphitobius diaperinus*.

The black fungus beetle, *Alphitobius laevigatus* (F.) resembles *A. diaperinus* in form and habits. It is shiny black above, reddish brown beneath, and the antennae and legs are brownish. The surface of the thorax is coarsely and profusely punctate, while that of *A. diaperinus* is finely and sparsely punctate. It is not so abundant as the lesser mealworm.

Other Tenebrionid Beetles

There are some related tenebrionids of minor importance. The black flour beetle, *Tribolium audax* Halstead, has long been known in the United States and Canada under the name *T. madens* (Charpentier), which is a European species not known to occur in North America. It is 2.8 to 4.5 mm long, and very dark brown. (The true *T. madens* is 3.8 to 5.1 mm long, among other characters. It is also a flour beetle, found occasionally in flour, meal, seeds, and grain, but usually it is found under the bark of trees.) *Tribolium audax* is most abundant in the Rocky Mountain areas of both the United States and Canada, but is also widely distributed elsewhere. It has occasionally been recorded from stored grain, meal, and flour, but is more often found outdoors under pine bark, in leaf mold, and inside beehives; the latter are common habitats for species of this genus. In Utah, large numbers were collected from the cells of the wild bee *Megachile rotundata* (F.), a highly successful pollinator of alfalfa (Halstead, 1969).

The false black flour beetle, *T. destructor* (Uytenboogaart), resembles *T. confusum*, but is slightly larger, and is black. In California, it is found in flour, grains, and mixed feeds (Strong and Okumura, 1958).

The longheaded flour beetle, *Latheticus oryzae* Waterhouse, is pale yellowish brown, slender, flattened, and less than 3 mm long. It resembles *T. confusum* in size and shape, but is lighter in color. It is widespread in the southern and midwestern states in rice and flour mills, causing the same type of damage as the confused flour beetle.

The broadhorned flour beetle, *Gnathocerus cornutus* (F.), derives its name from a pair of broad, stout horns on the mandibles. It is a stout, reddish-brown beetle, about 4 mm long. It is common in most of the United States except in the Great Plains area, where it is comparatively rare. It feeds principally on flour and meal. The slenderhorned flour beetle, *Gnathocerus maxillosus* (F.), resembles *G. cornutus*, but is a little smaller, and the horns on the mandibles are slender and incurved. It is less common than *G. cornutus*, and occurs principally in the southern states (USDA, 1968).

Flat Grain Beetle, *Cryptolestes pusillus* (Schönherr) (Cucujidae)

This beetle is widely distributed, and is one of the most common pests of stored grain, but it cannot attack sound, uninjured kernels. The larvae also feed on dead insects, and often subsist as scavengers.

Description. The small, flattened, reddish-brown adult (plate IV, 5; figure 183) is only about 2 mm long. The male's antennae are not quite as long as its body, and those of the female are about half as long. The elytra have 5 parallel ridges. The beetles can both jump and fly.

Life Cycle. The female oviposits in crevices in the grain, or in loose, farinaceous material. The larva is slender and pale, with a black head, and has a pair of slender, black, spinelike processes at its posterior end (Linsley and Michelbacher, 1943). It is particularly fond of the germ of the wheat, and many kernels are left uninjured except for the removal of the germ. The full-grown larva forms a cocoon of a gelatinous substance to which food particles adhere. Development from egg to adult may require only 5 weeks

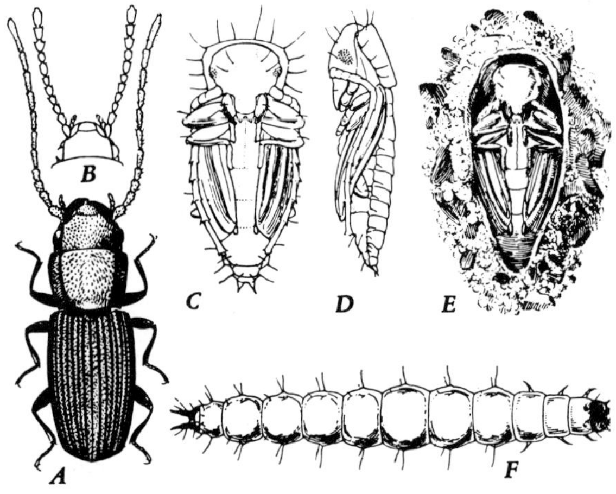

Fig. 183. Flat grain beetle, *Cryptolestes pusillus*. *A*, adult male; *B*, head and antennae of female; *C*, pupa, ventral view; *D*, pupa, lateral view; *E*, pupa case cut open, showing pupa within; *F*, larva. (From Linsley and Michelbacher, 1943.)

under favorable conditions, although the average in summer is about 9 weeks. The adult female may live a year (Payne, 1946; Cotton, 1963).

Other Cucujids

Another species of *Cryptolestes, C. turcicus* (Grouvelle), resembles *C. pusillus,* but the male's antenna is as long as or longer than its body. The

Fig. 184. Broadnosed grain weevil, *Caulophilus oryzae*. *A*, adult; *B*, adult and larvae in an avocado seed; *C*, injury to avocado seeds.

rusty grain beetle, *C. ferrugineus* (Stephens), is similar to *C. pusillus* in both appearance and habits, but the male's antennae are not more than half as long as the body. In the northern states and Canada, it is more commonly found in stored grain than the other 2 species (Cotton, 1963).

Siamese Grain Beetle, *Lophocateres pusillus* (Klug) (Trogositidae)

This cosmopolitan species attacks a wide variety of legumes, rice and rice products, other cereals, and various food products such as cocoa beans, coffee beans, cotton seeds, nutmeg, cassava, sweet potatoes, and dried fruits (Halstead, 1968). It is basically a tropical species, and is a relatively minor pest that is particularly associated with rice. Sound rice kernels are resistant to infestation, but they can be attacked if damaged (Corbett and Pagden, 1941).

Biology. The elongated, flat, reddish-brown female lays her eggs in crevices in fanshaped clusters of 11 to 14 eggs. They fail to hatch at temperatures below 15 or above 37° C (59 or 99° F), and the minimum incubation period is 5.8 days at 35° C (95° F). There are 4 larval instars. At 23° C (73° F) and 75% relative humidity, the larval stage requires an average of 53.8 days. The last-instar larva makes a cell in the food preceding pupation. The mean duration of the total life cycle (oviposition to adult) at 95% relative humidity is 179.7 days at 20° C (68° F) and 48.8 days at 35° C (95° F) (Halstead, 1968).

Broadnosed Grain Weevil, *Caulophilus oryzae* (Gyllenhal) [= *latinasus* (Say)] (Curculionidae)

The adult beetle is about 3 mm long. It has the typical prolonged snout of the weevils, and resembles the rice weevil (*Sitophilus oryzae*), having pitted elytra, but is darker brown. It is widespread in the southern United States. Unlike the rice and granary weevils, this species cannot feed on whole grain or any seed that is dry and hard. It is often associated with the rice weevil, and feeds on grain that the latter has damaged. In California, it can often be found in an orchard in avocado seeds on the ground (figure 184). It bores through the rotting flesh of fallen avocados and enters the seeds. It has been known to infest avocado seeds that had been planted in nursery beds.

Life Cycle. The female lays between 200 and 300 eggs. In summer, the period from egg to adult is about a month, and the adults may live for another 5 months (Cotton, 1963).

Mediterranean Flour Moth, *Anagasta kuehniella* (Zeller) (= *Ephestia*) (Pyralidae)

The Mediterranean flour moth owes its name to the belief by some authorities that this insect, now world-wide in distribution, originated in the Mediterranean region. It first appeared in the United States in California in 1892. The larvae spin silken threads that cause a matting of flour particles that may clog machinery in flour mills. Modern fumigation methods have greatly decreased the importance of this pest in mills, but it is still one of the important pantry pests in the home.

The mature larvae crawl around extensively, and may be found anywhere in the house. This moth infests flour, cereals, bran, biscuits, dog food, nuts, seeds, chocolate, dried fruits, and many other stored foods.

Description. The adult moth (plate V, 1; figure 185) is 7 to 12 mm long, and has a wingspread of 24 mm or less. The forewings are pale gray, transversely marked with 2 zigzag black lines. The hindwings are dirty-white. When at rest, the forepart of the body is arched in a characteristic posture.

Life Cycle. The female lays several hundred eggs in flour or other larval food, and they hatch in 3 to 5 days. The larvae (figure 185) confine themselves in silken tubes, and are full grown in 40 days, being about 15 mm long and whitish pink. They pupate in clean flour or other food material in silken cocoons—or may not spin cocoons if they pupate in cracks or crevices. In any case, they leave the mass of infested material at this period. The pupal stage lasts 8 to 12 days, and the entire life cycle usually takes 9 to 10 weeks. The rate of development is largely determined by temperature. Under favorable conditions, there may be at least 4 or 5 generations in a year.

Indian Meal Moth, *Plodia interpunctella* (Hübner) (Pyralidae)

This cosmopolitan insect prefers the coarser grades of flour, and is therefore the most common pest in wholewheat or graham flour and cornmeal (Indian meal). The common name was given by Asa Fitch, an early American entomologist, who

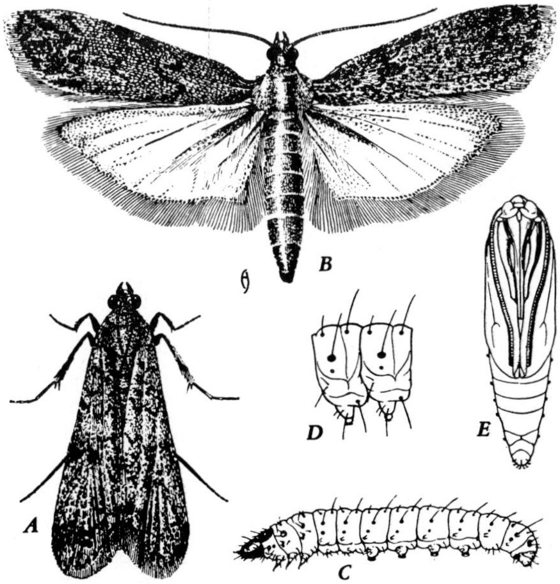

Fig. 185. Mediterranean flour moth, *Anagasta kuehniella*. *A*, adult in normal resting position; *B*, adult with wings spread; *C*, larva; *D*, enlarged drawing of 2 segments of larva in the region of the prolegs; *E*, pupa. (From Linsley and Michelbacher, 1943.)

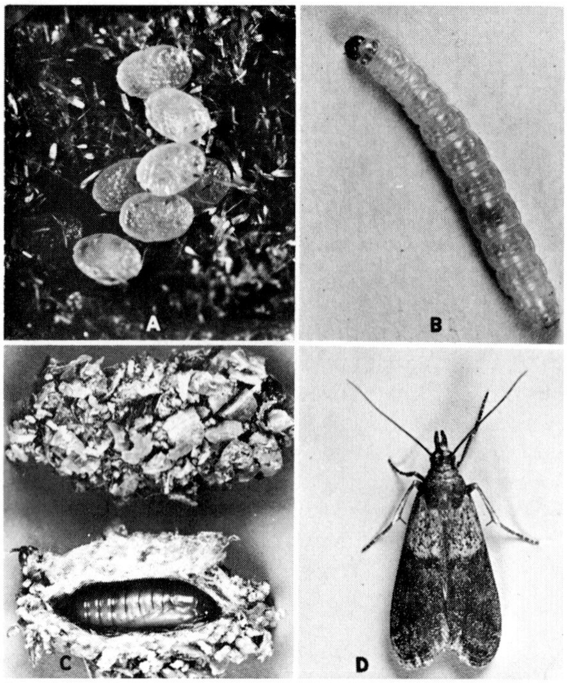

Fig. 186. Indian meal moth, *Plodia interpunctella*. *A*, eggs; *B*, larva; *C*, pupa case and pupa; *D*, adult. (Courtesy Stennett S. Heaton.)

found the larvae feeding on cornmeal. It also breeds successfully in shelled or ear corn, matting it over with silken threads. In the home, it infests broken grain and grain products, dried fruits, seeds, crackers, biscuits, nuts, powdered milk, chocolate, candy, red peppers, and dog food. By its extensive webbing, it spoils more food than it can consume.

Description. The adult Indian meal moth (plate V, 1; figure 186) has a wingspread of about 20 mm. The wings are pale gray, and the outer portion of the forewing is reddish brown, with a coppery luster. The adults are often seen flying about in the home, and are likely to be mistaken for clothes moths, but the wing color is very distinct from the uniform gray of the clothes moth's wings.

Life Cycle. Although generations overlap as the season progresses, oviposition usually starts in April, the female laying 200 to 400 eggs, singly or in groups, on the larval food over a period of up to 18 days. Egg-laying usually takes place at night. Shortly after hatching, the dirty-white larvae secrete themselves in crevices in the food medium, and feed in or near a tunnel-like case of silk with frass incorporated into it. The larva (figure 186, *B*) generally retains its dirty-white color, but may be yellow, pink, brown, or greenish, depending on its food. The head and prothoracic shield are brown. The average length is about 13 mm when full grown. The larval period is considerably variable, depending on food and temperature. A heavy webbing usually extends throughout the infested material. The larvae crawl out of the infested food to spin their cocoons and pupate. In heavy infestations, they frequently pupate far away from the original food source. Complete life cycles have been noted to range from 27 to 305 days (Hamlin *et al.*, 1931). There are generally 4 to 6 generations a year, but under favorable temperature conditions there may be 7 or 8.

The hymenopterous parasite *Bracon hebetor* Say is often seen in Indian meal moth infestations. According to Cotton (1963), outbreaks of the moth are nearly always terminated by this parasite, which can increase rapidly to enormous numbers.

Dermestid Beetles (Dermestidae)

The dermestid beetles are best known as scavengers and feeders on animal matter, but some species belonging to the genera *Trogoderma, Attagenus, Anthrenus,* and *Dermestes* can vary their diet by feeding at least in part on grain products. They are often found in warehouses, farm granaries, flour mills, and food-packaging plants, and particularly from the latter they may be transported to the home in infested food materials. They may also infest household food products after breeding in carpets, hides, bird or insect nests, dead rodents, or other sources of infestation within the house.

Trogoderma is the genus most commonly found in stored food, but not necessarily in the home. Some common species in the United States are *T. variabile* Ballion (=*parabile* Beal), *T. glabrum* (Herbst) (plate IV, 4), *T. inclusum* LeConte, *T. ornatum* (Say), *T. simplex* Jayne, *T. sternale* Jayne, and *T. grassmani* Beal. Their relative abundance is greatly affected by climatic factors. In California, *T. simplex* was found to be the most widely distributed species among 18 climatic zones (Partida and Strong, 1970).

The khapra beetle, *Trogoderma granarium* Everts (plate IV, 3), is a serious pest of cereals in India and other parts of the world. Effective eradication measures have been taken whenever it has become established in the United States. According to Hinton (1945), it normally was found to feed on vegetable products, particularly wheat, barley, rice, and malt, and was the only dermestid for which he had found no recorded instances of feeding on materials of animal origin, although it would do so experimentally.

In the United States as a whole, the black carpet beetle, *Attagenus megatoma* (F.), is probably the dermestid pest most often encountered in the pantry feeding on farinaceous materials, but it is seldom found in California. The description and biology of this species are given in chapter 8. Other dermestids described in that chapter could also become pests of farinaceous food products. The dermestids that are principally attracted to cured meats are discussed under "Insects Infesting Meats and Cheeses," later in the present chapter.

Dermestids as Museum Pests

These beetles are notorious pests of dried plant and animal museum specimens. Dried insects are particularly susceptible to attack, and at least 22 species of dermestids have been recorded damaging insect collections in various parts of the world, apparently more than all other insect species combined. The 4 most important museum pests are: *Anthrenus verbasci* (L.), *A. museorum* (L.), *A. scrophulariae* (L.), and *Trogoderma versicolor* (Creutzer).

Allergy and Canthariasis

Dermestids can cause severe allergic reactions. A man doing research on the control of dermestid beetles and having "close contact" with the insects suffered severe reactions, including extreme difficulty in breathing. The condition became gradually worse over a 2-year period. Species principally involved were in the genus *Trogoderma*. A baby suffered digestive distress and pain (canthariasis) after eating cereal containing *T. ornatum* fragments. The irritation was caused by larval setae. Although approximately 90% of the *Trogoderma* species received by the California Department of Agriculture are *T. variabile,* that species has not been recorded as causing canthariasis (Okumura, 1967).

After 3 years of rearing and otherwise working with carpet beetles (*Anthrenus flavipes* LeConte and *Attagenus megatoma*), a technician in our entomology laboratory began sneezing whenever he worked with the insects. After 10 years of such contacts with the beetles, he experienced severe itching and reddening of the eyeballs. The allergic reactions were apparently caused by dislodged hairs and fragments of cast skins.

Spider Beetles (Ptinidae)

Spider beetles are found throughout the world. They can remain active in freezing temperatures, and can therefore be pests in certain climates and circumstances in which other insects cannot. The farther north, the greater the relative importance of spider beetles. In North America, they attain their greatest relative importance in the northern U.S. and in Canada, where they can survive the winter in unheated warehouses and seriously infest cereal products during the spring and summer (Smallman, 1948). Because spider beetles can live as general scavengers, they can survive on rat droppings and miscellaneous debris in buildings such as empty warehouses. The source of infestation is sometimes difficult to locate.

Spider beetles tend to appear late in the succession of grain-infesting beetles. For example, in one investigation the succession of dominance of stored-grain beetles was found to be *Sitophilus granarius,* followed by *Ptinus ocellus* (= *tectus*), and finally by the dermestid *Attagenus pellio. Sitophilus granarius* transformed whole wheat kernels into husks (empty grain) and frass (a fine powder, mostly feces), and then this species declined. The developmental rate of *P. ocellus* was most rapid on crushed, dead *S. granarius,* but was also faster on frass than on wholemeal flour (Coombs and Woodroffe, 1963, 1970).

In a large food-processing plant in Wisconsin, the principal breeding area for spider beetles was found to be the void of a dropped ceiling into which food particles sifted through cracks in the floor above. The silica aerogel Dri-die 67® was blown into the infested areas with a water-type fire-extinguisher (figure 33, chapter 3), eliminating them as sources of infestation by spider beetles and dermestids.

Biology. Spider beetles are small, 1.5 to 4.5 mm long, oval or cylindrical, and can be easily recognized by their long legs and the constriction of the prothorax near its base. As in the closely related family Anobiidae (see chapter 5), the prothorax extends over the head like a cowl. When the insect is viewed from above, the head is not visible, and only the long filiform or moniliform antennae can be seen. In figure 187, the specimens were drawn when tilted back at a 35° angle in order to make the heads visible. If they had been on a horizontal plane, the heads could not have been shown. Except for the antennae, these features combine to give the insects a fancied resemblance to small spiders, hence the common name.

The larvae are white, fleshy, and scarabaeiform (curved). There are usually 3 instars. They spin some silk, often in the form of a feeding cocoon. A cocoon is also spun in which to pupate, and the adult spends considerable time in it before emerging. According to König (1936), this cocoon is constructed from an anal secretion applied by the mouthparts, to which debris and particles of foodstuff adhere. Howe and Burges (1953) quoted Vajropala as stating that the silk of ptinids was secreted by the Malpighian tubules and mesenteron. The adult life is longer than the developmental period, and the female oviposits during most of it. However, few eggs are laid when larval food is unavailable, and readily obtainable drinking water is required to ensure maximum reproduction (Hickin, 1942; Ewer and Ewer, 1942; Howe and Burges, 1952).

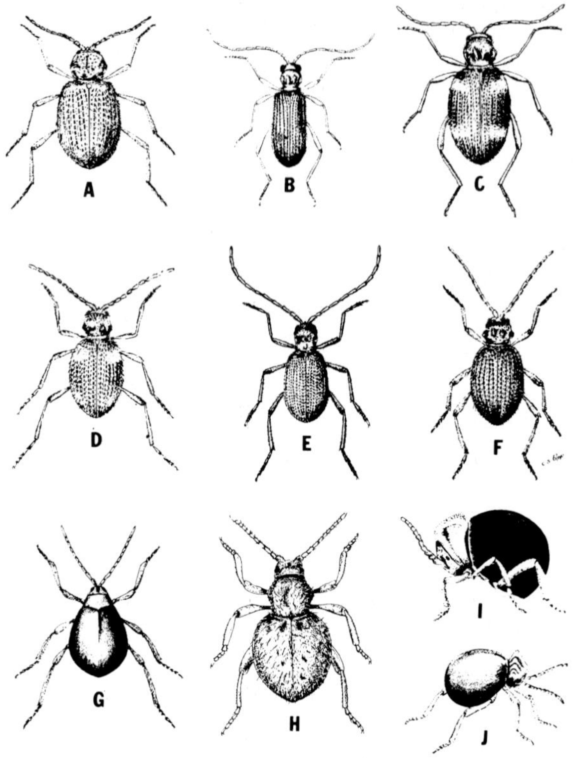

Fig. 187. Spider beetles. ***A, Ptinus ocellus; B, Ptinus fur*** **(male);** ***C, Ptinus fur*** **(female);** ***D, Ptinus villiger; E, Ptinus clavipes*** **(male);** ***F, Ptinus clavipes*** **(female);** ***G, Gibbium psylloides; H, Trigonogenius globulum; I, Mezium americanum; J, Mezium affine.*** **(Courtesy C. S. Papp.)**

Economic Importance. Most species of spider beetles are cosmopolitan, and are inclined to be omnivorous, feeding on broken grain or grain products, seeds, dried fruits or meats, wool, hair, feathers, rat and mouse droppings, insect and other animal remains, and plant and animal museum specimens. Howe (1959) pointed out that spider beetles had been observed in nature in dead or decaying wood, and that they also fed on mammal and bird dung. He made the interesting suggestion that some species were probably first introduced into primitive food-storage devices in the wood used to construct them. Some of the mammals and birds attracted to the food would have nested in the storage area, and their droppings would have provided "extra amenities" for spider beetles.

Spider beetles can be pests in warehouses, grain mills, museums, and homes, and appear to thrive best in old wooden buildings. In such structures, a typical scarring of the wood can sometimes be found where the insects have formed pupal cells. In California, one species (*Ptinus ocellus*) severely riddled the timbers of a grain elevator (Mackie, 1932). When preparing to pupate, this species also commonly damages containers made of cardboard, sacking, or wood. In warehouses, spider beetles are primarily pests of cereal products, particularly those that have long remained in storage. Their principal damage is the lowering of the quality of the stored food products because of the presence of dead insects and the silk and cocoons spun by the larvae. Whole grain is seldom seriously attacked, and any small damage that occurs is easily recognizable: the bran covering and the endosperm immediately beneath it are eaten off unevenly. The infestation is usually restricted to the peripheral few inches of a pile of stored grain (Howe and Burges, 1953).

One thing that increases the importance of spider beetles as warehouse pests is their habit of ovipositing through the meshes of cotton flour sacks. The adult beetles should be eliminated before they reach the sacks to oviposit (Gray, 1942; Smallman, 1948).

Principal Species

Included in the specimens identified by entomologists of the California State Bureau of Entomology from among the insects collected by county, state, and federal inspectors in a wide variety of foods and seeds in storage were 8 species of spider beetles. These, and the number of counties in which each species was collected, were as follows: *Ptinus clavipes* Panzer, 37; *P. ocellus* Brown, 18; *P. fur* (L.), 3; *P. gandolphei* Pic, 3; *P. villiger* (Reitter), 1; *Gibbium psylloides* (Czenpinski), 3; *Mezium americanum* (Laporte), 2; and *Trigonogenius globulum* Solier, 15 (Strong and

Okumura, 1958). Among the species Strong (1970) collected in California in his mixed food baits, *Ptinus clavipes* was again the most abundant, but a few of the other species just named were not collected (table 6). However, 5 species not found in the 1958 survey *were* collected: *Ptinus cognatus* Fall, *P. californicus* Pic, *P. verticalis* LeConte, *P. extendus* Smith, and *Niptus ventriculus* LeConte.

These surveys may have indicated the relative importance of the various species as "pantry pests" in California, but their potentials for household damage might be different in other geographical areas. The least widely distributed species in California (*P. villiger*) is the most important one in Canada.

Australian Spider Beetle, *Ptinus ocellus* Brown (= *tectus* Boieldieu)

This is a cosmopolitan species, second only to *P. clavipes* in California in its importance as a pest (Strong and Okumura, 1958; Strong, 1970). Although it was not recorded in England until 1892, it has become one of the most widely distributed pests of stored food products in that country (Hickin, 1964).

Description. *Ptinus ocellus* (figure 187, *A*) is dull reddish brown, 3.5 to 4 mm long, with golden-brown or yellowish hair covering its elytra. This hair hides a series of longitudinal rows of small pits that are visible only when the hair is rubbed or worn off. This species can be confused with the male of *P. fur*.

Life Cycle. The female lays about 100 opalescent, sticky eggs, singly or in small groups, over a period of 3 to 4 weeks. Food and debris adhere to the eggs. The fleshy larva, covered with fine hairs, is strongly curved, and rolls up into a tight ball when disturbed. It can chew its way through sacking, cellophane, or cardboard, and can hollow out a chamber in adjacent woodwork when making a place in which to pupate. Pupation takes place in a tough, spherical, thin-walled cocoon. The adult may remain in this cocoon as long as 3 weeks after emergence (Hickin, 1964). Howe (1943) found that at 21° C (70° F) and 70% relative humidity, the eggs hatched in 8.6 days; the 3 larval instars required 59.9 days; the pupal stage, 15.9 days; the pre-emergence period (adult in cocoon), 9.5 days; and the total life cycle, from egg to the adult leaving the cocoon, 93.9 days.

Habits and Food Preferences. The adults are most active during darkness, emerging from their hiding places at dusk and returning at dawn. The hiding places are usually cracks and small crevices, and the larvae are inclined to pupate in them (Howe, 1950; Howe and Burges, 1953). This species can live and develop in accumulations of organic dust or in rodent droppings, and is often found in the nests of pigeons and other birds, even though it cannot fly (Howe and Burges, 1953). Large numbers of *Ptinus ocellus* were found in the nests of house sparrows in a warehouse, and the beetles were most numerous on or around the stacks of flour sacks nearest the nests (Cutler and Hosie, 1966).

The amazingly omnivorous nature of *P. ocellus* is typical of spider beetles in general, and is indicated by the following list of foods and other materials that it has been known to consume or bore into: almonds, beans, cacao, carpets, casein, cayenne pepper, chocolate powder, corn, crab meat, desiccated soup, dried fruits, dried insects, figs, fish food, fishmeal, furs (completely destroyed), ginger, lead cable, nutmegs, packaging materials, paprika, poultry food, raisins, rye, stored hops, and wood timbers. This beetle thrives best when the vitamin B content of its food is high, but can complete its development on materials of relatively low nutritive value, such as casein and pure starch. Its longevity and egg-laying capacity are greatly increased if it has access to water at least once a week (Coombs and Woodroffe, 1965).

Brown Spider Beetle, *Ptinus clavipes* Panzer (= *hirtellus* Sturm and *brunneus* Duftschmid)

Mallis (1969) stated that this insect was a scavenger in cellars, attics, storehouses, henhouses, and the like, and fed on "books, feathers, skins, dried mushrooms, the excrement of rats and other animals, the powdered leaves of such drugs as senna and jaborandi [*Pilocarpus:* Rutaceae], dried fruit, grains, and sugar." It was found to be more widely distributed and was taken from a greater variety of grains, legumes, various seeds, animal feeds, and mixed-food baits than any other spider beetle in California (Strong and Okumura, 1958; Strong, 1970).

Description. This beetle (figure 187, *E, F*) is uniformly pale brown in both sexes, as a rule, but may vary from dark brown in the male to blackish brown in the female; it ranges from 2.3 to 3.2 mm in length.

Life Cycle. Howe (1956*a*) gave the following data: The eggs hatched in 13 days at 22.5° C (about 72° F) and 70% relative humidity. The complete life cycle required 6 to 9 months. However, about one-third of the larvae entered diapause (see chapter 4), and remained larvae more than 10 months. *Ptinus clavipes* was able to complete a generation in 5 months at temperatures in which *P. ocellus* could not have survived.

Hairy Spider Beetle, *Ptinus villiger* (Reitter)

Smallman (1948) stated that this species was the spider beetle most commonly found in grain warehouses in Canada, with *P. fur* and *P. raptor* also occurring.

Description. The adult beetles (figure 187, *D*) are reddish brown. The elytra of both males and females have irregular white patches anteriorly and posteriorly, and those of the females also have several longitudinal rows of shallow pits. The length varies from 2.2 to 4 mm.

Life Cycle. The female may lay up to 40 pearly, spindleshaped eggs, 0.6 mm long. They are usually laid on the outsides of bags, through the meshes of sacks, and in flour debris in cracks and corners. They hatch into cream-colored larvae with brown head capsules, that complete their development in 3 months in warm weather, and become about 4 mm long when full grown. Pupation takes place in a silken cell coated with food material. The larva often overwinters in this pupal cell, and does not pupate until the following spring (Gray, 1934).

Whitemarked Spider Beetle, *Ptinus fur* (L.)

Description. The beetles are reddish brown, 2 to 4.3 mm long, and are covered with yellow hairs. The females (figure 187, *C*) are larger and more nearly globular than the males (figure 187, *B*), and unlike them, have 2 white patches anteriorly and posteriorly on each elytron. In some specimens, the patches are joined to form white stripes. This species closely resembles *P. villiger* in appearance and habits.

Life Cycle. The larvae, which superficially resemble those of the drugstore beetle (*Stegobium paniceum*), feed in a globular cell formed from their foodstuff. The complete life cycle varies from 1 to 3.5 months, depending on temperature and other factors.

Golden Spider Beetle, *Niptus hololeucus* (Faldermann)

In addition to the usual wide variety of products that are generally attacked by spider beetles, the adults and larvae of this species feed on woolens, linens, and natural silks, particularly if the fabrics are soiled with grease. The beetles are most commonly found in vegetable and animal debris in warehouses, poorly kept storerooms, cellars, and old houses, and sometimes occur in considerable numbers. They shun light, but move actively in the dark (Hickin, 1964). The nests of bees, wasps, and birds provide their natural habitats (Howe and Burges, 1952).

Description. The adults are 3 to 4.5 mm long and golden yellow, with long, silky hairs covering the fused elytra. Flight wings are absent.

Life Cycle. Oviposition and the appearance of the eggs and larvae are similar to their equivalents in *Ptinus ocellus*. The eggs hatch in 11 to 20 days at 64 to 68° F (18 to 20° C); the larval period lasts about 150 days; and the pupal period, 18 to 26 days. The adults may live as long as 250 days (Hickin, 1964).

American Spider Beetle, *Mezium americanum* (Laporte)

This species has been noted to occur in dwellings, warehouses, and mills in California, and has been taken from such materials as dried animal products, melon seeds, mixed feed, and woolen carpet (Strong and Okumura, 1958), and in grain, tobacco, cayenne pepper, and opium (Papp and Okumura, 1959). In a San Francisco residence, a continual renewal of infestation after treatment led to the discovery that spider beetles of this species were coming from an old rat nest under the bathroom. The nest contained dead rats, rat droppings, old bones, and other materials on which the insects could subsist. The otherwise inaccessible area could be reached with an insecticide dust, and control was obtained (Montgomery, 1936).

Description. *Mezium americanum* has a particularly striking appearance. It is 1.5 to 3.5 mm long, broadly oval, and the elytra are strongly convex (figure 187, *I*). It is shining dark reddish brown to nearly black. The "humped" elytra give this, as well as the following species, a conspicuous structural feature for easy identification.

Shiny Spider Beetle, *Gibbium psylloides* (Czenpinski)

This is a cosmopolitan species that is well established in California and other western states, where it has been recorded as feeding on woolens and various dried animal products.

Description. The shiny spider beetle (plate IV, 5; figure 187, *G*) is red to nearly black, with dense, short, yellow hairs beneath, and is 1.7 to 3.2 mm long. It closely resembles *Mezium americanum* in appearance and habits, but the head and thorax are entirely bare, while in *M. americanum* they are densely covered with small scales and scalelike hairs. Both species are striking in their "humped" appearance, and are easily distinguished from other spider beetles.

Life Cycle. Relatively few eggs are laid when compared with some species, such as *Ptinus ocellus.* The immature stages are similar in appearance to those of *P. ocellus.* The larvae do not wander away from the foodstuff for pupation, which takes place in a spherical cocoon. *Gibbium psylloides* appears to thrive under drier conditions than other spider beetles. The life cycle requires 22 to 42 weeks, and the adult usually lives another 30 to 40 weeks, but an adult life of 18.5 months at 25° C (77° F) has been recorded (Hickin, 1964).

Economic Importance. This beetle has been found in houses, hotels, warehouses, mills, granaries, bakeries, and latrines, and on sponge-rubber bath mats, woolens, and towels. It is recorded as feeding on cereals, wheat, bran, baby food, stored seeds, stale bread, dog biscuits, decaying animal and vegetable refuse, cayenne pepper, and opium cake (Hinton, 1940; Curran, 1946*a*). In England, it has long been known to occur in large numbers in old houses, damaging woolens and paper. It infests tallow, and has the ability to thrive on drugs, including opium (Hickin, 1964).

Globular Spider Beetle, *Trigonogenius globulum* Solier

Essig (1926) recorded this species from dried animal products, herbarium and insect specimens, seeds, red pepper, and cereals. He kept a living colony in a jar of manure for 8 years. Papp and Okumura (1959) recorded the beetle as breeding in large numbers in a manure pile and being found in burlap sacks and old clothing, as well as in grains and feeds.

Description. This spider beetle (figure 187, *H*) is 2.5 mm long, with both prothorax and abdomen nearly globular, the latter being twice as wide as the former. The adult is pale or dark brown, and is densely clothed above and beneath with pale yellowish-brown scales that are irregularly dispersed with darker brown on the elytra, and with a small, elongate, blackish-brown spot on each side of the suture at the base of the elytra. There are no scales on the prothorax, but it bears a dense mat of interlacing hairs.

INSECTS INFESTING LEGUMES

Bean Weevil, *Acanthoscelides obtectus* (Say) (Bruchidae)

This beetle, thought to be indigenous to North America, is the principal pest of stored legumes (kidney beans, peas, lentils) and certain other seeds, attacking them both in the field and in storage if they are stored in a warm place. If the field-infested material is brought into storage, the insects can reinfest the dried seeds. The bean weevil can be eliminated by destroying infested legumes because it does not attack grains, cereals, or other stored food products.

Description. Despite its name, this pest is not a true weevil; its head is not prolonged into a beak (plate IV, 8; figure 188). The adults are 2 to 3 mm long, robust, somewhat triangular (cut off squarely at the rear and narrowing toward the front), almost flat above, and are velvety gray or brown, with pale, linear markings on the elytra, which do not cover the tip of the abdomen. The thorax is covered with fine, yellowish-orange hairs, the legs are reddish yellow, and the antennae have serrate (sawtoothlike) segments.

Life Cycle. The female lays an average of 75 eggs, singly, on or near the beans or related seeds. The eggs hatch in 5 to 20 days, and the tiny, legless larvae proceed to bore into the seed; several larvae can develop in a single bean. The life cycle requires 21 to 80 days, and the newly emerged adult makes an exit hole and leaves the seed. The adults do not eat legume seeds. They fly actively during the day at 70° F or more.

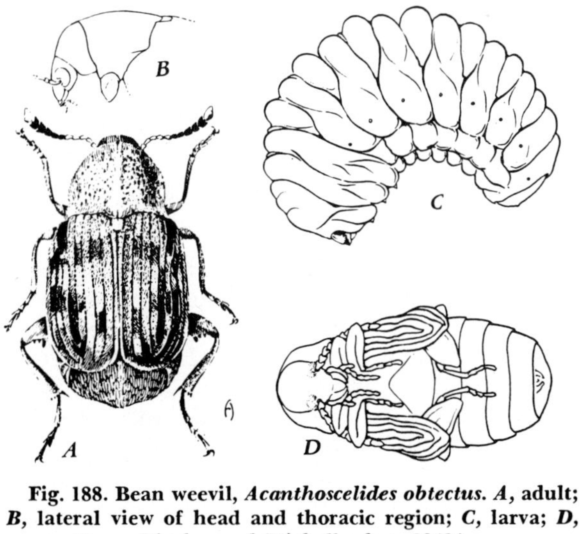

Fig. 188. Bean weevil, *Acanthoscelides obtectus. A,* adult; *B,* lateral view of head and thoracic region; *C,* larva; *D,* pupa. (From Linsley and Michelbacher, 1943.)

Cowpea Weevil, *Callosobruchus maculatus* (F.)

This insect is similar in size and habits to the bean weevil, but infests only cowpeas. The elytra are reddish brown, and each has 2 large, red spots. The thorax is covered with fine, white hairs. Whereas in the bean weevil the basal segment and the last segment of the antenna are reddish yellow, in the cowpea weevil only the basal segments are that color and the remainder of the antenna is dark.

Pea Weevil, *Bruchus pisorum* (L.)

Pea weevils are 4 to 5 mm long, gray to brownish gray, with small, distinct, white spots on the elytra. They attack peas in the field—one beetle per pea. The females must lay their eggs on the pods of growing pea plants in the spring or perish without ovipositing. Infested peas may be brought into storage, but eggs are never laid on dried peas, and there is no reinfestation in storage. Therefore, it is a relatively unimportant pantry pest compared with the bean weevil.

Broadbean Weevil, *Bruchus rufimanus* Boheman

This weevil is a pest only in California. It is only about two-thirds as large as the pea weevil, but is otherwise almost identical in appearance and habits. It prefers broad "beans" (*Vicia faba*), but also attacks other peas and vetches. Several weevils may be found in a single seed.

INSECTS INFESTING MEATS AND CHEESES

Redlegged Ham Beetle, *Necrobia rufipes* (De Geer) (Cleridae)

This beetle is the most important of the insects that infest dried or smoked meats. It can be a serious pest where smoked meats are stored for long periods, and is the principal reason for the wrappings, sacks, and washes used for their protection. The larvae boring into the meat, particularly the fat parts, do most of the damage; the adults are surface feeders. The redlegged ham beetle has also been recorded attacking cheese, bones, hides, drying carrion, copra, salt fish, herring, dried egg yolks, dried figs, "guano," bonemeal, palm-nut kernels, and Egyptian mummies. Substances infested but not fed upon have been silk, baled cotton, and woolen goods. These beetles belong to a family that is principally known as predaceous, and it is not surprising that they feed on the larvae of blow flies and cheese skippers that infest meat. When redlegged ham beetles become very numerous, their numbers are reduced by cannibalism (Simmons and Ellington, 1925). Infestation has been prevented by installing tight-fitting doors and careful screening with fine (30-mesh) wire cloth.

Description. The adult beetle (plate IV, 3; figure 189) is shiny green or greenish blue, with legs and bases of the antennae reddish brown, and is 3.5 to 7 mm long. It is convex, straightsided, and noticeably "punctured."

Life Cycle. The smooth, shiny, translucent eggs, 1 mm long, are glued in place, usually in clusters, on the surface of the foodstuff; during the adult life of 14 months or more, as many as 2,100 eggs may be laid. They hatch in 4 or 5 days in warm weather. The full-grown larvae are about 10 mm long, slender, tapering toward the head, and are purplish in color. They are repelled by light. There are 3 or 4 larval instars. The larva uses a frothy oral secretion for the construction of a white cocoon for pupation. The life cycle varies from 36 to 150 days or more, depending on the type of food and the temperature (Simmons and Ellington, 1925).

Related Species

Two other species of *Necrobia* with habits similar to those of *N. rufipes,* but seldom recorded as pests, are the redshouldered ham beetle. *N. ruficollis* (F.), and *N. violacea* (L.) (figure 189). *Necrobia ruficollis* is 6.2 mm long. The front of the head and the apical three-fourths of the elytra are metallic blue; the ventral surface of the head, the prothorax, the bases of the elytra, the meso- and metasternum, and the legs are brownish red; and the antennae and abdomen are dark brown. *Necrobia violacea* is 4.5 mm long, and metallic dark blue or dark green, with very dark brown or entirely black antennae and legs. Both these species are cosmopolitan, and are commonly found on the skin and bones of dead animals and on dead fish. *Necrobia violacea* has been mentioned as a predator of *Dermestes* larvae, and may be more beneficial than harmful (Papp, 1959).

Larder Beetle, *Dermestes lardarius* L. (Dermestidae)

Refrigeration, the purchase of meat in small quantities, and almost complete discontinuance of the home curing of meat have resulted in the larder beetle becoming less important as a pest in modern times. It is now more likely to be introduced into the home with dog and cat food (Gray, 1947).

Description. This dark-brown beetle is about 8 mm long. On the basal third of each elytron there is a wide, yellow band with 3 black spots. The ventral surface of the body and the legs are covered with fine, yellow hairs, and the antennae are distinctly clubbed. The larva is brown and very hairy, as is typical of dermestids, and is 11 to 13 mm long. There are 2 curved dorsal spines (urogomphi) on the next-to-last abdominal segment.

Life Cycle. The adults overwinter in crevices of bark, and enter buildings in spring and early summer. They seek food such as ham, bacon, dried beef, dried fish, cheese, feathers, horn, skins, or hair on which to oviposit, but will lay eggs in cracks and crevices if no food is available. The female lays 100 or more eggs, and they hatch in 12 days or less. The larvae may burrow into ham, bacon, or anything available in the vicinity, such as wood or even lead. The male larvae molt 5 times and the females, 6, before seeking places in which to pupate. There is usually a single generation per year, but as many as 5 or 6 have been observed (Kreyenberg, 1928; Hinton, 1945).

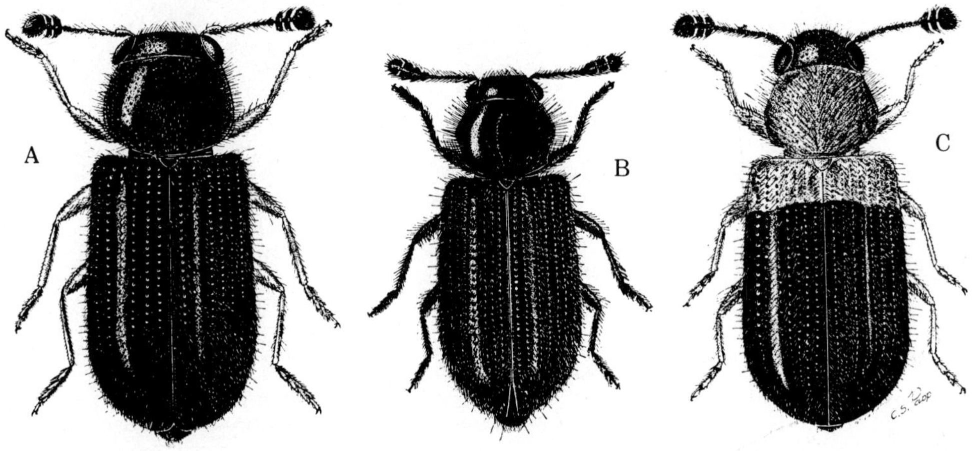

Fig. 189. Three species of *Necrobia* that infest dried and smoked meats. *A*, redlegged ham beetle, *Necrobia rufipes*; *B*, *Necrobia violacea*; *C*, *Necrobia ruficollis*. (From Papp, 1959.)

Black Larder Beetle, *Dermestes ater* De Geer (= *cadaverinus* F.)

This species, in common with *Dermestes lardarius,* is an occasional pest of cured meats, cheese, and other foods of animal origin, and is sometimes found in packaged pet foods. Johnson (1945) stated that this species was commonly found in incinerators in Connecticut. In one apartment building, for example, it was found flying about, apparently having bred in dead mice in the building. The adults were often observed coming out of crevices and from behind baseboards and moldings. They appeared to have dispersed via wall voids from the incinerator room in the basement. Complete combustion of all waste material in the incinerators was recommended because odors from unburned garbage attracted the adult beetles. It was also suggested that all ashes and unburned material be removed every 2 weeks.

Biology. *Dermestes ater* is a black beetle, about the size and shape of *D. lardarius,* and with a yellowish-gray pubescence. It can be easily distinguished from *D. lardarius* because it lacks the dull-yellow band across the base of the elytra of the latter. The female may lay up to 400 eggs over a period of about 2 months. At 80 to 82° F (27 to 28° C), the period from egg to adult requires 6 weeks (Roth and Willis, 1950).

Hide Beetle, *Dermestes maculatus* De Geer (= *vulpinus* F.)

The hide beetle prefers to feed on hides and skins, and is less likely than the larder beetle to be a pest of stored food products. This species can damage such hard substances as hair brushes, cork, tea chests, and woodwork, which the mature larvae bore into in order to pupate. Severe damage to rafters of Douglas fir (Oregon pine) was caused by *D. maculatus* in a mill in which bones and offal were received (Brimblecombe, 1938).

Description. This species (plate IV, 3) is similar to the larder beetle in size, shape, and habits. It may be distinguished from that species by the fact that the elytra are uniformly dark and the undersurface is white. Each elytron terminates in a sharp point where the elytra join. The apical margins of the elytra are serrate.

Life Cycle. In one investigation, it was noted that the female oviposited in any available crack or crevice, and that the eggs hatched in 3 days. After 6 molts, the larvae sought places to pupate, boring into any available material. The pupal period lasted 10 to 11 days. The life cycle required 64 days (Illingworth, 1918).

Dermestes frischii Kugelann

Description. This species is similar to *Dermestes maculatus* in appearance and feeding habits, but may be distinguished by the fact that the apical margins of the elytra are not serrate.

Life Cycle. Amos (1968) determined that the eggs hatched in periods ranging from 2 to 3 days up to 11 to 15 days in temperatures ranging from 40 to 15° C (104 to 59° F). The average number of eggs laid was 215. The larval stage lasted from a mean of 19.6 days at 35° C (95° F) and 90% relative humidity to 170.5 days at 20° C (68° F) and 45% RH. The investigation demonstrated the striking effect that humidity may have on the developmental periods of some insects. With this species, variation in relative humidity had about as great an effect as variation in temperature. Thus, at 25° C (77° F), the larval period averaged 40.5 days at 90%, 54.9 days at 75%, 74.6 days at 60%, 86.3 days at 45%, and 130.8 days at 30% RH. About the same range in the duration of the mean larval period was found at different temperatures at a constant RH of 60%: 35 days at 35° C, 33.7 days at 30°, 74.6 days at 25°, and 132.8 days at 20° (95, 86, 77, and 68° F).

Of equal interest in Amos' investigation was the finding that the presence of salt in fishmeal increased the developmental period and reduced the oviposition and rate of increase of *Dermestes frischii*. His data supported earlier field observations on the value of thorough salting and drying of fish for the control of this pest.

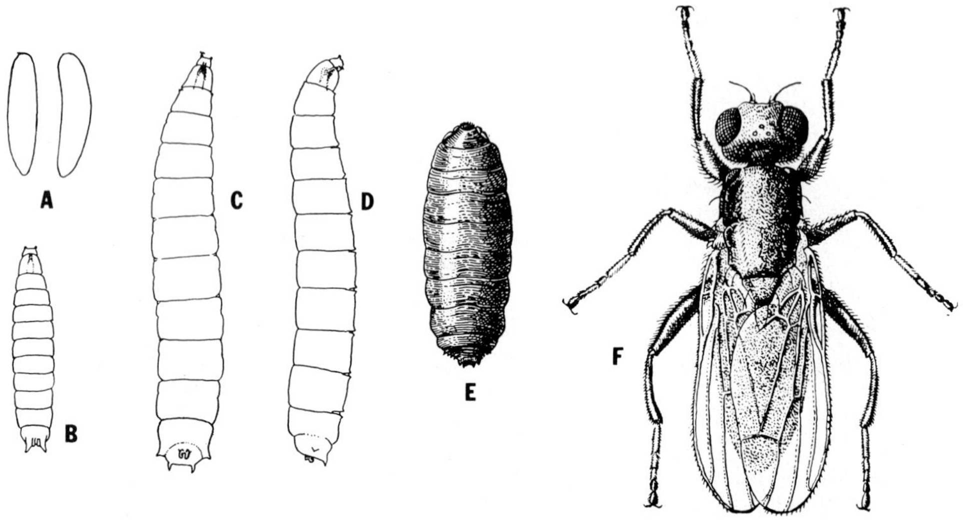

Fig. 190. Cheese skipper, *Piophila casei*. *A*, eggs; *B*, first-instar larva; *C*, third-instar larva (dorsal); *D*, third-instar larva (lateral); *E*, puparium; *F*, adult. (From Fulton, 1953.)

Cheese Skipper, *Piophila casei* (L.) (Diptera: Piophilidae)

The pests already described that infest meats and cheeses have all been beetles. There is also a fly of considerable importance—the cheese skipper. It occurs throughout the United States and most of the world. It is an important pest of cheese, but can also be destructive to cured pork. Other materials reported as foods of the cheese skipper include salted beef, smoked fish, brine-cured fish, bones, animal carcasses, human cadavers, and various other materials consisting largely of animal proteins. When they are accidentally eaten with infested cheese or other cold, infested food, live maggots can pass into the intestines, where they may cause irritation, lesions, and diseases. Unspoiled, salvaged meat can be eaten safely if it is thoroughly cooked (Fulton, 1953).

Description. The adult (figure 190) is a small, black or bluish-black fly, with bronze tints on the thorax, yellow face, mouthparts, and antennae, reddish-brown eyes, and iridescent wings that lie flat over the body when at rest.

Life Cycle. The female lays an average of about 130 eggs on the surface of the food substance. The maggot travels by peristaltic movements of its body or, when full grown, by "skipping," a movement by which it can propel itself as much as 25 cm horizontally or 15 cm vertically. It bores into cheese or meat for the most part. When about to pupate, the larva (figure 190) migrates to some dark, dry crevice. The life cycle may occupy as little as 12 days. The adults feed on the juices of the larval food, but live only 3 or 4 days, and appear only in the warm season (Simmons, 1927).

Control methods for the cheese skipper are given under "Control of Pantry Pests" at the end of this chapter.

GENERAL FEEDERS ON STORED FOODS

Sawtoothed Grain Beetle, *Oryzaephilus surinamensis* (L.) (Cucujidae)

This beetle often does considerable damage to grains, although its attack generally follows that of other insects, such as the granary or rice weevils; it cannot attack sound kernels. Its flat

Fig. 191. Sawtoothed grain beetle, *Oryzaephilus surinamensis.*

form allows it to penetrate into very narrow cracks, crevices, and imperfectly sealed packages. It is almost omnivorous, and attacks such foods as cereals, bread, macaroni, dried fruits, nuts, dried meats, sugar, and many other products. It is one of the most common pantry pests.

Description. The adult (plate IV, 5; figure 191) is a small, slender, flattened, brown beetle, about 3 mm long. It can be easily identified by the 6 sawlike "teeth" on each side of the thorax. The posterior femora of the males are armed with a tooth, while those of the females are not.

Life Cycle. The female lays 45 to 285 white, shiny eggs over a period of 2 to 5 months. They are deposited singly or in clusters in crevices in the foodstuff. In warm summer weather, at 80 to 85° F (27 to 30° C), they hatch in 3 to 5 days, and in cool spring and fall weather, at 68 to 73° F (20 to 23° C), in 8 to 17 days. The larvae molt 2 to 4 times (usually 3). When mature, they are yellowish white, and less than 3 mm long. Before pupation, they generally construct crude pupal cells or cocoons from particles of seeds or other foodstuffs, fastening them together with an oral secretion. Whether this cell is made or not, the larva attaches itself by its anal end to some solid object to pupate. The complete life cycle (egg to egg) may range from 27 to 375 days. The adult life may vary from a few days to more than 3 years (Back and Cotton, 1926). The adults have apparently never been seen in flight, although they have sometimes been captured in flight traps (Mallis, 1969).

Merchant Grain Beetle, *Oryzaephilus mercator* (Fauvel)

This species closely resembles *Oryzaephilus surinamensis,* and for many years the two were considered to be synonymous. Eventually, cross-breeding experiments proved that they were distinct species (Howe, 1956*b*). Also, a morpholog-

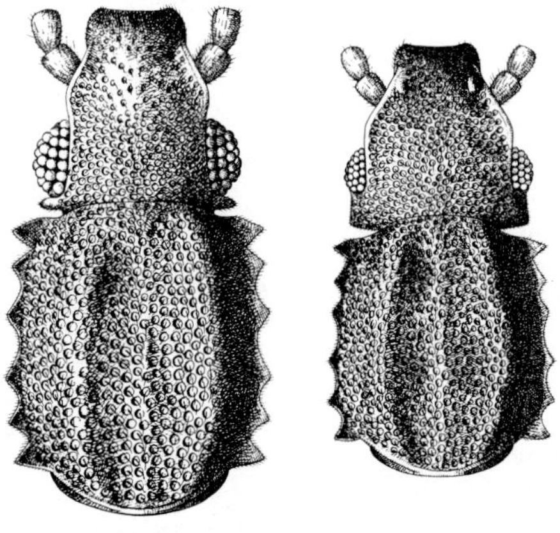

Fig. 192. Head and thorax of *Oryzaephilus mercator* (*left*) and *Oryzaephilus surinamensis,* showing difference in lengths of temple regions (seen directly behind the eye). (Courtesy H. E. Hinton.)

ical difference can be seen with the aid of a hand lens. In *O. mercator*, the length of the temple (the region directly behind the eye) is less than half the vertical diameter of the eye, while in *O. surinamensis*, it is greater than half that dimension (figure 192) (Howe, 1956*b*; Hinton and Corbet, 1963). An extensive survey made in Canada in 1966 revealed that *O. mercator* was widely distributed in that country, but mostly in processed cereals (rolled oats, rice, flour, cake mixes, macaroni, and cookies). In addition, it was found infesting nuts, coconut, and candy bars made with peanuts and puffed rice. It was found most frequently in kitchens in both apartment buildings and private residences. In imported commodities, those most often infested were copra, peanuts, and cashews from Asian countries, confirming Howe's belief that *O. mercator* was chiefly associated with oilbearing seeds. This species has not been recorded in Canada from grain stored in either farm granaries or grain elevators. On the other hand, *Oryzaephilus surinamensis* is a pest of both stored grain and processed cereal products. It can apparently endure the low Canadian temperatures in unheated granaries, whereas *O. mercator* cannot (Loschiavo and Smith, 1970).

Cigarette Beetle, *Lasioderma serricorne* (F.) (Anobiidae)

This beetle is considered to be the most destructive pest found in stored tobacco, but it also has the reputation of attacking a wider range of foods

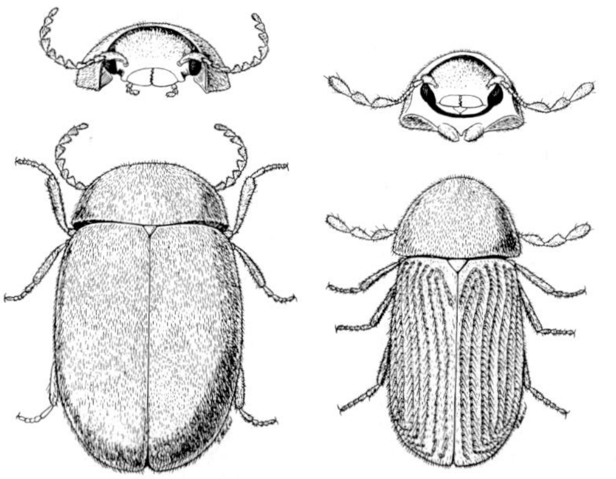

Fig. 193. Cigarette beetle, *Lasioderma serricorne* (*left*), and drugstore beetle, *Stegobium paniceum*. (Courtesy R. R. Snelling.)

than any other storage pest. It can live and fly outdoors in the tropics and subtropics, but can survive only in warm buildings in temperate regions. It is distributed throughout the United States and southern Canada.

Description. The adult beetle (plate IV, 7; figure 193) is light brown, small, oval, and about 3 mm long. In common with other anobiids, the ptinids, and the bostrichids, the head and prothorax are bent downward so as to give the insect

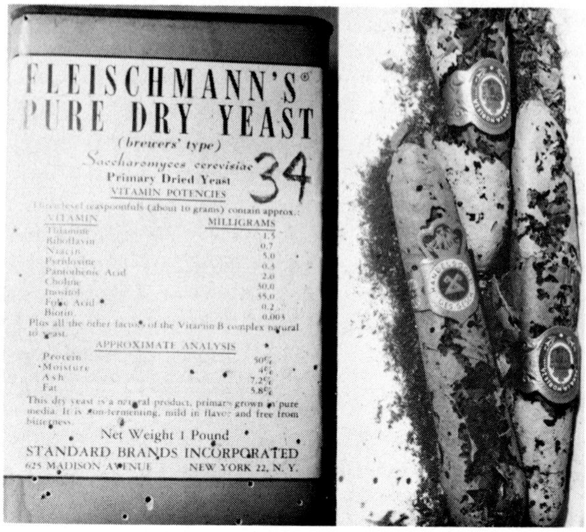

Fig. 194. Damage by cigarette beetles to a can of yeast (*left*) and to cigars.

a strongly humped appearance. It has occasionally been confused with the drugstore beetle (*Stegobium paniceum*), but *L. serricorne* has serrate antennae and smooth elytra, whereas in *S. paniceum* the last 3 antennal segments are long and broad, forming a distinct "club," and the elytra are striate. The larvae of the two species are also similar (plate IV, 7), except that the larva of *L. serricorne* has more and longer hairs, giving it a more fuzzy appearance.

Life Cycle. The female lays her oval, whitish eggs in and about the food materials, and they hatch in 6 to 10 days in warm weather. There are 4 to 6 larval instars. The full-grown larvae are about 4 mm long, curved, hairy, and pupate in silken cocoons covered with bits of foodstuffs. The life cycle may require 40 to 50 days, and there are usually 3 to 6 generations per year. The developmental period is affected by both humidity and temperature. Howe (1957) determined that at 30° C (86° F), the period required for the complete life cycle varied from a mean of 33.9 days at 70% relative humidity to 78.8 days at 40%, and that natural mortality increased from 50 to 73%, respectively. At 30% RH, there was 100% mortality. The minimal temperature for development was about 18° C (65° F), and food stored below that temperature was found to be safe from infestation. Swingle (1938) found that all stages of the beetle were killed at 36° F (2.2° C) for 16 days and 25° F (–3.8° C) for 7 days.

Food and Other Materials Attacked. Based on the records of many investigators, plant materials attacked by the cigarette beetle include: aniseed, areca nuts, atta (a wheat product in India), bamboo, beans, biscuits, cassava, chickpeas, cigars (figure 194), cigarettes, cocoa beans, coffee beans, copra, coriander, cottonseed (before and after harvest), cottonseed meal, cumin, dates, dried banana, dried cabbage, dried carrot, dried fruits, drugs, flax tow, flour, ginger, grain, herbs, herbarium specimens, insecticides containing pyrethrum, juniper seed, licorice root, paprika, peanuts, rhubarb, rice, seeds of various trees and plants, spices, and yeast (figure 194). It also breeds in animal matter, such as dried insects and dried fish, fishmeal, and meatmeal, and has been recorded attacking leather and the stored wax of *Cocos coronata*. This insect also feeds on furniture stuffing and bookbinders' paste, and may incidentally damage cloth, upholstery, paper, and books (Howe, 1957). Paprika and dog food are commonly infested in the home.

Drugstore Beetle, *Stegobium paniceum* (L.) (Anobiidae)

This cosmopolitan species resembles the cigarette beetle, and rivals it as a pantry pest. It feeds on any of the household foods and spices, as well as wool, hair, leather, horn, museum specimens, and drugs. It has been known to perforate books and wooden objects, and even tin or aluminum foil and lead.

Description. The adult is cylindrical, and 2.25 to 3.5 mm long (plate IV, 7; figures 193 and 179 [larva]). The color is uniformly reddish to reddish brown. Very fine hairs are arranged in longitudinal rows on the elytra. The drugstore beetle can be distinguished from the cigarette beetle by the distinctly striate elytra, the 3 enlarged segments at the tips of the antennae, and by the less hairy larva. When comparing the 2 species in figure 193, note that the sides of the pronotum of the cigarette beetle are more uniformly rounded than those of the drugstore beetle. The drugstore beetle resembles even more closely another anobiid, the furniture beetle, *Anobium punctatum* (figure 115, chapter 5), which also has striate elytra and a 3-segmented antennal club. However, the front margin of the pronotum of the drugstore beetle is much more rounded than that of the furniture beetle, which is more squared off (truncate). There is also a close resemblance in the appearance of the larvae, although that of the drugstore beetle is less hairy.

Life Cycle. The life cycle resembles that of the cigarette beetle. There may be from 1 generation to as many as 4, depending on the temperature. The drugstore beetle possesses symbiotic organisms that resemble yeasts and produce vitamins of the "B" group, and is therefore not dependent on the presence of this important nutritional factor in its foods. For this reason, it may subsist on foods of very low quality (Hickin, 1964).

Tobacco Moth, *Ephestia elutella* (Hübner) (Pyralidae)

This cosmopolitan pest of stored tobacco is also a pest of cereals, chocolate, cocoa beans, coffee, cottonseed, dried fruits, flour, nuts, seeds (various), and spices.

Description. The adult moth has a wingspread of 15 to 16 mm. The forewings are brownish gray, crossed with 2 oblique, light-colored bands, while the hindwings are uniformly gray. The mature larva is 10 to 15 mm long, and creamy white, tinged with yellow, brown, or pink. The light-brown pupa turns nearly black just before the emergence of the adult.

Almond Moth, *Cadra cautella* (Walker) (Pyralidae)

The larvae of this cosmopolitan insect feed on cereals, cocoa beans, dried fruits, flour, grain, peanuts, seeds (various), and shelled nuts. For some unknown reason, for many years the moth seldom infested dried fruit in California, whereas it evidently did so in other parts of the world (Donohoe, 1946). However, it has finally become a pest of some importance in California on both dried fruits and almonds (H. D. Nelson, correspondence). *Cadra cautella* is second only to the tenebrionid beetle *Tribolium castaneum* in its abundance in foodstuffs imported into Great Britain (Freeman, 1964).[16]

Description. The almond moth resembles the Mediterranean flour moth, *Anagasta kuehniella*. It has a wing expanse of 14 to 20 mm. The forewings are mottled gray on most individuals, but on some they are strongly suffused with fawn-

[16] The most common insects in the descending order of frequency in imported stored foodstuffs were: *Tribolium castaneum, Cadra cautella, Oryzaephilus mercator, Necrobia rufipes, Corcyra cephalonica, Oryzaephilus surinamensis, Lasioderma serricorne, Alphitobius diaperinus, Sitophilus oryzae, Trogoderma granarium, Plodia interpunctella,* and *Tenebroides mauritanicus*.

colored scales. The moths fly with a conspicuously rapid wing vibration, and dart about quickly. The larvae are so similar to those of the raisin moth, *Cadra figulilella* (described in the next section), that the two can be distinguished only by an expert (Okumura, 1955).

Whiteshouldered House Moth, *Endrosis sarcitrella* (L.) [= *lactella* (Dennis and Shiffermuller)] (Oecophoridae)

This widely distributed insect feeds on a great variety of plant material, such as dry seeds, seed potatoes, rubbish in bird nests, thatch on roofs, fungi on trees, and miscellaneous dry vegetable refuse. It is known to do serious damage in wine cellars by boring into corks, and is also said to infest homes, feeding on dry seeds, meal, and carpets (Austen *et al.,* 1935). In Great Britain, where the oecophorids appear to be more important pests of stored products than they are in the United States, *Endrosis sarcitrella* is almost certain to be found wherever broken grain, flour, or other vegetable debris is allowed to accumulate undisturbed (Munro, 1966). This species is widely distributed in California, and in a survey of insects on stored foods and seeds, it was recorded from corn and other grains, mixed feeds, and wheat bran (Strong and Okumura, 1958).

Description. The male and female average 6.3 and 10.5 mm long, respectively. Recently emerged adults can be readily recognized by their white shoulders and prothorax, contrasting with the grayish-white forewings that are marked by dark patches. The hindwings are narrowed toward the tips, somewhat like those of the Angoumois grain moth (*Sitotroga cerealella*), but not so acutely.

Life Cycle. Egg incubation requires 10 to 58 days, the larval period is 38 to 133 days, and the period from egg to adult is 62 to 235 days. The full-grown larva is about 12 mm long, and white, with a brownish head (Austen *et al.,* 1935; Woodroffe, 1951; Munro, 1966).

Brown House Moth, *Hofmannophila pseudospretella* (Stainton) (Oecophoridae)

This is another cosmopolitan species that appears to be especially important as a pest in Great Britain. The larvae feed on book bindings, cereals and cereal products, clothing, corks, dried fruits, dried plants and insects, furniture stuffed with horsehair, furs, grains, herbs, leather, legumes, mohair furniture, paper, seeds (various), and skins. They can cause damage by boring into the corks of wine bottles, causing them to leak. They can also eat their way through nylon, plastics, and the insulation of cables, and are known to infest bird nests (Austen *et al.,* 1935; Bender, 1941; Woodroffe, 1951). Cole (1962) concluded that the brown house moth in "attics and roof spaces" fed on bird and animal droppings, dried corpses, and other dry organic debris. In his experience, carpets were the household articles most often damaged, especially the edges of fitted carpets when they were folded under.

In California, this species is widely distributed, but is not an important pest and is not commonly found in homes. In a survey of pests of stored products, it was reported from cantaloupe seeds, celery seeds, fishmeal, grain, and mixed feeds (Strong and Okumura, 1958). In warehouses in Britain, it is usually an omnivorous scavenger in spilled cereals and flour, but sometimes attacks bulk wheat, bagged flour, and other commodities, and is also found in homes. This moth is believed to owe its persistent survival to its omnivorous habits, its high reproductive capacity, and the resistance to adverse conditions of all stages except the growing larvae, which thrive only in high humidity. However, the larvae can undergo prolonged diapause, during which they are resistant to desiccation (Woodroffe, 1951; Munro, 1966).

Description. The adult moth is bronze-brown, with dark-brown to black flecks on the forewings. The male and female average 8.5 and 14.5 mm in length, respectively. The full-grown larva is white, with a tan head, and is about 6 mm long.

Life Cycle. Egg incubation requires 8.5 to 110 days, and the larval stage lasts 71 to 145 days, both varying greatly with the temperature. The larva spins a cocoon in which it incorporates its foodstuff. The life cycle varies from 192 to 440 days in the laboratory, and takes about 12 months in nature (Woodroffe, 1951; Munro, 1966).

INSECT PESTS ATTACKING DRIED FRUITS AND NUTS

A number of moths and beetles are important pests of dried fruits, although some of them may be equally or even more important as pests of nuts, cereals, and other stored products. Of particular importance are the larvae of certain pyralid moths and the larvae and adults of nitidulid beetles. Some species infest fruit before it is harvested or while it is drying on trays. Others infest fruit principally or even entirely after it is completely dry and in storage or in packages.

(A) Insect Pests of Ripening or Partially Dried Fruits

Raisin Moth, *Cadra figulilella* (Gregson) (Pyralidae)

This moth is world-wide in distribution, and is one of the most common pests of dried fruits, but it occasionally feeds on cereals and nuts, and also bores holes in oranges on low-hanging branches in Tulare and Fresno counties in California. Damage to dried fruits, particularly raisins, is caused by the larvae feeding on the surface of the fruit and by the resulting pellets and webbing. This species primarily infests fruit in orchards or vineyards or while it is drying. Sound grapes may be attacked, but the larvae usually infest only grapes that have been injured by other causes, chiefly the cracking and crushing of the berries in tight bunches. After the fruit is dry, there is little danger of infestation. Processed fruits in storage, if less than a year old, may sometimes be infested; if older, this rarely occurs, and then only very lightly (Donohoe, 1946).

Description. The adult moth (plate V, 2) is about 10 mm long when at rest with its wings folded. The forewings are grayish and the hindwings are satiny white. Both pairs have a prominent, satiny fringe. The full-grown larvae are about 13 mm long, white, with 4 rows of purple spots along their backs.

Life Cycle. The adults, which live only about 2 weeks, stay in shaded and protected places during the day and are active chiefly in the early evening. The female lays about 350 minute, white eggs, scattering them over the surface of the host fruit. They hatch in about 4 days, and the larvae feed on the fruit for about a month, starting with the ripening of the earliest kinds, such as mulberries. They then enter the soil or some dark retreat to pupate. Breeding may continue from April or May until November.

Driedfruit Beetle, *Carpophilus hemipterus* (L.) (Nitidulidae)

This beetle is among the most important insects that damage fruit, such as figs, dates, and less often raisins, before they are completely dried and stored. Infestation takes place mainly on cracked or fermenting fruits; sound ones are rarely attacked.

Description. The adults (plate V, 4) are 3 mm long, dull or shiny black, with 2 conspicuous, amber-brown spots at the posterior tips of the elytra and 2 smaller, more obscure spots of the same color near the bases of the lateral margins of the elytra; the antennae and legs are amber; and the surface of the body is finely punctate, each pit giving rise to a hair. In common with most nitidulids, the elytra are short and truncate, exposing the 2 posterior abdominal segments. The antennae are distinctly knobbed.

Life Cycle. The female lays an average of over 1,000 small, white eggs in ripe figs, or on piles of fermenting fruits in orchards or refuse dumps. The incubation period is 1 to 7 days, averaging 2.2. The full-grown larvae are 6 to 7 mm long, white or yellowish, with the head and posterior tip of the body amber-brown. The body is clothed with hairs, and has 2 pairs of large, dorsal tubercles and 1 pair of small, lateral ones at the posterior end of the abdomen. All larval stages are very active, and try to hide quickly when disturbed. Larval development takes 6 to 14 days. The pupae are robust, oval, somewhat spiny, about 3 mm long, and white or pale yellow, becoming amber when nearly mature. The pupal stage lasts 5 to 11 days. The complete life cycle may vary from a minimum of 15 days in summer to several months in winter. In winter, the adults survive in decaying melons, cull figs, waste dates, and moist raisins; the larvae may be found in such favorable locations as moist cull figs or dates. The pupae generally overwinter in the soil (Simmons *et al.*, 1931; Stickney *et al.*, 1950; Lindgren and Vincent, 1953).

Other Nitidulid Species

Three other species of nitidulid beetles are associated with *Carpophilus hemipterus* as pests of dried fruits. The corn sap beetle, *Carpophilus dimidiatus* (F.) (plate V, 4), is about 5 mm long, pitch-black to dull yellow-brown, with the elytra somewhat paler yellow-brown. *Haptonchus luteolus* (Erichson) is one of the smallest, averaging about 3.5 mm in length. It is yellowish brown,

Fig. 195. Pineapple beetle, *Carpophilus humeralis*.

but considerably lighter in color than the preceding species. The pineapple beetle, *Carpophilus humeralis* (F.) (= *Urophilus*) (figure 195), is entirely black, and is one of the largest of the 4 species found in dried fruits, being 4 to 5 mm long.

(B) Insect Pests of Dried Fruits During Storage

Unlike the pests already discussed, those in this section generally begin their infestations after the fruit has been harvested and dried (usually during storage), particularly if the infested products have been stored for long periods. Certain species that are important pests of dried fruits in storage are not discussed fully in this section, since they are even more harmful to other stored foods. Their habits and biological data are given under more appropriate headings. The acarid mites are important pests of dried fruits, but are considered later under a separate heading.

Indian Meal Moth, *Plodia interpunctella* (Hübner) (Pyralidae)

In California, this is the most important moth attacking dried fruits in storage and in the home (figure 186). Although it is principally a storage pest, it can also attack fruit before it is dried. Infestations may begin on a ranch where the fruit has been dried and temporarily stored in boxes, and the insects will continue to develop in packing houses unless the fruit is fumigated and stored in insectproof enclosures (Simmons *et al.*, 1931). Because it attacks a wide variety of foods, the Indian meal moth is one of the most common species found in the home. The "white worms" seen in packaged dried fruits are nearly always the larvae of this moth (Donohoe, 1946). Abundant webbing in infested materials is characteristic of infestations by the Indian meal moth. Its larvae are often found far from infested foods because they usually crawl away from their foods to construct silken cocoons in which to pupate. (The description and life cycle of the Indian meal moth have been discussed earlier in this chapter.)

Mediterranean Flour Moth, *Anagasta kuehniella* (Zeller) (= *Ephestia*) (Pyralidae)

The larvae of this species infest dried fruits, and leave their webbing. As with the Indian meal moth, they crawl away from the food to construct pupal cocoons, but usually near the infested product. They may also migrate to other parts of the pantry or kitchen, or even elsewhere in the house.

Almond Moth, *Cadra cautella* (Walker) (Pyralidae)

This cosmopolitan insect, described earlier in the present chapter, was once referred to as the "fig moth" because early investigators in California believed it to be an important pest of dried fruits, particularly figs. However, despite being widely recorded in the world's literature as a major pest of dried fruits in storage, Donohoe (1946) and other investigators, after examining thousands of moths reared from dried fruits from field and storage areas throughout California and parts of Arizona, found no instance of infestation by this particular species. In fact, the moths were not even found in dried fruits when they were stored close to almonds, the favored hosts. The earlier records of the moths on dried fruits in California were believed to be results of misidentification. Their absence during a long period was considered to be a biological mystery. As already stated, *Cadra cautella* is again assuming some importance as a pest. According to the USDA Dried Fruit and Tree Nuts Insect Investigation Laboratory in Fresno, *C. cautella* is infrequently observed infesting raisins or other dried fruits and nuts in California (H. D. Nelson, correspondence).

Driedfruit Moth, *Vitula edmandsae serratilineella* Ragonot (Pyralidae)

This insect develops readily in dried fruits, preferably slightly fermented ones, as well as in rotting bee combs. It has also been found in the cells of a carpenter bee, *Xylocopa tabaniformis orpifex*, where the larvae had probably fed on pollen (Linsley, 1943*a*). The adults are likely to be found in the dark corners of storage bins of dried fruit or between rows of stacked boxes.

Description. The mottled-gray moths are nearly twice as large as the Indian meal moth. The larvae, 18 mm long, are white or pinkish except for the head, prothoracic shield, and anal plate, which are brown. They are larger than those of the Indian meal moth, but have the same habit of prepupal migration (Essig, 1926; Donohoe, 1946). They closely resemble the larvae of the raisin moth, *Cadra figulilella* (Okumura, 1955).

Stored Nut Moth, *Aphomia gularis* (Zeller) (Pyralidae)

This moth occurs principally in India, China, and Japan, and is found occasionally in Britain, continental Europe, and North America. It was first reported from North America in 1919 in a shipment of peanuts to California that originated in China. It was reported again in 1930 and 1931 from San Jose, California, under conditions that indicated it was infesting prunes. Its larval food appears to be limited to seeds, flour, nuts, and dried fruits. It is found in California somewhat rarely. It has been reared from walnuts in Montreal, Canada. (It is included in this section because nuts do not form its only diet.)

Description. The forewings are 10.5 mm long and 3 mm broad, and the length from head to tip of abdomen is about 10.5 mm. The forewings are light yellowish brown, and the hindwings are lighter. The thorax, abdomen, and legs are of the same basic color as the forewings.

Sawtoothed Grain Beetle, *Oryzaephilus surinamensis* (L.), and Merchant Grain Beetle, *O. mercator* (Fauvel) (Cucujidae)

These beetles (figures 191 and 192) infest dried fruits, usually after they have been stored for long periods. (They have been described under "General Feeders" earlier in this chapter.) The lower the moisture content of the fruit, the more severe the infestation can be. Without adequate protection, dried fruits, particularly raisins, can be so severely damaged by these pests as to be virtually worthless. They readily penetrate the packages in which foods are kept. The small, brown, flattened adults are only about 3 mm long, and the prothorax bears 6 sawlike teeth on each side (see figure 192). It is difficult to distinguish one species from the other, and the relative importance of the two has not yet been determined in regard to dried fruits.

Flour Beetles, *Tribolium* spp. (Tenebrionidae)

These beetles, *Tribolium confusum* Jacquelin duVal and *T. castaneum* (Herbst), described earlier in this chapter, rarely infest dried fruits severely, but both species can survive in them, and care should be taken to prevent infestation, particularly when dried fruit is stored or shipped in close proximity to products that are even more subject to infestation.

Rusty Grain Beetle, *Cryptolestes ferrugineus* (Stephens) (= *Laemophloeus*) (Cucujidae)

This very small cosmopolitan beetle (also mentioned earlier) is usually found in stored grain and other foodstuffs and under the bark of trees. It can also be found in stored fruits, and is sometimes the dominant species in moldy ones, especially in winter. The adults frequently infest fallen, moldy, drying fruits, particularly figs.

(C) Moths Attacking Walnuts and Almonds in the Orchard

Persian walnuts and almonds are grown commercially in the United States, principally in areas on the Pacific slope in California. The nuts can become infested in the orchard with the larvae of certain moths. The same species may also infest stored nuts, and can be accompanied by others that attack the nuts only after they are in storage. Insects found infesting walnuts or almonds in the home usually come from nuts obtained from dooryard trees and stored indoors. Also, moths emerging from the nuts outdoors sometimes become accidental intruders in the home.

Codling Moth, *Laspeyresia pomonella* (L.) (= *Carpocapsa*) (Olethreutidae)

This is an age-old cosmopolitan pest of apples. It was not found in walnuts until 1909, and had become a serious pest in a limited area of southern California by 1918. It is now widely distributed on walnuts, and annual treatment with pesticides is required.

Description. The adult moth (figure 196) has a wingspan of about 20 mm. The forewings are

Fig. 196. Codling moth, *Laspeyresia pomonella*. *Top*, female; *bottom*, overwintering larva in a silken cocoon.

brownish, with several gray or paler cross lines and a dark-golden and coppery patch at the tips. The hindwings are paler. The outer margins of the wings are fringed with hairs.

Life Cycle. The female lays flat, white, discoidal eggs, about 1 mm in diameter, singly on foliage or nuts. The full-grown larvae are about 20 mm long, yellowish or pinkish, with the heads and thoracic shields brown. They pupate in any hidden retreats in white, felty cocoons. The pupae are light amber to dark brown, and are about 13 mm long. In the larval stage, the winter is passed in cocoons (figure 196) under loose flakes of bark, in old pruning cuts, tree crotches, in litter on the ground, or in other protected places. Pupation occurs in early spring, and the first brood of moths appears in April or May. There may be a complete second generation within any year, depending on seasonal temperatures, and in some years even a partial third.

Cultural Control. It is helpful to destroy the hiding places of the hibernating larvae by scraping the loose bark off old trees and removing prunings and other rubbish from the ground.

Filbertworm, *Melissopus latiferreanus* (Walsingham) (Olethreutidae)

This pest, also known as the Catalina cherry moth, was formerly known to infest only the Catalina cherry (*Prunus lyonii*) in southern California, and the large, green galls of the blue oak (*Quercus douglasii*) in central and northern California. Investigations conducted in 1944 showed that this insect was an important pest of walnuts in the Sacramento Valley (Bacon *et al.*, 1948), and it now attacks most varieties of walnuts throughout the state. It may also attack acorns and hazelnuts (filberts). It is likely to become destructive only in areas where one of its native hosts, the oak-apple gall, is produced in abundance; the moths migrate from oaks to walnuts.

In 1968, the author received many larvae and cocoons of the filbertworm collected from a deep-piled woolen rug in a home in San Marino, California. The larvae had not been feeding on the rug, but appeared to have originated from infested walnuts in or outside the house. There are many oak trees near-by.

Description. The adult moth (figure 197) has a wingspread of about 13 mm. It is pale or dusky bronze, and has a complete, broad, coppery median band and a broken postmedian coppery band on each forewing.

Life Cycle. The whitish, disklike eggs are laid in the fruit of the host plant or on oak galls. The full-grown larvae are about 20 mm long. They resemble the larvae of the codling moth, but in the later instars they become uniformly creamy white, while the codling moth larvae become

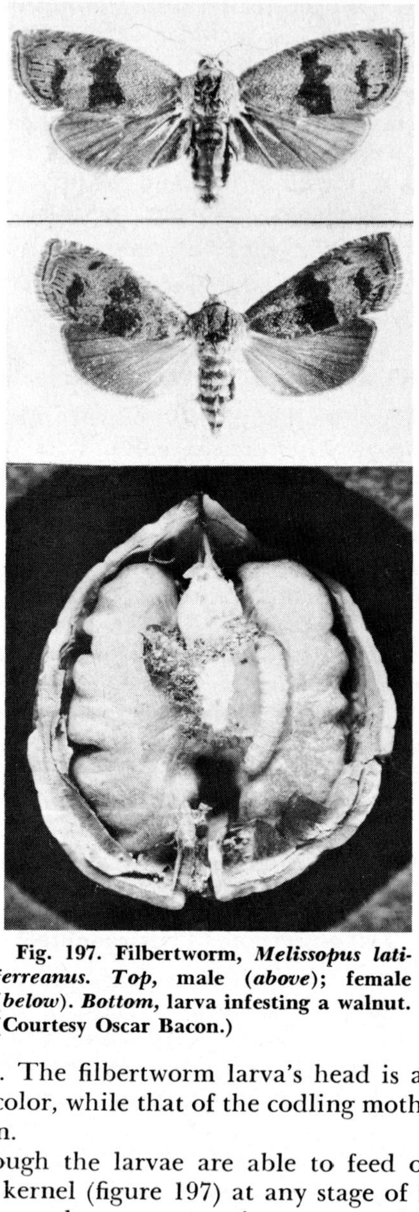

Fig. 197. Filbertworm, *Melissopus latiferreanus*. *Top*, male (*above*); female (*below*). *Bottom*, larva infesting a walnut. (Courtesy Oscar Bacon.)

pinkish. The filbertworm larva's head is a clear amber color, while that of the codling moth larva is brown.

Although the larvae are able to feed on the walnut kernel (figure 197) at any stage of its development, they cannot gain access to it until cracks appear in the husk at the time of maturity. Therefore, the nuts are not damaged until late August or September. If a continuous food supply is present, as when walnuts are grown near oak trees, there are 2 broods per year and a partial third brood (Bacon *et al.*, 1948).

Navel Orangeworm, *Paramyelois transitella* (Walker) (= *Myelois venipars* Dyar) (Pyralidae)

This insect was known for many years as a scavenger in cracks or wounds of navel oranges. Beginning about 1943, it began to attack walnuts, depositing its eggs on the green husks, and it now occurs throughout most of the nut- or fruit-growing areas of California. It is also found in many varieties of fruits as a scavenger, developing in those that are mature but injured or mummified. It is considered to be a primary pest of walnuts and almonds. The larvae can enter the sound nuts after their hulls have cracked, thus causing the damage a few weeks before harvest. Some larvae only rasp the surface, while others bore deeply into the nut meat. Webbing produced by the larvae is sometimes so profuse as to hide any unconsumed nut meat (Wade, 1961). When almonds are infested and damaged in the field by the navel orangeworm, peach twig borer, or codling moth, they are more subject to attack in storage by the merchant grain beetle (Spitler and Hartsell, 1967).

A Related Species. Another moth, *Ectomyelois ceratoniae* (Zeller), with similar habits, is the most important pest of stored almonds in Israel. It also attacks other nuts, dried fruits, stored carob pods (*Ceratonia siliqua*), and stored seeds in Israel, as well as walnuts in South Africa (Calderon *et al.*, 1969). It is said to infest dried fruits, seeds, and nuts from widely separated areas in Europe, Africa, and "America" (Corbet and Tams, 1943).

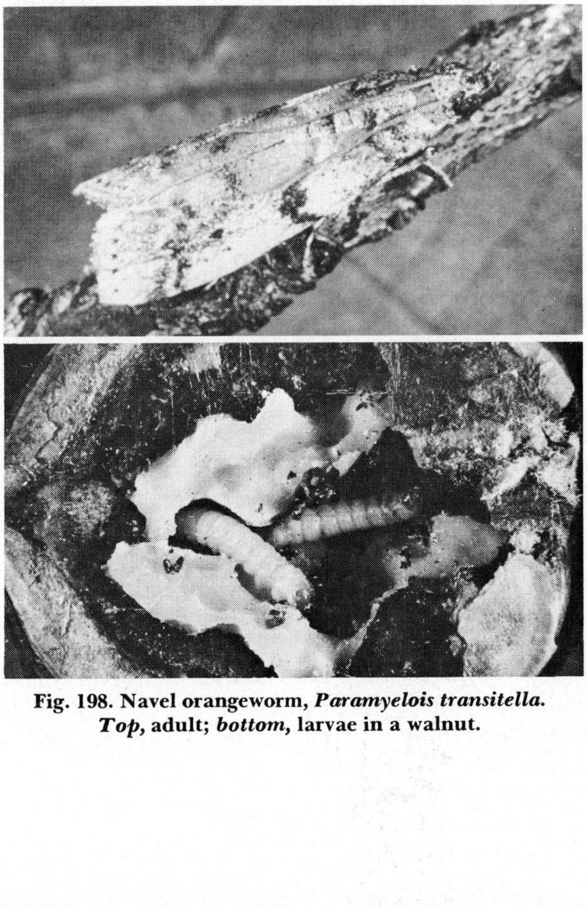

Fig. 198. Navel orangeworm, *Paramyelois transitella*. *Top*, adult; *bottom*, larvae in a walnut.

Description of the Navel Orangeworm. The adult (plate V, 3; figure 198) is silvery gray, with irregularly narrow to broadly wavy lines on the forewings. The hindwings are uniformly light, except for a darkening at the tips and along the veins. The females average 10.9 mm in over-all length; the males, 9.7 mm. The larvae (figure 198) are at first reddish orange and, after the first molt, pinkish orange to cream-color. As they develop, they may be entirely cream-colored if they are reared on oranges. If the food is contaminated with spores of black mold, the intestinal tract will appear black. Full-grown larvae range from 13 to 19 mm in length. The pupae are dark brown, with sutures clearly evident. They range from 7.25 to 12 mm in length, those of the females averaging larger (Wade, 1961).

Life Cycle. The larvae and pupae overwinter in old nuts left on the trees. The adults emerge in the spring, and the female lays her first eggs in April on old, sticktight nuts remaining on the trees or on twigs fairly close to the old nuts. The oval, flattened egg has many ridgelike marks on the surface. It is at first lustrous creamy white, but changes to pink and finally to reddish orange. Infestation becomes evident sometime in August, for the current crop is attacked only after the hulls have split. The infestation increases until the nuts are harvested (Wade, 1961).

Cultural Control. Early and complete harvest of nuts is desirable in order to minimize the possibility of their becoming infested. Sticktights should be removed from the trees and destroyed, as well as nuts left lying on the ground or in tree crotches. Good codling moth control is an aid in the control of the navel orangeworm, which often enters nuts through holes made by codling moth larvae.

There is no known effective chemical control for the navel orangeworm.

Almond Moth, *Cadra cautella* (Walker) (Pyralidae)

This moth has already been discussed as a pest of dried fruits. It derived its common name from its former importance as a pest of almonds in California. Then for many years it was seldom seen, but is now observed in increasing numbers on almonds as well as dried fruits.

Peach Twig Borer, *Anarsia lineatella* Zeller (Gelechiidae)

This moth is a serious pest of peaches, almonds, apricots, nectarines, plums, and prunes. The larvae injure the buds and young twigs and also enter the fruits, particularly those of the almond. They are also known to infest walnuts. The full-grown larvae average 10 mm in length, and are reddish brown except for the yellowish-white intersegmental membranes and the black head. The pupae are 6 mm long, naked, smooth, and dark brown. The adults are about 8 mm long, steel-gray, with white and dark scales. The wings are narrow and fringed, with an expanse of 12 to 15 mm, and are held rooflike when at rest.

Fruittree Leafroller, *Archips argyrospilus* (Walker) (Tortricidae)

This tortricid occurs on many varieties of fruit trees, feeding principally on foliage, but the larvae are occasionally found in walnuts. They are about the same size as those of the codling moth, but are greenish, while the codling moth larvae are often pink.

(D) Storage Pests of Walnuts and Almonds

Unless the field-insect pests of walnuts and almonds just discussed are controlled, they will continue to feed within the shells, and may completely destroy the nut meats. In addition, these insects create conditions favorable to the establishment of many secondary feeders among storage pests, such as *Tribolium confusum, T. castaneum, Oryzaephilus surinamensis,* and *O. mercator.* The Indian meal moth, *Plodia interpunctella,* is a primary feeder among the storage pests, and can become a problem if the nuts are stored more than 30 days (Spitler and Hartsell, 1969). (See figures 186, 191, and 192.)

ACARID MITES (ACARIDAE) (= TYROGLYPHIDAE)

There are several species in this cosmopolitan family that are of importance chiefly as pests of stored foods, flour, grain, seeds, and bulbs, feeding on these products or the fungi that develop on them. These mites are mostly very minute, almost colorless, with the body bearing a few long setae, and with the legs terminating in 1 claw and usually a sucker. When very abundant, the mites and their cast skins form a buff-colored dust and give rise to a characteristic odor. The living mites move slowly away from a source of light. An interesting characteristic of these mites is a nymphal resting stage (hypopus), during which they take no food but attach themselves to other mites, insects, and humans by means of suckers, and are thereby transported to other suitable places for development, where they drop off, molt, feed, and develop to the adult stage. The hypopus may be resistant to fumigants. However, the "cheese mites," *Tyrophagus casei* and *T. longior,* do not have a hypopial stage. An exhaustive treatise on the morphology and biology of the mites that may be found in stored foods was written by Mrs. A. M. Hughes (1961).

Prolonged storage of products subject to infestation without cleaning of pantries and shelves where such materials are placed may result in serious infestations, for these mites can subsist on organic debris in cracks, crevices, and corners, and are readily spread to other areas.

Control methods for acarid mites are given under "Control of Pantry Pests" at the end of this chapter.

Grain Mite, *Acarus siro* L. (= *Tyroglyphus farinae* De Geer)

The grain mite (figure 199) is the most important and best known of the mites infesting grain and flour, but can also infest other stored food products. In severe infestations, grain mites are responsible for the "mite dust" on the surface of cheese or on the floor at the bases of stacks of flour or wheat sacks. Although the bodies of the mites are almost colorless, their legs are brown, so the "mite dust" has a brownish tinge. If the mites are crushed, their "minty" odor is easily recognized. Whereas *Acarus siro* is widely distributed in temperate zones, it occurs in the tropics only through repeated introductions with grains and foodstuffs. It is not an important pest in hot climates, if it persists at all (Solomon, 1962).

Description. This mite is pearly white, with legs varying in color from pale yellow to reddish brown. The male is 0.32 to 0.42 mm long, and the female, 0.35 to 0.65 mm. The mite can be distinguished from members of other genera because of the femur of the male's foreleg being enlarged and bearing a stout spine ventrally, a characteristic of the genus *Acarus* (A. M. Hughes, 1961).

Life Cycle. In a day, a female may lay 20 to 30 oval, smooth, white eggs, 0.12 mm long, singly or at random on the food material. Under favorable conditions, the female can lay up to 800 eggs in her lifetime (about 40 days). The larvae feed and grow rapidly for 3 days, then lie motionless for 2 days before molting. The life cycle may require 17 days at a favorable temperature ranging from 64 to 71° F (18 to 22° C), but during the winter at 50 to 60° F (10 to 16° C), it requires 28 days (A. M. Hughes, 1961). *Acarus siro* does not have

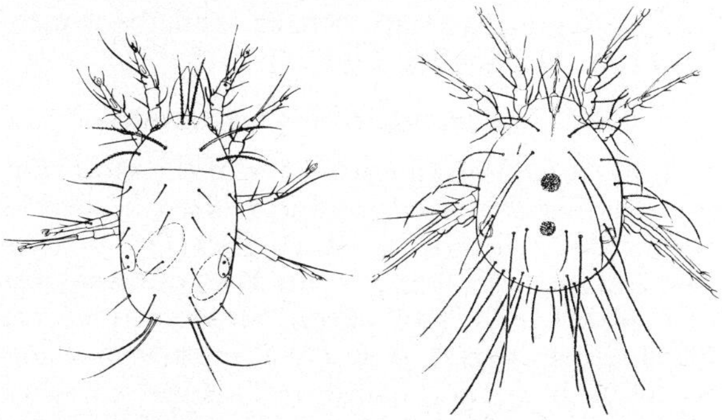

Fig. 199. Mite pests of stored products. Grain mite, *Acarus siro* (*left*), and grocer's itch mite, *Glycyphagus domesticus*. (From A. M. Hughes, 1961.)

tracheae, so the cuticle must be permeable to gases and therefore also to water vapor. Although the epidermis and cuticle obviously have some ability to regulate water balance, this does not suffice to prevent a lethal rate of water loss below 60% relative humidity, and the mites cannot survive at such humidities (Solomon, 1962; Munro, 1966). The distribution of the grain mite in stored grain is conditioned by the relative humidity of the air between the kernels, which in turn depends on the water content of the grain. The mite avoids relative humidities below 75% and above 85% (Knülle, 1963*a*).

Curry (1971) found wholewheat flour to be the best substrate for rearing this species. In 5 plastic vials, each containing 3 grams of foodstuff plus mold inhibitor, he reared a total of 26,055 mites in wholewheat flour, 6,140 in cornmeal, 1,395 in polished rice, 68 in white flour, and 48 in enriched white flour in a period of 30 days. He concluded that white flour, in the absence of mold, was not a suitable food substrate for *A. siro*. Fungi not only attract these mites to a food, but are themselves good food sources. On unbroken grain, fungi serve another function, for the mites cannot feed on unbroken kernels. The fungi weaken the hull, permitting access to the germ, where the mites can feed and reproduce (Thomas and Dicke, 1971).

The grain mite is commonly found in bird nests around houses and stores, and it is believed that birds aid in its dispersal. Rodents and other mammals also probably contribute to dispersal, as well as humans on their shoes and clothing (Woodroffe, 1953). The mites are common in grocery stores, and are usually the culprits responsible for a severe dermatitis known as "grocer's itch." In fact, probably any acarid mites could cause dermatitis on the body of a sensitive person (Pratt, 1963; Busvine, 1966).

Cheese Mite, *Tyrophagus casei* (Oudeman)

These cosmopolitan mites are common in stored food, cheese, grain, damp flour, old honeycombs, and insect collections. They can eat out small holes in cheese, and have been cultured for addition to Altenburger cheese to impart a characteristic "piquant" taste. When the cheese is covered with a grayish powder, consisting of enormous numbers of living and dead mites, cast skins, and feces, it is considered by some people to be "ripe" and particularly delectable (A. M. Hughes, 1961; Busvine, 1966).

Biology. This species, like others of the genus, has at least 4 pairs of long bristles trailing from the hind end of the body. It is translucent, but has tan-colored legs and mouthparts. The males are 0.45 to 0.55 mm long, and the females, 0.50 to 0.70 mm. The life cycle requires 15 to 18 days at 23° C (73° F) and 87% relative humidity.

Tyrophagus longior (Gervais)

These mites are commonly associated with grain, hay, and straw, but are also found on cheese. In the United States, they are not so serious a problem as *Acarus siro* (A. M. Hughes, 1961). *Tyrophagus longior* has been implicated in human pulmonary, urinary, and intestinal acariasis (Baker *et al.*, 1956).

Biology. The mites are translucent like the foregoing species, but the legs and chelicerae are more deeply tan. The males are 0.33 to 0.53 mm long, and the females, 0.53 to 0.67 mm. The life cycle requires 14 to 21 days.

Mold Mite, *Tyrophagus putrescentiae* (Schrank) (= *lintneri* Osborn)

This is one of the most destructive pests of stored products. According to A. M. Hughes (1961), the mold mite is found most frequently on products having a relatively high fat content, such as linseed, dried egg yolk, peanuts, cheese, ham, herringmeal, and copra, but occurs on dried banana, spilled wheat, oats, barley, and flour, and is also a pest on mushrooms. Linsley and Michelbacher (1943) recorded this mite in California, under the synonym *Tyroglyphus americanus* Banks, as occurring in enormous numbers in cereals and cereal products. They stated that the mites sometimes caused considerable concern and annoyance when infesting stored products in homes, and were responsible for the severe dermatitis called "grocer's itch" when attacking humans who handled infested materials. The mites also cause "copra itch," familiar to persons handling infested copra dust (A. M. Hughes, 1961). As already stated under *Acarus siro,* itches can also be caused by other acarid mites, among which *A. siro* is very notable.

Biology. This is a small, translucent mite, with almost colorless chelicerae and limbs. The males are 0.28 to 0.35 mm long, and the females, 0.32 to 0.41 mm. It is more slender than other acarids, and the long setae do not project so stiffly from the surface of the body. The life cycle requires 2 to 3 weeks at 23° C (73° F) and 87% relative humidity.

Driedfruit Mite, *Carpoglyphus lactis* (L.) (= *Acarus passularum* Hering)

This species is found principally on dried fruits, jams, and products containing lactic, acetic, or succinic acids. It has been found in honeycombs, fruit drink residue, senna confections, wine clinging to pieces of cork or forming a floating scum on the surface, rotting potatoes, flour, dried-milk powder, and caramel used in manufacturing sweets.

It seems to be particularly attracted to substances that are starting to ferment (A. M. Hughes, 1961). In California, this species is sometimes found in enormous numbers on a variety of dried fruits, imparting a disagreeable odor. It has also been observed on cured hams and the pollen paste of bees (Essig, 1926; Simmons *et al.,* 1931; Donohoe, 1946). *Carpoglyphus lactis* can cause "driedfruit mite dermatitis" (Baker *et al.,* 1956).

Biology. This oval, slightly flattened mite is translucent, except for the color of the contents of the alimentary canal, which can be seen through the body wall. The legs and mouthparts are pinkish. There is no transverse suture dividing 2 body regions, as with certain other species. The 2 sexes are 0.38 to 0.40 mm long, and are similar in appearance. The females can lay 25 to 30 eggs in a week, attaching them to the substrate by means of a stalk, and can then survive another 35 to 45 days.

Thyreophagus entomophagus (Laboulbene)

This mite, found on dried fruits and jams, has an elongate and oval body, a colorless, shiny cuticle, and pale-brown, stumpy legs. The males are 0.30 to 0.45 mm long, and the females, 0.45 to 0.60 mm. It has been recorded from England, France, Germany, Italy, Russia, and the United States (A. M. Hughes, 1961).

Glycyphagus destructor (Schrank)

This cosmopolitan mite is one of the most common species on stored products, often in association with *Acarus siro.* A. M. Hughes (1961) stated: "It lives in stores of grain, any kind of cereal, linseed, dried fruit, hay, cheese, straw, rodent and bumble bee nests." [The last 2 records probably represent its natural habitat.] This species is considered to be strictly mycophagous (fungus-eating), but is also attracted to sugar, and will live on fresh wheat germ.

Biology. The body is pearshaped, being somewhat constricted behind the fourth pair of legs. Minute papillae give the cuticle a dull-whitish coloration. The males are 0.35 to 0.50 mm long, and the females, 0.40 to 0.56 mm. Pectinate (feathery) hairs are characteristic, and are particularly long on the posterior third of the body. The legs are long and thin. The mite moves with a characteristic jerking gait.

Grocer's Itch Mite, *Glycyphagus domesticus* (De Geer)

This mite (figure 199), also called "house" or "furniture" mite, is world-wide in distribution, being found in "flour, wheat, hay, linseed, tobacco, sugar, cheese, and bee frames, in bees' and birds' nests, and in the sweepings from barns. It will increase to enormous numbers on rush furniture and on upholstered chairs stuffed with green Algerian fibre" (A. M. Hughes, 1961). As the common name implies, the mites often cause "grocer's itch," in common with some other acarids just noted. (See also, chapter 10.)

Biology. Live mites of this species can be distinguished from *Glycyphagus destructor* by having a more rounded body and by a uniform, unhurried gait in contrast with the jerking progress of the latter. The long hairs on the body and legs are less coarsely pectinate than those of *G. destructor,* and in life they radiate stiffly from the surface. Like the related species, this mite has long legs with tapering segments. The males are 0.32 to 0.40 mm long, and the females, 0.40 to 0.75 mm (A. M. Hughes, 1961).

PREDACEOUS MITES

Together with the acarid mites in stored products, one can often find their predators. Under conditions of proper temperature and humidity, these predators can greatly suppress acarid mite infestations, but in the meantime their presence in stored food is likewise a nuisance, and some species also cause dermatitis in humans. Prominent among the predaceous mites are those of the family Cheyletidae, an extensive review of which was prepared by Summers and Price (1970). The family Pyemotidae also figures prominently.

Cheyletus eruditus (Schrank) (Cheyletidae)

This is one of the most common predaceous mites. Like most predators, it is a solitary animal, and doesn't gather in masses like its victims. It moves about rapidly. A striking structural character is the pair of stout, pincerlike pedipalps used to seize its prey. It probably injects venom into its victim, for the latter is soon paralyzed. It then takes about a half hour to suck the disabled prey dry. Male mites are rare; reproduction is mainly parthenogenetic. The females are 0.45 to 0.62 mm long. The efficacy of this species as a predator is limited by the fact that it requires moderately high temperatures. It reduced the grain mite, *Acarus siro,* practically to extinction in grain at 25° C (77° F) and 70 or 90% relative humidity, but was unable to prevent its increase at 5 to 10° C (41 to 50° F). *Acarus siro* can increase in temperatures at which *Cheyletus eruditus* is static. However, the latter can live when the moisture content of the grain mite's food is too low to allow reproduction. Under such a condition, the predator may completely suppress *A. siro.* In summer, *C. eruditus* tends to be dominant, but *A. siro* is dominant under the moister and cooler conditions of winter (Solomon, 1962; Knülle, 1963*b*). Some species of *Cheyletus* may infest man and cause localized skin troubles (Horsfall, 1962).

Acarophenax tribolii Newstead and Duvall (Scutacaridae)

The females of this mite are parasites of flour beetles (*Tribolium* spp.), as the specific name indicates. They are 0.16 to 0.19 mm long before becoming gravid and 0.24 to 0.29 mm in diameter when gravid. The nongravid female is round and arched, tapers slightly toward the posterior end, and is yellow. With their styletshaped chelicerae, the mites pierce the cuticle of their host where it is soft, beneath the second pair of wings, and suck the body contents. Then, with bodies distended with embryos, they leave the victim. About 14 or 15 mites develop within the female's body, and only one of these is a male, which never leaves. Therefore, the females are probably all fertilized before leaving the enlarged genital opening. The mother dies as soon as the young have emerged (A. M. Hughes, 1961).

Straw Itch Mite, *Pyemotes ventricosus* (Newport) (Pyemotidae)

The females of this mite are ectoparasites on the immature stages of many arthropods, including pests of stored products. This species is discussed more fully in chapter 9, under "Mites," because of its primary importance as a cause of human dermatitis.

The scutacarids and pyemotids are closely related. *Pyemotes ventricosus* rivals *Acarophenax tribolii* in the peculiarity of its biology. In both species, the mites reach maturity within the mother's body, but in the case of *P. ventricosus,* the male emerges to fertilize the females after they have left the body (A. M. Hughes, 1961; Munro, 1966).

INSECT CONTAMINATION OF MILLED CEREAL PRODUCTS

From the great number of storage pests discussed in this chapter, the reader may conclude that the identification of all species would be a truly formidable task. Then, how much more difficult would be the identification of insect fragments in canned, milled, or otherwise processed foods! However, specialized methods have been found to accomplish this task (Harris, 1946, 1955; Harris *et al.,* 1952, 1953, 1969).

Detection of contamination by rodents, insects, mold, and sometimes yeasts and bacteria can often be accomplished by simple examination, preferably aided by the use of a hand lens or microscope. For example, live insects or the webbing of moths are obvious evidences of insect infestation. Insects may also leave characteristic feeding marks on the various food products, or leave their frass (excreta, cast skins, or particles of food) on them. The fecal pellets of rodents are also easily identified. If flour is sifted through a No. 64 wire screen, most of the eggs, larvae, and adults of any insects that may be present can be removed. Photographs of the hundreds of types of insect fragments from the many species of all orders and families of insects that commonly contaminate stored foods may be found in a monumental treatise on the subject by Kurtz and Harris (1962).

Insect mandibles are resistant to grinding, and may be recovered from ground food products and examined microscopically. Kurtz *et al.* (1952) published a key for the identification of the mandibles of common insects infesting stored grain, recovered from various food products. Heuermann and Kurtz (1955) published a paper on the identification, on the basis of elytral patterns, of 20 insects associated with food products. Even small pieces of elytra can be used to identify insect species.

If food contamination is severe, the nature of the contaminant can sometimes be determined by its odor. *Tribolium* and *Rhyzopertha* produce a musty, penetrating odor, and tyroglyphid (acarid) mites produce a distinctive, sweet, "minty" odor.

With low-voltage X rays, it is possible to reproduce an image of the amount of tissue removed from seeds by insects. Live insect larvae can be detected in a radiograph because larvae contain much water and X rays do not readily penetrate it. Pupae are less readily visible, and adult insects are practically invisible. Dead larvae can occasionally be seen, but not dead pupae or adults.

Various methods have been devised to isolate the contaminant. Some are based on the insolubility of the contaminant (or a product thereof) in solvents, chemical reagents, and enzymes that can dissolve the food product. Some are based on differences in size of the food and the contaminant. Others are based on differences in specific gravity or on selective wetting. The Waldman trays procedure, for example, is based on the fact that animal hairs, insects, and insect fragments are wet by oil, and plant material is wet by water, and the specific gravity of oil is less than that of an aqueous medium. Several of these principles of isolation are usually combined in an isolation and detection method (Harris *et al.*, 1969).

CONTROL OF PANTRY PESTS

For all pests, control begins in the field, but this is particularly true of the pests of dried fruits. The pests that infest injured fruits on the tree or vine or while the fruit is on drying trays can be partially controlled by eliminating near-by breeding places, such as waste fruits, or by protecting drying trays from infestation. Raisins dried on trays containing about 100 mg of malathion per sq ft (929 sq cm) are protected from insects during drying and storage (Nelson, 1967; Nelson *et al.*, 1967).

Figs are fumigated with methyl bromide for the control of field pests at the beginning of storage, and again every 30 to 60 days to control storage pests. Many pests of dried fruit are removed during processing, sorting, and cleaning before it is packaged. Malathion dust has proved to be effective in protecting almonds (Spitler and Hartsell, 1967), walnuts (Spitler and Hartsell, 1969), dried figs (Spitler, 1969), raisins (Spitler and Hartsell, 1970), and prunes (Spitler and Clark, 1970) for as long as a year in storage. Phostoxin tablets, placed on paper plates on large stacks of natural raisins in sweat boxes, have been found to give complete kills of merchant grain beetles (*Oryzaephilus mercator*), Indian meal moth larvae (*Plodia interpunctella*), and raisin moth larvae (*Cadra figulilella*) at dosages as low as 5 tablets per 1,000 cu ft (28 cu m) when the gas was allowed to remain for 6-day periods. Effective treatments leave residues of less than the 0.01 ppm federal tolerance.

Some pests, such as the sawtoothed grain beetle and Indian meal moth, attack dried fruits only after they are in storage or in the home pantry. Regardless of the source of infestation, the homeowner's efforts must be directed toward the detection and removal of infested material.

Methyl bromide fumigation is used to control both field and storage insects in the "in-shell" nuts. However, there is some danger that with repeated fumigations, the tolerance of 200 ppm of methyl bromide may be exceeded. Recent preliminary investigations with malathion, particularly as a dust in a "tumbling" method of treatment, indicated that this insecticide, applied to the surface of the walnut shell, would protect the nut meat from insect attack. The malathion caused no differences in flavor or odor, but the heaviest dust treatment resulted in a lighter color of the shell (Spitler and Hartsell, 1969).

In January, 1972, the Environmental Protection Agency's Pesticides Regulation Division registered the use of malathion 57% E for the protection of stored in-shell almonds from the Indian meal moth and merchant grain beetle. A suitable mechanical spray applicator should be used that regulates the rate of application of the insecticide to the flow of the almonds. The spray nozzle should be shielded against wind and air currents. A mist so fine that it drifts away should not be used. Four fluid ounces of the emulsifiable concentrate are enough to treat 5 tons of almonds (about 120 ml for 4,500 kg).

Some infestations of packaged food originate in the food-processing plant or warehouse. Broken packages should not be purchased, or should be exchanged for unbroken packages when discovered, for the chance of these being infested is greater than for perfectly sealed ones. Even sealed packages can be penetrated by some species of insects. The Food and Drug Administration currently approves only synergized pyrethrins for the treatment of packages to make them insect-repellent.

Opportunities for packaged foods to become infested in the home are numerous. Pantry pests, such as various species of beetles, moths, psocids, and mites, may gain access to a home in old furniture, rugs, drapes, bedding, and other products of plant or animal origin. Some pantry pests breed in the nests of rodents or insects, or in dumps or other places where by-products and waste from processing plants are deposited. They may enter the home under doors, through broken or loose-fitting screens, through fireplace flues, or around utility pipes. Some may breed in spilled food products on pantry shelves, particularly material in cracks, crevices, and corners that may have been overlooked for long periods. Periodic cleaning of the shelves helps to prevent infestation of stored food products by pantry pests.

Infestations are most likely to occur in packages that have been opened for the removal of a portion of the contents and then left unsealed for long periods. Some of the pests may find their way into other food packages, but even those in a single package may become so numerous that large numbers may find their way into every suitable material in the home, and will eventually crawl over floors, climb up walls, and gather about windows. Control of pantry pests requires that all sources of infestation, usually opened or damaged packages of such things as cereals, dried fruits, nut meats, flour, meal, cornstarch, crackers, breakfast foods, dehydrated foods, macaroni, chocolate, cocoa, dry soup mixes, spices of all kinds, dry dog food, or birdseed be found and destroyed. If the pests have already begun to migrate from the original and most obvious source of infestation, all exposed foodstuffs should be thoroughly examined. If there is any doubt as to whether or not they are infested, they should be destroyed. If insects are suspected in packaged food products but cannot be found, the packages can be placed in plastic bags for a few days. Occasional inspection of the transparent bags will reveal any insects that may have escaped.

Materials believed to be uninfested should be kept in tight containers, such as jars or tins with tight-fitting lids. Then for several weeks, food products subject to insect attack should be purchased only in quantities that can be consumed quickly without prolonged storage. Cheeses have been dipped in wax to provide a complete protective coating against mites. If only a small, incipient infestation of insects has been noted, the removal of the source of infestation may be all that is required for control. If the insects have had the opportunity to become widely distributed, it may be desirable to remove all the pantry contents, clean the shelves, and spray them with a safe and approved insecticide.

Residual sprays of 2% chlordane, 2% malathion, 1% diazinon, or 0.5% lindane are generally effective when applied on shelf surfaces, thoroughly treating all cracks and crevices. Malathion is sometimes used because of its low mammalian toxicity, but even "odorless" malathion leaves an odor that many people consider to be objectionable. Space sprays containing pyrethrins are useful for the rapid elimination of flying moths, but they have very short residual activity. Resmethrin (SBP-1382®), synergized with piperonyl butoxide, has been found to be effective against a strain of red flour beetles (*Tribolium castaneum*) that was highly resistant to pyrethrins.

Ham and cheese may be protected against cheese skipper infestation by sanitation and fly-proofing. The latter requires tight-fitting doors and 30-mesh wire or plastic screen, as well as residual sprays applied to walls and floors. Certain types of bags have been found to protect hams from infestation (Fulton, 1953). Cheese can be protected with cheesecloth and paraffin.

Infestations of acarid mites can be prevented by keeping their potential environment moderately dry. The mites are destroyed by a temperature of 130° F (54° C) for several hours (Linsley and Michelbacher, 1943). When chemical control is attempted, infested foodstuffs should first be removed and destroyed and the shelves cleaned. Sprays of pyrethrum and piperonyl butoxide or other suitable synergist have been successfully used against the cheese mite. Ethylene dibromide was found to be outstandingly effective as a fumigant against 2 species of mites found infesting packages of *chapatties,* a common baked wheat preparation in India. It was 46 and 17 times more effective in controlling *Tyrophagus putrescentiae* (Schrank) than methyl bromide and phosphine (normally generated from Phostoxin tablets), respectively, and 4 and 6 times more effective against *Caloglyphus krameri* (Berlese) than methyl bromide and phosphine, respectively (Muthu *et al.,* 1970).

Chapter 8 · *Pests of Fabrics and Paper*

THE WIDESPREAD USE of wool and other animal hair in clothing, carpeting, furniture, and other household materials gives insects the potential for being of great economic importance if they are able to feed on keratin, a proteinaceous constituent of these materials. There are approximately 30 species of moth larvae, 15 species of beetle larvae, and hundreds of species of bird-infesting biting lice (Mallophaga) known or suspected to be able to digest keratin, and they are the only animals that *can* digest it (Waterhouse, 1958). The pests of economic importance are found mainly in 2 groups: the carpet beetles and the clothes moths. Clothes moths are the more important of these groups in the southern states, but carpet beetles are more important in the remainder of the country. In either group, it is only the larvae that cause damage. They will feed on or damage many other substances of animal or vegetable origin besides fabrics and paper.

Although keratin-containing substances (wool, hair, and feathers) are their preferred food materials, carpet beetles and clothes moths will also attack other fabrics, such as cotton, linen, silk, and synthetics if the fabrics contain contaminants of nutritional value, such as urine, perspiration, beer, milk, or fruit juice. Merely handling a fabric with the bare hands will impart some nutritional factors to it, and even fungus spores settling from the air will add some nutrition (Pence, 1958). Clean, processed wool cannot support the normal life cycle of the clothes moth unless it is contaminated with certain nutritional supplements.

In one investigation, it appeared that minerals, proteins, and B vitamins were the attractant substances in fabric stains for the larvae of the webbing clothes moth, *Tineola bisselliella,* and the furniture carpet beetle, *Anthrenus flavipes.* Of the salts tested, K_2HPO_4, KCl, NaCl, and Na_2HPO_4 were the most readily eaten, and appeared to be the substances that would make a stain such as tomato juice attractive to these insects. However, the larva of the black carpet beetle, *Attagenus megatoma,* was not especially attracted to these salts or to the more common stains (Mallis *et al.,* 1962).

There are many species of insects that are unable to digest keratin, but nevertheless can cause damage by chewing through keratin-containing fabric. Termites, crickets, cockroaches, silverfish, psocids, and even some dermestids may be included among such species.

Wherever they occur, the fabric-infesting insects are potentially highly destructive, but the substitution of synthetics for many of the woolen fabrics and the treatment of rugs, carpets, and other woolen fabrics with insectproofing agents in their places of manufacture have led to a steady decline of the over-all importance of this group of household pests. However, the hazard remains undiminished for unprotected furniture and fabrics imported from areas of the world where they are not treated. Even initially protected materials can lose their immunity to attack through cleaning and aging.

CARPET BEETLES (DERMESTIDAE)

Four species of carpet beetles comprise not only the most important group of fabric pests, but also the group that is most difficult to control. The adults feed largely on pollen and nectar, and may enter homes in spring and early summer. All damage (figure 200) is done by the larvae, which develop in dark, undisturbed locations. Unlike clothes moth larvae, they spin no webbing, but their hairy cast skins and their sandlike pellets (shown in the figure; often the color of the fabric eaten) are evidences of infestation. The cast skins look much like live larvae, and may give the casual observer the impression that there is a greater infestation than is actually present. Pupation takes place in the last larval skin, and the adult may remain in the partially shed pupal skin for as long as 3 weeks. Evidence of a carpet beetle infestation may be the presence of the small, adult beetles flying to windows or larvae wandering from room to room. The adults resemble lady beetles in shape.

The source of a carpet beetle infestation is sometimes difficult to find. For example, one pest control operator treated an office building 3 times, each time failing to find the source of the beetles seen by the occupants. On the fourth attempt, he traced the beetles to a telephone cable in the wall, where the insects were discovered to be feeding on the insulation.

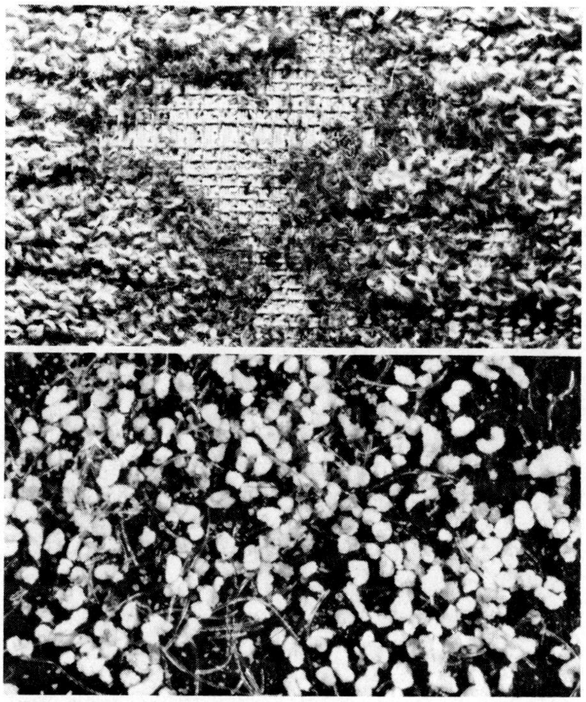

Fig. 200. Furniture carpet beetle, *Anthrenus flavipes*. *Top*, damage to a woolen carpet; *bottom*, fecal pellets, greatly magnified.

Whereas clothes moth larvae usually are found on their food material, carpet beetle larvae crawl about for considerable distances, and may be found on cotton goods or other materials on which they do not feed. They may often be found behind baseboards and moldings, in cracks in floors, behind radiators, in the air ducts of heater systems, and other secluded areas. If found in one area of a house, such as in the carpeting of a hallway, they can be expected to be found in adjoining rooms also. The adults and larvae of carpet beetles may also be seen outdoors in and under dead animals, in the hides and horns of cattle, or in the nests of birds, rodents, wasps, or bees (including the honeycomb). Beetles or clothes moths originating in such materials under floors or in wall voids or attics sometimes invade the living space, and the originally infested materials may continue to be sources of infestation after carpets and other infested articles have been treated (Linsley, 1944). The cast skins and fecal pellets of such insects in the attic sometimes drop into the rooms below via apertures around beams, through the holes in perforated ceiling tile, and around light fixtures. This material is sometimes mistaken for the frass of woodborers, and may lead to a mistaken diagnosis of the problem.

Black Carpet Beetle, *Attagenus megatoma* (F.) [=*piceus* (Olivier)]

Except in the arid western states, the black carpet beetle is the most important species. In California, it is of little importance as a fabric pest. It is most likely to be found infesting stored food products, or when it is found on woolens, the infested items have usually arrived recently from the eastern United States. Among infested materials found in federal quarantine were "woolen rugs and clothing, silk clothing, carpeting, felts, furs, skins, yarn, upholstered furniture, wool-filled blankets, house insulations containing sheep wool and cattle hair or binders of farinaceous materials, meat and insect meal [rearing mixtures], kid leather, milk powders, casein, books, birds' nests, cayenne pepper, beans, peas, corn, wheat, rice, and a long list of seeds and seed products" (Back and Cotton, 1938). In a survey of 118 grain-storage and feed mills in Ohio, the black carpet beetle was found in 110 of the buildings (Triplehorn, 1965). Although not a serious grain pest, it was the most widely distributed insect observed.

Although the larvae can eat large holes through any suitable food material, they tend to eat the nap from the fabric and shun the base threads. In furs, they tend to cut hairs at the base with no injury to the hide, but they leave it bare in

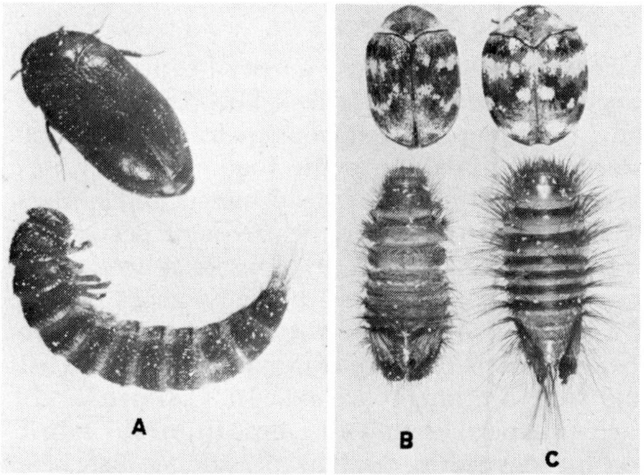

Fig. 201. Adults and larvae of the principal species of carpet beetles. *A*, black carpet beetle, *Attagenus megatoma*; *B*, varied carpet beetle, *Anthrenus varbasci*; *C*, furniture carpet beetle, *Anthrenus flavipes*.

appearance. The larvae may also burrow through packaging material to obtain food, thus providing access for other species of insects (Truman and Butts, 1967).

Description. The adults (plate VI, 2; figure 201, *A*) are 2.8 to 5 mm long, shiny black and dark brown, with brownish legs. The full-grown larva is narrow, up to 8 mm in length, tapers toward the rear (carrotshaped), terminates in a tuft of long hairs, and has short, stiff hairs covering the body. It varies from light brown to almost black. The cuticle is hard, smooth, and shiny, resembling that of a wireworm. It is relatively resistant to pesticides.

Life Cycle. The number of days required by the various stages are: egg, 6 to 16; larva, 166 to 330; and pupa, 8 to 14. The period from egg to adult is 180 to 360 days, and the adult may live another 30 to 60 days (Laudani, 1961).

The female lays about 90 eggs in the lint under and behind baseboards, in cracks, in ducts of hot-air furnaces, and other dark and protected places. There may be 5 to 11 larval instars, or even as many as 20 under particularly favorable conditions.

A Related Species

A black, indigenous dermestid, *Attagenus alfierii* Pic, resembling the black carpet beetle, has been reported from southern California in bird nests. It infests barley, cottonseed meal cake, and mixed feeds, and sometimes attacks fabrics in homes (Strong and Okumura, 1948).

Varied Carpet Beetle, *Anthrenus verbasci* (L.)

This species infests the usual range of household articles attacked by carpet beetles, such as carpets, woolen goods, silks, skins, furs, feathers, hair, horn, cereals, red pepper, fishmeal, or any processed animal or plant food. However, the favored larval foods are dead insects and spiders, which may account for the frequent occurrence of the varied carpet beetle in insect collections in museums. In common with other carpet beetles, this species will feed on cottons, linens, and synthetic fibers if they are contaminated, but it will not feed on rayon acetate. In California, the varied carpet beetle is the most injurious fabric-feeding dermestid, and its prevalence has been attributed to the wasp nests around homes, which these beetles also infest (Linsley, 1944).

Another Related Species

A newly introduced dermestid, *Anthrenus coloratus* Reitter, closely resembles the varied carpet beetle. It was originally known from Asia, Africa, and Europe, and has been found in Washington, D.C., Maryland, Illinois, and California. Its habits are apparently similar to those of other carpet beetles.

Description of *Anthrenus verbasci*. This species has a number of varieties. They differ in shape, size, color, and pattern of the scales, but in general the adults are 2 to 3 mm long, and have a varied pattern of white, brownish, and yellowish scales on the back and fine, long, grayish-yellow scales below (plate VI, 3; figure 201 *B*). Where the elytra terminate, they do not form a cleft as do those of the furniture carpet beetle, *A. flavipes*. The mature larva (plate VI, 2) is 4 to 5 mm long, and has a series of light- and dark-brown transverse stripes. If it is suddenly alarmed, the larva erects the 3 dense tufts of bristles and hair located on each side of the rear end of the body, spreading them out to form beautiful, round plumes (Back, 1938). The larva is broadest near the rear, and becomes narrower toward the front end, unlike other carpet beetles.

Life Cycle. The female lays about 40 eggs. There are usually 7 to 8 larval instars, but the number may vary from 5 to 16. The numbers of days for the various stages are as follows: egg, 17 to 18; larva, 222 to 323; and pupa, 10 to 13. The period from egg to adult is 249 to 354 days, and the adult may live another 14 to 44 days (Laudani, 1961).

Furniture Carpet Beetle, *Anthrenus flavipes* LeConte

As the name implies, furniture carpet beetles are best known for their attacks on upholstered furniture, feeding on hair padding, feathers, and woolen upholstery; but they also attack the usual food materials that are eaten by other species, including the glue of bookbindings and natural-fiber brushes. This species is distributed throughout the United States, and is important in California, where it thrives in a slightly drier environment than the varied carpet beetle.

Description. The adults (plate VI, 1; figure 201, *C*) are 2 to 3.5 mm long, spotted yellow, white, and black above and white beneath. They are slightly larger and more rounded than the varied carpet beetle, and have a cleft where the elytra meet posteriorly. The adult spends a quiescent period within the old larval skin before becoming active. Recently hatched larvae are white, but turn light yellow to dark red or chestnut brown as they develop. They are about 5 mm long when full grown. They are widest in front and become narrower toward the rear. Mature larvae are darker than those of the varied carpet beetle, and are able to run swiftly.

Life Cycle. The female lays about 60 eggs in from 1 to 3 clutches, generally in such places as the pile of mohair chairs, in the nap or on the surface of clothes, and in cracks and crevices. The numbers of days for the various stages are as follows: egg, 9 to 16; larva, 70 to 94; and pupa, 14 to 17. The period from egg to adult is 93 to 126 days, and the adult may live another 30 to 60 days (Laudani, 1961).

Common Carpet Beetle, *Anthrenus scrophulariae* (L.)

This species occurs throughout much of the world, but in the United States it is most abundant in the northern area. It is of little importance in California. One unusual feeding habit of this carpet beetle is its tendency to eat slits in carpets that correspond in position with the floor cracks (Back, 1938).

Description. The adult is 3 mm long, gray to black, with a varied pattern of whitish and orange scales on the back, and with whitish and orange-red scales around the eyes and on the clypeus. The mature larvae are about 3 mm long, reddish brown, and have many black or brown hairs. They move about actively.

Life Cycle. The female lays about 60 eggs on textiles or other larval food sources. The number of days for various stages are: egg, 10 to 18; larva, 60 to 80; pupa, 7 to 12. The period from egg to adult is 77 to 110 days, and the adult may live another 20 to 30 days (Laudani, 1961).

Birdnest Carpet Beetle, *Anthrenus pimpinellae lepidus* LeConte

This species is of interest as a fabric pest mainly in California, where several races or varieties are indigenous. The adults are mottled brown and white, and the black larvae are similar to those of the black and furniture carpet beetles. In California, this species is common in the nests of the house sparrow, house finch, black phoebe, barn swallow, and cliff swallow (Linsley, 1944). The adults emerge outdoors in March and April and fly to flowers for feeding and mating, then usually seek out bird nests for oviposition but may find their way into houses. The larvae feed primarily on feathers and animal hairs. They pupate in late fall, and the adults overwinter in the last larval skin.

Odd Beetle, *Thylodrias contractus* Motschulsky

The odd beetle, also called "tissuepaper beetle," is related to the carpet beetles despite their mutual lack of resemblance in general appearance. It is widely distributed in the United States, but not in large numbers, and is seldom a pest. A striking feature of the species is the great difference in appearance between male and female. The male is 2 to 3 mm long, slender, yellowish brown, with disproportionately long legs and antennae. When at rest, the elytra are contiguous for only about one-third of their length, and then separate to reveal the abdomen. The wingless, larviform females (plate V, 8) are also brownish in color, but differ from the males so strikingly that they would hardly be suspected to be of the same species. Both sexes have a median ocellus between the compound eyes.

The larvae somewhat resemble those of the common carpet beetle, but lack the tuft of hair at the rear end and the long hairs on the dorsal surface. Instead, there is a transverse row of stout bristles at the rear edge of each body segment, those on the prothorax being clubshaped. The larva can roll itself up into a ball when disturbed. The larvae feed on dry animal matter and woolen, silk, or muslin cloth, even when it is unsoiled. They can penetrate tissue paper used to wrap their normal food materials, but the common name "tissuepaper beetle" does not appear to be appropriate.

CLOTHES MOTHS (TINEIDAE)

Clothes moths are small, yellowish or golden insects, with narrow wings fringed with long hairs. They are not attracted to light, and attempt to hide when disturbed, so the housewife is not likely to see them. However, they are occasionally seen flying in subdued light. Most moths suspected by the layman to be clothes moths are other species, usually harmless accidental intruders. Clothes moths are most likely to become evident from damaged fabrics and by the presence of silken webs spun by the larvae. The larvae can feed on clothing, carpets, rugs, furs, fabrics, blankets, stored wool products, upholstery, piano felts, and brush bristles. They may feed on fabrics of vegetable origin or synthetics, if the fabrics are mixed with wool, or may use such materials to construct their cocoons.

Webbing Clothes Moth, *Tineola bisselliella* (Hummel)

This cosmopolitan species is the most common clothes moth in the United States. It is very destructive. Damage to clothing occurs most commonly in dark, hidden areas, as under collars and cuffs, but sometimes the larvae are very active and may be seen crawling on clothes or on the

Fig. 202. Webbing clothes moth, *Tineola bisselliella*, adult and larvae. *Inset*, eggs.

floor beneath badly infested furniture. When feeding between the carpet and the floor, the larvae may extend their silken feeding tubes along the floor cracks, and the feeding pattern is indicated by the webbing beneath the carpet (Curran, 1949).

Besides feeling on clothes, carpets, rugs, and upholstered furniture, the webbing clothes moth feeds on furs, stored wool, and such miscellaneous articles as the animal bristles of brushes and the felts in pianos. In nature, it feeds on pollen, hair, feathers, wool, fur, dead insects, and dried animal remains.

Description. The body and wings of the adult are uniformly golden colored, except for reddish-golden hairs on the top of the head (plate VI, 4; figure 202). The wingspread of the female is about 11 mm and that of the male is somewhat less.

Life Cycle and Habits. The adults are very active, can penetrate through surprisingly narrow cracks, and can fly considerable distances (Griswold, 1944). However, they are not attracted to light, and gravid females are weak fliers. The female dies after attaching about 40 to 50 eggs, singly or in groups of 2 or more, to the threads of infested clothes over a period of 2 to 3 weeks. The eggs (figure 202, *inset*) hatch within an average of 4 to 10 days in summer, but take as long as 3 weeks in winter.

The newly hatched larvae are only about 1 mm long, and translucent white. Some larvae may spin a small, frail, silken tube or tunnel, incorporating into the silk some fibers, excrement, or cast skins. They then feed within the confines of the tube. Others may merely spin flat mats under which they crawl about, or remain naked for several days before they spin any webbing. Some larvae leave the webbing and crawl about unprotected. The feeding tubes and silken mats make up the webbing that characterizes an infestation. The number of larval molts can vary from 5 to 45, and the period required for larval development can vary from 35 days to 2.5 years, depending on the availability of food as well as relative humidity and temperature. The full-grown larva is shiny, creamy white, and about 12 mm long. When preparing to pupate, it spins a pupal case of silk about 8 mm long, again incorporating textile fibers and excrement. The period required for pupation varies with temperature, but can be as brief of 8 or 10 days in summer or as long as 3 or 4 weeks in winter. The length of the life cycle varies from 50 to 90 days, but can be extended to as long as 4 years under unfavorable conditions (Back, 1923).

Casemaking Clothes Moth, *Tinea pellionella* (L.)

In the United States, the casemaking clothes moth is much less important than the webbing clothes moth as a pest. It can be found throughout the country, but is most common in the southern states. It is particularly capable of damaging hair and feathers, but will also feed on spices, tobacco, hemp, and skins. The moth derives its common name from the small silken case (figure 203) that the larva spins about its body and carries about wherever it feeds, thrusting its head and legs out

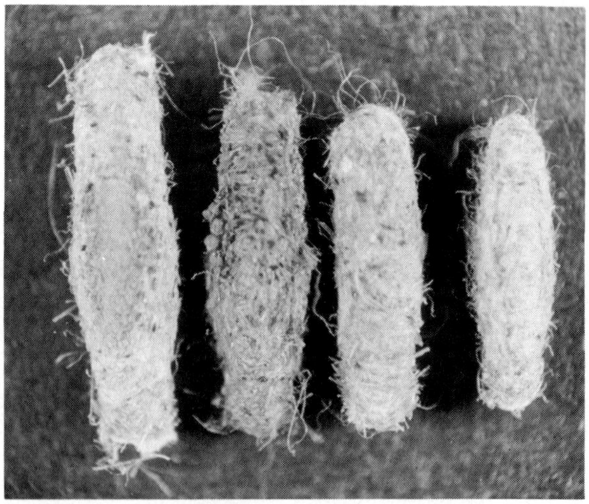

Fig. 203. Silken cases of the casemaking clothes moth, *Tinea pellionella*.

in front. Sometimes in a severe infestation, larvae may crawl up on a wall in large numbers, dragging their cases behind them.

Description. The adult is somewhat smaller and more brownish than the webbing clothes moth, and has 3 dark spots on the wings, but the spots become less discernible if the wing scales are worn off. The hindwings are yellowish brown. The males are smaller and lighter in color than the females, and are active fliers. The females are sluggish, and fly only for short distances. The first thoracic segment of the larva, at first brown, later becomes black, and is divided by a longitudinal band.

Biology. The larva can turn within its case and feed on food material at either end without altering the position of the case. If the case is removed from the larva when it is very near pupation, the larva will die. Rarely will the larva spin a web directly on the material on which it is feeding, but will usually attach its case to the material by means of silken threads. Pupation takes place within the case after both ends have been sealed with silk. There were found to be 3 or 4 generations a year at 26° ± 8° C (79° F) and 82% ± 10% relative humidity when larvae were fed on woolen fabrics impregnated with 5% yeast (Cheema, 1956).

Carpet Moth, *Trichophaga tapetzella* (L.)

This moth is rare in the United States. It is larger than the webbing or casemaking clothes moths, having a wingspread of about 19 mm. Its head and the basal third of the forewings are black, the outer two-thirds of the forewings are creamy white, and the hindwings are uniformly pale gray. The larvae prefer coarser and heavier fabrics than those of the 2 species just described, and they construct burrows or silk-lined galleries in all directions throughout the infested materials, such as carpets, tapestries, feltings, furs, and skins. This species has also been reported to be destructive to wallpaper (Back, 1923).

Control of Carpet Beetles and Clothes Moths

The best way to combat carpet beetles and clothes moths is to prevent them from becoming established in the home. The principal weapon is the vacuum cleaner and its brush attachment. Rooms should be cleaned often enough to prevent the accumulation of lint, hair, and other carpet beetle and clothes moth food materials. Close attention should be given to rugs and carpets, draperies, upholstered furniture, closets (especially those containing woolens and furs), radiators and heaters, corners, cracks, baseboards, moldings, and other hard-to-reach places. If an infestation is known or suspected, the sweepings of the vacuum cleaner should be disposed of immediately in such a way as to destroy the insects, to prevent a transfer of infestation from one part of the house to another (Pence and Davis, 1969).

Abandoned nests of birds, rodents, and insects (particularly bees and wasps) that are in or near the house should be removed, for the larvae of both carpet beetles and moths may feed on insect remains that they may contain. Bedding places of pets should be kept clean. Mounted animal specimens or trophies (or even fur-covered toys), insect collections, stored woolens, carpeting, clothing, feathers, furs, old spices, cereals, or seeds should be examined for signs of infestation. The attic and garage should be included in the inspection. Avoid bringing carpet beetle adults into the house on cut flowers, where they are sometimes found feeding on pollen.

Dry-cleaning kills all stages of clothes moths and carpet beetles, but gives no protection against reinfestation. Many cleaning establishments and pest control firms can apply protective treatments. Woolen garments or materials that have been stored for a long time should be occasionally shaken and aired. Brushing can crush most eggs, particularly the fragile eggs of carpet beetles. If they cannot find protection from light, many larvae that are not removed by the brushing will fall from any garments hung in the sun, as on a clothesline. (It is unwise to use such treatment for furs, since they are subject to fading, drying, or even theft.)

Woolen clothing and blankets can be protected against feeding damage by carpet beetles and clothes moths by spraying them lightly with oil solutions of methoxychlor or allethrin (Bry *et al.*, 1968); Gardona® (Bry *et al.*, 1969); Strobane® or Perthane® (USDA, 1966); or other recommended pesticides. A household hand sprayer may be used, or a pesticide may be purchased in a ready-to-use pressurized container that delivers a coarse spray. The clothing or blankets should be hung on a line and sprayed lightly and uniformly until their surfaces are moist (USDA, 1966). Infants' apparel and blankets should be cleaned or washed, and sprayed only if they are to be stored. They should be laundered and dry-cleaned before further use.

Woolens stored in a container, trunk, or closet may be fumigated with paradichlorobenzene

(PDB) crystals and naphthalene flakes or balls (mothballs). The storage container should be tightly sealed. Loose-fitting containers can be lined with heavy paper and sealed with tape. One lb (0.45 kg) of either fumigant is enough to treat 20 cu ft (0.57 cu m) of storage space. Scatter the fumigant between the layers of paper separating the articles to be treated. In a reasonably tight container, 1 application a year should suffice. In closets, apply the fumigant to the floor and shelving, and replenish the supply after it evaporates. All plastic buttons should be removed from clothing before treating, and plastic hangers should not be used, for PDB and mothballs are destructive to plastics and may fuse them with the fabrics.

The 20% Vapona Resin Strip (polyvinyl chloride resin impregnated with dichlorvos) is worthy of trial in almost any situation in which the pests are in a confined area. The strip slowly gives off vapors of dichlorvos, which is much more toxic to insects than PDB or naphthalene, less toxic to man when used as directed in limited storage areas, and is also less expensive. Either the 6-in. (15-cm) strip or the 2-in. (5-cm) "ministrip" can be subdivided to the appropriate size for the space to be fumigated. In a small, tightly sealed container, such as the Schmidt box in which pinned insects are kept, vapors leave a piece of Vapona Resin Strip very slowly because of vapor pressure built up in the box. As little as 1 sq in. (6.45 cm^2) of resin strip is 32 times the required amount to protect a Schmidt box of 150 cu in. (2,458 cm^3) from museum pests (Ryckman, 1969).

Woolens wrapped in heavy paper or enclosed in a cardboard box with sealed edges are protected against fabric pests, provided the woolens are not already infested. Spraying of furs is not recommended, for they can be protected with PDB, mothballs, or dichlorvos resin strips as recommended for woolens, or they can be frequently shaken and aired. If living or dead insects, fecal pellets, cast skins, or fiber particles are not present, as they would not be after the fur is brushed, shaken, or cleaned, mandibular scars on the individual fibers are unmistakable clues to insect damage. An illustrated guide for the diagnosis of fur damage with the aid of a microscope is available (Pence, 1966).

The edges of rugs or carpets can be pulled up and the common household sprays can usually be applied on both sides, as well as on both sides of the pad. Treatment is particularly important under heavy furniture that is seldom moved. Upholstery and draperies can be sprayed, following the same directions discussed for clothing and blankets.

Pesticide sprays may be applied to any surfaces upon which fabric-infesting insects are likely to crawl, such as along the edges of wall-to-wall carpeting, in closets, behind radiators, and in corners, cracks, baseboards, moldings, and other hard-to-clean places. Closets, particularly, should be thoroughly sprayed after removing the contents. Chlordane 2%, premium grade malathion or ronnel 5%, and lindane or diazinon 0.5% have been recommended as insecticides (USDA, 1966). Insecticide dusts are useful to blow into attics, basements, wall voids, or other areas difficult to reach with a spray, or dusts may be sprinkled on the floor and swept into cracks, after which the rug can be put back into place.

Deodorized kerosene (base oil) with dissolved insecticides, when applied as a mist or fog, remains as an oily deposit on the surface of a carpet. The addition of 30 to 50% of isopropyl alcohol (rubbing alcohol) results in good penetration of the insecticide to the bases of the fibers, where fabric pests feed (Pence and Viray, 1963). Oils can cause the nap to separate from the backing, and then the carpet buckles. By improving penetration, the isopropyl alcohol reduces the amount of oil required and decreases the possibility of damage to the carpet. Only enough spray should be used to reach the bases of the fibers, and any excess should be avoided. A kerosene–isopropyl alcohol solution should be used only for treatment of carpets and rugs—not as a space spray. Windows and doors should be left open during application. Pilot lights should be extinguished before treating near floor or wall furnaces, hot-water heaters, or other natural-gas appliances.

Water-base sprays can be effectively used if a wetting agent is added. Such sprays should not contact nonwashable wallpaper or other materials that will color-run or stain.

Fumigation is an effective method for the control of fabric pests. If the pests are widespread in a house, the entire building can be fumigated by a pest control operator. Fumigation is expensive, and requires that the house be vacated for a day or two. It provides no residual protection against reinfestation. Some pest control operators and storage firms have fumigation vaults where infested pillows, mattresses, upholstered furniture, and similar small articles can be fumigated when the infestation is localized. Methyl bromide, the most commonly used gas for fumigating buildings, is known to be a good ovicide. It was found to be effective against all stages of carpet beetles (Pence and Morganroth, 1962).

Insectproofing Agents[17]

In the United States, the 2 most commonly used insectproofing agents have been DDT and dieldrin. Since the recommendation by the Life Sciences Panel of the President's Science Advisory Committee in 1963 that more specific mothproofing treatments be developed, entomologists and chemists of the Savannah Laboratory of the USDA Stored-Product Insects Research Branch have searched for substitutes for the organochlorine compounds. They found that Ciba-Geigy C-9491 or iodofenphos [*O*-(2,5-dichloro-4-iodophenyl) *O,O*-dimethyl phosphorothioate], when applied to woolen cloth from emulsion bath treatments, protected wool fabric from feeding damage by larvae of the black carpet beetle, *Attagenus megatoma,* and the webbing clothes moth, *Tineola bisselliella,* at deposits as low as 0.05 and 0.03% by weight of the cloth, respectively (Bry *et al.,* 1971, 1972). The oral LD_{50} of this compound to rats is reported to be 2,000 mg/kg, approximately 8-fold greater than the LD_{50} for DDT and 40-fold greater than that for dieldrin. However, the registration of dieldrin is still valid for "mothproofing by those manufacturing processes which utilize the pesticide in a closed system" (EPA Determination and Order of June 25, 1972).

According to a more recent report from the same laboratory, when resmethrin was applied to woolen cloth as an aerosol, it protected the cloth against feeding by black carpet beetles and webbing clothes moths for as long as 6 months (Bry *et al.,* 1973*a*). These 2 species were also killed when hit by the resmethrin aerosol. This versatile and relatively safe synthetic pyrethroid shows promise as a replacement for some insecticides now used in the home against carpet beetles and clothes moths (*Bry et al.,* 1973*b*).

Fan Palm Caterpillar, *Litoprosopus coachellae* Hill (Noctuidae)

This insect is an occasional pest in southern California and Arizona. The adult is a light-tan moth, with dark lines on the forewings (figure 204). The full-grown larva is approximately the size of a mature cutworm larva—about 40 mm long—and when mature usually has a pinkish or pinkish-brown color. The larvae normally feed on the flowers, flower shoots, and to some extent on the fruits of fan palms. Large numbers of the more mature larvae sometimes crawl to the ground, or may be blown down by strong winds, and then sometimes enter homes, where they may make their cocoons from material obtained from rugs, drapes, or other household fabrics. Large patches of pile from rugs, drapes, or even velvet can be removed by the larvae. They are attracted to light, and therefore any areas of ingress into lighted homes near palm trees should be closed at night. The application of DDT spray (no longer available) to the crowns of palm trees in June or July and the application of lead arsenate to the flowering shoots early in the season were found to be effective controls (Flock, 1951). Tests have not been made with currently available insecticides.

Fig. 204. Fan palm caterpillar, *Litoprosopus coachellae. Left,* larva and palm flowers on which it sometimes feeds; *right,* adult female.

SILVERFISH (THYSANURA: LEPISMATIDAE)

Silverfish are primitively wingless insects, without metamorphosis (ametabolous), and have flattened, slender, scale-covered bodies that are sometimes silvery, usually with a metallic sheen. The body tapers gradually to the rear, giving the insect a fishlike appearance, suggesting the common names "silverfish" or "fishmoths" or the German *Fischen.* They have long, slender antennae and 3 long, slender appendages at the rear of the body, suggesting another common name—"bristletails."

[17] The term "insectproofing" is used here, even though the term "mothproofing" is firmly established in the literature. There appears to be no reason to believe that moths are more important than carpet beetles as destroyers of fabrics, and the beetles are much more difficult to control.

Silverfish are very long-lived insects, and are unique in continuing to molt their skins for years after they become adults. The female usually lays fewer than 100 eggs, singly or in small groups, but ranging up to as high as groups of 45 for firebrats, in cracks and crevices, or behind baseboards. The eggs are elliptical, and are only about 1 mm long. They hatch in from 2 weeks to 2 months or more, depending on conditions. The young closely resemble the adults, except for size. In their third or fourth molt, the young develop the scales that thereafter give the insects their characteristic coloring.

Silverfish are nocturnal insects that may occur almost anywhere in a house, including attics, wall voids, and subfloor areas. They are very fast-moving, and sometimes are seen only when they are trapped in such places as washbasins and bathtubs, where they remain because they cannot climb smooth, vertical surfaces.

Silverfish feed on any human food, and in addition they may feed on starch, paste, glue (as in bookbindings), starched cotton, linen, silk, or certain synthetic fibers and paper products (figure 205), to which they are attracted by sizing or, as in the case of wallpaper, by paste. Sometimes they cause wallpaper to flake off by removing the paste. Possibly, the most common evidence of the presence of silverfish is paper with the glaze removed in an irregular fashion, with irregular holes, or with the edges notched. Scales, excrement, or yellowish stains may be seen on paper or fabric that has been infested. Silverfish may also feed on dead animals, including dead or injured individuals or cast skins of their own species.

As might be expected, silverfish are serious pests in libraries, where they attack bookbindings and heavily sized paper. They seldom injure wool, hair, or other fibers that are of animal origin. Like termites, they are said to harbor organisms that aid them in the digestion of cellulose materials. A silverfish infestation develops slowly, so when these cryptobiotic insects become numerous enough to be seen occasionally in a home, this indicates that the infestation is probably an old one, or that large numbers were brought into the home as eggs, active stages, or both, in cardboard cartons, books and papers, and other household materials. All the species described in this chapter are in the family Lepismatidae.

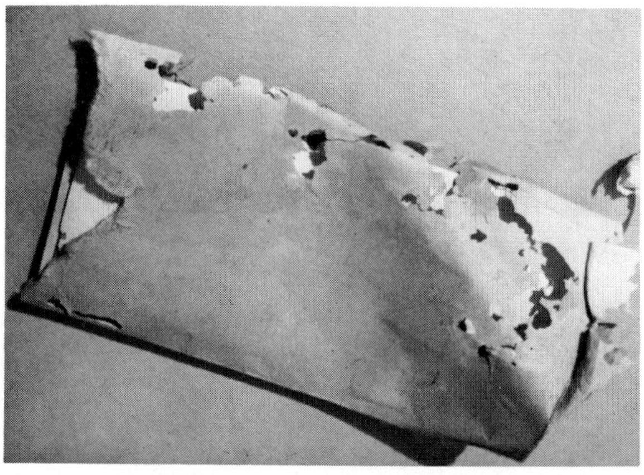

Fig. 205. Paper damaged by silverfish.

Fig. 206. Two common thysanurans. *Left,* firebrat, *Thermobia domestica; right,* gray silverfish, *Ctenolepisma longicaudata.*

Silverfish, *Lepisma saccharina* L.

This common household pest is believed to have come originally from the tropics but, like so many other household pests, it has been able to extend its range to temperate countries by dwelling in damp and warm locations in the homes and other buildings of man. Silverfish or their eggs can be unknowingly transported from one building to another in cardboard cartons, books, papers, and many other carbohydrate substances.

Description. This is an elongated silvery-gray insect with a metallic sheen. It attains a length of about 12 mm, not including the very long appendages.

Life Cycle. According to Sweetman (1939), females might lay 1 to 3 eggs per day on a number of successive days or at intervals of several days or even weeks. Wigglesworth (1964) stated that the female alternately molted and laid eggs, and might molt up to 50 times before becoming an adult. He found that the eggs hatched in 50% relative humidity at 22 and 27° C (71 and 80° F), but only above 75% RH at 29 and 32° C (84 and 90° F). He noted that the period from egg to adult could be as long as 2 or 3 years or, under favorable conditions, only 3 or 4 months. Development was favored by warmth and high humidity. Sweetman also observed that at temperatures ranging from 22 to 32° C (71 to 90° F), nearly all reproduction occurred at above 75% RH, and the highest percentage of oviposition took place at 84 to 100% RH. This insect may continue to grow for well over 3 years, molting every 2 or 3 weeks.

Firebrat, *Thermobia domestica* (Packard)

As the common name implies, firebrats are found in locations where temperatures are high, such as around ovens, heating units, fireplaces, and hot-water pipes, provided these places are not too dry, for they thrive best where it is both warm and damp. They can be serious pests in bakeries. They are cosmopolitan in distribution, but prefer temperatures above 90° F (32° C), with an optimum of 98 to 102° F (37 to 39° C), but nymphs and adults can survive at temperatures between 32 and 112° F (zero and 44° C). Eggs fail to hatch below about 70° F (20 to 22° C) (Sweetman, 1938).

Description. This insect (Plate V; figure 206) is silvery, about 14 mm long, but usually with transverse gray areas that give it a mottled appearance best shown on page 108, figure 54. When its scales are rubbed off, the light-yellow color of the body can be seen.

Life Cycle. At temperatures of 90 to 106° F (32 to 41° C), the female may oviposit when she is 45 to 135 days old, depositing her eggs in crevices, an average of about 50 in a lifetime. Like the silverfish, the firebrat continues to molt during its adult life. Only 1 clutch of eggs is laid between molts, and fertilization must take place before each clutch of eggs is laid. The elliptical, nearly white eggs are about 1 mm long and 0.7 mm wide in their greatest dimensions. Optimum conditions for incubation of eggs are 99° F (37° C) and 76 to 85% relative humidity, and eggs kept at these conditions will hatch in 14 to 18 days. Under optimum conditions, only 1 day or less is spent in the first instar, 4 in the second, about 6 in the third and fourth, and about 8 in the fifth to tenth. This period gradually increases to 12 or 13 days in further instars, and the insect may pass through 45 to 60 instars before death (Sweetman, 1938). Brett (1962) found that females started to oviposit in the fourteenth instar, and that molting of the growing firebrats appeared to be continuous, with very little change in size occurring after the thirty-fifth instar.

A curious and noteworthy thing about firebrats is their "love dance," described by Brett as follows:

> The courtship of firebrats has been described as a "love dance" wherein the male constantly approaches the female and repeatedly contacts her antennae, mouthparts, and legs. As the male whirls about, he curves his abdomen and deposits a sperm bundle about one-half inch in front of the female. He then contacts her head and legs and comes to rest, apparently losing all interest in her. After his final contact, the female moves forward, straddles the sperm bundle, and secures it to certain reproductive structures.

At 37° C (99° F) and 50 to 70% relative humidity, firebrats have been reared both with and without liquid water to imbibe, but in one experiment the insects reared with water weighed 50% more than those receiving none. Firebrats can obtain water from wet cotton wicks, but they avoid actual drops of water on the wick. Without access to water, development is slower, and oviposition is retarded and decreased (Adams, 1933). At ordinary room temperatures, firebrats lose water from their bodies only below 45% RH; above that, they absorb water via their cuticles (Noble Nesbitt, 1969).

Lepisma saccharina and *Thermobia domestica* are seldom seen outdoors in temperate regions, but have been occasionally found under bark, in bird, mammal, and insect nests, and in debris (Linsley, 1944).

Fourlined Silverfish, *Ctenolepisma lineata* (F.) (= *quadriseriata* Packard)

This species occurs in the eastern United States as far south as Georgia and Arkansas and also in California. This and the following species are not so limited in distribution in a building by temperature and moisture conditions as the silverfish already mentioned and the firebrat. Consequently, *C. lineata* can be found throughout the house and in the basement, in wall voids, and in the attic, where it is often seen in large numbers, particularly in houses with roofs of wooden shingles. It occurs in the mulch of flower beds around the foundation, and also in the garage, even if it is not attached to the house (Zeigler, 1955).

Description. The fourlined silverfish is about 15 mm long, tannish gray, and has 4 dark lines extending down the length of its back. The young are light brown, and are often tinged with pink until the fourth molt, which occurs a month or so after hatching. The subspecies *Ctenolepisma lineata pilifera* (Lucas) frequently enters homes in rural areas of northern California (Smith, 1970).

Gray Silverfish, *Ctenolepisma longicaudata* Escherich (= *urbana* Slabaugh)

This species is similar in size and habits to *C. lineata*, but differs somewhat in color, being uniformly light to dark gray (plate V, 5; figure 206). It has been reported from a few eastern and southern states and from California and Hawaii. In laboratory tests, it was found to do much more damage to paper and fabrics than *Lepisma saccharina* or *Thermobia domestica* (Sweetman and Kulash, 1944). Mallis (1941*b*) found *C. longicaudata* to be an important pest in southern California, being distributed throughout the home from basement to attic in both old and new houses, but never outdoors. He observed that ventilators and heat conduits originating in the basement facilitated the spread of this insect throughout the building.

Dietary Factors. This species is the most widely distributed and abundant silverfish in Australia, where its biology was intensively investigated by Lindsay (1940). In wallpaper, the palatable materials were the starch and dextrin sizes on the surface. In writing paper, the palatable material was the "chemical pulp" containing degraded celluloses. Papers containing more than 45% "mechanical pulp" were not eaten; the unpalatable materials were associated with the "ether extract" fraction. Cellulose-digesting bacteria, and enzymes that passed forward from the midintestine, supplemented the action of the gizzard in the digestive process.

Control of Silverfish

Residual liquid sprays have long been used for control, including 2% chlordane, 0.5% diazinon, 0.5% lindane, and 0.5% propoxur. Lindane possibly has special merit because of its vapors, which may penetrate into cracks and crevices not reached by the spray. Dichlorvos added to any spray should provide a similar action. Chlordane 5% dust and the silica aerogel Dri-die 67® have been used effectively. A widely distributed infestation should be treated with the same thoroughness and in very much the same out-of-sight places as suggested for the control of German or brownbanded cockroaches in chapter 6.

In recent years, pest control operators have had increasingly erratic results with chlordane against the firebrat. In laboratory tests, residues of chlordane gave less than 100% mortality on a porous surface (unfinished plywood) in 144 hours when used at less than 4% concentration, whereas diazinon, dichlorvos, malathion, and propoxur gave 100% kills in 24 hours used at 0.25, 0.25, 1.25, and 0.55% concentration, respectively. By far the most effective insecticide tested in both topical and residual treatments was chlorpyrifos (Dursban®). This insecticide gave a 100% kill on unfinished plywood in 12 hours at 0.20% concentration (Luke and Snetsinger, 1972).

In the author's experience, when treating mixed populations of cockroaches and silverfish, a thorough cockroach control has invariably eliminated silverfish also. However, *Lepisma saccharina* and particularly *Thermobia domestica* may occur in very localized infestations, and treatment can be confined to the areas where they are known to occur. For firebrats, emphasis is placed on treatment of areas of high temperatures, such as around heating units, heating pipes or conduits, steam or water pipes, and fireplaces (Zeigler, 1955). *Ctenolepisma lineata* is more difficult to control because it is so widespread both inside and outside the house.

The fourlined silverfish particularly, and sometimes other species, may be found in attics. Dri-die is so light that it can be uniformly blown throughout the attic from a single crawl hole. Because it is inorganic, it affords protection against silverfish and other attic-infesting pests for much longer periods than the conventional organic pesticides.

Most of the various insecticide dusts can be used effectively to eliminate existing infestations. For attic dusting, electric blowers (figure 32, chapter 3) or water-type fire extinguishers (figure 33, chapter 3) can be used, and the latter, because of the narrow discharge orifice, can also be used for blowing dust into wall voids or under cabinets through existing apertures or through holes that may be drilled in appropriate places.

Among fabrics, silverfish do most damage to rayon. Zeigler (1955) noted in several cases of damage to rayon draperies that most of the holes made by the fourlined silverfish were made at windowsill level. He noticed that the insects were often found between the drapery and its lining. A light mist of residual pesticidal spray can be applied to cornices and both sides of the infested draperies.

Poison Baits. Adams (1933) found that the odor of ground rolled oats, which was generally considered to be a favored food of firebrats, did not attract them at a distance of 8 cm, even after the insects had been denied food for 2 days. Direct antennal contacts were required for strong, positive responses to dry food particles. This probably accounts for the very poor results the author has had in attracting firebrats to currently popular poison baits when the baits were placed in traps on the centers of the floors of enclosed wooden boxes 29 cm long, 14 cm wide, and 9 cm deep, even when other food sources were absent. A few insects would invariably live longer than 25 days without being attracted to the bait. Many firebrats congregated on the vertical walls of the boxes, and were seldom seen crawling over the floors. Contacts with baits were probably by chance. This was in accord with the observation of Berger (1945) that the location of a baited trap, and not the bait, determined the number of firebrats caught. The same is probably also true of other silverfish species. When placing poison baits in areas where silverfish normally crawled, Berger found that among equally available foods, the firebrats showed preferences. He concluded that because of the ease of mixing and their availability, wheat flour 85% and powdered sugar 15% were about the most practical food combination for bait. However, another factor to be considered is the repellency of the insecticide in the bait. For example, Mallis (1941*b*) found that *Ctenolepisma longicaudata* was repelled by white arsenic, sodium arsenate, sodium fluosilicate, and sodium fluoride in flour pastes, but not by barium fluosilicate or barium carbonate.

Traps. Where the application of insecticides is undesirable, silverfish can be trapped. The outer surface of a small jar, such as a 1-oz ointment jar, can be covered with masking tape to enable silverfish to climb up the outside of it. The insects fall into the jar, and cannot escape because they are unable to climb its smooth, vertical walls. The traps should be placed in paths normally used by silverfish, as in the intersections and corners of a pantry or bookcase or next to the baseboard on the floor (in the warmest locations, in the case of firebrats). The first to advocate a trapping procedure for firebrats was Mallis (1941*b*), who placed a teaspoonful of white wheat flour in the jar. In our replicated laboratory experiments with large numbers of firebrats, we found that the trap jars had to be judiciously placed, as in corners of the experimental boxes, but that the kind of bait used had no influence on the number trapped. Empty jars trapped as many firebrats as jars with wheat flour and sugar or wheat flour, sugar, and chipped beef.

Comparison of Destructiveness of Fabric-Feeding Insects

Mallis *et al.* (1958) made a comparison of the amount of feeding done by the larvae of the webbing clothes moth (*Tineola bisselliella*), the black carpet beetle (*Attagenus megatoma*), the furniture carpet beetle (*Anthrenus flavipes*), and the adults of the firebrat (*Thermobia domestica*) to 3 classes of fabrics: (1) synthetic (nylon, dynel, dacron, orlon, vicara, and acetate rayon); (2) natural (silk crepe, wool [AATC], linen, and cotton percale); and (3) a combination of wool and synthetic fabrics (86% wool and 14% nylon, 45% wool and 55% orlon, and 45% wool and 55% viscose rayon). The clothes moth and the carpet beetle did extensive damage to wool and the combination of wool and synthetic fabrics. The furniture carpet beetle fed extensively on acetate rayon fabric, but the black carpet beetle did not. The firebrat was the only species that fed extensively on viscose rayon. It also fed slightly on silk crepe and linen. It is evident from the feces of silverfish that they digest silk, whereas if carpet beetles feed on silk, an examination of the feces shows that the material remains undigested (Hartnack, 1943).

Mallis *et al.* (1958) observed that the carpet beetles fed slightly on nylon, but the clothes moth and the firebrat did not. None of the 4 species they tested did any damage to dynel, dacron, orlon, vicara, or cotton percale. However, any fabric was attacked when contaminated with nutrient material, such as starch, beer, urine, fats, and molds. Stains containing minerals, proteins, and B vitamins caused fabrics to be more attractive, even those that would otherwise not be infested. The predilection of firebrats for viscose rayon noted by Mallis *et al.* had also been demonstrated by Wall (1953) for both the firebrat and another species of silverfish, *Ctenolepisma lineata*.

MISCELLANEOUS PESTS

Many species of insects have been implicated in damage to fabric. In some cases, stains or contaminants of some nutritional value have initiated the attack. This appeared to be the explanation for some damage to a nylon viscose rug by southern fire ants, *Solenopsis xyloni* McCook, that was similar to damage caused by carpet beetles (Smith, 1965). Finley *et al.* (1968) exposed a number of cotton and synthetic fabrics to the American cockroach (*Periplaneta americana*), the house cricket (*Acheta domesticus*), and the striped earwig (*Labidura riparia*). Half of the 144 pieces of fabric were stained by dipping them into hot animal fat. The amount of visible damage from cockroaches to stained and unstained cotton fabrics (unfinished and resin-finished fabrics) and to unstained synthetic fabrics was negligible, but damage to stained acetate, viscose, and triacetate was extensive. Crickets preferred the same 3 synthetic fabrics as the cockroaches, and damaged both stained and unstained fabrics. Stained and unstained polyester, nylon, and acrylic fabrics were attacked to a lesser extent, and cotton was undamaged. Earwigs caused a slight "picked appearance" on the stained nylon.

Most of the termites that normally feed on wood eat fabrics of cotton, linen, jute, silk, rayon, and leather. Termites coming up through the floorboards and joists can damage carpets (Harris, 1961). Termites may destroy materials they do not use as food but which are damaged when the insects bore through them or pack earth against them. The termite *Cryptotermes brevis* readily attacks cellulose fabrics (Light, 1934*a, b*). Termites, crickets, and silverfish will attack cotton, especially if it is starched or soiled (Labarthe, 1964). Sometimes, the injurious insect destroys the fabric to gain access to fibers in building a nest or cocoon, e.g., damage from the fan palm caterpillar, *Litoprosopus coachellae,* as already described (figure 204).

There are many cases of damage to fabrics merely because they are in the way of insects. An example would be carpeting damaged by termites and the adults of woodwasps and certain woodboring beetles, when it happens to be in the way of insects emerging from the wood floor. The insect must bore through the fabric in order to escape. A rare and curious type of damage was reported from a home in which insect damage to a man's suits had amounted to about $1,000. Drugstore beetles (*Stegobium paniceum*) (figure 193, chapter 7) infesting cayenne pepper in the kitchen pantry had apparently found their way to a clothes closet, via the wall voids, and emerged in the closet to gather on a shelf. They dropped from the shelf onto the right shoulder of each suit, cut their way through the fabric, and dropped to the floor. Removal of the infested spices and spraying of the closet prevented further damage (Cushing, 1970).

Chapter 9 · *Pests Attacking Man and His Pets*

Many arthropods are pests because they bite, sting, or otherwise annoy humans and their pets. Some envenomate their victims, some transmit disease, and some, both. Others do not envenomate or transmit disease, but are merely annoying. All such pests are discussed in this chapter. Not included here are certain pests that are annoying and are also mechanical carriers of disease germs, such as the house fly and cockroach. These happen to be even more important in other ways (e.g., as food pests), and are more appropriately discussed under other chapter headings.

A distinction should be made between venomous (phanerotoxic) and poisonous (cryptotoxic) animals. The 2 terms are often used synonymously, but biologists generally consider venomous animals to be those that have a gland or group of highly specialized secretory cells, a venom duct (not always found), and a structure for delivering the venom. The *venom apparatus* is most commonly considered to be not only the stinger or fang, but also the associated gland and duct. Poisonous animals, as distinguished from venomous ones, have no venom apparatus, and usually poison their victims by being ingested (Russell, 1967*b*). There are at least 700 species of venomous arthropods, but only a few poisonous species. The latter would be hazardous to humans only if they were ingested. Among arthropods, only a few crustaceans are universally consumed. Members of a few primitive tribes eat insects, but they are well aware of the poisonous kinds. Therefore, poisonous arthropods are not discussed in this chapter.

Arthropods that attack humans and pets can be divided into 4 groups: (1) species that inject their venom by means of fangs or stingers; (2) species that inject their venom along with their saliva; (3) species that are sources of allergenic emanations and inhalant allergies; and (4) species that do not envenomate, but are merely pestiferous. The indictment against all but the fourth group of arthropods is that each species has the ability either to envenomate its victims or cause allergic reactions, or both. Some of the same species are also vectors of disease organisms. Therefore, a discussion of envenomation, allergy, and disease transmission by arthropods is appropriate at this point. Venomous snakes are also included in this chapter because for practical purposes it is desirable to discuss all the principal venomers of the United States together.

ENVENOMATION

Every year there are many thousands of arthropod and reptile envenomations in the United States, and during the period 1950 through 1959, they caused several hundred deaths. Hymenopterous insects were responsible for 229 of these deaths: bees, 124; wasps (yellowjackets and hornets), 101; and ants, 4. Spiders caused 65 deaths, and scorpions, 8. Snakes caused 138 deaths: rattlesnakes, 94; cottonmouth moccasins, 8; coral snakes, 2; snakes not native to the United States, 3; and unidentified snakes, 31. Thus, the hymenopterous insects caused about half of the deaths from venomous animals during the 10-year period. Bees and wasps were each responsible for more deaths than rattlesnakes, which caused 20.4% of the total number of deaths (Parrish, 1963).

People who succumb to envenomation by Hymenoptera usually die with alarming suddenness, often within 15 to 30 minutes. Of 208 deaths, 80% occurred in less than an hour. Only about 8 deaths resulted from overwhelming envenomation caused by hundreds of stings; the remainder resulted from insect allergy. This points up the importance of identifying and desensitizing allergic patients with appropriate insect antigens and instructing hypersensitive patients to carry insect allergy first-aid kits with them. In striking contrast to stings by Hymenoptera, only 11% of 54 victims of spider bite, for whom the period between bite and death was stated, died in the first 12 hours. Following the bite, 41% died in 12 to 48 hours, 26% from the third to the seventh day, and 22% after a week (Parrish, 1963).

Statistics also revealed that only 17% of snakebite victims died in less than 6 hours and 64% died in 6 to 48 hours after the bite. It usually requires several hours for the full toxic effects of the venom of American pit vipers to become evident, and there is a period of an hour or two following most snake bites when first aid and specific antivenin therapy may be used to best advantage. One of the most important causes of death from snake bite is said to be the failure to seek prompt and vigorous medical treatment (Parrish, 1963). Other things being equal, usually the smaller the victim, the greater the reaction to a given amount of venom.

The toxic effects of a venom result from the combined action of its components. However, additional effects may be attributed to metabolites formed when the venom components react with the constituents of the envenomated organism (Neumann and Habermann, 1956). The effects of autopharmacological substances such as histamine, bradykinin, and adeposine may be far more deleterious than those of the venom. In this chapter, the discussion of each arthropod will be accompanied by a discussion of the type of envenomation or allergic reaction produced by that species, whenever data are available.

Treatment of Arthropod Bites and Stings

Localized allergic reaction is evidenced by intense itching (pruritus), erythema, or scattered small urticarial papules. There are several effective anti-pruritus agents. Two are Calamine lotion and colloidal oatmeal bath, using Quaker Oats or a similar product. Antihistamine topical ointments such as tripelennamine hydrochloride 2% may be useful, but can cause sedation as well as allergic dermatitis (H. C. Spencer, correspondence). Pruritus may also be treated with an oral antihistamine and, on a physician's advice, by the local application of a combination steroid–analgesic preparation. Treatment should include thorough cleansing of the area around the wound to avoid secondary infection. Insects such as wasps that carry animal food or carrion to their nests are more likely to have infective stings than are bees. The above treatments may suffice, but further first-aid and medical measures and even hospitalization are occasionally required. If the bite or sting occurs on an extremity, one should avoid exercising it or placing pressure on it. Cold compresses may be applied to areas of localized swelling. A constriction bandage may be useful proximal to the sting site. It should not be applied so tightly as to cut off arterial pressure, and it should be loosened every 3 to 5 minutes. In cases of severe envenomation, as from the bite of the black widow spider, the sting of the deadly scorpion of Arizona (*Centruroides sculpturatus* Ewing), or multiple stings of bees and wasps, hospitalization may be required. Russell (1969) advised that all patients under 16 or over 60 years of age or with hypertensive heart disease receive antivenin following black widow spider envenomation.

For the reaction-prone person, an insect-sting kit is available that can be used during the period required to obtain a physician's help. The kit should contain a tuberculin syringe with a 26-gauge needle, an ampule of 1:1,000 aqueous epinephrine, and instructions for dosage and use. The recommended dose is 0.01 ml/kg of 1:1,000 solution subcutaneously. For oral medication, isoproterenol linguets, 10 or 15 mg, can be included (Zwemer & Handelman, 1971). An emergency kit should contain a constriction bandage, a sterile alcohol pad, suggestions for avoidance of hymenopterous insect stings, and Medic-Alert literature dealing with treatment.

Arthropod-Related Allergies

"Allergy" is thus defined by Pennak (1964) in the *Collegiate Dictionary of Zoology:*

> Excessive sensitivity to a common substance such as pollen, many foods, natural oils, hair, dust, feathers, medicines, etc. The sensitivity is manifested in many ways, including hives, rash, asthmatic attacks, higher blood pressure, headaches, and nasal congestion; such symptoms are external manifestations of antigen–antibody reactions, the antigens being substances mentioned above, and the antibodies being their specific antagonistic materials in the tissues and blood plasma.

Although "allergy" was originally meant to include all forms of human hypersensitivity, the term has later been generally applied to a group of diseases such as hay fever, asthma, urticaria, and eczema, all the consequences of an immune response to an exogenous factor. They characteristically involve production of an unusual kind of a reagin or skin-sensitizing antibody. This antibody reacts with the antigen or allergen to produce tissue damage. This phenomenon can be conveniently demonstrated by intradermal injections of various dilutions of the offending allergen, which result in a swelling or reddening at the site of injection. A clinically evident hypersensitivity having a basis of hereditary predisposition is called *atopy*. While allergic conditions develop spontaneously and only in some individuals, presumably from mild exposure to environmental agents, a more intense reaction, called *anaphylaxis*, requires intense artificial exposure, usually by injection. The allergic response can develop only in some individuals, but anaphylactic response can be obtained in all individuals of a species. An anaphylaxis-like reaction can occur in a person from an insect sting or following administration of drugs, foreign serum, vaccines, or from other allergenic exposures. Such a response has been termed "anaphylactoid" to distinguish it from experimental anaphylaxis, although this distinction is not always made with precision in published reports (Arbesman, 1965; Shulman, 1967).

Man's sensitizing agents are made up of the various antibodies which a number of body tissues can produce to react with the foreign matter. The antibodies include specific immune substances against living viral or bacterial organisms or the foreign protein of allergens. The antibodies produced by the body may be of the wrong kind, produced in excessive amounts, or localized in the wrong places, or there may be combinations of these circumstances. In these situations, the antibodies go beyond the point of being purely defensive and become pathologic (Kern, 1962). Kern discussed in detail the relationship of sensitization to numerous environmental allergens that are ubiquitous, and in many cases are subtle and difficult to diagnose and define. Prolonged contact with the allergen may be required before symptoms become evident.

Allergies from Bites or Stings

Arthropods can be the sources of *injectant* or *inhalant* allergies. The arthropod may inject venom by means of (1) a stinger in the abdomen, e.g., bees, wasps, ants, and scorpions; or (2) by means of the mouthparts, e.g., sucking lice, bed bugs, reduviids, mosquitoes, gnats, fleas, spiders, ticks, and mites. Allergic reaction may result from venoms secreted in venom glands or from salivary secretions that some arthropods inject subcutaneously.

The Insect Allergy Committee of the American Academy of Allergy studied over 3,000 completed questionnaires from persons experiencing allergic reactions to the stings of bees and wasps, of which 2,606 were recorded and analyzed (IAC, 1965). Of these, 13.3% reported only "local" reactions; 16.1% reported "slight general" reactions, in which there might be such symptoms as a few hives or itching beyond that which local swelling and pain might be expected to produce; 43.6% reported "moderate general" reactions; 24.2% reported life-threatening "severe general" reactions; and 2.8% reported "delayed" reactions, in which the time of onset of reactions was an hour or more after the sting. Symptoms indicating "severe general" reactions were dyspnea (difficult respiration), swelling in the throat, shock, and unconsciousness, the latter affecting 62.2% of the persons in the "severe general" reaction group. A sharp rise in the proportion of serious reactions in both sexes after age 30 suggested increasing sensitivity as the total number of stings received would mount over the years. A particularly disquieting finding was that responses to stings might be completely normal before the occurrence of a particular sting that produced a life-threatening allergic response.

An interesting aspect of the allergy problem is the fact that various biting or stinging arthropods often prefer to attack certain individuals in a family or group rather than others. The variations in the attractiveness of persons to insects result from differences related to temperature, moisture, age, sex, color, or physiological conditions of the person. Hormonal factors have been found to account for host selection by various species of mosquitoes (Brown, 1966) and fleas (Rothschild, 1965).

The clinical severity of reaction to insect bites and stings has been classified by Mueller (1959*a*, *b*), based on a study of 84 patients stung by bees or wasps, as follows:

1. *Slight General Reaction.* Generalized urticaria (inflammation, wheals, itching, malaise and anxiety).
2. *General Reaction.* A slight general reaction plus 2 or more of the following: generalized edema (swelling), constriction in chest, wheezing, abdominal pain, nausea and vomiting, or dizziness.
3. *Severe General Reaction.* Any of the above reactions plus 2 or more of the following: dyspnea (difficulty in breathing), dysphagia (difficulty in swallowing), hoarseness or thickened speech, confusion, or feeling of impending disaster.
4. *Shock Reaction.* Any of the above reactions plus 2 or more of the following: cyanosis (blue color because of insufficient oxygenation of blood), fall in blood pressure, collapse, incontinence (inability to restrain natural evacuations), and unconsciousness.

In the above investigation, there were a few patients in each group who reported a very rapid onset of symptoms after being bitten—as brief a period as 2 minutes. The average time reported declined from 24 minutes for group 1 to only 5 minutes for group 4.

Mueller reported that of the 84 patients studied, all of whom suffered at least a slight general reaction, 28 had personal histories of other allergies as well (close to the 30% found by the Insect Allergy Committee). There were 63 patients (75%) that had family histories of allergy, 6 having histories of severe general reactions or shock reactions to insect stings. In 18 patients (21%) in which there was no family or personal history of other allergies, the reactions were generally mild, and there was relatively little skin sensitivity.

Repeated exposures to bites of a hematophagous insect generally result in changes in skin reactivity of the host that follow a definite sequence of 5 stages: (1) no observable skin reactions during the period of induction of hypersensitivity; (2) delayed skin reactions; (3) immediate skin reactions followed by delayed reactions; (4) immediate but no delayed reactions; and (5) no reactivity (Feingold *et al.,* 1968). This sequence of stages of skin reactivity, as it applied to mosquito bites in an experiment on 25 human volunteers, is described on page 424.

Immunization or Desensitization

Mueller (1959*b*) presented extensive evidence that people suffering more than local reactions to hymenopterous insect stings could be successfully inoculated with mixed whole-insect extract of equal parts of bee, wasp, hornet, and yellowjacket ("Mixed Stinging-Insect Extract," Hollister-Stier Laboratories). The evidence seems to be overwhelming that the antigen that causes allergic reactions is present throughout the insect's body, independently of venom, but that the venom contains a small amount of this antigen (Hecht, 1929; Benson and Semenov, 1930; Ellis and Ahrens, 1932; Benson, 1936, 1939; Rockwell, 1952; E. Perlman, 1955; F. Perlman, 1962). At least 6 antigenic fractions are found in bee and wasp extracts. Some are species-specific, but others cross-react with those of other Hymenoptera (Arbesman *et al.,* 1966). Patients generally are extremely poor observers concerning the true identity of the offending insect, and polyvalent therapy—long-term treatment with polyvalent bee and wasp antigens—is usually suggested (Torsney, 1968).

Mueller (1959*a, b*) concluded that testing a patient for sensitivity or treatment should always be started at a dilution of 1:100,000,000. Scratch testing was found to be an unreliable index to the degree of skin sensitivity. He stated that treatment is started at 0.05 ml of the dilution selected according to the patient's "initial positive test." Weekly increments in dosage are then given until 0.2 to 0.3 ml of a 1:100 dilution, the "maintenance dose," is reached, "or until a local reaction, larger than a silver dollar, is encountered twice with the same dose, after having dropped back and approached that dose a second time" (Mueller, 1959*b*). This is similar to one of the methods employed for pollen desensitization.

Mueller's patients were given their maintenance doses approximately every 4 weeks in the insect season and every 6 weeks through the winter for a period of 3 years. Subsequent stings were suffered by 40% of them. With 1 exception, none had resulting systemic symptoms (Mueller, 1959*b*). Patients treated by Frazier (1969) were given their maintenance doses about every 2 weeks in the insect season and every 3 weeks during the winter. He suggested that the mild-reaction patients should be discharged after 3 years of treatment, that the severe or shock-pattern reactors should be maintained in therapy indefinitely, but that time intervals between treatments should be increased where possible. Levine (1971) stated that the duration of immunotherapy should be at least 3 years, and perhaps should be continued indefinitely, but that there was no agreement on that point.

It may appear anomalous that repeated stings induce antibodies responsible for allergic reactions, and yet a large enough dose of the same venom (arrived at in gradual increments) will induce immunity. In the latter case, the "blocking antibody" differs from the allergic antibody in its molecular size, electric charge, capacity to withstand heat, and most importantly, in its effect on the host when in combination with its allergen (Frazier, 1969).

The questionnaires of the Insect Allergy Committee revealed an expectancy of progressively severe reactions in about 65% of persons not hyposensitized, whereas following hyposensitization, reactions to subsequent stings were reduced in about 90% of treated persons. The study showed that this protection by hyposensitization may be maintained for years, or may be lost in less than a year. The committee recommended hyposensitization for those persons who demonstrated any degree of systemic sensitivity following an insect sting (IAC, 1965).

Inhalant Allergies Attributable to Insects

The enormous numbers and wide distribution of insects and insect parts in the atmosphere give insects, like pollens, great potential as producers of allergic responses, particularly inhalant or respiratory allergies—that is, reactions similar to asthma. Small arthropods, and weak fliers in particular, may be carried for great distances from their breeding grounds by convection currents and upward in airstreams to as high as 10,000 ft (about 3,000 m). The dust and debris of decomposed arthropods become distributed in the same way, and these minute particles are important causes of inhalant allergies.

Inhalant allergy has long been known and can be readily demonstrated by skin tests. One of the earliest published reports on inhaled insect allergens was made by Wilson (1913), and concerned swarms of mayflies (Ephemerida). By placing a drop of a mixture of mayfly parts in 2 ml of sterile saline solution to one eye of a susceptible person, a marked conjunctival redness resembling that seen in allergy patients could be produced, whereas the other eye remained normal. Mayflies continue to be important sources of inhalant allergy, and caddisflies (Trichoptera) are other common sources, apparently first revealed by Parlato (1929). The hairs and wing fringes are easily dislodged and become windborne. Aphids, being extremely numerous and widely distributed, are also important sources of inhalant allergy. However, insects from most orders have been implicated. In an investigation of 222 allergy patients, 121 were sensitive to more than 1 insect, 25 were sensitive to only a single insect, and 76 were negative to all insects. Sensitivity to multiple insect antigens appeared to be the rule (Frazier, 1969).

Respiratory insect allergy has resulted from heavy occupational exposures. Examples are mushroom fly allergy in persons who grow mushrooms; beetle allergy in a museum curator; moth allergy in an entomologist in continuous contact with larvae and eggs of the range moth caterpillar; and also in an employee who had worked for 8 years with the larvae of the greater wax moth (bee moth), *Galleria mellonella* (L.) (Stevenson and Mathews, 1967). An employee of the then Federal Bureau of Entomology, after 2 years of employment, found that her hands and forearms itched severely when she worked with the Madeira cockroach, *Leucophaea maderae* (F). (Bernton and Brown, 1964). A man doing research on the control of dermestid beetles suffered severe allergenic reactions which became gradually more severe over a 2-year period (Okumura, 1967).

Thirteen cases of bronchial asthma were reported among workers at a sewage works in the Transvaal, South Africa, all the result of allergic response to the filter fly, *Psychoda alternata*. The inhalant allergen was demonstrated to be the dust resulting from the disintegration of the bodies of the flies. One individual was found to have no family history of allergy, and was not found sensitive to any of the more common causes of allergy, such as feathers, dusts, and pollens (Ordman, 1946).

Cross-Antigenicity. An interesting aspect of the problem of inhalant allergy has been the study of the cross-antigenicity between different types of insects. One antigen has been found to be common to a fly, a cockroach, and a cricket, another to a fly and a cockroach, and still another to a cricket and a cockroach (Pruzansky *et al.*, 1958). Common antigenicity has also been found between crickets, grasshoppers, moths, and butterflies on the one hand and caddisflies on the other (Langlois *et al.*, 1963).

Inhalant Allergy Attributable to Mites

The importance of house dust in inhalant or respiratory allergy was apparently first discussed by Kern (1921). He considered dust to be the most common cause of bronchial asthma. In 1922, mites were first mentioned as possible antigens in house dusts (Storm van Leeuwen, 1922). Inhalant allergy is attributable to many orders of arthropods, but in house dust certain species of mites appear to be the most frequent causes. Spieksma and Spieksma-Boezeman (1967) believed that no house dust allergen is formed in the absence of living mites, and proposed the name "house dust mite" for the leading European species, *Dermatophagoides pteronyssinus* (Trouessart). In Ohio, the predominant species of house dust mite is *D. farinae* Hughes (Mitchell *et al.*, 1969), for which the common name "North American house dust mite" has been suggested (Wharton, 1970). In California, these 2 species are common, and may coexist in certain locations (Furumizo and Mulla, 1971). (See also the section on house dust mites at the end of this chapter.)

DISEASE TRANSMISSION BY ARTHROPODS

Many species of insects, particularly those that feed on or breed in excrement, are mechanical carriers of various "filth diseases," such as typhoid fever, cholera, and amoebiasis. The causal organisms may adhere to the tarsi, mouthparts, and other body parts of such insects, notably house flies and cockroaches, and may be deposited on food and on areas where food may be placed, as on sink tops, breadboards, or tables. The pathogenic organisms (bacteria, protozoa, helminthic ova, etc.) may also pass uninjured through the alimentary tract of the insect and be deposited on food via its excrement. Diseases like yaws (a spirochete infection) and certain eye infections such as "pinkeye" can also be transmitted mechanically by some species of flies. Some bacteria, such as anthrax (*Bacillus anthracis*) may be disseminated by flesh flies bred in carcasses of animals that have succumbed to the disease. The bacteria consumed by the fly larvae may survive through the pupal stage, and the adult flies may be infected and become highly mobile carriers of the disease. The piercing mouthparts of some insects like the horse flies (Tabanidae) may become contaminated by blood-inhabiting pathogenic organisms, and the insect may then spread the disease by simple mechanical means.

In all the foregoing cases of mechanical disease transmission, the pathogenic organisms undergo no developmental changes. Of much greater biological complexity and interest is transmission of pathogenic organisms that pass through cyclical changes and/or multiply in the body of the carrier. Huff (1931) suggested the following classifications for such "biological" transmissions:

I. *Cyclo-Propagative.* The organisms undergo cyclical change, and multiply.

II. *Cyclo-Developmental.* The organisms undergo cyclical change, but do not multiply.

III. *Propagative.* The organisms undergo no cyclical change, but they multiply.

An example of *cyclo-propagative* transmission is the transmission of the malarial parasite *Plasmodium falciparum* to man by anopheline mosquitoes which are then called *vectors.* Another example is the transmission of the protozoan *Babesia bigemina,* the causal agent of "Texas fever" of cattle, by another arthropod vector, the Texas cattle tick, *Boophilus annulatus* (Say). An example of *cyclo-developmental* transmission is filariasis, in which the filarial worm is transmitted by certain mosquitoes. Probable examples of *propagative* transmission are the transmission of the bubonic plague bacillus, *Yersinia pestis* (= *Pasteurella*), which can multiply in the foregut of the flea, and the transmission of relapsing fever by ticks. Huff (1931) called attention to the fact that the most important types of transmission are I and III of the biological types, in which the disease organisms multiply within their arthropod hosts. Multiplication of the organism does not take place in mechanical transmission or in biological transmission type II.

It is probably an unfortunate circumstance that insect species with benign bites, and therefore those less likely to be shunned by humans, are the most potent vectors of disease. Among the Diptera, for example, *Anopheles maculipennis,* which does not inflict a painful bite, is a disease vector, whereas other species with painful bites, such as the field mosquito, *Aedes dorsalis,* and the stable fly, *Stomoxys calcitrans,* are seldom vectors of disease. If an insect is to be a successful vector, it appears to be generally necessary that it must be able to decrease the pain caused by its bite, as by local anesthesia.

Some arthropod vectors can transmit disease via their feces, such as flies transmitting plague, conenose bugs (*Triatoma*) transmitting Chagas' disease, or the mosquito *Aedes aegypti* transmitting yellow fever.

Disease organisms can also be transmitted via the egg of an infected arthropod (transovarian transmission). Examples are rickettsiae, bacteria, and spirochetes transmitted through the eggs of ticks that cause Rocky Mountain spotted fever, tularemia, and relapsing fever, respectively, and sand fly (pappataci) fever caused by the bites of the psychodid *Phlebotomus papatasii* reared from the eggs of infectious flies.

An important factor in the transmission of diseases by insect vectors is the fact that many wild animals serve as reservoirs for disease organisms. They suffer few or no ill effects from the diseases, but insect vectors can transmit the disease agents from these animals to man or domestic animals. An example is the rabbit as a reservoir for Rocky Mountain spotted fever and tularemia. Tularemia also exists in many other animals, among them meadow mice, ground squirrels, coyotes, sheep, and quail. Wild rats, particularly the roof rat, are reservoirs for human plague and typhus, for which fleas are the vectors. Care should be taken to avoid picking up fleas from wild animals.

(A) PESTS THAT INJECT VENOM BY FANGS OR STINGERS

This group of pests includes all the arthropods that cause the annual death toll in the United States, as discussed earlier in this chapter under "Envenomation": ants, wasps, bees, spiders, and scorpions. This is not meant to imply that other envenomating arthropods are not important. In fact, such pests as mosquitoes and gnats are probably of greater economic importance. Their bites are painful enough to cause great annoyance, and their numbers can be large enough to constitute the determining factor as to whether a district proposed for farming or residential development is inhabitable from a practical standpoint, or whether a proposed recreational area will be visited and enjoyed by the public. Mosquitoes and certain other insects that fall into the second group listed in the introductory part of this chapter, particularly fleas and lice, are also important disease vectors. Nevertheless, the pests that envenomate by injecting their venom should be recognized and understood, for their envenomation, while less common, can lead to serious consequences. As already stated, reptiles are included in this section so that all the important venomous animals can be discussed in one chapter.

SPIDERS (ARANEIDA)

All spiders are predators, feeding primarily on insects and other arthropods. When feeding, spiders inject a predigestive liquid into the wounds of their prey and then suck up the digested food. They can survive for very long periods without food. Some South American brown spiders, *Loxosceles laeta,* have been kept alive for over 2 years with no food or water (M. Madon, personal communication). A concisely informative and well-illustrated (mostly in color) little book on spiders and related arthropods is *Spiders and Their Kin,* by Levi *et al.* (1968). Illustrated keys for 44 of the families of spiders of the United States and Canada, as well as descriptions, illustrations, and other information on 398 species in 212 genera, are given by Kaston (1973).

Most spiders are nocturnal, unobtrusive, often feigning death when molested, and those frequenting human habitations will usually remain in undisturbed, dark, or dimly lighted, cool places. People are most likely to be bitten by spiders while cleaning out dark, neglected places such as basements, garages, or barns, or when gathering or cutting wood. Children may be bitten when playing in old, unused buildings. Bites may occur from spiders that crawl into a victim's clothing or shoes during the night or hide in bedding during the day.

Most spiders never attempt to bite without the greatest provocation, such as being squeezed or held. Many must be forced to bite when their venom is required for an experiment. However, only the larger spiders can penetrate the skin of a human with their fangs. Almost all spiders possess venom, but relatively few are dangerous to man. In the mainland United States, the black widow and the brown or violin spiders are the most dangerous, but moderate to severe symptoms can result from the bites of other species.

Biting Apparatus and Venoms

Spiders have an anterior pair of pedipalps and a pair of chelicerae with horny fangs at their tips (figure 207). Venom from modified salivary glands is injected through the hollow tips of the fangs (figure 208, *D*). It happens that the most common spiders (suborder Labidognatha) have their chelicerae (jaws) attached below the head. They are horizontal, open sideways, and oppose each other (figure 209, *left*). On the other hand, mygalomorphs (suborder Orthognatha), the trapdoor and hunting spiders, including the tarantulas, have chelicerae that are attached on the front of the head and move up and down, opening parallel with the long axis of the body (figure 209, *right*) (Levi *et al.,* 1968).

Chemical and Pharmacological Properties of Spider Venoms

In a study made of the venoms of spiders in 4 genera, including *Latrodectus,* McCrone (1969) found that they contained many biologically active substances, ranging from compounds of relatively low molecular weight, such as γ-aminobutyric acid, to those of relatively high molecular weight, such as enzymes and lethal polypeptides. The lethal effects of the venoms on both arthropods and man appear to result from the nonenzymatic peptides. Most of the other substances appear to influence primarily the distribution and action of the major lethal components.

Most venomologists feel that it has not yet been possible to classify spider venoms satisfactorily on the basis of their chemical or pharmacological properties. Such classifications as "neurotoxins," "hemotoxins," and "cardiotoxins" are misleading, and may give rise to unfortunate clinical decisions (Russell, 1967*b*). Black widow spider venom provokes a number of responses in hu-

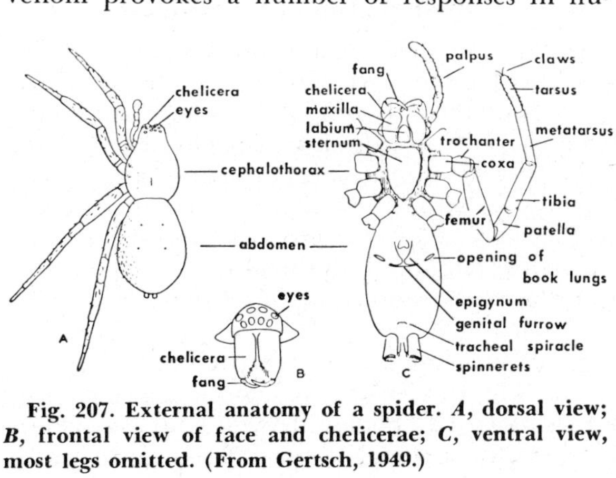

Fig. 207. External anatomy of a spider. *A*, dorsal view; *B*, frontal view of face and chelicerae; *C*, ventral view, most legs omitted. (From Gertsch, 1949.)

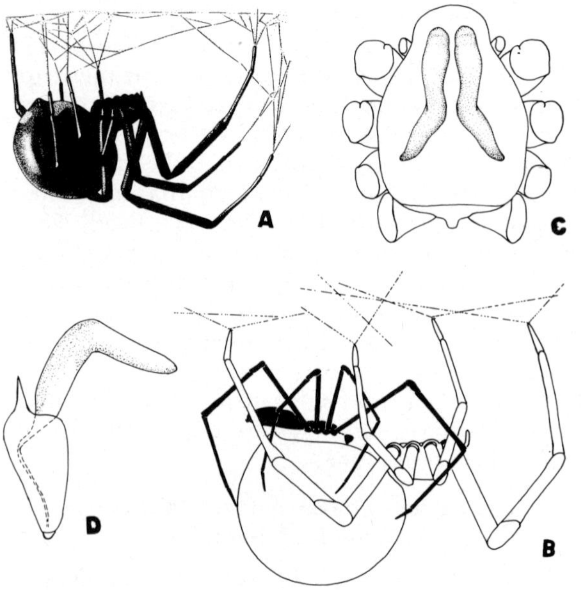

Fig. 208. Western black widow spider, *Latrodectus hesperus*. *A*, female in normal position, hanging in web; *B*, mating position, with female represented in outline and male blackened; *C*, venom glands as seen from above in relation to entire cephalothorax; *D*, left venom gland, with its duct and left chelicera. (from Kaston, 1970.)

mans, including hypertension, muscle spasms, weakness, and even paralysis. The bites of some spiders, such as *Phidippus formosus* (Peckham) (Salticidae), produce sharp pain, a wheal, and various local tissue reactions, including swelling and edema (Russell, 1970). The venom of *Loxosceles* sp. produces an ulcerative lesion which may become very large, but it is not yet clear whether the effects are caused directly by the venom or by some autopharmacological response provoked by the venom (F. E. Russell, correspondence).

Black Widow Spiders, (*Latrodectus* spp.) (Theridiidae)

In the United States, published reports of bites attributed to the black widow spider began to appear simultaneously from the southern states and from California beginning in 1889 (Herms *et al.*, 1935). Black widow spiders (also known as comb-footed spiders) are now the cause of the majority of the cases of serious arachnidism. When considered in relation to the total population of the country, black widow bites are infrequent.

The genus *Latrodectus* is cosmopolitan (Levi, 1958, 1959), and 5 species occur in the United States. The species are distinguished principally on the basis of differences in male genitalia. There are usually accompanying differences in gross characters, such as color and markings, but they cannot be relied upon to distinguish a species in all cases.

There are 2 species restricted to southern Florida, the "brown widow," *L. geometricus* C. L. Koch, and the "red widow," *L. bishopi* Kaston (McCrone and Levi, 1964). The brown widow prefers to live in human habitations, and is generally brownish, although there are also many black individuals. The cephalothorax of *L. bishopi* is usually bright orange, but is yellow or brick red in some specimens. The webs are not built under stones and debris, as with other species of *Latrodectus,* but are built 3 to more than 4 ft (about 1 to 1.5 m) from the ground, usually stretched between palmetto trees (Kaston, 1937).

Latrodectus mactans (F.) is the species most generally associated with the common name "black widow spider" in the United States. It occurs in the southern states, and its range overlaps that of the northern widow, *L. variolus* Walckenaer in the southern New England states, where the latter species is more common. *L. variolus* is also common throughout the northern states and Canada. Black widow spiders from the western United States and western Canada are *L. hesperus* Chamberlin and Ivie. The males of

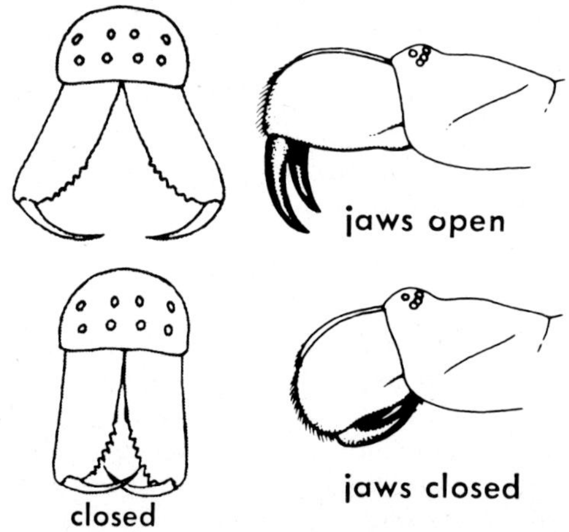

Fig. 209. Jaws of most common spiders, suborder Labidognatha (*left*) and jaws of mygalomorphs, suborder Orthognatha (*right*). (From Levi *et al.*, 1968.)

this species are usually light brown, whereas those of *L. mactans* and *L. variolus* are usually black (Kaston, 1968). Most investigators have found that the sex ratio of *Latrodectus* spp. is approximately 1:1. One investigator found that when the spiderlings were underfed, a higher percentage of males matured (Kaston, 1968, 1970).

Description. Female black widow spiders (plate VI, 5, figure 208) are black, with a body about 12 mm long and a nearly globular abdomen 7.2 to 9.6 mm in diameter. The over-all length, including the legs, is 38 to 43 mm (Baerg, 1936). According to Kaston (1970), the body lengths are as follows:

Of the three species, *Latrodectus mactans* averages smallest for both sexes. Thirty-seven adult males ranged from 2.9 to 5.1 mm in length, with most between 3.2 and 4 mm; 52 females ranged from 5 to 13.5 mm, with most between 8 and 10 mm. *Latrodectus variolus* has the largest males, mostly between 5.5 and 6.5 mm, with a range for 34 specimens of 4.5 to 8.3 mm. Females of *L. variolus* are mostly 9 to 11 mm in length, with a range for 32 specimens of 7.4 to 13 mm. *Latrodectus hesperus* has the largest females, 59 specimens ranging from 8 to 15.5 mm, with most from 10.5 to 13 mm. Sixty-three males ranged from 3 to 6.5 mm, with most between 3.8 and 4.5 mm.

The great difference in sizes of males and females (figure 208) is even more apparent when weights are compared. Most males of *L. hesperus* weigh between 8 and 18 mg, while most of the adult females weigh between 120 and 400 mg (Kaston, 1970).

The black widow can be distinguished from all other spiders by the red hourglass figure on the underside of the abdomen (plate VI, 5). In many specimens, the pattern varies somewhat from the hourglass figure, or may be reduced in size, or in rare cases it may be absent. According to Sauer (1970), *L. variolus* has 2 red triangular marks ("split hourglass") instead of the typical hourglass of *L. mactans.*

Male black widows have relatively longer legs than the females, and have a lower and narrower abdomen that appears somewhat ellipsoidal (plate VI, 5; figure 208).

In addition the males are commonly more brightly colored. The mature male resembles the fifth-instar female in body markings, but it is much smaller. The male also possesses the ventral red hourglass figure on the abdomen. Dorsally, the pattern is variable, but usually consists of a median row of red spots with white lines radiating out to the sides. The pedipalps are large and bulbous. The venom sac of the male is small, and is not functional after he matures.

Kaston (1968) pointed out that many specimens of the female black widow spider were noted to be purplish brown, chestnut brown, or light chocolate brown rather than black, and that they might undergo changes from one of these colors to another.

The Black Widow's Web

The web of the black widow spider is an irregular mesh in which the spider hangs in an inverted position (figure 208). It is used to ensnare insects and other small prey, and consists of a network of silken strands that are stronger than those of most spiders. There is usually a small, central pocket or retreat to which the spider resorts when frightened. The prey sets up vibrations in the thread that announce its presence to the spider. Baerg (1936) noted that when black widow spiders lived under stones, they had relatively small webs or none at all. Presumably, this particular ecological niche attracts plenty of victims, and the spider is aware of their presence when they contact the few threads spun around her body. The webs usually found in old barns, sheds, crawl spaces under houses, or in little-used parts of garages and basements are likely to be about 30 cm across and equally as high, but they can sometimes be larger. Kaston (1970) described a web of *Latrodectus hesperus* in an unused wooden shed that had a "catching portion" about 30 in. (75 cm) above the ground level, with threads extending to a retreat in the rafters about 12 ft (4 m) above the ground. The female avoids light, and during the day may stay under a piece of overhanging board or a clod of earth, or may remain back in her retreat. She may move out over her snare at night and assume a position several inches in front of the overhang of the object under which she was hiding.

Life Cycle. The eggs of black widow spiders are deposited in silken egg sacs (figure 210) that can be constructed by the female in 1 to 3 hours. According to Kaston (1970), the egg sacs of *Latrodectus mactans* average about 9.5 mm in diameter, are spherical, with a conspicuous nipple on top, and gray. The egg sacs of *L. hesperus* are about 1.1 cm in diameter, pyriform, and tan, and those of *L. variolus* are 1.2 cm in diameter, pyriform, and gray. Kaston observed as many as 6 egg sacs per female for *L. variolus,* as many as 10 for *L. mactans,* and as many as 21 for *L. hesperus.* The mean number of eggs in 185 sacs of *L. mactans* was 255, and for 464 sacs of *L. hesperus* it was 196. The outer covering of the egg sac is tough and closely woven, differing in this respect from that of the brown recluse spider, which has an outer covering of loose threads. Baerg (1936) frequently saw a female with 4 egg sacs, and once saw one with 6. Under laboratory conditions, they have been observed to construct as many as 9 sacs in succession.

An egg sac will occasionally be found to be empty. The eggs of all 3 species are usually creamy white to yellow. The incubation period was noted to be 14 to 30 days in California by Herms *et al.* (1935), and from 15 to 21 or 30 days in Kansas by Lawson (1933). In the laboratory at 25° C (77° F),

Fig. 210. Black widow spider, *Latrodectus mactans*, with egg sac.

Kaston (1970) found the average incubation period (in days) to be 13.4 ± 2.0 for *L. variolus,* 14.2 ± 1.4 for *L. mactans,* and 14.6 ± 2.0 for *L. hesperus.*

The Spiderlings. When newly hatched, the tiny spiders undergo their first molt inside the egg sac about 3 or 4 days after hatching. They emerge from the sac in about 26 to 30 days after oviposition. There is usually a single emergence hole about 1 mm in diameter cut out by 1 or 2 spiderlings, but sometimes there may be 2, and rarely 3 holes (Kaston, 1970). The newly hatched spiders are pale reddish brown, with light and dark stripes on the abdomens and legs. They move about in the vicinity of the nest for several weeks, and many are destroyed by other spiders in search of prey, including their own kind. The mother will not eat them, even when she is starved.

Spiderlings are poisonous (when ingested) before emerging from the egg sac and until they are about 18 days old, after which the poison disappears. It is said to be "dramatically different" from the venom of the adult spider. Extracts of the eggs and spiderlings, when given to cats, caused a precipitous and often fatal hypotension, whereas the venom of the adults caused an acute and persistent hypertension (Buffkin *et al.,* 1971).

Description of Instars. There may be as many as 9 instars, but the spiders can mature in fewer instars. In a study of a total of 544 spiders, including males and females of 2 species, Kaston (1970) found that the mean periods required for them to become mature varied as follows: Males of *L. hesperus* matured in 4 to 7 instars in 62 to 151 days. The periods required for maturity increased with increasing numbers of instars. Females matured in 6 to 9 instars in 137 to 242 days. *Latrodectus mactans* males matured after 4 to 7 instars in 54 to 88 days, and females after 6 to 8 instars in 112 to 140 days. Only 1 male had 8 instars (in 166 days), and only 1 female had 9 (in 107 days).

Kaston (1970) described each instar of the 3 species of black widows. These vary greatly in appearance from one another and from the adults, and this in part accounts for the long list of synonyms. Figure 211 shows some dorsal and ventral patterns of late instars of *L. hesperus.* Kaston described all the instars in detail, but space permits only the inclusion of his description of the sixth instar (figure 211, *A, B, C*), which could be the penultimate (next to last) one, for females can become mature in the seventh, eighth, or ninth instar.

The darker specimens show more pigment on the carapace and have the dark areas more extensive than previously. There remains on the sternum only a narrow central light band. On the legs the dark areas have increased in size.

The abdominal dorsum is mostly covered with dark pigment now, with the only light areas reduced to a basal transverse band, a row of spots along the midline, and 2 pairs of diagonal stripes extending down the sides to the rear. These latter are the areas that had previously been narrow. Each of the light spots along the midline encloses a reddish spot. The hourglass mark is becoming more constricted at the middle and has more red pigment.

Some individuals, showing a more or less similar arrangement of spots, have the pigmented areas lighter. Also, the light diagonal bands extend farther down on the sides, and the dorsal spots are more orange than red.

The figure shows that the dorsal marking of the penultimate instar of the *Latrodectus hesperus* male is remarkably similar to that of the fifth-instar female. There is subsequently little if any difference in the markings in the male, even if it has as many as 7 instars; they continue to resemble females of the "light variety" of *L. hesperus* in the fifth or sixth instars.

Longevity of Adults. In California, most of the spiders overwinter as immature individuals, maturing sometime in the spring (Herms *et al.*, 1935). In Arkansas, the young spiders generally mature in May or June (nearly a year from egg to adult), and begin to die in considerable numbers in late July. They can be kept alive in the laboratory for 2 years or more (Baerg, 1936). The maximum number of days of survival of males after maturity was found by Kaston (1970) to be as follows: *Latrodectus mactans,* 127; *L. variolus,* 155; and *L. hesperus,* 196. The corresponding figures for females were: *L. mactans,* 849; *L. variolus,* 822; and *L. hesperus,* 952. Spiders are known for their ability to live for long periods without food. Kaston kept 37 black widow females without food for periods ranging from 36 to 193 days, with an average of 89.3 days.

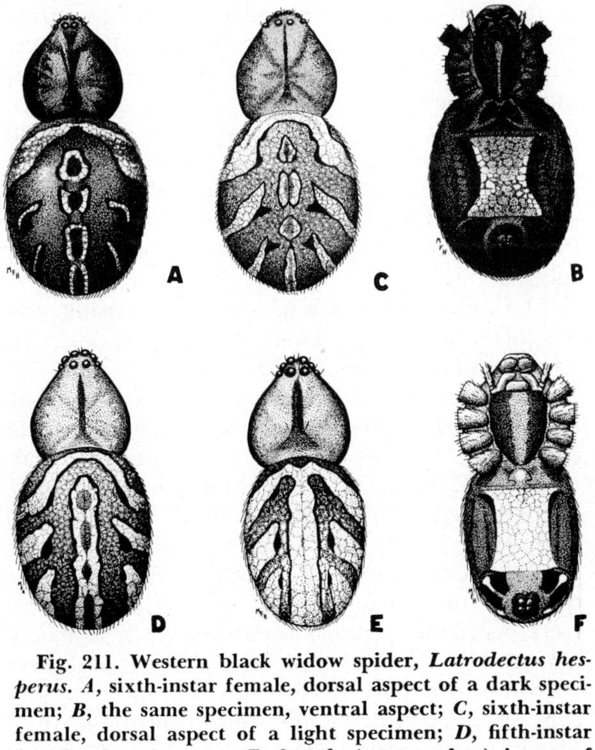

Fig. 211. Western black widow spider, *Latrodectus hesperus. A,* sixth-instar female, dorsal aspect of a dark specimen; *B,* the same specimen, ventral aspect; *C,* sixth-instar female, dorsal aspect of a light specimen; *D,* fifth-instar female, dorsal aspect; *E,* fourth (next to last) instar of male, dorsal aspect; *F,* the same, ventral aspect. (From Kaston, 1970.)

Habits. In the laboratory, the reaction of spiderlings to a fan-produced breeze was to climb upon whatever supports were provided, to distances at which the air currents were most suitable, and to spin webs and let themselves float like kites, the silk functioning as the tail of the kite. If the wind is in the right direction, a house may be showered with spiderlings, and many will find suitable locations, establish their webs, and grow to maturity. "Ballooning" of black widow spiderlings provides for general distribution of the species in upland, lowland, wild, and rural areas, towns, and cities (Baerg, 1936).

Outdoors, black widow spiders commonly live under and among stones, under pieces of wood, in hollow stumps, in rodent burrows, and less commonly among the leaves of plants and in low shrubbery. Sometimes the black widow becomes so abundant in the vines of such crops as tomatoes and grapes that harvesting may be hazardous and gloves should be worn. However, these spiders are most commonly encountered in dry and sheltered man-made structures, such as outdoor privies, barns, henhouses, garages, cellars, furniture, in water-, gas-, and light-meter boxes, and in woodpiles or piles of rubbish. One often sees the egg sacs of the black widow spider, but not the spider itself, when crawling about extensively under a house, as when inspecting for termites (figure 72, chapter 5). There, the black widows are preyed upon by ground beetles, scorpions, wolf spiders (Lycosidae), funnelweb weavers (Agelenidae), and mud daubers, as well as by others of their own kind. However, their predators may sometimes themselves be vanquished in the encounters and, along with other small arthropods, may serve as food for the black widow.

According to Sauer (1970), the northern widow may be found throughout the lower peninsula of Michigan, where it "occurs in marginal land where the vegetation is rather sparse, often hiding in old stumps, hollow logs, under fallen fenceposts, in abandoned animal burrows or piles of dead tree branches, and other debris. It prefers these outdoor situations to buildings, whereas the black widow of the southern United States can be found in both habitats."

Natural Enemies. A dipterous egg predator of black widows, *Pseudogaurax signatus* (Loew) (Chloropidae), has a sparse distribution. The most common hymenopterous parasite is a scelionid, *Baeus latrodecti* Dozier, reported from egg sacs by Pierce (1942). Other hymenopterous egg parasites are a eulophid and a *Eurytoma.*

Another spider, *Steatoda grossa* (C. L. Koch) (figure 212), in the same family (Theridiidae) as the black widow, has been observed feeding on the latter. This species is mentioned in chapter 6 as one of the predators of cockroaches. Other black widow predators are species of *Mimetus,*

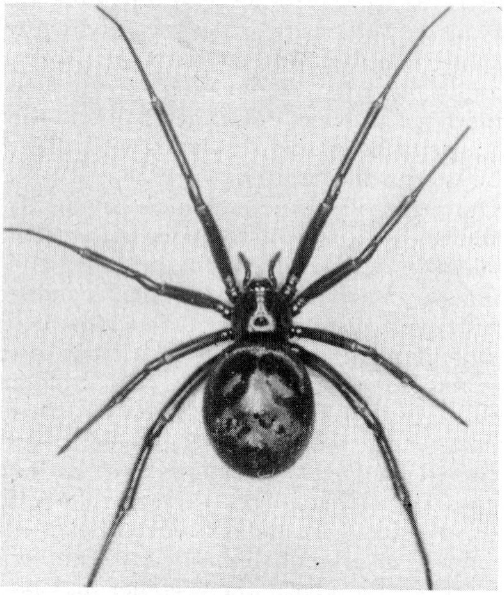

Fig. 212. A theridiid spider, *Steatoda grossa,* a natural enemy of cockroaches as well as of the black widow spider.

the pirate spider. The most widespread and possibly the most effective predator is the blue burglar wasp, *Chalybion californicum* (Saussure). It will provision its cells with black widows in preference to other spiders. The San Diego alligator lizard has also been found to be an effective predator (Kaston, 1970).

The Bite of the Black Widow

The bite of the black widow spider is not always felt, and often there is little evidence of a lesion. There is usually a slight local swelling and redness, and 2 tiny red spots may appear. It is important for a physician to take note of these spots, for they indicate the 2 entrance points of the spider's fangs. A single entry mark might indicate the bite of a conenose bug or some other insect, and the administration of an antitoxin for black widow spider venom would not only be useless but could actually be harmful. Pain is felt almost immediately, reaches its maximum in 1 to 3 hours, and continues for 12 to 48 hours, gradually subsiding. In severe cases, there is rigidity and spasm of most of the larger muscles, particularly of the abdomen, which becomes "boardlike." There is a rise in body temperature, increased blood pressure, profuse perspiration, and a tendency to nausea. Other symptoms, such as chills, urinary retention, constipation, hyperactive reflexes, and a burning sensation of the skin, are frequently reported.

The black widow can control the amount of venom ejected from her venom glands. For example, when attacking insects on which she feeds, she injects a dose commensurate with the size of the victim. When biting a human being, she attempts to inject her entire supply of venom. The amount of venom held in her sac at the time of the bite depends on how recently she has used her fangs. These do not always break through tough skin, since they are less than a millimeter in length. If she drops on the hand or arm, she should be brushed off rather than swatted, in order to avoid pushing the fangs into the skin. It is fortunate that the black widow can inject only a relatively small amount of venom, for the venom is 15 times more powerful than that of a rattlesnake. Nevertheless, the bite of a rattlesnake is considered to be more dangerous because of the much greater dose of venom injected. There is about 15 to 25% mortality of persons bitten by rattlesnakes. According to Wingo (1960), there is less than 1% mortality of persons bitten by black widow spiders.

The black widow will attack any object that touches her web, but she feeds mainly on insects. Insect remains are generally not seen on the web, suggesting that the black widow may remove them from the web sooner than do other spider species.

The web of the black widow is very strong, and has been known to snare animals as large as lizards and mice (Kaston, 1970).

Approximately 300 spider bites, of all kinds, are reported to physicians in southern California each year, but fewer than 5 persons per year, usually children, die from spider venom poisoning (Russell, 1969). Yet a child weighing 30 pounds or less would have only about a 50 per cent chance of surviving if he received the entire quantity of a black widow spider's venom and were given no medical attention (F. E. Russell, personal communication). The physical and mental condition of the victim of a black widow spider can, of course, have a great effect on the seriousness of the reaction. However, the bite of a black widow spider need never be fatal if treated promptly by a physician.

Brown Recluse Spider, *Loxosceles reclusa* Gertsch and Mulaik (Loxoscelidae)

Montgomery (1908) found that the brown recluse spider was common in its silken hibernaculae under stones in a Texas area during winter. It was described later by Gertsch and Mulaik (1940). What appeared to be its first proven bite occurred in Kansas in 1928 (Schmauss, 1929), but not until 1957 was there circumstantial evidence that associated this species with the typical sloughing bites that physicians had been seeing (Atkins *et al.,* 1957). The following year, the sequence of results from the bite on experimental animals was recorded (Atkins *et al.,* 1958).

The brown recluse occurs in the United States in the southern parts of Ohio, Indiana, and Illinois, in Missouri, Kansas, and parts of Nebraska, and southward to the Gulf of Mexico. This species and its near relatives may now be included with the black widows (*Latrodectus* spp.) as being among the few spiders in the country for which it can be said that their bites, although uncommon, are causes for justifiable concern.

Description. The adult females (plate VI, 6; figure 213) vary from 7 to 12 mm in length, averaging about 9 mm, and the males are a little smaller, averaging about 8 mm (Gertsch, 1958).

Fig. 213. Brown recluse spider, *Loxosceles reclusa* (female). (U. S. Army photo, from Glick, 1969.)

The males are readily distinguished by their bulbous pedipalps. The legs of *L. reclusa* are long, and are covered with minute brown hairs, but appear almost bare to the unaided eye. The body color varies from light fawn to dark brown. Recently molted individuals have a lighter color. They have 6 eyes, arranged in pairs in a semicircle unlike most spiders, which generally have 8 eyes. Immediately behind the eyes is a violinshaped marking on the anterior portion of the cephalothorax, with the neck of the violin pointing backward and narrowing down to a center line extending almost to the abdomen. This is a conspicuous mark, and has resulted in this and related species being called "violin spiders." The immature stages closely resemble the adults except for size and generally having a slightly lighter color.

The Web. The medium-sized, irregular web is made up of a maze of very viscid threads that extend in all directions without a definite pattern. It has been described as an off-white or grayish, nondescript, "cobweb-type" webbing. The web is not used specifically for trapping insects and other prey, but mostly as a retreat for the spider. In its outdoor habitat, the brown recluse spider spins a tube of thick silk as a retreat in winter. Likewise, in the laboratory the spider often constructs a retreat of loose silk in 1 part of its web (Hite *et al.,* 1966).

Life Cycle. The egg sac consists of a circular base of densely woven silk threads upon which the eggs are laid. The female then spins a dense cover over the eggs. After resting, she then spins a more loose and flocculent cover. The completed egg sac is white, flat beneath, and convex above. It averages 3.7 mm in height and 17 mm in diameter. From 1 to 5 egg sacs are produced. The young emerge from the eggs in 25 to 39 days (a mean of 32.6) (Hite *et al.,* 1966).

In a laboratory, egg production varied from 31 to 300 per female. From 146 egg cases containing 7,374 eggs, 3,576 young emerged. Sometimes females fed on eggs, and sometimes the young from a preceding egg sac fed on later eggs. There were 8 instars, and the period from egg to adult varied from 266 to 444 days, with a mean of 336 days. Males lived an average of 543 days, and females, 628 days (Hite *et al.,* 1966). On the other hand, 30 females, assumed to be about 540 days old when they were collected, were kept at normal room temperature in summer and a near-natural environment in winter. They were reared another 880 days, and were then in their fifth overwintering hibernacula (Horner and Stewart, 1967).

In Arkansas and surrounding states, mating occurs from early February to early October, the most active period being in June and July. No egg-laying was observed in October to January, inclusive (Glick, 1969).

The Brown Recluse's Habitat. Hite *et al.* (1966) recorded the location of 298 brown recluse spiders, collected in homes in Arkansas, as follows: in boxes, 156; among papers, 39; in bedrooms, 29; attics, 22; halls, 14; utility rooms, 11; kitchens, 11; livingrooms, 5; bathrooms, 4; front porches, 3; window wells, 2; cellar, 1; and basement, 1. In school buildings, among 74 spiders, 45 were found in schoolrooms, 26 in basements, 2 in attics, and 1 in a storeroom. In other sites, among 58 spiders, 21 were found in the loft of a feed mill, 11 in a broiler-house storeroom, 10 in storehouses, 8 in cabinet shops, 6 in garages, and 2 in sheds. Outdoors, of 196 spiders collected, 158 were found under rocks, 34 under piles of inner tubes, 2 under houses, and 2 under bark.

In homes the brown recluse spider may be found in old clothes, on the undersides of tables and chairs, behind baseboards and door casings, or in corners and crevices. It usually runs for cover when disturbed, accounting for its common name. People are most likely to be bitten when putting on old clothes or shoes that may not have been worn for long periods, or by rolling on the spider in bed.

Symptoms of the Brown Recluse's Bite

Venom was extracted from 3,000 *Loxosceles reclusa* spiders by electrical shocks and was collected in microcapillary tubes. The quantity of venom collected per spider ranged from 0.25 to 0.62 μl, with an average of 0.36 μl. Male spiders gave less than half as much venom as females, but it had the same toxic properties (Morgan, 1969). Both sexes are able to inflict venomous bites to mammals. The following is a description in nonmedical terms of the reaction to the bite in man, as quoted from Wingo (1960):

> The typical reaction in man following a bite by the brown recluse is necrosis (killed tissue) at the site of the bite. The victim may not be aware of being bitten for 2 or 3 hours, or a painful reaction may occur immediately. A stinging sensation is usually followed by intense pain. A small blister usually rises, and a large area around the bite becomes congested and swollen. The patient may become restless, feverish, and have difficulty in sleeping. The local pain is frequently quite intense, and the area surrounding the bite remains congested and hard to touch for some time. The tissue affected locally by the venom is killed and gradually sloughs away, exposing the underlying muscles. The edges of the wound thicken and are raised, while the central area is filled by dense scar tissue. Healing takes place quite slowly, and may take 6 to 8 weeks. The end result is a sunken scar which has been described as resembling "a hole punched or scooped from the body." Scars ranging from the size of a penny to half-dollar have been reported.
>
> The necrotic condition described above is typical of all bites of the brown recluse. However, in some cases a general systemic reaction has also occurred. In one case, the patient broke out with a rash resembling that of scarlet fever. In another case, the kidneys were apparently affected, causing bloody urine to be passed. These systemic disturbances probably occur infrequently, and are the results of a "full" bite (i.e., the injection of a maximum amount of venom) or extreme sensitivity to the venom. This general reaction to the bite of the brown recluse is certainly a serious condition, and hospitalization of the patient is usually required. Those in poor general physical condition, young children, and older people are more apt to be affected seriously by the bite of the brown recluse.

The principal injury inflicted by the brown recluse is the severe tissue damage, which is deep and heals slowly. Death rarely results from the bite of this spider. Successful therapy and prevention of local necrosis and severe systemic toxicity depend on early recognition of the symptoms followed by the application of appropriate, vigorous therapy (Bolton, 1970). Good results have been obtained with a course of treatment employing certain corticosteroids, but treatment initiated 48 hours or more after the bite is unlikely to have any effect (Dillaha *et al.*, 1964).

Figure 214 shows the healing sequence of a lesion resulting from the bite of a brown recluse spider that was first photographed on June 24, 1968, about 48 hours after the bite occurred, and for which photographs were made at intervals until August 23, 1968, a period of 58 days. Debridement (surgical removal of lacerated, devitalized, or contaminated tissue) was done at necessary intervals as the tissue sloughed off. A bleb (small blister) recurred at the initial site of the bite almost 6 weeks following the original blister, and this was found to be typical of the lesions observed (Glick, 1969). Before 1968, there had been at least 126 cases of this type of necrotic spider bite in the United States (and 6 deaths), and about 400 cases in South America (and at least 35 deaths) (Gorham, 1968).

Some Relatives of the Brown Recluse, Native and Introduced

There are about 50 species of *Loxosceles* in the Western Hemisphere, and according to W. J. Gertsch (correspondence), identification depends almost exclusively on a study of the genitalia. In immature stages where the female genital receptacles are not present, it is virtually impossible to identify the specimens. Any species within the United States can be distinguished by superficial observation, particularly in view of the relatively small overlap in geographical ranges. There are several other *Loxosceles* spiders in the country that are similar in size, shape, and coloration to *L. reclusa,* but their violinshaped markings are sometimes less distinct, and may be somewhat faint in some species.

Loxosceles unicolor Keyserling

This is the only species of *Loxosceles* that is indigenous to California. It is distributed in parts of California, Arizona, New Mexico, Nevada, Utah, and Texas, in Mexico at least to southern Sonora, and in Baja California (Gertsch, 1958). Specimens collected by Russell *et al.* (1969) were generally found near human habitations, from beneath ground cover such as lumber, rocks, or debris, and also inside houses. These chiefly nocturnal spiders live a solitary type of life and move quickly when attempting to escape.

Description. The females (plate VI, 6) are from 6 to 9 mm long and average about 7.5 mm (Gertsch, 1958; Russell *et al.,* 1969). The males are slightly smaller. Thus, *L. unicolor* has a smaller body than *L. reclusa,* but has longer legs. According to Russell and associates, *L. unicolor* is light brown, sometimes with a shading toward beige or yellow. Ennik (1971) found that the color and to some extent the size of the abdomen of *L. unicolor* depended on the quality and kind of prey it ate. *Drosophila* flies as food caused their abdomens to turn orange-red, but house flies caused them to turn gray. Grasshopper nymphs caused a light green and German cockroaches a violet color. The "violin" on the cephalothorax of *L. unicolor* is not so distinct as in *L. reclusa.* Males can be recognized in the penultimate stage by the enlargement of the pedipalpal tarsus. Females can be recognized in the same stage by the reddish color of the tibiae and tarsi of the pedipalps and of the genital furrow border (Ennik, 1971).

Life Cycle. Spiderlings hatch 15 to 19 days after oviposition. Then follows the first post-embryo, confined in a membranous sac; a second post-embryo, when the sac is ruptured; and the first molt 11 to 16 days later. Although the spiders may mature in any stadium after the sixth molt, the

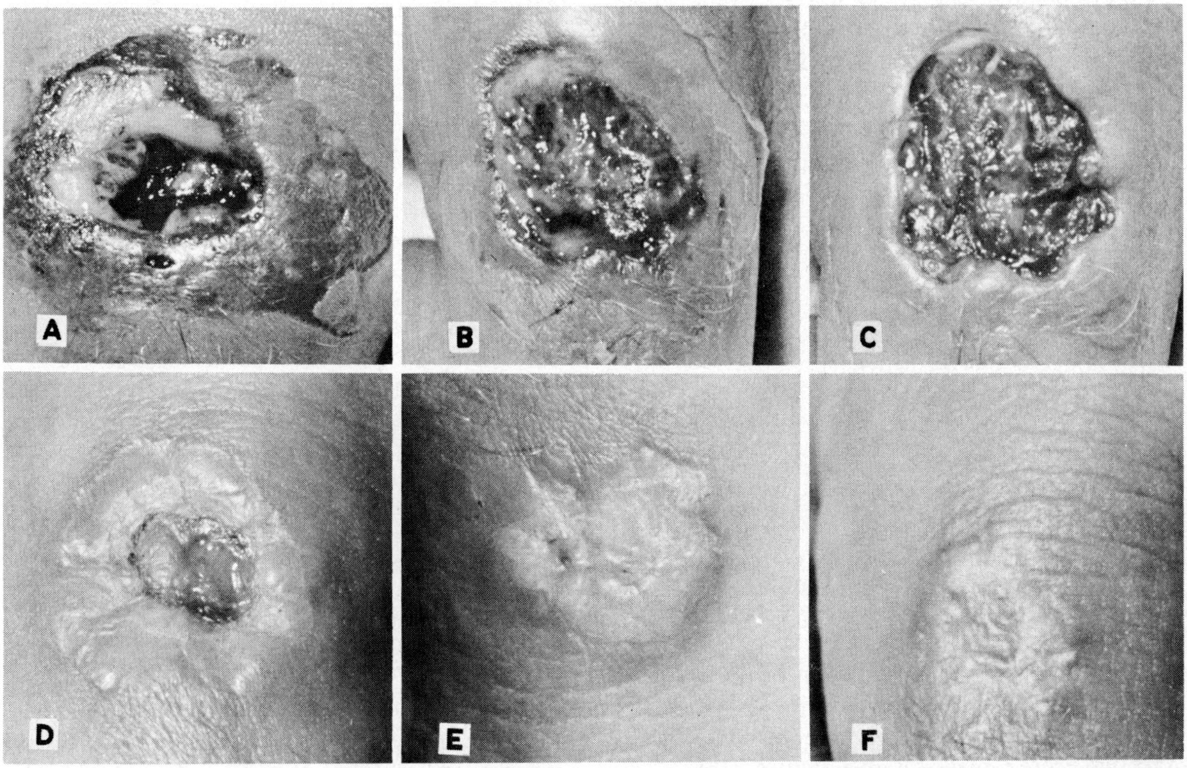

Fig. 214. Healing sequence of a lesion on the upper joint of an index finger caused by the bite of a brown recluse spider. The sequence covered a period of 62 days. *A,* 2 days after bite; *B,* 4 days; *C,* 9 days; *D,* 27 days; *E,* 38 days; *F,* 62 days. (Courtesy Mrs. Lena Glick, R.N.)

majority of males become adult in the seventh and eighth molts, and the majority of females in the sixth and seventh molts. The developmental periods of the sexes were found to be as follows: males, 290 to 680 days (mean of 464); females, 276 to 562 days (mean of 399) (Ennik, 1971).

Bite Symptoms. *Loxosceles unicolor* can inflict a bite that causes pain, bleb formation, erythema, and necrosis leading to ulceration, but the skin responses and systemic changes are not usually so severe as those caused by *L. reclusa*. Apparently, only 3 cases of necrotic arachnidism in which *L. unicolor* was positively identified as the cause have come to the attention of medical authorities in California. However, bites by *L. unicolor* appear to be more common than had formerly been expected; descriptions of some of the lesions reported by physicians appear to be similar to some that have been definitely connected with this spider (Russell *et al.*, 1969).

Loxosceles arizonica Gertsch and Mulaik

This species occurs in Arizona, New Mexico, western Texas, and southward into Mexico from Coahuila to Baja California. The females are 6 to 10 mm in length, averaging 8 mm (Gertsch, 1958). The males are slightly smaller. These spiders have been described as being darker in color than *L. unicolor*, but Russell *et al.* (1969) found no such difference in specimens they examined. Specimens from arid regions tended to be paler. The violin markings were always distinct (figure 215).

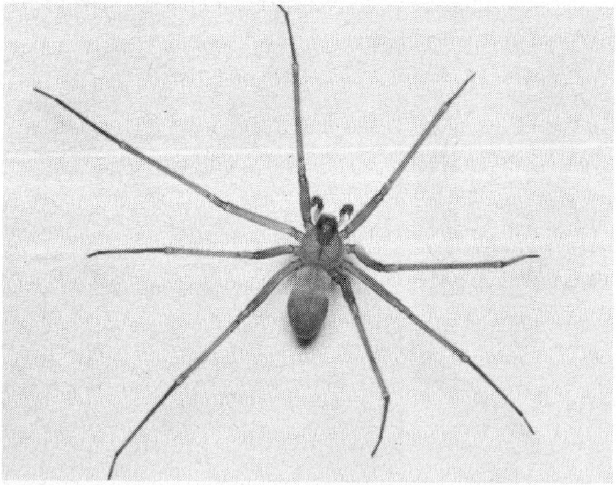

Fig. 215. A brown or violin spider, *Loxosceles arizonica*, of the southwestern United States. (From Russell *et al.*, 1969, courtesy F. E. Russell, M.D.)

As with *L. unicolor, L. arizonica* can now be considered to produce lesions. Many of the lesions that have been recorded as "necrotic arachnidism" during the past several years in areas where *L. arizonica* is indigenous have probably been caused by that spider. Experimental envenomation of hairless mice by *L. arizonica* and *L. devia* (see below) has been demonstrated (Cutler and Cutler, 1971).

Loxosceles devia Gertsch and Mulaik

This species ranges through southern Texas and the adjacent states of eastern Mexico. The area of distribution of *L. reclusa* slightly overlaps that of *L. devia* in Texas. The latter is the smaller of the 2 species, with both males and females averaging about 7.5 mm in length. It also has a paler carapace, and the typical dorsal pattern ("violin" marking) of *L. reclusa* is mostly lacking (Gertsch, 1958).

South American Brown Spider, *Loxosceles laeta* (Nicolet)

This spider has been associated with the "gangrenous spot" syndrome in Chile since at least as early as 1937 (Macchiavello, 1937). The lesion is similar to the one produced by *L. reclusa*. An antivenin has been prepared against *L. laeta* for the treatment of both the cutaneous lesions and the systemic manifestations of that spider's toxin, but at least by May, 1970, the antivenin was not available in the United States, possibly because it had been found to be unsuccessful against the toxin of *L. reclusa* (Bolton, 1970).

Beginning in 1960, *L. laeta* was found in basement areas in a museum and in a terrarium at Harvard University, Cambridge, Massachusetts (Levi and Spielman, 1964). Eventually, more than 50 spiders were collected and many more were observed. One building occupant stated that the spiders had been there for at least 20 years. A further observation was made that it had been about 20 years since firebrats (*Thermobia domestica*) had had to be controlled in the infested area. *Loxosceles laeta* apparently did not spread beyond the infestation at Harvard, and is not known to occur elsewhere in the New England states (Spielman and Levi, 1970).

In 1969, this species was found in Sierra Madre, California, a suburb of Los Angeles, where about 190 spiders were collected in an 8-block area. The infestation was believed to be at least 2 years old (Hawthorne, 1969; Waldron, 1969*a*). During the same year, an even greater area of infestation and greater numbers of spiders were found 11 miles (18 km) away, in the city of Alhambra. It seemed apparent that the spiders had been in the Alhambra area 5 years or longer. The extreme shyness of *Loxosceles laeta* and its rapid movements may have helped to keep it from being detected sooner (Waldron, 1969*a*).

Description. *Loxosceles laeta* is similar to *L. reclusa* in appearance. It is the largest species in the genus. The living specimens (plate VI, 7) are tawny to brown or blackish, and the dead specimens generally seen in collections are yellow to reddish brown. The cephalothorax and legs are more reddish than in *L. reclusa*. The "violin" on the cephalothorax is not quite so conspicuous as in *L. reclusa,* but is readily discernible. W. J. Gertsch (correspondence) pointed out that the palpal tarsus of the male *L. laeta* is longer than wide, whereas in *L. reclusa,* it is wider than long. In *laeta,* the palpal tibia is narrow, and in *reclusa,* it is shorter and broader. In *laeta* females, the fourth leg is longer than the others, while in *reclusa,* the second leg is longest in both sexes. The palpi of the *laeta* males have a distinctive form. Of 5,449 *L. laeta* collected in houses in central Chile, the ratio of males to females was 1:8 (Schenone *et al.,* 1970).

Habitat. In the Harvard infestation, the spiders were usually seen on floors and baseboards, only rarely on desks and tables, and never on walls. It was noted that *L. laeta* could become very numerous itself, while appearing virtually to eliminate all other arthropods. Webs sometimes reaching 30 cm in diameter were typically found in corners, as noted by Waldron (1969*a*) in the California infestation. This was in keeping with a common name for *L. laeta* in South America, *araña de los rincones* (corner spider). The loosely woven egg case, containing about 50 large, white eggs, was always placed in a dense portion of the multilayered web. Males were frequently found without webs. Lindane proved to be the best insecticide in control tests at Harvard.

Attempts to Provoke Biting. Six females of *L. laeta* were confined to a guinea pig's flank, and although they were pressed against it and provoked, none bit the animal (Levi and Spielman, 1964). This was contrary to tests with *L. reclusa* in Arkansas, in which the spider could easily be provoked to bite under similar conditions (Atkins *et al.,* 1958).

In another similar experiment *L. laeta* was provoked to bite rabbits. The reactions of the rabbits to the bites of *Loxosceles reclusa, L. laeta,* and *L. rufescens* (mentioned following) were similar, and the reactions to males and females of the 3 species were also similar. The spiders probably did not usually deposit all their venom in a single bite; a second bite frequently caused a lesion at least half the size of the first (Smith and Micks, 1968).

Loxosceles rufescens (Dufour)

This species, apparently imported from Europe, now occurs in a number of localities, from New York to Illinois and throughout the southeastern states. It is similar to our indigenous species in general appearance. The female is 7.5 mm long and the male, 7 mm (Kaston, 1973).

Chiracanthium inclusum Hentz (Clubionidae)

This is a pale-yellow spider, with a body about 5 to 6 mm long, that is widely distributed throughout North and South America and frequently appears in houses.

In California, a man awakened by a sharp, penetrating pain below the outer margin of the right eye found that he had been bitten by an immature male of *C. inclusum.* Cold compresses applied over the area of the bite for an hour and taking 2 grams of aspirin over a 4-hour period did not alleviate the pain. Within 3 hours after the bite, a throbbing pain had spread over the cheek; after 7 hours, the pain was still intense, but seemed to be abating; and after 12 hours, it was gone. The venom appeared to have only a local neurotoxic effect.

Later, 14 more specimens were found in the same house, apparently having come from the adjacent shrubbery. Subsequently, an active specimen could not be induced to bite a man when placed on an area of thin skin, indicating that this spider is not pugnacious. However, a number of persons have been bitten by it in Hawaii, suffering moderate to grave symptoms (Furman and Reeves, 1957).

Another species, *C. mildei* C. L. Koch, indigenous to Europe, north Africa, and the Middle East, was considered to be probably responsible for necrotizing skin lesions seen in 5 persons in the vicinity of Boston, Massachusetts (Spielman and Levi, 1970). (*Loxosceles reclusa,* also responsible for necrotizing lesions, does not occur in the New England states.) *Chiracanthium mildei* was introduced into the United States, and has spread from New England south to Alabama and west to Utah, also occurring in California (Bryant, 1951; Spielman and Levi, 1970). This species is 7 to 10 mm in body length, and has a pale-brown cephalothorax and white abdomen. It inhabits thickets and hedgerows, but adapts well to life in homes and other occupied buildings. Minton (1972) described symptoms and treatment of a woman bitten by this spider while swimming in a heated indoor pool in Indiana.

Tarantulas (Theraphosidae)

The tarantula is the largest spider in the United States. It is not ordinarily a household spider, and in fact the encroachment of civilization greatly decreases its numbers in an area. Nevertheless, as more and more people build their homes in formerly wild foothill or desert areas, we receive an ever larger number of reports of tarantulas being found in houses.

The tarantula is an interesting spider because of its huge size (up to a 5-in. [13-cm] leg span) and forbidding, hairy appearance, which have given it an undeservedly sinister reputation. Baerg (1922) allowed himself to be bitten twice, and stated that the bites felt like pinpricks, with mild pain lasting only 15 to 30 minutes, not accompanied by inflammation or swelling. Tarantulas are very sluggish, and will not bite unless provoked. They are unlikely to bite even when picked up if they are handled gently. Some people keep them as pets; a good place for this is in a small, empty aquarium with a sandy bottom. A small clay flower pot, laid on its side and with sand at the bottom, may be put at one end as a hiding place. A jar lid can be sunk into the sand near-by to serve as a container for water. Tarantulas should be fed live insects. It is convenient to use live crickets, which can be purchased from some suppliers of fish bait. The female spider can live as long as 20 years in captivity, and even longer in the wild state.

There are some 30 species in the United States, mostly in the southwestern area. The light- to dark-brown *Aphonopelma californica* (Ausserer) (= *Eurypelma* = *Dugesiella*) (plate VI, 8; figure 216) is the principal species, usually found on the dry, southern slopes of hills. It is the largest of the United States spiders, being up to 5 in. (13 cm) in total length. It is covered with velvety wool and long, silken hairs. The male may be black or reddish brown, and is usually much darker and has a smaller abdomen than the brownish female. Another tarantula, *Dugesiella hentzi* (Girard), more common in Arkansas, Oklahoma, and Texas, has often been called *Aphonopelma californica*.

Tarantulas either burrow into the ground or utilize small holes of suitable diameter near rocks, roots, or other objects. They line their tunnels with silk, and form a neat, webbed rim at the entrance or conceal it entirely. They generally travel at night, and usually those seen wandering about in the open are adult males. It requires 10 years for tarantulas to reach maturity. The males enter the burrows only in mating season, from July to November. Many are killed by the females, and

Fig. 216. A tarantula, *Aphonopelma californica*.

others die from many hazards, including man. On the other hand, the females may live for 25 years or more, the longest-lived spiders known. The females molt after maturity, but the males do not (Levi *et al.*, 1968)

Other Species of Venomous Spiders

Although all but 2 small groups of spiders are venomous, the fangs of most species are too short or too fragile to penetrate skin. Nevertheless, in the United States at least 50 species have been implicated in bites on humans. Aside from the species already discussed, those which have produced some local or systemic effects include the following:

Theraphosidae: *Pamphobeteus* spp.

Trapdoor spiders (Ctenizidae): *Bothriocyrtum californicum* (O. P.-Cambridge), and *Ummidia audouini* (Lucas).

Cobweb spiders (Theridiidae). *Steatoda grossa* (C. L. Koch) (possibly).

Orb weavers (Araneidae): *Araneus* spp., *Argiope aurantia* Lucas, and *Neoscona vertebrata* McCook.

Lynx spiders (Oxyopidae): *Peucetia viridans* (Hentz).

Wolf spiders (Lycosidae): *Lycosa* spp.

Gnaphosidae: *Drassodes* spp.

Sac spiders (Clubionidae): *Liocranoides* spp.

Wandering spiders (Ctenidae): *Phoneutria* spp. and *Cupiennus salei* (Keyserling).

Crab spiders (Thomisidae): *Misumenoides aleatorius* (Hentz).

Jumping spiders [Salticidae (= Attidae)]: *Phidippus* spp.

Pamphobeteus, Phoneutria, and *Cupiennus salei* are not native to the United States, but are sometimes introduced in produce or other materials (Anonymous, 1970*g;* Russell, 1969).

Of the 5 species of spiders other than *Latrodectus* spp. and *Loxosceles* spp. that Waldron (1965) was able to identify with cases of arachnidism (poisoning caused by the bite of an arachnid) in southern California, only a single species, *Neoscona vertebrata,* which bit the back of the hand of an adult man, caused an ulceration 1 in. (2.5 cm) in diameter. The others were *Phidippus formosus* (Peckham) (figure 217), (3 cases), an immature *Misumenoides aleatorius, Chiracanthium inclusum,* and *Araneus* sp. The symptoms in these cases included initial pain, redness, and sometimes swelling. Marks caused by the spider's chelicerae were observed in all cases. A woman bitten by *Phidippus formosus* reported that the bite area on her leg was sore and tender for almost 6 weeks. A 16-year-old boy bitten by *Araneus* sp. noted no swelling, but the pain lasted about 4 days.

Phidippus formosus bites aggressively and, unlike *Latrodectus* and *Loxosceles,* it does not have to be stimulated to bite. While photographing the specimen shown in figure 217, the photographer was bitten twice on the back of his left hand. The swelling gradually increased until in 4 days it involved the entire back of the hand and the fingers. The back of the hand was sensitive to touch and pressure, and itched severely. The swelling and a dull pain lasted for 9 days (Russell, 1970).

Importance of Submitting Captured Arthropod When Bitten

It is important to try to capture the arthropod that does the biting and submit it to the proper authority for identification. The stings and bites of bees, wasps, mosquitoes, biting flies, fleas, bed bugs, lice, ticks, and mites are far more common than spider bites, and some of these arthropods can produce bulbous lesions that may rupture and ulcerate. Many reports of necrotic or gangrenous arachnidism currently attributed to *Loxosceles reclusa,* especially in areas where it is not likely to be found, may be traceable to other species of spiders. Bites from some of the spiders just listed, while not causing as severe symptoms as those caused by *L. reclusa,* can be serious enough to warrant hospitalization and intensive care (Russell *et al.,* 1972).

Fig. 217. A jumping spider, *Phidippus formosus.* (From Russell, 1970.)

Although most bites or stings by arthropods are not likely to be dangerous, medical care or advice should be sought in each case. Regardless of the offending animal, rest or decreased activity is always indicated, and reassurance should be given to the victim. According to F. E. Russell (personal communication), of the millions of stings and bites by arthropods each year in the United States, fewer than 10 persons die as a direct effect of the venom. More than 50 deaths per year are attributable to the sensitivity reactions to venom proteins, usually to those of Hymenoptera. Irrespective of the arthropod involved, whenever possible it should be sent with the victim to the physician or medical facility so that proper identification can be made. This is particularly important in the case of a spider or tick bite or a scorpion sting.

Treatment for Spider Bites

There are no first-aid measures of value for bites by the black widow spider. If the pain is severe, an ice cube may be placed on the bitten spot to reduce pain. It should be removed when antivenin is injected. The same measures recommended for scorpion stings—immediate immobilization of the affected part, decreased activity or complete rest, and reassurance—are indicated. Hospitalization is usually advised for all patients under 16 or over 60 years of age, those with hypertensive heart disease, or symptoms and signs of severe envenomation. (Such symptoms and signs for the black widow spider and for the brown recluse are described on preceding pages.) In case of a black widow spider bite, 1 ampule of *Latrodectus mactans* antivenin should be given as soon as possible at a medical facility following the appropriate skin or eye test. Patients with a history of hypertensive heart disease should receive mild sedation, and appropriate measures for reducing the blood pressure should be instituted, if necessary. For adults, treatment usually consists of the administration of muscle relaxants or calcium gluconate, hot baths, aspirin, and mild sedation.

In all cases, the patient should be advised to rest for 12 hours. The symptoms and signs of black widow poisoning rarely persist, even in an untreated patient, for more than 2 days (Russell, 1969).

Treatment for *Loxosceles* bites is largely empirical. Most authorities agree that in order to be effective, treatment must begin within the first 24 hours following the bite. Local excision of the wound appears to be the method of choice when the case is seen within 12 hours. Thereafter, steroids appear to be the drugs of choice, but even these must be administered early to obtain a good result. Antihistamines are of questionable value (Russell *et al.,* 1969; Asel, 1969; Russell *et al.,* 1972).

Although *Loxosceles* is the only spider at present implicated in the syndrome known as "necrotic arachnidism," many arachnologists and physicians feel that it is quite possible that other genera might be implicated in this syndrome in the future. A number of new drugs are now being tried for the treatment of the necrotic lesions produced by spiders, and some show promise.

Control of Spiders

The habitats of the black widow and brown recluse spiders in buildings are similar to those of any building-invading spiders—in corners, under and behind furniture and household appliances, in undisturbed clutter in the garage or workshop, piles of lumber, boxes, and debris, and in the crawl spaces under buildings. Premises should be cleared of rubble, scrap, and lumber piles, but care should be taken not to accidentally imprison or crush spiders with the fingers or other body surfaces. The female black widow spider may rush out and bite when her web is disturbed. Most spiders will bite, or attempt to bite, when they are accidentally trapped in clothing or shoes.

Any secluded and seldom-disturbed location may provide harborage for spiders. Webs should be destroyed, or may be sprayed or dusted. Suspect areas may be sprayed or dusted with insecticides such as lindane, chlordane, diazinon, or chlorpyrifos, using concentrations recommended on the package labels. The crawl space can be most effectively treated with an insecticide dust, which can be applied from the crawl hole and foundation vents. Large hand dusters may suffice for dusting extensive areas, but equipment that can apply a dust with greater velocity, such as an electric blower or a modified water-type fire-extinguisher, is desirable. The silica aerogel Drie 67, available only in 15-lb. (7-kg) bags, is useful for attics, wall voids, and other enclosed or unused places, for it leaves a permanent film of protection against reinvasion by spiders as well as other cryptobiotic arthropods.

Lindane has long been considered to be the most effective insecticide for control of the brown recluse spider (Wingo, 1964; Hite *et al.,* 1966; Norment and Pate, 1968), but more recently resmethrin, closely related to the natural pyrethrins, was found to be just as effective in topical application (Gladney and Dawkins, 1972). On the other hand, dichlorvos was relatively ineffective, even by topical application, and the fumigant action of dichlorvos strips was practically worthless; out of 30 spiders of all ages, the first adult male died after 33 days and no adult female died after 60 days of exposure.

Spiders feed principally on insects, and any measures taken to decrease the number of insects in a building will also decrease the number of spiders. Spraying for spiders also decreases the number of insects in the vicinity upon which the spiders feed. Newly hatched spiders can enter the home through screens or around loose-fitting windows and doors, but the older ones can be kept out by careful screening. Spiders of all stages can be introduced into homes with firewood, plants, boxes, and many other items. Within the home, the white egg cases can be eliminated with a vacuum cleaner. The bag of the vacuum cleaner should then be emptied before the eggs have had a chance to hatch.

Particularly in the warmer parts of the country, where control is necessary, it is advisable to treat for spiders outdoors as well as in the house. A residual-type insecticide should be applied to coves, eaves of porches, low roofs, window wells, around door and window frames, and to a 3- to 5-ft (1- to 1.5-m) area around the foundation, as well as to the foundation itself. It is also desirable to treat the garage, shed, or any other outbuilding on the premises, as well as piles of old lumber, woodpiles, weedy areas, and fences. Check the label on the insecticide package to make sure it will not be injurious to any of the plants and shrubs in the area to be treated. Dusts or liquid sprays of chlordane, diazinon, or propoxur (Baygon) have been effective against spiders outdoors. A dust is easier to apply and will penetrate more effectively into woodpiles, dense weed patches, or inaccessible areas in the garage. The spiders will succumb after they crawl over the residue, either on their webs or in adjacent areas.

As is the case with other pests already mentioned, a treatment may be adversely affected if the toxicant repels the spider from the spray residue. Malathion and propoxur, for example, have been found to be highly repellent to brown recluse spiders, at least for a period of 2 weeks, and sometimes longer (Sterling, 1970). To overcome this effect, an attempt might be made to contact the spider with the spray or to treat a large enough area so that it is forced to pick up a lethal quantity of the toxicant while attempting to escape.

SCORPIONS (SCORPIONIDA)

Scorpions are most abundant in warm parts of the world, and particularly in semiarid regions (see chapter 4). In areas that occasionally have subfreezing temperatures, they remain in the warmer crevices of cliffs and piles of rock, and lie dormant during cold periods. Despite their usual occurrence in dry climates, scorpions need access to moisture, for they drink water. Their nocturnal habit and tendency to inhabit shady crevices reduce the evaporation of their body fluids. The crevices may also provide moist microclimates, or they may be extended and connected by the scorpions to moist areas in the soil. In the United States, scorpions are most abundant in the arid Southwest, and they are rare north of a line traced from Baltimore to St. Louis, Salt Lake City, and San Francisco (Scott and Pratt, 1957).

Like spiders and ticks, scorpions can withstand starvation for a long period, 4 or 5 months being common. Scorpions emerge from their hiding places at night and prey upon ground-inhabiting insects and other small animals, including small mice. Even in the dark, the scorpion is aware of approaching prey because it can detect slight air movements and minute vibrations. It blindly grasps its prey with the chelae of its large pedipalps, which are always kept in a forward position, and stings it. Normally, the "tail" of a scorpion is extended horizontally, and the stinger, situated in the bulbous last segment, curves downward. However, when the scorpion readies itself for attack or defense, the entire tail-like postabdomen is curved dorsally and forward, the stinger then curves upward (figure 36, chapter 4), and the victim is stung quickly and repeatedly.

Around the home, scorpions are most commonly found in the crawl space under the house and in the attic, which they usually enter by way of the wall voids. They seek water, and sometimes may be seen in washrooms, kitchens, and bathrooms where water is available, although during the day they are more likely to be seen in closets, shoes, folded blankets, or similar retreats. Ground scorpions (all scorpions other than *Centruroides* spp.) frequently burrow into children's sand boxes or gravel banks, where they can remain buried without food or water as long as 6 months (Stahnke, 1966).

Sculptured Scorpion, *Centruroides sculpturatus* Ewing (Buthidae)

Few species of scorpions are deadly to humans, but they are all venomous, and envenomation by arachnids should always be considered as potentially dangerous. The most venomous species in

Fig. 218. The sculptured scorpion, *Centruroides sculpturatus*, a very venomous Arizona species.

the United States is the deadly sculptured scorpion, *Centruroides sculpturatus* (plate VII, I; figure 218). It is comparatively small and slender, and may be yellowish or straw-colored over the entire body, or may have 2 irregular, blackish stripes lengthwise along the dorsal surface. This striped form is a color phase formerly considered to be a distinct species, *C. gertschi* Stahnke. In fact, 4 color phases of *C. sculpturatus* have been recognized by Stahnke (1971). He stated:

> *Centruroides sculpturatus* is to be recognized as a species ranging from a primarily concolorous condition to 3 color phases bearing various interstices of fuscous stripes along the trunk dorsum, on the ventral surfaces of the caudal segments, and some fuscous marbling on the lateral surfaces of the legs. All evidence is unfavorable for recognizing subspecies taxa.

This species ranges from about 12 mm in length when the young first leave the mother's back to about 7 cm when full grown, the males being generally longer and more slender. The adult female's body is not more than 10 mm in width, and the male's is 7 mm at the greatest dimension. The "tail" of the female may be a little over 2 mm in diameter, and that of the male even less. The pincers are very slender—about 6 times as long as the greatest width. At the base of the stinger is a little blunt thorn that is lacking on all other known Arizona scorpions (Stahnke, 1948). *Centruroides sculpturatus* is principally an Arizona species, ranging only a short distance into New Mexico and Mexico.

The sculptured scorpion caused 75 deaths in Arizona from 1926 to 1965, twice as many as from all other venomous animals, including Arizona's many species and subspecies of rattlesnakes and the Sonoran coral snake. Most of the deaths were those of children under 16 years of age (Stahnke, 1966).

Three Related Forms

In Big Bend National Park, Texas, there is a scorpion, *Centruroides pantheriensis* Stahnke, that superficially closely resembles *C. sculpturatus,* but has venom more like that of the Big Bend scorpion, *C. chisosarius* Gertsch, and the common striped scorpion, *C. vittatus* (Say), which ranges from South Carolina to Kentucky and west to New Mexico and Mexico (Stahnke, 1956*a*). Preliminary evidence indicates that these 3 are all the same species (Stahnke, 1971).

Envenomation

Stings by most species of scorpions are not likely to be serious, usually resulting in localized pain and paresthesia, some swelling and tenderness and, on occasion, localized ecchymosis (black-and-blue areas) and the formation of vesicles (Stahnke, 1956*b;* Russell *et al.,* 1968). However, stings by certain *Centruroides* species common to parts of Arizona and New Mexico may be fatal, as already stated. Stings by *Centruroides* are often very painful and give rise to immediate distress, including numbness about the wound which may rapidly spread to involve the entire extremity, weakness or even paralysis of the injured part, hyperactivity and anxiety, profuse salivation, dizziness, difficulties in speaking and swallowing, respiratory distress, and in some cases convulsions (Shannon, 1965). The site of the sting does not swell or become discolored as it does with other less dangerous species (Scott, 1966*b*).

Centruroides sculpturatus venom administered to anesthetized dogs and cats caused hypertension, respiratory failure, and skeletal muscle stimulation. The animals developed an intermittent gasping form of respiration which progressively diminished in amplitude and frequency until death resulted from anoxia (deficiency of oxygen reaching tissues of the body) (Patterson, 1960).

Preventive Behavior to Avoid Bites and Stings

According to Stahnke (1966), 2 principles of action would make an individual almost free from the danger of envenomation:

> (1) Never thrust your hand or kick your foot where the eye cannot see.
>
> (2) If you feel something crawling on your body during the night, or on the back of your neck, etc., do not swat, *brush!* If it is a stinging or biting arthropod, the swatting action will guarantee envenomation, but a quick brush will catch the invader by surprise and remove it before damage to the host can occur.
>
> The first principle applies to vertebrate as well as invertebrate venomers.

Treatment if Bitten or Stung

The first thing to do is to summon a physician and keep the patient warm. Stahnke (1965, 1966) emphasized the importance of protecting the bitten or stung patient as much as possible from emotional concern and stress, stating that stress might more than double the toxicity of rattlesnake or scorpion venoms, and that the increased toxicity appeared to be caused in part by the synergistic action between the venom and epinephrine which was discharged into the bloodstream under conditions of emotional stress.

In scorpion stings, if available, an ice cube can be placed over the wound and the part immobilized. The ice can be left in place until the patient is seen by a physician. The cooling reduces pain, but prolonged exposure to the ice should be avoided. Incision and suction are not indicated, and the use of a tourniquet should be avoided except when the symptoms indicate anaphylactic shock. In the event that ice is not available, the area should merely be cleansed, the injured part immobilized, and the patient taken as quickly as possible to the nearest medical facility. Exertion should always be kept at a minimum. If there is a long delay in transit, respiratory distress may develop, and it may be necessary to give artificial respiration. Protective measures may also be required in case of convulsions. Antivenin is available for *Centruroides* stings, and can be obtained by a physician or hospital when needed. The location of the nearest source of antivenin can be obtained from the district Poison Control Center. Under no circumstances should a lay person administer antivenin in the field. Supportive measures for stings by *Centruroides* include vasopressor agents, atropine, oxygen, and pentobarbital in cases where convulsions occur (Russell, 1969).

The "ligature and cryotherapy" (L-C) technique in the treatment of snake bite or scorpion sting has long been advocated by some authorities, and therefore some mention of it should be made at this point. This is the treatment that involves placing the bitten member in iced water to a point well above the wound. The treatment is considered to be dangerous by most physicians. Many people on whom this treatment has been practiced have had to have the affected arm or leg amputated.

Control of Scorpions

As in the control of spiders, the elimination of harborage places, such as piles of wood and trash, is important. Lumber and firewood should be stored off the ground. Spraying or dusting of the areas surrounding the house with insecticides of long residual action is useful for killing scorpions and also for reducing the supply of insects upon which they feed. Scorpions in a house are found most commonly under the building (crawl space) and in the attic. However, they shun temperatures above 100° F (38° C), and when the attic gets too hot, they will move downward into the living space where they are attracted to the kitchen, bathroom, and other rooms where water is available.

Insecticide dusts, such as 5% chlordane, are preferred for treatment in the house because they can be blown into wall voids and attics. A long-lasting insecticide is desirable, for scorpions may remain in such places for a long time. Sprays of chlordane or diazinon are also effective when applied to baseboards, under furniture, and other hiding places (Stahnke, 1948; Homer, 1969).

CENTIPEDES (CHILOPODA)

Centipedes are long, flattened, many-segmented arthropods. Most of the segments have a pair of legs (plate VII, 2). They may be found in damp locations under leaves, rotten logs, stones, or boards, and run swiftly when disturbed. They prey on many species of arthropods, earthworms, snails, and other small animals, killing them with their maxillipeds or toxicognaths, appendages on the first segment behind the head. These have strong, piercing, terminal segments with orifices of ducts that lead to venom glands (figure 219).

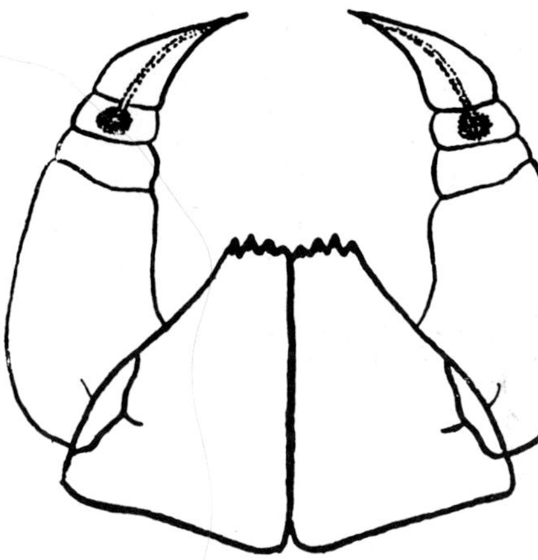

Fig. 219. Venom glands and their ducts on centipede claws. (From Cloudsley-Thompson, 1968.)

[Spiders are the only arthropods in which the venom glands open through the chelicerae (figure 207).] Centipedes are long-lived, some species living as long as 5 or 6 years. Many species of centipedes can inflict venomous wounds on man, but these rarely result in death. Wounds caused by a centipede's claws (maxillipeds) should be disinfected and a physician should be consulted (Scott, 1966*b*).

House Centipede, *Scutigera coleoptrata* (L.)

The house centipede came from Mexico, but is now distributed throughout the United States. It often invades houses, particularly if subfloor areas or basements are damp. It searches for insects at night. Unlike other centipedes in the United States, it is able to reproduce within the house. It occurs both indoors and outdoors in the warmer regions of the country, but only indoors in the colder regions. It runs rapidly, holding its body well elevated by means of its conspicuously long legs, then suddenly stops, remains motionless, and with equal agility runs again to a place of concealment (Back, 1939). There are only a few records of this species wounding man, resulting in pain no more severe than that of a bee sting (Curran, 1946*b*).

Description. The house centipede (plate VII, 2; figure 220) has a very strange appearance. It is

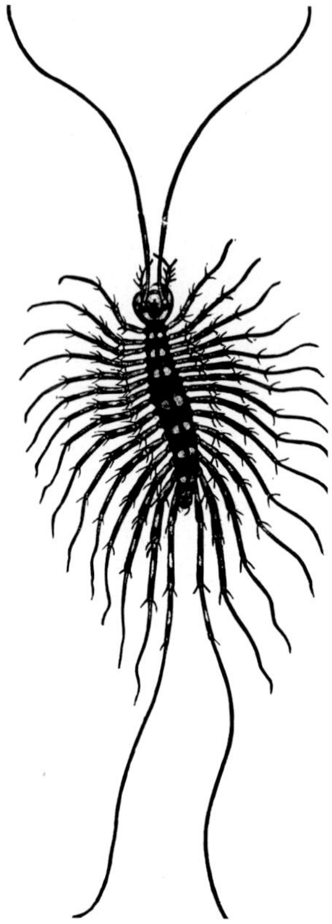

Fig. 220. House centipede, *Scutigera coleoptrata*. (From Back, 1939.)

2.5 to 4 cm long, with very long, slender antennae, and with 15 pairs of legs, also very long. In the female, the last pair of legs is more than twice the length of the body. The grayish-yellow body has 3 longitudinal dorsal stripes, and the legs are banded with white (Back, 1939). This species is in an order (Scutigeromorpha) in which all species have 15 pairs of long legs and long antennae, and are the only centipedes with compound eyes, the others having clusters of ocelli. They can readily detach their legs if they are grasped by an enemy.

Life Cycle. When the larvae hatch from the egg, they have 4 pairs of legs. There are then 5 more larval stages, with 5, 7, 9, 11, and 13 pairs of legs, respectively. Thereafter, there are 4 adolescent stages, each with 15 pairs of legs (Cloudsley-Thompson, 1968).

Other Western Species

In the West, the large *Scolopendra polymorpha* Wood (plate VII, 2) can attain a length of 15 cm, and its venom-bearing claws can cause severe pain and swelling. Its claws can scratch the skin, and such scratches should be disinfected to prevent secondary infection. The large, greenish *Scolopendra heros* Girard, which occurs in southern California, can not only cause severe pain and swelling by its venom, but can produce a reddish streak where it has crawled upon the body. Its numerous feet make many incisions into which it drops venom that causes intense irritation and leaves 2 rows of white punctures (Cloudsley-Thompson, 1968).

Control of Centipedes

For control of centipedes outdoors, the removal of damp harborage places, such as piles of trash, rocks, boards, leaves, grass, and compost, is desirable. In addition, it may be necessary to spray or dust with such insecticides as chlordane, dieldrin, or carbaryl at the concentrations used for most insects (as stated on the package labels) on and around foundations and into the subfloor crawl space of the house.

For control of the house centipede, *Scutigera coleoptrata*, sprays of 2% malathion, 0.5% diazinon, or 0.5% lindane have been found to be effective when applied to baseboard crevices, cracks, openings in concrete slabs, or other hiding places. Dusts of the same insecticides are useful to blow into wall voids and crawl spaces.

ANTS (FORMICIDAE)

The common household ants have been discussed in some detail in chapter 6. At this point, only certain species will be discussed that do not commonly invade the home, but occur outdoors and inflict especially painful bites or stings. They are pests principally in the yard and lawn and in recreational areas. The best-developed stingers and a potent venom are possessed by species in the subfamily Ponerinae, considered to be the most primitive ants. However, many species of ants in other subfamilies sting. Other species do not possess a stinger and merely bite, but some of these inject a venom into the wound, causing considerable pain (Snelling, 1970).

In the more primitive subfamilies of ants, the venoms are proteinaceous, corresponding to those of wasps and bees. In the fire ant (*Solenopsis*), the venom is an alkaloid, noted at the end of this section. In the highly evolved subfamily Formicinae, containing such genera as *Formica, Prenolepis, Lasius,* and *Camponotus,* the venom glands secrete large quantities of formic acid, believed to be a remarkable chemical simplification in the evolutionary pattern (Cavil and Robertson, 1965). The formic acid reaches intradermal or subdermal tissues only through wounds made by the mandibles. The tip of the abdomen is twisted forward, and the formic acid is sprayed into the wound from glands in the anal region. Envenomation from ant, wasp, or bee stings is discussed under "Hymenopterism" in this chapter, preceding the section on venomous snakes.

Southern Fire Ant, *Solenopsis xyloni* McCook

Fire ants get their name from the severe reaction caused by their stings. The southern fire ant ranges from California to southern South Carolina and the northwest corner of Florida. In California, it occurs at lower altitudes from southern California up through the Sacramento Valley, but seldom or never along the coast in the northern or central sections. In the literature, the California form is usually designated as a subspecies (*S. x. maniosa* Wheeler), but ant taxonomists now believe that this is a synonym.

This species is probably our most common native ant, but like some other ants, both its actual and relative importance are declining, particularly in urban areas, as the Argentine ant increases in numbers and range of distribution. However, wherever the Argentine ant is successfully controlled, the fire ant make a rapid resurgence. The workers are ordinarily slow-moving when compared with Argentine ants. However, they are sensitive to vibrations or jarring. If their nest is stepped on, the workers rush out and sting the feet and legs of the intruder. There is a great variation in the way people react to ant stings, depending on their degrees of allergy, but an infant was reported to have been stung to death by southern fire ants (Smith, 1965).

Description. The worker (figures 221, *A,* and 222, *top*) is 1.6 to 5.8 mm long, with yellowish-red head and thorax and black abdomen. There are 2 nodes in the petiole and 2 segments in the antennal club. The body, especially the abdomen, is usually very hairy. The workers are polymorphic; the great difference in size is the result of there being 2 forms: the worker minor, which is generally the form that stings, and the worker major, which rarely stings. Both forms bite. The females average about 6.6 mm long, and are light reddish brown except for the abdomen, which is black. The males are reddish black, with reddish-yellow legs. Winged forms are often observed in large numbers in store windows (Eckert and Mallis, 1941; Smith, 1965).

Biology. The southern fire ant is largely a ground-nesting species, with nests either exposed or under cover of stones, boards, and other objects. When exposed, the excavated earth is usually placed in irregular, cratershaped masses of loose soil. Mounds in lawns cause much damage. Fertilized females have often been seen establishing nests alone. However, more than 1 female is involved in founding a new colony; there may be as many as 12 or more reproductive females (R. R. Snelling, personal communication). The biology has not been completely studied.

Habits. Besides nesting in earth, *S. xyloni* also nests in wood, sometimes in the wood or masonry of houses. Nests also occur in cracks in soil or in sidewalks. When the ants are outside in the vicinity of the house, they are usually near the kitchen. They travel in trails, and are relatively slow-moving when compared with most other ants. Sometimes the nests are in wall voids, and the ants may then emerge from crevices and between sink tiles (Mallis, 1938).

According to Smith (1965), the southern fire ant is practically omnivorous, feeding on seeds, honeydew, flesh, the juices or sap of fruits and plants, and household foods such as nuts, cereals, cookies, butter, grease, meats, and fruits. It takes seeds from seedbeds, kills young or newly hatched poultry and wild birds, girdles nursery stock such as citrus and pecans, destroys fruits and vegetables, tends honeydew-excreting insects, removes the rubber insulation from telephone wires or electrical wires of traffic signals, and gnaws clothes, silks, woolens, linens, and cotton articles. A pest control operator reported finding fire ants in a nylon viscose rug, where they had caused damage similar to that of carpet beetles. The rug had probably been contaminated with some food substance. The southern fire ant is aggressively predaceous on live insects, but will eat dead ones.

Fire Ant, *Solenopsis geminata* (F.)

This species (figure 221, *B*) ranges from Texas to South Carolina and Florida and south to Brazil and Peru. It does not occur in California. It resembles the southern fire ant, except that its head is much larger and the petiolar node is higher, narrower and, in profile, bladelike at the summit. It also closely resembles that species in biology and economic importance (Smith, 1965).

Red Imported Fire Ant, *Solenopsis invicta* Buren

This fire ant (figure 221, *C*), apparently indigenous to central Brazil (Buren, 1972), is now a serious pest throughout much of the southeastern United States, and occurs as far west as Texas. It infests almost any type of land except swampy areas and dense forests, constructing large, hard, earthen mounds that interfere with cultivating and harvesting operations and cause damage to farm machinery. The ants are very pugnacious, and painfully sting people and farm animals. More than a million persons are estimated to be stung each year by fire ants in the southeastern states. They sometimes nest in houses. Without continued control measures, the red imported fire ant could spread westward to the Pacific and as far north as Oregon and, in the East, to as far north as New Jersey (Anonymous, 1973*b*).

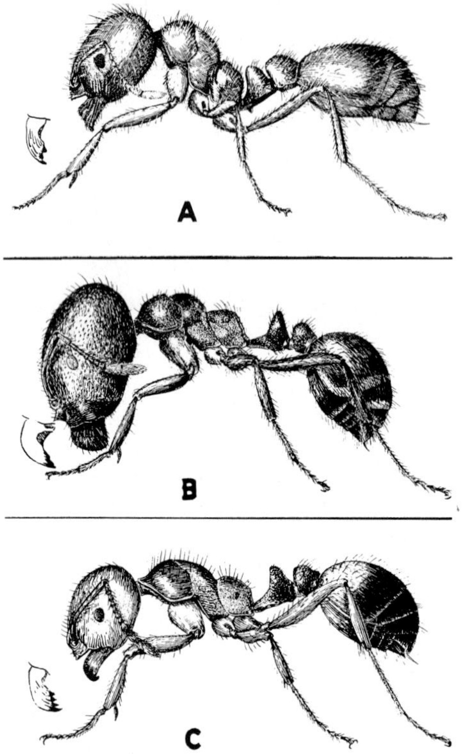

Fig. 221. Fire ants. *A, Solenopsis xyloni; B, Solenopsis geminata; C, Solenopsis invicta.* (From Smith, 1965.)

The literature before 1972 refers to the imported fire ant as *Solenopsis saevissima richteri* Forel, which was believed to comprise a red and a black form. The black form was known to have been in the Mobile, Alabama, area since 1918, and was probably introduced from Argentina or Uruguay. The red form, probably introduced from the Matto Grosso of Brazil, was also first noted in the Mobile area, but not until the early 1940's. It is more vigorous than the black form and has largely replaced it, marking the beginning of the explosive increase in the imported fire ant population in the United States. The black form is now found only in a small area of northern Mississippi. Buren (1972) considered the black form to be a distinct species, *Solenopsis richteri* Forel, and the red form to be a new species which he named *S. invicta*. These 2 species are currently being referred to in the literature as the black imported fire ant and the red imported fire ant, respectively (e.g., Banks *et al.*, 1973). The

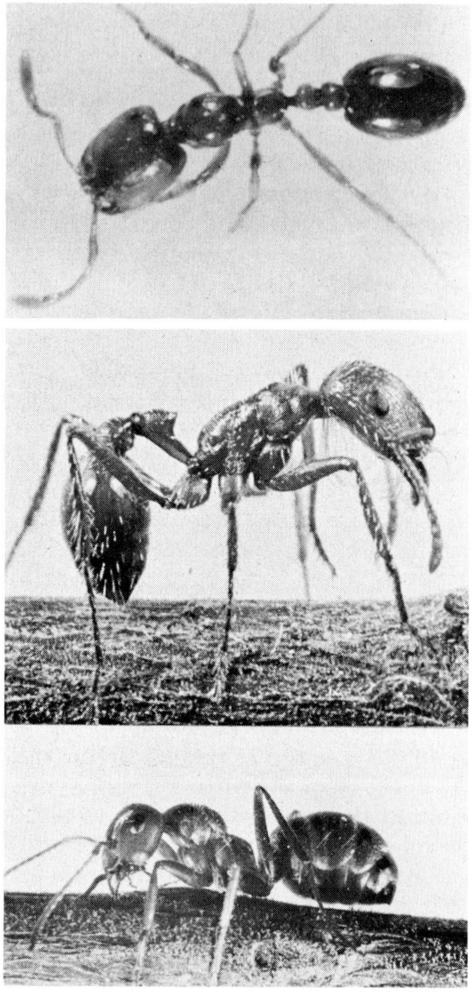

Fig. 222. Some ants with painful stings or bites. *Top,* southern fire ant, *Solenopsis xyloni; center,* California harvester ant, *Pogonomyrmex californicus; bottom,* a field ant, *Formica pilicornis.*

voluminous literature on the biology and control of "*S. saevissima richteri*," earlier than 1972, refers almost exclusively to what is now known as *S. invicta*.

Description. The red imported fire ant (figure 221, *C*) closely resembles the southern fire ant (figure 221, *A*), but can be distinguished from it by the 4 teeth along the biting edge of the mandible, as shown in the figure; the southern fire ant has only 3 such teeth.

Biology of the Imported Fire Ant. Based on observations made along the Gulf coast in Mississippi, alates are most abundant in early spring, and may comprise as much as 9% of the colony in June. Then, a decreasing number of alates is produced, until production of this caste ceases completely in September. During the 8-month developmental period (April–November), an average colony may produce an estimated 1,500 females and 2,500 males (Markin and Dillier, 1971). Mating flights of male and female ants will occur from large mounds during warm days of most months of the year (Green, 1962). The production of workers begins in March, and peaks of production occur in May and October. Colonies producing mounds between 25 and 65 cm in diameter might contain between 30,000 and 100,000 workers, and over the period of a year they were found to average 76.3% of the colony (Markin and Dillier, 1971).

As many as 1,000 colonies of the imported fire ant have been observed per acre (0.405 ha) under ideal conditions. However, 20 to 30 mounds per acre is the more common number, and the mounds gradually increase to about 45 to 90 cm (18 to 36 in.) in diameter. Sometimes, "giant mounds" 30 to 90 cm (1 to 3 ft) high and 150 to 240 cm (5 to 8 ft) in diameter are formed, possibly as the result of a number of colonies moving together under adverse soil conditions. Adjoining areas are uninfested. The giant colonies may disperse again to form the characteristically smaller ones (Green, 1962).

There is normally little or no movement or communication between adjacent colonies, and discrete territorial boundaries are observed by the ants. As a result, if ants in a colony have access to mirex bait, for example, they will all be dead in 1 to 3 weeks, whereas those in neighboring nests will still be very active (Wilson *et al.*, 1971).

Habits. Imported fire ants gnaw into the roots, stems, buds, and fruit of plants, such as cabbages, collards, okra, eggplant, and field peas, or seriously injure young trees by girdling or removing the outer bark from roots or trunks. Citrus nursery stock is especially subject to attack (Smith, 1965). The ants are often so numerous that the gathering of vegetables and other crops is almost impossible. They are also a menace to wild animals and to the eggs and young offspring of ground-nesting birds. However, imported fire ants are principally predators against other insects (Wilson and Oliver, 1969).

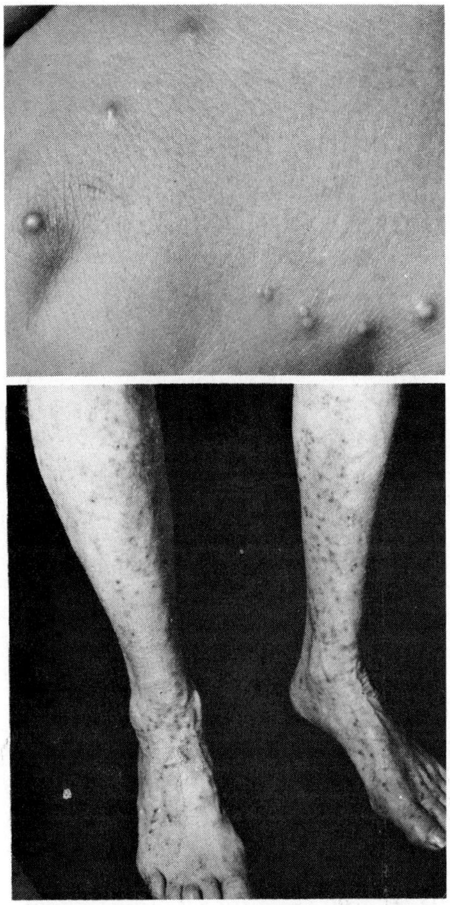

Fig. 223. Results of stings by imported fire ants. *Top*, pustules 2 to 3 mm in diameter that appear within 24 hours; *bottom*, pink areas and scar tissue that appear after the purulent material in the pustules is absorbed or sloughed off. (Courtesy USDA, Entomology Research Service.)

Envenomation. The structures of the stinger and poison sac of the imported fire ant were studied and illustrated by Caro *et al.* (1957). Also, the histopathology of biopsy specimens of volunteers who allowed themselves to be stung was investigated. The ant first pinches up the skin with its mandibles, raising it slightly, and this in itself causes pain. The ant then arches its back at the petiole and inserts its stinger, usually maintaining this position for 20 to 25 seconds. Then, using its head as a pivot, it may rotate and insert its stinger in 2 or 3 additional places, causing a clustering of sting sites that is useful in diagnosis. Sometimes 2 minute, bright-red, hemorrhagic punctures may be seen at the point where the mandibles entered. Within several minutes, a wheal 4 to 8 mm in diameter appears, and by that time the stinging sensation has subsided. Wthin 24 hours, a pustule 2 to 3 mm in diameter appears (figure 223, *top*). In 3 to 8 days, the purulent material is absorbed or sloughed off, and leaves a smooth, pink area 2 to 3 mm in diameter. This persists for several weeks, and then scar tissue gradually develops (figure 223, *bottom*). The pustular lesions are often arranged in a circular pattern. Thousands of the lesions were found on inebriated persons who inadvertently disturbed an imported fire ant mound (Smith and Smith, 1971).

Whereas the venoms of stinging ants are typically proteinaceous, the venom of the imported fire ant was found to contain an alkaloid, *trans*-2-methyl-6-*n*-undecylpiperidine (solenopsin *A*) or its mirror image, and the structural formula for the compound was determined. This is apparently the first record of an alkylated piperidine being described from a venom of animal origin. The venom is soluble in organic solvents but insoluble in water. It is a potent necrotoxin, and also has pronounced hemolytic, phytotoxic, insecticidal, and antibiotic activities (MacConnell *et al.,* 1970). Five unique alkaloids were finally identified (MacConnell *et al.,* 1971).

Harvester Ants

Harvester ants, so called because they gather seeds and grasses, are large red to dark-brown ants, usually 5 to 6 mm or more in length. All but 1 of the 22 species of the United States occur west of the Mississippi River. Most species have 4 fringes of long hairs on the posterior surface of the head, and these comprise the *psammophore.* With the psammophore they clean their legs and antennae, carry water, and remove sand during the excavation of their nests. They can carry 4 times more very fine sand than they could with the mandibles alone. Some species are also characterized by many small, parallel ridges on the head.

Harvester ants do not invade the home, but their nests may be found in areas surrounding the house. When houses are built in areas infested by them, these insects can constitute a severe problem. They build their nests, with big craters, in lawns and yards. The nests of California species consist of 1 or more holes with a surrounding low, flat crater about 6 in. (15 cm) across and surrounded by large, vegetation-free areas. Some ants bring seeds into the nest, often with husks and air floats attached. The air floats and refuse are then taken out of the nest by other ants, which deposit the material on the outer edge of the crater. If the nest is disturbed, these ants bite the intruder and hang on tenaciously while inflicting a very painful sting.

Harvester ants are considered to be the most aggressive and pugnacious ants in the United States. The reaction to their stings is not localized, but spreads along the lymph channels. Long after the original pain of the sting has ceased, intense discomfort may be felt in the lymph glands of the axil or groin (Creighton, 1950).

California Harvester Ant, *Pogonomyrmex californicus* Buckley

This species (plate II, 6: figure 222, *center;* figure 224, *top*) is 5.5 to 6 mm long, light rusty red with somewhat lighter legs, and the thorax lacks spines. In southern California, the nests are closed continuously during the winter, and every night for the remainder of the year. At 70° F (21° C), the ants are sluggish, and maximum foraging activity occurs when the temperature at the ground's surface is 90 to 115° F (32 to 46° C). Swarming occurs during certain clear, hot days in June and July (Michener, 1942).

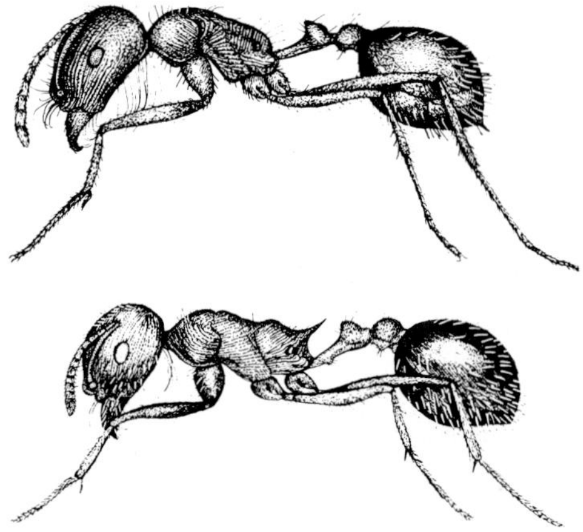

Fig. 224. Harvester ants. *Top, Pogonomyrmex californicus; bottom, Pogonomyrmex rugosus.* (Courtesy R. R. Snelling.)

Wheeler and Wheeler (1973) state that *P. californicus,* in well-established colonies, "is undoubtedly the fiercest, the boldest, and the most irascible ant of the Sonoran desert," the quickest to sting, and the effects of the sting are the most painful. Like the sting of the honey bee, but unlike the sting of most ants, its stinger easily becomes detached and remains in the wound. The Wheelers give a number of personal case histories of severe reactions to the sting of *P. californicus,* describing symptoms that persisted in one form or another for as long as 37 days.

Pogonomyrmex rugosus Emery

This harvester ant (figure 224, *bottom*) is 4.9 to 7.2 mm long, and varies from brownish black to various combinations of brownish, yellowish, and brownish black. The petiole always has a few hairs underneath. This species occurs from California to western Texas, in the chaparral belts and the upper desert mountains. It is very aggressive, and inflicts a painful sting (R. R. Snelling, correspondence).

Western Harvester Ant, *Pogonomyrmex occidentalis* (Cresson)

This is a red ant, 10 mm long, that builds large mounds 4 to 12 in. (10 to 30 cm) high and 24 to 36 in. (60 to 90 cm) in diameter. The galleries honeycomb the soil beneath to a depth as great as 9 ft (about 3 m), and most of them reach a depth of over 6 ft (2 m). There are chambers (granaries) for storing seeds, and others for storing "trash." A chamber containing the queen may range from 10 to 70 in. (25 to 180 cm) in depth. Many workers (250 to 1,200) share the queen's chamber. The number of workers found in 33 excavated colonies ranged from 412 to 8,796 (Lavigne, 1969).

This species occurs at 6,000 ft (1,800 m) or more elevation in Kansas, Colorado, Wyoming, Idaho, New Mexico, Arizona, and Nevada. It is uncommon in California. In one survey in Wyoming, the cleared area around the mounds averaged 88 sq ft (8.2 sq m). Thousands of acres of rangeland had been denuded. The ants had also caused failure of experimental range seedings in one area by gathering seeds or by cutting off young grass plants as they emerged (Lavigne, 1966).

Field Ants

These are medium-sized ants that occur throughout the United States and are quite common. They may be brown, black, red, or various combinations of these colors. The females may be winged or wingless, and usually have a long gaster. The petiole has 1 node. The workers are attracted to insects and honeydew, and are fierce predators of other insects. Many species build nests consisting of large mounds of leaves and sticks around rocks or small trees and shrubs. These ants can be troublesome pests in mountain recreational areas. They bite, and then curve their abdomens forward to inject a secretion into the wound; their bites are therefore very painful.

California Red-and-Black Field Ant, *Formica occidua* Wheeler

These ants (figure 225, *A*) are 4 to 7.5 mm long, have a red thorax, silvery-black abdomen, and the top of the head is dusky brown. They have a characteristic, jerky method of running and stopping. Nests may be found in nature beneath rocks in open, unshaded areas. In urban areas, the nests are often located in cracks in sidewalks, along the sides of buildings, and at the bases of trees. The ants may forage in houses and seek honeydew on trees and shrubs. They have been observed carrying great numbers of workers and sexual forms of subterranean termites toward their nests (Mallis, 1941*a*).

Western Thatching Ant, *Formica obscuripes* Forel

These ants are 3.8 to 8 mm long, and have a red head and thorax and brownish-black abdomen. They occur from British Columbia to California, at elevations of 5,000 to 8,000 ft (1,500 to 2,400 m). They construct mounds from small sticks, leaves, and pine needles, and often are injurious to seedling trees and native plants near the mounds. They can become pests in mountain cabins.

Formica haemorrhoidalis Emery

This species (figure 225, *B*) is similar to *F. obscuripes* in appearance, except that pubescence is almost totally absent from its head and thorax. It also has a similar distribution range and habits, except that it is more completely confined to forested areas.

Formica pilicornis Emery

These field ants (figures 222, *bottom,* and 225, *C*) are 3 to 7 mm long, and have a blackish head and gaster and usually a reddish thorax. They have many erect hairs on the legs, scapes, and eyes, and the abdomen appears grayish because of dense, appressed hairs. They may be found along the coast and throughout the chaparral belt of southern California, nesting in the soil, often under cover, and are frequently very abundant.

They often nest in cracks in sidewalks and along sides of buildings. They travel in long trails, and sometimes invade the home. They are diurnal scavengers, and tend aphids and mealybugs (R. R. Snelling, correspondence).

Silky Ant, *Formica fusca* L.

Formica fusca (figure 225, *D*) occurs over much of the United States, but only in the mountains in southern areas (in southern California, above 4,500 ft (1,450 m). These ants are 4 to 6.5 mm long, uniformly blackish, often with bronze tints on the head, and the legs are sometimes reddish. The body is almost devoid of erect hairs such as may be found in *F. pilicornis,* but the silky pubescence (fine hairs) of the abdomen give this species its common name. The clypeus has a sharp median carina. They nest in the soil, throwing up messy tumuli. They are diurnal, tend aphids, and are general scavenger-predators. They may be attracted to meats, but are much more timid than is the case with *F. pilicornis.*

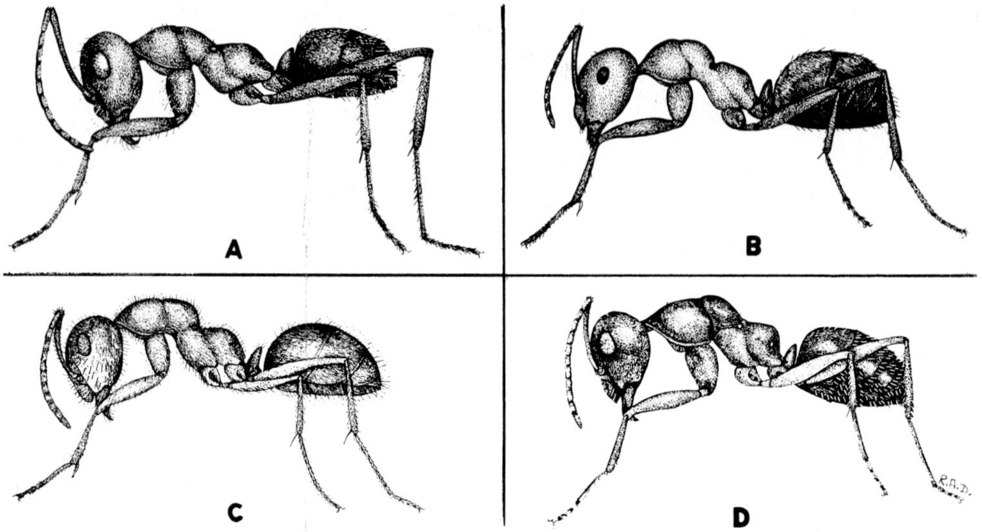

Fig. 225. Field ants. *A, Formica occidua; B, Formica haemorrhoidalis; C, Formica pilicornis; D, Formica fusca.* (Courtesy R. R. Snelling.)

Allegheny Mound Ant, *Formica exsectoides* Forel

This common eastern species ranges from Nova Scotia to Georgia and from the Atlantic Coast as far west as Ontario, Wisconsin, and Iowa. The head and thorax are rusty red and the legs and abdomen are blackish brown. The workers vary greatly in size, from 3.2 to 6.3 mm, the largest being about the size of the soldiers. The Allegheny mound ant builds the largest anthills of any of the North American moundbuilders. They are rounded cones, 12 to 36 in. (30 to 90 cm) high and often up to 2 m in diameter. A mound 19 in. (50 cm) high was found to contain 237,103 workers and 1,407 queens. One 10-acre area (about 4 ha) averaging 7 mounds per acre was estimated to contain about 12 million ants. Inside the mounds are many approximately cylindrical passageways about 12 mm in diameter. Passageways also extend down into the ground beneath the mound for a depth of 30 to 36 in. (75 to 90 cm). The ants spread in the spring by swarming and starting new mounds, a moderate distance up to 30 m from the old mounds (Haviland, 1947).

This species girdles small trees and shrubs by injecting formic acid into wounds it makes in the bark with its jaws. It is not a household pest, but can be one in recreational areas (Mallis, 1969).

Control of Ants

Although chlordane is generally used by the homeowner for ant control, good results can be obtained with other commonly used household and garden sprays or dusts at the usually recommended concentrations. In field applications, heptachlor and Nonachlor (related to chlordane) at dosages as low as 0.25 lb (0.113 kg) per acre of actual toxicant gave good control of the red imported fire ant (Banks *et al.,* 1966). Mirex and Kepone baits have been suggested where large acreages are treated because less weight of material needs to be transported compared with water- or oil-based insecticides (Lavigne, 1966).

In connection with baiting, an interesting observation was made by Haviland (1947) with regard to *Formica exsectoides.* After feeding on a sodium arsenite–sugar solution bait for several hours, during which several hundred ants died, the remainder stopped feeding, even though no other source of food was available. For 2 weeks they ignored any food placed on the spot where the poison had been placed. It was 6 weeks before they would again feed on a sugar solution, and they never again fed on a poisoned solution. A similar observation was made on red imported fire ants, *Solenopsis invicta,* by Rhoades and Davis (1967). These ants ceased feeding on a bait when they became aware that it contained a poison, and they moved the colony to another location.

A recent innovation in baits consists of the encapsulation of microdroplets of vegetable oil containing 2% mirex within thin gelatin or plastic coats. These microencapsulated oil baits have been found to be very attractive to the imported fire ant, and give excellent control. The bait can withstand weathering, making it superior to baits used in the past in which mirex has been contained in corncob grits. Lacking a bulky carrier, this type of bait is particularly suitable for application by aircraft (Markin and Hill, 1971).

The first full-scale attempts to totally eliminate the red and the black imported fire ants (*S. invicta* and *S. richteri,* respectively) from large areas by aircraft application of mirex bait, using 3 applications at 1.25 to 2.5 lb/acre, indicated that total elimination of the ants from large, isolated areas might be technically possible. The bait consisted of 85% corncob grits, 14.7% soybean oil, and 0.3% mirex (Banks *et al.,* 1973).

J. N. Roney (correspondence) states that, in the control of agricultural ants in Arizona, if insecticide dusts or liquids are applied in the opening of the hill, the ants will make an opening in some other part of the hill. A narrow band of dust around the hill and about 24 in. (60 cm) from the opening is effective. However, dust may cake over after 5 days or so, or if it gets wet and subsequently dries, and the ants can then crawl over the deposit. For this reason, granular formulations of the same insecticide are more effective. A very large anthill may require 2 applications at intervals of 6 to 8 weeks for control. Liquid formulations must be used in large quantities, and even then results are usually poor. R. R. Snelling (personal communication) has injected calcium cyanide into the soil around the mounds of fire ants and harvester ants with great success.

WASPS (VESPOIDEA, SPHECOIDEA, BETHYLOIDEA, AND SCOLIOIDEA)

In this section, wasps of varying degrees of importance because of their stings will be discussed. They are contained in 4 superfamilies. Best known is the superfamily Vespoidea, which includes the social wasps (yellowjackets, hornets, umbrella wasps, and the genus *Mischocyttarus*), all in the family Vespidae. The vespids construct a nest of paperlike material, consisting of wood, foliage, or cellulosic debris, chewed up and elaborated by the insects. The nests may be of various sizes, and may be constructed above or below ground, depending on the species. The family Vespidae also includes species of solitary wasps in the subfamilies Masarinae and Eumeninae. The superfamilies Sphecoidea, Bethyloidea, and Scolioidea contain only solitary wasps.

SOCIAL WASPS (VESPIDAE)

Millions of people are stung by venomous arthropods each year, and many thousands of these are seriously envenomated. Social wasps are only slightly less incriminated than bees in the number of fatalities for which they are responsible, and each of these 2 groups is more important than spiders, scorpions, and ants combined in this respect. Besides the approximately 22 deaths per year recorded as being directly caused by bees and wasps, there is no way of knowing how many other deaths reported to be caused by "heart attacks" and "heat strokes" may actually have resulted from the sting of a venomous insect (Parrish, 1959, 1963), or how many deaths attributed to "automobile accidents" may have resulted when bees or wasps caused the accidents, either by their stings or by the consternation and loss of vehicle control resulting from their mere presence (Miller, 1956).

Some social wasps, particularly yellowjackets, are greatly benefited by association with man, for in his garbage cans, discarded food materials, and trash and garbage dumpsites he offers much greater sustenance to scavenging species than they can find in nature. Predatory species are attracted to flies that abound in such areas. Therefore, in wilderness regions, wasps are often abundant around human habitations and camps, but are rarely seen elsewhere. The relative importance of wasps in relation to other venomous arthropods will probably increase with the growth of human population, and particularly with the increase in the number of parks, camps, recreation areas, and suburban developments in formerly uninhabited places. Wasps decrease the desirability of recreation areas for public use.

In most of the country, wasps vary greatly in abundance from year to year. The number found in late summer depends to some extent on the weather during the preceding spring. If an early spring prompts queens to establish their nests and the weather then becomes inclement for a prolonged period, the wasp population can be greatly diminished (Andrewes, 1969). In California, the wasp population is far less variable from year to year than it is in parts of the country with more severe changes in spring weather.

Description. The vespid wasps are medium-sized, and usually black and yellow in color. Species of the most common genus, *Vespula,* are readily distinguished by the jagged bands of black and bright yellow on the abdomen (plate VII, 3, 4; figures 226, 227). The wings are folded longitudinally. The antennae are elbowed, but not so abruptly as in ants, and the scape (basal segment) is long. The mouthparts are of the "chewing–lapping" type (figure 49, chapter 4). The middle tibiae have 2 apical spurs and the claws are simple. Generally there are 3 castes: the queen, a mated female that has hibernated over the winter before starting the nest; the males, produced only in the late summer or fall from unfertilized eggs; and the workers, infertile females. In *Vespula,* the queen is usually about 25% longer than the workers, whereas in *Polistes* and *Mischocyttarus,* queens, males, and workers are about the same size.

Biology. The queen overwinters under loose bark, in cracks and crevices, and occasionally in attics or similar sheltered locations. In the spring, she builds a few "paper" cells in a suitable location and lays a single egg in each cell. She feeds the legless, white larvae on meat or other proteinaceous food fragments, which she first macerates with her mandibles. These larvae become workers, and take over all the duties of the community, such as nest enlargement, providing food, and feeding the continuous broods of larvae. Throughout the summer the nest is enlarged, and the wasp population increases to a maximum that ranges from about 200 in the nests of *Polistes* and *Mischocyttarus* at any one time, to as many as 15,000 individuals in the nests of *Vespula* (Bohart and Bechtel, 1957). There are great differences in the numbers attained by different species, and even within a species nests vary greatly in size. Mature *Vespula pensylvanica* nests in southern California appear to have an average of about

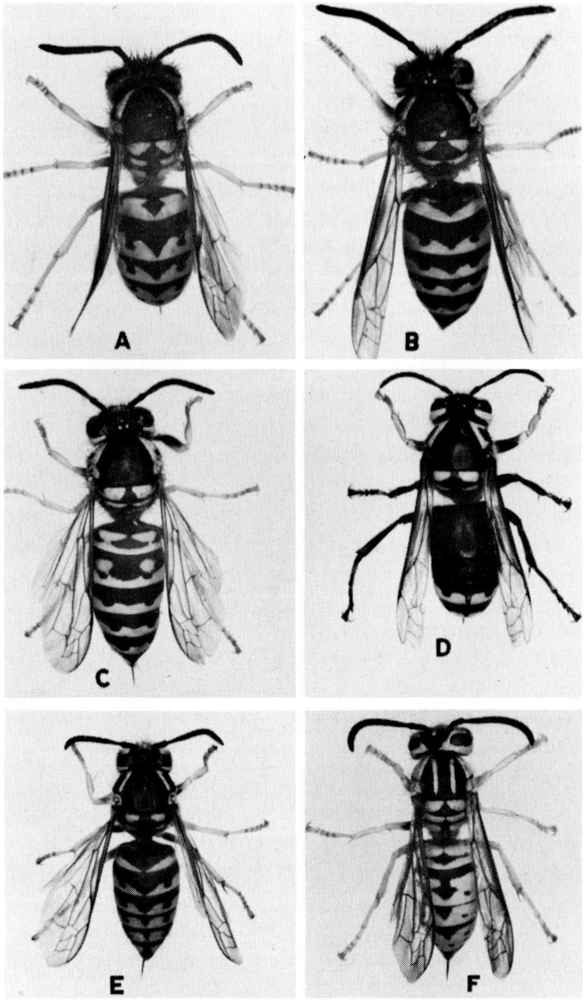

Fig. 226. Western yellowjackets. ***A, Vespula pensylvanica; B, Vespula vulgaris; C, Vespula atropilosa; D, Vespula maculata; E, Vespula arenaria; F, Vespula sulphurea.*** **(All species have yellow markings on a black background except *Vespula maculata,* which has white markings.)**

3,000 individuals. Mature nests of the most common species of *Polistes* (*P. fuscatus* and *P. apachus*) generally have from 20 to 30 individuals, and the largest one found by R. E. Wagner (personal communication), a nest of *P. f. aurifer,* had 72 individuals.

Nests decline in summer or fall, and males and new queens are produced. The new queens of many species (most *Vespula*) are reared in specially constructed cells that are much larger than those in which the workers and males develop. The new males and queens leave the nest, and after mating the queens go into hibernation until the next spring. The nests of social wasps are used for only 1 year.

The adult social wasps feed on nectar, honeydew, fruit juices, sap, or similar substances. The proteinaceous food for the larvae consists of other insects and spiders if the wasps are predators, or bits of meat from any source if they are scavengers. After the workers take over the work of feeding the larvae, they continue to do so during their entire larval development. While the worker is feeding the larva, the larva exudes a drop of *trophallactic fluid* from its mouthparts. This fluid is ingested by the worker and provides its required proteins. Queens feed on nectar in the spring, and the workers and males feed on it in the fall, when there are no more larvae in the nest. Apparently the new queens, after mating, rapidly seek overwintering quarters, for they are not seen feeding (R. E. Wagner, personal communication).

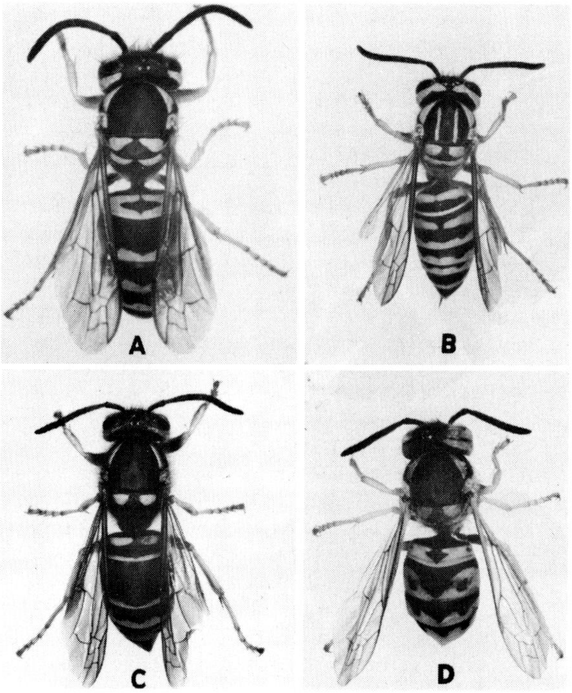

Fig. 227. Eastern yellowjackets. *A, Vespula maculifrons; B, Vespula squamosa; C, Vespula vidua; D, Vespula germanica.* (All species have yellow markings on a black background.)

The Stinger of the Wasp

The stinging apparatus of the wasp is not fundamentally different from that of the honey bee (see figure 264, later in this chapter), but the stinger is larger and is semicylindrical in shape. The wasp can easily withdraw its stinger from human flesh to reinsert it or to escape, because the lancets of most wasp stingers, unlike those of the bee, are not barbed. Envenomation from the stings of hymenopterous insects is discussed under the heading "Hymenopterism" later in this chapter. Social wasps are more hazardous to humans than solitary wasps, not only because of their greater abundance but because the victim may be stung by many wasps at one time if he happens to disturb their nest (Fluno, 1961).

VESPINE WASPS (VESPIDAE: VESPINAE)

This is the subfamily that contains the typical yellowjackets, so called because of the black-and-yellow pattern of the majority of species, although in some species a very pale yellow or "white" is substituted for the deeper yellow. In the *Vespula* species, the ground color of the integument is black. The yellow (or white) is present in a series of maculations that are quite variable but nevertheless present a distinct pattern, most strikingly exhibited on the abdomen. Therefore, the color pattern of the abdomen is the character most easily used to distinguish species. In figure 226, the typical abdominal color patterns are shown for 6 *Vespula* species found only in the West (*V.*

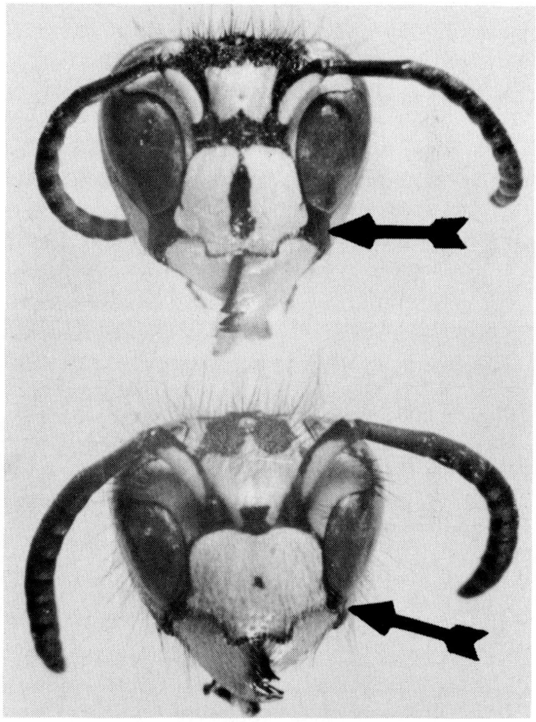

Fig. 228. Faces of 2 yellowjackets. *Top, Vespula maculata; bottom, Vespula pensylvanica.* (The arrows indicate ocular–malar areas, which are long in the subgenus *Dolichovespula* (*top*) and short in the subgenus *Vespula*.)

pensylvanica, V. atropilosa, and *V. sulphurea)* or widely distributed throughout the North American continent, including the western states (*V. vulgaris, V. arenaria,* and *V. maculata*). In figure 227, the patterns are shown for the most important species confined to the East.

The facial morphology (figure 228) provides the taxonomic character that distinguishes the 2 subgenera. In the subgenus *Vespula,* the oculo–malar distance (distance between the eyes and mandibles) is always less than half the length of the penultimate (next to last) antennal segment. In the subgenus *Dolichovespula,* the oculo–malar distance is greater than half, and is usually as long as or longer than the penultimate antennal segment. The subgenus *Dolichovespula* is represented by *V. arenaria* and *V. maculata* in figure 226.

A key to the *Vespula* species of the Pacific Coast states was prepared by Bohart and Bechtel (1957). The accompanying practical field key for all presently known North American *Vespula* species was prepared by the author's co-workers, R. E. Wagner and D. A. Reierson. When using this key, it is important to examine as many specimens as possible because atypical individuals sometimes exist which can only be correctly determined by using secondary characters not specified in the key.

KEY FOR FIELD IDENTIFICATION OF THE WORKERS OF YELLOWJACKETS AND HORNETS OF NORTH AMERICA

1. Top and sides of head brownish .... *Vespa crabro germana*
   Top and sides of head black, marked with yellow or white .... (genus *Vespula*) 2
2. Eyes nearly touching the upper edge of the jaws (figure 228) .... (subgenus *Vespula*) 3
   Eyes separated from the upper edge of the jaws .... (subgenus *Dolichovespula*) 13
3. Top of thorax with 2 prominent longitudinal yellow stripes 4
   Top of thorax without such stripes 5
4. Undersurface of abdomen bright yellow ...... *V. sulphurea*
   Undersurface of abdomen conspicuously marked with black .... *V. squamosa*
5. Pale markings whitish 6
   Pale markings yellow 7
6. Reddish markings present on sides of abdomen .... *V. intermedia*
   Abdomen without reddish markings *V. consobrina*
7. Eyes encircled with yellow ...... *V. pensylvanica*
   Eyes with some black contacting them at top . 8
8. Long basal segment (scape) of antennae black, or with only a short yellow marking near its center 9
   Scape black with yellow marking for most of its length 11
9. Yellow stripe behind eyes, usually with dark or cloudy marking on it; clypeus with a prominent vertical black bar .... *V. vulgaris*
   Stripe behind eyes entirely yellow 10
10. Black spots present on top of at least 1 abdominal tergite or on the second abdominal sternite .... *V. germanica*
    Black spots not present as above .. *V. maculifrons*
11. Markings on second abdominal tergite variable, ranging from predominantly yellow, with a prominent central black marking accompanied by a pair of lateral black spots, to predominantly black, with the posterior yellow band similar in width to those on the succeeding segments 12
    Markings on second abdominal tergite predominantly black, with a yellow posterior band much narrower than those on tergites 3 through 5; free black spots usually present on tergites 4 and 5 .... *V. vidua*
12. Clypeus with 3 black spots or rarely an anchor-shaped marking; stripe behind eyes entirely yellow or with a small black interruption near its center; second abdominal sternite with free black spots .... *V. atropilosa*
    Clypeus with a longitudinal central black bar; stripe behind eyes broadly interrupted by black. Second abdominal sternite rarely with free black spots .... *V. acadica*
13. Pale markings yellow 14
    Pale markings white 15
14. Yellow band on first abdominal tergite interrupted by black. Rear edge of abdominal black bands with 3 points or 1 central point, flanked by a pair of black spots .... *V. arenaria*
    Yellow band on first abdominal tergite continuous. Black abdominal bands wide, with only a single faint point on the rear edge of each .... *V. norvegicoides*
15. First and second abdominal tergites entirely black .... *V. maculata*
    First and second abdominal tergites with a narrow whitish band at the rear and reddish marks at the sides .... *V. albida*

Persons using the key are reminded that it is accurate only for the worker caste. Queens are usually much larger than workers, and commonly have somewhat different markings. Males, which are similar in size and markings to workers, have longer antennae, and are best determined by structural differences in genitalia. The scape (basal segment) of the antenna and the "stripe behind eyes" referred to in the key are easily recognized in figure 229. The terms "tergite" and "sternite" used in the key are defined in chapter 4, in connection with figure 42.

The nesting behavior of *Vespula* species is widely variable. Certain species, such as *V. maculata* and *V. arenaria,* build aerial nests in trees, shrubs, and under eaves. The other members of the subgenus *Dolichovespula, V. norvegicoides* and *V. albida,* also sometimes build aerial nests in shrubs, but they commonly construct them on the ground in such materials as leaf litter and sphagnum moss clumps. These supraterrestrial nests differ from true ground nests in that the wasps do not excavate a soil cavity in which the nest is constructed.

Among the species in the subgenus *Vespula,* the most common nesting habit is terrestrial. Most of them, however, show remarkable adaptability, and supraterrestrial nests, nests in the voids of buildings, and even truly aerial nests are occasionally constructed. Some species (e.g., *V. austriaca* and *V. arctica,* both widely distributed in the United States and Canada) do not build their own nests, but are social parasites in the nests of other species. They do not produce a neuter (worker) form.

The color patterns actually seen when examining the species of *Vespula* depicted in figures 226 and 227 will not always be exactly like those shown, but will tend to approximate them, at least to the extent that there should be little difficulty in identifying a wasp of any of the species represented.

The Subgenus *Vespula*

Western Yellowjacket, *Vespula pensylvanica* (Saussure)

This is the most common and most pestiferous yellowjacket in California. The queens are about 20 mm long. The workers are about 16 mm long, and the typical pattern of their black-and-yellow markings distinguishes them from other species (plate VII, 3; figure 226, *A,* and 229). It is a pest because it is abundant and is often a scavenger. Although it has been observed to be predatory when prey species are abundant, it readily becomes a scavenging pest if native food supplies are inadequate. Completely predatory species of wasps are seldom important as pests, even when they are abundant. *Vespula pensylvanica* occurs throughout the western states and adjacent Canadian provinces. In Mexico, it has been taken in the state of Michoacán at an elevation of 6,600 ft (2,000 m). It usually constructs subterranean nests, but occasionally builds them in wall voids, attics, and other aboveground sites. The biology and habits of *V. pensylvanica* were thoroughly investigated by Duncan (1939).

In certain years, *V. pensylvanica* appears in unusually great numbers in public camps in mountain areas, and sometimes surpasses mosquitoes and flies as an annoyance to campers. These wasp

Fig. 229. The most common western yellowjacket, *Vespula pensylvanica.*

outbreaks are believed to result, at least in part, from a relatively mild winter and early spring (Bohart, 1941). Regional fluctuations in wasp populations are more likely to occur in the mountains than in the lowlands, for the insects are more adversely affected by severe spring weather. Campers sometimes leave their vacation places in large numbers because of numerous stings and the necessity of having to compete with yellowjackets for every mouthful of food eaten. In hot weather, the wasps seek water wherever it is spilled or in the wet sand of bathing beaches along the shores of mountain lakes, thus annoying bathers. In a discussion of one such outbreak of *V. pensylvanica* near Lake Tahoe, California, Bohart (1941) stated that persons stung were mostly those who hit or squashed the yellowjackets. A sting often results in a painful swelling. After the pain has subsided, there follows a period of itching. An ice cube or ice pack on the site of the sting will reduce both the pain and the swelling.

Vespula vulgaris (L.)

This is a common species, similar to *V. pensylvanica* in appearance (figure 226, *B*), and occurs in all the western states, throughout the northern states of the East and the adjoining areas of Canada, throughout western Canada, and in Alaska. In California, it is usually found north of Fresno. It has been reported from the state of Michoacán, Mexico, at an elevation of 6,600 ft (2,000 m) and on the western slope of Mt. Popocatepetl at 10,000 ft (3,000 m). As with *V. pensylvanica,* it is predatory, but often becomes a scavenger and is second only to that species as a pest in the western United States. The nest is almost always located near coniferous trees in partially rotted stumps, partially buried logs, or soil rich in organic matter in damp and shady locations.

Vespula atropilosa (Sladen)

This species occurs throughout the western states, northward into the southern halves of British Columbia and Alberta, and southward to Tijuana, Mexico. Figure 226, *C,* shows that *V. atropilosa,* as well as *V. sulphurea* (figure 226, *F*), usually has more yellow on its abdomen than the 2 preceding species. The nest is terrestrial.

Vespula sulphurea (Saussure)

This wasp is easily recognized by the 2 longitudinal yellow bars on the thorax (plate VII, 4; figure 226, *F*). It has been collected in many localities in most of California, but is most common in the northeastern area. It has also been found in southern Oregon, western Nevada, the Huachuca Mountains of southeast Arizona, and in northern Baja California. The nest is terrestrial.

Vespula maculifrons (Buysson)

This species (figure 227, *A*) closely resembles *V. vulgaris,* and is also ground-nesting, but does not occur in the West. It is found principally in the central Atlantic Coast states, and is common in deciduous forests, where it builds nests.

Vespula squamosa (Drury)

This is considered to be the most common ground-nesting yellowjacket in the southeastern states. It is easily identified by the pair of longitudinal, parallel, yellow stripes on the dorsal surface of the worker's thorax (figure 227, *B*). Although these stripes are also prominent on the queen, she is unique among Nearctic *Vespula* species by being notably different from her progeny in color and markings. Her abdomen is amber-colored, with some black markings, in contrast to the other castes in which the abdominal segments are black, bordered with bright yellow. This difference is so great that the queen was originally described as a separate species. *Vespula squamosa* has the ventral surface of the abdomen conspicuously marked with black, while the corresponding surface of its most closely related species, *V. sulphurea,* is bright yellow.

Vespula vidua (Saussure)

This species (figure 227, *C*) is a relatively common ground-nesting yellowjacket in the northeastern states. Note that it has much less yellow marking than the other species shown in figure 227. It has been found nesting within the walls of buildings on several occasions.

Vespula germanica (F.)

This European yellowjacket has become well established in the eastern United States. It somewhat resembles *V. maculifrons,* but is generally larger and has a greater proportion of yellow (figure 227, *D*). Also, *V. germanica* has black spots on at least 1 abdominal tergite, whereas *V. maculifrons* does not. Although this species is primarily a ground-nesting yellowjacket in Europe, it commonly nests in wall voids and other protected places in buildings in the United States. Its nests tend to be very large.

The Subgenus *Dolichovespula*

Vespula arenaria (F.)

This is one of the commonest yellowjackets in North America (figure 226, *E*). It occurs throughout the western states as well as the northeastern ones, and in most of Canada and Alaska. It is not found in the Plains states, and is absent in all but the northernmost areas of the southern states. The nest is aerial, and is often built under the eaves of houses (figure 235).

Baldfaced Hornet, *Vespula maculata* (L.)

The baldfaced hornet (figure 226, *D*) is the only black-and-white species among those shown in the figure. It is morphologically more similar to the yellowjackets than to the hornets. However, it resembles the hornets in behavior as well as in size. The worker caste cannot be separated from the queen caste on the basis of external morphology. There is a broad range of sizes in these females, but they are sometimes as large as the queen.

The white maculations are confined to the face, the tip of the abdomen, and to 2 stripes, 1 on each anteriolateral margin of the thorax. The species occurs throughout the United States, but is scarce in the southern half of the country. It is abundant in the northern half of the United States and the southern portions of the adjacent Canadian provinces from the Pacific to the Atlantic, and extends north into Yukon Territory (Miller, 1961), and also into Alaska. In California, this species occurs

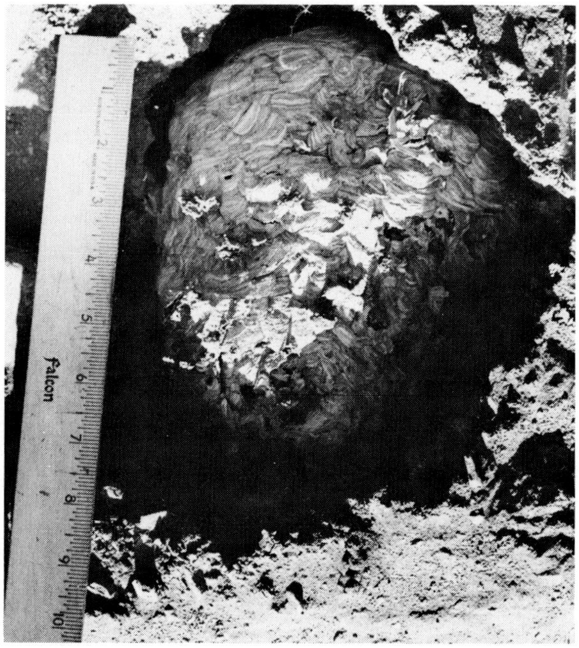

Fig. 230. Ground nest of *Vespula pensylvanica.*

primarily in the north coast mountain ranges and at moderate elevations in the Sierra Nevada. It is not found in southern California. Nests of *V. maculata* are aerial, sometimes high off the ground, and are often 12 in. (30 cm) or more in diameter (Bohart and Bechtel, 1957).

Balduf (1954) had the rare opportunity to observe the construction of a nest of *V. maculata* against a windowpane in northern Minnesota. The wasps allowed the windowpane to serve in place of the usual "paper" envelope for one side of the nest. The activities of the wasps could be observed, and construction stages of the nest were photographed from the time its dimensions were 3.5 × 3 in. (9 × 7.5 cm) to the time the nest was completed at 8.4 × 6.8 in. (21.3 × 17.3 cm). Another such nest was built against a windowpane of a farm house in Illinois, and an account of the nest, with a photograph, appeared in the "State Edition" of the Chicago *News* on September 28, 1948. At its maximum, the nest was 24 in. (60 cm) high and 18 in. (45 cm) wide. Balduf observed that the wasps, like bees, manifested irritable and combative behavior on rainy days, and that food gathering was greatly curtailed on such days.

The Terrestrial Wasp Nest

The raw materials for constructing the "paper" from which the nest is made are vegetable fibers and salivary secretion. The fibers may be from the surface of rotten or weatherworn wood, the cortex of dead plants, fragments of caps of cocoons within the nest, cut away by young wasps when emerging, and from artificial products such as newspapers, paper bags, or cardboard signs.

The collected fibers are chewed and mixed with a salivary secretion to make a doughy mass. The queen selects the site for the nest and builds the first cells. Whether the nest is aerial or underground, the first structure to be built is generally a pedicel or suspensorium for the support of a small comb of cells. The pedicel is expanded into the first cell of the nest. Either simultaneously or after the first small comb is completed, a paper-like envelope is built around it, leaving just enough space between the envelope and the comb to allow for the movement of the wasps. An opening is left at the bottom of the envelope to allow for the entry and exit of the queen. Nestbuilding is continuous, from the time the first workers appear until the colony declines. Eight to ten or more tiers of combs can be built into a nest, all encased in an envelope that is enlarged as the combs are constructed (figure 230). As additional combs are built, many secondary suspensoria are attached to the nest envelope and comb above the one being constructed (figure 231).

Fig. 231. Nest of *Vespula pensylvanica* with outer paper envelope removed to show combs and suspensoria. *Inset:* Some suspensoria (enlarged).

Some entomologists, such as Andrewes (1969), believe that underground nests must have a root or stone in the roof from which the nest may be suspended, but excavations of nests of *Vespula pensylvanica* have revealed that this is not necessarily true. By the time the roof of the hollow has been wet by the queen and subsequently dried, it is quite hard and firm. It often contains gravel. The base of the pedicel is expanded to form a broad area of attachment to the roof. As the nest increases in size, many suspensoria are built from the "paper" envelope to the surface of the hollow, so the nest does not depend solely on the primary pedicel for its support.

Fig. 232. Entrance hole of *Vespula pensylvanica* nest, with typical cleared area surrounding the hole.

Wasps that build subterranean nests do much more work on their nest sites than those that build aerial nests. They clear away leaves, grass, and other litter surrounding the entrance to the burrow (figure 232). They also do much excavating to enlarge the cavity in which the nest is built. Yellowjacket nests may be over 30 cm in diameter. The excavation proceeds at a rate that depends on the speed with which the colony itself is developing. The clearance between the outer envelope of the nest and the wall of the cavity averages about 12 mm.

Whether the nests are excavated in naturally moist or in dry soil, the pellets of soil carried by yellowjackets when excavating them are either moist or appear to have been recently moist. Also, the interior of the nest cavity has the appearance of a dried mud surface. In the season of active excavation, worker wasps are abundant at sources of water. Duncan (1939) once observed a *V. sulphurea* worker returning at intervals of about 5 minutes to a particular spot at the edge of a small pond, each time drinking deeply before leaving. Wasps of other species were likewise observed doing the same thing, some drinking so much water that they experienced difficulty in taking flight. Water is apparently carried to the nest cavity, regurgitated on its walls, and the mud is then scraped off in pellets and carried away.

Much of the excavated earth is carried outside the burrow and dropped. Pebbles that are small enough to be moved but too large to carry very far may be dragged to the mouth of the burrow and left near-by. Duncan (1939) observed a *V. pensylvanica* worker carrying a pellet, in flight, at least 70 ft (21 m) from the burrow opening before dropping it. Some of the excavated mud is used to fill cavities leading out from the nest chamber, such as lateral rodent burrows. During the excavation of many *V. pensylvanica* nests in southern California, much of the excavated soil was found to have been used to reduce the diameter of rodent burrows in which the nests were constructed, where the burrows formed the entrance tunnels. Duncan took 3 qt (about 3 l) of dried pellets from a cavity left by a decayed root, which extended out from a nest cavity constructed by a colony of these yellowjackets. Sometimes there are 2 entrances to an underground nest, but the wasps use both entrances for ingress and exit.

As the end of the season of activity approaches, workers of most species begin to build cells of a larger size for the rearing of new queens. Queens are always reared in large cells and males and workers in small cells. Workers of *V. pensylvanica* may be reared in used cells. The number of adults successively reared in the same cell can be determined from the number of fecal pellets at the bottom of the cell (R. E. Wagner and D. A. Reierson, personal communication).

In extremely rare instances, wasps continue to live in and develop a nest into a second season in California. One such nest of *V. pensylvanica* measured 46 in. (115 cm) in one diameter and 40 in. (100 cm) in the other, and was 30 in. (75 cm) long. Only about half the height of the nest was occupied by combs, but there were 21 comb levels. The nest contained 4 gal (15 l) of wasps, including some newly transformed males and queens and 22 functional queen mothers, recognizable because of their worn condition.

Species that generally build their nests underground need not necessarily do so if adequately protected nest sites can be found aboveground. Figure 233, *top*, shows a *V. pensylvanica* nest that had been constructed in an attic. The lower figure shows the exposed combs of a nest that had been constructed between 2 studs in a house. The wasps in this nest were exterminated with a resmethrin aerosol, and then a considerable section of the outer wall had to be taken off in order to remove the extensive nest. In figure 234, the screwdriver shows the narrow entrance to a nest site chosen by a *V. vulgaris* queen in an old tree stump in Lassen National Park, California. Removal of some surface layers of wood revealed the nest in a cavity that had been excavated by the yellowjackets.

Fig. 233. Nests of *Vespula pensylvanica* in houses. In an attic and in a wall void between 2 studs.

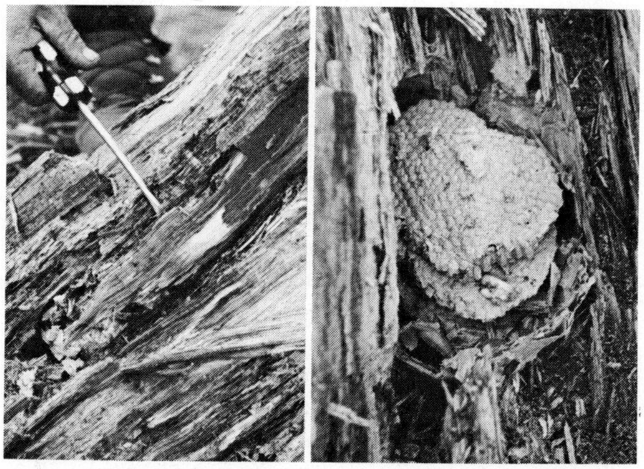

Fig. 234. Entrance to nest of *Vespula vulgaris* (*left*) and exposed nest *in situ*.

The Aerial Wasp Nest

The subgenus *Dolichovespula* contains the 2 species (*Vespula arenaria* and *V. maculata*) that are the most common aerial nestbuilders (figure 235). They often construct their nests on buildings, especially under the eaves (Wagner and Reierson, 1971). Most people are familiar only with aerial nests, although the majority of *Vespula* species build their nests underground. Wasps of the subgenus *Dolichovespula* tend to have smaller nests than those of the subgenus *Vespula*. The largest nest among the *Dolichovespula* ever observed by Duncan (1939) was one of *V. arenaria* a little over 30 cm in diameter and 35 cm long. On the other hand, a nest of *V. pensylvanica* with 5 comb levels that measured 25 and 40 cm in its diameters and 44 cm long was only a little larger than the average for the species. The *Dolichovespula* also build a smaller number of cells in their nests, but a larger proportion of queen

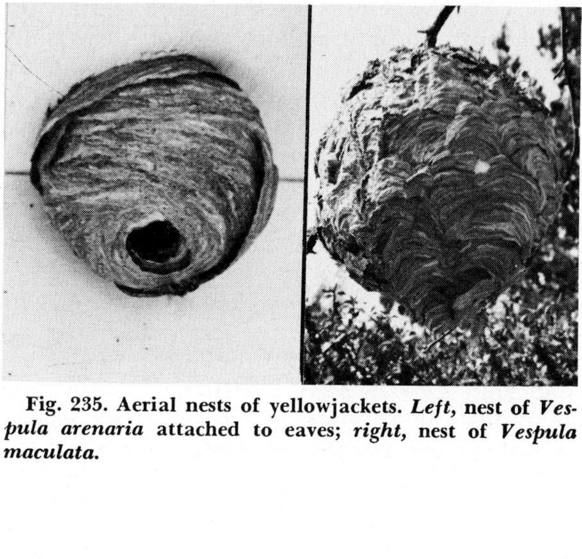

Fig. 235. Aerial nests of yellowjackets. *Left*, nest of *Vespula arenaria* attached to eaves; *right*, nest of *Vespula maculata*.

(large) cells. For example, a completed nest of *V. arenaria* contained 1,199 small cells and 1,290 large cells, distributed among 5 combs. A much larger nest of *V. pensylvanica* at the height of its activity in mid-August contained 7 combs and 10,456 cells, all small. The colony had about a month longer to develop (Duncan, 1939). In nests of *V. pensylvanica* evacuated during the fall of 1970, Wagner and Reierson (1971) found an average of 15% as many queen cells as worker cells in mature nests.

Duncan (1939) observed that workers of *V. arenaria* (*Dolichovespula*) began to rear new queens as early as late June in central California, whereas none of the species in the subgenus *Vespula* in that area normally produced queens earlier than September and some not until late October.

The nests of *Dolichovespula* wasps show that the change from worker- to queen-cell construction is gradual, whereas with the subgenus *Vespula* it is abrupt.

Life History of Yellowjackets

In the early spring, the mated queens emerge from protected harborages in which they have spent the winter months. They feed on nectar, honeydew, and other sweets while flying about singly in search of a suitable site for a nest in which to start a new colony. These queens usually resemble the workers of their own kind, but are considerably larger in most species.

Egg. The queen of a vigorous colony may lay as many as 25,000 to 30,000 eggs. In the process, her wings may become frayed, and are sometimes worn down to mere stubs and most of her body setae are broken off as the result of the friction in inserting her body into the "paper" cells. The same cell may be used a second or third time for egg-laying. The milky-white eggs of the vespine wasps are about 4 times as long as broad, slightly curved, slightly narrowed and rounded at the anterior end, and more narrowed and less blunt at the posterior end. The posterior end of the egg is glued to the wall of the cell (Duncan, 1939). As with many Hymenoptera, the egg membrane is not shell-like, and does not maintain its form after the larva emerges.

Larva. At first, the larva is held in place at its posterior end by a mucus it secretes. It later becomes large enough to fill the lumen of the cell and hold itself in place by friction. It can then be pulled out without injury, and if the colony is asphyxiated, some of the larger larvae fall out of their cells. The mature larva is elongate and flattened dorsoventrally (plate VII, 3). The first abdominal segment is the broadest part of the body, which tapers abruptly anteriorly and more gradually posteriorly. Only about one-fifth of the total length of the larva is made up of head and thorax, compared with over a third in the adult. The full-grown larva is creamy white, except that the posterior section may appear somewhat blackish because of the dark intestinal contents.

Pupa. The mature larva spins a cell cap, and then voids from its intestine a blackish mass of accumulated intestinal wastes, called the *meconium.* This is attached to the bottom of the cocoon, or the bottom of the cell in the case of those species in which the base of the cocoon is incomplete. It dries into a brownish pellet, and the number of pellets at the bottom of a cell indicates the number of times it has been used for larval rearing.

Cocoons of the larvae that are destined to become queens are larger than the others and extend somewhat farther out of the cells. Pupation takes place in the cocoon. The legs and wings of the pupa are free from its body and, except for the pupa's whiteness and delicacy, its appearance is similar to that of an adult. The queens are larger than the workers, and the males have long antennae.

The total period required for larval and pupal development is about 1 month; about two-thirds of this period is required for the larval stage (Duncan, 1939).

Teneral Adults. The new adults, called "teneral" (soft and unpigmented), cut their way out of the cocoons, seek out the larvae in the comb, and bite the heads of these larvae to induce the secretion of a salivary liquid, a process called *trophallaxis.* This is the first food received by the newly emerged wasps. The larvae are said to make a rustling sound by scratching the walls of their cells in order to attract the feeding wasps (Andrewes, 1969). Soon afterward. the larvae are fed by the older workers in the nest on regurgitated sweets or bits of macerated and malaxated insects from which they suck the liquid contents. The teneral adults remain in the nest for a few days. If they are workers, they gradually assume the duties of nest-building and brood care. If they are queens, they are the founders of the next year's colonies. The males do much flying about after leaving the nest, visiting flowers for nectar, whereas the queens go into hibernation almost immediately after mating.

Control of Wasps

Wasps may be controlled by spraying directly into the nests if they are aboveground, or by pouring insecticide liquid into the openings of the nests of subterranean species. This type of treatment must be done carefully, because the wasps defend their nests ferociously. A safer treatment is the use of poison baits in "bait stations" for the few species for which baits have been developed. This method of treatment also obviates the necessity for finding the wasp nests.

Sprays and Dusts

For the control of wasps that build aerial nests, various insecticide formulations have been made available in pressurized containers that shoot out a long, narrow stream of spray (Anonymous, 1962, 1970*f*). The author's associates, R. E. Wagner and D. A. Reierson, found resmethrin (SBP-1382®) to be spectacularly effective and superior to all other commercially available sprays of this type, as well as the safest. The liquid should first be directed into the opening of the nest to knock down insects as they attempt to leave, and then the entire nest should be sprayed.

For species that build their nests in the ground, the colony should be located during the day and then, for treatment, should be approached only in darkness, using a flashlight covered with red cellophane for light. Anonymous (1962, 1970*f*) recommend that 1 quart (1 l) of insecticide liquid be poured into the nest entrance as quickly as possible. Dichlorvos 1%, chlordane 2%, or a suspension of 2 tablespoonfuls of carbaryl 50% wettable powder in a quart of water are recommended. If the nest is still active after 48 hours, pour in another quart of the same mixture. A 5% carbaryl dust blown into the nest opening has also been found to be effective.

Pouring insecticide liquid into a nest is accompanied by a certain degree of risk that can be avoided by wearing protective clothing or by directing a stream of spray liquid into the nest entrance from a distance of 5 or 6 m by means of a power sprayer. A quart (liter) of kerosene poured into the entrance will provide almost immediate extermination of all the adult yellowjackets that are in the nest (Wagner and Reierson, personal communication).

Cyanogas A dust, applied with a plunger duster, is very effective, almost instantly disabling all the wasps within the nest. Cyanide dust is extremely toxic, and should be used only by professional pest control personnel.

Wagner (1961) worked out a control method for ground-nesting yellowjackets (*Vespula pensylvanica*) in a zoological garden and an adjacent picnic ground where they had been severe pests. He recommended that once a week, immediately after trash barrels were emptied, the inner surfaces, particularly near the rims, be sprayed with 0.75% dichlorvos. Since poor sanitation attracted the wasps, he stated that the trash barrels should be emptied as often as necessary to avoid filling them more than half full. Keeping the trash level in the barrels low exposed the toxicant residue to the wasps, and also tended to reduce the amount of trash dropped indiscriminately by the public. He also recommended that such a control should be begun early in the year before yellowjackets become numerous. The spray program should be continued through the summer season of greatest yellowjacket activity, thus insuring maximum control of these insects as well as flies and honey bees. If control is not satisfactory the spray program can be supplemented by searching the surrounding area for nests. They can be destroyed by means previously mentioned.

Bait Stations

A more effective and far more convenient method has been developed for the control of yellowjackets. Keh *et al.* (1968) were able to attract *Vespula pensylvanica* to poison baits of mirex in beef liver or undiluted fruit-drink concentrate. Yellowjacket workers carried the bait to their underground nests, and the mirex was detected in dead workers, males, queen, and larvae. Wagner and Reierson (1969, 1971) worked out a control program against the same yellowjacket species, using a bait consisting of fish-flavored cat food with 0.5% mirex. Results of tests with different pet-food bases disclosed that yellowjackets were very discriminating, and preferred Pacific mackerel over several other species of fish used in cat foods.

Mirex was the only insecticide of the 12 investigated that was not repellent to the yellowjackets. Using the figure "100" as the amount of cat food removed by the yellowjackets when no insecticide was added, the relative attractancy of the cat-food–insecticide baits ranged from 1 to 121. The latter figure indicated the relative attractancy of the mirex bait, which was the only bait more attractive than cat food alone. The workers carried the bait to their subterranean nests, killing the entire colony. To assist the foraging wasps in locating the bait station, an "attractant generator" was placed in it. This consisted of 25 ml of heptyl butyrate (Davis *et al.,* 1967) in a 60-ml ointment jar with a cotton dental-roll wick protruding 1 cm through the lid. This ester was easily synthesized. It tripled the number of yellowjackets attracted to the bait station. A proprietary bait station based on these findings became available in the summer of 1971.

The first yellowjackets to be seen in the spring are the overwintering queens. They feed on nectar, honeydew, and other carbohydrates. Carbohydrate baits were not tested, since they might have the disadvantage of attracting honey bees and other nontarget insects. When the workers appear, they search for proteinaceous food to feed the developing larvae in the nest. There follows a long period when a protein bait can be used that attracts yellowjackets and practically no other insects. The virgin queens that appear in the fall search for overwintering sites at the time the colonies are rapidly declining, and only the newly mated queens survive the winter.

When protein foods are being scavenged by yellowjackets, the baiting method is safe, convenient, adds practically no insecticide to the environment, and is harmless to all species but *V. pensylvanica* and *V. vulgaris.* The suppression of yellowjacket populations does not upset the "balance of nature." Only the nests within flight range of baited residential areas, parks, zoos, picnic grounds, camps, and recreation areas are eliminated. Most yellowjacket species are beneficial predators, and are not attracted to protein-based baits.

Hornets

The genus *Vespa* is represented in North America only by the giant hornet, *V. crabro germana* (Christ), also sometimes called the brown hornet or European hornet. This large, robust, brown-and-yellow wasp (plate VII, 5), about 22 mm long, was introduced from Europe, and now occurs sparingly in the Atlantic seaboard states. According to Duncan (1939), *Vespa* spp. are the "true hornets," and are characterized by having the head swollen behind the eyes and the ocelli remote from the margin of the head. The nests of the giant hornet may be in hollow tree trunks, in buildings, or in other sheltered places, but occasionally in the ground or in aerial situations. *Vespa crabro* is the only wasp in the United States that is known to use fibers from the cortex of living plants as materials for nestbuilding, and it has caused injuries to dahlias and various shrubs when obtaining these fibers. When this hornet builds its nest in the open or in cavities in soft earth, it constructs an envelope around the combs like that found in yellowjacket nests. Its subterranean nests lack envelopes only when they are built in very hard and stony ground in which the stones prevent their normal enlargement. In hollow trees, there may be no envelopes at all, except perhaps for a few rudiments above the first comb and to partly close the access hole to the hollow interior.

The term "hornet" is also applied to *Vespula maculata* (described earlier) because it is the only species of *Vespula* in North America that resembles the giant hornet in size and behavior. Its white markings make the term "yellowjacket" inappropriate.

POLISTINE UMBRELLA WASPS (VESPIDAE: POLISTINAE)

The wasps of the genus *Polistes* are sometimes called "paper wasps" because of the paperlike material from which their nests are made, but that name is equally appropriate for other vespids. The author suggests that they be called "umbrella wasps," based on the shape of their nests. The closely related but relatively unimportant wasps of the genus *Mischocyttarus* should then also be called "umbrella wasps." The genus *Polistes* is world-wide in distribution, and in Europe and North America its colonies outnumber those of all other social wasps combined. *Polistes* may be distinguished from *Vespula* by the presence of a distinct lobe on the hindwings and by the spindleshaped abdomen, which tapers at both ends and is never truncate at the base. A key to the California species of *Polistes* was prepared by Bohart and Bechtel (1957).

Whereas the *Vespula* species are best known as pests in resort areas, parks, picnic grounds, and zoological gardens, *Polistes* are better known as pests in orchards and vineyards. They are often seen on or near their nests in fruit trees, feeding on the juices of injured fruits, or on flowers.

As with *Vespula,* the nests of *Polistes* are begun in the spring by fertile, overwintering females. The females are sometimes joined by one or more other fertile queens that assist in the nest's construction and maintenance. These additional reproductives are relegated to more or less worker status, for the founding queen maintains dominance and is the only egg-laying individual. In such cases, should the dominant queen die, one of the other fertile females can take over her duties and the nest will survive. There is no appreciable difference in the size and appearance of egg-laying queens and the workers.

The nests of *Polistes* differ from those of *Vespula* in that they are single-layered with the cells exposed. The nests are umbrella-shaped, and are generally supported by a longer pedicel (figure 242) than the nests of vespids. They are never completely enveloped in paper. Although they are commonly suspended from the branches and twigs of trees, they are also often suspended from the eaves and attics of houses or the ceilings of barns and sheds, more so than the aerial nests of *Vespula*. *Polistes* nests, just as those of *Vespula,* may be found in almost any imaginable location. A nest of *P. exclamans exclamans* was transported from Pennsylvania to Germany inside the bumper of an automobile. Nineteen wasps emerged from the nest, along with 15 specimens of a hymenopterous parasite, *Elasmus polisti* Gauss (Gauss, 1972). Eberhard (1969) suggested that the pedicel between the nest and the attachment surface might lessen the probability of discovery of the exposed brood by predators, and that a single narrow access to the nest might result in more effective defense against them. She observed ants being effectively repelled by umbrella wasps when the ants attempted to cross the narrow pedicel.

Polistes fuscatus (F.)

This is the most common North American species of *Polistes,* and occurs with its varieties throughout the United States. It also extends into southern Canada, Mexico, and Central America, and has been accidentally introduced by man into the Bermudas, Jamaica, Barbados, Society Islands, and Hawaii (Bequaert, 1940). It is the most abundant and widespread umbrella wasp in Califor-

Fig. 236. *Polistes fuscatus aurifer.*

Fig. 237. Nests of *Polistes fuscatus aurifer.*

nia, having been recorded in 51 of the state's 60 counties and from sea level to over 8,000 ft (about 2,400 m) elevation. The most widespread subspecies, *P. f. aurifer* Saussure, ranges from British Columbia to southern California, where it merges with *P. f. centralis* Hayward. The latter extends south into Baja California and east and northeast as far as Texas, New Mexico, and Colorado. At the eastern extremities of its range, *P. f. aurifer* intergrades with *P. f. variatus* Cresson in northern Washington and Idaho and with *P. f. utahensis* Hayward in the foothills of the Rockies, particularly in Utah (Bohart and Bechtel, 1957).

Polistes fuscatus aurifer is about 18 mm long. The face and the bulk of the abdomen are bright yellow (plate VII, 6; figure 236). Only the first 2 anterior bands of black are conspicuous on the abdomen. The thorax is black except for 6 narrow, yellow stripes: 2 marginal stripes on the pronotum, 2 transverse stripes on the mesonotum, and 2 lateral stripes on the metanotum. The upper surfaces of the coxae and femora are brownish, but the remaining surfaces of the legs are yellow.

The nests of *P. f. aurifer* (figure 237) are often found attached under eaves and even inside attics. Umbrella wasps in general are more apt to seek man-made structures for nestbuilding than are the species of *Vespula* that build their nests aboveground. This is to be expected, in view of the fact that *Polistes* nests do not have a strong and complete envelope, such as that possessed by *Vespula* nests, to protect them from the elements.

Polistes fuscatus centralis (figure 238) ranges in color from the red and yellow of the more typical forms to nearly all yellow. In California, it occurs in the southeastern area as an inhabitant of creosote bush (*Larrea divaricata*) and prairie, plains, and desert grassland areas (Bohart and Bechtel, 1957). In Baja California, these wasps have at times been so numerous in cotton fields that they

Fig. 238. *Polistes fuscatus centralis.*

destroyed populations of lepidopterous larvae before the insecticides applied had time to take effect (Snelling, 1954).

Polistes apachus Saussure

This wasp (figure 239) has been referred to as the "Apache wasp" (Simmons *et al.,* 1948). It occurs in the central valleys and in southern California, and also in Texas and adjacent areas to the north, west, and south. It appears to have been introduced into California from Texas about 1920, and has been gradually extending its range there since that time (Bohart and Bechtel, 1957). It has become established in some urban areas of southern California as the primary polistine species nesting under the eaves of buildings. It has a tendency to build its nest close to that of the previous year, or at least in the same building. In one case, this habit was observed for 10 years, and during that period the wasps shunned adjacent buildings of similar construction (Gibo, 1972). This habit accounts for the occasional finding of as many as 12 nests in the attics and/or eaves of some buildings.

In Texas and adjacent regions, *P. apachus* is primarily associated with mesquite and grassland rather than with wooded areas. Nests are large, circular, and contain about 150 cells. The frequency with which this wasp builds its nest in fruit trees and vines, and its pugnacity when disturbed, have resulted in its reputation as an agricultural pest.

Description. *Polistes apachus* attains a length of as much as 20 mm. The basic color is golden brown. The pronotum is bordered with a thin, yellow stripe. The mesonotum has 2 transverse stripes of yellow—the anterior stripe narrow and the posterior stripe broad. The abdomen has alternate stripes of golden brown and yellow. On some wasps, the 2 colors are about equal in extent, and in others, one or the other color may predominate.

Fig. 239. Apache wasp, *Polistes apachus* (worker).

Biology and Habits. Simmons *et al.* (1948) studied the biology and habits of *P. apachus* in fig orchards. They found that in the spring, 2 to 6 wasps cooperated in the building of a single nest, and that such nests initially harbored from 1 or 2 to as many as 6 or 7 queens. They found that often new nests were rebuilt within a week in fig trees in which they had been removed. Nests in fig orchards were said to average about 13 cm in diameter and to contain 320 cells, over twice the number reported by Bohart and Bechtel (1957). Combs as large as 15×20 cm were found.

Polistes dorsalis californicus R. Bohart

This is the dominant subspecies of *P. dorsalis* (F.). It has light-red and yellow markings, and practically no black (figure 240). This form most commonly hibernates under loose bark of cottonwood or eucalyptus and in company with *P. fus-*

Fig. 240. *Polistes dorsalis californicus.*

catus aurifer. Although its nests are usually small, Bohart (1949) collected an elongate nest with a double pedicel near the middle and containing about 250 cells. The typical species occurs at low to moderate elevations in California, from Shasta County south, and also occurs in southeastern Arizona. In southern California and Arizona, this subspecies overlaps with it.

Polistes exclamans arizonensis Snelling

This subspecies ranges from the state of Sonora, Mexico, into Arizona as far north as Phoenix and into the extreme southeast of California. According to Bohart and Bechtel (1957), it can be recognized by its "brownish-red color, moderate size, and spotted antennae in the female" (figure 241). This wasp is a nuisance in citrus orchards. Laborers have refused to work in infested orchards, particularly to pick fruit. The stings are very painful, and some workers who were stung have had to be hospitalized. In Mexico, some orchards were

Fig. 241. *Above, Polistes exclamans arizonensis* on a portion of its nest; *below,* the entire nest in its characteristic vertical position inside an orange tree.

abandoned because the wasps became so abundant. In the Salt River Valley of Arizona, in some cases up to 20% of the fruit has remained unpicked because of umbrella wasp infestations. From 1 to 26 nests have been observed per tree. The nests (figure 241) occur inside the foliage

Fig. 242. *Polistes exclamans exclamans* worker and nest—typical of the nests of most *Polistes* species. (From Wagner and Reierson, 1971.)

canopy of a citrus tree. They are elongate, 15 to 35 cm long and 5 to 15 cm wide. They are positioned vertically, instead of horizontally as with most other *Polistes* species, and face the interior of the tree. There may be from 100 to more than 200 wasps per nest. Numbers of them will attack if the nest is disturbed (Hopkins, 1955).

Polistes exclamans exclamans Viereck

This subspecies (figure 242) is common in the southeastern and Atlantic seaboard states. It is not very pugnacious, but inflicts a potent sting. Although variable in color, it is always well marked, sometimes profusely, with yellow. The pendulant, umbrella-shaped nest (figure 242) of this wasp was removed from the eaves of a house in New Jersey. It has the typical form of the nests of most species of umbrella wasps—single-layered and never completely covered with a "paper" envelope, as are the nests of *Vespula.*

Control of Umbrella Wasps

In attempts to control *Polistes apachus* in fig trees, Simmons *et al.* (1948) tried poisons with beef liver and with honey, as well as proprietary ant poisons, but without success. Dusting the nests with 10% DDT dust (available at that time) proved to be successful.

Control of *Polistes exclamans arizonensis* in citrus trees was obtained with a spray of 0.5 pt (about 0.25 l) of 100% Thanite® to 5 gal (19 l) of spray oil. This was applied to the wasps and their nests with a 2- or 3-gal (7.5- or 11-l) compressed-air sprayer (LaFollette, 1952). Also, 1% Thanite in base oil (deodorized kerosene) may be used (Hopkins, 1955). Later on, DDT, lindane, and chlordane were used to control this species, but they killed the wasps too slowly. After being hit by the sprays, they became highly agitated for an hour or so before paralysis ensued. During this period, they flew around rapidly and attacked anyone within 30 m of the sprayed nest.

Wene (1969) recommended the substitution of some pesticide that would paralyze the wasps more rapidly. He compared 4 preparations, 3 used at 1 lb per 100 gal (0.45 kg per 378 l) and another (chlordane) used at 1.5 lb per 100 gal. When sprayed with propoxur (Baygon), the wasps started dropping after a minute, and one could pick them up without being stung. After 5 minutes, 80% were completely paralyzed, and after 15 minutes, all were apparently dead. After spraying with dichlorvos, the wasps started dropping in 2 minutes. Once on the ground, they crawled around without making any attempt to fly. In an hour, all were apparently dead. With Gardona, the wasps started dropping to the ground within a few minutes, then crawled around without attempting to fly. In an hour and a half, 32% were dead and the remainder were paralyzed. Gardona is one of the safest insecticides to humans, with an acute oral toxicity of 4,000 to 5,000 (tables 1 and 2, chapter 2). When sprayed with chlordane, the wasps flew around vigorously or, if they dropped, they were nevertheless quite active, and in fact they flew away within an hour; 2 days after application, no live or dead wasps were seen. When propoxur, dichlorvos, or Gardona were used, many dead wasps were seen on the ground.

Most commercially available aerosol sprays that contain pyrethrins are extremely effective in knocking down umbrella wasps within a few seconds. Also, synthetic pyrethrins have been used with good results. Pyrethrins and pyrethroids appear to be the best insecticides for control of umbrella wasps. This is in sharp contrast to the relative ineffectiveness of most pyrethroids against yellowjackets and hornets, although resmethrin is effective. One is likely to arouse the pugnacity of yellowjackets with pyrethrins, and to be stung by these wasps before they die (R. E. Wagner, personal communication). The best time to treat wasp nests is in the early morning or at night, when most of the workers are in the nest. The nest should then be knocked down in order to prevent its re-use (Wagner and Reierson, 1971).

POLYBIINE UMBRELLA WASPS (VESPIDAE: POLYBIINAE)

In this subfamily, *Mischocyttarus* is the best-known genus. This genus can be distinguished from *Vespula* and *Polistes* by the slender, petiolate basal abdominal segment and the shape of the gaster (plate VII, 7; figure 243). There are at

Fig. 243. *Mischocyttarus flavitarsis flavitarsis.*

least 183 species in *Mischocyttarus,* and it ranges from northern Argentina northward to British Columbia in the West and through the West Indies to Florida in the East (Wilson, 1971), but there are only 2 species in the United States.

As in the case of the pedicels of *Polistes* nests, as observed by Eberhard (1969), the pedicels of *Mischocyttarus* nests may contain a repellent to keep scouting ants from gaining access to the nest and discovering the brood. The Neotropical *M. drewseni* Saussure has a gland at the base of the terminal sternite that excretes the repellent onto a tuft of hair. The tuft is rubbed against the pedicel to deposit the repellent substance (Jeanne, 1972).

Mischocyttarus flavitarsis flavitarsis (Saussure) (figure 243) is the species found in California, where it is common, widespread, and found from sea level to 9,000 ft (2,700 m) elevation. It is 16 to 20 mm long, and is colored orange-yellow, reddish, and black. Females are sometimes found hibernating under bark together with *Polistes fuscatus aurifer* and *P. dorsalis californicus* (Bohart and Bechtel, 1957), and have also been found in an abandoned nest of *Vespula arenaria* (Smith, 1944). The nests of *M. f. flavitarsis* may be found suspended from the eaves of houses, branches of trees, and similar supports, and are almost indistinguishable from the nests of *Polistes,* but are usually somewhat smaller.

SOLITARY WASPS

There are many species of wasps commonly referred to as "solitary wasps," although among these, in the subfamily Eumeninae, there are species with subsocial behavior (Wilson, 1971). In general, however, the solitary wasps do not have a socially organized life, each female making a burrow or other form of nest for herself and caring for her own offspring. She stings her prey (insect or spider) to paralyze it, and usually brings it back to a cell in her nest, oviposits on it, and seals the cell. Most species will repeat the process until they have prepared a number of cells in the same nest, each with food for larvae. Some "spider wasps" (Pompilidae) will attack and paralyze a spider in its own cell or burrow, oviposit on it, and leave it in place.

There is a great variety of forms and sizes of solitary wasps, ranging from about 6 to 38 mm in length. They vary in color from dull black or brown to brilliant red, yellow, or blue. They have stingers, but their venom is generally not so potent as that of the social wasps, although the mutillids, psammocharids, and some others can inflict a very painful sting. As stated earlier, there are a few species of solitary wasps in the Vespidae, although its species are mainly social.

Fig. 244. A solitary wasp, *Pseudomasaris wheeleri*. *Top*, an adult emerging from one of the cells in a nest of mud and sand attached to a twig; *bottom*, a female covering the last cell in a nest in which all cells have been provisioned with pollen and nectar. (Courtesy L. R. Brown.)

Genus *Pseudomasaris* (Vespidae: Masarinae)

Wasps of the genus *Pseudomasaris* build hard nests of mud, or sand and mud, attached to twigs or rocks. Each nest is elongate or globular, and contains several cylindrical cells (figure 244). Unlike other wasps, which provision their cells with meat, the species of *Pseudomasaris* provision them with pollen and nectar as do the bees (Hicks, 1927; Bequaert, 1929). They look like vespine wasps in having a black body color marked with yellow. *Pseudomasaris vespoides* (Cresson), the largest species, occurs throughout the West, and has been recorded as far east as South Dakota. It attaches its nests to twigs or rocks in California, but only to rocks in Colorado (Hicks, 1927). Adults may be seen on *Penstemon, Phacelia,* and other flowers. *Pseudomasaris edwardsii* (Cresson) also is found throughout the West, and is the species most commonly seen in southern California, usually on *Phacelia* flowers. Less commonly found is *P. wheeleri* Bequaert. The adults of *P. wheeleri* are 13 to 17 mm long, and black, with numerous and rather variable sulfur-yellow markings. The mesonotum has 2 complete or interrupted longitudinal yellow stripes (figure 244). The eyes are bulging in the male. The yellow markings are more extensive and more variable in the female than in the male (Bequaert, 1929). They visit *Eriodictyon* and *Penstemon* flowers.

Potter and Mason Wasps (Vespidae: Eumeninae)

Most solitary vespids are in the subfamily Eumeninae, the potter and mason wasps (Muesebeck *et al.*, 1951). These wasps vary from 9 to 20 mm in length, and are dark, with white, yellow, orange, or red markings. The females have 12-segmented antennae, and those of the males have 13 segments. The wings are folded longitudinally when at rest, and the middle tibiae have 1 apical spur, as in all other vespids. The affinity of these wasps to other vespids can be seen in some aspects of their behavior as well as in morphology. They always lay their single egg in an empty cell, often suspending it from the wall by a slender filament, whereas other mud-users lay their egg on the prey. The potter and mason wasps place from 1 to 12 caterpillars in their nests, soon after the egg is laid, and then seal them (Evans, 1963).

The best-known Eumeninae are the potter wasps (*Eumenes*). They differ from most other genera in the subfamily by having a very narrow, elongated first segment of the abdomen. *Eumenes iturbide pedalis* Fox (figure 245) is 13 or 14 mm long and predominantly black, but with prominent yellow markings, particularly on the abdomen, and with the tibiae, tarsi, and apices of the femora dull red (Fox, 1894). It is distributed throughout the western states and in western Canada and Mexico (Muesebeck *et al.*, 1951).

Fig. 245. Potter wasp, *Eumenes iturbide pedalis*, emerging from a pot on the stem of an *Epidendrum* orchid.

All species of *Eumenes* build little clay pots for nests, and usually attach these to twigs, as shown in the figure. They provision the nests with caterpillars. The pots are globular, and have thin, short necks with expanded lips, resembling miniature jugs. Evans (1963) stated that they were among the finest examples of insect architecture. This cannot be said of the work of some other genera of the subfamily, such as *Odynerus, Rygchium,* and *Ancistrocerus,* which may use mud merely as a partition between cells in a hollow twig, abandoned ground-bee burrow, or *Polistes* cell (some species make their own ground burrow), or they may make mud cells on rocks.

Spider Wasps (Pompilidae)

These are medium-sized to large, metallic dark-blue or reddish wasps, with iridescent wings and long, spiny legs. They all provision their nests with spiders, but unlike the mud daubers, they use only 1 spider in each nest. Unlike other solitary wasps, the pompilids catch their prey before preparing their nests. The spider is generally paralyzed by a single sting into the underside of the body in the region in which the nerve centers are clustered together. However, some spider wasps give the victim numerous stings at random to slow it down before administering the principal ventral sting. Unlike most solitary wasps, the stings of which are not very painful to humans, the sting of a spider wasp is very painful. The wasp drags the paralyzed spider while she backs up to a location suitable for digging a nest hole, hides the spider, excavates the nest, and then pulls her prey into the hole and lays an egg, usually on its abdomen. After the egg hatches, the larva gradually devours and finally kills its host. The habit of capturing prey before preparing the nest has its disadvantages, for the paralyzed prey may be stolen by some other insect (such as ants or another pompilid) while its captor is digging. One species places its prey in the crotch of a plant while digging, and another hides it temporarily in a shallow grave to decrease the possibility of theft. After oviposition, the burrow is closed with earth, and the entrance is tamped down with the apex of the wasp's abdomen. Some species disguise the entrance by placing pebbles or other objects on and around it (Andrewes, 1969).

Tarantula Hawks, *Pepsis* spp. (Pompilidae)

These are very large wasps, metallic velvety blue-black throughout, with violaceous, amber, or red wings. They prey on trapdoor spiders and tarantulas. They have long been well known in the southwestern United States, and were further popularized in Disney's remarkable film about life on the desert, in which a fierce struggle between a tarantula hawk and a tarantula was one of the exciting episodes. When the spider rises on its hind legs in defense, it exposes its vulnerable ventral area, and the wasp disables it with a rapid thrust of its stinger.

Pepsis formosa (Say) is one of the largest and best-known species of tarantula hawks. It is 25 to 45 mm long, steel-blue and black throughout, with fiery-red wings. Essig (1926) stated that *P. formosa* occurred in "Mexico, Texas, New Mexico, Arizona, and southern California" (it was originally described from Arkansas), but it is now known that the species does not occur in California. *Pepsis thisbe* Lucas (= *sherillae* Hurd), a species closely resembling *P. formosa,* occurs in all but the northern third of California. Other species of the genus found in California are *chryso-*

Fig. 246. One of the tarantula hawks, *Pepsis mildei*.

themis Lucas, *pallidolimbata pallidolimbata* Lucas, *pallidolimbata smithi* Hurd, *elegans* Lepeletier, *pattoni* Banks, *mexicana* Lucas, and *mildei* Stål. The latter species (figure 246) may be distinguished from all the other known species of *Pepsis* in California by the antennae having at least the apical segments, but usually the apical three-fourths of the antennae, orange or reddish orange, and by the yellowish-brown wings, with bands of dark brown (Hurd, 1948).

A curious characteristic of the genus is that the antennae of the females are curled, whereas those of the males are straight. This difference is accentuated in mounted specimens, in which the antennae of the females form a complete loop.

Other Pompilids

Aside from the tarantula hawks, a few other species of pompilids have attracted attention in the West. *Pompilus luctuosus* Cresson is 10 to 12 mm long, metallic blue-black, with smoky wings having a violet iridescence. It occurs throughout boreal North America. Another species, *Psorthaspis planata* (Fox), has been reared from the nests of a trapdoor spider (Davidson, 1905).

Mud Daubers (Sphecidae)

Mud daubers are medium-sized to large wasps, with petiolate abdomens, the petioles long and slender as is characteristic of the Sphecidae. They are sometimes called threadwaisted wasps. Probably the most familiar of their nests are those made of mud or clay and attached under bridges, against houses, or under eaves, in garages that are commonly left open, or in barns or sheds. Cylindrical cells of mud are built side by side until they make a mass that may be the size of a fist. The entire structure is plastered over to make a relatively smooth and uniform outer surface. The cells within the nest are provisioned with insect larvae or spiders which have been paralyzed by the female's venom, and a single egg is laid in each cell. The larvae feed on the paralyzed prey. A larva can mature in 3 weeks and then spins a cocoon, but does not actually pupate until the following spring. Mud daubers rarely sting, and do not defend their nests. They can be eliminated by tearing down their nests.

Black-and-Yellow Mud Dauber, *Sceliphron caementarium* (Drury)

This species (figure 247) occurs throughout boreal North America. It is 15 to 17 mm long, dull black, with the bases of the antennae, collar, scutellum, a portion of the metathorax, petiole, and portions of the legs bright yellow. Its nests

Fig. 247. Black-and-yellow mud dauber, *Sceliphron caementarium*, and mud nest attached at the intersection of a plaster ceiling and a rafter.

(like the one shown in the figure) are common in attics, porches, and outbuildings. The cells are 25 to 36 mm long, and there may be 6 or more in a nest. The whitish larvae are 10 to 14 mm long.

Biology. This has been investigated in great detail by Rau and Rau (1918) and Shafer (1949). A single cell is provisioned with several spiders. In one cell containing 8 spiders, 2 died in 8 days, 1 on the tenth day, 3 on the fourteenth, 1 on the fifteenth, and 1 on the thirty-second day. Crab spiders are preferred, but black widow spiders are sometimes used to provision the cells.

Chalybion californicum (Saussure)

This beautiful wasp (figure 248) is 12 to 18 mm long, metallic blue, blue-green, or blackish, with blue wings. It occurs throughout the United States. It has been reported that this species does not build its own nests, but occupies those built

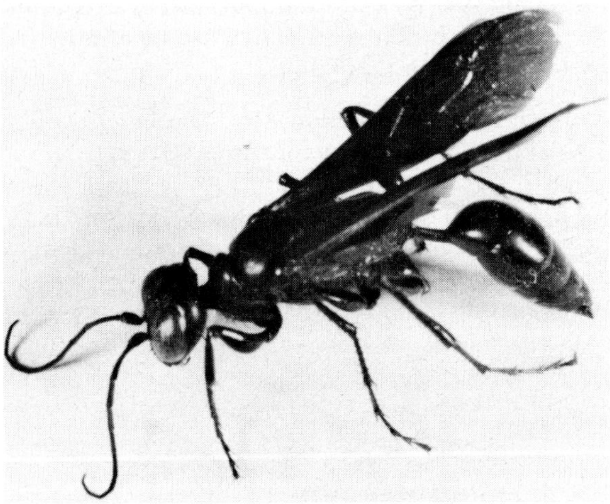

Fig. 248. A beautiful metallic-blue wasp (*Chalybion californicum*) that occupies the nests built by *Sceliphron caementarium*.

by *Sceliphron caementarium*. It carries water to the nests of the latter to soften the mud at the cell entrances, and opens them with its mandibles. It then empties the cells of all the spiders its host stored in them, and replaces them with spiders of its own. An egg is laid on the first paralyzed, replaced spider, others are placed on top of the first one, and the cell is sealed (Muesebeck *et al.*, 1951; Matheson, 1951).

Digger Wasp, *Ammophila murrayi* Menke (Sphecidae)

This ground-nesting wasp (figure 249) has habits characteristic of the genus, and is commonly seen in southern California at moderate

Fig. 249. A digger wasp, *Ammophila murrayi*.

elevations. It is 16 or 17 mm long. The head and thorax are black. The first 2 abdominal segments are long and petiolate, the first black and the second rusty red-brown. Dorsally, the first 2 gastral segments are red-brown and the others are black. (In some species, the abdomen is orange or yellowish.)

Habits and Biology. The members of the genus *Ammophila* are very slender, with petiolate abdomens. They drag caterpillars into burrows they have dug in the ground, and their larvae feed on the paralyzed prey. These wasps were made famous by Peckham and Peckham (1905), who described their habits in great detail, including their use of a "tool." They temporarily close the tunnel leading to their nests while they are out hunting caterpillars, using a pebble or a piece of twig as an implement. Also, after the nest is provisioned and the tunnel is blocked with soil, the wasp will seize a pebble in her jaws and use it to tamp down the soil at the entrance.

Cicada Killer, *Sphecius speciosus* (Drury) (Sphecidae)

This is a large insect, up to 40 mm long. It has a black body marked with yellow, like the yellowjackets, but is much larger (figure 250). It occurs in the United States east of the Rocky Mountains and south into Mexico, preying on many species of cicadas (Dow, 1942; Damback and Good, 1943). After paralyzing her prey with her stinger, the female places it at the bottom of a long burrow she has dug and lays an egg on it. The burrow is about 12 mm in diameter, so a considerable amount of earth is removed and thrown out to form a mound. When this happens in a lawn, the wasp becomes a pest. The female will not sting unless she is handled, but sometimes stings people painfully who are working on lawns, or barefooted children who are playing on them. The males only buzz around people, but do not sting.

Other species of *Sphecius* found in the United States are *S. convallis* Patton, ranging from Kansas to California and Baja California, and known

Fig. 250. Cicada killer, *Sphecius speciosus*.

to prey on cicadas; *S. grandis* (Say), recorded from Tennessee, Missouri, Arkansas, and throughout the West; and *S. hogardii* (Latreille), that occurs in the southern tip of Florida and in the West Indies (Muesebeck *et al.*, 1951).

Control of Digger Wasps

For the control of cicada killers and other solitary, ground-nesting wasps, sprinkle 10% chlordane dust at the entrances to their burrows (Anonymous, 1970*f*).

Sand Wasps (Bembicidae)

Bembicids have a beelike form, but are solitary in habit, and nest in the ground. They are large and robust, 11 to 20 mm long, smooth or finely pubescent, and are black or grayish, with greenish-yellow, orange, or red markings and abdominal bands. They have relatively large heads and especially large eyes. The abdomen is not petiolate, but has a broad base and tapers to a point posteriorly (Essig, 1942; Andrewes, 1969). These insects build their nests in sunny, sandy places. Their forelegs are fossorial, and have a fringe of long, inner hairs for moving the sand. The males (sometimes the females also) may dig shallow burrows in the sand in which to spend periods of inactivity. The nesting burrows are 12 to 24 in. (30 to 60 cm) in depth, and each contains 5 or 6 larvae in a single cell. These are fed daily, nearly all species receiving masticated flies. They hibernate in thick, oval cocoons, and the adults emerge in spring and early summer. Rau and Rau (1918) described the sudden emergence of thousands of sand wasps (*Bembix nubilipennis* Cresson) in a "sand-lot" baseball field in St. Louis (Missouri) to participate in a "dance" 6 to 20 in. (15 to 50 cm) from the ground. They counted 44 emergence holes in an area about a meter square where none had been seen before the simultaneous emergence. This species brings small flies to its nest while the developing larvae are small, and larger ones as they grow larger (Muesebeck *et al.*, 1951).

The 2 most important genera are *Bembix* and *Microbembex*.[18] These are black, usually with yellow markings similar to those of yellowjackets. The *Bembix* species provide their nests with flies, many of which are tabanids and muscoid species. In California, *Bembix americana* F., *B. amoena* Handlirsch, and *B. occidentalis* (Fox) are the most abundant and widespread, although Bohart and Horning (1971) included 8 others, as well as some varieties, in their keys to California species. These are well known to collectors of wasps, possibly because of their large size and bright markings. *Microbembex* species are relatively small-sized, averaging 11 to 13 mm long, whereas *Bembix* species average 15 to 20 mm. They often occur in large colonies in areas of fine sand. They do not capture live prey as do other sand wasps, but provision their burrows with a wide variety of dead or moribund arthropods. *Microbembex monodonta* (Say) is the best-known species. It ranges from Canada to Panama, east of the Continental Divide. A closely related species, *M. californica* R. Bohart, occurs west of the Continental Divide in the United States and south to Sonora and Baja California in Mexico. The markings are yellow in the male and whitish to yellow in the female. Bohart and Horning (1971) included 4 other species of *Microbembex* in their keys.

Bethylid Wasps (Bethylidae)

These are small to medium-sized, usually metallic-bronze wasps. Either or both sexes of a species may include winged and wingless individuals.

[18] The generic names *Bembix* and *Microbembex* appear to be inconsistent in their spelling. They are given here just as they were originally named, and as they are recognized by entomologists.

The larvae are commonly endo- or ectoparasitic on the paralyzed or dead larvae of beetles, moths, and butterflies.

In 1927, a tiny black wasp of the genus *Epyris,* which inflicts a severe sting, was reported from Yolo County, California (von Geldern, 1927). These wasps were said to appear in large numbers in the fall after a warm spell, invade houses, and get into bedding and clothing, where they stung when crushed against the skin. Individuals within a family varied greatly in their reactions, the most sensitive suffering general systemic disorders over a period of about a half hour, including itching, cramps, diarrhea, marked prostration, weakness, and sweating. The wasp was later identified as *Epyris californicus* (Ashmead). It is occasionally reported to sting people in the lowland delta area in northern California. It is believed to parasitize the larvae of tenebrionid beetles (Essig, 1932, 1942). Muesebeck *et al.* (1951) recorded the species as occurring in California and Arizona.

Cephalonomia gallicola (Ashmead) occurs in the eastern United States, Europe, and Asia, and is probably cosmopolitan. Certain beetles, such as *Stegobium paniceum* and *Lasioderma serricorne,* which commonly infest stored products in the household, are hosts of this wasp. Adult wasps are likely to become abundant in houses with a persistent infestation of the beetles, and there have been many reports of the human occupants being stung. Other bethylid wasps of the genus *Sclerodermus* parasitize cerambycid beetle larvae infesting the woodwork of houses, and the adult wasps have often been reported to sting people (Muesebeck *et al.,* 1951).

Velvetants (Mutillidae)

Mutillids or velvetants are medium-sized, brightly colored insects that resemble ants, but lack the node that is characteristic of the petiole of the ant. There are many species, and most occur in the West and South. Their bodies are black, but are densely clothed with white, yellow, golden, orange, or bright-red hairs that give the insects a velvety appearance. The females inflict very potent stings, and can make a queer squeaking noise when captured. The males usually have wings, and have straight, 13-segmented antennae; the females are wingless, and have curled, 12-segmented antennae.

Biology. Unfortunately, little is known about the biology of these interesting insects. The larvae are external parasites on larvae and pupae, principally of wasps and bees, but a few species

Fig. 251. Mutillid or velvetant, *Dasymutilla satanas* (female).

also feed on certain beetles and flies. The female oviposits on the prepupal or pupal stage of its host, the larva devours its host in a few days, and the cocoon is spun within the pupal case of its host. Some species are believed to dig nests in the soil and provision them with insects, but more are believed to be parasitic in the nests of wasps (e.g., *Chalybion*) and ground-inhabiting bees (*Bombus* and *Nomia*) (Essig, 1926, 1942). Some species are known to prey on honey bees (Matheson, 1951).

Descriptions. Following are descriptions of some of the California mutillids. The colors given always refer to the dense covering of hairs, for the body itself is always black or nearly so. The female of *Dasymutilla satanas* Mickel (figure 251) is about 20 mm long, and is clothed with long, erect, pale-ochraceous hairs, except that those on the legs are black. In the male, which is 13 mm long, tergites 3 to 6 are clothed with light-fulvous pubescence. The male was once believed to be a distinct species (*Dasymutilla mimula* Mickel) (Mickel, 1928) until it was found mating with *D. satanas* females by Barr and Hurd (1947). *Dasymutilla aureola* (Cresson), 12 to 13 mm long, varies from almost white in Inyo and Tulare counties to orange or almost scarlet in coastal California as far north as San Francisco (Hurd, 1951). In *D. californica* (Radoszkowski), the female is 8 to 14 mm long, and brick-red. It has been recorded from New Mexico, Colorado, Utah, and California. In *D. gloriosa* (Saussure), the female is 13 to 16 mm long, has long, white hairs, and is commonly known as the "thistledown mutillid." It occurs in central and southern California and eastward through Arizona, New Mexico, and Utah. In *D. sackenii* (Cresson) in southern California, approximately the anterior half of the male's abdomen is black and the remainder is white. In the female, approximately the anterior fourth of the abdomen is black and the remainder is white. Both sexes have white hairs on the dorsum. In northern California, the white dorsal vestiture changes to yellow. In *D. flammifera* Mickel, individual specimens vary from bright scarlet to nearly ocher, some of the latter being difficult to distinguish from the yellow females of *D. sackenii* by the nonspecialist (Hurd, 1951). The male of *Pseudomethoca propinqua* (Cresson) is 9 mm long, the dorsum of the thorax is black, the vertex and apical margins of the abdominal segments are golden, and the wings are dark. The female is similarly colored, and is 8 to 12 mm long. This species has been recorded from Minnesota south to Texas and west to Alberta and California (Essig, 1926; Muesebeck *et al.*, 1951).

Scoliid Wasps (Scoliidae)

The scoliids are large wasps, with large heads and short antennae. The legs are stout, and the tibiae of the middle pair have only 1 spur, whereas those of vespine wasps have 2. The general background color is black, but with yellow, orange, or red bands or patches, principally on the abdomen. The females are generally larger and more brightly maculated than the males. They do not dig their own burrows, but attach their eggs to a scarab larva and construct a crude cell around it (Essig, 1926; Hurd, 1952).

Campsomeris tolteca (Saussure) appears to be the dominant scoliid in southern California, and is commonly seen in lawns infested with scarabaeid larvae. Its geographic range includes California, Arizona, Texas, Mexico, and Haiti. The head is black, and the thorax is black, with whitish hairs anteriorly, as shown in figure 252. The wings are colorless. The abdomen is rufous, with black markings. The legs are black, but are sparsely clothed with white hairs. These wasps are large and robust, and are about 25 mm long.

Fig. 252. Scoliid wasp, *Campsomeris tolteca* (female).

BEES (APOIDEA)

Of the approximately 20,000 known species of bees, only about 5% are social insects. These are included in the family Apidae, subfamilies Apinae (honey bees and stingless bees), Bombinae (bumble bees), Halictinae (sweat bees), and Xylocopinae (xylocopine bees) (Michener, 1969).

Bees are closely related to wasps, but in bees at least some of the body hairs, particularly those on the thorax, are branched and plumose. The body hairs of wasps are simple and unbranched, and many wasps have relatively little body hair. Bees also differ from wasps in that bees provision their nests with pollen and honey, whereas wasps generally feed their young on meat. [Some masarid wasps (*Pseudomasaris* spp.) (figure 244) provision their nests with pollen and nectar.]

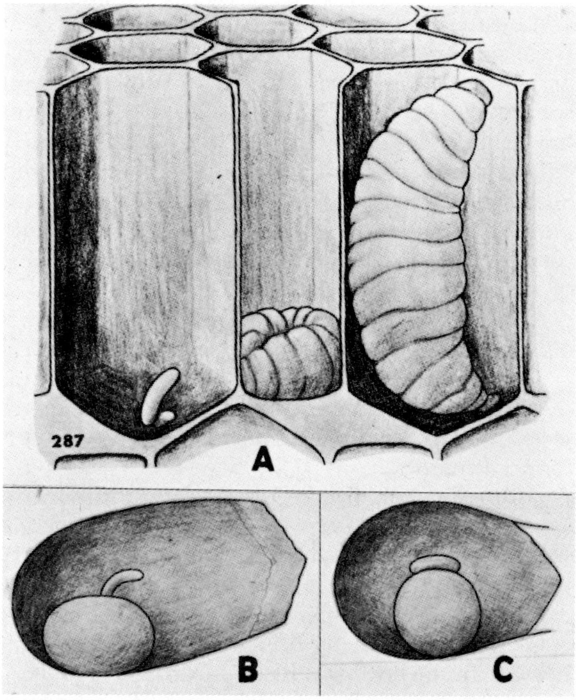

Fig. 253. Food for the young of bees. *A*, egg of *Apis mellifera* in unprovisioned cell; young larva; mature larva. (The larvae are fed by workers.) *B* and *C*, eggs of solitary bees, *Melitturga clavicornis* and *Systropha planidens*, respectively, attached to pollen balls that will serve as food for larvae. (From Stephen *et al.*, 1969.)

Honey bees and bumble bees differ from most nonparasitic bees in laying their eggs in unprovisioned cells (figure 253), and even the bumble bees place the eggs of the first generation in a bed of pollen. Nearly all other species of nonparasitic bees bring food to the nest before any eggs are laid. The honey bee and other species of *Apis* feed their larvae from the time they are hatched until they are ready to pupate (Stephen *et al.*, 1969).

Bees are best known as providers of honey and wax, but they are of much greater importance as pollinators. Their branched and plumose hairs entangle and hold pollen, and this, along with their continual contact with pollenbearing flowers, results in their being particularly efficient pollinators. Pollen is carried in large masses in hair brushes located on the hind legs of most bees and on the ventral side of the abdomen in a family of solitary bees, the Megachilidae. Honey bees and bumble bees carry pollen in moist balls in special pollen baskets on their hind tibiae, and have the first segment of the hind tarsus greatly enlarged and more or less flattened. Eckert and Shaw (1960) compiled a list of the major nectar and pollen plants of the United States, their major distribution as sources of nectar, and the color of the honey produced. Seventy groups (families, genera, or species) of plants were listed.

In the United States, about 50 agricultural crops either depend on honey bees for pollination or are at least somewhat benefited by these insects. Many species of wild bees are also important pollinators.

Unfortunately, bees are also the most important sources of insect stings, the honey bee being the worst offender. Both bees and wasps frequently find their way into moving vehicles, and have caused many accidents. Honey bees can also become pests by locating their nests in attics, chimneys, or in the hollow walls of buildings. The bees themselves may be annoying, and in warm weather or near sources of heat, such as stoves, fireplaces, or furnaces, combs filled with honey may melt and saturate ceiling or wall plaster. The honey may drip through the ceiling plaster to rooms below.

Honey Bee, *Apis mellifera* L. (Apidae)

Honey bees are not indigenous to North America north of Mexico, but were introduced by early immigrants. Any "wild" honey bee colonies are the progeny of those that have escaped from apiaries and settled in hollow trees and other suitable nesting sites, where they may establish large colonies of 20,000 to 60,000 individuals.

Description. There are 3 castes: the workers, the queen, and the male or drone. Most people see only the worker caste. The workers are 11 to 15 mm long, mostly various shades of brown and black, with the head, antennae, legs, and portions of the abdomen dark (figure 254). A dense, short, buff or pale pile clothes most of the body, being thickest on the thorax and thinnest on the dorsum of the abdomen. The queens are large, and 15 to 20 mm long. The queen's pointed abdomen extends a considerable distance beyond the tips of the wings. The drones are robust, and 15 to 17 mm long. A particularly significant anatomical feature of bees is the elongate "tongue," formed by the maxillae and labium. The tongue enables bees to reach the nectar in flowers that have moderately deep corollas.

Biology. The queen may live as long as five years. She can lay as many as 1,500 to 2,000 eggs a day, placing 1 egg in each cell of the *brood comb*. However, queens are seldom profitable after the second year, and most beekeepers believe it is profitable to "requeen" yearly (Root and Root, 1935). The queen bee produces many pheromones, mostly in her mandibular glands. The pheromones include the "queen substance" (*trans* 9-keto-2-decenoic acid), which can be made synthetically. It is passed from worker to worker orally. Among its functions is the inhibition of the development of ovaries in worker bees. Its absence induces the workers to construct new queen cells. When the concentration of queen substance in the hive falls below a certain level, new queens are produced. This pheromone is also a sex attractant for drones in the mating flight (Blum, 1970).

The workers are abortive or neuter females that arise from fertile eggs. Their reproductive organs are incompletely developed. The young workers

Fig. 254. Honey bee crawling away from her stinger left in the arm of a victim. (Courtesy Treat Davidson, from National Audubon Society.)

take care of the young brood, build the comb, and protect the entrance to the colony against invaders. Older workers collect pollen, nectar, and propolis. Propolis, sometimes called "bee glue," is a brownish, resinous material of waxy consistency collected from the buds of trees and used as a cement for closing small crevices in the hive. During the summer months, the worker bees live only 6 or 7 weeks, but those hatching in the fall survive the winter, feeding on honey stored up through the summer, and then perform the work of early spring.

The comb consists of 2 layers, back to back, of 6-sided cells constructed of wax. The cells around the top and sides of the comb are for storing honey, and are capped with wax. The lower and central portions of the comb comprise the brood comb. Most of the cells of the brood comb are similar in size to those used for storing honey, but the development of worker bees takes place in most of these uncapped cells. The wax used to construct the comb is secreted in glands on the underside of the abdomen, and is masticated and mixed with cephalic gland fluids. If the colony increases in size to the extent that the founding of a new one becomes necessary, the workers construct special cells for the rearing of queens and males. Males (drones) are reared in larger cells than those of the workers, often located in the lower corners of the comb.

To prepare a cell for rearing a queen, the workers remove the partitions between 3 adjacent cells containing eggs, and destroy 2 of the eggs. The remaining egg hatches, and the larva is fed the highly nutritious *royal jelly* during its entire larval life. The sequence of events leading to royal jelly is as follows: After nectar is brought back to the hive in the honey stomach of a forager, it is transferred to other workers (hive bees). The latter "ripen" it by forming a droplet beneath their mouthparts and expanding and contracting this droplet, thereby evaporating excess water to about 20% of the total solution. The resulting 80% sugar solution, with small percentages of pollen, is called honey, and is stored in cells as adult food. Later, nurse bees (young adult workers) consume pollen and some honey to manufacture royal jelly, and others mix pollen, nectar, and secretions to form *worker jelly* or *bee bread* to feed the brood. Several queens may be produced simultaneously within the hive. All other larvae are fed royal jelly for only 3 days, after which they are provided with worker jelly and other food.

The old queen tries to destroy the new virgin queens, but the workers protect them. Then the old queen and a large number of workers leave the hive and "swarm" to seek a suitable place for a new colony. A struggle among the virgin queens then takes place, and only the most vigorous one survives. Ordinarily, this is the only time a queen uses her stinger. Several days later, the surviving queen makes her mating flight. Mating takes place while in flight, and it may be with males from her own hive or any other. The queen mates only once during her life. Egg-laying ordinarily diminishes by the first or second year of the queen's adult life. In any case, the supply of spermatozoa in her spermatheca becomes exhausted, and the eggs are no longer impregnated. Even though she may continue to lay eggs in worker cells, only drones are produced (Root and Root, 1935). After the swarming season sometime during the first warm days of spring, the males are driven from the hive or killed by the workers.

Under ordinary conditions, the developmental period for the worker is 21 days; for the drone, 24 days; and for the queen, 15 or 16 days. About the first half of the worker's life is normally devoted to hive duties, and the last half to field duties. The average life of the worker during the summer is not over 3 months and, over the height of the honey flow, perhaps an average of 5 weeks. If drones were kept in a queenless hive, they might live for 3 or 4 months, but ordinarily their lives are even shorter than those of workers. However, the queen may live several years.

Among the races of *Apis mellifera,* the Italian race is superior. It is gentle, is the best producer of honey, builds few queen cells, does not run over the comb, keeps out the wax moth (*Galleria mellonella*), and is the most resistant to European foul brood.

The Language of Bees

One of the most amazing phenomena in biology is the bee dance, a form of communication unique in the animal kingdom. We owe the revelation of the full significance of the bee dance, and of the physical and biological phenomena involved, principally to the ingenuity and perseverance of the great entomologist and Nobel laureate Karl von Frisch[19] and his associates in a series of thor-

[19] Unknown to von Frisch in the early years of his experimentation was the fact that, in 1788, Ernst Spitzner had observed that a scout bee, "full of joy," twirled in circles in the hive and was followed by other bees when she returned to the source of the food, which in Spitzner's experiment was honey (von Frisch, 1967).

ough and painstaking experiments. A scout bee is able to communicate the location of a food source to other bees in the darkness of the hive and on the vertical plane of the honeycomb. Upon returning from a good source of food, she may perform a *round dance* if the food is less than 10 m away. She runs rapidly in a small circle on the comb, first in one direction then in the other, followed by other bees which have received the message that food is near-by. If the food is nectar or pollen from flowers, their fragrance will remain on the scout and give the other bees a clue to the source. If the food source is farther than 100 m from the hive, the scout performs a different kind of dance, called the *tail-wagging dance*. (Distances between 10 and 100 m are indicated by transitions between the round and tail-wagging dances, with a gradually increasing preponderance of the latter with increasing distance.) In the tail-wagging dance, the scout imparts not only the information given in the round dance, but also the direction and distance of the food source. She runs through a short, straight stretch and then returns in a semicircle, alternately to the right and to the left to the starting point, completing a figure 8. On the straight portion of her course, she makes lateral waggling motions of the body—about 13 to 15 waggles per second. The farther away the source of honey, the fewer the cycles of the dance in a given time. The direction of the straight (tail-wagging) part of the dance indicates the direction of the food source, the sun being used as a reference point. If the scout moves upward in the tail-wagging part of her dance, this indicates that the food source is in the direction of the sun; if she moves downward, this indicates that the food source is in the direction opposite to that of the sun. If the tail-wagging stretch of her run is to the left (or right) of vertical, this indicates to the bees following her that the food source is left (or right) of a direct line to the sun. The angle of the line from the vertical indicates the angle from the direction of the sun the bees must take to reach the food source. Since all this activity is taking place in the dark except when observed under glass by the investigator, the bees orient the straight portion of the dance at the same angle to the force of gravity as the angle they have flown with respect to the sun during the flight from hive to the food source (von Frisch, 1950, 1967).

The amount and quality of the food are indicated by the duration and vivacity of the dance, and the soliciting dances last only as long as the food is in abundance. The bees directed to the food source by the original discoverer likewise become messengers via the bee dance, and soon a large portion of the colony is engaged in the exploitation of the newly discovered food source (Lindauer, 1960; von Frisch, 1967).

In their waggle dances, bees indicate a greater distance if the flight to the food source is to be against a head wind than if it is to be in a quiet atmosphere. Also, a shorter distance is indicated if a tail wind is to be encountered (von Frisch, 1948; von Frisch and Lindauer, 1955). The dancers even indicate a longer distance if the flight is to be uphill than if it is to be downhill (Heran and Wanke, 1952). This implies that distance information is based on the amount of energy needed during the flight rather than the flying time (Lindauer, 1960).

The sun is not always in sight as a compass for orientation. It may disappear behind clouds, mountains, or the edge of a crest, but the bees still orient their hivemates with their waggle dances under such conditions. A small patch of blue sky is sufficient to reveal the sun's position to bees, for they are able to recognize the plane of vibration of polarized light that comes to earth through this clear patch of sky. The direction of vibration of polarized light, together with the percentage of light that is polarized, determines the sun's position in the sky. Provided the clouds are not too thick, bees can also determine the position of the sun when the sky is completely overcast, for 5% more ultraviolet light penetrates through clouds directly in front of the sun than elsewhere, and bees are particularly sensitive to ultraviolet light (von Frisch *et al.*, 1960).

Mautz (1971) determined that all bees that had been successful in finding an artificial food source 400 m from the hive were individuals that had followed 5 or more tail-wagging runs of the dancing bee. Those following fewer than 5 runs might not find the food source.

Forager bees can become househunters, and a few days before swarming occurs, they search for a suitable home and impart the distance and direction by the same kind of waggle dance used for informing their hivemates about a food source. In either case, the bees' language can be interpreted by man, who has on a number of occasions obtained the message of the waggle dance and awaited the arrival of the first bees at a new food source or future home. One can readily imagine this to be the "unforgettable experience" described by Lindauer (1967).

Honey Bees on Residential Properties: Outdoors

Honey bees may form a "colony" or "swarm" on the limb of a tree or in a shrub (figure 255) on a residential property. This happens when a queen becomes tired and settles on a place to rest. The location where she settles is not necessarily suitable for a permanent colony. In that case, scout bees seek a suitable location, and the colony will follow the scout that gives the most enthusiastic "dance" to indicate the discovery of a suitable location for a permanent colony. If the homeowner will wait 3 or 4 days, the bees will usually leave of their own accord. In the meantime, it is desirable to cover chimney openings or any hole leading to an attic, wall void, or other enclosed place that may be found by the bee scouts. However, in areas of prolonged dry seasons, as in the Southwest, if the colony happens to become established under a dense canopy of foliage (figure 255) or in a thicket, the bees may remain there until the first rains of late fall or winter before moving on to a better location. If a garden sprinkler is set so that it allows a "rain" to fall on the colony, the bees may move immediately. A strong stream of water can dislodge bees, but will infuriate them and cause them to sting.

Sometimes, beekeepers are willing to come and pick up bee colonies. But in the Los Angeles area,

Fig. 255. Swarm of honey bees attached to the limb of a shrub.

the experience has been that those beekeepers who have left their telephone numbers at fire stations or other appropriate public places for the benefit of persons who wish to have bee colonies removed from their premises soon ask to have their names removed because of the overwhelming number of such requests.

One can get rid of bees that have colonized on a tree, shrub, under the eaves of the house, or in some other suitable location, by the use of insecticides. Using a "hose-attachment" garden sprayer and a thin stream of spray, the colony can be treated from a safe distance with potent insecticides such as dichlorvos or lindane. The cluster of bees should be completely covered with spray. The person applying it should then run into the house. By the next morning, the bees should all be dead or gone, despite the fact that many may not have been wet by the spray.

Most people are reluctant to kill bees or to ask pest control operators to do so because bees are thought to be beneficial as pollinators. A simple and effective way of removing a colony hanging free on a limb is to place beneath it an open cardboard box with the cover flaps bent outward. The box should have a slot about 5 in. (13 cm) long and 2 in. (5 cm) wide cut out of one side near the bottom in such a way that the part cut out can be bent down to form a runway for the bees. (The reason for the slot near the bottom of the box is to encourage the bees to use it instead of the top.) Then, strike the top of the limb with a pole, which will cause the bees to fall into the box. The limb should be tapped until all the bees have fallen. This will not arouse their anger, for they prefer to have their colony in an enclosed space, and will be "glad" to find themselves in the more suitable location. The best time to do this is during the warmest period of the day, for the bees are then at their peak of activity and are therefore the most gentle. Although homeowners have done this without previous experience, it is probably prudent to get someone with more experience or with protective clothing and a bee veil to do the job, for it is possible that a person could be stung, and the reactions of some individuals to bee stings are very severe. Later, the flaps on top of the box can be closed, using a long pole, and a board placed across the top. All motions should be slow and deliberate, and it is helpful to wear clean, white clothes, free of perspiration. At night, when the bees' activity in leaving and entering the box has practically ceased, the flaps and the slot near the bottom can be sealed with masking tape. The colony can then be taken away and released in some appropriate area.

The rule for safety in the presence of bees is slow, deliberate movement. By observing this rule, it is possible to open the hive and even tear the comb to bits.

Honey bee colonies situated in the sun are said to be "good-natured," while those in dense shade are usually the reverse. Bees are generally very gentle in the middle of the day when the air is warm and they are going out into the field. If a sudden rainstorm develops, shutting off their supply of nectar, they will usually become quite irritable, and this "bad temper" will last until the normal supply of nectar is again available. Bees will become "angry" if one offers them a bit of food, such as honey, and then suddenly withdraws it. They are also likely to be belligerent toward nightfall on chilly days. Root and Root (1935) stated: "When all are at home, and the hives are opened unceremoniously, they may resent the intrusion. It is then that beginners discover, much to their sorrow, that bees should not be handled during cool or chilly weather, right after a rain, or at night."

Bees never injure sound fruit, no matter how soft and juicy it is or how thin the skin. Even fruits such as grapes, peaches, and apricots that were placed in hives of bees that had been purposely starved were not molested (Root and Root, 1935). If they are seen sipping fruit juice, it is an indication that the skin or peel has been broken by some other marauder, such as a yellowjacket or a bird.

Fig. 256. Honey bee colony in an attic.

Honey Bees on Residential Properties: Inside a Structure

If bees colonize in the attic (figure 256) or wall voids of a house, they present a potentially serious problem. The wise procedure is to get rid of them as soon as possible. Large quantities of honey can be produced within a few months, and the combs and honey can melt and seep through plaster, particularly near a kitchen stove, furnace, or fireplace. The odors of dead brood and rancid honey are other adverse results if removal of the bees is delayed too long (Pence, 1955). During warm weather, the hive is constantly cooled by the fanning of thousands of wings. If the colony is destroyed, the cooling process ends and the wax combs may melt, releasing the honey. Bees also remove excessive moisture, and when this process stops, the honey may burst the wax cappings and drain into the wall spaces. If the bees are killed, ants, wax moths, carpet beetles, soldier flies, and cockroaches may find their way to the comb to feed and breed.

In one way, infestation by wax moths (*Galleria mellonella*) after the bees are gone can be beneficial, for their larvae can eventually clear away the old combs in wall voids and discourage further occupancy by bees. On the other hand, if combs containing honey are dislodged and fall to a fire block or floor plate, the honey may soak through plaster, wallpaper, or wood interstices and cause damage. Wax moth larvae will sometimes move to the edges of the comb area and gnaw out oval depressions in which to pupate. If they crawl through a crack or opening in the wall plaster and pupate beneath the wallpaper, a lump beneath the paper may result (Happ, 1966).

If Argentine ants find an inaccessible hive, their trails may be found leading to points of entry into the wall void, chimney, or other closed space where the hive is located. The ants can be allowed to remove the store of honey, but extra care must be exercised to avoid leaving any exposed sweet material in the living space of the house that will attract their foraging scouts.

Control with Insecticides

Nearly any insecticide sprayed or blown into an enclosed space such as a wall void will kill bees, particularly when applied as a dust, which penetrates well. With a volatile insecticide, such as dichlorvos or lindane (particularly the former), one can use a hose-attachment sprayer and with a thin stream sometimes reach the bees' entrance hole from the ground. With such volatilizing insecticides, a large quantity and thorough coverage of the comb are unnecessary. A 5% chlordane or carbaryl (Sevin®) dust is commonly used. Dusts can easily be applied with a plunger-type garden duster with the baffle removed, using about 0.5 lb (0.23 kg) of dust per colony. Among toxic insecticides, carbaryl dust appears to have an important advantage in addition to its efficacy in killing bees when exterminating them in wall voids and attics. It kills the bees slowly, and they tend to leave the building before they die. The odor resulting from dead bees is avoided.

One can also drop small pieces of dichlorvos "resin strip" into a bee-infested cavity. This, as well as dusting and most types of spraying, requires that a person climb up to the hole to introduce the insecticide. This should be done at night when there is minimal traffic in and out of the entrance hole. Generally after the colony has been treated, one side of the double wall must be taken down and the comb and honey removed.

One pest control operator set up a chlordane fogger outside a house to prevent bees from attacking him while he drilled 2 holes in the siding. He inserted 2 plastic void-injector tubes in the holes and attached the tubes to 3-lb (1.36-kg) pyrethrin aerosol bombs to kill the bees. When the siding was removed, he found the combs were as wide as 18 in. (45 cm) (the distance between 2 studs) and as long as 7 ft (over 2 m). He removed the combs and honey, cleaned the studs, applied 2% chlordane spray to the infested area, and replaced the siding (Goforth, 1964). A particularly useful device for dispensing aerosols in the control of bees, as well as wasps, is the Hi-Spray Aerosol Dispenser,[20] on which the canister is mounted at the end of a pole. With this equipment, one can reach a distance of 13 ft (4 m) without an extension and 18 ft (5.5 m) with an extension pole. The aerosol is activated by pressing a button inside the base of the handle. The equipment weighs less than 3 lb (1.36 kg) with the can.

Removal of a Living Colony

Even when bees have located their colony within a wall void, many people are reluctant to destroy them. A method of removing them from a wall void has been devised which does not require killing the bees, and in fact after they have been relocated outside the house, they will remove the honey from the wall void. Building repairs are avoided and a useful colony of bees is made available. The job can be done best by an experienced beekeeper, although pest control operators and even the homeowner may be tempted to try, with the aid of an illustration and a description of the simple procedure. The instructions are copied verbatim from Stanger (1967):

> Fold a piece of window screen to make a cone 18 in. [45 cm] long and wide enough at the bottom to cover completely the entrance used by the bees. Bend the cone slightly upward toward the smaller opening, which should be about 0.5 in. [12 mm] in diameter (figure 257).

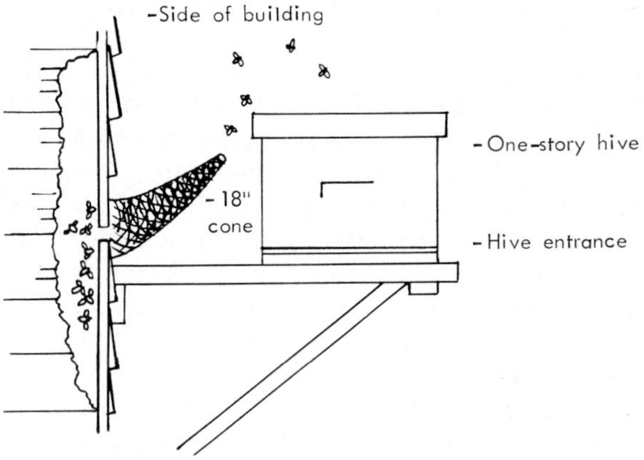

Fig. 257. A window-screen cone placed over the entrance hole of a honey bee colony in a wall void. The escaping bees cannot return to the wall void, and will form a colony in the adjacent hive, which contains a queen. (From Stanger, 1967.)

> Protect yourself with a bee veil and use a bee smoker to smoke the bees' entrance to the building. Then tack the screen cone tightly over the entrance.
>
> Plug all other holes where the bees may re-enter the building.
>
> Near the entrance, place a 1-story hive containing a queen bee, 1 frame of unsealed brood covered with bees from an established colony, and new frames. (A frame with brood and a queen can be obtained from a bee supply house or local beekeeper. It can be ordered by mail from the bee supply house.) The hive may be on a separate stand or on brackets nailed to the building (figure 257).
>
> The bees will emerge from the screen cone and fly to the outside, but will be unable to find their way back into the old nest through the opening in the cone. Instead, they will enter and become established in the new hive near their old entrance. As the bees move out of the building into the hive, the old colony will grow weaker and be unable to maintain itself in the building. In about 4 weeks, remove the screen from the entrance. The bees in the hive will enter the old nest and transfer the honey to the new hive. When the bees have moved from the building, close all holes and cracks to prevent another swarm from entering.

[20] Product of Ashcombe Products Company, Route 5, Box 75, Dover, Pennsylvania 17315.

The African or Brazilian Bee

The African bee is a yellow strain of the subspecies *Apis mellifera adonsonii* Latreille, said by some observers to be slightly smaller than the common honey bee of the United States, but otherwise indistinguishable by casual observation. Two dozen swarms of this strain of bees were inadvertently permitted to escape into Brazil in 1956, and have subsequently multiplied rapidly and are reported to have spread over most of Brazil and into Paraguay, Uruguay, Argentina, and Peru. Some investigators believe that they are now moving northward at the rate of about 200 mi (320 km) a year, and could reach the United States in 10 years, or sooner if accidentally introduced. The mixture of the African bee with the European bees of Brazil has resulted in a strain that can appropriately be called the "Brazilian bee." Some investigators have reported the Brazilian bee to have a much different behavior than the European bees of the United States, mainly the German race, *Apis mellifera mellifera* L., among the wild bees and mainly the Italian race, *Apis mellifera ligustica* Spinola, in the apiaries. They state that the sting of the African bee is no more painful than that of the bees of the United States, but that it is more aggressive and stings more readily. They report a common belief that Brazilian bees swarm and migrate oftener, that the queen lays more eggs per day than the queen of the United States bee, and that as the Brazilian bees invade an area they replace other bees (McGregor, 1970; Gary, 1971).

The initial apprehension concerning the Brazilian bee is now subsiding. Most of the adverse claims concerning the Brazilian bee could not be verified by the Committee on the African Honey Bee, consisting of 9 members from the United States, Canada, and Mexico under the auspices of the Division of Biology and Agriculture, National Research Council, National Academy of Sciences, Washington, D.C. In the early 1970's, the African bee was introduced into Poland and built up to 30 colonies. These bees were reported to be more aggressive than the local ones, but no more so than a race (*A. m. silvarum*) from northeast Poland. The African bees did not swarm, and their honey production did not differ from that of domestic colonies. These bees did not spread and become a problem throughout Europe. Also, African bee semen was used to inseminate queens of domestic bees through 3 generations at Baton Rouge, Louisiana, to produce a strain over 90% African, but the resultant offspring presented no special problems. In any case, colonies with aggressive bees can easily be requeened to change their temperament (Cantwell, 1974).

In southern Brazil, where European bees were kept in considerable numbers before the arrival of the Brazilian bee, the new race of honey bees is gentler than it is in northern Brazil. Presumably, there has been a dilution of African traits through interbreeding. Also, beekeepers have selected for gentler colonies. The new Brazilian strain is reported to be producing more honey per colony than the Italian bee under the same circumstances, and beekeepers in southern Brazil have learned to handle the new race and to take advantage of its productivity.

The large number of European honey bees now in Mexico may considerably dilute the characteristics of the bees now in northern Brazil, and the strain that eventually reaches the United States may be considerably different from the one now spreading northward in Brazil (Michener, 1973).

Bumble Bees, *Bombus* spp. (Apidae)

Bumble bees are almost world-wide in distribution, and are found even within the Arctic regions and up to 13,000 ft (about 3,900 m) elevation in temperate regions. Unlike most social insects, they are more numerous in temperate and even in Arctic zones than in the tropics. They occur throughout the United States, but are most abundant in the Sierra Nevada, Cascade, and Rocky mountains and in the northernmost states. All are nestbuilders, and the colony consists of queens, males, and workers. Bumble bees are medium- to large-sized, robust, and densely haired. The hair is usually black, with areas of orange or bright yellow. Queens and workers have pollen baskets on the hind legs. Unlike honey bees, they have terminal spurs on the hind tibiae.

Biology and Habits. Fertilized bumble bee queens must hibernate. Most of them prepare for hibernation by digging a small chamber or hibernaculum in the soil, frequently near the entrance to the nest in which they were reared. Some hibernate in hollow trees and other protected places. In the spring, they appear and feed on nectar when the early plants bloom. A queen will then seek a suitable nesting place in a clump of dry grass, an old bird nest, a deserted mouse nest aboveground, in a burrow abandoned by a rodent, or sometimes even in a building. When the nest has been established, she proceeds to build a "honey pot" (figure 258, *F, L*) from the wax that extrudes from between the sclerites of both dorsal and ventral abdominal surfaces. She fills the honey pot with nectar, and builds a cell, usually from a mixture of wax and pollen, for her eggs. The eggs of the first brood vary in number, from 1 per cell in the subgenus *Bombias* to as many as 10 or 12. She constructs the cell on top of a mass of paste made of pollen and honey, covers it with a wax cap, and broods over the eggs to incubate them. The larvae hatch in 3 to 5 days, and feed on their bed of pollen-and-honey paste. The queen continues to care for her young. Some queens gnaw a hole in the lid of the cell and supply the larvae with regurgitated honey and pollen, while others push pollen in under the developing brood.

About 3 weeks are required for larval and pupal development, and then the first brood of small workers appears. These gather food and construct brood cells, relieving the queen of all duties except egg-laying. While the colony is in the early stages of development, some of the queen's eggs are never laid, but are reabsorbed. As the number of workers increases, the queen usually lays more eggs. If larvae receive ample food, they become queens, but if partially starved, they become workers (Free and Butler, 1959). A large community of bees eventually occupies the nest, which may attain a diameter of 20 to 30 cm and be lined with straw or other dry plant material. The nest may have several entrances. The upper and most recent cells contain brood (figure 258, *D, E*), whereas the lower cells, from which brood has emerged, contain pollen and honey. In the later broods the workers are larger, some approaching the size of the queen. With the appearance of the new queens, unfertilized (male) eggs are laid. When the males mature, mating takes place. The fertilized queens seek hibernating quarters, and the workers, old queen, and males soon perish.

Bumble bee colonies generally contain a few hundred individuals, although the population of a thriving colony may reach 1 or 2 thousand (Michener and Michener, 1951). There is no strict

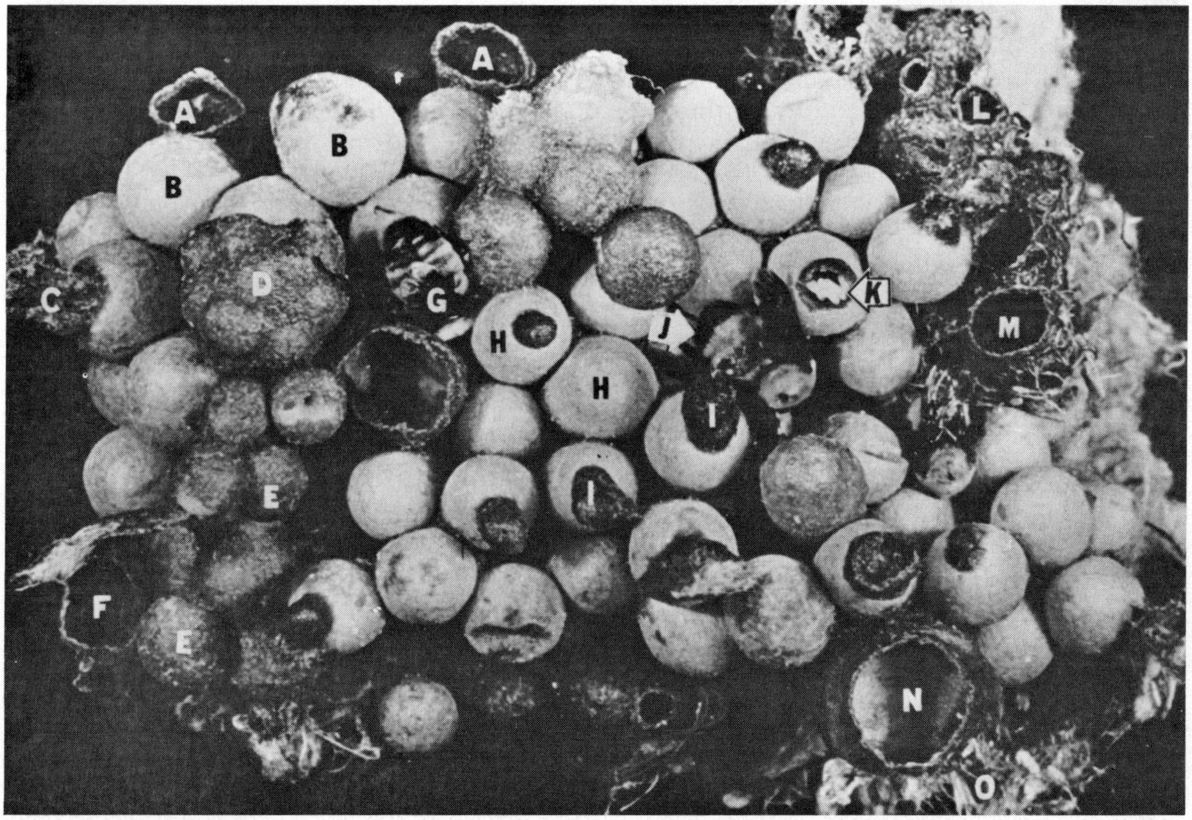

Fig. 258. Nest of bumble bees, *Bombus morrisoni*. *A*, old cocoons used for nectar storage; *B*, new queen cocoons; *C*, egg baskets on queen cocoons; *D*, cluster of young, developing brood; *E*, older developing brood; *F*, an original honey pot; *G*, young male bee; *H*, worker or male cocoon; *I*, egg basket on worker or male cocoon; *J*, worker bee; *K*, egg basket opened to show eggs; *L*, honey pot; *M*, probably a pollen pot; *N*, pollen pot; chicken feathers (the nest was in a chickencoop). (Courtesy W. P. Nye.)

division of labor within the nest, and the large workers are more inclined to go on foraging trips. These large bees are probably stronger fliers, and can collect larger loads of nectar and pollen (Free and Butler, 1959).

Bumble Bees as Pollinators

Certain kinds of clover seed cannot be grown where bumble bees or other long-tongued bees do not occur. The tubular part of the flower of the red clover, *Trifolium pratense,* may be as long as 9 to 10 mm (as short as 5 or 6 mm under arid conditions), and the tongues of some bumble bees are long enough to enable them to reach the nectar at the bottom of the longest tubes. In doing so, they pollinate the flower, visiting as many as 25 to 35 flowers a minute. In his *Origin of Species,* Darwin (1859) recorded that 100 heads of red clover visited by bees produced 2,700 seeds, whereas another 100 heads from which bees were excluded produced none. In New Zealand, where the bumble bee is not native, a bumble bee (*Bombus terrestris*) had to be introduced before red clover began to produce seed. In cool, humid areas, such as parts of Canada and the northern United States, bumble bees and leafcutting bees (*Megachile*) are more important than honey bees for the pollination of alfalfa (Bohart, 1958). With increasing knowledge of the biology of bumble bees, their colonies might be managed for the pollination of many important crops, particularly legumes (Medler, 1958; Holm, 1966).

Common California Species of Bumble Bees

A taxonomic study of 27 species of bumble bees of western America was made by Stephen (1957). It includes keys to queens, workers, and males, species descriptions, and notes on variations within species, biology, distribution, and flight records. The color patterns of most bumble bees are very variable, particularly for males, and structural characters used by taxonomists are more reliable for identification. However, they would be of little value to the average reader. The following descriptions are brief and nontechnical, adapted mainly from Stephen's work, of some species that are common in California.

Yellowfaced Bumble Bee, *Bombus (Pyrobombus) vosnesenskii* Radoszkowski

This is the most common species in many parts of California. It ranges throughout the Pacific Coast states from Washington southward, and eastward to western Nevada. Its appearance is remarkably constant throughout its range. The queens are 14 to 19 mm long, the males 10 to 14 mm, and the workers 9 to 14 mm. The color is black, except for the face, occiput, pronotum, and mesonotum in front of the wings, and the dorsum (tergite) of the fourth abdominal segment, which are yellow.

Bombus (Pyrobombus) caliginosus (Frison)

This species is found in coastal regions of the Pacific Northwest, and in California as far south as the San Jacinto Mountains. It closely resembles the foregoing species, and it is hard to distinguish the workers and queens of the two; their color patterns are almost identical. The male genitalia offer the only certain characters for determination.

Van Dyke's Bumble Bee, *Bombus (Pyrobombus) vandykei* (Frison)

This bumble bee ranges throughout the Pacific Coast states, and appears to be most closely related to *B. caliginosus.* The females of *B. vandykei* from southern Oregon and California resemble those of *B. caliginosus* superficially, but in *caliginosus* the third abdominal tergum has black hairs and the fourth has mostly yellow hairs, while in *vandykei* the third abdominal tergum has yellow hairs apically and the fourth has black hairs. *Bombus vandykei* has been most abundant in collections from the transverse mountain ranges of southern California and along the western foothills of the Sierra Nevada below 6,000 ft (1,800 m) elevation, but is uncommon in coastal areas where *B. caliginosus* occurs (Thorp, 1969).

Edwards' Bumble Bee, *Bombus (Pyrobombus) edwardsii* Cresson

These are relatively small bumble bees, the queens ranging from 15 to 17 mm long and the workers, 11 to 12 mm. The queen has 2 yellow stripes on the thorax, separated by a wide, median, transverse band of black between the bases of the wings. The yellow pile comes down the side of the thorax and covers the area from legs to wings. It is a less intense yellow color than in the preceding species. There is also yellow pile on the face, as well as yellow bands on the first, fourth, and fifth abdominal tergites. The workers have the same general yellow-black pattern as the queens.

Bombus (Protobombus) bifarius bifarius Cresson and *B.* (*P.*) *b. nearcticus* Handlirsch

Bombus bifarius bifarius and *B. b. nearcticus* are common and highly variable species, found from Alaska to California and southern Utah and from sea level to 10,500 ft (3,200 m) elevation. Their range does not extend as far south as southern California. *Bombus b. nearcticus* is by far the more abundant of the 2 subspecies, being found in great numbers throughout eastern Washington, Oregon, and in the mountains of northern California. Both subspecies have the first and fourth abdominal tergum yellow and the fifth and sixth black. *Bombus b. bifarius* has red-orange pile predominating on the third abdominal tergum and traces of it on the second. *Bombus b. nearcticus* has a small amount of orange-red pile only on the third tergum. There is an intergradation between the two.

California or Blackfaced Bumble Bee, *Bombus* (*Fervidobombus*) *californicus* Smith

This species ranges from the eastern slopes of the Rocky Mountains to the Pacific and from British Columbia to Baja California. The queens are 20 to 24 mm long, and have yellow pile on the anterior half of the thorax and a single yellow stripe across the fourth abdominal segment. The workers have the same general yellow markings. They vary greatly in length, from 12 to 22 mm. Neither queens nor workers have yellow pile on the head. This species has been confused with *B. vosnesenskii*, but has black pile on the face.

Fig. 259. Sonora bumble bee, *Bombus sonorus* (worker).

Sonora Bumble Bee, *Bombus* (*Fervidobombus*) *sonorus* Say

This species is one of the larger bumble bees of western America. It appears to be the most common one in southern California. The queen is about 30 mm long, and has yellow pile on the thorax, but usually has a black dividing band across the thorax from wing to wing (figure 259), and the sides are black. The first 3 abdominal tergites have yellow pile. The workers are much smaller (15 to 20 mm), but have a similar color pattern. Throughout the range of the species in western America may be found specimens with the dorsum of the thorax almost or completely covered with yellow pile.

On the UCLA campus, *B. sonorus* had a nest in the ground at the base of a high fence heavily covered with ivy. The author was stung on the head on 2 occasions while walking alongside the fence, but the workers are normally more likely to merely threaten an intruder by fast, circular, vertical flights in the region of the head and shoulders. Protective clothing was used when digging up the nest in 1971, and nearly all the bees were captured. There were about 450 bees in the colony.

Western Bumble Bee, *Bombus* (*Bombus*) *occidentalis* (Greene)

This species occurs along the eastern slopes and plains of the Rocky Mountains, westward to the Pacific Ocean, and northward into Canada and Alaska. The queens are 15 to 18 mm long, the males 11.5 to 15.5 mm, and the workers 9.5 to 15 mm. In the Rocky Mountain form, the face has yellow pile mixed with black; the anterior dorsum of the thorax is yellow; a broad black band appears between the bases of the wings; the scutellum, has yellow and black pile; the legs are black, the tarsi brown, and the pollen basket is rusty red-brown. The abdomen is always black at the base, yellow on the dorsum (tergite) of the third segment, and whitish-ferruginous at the posterior 2 segments. The workers and queens are similarly colored, but the males have more yellow and whitish pile. In northern California, *B. occidentalis* has predominantly black hair, with yellow and black intermixed in the face, yellow on the thorax in front of the wings only, and white on the terminal 2 segments (fifth and sixth), and often the apex of the fourth tergum. This form grades into the one in the San Francisco Bay area (subsp. *nigroscutatus*), in which the terminal white becomes buff, with yellow on the third tergite and apicolaterally on the second tergum.

Crotch's Bumble Bee, *Bombus* (*Crotchiibombus*) *crotchii crotchii* Cresson

This large and rather striking species occurs in California and northern Mexico. Three queens measured were 23 and 24 mm long, and 4 workers were 14 to 18 mm long. There is little difference in appearance between queens and workers. These bees have a little yellow pile on the head and a yellowish golden pile on the anterior half of the thorax. The abdomen is black, except that the wide second abdominal tergite is yellow, and segments 4, 5, and 6 may have much brick-red or rusty red-brown pile. The yellow shades on *B. c. crotchii* vary from light to bright yellow.

Morrison's Bumble Bee, *Bombus* (*Separatobombus*) *morrisoni* Cresson

This species is almost entirely distributed east of the Cascade–Sierra chain of mountains, becoming rather common in southern Idaho and Utah and one of the most common species in New Mexico and Arizona. In the workers, almost the entire dorsal surface is covered with bright-yellow pile, only the terminal part of the abdomen being black (figure 260). In the queen, the yellow pile does not extend so far posteriorly as in the workers. The yellow is paler in the north and at higher altitudes. In California, this species occurs only east of the Sierra and apparently not south of Whitney Portal in Inyo County.

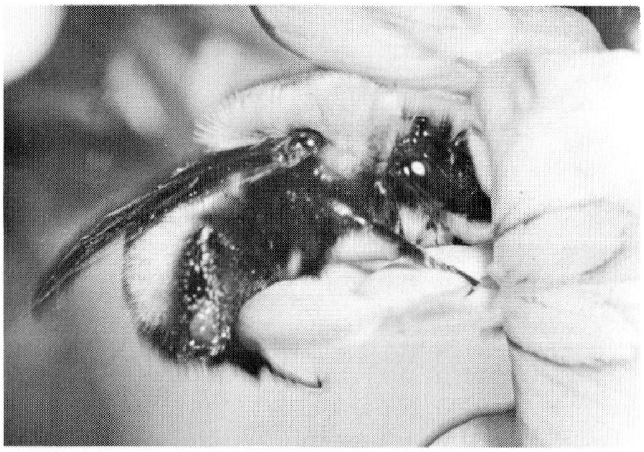

Fig. 260. A worker of Morrison's bumble bee, *Bombus morrisoni*, collecting nectar from bladder senna, *Colutea arborescens*. (Courtesy W. P. Nye.)

Control of Bumble Bees

Bumble bees can be destroyed by spraying or dusting an insecticide into their nests, as for the control of honey bees and wasps.

Cuckoo Bumble Bees

Cuckoo bumble bees (*Psithyrus*) are inquilines in one of the accepted definitions: "Any animal living habitually in the nest or burrow of another animal" (Pennak, 1964). They closely resemble *Bombus* spp. except that they do not have pollen baskets on the hind legs. They live as inquilines or uninvited guests in the nests of true bumble bees. The females may be seen in the spring a few weeks after the queen bumble bees appear, and they lay eggs in the nests of the latter. If not promptly ejected or killed, they become accepted into the bumble bee colonies. *Psithyrus* has no worker caste, and the young are fed and cared for by the bumble bee workers.

According to Free and Butler (1959): "It is perhaps significant that when a *Psithyrus* female first enters a nest, she often hides under a comb or nesting material for some time, for by so doing she may, in fact, be acquiring an odor similar to that of the colony's inhabitants." The cuticle of *Psithyrus* is thicker and tougher than that of *Bombus,* and her stinger is longer. She is a formidable adversary when fighting is required, and in one nest, the bodies of no fewer than 20 *Bombus* queens were found after such a battle. There are some exceptions, but the cuckoo bumble bee in most cases strongly resembles her principal host species in color and markings.

There are 2 species of *Psithyrus* presently recorded for California: *P. crawfordi* Franklin and *P. fernaldae wheeleri* Bequaert and Plath. Both species are represented in California by color forms that mimic *B. vosnesenskii* and other bumble bees of this color pattern, especially *B. caliginosus, B. vandykei* (var.), and *B. californicus.* The females of both species of *Psithyrus* are predominantly black, with yellow hair on the vertex, the thoracic dorsum in front of the wings, and the fourth abdominal tergite. In addition, *P. crawfordi* has yellow on the face above and sometimes below the antennal bases and laterally on the fifth tergite. Another striking difference is that *P. fernaldae* has the tip of the abdomen strongly recurved beneath, with the last sternite projecting beyond the apex of the last tergite (R. W. Thorp, correspondence).

SOLITARY BEES

As stated at the beginning of this section, about 95% of the bees are solitary in habit. The female prepares her own nest, provisions the cells of her nest with nectar and pollen, oviposits, but manifests no further interest in the welfare of her offspring and survives for only a single season. The solitary bees are rather closely related to the sphecoid wasps and have similar nesting habits, except that they provision their cells with pollen and honey instead of meat. Unlike honey bees and bumble bees, solitary bees do not have pollen baskets on the hind tibiae. Pollen is carried on pollen brushes on the legs or abdomen (Eckert and Shaw, 1960).

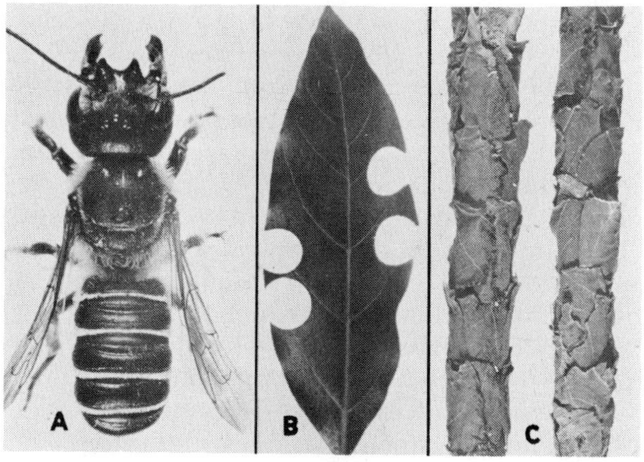

Fig. 261. Leafcutter bee, *Megachile davidsoni*, and its work. *A*, adult; *B*, holes left where the bee cut disks from an avocado leaf; *C*, two columns of nests.

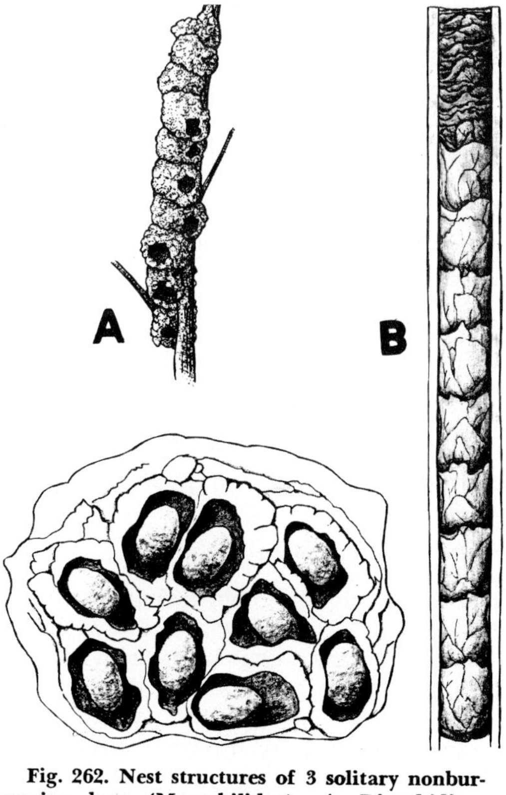

Fig. 262. Nest structures of 3 solitary nonburrowing bees (Megachilidae). *A, Dianthidium concinnum; B, Megachile rotundata; lower left, Osmia* species. [The first is from Stephen *et al.* (1969), after Fischer (1951); the second and third are from Stephen *et al.* (1969).]

The number of solitary bee species is particularly great in the warm, arid, or semiarid deserts of the world, such as in the southwestern United States. Here there are species that fly in the spring and obtain nectar and pollen from herbaceous mesophytes (plants that exist under medium conditions of moisture) that grow and bloom during the brief period when the desert is relatively moist. Other species depend on xerophytes (plants structurally adapted to resist drought) that bloom not only in the spring, but also later in the year. Even entire genera tend to fall into one or the other of these categories. An abundant solitary bee fauna may also be found in mountainous areas with moderate rainfall, varied floras, and well-drained soils favorable for nesting. Agriculture may be beneficial to native bee species that are preadapted to make use of its cultivated flowers, such as cucurbits and alfalfa, but many agricultural practices have reduced the ranges of some native species by destroying their nesting sites, for most of them are ground-nesters (Linsley, 1958).

Some species of solitary bees are important pollinators, at least of certain crops. They have stingers, and sometimes sting people, but the reaction to the stings of most species is mild in comparison with the usual one to the stings of social bees, except in rare cases of specific allergies. One exception is the extremely painful sting of certain solitary bees of the genus *Eulaema,* which are the size of bumble bees but do not occur in the United States (Stephen *et al.,* 1969).

Family Megachilidae

This family includes the leafcutter bees (*Megachile*), mason bees (*Osmia*), and their allies. They are medium-sized, and are usually black, brown, or metallic green or blue in color. The nonparasitic females of the megachilids can be identified by their pollen-collecting hairs (scopae) being located only on the ventral surface of the abdomen (Hurd and Michener, 1955). In other pollen-collecting bees, such hairs are largely on the hind legs or, in a few species, on both the legs and the ventral surface of the abdomen.

Most of the species of leafcutter bees in North America are about the size of the honey bee. To build nests, some females cut out tunnels in rotten or solid wood or in the ground, but most use the hollow stem of a plant, and line the bottom and sides of the nest with circular or oblong pieces of leaves that they have cut neatly and with geometric precision (figure 261). The female then places a paste of pollen and honey in the nest, and lays an egg. This is the first of what is usually a series of cells that are separated by leaf disks (figures 261 and 262). The female of a common southern California species, *Megachile davidsoni* Cockerell, and her work are shown in figure 261.

The circular leaf disks which this solitary bee had cut from plants were neatly packed into the tube to form a series of cells. The rounded base of each cell was inserted into the concavity of the cell below.

The mason bees (*Osmia*) are medium-sized, robust bees, and have a uniformly dark, metallic (rarely black) color. The females construct their nests of sand, clay, or soil, bound together with a glutinous substance, in any available cavity, such as "old snail shells, old galls, the keyholes of doors, apertures in wood, stems, or mortar, or under stones" (Matheson, 1951). Some species build their nests on surfaces, as on a wall or a rock, build a cell (figure 262, *C*) in which to oviposit, and cover the cell with a mixture of clay and saliva. Except for the fact that they substitute clay for leaves, the habits of mason bees are similar to those of leafcutter bees.

Family Colletidae

This is a family of ancient origin, from which all other bees could have been derived. *Colletes* spp. are moderate-sized bees, with an entirely black integument and usually with well-defined hair bands. They are ground-nesters, forming transparent, cellophanelike, broad cells (figure 263, *F*), which Torchio (1965) believed could be appropriately called "cocoons." They are formed from a clear, filamentous, gelatinous material secreted by the Malpighian tubules and excreted through the anus. Species of *Hylaeus* are so distinct from other North American bees that they are easily distinguished. The body is jet black and nearly naked, but has light yellow maculations in nearly all species. The face is triangular, with a long clypeus (Stephen *et al.*, 1969).

Family Halictidae

The halictids are called "miner bees." They are small to medium-sized bees, mostly black, but some with a metallic blue, green, brazen, or cupreous iridescence. A few are mostly or partially red or yellow. Some are communal, in that a few bees work together to dig out an underground tunnel (figure 263, *A, E, H*). However, each female then digs a short lateral tunnel, into which she brings pollen and nectar before ovipositing. Often, many bees nest close together and use a common passageway to the outside. The nests are frequently made in banks along washes or ravines. Other species are semi-social or primitively social.

The alkali bee, *Nomia melanderi* Cockerell, is the most important pollinator of alfalfa in the Pacific Northwest, where it can be distinguished from all other bees by the iridescent, yellow, integumental bands present on the abdominal tergites of both sexes (Stephen *et al.*, 1969). Batra (1970) investigated the behavior of the alkali bee in burrow and cell construction, provisioning, oviposition, and larval development by observing 23 nests in 3 glass-sided chambers. Females, beginning between 3 and 6 p.m., excavated their vertical main burrows, 30 to 40 cm long, by midnight, and the lateral burrows by morning. The burrows were about twice as long as they would be in the harder soils of natural environments. The lateral burrows terminated in cells that were made during the second or subsequent nights at irregular intervals. They were lined with a waterproof oral secretion. The cells were provisioned, usually the following morning, with pollen balls containing 8 to 10 loads of pollen, the first 3 or 4 levels being mixed with nectar. The bee then laid an egg on the pollen ball and closed the cell opening with earth. Eggs hatched in 3 or 4 days, the larva consumed the pollen ball within 5 to 7 days, and the prepupa was formed within 11 to 13 days, compared with 10 to 12 days in the field.

The sweat bees (*Halictus*) are small insects that are sometimes bothersome because they are attracted to perspiration in attempting to obtain moisture. Occasionally, a mild sting may result when attempts are made to brush them away. Even more frequently encountered are tiny bees belonging to the genus *Dialictus*.

Family Andrenidae

The andrenids are sometimes called "mining" or "burrower" bees. They are always somewhat flattened dorsoventrally, and have an oval rather than a parallel-sided abdomen (Stephen *et al.*, 1969). These are small bees, mostly metallic black, but some are reddish or brownish, and all have a dense pubescence on the head and thorax. The females construct their own burrows and cells in the ground (figure 263, *B, C, D*) and provision them with nectar and pollen. Both sexes overwinter in the nests and emerge in the spring.

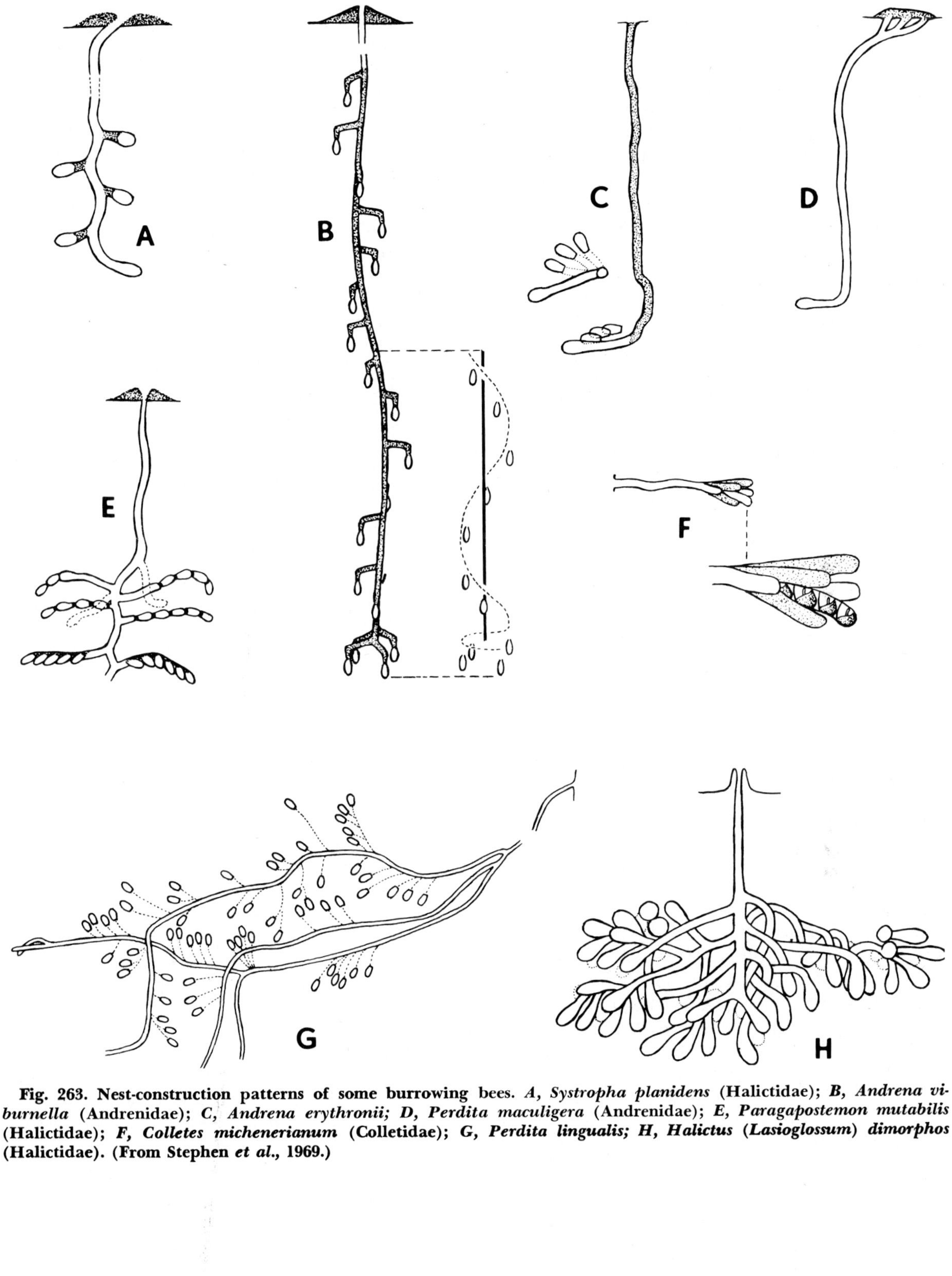

Fig. 263. Nest-construction patterns of some burrowing bees. *A, Systropha planidens* (Halictidae); *B, Andrena viburnella* (Andrenidae); *C, Andrena erythronii; D, Perdita maculigera* (Andrenidae); *E, Paragapostemon mutabilis* (Halictidae); *F, Colletes michenerianum* (Colletidae); *G, Perdita lingualis; H, Halictus* (*Lasioglossum*) *dimorphos* (Halictidae). (From Stephen *et al.*, 1969.)

Family Nomadidae (Cuckoo Bees)

The cuckoo bees are small (10 to 14 mm long), wholly or partially black or reddish, or bright, metallic blue or green, marked with yellow. They resemble wasps in having a wasplike color pattern and sparse pubescence. Most species carry the abdomen in a high or arched position, hold the wings low, and tip the face upward, although these characteristics cannot always be seen on pinned specimens (Stephen *et al.,* 1969).

The cuckoo bees are inquilines in the nests of other bees, often those of the andrenids. In the spring, the females lay eggs in the provisioned nests of their hosts. They lurk about the nest of the host and slip in while she is out foraging for pollen and nectar. The larva of the cuckoo bee is the first to hatch, feeds more rapidly then the host, and may cause the latter to starve, but sometimes there is enough food for both. The inquiline larvae may consume their host's eggs or its larvae.

Cuckoo bees are considered to be harmful because they can destroy pollinating bee species.

Family Anthophoridae

The anthophorids comprise a large family of bees that are intermediate in size between the honey bees and the bumble bees, often black, yellowish brown, or gray, and with thick pile. They nest in burrows in the ground, in banks or, in the case of carpenter bees, in galleries made in wood timbers, provisioning their cells with nectar and pollen. The cells are lined with a thin wax or varnishlike substance.

Carpenter Bees. Formerly considered to be in the family Apidae, the carpenter bees are now regarded as a subfamily of Anthophoridae (Hurd and Moure, 1963). These bees are discussed more fully in chapter 5, because their role as pests, though rather limited, is in connection with their nestbuilding in wooden structures. They merit brief mention at this point principally because they are so easily confused with bumble bees, having the large, robust appearance of the latter. However, except for the males of certain species, they generally do not have the yellow markings on otherwise black bodies that are so characteristic of the bumble bees.

Carpenter bees do not have the pollen baskets of honey bees, but the posterior tibiae and basitarsi are covered with a dense mass of long, simple or branched hairs that are similar to those of other Anthophoridae. Some carpenter bees are said to be efficient alfalfa pollinators (Bohart, 1958). The females have stingers, but will not use them, even around their nests, unless they are handled or accidentally collide with a person.

The Stinger of the Hymenoptera

In some of the stinging Hymenoptera, the tubular ovipositor has lost its original function and has become a stinging apparatus, or "stinger." This is called the "sting" in most entomological literature, but it seems logical to use a term for the stinging apparatus that is different from the verb or the term used for the *effect* produced. The stinger is situated at the tip of the abdomen, and is withdrawn into the body when not in use. It can be used to identify the females, for males do not sting. The stinger consists of 3 parts; the piercing apparatus, the lateral plate and appendages, and the venom sac and glands, comprising a "hypodermic syringe" type of mechanism.

There are Hymenoptera that bite and others that sting. Among the stinging or aculeate Hymenoptera, there are those that sting to paralyze their insect or other prey so that it may be easily transported to the nest and will not "spoil" while the larva is devouring the nonvital parts. Other species sting their prey in order to kill it, and their stinging apparatus is more complex. A combination of acid and alkaline fluids, each from a separate gland, forms the lethal venom.

The Stinger of the Honey Bee

In the honey bee, the stinger is in a chamber at the end of the abdomen, and only the tapering, sharp-pointed *shaft* protrudes (figure 264, *A*). The shaft is composed of 3 parts, the one above termed the *stylet* and the two below, the *lancets*. The lancets are held firmly in place, but can slide freely back and forth. The stylet and lancets form a channel, the *venom canal,* leading from the bulb to the tip of the shaft. The lancets each have a series of 9 or 10 recurved barbs. At *D* and *E* of the figure, the stinger is shown in positions of repose and protraction, respectively. To insert its stinger into the victim, the worker bee bends its abdomen downward and inserts the tip of its outthrust shaft with a sudden jab. The lancets are alternately thrust deeper and deeper into the flesh, an action made possible by the recurved points (barbs) along their sides, which keep them from slipping back (Snodgrass, 1949, 1956). A honey bee can withdraw its stinger from an insect, but in the human skin, the lancets become so firmly anchored by their barbs that they cannot be withdrawn (figure 264, *B*). The bee gets away, but the entire stinging apparatus is left in the skin, as shown in figure 254, and the injured bee soon dies. Reflex action of the stinging apparatus drives the lancets ever deeper into the flesh, and venom continues to be driven out of the sac.

As shown in figure 264, *F,* the acid issues from the gland labeled *PsnGld* and the alkaline part from the gland labeled *BGld.* The mixture of acid and alkaline fluids is a potent venom used in the combat of queen bees or the defense of the hive by workers. A similar venom apparatus enables the predatory Hymenoptera to kill their prey. The queen has a longer stinger than the worker. Her venom glands are well developed, and the venom sac is very large, but the lancets have fewer and smaller barbs. A queen rarely uses her stinger, even under the greatest provocation, except to attack rival queens.

Before a bee can sting, it must first take a firm hold with its claws. An experienced person is often able to feel this, and can brush the bee off before its stinger penetrates the skin. If it has already been inserted, it should be quickly scraped off with a fingernail or knife. No attempt should be made to pick it off with the fingers, for this will result in squeezing the venom sac, causing the venom to be injected into the wound. If itching and swelling result, one should not rub the affected area, for this will only cause increased pain and swelling. Application of cold cloths or ice packs (mentioned farther on) is recommended.

The Stinger of the Bumble Bee

The stinger of the bumble bee does not remain in the flesh of a vertebrate as does that of the honey bee. The female bumble bee can use her stinger repeatedly and effectively. She may be aggressive in defense of the nest, but most species soon cease to attack. Away from the nest, the female is tolerant and timid. Only 1 victim in 120 cases of Hymenoptera stings studied by Mueller (1959*b*) was stung by a bumble bee. Bumble bee attacks are most likely to occur if a person inadvertently steps on a nest. The species of the

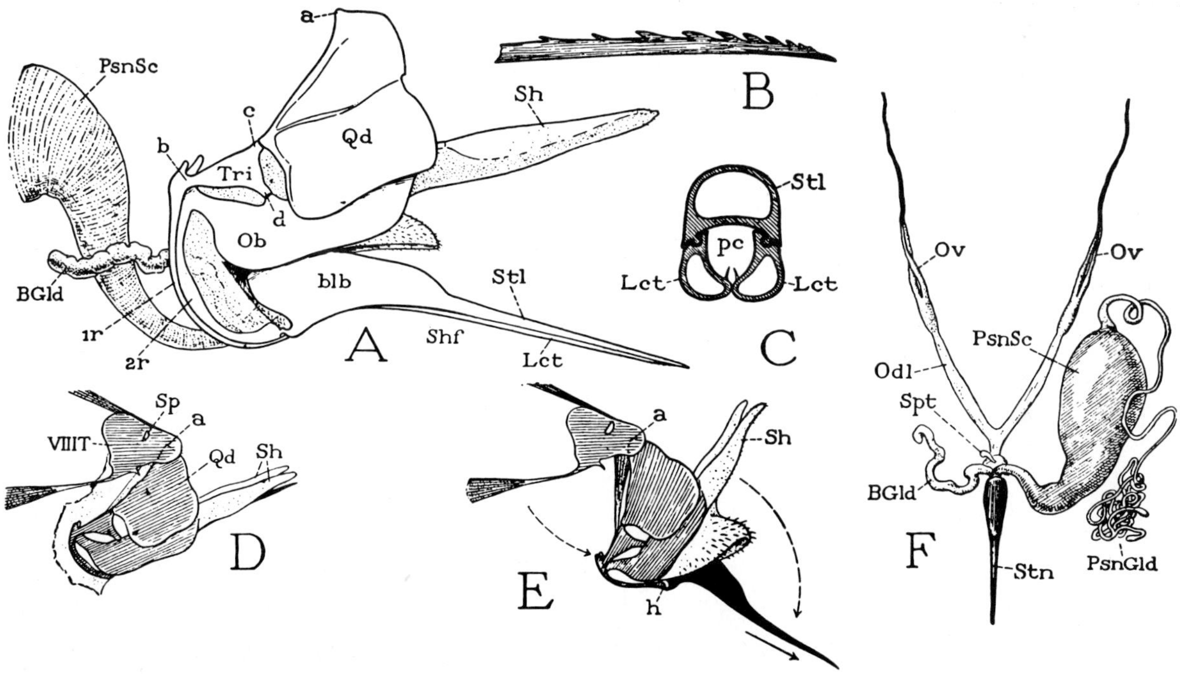

Fig. 264. The stinger of a worker honey bee. *A,* entire stinging apparatus, left side; *B,* end of a lancet; *C,* cross section of shaft of stinger; *D,* stinger in position of repose, suspended from wall of stinger chamber between spiracular plates (*VIII T*); *E,* stinger in position of protraction (arrows indicate the 2 essential movements); *F,* reproductive organs of a worker, together with shaft of stinger, stinger glands, and venom sac; *a,* attachment of quadrate plate with spiracular plate; *b,* apex of triangular plate continuous with lancet; *BGld,* "alkaline" gland of stinger; *blb,* bulb of stylet; *c,* hinge of triangular plate on quadrate plate; *d,* hinge of triangular plate on oblong plate; *h,* hinge of bulb; *Lct,* lancet; *Ob,* oblong plate; *Odl,* lateral (paired) oviduct; *Ov,* ovary; *pc,* venom canal; *PsnGld,* venom gland of stinger; *PsnSc,* venom sac; *Qd,* quadrate plate; *1r,* basal arm (ramus) of lancet; *2r,* basal arm of bulb and stylet; *Sh,* sheath lobes; *Shf,* shaft of stinger; *Sp,* spiracle; *Spt,* spermatheca; *Stl,* stylet; *Stn,* stinger; *Tri,* triangular plate; *VIII T,* spiracular plate associated with stinger base. (From Grout, 1949.)

subgenus *Fervidobombus,* as well as the species *Bombus crotchii,* are said to be somewhat more persistent in their attacks than other California bumble bees.

Two days after being stung twice by *Bombus sonorus,* the author was again stung 6 times on exposed areas on an arm and a leg while attempting to find a nest, using a bee suit for protection. Within a half hour, a swollen, red area 4 to 5 cm in diameter had developed around each dark-red spot that indicated a point of entry of a stinger. By evening, the swollen areas had become very sensitive to pressure, and fatigue and loss of appetite appeared to be associated with the envenomation. By morning, the red areas had spread even farther, but the swelling and sensitivity to pressure had largely subsided. The shoulder muscles of the arm that had been stung twice in the forearm area were slightly sore when exerted. A colleague, stung on the ankle for the first time, reported increasing swelling over a 2-day period; his ankle finally became very badly swollen and itched severely. Two weeks earlier, the author had been stung about a dozen times by yellowjackets (*Vespula pensylvanica* and *V. atropilosa*) over a 5-day period, suffering only minor reactions after the sharp initial pain. The colleague just mentioned had likewise received multiple stings of yellowjackets and hornets during the preceding months, but their effects were relatively mild compared with the sting of *Bombus sonorus.*

Hymenopterism

Since wasps and bees are responsible for about 75% of the deaths that occur each year in the United States from arthropod envenomations, it may be appropriate at this point to quote from the classification by Fluno (1961) of 3 types of reactions of differing severity resulting from stings of these insects. These are named "hymenopterism," from the order Hymenoptera, to which ants, wasps, and bees belong.

Hymenopterism vulgaris. In the majority of instances, the effects of a wasp sting are painful, but not serious or lethal, except when the driver of a vehicle loses control in sudden panic. Actual pain felt varies with the individual victim and with the genus or species of wasp, but generally an intense, immediate pain is felt at the site of the wound. Localized reddening and swelling also almost always occur. Depending upon the site affected, there may be interference with normal function; for example, a sting at a finger joint often results in stiffness, with consequent impairment of movement. The pain, swelling, and stiffness may last for only a few minutes or for 1 or more days.

Hymenopterism intermedia. In this arbitrarily classified type of reaction, swelling includes considerable tissue beyond the actual site of the sting, for example, an entire arm or leg. This intermediate reaction may last for several days, but it is not necessarily lethal. However, in this category should be placed cases in which the site of the sting is the tongue, neck, or throat, with occasional involvement of the glottis and consequent impairment of swallowing and breathing.

Hymenopterism ultima. This is the lethal or near-lethal reaction, quickly diagnosed by the alert physician as anaphylactic shock. Generally, the patient goes into shock within 10 to 20 minutes after receiving the sting. Hives may be present, breathing is shallow, pulse and heartbeat are almost undetectable. Profuse sweating generally occurs, and the patient becomes unconscious and obviously in need of medical assistance.

Of the 3 types of hymenopterism, the "vulgaris" type seldom prompts the victim to consult a physician, and then only for relief from pain. He may be reacting to only a part of the venom. The victim with an "intermedia" type may be reacting to more of the venom components (enzymatic materials, proteins and proteinlike materials, histamine, 1 or more spreading constituents, and possibly other compounds).[21] Many persons with the intermedia-type reactions do not consult a physician, but they should, for their sensitivity is indicated. Sometime in the future, and possibly with the very next sting, such a person is likely to demonstrate hymenopterism ultima.

Based on published reports, Barr (1971) concluded that the causes of death from hymenoptera-sting allergy, in descending order of frequency, had been obstruction of the respiratory tract, acute vascular collapse, and vascular or hemorrhagic reactions, and that death had also been caused by various types of delayed reactions, frequently involving the central nervous system. Also, that in many older persons, an associated heart ailment had been a complicating factor.

[21] At least 8 fractions have been isolated from bee venom. It contains histamine, but this is not considered to be the major pharmacological component. Two fractions, F_1 and F_2, are basic protein components, with probable molecular weights of 35,000 and 44,000, respectively. Melittin (F_1) accounts for most of the local and general toxicity, although the total effect is from the combined action of all components of the venom (Marshall, 1957). Melittin contains 13 amino acids, none of which contain sulfur. F_2 is a slower-migrating fraction than F_1, but supplements the effects of the latter, mainly through the 2 enzyme components. Habermann (1972) listed histamine and dopamine as the biogenic amines contained in bee venom, and the same compounds plus serotonin in wasp venom. He listed melittin, apamin, MCD-peptide, and minimine as the protein and polypeptide toxins (nonenzymatic) in bee venom, and "wasp kinin" in wasp venom. Among the enzymes, he listed phospholipase *A*, phospholipase *B*, and hyaluronidase as occurring in both bee and wasp venoms.

Hymenoptera venom generally contains histamine, serotonin, acetylcholine, kinins, a hemolysin, a neurotoxin, a convulsant factor, hyaluronidase, other enzymes, and amino acids. Formic acid presumably accounts for most of the pain (Mueller, 1965). The LD_{50} of bee venom for 2 strains of mice was 1.75 mg/kg (intravenous) and 3.50 mg/kg (subcutaneous). Intravenous injection of dried bee venom in a saline solution produced a fall of blood pressure in dogs and monkeys in 35 seconds. Animals that survived returned to control levels within 5 to 30 minutes. Death resulted primarily from respiratory failure associated with some cardiotoxic action (Brooks and Vick, 1972). The most toxic component of bee venom is phospholipase *A*, which caused the death of a dog by respiratory paralysis in 10 minutes. It is similar to the phospholipase *A* obtained from Indian cobra venom, known to be a neurotoxin. Bee venom also contains melittin and apamin, which increase plasma cortisol. This could explain, in part, the observation that bee venom or some of its components alleviate the symptoms of arthritic-like conditions. This is a finding of special significance in that it suggests the possibility that fractions of bee venom might be used to control some of the symptoms of arthritis and thereby circumvent the high toxicity of whole bee venom (Vick and Brooks, 1972).

The immediate pain from a bee sting is of short duration; the swelling and itching that follow are the most disturbing. In some persons this does not occur, but susceptible or sensitized persons are so severely affected that they may be confined to bed for days. The combined stings of many bees, or the sting of a single bee in a sensitized person, may result in red blotches on the skin, shortness of breath, nausea, fainting, and even death. Even though bee and wasp venoms ordinarily have a relatively low order of lethality, they are great hazards to sensitized persons. This is a good example of severe autopharmacological action. The variability in the reactions of different people to insect stings is enormous. A beekeeper, whose wife and son suffered anaphylaxis whenever stung by bees, stated that he was stung without symptoms 200 to 300 times every summer. He rarely suffered generalized symptoms, and these only when stung on the scalp or neck. He would then develop generalized redness, pruritis, and dizziness. He would be well after lying down in the field for a half hour. He suffered no reaction when stung on the extremities, and was convinced that the development or absence of symptoms depended on the site of the sting (Brown and Bernton, 1971).

The potency of honey bee stings varies with the time of year and with variation in the plants the bees are visiting. In California, the sting is most potent when bees are visiting California buckeye, *Aesculus californica*, the pollen and nectar of which are deleterious to both adults and brood (Eckert, 1940).

The Stinger Pheromone

When the bee stinger is protruded, it gives off a sweet scent reminiscent of the odor of banana oil. When the stinger is detached, as is ordinarily the case after a vertebrate animal is attacked, this odor is much greater. It is absent from the stingers of very young workers or queens. Because of the alarm and attack behavior which it releases in bees, the odor is considered to be a pheromone (Karlson and Butenandt, 1959). One of the constituents was found to be isoamyl acetate (banana oil), present in the amount of approximately 1 μg per stinger. When mixed with paraffin and absorbed in cotton balls, the isoamyl acetate alerted and agitated guard bees. Cotton balls that contained odoriferous stingers not only agitated guard bees, but incited them to attack, an additional reason for removing the stinging apparatus from the skin as rapidly as possible (Boch *et al.*, 1962).

First Aid for Stings by Hymenoptera

When stung by a bee, remove the stinger, carefully scraping it off with a fingernail or knife blade to avoid squeezing more venom into the wound. When stung by any Hymenoptera, wash the site of the sting with soap and water and apply an ice pack to lessen local swelling. If infection, gangrene, or neuritis occur as complicating factors, additional appropriate local care is required. Hypersensitive persons should carry an emergency kit that can be used in the event of a sting and subsequent reaction while seeking professional medical aid. (See under "Treatment of Arthropod Bites and Stings" at the beginning of this chapter.)

VENOMOUS SNAKES

As stated earlier, although this chapter deals principally with venomous arthropods, snakes are included because there are practical advantages in discussing all venomers in the same chapter. Snakes are the last of the animals to be discussed in this chapter that inject their venom by means of fangs or stingers.

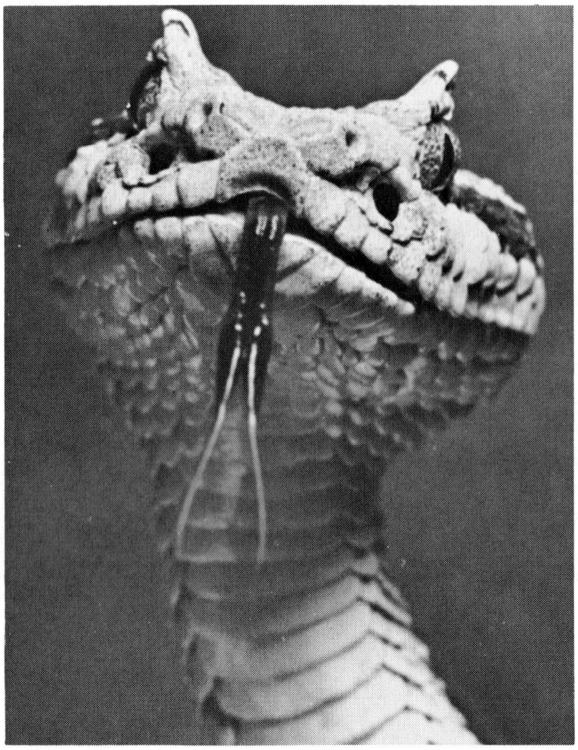

Fig. 265. Head of the sidewinder, *Crotalus cerastes*, showing the facial characters of the pit vipers and the "horns" of the sidewinders.

There are some dangerously venomous snakes throughout the United States, although they are extremely rare in Maine, New Hampshire, and most of Vermont, where there are only occasional reports of rattlesnakes in the southern part of the state. In North America north of Mexico, the pit vipers, which include the cottonmouth moccasin, the copperhead, and about 20 species of rattlesnakes, are the principal venomous snakes. There are also 2 species of coral snakes.

There are certain snakes that are venomous, but harmless to man. Their venom glands and fangs are located in the back of the upper jaw, instead of in the front as in pit vipers. Also, the fangs are short and grooved instead of long and hollow, and they must be embedded in small prey. The prey must be chewed instead of bitten, to get the venom into the wound. These "back-fanged" species occur in the desert areas of the Southwest.

There are also nonvenomous snakes that will bite. A gopher or bull snake, for example, coils and vibrates its tail and hisses loudly when cornered, and repeatedly lunges at any intruder. Many are killed in the mistaken belief that they are dangerous; they are actually beneficial, being effective predators against rodents. Klauber (1956) cited an instance of a man being bitten by a gopher snake (*Pituophis catenifer annectans*) which he thought was a rattler, and almost dying from shock. Many similar stories of near death from fright are recounted by various authors.

Pit vipers are so called because on each side of the head, between the eye and the nostril, there is a deep pit (figure 265), a heat-sensitive structure, that is more prominent than the nostril itself. The pupil of the eye is elliptical. The tongue of the pit viper and other snakes is narrow, flexible, ribbonlike, and has a forked tip, as shown in the figure. Even when the mouth is closed, the tongue can protrude through a notch in the lower jaw, and aid in smelling by bringing chemical stimuli to Jacobson's organs in the nasal cavity.

The pit vipers have a pair of long, hollow fangs anteriorly in the upper jaw. The fangs have an orifice at the tip, like a hypodermic needle. By a combination strike and bite, or by a deliberate bite alone, the fangs are driven into the flesh of the snake's victim. Then, a muscle over each venom gland (figure 266) forces venom out of these hollow fangs into the dual wounds produced. Pit vipers strike very rapidly from an S-shaped posture of the body, and instantly return to this defensive position. In order to thrust the fangs forward into the flesh of its victim, the snake must open its mouth as widely as possible (figure 267). When the jaws are closed, the fangs

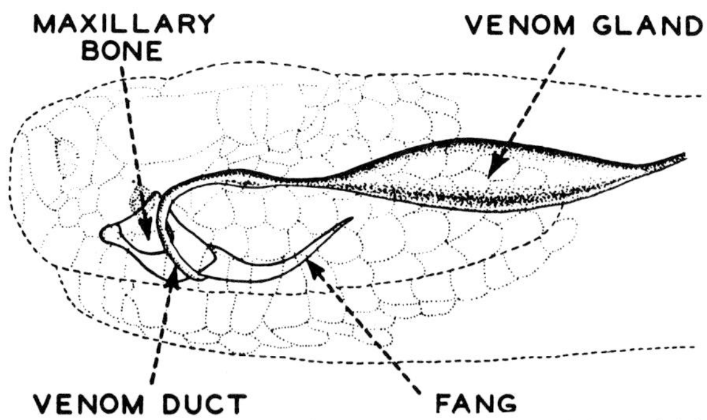

Fig. 266. Relative positions of venom gland, duct, and fang in the head of a rattlesnake. (From Klauber, 1956.)

fold back against the roof of the mouth (Klauber, 1956). Pit vipers can strike while crawling, for if they are alarmed the neck is usually kept laterally looped, and they can also turn and bite without darting the head from a loop.

The teeth of snakes are too thin and pointed to tear, shred, and masticate their food. Snakes

Fig. 267. *Crotalus atrox* with the mouth widely opened and the fangs advanced, as at the end of the forward drive of a strike. (From Klauber, 1956.)

swallow their food without mastication. The venom is of importance not only for procuring prey, but also for digesting it, and may be thought of as a venomous saliva (Klauber, 1956).

Snakes, like amphibians, arthropods, and many other animals, are poikilotherms (coldblooded animals), that is, they have a variable body temperature, usually slightly higher than the temperature of their environment. They cannot regulate their body temperature as do the homoiotherms (warmblooded animals), which have a unform body temperature maintained almost independently of the environmental temperature. If it is not too hot, rattlesnakes will hunt during the daytime, but if it is very hot, as in desert areas during the summer months, they hunt at night. If rattlesnakes hide during the daytime to protect themselves from the weather, the danger from bites is not eliminated, for bites are often received when people reach behind or under objects such as rocks, logs, or shrubs where the snakes hide.

Rattlesnakes (Crotalidae)

Rattlesnakes are found only in the Western Hemisphere. The 2 genera, *Crotalus* and *Sistrurus,* both occur in the United States, and are distinguished by the size and arrangement of the scales that cover the forward half of the top of the head (figure 268). In both genera, the head is broad and distinct from the narrow neck. Rattlesnakes may mate once a year or every second year. The females do not lay eggs, but give birth to active young which generally appear in late summer or early fall, depending on the species and on temperature. Rattlesnakes may attain an age in excess of 20 years (Klauber, 1956). *Crotalus* contains the most species and the largest and most dangerous snakes, and these range over a much greater territory. Species of *Sistrurus,* called massasaugas and pygmy rattlesnakes, do not occur in the Pacific Coast states.

Rattlesnakes are distinguished from other pit vipers, and in fact from all other snakes, by their possession of the rattle, a loosely articulated but interlocking series of horny rings at the end of the tail. The rattle makes a hissing sound when it is used. The flat side of the rattle is perpendicular to the ground, not parallel as is often believed. A slight upward tilt of the end of the tail keeps the rattle clear of the ground when the snake is crawling. The sound of the rattle is caused by the clashing of the successive lobes against each other. On the coiled snake, the rattle string is pointed almost vertically upward—it is tipped backward about 10° from the vertical. The vibration is always transverse or lateral and to the flat or side face of the rattle (Klauber, 1956). The average vibration frequency for 16 prairie rattlesnakes (*Crotalus viridis viridis*), as determined by a mercury-arc stroboscope, varied from 21.1 cycles per second at 10° C (50° F) to 98.6 cycles per second at 40° C (104° F) (Chadwick and Rahn, 1954). Rattlesnakes sound their rattles when disturbed or frightened, as by some movement or the approach of an intruder.

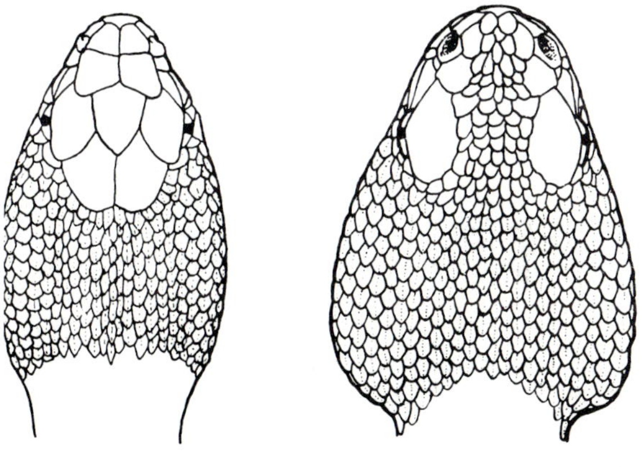

Fig. 268. Size and arrangement of the scales that cover the forward half of the head of the massasauga, *Sistrurus catenatus* (*left*), and the western diamondback rattlesnake, *Crotalus atrox.* (From Klauber, 1956.)

There is no basis for the common belief that the age of a rattlesnake is indicated by the number of segments on the rattle. A new rattle segment is added with each shedding of the skin, which for adult snakes in nature may be from 1 to 3 times per year, depending on the climate. However, segments can be accidentally broken off, particularly in nature. From 5 to 8 segments are commonly seen on wild rattlesnakes. Complete strings of 15 or 16 segments are not exceptional among caged rattlers, but the greatest number Klauber ever found on a wild rattlesnake was 13 (on 2 specimens), and he was also given a wild eastern diamondback rattler with a rattle of 17 segments. When first born, rattlers possess a "button" instead of a rattle. These young rattlers are venomous.

Another erroneous belief about rattlesnakes is that they lay eggs. Some snakes lay eggs, but rattlesnakes give birth to active young. There are many myths about snakes, including rattlers, among them the belief that an injured snake cannot die before sundown, or that the head, if cut off, will live for 10 days or more. However, reflex body movements do continue long after a snake has been fatally injured or even decapitated. Heartbeats in different species of rattlesnakes, when decapitated, have been recorded for periods ranging from 15 to 59 hours (Klauber, 1956). The severed head can bite by reflex action, and should be handled with care. Klauber experimented with severed heads of 13 rattlesnakes of 6 different forms. He found them to be dangerous for at least 20 minutes and sometimes for almost an hour. Continued life was manifested in such ways as opening the mouth at the approach of a hand or other object or by biting a stick thrust into the mouth. Serious accidents have been caused by careless handling of severed rattlesnake heads.

There appears to be some scientific basis for the belief that horsehair ropes repel rattlesnakes. These reptiles manifest a strong fear reaction to *N*-butyl mercaptan, which simulates the human body odor, and any odor-absorbing rope, or even clothing that has been worn, should serve as a barrier (Cowles and Phelan, 1958).

There are 59 named forms of the genus *Crotalus,* 26 of which occur in the United States, grouped into 13 species (Cockran and Goin, 1970). Experienced herpetologists can distinguish the various species and subspecies of rattlesnakes with moderate success on the basis of patterns and colors, but there are variables even within a subspecies that may, for example, be gray in one part of its range and pink in another. Scale arrangements are more consistent, and are the principal bases for keys to species and subspecies. Only a superficial discussion of some of the better-known rattlesnakes of the United States will be attempted here. For keys and detailed information, Klauber's (1956) monumental 2-volume treatise on rattlesnakes, published by the University of California Press, is recommended. A pictorial key to the venomous snakes of the United States has been developed as a part of the CDC Training Program of the U. S. Public Health Service (CDC, 1966).

Species of *Crotalus*

Best known among western rattlesnakes is the western diamondback, *Crotalus atrox* Baird and Girard. It ranges from Arkansas and Oklahoma south to central Mexico and west to California. Many adults are between 3 and 5.5 ft long, with a maximum of 7 ft (90, 170, and 215 cm). They often may attain a length of 6 ft (185 cm) and a weight of 15 lb (7 kg), second only to the eastern diamondback in size. The western diamondback causes more fatalities than any other snake in the United States, owing to its abundance over a wide area. It usually avoids dense forests, swamps, and elevations above 5,000 ft (1,500 m) in the United

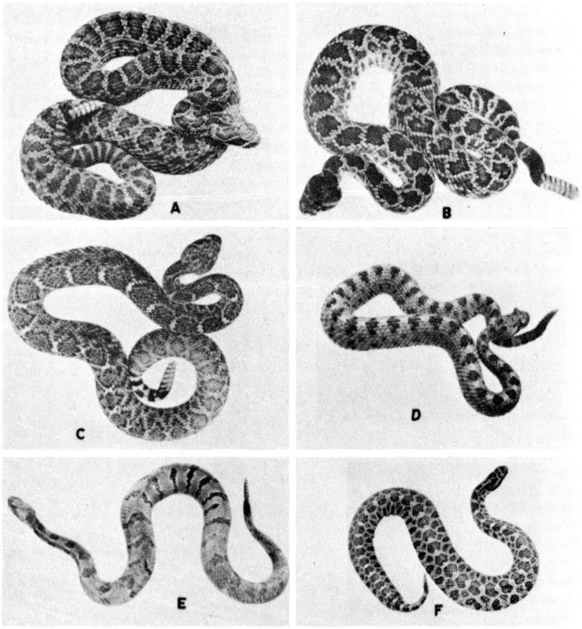

Fig. 269. Some rattlesnakes of the United States. *A*, prairie rattlesnake, *Crotalus viridis viridis; B*, southern Pacific rattlesnake, *Crotalus viridis helleri; C*, western diamondback rattlesnake, *Crotalus atrox; D*, Sonoran desert sidewinder, *Crotalus cerastes cerastes; E*, timber rattlesnake, *Crotalus horridus horridus; F*, eastern massasauga, *Sistrurus catenatus catenatus*. (From Klauber, 1956.)

States, but has been found up to 8,000 ft (2,400 m) elevation in Mexico. In the northern part of their range, the western diamondbacks congregate in large numbers during hibernation, as do some other rattlesnakes, such as the subspecies of *C. viridis*. The western diamondback is inclined to prowl into open places, and is particularly dangerous in its abrupt defense when approached. If suddenly surprised, it quickly throws its body into a coil, sounds the rattle, and strikes.

This species has a chain of symmetrical blotches (rhombs) on a grayish ground color (figure 269, *C*.) In barren areas, it is pallid, with obscure blotches, and varies from gray to yellow or even pinkish. In common with several closely allied species of the Southwest, its tail is vividly marked with black-and-white rings.

The red diamondback rattlesnake, *Crotalus ruber ruber* Cope, ranges throughout certain areas of southwestern California, and is largely confined to rocky hillsides with scrubby vegetation. It is 4 to 5 ft (120 to 150 cm) long, and has a distinctly reddish coloration, with rhomblike markings (figure 270, *A*). It is an unusually gentle species, and shows little excitement if disturbed in protected places, but if found wandering in the open, it will quickly coil and assume a threatening attitude. The venom of the red diamondback is much less toxic than that of the eastern and western diamondbacks, and fatalities from its venom are rare (Navy, 1965).

The eastern diamondback rattlesnake, *C. adamanteus* Beauvois, is a lowland species, ranging through the coastal plains of North Carolina south of Albemarle Sound, South Carolina, Georgia, Florida, Alabama, Mississippi, and southeastern Louisiana. It is the largest of the rattlesnakes. Most adult specimens are 3.5 to 5.5 ft long, but they commonly attain a length of 6 ft; Ditmars (1931) measured specimens 8.3 and 8.5 ft long (110, 170, 185, and about 260 cm). The venom glands of the eastern diamondback hold more venom than those of any other rattlesnake, and probably more than those of any other snake in the world. This species is the only rattlesnake in its range that has distinct, diagonal, whitish stripes on the side of its head. It has a bold and constant pattern in the form of a chain of symmetrical blackish rhombs (plate VII, 8), with light centers of "diamonds" edged with paler hue against a background body hue of dark olive or olive-brown. Ditmars also stated that the food of the adult eastern diamondback rattler consisted largely of rabbits, but that they also ate quail,

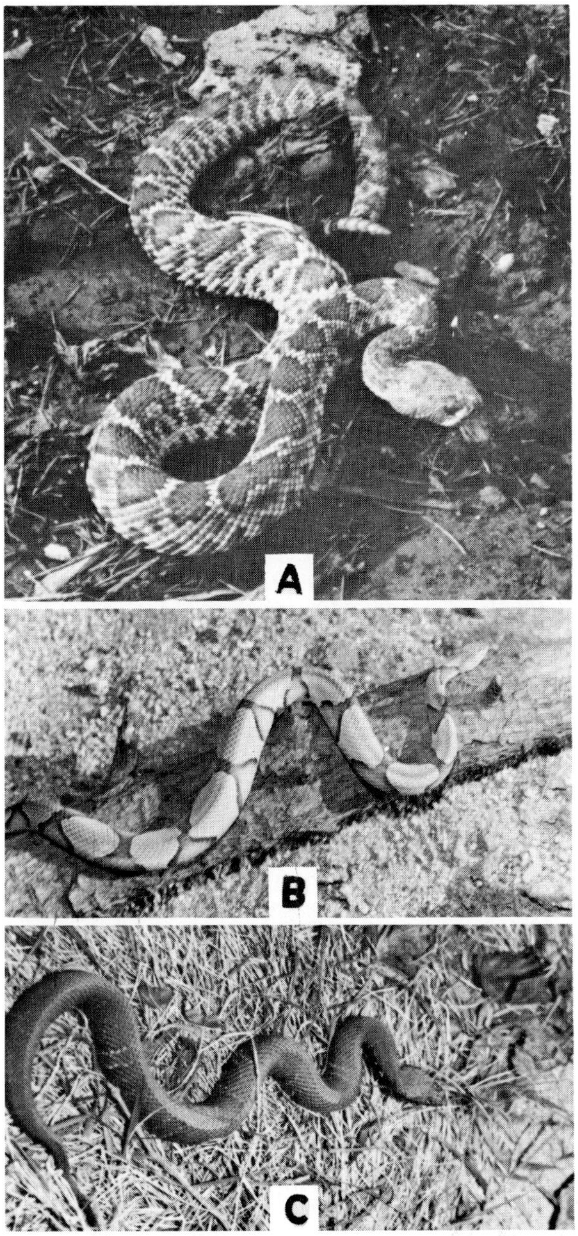

Fig. 270. Some pit vipers of the United States. *A*, red diamondback rattlesnake, *Crotalus ruber ruber; B*, copperhead, *Agkistrodon contortrix; C*, eastern cottonmouth moccasin, *Agkistrodon piscivorus piscivorus*. (Courtesy F. E. Russell, M.D.)

and that they gave birth to young in August and September, with 8 to 12 in a litter. The young of a captive female were 35 cm long.

The prairie rattlesnake, *C. viridis viridis* (Rafinesque) and its related subspecies, have the widest range of any North American rattlesnakes. The subspecies *viridis* (figure 269, *A*) ranges throughout the Great Plains, from longitude 96° W. to the Rocky Mountains and from southern Canada to northern Mexico. This yellowish to brown snake is usually less irritable than the diamondbacks, and is less likely to make a determined defense. The subspecies *C. v. oregonus* Holbrook ranges throughout the Pacific Slope, from British Columbia to San Luis Obispo and Kern counties in California. In California mountains, it occurs up to 11,000 ft (3,300 m) elevation. It varies in patterns and hues, from brown to grayish or greenish, or even to blackish specimens with no markings but the pale edges of what would have been dorsal blotches on lighter specimens. In its southern range, *C. v. oregonus* intergrades with the closely related *C. v. helleri* Meek (figure 269, *B*), which ranges throughout southern California and northern Baja California west of the desert, and on Santa Catalina Island.

The southwestern speckled rattlesnake, *C. mitchelli pyrrhus* (Cope), ranges through southern California, southern Nevada, and western Arizona, but is absent from the San Gabriel Mountains and the Los Angeles coastal plain. Its usual hues are yellowish, gray, or even pinkish, but some specimens are almost white.

The Mojave rattlesnake, *C. scutulatus* (Kennicott) ranges from west Texas northwestward to the Mojave Desert of California, and inhabits principally the desert and the prairie–desert transition zone. It is similar to the western diamondback and the prairie rattlesnakes in pattern and general appearance. Its general color is often greenish or olive. The adult is smaller than the species discussed earlier, generally 2.5 to 3.3 ft long, with a maximum length of about 4 ft (75, 100, and 120 cm). The approximate yield of venom (on a dry-venom basis) is 50 to 90 mg compared with 175 to 320 mg for the western diamondback. However, the LD_{50} of the venom is 0.23 mg/kg compared with 3.71 mg/kg for the western diamondback, i.e., about 16 times more toxic. (Navy, 1965; Russell, 1967*b;* Russell and Puffer, 1970). In fact, its venom is more toxic and has a greater effect on respiration than that of any other North American rattlesnake. There is frequently little local reaction to the bite, so it may not be considered serious until the victim suffers severe respiratory difficulty (Navy, 1965).

Sidewinders are subspecies of *C. cerastes* Hallowell. They may be found in the deserts of southeastern California, southern Nevada, southwestern Utah, and western Arizona, usually in sandy flats and dunes with sparse vegetation, but sometimes on arid, rocky hillsides. They are active at night, and during the day they may be resting with part of the body under sand. Their common name derives from their sidewinding type of movement, resulting in characteristic tracks in the sand (figure 271). Sidewinders are usually 1.5 to 2 ft long, but may attain a length of about 2.5 ft (45, 60, and 75 cm). The background color is cream, tan, gray, light brown, or pinkish, and there are also rows of darker spots (Navy, 1965). They may be identified by the elevated "horns" above the eyes (figures 265; 269, *D*), which are enlarged supraocular scales that appear to function as "eyelids." Pressure on the horns causes them to fold down in a perfect fit over the eyes, probably protecting them from abrasion by such objects as creosote bush roots, rocks, and gravel as the snake passes through burrows (Cohen and Myres, 1970). Increasing numbers of bites by sidewinders are being recorded coincidentally with the ever-growing use of desert areas for residential and recreational purposes. However, there are few fatalities, because the quantity of venom is small (18 to 40 mg of dry venom).

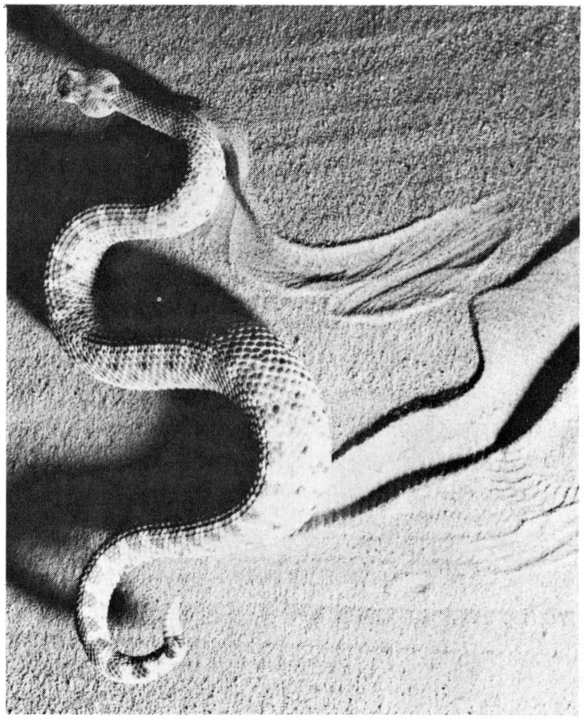

Fig. 271. A sidewinder, *Crotalus cerastes*, showing typical crawling posture and tracks left in sand.

The timber rattlesnake, *C. horridus horridus* L., ranges throughout the northeastern and north central United States, south as far as northern

Georgia and Alabama, and west as far as southeastern Nebraska to northeastern Texas. However, its distribution is intermittent, for it has been exterminated in many areas over which it once ranged, particularly in industrial and agricultural districts, but persists where rocky or wooded retreats are available in adjacent mountains and forests (Klauber, 1956). It averages slightly under 4 ft in length, although specimens around 6 ft have been found (120 and 180 cm). It is typically yellow or tan in color, with wavy crossbands of dark brown or black (figure 269, *E*), although many specimens have a noticeable blackish hue, and some from upland areas are almost entirely black. In the flat, forested country back of the central coast, often damp and with large sections covered with sphagnum moss, the timber rattler is a little larger than the mountain type. It also has a somewhat different color pattern, being grayish, with strongly contrasted black bands, and a faint rusty, dorsal stripe, several scales wide. It resembles the canebrake rattlesnake, *C. h. atricaudatus* Latreille, which inhabits the lowlands of the south Atlantic and Gulf states and the lower Mississippi Valley. The latter species can attain a length of close to 8 ft (240 cm), and has a distinct reddish band along the back (Ditmars, 1931). The timber rattlesnake tends to congregate to bask and hibernate on rocky bluffs and ledges, and this has facilitated the extermination of the species in populated areas. These snakes are mild-tempered, and do much preliminary rattling and feinting before striking. Along with the copperheads, they have been used in rituals by the snake-handling cultists of the southern mountains, who are often bitten and usually obtain no medical care. By 1965, there had been at least 20 fatalities among these snake-handlers (Navy, 1965).

Species of *Sistrurus*

The eastern massasauga, *Sistrurus catenatus catenatus* (Rafinesque), is one of two species of the genus in North America. The tops of the heads of *Sistrurus* species have large, symmetrical shields (figure 268) like the moccasins. Unlike the species of *Crotalus,* they frequently feed on frogs, as do the moccasins. The eastern massasauga (figure 269, *F*) ranges from central New York westward to Nebraska and Kansas. It is a small, brownish or grayish rattler, with chestnut-brown blotches on the back and a similar row on each side. It is 2 to 3 ft (60 to 90 cm) long. It frequents the vicinity of swamps where frogs abound, but shuns actually wet places. *Sistrurus c. catenatus* merges with the western massasauga, *S. c. tergeminus* (Say), in southeastern Nebraska and northeastern Kansas, and the latter ranges throughout the southwestern plains as far as extreme southeastern Arizona. The massasaugas usually remain quiet or try to escape, but bite readily when angered. There have been well-authenticated cases of bites that were fatal to man.

The pygmy rattlesnakes, *Sistrurus miliarius* (L.), may be found in pine woods and grassy marshes in the southern part of their range, and in low, rocky, wooded hills in the northwest part. There are 3 subspecies, ranging from North Carolina to Florida and westward to southern Missouri, Arkansas, southeastern Oklahoma, and eastern Texas (Klauber, 1956). The pygmies are seldom more than 20 in. (50 cm) long, and are grayish, with a series of rather widely separated, rounded blotches. A reddish band along the back becomes increasingly vivid anteriorly. The tiny rattle cannot be heard more than 2 m away. These snakes are alert and bad-tempered. Their bite causes severe pain and extensive swelling, even from specimens only 6 to 9 in. (15 to 23 cm) long, but there are no well-documented cases of fatalities (Navy, 1965).

Moccasins (Crotalidae)

Among the pit vipers are 2 species of moccasins, the copperhead and the cottonmouth or water moccasin. Like *Sistrurus,* they have large plates on the crown of the head, but they have no rattles and have tapering tails.

The 4 subspecies of the copperhead, *Agkistrodon contortrix* (L.), may be found from Massachusetts to northern Florida, west to eastern Kansas and Nebraska, and south to beyond the Pecos River in Texas. Copperheads are pale brown, pinkish, or light reddish brown, with a series of large, chestnut-brown blotches on the sides, somewhat like inverted Y's (figure 270, *B*). Some of the markings unite dorsally, producing a continuous, transverse pattern, with the central portion narrow and broadening on each side in hourglass fashion. The top of the head is without marking, and is often slightly paler than the body. This species usually has a pale pinkish-brown undersurface, with a row of dark spots on each side. A moccasin 3 ft (90 cm) long is considered to be a large specimen, but occasionally 4-ft (120-cm) specimens are seen (Ditmars, 1931). Copperheads may survive in suburban areas where larger snakes have been killed off (Minton, 1970).

The young are born during August and early September, 6 to 9 in a litter. Their tails are bright sulfur-yellow. In the broadbanded copperhead, found from Texas to Kansas, the tail retains its yellow color. This subspecies has fewer and wider bands, not narrowed at the top as in the eastern moccasins.

When the copperhead fears it is about to be attacked, it will rapidly vibrate its tail which, if the snake is among dried leaves, produces a distinct buzzing sound readily heard for about 5 m. It will seldom attempt to strike unless annoyed, attacked, or stepped on. Most cases of copperhead bites occur when persons wearing low shoes step on the snake and are bitten on or near the ankles. Copperheads are said to account for most of the snake bites in the eastern United States, exclusive of Florida and the Mississippi delta, but fatalities are almost unknown (Navy, 1965).

The eastern cottonmouth moccasin, *Agkistrodon piscivorus piscivorus* (Lacepède), ranges from southeastern Virginia to southern Florida, and the western cottonmouth moccasin, *A. p. leucostoma* (Troost) ranges through southern Texas and the Mississippi Valley north to southern Illinois (Klauber, 1956). The cottonmouth moccasin is also called water moccasin because of its commonly aquatic habits. Over much of its range it lives along streams and lakes, in swamps, or along rice ditches. The term "cottonmouth" derives from the whitish mouthparts, seen when the snake opens its mouth in threatening fashion. The cottonmouth is a much larger, more pugnacious, and more dangerous snake than the copperhead. It is usually 3 to 4 ft long, but may attain a length of 6 ft (90, 120, or 180 cm). It is dull brown or olive in color, crossed with darker, usually obscure, bands (figure 270, *C*). The young are more brightly colored, with transverse blotches on a reddish background.

The cottonmouth is often confused with nonpoisonous snakes of the genus *Natrix* that likewise have semiaquatic habits. However, the cottonmouth has a much heavier head, and its eyes are less prominent. The cottonmouth is frequently belligerent, and throws back its head with mouth widely open and vibrates its tail. Nonvenomous snakes almost always try to escape when alarmed (Navy, 1965).

Although bites by cottonmouths are fairly frequent in the Mississippi Valley and along the Gulf Coast, deaths are rare. However, the venom has a strong proteolytic activity, and may cause severe tissue destruction. No species-specific antivenins are available for the cottonmouth and copperhead, but Polyvalent Crotalid Antivenin is recommended (Navy, 1965).

Coral Snakes (Elapidae)

The coral snakes of North America are related to the cobras, kraits, mambas, and dangerous Australian species, all in the family Elapidae. The elapids have short, rigid, and permanently erect fangs in the forward part of the upper jaw. The head is small, and is not distinct from the neck, and the snout is rounded. In the New World, there is only a single genus, *Micrurus*. In the United States, there are 4 subspecies of *M. fulvius*, ranging from North Carolina to Florida and as far west as southern Arizona (Klauber, 1956). Coral snakes are usually less than 3 ft (90 cm) long, slender, not tapered, with smooth and glossy scales. They are brilliantly colored, with transverse rings of red, black, and yellow, each black ring being bordered with yellow (plate VIII, 1). There are some nonvenomous snakes that closely resemble coral snakes, but these have pairs of black rings bordering a yellow one, the reverse of the coral snakes, in which the red and yellow rings touch. Navy (1965) gave these guidelines:

> If the red and black rings touch each other, if the end of the snout is red, whitish, or speckled, and if the colors fail to encircle the body, the snake is not a North American coral snake.
>
> "Red and yellow, kill a fellow;
> Red and black, friend of Jack."

The coral snake feeds largely on the young of other snakes and on lizards. It does not strike like the pit vipers, but when stepped on or touched, it will turn and bite, retaining its hold. Compared with those of rattlesnakes and moccasins, bites from coral snakes are relatively rare, but fatal cases have been reported (Ditmars, 1931). This is of great interest in view of the small amount of venom they inject.

Protective Clothing

Authorities are all in agreement on the importance of protective clothing, particularly foot and leg coverings. In countries where hikers are well shod, there are about 40% more bites on the lower limbs than on the upper ones, whereas in countries where most people go barefoot, bites on the legs exceed those on the arms by three or four to one. Boots, leggings, or a shoe–puttee combination give good protection. Klauber (1956) provoked a 130-cm western diamondback rattlesnake with fangs 12 mm long to strike a man's oxford shoe stuffed with cloth. In none of several strikes did the fangs penetrate the leather. He concluded that leather of moderate thickness could not be penetrated by a large rattler, and that heavy boots or a heavy shoe–puttee combination would afford adequate protection to the part of the leg covered. The tongue of a boot might be its weakest point.

Trousers should be worn outside boots instead of being tucked inside. A snake may close its mouth during an attempt to bite when it feels cloth. Klauber cited the case of a boy who was bitten several times by a rattler, but the venom was all absorbed by intervening flannel and no harm was done. He stated:

> In considering the protective value of boots or clothing, it should be remembered that the point of a fang is solid; the discharge orifice is above the point and may be blocked by leather, or the venom may be absorbed by cloth, even though the point of the fang may have penetrated the skin of the victim. There is no doubt that heavy, gauntlet-type gloves are of value for anyone doing much climbing or working among rocks.

Cultivated Fear of Snakes

Klauber (1956) considered that the inculcation of a moderate fear of snakes in children is probably justified, since most children cannot distinguish between venomous and nonvenomous snakes. But he believed that it is a mistake to instill in them such a horror and revulsion at the sight of a snake that, even when they have become adults, they cannot bear to look at a snake in a cage, and that such insensate fear of snakes is not instinctive. He stated:

> The cultivated fear of snakes has had results quite the opposite of those desired. It causes people to become so paralyzed upon encountering a rattler in the field that they cannot take the most elementary safety precautions. If bitten by a snake, they are in no condition to judge whether it was venomous or harmless. This is a matter of importance in deciding whether painful and even dangerous remedial measures are necessary, for the bite of a harmless snake requires only a touch of antiseptic, as would a scratch or a splinter. Even in areas of the United States—and they are extensive—where rattlers are the only venomous snakes, victims usually cannot even report with certainty whether or not the culprit had rattles on its tail. These are the people who destroy all harmless snakes—the natural competitors, and even the destroyers of rattlesnakes—thus aiding in the protection and the increase of the dangerous snakes they so greatly fear. So children should be taught to avoid snakes, not to be terrified by them, and eventually we may have a more understanding adult population.

Envenomation by Snakes

Reasonably current information on the biochemistry and pharmacology of snake venoms has been reviewed by Jiménez-Porras (1970) and Russell and Puffer (1970). In the preceding 3 years, about 300 papers had been published on the subject (Jiménez-Porras, 1970). Few poisons are so complicated and diversified, and collectively they appear to have an effect on every organ system and perhaps every cell. Their complexity is indicated by the fact that a single snake venom may contain 5 to 15 enzymes, 3 to 12 nonenzymatic proteins and peptides, and at least a half dozen other substances. They cannot be classified simply as neurotoxic or shock-producing or hemotoxic and hemorrhagic. However, envenomation is usually directed against the cardiovascular and blood systems, the nervous system, and the respiratory system. Further complications arise as the result of autopharmacological responses, such as the production and release of histamine, bradykinin, and adenosine in the envenomated organism. These may complicate the poisoning, and may in themselves sometimes produce more serious consequences than does the venom.

It accordingly appears that brief descriptions often encountered in the literature of the differences in the actions of different snake venoms, such as those in the effects of rattlesnake and coral snake bites, are likely to be oversimplified and misleading. A generalization that appears to be justified is that all crotalid (rattlesnake, copperhead, cottonmouth) venoms so far examined appear to be rich in proteolytic enzyme activity, which causes tissue changes and destruction. On the other hand, the venoms of the elapids (in the United States, the coral snakes) either have no proteolytic enzyme activity or very little. Elapid venoms have a curare-like action—a marked effect on neuromuscular activity.

Aside from the nature of the venom, which can differ even within a single genus of snakes, the amount of venom injected is, of course, an important factor influencing the reaction suffered by a snakebite victim. The approximate yield of dry venom will vary from 370 to 700 mg for the eastern diamondback rattlesnake to 2 to 6 mg for the eastern coral snake. The latter has a very potent venom, however, with an intraperitoneal LD_{50} of 0.97 mg/kg. All the LD_{50}'s given here were obtained from tables presented by Russell (1967*b*) and again by Russell and Puffer (1970). The LD_{50}'s were the doses per kg of body weight that gave a 50% kill of mice; the lower the LD_{50}, the more toxic the venom. The dreaded cobra (*Naja naja*) has less venom (170–325 mg) than the eastern diamondback rattlesnake, but the LD_{50} of its venom is only 0.40 mg/kg. [The tiger snake of Australia (*Notechis scutatus*) yields 30–70 mg of venom, approximately the same amount as the copperhead in the United States, but the intraperitoneal LD_{50} is the lowest for any listed by Russell (1967*b*)—0.04 mg/kg!]

The eastern diamondback not only has a very high yield of venom, but also a relatively low intraperitoneal LD_{50} (1.89 mg/kg). The western diamondback is second among rattlesnakes in venom yield (175–320 mg), and the LD_{50} of the venom is 3.71 mg/kg, about twice that of the eastern diamondback. The Mojave rattlesnake yields only 50 to 90 mg of venom, but the LD_{50} of its venom (0.23 mg/kg) is by far the lowest of all the rattlesnakes. Its venom is strongly neurotoxic, in contrast to the much less neurotoxic but more necrotizing venom of the western diamondback (Minton, 1970). Among the moccasins, the cottonmouth yields 90 to 145 mg of venom of relatively high LD_{50} (5.11 mg/kg), but the copperhead yields less than half as much venom (40–70 mg), with an LD_{50} more than twice as high (10.50 mg/kg), and is apparently a much less venomous snake (Navy, 1965; Russell, 1967*b;* Russell and Puffer, 1970). The latter authors had the following to say about snake venoms in general:

> Changes in the dynamics of the heart; in the resistances and permeability of the blood vessels, particularly in those of the pulmonary system; in the integrity and viscosity of the blood cells; in conduction along nerves, at the neuromuscular junction, at the muscle, or in the central nervous system or spinal cord; and changes in the respiratory system at various levels have all been observed following poisoning with snake venoms.
>
> For this reason, it is best to consider all snake venoms as capable of producing 1 or more deleterious changes in several organ systems and to sometimes provoke these concurrently. No venom should ever be classified as a "neurotoxin," a "cardiotoxin," or a "hemotoxin" and have its other pharmacological properties minimized or dismissed. It has been shown that neurotoxins can and do have cardiotoxic or hemotoxic activities or both; cardiotoxins may have neurotoxic or hemotoxic activities or both; and the activity of hemotoxins may not be restricted to the blood. More errors in clinical judgment have been made on the basis of oversimplified classifications of the physiopharmacological properties of snake venoms than are generally appreciated. It is our own feeling that the physician faced with a case of snake-venom poisoning is confronted with 2 or perhaps 3 or 4 serious disease states. He must therefore give prompt and careful attention to an evaluation of a number of important physiological parameters in the treatment of his patient.

Symptoms of Crotalid Envenomation

Nearly all the seriously venomous snakes of the United States are crotalids (rattlesnakes and moccasins), and the symptoms are well known. Pain is immediate, and is most severe from bites of the eastern or western diamondbacks and the timber rattler. Many pain-producing venoms form the protein-splitting enzyme bradykinin from plasma. Plasma kinins in the victim's body contribute to the production of pain. *Crotalus atrox* forms much bradykinin and causes much pain, whereas the South American *C. terrificus* forms very little bradykinin and causes practically no local reaction (Keele, 1963). Swelling and edema are usually noted in the bitten area within 10 minutes. In the case of most rattlesnakes, without treatment the swelling may involve the entire injured extremity within an hour. Swelling and edema are less marked in the case of the American moccasins, and are not symptoms of the bites of the Mojave rattlesnake. Within several hours, the skin in the area of the bite may become discolored, appear tense and shiny, and vesicles may form. Vesicles are usually present by the end of 24 hours (Navy, 1965).

Common symptoms of snake bite are weakness, sweating, faintness, and nausea. Regional lymph nodes may become enlarged and painful, and some rattlesnake bites result in tingling or numbness over the tongue and mouth or scalp. Paresthesia (a prickling, tingling, or creeping sensation) is sometimes felt in the area of the wound (Navy, 1965).

Reptile venoms are almost innocuous when taken by mouth, as is sometimes demonstrated by showmen. The amount of venom required to kill mice, when it is administered by stomach tube, is about 750,000 times the lethal dose by injection in the case of rattlesnake venom, but only 80 to 100 times the lethal dose in the cases of cobra or krait venoms (Macht and Kehoe, 1943; Minton and Minton, 1969).

First Aid for Snake Venom Poisoning

In the foregoing discussions and accompanying illustrations, descriptions of the dangerously venomous snakes of the United States have been given. Authorities are unanimous in their advice to identify the snake, if possible, before any treatment is made. Russell (1967*a*) stated: "It should be borne in mind that some persons bitten by nonvenomous snakes become excited and even hysterical, and that these emotions may give rise to disorientation, faintness, dizziness, rapid respiration or hyperventilation, a rapid pulse, and even primary shock." If the snake is not identified at the time of the bite, it can be found quickly in the immediate area of the accident. Extreme caution should be exercised in hunting for the snake, for it can strike again. It can be killed by a sharp blow to the head, but it should not be handled, for it can bite by reflex action. If it cannot be positively identified, it should be carried on a stick or in a cloth bag. If the snake is venomous, first-aid measures should be begun immediately. The amount of pain or absence of pain should not be depended upon as a criterion of whether or not the snake was venomous. At the doctor's office or hospital, the doctor should either be informed of the genus of the snake, or else the dead, unidentified reptile should be turned over to him. He should also be told the approximate period between bite and arrival and of what first-aid measures have already been given. Russell (1967*a*) named 5 ideal objectives of first-aid measures:

(1) To retard absorption of the venom.
(2) To remove as much venom as possible from the wound.
(3) To neutralize the venom.
(4) To mitigate the effects of the venom.
(5) To prevent complications, including secondary infection.

The large quantity of venom injected by rattlesnakes and moccasins justifies first-aid procedures that are more severe and painful than those recommended for spider and scorpion envenomation, and they are also attended by a certain amount of risk to the patient if not properly applied. Therefore, it is important for the person bitten, if he is alone or if only he and the person applying first aid are present, to remain cool and follow directions carefully.

An important procedure in first aid for snake bite is to use the constriction band. Immediately after the bite, it should be placed above the first joint proximal (on the body side) or 2 to 4 in. (5 to 10 cm) proximal to the bite, and it should be tight enough to occlude the superficial venous and lymphatic returns, but not the arterial flow. Every 10 minutes, it should be released for 90 seconds, and can be moved in advance of the progressive swelling. As soon as antivenin has been started or suction discontinued, the constriction band should be removed. It is of little value if applied later than 30 minutes following the bite, and it should not be used for more than 2 hours. The constriction band is of questionable value following envenomation by a coral snake (Russell, 1967*a*).

In pit viper bites, the constriction band should be used in conjunction with incision and suction, but these are probably of little value following envenomation by a coral snake or other elapid. In pit viper bites, excluding those of small copperheads, cross-shaped or longitudinal incisions 0.125 to 0.25 in. (3 to 6 mm) long should be made through the fang marks. Most rattlesnake bites barely penetrate the skin, and incisions should be made to an accordingly limited depth. When making the incisions, the direction of the fangs' penetration should be taken into consideration. Oral suction should then be applied and continued for the first hour following the bite, but not if other means of suction are available. Without suction, incisions through the fang marks are of questionable value, and should be avoided. In sucking out the venom by mouth, a slow, steady pull is exerted and, of course, the venom is expelled from the mouth. Russell (1967*a*) stated that in treating 104 rattlesnake bites over a period of 12 years, most victims having arrived at the hospital following the initiation of some first-aid measures, he observed that those patients who had properly applied a constriction band, incised the fang marks, and applied suction fared better than in seemingly similar cases where these measures had not been used. He reported being able to kill mice with the exudate removed from incisions by suction cups.

Russell likewise warned that under no conditions should antivenin be administered by an untrained individual, for injection of antivenin into persons sensitive to horse serum might be fatal. However, antivenin might be required when severe symptoms developed early in the course of the illness, or where a delay was foreseen of 4 hours or more after envenomation by pit vipers, or of 2 hours or more after envenomation by elapids. Under such circumstances, Russell's advice was as follows:

> In such cases, the antivenin should be given intramuscularly following appropriate skin or conjunctival tests. The antivenin should be given intramuscularly at a site distant from the wound. The antivenin should never be injected into a finger or toe, and it should be administered intravenously only by qualified personnel. As the amount of antivenin available will more

than likely be limited, 1 unit (vial or package) will probably be all that can be administered. The earlier this is injected, the better the results that can be expected. Remember, if the victim is in shock, the antivenin will be absorbed slowly from an intramuscular site. In such cases, the antivenin may need to be given intravenously, but *only by a qualified person.*

As might be expected, complications may arise on the way to the doctor's office or hospital, particularly on a long trip. In his paper titled *First Aid for Snake Venom Poisoning,* Russell (1967*a*) detailed other supportive measures that might be utilized in case of such sequelae to snake bite as shock, respiratory distress, vomiting, excessive salivation, or convulsions. "Cryotherapy," which involves placing the bitten member in iced water to well above the wound, is considered to be a dangerous treatment by most physicians. An ice bag may be applied directly over the wound site following incision and suction, or until antivenin is given, but the bitten arm or leg should *never* be packed in ice.

Russell's paper and *Poisonous Snakes of the World,* issued by the United States Bureau of Medicine and Surgery, have formed the basis of a pocket-sized booklet, prepared by the Entomology Section of the Bureau of Environmental Sanitation of the Los Angeles County (California) Health Department, titled *Emergency Treatment for Rattlesnake Bite.* The booklet is designed and organized for rapidly finding brief, precise, and itemized instructions on procedures to follow when alone or with others present, with or without a snakebite kit, and after the initial treatment. Precautions to follow for avoidance of snake bite are also itemized. With the permission of the Los Angeles County Health Department, the instructions in the booklet are quoted here. It should be emphasized that the victim must receive medical assistance as soon as possible, to prevent the serious complications that can arise following rattlesnake bites. These bites are rarely fatal when adequate first-aid measures and subsequent medical treatment are given. The following information should aid you when treating yourself or others.

Treatment When Alone

Without a Snakebite Kit:

1. Keep calm. *Immobilize* the affected part if possible.
2. Tie a *constriction band* (tie, belt, or handkerchief) above the fang marks or swelling. *Do not* tie too tight. Do not restrict arterial blood flow. Release every 10 minutes for 90 seconds.
3. If available, apply antiseptic to fang marks.
4. Cut with sterile knife. Flame the knife if necessary. Make incisions 0.125 to 0.25 in. [3 to 6 mm] long through fang marks—just through the skin.
5. Initiate suction as soon as possible. If you have no sores in your mouth, or an ulcer, you may suck out the wound. Apply a slow, steady pull. Spit the venom out. Continue the suction for about an hour.
6. Never attempt to walk for help unless it is quite evident that failure to do so would jeopardize survival more than the venom itself.

With a Snakebite Kit:

1. Keep calm. *Immobilize* the affected part if possible.
2. Use the equipment in the snakebite kit. Follow the directions.
   (*a*) Tie the *constriction band* above the fang marks or swelling. *Do not* tie too tight. Do not restrict arterial blood flow. Release every 10 minutes for 90 seconds.
   (*b*) Use the antiseptic supplied in the kit. Apply to fang marks.
   (*c*) Cut with sterile knife. Flame the knife if necessary. Make incisions 0.125 to 0.25 in. [3 to 6 mm] long through fang marks—just through the skin.
   (*d*) Apply suction and continue for 1 hour.
3. Never attempt to walk to help unless it is quite evident that failure to do so would jeopardize survival more than the venom itself.

Treatment When Others Are Present

Without a Snakebite Kit:

1. Keep the victim warm, calm, and comfortable. *Immobilize* the affected part. Keep the patient *quiet.*
2. Tie a *constriction band* (tie, belt, or handkerchief) above the fang marks or swelling. *Do not* tie too tight. Do not restrict arterial blood flow. Release every 10 minutes for 90 seconds.
3. If available, apply antiseptic to fang marks.
4. Cut with sterile knife. Flame the knife if necessary. Make incisions 0.125 to 0.25 in. [3 to 6 mm] long through fang marks—just through the skin.
5. Initiate suction as soon as possible. If you have no sores in your mouth, or an ulcer, you may suck out the wound. Apply a slow, steady pull. Spit the venom out. Continue the suction for about an hour.
6. Send for medical help, or carry the person to medical assistance.

With a Snakebite Kit:

1. Keep the victim warm, calm, and comfortable. *Immobilize* the affected part. Keep the patient *quiet.*
2. Use the equipment in the snakebite kit. Follow the directions.
   (*a*) Tie the *constriction band* above the fang marks or swelling. *Do not* tie too tight. Do not restrict arterial blood flow. Release every 10 minutes for 90 seconds.
   (*b*) Use the antiseptic supplied in the kit. Apply to fang marks.

(*c*) Cut with sterile knife. Flame the knife if necessary. Make incisions 0.125 to 0.25 in. [3 to 6 mm] long through fang marks—just through the skin.

(*d*) Apply suction and continue for 1 hour.

3. Send for medical help, or carry the person to medical assistance.

Following Initial Treatment

1. The patient should always be seen by a physician as quickly as possible, so that appropriate antivenin can be given if necessary.
2. To be effective, sucking out of the wound should be continuous. However, discontinue the suction after 1 hour, for it is of no value after that period.
3. Keep the injured extremity as cool and dry as possible.
4. It is best to leave the *constriction band* in place until medical assistance is obtained. *Remember* to release the constriction band every 10 minutes for 90 seconds.
5. Keep the bitten part lower than the rest of the body.
6. Do not let patient take any alcoholic stimulants. He may drink water if necessary.

Precautions

1. Do not go into snake-infested areas alone. It is very difficult to administer adequate first-aid measures to yourself and also obtain the necessary medical assistance.
2. Rattlesnakes are active mostly at night in warm weather. They hide during the hottest part of the day or when not hunting. Keep hands away from areas which could hide rattlesnakes, such as logs, rocks, underbrush, or rodent burrows.
3. Use a stick if you have to move or transport a freshly killed snake, because it may bite by reflex action.
4. Carry an adequate snakebite kit along with you. Know how to use it.
5. Check the immediate area before sitting down.
6. Camp in open areas away from places where snakes could hide.
7. Be careful when gathering material such as firewood—day or night.
8. It is advisable to tell someone of your route and your general intentions. Help could be forthcoming much more rapidly in case an emergency should arise.

In addition to the precautions just quoted, it is well to remember to step ON to logs and large rocks, not OVER them. A snake may be lying on the other side. Keep in mind the value of protective clothing, as discussed a little earlier in this chapter—particularly, good leather shoes with high tops.

Finally, obtain a good snake-identification book and make sure the entire family or party can identify the venomous species.

Control of Rattlesnakes

One way of reducing the rattlesnake population of an area is to eliminate habitats such as potential dens (rock crevices or caves), as well as brush or piles of debris that might afford temporary shelter.

All openings in or under buildings should be sealed, and foundation vents should be covered with 0.25-in. (6-mm) hardware cloth. Shrubs should be trimmed to provide adequate ground clearance. Gopher and squirrel holes in the yard should be filled with stones or dirt, for snakes use them as passageways.

If it is not too hot, rattlesnakes will hunt during the daytime, but if it is very hot during the day, as in desert areas, they hunt at night. They avoid hot, bare ground or areas where there is no harborage.

Rattlesnakes remain active from the first warm days of spring until the onset of cold weather. They hibernate during the winter. They feed on rodents and other small mammals, amphibians, and birds, using their keen sense of smell and heat receptors—deep pits, one on each side of the head—that sense a passing animal.

Rodents may be eliminated as food sources for snakes by trapping or poisoning them (chapter 13). Also, rodent harborages can be removed by piling wood 18 in. (45 cm) above the ground, keeping all rubbish and garbage cleared away, and removing grass cuttings, weeds, tree trimmings, and similar rubbish where rats and mice may nest.

Squirrels and gophers are other rodents that also attract snakes.

A 50% DDT wettable powder was formerly sprayed or dusted in a strip around the area to be protected, as a repellent, or over the entire area to kill the snakes. However, DDT is no longer available to the public. Nicotine sulfate in drinking water set out for snakes has been recommended. The water is contained in a shallow pan covered with a screen in such a way as to allow the entry of snakes, but not pets, birds, or livestock.

The nicotine sulfate is used at the rate of 1 part to 250 parts of water. Burrows and dens may be fumigated with calcium cyanide, using 2 to 4 oz (60 to 120 g) for small burrows or dens and a pound (0.45 kg) or more for large dens, then sealing them tightly with soil. Hunting and clubbing of snakes is considered to have only very limited value as an area control measure (Waldron, 1969*b*).

(B) ARTHROPODS THAT INJECT VENOM WITH THEIR SALIVA

Bloodsucking arthropods are found among the arachnids and, among insects, almost exclusively in the Anoplura, Hemiptera, Diptera, and Siphonaptera. The bloodfeeding apparatus of the arachnids is a "one-way" channel that alternately conveys the salivary secretion outward from the mouth and into the tissues of the host, and the blood meal inward. This phenomenon is best observed with pseudoscorpions. For example, while the British species *Dactylochelifer latreillei* is feeding on a soft-skinned insect prey, the latter becomes alternately flaccid from having its liquid body contents sucked out, and then again turgid as the predator exudes a liquid that presumably contains digestive enzymes. The meal may last an hour or more before the pseudoscorpion discards its prey (Gilbert, 1951).

The bloodfeeding apparatus of the insect groups mentioned above consists of 2 channels extending from the mouth: a small dorsal channel through which blood is ingested, and a still smaller ventral one, usually within the hypopharynx, through which the saliva is conveyed outward into the host (Hocking, 1971). Vertebrates bitten by bloodsucking arthropods are usually more adversely affected by sensitization to saliva than by loss of blood.

Some arthropods do not envenomate their victims by injecting the irritating substance through ducts in their chelicerae or stingers, but by incorporating it with their saliva. The saliva contains an irritating substance that is hemorrhagic, increasing blood flow in the host. The bite of the arthropod may lead to swelling, which may sometimes be so extreme as to double the size of the affected arm or leg. The hemorrhagic substance prevents clotting of the blood while the arthropod is feeding. The bite may be immediately painful or the pain may be delayed, depending on the species of the attacker. If the pain is delayed, the presence of an anesthetic in the saliva may be indicated. Multiple bites may lead to a generalized toxic reaction and severe illness. The resultant itching may cause the victim to scratch the site of the wound, resulting in secondary infection.

If a person lives in a part of the world where the arthropods discussed under the above heading are considered as pests mainly because of the pain and discomfort caused by their bites, because of their effect on tourism, or because of the depreciation of real estate values, he is relatively fortunate, for this group of arthropods includes the principal vectors of some of the world's most serious diseases, such as malaria, yellow fever, sleeping sickness, elephantiasis, typhus, bubonic plague, cholera, dengue, Rocky Mountain spotted fever, and others.

Among the arthropods discussed in this section, the fleas and lice probably could at one time have been considered to be the most important, but owing to a world-wide improvement in sanitary conditions and personal hygiene, and particularly because of improved insecticides and more universal insecticide usage, their importance as pests has declined, and mosquitoes are now of greater relative importance.

MOSQUITOES (CULICIDAE)

Predominant among the arthropods that are pests of man and domestic animals are certain insects in the order Diptera (flies), and particularly those included in the family Culicidae. The insects in this family, with the exception of 2 small subfamilies, Chaoborinae (phantom midges) and Dixinae (dixa midges), are known as mosquitoes, from the Spanish word meaning "little fly." Mosquitoes are small, delicate insects, most of them from 3 to 6 mm in body length. The scales on their bodies and on the veins and posterior

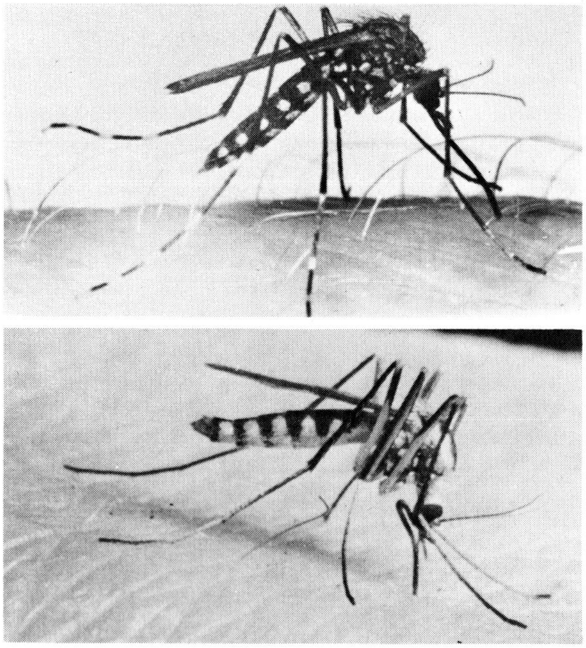

Fig. 272. Mosquitoes in the act of biting man. *Top,* western treehole mosquito, *Aedes sierrensis* (courtesy E. S. Ross); *bottom,* eastern treehole mosquito, *Aedes triseriatus* (courtesy John H. Gerard, from National Audubon Society).

margins of the wings and the long, slender proboscis are the most obvious characteristics separating mosquitoes from other Diptera.

Adult females of most species of mosquitoes are well known for their propensity for sucking blood. For such species, a blood meal is usually

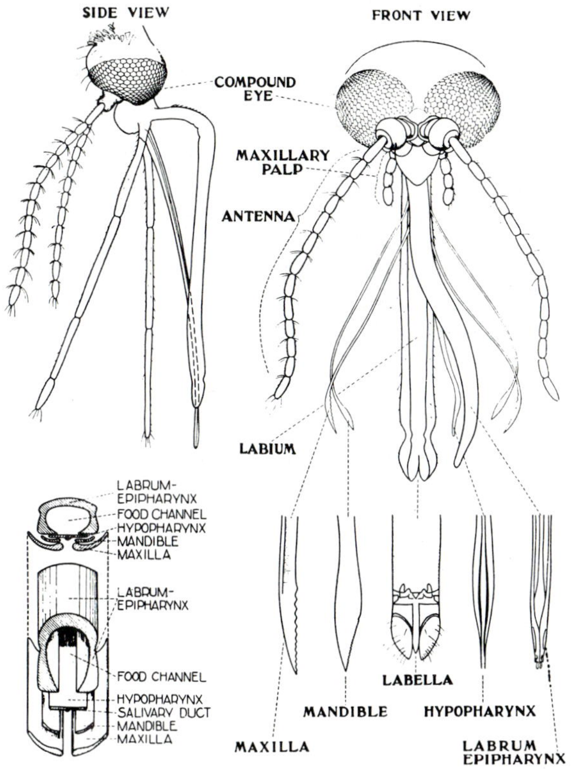

Fig. 273. Heads and mouthparts of 2 mosquitoes. Side view of *Anopheles* (*left*), and front view of *Culex*, with piercing stylets exposed. (From Herms, 1939.)

required for egg production. Adult females of *Culex tarsalis* averaging 2.7 mg in weight, after being deprived of liquid for 48 hours, weighed 6.5 mg after ingesting the blood of chicks—a weight gain of 140% (Thomas, 1963). Adult females of a few species (*Toxorhynchites* spp.) and males of all species feed only on nectar, honeydew, fruit juices, and "oozes" from injured or diseased plants. Even the females of most bloodsucking species obtain sugar from such liquids to supply the energy for the activity of daily life, particularly flight. The clear liquid can be easily seen in the swollen abdomens of these mosquitoes while they are resting on the vegetation from which the liquid is obtained (Haeger, 1955).

Mosquitoes are vectors of such human diseases as malaria, filariasis, and many of the arboviruses (arthropod-borne viruses), which include yellow fever, dengue fever, and the various strains of encephalitis. They are also vectors of many diseases of domestic and wild animals. Before the establishment of mosquito-abatement districts in 1915, there were 5,000 to 6,000 cases of malaria annually in California. In the United States, over 100,000 cases of malaria were reported annually as late as the early 1930's. The incidence of malaria then gradually decreased because of drainage and other successful source-reduction programs. Then, with the advent of DDT, progress in malaria eradication in the United States and throughout the world was spectacular.

A good general treatise on mosquitoes is W. R. Horsfall's *Mosquitoes: Their Bionomics and*

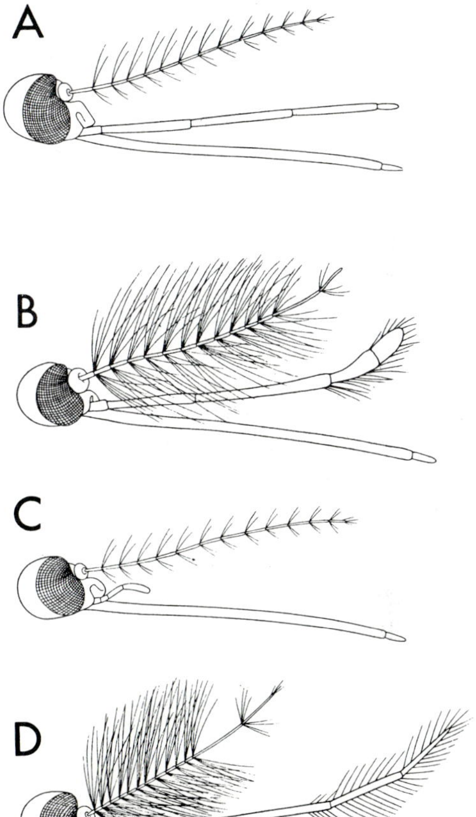

Fig. 274. Heads and appendages of mosquitoes. *A*, female *Anopheles; B*, male *Anopheles; C*, female culicine; *D*, male culicine. (From Carpenter and LaCasse, 1952.)

Relation to Disease, published in 1955 and republished in 1972. So much information on mosquitoes has been accumulated in the intervening years that a complete revision of the book, or a newer general treatise, has been much needed.

J. D. Gillett's *Mosquitos,* published in 1971, has therefore been a welcome addition to the literature.

General Morphology of Adults. Mosquitoes are slender, long-legged insects (figure 272). The head is globular, the thorax laterally compressed, and the abdomen long and cylindrical. The wings are long and narrow, lie flat on the back when at rest, and have 6 longitudinal veins. The second, fourth, and fifth veins are forked. The mosquito differs from nearly all other flies in having scales on its wings and on parts of its head, body, and appendages.

Mosquitoes have a highly modified assemblage of piercing–sucking mouthparts that is prolonged into a beak. The mouthparts consist of long, bladelike stylets (mandibles and maxillae) within a sucking tube formed by the labrum and hypopharynx, and all these elements are carried in the U-shaped trough of the elongate labium. The wound made by the stylets is not entered by the labium. As the piercing mouthparts penetrate the skin of the host, the labium gradually bends away from them, but loops around to guide them at the point of insertion. This loop is clearly seen in figure 272, *bottom.* A smaller duct in the hypopharynx provides for the outward passage of the secretion of the salivary glands. It is this secretion that causes itching.

As seen in figures 273 and 274, the mouthparts exhibit a sexual dimorphism that is sometimes useful in classification. Note also the striking differences in palps between the anopheline and culicine mosquitoes. In the anophelines, the palps are as long as the proboscis, whereas in the culicines, the females have short palps. The 15-segmented antennae of both sexes possess whorls of hair, but the hairs on the antennae of the males of most species are more abundant and much longer, presenting a plumose appearance (figure 274). It is the antennae, not the eyes, that are indispensable organs to enable mosquitoes to find their hosts (Roth, 1951).

General Morphology of Aquatic Stages. All mosquito larvae (figure 275) are aquatic. They possess a well-developed head followed by a swollen, unsegmented thorax and an 8-segmented abdomen. They breathe through a pair of spiracles (breathing siphons) at the hind end of the body, which they extend through the surface film of the water. They spend most of their time in that position. Tufts of bristles arise from many of the body segments. As shown in the figure, at the posterior end of each larva there are "gills" (on the left side) and, in *B, C,* and *D,* a tubelike projection called a "siphon." Mosquito pupae (figure 276) are also aquatic, but differ greatly from the larvae in appearance. The slender abdomen hangs down from a large, oval mass encompassing in rudimentary form the head, head appendages, thorax, legs, and wings of the adult. The breathing siphons of the larvae are replaced by 2 large, trumpetlike breathing tubes arising from the anterior part of the thorax. The thorax contains air sacs that cause the pupa to be lighter than water. This allows the adult to issue from the pupal skin

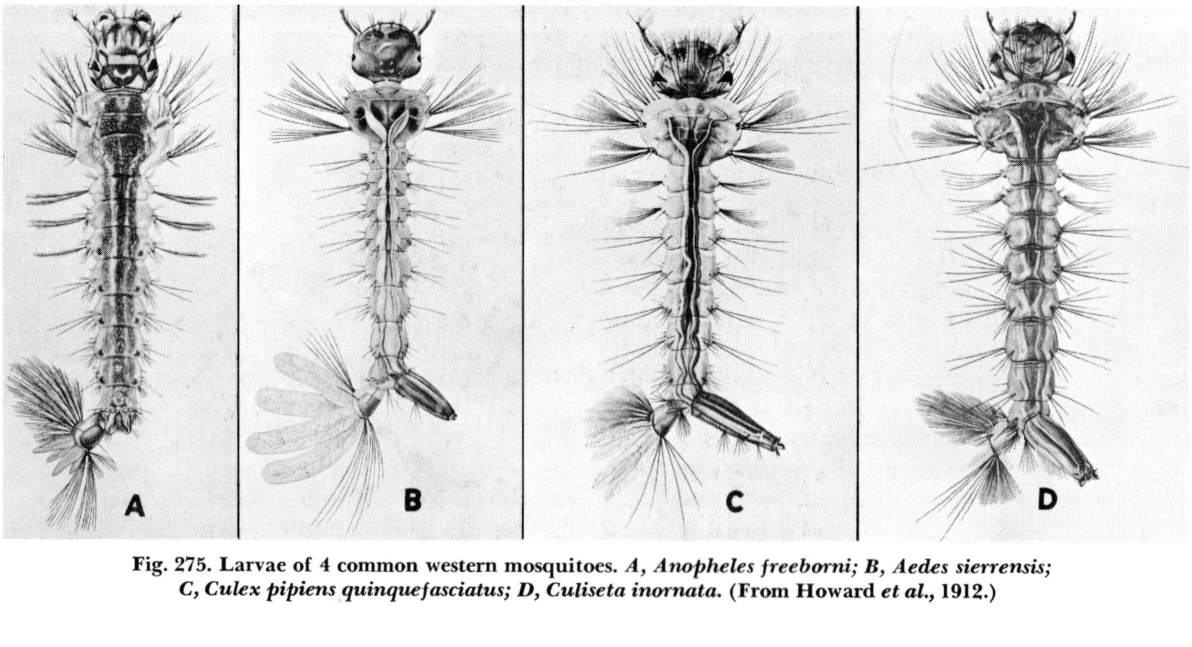

Fig. 275. Larvae of 4 common western mosquitoes. ***A, Anopheles freeborni; B, Aedes sierrensis; C, Culex pipiens quinquefasciatus; D, Culiseta inornata.*** **(From Howard *et al.*, 1912.)**

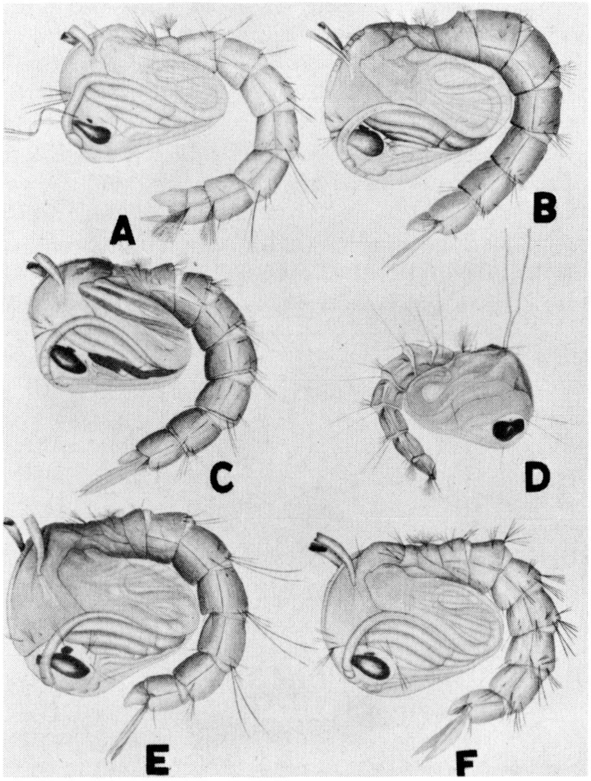

Fig. 276. Some representative forms of mosquito pupae. ***A, Wyeomyia smithii*** **(Coquillett);** ***B. Culex territans*** **Walker;** ***C, Aedes sylvestris*** **Theobald;** ***D, Wyeomyia circumcincta*** **Dyar and Knab;** ***E, Culex restuans; F, Culex pipiens quinquefasciatus.*** **(From Howard** ***et al.,*** **1912.)**

through a split in the back while it is floating on the water (Snodgrass, 1967).

Distribution and Habitat. Mosquitoes are distributed over the entire surface of the earth, from the tropics to the extremities of the continents in Arctic and Antarctic regions. Some species live at altitudes of over 12,000 ft (3,600 m). The larvae of all species are aquatic, and can be found in any small or medium-sized body of water—fresh, saline, or foul. Poorly drained land (including carelessly developed agricultural land), tidal marshes, flood plains of rivers, and ponded areas of prairies and muskeg support enormous mosquito populations. Permanent bodies of shallow water are best suited for high mosquito production if they are broken by plants or plant parts, for open water is unfavorable to mosquito development. Water held behind the axils of leaves (as in the bromeliads of the tropics), in broken bamboo stems, or in treeholes may be a common and important habitat for mosquito larvae, to cite some lesser examples.

Flight Range. The flight range of mosquitoes is of interest to city-dwellers because it indicates the potential hazard from surrounding rural areas having natural or artificial sources of infestation. Flight range is also of interest to mosquito-abatement districts because it indicates the extent to which a district where mosquitoes are under control might become infested from areas where they are *not* under control. The flight range varies with the species, time of year, wind direction, and also with the distance the mosquitoes must fly from their sources to where there are warmblooded animals upon which to feed. There are records in the literature of flights of marked specimens of female *Aedes* mosquitoes: *Ae. nigromaculis* as far as 28 mi (45 km) downwind in the lower San Joaquin Valley of California; *Ae. solicitans* (Walker) 110 mi (177 km) at sea off the coast of North Carolina; *Ae. dorsalis* and *Ae. squamiger* as far as 38 mi (61 km) downwind in the San Francisco Bay area; and *Ae. taeniorhynchus* 25 mi (40 km) downwind in southwestern Florida and 18 mi (29 km) in the coastal area of Georgia. The flight of males was always relatively short (Bailey *et al.,* 1965).

Bailey and his co-workers released approximately 253,000 adult *Culex tarsalis* mosquitoes marked with fluorescent Helecon products originally used by Reeves *et al.* (1948), and were able to trap 0.23% of them at various distances from the point of release, using CO_2-baited traps, which attracted only the bloodseeking females. This work was done in a large rice-growing area in Yolo County, California, for rice culture assures a high mosquito population. Figure 277 shows the pattern of dispersal of the trapped mosquitoes in an experiment during a 3-year period. The wind pattern was quite consistent. Their conclusions on the effects of prevailing summer evening winds in this experiment were as follows:

1. At low wind velocities up to 2 mph [3.2 km], dispersal takes place in all directions. The greatest upwind distance of recapture was 2.75 mi [4.43 km] on August 30, 1961, against a wind of 0 to 2.9 mph [4.67 km].

2. At least 10% of the mosquitoes from any particular release may disperse laterally, i.e., across the direction of the wind.

3. Above a limiting wind velocity of about 4 mph [6.4 km], the general direction of dispersal is downwind. There is only very limited movement against or across a wind as high as 4 mph. No recaptures were made upwind when velocities were 5.4 mph [8.69 km] or higher.

4. Wind velocities above 6 mph [9.7 km] probably would not extend the distance of effective dispersal—

i.e., the movement of significant numbers, which might become a nuisance or spread disease—because such high winds appear to discourage flight.

5. The swarms of males of this species exhibited very little horizontal movement, and moved vertically only enough to remain in a breeze of slightly more than 2 mph [3.2 km]. When the wind increased to 9 mph [14.5 km], even momentarily, the males dropped down behind trees, tall grass, levees, or even a parked automobile. When there was little wind, the swarms were observed sometimes as high as 30 ft [about 9 m] above the ground.

Culex tarsalis tends to head into the wind whenever possible, for wind brings the scent of near-by hosts. Odor trails cause some individuals to move across the prevailing wind or to fly against winds of low velocity. Nevertheless, the dispersal shown in figure 277 indicates movement predominantly downwind, and 1 individual was recovered 15.75 mi (25.36 km) from the release site. This mosquito can be caught in wind currents and be lifted as high as 2,000 ft (about 600 m) (Glick and Noble, 1961), but the numbers of

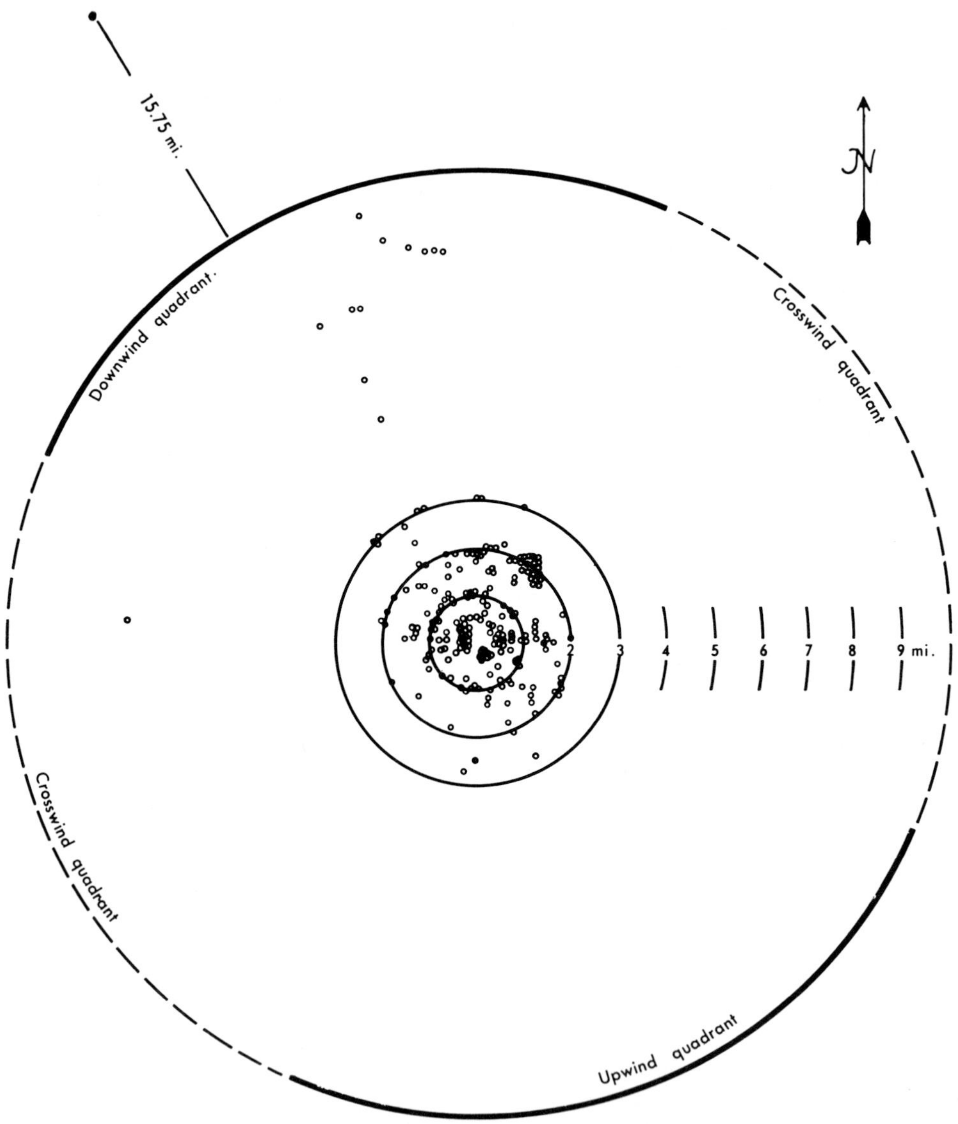

Fig. 277. Composite diagram of 18 release experiments with *Culex tarsalis* over a 3-year period, showing direction and distance of flights to points of recapture in traps baited with carbon dioxide. The prevailing winds on most summer evenings were toward the north or northwest. (From Bailey *et al.*, 1965.)

mosquitoes in the upper air are very small when compared with other insects. Bailey *et al.* (1965) believed that this species usually flew at heights between 5 and 50 ft (1.5 and 15 m), where birds, their preferred hosts, roosted in trees at night, and that they normally did not fly above the upper story of the biosphere. Horsfall (1942) reached the same conclusion concerning other mosquito species.

A similar study was made with marked adults of *Anopheles freeborni,* the only North American anopheline mosquito with a definite fall migratory phase. Although the great majority of these mosquitoes found adequate overwintering quarters within 5 mi (8 km) of their breeding sites in most farming areas of the Sacramento Valley of California, flights as far as 17.5 mi (28 km) were recorded. A small part of the "prehibernation" female population migrated into the foothills to a distance of at least 10 mi (16 km) and to an elevation of about 400 ft (120 m). A few males also migrated into the foothills up to 6 mi (9.7 km).

Culex tarsalis and *Anopheles freeborni* fly for great distances, with different flight patterns in summer and fall. Therefore, control in rice-growing areas is particularly difficult when organized on the basis of "political boundaries" (Bailey and Baerg, 1967).

Larval Feeding. Food for both the larvae and adults must be present near larval sites. Larvae ingest some dissolved substances, but depend for their sustenance principally on particulate matter, ranging in size from bacteria to clearly visible pieces collected by vibrating the labial brushes. Anopheline larvae, which lie horizontally just beneath the surface of the water and maintain a horizontal position by means of palmate hairs, can rotate the head through 180° while feeding against the surface film, which contains bacteria and other microorganisms. Culicine larvae, on the other hand, commonly observed in rain barrels, watering troughs, neglected tin cans, or similar containers, hang suspended diagonally from the surface by means of a prominent breathing siphon, with the head downward. The breathing siphons of mosquito larvae make them independent of the aeration of the water they inhabit. Large numbers can thrive in a small quantity of stagnant water, provided it contains sufficient food material.

Particulate matter may be floating in the surface film, suspended throughout the water, or clinging to submerged surfaces, or it may be in the form of other animals that can be as large as the larvae themselves (Horsfall, 1962). A few species of mosquitoes have larvae that are predaceous, and feed only on other mosquito larvae.

The Mosquito Life Cycle. Where the various species of mosquito larvae are found depends, of course, on the oviposition instincts of the adults,

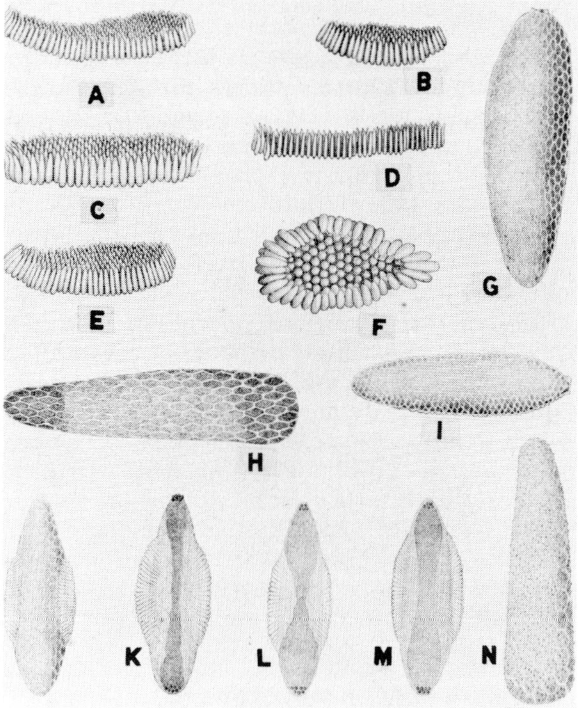

Fig. 278. Some representative forms of egg rafts and individual eggs of mosquitoes. *A, Culex pipiens quinquefasciatus; B, Culex salinarius* Coquillett; *C, Culex territans; D, Culex erraticus* (Dyar and Knab); *E, Culex restuans; F, Culiseta inornata; G, Aedes grossbecki* Dyar and Knab; *H, Culiseta inornata* (single egg); *I, Aedes aegypti; J, Anopheles punctipennis* (dorsal); *K, Anopheles punctipennis* (ventral); *L, Anopheles quadrimaculatus* (ventral); *M, Anopheles crucians* Wiedemann (ventral); *N, Coquillettidia perturbans.* (From Howard *et al.*, 1912.)

which are highly specific in this respect. Mosquitoes that deposit eggs in rafts (figure 278) and some that deposit their eggs singly require standing water for oviposition. Floodwater species require water present as moisture in the soil surface. Other species deposit their eggs on or barely beneath the surface of standing water, but attach them to masses of vegetation. Still others will oviposit in moist debris or depressions subject to inundation, or may even deposit their eggs in fairly dry situations where pools will be formed later by rains or tidal action. The eggs of some species, notably those of the floodwater mosquitoes, will hatch after being dry as long as 2 years or more. The incubation periods vary greatly, depending on species and temperature, but for many kinds the period is 16 to 24 hours.

The larvae molt 4 times, the last molt resulting in the pupa. Under optimum conditions, the larval stage of some culicines may require only a week, but that of the anophelines takes much longer. The pupa or "tumbler" does not feed, but is nevertheless very active. It breathes by means of a pair of "trumpets" thrust through the surface of the water, but can let go and dart through the water with a tumbling motion to seek shelter. After a few moments, it returns to the surface with little effort and re-establishes contact with the air. The pupal stage requires only 2 or 3 days.

The several minutes required for the emergence of an aquatic pupa to an aerial adult constitute a critical period in the metamorphosis of a mosquito. During this interval, the water/air interface must remain stable, and for this reason emergence usually takes place at night, the time of greatest water stability. Pupae of species that develop in large bodies of water move to areas where the surface is undisturbed for adult emergence.

Active females generally live 2 or more weeks in natural sites, except that hibernating species may remain dormant for several months. The life span of the adult male is even shorter than that of the female.

Recognition of Genera and Species

An illustrated field guide to California mosquitoes was prepared by Loomis (1967) in collaboration with the Entomology Committee of the California Mosquito Control Association. This field guide should aid greatly in identification of the principal pest mosquitoes as well as to contribute introductory knowledge of the species. Detailed pictorial keys to aid in the identification of the medically important mosquitoes of various geographic regions of the world, as well as much other useful information on mosquitoes and mosquitoborne diseases, are contained in *Mosquitoes of Medical Importance* (Foote and Cook, 1959). Current concepts of mosquito taxa from a world standpoint may be found in *A Synoptic Catalog of the Mosquitoes of the World,* by Stone *et al.* (1959), and subsequent supplements.

Most species of mosquitoes are included in the tribe Culicini. In this group may be found species that transmit yellow fever, filariasis, dengue, and other diseases, but not malaria. The malaria-transmitting species are in the tribe Anophelini, made up principally by the genus *Anopheles.* It is important to be able to distinguish between anopheline and culicine mosquitoes, and fortunately the groups are easily separated on the basis of data in the accompanying tabulation, taken with permission from the work by J. R. Busvine, entitled *Insects and Hygiene,* published in 1966.

The pointed shapes of the terminal abdominal segments of *Aedes* and *Psorophora* females generally can be used to separate them from those of the genera *Culex* and *Culiseta,* the abdomens of which are normally rounded or blunt. The larvae of the various genera usually can be separated on the basis of the number and position of hair tufts on the siphons (figure 275).

	Anophelines	Culicines
Eggs	1. Provided with floats.	1. No air floats.
	2. Laid separately.	2. Sometimes stacked together in egg rafts.
Larva	1. Hind spiracles on the body surface (eighth abdominal segment).	1. Hind spiracles at the end of a tail-like tube or siphon (projecting from the eighth abdominal segment).
	2. Larva held up horizontally beneath water film by float hairs, etc.	2. Larva hangs down at an angle, with only the tip of the siphon in the surface film.
Pupa	Respiratory trumpets more conical.	Respiratory trumpets more cylindrical.
Adult	1. Females with palps as long as proboscis.	1. Females with short palps.
	2. No scales on abdomen.	2. Abdomen covered with scales.
	3. Rests with proboscis and abdomen in a line, at an angle to surface.	3. Rests with proboscis and abdomen forming obtuse angles, the abdomen more or less parallel with surface.

The Genus *Anopheles*

The members of this genus are called "malaria mosquitoes," owing to the part played by certain species in transmitting the disease. For this reason it is important, in areas of the world where malaria is present, to be able to distinguish them from other genera, and particularly to be able to identify the vector species. All species of *Anopheles* are believed to be potential vectors of malaria, but as far as is known, only 2 species in the United States have transmitted it, provided, of course, that human reservoirs of the causal organism were present. These are *Anopheles freeborni* (figure 279, *A*) in the West and *A. quadrimaculatus* in the East.

Anopheles larvae are generally found in the larger permanent bodies of fresh water, but may occur in many other types of water. They are often seen in marshes, ground pools, and places where there is considerable vegetation. A good growth of green algae in shallow water provides an especially favorable environment. Some species inhabit cavities or artificial containers, but not so frequently as the culicine mosquitoes. Some live in recesses among epiphytes. Outbreaks of *Anopheles* commonly follow a change from dry-land agriculture to the use of irrigation (Horsfall, 1955). In North America, only 2 species, *A. atropos* Dyar and Knab and *A. bradleyi* King, breed in salt or brackish water (Pratt *et al.*, 1963).

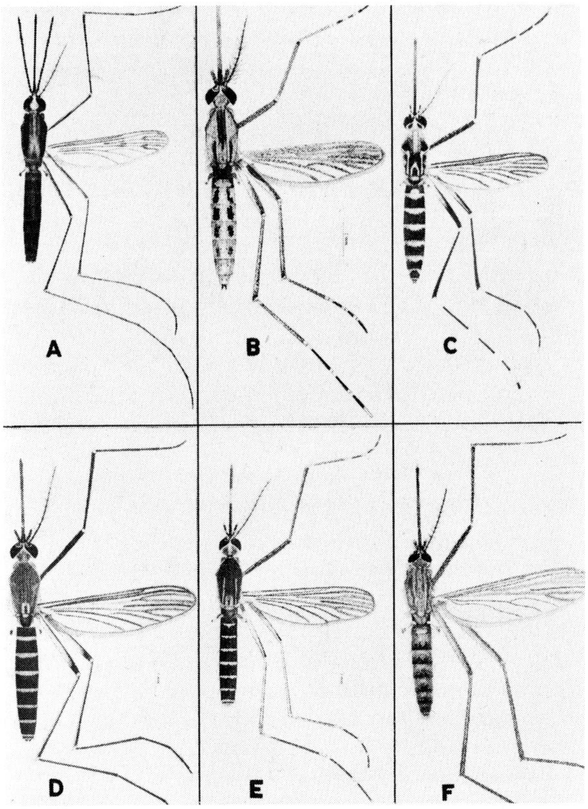

Fig. 279. Some common species of western mosquitoes (all females). *A, Anopheles freeborni; B, Aedes dorsalis; C, Aedes sierrensis; D, Culex pipiens quinquefasciatus; E, Culex tarsalis; F, Culiseta inornata.* (From Yamaguti, 1952.)

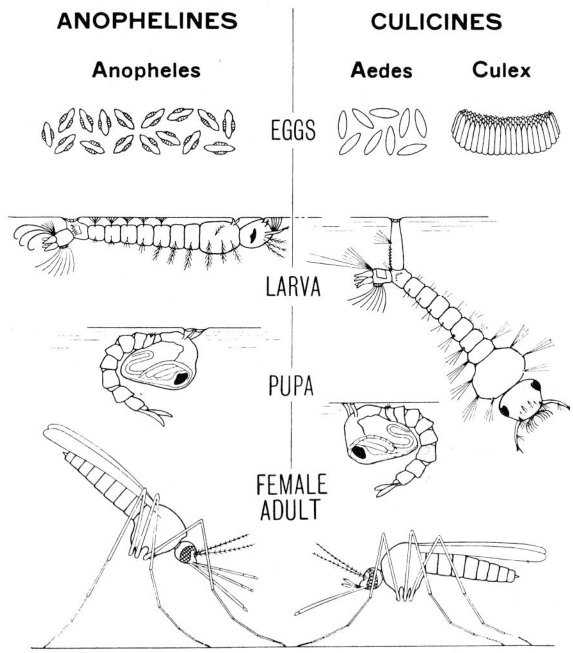

Fig. 280. Principal differences in the 4 stages of anophelines and culicines. (From Marshall, 1938.)

In this genus, the palpi of both sexes are usually about as long as the proboscis (figure 279,*A*); the mandibles and the maxillae of the females are well developed and toothed; the scutellum is evenly rounded, whereas in nearly all other mosquitoes it is trilobed, with a tuft of hair in each lobe; the legs are very long and slender, with no distinct tibial bristles and no pulvilli; the abdomen lacks scales, or at least the sternites are largely bare; and the wings usually have distinct markings, whereas in *Culex, Aedes,* and most other genera, they do not.

Anopheline mosquitoes rest with the head, thorax, and abdomen in a straight line, at an angle from 40 to 90° to the surface, while the culicine mosquitoes rest with their bodies nearly parallel to the surface. Characteristics of structure and habits that may be used to separate *Anopheles* eggs, larvae, pupae, and adults from those of the other 4 genera (culicines) to be discussed are listed in the box and illustrated in figure 280.

Anopheline mosquitoes spend the winter as fertilized females. They emerge from their nesting places in the spring to feed and oviposit, and then die. Breeding is continuous during the warm part of the year, and the eggs are laid in batches of 100 or more, each female laying an average of 400 to 500 eggs. The eggs float singly on the water, remaining properly oriented by means of an entrapment of air in "floats" on the sides of each egg, as well as by the combination of hydrophilic and hydrophobic areas. The eggs tend to gravitate toward one another or toward reeds, stems, sticks, or any objects that break the surface film. They usually hatch 1 to 3 days after being laid.

Western Malaria Mosquito, *Anopheles freeborni* Aitken

This species ranges over most of the area west of the Rockies, from southern British Columbia to Baja California, but is also found in southern Colorado, New Mexico, and extreme western Texas (Carpenter and LaCasse, 1952). In California, it occurs commonly wherever water is available, except in the cool coastal belt, which is inhabited by *A. occidentalis,* and above the digger pine belt in the Sierra. *Anopheles freeborni* was once greatly feared as a carrier of malaria, but with most human sources of the malaria organism eliminated, it is primarily a "pest mosquito" (a pest only because of its bites). The suggestion has been made that it could more appropriately be named the "ricefield mosquito" because of the great numbers that develop in rice fields in northern California (Bailey *et al.,* 1954).

Description. This is a large, dark-brown species, the wings having 4 indistinct, dark concentrations of scales (figure 279, *A*), which separate it from any other California anopheline mosquito except *A. occidentalis.* The latter, however, bears light scales on its wingtips.

Biology. During the first warm days of early spring, the mosquitoes emerge from their winter hibernating places (caves, hollow trees, basements, and other protected spots) and, after getting a meal of blood from a human or animal, the female retires to some secluded place to mature her eggs. A few weeks later, she deposits her matured eggs in collections of rainwater along roadsides, in old irrigation ditches, road ruts, hoofprints, or any ground pool where green algae can develop. The eggs float singly on the surface, and are properly oriented by "floats." They are identical in appearance to those of *A. quadrimaculatus* (figure 278, *L*), and tend to gravitate together or around reed stalks and other objects that break the surface. Incubation requires 2 to 6 days, and the larvae (figure 275, *A*) take about 2 weeks to develop.

The first summer brood of adults oviposits in late rain pools, drainage ditches, or swampy areas. Succeeding broods move into rice fields if they are available, and enormous mosquito populations can develop by the time the fields are drained in September if no control measures are practiced (Bailey *et al.,* 1954).

Habitat. This species prefers clear, clean, slightly alkaline water, particularly sunlit pools where algae and emergent vegetation are present, and normally avoids water that is polluted with sewage or other organic materials. The larvae are commonly found along the edges of irrigation ditches where weeds are abundant, and in residual pools along receding streams. The really typical aquatic sources for *A. freeborni* in California seem to be the shallow seepages resulting from careless use of irrigation water. In contrast to the common species of *Culex and Culiseta,* this species seldom, if ever, is found in artificial containers such as barrels or tin cans (BVC, 1951; Pratt *et al.,* 1963).

Females from the final generation in late fall will migrate long distances—as far as 10 or 12 miles (16 or 19 km)—to find shelter in homes, cellars, outbuildings, or caves and enter into a state of semihibernation. These emerge on warm days and nights for feeding, but do not bite viciously, and the winter feedings usually do not result in egg development. In summer, however, *A. freeborni* enters houses freely and is a vicious (but not painful) indoor biter, a characteristic that accounts for its potential as a malaria vector.

Common Malaria Mosquito, *Anopheles quadrimaculatus* Say

This eastern counterpart of *A. freeborni* has an equal potential for transmitting malaria. It occurs in the United States from Mexico to Canada, throughout the Mississippi Valley, and east to the Atlantic. It has been the chief vector of malaria in that extensive area. It is similar to *A. freeborni* in appearance and habits and, in fact, when the latter was described in 1939, it was considered to be a subspecies of *A. quadrimaculatus.*

Anopheles punctipennis (Say)

This mosquito occurs throughout most of the United States, but is not adapted to semiarid conditions, and its distribution in California appears to be limited to the central valley and north coast. It bites readily outdoors but seldom enters houses (Freeborn and Bohart, 1951).

Description. This is a medium-sized brown mosquito with a hairy mesonotum, the integument of which is whitish in the middle and dark brown on the sides. The long, black legs have small, white spots on the tips of the femora and tibiae. The palpi are unbanded. The wings have patches of light scales, unlike the wings of *A. freeborni* and *A. occidentalis.* Orange and yellow markings on the wings and the solidly black palpi separate this species from *A. freeborni.*

Biology. The larvae can be found in small, temporary pools and the edges of slow-moving streams, preferring partially shaded, cool, clear water. Marginal fringes of green, filamentous algae provide a very favorable environment. The developmental stages of *A. punctipennis* were described in detail by Smith (1914). In midsummer in Ithaca, New York, she found the entire developmental cycle from egg to adult to require 24 days in a laboratory culture. A larva found frozen solid in a pool on December 3 was thawed out on December 4 and was apparently normal in an hour.

Another very interesting observation regarding the biology of *A. punctipennis* was credited to a Dr. Dupree, who found that laboratory-reared mosquitoes supplied with blood oviposited 6 or 7 times (even 9 times in 1 case), depositing 100 to 300 eggs at a time, for a total of as many as 2,000 or more.

Anopheles pseudopunctipennis (Theobald)

This is a common species that ranges throughout the central United States and on into Mexico and Central and South America, but does not occur in California. It is similar in appearance to *A. punctipennis,* except that the palpi are banded with white. The larvae are often found in water too warm for other anophelines. It is an important malaria vector in some countries.

Anopheles franciscanus McCracken

This mosquito occurs along the Pacific Coast, from southern Oregon into Mexico and eastward into Nevada, Arizona, and New Mexico. It is similar to *A. pseudopunctipennis* in appearance and habits. It breeds in almost any type of available water, but reaches highest densities in drying pools with dense algal growth. A 6-acre lake in the vicinity of Los Angeles produced an estimated 500,000 adults per day.

Adults become active at twilight, do no blood feeding during the day, but begin to feed in the evening when the light intensity falls below 50 Weston units. *Anopheles franciscanus* is only moderately anthropophilic, feeding in preference on the larger mammals (horse, cow, sheep) and in fewer numbers on smaller animals (duck, turkey, rabbit, guinea pig). In central and northern California, it does not bite man, but specimens from the north and south are not otherwise distinguishable (Belkin *et al.,* 1951).

The Genus *Aedes*

This genus contains over 500 species of mosquitoes that occur in great numbers in tropical, temperate, and Arctic regions throughout the world, with about 60 species distributed throughout the United States. They are important pest mosquitoes, but are also important vectors of diseases, such as yellow fever, dengue fever, and filariasis. Nearly all species of *Aedes* suck blood, and many are vicious biters. They usually bite during the evening, but some species bite only during the day and others bite either by day or night (Horsfall, 1955). The daytime biters are important pests in the dairy and beef industries, fruit- and vegetable-growing regions, and in recreation and resort areas (BVC, 1951).

Most *Aedes* adults are small. They are not easily separated from those of the genus *Culex.* However, in *Aedes* adults, postspiracular bristles are present, whereas they are absent in *Culex.* In *Aedes,* the tip of the female's abdomen is usually pointed, and the cerci protrude. In *Culex,* the tip of the female's abdomen is usually blunt, with the cerci retracted. *Aedes* larvae have only a single pair of hair tufts on the breathing siphon, and it is relatively short and stout (figure 275, *B*). *Culex* larvae have more than 1 pair of hair tufts on the breathing siphon, and it is relatively long and slender (figure 275, *C*).

The adult females lay their black, spindle-shaped eggs (figure 278, *G* and *I*) singly on surfaces that are subject to flooding but are above the water table at the time of deposition. The females die soon after depositing their eggs, and the winter is passed in the egg stage in all regions having cold winters. Some species of *Aedes* breed in coastal salt marshes that are flooded by unusually high tides, others breed in the vicinity of rivers and streams that annually overflow their banks, and still others breed in irrigated fields or pastures. They may oviposit in debris or on the surface of the soil when it is moist but not waterlogged. Some species oviposit in treeholes, leaf axils, or artificial containers, attaching their eggs to the walls just above the receding meniscus. Their eggs can remain exposed above water for up to a year or more without desiccating. They submerge and hatch later after rains have caused streams and rivers to overflow their banks, or have filled treeholes, or after irrigation waters have filled depressions in which eggs were laid.

Eggs of the "snow mosquitoes," feared by early visitors to the high mountains, hatch only in the cold water of early spring, and there is only 1 generation per year. There are about 6 species involved. With rapidly increasing numbers of people in mountain resorts and recreational areas, their relative importance is likewise increasing. The snow mosquitoes are also a nuisance to farmers who pasture their cattle in mountain valleys.

Gjullin *et al.* (1930) found that the eggs of the 2 "floodwater" species they investigated, *Ae. vexans* and *Ae. sticticus,* hatched only in water of low oxygen content. Newly hatched larvae were found only in shallow, quiet waters where conditions were most favorable for reduction of oxygen by bacteria and organic material. Thus, the eggs of these mosquitoes hatch only when food material for the larvae is abundant. In the absence of flooding, eggs of *Ae. vexans* and *Ae. sticticus* were found to survive for 2 or 3 years, and small numbers even for 4 years. Despite the long periods that the eggs of floodwater mosquitoes can survive desiccation, the adults oviposit only on moist soil. The explanation seems to be that (1) a damp oviposition site is likely to be reflooded before the near-by dry areas, and (2) newly laid eggs will remain viable only in a very humid environment (Telford, 1957).

Western Treehole Mosquito, *Aedes sierrensis* Ludlow

This species is widely distributed along the Pacific Coast, from British Columbia to southern California and east to the summit of the Sierra. It also occurs along the desert margins in western Nevada, southern Utah, and southern California. In the literature earlier than 1957, it is referred to as *Aedes varipalpus* (Coquillett), but is now considered to be one of the 4 species of the "*varipalpus* complex." The other 3 species, *Ae. varipalpus, Ae. monticola* Belkin and McDonald, and *Ae. deserticola* Zavortink, are relict species, with relatively limited distributions in the interior portions of the southwestern United States and in Baja California. Along the desert edges of southern California, *Ae. sierrensis* and *Ae. deserticola* have sometimes been found together in the same treehole (Zavortink, 1969).

Aedes sierrensis bites fiercely, and can become a severe pest during the warm days of late spring. Its importance as a pest is increasing as residential subdivisions are being developed adjacent to metropolitan areas. When the larvae are too crowded, the adults may be so small that they can pass through ordinary fly screens, but they do less biting indoors than outside (Freeborn and Bohart, 1951).

Description. This is an attractive, small, black mosquito, with bright, white rings at both the bases and apices of the tarsal segments, and with many white or silvery scales distributed over the body to give a silver-mottled appearance (figure 272). There are somewhat triangular white bands at the bases of the abdominal segments (figure 279, *C*).

Biology. *Aedes sierrensis* usually deposits its eggs on the sides of treeholes. The eggs hatch whenever they are wet by the rising water following periods of heavy rainfall. The larvae are straw-colored, with brown heads and enormously developed gills (figure 275, *B*). They spend most of their time with their heads and thoraxes buried in the silty deposits at the bottoms of the treeholes.

The period of development is extremely long, lasting from 1 to 7 months. There are 2 pronounced peaks in the broods, 1 in the early summer and another in the fall. Eggs deposited by the fall adults hatch into larvae that overwinter (Freeborn, 1926).

Habitat. The western treehole mosquito most commonly breeds in holes in live oaks, valley oaks, and California laurel, but can also breed in black walnut, maple, sycamore, eucalyptus, and orchard and dooryard trees. However, it has also been found in rain barrels that contained a heavy sediment of decaying leaves (Freeborn and Bohart, 1951).

An interesting ecological niche is provided by the unique ability of redwood trees (*Sequoia sempervirens*) to develop new growth after they have been cut. The new growth provides shade for the cavity formed on top of the stump when a portion of the central core is pulled out as the cut tree falls over. The shade retards evaporation of rainwater in the cavity, allowing it to remain long enough for completion of the aquatic stages of *A. sierrensis* (Strickland, 1969).

Eastern Treehole Mosquito, *Aedes triseriatus* (Say)

This is a common treehole mosquito in the eastern United States, west to Montana, and south to Texas. It is an aggressive biter, and its bite is painful. The mesonotum of the female is black, and is densely clothed with narrow scales that are dark brown on the dorsum and silvery white laterally and on the anterior margin of that body region. The abdomen is black, and there is a large, triangular patch of white scales on the sides of each segment (figure 272). The abdomen is white-scaled ventrally, with the apices of the sixth and seventh segments bluish black. The wing scales are brown, and the legs are black, with tips of the femora white (Matheson, 1944).

Field Mosquitoes, *Aedes dorsalis* (Meigen) and *Ae. melanimon* Dyar

Aedes dorsalis occurs throughout the United States, southern Canada, and in Europe and Asia. The females are vicious biters, either by day or night, but are particularly active in the evening or on calm, cloudy days. They are stronger fliers than most mosquitoes, and are commonly found 10 miles (16 km) or more from their breeding places. They commonly breed in the salt marshes of coastal areas, but a fresh-water variety is often associated with lakes in the Sierra Nevada and in the southeastern and far northern regions of California (Loomis *et al.,* 1956). *Aedes dorsalis* is particularly bad in the cattle ranges of the east slope of the Sierra, up to 7,000 ft (about 2,100 m) elevation (Freeborn and Bohart, 1951). What was once considered to be *Ae. dorsalis* in irrigated pastures and waste-water pools in the interior valleys of California is now known to be *Ae. melanimon.* Mixed collections of the 2 species have been made only in brackish water, delta areas of Solano County, along the west edge of the San Joaquin Valley, and in Owens Valley (Loomis *et al.,* 1956). In irrigated areas, several broods of *Ae. melanimon* are produced during each year, one following each flooding.

Description. The field mosquitoes are medium-sized insects, varying in color from dark brown to a dusty straw color. The abdomen is marked above with a longitudinal stripe of pale scales, and the hind tarsi are banded with yellowish scales at both ends of the segments (figure 279, *B*), differing in this respect from *Ae. nigromaculis,* on which the tarsal segments are banded only at the basal ends. *Aedes melanimon* is difficult to distinguish from *Ae. dorsalis* except on the basis of microscopic examination of diagnostic characters of the males and larvae (Loomis *et al.,* 1956).

Irrigated-Pasture Mosquito, *Aedes nigromaculis* (Ludlow)

This is an important pest mosquito throughout the western plains, from Minnesota to Manitoba and Washington and south to Texas and Mexico. It was first taken in California in 1937, but now occurs throughout the state in open, sunlit pools of waste and intermittent water. It is now the principal mosquito pest in most areas of the San Joaquin Valley, attacking people and animals and inflicting a painful bite. It will bite during the day, but is most active in the evening (Pratt *et al.,* 1963).

Description. This is a medium-sized, dark-colored mosquito, with a longitudinal line of yellowish-white scales on the upper surface of the abdomen. It has a band of white scales at the base of each tarsal segment, but not at the apex, differing in this respect from *Ae. dorsalis* and *Ae. melanimon,* with which it is often associated. The latter 2 species have both base and apex ringed with yellowish bands. It also generally has a white-banded proboscis, while the others do not.

Biology. Although *Ae. nigromaculis* probably has only 1 or at most 2 or 3 broods per year in the arid sections of the western plains, in an environment of permanent pastures and abundant waste irrigation water its biotic potential is enormous. In the central valley of California, where pastures are irrigated every 8 to 12 days, a brood may be produced in 5 days, and there may be as many as 20 broods in a single season. It is especially well adapted to pasture irrigation because its eggs hatch in 2 to 6 days after oviposition. Considerable basic research has been done on the biology of this and other floodwater mosquitoes in recent years, particularly with reference to the influence of photoperiod and temperature on diapause in eggs (Telford, 1958, 1963; McHaffey, 1972).

Aedes vexans (Meigen)

This is probably the most widespread mosquito in the United States, and is also the most abundant and troublesome in many areas. It is a vicious biter outdoors, and is especially annoying at dusk and later. It is widespread in California, where it is but 1 of the 3 principal species of "floodwater mosquitoes," the others being *Ae. increpitus* and *Ae. sticticus.* When abatement work was begun in northern San Joaquin County

(California) in 1945, these 3 species in the jungle-covered river bottom of the Mokelumne River were the principal mosquito problems. They were even more of a problem than the pests of irrigated pastures, *Ae. nigromaculis, Ae. melanimon,* and *Culex tarsalis.* Ten years later, in the same river-bottom country, it was difficult to find specimens for a collection, indicating the outstanding success of the abatement program (Peters, 1957).

Description. *Aedes vexans* is a medium-sized brown mosquito. It is easily identified by the band of white scales at the base of each abdominal segment, indented with a central V-shaped notch that often separates the band into 2 parts. There are also basal tarsal bands, creamy to bronze in color, often so dark that they are easily overlooked.

Biology. In California, *Ae. vexans* is primarily a floodwater species in woodland water courses, but it is also occasionally found in open pasture land in intermittent pools, along with *nigromaculis* and *dorsalis.* It can become a problem in irrigation ditches and drainage sloughs where there is enough shade. In this respect, it differs from its eastern populations, which breed in open, sunny marshes and rain-filled road ruts. Eggs are deposited in shallow depressions subject to inundation. No adverse effect to eggs in natural sites results from prolonged freezing. There may be 1 or more generations per year, depending on the number of inundations of oviposition sites. The adults rest during the day in grass and other vegetation, but seldom in shelters (Freeborn and Bohart, 1951; Horsfall, 1955).

Aedes increpitus Dyar

This is one of a "group" of 4 loosely related species, the others being *Ae. stimulans* (Walker), *Ae. excrucians* (Walker), and *Ae. fitchii* (Felt and Young), and is the only one of these species confined to the western states. The others are among the most abundant and annoying of the "woodland" mosquitoes in many of the northern states, and bite readily in the daytime. *Aedes increpitus* ranges between the Rocky Mountains and the Pacific Coast as far south as Arizona and southern California. In California, it was found originally only in the Sierra and Coast Range at altitudes of 4,000 to 5,000 ft (1,200 to 1,500 m), and was the species that once made Yosemite Valley almost uninhabitable, but it was eradicated there. It now occurs in the central valleys, breeding in overflow pools in wooded areas in company with *Ae. vexans* (Freeborn and Brookman, 1943; Freeborn and Bohart, 1951).

Description. *Aedes increpitus* has basal abdominal white bands, broadest in the middle, and a quadrate patch of white scales at the sides that are so much whiter than those in the band that they can be used effectively for field identification, distinguishing *Ae. increpitus* from other floodwater species. White basal bands occur on the tarsi (BVC, 1951; Freeborn and Bohart, 1951).

Biology. The winter is passed in the egg stage. The adults may live most of the summer, there being only 1 generation per year.

Floodwater Mosquito, *Aedes sticticus* (Meigen)

This is a floodwater species that occurs throughout most of the United States, but is most abundant in the northern states. In California, it is found throughout the central valleys as far south as the Kings River. It prefers comparatively large bodies of flood water rather than the small, detached pools favored by *Ae. vexans* (Freeborn and Bohart, 1951). Following floods, the adults can be very abundant. They bite fiercely during the evening, or during the day when it is cloudy or when they are in shady places.

Description. *Aedes sticticus* is a medium-sized species. The thorax is dotted with pale scales, and the legs are speckled with white scales, but not banded. The adults can be separated from those of *Ae. vexans* by their all-black tarsi, and the larvae by the relatively small anal saddle (Freeborn and Bohart, 1951).

Biology. The female prefers to oviposit in a loam soil with dead or live vegetation, or both, rather than in bare areas exposed to sun and wind. The eggs overwinter and hatch in spring or summer, there being only 1 generation per year. The eggs can survive for 2 or 3 years if flooding does not occur. The females may live as long as 3 months, and have a flight range as far as 25 or 30 mi (40 or 48 km) (Pratt *et al.,* 1963).

California Saltmarsh Mosquito, *Aedes squamiger* (Coquillett)

This mosquito is restricted to salt-water breeding, and ranges from Sonoma County, California, south into Baja California. It can migrate for long distances inland, particularly along wooded streambeds, and can invade residential areas and become a troublesome pest—it is a vicious daytime biter and is even worse at dusk. It has a very feeble hum, and prefers to attack under cover, as about the ankles, under coatsleeves, and on the back of the neck.

Description. In late spring, the large, brownish-gray adults fly in company with *Ae. dorsalis,* but can be distinguished by their larger size, dark color, and speckled wings. They also lack the pale longitudinal band on the abdomen and apical bands on the tarsal segments, such as possessed by *Ae. dorsalis.*

Biology. This species breeds in saltmarsh pools that are fed by fresh water from rains. The eggs are laid in the marsh, and even though flooded by high tides, they do not hatch until the first rains of late fall or winter (Bailey *et al.,* 1954).

Dark Brown Saltmarsh Mosquito, *Aedes taeniorhynchus* (Wiedemann)

This species ranges along coastal plains from Massachusetts to Texas and as far as Brazil in South America. In California, it replaces *Ae. dorsalis* along the coast south of Monterey, and its range extends down to Peru. The adults are strong fliers and fierce biters. They are active principally at night, but can be very annoying in the shade in the daytime.

Description. This is the only California member of the genus that has a white band on its proboscis. *Aedes nigromaculis* sometimes has a concentration of pale scales midway on its proboscis, but its palpus is solid black, while that of *Ae. taeniorhynchus* is tipped with white scales. South of Monterey, the adults of *taeniorhynchus, squamiger,* and *dorsalis* may be found together. However, the dark brown of *taeniorhynchus* can be readily distinguished from the grayish and black color of *squamiger* and the light brown to straw color of *dorsalis.*

Biology. *Aedes taeniorhynchus* larvae can be found in the same pools as those of *Ae. squamiger,* but can be easily distinguished because the anal segment is completely ringed by the "saddle," while the anal segment of *Ae. squamiger* is only partly encircled. In California, *Ae. taeniorhynchus* occasionally breeds in fresh water adjacent to salt marshes, but unlike *Ae. dorsalis,* it does not breed in inland bodies of fresh water. On the other hand, there are some areas in the southwestern United States where it breeds in fresh water far from salt marshes (Freeborn and Bohart, 1951).

At least along the warmer coastal marshes of the United States, there are multiple and continuous generations of *Ae. taeniorhynchus.* When the eggs are ready, only rainfall or other sources of water are required to cause them to hatch. Development is then very rapid. In Florida, for example, the aquatic stages require only 5 to 7 days in summer and 12 to 15 days in winter (Nielsen and Nielsen, 1953).

Yellowfever Mosquito, *Aedes aegypti* (L.)

This species is widely distributed throughout the tropical and subtropical regions of the world, but does not occur in areas where nights are cool (68° F [20° C] and below), even in areas considered to be subtropical, as in California. It is common in the southern states, and is important as a potential carrier of yellow fever, dengue fever, and experimentally, eastern equine encephalitis.

Description. The adult is beautifully and conspicuously marked with silvery-white or yellowish-white bands and stripes on a nearly black background (figure 281). Dorsally on its thorax it has 2 outer, curved, yellowish-white lines and 2 medium parallel lines, creating a pattern that has been called "lyrelike." The tarsi are banded with white.

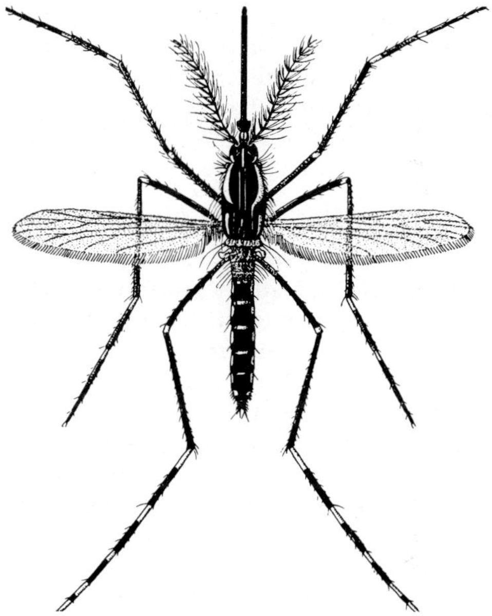

Fig. 281. Yellowfever mosquito, *Aedes aegypti.* (Courtesy C. S. Papp.)

Biology. Mating takes place while the adults are in flight. The females can live for a long time on honey or nectar, but must have a blood meal to initiate ovarian development. A female can lay 50 to 150 eggs during her lifetime. These can remain viable out of water for at least 5 months, and the dry season may be passed in the egg stage. This species has been known to develop from egg to adult in 9.5 days, but it generally takes 11 to

18 days. The female can live 4 months or more. The mature larvae are rather stout, and have a comparatively short and somewhat pointed siphon. The slightest disturbance, even a passing shadow, causes them to dart to the bottom of the water. They press themselves close to the bottom, and are not easily dislodged (Matheson, 1951).

Habitat. This is probably one of the most "domestic" of all mosquitoes, being usually found near buildings, often within them, and prefers to bite humans. Its breeding sites were once believed to be almost entirely artificial—abandoned automobile bodies, refrigerators, and other appliances, as well as tin cans, bottles, jars, and the like. It is known that *Ae. aegypti* will breed in artificial containers at distances of up to a mile (1.6 km) from the nearest human habitation, as well as in treeholes and other natural water containers (Schliessmann, 1966). Treeholes are believed to have been the original breeding sites for *Ae. aegypti.* Its oviposition habits in both treeholes and artificial containers are similar to those described for the western treehole mosquito, *Ae. sierrensis.*

The Genus *Culex*

The genus *Culex* contains many species, especially in the tropical and warmer temperate regions, but some species occur in the colder northern regions. The adults are usually the most common "pest mosquitoes," particularly in urban areas. Some species are vectors of arboviruses and avian malaria. The overwintering females oviposit on the surface of the water. The eggs are spindleshaped, about 2 mm long, and are stacked together as they are laid, each one in a vertical position. They adhere to one another, forming raftlike masses, commonly of several hundred eggs.

Only the lower half of each egg is wet. The compact egg rafts collectively appear like masses of soot floating on the water. The larvae of some important species of *Culex* live in water that is either foul or supports microbial activity. They feed on the flocculent detritus, and often breed in water that contains sewage.

Southern House Mosquito, *Culex pipiens quinquefasciatus* Say

This species occurs in all the southern states from coast to coast, and ranges northward into Nebraska, Iowa, Illinois, and Ohio. The closely related northern house mosquito, *C. pipiens pipiens* L., difficult to distinguish from *C. p. quinquefasciatus,* is found throughout the northern United States and ranges as far south as Georgia and Oklahoma. Probably one or both of these subspecies occurs in every one of the states (Pratt *et al.,* 1963). Both are found in California. As the common name implies, this mosquito invades houses freely. It usually enters through defective screens, cracks around ill-fitting doors, floor furnaces, or by way of the furnace flues. Its high-pitched hum, continued late into the night, and its vicious bites make it one of the most annoying of the "pest mosquitoes."

Description. This is a medium-sized, dull-brownish or yellowish mosquito, with prominent, uniform, white bands at the bases of the abdominal segments (figure 279, *D*).

Biology. The egg rafts (figure 278, *A*) contain 50 to 200 eggs. They hatch in a day or two in warm weather, and the larval (figure 275, *C*) and pupal (figure 276, *F*) stages are completed in another 8 or 10 days, although 2 weeks or more may be required in cooler weather. In any case, since there are quite a few generations per year, the potential for enormous mosquito infestations in a single season is obvious.

Habitat. During the day, the adults rest in and around homes or other buildings and various shelters, such as culverts or bridges, near their breeding places. They are active only at night, and are readily attracted to lights. The larvae can develop in ponds, dairy drains, tanks, old wells, rain barrels, cans, and other artificial receptacles.

Encephalitis Mosquito, *Culex tarsalis* Coquillett

This species ranges throughout the United States west of the Mississippi River and into Canada and Mexico. In California, it occurs from sea level up to 7,600 ft (2,300 m) in the Sierra, but is principally a rural mosquito. It is the most ubiquitous and one of the most abundant of California mosquitoes. Investigations made in Fresno County, California, over a 5-year period have indicated that *C. tarsalis* moves into foothill areas in the fall, and that the females overwinter there in culverts, under bridges, in barns or sheds, and in other sheltered places (Kliewer *et al.,* 1969). After breeding in January, they move out of the foothill area to the San Joaquin Valley without breeding significantly in the foothills. In the Imperial Valley of California, diapause is so weak that this species remains active all winter, and would appear to be a good candidate for overwintering arboviruses, although no direct evidence of this is available (Nelson, 1971).

Description. *Culex tarsalis* is a medium-sized dark mosquito, with a broad, white band at the middle of the proboscis. This separates it from nearly all other species of *Culex* in North America north of Mexico (Freeborn and Bohart, 1951). It also has white bands at each end of the tarsal segments (figure 279, *E*).

Biology. The egg rafts contain 200 to 300 eggs, hatching normally occurs in 2 or 3 days, larval and pupal stages develop rapidly, and breeding continues from spring until late fall. In the laboratory at 74 to 75° F (23 to 24° C), development from egg to adult required approximately 18 days (eggs 3, larvae 12, and pupae 3 days) (Thomas, 1963).

Habitat. These mosquitoes are most active soon after dusk, and may enter buildings in search of a blood meal. During the day, the adults may rest in secluded spots, such as on porches, on the shaded sides of buildings, or under bridges, but the majority rest in grass and shrubs or along cut banks of streams (Pratt *et al.*, 1963). In southern California, females have been found during the winter in nests of the wood rat (*Neotoma*). In arid and semiarid regions, *C. tarsalis* is most frequently found in temporary or semipermanent bodies of water associated with irrigation or sustained wetness, including canals, ditches, burrow pits, impoundments, ground pools, and hoofprints. However, it can also breed in the effluent from cesspools and in artificial containers such as cans, jars, barrels, drinking troughs, ornamental ponds, and catch basins. In California, the form occurring near sea level develops primarily in foul water, but the one found in mountains often occurs in clear, cold, unpolluted water (Jenkins, 1950).

Encephalitis. *Culex tarsalis* commonly feeds on the blood of cows, horses, and humans, but shows some preference for birds. It is believed to be the most important vector of western equine encephalitis to man and horses in the western states. This and other encephalitides are discussed later in this chapter.

Culex peus Speiser (=*stigmatosoma* Dyar)

This is one of the most common and widespread species in California, and also occurs in Oregon, Utah, and south through Texas to Mexico and northern South America. It apparently does not bite man, and probably confines its feeding to birds.

Description. *Culex peus* is similar in appearance to *C. tarsalis,* and the 2 species are the only ones in the genus in California that have a white-banded proboscis. However, *C. peus* has a median black spot on each abdominal segment in a field of white scales instead of the inverted "V" of *C. tarsalis,* but this difference is not well marked in all specimens (Freeborn and Brookman, 1943).

Habitat. This species breeds in almost all types of nonsaline water, in ground pools and artificial containers, and is the dominant mosquito in the oxidation ponds of sewer farms in some areas.

The Genus *Culiseta*

The genus *Culiseta* contains a few species of large mosquitoes inhabiting temperate regions. *Culiseta* adults can be distinguished from those of *Culex* by the presence of a patch of bristles on the lower surface of the wing, on the subcostal vein. Also, *Culiseta* has characteristic bristles in front of the thoracic spiracle, whereas *Culex* does not. In contrast to *Culex* mosquitoes, the adults of *Culiseta* are sluggish biters, attacking humans, horses, and cattle. There are only 10 species in the United States, of which 5 are fairly widespread. They are somewhat similar in appearance and habits to *Culex,* but are relatively unimportant as pests. The females lay their eggs (figure 278, *F* and *H*) in rafts, like *Culex.* They are typically winter mosquitoes, all stages being found from September through May or June, but in coastal areas some species may undergo development throughout the year. The larvae are generally found in permanent bodies of water, but a few develop in artificial receptacles. They may be found in foul water or in clear water over foul sediment. Species of *Culiseta* may occur abundantly in drainage from stables that contains urine, in pools with malodorous sediment at the bottom, or in water beneath roosts of blackbirds and waterfowl in cattail marshes (Horsfall, 1962).

Broadwinged Mosquito, *Culiseta inornata* (Williston)

This species ranges over the entire Pacific area, including southern Canada, but occurs in greatest numbers in southern California. It has been reported from almost all the states except in upper New England. It does not readily attack man, but can be a great annoyance to livestock.

Description. This is a large, grayish-brown species, with brown mouthparts, 2 bare, pale-brown lines on the mesonotum, pale-yellow, irregular bands at the base of each abdominal segment, and broad, lightly scaled wings (plate VIII, 2; figure 279, *F*). The broad, unspotted wings and unbanded tarsi separate it from all other species of *Culiseta* except the rare *C. impatiens* (Walker).

Biology. *Culiseta inornata* breeds in woodland pools by choice, sometimes in very cold water, but its larvae (figure 275, *D*) can also be found in any stagnant water, including artificial receptacles, and they are able to tolerate a considerable degree of salinity.

There are multiple generations, with their activity varying from the fall, winter, and spring in the southern part of its range to summer only in the northern part. It estivates in the southern area.

Rainbarrel Mosquito, *Culiseta incidens* (Thomson)

This mosquito occurs in Texas, Oklahoma, Nebraska, and all states to the west and north up to Alaska. It is one of the most widely distributed mosquitoes in California, and breeds from below sea level to elevations of at least 9,500 ft (2,900 m), in brackish water on the coast and spring water and snow pools in the mountains. The larvae are more partial to shade and cool water and less tolerant of salinity than those of *Culiseta inornata*. They may be found in water collected in discarded tires, rain barrels, reservoirs, and hoofprints.

There could be physiological strains of this species, for in some areas they bite humans and are extremely annoying, while in others they can be induced to bite only with the greatest difficulty. In the latter situations, they feed more frequently on domestic animals than on man. For example, a horse may be bitten while his rider remains unmolested (Freeborn and Bohart, 1951).

Description. This is a large, dark mosquito, with spotted wings. The abdomen is black, with basal segmental white bands. The wings have patches of dark scales clumped together, and the hind tarsi have narrow, white bands. In both these characteristics, *Culiseta incidens* differs from *C. inornata*.

The Genus *Psorophora*

Adults of this genus can be distinguished from those of the closely related genus *Aedes* by the presence of spiracular bristles. The larvae are characterized by "the single pair of hair tufts on a swollen air tube and by the comb scales of the eighth segment arising from a plate" (Freeborn and Brookman, 1943). There are 13 known species of *Psorophora* in the United States, 10 of them rather widely distributed in the southern and eastern states, They behave like the floodwater species of *Aedes*.

Dark Ricefield Mosquito, *Psorophora confinnis* (Lynch-Arribalzaga)

This species ranges from the southeast corner of California (Coachella and Imperial valleys) east throughout the southern states and north to Nebraska, Iowa, New York, and Massachusetts. It also occurs in Central and South America. In Florida it is known as the glades mosquito, and in the southeast California desert area it is called the Coachella Valley mosquito. These insects are attracted to lights. They are vicious biters, day or night, and it is sometimes almost unbearable to remain outdoors in infested areas. Control is closely associated with proper irrigation (Bailey *et al.*, 1954).

Description. This is a medium to large, dark species (figure 282), with speckled wings and with the proboscis, tarsi, and abdomen banded with white. It can be readily separated from the similarly marked *Coquillettidia perturbans* by the abdominal light banding being on the apices of the segments and by the narrower scales on the wings.

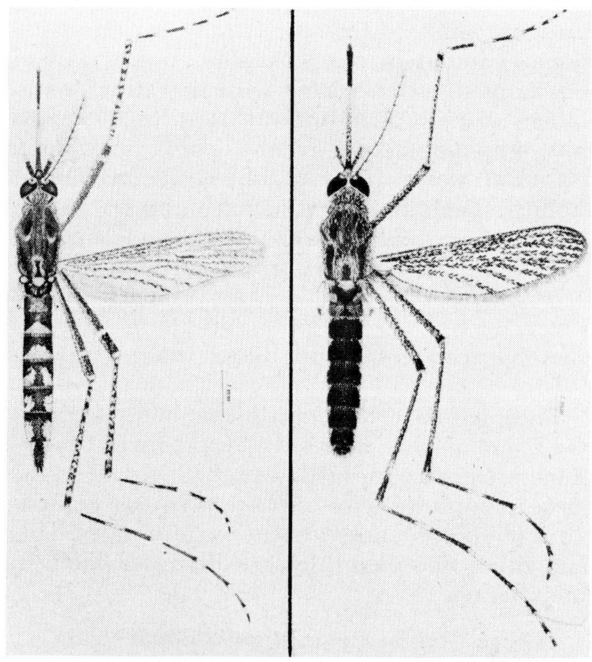

Fig. 282. ***Psorophora confinnis (left)* and *Coquillettidia perturbans.* (From Yamaguti, 1952.)**

Biology. The eggs are laid above the existing water level on damp ground, and are able to withstand a long period of desiccation. Drained rice and alfalfa fields are among the most favorable sites for oviposition. The eggs will hatch in 4 or 5 days if submerged in rising water immediately after oviposition. Once they have remained on the surface of the soil 2 or 3 weeks or longer, hatching may begin within a few minutes after they are wet. In midsummer, the larval period may be as brief as 4 days, and the pupal period, 1 or 2 days. Great swarms of adults can originate from pools of water that last a few days under the hot desert sun. A brood can develop with each flooding of a field. The adults live 1 or 2 months, and have a flight range of at least 10 mi (16 km).

Psorophora ciliata (F.)

These vicious biters are widely distributed in the eastern United States east of the plains states from Canada to Mexico, and also occur in parts of South America. The adults are very large (9 mm long), yellowish, and have shaggy legs (figure 283), which separate this species from others of the genus. The larvae are predaceous, feeding on those of other mosquitoes that breed in the temporary ground puddles inhabited by *P. ciliata.* The larval and pupal lives of this species are short, and hibernation takes place in the egg stage.

Fig. 283 ***Psorophora ciliata.***
(Courtesy C. S. Papp.)

The Genus *Coquillettidia*

Coquillettidia, formerly considered to be a subgenus of *Mansonia,* is now recognized as a separate genus, and is represented by a single species that is widespread in the United States.

Coquillettidia perturbans (Walker)

In California, these mosquitoes occur in various localities in the San Joaquin Valley, and can be severe pests near their breeding places. They breed in marshes, ponds, and lakes that have a thick growth of aquatic vegetation. The adults are attracted to lights. They are fierce biters, active principally in the evening and early part of the night, but will bite during the day in shady, humid places (Pratt *et al.,* 1963).

Description. This is a rather large, speckled, brown-and-white mosquito, with a rough, fuzzy appearance and with very broad wing scales, banded beak and tarsi, and pale basal abdominal bands (figure 282). The larvae are characterized by a peculiarly shaped siphon and long antennae.

Biology. The larvae are unusual in that they attach themselves to the roots of aquatic vegetation by means of their heavily sclerotized and serrate breathing siphons. Neither larvae nor pupae need to rise to the surface; their siphons and pointed trumpets, respectively, penetrate air cells of the roots to which they are attached; they can detect roots in which these air cells are abundant. Unlike most mosquitoes, larval development is quite slow; larvae that hatch during one season ordinarily complete their development the following spring. However, the pupal period is only 5 or 6 days, and the adults emerge in late spring or early summer (Freeborn and Brookman, 1943; Freeborn and Bohart, 1951; Jamnback, 1969). Overwintering larvae are killed if their ponds are drained. It is difficult to kill the larvae with insecticides because they do not come to the surface of the water. Granular formulations that sink to the bottom before releasing the toxicant have given the best results (Jamnback, 1969).

Attraction of Mosquitoes to Hosts

The differences in the attractancy of different individuals to insects are so conspicuous that they have become well known to most people. Brown (1966) discussed the factors determining the attraction of mosquitoes to their hosts. Among the airborne factors, moisture appears to be the most important, provided 100% relative humidity is not reached. Warmth is considered to be next in importance, and the combination of warmth and moisture is highly attractive. Carbon dioxide, as an airborne factor, is thought by some investigators to be an activator rather than an attractant, but on the other hand, carbon dioxide was attractive to *Aedes aegypti* when simply added to a dry airstream, and dry ice provided an effective field trap for *Culex tarsalis.*

Dark colors have been found to be more attractive than light ones for some species, but not for others. Dark hands were found to be more attractive than pale hands, and warm hands more attractive than cool ones. Men were more attractive than women to *Aedes aegypti.* The attractiveness of women during the menstrual cycle was directly correlated with the urinary excretion, and presumably blood concentration, of estrogens. Also, the comparative attractiveness of two men was found to be proportional to their comparative blood and urinary estrogen levels. Certain amino acids, such as alanine, arginine, cystine, tyrosine, and threonine, were attractive, some with and some without the addition of carbon dioxide. Brown arrived at this conclusion:

> The experimental evidence suggests that amino acids and estrogens each constitute a group of compounds capable of enhancing the base attractiveness of moisture, warmth, and carbon dioxide, and that the spectrum of amino acids and estrogens put out by different host types might largely determine the host selection exhibited by various species and races of mosquitoes, and the difference in attractiveness between the sexes and individual humans.

Toxicogenic Properties of Secretions

Despite the fact that we hear or read about the "bites" of mosquitoes, black flies, deer flies, midges, and related insects, it is a piercing and bloodsucking action that is ordinarily involved (figure 272). The initial wound can range in severity from that caused by the delicate mouthparts of midges to that caused by the large, bladelike cutting instruments of the horse flies, the latter being very painful. However, the bites of some gnats with small mouthparts, such as certain species of *Culicoides* and *Simulium* are also painful. The wound trauma is generally inconsequential, but there can be accompanying allergic reactions, intense itching, and even anaphylactic shock. Sensitization can result from successive bites of mosquitoes, black flies, biting midges, and deer flies.

The wheals and erythema resulting from mosquito bites have been attributed to mechanical damage or irritation of the skin caused by specific toxins or histamine injected by the insect. It has been suggested that the specific allergens may be mucosubstances, proteins, or peptides. Bites of *Aedes stimulans* did not produce skin reactions (wheals) when its salivary glands were cut. However, the pain from the bite itself was greater, indicating the presence of an anesthetic in the saliva (Hudson *et al.*, 1960*a*).

Oral secretions have been obtained from yellowfever mosquitoes, *Aedes aegypti,* by inducing them to probe membranes containing distilled water. Rabbits and guinea pigs were sensitized by injecting them with these secretions. Nearly all animals then showed positive reactions when bitten by *Aedes aegypti,* but not when bitten by *Anopheles quadrimaculatus.* Hypersensitivity could be transferred from a sensitized to a normal animal by injecting serum or lymphoid cells from the former into the latter. An "immediate" type of sensitivity was transferred by the serum and a "delayed" type by the cells, but not by the serum. The experiments demonstrated for the first time that antibodies were involved in the reactions of animals to mosquito bites (Anonymous, 1964*c*).

Using 25 human volunteers who had never been exposed to *Aedes aegypti* mosquitoes, Mellanby (1946) studied their reactions to bites by that species on several occasions over an extended period. He described their reaction to the first bite as follows:

> There was no immediate cutaneous response, other than a tiny red spot about 1 mm in diameter at the site of the bite, and no itching was observed. After a variable period, however, usually between 20 and 24 hours, a marked delayed reaction occurred. A red patch about 3 cm in diameter surrounded the bite, and the central 1 cm was seen to form a definite wheal. This condition was observed over several days, the itching and other symptoms waxing and waning several times.

A month later, after several more bites, the reactions were markedly different and were described as follows:

> An immediate reaction had developed: As soon as the mosquito had fed, a wheal developed at the site of the bite, an area of erythema appeared surrounding the wheal, and the skin itched. Within 2 hours, all these symptoms disappeared completely, but after 20 to 24 hours, the same delayed reaction noticed on the occasion of the earlier exposures appeared.
>
> After a further period of exposure, there was another modification. The immediate reaction persisted unchanged, but the delayed reaction got gradually less severe and eventually disappeared. This meant that the bites were less troublesome to the victim.

No further experimentation was done with the 25 volunteers, but in further tests with persons exposed to thousands of bites, Mellanby observed that the immediate reaction also eventually disappeared, i.e., the victim suffered neither immediate nor delayed reaction.

Reactions to mosquito bites are usually less severe than to bee stings. However, the vesicles, papules, bullae, edema, and pruritus of the average immediate or delayed reaction are occasionally complicated, in some particularly allergic individuals, by severe reactions, such as necrosis or scarring, hivelike lesions, hemorrhagic spots, urticarial lesions, nausea, vomiting, and high temperatures. Case histories of such involvements were presented by Frazier (1969).

Desensitization to one genus of mosquitoes does not necessarily involve desensitization to another. Some travelers to tropical regions have been so impressed by the widespread immunity of the native populations to stinging insects as to conclude that it was a racial characteristic. However, the immunity is considered by experts to be the result of regular exposure of the native people to the bites of the insects from the time of their birth (Gordon and Crewe, 1948).

The subject of systematic desensitization by hypodermic injection of graduated quantities of venom is discussed near the beginning of this chapter.

Mosquitoborne Diseases and Their Vectors in the United States

Malaria

Malaria is a widely distributed disease, occurring to some extent in every continent and on many islands of the oceans. It is an acute, intermittent, febrile ailment, prevalent between 45° N. and 40° S. around the world. As might be expected from the importance of the subject, the literature on malaria is overwhelmingly voluminous, yet all aspects of the multifaceted malaria problem from a global standpoint are discussed in a monumental treatise of 2 volumes titled *Malariology,* with chapters by 65 scientists, edited by M. F. Boyd (1949). Most of what is currently known about the epidemiology of malaria was already known in 1949, and Boyd's work continues to be a useful reference.

The disease is characterized by periodic attacks of chills and fever that coincide with mass destruction of blood cells and the release of toxic substances by certain protozoan parasites at the end of each of their reproductive cycles. It sometimes persists in individuals for several years. Although with modern methods of treatment the disease is seldom fatal, the large number of people contracting malaria in areas where control methods are inadequate results in the disease being very debilitating to morale and social and economic welfare.

In humans, malaria is caused by 4 species of protozoa belonging to the genus *Plasmodium* (*P. vivax, P. falciparum, P. malariae,* and *P. ovale*). Each causes somewhat different symptoms (Boyd, 1949; Foote and Cook, 1959). The parasites are invariably transmitted by *Anopheles* mosquitoes. When they feed on infected human blood, the female mosquitoes draw up the parasites along with the blood. These parasites enter the gut of the mosquito, undergo a series of developmental stages, eventually invade the salivary glands, and are injected by the mosquito at a subsequent feeding. Of approximately 400 species of *Anopheles* mosquitoes, 85 have been found to transmit malaria. Very little is known as to why only *Anopheles* can transmit malaria or why only certain species have been implicated (Foote and Cook, 1959).

In the United States, where 2 species of *Anopheles, A. quadrimaculatus* to the east of the Rockies and *A. freeborni* west of the Rockies, are potentially important vectors of malaria organisms, malaria was once a very serious problem. But all foci of natural infection by malaria in the United States have been eradicated, and the country is free of endemic malaria. An attempt has been made to prevent its re-establishment, and this attempt will continue as long as malaria occurs in other parts of the world. This is being accomplished by exhaustive surveillance by state epidemiologists in collaboration with the Malaria Surveillance Unit of the Center for Disease Control of the U. S. Public Health Service (Bradley, 1966.)

The spectacular results of DDT in mosquito control during and immediately following World War II resulted in the general belief that the malaria problem would be effectively and permanently resolved by vector control. Increasing resistance of *Anopheles* mosquitoes to pesticides and of plasmodia to drugs has dispelled such optimism, and current efforts are directed toward control and containment where possible. Ultimate solution of the malaria problem is believed to be dependent upon better knowledge of the parasite and its interaction with hosts, including subhuman primates, and vectors.[22]

Intensive biochemical investigations of malarial parasites are in process. Considerable progress has been made in chemoprophylactic medication, for example, the chloroquine–primaquine tablet (Winkler, 1970).

Of probably even greater importance for the future is the growing interest in the possibility of successful immunization to prevent malaria. Fortunately, basic research is again flourishing, as

[22] The problem in the United States has also been complicated in recent years by more than 3,000 cases of malaria annually reported that appeared to be transmitted by needles. Over 40 cases of malaria caused by *Plasmodium vivax* occurred in Kern County, California, in 1971. The infected persons were among a group of over 400 heroin users that included an infected Vietnam veteran. The last report of mosquitoborne malaria in Kern County was in 1938. Work of the Kern County Mosquito-Abatement District has resulted in the county's being an "unreceptive" area for mosquitoborne malaria, but there are many regions in the nation that still have enough anopheline mosquitoes for potential mosquito transmission of the disease (Reeves, 1971).

can be seen from the magnitude of the series of panel workshops that have been held at the Walter Reed Army Medical Center in Washington, D.C. Sixty-two scientists participated in the May, 1969, workshop, and their 64 papers were published together that year in a special 577-page issue of *Military Medicine*.

Yellow Fever

Yellow fever is a viral disease, of which there are 2 epidemiological types, called "urban" and "jungle." The former is transmitted from man to man by the yellowfever mosquito, *Aedes aegypti*. Urban yellow fever once occurred throughout much of South and Central America and Africa, and was repeatedly introduced into the United States. However, it has not occurred since 1905, and there have been no major epidemics in the Americas since 1942 (Pratt *et al.*, 1963).

Dengue Fever

Aedes aegypti also transmits from man to man an acute, but rarely fatal, viral disease known as dengue or breakbone fever. The disease attacks suddenly, causing severe rheumatic pains in the joints and limbs, accompanied by headache and high fever. This lasts about 3 days, followed by a great diminution of intensity of the disease for 2 days. The second attack lasts about a day, and is accompanied by a rapidly spreading rash.

Encephalitides

Encephalitis is the only mosquitoborne disease of man now known to occur in the United States. It is a viral (arbovirus) disease that affects the central nervous system and causes an encephalitis or inflammation of the brain. The various encephalitides (types of the disease) are caused by different viruses that are normally transmitted from bird to bird and less commonly from bird to man or his domestic animals. The primary cycle for the virus involves *Culex tarsalis* as the vector and wild birds and domestic fowl as reservoir hosts (Hammon *et al.*, 1951; Reeves and Hammon, 1962). Humans, horses, and other mammals apparently do not circulate the virus in high enough concentrations to infect mosquitoes. When all types of virus are considered, encephalitis is distributed throughout the United States except for northern New England.

In California, there were 2,233 cases of encephalitis reported in the period 1945 through 1953, and 266 deaths—a fatality rate of 11.9%. The largest number of cases were reported in 1952, when there were 813: 375 cases of western encephalitis, 45 of St. Louis encephalitis, and 393 of undetermined etiology. There were 58 deaths from encephalitis in 1952, a fatality rate of 7.1% (Longshore *et al.*, 1956).

Most insectborne diseases, such as malaria, plague, tularemia, Rocky Mountain spotted fever, relapsing fever, murine typhus, etc., while still important, have dramatically diminished in importance in recent decades through a combination of newer therapeutics, suppressives, vaccines, and improved arthropod control measures. Because of the absence of specific therapeutic or suppressive measures or even a human vaccine, encephalitis has been a notable exception to this trend (Peters, 1959).

Western Encephalitis. This encephalitide occurs west of the Mississippi River and in Wisconsin and Illinois. An extensive interdisciplinary ecologic and epidemiologic investigation of western encephalitis (WE) or western equine encephalitis (WEE) in the period 1943–1952 was undertaken by W. C. Reeves and W. McD. Hammon (1962) and collaborators in Kern County, California, where encephalitis had occurred in man for years at high endemic levels. WE causes a dengue-like disease in man and an epidemic neurotropic disease in horses. It is generally not fatal to man, but causes a high mortality in horses. Nevertheless, from 1938 to 1959 there were 64 human deaths from WEE in California. *Culex tarsalis* is the most important vector. It apparently obtains the virus from feeding upon birds and later transmits it to other birds and to man and horses. The mosquito itself is not affected. Birds have been implicated as overwintering hosts of the virus (Ferguson, 1954; Kissling *et al.*, 1957; Reeves and Hammon, 1962). Possibly, it is also carried through the winter by *C. tarsalis* females (Hammon *et al.*, 1951; Blackmore and Winn, 1956; Kliewer *et al.*, 1969). The possibility that snakes may also be overwintering hosts for the virus has been suggested (Gebhardt *et al.*, 1964).

The important foci of infection on the West Coast occur in irrigated areas adjacent to wasteland, where various migratory waterfowl, blackbirds, and sparrows, together with *Culex tarsalis*, comprise the basic reservoir. Gray squirrels (*Sciurus griseus*) and ground squirrels (*Citellus beecheyi*) may be important reservoirs, at least in California. Domestic reservoirs of the encephalitis virus include the chicken, duck, pig, horse, and house (English) sparrow. The sparrow ranges between sylvatic and domestic foci, and could take the virus from one to the other. While domestic fowls can act as effective hosts of the virus, they are not necessarily essential to the maintenance of infection, for if there were no domestic fowls, the virus could be maintained in wild birds (Reeves and Hammon, 1962).

Unusually high rainfall and heavy snowpack can lead to a markedly increased population of *Culex tarsalis,* and can contribute to the development of encephalitis epidemics in California (Reeves, 1965, 1967). Although wild birds are the principal reservoirs for the virus, rodents, rabbits, and deer (Kandle, 1964) and cats and dogs (Goldfield *et al.,* 1966) may also be carriers.

The WE virus was once believed to be limited to the western United States, but has been found along the East Coast and in Canada, Brazil, Argentina, and British Guiana. In the East, where *Culex tarsalis* is rare, western encephalitis has most often been isolated from *Culiseta melanura* (Coquillett), but also from some other mosquitoes (Reeves, 1965).

California Encephalitis. California encephalitis (CE) first isolated in 1943 (Hammon and Reeves, 1945), was considered rare and confined to a single area in California (Hammon and Reeves, 1952) until 1963, when serologic evidence of past infection with CE virus in many adults in Wisconsin was reported (Thompson *et al.,* 1963). Subsequently, many cases of childhood encephalitis associated with CE virus infection were recognized, mostly in midwestern states and particularly in Ohio (100 cases during 1964–66) (Johnson *et al.,* 1968; Masterson *et al.,* 1971).

More than 50% of the cases of arthropod-borne encephalitis reported to the Center for Disease Control from 1967 to 1969 were caused by the CE virus, mostly in midwestern states, but some in southern states (CDC, 1969*b*). In 1966, 1 case and in 1969 2 additional cases of CE virus were found in New York state. The similarity in ecologic characteristics and in mosquito and wildlife populations throughout the northeastern states indicates that central nervous system diseases due to CE virus can occur throughout that region (Vianna *et al.,* 1971).

Twelve different strains of the CE virus were recovered from *Aedes canadensis* (Theobald) in the Yukon Territory at latitude 61° N. during the summer of 1971. This confirms the endemic prevalence of CE virus throughout western North America from California through Montana, Alberta, and British Columbia to Alaska (McLean and Wilson, 1972).

St. Louis Encephalitis. This encephalitide occurs principally in all states west of the Mississippi and in the Ohio River valley. The chief urban vectors are in the *Culex pipiens* complex. *Culex tarsalis* is the chief vector in some western states, and *C. nigripalpus* Theobald has probably been a vector in Florida (Pratt *et al.,* 1963). The virus has an active mosquito–bird cycle, and possibly wild rodents are of some importance (Henderson *et al.,* 1962). Epidemics of St. Louis encephalitis (SLE) are most common following unusually warm spring weather. Most symptoms of the disease in humans are mild, generally a few days of fever and severe headache followed by complete recovery. In the more severe cases, there is usually a sudden malaise, chills, and rapid rise in temperature, with severe headache, nausea, convulsions, and speech and visual difficulties. Recovery is often sudden and without complications, although weakness, dulled mentality, and paralysis may follow. The largest epidemic of the disease occurred in St. Louis in 1933, when there were 1,100 cases and more than 200 deaths. SLE is still the most important mosquitoborne disease remaining in the United States.

Eastern Encephalitis. This encephalitide is found along the Atlantic to the Dominican Republic and Jamaica, and on down to Argentina. In the eastern United States it is most commonly isolated from *Culiseta melanura,* but has been found in other mosquitoes and various insects. Eastern encephalitis (EE) or eastern equine encephalitis (EEE) is one of the most dangerous of all the encephalitides. Of 34 cases in Massachusetts in 1938, 25 were fatal and, in the same state, 10 out of 13 cases were fatal in 1956 (Anonymous, 1971*c*). In infants and children, there is usually a sudden high fever, vomiting, drowsiness or coma, twitching, and convulsions. Death may occur in 3 to 5 days. Survivors under 5 years of age are often left with mental retardation, convulsions, and paralysis, but survivors over 40 generally recover completely. The disease also occurs in horses and Chinese ringneck pheasants.

Venezuelan Equine Encephalitis. An outbreak of the type 1B epidemic strain of Venezuelan equine encephalitis (VEE-1B) occurred in Ecuador in the spring of 1969, with an estimated 20,000 human cases. The usual symptoms of fever, severe headache, myalgia, and some vomiting occurred, but with a mortality rate of less than 1%. There were between 30,000 and 60,000 equine deaths, a mortality rate varying from 50 to 90%. The same disease was recognized in El Salvador and Guatemala in July, 1969; in Mexico in March, 1971; and the first equine deaths in Texas occurred during the first week of July, 1971. By late July, the USDA reported 1,937 sick horses and 1,505 dead ones in Texas and fewer than 100 human suspect cases, with no deaths. Equines with VEE-1B developed high virus levels in their blood, and large numbers of the many species of mosquitoes that fed upon them became infected. At least 11 species of mosquitoes became involved in the Texas outbreak (Sudia and Newhouse, 1971).

VEE-1B was the first strain of this encephalitide to be confirmed in the United States. VEE is not as well known as other encephalitides, but the cycle is believed to involve wild birds, rodents, bats, various other animals, and many species of mosquitoes. Equine symptoms include fever, weakness, weight loss, and death in 5 to 11 days unless recovery occurs. Symptoms of VEE in humans are usually mild, although death can occur (Anonymous, 1971*c*).

Horse quarantines and embargoes were employed in almost all southern and southwestern states, and in most of these states horses were vaccinated with an attenuated live VEE virus vaccine to reduce or eliminate the potential of their infecting vector mosquito species. Millions of acres of the Gulf Coast of Texas were sprayed with a ULV aerial application of technical malathion, and effective mosquito control was obtained everywhere except where wind conditions were unfavorable for that type of application (Sudia and Newhouse, 1971).

Relation of the Artificial Environment to the Mosquito Problem

The natural environment is not necessarily conducive to low mosquito populations, as anyone can determine by visiting wilderness areas where adequate rainfall and the proper topography provide the required habitats for these insects. However, man's activities can greatly increase the potential for mosquito production in both rural and urban areas, but particularly in arid and semiarid regions where irrigation is practiced and where catch basins and spreading grounds are provided to conserve the meager rainfall. Some aspects of the relation of artificial environments to mosquito populations will be mentioned, with particular regard to California conditions, but the points to be discussed are equally applicable to other arid and semiarid regions and to some extent to other mosquito-infested regions also. Mosquito populations have been drastically reduced and, in some areas, practically eliminated by certain cultural control measures and insecticide treatments, despite many of man's activities that greatly favor mosquito production. This is in large measure a tribute to the efficacy of the mosquito-abatement districts (treated a little farther on) and the ability and dedication of those who have been responsible for their success.

Artificial Rural Environments

In many areas, an intimate relationship exists between irrigation practices and mosquito production. This is a particularly important factor in California, which contains 25% of all the irrigated land in the United States and where the amount of such land is expected to double in 30 years. With species of *Aedes* and *Psorophora,* which lay their eggs on the ground and for which hatching depends on rising water, irrigation practices, particularly those depending on flooding, can result in many generations of mosquitoes per year where otherwise there might be only 1 or at most 2 or 3. Where there are 10 irrigations per year, for example, irrigation practice alone can account for that many generations of mosquitoes. Swarms of mosquitoes can develop that are so large and so dense that one's clothing becomes covered with a continuous layer of the insects, and livestock may be killed by their continuous attacks. This gives an idea of the potential threat from mosquitoes and the extent to which the public depends for its health and comfort on continued scientific and technological progress to solve the ever-recurring problems brought about by the resistance of mosquitoes to insecticides. It also emphasizes the urgency of continued public support for the organization and legislation that have in past years proved to be so effective in implementing the control programs that have been developed to meet continually changing conditions.

The ultimate fate of water applied to fields depends on (1) surface runoff from the field, (2) evaporation from water and soil surfaces during and after the irrigation application, (3) deep percolation of water below the plant root zone, and (4) transpiration by the plants. The first 3 factors contribute to mosquito production. The first 2 factors are local problems, and the third contributes to area-wide drainage problems, yet some deep percolation is required to provide leaching of salts from the soil. In some studies made in several irrigated pastures in Merced County, California, it was found that when "irrigation efficiencies" (relation of "consumptive use" to the amount of water applied) were about 60% or greater, no mosquitoes were produced. An irrigation efficiency as high as 86% allowed for ample leaching to prevent salt accumulation. Information now available on irrigation procedures can be used in educational programs to improve water management (MacGillivray, 1964).

Proper preparation of land, adequate surface drainage, and the correction of poor irrigation practices are among the principal measures that should be considered for abatement of mosquitoes in rural areas. These measures also lead to higher yields of agricultural crops at lower cost. The same may be said regarding irrigated pastures. In addition, a pasture should not be grazed when the soil is wet, for water that accumulates in hoofprints causes much of the mosquito problem with such species as *Aedes nigromaculis*. Also, the soil is compacted, reducing water penetration, and the planted crop is badly cut up. A pasture should be divided into a number of smaller fields so that animals can be moved from one field to another (Bailey *et al.*, 1954).

Land drainage can be improved by weed control in drainage ditches. Cleared ditches help to keep water levels low, minimize mosquito production, and bring benefits to agricultural land that may in themselves repay the cost. A mixture of 2,4-D, low-volatile ester, and diesel oil will control cattails (*Typha* spp.) and tules (*Scirpus* spp.) as well as broadleaved and surface water weeds in a single application. When the use of 2,4-D seems undesirable, dalapon and amino triazole may be used. Mechanical equipment may be needed when a drainage channel is completely overgrown and partly clogged by silt. Chemical treatment may follow this (Greenfield and Wood, 1959).

In addition to proper land usage and irrigation practice, the following measures for mosquito prevention that pertain particularly to rural areas have been suggested by Bailey *et al.* (1954):

> The overflow of water from a dairy should be properly drained and kept free of vegetation. In some instances, the collected water should be pumped into the irrigation system or the drainage should be run down furrows to percolate into the soil. If these measures are not possible, the dairy should be kept free of weeds and the accumulations of water should be sprayed with kerosene or diesel oil.
>
> Public health departments should be consulted regarding the construction (or repair or rebuilding) of cesspools and septic tanks. When approved from the standpoint of sanitation and public health standards, they are also least conducive to mosquito production.
>
> The area around a watering trough should be properly drained or paved, or the trough should be situated on a mound of gravel. Otherwise, water collects in the many hoofmarks around the trough, and can easily become a source of mosquitoes. Unnecessary water containers should be drained, overturned, or destroyed.
>
> In the late spring, when neither rainpools nor irrigation overflow pools are readily available, drying streams adjacent to farmlands create stagnant pools to sustain mosquito infestations. Ditching to connect and drain the isolated pools and removing adjacent vegetation will eliminate these sources of infestation.
>
> Coastal areas that are flooded by the tides, and where low spots retain water, can be breeding grounds for *Aedes dorsalis, Ae. squamiger*, and *Ae. taeniorhynchus*. Ditches that will drain off the water when the tide falls are sometimes useful. Otherwise, a levee with a tide gate may have to be constructed that prevents entrance of salt water during high tide, but allows fresh water to flow out at low tide.

Treeholes in native, shade, and orchard trees collect water in which several species of mosquitoes can breed, particularly *Aedes sierrensis*. The holes may be treated by a tree surgeon, or a hole can be bored through the trunk to drain the treehole. Otherwise, treeholes can be treated with an insecticide.

Artificial Urban Environments

The special problems associated with mosquito control in the Los Angeles metropolitan area were discussed by Bristow (1966). Some of these problems are common to all heavily populated areas, and others pertain particularly to the large cities of California and the arid Southwest. In the Los Angeles area, for example, much of the imported water and local precipitation is diverted into water-spreading basins totaling over 1,800 acres (730 ha) for percolation into underground storage. These basins result in mosquito and midge problems. A vast drainage network of underground and open storm-drainage channels, together with the associated structures of catch basins, cross drains, and siphons, continuously supplied with waste waters, provides additional breeding sources. During the dry, warm summers, lawns must be watered frequently; the water continuously maintained in gutters provides an ideal environment for *Culex pipiens quinquefasciatus* and *C. tarsalis*. Because of the shallow slope of the land, many areas are not adequately drained.

Industry provides the usual accumulations of water in which mosquitoes can breed, such as pits, sumps, water towers, vegetable- and fruit-washing operations, and yard flushing. In residential backyards, there are thousands of mosquito sources, such as swimming pools, ornamental pools, bird baths, cesspools, empty barrels, and tin cans or other discarded containers. An ideal environment can be provided for several species of mosquitoes by thousands of flower vases in cemeteries. The water hazards of golf courses and areas provided for waterfowl protection may also provide mosquito sources. Salt marshes further aggravate the problem.

In the environs of large cities, as agriculture becomes less extensive it becomes more intensive. There are many dairies in which enormous quantities of water are used daily for cow washing and cleanup. This water, heavily loaded with organic matter, is disposed of on ground that is often impervious. Crop operations, livestock-watering troughs, poultry-watering facilities, and *Tubifex* earthworm culture (for fish bait) all contribute to the mosquito population. The close proximity of people and animals to the areas where sprays or granules are applied results in the necessity for great care in the selection and use of insecticides and complicates control procedures.

The development of new irrigated areas in arid or semiarid regions has favored species of mosquitoes that oviposit in sites where egg hatch depends on flooding and for which the number of generations per year depends on the number of floodings (e.g., *Aedes nigromaculis* or *Psorophora confinnis*). Urbanization of an area favors species that do not depend on an intermittent rise and fall of the water level and breed readily in the accumulations of water in the multitude of containers that make up the residue of urban living. Species of *Culex* and some *Aedes* are likely to be favored by the typical accumulation of litter in affluent societies.

Urbanization results in changes in the relative importance of mosquito species, but the total problem can be greatly ameliorated by sound regional mosquito-abatement programs. For example, in Orange County, California, which has probably been subjected to the most rapid rate of urbanization in the state, when the county's Mosquito-Abatement District was established in 1947, the irrigated-pasture mosquito, *Aedes nigromaculis,* was the predominant daytime species. It was eventually confined to the few small remaining dairy-farm pastures. The saltmarsh mosquitoes, formerly pests in coastal communities, are practically extinct. The sources of infestation have been drained, filled, or converted into residential tracts and yacht harbors. *Aedes squamiger* and *Culiseta inornata* are still potential sources of infestation if late rainfalls keep rain-filled depressions from drying up. The western treehole mosquito, *Aedes sierrensis,* was once common in populated areas, but all known treeholes have been filled with sand, and the species has become rare. It still breeds in treeholes of the older coast live oaks, *Quercus agrifolia,* and mesa oaks, *Q. engelmannii,* in the foothill areas. The predominant mosquito species in populated areas are now *Culex peus, C. pipiens quinquefasciatus,* and *C. tarsalis,* but they are not serious problems (Kimball, 1965).

Aside from the elimination of accumulations of water that may be sources of mosquito infestation on his own premises, the screening of windows and doors, and the use of bed nets, repellents, insecticidal aerosols, "pest strips," and other formulations, there is little that the average homeowner can do except to be aware of and support community-wide mosquito-abatement activities, the chief promoter of which is usually the local mosquito-abatement district.

Mosquito-Abatement Districts

To be adequately organized and effective, area control of mosquitoes must be based on written authority if the state or lesser geographical area is to spend money for mosquito abatement. This authority is derived from what has been termed in the various states a "Mosquito-Control Ordinance" or "Mosquito-Abatement Act." In California, Professor W. B. Herms and an engineer, H. F. Gray, had begun educational campaigns for controlling malaria by combating anopheline mosquitoes in Penryn, Placer County, in 1909. In 1913, Herms tried, but failed, to persuade the State Legislature to pass a law enabling the state to control mosquitoes as a basis for controlling malaria. However, a real estate lobby group representing Marin and San Mateo counties added their influence to reinforce Herms' malaria-control crusade, even though their principal concerns were culicine "pest mosquitoes" rather than the anopheline malaria carriers. Programs directed against pest mosquitoes often reduce the number of *Anopheles* vectors of malaria and the encephalitis vector, *Culex tarsalis* (Mulhern, 1966). As a result of the pressure exerted by the real estate lobby, in 1914 the State Legislature added to the State Public Health Code certain sections now known as the "Mosquito-Abatement Districts Acts." Two districts were formed in 1915, 3 in 1916, 3 in 1917, and the number was continually increased until in 1970 there were 63. These districts are currently responsible for the abatement of mosquitoes and certain other pests over more than 40,000 sq mi (103,600 sq km), and they spend over 9 million dollars annually. A district may coincide with a political unit such as a county, but the Los Angeles metropolitan area alone has 5 mosquito-abatement districts. The code specifies only that "any territory in 1 or more counties having a population of not less than 100 inhabitants may be organized as a mosquito-abatement district."

The formation of a mosquito-control agency is generally initiated by local citizens after they have become sufficiently impressed by the severity of the mosquito problem. It is desirable for concerned individuals to work through groups, clubs, or other public or private organizations. Legal counsel should be obtained, and in fact in some areas the county or city employs an attorney who may be able to give the necessary legal guidance. The board of trustees in each Mosquito-Abatement District (MAD) has the following authority:

(*a*) To take necessary steps to exterminate mosquitoes in or within migrating distance of the district; (*b*) to abate as public nuisances artificially created mosquito breeding places; (*c*) to notify a property owner of the existence of a nuisance, hold hearings on the notice, determine thereafter whether abatement must be made, direct him to comply, or abate the nuisance when he fails, and initiate lien action against the property involved to enforce payment.

The California law also includes a penalty clause: "Any person who retards activities of the district is guilty of a misdemeanor" (Beadle, 1957)

Mosquito-abatement districts are empowered to control house flies and other insects besides mosquitoes. Those organized since 1931 can also control rats when this activity is supported by a separate tax. The Coachella Valley MAD was organized in 1928 to control eye gnats. Elsewhere in California, the Moorpark MAD was organized primarily for house fly control, and the Contra Costa and Delta MAD's also control flies. Many of the districts control midges, and several also control eye gnats (such as the one just noted), biting gnats, and yellowjackets.

The governing power of a MAD is vested in a board of trustees representing all cities as well as the county. The city council for each city and the board of supervisors for the county at large each appoint 1 representative to serve without pay for a 2-year term. The board of trustees establishes policy and budget requirements for the district. Funds supporting the budget are raised by a tax on the total secured valuation within the district. It is desirable that the people serving on the board of trustees should be capable and respected citizens from a variety of businesses and professions who are willing to serve without pay as a public service. In the initial period of organization, the help of an experienced consultant is advantageous, possibly for a few days each month. The principal personnel will be the manager, source-reduction man, and entomologist, and these must start by training subordinates, such as inspectors and foremen. The public should be informed that little or no observable relief from mosquito annoyance can be expected during the first year, although some localized improvement actually may result from the initial training process (Anonymous, 1961*a*).

As part of its mosquito-abatement program, a MAD should provide the following services to individuals, corporations, and public agencies whose operations may create potential mosquito-breeding conditions:

(1) Maintain surveillance of all potential mosquito-breeding sources, and inform the responsible party of any breeding that requires control to prevent a public nuisance.

(2) Furnish information and training on routine mosquito-control methods to the person in charge of the facility.

(3) Furnish and deliver whatever quantity of mosquito fish (*Gambusia*) is necessary for adequate control within suitable ponds and holding reservoirs.

(4) Furnish emergency mosquito-control services, using the district's special spraying equipment [figure 284] and mosquito larvicides to prevent a public nuisance whenever mosquito-breeding conditions are created by operational failures. (From Kimball, 1965.)

California's Health and Safety Code defines any breeding source for mosquitoes created by the improper use of the land as a "public nuisance," and requires the abatement or control of the nuisance. The board of trustees of a MAD will do well to place the responsibility for abatement on the owner (individual, corporation, or public agency). Experience has shown that the property owner, once he is aware of the benefits and the simplicity of sound mosquito control methods, prefers to do the job himself. The district should cooperate by providing technical information, training, and guidance, and should also provide emergency services in case of failure. The owner probably will not be charged for these services, but repeated violations may result in stronger notification to abate the public nuisance. Full use of the formal abatement procedure as provided

Fig. 284. Spraying and mist-blowing for control of mosquitoes and midges in Orange County, California. (Courtesy Orange County Mosquito-Abatement District.)

by the Health and Safety Code is generally not necessary (Kimball, 1965).

Rural areas provide the greatest scope for action by mosquito-abatement districts. An account of mosquito abatement in a predominantly rural section of the central valley of California was given by the manager of the Northern San Joaquin County Mosquito-Abatement District (Peters, 1957). In the decade following the formation of the district in 1945, tractors, scrapers, draglines, graders, and ditchers were utilized in the 409 sq mi of the district to clear and/or rough-level 1,150 acres of land, and to construct and maintain about 7.5 mi of levees, about 15.5 mi of drainage ditches, and 20 water-storage ponds with a total area of 105 acres (1,060 sq km; 466 ha; 12 km; 25 km; 43 ha).

The district has opened up new areas to agriculture, designed recirculation systems for pastures, developed drainage systems, and assisted in the reshaping and construction of recreation areas and small aquatic parks. The district's participation in such work requires only that an actual or potential mosquito problem be involved.

California law places the responsibility for mosquito control in "natural locations" on the district. Wherever the district can change mosquito sources from natural to artificial, the legal responsibility is switched to the owner of the reclaimed land, reducing the taxpayers' burden. Wherever mosquito control can be obtained by elimination of "sources" (breeding areas), eventual problems arising from resistance to insecticides are precluded (Peters, 1957).

MAD's and other agencies interested in mosquito abatement can become corporate members of the Mosquito-Control Association, of which there are 12 regional associations in the United States. (See chapter 1.)

Household Mosquito Control

In the foregoing discussion, the importance of community-wide programs of mosquito control, aimed principally at the elimination of sources of infestation, has been emphasized. The homeowner can participate in the community effort by eliminating unnecessary accumulations of water on his own property in which mosquitoes can breed, and by stocking bodies of water, such as ornamental ponds, with mosquito-eating fish (*Gambusia*). Following are additional measures that an individual can employ for personal protection against mosquitoes.

Screening

Screening is such a simple and obviously sensible measure for protection against mosquitoes that, in parts of the world where it is almost universally practiced, its major contribution to health and comfort can easily be overlooked. Screens are constructed of galvanized iron, copper, bronze, aluminum, and plastic, the 3 last being particularly popular in coastal areas where there are corrosive salt sprays. To be effective, screens ordinarily must be $16 \times 16 \times 20$-mesh, and must fit tightly and be kept in good repair. Screens of smaller mesh ($16 \times 20 \times 23$) are required for some mosquito species, such as *Aedes aegypti* and *Ae. taeniorhynchus* (Bidlingmayer, 1959) or *Ae. sierrensis* when the mosquitoes develop under crowded breeding conditions (Freeborn and Bohart, 1951). Screen doors should open outward, and be provided with closing springs or automatic closing devices.

Added protection can be provided by insecticides. Kerosene solutions of insecticides are preferred to xylene emulsions because rust problems on galvanized-iron screens, and possibly adverse effects on some plastic screens, may result from the xylene (Pratt *et al.*, 1963).

Mosquito Netting

Mosquito netting is seldom used in the United States because in most inhabited areas mosquitoes are under good control, and in any case people are usually lodged in buildings in which mosquitoes are effectively excluded. However, mosquito netting can be useful on camping trips and when traveling abroad in areas where buildings are not screened. Mosquito netting consists of cotton or nylon cloth with 23 to 26 meshes per inch (2.4 cm). From this material, "bed nets" can be made that are fitted over a frame that is high enough above the bed to permit a person to sit up. Before retiring, any mosquitoes inside the bed net can be killed by hand or with an aerosol bomb. If the lower edges of the net are tucked under the mattress, mosquitoes can be kept out indefinitely.

Repellents

Great progress has been made in the development of chemicals that can be applied to the skin and clothing to repel mosquitoes and other arthropods. A few drops applied to the neck, face, hands, and arms or sprayed onto clothes will effectively repel mosquitoes for periods of 2 hours to half a day, depending on the person and the abundance and species of the attacking mosquitoes. Some mosquito repellents may also be effective against other biting flies, fleas, chiggers, and ticks.

A popular repellent, and one commonly used in the former combat areas of Vietnam, is deet (diethyl toluamide), sold commercially as OFF or DEET (Gilbert *et al.,* 1957). It is available as a liquid in bottles or as a spray in pressurized cans. Deet is a standard with which newly developed repellents are compared by the USDA Entomology Research Division. Deet protected women for longer periods than it did men against the yellowfever mosquito, *Aedes aegypti,* in experiments conducted in Florida. Men were found to be significantly more attractive than women to the mosquitoes (Gilbert *et al.,* 1966). In a number of such investigations during the past decade, it has proved to be as effective, and usually more so, than new candidate materials (Gilbert *et al.,* 1970*a, b, c*). TMPD (2,2,4-trimethyl-1-1-3-pentanediol) was compared with deet used against several species of mosquitoes, as well as sand flies, black flies, chiggers, fleas, and ticks. At equal strength, it ranged from 66 to 105% as effective as deet (Gerberg, 1966).

Dimethyl phthalate is considered to be a particularly effective repellent against some malaria mosquitoes (*Anopheles quadrimaculatus* and *A. freeborni*) and sand flies, and is also one of the preferred repellents for chiggers. Rutgers 612 (2-ethyl hexanediol-1,3) and dimethyl carbate are more effective than dimethyl phthalate against *Aedes* and other pest mosquitoes. Indalone is especially effective against stable flies and ticks. As might be expected, a combination of these compounds will give better protection against a greater variety of pests than any one of them alone. A combination of 3 parts of dimethyl phthalate, 1 part of Indalone, and 1 part of Rutgers 612 (the original 6-2-2 formula) has long been used as a general insect repellent. Later, dimethyl carbate was substituted for Rutgers 612 in some formulations, including a cream containing 92.5% liquid repellent and 7.5% cream base. Care should be taken to avoid getting repellents into the eyes or on the lips or other mucous membranes. Many repellents are solvents for paints, varnishes, lacquers, plastics (such as synthetic watch crystals), and synthetic fibers, and should be kept off such materials. However, deet will not affect nylon (Pratt *et al.,* 1963; Mallis, 1969).

Hach and McDonald (1971) found that diethyl amide was an effective repellent for 50 days and that 2-butyl-2-ethyl-1,3-propanediol was effective for 196 days in keeping the yellowfever mosquito, *Aedes aegypti,* from arms covered with treated cotton stockings.

Aerosol Canister

The aerosol canister described in chapter 3 is convenient and effective for rapidly knocking down insects in flight in a room or resting or crawling about in exposed areas, as is likely to be the case with mosquitoes, gnats, and house flies. The label on the canister provides information as to how long the spray valve should remain open, in a room of a given volume, to ensure sufficient insecticide to rid the room of the specified insects. Best results are obtained if doors and windows are closed for a short time and then opened to ventilate the room. Aerosols are most effective in enclosed spaces, but are nevertheless able to provide temporary relief outdoors, such as at picnic tables. They are also excellent for use in closed tents or cabins.

The Dichlorvos Insecticide Resin Strip

A convenient and highly successful method of eliminating mosquitoes from the home is by using the dichlorvos resin strip. This is a solid polyvinyl chloride (PVC) resin formulation, impregnated with 20% dichlorvos and related products. It is available in a strip 2.5 in. (6 cm) wide and either 10 in. or 2.75 in. (25 or 7 cm) long, the latter known as the "ministrip." When suspended in a room, the strip gives off dichlorvos vapor at a slow and uniform rate that is sufficient for the control of most small, flying insects, whether in flight or resting on exposed surfaces. Dichlorvos is quickly hydrolyzed by moisture, including that in the atmosphere, and the manufacturer says it is safe to those occupying the treated room. In Italy, where dichlorvos resin strips were used in hospital wards, the hospital authorities concluded that they were clinically harmless to healthy pregnant women or to sick children and adults exposed for 24 hours to vapor concentrations of 0.1 μg per l of air.

Nevertheless, Anonymous (1970*a*) advised that in the United States, dispensers of this type should not be used where infants or aged persons were continually exposed to the vapor. Also, that dichlorvos resin strips were not supposed to be used in a food-preparation room.

The 10-in. (25-cm) strip has been found by various investigators to be safe and effective when suspended in 1,000 cu ft (28 cu m) of space for the control of a wide variety of household and farmyard pests. It is particularly suitable for the control of mosquitoes, midges, gnats, house flies, and other small, flying insects in homes, restaurants, dairy and other livestock barns, milk rooms, chickenhouses, or dog kennels, and when so used, it is said to be effective for about 3 months. The dichlorvos ministrip can be used in smaller enclosed spaces, such as closets and dresser drawers. A concentration of dichlorvos in an enclosed space at the extremely low concentration of 0.02 μg per l of air gives a complete mortality of adult mosquitoes.

The degree of vapor concentration in a room depends on the degree of ventilation—the less ventilation, the closer will the results approach the maximum obtainable.

Residual Sprays

If an insecticide with a long residual life is applied to the surfaces upon which house-frequenting mosquitoes characteristically rest, control can be obtained for long periods. The control of mosquitoes in developing countries has been based principally on this method, using DDT as the insecticide. Virtually all applications are made to the inner walls, and usually to the ceilings of houses, at a rate of about 200 mg of DDT per sq ft (929 sq cm). This type of treatment poses no threat to the occupants of the houses or to the environment.

The principal use of DDT continues to be in the campaign against malaria, about 90 million lb (40,860,000 kg) annually being used for this purpose. One or two treatments per year are usually sufficient. Constant protection against bites by infective mosquito vectors must be provided for several years in order to interrupt malaria transmission. In developing countries, the areas to be treated are so large and the remote rural areas are so inaccessible that frequent spraying of houses is impractical and excessively costly. The long residual efficacy of DDT is essential. The discontinuance of the DDT spray program in developing countries, without an effective and practical alternative, would lead to major catastrophe. Where DDT has been used only for mosquito control, there has been remarkably little indication of resistant strains of mosquitoes. Resistance is a problem principally in areas where there is also widespread use of DDT in agricultural pest control. In the United States, mosquitoes are everywhere resistant to DDT, and in any case, the insides of dwellings are not generally sprayed for mosquito control. However, the spraying of other types of buildings is sometimes practiced, particularly in rural areas. In a rice-growing region of Arkansas in 1970, insecticides were applied with a compressed-air sprayer to potential resting places of mosquitoes (*Anopheles quadrimaculatus*) inside small buildings, such as chickenhouses, barns, doghouses, pumphouses, storage sheds, and garages, all of which housed animals or had animals near-by. Propoxur was the only insecticide that caused a 100% reduction during the entire 13-week period of the test, although chlorpyrifos nearly always produced a 99 to 100% control (Gahan and Wilson, 1970).

Area Control of Mosquitoes

As already stated, mosquito control depends principally on control programs carried out by the community or by mosquito-abatement districts. However, the understanding and support of the public are needed for the continuance and success of these programs. In addition to the value of such programs for reducing the incidence of vectorborne diseases, they make possible the full utilization of residential, agricultural, industrial, recreational, and wilderness areas merely by freeing them from the discomfort and frustrations that characterize mosquito-infested localities.

Aside from the obvious human values, the economic significance of mosquito abatement is almost incalculable. Importantly, an intensive and well-organized mosquito-abatement program has another outstanding benefit: If outbreaks of vectorborne disease occur, such as the encephalitis epidemics in California in 1952 and in Texas in 1971, the major portion of abatement activities can be redirected to control of the vectors, and these operations can be supplemented by emergency funding. A surveillance and early-season warning system is useful in this connection, and has been used in California since the 1952 outbreak. Far less time is required to redirect existing programs to disease-vector control than to develop a new program (Mulhern, 1966).

Fogging and Misting

The measures by which elimination of mosquito sources is accomplished have been discussed already. These procedures have been effectively supplemented by insecticides, generally applied as fogs or mists. Fogging or misting is done in the late afternoon or early evening, at night, or early in the morning when the air is calm. Ground equipment is generally used for generating the fogs or mists. The machines travel at an average of 5 mi (8 km) per hour, and generally disperse 15 to 25 gal (57 to 95 l) per hour, but some may have an output of 40 gal (151 l) per hour. At the higher rate, using a recommended insecticide in fuel oil and covering a swath about 400 ft (120 m) wide, and with a vehicle speed of 5 mi (8 km) per hour, a dosage of about 0.15 lb per acre (0.07 kg per 0.4 ha) of insecticide might be applied. A satisfactory swath can be produced when wind velocities are 3 to 5 mi (5 to 8 km) per hour. This type of application is only for an immediate kill of mosquitoes, for the amount of insecticide applied is too little to have residual effect. Dusting is not so effective as fogging or misting, even at higher dosages (Pratt *et al.*, 1963). Fogging with No. 2 fuel oil alone has proved to be successful in controlling mosquitoes in backyards. The exhaust system of a tractor has been used to generate fuel-oil fogs for the protection of herds of cattle during heavy mosquito flights, and apparently has also controlled the horn fly (Thompson, 1968).

No differences in effectiveness were detected between thermal or nonthermal fogs or between oil solutions and water emulsions with malathion at 6 oz per gal or chlorpyrifos at 2 oz per gal (180 or 60 cc per 4 l) used against *Aedes aegypti, Ae. taeniorhynchus, Anopheles albimanus* Wiedemann, and *Culex pipiens quinquefasciatus* (Taylor and Schoof, 1968). When comparing malathion, naled, and pyrethrum individually and in combinations, the combination of malathion and naled, applied with a mist blower, was the most effective against caged mosquitoes (*Culex pipiens pipiens*) and house flies, and the least toxic to honey bees (Colburn and Langford, 1970). (It should be borne in mind that insecticides found to be more effective one year in a given area may become ineffective within a few more years because of the development of resistant strains of mosquitoes.)

Aircraft Application

An important factor in making insecticide treatment practicable has been the use of aircraft equipped with ultralow-volume (ULV) dispensers of the insecticide liquids. Ultralow-volume refers to any spray in which total volume applied is less than 0.5 gal per acre (2 l per 0.4 ha) (Messenger, 1963; Mount and Lofgren, 1967). The ULV method permits the aircraft to carry a much greater load of toxicant, reducing the cost of application. Using adult female *Aedes taeniorhynchus* mosquitoes exposed in 16-mesh screen cages, Mount and Pierce (1971) found that the use of ULV cold aerosols permitted a 50% decrease in doses of malathion, naled, and fenthion. Another advantage was that solvents, carriers, and additives were not required, and therefore the need for formulating facilities was eliminated. Also, traffic hazards were eliminated, for the dense fogs typical of thermal aerosols were obviated. A highly efficient rotary-cylinder atomizer was found to be particularly useful. Malathion, naled, propoxur, fenthion, and pyrethrins are currently registered for aircraft application for mosquito control. Also, pyrethrum is registered as an adulticide for the protection of workers from mosquitoes and gnats during the harvesting of agricultural croplands. It may be used at 1 oz to 4 gal of water per acre (30 cc to 15 l per 0.4 ha) when applied by means of fixed-wing aircraft.

Larvicides

Much mosquito control with pesticides is directed against the larvae. In heavily polluted water, as in log ponds, a treatment may become effective within 2 days. Where vegetative cover in the treated areas is heavy, the concentration of toxicant in the solution or the amount of solution applied must be increased to obtain satisfactory results. Also, under such conditions granular formulations often provide better control than emulsions or oil sprays. Granular formulations are a great convenience, and will probably speed up a treatment. Whereas a liquid-spray rig must often be driven back to a source of water to refill the spray tank, an operator can carry enough granules with him for an entire day of treatment without stopping. Special compressed-air granule guns are used for application (Anonymous, 1970*a*). Resistance of mosquitoes and midges to relatively safe insecticides such as DDT and malathion has led to the use of far more toxic compounds. Reduced hazard to insecticide applicators and beneficial wildlife, as well as reduced toxic residues on food and forage crops in agricultural areas, can result from switching from liquid and dust to granular formulations in vector control (Mulla, 1962).

In California, where public agencies have been treating nearly 1.5 million acres (607,000 ha) annually for mosquito control with about 1 gal (4 l) per acre of emulsifiable water-mixed spray, ULV application of larvicides by aircraft has created special interest. In the irrigated pastures of the San Joaquin Valley and the rice fields of the Sacramento Valley, experiments were made in which approximately 8 fluid oz (240 cc) of liquid were applied per acre (0.4 ha), resulting in dosages of 0.1 lb of fenthion or 0.05 lb of chlorpyrifos per acre (0.045 or 0.023 kg per 0.4 ha) (Burgoyne *et al.*, 1967). Good mosquito control was obtained with chlorpyrifos at 0.012 to 0.025 lb (0.005 to 0.011 kg) per acre, using carefully controlled droplet sizes of approximately 150 μ MMD (mass median diameter) when air drift was minimal (Akesson, 1970). (See chapter 3, under "Aerosol or Mist Applicators.")

In California, as part of an emergency encephalitis program in 1969, chlorpyrifos and fenthion were applied by the ULV technique at dosages of 3 to 6 fluid oz (90 to 180 cc) per acre with excellent results against the vector *Culex tarsalis*. However, in certain counties in California, resistance had reached such levels that the larvae of *C. tarsalis* could not be controlled by any organophosphorus or carbamate insecticide at legal dosages (Anonymous, 1970*a*).

Abate® is a relatively new insecticide, of great interest because of its low mammalian toxicity (acute oral, 1,000–3,000) and high effectiveness against mosquito larvae. (See table 2, chapter 2.) Applied in sand granules, 1 ppm in drinking water, bird baths, and animal-watering devices, it has been effective against *Aedes aegypti*. In toxicological studies on human volunteers, no adverse effects were found with dosages far in excess of those required for kills of *Ae. aegypti* (Anonymous, 1970*a*). In Orange County, California, where fenthion granules have long been used effectively as a larvicide at 0.2 lb (0.090 kg) per acre (0.405 ha), Abate granules at 0.1 lb (0.045 kg) of toxicant per acre have given equally good control. Low-cost "airboats" (figure 284) have been useful in applying larvicides. Amphibious track vehicles are also used effectively (Kimball and Thompson, 1967; Thompson, 1967).

There is currently much interest in "slow-release" mosquito larvicides. Mosquito larvae in temporary pools, potholes, and small, isolated swamps and salt marshes were effectively controlled with chlorpyrifos-impregnated plaster-of-paris briquettes at the rate of 0.75 to 1.5 lb (0.34 to 0.7 kg) of actual insecticide per acre (McDonald and Dickens, 1970). In experiments made in polyethylene-lined field pools, treatments of chlorpyrifos in polyethylene and in charcoal at rates of 0.5, 1.0, and 2.5 ppm provided effective control of the southern house mosquito, *Culex pipiens quinquefasciatus*, for 20 to 26 weeks. Fenthion in polyvinyl chloride at 2.5 ppm provided control for 20 week (Wilkinson *et al.*, 1971). An interesting procedure in mosquito control, but one that is not being widely practiced, is the use of dichlorvos resin strips immersed under water, as in the catch basins at Turlock, California. In one test, a 10-in. (25-cm) strip per catch basin resulted in a full season of control, whereas in the untreated basins, mosquitoes became so abundant that it was necessary to spray them in July and August (Silveira *et al.*, 1964).

Continued success with pre-DDT insecticides is of great interest. In the southeastern United States, paris green pellets have proved to be highly effective against saltmarsh mosquitoes at the rate of 15 lb (7 kg) of 5.0 to 7.5% granules per acre (0.405 ha), applied with ground or aerial equipment. In Georgia, such usage of paris green has resulted in no loss in susceptibility of mosquitoes over a 10-year period. Likewise, petroleum oil continues to be effective as a larvicide. Just as in agricultural pest control, no resistance to oil has ever been noted. In one experiment, after 60 generations of selection, *Culex pipiens quinquefasciatus* showed no change in susceptibility to petroleum derivatives (Micks and Berlin, 1970). At the rate of 2 to 3 gal (7.5 to 11 liters) per acre, fuel oil plus Triton X-100 has been effective in areas harboring valuable fish and wildlife (Anonymous, 1970*a*). In California, petroleum oil with biodegradable surfactant and cosurfactant gave good results with low dosages of oil. Also, botanical insecticides, such as pyrethrins and chemically similar synthetic compounds, gave good results with relatively small dosages when petroleum oil was used as an extender. One of the compounds investigated was the synthetic pyrethroid SBP-1382®, which had very low toxicity to mammals but was not a good larvicide (Mulla, 1970).

In laboratory tests, several aliphatic amines were found to be effective against mosquito larvae and pupae. They were dispersed as films on the surface of the water, and resulted in symptoms similar to those caused by petroleum oils. Peak mortalities were reached more rapidly than with conventional insecticides. *Anopheles* larvae were particularly susceptible, presumably because they remained at a near-horizontal position at the surface and thereby obtained a maximum of contact with the insecticide. Several aliphatic amines were effective ovicides as well as larvicides and pupicides (Mulla and Chaudhury, 1968). The amines can be successfully used against mosquitoes that are resistant to organophosphorus insecticides. The peculiar mode of action of these compounds gives some reason to believe that mosquitoes may not develop resistance to them. Unlike organophosphorus insecticides, they have few or no disruptive effects on predaceous and herbivorous aquatic insects (Mulla, 1967; Mulla and Chaudhury, 1968; Mulla *et al.,* 1970*a;* Mulla and Darwazeh, 1971).

Insecticide solutions or emulsions can be applied a drop at a time into moving or quiescent bodies of water by means of "drip cans." In experiments, effective and economical control of mosquito larvae was obtained by the drip method in flowing and quiescent waters by the application of chlorpyrifos, Abate, or fenthion at 0.2 ppm and 0.15 to 0.25 lb (0.07 to 0.11 kg) per acre (Sjogren and Mulla, 1968).

For *Aedes melanimon* and *Ae. nigromaculis* in irrigated pastures, the efficacy of a given insecticide was greater when applied by the drip method than when applied in the conventional way by aircraft (Mulla *et al.,* 1969). Mosquito larvicidal sprays, applied from levees surrounding sewage oxidation ponds, often fail to penetrate the heavy vegetation surrounding them. Both the drip method and that of directly pouring the emulsifiable concentrate into the flowing water for a period of 5 minutes were found to be particularly suitable (Mulla *et al.,* 1970*b*).

FLIT® MLO. This petroleum derivative is a colorless, low-viscosity hydrocarbon that has little odor, a high initial boiling point of 500° F (260° C) or higher (low fire hazard), and has a high unsulfonated residue of 90 vol./% or greater (low phytotoxicity). It is said to be harmless to fish and wildlife, and to be much more effective than No. 2 diesel oil, long used as a mosquito larvicide. Micks *et al.* (1968) made laboratory experiments on the spectrum of activity of FLIT MLO against all aquatic stages and adults of the yellowfever mosquito, *Aedes aegypti;* the common malaria mosquito, *Anopheles quadrimaculatus;* and the southern house mosquito, *Culex pipiens quinquefasciatus.* The petroleum derivative proved effective at dosages as low as 1 gal (about 4 l) per acre. It delayed egg hatch, interfered with the development of early larval instars, produced high mortality of fourth-instar larvae and of pupae, and high adult mortality during emergence. Insecticide-resistant mosquito strains were as susceptible as nonresistant strains. Heavy selection pressure with petroleum hydrocarbons for periods of up to 5 years failed to produce resistance, confirming all earlier experience concerning the inability of oils to produce resistance in insects (Micks and Berlin, 1970).

In small-plot and field tests in rice fields, Bearden and Steelman (1971) applied FLIT MLO against the dark ricefield mosquito, *Psorophora confinnis.* It was applied (1) as a spray, (2) dripped into the entrance of a plot during the time flood water was entering, (3) scattered by means of gelatin capsules containing 1 ml of the liquid, and (4) poured into the centers of the 3×6-m small plots or in a sufficient number of places to ensure a uniform film in the larger field plots. In the small-plot tests, FLIT MLO caused larval mortality of 96% or greater at 1 gal per acre and 97% or greater at 2 gal per acre (about 4 or 7.5 l per 0.4 ha), employing any of the 4 methods of application. In the larger field tests, 99% larval mortality was obtained with spray or capsule application at 0.5 gal per acre, but for the same degree of control with the "drip-bucket" or "pour-on" methods, a dosage of 1 gal per acre was required. FLIT MLO did not reduce the yield of rice, regardless of the method of application, and the petroleum derivative was not detected in rice grain or in straw from plots sprayed with 2 gal per acre.

In California, FLIT MLO has been used effectively against the irrigated-pasture mosquito, *Aedes nigromaculis,* which has become resistant to chlorinated hydrocarbon and organophosphorus larvicides (Murray, 1968; Gillies and Womeldorf, 1968; Mulla, 1970). The lowest rates giving acceptable control by hand and by aircraft application were 2 and 3 gal (7.5 and 11 l) per acre, respectively (Gillies and Womeldorf, 1968). Aircraft application has resulted in incomplete control, possibly because of insufficient coverage in areas with exceedingly dense vegetation. Because FLIT MLO is not applied with a diluent, it is not possible to increase the application rate from ULV to many gallons per acre by dilution, with the dosage level remaining constant, as can be done, for example, with organophosphorus insecticides (Gillies *et al.*, 1971).

In the Orange County (California) Mosquito-Abatement District, FLIT MLO has been directed against *Culex pipiens quinquefasciatus, C. peus,* and *Culiseta incidens.* It has given good control when applied with mechanized or hand-operated sprayers at the rate of 2 to 3 gal (7.5 to 11 l) per acre to breeding areas associated with urban community drainage facilities. The use of the compound has eliminated hazard to the public and to district employees, as well as to fish and wildlife, and has reduced the spray route man-hours and vehicle mileage by as much as 50% by increasing to 14 days the 7-day spray cycle required for emulsions or granular-type larvicides (Kimball and Perruzzi, 1970).

There has been a considerable difference in degrees of optimism about FLIT MLO expressed by different researchers. This insecticide is much less effective against first-, second-, and third-instar mosquito larvae than against later instars. For a given species, it might be advisable to allow larvae to attain the stage of development most conducive to good control. However, large fields are generally irrigated from one side to the other over the course of several days. This results in larvae of various stages of development on the day of treatment. Many fields that contain late-instar larvae and pupae of *Aedes nigromaculis* also contain many early-instar *Culex tarsalis* larvae. The latter are not adequately reduced in numbers and must be treated later (Schaefer and Ramke, 1971).

Insect-Development Inhibitors

Insect-development inhibitors are compounds with biological activity which mimics that of natural juvenile hormones. In California, aircraft applications of a slow-release formulation of ZR-515 (isopropyl 11-methoxy-3,7,11-trimethyl-2,4-dodecadienoate) were successful in controlling the larvae of strains of *Aedes nigromaculis* that were resistant to organophosphorus (OP) insecticides. Doses as low as 0.025 lb (0.011 kg) of actual insecticide per acre were found to be effective against *Ae. nigromaculis* and 0.05 lb (0.023 kg) per acre showed promise in the control of *Ae. melanimon* but not *Culex tarsalis.* Continual selective pressure with ZR-515 against OP-susceptible *C. pipiens quinquefasciatus* failed to produce evidence of development of resistance (Schaefer and Wilder, 1972, 1973). Dosages of the slow-release formulation of ZR-515 used for mosquito control had no visible effects on most of the nontarget organisms, but larvae of aquatic Diptera (Chironomidae, Ephydridae, Psychodidae) showed some sensitivity (Miura and Takahashi, 1973).

Insecticide–Fertilizer Mixtures

Fertilizer is routinely applied to flooded rice fields. The cost of insecticide application can be eliminated when insecticide can be effectively added to fertilizer.

In a recently conducted experiment, several concentrations of diazinon, chlorpyrifos, Abate, and fenthion were added to a granular fertilizer, and the insecticide–fertilizer formulations were applied at rates of 100 to 500 lb (45 to 225 kg) per acre. Effective control of the first-instar larvae of *Psorophora confinnis* was obtained with low concentrations of insecticide. Economically significant increases in rice yields were obtained by combining insecticides and fertilizer (Steelman and Poché, 1970).

The Resistance Problem

Although in many parts of the world DDT remains effective in mosquito control, it has long since lost its effectiveness in the United States, and is now entirely replaced by organophosphorus and carbamate insecticides. Resistance has also been developing to those compounds. In California, a particularly critical situation has developed because of the widespread resistance of *Culex tarsalis,* the carrier of encephalitis, to currently available larvicides.

Some adulticides have been determined to be sufficiently toxic to the mosquitoes, but the habits of *Culex* species result in adulticides being particularly ineffective against them. The adults emerge irregularly from the quiet, clear pools of standing water in which they breed, and treatments must be repeated often.

The irrigated-pasture mosquito, *Aedes nigromaculis,* bites more viciously than *Culex tarsalis,* but carries no serious disease. Against this species also, only an adulticide (propoxur) is now effective among the several cholinesterase-inhibiting insecticides.

Fortunately, the adult emergence of species of *Aedes* is relatively uniform, and occurs at predictable intervals after pastures have been irrigated, provided they are not irrigated a section at a time. Therefore, adulticides can be more effective than they are against *Culex.*

The greatest danger posed by insect resistance in California is that an outbreak of encephalitis might occur when there are no effective controls against *Culex tarsalis.* That situation developed in 1952, when this species was found to be resistant to DDT, leaving mosquito-control agencies helpless until effective organophosphorus insecticides were developed (Anonymous, 1970*b*).

Mosquito control is becoming increasingly complex and difficult because those entrusted with the job of control must deal not only with genetic resistance to toxic insecticides by mosquitoes, but also with political and social resistance by a public increasingly concerned with pesticide pollution of the environment. The use of FLIT MLO would obviate both problems, but, as already explained, this insecticide has proved to be unsatisfactory in large fields in which mosquito larvae are not in a uniform stage of development because of irrigation dates.

MOSQUITO-EATING FISHES

The usefulness of mosquito-eating fishes for mosquito control has been known since late in the nineteenth century. The fish most commonly used has been *Gambusia.* The famous American entomologist, L. O. Howard (1901), was among the first to advocate the use of *Gambusia,* as well as sticklebacks and sunfish, for the control of mosquito larvae. *Gambusia* stocks were transported from the southern states to New Jersey and Hawaii in 1905 and to California in 1922. In 1923, the California State Board of Health established 25 hatcheries, from Redding to Santa Ana, and *Gambusia affinis* stocks were distributed to 24 counties during antimalarial and mosquito-abatement work.

By the year 1968, 265 species of fishes had been listed as preying on mosquito larvae throughout the world (Mallars and Fowler, 1970). Sweetman (1958) discussed the most effective larvicidal fishes in still or moving waters of North America.

Mosquito Fish, *Gambusia affinis* (Baird and Girard)

The generic name is derived from *gambusino,* a colloquial expression among Italian-speaking people meaning "nothing" in a scoffing sense. If a fisherman catches nothing, he is said to have been fishing for *gambusino* (Mallars and Fowler, 1970). While *Gambusia affinis* is not an edible game fish, for the biological control of mosquitoes it surpasses any other fish species that has been tried for that purpose.

Description. The maximum size of these fishes depends upon the absence of obstructive plant life and the attendant food supply of the water in which they live. The females grow throughout life, and may reach a length of 60 mm. The males make little growth after the gonopodium (modified anal fin serving as the copulatory organ) is formed, and do not exceed 40 mm in length. The sexes also differ in form and structure (figure 285). The female is more robust, and has a small, rounded anal fin, with rays connected by a membrane. In the male, the third, fourth, and fifth anal fin rays are united and lengthened to form the gonopodium (Mallars and Fowler, 1970).

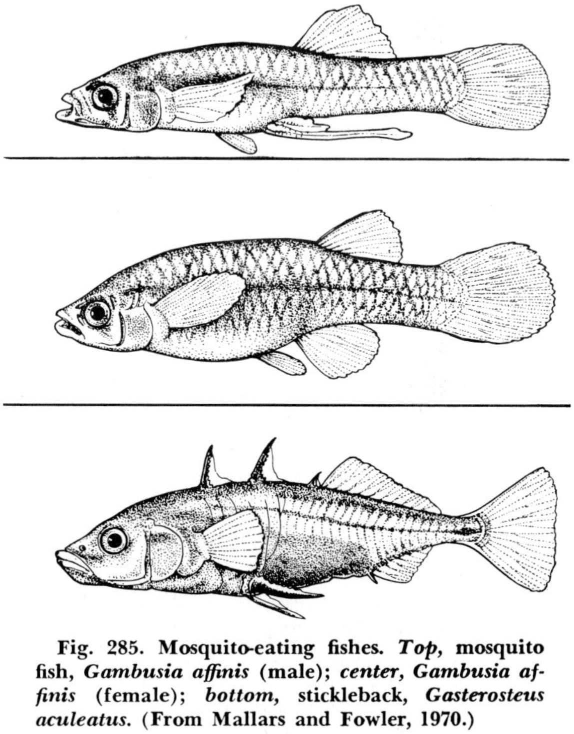

Fig. 285. Mosquito-eating fishes. *Top,* mosquito fish, *Gambusia affinis* (male); *center, Gambusia affinis* (female); *bottom,* stickleback, *Gasterosteus aculeatus.* (From Mallars and Fowler, 1970.)

Identifying Characters. Since *G. affinis* appears to be the only member of the genus introduced into California, one should be able to distinguish it from other poeciliids and the related cyprinodontids. The male is easily identified by the possession of a gonopodium and by his dull-olive coloration. In the female, the dorsal fin originates well behind the origin of the anal fin. In the female guppy, *Poecilia reticulata,* the dorsal fin originates directly above the origin of the anal fin. Otherwise, the females of these 2 species are similar in appearance. Overlap of species may occur in the salt marshes of coastal southern California, where the cyprinodont *Fundulus parvipinnis* Girard is quite widely distributed, and in desert areas where species of *Cyprinodon* occupy restricted environments. In certain drainage canals near the Salton Sea (California), a poeciliid, *Poecilia latipinna* (LeSueur), has become permanently established. From time to time, "aquarium" fishes related to *Gambusia,* such as the guppy, have been introduced.

Larvicidal Efficacy. Hildebrand (1919) considered *Gambusia affinis* to be especially effective against mosquito larvae for the following reasons:

(*a*) It seeks its food at the surface; (*b*) it is very prolific; (*c*) it gives birth to living young, therefore requiring no special environment for depositing and hatching the eggs; (*d*) it lives and thrives under a great variety of conditions, and frequents areas especially suitable for the support of mosquito larvae; and (*e*) it lives and multiplies in ponds stocked with predaceous fishes, provided it has very shallow water for refuge.

This species is especially effective where stock ponds, ditches, natural streams, lakes, ornamental garden pools, and animal troughs are free of excessive vegetation and algae. The feeding efficiency of *Gambusia* is somewhat reduced in the presence of aquatic vegetation. The fishes may not be able to penetrate the vegetation to get to the larvae.

When ponds and troughs have been located under certain types of trees, such as mulberry, eucalyptus, or walnut, their leaves fall into the water and provide hiding places for mosquito larvae. This does not mean that the *Gambusia* are rendered completely ineffective. In one investigation where ponds afforded good protection for immature mosquitoes, 57.6% of the anopheline larvae and 80.7% of the culicine larvae were eliminated. Pupae were reduced 74.2% in anophelines and 85.1% in culicines. In another investigation, anophelines were reduced 80.9 to 94.8% in ponds having a natural growth of aquatic vegetation. On the other hand, these fishes have been observed to be only 20% effective when heavy aquatic vegetation was present. Under some conditions, *Gambusia* may be eliminated by other fishes, such as sticklebacks (Mallars and Fowler, 1970).

There are nearly 400,000 acres (162,000 ha) of rice in California, all of it potential breeding ground for mosquitoes. Tests have shown that if rice paddies are stocked with 200 mature female *Gambusia affinis* per acre (0.4 ha), mosquito production can be decreased by 95% (Anonymous, 1971*a*). Experiments made in rice fields where nearly all mosquitoes were *Culex tarsalis* revealed that fewer fields exceeded the population of mosquitoes normally used as a criterion for successful insecticide treatment where *Gambusia* were stocked than in fields in which they were not stocked. No difference in results could be determined between a stocking rate of 100 and 200 fishes per acre, possibly because the numbers of fishes in a field reached an asymptote well before midseason (Hoy and Reed, 1971).

Gambusia for stocking are usually collected in 30- to 50-ft (9- to 15-m) seines. This procedure requires 2 men, and is very time-consuming. Many fishes escape. A box trap 2 ft (60 cm) square, made of a wooden frame and galvanized hardware cloth, set 2 to 3 ft (60 to 90 cm) deep in shallow water along the margins of ponds or slow-flowing streams, has proved to be highly successful (Caton and Sjogren, 1969). *Gambusia* could be transported in plastic bags containing 50 mature fish in a gallon of water, provided an anesthetic, methyl pentynol, was added to the water at 132 ppm (0.5 ml/gal). Mortality after 72 hours at 20 to 22° C (68 to 71° F) was only 4% compared with 79% in controls without anesthesia (Bay, 1970).

Ecological Considerations. *Gambusia* are voracious feeders, and will consume over 50 kinds of plant and animal life. About 20% of their diet is plant material (blue-green algae, diatoms, desmids, etc.) and, of the aquatic animal life besides mosquitoes, they consume cyclops, chironomid larvae, nematodes, and other planktonic forms. *Gambusia* will eat their own young as well as the young of native larvivorous and game fishes. The use of *Gambusia* at a black-bass hatchery had to be discontinued because of the considerable destruction of young bass. In native streams in the southern states, naturally evolved enemies of *Gambusia* have prevented excessive destruction of desirable fish species, and smaller fish have learned to hide. When transplanted to a new situation where natural checks do not occur and where native species have evolved no defenses, *Gambusia* may crowd out desirable native species

PLATE I

1 2

3 4

5 6

7 8

1. Western subterranean termite, *Reticulitermes hesperus,* alates and dealate. 2. *Reticulitermes hesperus,* queen, soldier, and workers. 3. *Reticulitermes hesperus,* soldier. 4. Western drywood termite, *Incisitermes minor,* alates. 5. *Incisitermes minor,* queen, soldier, and nymphs. 6. Powder-post termite, *Cryptotermes brevis,* alate, soldier, and nymphs. 7. Pacific dampwood termite, *Zootermopsis angusticollis,* alate. 8. *Zootermopsis angusticollis,* supplementary reproductive, nymphs, and eggs.

PLATE II

1 2

3 4

5 6

7 8

1. Southern lyctus beetle, *Lyctus planicollis (left)*, and a common buprestid beetle, *Dicerca horni*. 2. Blue woodwasp (horntail), *Sirex cyaneus*, female. 3. Carpenter ant, *Camponotus vicinus*, workers and winged male. 4. *Camponotus semitestaceus*, worker. 5. Argentine ant, *Iridomyrmex humilis, (left)*, and *Iridomyrmex pruinosus*. 6. Harvester ant, *Pogonomyrmex californicus*, worker. 7. Mountain carpenter bee, *Xylocopa tabaniformis orpifex*. Adult, balls of pollen, and larva in gallery system in a cedar shake. 8. Valley carpenter bee, *Xylocopa brasilianorum varipuncta*, male (*left*) and female.

PLATE III

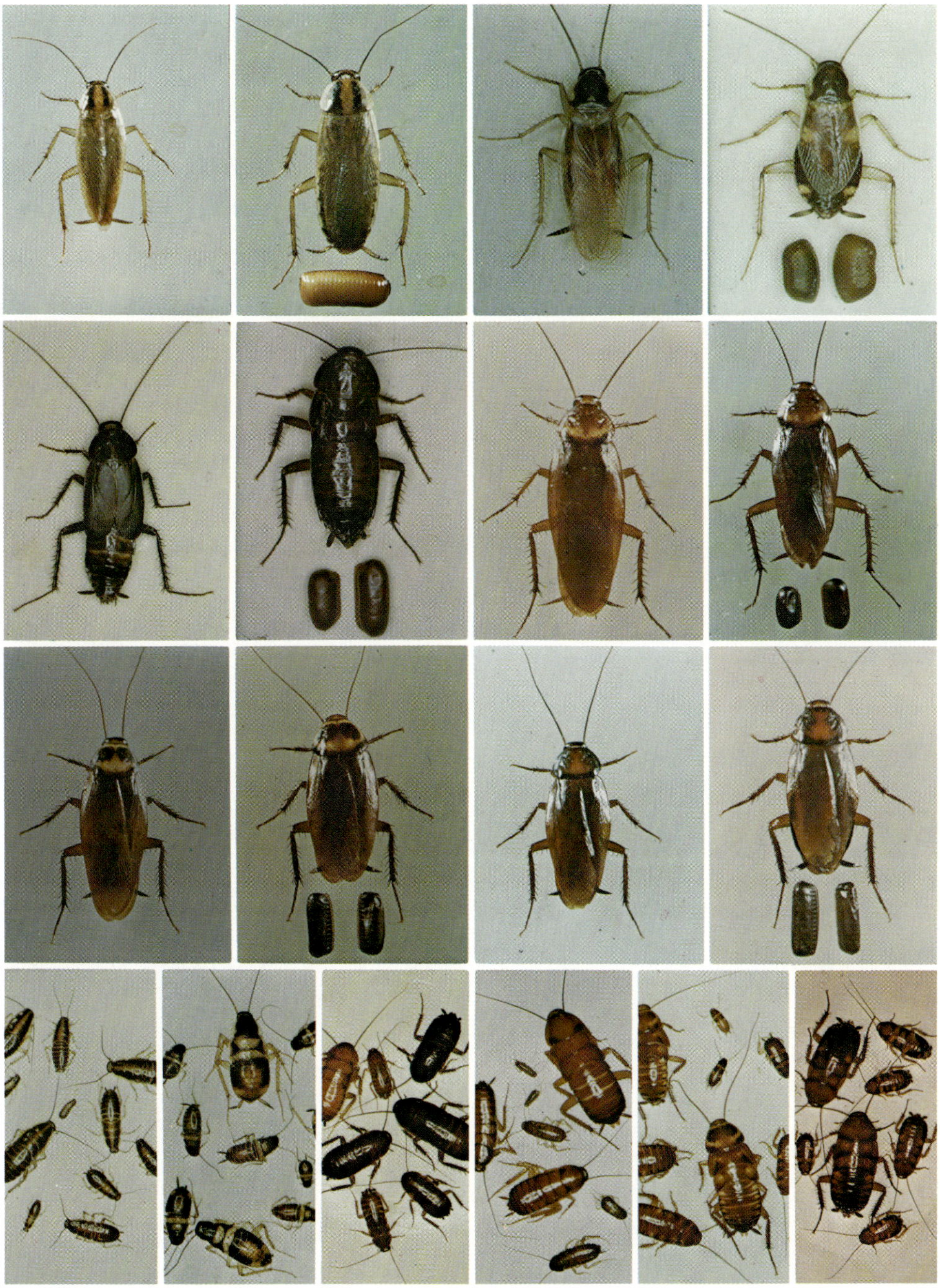

1. German cockroach, *Blattella germanica,* male (*left*) and female. 2. Brownbanded cockroach, *Supella longipalpa,* male (*left*) and female. 3. Oriental cockroach, *Blatta orientalis,* male (*left*) and female. 4. American cockroach, *Periplaneta americana,* male (*left*) and female. 5. Australian cockroach, *Periplaneta australasiae,* male (*left*) and female. 6. Brown cockroach, *Periplaneta brunnea,* male (*left*) and female. 7. Nymphs of German (*left*), brownbanded (*center*), and oriental (*right*) cockroaches. 8. Nymphs of American (*left*), Australian (*center*), and brown (*right*) cockroaches.

PLATE IV

1 2 3 4 5 6 7 8

1. House fly, *Musca domestica*. Eggs, larva, pupae, and metric scale (*left*), and adult male (*right*). 2. A bluebottle fly, *Calliphora terraenovae* (*top*), and a flesh fly, *Sarcophaga bullata*. 3. Redlegged ham beetle, *Necrobia rufipes*, and metric scale (*left*); hide beetle, *Dermestes maculatus* (*center*); khapra beetle, *Trogoderma granarium* (*right*). 4. A dermestid beetle, *Trogoderma glabrum*, adults, larvae, and a pupal skin. 5. Sawtoothed grain beetle, *Oryzaephilus surinamensis* (*left, top*); flat grain beetle, *Cryptolestes pusillus* (*left, bottom*); rice weevil, *Sitophilus oryzae* (*right, top*); shiny spider beetle, *Gibbium psylloides* (*right, bottom*). 6. Dark mealworm, *Tenebrio obscurus* (*left*), and red flour beetle, *Tribolium castaneum*. 7. Cadelle, *Tenebroides mauritanicus* (*left*); drugstore beetle, *Stegobium paniceum* (*center*); cigarette beetle, *Lasioderma serricorne* (*right, top*), courtesy Stennett Heaton; cigarette beetle larva (*right, bottom*). 8. Bean weevils, *Acanthoscelides obtectus*, on a lima bean.

PLATE V

1. Mediterranean flour moth, *Anagasta kuehniella* (*left*), and Indian meal moth, *Plodia interpunctella*. 2. Angoumois grain moth, *Sitotroga cerealella* (*left*), and raisin moth, *Cadra figulilella*. 3. Navel orangeworm, *Paramyelois transitella* (*top*), and larva. 4. Driedfruit beetle, *Carpophilus hemipterus* (*left*), and corn sap beetle, *Carpophilus dimidiatus*. 5. Gray silverfish, *Ctenolepisma longicaudata*. 6 and 7. Firebrat, *Thermobia domestica*. 8. Odd beetle, *Thylodrias contractus,* female.

PLATE VI

1 2

3 4

5 6

7 8

1. Furniture carpet beetle, *Anthrenus flavipes,* adults and larvae. 2. Black carpet beetle. *Attagenus megatoma* (*left*), and varied carpet beetle, *Anthrenus verbasci,* larva. 3. Varied carpet beetle, *Anthrenus verbasci,* adult, pupal case, and eggs. 4. Webbing clothes moth, *Tineola bisselliella.* 5. Western black widow spider, *Latrodectus hesperus,* female (*left*) and male. 6. Brown recluse spider, *Loxosceles reclusa* (*left, top*), and a southwestern brown spider, *Loxosceles unicolor* (*left, bottom,* and *right*). 7. South American brown spider, *Loxosceles laeta,* female (*left*) and male. 8. Tarantula, *Aphonopelma californica.*

PLATE VII

1. Deadly sculptured scorpion, *Centruroides sculpturatus,* female. 2. A centipede, *Scolopendra polymorpha* (*top*), and the house centipede, *Scutigera coleoptrata.* 3. A yellowjacket, *Vespula pensylvanica,* adult (*left*) and larva. 4. A yellowjacket, *Vespula sulphurea.* 5. Giant hornet, *Vespa crabro germana.* 6. A polistine wasp, *Polistes fuscatus aurifer.* 7. A polybiine wasp, *Mischocyttarus flavitarsis flavitarsis.* 8. Eastern diamondback rattlesnake, *Crotalus adamanteus,* courtesy F. E. Russell, M.D.

PLATE VIII

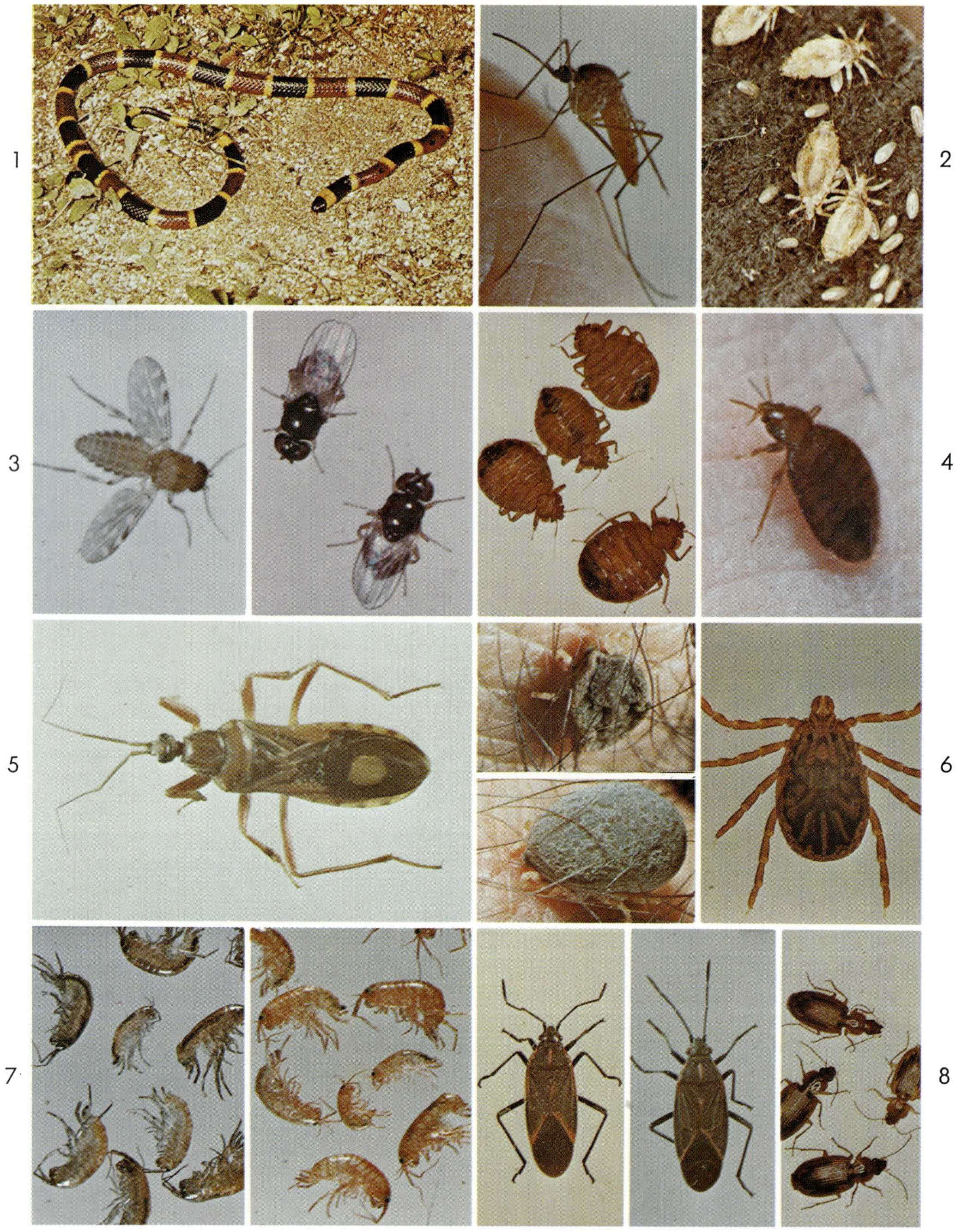

1. Coral snake, *Micrurus fulvius fulvius,* courtesy A. D. Cruickshank, from National Audubon Society. 2. Broadwinged mosquito, *Culiseta inornata* (*left*), and body lice, *Pediculus humanus humanus,* with eggs. 3. Biting midge, *Culicoides californiensis* (*left*), and eye gnats, *Hippelates collusor*. 4. Bed bugs, *Cimex lectularius (left),* and bed bug feeding. 5. Twospotted corsair, *Rasahus thoracicus*. 6. The pajaroello, *Ornithodoros coriaceus,* before feeding (*left, top*), and after 10 minutes of feeding (*left, bottom*); Rocky Mountain wood tick, *Dermacentor andersoni* (*right*). 7. An amphipod, *Talitroides sylvaticus,* alive *(left)* and dead. 8. Boxelder bug. *Leptocoris trivittatus* (*left*); western boxelder bug, *Leptocoris rubrolineatus* (*center*); tule beetles, *Agonum maculicolle* (*right*).

and become a pest (Myers, 1965). In fact, the possibility of extermination of native fishes, especially localized endemic species, accompanies every introduction of a new species (Deacon *et al.*, 1964). When discussing the problems connected with the introduction of new species, Bay (1965) stated:

> There is a universal reluctance among biologists, naturalists, and sportsmen to see any new species introduced into an area for fear that it will succeed too well and become a pest. These fears are based on such mistakes of the past as the introduction of the gypsy moth, the English sparrow, the European starling, and the carp into the United States. Similar cases of unwanted introductions can be cited throughout the world, and they have been perhaps most common in aquatic environments. For this reason, it is well for those responsible for mosquito control to work as closely as possible with those entrusted with the management of fresh-water resources. Carefully supervised ecological surveys of would-be larvivore release sites, followed by restricted, in-the-field larvivore evaluations, can benefit both mosquito-control and fisheries staffs. If the larvivore proves to be an effective mosquito predator without threatening existing fisheries, the sportsman can pursue his pleasure in greater comfort. If, on the other hand, it should transpire that a candidate larvivore has disadvantages that outweigh its mosquito-control potential, careful documentation of the situation is of value to all parties concerned.

The successful use of *Gambusia* and other mosquito-eating fishes depends on an effective integrated mosquito-control technology involving the use of selective insecticides that give adequate control of mosquitoes without adverse effects on the mosquito-fish population. Judging from the results of past investigations (Mulla *et al.*, 1963), it is reasonable to suppose that such an integrated-control program should continue to be feasible. On the other hand, researchers in the central valley of California are beginning to question the desirability of incorporating *Gambusia* in permanent and semipermanent mosquito-breeding habitats. They believe that the adverse effects of *Gambusia* on beneficial invertebrate predators (notonectids, dytiscids, hydrophilids, and others) may offset any advantage gained by predation against mosquito larvae. Gut analyses show that *Gambusia* apparently prefers to feed on the above invertebrate predators, and usually not more than 15% of its diet consists of mosquito larvae. *Gambusia* appears to be unable to graze as effectively as notonectids, dytiscids, and hydrophilids in emergent grass and other vegetation. When these predators come out into the open water, the mosquito fish consumes them, resulting in a gradual depletion of effective *Culex* predators. In view of these facts, as revealed by various competent and careful investigators, the widespread dissemination of *Gambusia* in permanent and semipermanent bodies of water, such as duck ponds and rice fields, would not appear to be justifiable. The introduction of *Gambusia* may be more logical in pastures, swimming pools, and temporary mosquito sources in urban areas (E. F. Legner, correspondence).

Guppy, *Poecilia reticulata* Peters

The guppy is a mosquito-eating fish, belonging to the same family (Poeciliidae) as *Gambusia.* It is a small, live-bearing top minnow from Central and South America. Females reach lengths up to 34 mm. Their body color may vary from yellow to yellow-green, with rows of blue scales. The males reach a length of up to 20 mm, and do not grow appreciably after attaining sexual maturity. Their bodies are often brightly colored with iridescent shades of red, blue, or green. There are large, black spots on the flanks, and the dorsal and caudal fins are often long and trailing. While the females of this species superficially resemble *Gambusia affinis,* the males of the 2 species differ greatly. In similar periods, guppies have developed in greater numbers in ponds than have *Gambusia.* The fishes are smaller, and are able to penetrate denser vegetation (Mallars and Fowler, 1970). These authors observed that in some polluted waters in which both *Poecilia reticulata* and *Gambusia affinis* were introduced, only the former survived and thrived. In the mild, nontropical climates of California, the high pollution tolerance and fecundity of *P. reticulata* may result in this species being useful for the control of mosquitoes in polluted waters, such as in "sewage oxidation ponds, dairy drains, ditches receiving organic industrial effluents, and temporary reservoirs that would be incapable of supporting other larvivorous fishes."

Sticklebacks, Genus *Gasterosteus*

Description. In California, the threespine stickleback, *Gasterosteus aculeatus* L., with 3 subspecies, is the most important of about 12 species. Its maximum standard length is 74 mm for females and 65 mm for males. ("Standard length" is the distance from the tip of the nose to the end of the hypural structure, or beginning of the tail fin.) Sticklebacks get their common name from the first 3 rays (rarely 2 or 4) of the dorsal fin that form a series of sharp, detached spines (figure 285, *bottom*). The elongate, compressed body tapers behind to a narrow caudal peduncle. Vertical bony plates cover the flanks, but other parts are free of any scaly covering. Both jaws have teeth, but they are absent from the tongue and palate. Both sexes are pale olive green.

Biology and Habits. When the mating season begins, as early as mid-February, the male's body changes to blue or greenish blue, the ventral surface becomes reddish, and the iris of the eye becomes a highly iridescent blue-green. There is little or no color change in the female (Mallars and Fowler, 1970). The female deposits 50 to 100 eggs in a small pit formed by the male in the top of a complex nest he has constructed from pieces of straw, sticks, and other vegetable matter. The parts of the nest are bound together with a silken fiber secreted by an organ closely connected with the male's kidneys (Vrat, 1949). Sticklebacks have high salinity tolerance, and can even move out into the ocean. They will prey on the smaller *Gambusia* when the 2 species are placed together in a restricted habitat.

Larvicidal Efficacy. Hubbs (1919) made a study of the stickleback in swamps, pools, and streams of coastal California, and wrote an enthusiastic report on the role of this species in mosquito control. The abundance of other food does not deter it from feeding on mosquito wrigglers, both larvae and pupae. It feeds at all water levels, and thereby captures mosquito wrigglers of different habits. Its spines and armor make it immune from attack by larger fishes. The species is widely distributed in the brackish waters of bays and estuaries, and in coastal streams as far up as the fishes can penetrate. Hubbs noted that mosquito wrigglers were abundant in streams in which sticklebacks were unable to penetrate because of the steep descent of their rocky beds.

Where the water in shallow ponds and reservoirs becomes warm in summer, the stickleback can survive when other species may die. It has even been found in the waters of hot springs. The streams of California provide an ample supply of this species for stocking artificial and natural pools, ponds, and reservoirs. It is easily handled, using open buckets or cans (Hubbs, 1919).

Goldfish, *Carassius auratus* (L.)

Goldfish resemble their relatives, the carp, in having long dorsal fins, but differ by lacking barbels. They usually keep a well-balanced garden pool free of culicine larvae, but are not so effective against anophelines. When overfed, they become lethargic and less effective. Under some conditions, *Gambusia* will attack goldfish and tear their fins. Also, their presence at goldfish spawning time is detrimental.

Annual Fishes

One of the possibilities that is currently being investigated is the introduction of "annual fishes" —species that have the ability to survive periods of desiccation in the egg stage. The desiccated eggs hatch when submerged in water. Such fishes might be particularly useful for controlling mosquitoes that also pass dry periods as eggs, such as species of *Aedes* and *Psorophora*.

The annual fishes are egg-laying tooth carps of the family Cyprinodontidae. The Argentine pearl fish, *Cynolebias bellottii,* has been shown to be adaptable to California conditions (Mallars and Fowler, 1970).

Regulations Regarding the Stocking of Fish

There are 5 regional offices of the California Department of Fish and Game to which inquiries about stocking fish in lakes and streams should be directed. Some regulations follow:

> Only trout and salmon (family Salmonidae); black bass, crappie, and sunfish (family Centrarchidae); catfish (family Ictaluridae); and minnows (family Cyprinidae) of species already present in California may be planted.
>
> The planting of mosquito fish *(Gambusia affinis)* for purposes of mosquito control is hereby authorized without first applying for and obtaining a permit under these regulations.

Mosquito-abatement districts appreciate any efforts that are made by citizens in either eliminating mosquito sources around their homes or in planting mosquito fish. Most abatement districts will provide fish for this purpose free of charge.

Other Promising Natural Control Agents

With the exception of mosquito-eating fishes, the chance of finding satisfactory natural enemies of mosquitoes has not been considered good until somewhat recently. Both nymphal and adult dragonflies are predaceous on mosquitoes, but their relatively small numbers preclude an important effect on control. While bats feed freely on the pests, they are likewise factors of little consequence.

A nematode, *Reesimermis nielsoni,* was found by USDA entomologists to be effective against mosquito larvae, and a fungus, *Lagenidium culi-*

cidum, was found by the Center for Disease Control to be lethal to a number of culicines, including *Culex tarsalis* (CDC, 1973). A bacterial pathogen has been isolated from moribund *C. tarsalis* larvae. This is a crystalliferous *Bacillus,* tentatively considered to be a new species. It is ineffective against *C. tarsalis,* but is a very effective pathogen against *Aedes* larvae (Reeves, 1970).

Another promising development in the field of natural control agents was based on the observation that when mosquito larvae came in contact with certain mucilaginous seeds, their oral brushes became irreversibly attached, causing the larvae to die from drowning, exhaustion, and other forms of stress. When the seeds of the cruciferous weeds *Descurainia sophia, D. pinnata,* or *Lepidium flavum* were placed in water, the dehydrated mucilage contained in the seed coats popped out in less than a minute. Figure 286 shows the remarkable efficiency of *D. sophia* in trapping second- and third-instar larvae of *Culex peus.* A single seed of *L. flavum* was observed to entrap 27 second-instar *Culex* larvae. There were determined to be over 2 million seeds of *L. flavum* per lb (0.45 kg), and these could entrap a maximum of 54 million larvae. The seeds of the cruciferous weeds could be heated to 121° C (250° F) for 20 minutes to prevent germination, if this

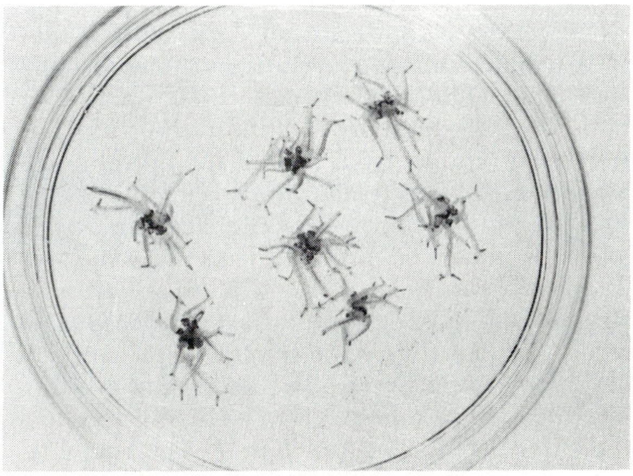

Fig. 286. Second- and third-instar larvae of *Culex peus* attached to seeds of *Descurainia sophia* by mucilaginous layer. (From Reeves and Garcia, 1969.)

was found to be desirable in order to keep the weeds from spreading into areas where they were not wanted. The mucilage formation was not adversely affected by heat treatment. The crucifers were found growing extensively along the borders of fields, in vacant lots, and along roadsides. If such sources were insufficient to meet the demand, they could presumably have been produced on marginal land that was unsuitable for the production of other crops (Reeves and Garcia, 1969).

Encouraging prospects have been reported by Reeves (1970) concerning the investigations of the Mosquito Pathology Laboratory at the University of California (Riverside). It should be emphasized that to this point these investigations have been successful only in their research phases, and practical results are yet to be demonstrated.

A natural low level of mosquito population along some streams in southern California has been correlated with the presence of the following 5 green filamentous algae isolated from mosquito-breeding habitats: *Cladophora glomerata, Schizothrix friesii, Chlorella ellipsoides, Chara fragilis,* and *C. elegans.* Hot methanol extracts from dehydrated algal material were purified, using column chromatography. Each of the 5 algae yielded a pure compound that was toxic to immature mosquitoes, whereas 2 inactive algae did not. Histopathological studies on larvae of *Aedes triseriatus* revealed at least 1 mode of action—a disruption of the epithelial lining of the gut wall and stoppage of gut contents. These findings indicate that strains of algae toxic to mosquitoes could be propagated in rice fields, ponds, and streams to suppress mosquito larvae. Also, the active compounds might be chemically characterized and commercially synthesized, along with their analogs, and applied like other chemical insecticides. No adverse effects on fish or aquatic insect predators have been observed, so the toxic compounds would be rather specific natural larvicidal agents. Being natural botanical larvicides, there would probably be less risk of mosquito larvae becoming resistant than with synthetic chemicals.

Crushed *Chara* released an odor resembling that of crushed garlic. An oil fraction of garlic (*Allium sativum*) proved to be highly larvicidal against the third instars of 5 species of *Culex* and *Aedes.* The LC_{50}'s were in the range of 2 to 6 ppm (Reeves, 1970). Crude methanolic extracts and an oil fraction from minced, dehydrated garlic were found to be larvicidal in tests made against the third-instar larvae of 6 species of mosquitoes. The partially purified oil fraction was 12.5 times more potent against *Culex tarsalis* than the crude extract (Amonkar and Reeves, 1970). The larvicidal principles of garlic have since been isolated, and were identified as diallyl disulfide and diallyl trisulfide. Natural and synthetic samples of the larvicides were fatal at 5 ppm to *Culex pipiens quinquefasciatus* (Amonkar and Banerji, 1971).

Natural pathogens, toxins, and hormones are under intensive investigation. The juvenile hormone mimic 2,6-di-*t*-butyl-4-(*a,a*-dimethylbenzyl) phenol, tested under the code designation MON-585, was used at a concentration of 0.1 ppm (1 part in 10 million parts of water), and was found to be able to block the metamorphoses of all instars of *Aedes aegypti, Anopheles albimanus,* and *Culex restuans* Theobald (Sacher, 1971). In another investigation, this compound gave at least 95% mortality of *Aedes aegypti* and *Anopheles stephensi* Liston, based on the emergence of "normal" adults in relation to the number of third-instar larvae exposed, when used at 0.25 ppm in water. Equally good results were obtained against *Culex peus, C. tarsalis, Anopheles albimanus,* and *Aedes taeniorhynchus* at 0.1 ppm (Jakob and Schoof, 1971, 1972). The work of Schaefer and Wilder (1972, 1973) with ZR-515, already discussed in this chapter, is another example of the potential existing in this type of compound for mosquito control. Currently, an increasing number of insect growth regulators (IGR's) show promise when used against the larvae of mosquitoes and midges. These compounds adversely affect insect development, but not necessarily in the same way as juvenile hormones and their analogs and mimics. Instead of preventing the molting process, they may cause mortality during it.

One of the most interesting developments of recent years was the discovery that third-instar larvae of *Culex pipiens quinquefasciatus* reared under overcrowded conditions (5 to 7 larvae per ml) produced toxic chemical factors that were ether-extractable. The most active materials showed rapid lethality and residual action against first-instar larvae, in some tests manifesting biological activity for as long as 22 days. Considerable activity was also manifested against first-instar larvae of 3 other mosquito species (Ikeshoji and Mulla, 1970). In a later investigation, the growth-retarding effects of the overcrowding factors and their bacteriostatic or bactericidal properties were investigated.

It is evident that there is much emphasis on mosquito control research that may help attain the goal designated as "integrated control."

All the phases of research toward this objective are obtaining increased support (Coppock, 1970). However, the objective is much easier to define than to accomplish. The complexity of the problems and also the fields of research that appear to offer the greatest opportunities were discussed in great detail by Mulla (1970).

MIDGES AND GNATS

In the United States, the terms "midge" and "gnat" refer to a number of families of small flies. There has been some tendency to refer to the more pestiferous of these small flies as gnats, but the literature in which the 2 terms have been used interchangeably is so extensive that any attempted distinction would probably cause even more confusion. A few midges and gnats are vectors of disease, some inflict painful bites, while others are annoying merely by their presence. This section will deal only with some of the envenomating species; others are discussed later in the present chapter under the heading "Pestiferous Arthropods That Do Not Envenomate."

Sometimes, regional mosquito-abatement programs will also result in the control of midges and gnats, but in other instances, a special program must be developed for a particular species. Partial measures that can be employed by the homeowner in preventing or combating mosquitoes, such as screening, sprays, aerosols, dichlorvos resin strips, and repellents, are equally applicable to midges and gnats, but for the latter, the screens must be of a smaller mesh than those used for mosquitoes, and a 60-mesh bolting cloth may be necessary to exclude these tiny insects.

Biting Midges (Ceratopogonidae)

The ceratopogonids are all dark-colored, very small, slender midges, rarely 5 mm long, usually 1 to 3 mm. The head is small, and rounded or hemispherical (flattened behind); the antennae are slender, and hairy in the females but plumose in the males. The females have sharp, piercing mouthparts. Most ceratopogonids feed on a wide variety of plant and animal juices, and some subsist entirely on nectar. Some females are ectoparasitic on larger insects, others are predaceous on insects smaller than themselves, others feed on coldblooded vertebrates, and only a few feed on warmblooded vertebrates. Species of only 4 genera (*Culicoides, Ceratopogon, Lasiohelea,* and *Leptoconops*) are known to attack man. The males do not bite. Keys to the Nearctic species of bloodsucking midges were prepared by Fox (1955).

The structure of the mouthparts of the biting midges is shown in figure 287, using those of *Culicoides furens* as an example. These mouthparts are tiny but vicious. The stylets are enclosed in the labial gutter, which is covered in front by the broad labrum. Each of the female mandibles is a thin, flat blade (figure 287, *B*), with a finely toothed distal margin. The labrum (*A*), hypopharynx (*C*) and the mandibles (*B*) together form the piercing organ (Snodgrass, 1944).

Biting midges are sometimes called "no-see-ums," "punkies," "sand flies," or "black gnats." They were once included in the family Chironomidae, but may be distinguished from chironomids by their shorter legs, broader wings, and shorter thorax, with the pronotum never prominent. Most importantly, some species have a vicious bite, whereas chironomids never bite. The more conspicuous "black flies" (simuliids) and mosquitoes are sometimes blamed for bites for which the biting midges are responsible.

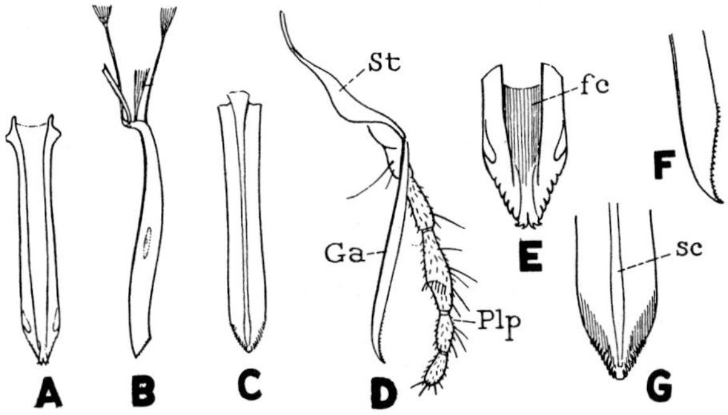

Fig. 287. Mouthparts of a biting midge, *Culicoides furens*. *A*, labrum; *B*, right mandible (anterior); *C*, hypopharynx; *D*, maxilla; *E*, end of labrum more enlarged, showing food canal on undersurface; *F*, end of galea; *G*, end of hypopharynx; *fc*, food canal; *Ga*, galea; *Plp*, palpus; *sc*, salivary canal; *St*, stipes. (From Snodgrass, 1944.)

The larvae are aquatic or semiaquatic. Various species may be found in mud, wet sand, decaying vegetation, damp, rotting wood, foul and stagnant water, brackish or salt water, fresh, flowing streams, in sap oozing from trees, or in treeholes. The larvae have oval, pale-brownish heads. Their long, cylindrical, dull-whitish or translucent bodies are devoid of appendages, lacking even the pseudopods of the chironomid larvae. They move about with a wriggling, eel-like action.

Bodega Black Gnat, *Leptoconops kerteszi* Kieffer

This minute, diurnal biting midge obtained its common name from Bodega Bay, California, where it was reported in 1934 and where its biology was investigated by Smith and Lowe (1948). The larvae breed in the surface layers of porous or sandy soils having adequate moisture in various areas of western North America. This species also occurs in the Mediterranean region of North Africa. In southern California, *L. kerteszi* is a serious pest in desert areas, breeding in damp sand along watercourses, around desert springs, in an extensive marsh along the north shore of the Salton Sea, and in the Santa Ana River basin in Riverside County. In the Salton Sea region, the expanding urban population is showing increasing interest in desert recreation areas, and there is a corresponding increase in the importance of *L. kerteszi* as a public health pest (Foulk, 1967; Sjogren and Foulk, 1967). It is also occasionally a pest in the Santa Monica Mountains, at least in years following unusually heavy winter rainfalls. In the spring and summer of 1973, following a winter of unusually heavy precipitation, a severe infestation of *L. kerteszi* was called to the author's attention in residential properties in Ramirez Canyon, about a mile from the coast in the Santa Monica Mountains. The creek had stopped running in that area in mid-June, but people were bitten all through the spring and summer months. While walking about in infested areas, they were bitten on the lower parts of their bodies and under their clothes. While sitting in chairs outdoors in the sunlight, they were bitten at the hairline, on the ears and neck, and around the eyes, frequently causing itching and painful, persistent swellings. The midges did not attack inside their homes.

Description. The adult female (figure 288) is 2.5 to 3 mm long, and black, except for the legs, which vary from dark brown at the base to light brown at the extremities. The transparent wings, when folded on the back and seen by reflected light, appear opaque white. The dark-brown antennae are 13-segmented, the basal segment flat

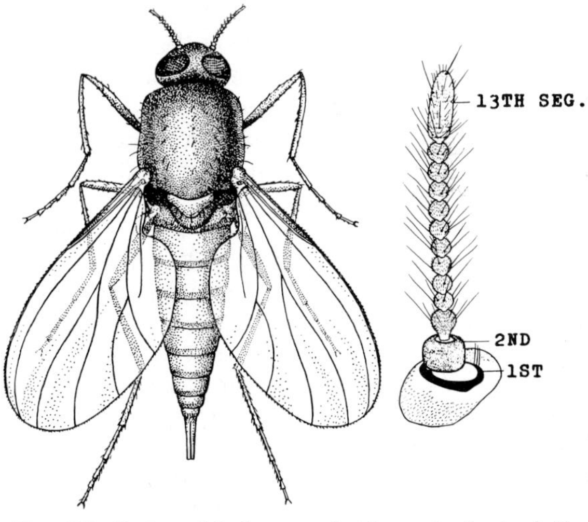

Fig. 288. Bodega black gnat, *Leptoconops kerteszi* (female). *Right*, detail of antenna. (Antenna from Smith and Lowe, 1948.)

and ringlike and having 3 bristles on the inner margin (figure 288). The male is darker than the female, and live specimens appear shiny black. It is slightly larger, being 3.5 to 4 mm long. Its antennae are 15-segmented.

Life Cycle. At Bodega Bay, Smith and Lowe (1948) noticed the first swarms of males and a few females in mid-April, and the residents started to complain about midge bites at that time. The pools in the area were without surface water for the first time since December, and the moist sand appeared to be right for egg-laying. Captured males died within 4 days. Captured females lived as long as 11 days, and took as many as 4 blood meals. The adult lives of the insects might have been longer in the field. The incubation period was at least 12 days. Two gravid females were dissected, and were found to contain 48 and 54 eggs. Eggs hatching from April to August produced larvae that pupated in March to August of the following year. Adults emerged almost continuously from early March until August or September, and the last small swarms, predominantly females, were seen in early October.

There were 4 larval instars, and the period required to reach the pupal stage was 8 to 10 months. The fourth-instar larvae were 4.2 to 5.6 mm long, and since they were legless and the thoracic segments had no special characteristics, the larvae appeared to have 23 segments. They lacked spines, hairs, or bristles, except for 2 bristles just anterior to the mandibles. The larvae of the first 2 instars were colorless and transparent, but became orange in color when they reached the third instar. The pupae were noted to be 2.7 mm long, had dark-brown thoraxes, and tan to dull-orange abdomens.

Smith and Lowe found larvae in water-saturated soil from 1 to 3 in. (2.5 to 7.5 cm) beneath the surface. In a block of soil 12 × 12 in. (30 × 30 cm), they found 166 larvae and 41 pupae in soil samples 1 to 3 in. in depth, but found only 2 larvae at 3 to 5 in. (7.5 to 12.5 cm) and none at 5 to 7 in. (12.5 to 18 cm). The pupae wriggled actively to the surface, and stood erect on the tips of their abdomens until the adults emerged. In the laboratory, the pupal stage averaged 8 days. Pupae were found in the field from early March to late August.

Valley Black Gnat, *Leptoconops torrens* Townsend

This small biting midge or gnat is widely distributed in California, and occurs in several other western states. In California, Smith and Lowe (1948) found this species along the western side of the Sacramento Valley and in isolated locations in the San Joaquin Valley. It has not been recorded from outside the United States. Smith and Lowe attempted to trap these midges from many types of environment, but found larvae emerging only from clay adobe soil, particularly in sinkhole areas that had contained water during the winter. It feeds on man, domestic animals, poultry, and other birds, causing in man a transient, inflammatory swelling and often an intense itching which may continue for about a week.

The midges tend to get under clothing, and their bites are concentrated in areas where the clothing fits tightly. Discomfort is most acute where bites are concentrated. Loose-fitting, short-sleeved garments prevent the concentration of bites, and may even reduce their total number. The midges have been observed to bite as late as 8 p.m. on warm, still evenings (Smith and Lowe, 1948). It is advisable to stay indoors during periods of maximum midge activity. Because many of the endemic habitats of the valley black gnat have been invaded by people, human exposure to the insects has greatly increased, together with public demand for gnat control (Whitsel and Schoeppner, 1966).

Description. The adults of this species closely resemble those of *L. kerteszi,* except that the antennae of the female contain 14 segments. The larvae of the 2 species are also similar, except that the mouthparts of the larvae of *L. torrens* are much more heavily chitinized and easily visible in the head capsule (Smith and Lowe, 1948).

Life Cycle. The flight period of the adults is from 4 to 6 weeks, usually beginning in mid-May. Emergence from any one spot probably does not extend over a period of more than 3 weeks. Females feed only once, and the males do not feed at all. Gravid females were found to have 60 to 70 eggs at various stages of development in their ovaries. Smith and Lowe believed that mated and fully engorged females entered cracks in clay adobe soil to lay their eggs, and did not again emerge. The eggs may hatch soon after fall rains have wetted the soil. The larvae spend at least 2 years in the soil, at a depth of 15 to 30 in. (38 to 76 cm), probably feeding on organic detritus with its associated microorganisms. They can also enter a period of diapause that can last as long as 3 years if the soil does not crack on schedule. Apparently, mature larvae orient themselves on the closed crack faces and, as the cracks open, pupation occurs. The adults emerge via the open cracks.

Envenomation. Smith and Lowe found valley black gnats to be voracious feeders that could not be frightened away from their hosts until after feeding. The mouthparts seem to saw for 1.5 minutes, and the abdomen is then quickly distended with ingested blood. No sensation is felt by the host for 30 to 60 seconds, then a slight tingling. A tiny hemorrhage remains under the skin where the host is bitten. In about 10 to 15 minutes, a hard, white, raised area appears, about 0.5 cm in diameter, along with severe itching. In 1.5 hours, the hard, white area is partially reduced and becomes inflamed. Then a diffused, red, blotchy area 25 to 50 mm across appears around the central swelling. This disappears about 3 hours after the bite, leaving a central 5-mm red area resembling a flea bite, and itching disappears. Eighteen hours after the bite, a hard, inflamed, conical swelling about 2.5 mm in diameter marks the site of the bite, and itching returns. On many persons, a small, clear blister forms on the second day and disappears a day or two later. For about 5 days, itching is intense, especially early in the morning.

Biting midges are so small that they are not noticed until they begin to pierce the skin. Unlike the black flies (Simuliidae), they appear to inject no anesthetic, and therefore the bites of most species are felt immediately. For some persons, reactions to biting midges are intensified by repeated bites, and include both an immediate and a delayed reaction. Blisters filled with clear, serous fluid may develop. The blisters may last for several days, itching and burning all the while, but there are no general systemic symptoms. Most people show no reaction except for the discomfort and annoyance provided by the initial attack (Hase, 1933; Gordon and Crewe, 1948; Areán and Fox, 1955).

Of the 2 North American biting genera of Ceratopogonidae, *Culicoides* and *Leptoconops,* the latter produces the more intense reaction. The initial transient swelling may become vesicular, and then rupture and produce an open, moist lesion which may exude for weeks. However, even *Culicoides* causes itching and, in sensitive individuals, it can cause welts and lesions that may persist for several days. The itching will cause the victim to scratch, and this may result in secondary infection (Frazier, 1969).

Culicoides furens (Poey)

This midge is a vicious biter that breeds in salt marshes and similar habitats along the Atlantic and Gulf coasts from Massachusetts to the West Indies and south to Brazil, and also along the Pacific Coast from Mexico to Ecuador. According to Jamnback (1969), it breeds all year round in tropical areas, and becomes a very serious pest. In New York state, it probably has only 1 extended generation each year, attacking man from June into August.

Culicoides sanguisuga (Coquillett)

This is said to be the most abundant and annoying midge of the northern woods (Jamnback and Wathews, 1963). Regarding forest-inhabiting biting midges, Jamnback (1969) pointed out that these species lose moisture readily, and are therefore inactive during the day except in shaded, humid woodlands or on wet, humid, and calm days in open areas. Therefore, they usually fly about in search of their blood meals at dusk, and will bite all night long. They bite people indoors during the evening, for they are attracted to light and are small enough to pass through ordinary window screens. Where biting midges are abundant, it may be desirable to draw the shades or close the windows on warm summer evenings.

Culicoides californiensis Wirth and Blanton

This species was described from specimens reared from cottonwood treeholes in Kern and San Bernardino counties, California. (See plate VIII, 3.)

Nematode Parasites

A perceptible degree of natural control of *Culicoides* species under favorable conditions was believed to have resulted from parasitism of larvae by merminthid nematodes (Smith and Perry, 1967). [Merminthid nematodes have also been given credit for holding down populations of simuliids in Wisconsin (Phelps and DeFoliart, 1964), and more than 25 species of mosquitoes were said to be parasitized by these nematodes (Welch, 1960).]

Midges or Gnats Affecting Tourism

In Florida, the Bahamas, and the Caribbean area, ceratopogonid midges or gnats are major factors affecting the growth of tourism. The 3 species of greatest importance are *Culicoides furens, C. barbosai* Wirth and Blanton, and *Leptoconops bequaerti* (Kieffer). They are generally called "sand flies," although of the important species, only *L. bequaerti* is normally found breeding in sand on the beaches, and then only under very special conditions. The important *Culicoides* species normally breed in swamps and wet soil, conditions commonly present near the beach areas where tourism is so intensively developed. If there are breeding areas within 5 mi (8 km) of a tourist development, regardless of wind direction, annoyance from "sand flies" will be certain. They can become so abundant that hotel guests may suffer multiple bites within an hour on exposed body surfaces, whereas for most people, 5 bites per hour is considered to be the tolerable limit. Developers have been advised to investigate the sand fly situation thoroughly before commencing their projects. Preventive measures taken before development and construction are more satisfactory and less expensive than control measures that may become necessary later (Linley and Davies, 1971).

Control of Midges or Gnats

The experience in California has been that irrigation water keeps the soil from cracking and hinders the emergence of adult biting midges, while disking the soil apparently discourages egg deposition. Pre-emergence treatments of resting places with various organophosphorus insecticides have greatly decreased *Leptoconops torrens* populations (Whitsel and Schoeppner, 1966). Control of *L. kerteszi* was obtained by reducing the moisture of breeding grounds to less than 5% by weight through drainage. Diazinon at 2 lb per acre (roughly, 2 kg per ha), applied by aircraft before the spring emergence periods, virtually eliminated the midges from a 20-acre (8-ha) breeding ground (Foulk, 1967). Because of its high degree of efficacy against *L. kerteszi* and relatively low mammalian toxicity, chlorpyrifos was considered to be a promising insecticide for spot treatment against this gnat in parks and other recreational areas (Georghiou *et al.*, 1972*b*). A formulation of deet and other isomers prevents *L. kerteszi* from landing on treated surfaces for 2 hours or longer (Sjogren, 1971).

In Florida, the principal problem has been in coastal areas where marshlands and mangrove swamps, with water levels rising and falling with the changing tides, have provided optimal breeding grounds for pestiferous "sand flies" (Linley and Davies, 1971). The most satisfactory control has been "impoundment," which was originally developed for controlling saltmarsh mosquitoes. The swamp or marsh is surrounded by a wide dike, constructed so as to retain water at a fairly constant level within the permanently flooded area. Water needs to cover the highest mudflats to a depth of only 2 to 3 in. (5 to 8 cm), provided the water level is sufficiently constant to ensure that the mud is never exposed. A constant water level requires that water be available from such sources as drainage from higher ground, rivers or streams, tides, direct rainfall, artesian wells or other underground sources, and pumping.

The effect of impoundment is a permanent ecological change, whereas the best that can be expected from insecticides is only temporary relief. A valuable bonus effect from midge control is source reduction of floodwater mosquitoes. A great variety of bird life is attracted to the area, and fish may also proliferate. The elimination, or great diminution, of insecticides is an important advantage from the ecological standpoint, and the odor associated with alternately flooded and dried mangrove swamps is eliminated. A potential advantage of impoundment is that a peripheral dike around a swamp or marsh is the first step in reclamation by the dredge-and-fill method where the sale value of new land is sufficiently high to justify the expense. If earth or marl is used as fill, provision should be made for adequate drainage. Sometimes a swamp is filled by first cutting down the vegetation and then dredging sand offshore and pumping it into the swamp. If *Leptoconops bequaerti* is present, it is essential to stabilize the entire filled area at a minimum of 30 in. (75 cm) above the level of the free soil water to prevent breeding of this gnat. A 6-in. (15-cm) layer of earth or marl may be spread over the sand as an additional safeguard. It is well to encourage a rapid growth of grass and other vegetation, for dense vegetation deters oviposition of *L. bequaerti*. Constant surveillance and maintenance are required to avoid new depressions, such as could be caused by flood rains or rough seas that would erode the front of a beach and create conditions suitable for insect breeding (Linley and Davies, 1971).

Black Flies, Buffalo Gnats, Turkey Gnats (Simuliidae)

Both sexes of all species of simuliids feed on the nectar of flowers, but the females of most species suck blood. They bite on any exposed part of the body, and also get into the hair and under clothing. They tend to bite where clothing fits snugly against the body, and therefore often leave a ring of bites just above or below the belt line. Children are particularly susceptible, and may be severely bitten while adults in the same area are scarcely aware of the flies (Jamnback, 1969). Bleeding continues for some time after the fly finishes feeding. The site of a bite is typically a small, red, central spot, surrounded by a slightly reddened, swollen area. The bites become increasingly itching, swollen, and irritating for several days. Envenomation causes a specific effect on glands about the ears and neck, causing symptoms similar to mastoiditis. Some simuliids readily attack people, but others prefer domestic animals or birds, feeding during the daylight hours and sometimes into the night. They are sometimes so abundant as to be drawn into the air passages of poultry and larger animals, and have been known to smother them. Livestock can be killed in a few hours by their large-scale attacks. In some parts of the world, black flies transmit diseases to man and animals, including protozoan diseases to poultry in the United States.

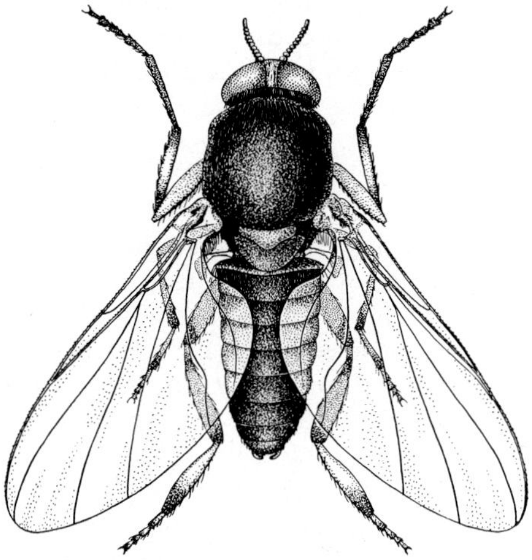

Fig. 289. A western species of black fly, *Simulium virgatum*.

Description. The simuliids are minute but heavy-set, humpbacked gnats, and are usually black or gray, but some are orange or yellowish. They vary from 1 to 5 mm in length. Their antennae are short and 11-segmented and, unlike those of the chironomids and ceratopogonids, are bare. They have large, round eyes, holoptic (fused together) in the males. The wings are broad and short, lack scales and hairs, and the heavy veins are concentrated near the anterior margins. A medium-sized western black-and-gray species, *Simulium virgatum* Coquillett, 3 mm long is shown in figure 289 as an example of the general appearance of the simuliids. Only the females are bloodsuckers, possessing strong, bladelike, piercing mouthparts; those of the males are more or less rudimentary (Malloch, 1914; Dyar and Shannon, 1927; Wirth and Stone, 1956).

Biology and Habits. Simuliid eggs may be deposited in great numbers in streams or rivers, on rocks and other objects washed by spray or turbulent water, as in rapids, cascades, and under waterfalls. Some eggs are deposited on the bottom, where they may lie dormant under stones and other objects throughout the summer. Some species breed in irrigation canals and ditches. However, in most species the immature stages are passed in swiftly flowing streams, differing from other bloodsucking flies in this respect. That is the reason they are abundant in hilly or mountainous areas. They usually deposit their eggs in masses of 200 to 500 on partly submerged vegetation or stones in running water. Very large, communal masses may be laid by some species. However, other kinds lay their eggs in small masses of a few eggs, and others drop them into the stream. Incubation may take 3 to 7 days in warm weather, although eggs laid in late summer or early fall do not hatch until the following spring. Species having only a single generation per year lay their eggs in late spring, and they hatch the following autumn or winter. Eggs of the important vector and pest species will not mature unless the female has had a blood meal before ovipositing (Jamnback, 1973).

The whitish to brown aquatic larvae are cylindrical, 12-segmented, narrowest in the midregion, and attain a length of 10 to 15 mm. They are able to move about with a looping motion by means of an anterior proleg that is modified into a prehensile, toothed disk, and also by using a posterior, toothed, disklike sucker. The larvae attach themselves to rocks or other objects in the water after covering the place of attachment with silken threads produced by secretions from the salivary glands. The hooks of the posterior sucker are inserted into the network of silken threads.

The larvae have a well-developed tracheal system and 9 pairs of closed spiracles. Respiration is accomplished by means of branched, retractile gill structures located dorsally on the last abdominal segment. They have elaborately modified mouthparts, similar to those of the mosquitoes, and feed on particulate material screened from the water. The larva lies on its dorsum and holds its 2 labral brushes open to form palmate sieves. By means of these sieves, the larva strains food particles out of the swiftly flowing water. The current of water keeps the larval mouth brushes open. At intervals, the brushes are clapped over the mouth opening and combed free of detritus. The mass of food particles is sucked into the pharynx, and the water is expelled through a complex mass of combs within it. Anything small enough to be swallowed is taken in: diatoms, algal filaments, flocculent organic matter, animal plankton, and silt. Under favorable conditions, the larva will pass through 6 or 7 instars in a week or 10 days.

The larva spins a crude, pocketlike cocoon, open in front and facing downstream, in which to pupate. Cocoons are attached to the sides of rigid supports found at larval feeding sites. The pupae possess respiratory filaments attached anteriodorsally to the thorax. The emerging adult has a hydrofuge surface that is surrounded by air bubbles. The bubbles carry the adult upward, and the insect is virtually catapulted into the air and immediately flies away as the bubbles burst at the surface of the water. It does not require the period for drying and hardening that is required by many aquatic insects. The males usually emerge first, and mate with the females that emerge soon after. The adult life of simuliids is brief, usually 3 weeks or less (Jamnback, 1973). Most species will travel considerable distances from their larval sources.

The number of generations per year depends not only on latitude, but also on whether the fly is a "riverine" or a "cold-water" species, and may vary from 1 to 4 or 5.

Unlike most mosquitoes and biting midges (ceratopogonids), black flies are active during the day; they do not bite late at night outdoors, or even in the daytime within buildings. They enter buildings accidentally and fly to windows where they crawl about on the panes until they die. They can be blown by the wind for miles, and can be nuisances far from any stream (Jamnback, 1969).

Some Common Species

The family Simuliidae contains relatively few species and only 3 important genera: *Cnephia, Prosimulium,* and *Simulium,* the last-named having the most species. Wirth and Stone (1956) prepared keys for males, females, and pupae of 16 California species.

Simulium vittatum Zetterstedt, which is 2 to 3 mm long and dark gray to velvety black, is the species most widely distributed in North America, and it is also common in Europe. It attacks domestic animals, but does not appear to be as bad a pest as some of the other species.

Simulium venustum Say, apparently the original "black fly," is also widespread, and is readily recognized by its white-marked tibiae. It has been reported from Plumas to El Dorado counties in California (Wirth and Stone, 1956). In the New England states and Canada, this species and the "Adirondack black flies," a complex referred to in the older literature as *"Prosimulium hirtipes* (Fries)," are the ones particularly feared by fishermen and campers. (The true *P. hirtipes* appears to be a European species.) *Prosimulium mixtum* Syme and Davies is the most important pest of the "*hirtipes* complex." In common with other species of the complex, the adults emerge in early or mid-May, and attack people into early June. *Simulium venustum* is annoying from the end of May into early July. Black flies are not troublesome late in the summer because stream levels are subsiding and the streams are becoming unsuitable for the larvae. Campers in the Adirondacks can avoid black fly attacks by making their trips after mid-July. *Prosimulium mixtum* and *Simulium venustum* can be distinguished by the uniformly brown legs of the former in contrast to the white bands on the legs of the latter (Jamnback, 1969).

Cnephia pecuarum Riley, the southern buffalo gnat, is a severe pest to both people and livestock in the Mississippi Valley. The gnats become most abundant in the early spring. Annoyance to horses is often so great that work on farms has to be discontinued.

Simulium meridionale Riley, the turkey gnat, is 2 mm long and grayish or brownish black. It is distributed throughout the United States, including California, but is particularly common in the southern states. In late spring following the southern buffalo gnat, it attacks poultry, biting the combs and wattles. This species has been called the "cholera gnat" because its attacks on poultry may cause "cholera-like" symptoms.

A Few Others. Among other better-known species may be included *Simulium bivittatum* Malloch, *S. argus* Williston, *S. tescorum* Stone and Boreham, the orange-colored *Prosimulium fulvum* (Coquillett), *Cnephia mutata* (Malloch), and the black *C. minus* (Dyar and Shannon).

Envenomation by Black Flies

Black fly bites are characterized by extreme pain, intense itching, and a localized swelling that is often accompanied by complications that indicate the presence of an active toxin. A small, red, hemorrhagic spot develops at the site of each bite, an itching wheal soon develops, and the entire area may become hot and swollen. Headache, fever, nausea, and adenitis (inflammation of 1 or more lymph nodes) are common symptoms, and constitute what is known as "black fly fever." The glandular enlargement slowly subsides, but itching may last for weeks. Bites around the eyes cause the surrounding tissues to swell to such an extent that they sometimes obscure vision (Jamnback, 1973).

The severity of symptoms that can result from black fly attacks is indicated in a report by Gudgal and Grauer (1954) concerning a patient who did not notice bites (probably by *Simulium venustum*) on her feet, ankles, and lower legs for several hours. Then, small wheals with hemorrhagic centers appeared, and these rapidly enlarged until the lower legs were twice their normal size and so acutely tender that walking was difficult. She developed a fever of 101° F (38.3° C) and enlarged, tender, femoral lymph nodes. The symptoms disappeared after several days, but some of the bites remained as small papules for 7 to 10 days.

Control of Black Flies

Control with insecticides generally requires large-scale efforts by a governmental agency. Differences in the life cycles of the various species must be taken into consideration. Control has sometimes been directed against adults along stream courses, using either aircraft or ground equipment to apply fogs and mists. Such treatments are usually repeated every few days. Insecticides at dosages below the toxic levels for fish have been applied to streams for larvicidal effect. After the banning of DDT, methoxychlor appeared to be a more promising candidate because it was an outstanding larva-detaching agent, was good as a killing agent, had an acceptably low vertebrate toxicity, and was more readily biodegradable than DDT. However, in simulated stream tests with larvae of *Simulium pictipes* Hagen, chlorpyrifos was at least 10 times more effective as a killing agent than methoxychlor, as well as 2 other insecticides of low mammalian toxicity that were tested (Abate and ronnel) (Travis and Schuchman, 1968). In southern California, Abate was sprayed into streams for periods of 60 minutes to achieve a final concentration of 0.5 ppm. Suitable control of larvae (*Simulium argus, S. vittatum, S. virgatum,* and *S. tescorum*) was obtained without affecting nontarget organisms (Pelsue *et al.*, 1970).

To control black flies in homes, cabins, tents, or other confined or limited areas, screens (60-mesh bolting cloth), household sprays and aerosols, and repellents may be used. Travis *et al.* (1951) found dimethyl phthalate, Rutgers 612, Indalone, and a 6–2–2 mixture of these materials to be effective repellents against "*Prosimulium hirtipes* (Fries)" and *Simulium venustum* in the Adirondack Mountains, New York state. In Colombia, Rutgers 612 and dimethyl phthalate gave almost complete protection against black flies for 2 to 2.5 hours, but Indalone was less effective.

Hilton (1970) found unpainted aluminum "hard hats" to be very attractive to simuliids in northern Alberta. A thin film of fuel oil smeared over the outer surface of a hat resulted in these flies becoming trapped in the oil, and reduced to almost zero the number attacking and biting the wearer. No protection was afforded against mosquitoes by this method.

HORSE FLIES, DEER FLIES, AND SNIPE FLIES

These are diurnal bloodsucking Diptera of moderate to large size, in 2 families: Tabanidae (horse flies and deer flies) and Rhagionidae (snipe flies), both in the suborder Brachycera. The species in both families have very short, 3-segmented antennae, the terminal segment (flagellum) being elongate and annulate. Biologies are similar.

Horse Flies and Deer Flies (Tabanidae)

The females of most tabanid species suck blood, while the males feed on plant juices, nectar, the excreta of certain other insects, or any liquids that contain nutritive material. Tabanids are diurnal in their habits.

Horse flies are cosmopolitan in distribution, and are especially abundant in moist, wooded areas up to several thousand feet elevation. They are notorious pests of horses, mules, cattle, hogs, dogs, deer, and other mammals, and also attack man. Their bite is painful, and they make a deep wound, causing a considerable flow of blood. The horse fly's stylets are large and easily distinguished. The labrum (figure 290) is not an effective piercing organ; its undersurface is deeply excavated to form the food canal. The mandibles (*B*) of the female are large, sharp-pointed blades. The puncture is formed by the mandibles and the maxillary galeae (*Ga*). The labial labella (*Lbl*) are used in the manner of nonpiercing flies, such as house flies and blow flies (figure 46, *I, J,* chapter 4), that feed on exposed liquids. The soft undersurfaces of the labella are transversed crosswise by numerous fine, closely set channels ("pseudotracheae") that lead to a pair of median, lengthwise channels (Snodgrass, 1944). Animals can lose an estimated 300 ml of blood per day as the result of horse fly attack, and this does not include the

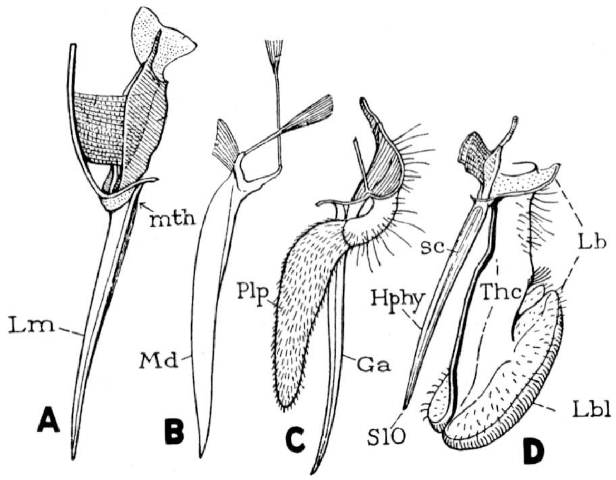

Fig. 290. Mouthparts of a horse fly, *Tabanus atratus*. *A*, labrum and sucking apparatus; *B*, mandible; *C*, left maxilla; *D*, hypopharynx, labium, and salivary pump (left side); *Md*, mandible; *mth*, mouth; *Ga*, galea; *Hphy*, hypopharynx; *Lb*, labium; *Lbl*, labellum; *Lm*, labrum; *Plp*, palpus; *sc*, salivary canal; *SlO*, salivary orifice; *Thc*, theca. (From Snodgrass, 1944.)

blood that exudes after the flies leave (Philip, 1928). Where horse flies are abundant, animals in harness often become unmanageable and bolt.

A deer fly, *Chrysops discalis,* transmits tularemia to man, principally from hares, rabbits, and other rodents. It is the only species of deer fly known to be a vector. The fly becomes a carrier by biting a tularemic animal, and can transmit the disease for about 5 days. Transmission is believed to be mechanical (McDowell *et al.,* 1964). Ticks appear to be even more important vectors, and the disease is discussed more fully in the section on ticks later in this chapter.

Description. Horse flies and deer flies are large, robust, pubescent flies, 10 to 25 mm long. The eyes are large, and are usually holoptic (contiguous) in the males. The proboscis projects forward, and is sometimes longer than the body. The mouthparts of the females are bladelike, and are adapted for piercing. Deer flies can be distinguished from horse flies by the apical spurs on the hind tibiae. Also, deer flies have wings with dark markings, whereas horse flies usually have clear wings. Keys to the adult horse flies and deer flies of California were prepared by Middlekauff (1950).

The tabanid larva is cylindrical, tapering toward both ends. Its body is tough and leathery, usually white, and has a ring of tubercles around each segment. The head can be completely retracted into the thorax. A single posterior siphon can be telescoped within the anal segment, and bears the openings of the respiratory system on its tip.

Biology. Tabanids may deposit their long, spindleshaped eggs on foliage or other objects over shallow, quiet water at the edges of pools or lakes, or in relatively deep water at some distance from shore; or on ledges or rocks over deep water. Some oviposit on rocks or other objects over flowing streams; or on vegetation, such as the leaves or trunks of trees; or over moist or even quite dry soil (Stone, 1930). The eggs are deposited in masses of 100 to 800. The egg masses are protected by a gluey, waterproof covering. The horse flies (*Tabanus, Hybomitra, Atylotus,* and *Haematopota*) generally lay their eggs in several layers, and the deer flies (*Chrysops*), usually in a single layer. The eggs normally hatch in 5 to 7 days, and the larvae drop to the water or ground. The eggs all hatch at about the same time, and the larvae fall in clumps to the water, mud, or damp earth and begin feeding on organic matter (Stone, 1930; Matheson, 1951).

The larvae are primarily carnivorous and cannibalistic, preying on other insect larvae, snails, earthworms, and other soft-bodied animals. Some *Chrysops* are considered to be saprophagous (feeding on decaying matter). If the larvae are under water, they stay in the bottom mire, and are rarely seen swimming in clear water. They move above the water level to pupate. The cylindrical, brown pupae are normally found in a damp mass of peat or other dead vegetation.

Some horse flies are slow to mature and require a year or more to complete their development, but a few species have 2 or more generations per year. Deer flies usually need less time for development than horse flies.

Some Western Species of Horse Flies

Tabanus punctifer Osten Sacken, the western horse fly, is the largest and best-known species in this group, especially along the Pacific Coast. It is 25 mm long, black and white, with whitish or golden pile on the thorax, and usually has a small, dark spot near the wingtip (figure 291).

Hybomitra californica (Marten) is 17 mm long, and the first 4 abdominal segments are tawny.

Atylotus incisuralis (Macquart) is 10 mm long, with only the sides of the first 4 abdominal segments tawny. It is a serious pest to horses and cattle, attacking chiefly the abdomen.

Haematopota species are usually small, slender, and grayish in color, with wings beautifully maculated in brown and gray. The most widely distributed western species is *H. americana* Osten Sacken. It ranges from Hudson Bay to Alaska and south to northern California and New Mexico.

Some Eastern Species of Horse Flies

Tabanus stygius Say, another black-and-white species, is 20 to 22 mm long, and is distributed east of the Rocky Mountains.

Tabanus atratus F., the black horse fly, is 16 to 28 mm long and is uniformly black. It occurs over most of the United States east of the Rockies and southward into Mexico. The thorax and abdomen are covered with a whitish dust, and if this is not rubbed off, the abdomen appears bluish in certain lights. C. H. Curran believed this to be the "blue-tailed fly" of the well-known ballad.

Tabanus lineola F., the striped horse fly, is 12 to 15 mm long. The thorax is striped brown and gray, and the abdomen is brown or black above, with 3 prominent gray stripes. It is an important livestock pest, ranging throughout eastern, central, and southern North America (James and Harwood, 1969).

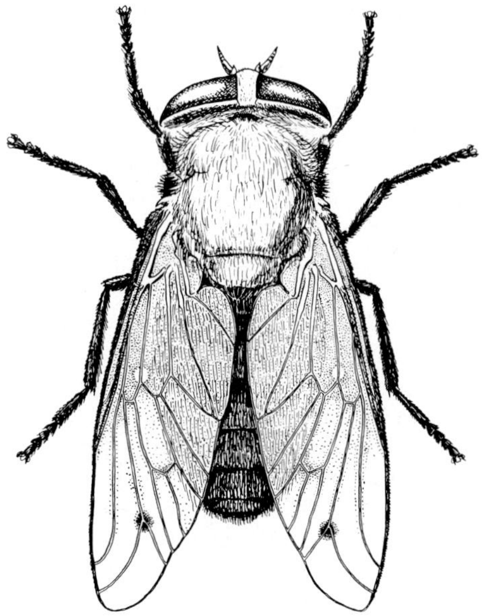

Fig. 291. Western horse fly, *Tabanus punctifer*.

Some Deer Flies

Chrysops discalis Williston is a western species that has already been mentioned as a vector of tularemia. It is 8 to 10.5 mm long. The female is gray to yellow-gray, with black spots on the abdomen, while the male is mostly black, with yellow-gray spots on the abdomen (figure 292).

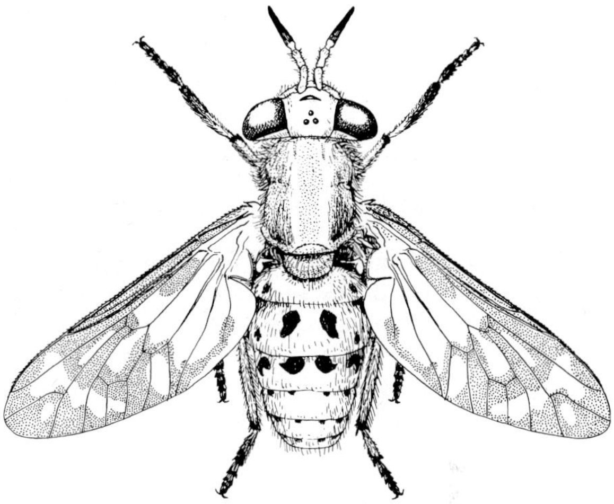

Fig. 292. A western deer fly, *Chrysops discalis*.

Chrysops noctifer Osten Sacken is one of the most common western species. It is 9 mm long and shiny black, with 2 large, basal orange spots and heavily clouded wings. These deer flies seriously bite the flanks of horses and cattle during the spring months. *Chrysops proclivis* Osten Sacken is similar in size, appearance, and habits.

Chrysops pachycerus Williston is 8 mm long. The thorax is gray, lined with black, and the abdomen is yellow, with large, median, basal and many smaller apical black spots. The wings are smoky gray-brown to the tips, differing in this respect from many other deer flies.

Envenomation by Tabanids

Tabanid bites are painful, and some people who are attacked suffer severe lesions, high fever, and even general disability. These symptoms are the allergic responses to the hemorrhagic saliva that is poured into the wound to prevent clotting of the blood while the fly is feeding (McDowell *et al.*, 1964). A person can become increasingly sensitive to repeated bites. In Florida, a man had been occasionally bitten by deer flies while working in orange groves over a period of 12 years. The flies later became more attracted to him than to other workers in the groves. He protected his hands with gloves and his face and neck with mosquito netting, but was bitten on the wrists, causing his arms to swell and become very painful. Finally, they swelled to twice normal size. His wrists had areas of erythema, and red streaks radiated upward on the forearm, which was very painful. He collected 35 deer flies, from which material for injection was prepared. Injection resulted in rapid hyposensitization lasting at least 4 years, and in addition the man reported that he believed he no longer attracted deer flies more than the average person (Mease, 1943).

Snipe Flies (Rhagionidae)

Snipe flies occur only at higher altitudes. They are common in woods or chaparral, especially near moist places, and may be found in foliage, in long grass, and in hollow tree trunks. Unlike horse flies, they are sluggish and easily captured. Both larvae and adults are predaceous. The adults inflict a painful bite. Deer are apparently the principal victims, but man and livestock animals are subject to attack. The biology and habits of most species are not well known, but as foothills and mountains become more important, not only as recreational areas but also as building sites, snipe flies will probably increase in importance and become better known.

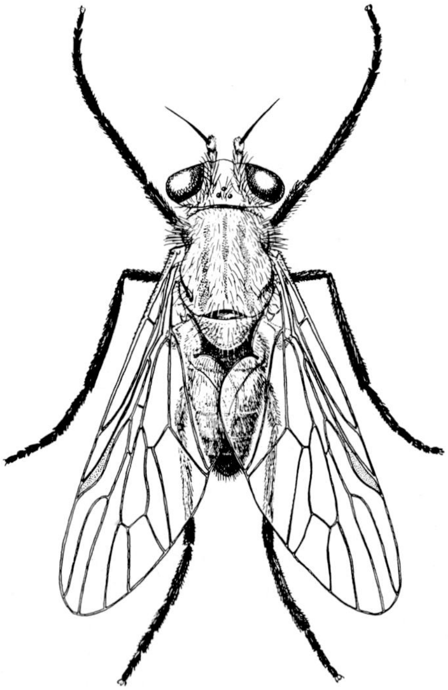

Fig. 293. A small female snipe fly, *Symphoromyia* sp.

Description. Snipe flies are moderate- to large-sized, and slender or rather stout, but in either case with the body tapering posteriorly (figure 293). They are long-legged, and naked or pubescent. The eyes of the males are holoptic (contiguous), or may be narrowly separated. The antennae are 3-segmented, the third segment bearing a terminal or dorsal arista (slender style). The mouthparts are of the tabanid type; strongly sclerotized piercing and cutting structures are enclosed in a retractile labial sheath, and blood is lapped up by the labella (Ross, 1940).

Biology. The biology of snipe flies is similar to that of the tabanids. The females of some species lay masses of eggs on tree branches overhanging streams. The newly hatched larvae drop into the water. The larvae of aquatic species have 2 rounded, protrusive blood gills on the apical segment below the long, terminal filaments, while the terrestrial species have small spiracles. The latter are found in rotten wood, leaves, or other decaying vegetation, where they are predaceous on insects and other small animals.

The Genus *Symphoromyia*

This genus contains the most important biting species, although hematophagy by adult snipe flies is widespread and well documented for several other genera. The flies alight silently on the exposed parts of the body, and inflict a sudden, painful bite, somewhat like that of certain deer flies, before their presence is known. Their bite is said to resemble "the sensation produced by a needle being poked into the skin" (Hoy and Anderson, 1966).

Symphoromyia hirta Johnson was reported by Mills (1943) to have been seen in numbers up to 50 on the back of a fisherman in Montana.

Symphoromyia atripes Bigot was observed by Ross (1940) in the Cascade Mountains of British Columbia at elevations of 3,000 to 5,000 ft (900 to 1,500 m). He stated that its bite was very painful, causing profuse bleeding and some swelling, adding that the discomforts caused by this fly were "equaled only by those of mosquitoes." However, it was almost completely absent on cloudy days or in dense shade, conditions that favored mosquitoes; it apparently attacked only on sunny days, and only in the open. It could be easily swatted, and in fact could be picked up with the fingers. Aldrich (1916) reported that when *S. atripes* attacked horses, blood trickled from the wounds, and that he had never seen horses attacked so severely, even by actual horse flies.

California appears to have 6 species of *Symphoromyia* that are limited to the Coast Range, 6 to the Sierra Nevada, and 5 shared by both ranges. At least 7 species indigenous to California are known to take blood from humans: *Symphoromyia atripes, fulvipes, kincaidi, limata, pachycerus,* several species of the *pachycerus* complex, and at least 1 of the *johnsoni-inurbana* complex. In many areas of the western United States and Canada, including California, these flies can be very abundant locally during certain years (Hoy and Anderson, 1966). Some species also occur in eastern North America and in Europe.

LICE (PHTHIRAPTERA: PEDICULIDAE)

Lice are external parasites on birds and mammals. They cannot survive very long when removed from their hosts. They are small, flattened, wingless insects with short, 3- to 5-segmented antennae; with eyes either very small or absent; with no ocelli; and with mouthparts greatly modified for chewing or sucking. The thorax is small, with segments more or less fused and sometimes indistinct; the abdomen has 5 to 8 distinct segments; the legs are short but stout, with 1- to 2-segmented tarsi, and are well adapted for clinging to the host.

There are species of lice that have biting and chewing mouthparts and others that have piercing and sucking mouthparts. All the biting lice were at one time included in the order Mallophaga and the sucking lice in the order Anoplura, but these are now generally considered to be suborders. There is also the suborder Rhynchophthirina, which contains only 1 species, occurring on the Indian elephant. This can be distinguished from other lice by the front of the head being narrowed down to a beak that is longer than the rest of the head.

Lice are most likely to come to public attention during periods of war, famine, or widespread disaster when it is often not possible to bathe and wash clothing frequently and people may be crowded in confined quarters. Sucking lice (Anoplura) then generally become abundant, and the irritating saliva they inject into the skin while feeding causes itching, which may in turn result in scratching, infections, and scarred, hardened, or pigmented skin, known as pediculosis.

There are 2 groups of microbial agents pathogenic to humans that are adapted biologically to lice, and are therefore effectively transmitted by them to man: rickettsiae, causing typhus and trench fever; and spirochetes, causing relapsing fever.

"Louseborne" or "epidemic" typhus is a particularly dreadful disease that has played an important role in history, for until World War II it was often a more important factor than the armed might of contending armies in deciding the course of battle. The introduction of DDT and other organochlorine insecticides during World War II provided a means of reducing the ancient scourge to manageable proportions.

Suborder Mallophaga (Biting Lice)

Mallophaga, the more primitive of the 2 major suborders, contains 6 families with several hundred species in North America. Of the 6 families, 5 are parasites of wild or domestic fowl. Only the species of the small family Trichodectidae infest mammals. The biting lice feed on scaly skin, bits of feathers, hair, clotted blood, and surface debris. They glue their eggs to the feathers or hair of their hosts. The biting lice cannot be properly considered as household pests. The remainder of this section will be devoted to the sucking lice.

Suborder Anoplura (Sucking Lice)

Because of the uniformity of temperature and the constant food supply afforded by human skin, lice can thrive wherever man exists. Buxton (1947) stated that he knew of no part of the world or of any type of climate in which sucking lice could not live and multiply. In North America, there are about 20 genera of sucking lice and about 100 species. They range in length from 2 to 5 mm. They are all external parasites of mammals, feeding on blood which is sucked through a tube formed by an eversible set of fine stylets. Lice glue their eggs to hairs, and spend their entire life cycles on the host. The egg has a cap or operculum which comes off intact when the egg hatches, 5 to 10 days after it is laid. Since lice develop by gradual metamorphosis, the newly hatched nymphs are similar to the adults in appearance and habits. An even temperature is maintained by the body of the host, so breeding of lice continues throughout the year. Lice are strongly negatively phototropic, and will quickly seek cover when exposed. This must be kept in mind when searching for them.

Lice superficially resemble psocids or "booklice" (Psocoptera) and the workers of the subterranean termites, indicating a common ancestry. The most commonly accepted theory is that cockroaches, termites, psocids, and biting lice are all derived from an ancient group of pre-cockroaches (Protoblattoidea), fossil forms of which may be found in strata of the upper Carboniferous period, about 250,000,000 years old. Probably the Anoplura evolved from the Mallophaga (Zinsser, 1935).

Among the Anoplura, the family Pediculidae contains the species that occur on man and monkeys. All the Pediculidae have eyes, a character that distinguishes them from all other species of sucking lice. The family contains 3 genera, *Pedicinus, Pediculus,* and *Pthirus. Pedicinus* is found on Old World monkeys. *Pediculus* and *Pthirus* are found on the higher apes and on man.

Mammals that are closely related to one another tend to be infested by closely related or identical species of lice. In the course of evolution, mammals have often come to differ from one another more than the lice that parasitize them, so the lice can sometimes indicate evolutionary relationships between mammal species that diverged (Buxton, 1947). Lice can also be used to trace the prehistoric migration of man. Distinguishable strains of the human louse developed in different parts of the world. The lice collected from prehistoric Indian mummy scalps were found to be similar to the Chinese head lice and to those found on Aleutian Eskimos—another evidence of the migration of human populations across the Bering Strait. Several species of *Pediculus* occur only on monkeys of tropical America, particularly the spider monkey (*Ateles*). These lice are believed to have been transferred to the New World monkeys from human beings in the remote past (Zinsser, 1935).

Head and Body Lice, *Pediculus humanus* vars.

Humans are infested by 2 species of lice: *Pediculus humanus* and *Pthirus pubis. Pediculus humanus* has 2 varieties, *P. humanus capitis* De Geer, the head louse, and *P. humanus humanus* L., the body louse (plate VIII, 2). Anatomically, the 2 varieties differ principally in average size and bodily proportions, although the extremes overlap in this regard. Body lice are 3 to 3.5 mm long. They are generally about 10 to 20% larger than head lice. Head lice tend to have relatively thicker antennae, deeper indentations between abdominal segments, and less well-developed abdominal musculature. The most important differences between the 2 varieties or forms are in their habits. The head louse principally infests the head, but may occasionally be found on hairs situated in other parts of the body. The body louse occurs chiefly on the clothing, and moves to adjacent body areas to feed (Nuttall, 1917*b*). *Pediculus humanus capitis* is generally believed to be the form that originally became adapted to man before he began to wear clothes, and *P. h. humanus* is thought to be a recent form that became adapted to the more difficult life in movable and removable clothing (Busvine, 1966).

Head lice and body lice produce fertile offspring when crossed (Bacot, 1917). If head lice are reared for a few generations in the usual habitat of the body louse, no differences can be ascertained between the 2 varieties (Keilin and Nuttall, 1919). In attempts to rear head lice on the body, they had a strong tendency to migrate back to the head, and had to be put back onto the body at intervals. In a later experiment, many of their offspring (F_1 generation) showed no tendency to migrate toward the head. Most of them laid their eggs on clothing, and only a few on the hairs of the body. In the F_2 generation, only a few individuals demonstrated the tendency to migrate, and only feebly so (Howlett, 1917). The imperfect isolation of the 2 varieties is probably the factor that has prevented them from diverging into distinct species (Busvine, 1966).

Head Louse—Description. The color of adult head lice varies from dirty-white to grayish black, the color of the adult being largely dependent on the color of the substrate on which the nymph develops. Lice from blonds are paler than those from people with black or brown hair. The male is generally smaller than the female, has a posteriorly rounded abdomen, and is relatively narrower. The abdomen of the female terminates in 2 lobes (figure 294). The mouth of the head louse (figure 46, chapter 4) bears recurved "teeth" which can be pressed into the host's skin to anchor the head in position, allowing the piercing mouthparts to enter easily. The 3 pairs of legs are similar. The short tibia bears a blunt "thumb" on the inner side, close to the apex. The thumb bears spines and hairs. The short, 1-jointed tarsus has a single claw that is rough along its concavity. The anterior legs of the male are stouter than those of the female, and are provided with larger tibial thumbs and tarsal claws. The louse grasps the host's hair between the thumb and the claw (Buxton, 1947).

Head Louse—Biology. The adult female lives for a month or more, laying 50 to 100 eggs that hatch in 5 to 10 days. She cements her oval, rather large eggs on a hair of the head near the scalp. The eggs are yellowish-opalescent, and are about 0.8 mm long by 0.3 mm broad. At one end is a cap (figure 294) to admit air and facilitate the escape of the young nymph. The eggs ("nits") remain attached for a long time, and the hair of the infested person will eventually contain many of them, some viable and others empty. The cement cannot be dissolved without harming the hair or scalp. Buxton (1947) noted that there were

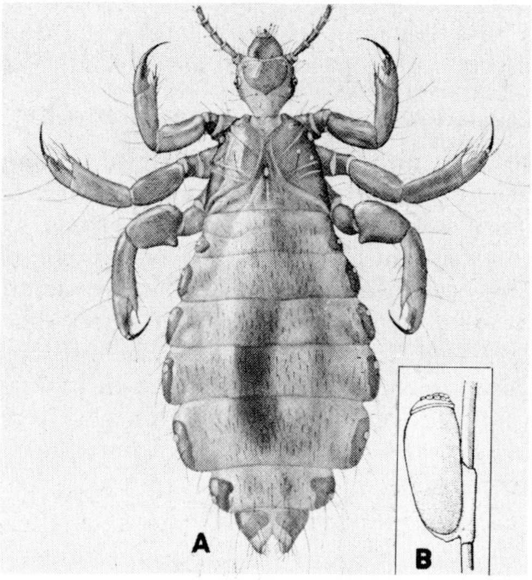

Fig. 294. Body louse, *Pediculus humanus humanus. A*, adult; *B*, egg or "nit" of head louse attached to human hair. The head louse (*Pediculus humanus capitis*) is morphologically similar, but smaller. (From Smart, 1948, with permission of the trustees of the British Museum [Natural History].)

3 nymphal instars. The first was straw-colored, but the first blood meal, visible through the cuticle, gave the insects the appearance of "tiny rubies." The gut of the older lice was usually purplish black. The entire life cycle, from egg to egg, normally required about 3 weeks. The ratio of males to females varied from 43.5:100, when there were only 1 or 2 adult lice per head, to 150:100, when there were more than 100 per head. The longest adult lives of body lice confined in cages at the wrist were 28 days for a male and 29 days for a female. The corresponding periods for adult head lice were 23 and 22 days, respectively (Nuttall, 1917*b*).

Body Louse—Biology. The biology of the body louse is similar to that of the head louse, but its eggs are generally glued to fibers of clothing, and only occasionally to hair. The body louse tends to lay its eggs in large numbers along the seams inside of underwear and in places where the clothing contacts the body, such as the neck, shoulder, armpit, waist, and crotch of the trousers. When clothes are removed, the eggs are likewise removed from the warmth of the body surface. The lowest constant temperature at which eggs of the body louse will hatch is 24° C (75° F), and they will then begin to hatch on the seventeenth day and continue hatching until the twenty-first day. The highest constant temperature at which the eggs will hatch is 37° C (99° F), and they will then hatch on the sixth and seventh days. The eggs hatch most rapidly (in 5 days) at 35° C (95° F). The maximum percentage of hatch occurs between 29 and 32° C (84 and 90° F), when the incubation period is from 7 to 11 days. Newly deposited eggs will not develop if kept at 23° C (73° F) for 14 days or 8° C (46° F) for 7 days (Leeson, 1941).

The females are likely to be found along the seams of clothing, where oviposition occurs, while the males tend to be more generally matted over the surface of the clothing. Adult lice migrate farther than nymphs; therefore, the proportion of adults is higher in layers of clothing farther from the skin. However, an individual must be very heavily infested before lice can be seen crawling outside the outer clothing. The body louse prefers wool clothing, and spends most of its life in the garments except for the brief periods when it is feeding on the skin. It usually feeds where the skin is soft or folded, as at the joints. Even while feeding, it retains its hold on the adjoining fabric (Eddy and Bushland, 1946; Buxton, 1947).

The period required for the different stages to develop is approximately the same for both head and body lice. If the clothing is not removed, the nymphal stages are completed in 8 to 9 days, but may require 2 to 4 weeks when the clothing is habitually removed at night, for the lice remain on it and are subjected to lower temperatures. If the infested clothes are not worn for several days, all the lice will usually succumb (Pratt and Littig, 1961). The active stages of lice were killed much more rapidly than were the eggs at temperatures below freezing. At –20° C (–4° F), eggs were killed in 4 or 5 hours and adults in 0.5 to 0.75 hour. At –5° C (23° F), eggs were killed in about 75 hours and adults in about 20 hours (Nuttall, 1917*b;* Busvine, 1944).

At least when reared in "breeding boxes" worn on the human body, body lice show less mortality during development than do head lice. They also have a longer adult life, more eggs laid per female (average, 110, compared with 57 for head lice), a higher percentage of hatch, and greater resistance to starvation. The latter advantage is particularly important to the body louse because it can feed only when the infested person is resting. Also, it may be separated from the host, at least overnight, when he removes his clothing. The reason the body louse is larger than the head louse is possibly because it feeds less often, and so must take larger meals (Busvine, 1966).

Crab Louse, *Pthirus pubis* (L.)

In common with the head and body lice, crab lice occur only on human beings. They are most commonly found in the hair of the pubic and perianal regions, and rarely elsewhere. Small children are rarely infested, but if crab lice are present, they are most likely to be found in the eyelashes. Infants may become infested from their mothers (Roy and Ghosh, 1945).

Description. As the common name implies, crab lice are crablike in appearance (figure 295). They are smaller than the head and body lice, being 1.5 to 2 mm long, and look like small, gray specks to the unaided eye. Upon magnification, it is noted that the first leg and its claw are much more slender than the second and third. The abdomen is broader than long, with lobelike protuberances at the sides.

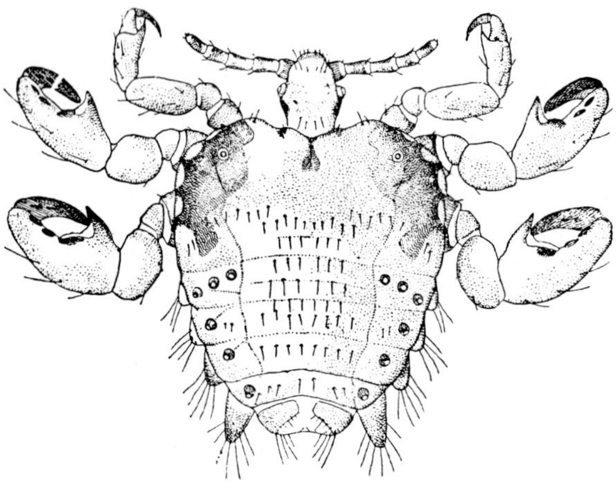

Fig. 295. Crab louse, *Pthirus pubis*. (From Smart, 1948, with permission of the trustees of the British Museum [Natural History].)

Biology. Little is known about the biology of crab lice. Only 3 or 4% of people infested with lice have crab lice. In one observation, a single female laid 26 eggs, an average of 3 per day. At skin temperature, the eggs hatched in 7 to 8 days, nymphal development required 13 to 17 days, and adult life appeared to be less than a month. Both adults and nymphs tend to settle in one spot, and move about very little. The insect grasps hair with the legs on both sides of its body. The reach of 2 extended legs is approximately the distance apart of the coarse hairs on the parts of the body the crab louse infests. Probably, it rarely infests the scalp because the hairs there are too close together. After grasping the hairs, the louse takes blood intermittently for many hours at a time. When 200 crab lice were taken from a man, only 1 was alive less than 24 hours later, even though the insects were kept in moist air (Nuttall, 1918).

Conditions Favoring Infestation by Lice

Since lice cannot survive for long periods away from their hosts, infestations usually result from regular association with infested persons. Children may spread head lice at school during play. Lice are readily spread among family members, and as a result, the measures that school officials may take to disinfest children may be nullified by the close association that the children have with their families. On the other hand, children acquiring lice at school may infest a previously uninfested family.

Head Lice

Transmission of head lice is greatest where sleeping quarters are crowded and where facilities for bathing are inadequate. In slum areas, over 50% of the children may be infested. In both sexes, the number of lice per head may be correlated with hair weight, which is the best measure of hair length. In a group of prison inmates, among those having 10 grams of hair or less per head, only about 14% were infested with head lice; whereas among those with 20 grams or more, over 55% were infested (Buxton, 1941*a*), which would indicate a correlation of head louse infestation with approximately the square of hair length. Whether the sharp increase in the incidence of head lice in recent years was affected by the current long-hair style in both sexes can only be a matter of conjecture. In any case, long hair increases the difficulty of control. Also, it is well to bear in mind that any life style featuring lax personal hygiene, as in hippie-type culture groups, would tend to increase insecticide usage for head lice or other species of lice and would also increase the rate of development of strains of lice resistant to the currently available insecticides.

In the United States, the head louse problem has become particularly serious among schoolchildren. The problem is compounded because of the fact that the great cost of developing a new insecticide formulation to the point of registration for use is discouraging investment by industry in "lousicides" and other medically important formulations with a small potential market from the standpoint of the volume of pesticide required.

Body Lice

Body lice are readily transmitted under crowded and unsanitary conditions. Where several layers of wool clothing are worn more or less continuously, infestations reach a peak in winter. In the layers near the skin, the temperature is high and the humidity is low. In the outer layers of clothing, the relative humidity is higher because the moisture evaporating from the skin is cooled there. In the summer, the moisture content of both skin and clothing may be greater than lice can tolerate. Under such conditions of high humidity, they are forced to move outside the clothing, where they may die and their eggs cannot survive. Likewise, a change in outer garments kills many lice. In warm climates where wool clothing is not worn in summer, body lice are said to be absent whether people bathe regularly or not (Nuttall, 1917*b*).

The rapid spread of body lice is favored where people sleep in their clothing and huddle together for warmth, as in very cold climates. Bedding and furniture may be sources of infestation, although lice cannot live very long after they leave their hosts. According to Leeson (1941), separation of nymphs or adults from their hosts can be tolerated for as many as 5 days at 24° C (75° F) and 10 days at 15° C (59° F). However, it is doubtful if they would be able to feed and thrive if starved for more than 2 days (M. M. Cole, correspondence). Eggs remain viable for up to 14 days at 24° C.

Crab Lice

Like the head and body lice, crab lice usually occur only on man. Infestation is said to result most often from contact during sexual intercourse. A great increase in the incidence of crab louse infestation, as well as venereal diseases, appears to have resulted from the greater sexual promiscuity that has accompanied the increased permissiveness and changes in life style in recent years (Ackerman, 1968). However, there appears to be little doubt that crab lice can be transferred when clothing is crowded in locker rooms and gymnasiums, or by such items as towels, bedding, and toilets, or by loose hairs (with attached eggs) dropped by infested persons. Infestation of the scalp by crab lice was formerly extremely rare, but in recent years it has apparently become more common. Kwell shampoo has been used successfully for treatment (Elgart and Higdon, 1973).

Irritation and Disease

Louse infestation, called pediculosis, can cause severe skin irritation and sometimes a generalized toxic reaction. Generally, red papules develop at the site of each feeding puncture, and the affected area may swell. If the infestation is allowed to continue, the infested person may become sensitized, suffering even more acute symptoms.

Moore (1918) discussed his reaction when allowing about 700 to 800 body lice to feed on him twice a day. He almost immediately experienced a general tired feeling in the legs and later an illness that was at first similar to grippe, with a rash similar to German measles, particularly over the shoulders and abdomen. After remaining in bed for several days, he again allowed the lice to feed on him and again experienced the same symptoms with increased intensity. The rash was then diagnosed as typical of German measles, but other symptoms of measles were absent. He experienced a severe headache and pains in the legs, and in all joints of the body. A third trial led to the same symptoms. One doctor considered the symptoms to be similar to those of trench fever. An assistant who started with feeding a small number of lice, which was gradually increased, developed similar symptoms, along with chills and fever.

Most of Moore's symptoms disappeared within 2 days after he quit feeding the lice, but a peculiar tired feeling in his legs and feet persisted for several days longer. Normal health was restored within 2 weeks. He concluded that his illnesses appeared to be the result of an intoxication of the system with some toxin secreted by the lice while they were feeding. He believed that sometimes symptoms ordinarily associated with trench fever might be caused by certain toxins introduced by lice, rather than by louseborne rickettsiae.

The infested person usually scratches, and may abrade the skin. This may lead to impetigo, furunculosis, or eczema, caused by *Staphylococcus* bacteria and other organisms that normally occur on the skin. A long, continuous period of infestation may result in a thickening and pigmentation of the skin, known as "vagabond's disease" or "morbus errorum." In some severe head louse infestations, a fetid mass may develop in the scalp, forming a sort of carapace in which fungus can develop. The carapace can sometimes be lifted up, revealing myriads of lice (Nuttall, 1917*a*).

Often a few hours after a crab louse bite, characteristic blue spots about 0.2 to 3.0 cm in diameter and with an irregular outline appear. They are painless, and appear to be in the deeper tissues. The blue spots do not always develop, but when they do they are a characteristic of crab louse rather than body louse infestation (Nuttall, 1918). They may persist for a number of days (Ackerman, 1968). In some parts of the world, body lice (but not head or crab lice) transmit epidemic typhus, trench fever, and relapsing fever. Louse bites should be disinfected immediately, and in case of severe reaction, a physician should be consulted (Scott, 1963*a*).

Epidemic Typhus

The most important of the several typhus fevers is "classical," "louseborne," or "epidemic" typhus. This disease, caused by *Rickettsia prowazeki,* is transmitted by the body louse, occurs only in man, is highly fatal during epidemics, and is chiefly a disease of winter and spring. The extensive use of DDT against body lice during the outbreak of typhus in Naples in the winter of 1943–44 resulted in the first successful control of epidemic typhus in winter.

Typhus continues to be endemic where people live in filthy, crowded conditions. Epidemic typhus has long been suspected of being associated with malnutrition. Credence for this belief was increased by laboratory experiments with rats, which showed that their susceptibility to the disease increased with various protein and/or vitamin deficiencies (Fitzpatrick, 1948). The symptoms of epidemic typhus are high fever which continues for about 2 weeks, backache, intense headache, bronchial disturbances, mental confusion, stupor, and a congested face. On the fifth or sixth day, a brick-red macular eruption appears on the chest and abdomen and later spreads to other parts of the body. The typical mottled appearance of the victim initially led to the belief that the disease was identical with the tickborne Rocky Mountain spotted fever, but Ricketts and Wilder (1910) proved that typhus was transmitted by lice and proved that bed bugs and fleas did not transmit the disease.

A louse cannot transmit *Rickettsia prowazeki* to its offspring through the egg, so each louse must acquire the organism by feeding on an infected person. The rickettsia multiplies rapidly in the epithelial cells of the midgut of the louse until the cells rupture, releasing enormous numbers of the organism into the gut contents and feces of the insect. The louse ordinarily dies from the infection in a week or 10 days, but occasionally it survives, and is then infective for life.

Man usually becomes infected through fecal contamination or through the crushed body contents of the louse; the bite of the louse is not directly involved. At ordinary room temperatures, the rickettsiae may remain alive and virulent in the louse's feces for more than 60 days. Although normally the rickettsiae are transmitted through scratches and through contact with the conjuctivae or mucous membranes by fingers contaminated with louse feces, one may also become infected by inhaling minute particles of louse excrement. When the typhus patient develops a high fever, the lice frequently leave and spread the disease to other persons. Lice can become infected from human carriers of the disease for many years, even though the carriers may be without any symptoms of the disease (Wilcocks, 1959; Pratt and Littig, 1961). Modern vaccines give good protection against epidemic typhus. Also, drugs such as aureomycin, terramycin, and chloromycetin can be used for treatment. Modern synthetic insecticides are providing excellent control of the louse itself.

Trench Fever

Trench fever, also known as "five-day fever" and "shank fever," is a nonfatal, louse-transmitted disease first noticed during World War I under conditions of trench warfare when over a million cases were reported. The disease reappeared in World War II in Yugoslavia and the Ukraine. As in the case of typhus, human epidemics depend on heavy louse infestations in a human population living under crowded, filthy conditions. The symptoms of trench fever include a sudden onset of fever, headache, dizziness, pains in the muscles and bones, particularly in the legs, with special tenderness of the shins. This stage of the disease lasts 24 to 48 hours or longer. Other attacks of fever of diminishing severity occur at intervals of about 5 days. Trench fever is caused by *Rickettsia quintana.* The organisms are found in the louse's midgut, as are other rickettsiae, but are not pathogenic to the insect. The louse remains infective for life. The louse obtains the rickettsiae by sucking the blood of infected humans, and passes it on to others through the feces or the body contents of a crushed louse. The infective organism is retained in dry louse feces for a long time, and may in this way result in delayed transfer to man. Man is the only known reservoir, and may be a carrier of the disease, without symptoms, for as many as 5 to 8 years, and can infect lice during this entire period.

Louseborne (Epidemic) Relapsing Fever

Relapsing fevers are closely related diseases caused by spirochetes and characterized by periods of fever that occur at regular intervals. Louseborne relapsing fever may have arisen from the more widespread relapsing fevers transmitted by ticks of the genus *Ornithodoros*. Louseborne relapsing fever occurs in many parts of the world, and has been a particularly serious problem during the great wars and their aftermaths, including World War II.

During the years 1942 through 1944, an epidemic of relapsing fever in North Africa resulted in an estimated 1,000,000 persons becoming ill, of whom about 50,000 died (Gaud and Morgan, 1947). Relapsing fever is characteristically epidemic where lice are the vectors and endemic where argasid ticks (principally *Ornithodoros turicata* and *O. hermsi* in the United States) are the vectors.

Relapsing fever is characterized by a sudden onset of headache, chills, fever, and generalized pain. The symptoms appear from 3 to 10 days after infection. The fever remains high for an average of 4 days, subsides abruptly for a period of 3 to 10 days, and is followed by 1 or more relapses. Fatalities are generally low (2 to 5%) among well-nourished persons, but soar to 50% or more among undernourished populations.

Epidemic relapsing fever is caused by the spirochete *Borrelia recurrentis*. The human body louse can acquire the spirochete by a single feeding on an infected person. The louse sucks out about a milligram of blood, which contains many spirochetes, but by the end of the first day all live spirochetes disappear from the gut and reappear in great numbers in the insect's hemolymph. The louse is not adversely affected by the spirochetes, and remains infected for life. The disease is not transmitted to humans by the bite or the feces of the louse; it is acquired by crushing the infected insect, which allows the hemolymph to contaminate skin or mucous membranes and enter through minor abrasions. Scratching the skin results in an occasional engorged louse being crushed and in abrasions being formed in the skin through which the louseborne spirochetes can enter (Sparrow, 1955; Varma, 1962; Dubos and Hirsch, 1965; Burgdorfer and Varma, 1967).

Other Louseborne Pathogenic Organisms

Other than rickettsiae and spirochetes, bacteria are considered to be the pathogenic organisms most likely to be transmitted by lice. *Salmonella* bacilli were observed attacking the midgut epithelia of lice that had fed on infected rabbits, and the bacilli remained infective in the feces of the lice for more than a year (Milner *et al.*, 1957). Other researchers have obtained similar results with typhoid, paratyphoid, and colon bacilli. Multiplication of the tularemia bacillus in body lice has been observed after experimental infection (Price, 1956).

Control of Lice

Although the opportunities for cleanliness for the great majority of people in much of the world have greatly improved in recent decades, and the efficacy of insecticides even more so, it is nevertheless desirable to understand the methods of louse prevention and control. This is particularly important in areas where louseborne diseases remain a threat to humans. To avoid becoming infested with body lice, one should avoid contact with people infested with lice, at least during the night. One is not likely to acquire body lice merely from contact with people during the day. Buxton (1947) stated that he never picked up a stray louse, even though he examined many infested garments and counted the lice. He believed the reason was that in daylight hours, a louse either crouched motionless or quickly found a fold or seam in which to remain.

Lice are most likely to spread under unsanitary, crowded conditions. One should bathe frequently with hot water and soap and rub dry with a rough towel. Underwear should be changed often to lessen the chance of infestation by lice. Adults, nymphs, and nits are killed by laundering with hot water or by dry-cleaning. If underwear is dispensed with completely, the chance of becoming infested is increased, for underwear is laundered more frequently than outer garments. Avoid sleeping between blankets that have been used by persons suspected of being infested with lice, and make sure that the bed linen is really fresh. Frequent inspection of the body and clothing is essential if there is any likelihood or suspicion of infestation.

Particularly since the advent of DDT during World War II, louse powders applied to the inside surface of the clothing, particularly underwear, and with special attention to the treatment of seams and folds, have had spectacular success against the body louse. For some years, DDT played an important role in combating typhus after it was demonstrated in Italy in late 1943 that 10% DDT powder applied at the rate of 30 cc (1 oz) per person completely controlled body lice and halted an incipient typhus epidemic (Simmons and Upholt, 1951). An ounce of any effective powder, applied either to the clothed individual or to the garments, is usually sufficient for effective treatment. Dusts of 10% DDT, 1% lindane, or 1% malathion have been effective for mass treatments under epidemic conditions. Strains of lice that are resistant to DDT and lindane have developed in various parts of the world, and in those areas these insecticides have been replaced by malathion, which is just as effective, and will control lice for 3 or 4 weeks with 1 application (Anonymous, 1970*a*). The efficacy of malathion dust was again demonstrated, for other louse species, by the excellent results obtained when it was applied in dust-bath wallows used by game birds (partridge, quail, and pheasant) for the control of wing lice (*Lipeurus* sp.) and body lice (*Goniocotes* and *Goniodes* spp.) (Philopteridae). The birds remained free of lice for more than 3 months (Rodriguez and Riehl, 1959*b*).

Powders of 0.2% pyrethrum or 0.3% allethrin, synergized with sulfoxide at a ratio of 1:10, are effective against body lice, but do not provide as long a control as DDT or malathion. They also tend to clog mechanical equipment, and are therefore more satisfactory for manually operated dusters than for the mechanically operated equipment generally used for mass treatments. It is usually necessary to repeat the dusting in 7 to 10 days to kill newly hatched lice. Tests with Abate®, Mobam® (4-benzothienyl methylcarbamate) (a product that has never been commercialized), and an experimental insecticide, Geigy GS-12968, all used as dusts, indicated that they were considerably more effective in controlling lice than were malathion, lindane, or carbaryl plus piperonyl butoxide (Cole *et al.*, 1969*b*; Anonymous, 1970*a*). In an experiment in Korea in which 100 volunteers were dusted with 5% Mobam in pyrax diluent with an average of 34 g per man and while the men were clothed and in a standing position, a period of 4 weeks of protection was provided without medical contraindications (Steinberg *et al.*, 1971). In a later experiment in Korea with groups of 100 volunteers, dusts of 2% Mobam, 1% Abate, and 2% malathion were equally effective in controlling louse populations for at least 25 days at a dosage of 56.5 g per man. Most of the dust was applied in thorough treatment of clothing, but some was applied to individual bedding (Steinberg *et al.*, 1971, 1972).

Dust applications as used for body lice will control crab lice. Persisting infestations should be retreated at 1- to 2-week intervals. To obtain the full benefit of a treatment for head lice, the insecticide should remain in the hair for 24 hours. For this reason, people are inclined to object to a dust, which is highly visible and indicates that the person is lousy. Liquid shampoos containing 0.2% lindane or 0.5% malathion are generally more acceptable than dusts (Anonymous, 1970*a*).

Keh and Poorbaugh (1971) listed some proprietary products that were available in California for control of head lice at that time. Not requiring a prescription were A-200 Pyrinate®, a shampoo consisting of 0.165% pyrethrins, 2% piperonyl butoxide, and 5% deodorized kerosene; Cuprex®, a lotion consisting of 30.97% tetrahydronaphthalene and 0.03% copper oleate; and Bomate®, a lotion consisting of 0.5% isobornyl thiocyanoacetate and 0.6% dioctyl sodium sulfosuccinate. Available by prescription were Topocide®, a lotion consisting of 12% benzyl benzoate, 2% benzocaine, and 1% DDT; and Kwell® (shampoo, lotion, or cream). The directions for application and periods recommended to retain the preparations on the treated parts vary greatly, and should be followed carefully.

In East London, the rate of infestation in 2 primary schools reached 25% of the school population, and partial control was achieved only by the use of fine-toothed combs. Proprietary insecticidal preparations failed completely to control the lice. Lice collected in various parts of London were shown to be resistant to organochlorine insecticides but highly susceptible to propoxur, malathion, and carbaryl (Maunder, 1971*a*). The treatment of nearly 3,000 people with 0.5% malathion lotion resulted in no head lice being recovered the following day, a complete kill of eggs, considerable residual effect, and no observable side effects except for a slight temporary increase in dandruff, in persons prone to that condition, which could have been caused by the solvent rather than the insecticide (Maunder, 1971*b*). Laboratory colonies of body lice were selected for malathion resistance for over 44 generations with no sign of the development of resistance (Cole *et al.*, 1969*a*). However, in previous selection experiments, body lice had become resistant to carbaryl (Clark and Cole, 1967). The means to virtually exterminate lice appear to be available, and there appears to be a need for "an end to half-hearted measures" (Maunder, 1971*b*).

The eggs or "nits" of head lice are so firmly attached to hair (figure 294) that they remain for long periods after treatment, even if they are killed. The cement that binds them to the hair is not dissolved by any known treatment. They may eventually be removed by thorough combing with a fine-toothed metal comb. Lice, like female mosquitoes, are attracted to warm objects. If a warm comb, "pleasantly hot to the hand," is passed through the hair, lice become excited, leave their usual places among the bases of the hairs, and are much more easily combed out. Although this procedure has not been recommended by entomologists as a control measure, there have been enthusiastic testimonials as to its efficacy (Howlett, 1917; Nuttall, 1917*b*).

To eliminate lice and other pests, the armed services have fumigated clothing with chloropicrin or methyl bromide in ampules broken within tightly tied duffle bags. In the rare cases of crab lice in the eyelashes of small children, the insects can be eliminated by the application of 0.25% physostigmine ophthalmic ointment with a cotton-tipped applicator (Ackerman, 1968).

BED BUGS AND ALLIES (CIMICIDAE)

Bed bugs are pests of man and his domestic animals, as well as of bats and wild birds. Thirteen genera of them are associated exclusively with bats, and 9 with birds. Only the genus *Cimex* contains some species that attack bats and others that attack birds. Half of the 74 known species of bed bugs were discovered during the 10 years preceding the publication of the late R. L. Usinger's extensive work, *Monograph of Cimicidae,* in 1966.

There are 3 species of bed bugs that commonly attack man. One of these, *Cimex lectularius,* is associated with man, bats, chickens, and occasionally with other domestic animals. It is widely distributed over most of the world, and is the species that occurs in north temperate regions. The oily secretion from human skin is strongly attractive to it, but pure perspiration alone is repellent (Aboul-Nasr and Erakey, 1958).

The tropical bed bug, *C. hemipterus,* attacks man, chickens, and rarely bats throughout the tropical regions of the world. It has been found in Florida homes since 1938, and has become widely distributed in that state (Hixon, 1943). It has apparently not been able to adapt itself to colder climates, and is not likely to spread to north temperate regions. *Leptocimex boueti* (Brumpt) attacks both bats and humans in West Africa.

Bed bugs have probably been associated with man since the earliest recorded history, and possibly since the days of the cave man. This is suggested by the fact that some cimicids are parasites of animals that frequent caves, such as bats, swallows, and martins (Sailer, 1952; Usinger, 1966). Usinger stated:

> It is of course possible that the bed bug was associated with man throughout his early evolution in the Pleistocene and on through the Paleolithic and Neolithic periods. However, it seems more likely that it became permanently associated with man during the period of movement from cave to village to city. As an alternative hypothesis, the bugs may have turned to man on many occasions while their bat hosts roosted in his houses. *Cimex lectularius* occurs in Europe today in bat roosts in churches and other man-made structures and in treeholes, and a colony was discovered by Povolny in a bat cave near Kabul, Afghanistan.

Bed bugs were abundant in the Mediterranean region in ancient times. The generic name *Cimex* was derived from the common name used by the Romans for the insect. Bed bugs have been the subjects of much folklore. For example, it was believed that when consumed, usually with wine, they were a cure for many diseases and for snake bites. The northward spread of bed bugs was remarkably delayed. They reached Germany in the eleventh century, England in the sixteenth, and Sweden early in the nineteenth century (Busvine, 1966). They came to the United States and Canada with the early colonists (Marlatt, 1916). Although it was late in invading north temperate regions, *Cimex lectularius* thrives best in such areas. While it usually succumbs to temperatures of 96 to 100° F (36 to 38° C) when accompanied by high humidity, at temperatures of 60° F (16° C) or less it enters semihibernation, and can overwinter successfully in unheated buildings. The great increase in infestations in northern European cities in the early twentieth century was believed to have been favored by the increased use of central heating (Johnson, 1942).

Mallis (1969) prepared a key for 8 species of bed bugs that are most likely to become pests in the United States, although some of them are comparatively rare. In the genus *Cimex,* the 2 species infesting man, *C. lectularius* and *C. hemipterus,* have short body hairs, and the front margin of the pronotum is deeply concave. *Cimex lectularius* has winglike expansions on the pronotum (figure 296), and the hairs fringing the pronotum are relatively straight. In *C. hemipterus,* the pronotum is not winglike, and the hairs fringing it curve backward. The 2 other species of *Cimex* in Mallis's key have long body hairs, and

Fig. 296. Bed bug, *Cimex lectularius.*

the bat is their usual host. The bat bug, *Cimex pilosellus,* occurs in the western United States, and *C. adjunctus* (Barber) is found in the eastern states. Two other cimicids that occur in the United States, the swallow bug, *Oeciacus vicarius,* and the poultry bug, *Haematosiphon inodorus,* are occasionally pests in human habitations.

In some parts of the country, there is a belief that bed bugs can live in logs or dead trees for long periods. This is probably because they are confused with flat bugs, *Aradus* spp. (Wright, 1944). Flat bugs are also small, flat, oval, dark-colored bugs and, according to Essig (1926), they are often mistaken for bed bugs. They live in cracks and crevices or under the bark of living or dead trees. Some flat bugs are predaceous, while others subsist on plant material. Although bed bugs look very much like flat bugs, they are phylogenetically closer to the Anthocoridae, that prey on insects and mites and only occasionally bite warmblooded animals, than they are to flat bugs (Usinger, 1966). Anthocorids are also flat, oval insects that have 4 antennal segments, and their reddish nymphs superficially resemble bed bugs.

Bed Bug, *Cimex lectularius* L.

A half-century or more ago, the bed bug (figure 296) was a common household pest, and was the bane of travelers, for it was common even in "respectable" hotels. The importance of this insect gradually diminished with general improvements in household and personal cleanliness, as well as with the increased availability and use of some reasonably effective insecticides, such as pyrethrum. The widespread use of DDT after World War II resulted in the bed bug's finally being a minor pest. Bed bugs may still be problems where there are primitive and unsanitary conditions. They can also present at least temporary problems when they are introduced by people who have lived under such conditions and then move into living quarters where one would least expect bed bugs, such as in well-kept university housing. Bed bugs can migrate from one apartment into another, and if an infested apartment is vacated, the rate of their dispersal to other apartments is frequently increased.

Except for the tropical bed bug, *Cimex hemipterus,* in Florida, *C. lectularius* is the only cimicid in the United States that should be called a household pest, although there are other species that sometimes are brought into houses via bat roosts or the nests of swallows or chimney swifts.

Description. The adults are 4 to 5 mm long, oval, flat, and usually a rusty-red or mahogany color (plate VIII, 4), but become brighter red after feeding. They are thin when starved, but become plump and elongate after feeding. The body is covered with microscopic hairs, resulting in a banded appearance when engorged, revealing hairless areas—the conjunctivae between the segments. The 4-segmented antennae arise from between the pair of compound eyes. Bed bugs have a jointed beak or proboscis (figure 46, *B,* chapter 4), situated beneath the head and passing backward to between the coxae of the forelegs. The beak consists of 2 pairs of stylets. The outer pair, a modified form of the mandibles of biting insects (figure 46, *Mds*), is provided with barbs that exert a sawing action and facilitate the penetration of the host's skin. The inner pair (figure 46, *MxS*), a modified form of the maxillae of biting insects, forms 2 tubes. One tube sucks up the blood, and the other is the channel through which the bug injects a salivary secretion into the wound. The secretion contains a strong anticoagulant that allows the bug to withdraw the blood, which it could not do if the blood were allowed to coagulate. The salivary secretion is the cause of the irritation that most people feel after being bitten by cimicids (Snodgrass, 1944; Wright, 1944).

Life Cycle and Habits. Female bed bugs deposit 10 to 50 eggs at a time in irregular masses, and may eventually lay from 200 to 500 eggs. For the size of the insect, the eggs are rather large, being about 1 mm long. They are elongated, pearly to yellowish white, and are laid in secluded places in cracks and crevices or on rough surfaces, covered with a transparent cement. The cement attaches them firmly to the substrate, and the empty shells remain in place long after the eggs hatch. Wright (1944) stated that eggs were seldom deposited at temperatures below 48° F and very few between 48 and 60° (9 and 16° C). If the female was well fed, her maximum number of eggs was laid between 70 and 82° F (21 and 28° C). Under experimental conditions, eggs did not hatch at above 98.5° F (37° C) or below 55.5° F (13° C). At 70° F (21° C), the eggs hatched in 6 to 17 days, but at 57° F (14° C), hatching was delayed as long as 30 to 40 days. Bed bug eggs could not live longer than 3 months, and therefore did not survive the winter in most unheated houses in cold climates (Anonymous, 1949).

Bed bugs tend to congregate on rough surfaces that are dry and at least partially dark, and may be found in clusters in such locations (Usinger, 1966). Here, they oviposit and return after each meal, and eventually there is an accumulation of much fecal matter, eggshells, and exuviae in the area. The bugs prefer wood and paper surfaces to stone, plaster, metal, or even textiles for such harborages (Kemper, 1936). Herms (1939) noted that, where old-fashioned wooden bedsteads were used, bed bugs stayed close to the points of attack, but that iron bedsteads did not afford convenient hiding places. This does not mean that the insects will not reach their sleeping victims, for they are very active at night, and can crawl for considerable distances to find their hosts. In one instance, when the legs of beds were placed in cans of oil, people were disturbed by the bugs dropping from the ceiling! The pests can also migrate from house to house (Anonymous, 1949).

There are 5 nymphal instars. Ordinarily, only 1 meal is taken between each molt, and 1 before each egg deposition (Marlatt, 1916). The nymphs look like the adults, but the cuticle of the nymphal abdomen is thinner than that of the adult, and therefore more distinctly reveals the color of the partly digested blood inside. The cast skins of the various instars are thin, white, and semitranslucent, and have the conformation of the nymph. The nymphal period may vary from 14 days at 82° F (28° C) to 118 days at 59° F (15° C). Wright (1944) concluded that the period from egg to egg would be 5 to 8 weeks during the hotter parts of the summer in temperate latitudes, and 7 to 12 weeks in centrally heated houses during the winter. Under favorable conditions, there might be 3, 4, or even more generations per year.

Because of variations in developmental period, the generations overlap, so that all instars and stages may be found together, except perhaps in unheated rooms in which development is arrested in the winter months (Wright, 1944). Johnson (1942) calculated that in an unheated bedroom in London, *Cimex lectularius* could have only 1 full generation and an incomplete second one in a year.

Factors Affecting Longevity. Starvation and low temperatures contribute to the longevity of bed bugs, for both reduce activity to a low level and force the insects into a state of semihibernation. The bed bug is a hardy and long-lived insect. Wright (1944) recorded the following observations: At 44.6° F (7° C) and 90% relative humidity, unmated females survived for 550 days when kept without food. Survival was not as long at lower humidity, for water loss decreased longevity. At 73.4° F (23° C) and 90% RH, unmated females survived only 181 days. With normal feeding and reproductive cycles, individual bugs survived for 316 days. If eggs and newly hatched nymphs were kept at temperatures below freezing for 15 to 30 days, they succumbed. Eggs were killed after an exposure of 1 hour to 113° F (45° C), and first-instar nymphs, after 15 minutes at the same temperature. For adult bugs, the thermal death point was about 1.8° F below that for the eggs.

Once adult bed bugs have fed, their longevity is greatest at low temperatures; this accounts for their survival in unheated rooms in northern climates (Usinger, 1966).

Bed Bugs as Pests

Bed bugs share with lice, fleas, and cockroaches the characteristic of being especially abhorrent to most people, probably because their presence is associated with poor housekeeping and low hygienic standards. The bugs can emit odors from scent glands situated anteriorly on top of the abdomen of the nymph and under the thorax at the base of the abdomen of the adult. Although it is reputed to have a disagreeable odor, those who have smelled the isolated, clear, oily, volatile liquid that is secreted have described it as similar to "fresh, red raspberries" (Mallis, 1969), "an obnoxious sweetness" (Kemper, 1936), and "not particularly unpleasant" (Busvine, 1966). The odor is caused principally by carbonyl compounds (Schildknecht, 1964; Collins, 1968). Busvine (1966) stated that although it was sometimes claimed that houses infested with bed bugs could be recognized by the "bed bug smell," in his own experience most badly infested houses were likely to have a mixture of disgusting odors, characteristic of lack of hygiene.

If social stigma does not serve to motivate the occupant of infested premises to exterminate bed bugs, there remains an even more compelling reason—the bite of the bug. The bite itself is painless, and Kemper (1936) stated that 20% of the persons he tested showed no subsequent reactions. Most people, however, develop an allergic reaction, stimulated by the saliva injected by the bug as it feeds (Hecht, 1930). The bite ordinarily produces a lump or swelling (figure 297), with no red spot or other distinguishing characteristics such as are produced by the bites of black flies, fleas, and some other insects. When many bugs feed on a small area, reddish spots caused by hemorrhage may appear after the swelling has disappeared. Swelling is severe with some highly allergic persons, and may not be confined to the immediate area that was bitten. Of 3 persons who allowed themselves to be bitten by many bed bugs over periods of up to 7 years, 2 reported no immunity to allergic reaction (Kemper, 1936;

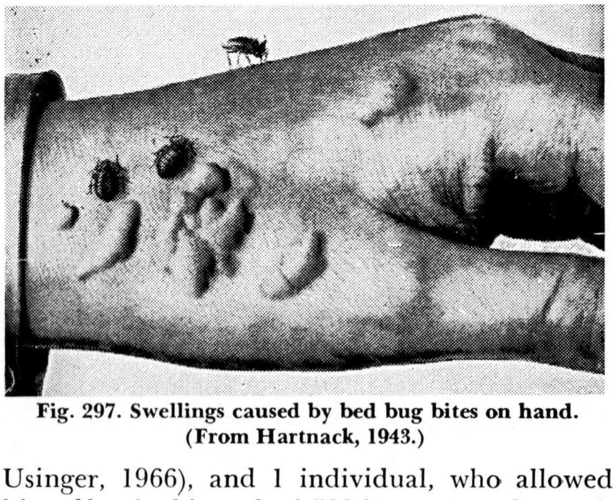

Fig. 297. Swellings caused by bed bug bites on hand. (From Hartnack, 1943.)

Usinger, 1966), and 1 individual, who allowed himself to be bitten by 2,500 bugs over a 9-month period, reported the development of immunity (Hartnack, 1939).

The period required for a bed bug to obtain a meal was stated by Wright (1944) to be 3 to 5 minutes, and by Kemper (1936) to be 5 to 10 minutes. The bugs may defecate while they are feeding, and their movements may result in the wound being contaminated by the excrement. Some individuals are unable to get enough sleep while the infestation continues, and their general health may suffer. Cases of iron deficiency caused by excessive feeding of bed bugs on infants in India have been reported (Venkatachalam and Belavady, 1962). Usinger (1966) allowed a colony of bed bugs to feed on him weekly for 7 years, and stated that his hemoglobin decreased from a normal 14.5 g per 100 ml of blood in 1958 to 11.5 g in 1963 and 1964, and that the hemoglobin count could not be increased with supplemental iron taken orally or by injection. However, after he stopped feeding the bugs, his hemoglobin rose to 13.2 g per 100 ml in a few months.

Bed bugs have been found to be infected in nature with various disease organisms. Burton (1963) prepared a table showing the maximum survival time, in days, of the causative organisms of 25 human diseases in *Cimex lectularius* and of 5 in *C. hemipterus*. Particularly long survival times (147 to 285 days) were reported for the organisms of plague, relapsing fever, tularemia, and Q fever.

However, after making an extensive review of the literature of the present century, some leading authorities have concluded that although the cimicids have been suspected in the transmission of many diseases or disease organisms of man and bats, in most cases conclusive evidence has been lacking (Nauk and Zumpt, 1940; Burton, 1963; Usinger, 1966).

Natural Enemies

Usinger (1966) listed a number of natural enemies of bed bugs. Several species of ants and reduviid bugs have proved at times to be strikingly effective. Solpugids, pseudoscorpions, centipedes, and spiders are known to be effective predators. Kemper (1936) reported complete destruction of bed bug colonies in 3 weeks by the straw itch mite, *Pyemotes ventricosus*. Whether or not bed bugs are eaten by cockroaches has long been a matter of controversy. Mallis (1971) provided new information on this study in an experiment in which he showed that neither American nor German cockroaches fed on bed bugs when Purina Dog Chow was available to them as food. In a later test with American cockroaches only, he found that the insects would eat bed bugs when no other food was to be had.

Control of Bed Bugs

Until the advent of DDT, little improvement had been made over the ancient recommendations to hang the feet of a hare or of a stag at the foot of the bed. Pyrethrum products became widely distributed about the middle of the nineteenth century, and at least temporary control could be obtained with them, so that one could have a night of undisturbed sleep by the thorough spraying of the bedroom. HCN fumigation was occasionally practiced, but it was dangerous and expensive, and reinfestation was rapid. DDT proved to be highly effective. It provided long, residual action, and pyrethrins could be added if rapid knockdown of the bugs was desired. Although DDT was developed during World War II, resistance to it by bed bugs was reported at Pearl Harbor by 1947 (Johnson and Hill, 1948). Other reports of resistance came from many parts of the world. One kind involved DDT, methoxychlor, and their analogs, and the other kind involved benzene hexachloride, dieldrin, and other chlorinated cyclodiene compounds. Resistance by bed bugs to one insecticide in each general group resulted in resistance to all the others. Strains showing the highest resistance were almost completely immune (Busvine, 1958). The use of DDT inside houses in malaria-eradication projects resulted in wide-scale buildup of resistance of bed bugs to organochlorine compounds, leading to resurgences of the pests within a few years after initial treatment (Rafatjah, 1971). Resistant bed bugs were shown to have retained their resistance when no treatment had been applied to their descendants for a period of up to 5 years (Mallis and Miller, 1964). Malathion proved to be an effective alternative treatment. Some strains resistant to DDT were also about 10-fold resistant to pyrethrins. DDT and malathion were used for some years, and diazinon at 0.5% concentration was used where the bed bugs became resistant to DDT and malathion. In 1964, organophosphorus resistance in bed bugs was reported from Israel (Barkai, 1964). At late as 1970, the entomologists of the Center for Disease Control of the U. S. Public Health Service stated that resistance to organochlorine compounds in bed bugs was not a problem in the United States. Their recommendation for bed bug control (Anonymous, 1970*a*) was as follows:

> A single residual application of insecticide emulsion or solution to baseboards, wall crevices, bedsteads, and mattresses of infested premises will usually control a bed bug infestation. Sprays containing lindane (0.1% for beds and mattresses and 0.5% elsewhere), trichlorfon (0.1%), ronnel (1%), dichlorvos (0.5%), or malathion (0.5 to 1%) are commonly used. Special care should be exercised in the treatment of mattresses and upholstery so that only a light application is made. *Under no circumstances should mattresses be soaked with spray. Infant bedding, including the crib, should not be treated. With dichlorvos or trichlorfon, spray only tufts and seams of mattresses, and air until dry—at least 4 or 8 hours, respectively—before re-use.* If infestation persists, re-treat at not less than 2-week intervals.

The addition of pyrethrins at 0.1 to 0.2% to the foregoing formulations will increase the effectiveness of the treatments by flushing the bed bugs out of their harborages. These may include the "crevices and corners in wooden bedboards, behind loose wallpaper, in nail holes and cracks in the walls, behind pictures, conduits, light switches, wall panels, baseboards, door and window frames, floor crevices, and in furniture and mattresses" (Usinger, 1966).

Bed bugs are sometimes abundant in attics, wall voids, and other enclosed places. In such cases, the application of the silica aerogel Dri-die 67 is appropriate, for it is highly effective against these insects and, being inorganic, should retain its insecticidal effect indefinitely.

Bat Bug, *Cimex pilosellus* Horvath

The bat bug occasionally becomes a household pest in the western United States, particularly in summer cabins that are left unoccupied in winter (Wright, 1944). The prothorax is relatively large, and extends forward on either side of the head, forming the "winglike expansions" that distinguish this species from *Cimex hemipterus.* The visible portion of the mesothorax is short and narrow. Two small pads, the rudimentary forewings, hide the metathorax from view. The abdomen has 9 segments, the first one so reduced in size as to be inconspicuous. The female's abdomen is broad and bluntly rounded, while the male's is narrower and more pointed posteriorly.

Swallow Bug, *Oeciacus vicarius* Horvath

Among the cimicids for which man is not the primary host, the swallow bug apparently is the one most commonly reported as a pest. The front margin of this cimicid is much less concave than that of the bed bug, and its body has long hairs. When about 200 swallow nests were knocked off a house in Banff, Alberta, the bugs invaded it in such numbers that the occupants had to leave it and sleep in tents (Spencer, 1930). In the same city, when swallow nests under the eaves of a high-school building were destroyed, swallow bugs soon invaded every room. Many students were bitten, and developed reddened, itchy areas 2 or 3 in. (5 or 8 cm) wide. The building was fumigated with HCN gas (Mail, 1940). Mallis (1969) reported an infestation of swallow bugs in Los Angeles, California, where swallows had been nesting under the eaves of a house. On several occasions, one of the occupants developed a swollen mouth, arm, and face. The cause was found to be bites of the bugs, which were later detected in cracks and crevices on the second floor. They were eliminated with pyrethrum fly spray.

Usinger (1966) stated that every cliff swallow nest he examined in North America and Europe had been infested with swallow bugs, and that obviously they could be carried to new nesting sites by the birds. However, it is difficult to envisage how they were carried, for unlike bird lice and fleas, cimicids have no special adaptations for clinging to the feathers or body of a bird while in flight.

Poultry Bug, *Haematosiphon inodorus* (Dugès)

The poultry bug occurs in the hot, dry regions of the southwestern United States and Mexico. It has longer legs than the bed bug, and has a beak that reaches to the coxae of the hind legs, while all other species have beaks that reach only to the coxae of the forelegs. It is more active than the bed bug, and does not have an odor. Originally a pest of birds and bats, it has become a severe pest of poultry, sucking their blood at night. It may also invade homes, and can be a severe pest in them. Benzene hexachloride was found to provide longer control than DDT when applied in chickenhouses (Hall and Wehr, 1949).

CONENOSE BUGS, ASSASSIN BUGS, CORSAIRS, AND WHEEL BUGS (REDUVIIDAE)

Reduviids are predators that suck the blood and lymph of their victims. Although most of them feed on other insects and on one another, some attack humans, inflicting painful bites and/or transmitting disease. The reduviids are sometimes referred to as "kissing bugs," a name that had its origin in the summer of 1899, when many newspapers of the eastern United States created widespread popular alarm with the story that reduviids, principally *Reduvius personatus,* most commonly fed on the faces and especially the lips of human victims (Howard, 1900).

Reduviids are large, elongate insects, with long, freely movable heads, 4-segmented antennae, conspicuous eyes, and strong, 3-segmented beaks. S. F. Wood devised the adjacent simplified key for field identification, with the aid of a hand lens, of some common reduviids that can inflict painful bites and that invade or are occasionally carried into homes in the southwestern United States. [A more detailed and comprehensive key to the reduviids, but one requiring more careful examination under considerable magnification, will be found in Furman and Catts (1970).]

Key for Field Identification of Some Reduviid Bugs of the Southwestern United States

1. Beak stout and arched under head in side view . . 2
   Beak slender and extending straight back to "chest" under head, producing a coneshaped head, as seen from the side 3
2. Top of prothorax (pronotum) constricted and grooved behind middle; outer sides of upper part (femur) of first leg flat or concave; body mostly yellowish tan, with a circular yellowish-tan spot in the center of the membrane of each forewing; 19 mm long . . . . .Twospotted corsair, *Rasahus thoracicus*
   Pronotum constricted and grooved near or before the middle; outer sides of femur of first leg rounded outwardly or convex; 12 to 19 mm long . . . . .Assassin bugs (*Reduvius, Zelus*)
3. Body more than 25 mm long 4
   Body 25 mm or less in length 5
4. Body uniformly black, with pale yellowish-brown or reddish-brown margin around top edge of abdomen; 30 mm long; Arizona . . . . Longlegged conenose, *Triatoma recurva*
   Body black; top edges of abdomen with yellowish blotches on the posterior edges (one-third) of the abdominal segments; 25 to 30 mm long; New Mexico, Texas . . . . Blotchbordered conenose, *Triatoma gerstaeckeri*
5. Body pale, tan, or brown, covered with curved black hairs and appearing distinctly hairy when seen with a magnifying glass; top of head rounded; 12 to 15 mm long; California, Arizona . . . . Hairy conenose, *Paratriatoma hirsuta*
   Body not hairy, top of head nearly flat 6
6. Mostly black, with 6 red spots on edge of abdomen above and below; 18 to 24 mm long; southern United States from Atlantic to Arizona, north as far as southern Illinois . . . . Bloodsucking conenose, *Triatoma sanguisuga*
   Black or brownish black, sometimes with a narrow tan edging along the top rim of the abdomen; lacking 6 red spots noted above; 14 to 20 mm long; California, Nevada, Utah, Arizona, New Mexico . . . . . . Western bloodsucking conenose, *Triatoma protracta*

Brownish black, with a tan border along top edge of abdomen; 13 to 19 mm long; Arizona, New Mexico, Texas .... subspecies *woodi*
Black or brownish black; 20 to 23 mm long; Arizona, New Mexico, Utah, Colorado .... subspecies *navajoensis*
Pale-reddish markings on thorax and wings, reddish border blotches extend to lateral margins of the abdomen; 16 to 23 mm long; New Mexico, Arizona .... Redbordered conenose, *Triatoma rubida uhleri*

There are many subfamilies of Reduviidae, but only 4 contain species of medical importance. The 4 major medically important subfamilies are (1) Triatominae, or conenose bugs; (2) Reduviinae, or assassin bugs; (3) Piratinae, or corsairs; and (4) Harpactorinae (includes the wheel bugs).

Conenose Bugs (Subfamily Triatominae)

This subfamily includes the species that feed exclusively by sucking the blood of vertebrate animals. The beak is straight and folded back under the head, as is characteristic of bugs that extend their beaks and take blood. They are mostly confined to the Western Hemisphere. Their principal hosts are various species of wood rats or pack rats (*Neotoma*), which build large "lodges" of sticks and plant parts (figure 298). Conenose bugs can be found in the inner grass nests and crawling about on the sticks and twigs of wood rat lodges. They will not be found in large numbers, for wood rats feed on the bugs when they become too numerous. Despite their high reproductive capacity, the bugs are kept at a relatively low level in wood rat nests (Ryckman and Ryckman, 1967*b*). In suburban areas, conenose bugs may be found on warm summer evenings on warm concrete or asphalt under street lights.

Conenose bites are usually not painful. Usinger (1944) pointed out that this was characteristic of habitual bloodsuckers in contrast to the usually severe effects of the bites of occasional bloodsuckers. Wood (1942) allowed 8 species of *Triatoma*, and also *Paratriatoma hirsuta* Barber, to feed to capacity on his left fingers, wrist, and forearm at different times. From the *Triatoma* species, he never felt more than a slight "pricking" or "tickling" sensation, and there was no marked injury. Forty hours after 2 *Paratriatoma* had fed on his wrist, a severe itching developed, and the area around each of the 2 bites was swollen and inflamed. Twelve hours later, the itching was less intense, but there were round, swollen, reddened areas 4 mm in diameter at the sites of the bites. Visible evidences of the bites remained for 19 and 30 days. However, bites of triatomine bugs have sometimes resulted in severe systemic effects lasting several days or weeks. These drastic symptoms appear to be the anaphylactic reactions of hypersensitive persons. Wood (1950*b*) described the symptoms of 5 individuals with obviously high sensitivity who were bitten by *Triatoma protracta*. One of the victims, a healthy-appearing, middle-aged woman, who was bitten on the right forearm, was unaware of having been bitten, but found some of the bugs, full of blood, in her bed. After the usual initial painless period, her hands itched severely and a general edema (swelling) of the arms and hands spread gradually over her body, and was accompanied by severe dizziness and nausea lasting about 2 days. At the site of the bite was a "red pinpoint puncture surrounded by a white wheal enclosed by a reddish ring." Similar experiences were recorded by Shields and Walsh (1956) with bites of *Triatoma sanguisuga*—initial painlessness, followed by a period of swelling and itching, and an increased sensitivity with repeated bites.

Fig. 298. Wood rat (*Neotoma*) nest built among pendant branches of an elder (*Sambucus*) tree on the University of California (Los Angeles) campus.

A more serious indictment against the triatomine bugs in the Western Hemisphere south of the United States border is that they are vectors of *Trypanosoma cruzi,* the causal agent of Chagas' disease or American trypanosomiasis. Swelling of the eyelids and face, and a fever, usually accompanied by conjunctivitis, are the most apparent initial symptoms of the disease (Wood, 1950*a*).

Conenose bugs defecate while they feed, or soon afterward. If the feces are infected with trypanosomes, these organisms may be rubbed into the bite by scratching or rubbing. Infection may also take place in other wounds or scratches on the body, or through the conjuctivae of the eyes and mucous membranes of the mouth.

The following rates of infection of *Triatoma* species by trypanosomes have been recorded: (1) 41% of 957 *T. protracta* examined in California, Arizona, and New Mexico (Wood and Wood, 1941); (2) 26% of 1,297 *T. protracta* in California; (3) 25% for over 3,500 western and southwestern species of *Triatoma* (Olsen, 1964); (4) 77.9% of 698 *T. gerstaeckeri* collected in Texas; and (5) an average of approximately 20% for the entire United States—similar to the rate observed in South America, where the incidence of Chagas' disease is high (Yaeger, 1961). However, there have been only 2 recorded cases of Chagas' disease in the United States, both in Texas, so in this country the insects are primarily important because of their bites.

With regard to the 2 Texas species, *T. sanguisuga texana* Usinger and *T. gerstaeckeri* (Stål), Pippin (1970) noted that the bugs preferred to feed with at least their hind legs in contact with a surface other than the skin of the host, such as clothing. The bugs rarely moved about on the host's body except when extremely crowded. Transfer of trypanosomes by fecal contamination is undoubtedly less likely with these relatively stationary bugs than by species such as *Rhodnius prolixus* Stål, which is an effective vector of Chagas' disease in South America. It does not hesitate to crawl on the host's body in an effort to feed.

The increasing tendency of people to build their homes in chaparral areas surrounding the cities of southern California or in the foothills of the Sierra and the Coast Range has increased the possibility of home invasions by conenose bugs and the potential for the development of hypersensitivity to their bites. It also provides greater exposure of man and domestic animals to Chagas' trypanosome (Wood and Wood, 1964).

Biology. The biology of conenoses in California is summarized as follows: The eggs are laid during the summer and hatch in 3 to 5 weeks, depending on the species and the temperature. Each of the 5 nymphal instars must have a blood meal before it can molt. In California and Arizona, all stages of *Triatoma* can be found in nature at all seasons of the year. However, there is a tendency for some species to overwinter in the fifth instar and become adults in the spring; this is less so for *T. protracta* than for *T. rubida* (R. E. Ryckman, correspondence). Although the life cycle is normally 1 year, sometimes 2 years are required, in which case the third nymphal instar is reached by the end of the first year (Keh, 1956). According to Grundemann (1947), the life cycle of *Triatoma sanguisuga* required approximately 3 years under simulated natural conditions in Kansas.

In an extensive survey made by Walsh and Jones (1962), although nearly 100 adult bugs were found in homes, in only 2 cases were nymphs of *T. protracta* recorded inside. They believed that the insects probably went through their life cycle only in the nests of small animals, principally wood rats, and were unable to complete the cycle in a home. However, in areas where conenoses are abundant, both nymphs and adults frequently invade homes. In Mexico and the rest of Latin America, millions of triatomine bugs go through their complete life cycles in homes. R. E. Ryckman (correspondence) believed that they could do this in the United States also if the occupants of infested houses permitted it. In fact, there have been a number of reports of *T. sanguisuga* living and reproducing in houses in southern Illinois throughout the year, apparently feeding on pets.

Western Bloodsucking Conenose, *Triatoma protracta protracta* (Uhler)

This subspecies is widely distributed in the southwestern United States and northern Mexico. It ranges in length from 14 to 20 mm, and its general body color is dark brown to black (figure 299). It sometimes has a narrow tan edging along the lateral margin of the abdomen. This conenose occurs throughout California in the foothill areas inhabited by its principal hosts, the various species of wood rat or pack rat, *Neotoma*. It is also found in northern Baja California, Sonora, Nevada, western Utah, most of Arizona, and western New Mexico. The subspecies *Triatoma protracta woodi* Usinger occurs in southeastern Arizona, southern New Mexico, western Texas, and northeastern Mexico. It is distinguished by the light, longitudinal connexival band (prominent abdominal margin at the junction of dorsal and ventral sclerites) on the abdomen. The bug is 13 to 19 mm long. *Triatoma protracta navajoensis* Ryckman occurs in the high Colorado plateau (Navajo country) of eastern Utah, western Colorado,

Fig. 299. Western bloodsucking conenose, *Triatoma protracta protracta*.

northeastern Arizona, and northwestern New Mexico. It is the largest of the 3 subspecies in the United States, being 20 to 23 mm long (Ryckman, 1962).

Bloodsucking Conenose, *Triatoma sanguisuga* (LeConte)

Throughout the country as a whole, *Triatoma sanguisuga* is the best-known triatomine species. It ranges from the Atlantic seaboard to Arizona, occurring throughout the entire southern part of the United States and as far north as southern Illinois. It is also found for a few hundred miles along the western seaboard of Mexico. This is a dark-brown to black insect, with red markings about the periphery of the posterior two-thirds of the abdomen. It is 18 to 24 mm long (Grundemann, 1947). *Triatoma lecticularius* (Stål) ranges over most of the area in the United States that is infested by *T. sanguisuga*. *Triatoma gerstaeckeri* (Stål) is found in southern Texas and eastern New Mexico, and *T. recurva* (Stål) occurs in a limited area of Arizona (Usinger, 1944).

Envenomation by *Triatoma* Species

As already stated, most persons bitten by conenose bugs report either no sensation or a sensation varying from a slight tickling to a sharp pricking, and local reactions may vary from none to a local reddening, some swelling, and itching. Physicians' reports indicated that fewer than 5% of persons bitten by these insects exhibited severe systemic reactions from their first bite[23] (Walsh and Jones, 1962).

In one survey throughout an extensive area in the Sierra foothills of California, 110 case histories were taken from questionnaires sent to persons who had reacted severely to *Triatoma* bites. Out of 92 persons reacting systemically, 86% reported pruritus (severe itching). The itching might occur anywhere on the body, but was usually most severe on the scalp, palms, and soles of the feet, and lasted from 30 minutes to 5 hours. The next most frequent symptom (72%) was edema (swelling), occurring generally throughout the body, but especially around the eyes, tongue, larynx, and trachea. This usually made speaking, breathing, and swallowing difficult. Even without medication, the swelling began to recede within 10 to 12 hours after the bite. In over half of the cases, welts and rashes appeared, lasting several hours or until medication was administered. In 42% of the cases, the bites were followed by nausea, and in 38% of them, fainting occurred, invariably as the patients arose from bed. Other symptoms reported were pain, vomiting, fever, cramps, and diarrhea. Infrequent symptoms were dizziness without fainting, palpitation, temporary loss of memory, and "diffuse malaise with whole-body aches and weakness" (Walsh and Jones, 1962).

Of the people contacted in the survey who noted a difference in the severity of reaction following each successive bite, 63% remarked that they believed the reactions to be increasingly severe. Some of the patients who reported that the reactions became less severe questioned the validity of their own interpretations because they had sought medication more promptly after their initial experiences with the conenoses. Balazuc (1950) found that the reaction to *Triatoma* bites on his own body became increasingly severe with successive bites over a 2-year period. Also, the severity of the reactions increased with increasing

[23] However, it does not appear possible that 5% of the persons bitten by *Triatoma* could be hypersensitive to the first bite when hypersensitivity was developed by repeated bites or exposures. These people may have been bitten before. Dr. R. E. Ryckman has informed us that he let conenose bugs feed on hundreds of medical and dental students, and that none had a serious reaction to the bites. A few showed a reaction comparable to that from a mosquito bite. This was probably an immunological cross-reaction with some other insects by which they had been bitten.

numbers of bites at any one time. However, he continued to allow the bugs to feed on him, and eventually lost his sensitivity. This indicated that the symptoms were the results of sensitization rather than responses to the primary irritant.

The potential for serious reactions is indicated by the fact that some inhabitants of the Sierra foothills have had as many as 15 systemic allergic reactions in a single season. Nichols and Green (1963) found in a few cases that their patients who had been bitten by conenoses were in shock, with blood pressure of 40/0 mm of mercury (normal is 120/80 or higher) and that all the physical findings were consistent with shock. In general, they noted the following complaints, listed in the descending order of frequency:

Itching at the site; generalized itching; generalized hives; fainting; nausea and vomiting; swelling of the eyes; swelling of the tongue and throat, with difficulty in swallowing, speaking, and breathing, etc. Physical findings were usually consistent with the complaints, and included urticaria, angioneurotic edema, immediate and delayed local reaction at site of bite, hypotension, and edema of tongue and larynx.

Walsh and Jones (1962) reported that all but 1 of 110 persons questioned were in the house when they were bitten, and most of them found the engorged bugs in their beds. The bugs were likely to enter the home through openings resulting from shifts in the foundation, gaps around the doors, through windows, or from under the house. They were sometimes brought in on clothing or bedding that had been hanging outside and was later carried into the house. In addition to being found in beds, the bugs were found in bedroom closets, especially in clothing pockets, and in shoes, under rugs, behind draperies and pictures, and crawling around crevices and baseboards. Some were seen on walls or ceilings. Some of the victims were bitten while sitting on overstuffed furniture, where they then found the bugs under the cushions.

Control of Conenoses

S. F. Wood has kindly furnished the following recommendations for the control of these bugs by homeowners.

I. Structural Insectproofing

1. Destroy all wild-animal harborages, such as ground squirrel dens, wood rat lodges, etc., within 100 yards [90 meters] of the premises, to control insect-breeding sites close to man.
2. Check window and door screens, frames, and moldings for tight fit. Use weatherstripping if house flies can enter around the screens.
3. Keep fireplace flue shut when not using fireplace, especially in summer.
4. Check flooring and moldings, including sidings, for possible openings into the subfloor area under the house. Probe openings with stiff wire and fill with calking compound.
5. Seal all foundation cracks, floor-slab breaks, and other openings caused by settling or shifting of the foundation, to prevent bugs entering through these spaces.
6. Insectproof special dog or cat entranceways. Be sure to keep the bedding of pets clean, and check the animals frequently in warm weather for bugs in their fur.

II. Chemical Insectproofing

1. On permanently dry, unweathered surfaces, such as rough inside walls, attics, and subflooring, use insecticide dusts for residual and permanent killing effect.
2. On exposed or weathered surfaces, painted walls, draperies, and home furnishings, use liquid sprays in volatile solvents that are not odorous and will not stain or spot.
3. Your local agricultural inspector or agent, or agricultural, horticultural, or plant supply dealer can advise you on the use of insect sprays and dust formulations. *Read directions on the insecticide label.*
4. For the western bloodsucking conenose, *Triatoma protracta,* use currently registered residual insecticides in powder, solution, or emulsion formulations, depending on the type of exposure, surface texture, or time factor.

III. Household Precautions

To decrease the possibility of conenose bugs getting into the bed, where most people are bitten, the bed should be moved at least 1 foot [30 centimeters] away from walls and other objects, and a search should be made for the bugs in and around the bed before retiring. The legs of the bed should be wrapped with adhesive tape with the sticky side out.

An investigation was made by Sjogren and Ryckman (1966) of the attraction of *Triatoma protracta protracta* to black-light traps. Three such traps were placed near a residence in the chaparral-covered foothills east of Redlands, California, and 398 of the insects were caught in 1 year. Of these, 170 were females and 228 were males. The traps were believed to be potentially inexpensive means of reducing conenose bug annoyance to foothill residents.

Subfamily Reduviinae (Assassin Bugs)

The best-known species in this subfamily are the predaceous assassin bugs in the genus *Reduvius.* They feed on other insects, including *Triatoma,* and may be of some importance in the biological control of the latter and thus also play a role in the epizootiology of *Trypanosoma cruzi,* for which conenose bugs are vectors (Wood, 1954*a;* Ryckman and Ryckman, 1967*a*). Unlike that of the conenose bugs, the beak of *Reduvius* is curved away from the ventral surface of the

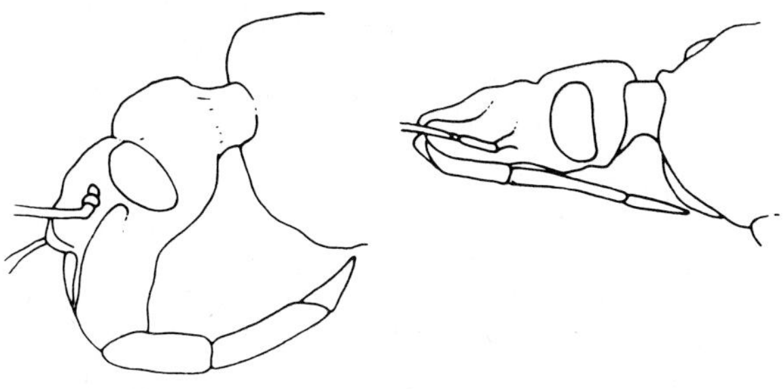

Fig. 300. Profiles of the head of a predaceous bug, *Reduvius sonoraensis* (*left*), and of a bloodsucking conenose, *Triatoma protracta protracta.* (From Ryckman and Ryckman, 1967*a*.)

head (figure 300), as is characteristic of raptorial predaceous bugs that must hold and pierce their prey.

According to Ryckman and Ryckman:

> This curved configuration is well suited for scraping or probing action on the far side of the rounded body of the prey; when a suture or intersegmental membrane is detected by the predator, the stylets are immediately inserted. If the predator's beak were straight, it would be much less effective in the rough-and-tumble battles which occur when the predator grapples with its prey.

Masked Hunter, *Reduvius personatus* (L.)

This species was introduced from Europe before the beginning of the century, and was known at first only in the eastern states. It has since extended its range across the country to Arizona and California. It is somewhat rare in California (Wygodzinsky and Usinger, 1964). In the northern part of the United States, *R. personatus* appears to live in houses and barns, and feeds on other insects, including flies and bed bugs. It is common in the eastern and northwestern states, where it is often found in homes. The adult is about 20 mm long, very active, and is attracted to lights. When it alights on the face or hands, it is often roughly handled, and will then bite readily and painfully. Assassin bugs and conenose bugs should be brushed away gently.

The nymphs are covered with a sticky substance, and camouflage themselves with dust, lint, and other small particles of debris; this has given rise to the common names, "masked hunter" or "masked bed bug hunter." The dust and debris are thrown onto the body by means of the hind legs, and may eventually amount to a fourth of the body weight of the insect (Hartnack, 1939).

In the dry climates of the West, this species is not confined to houses and barns. It has apparently been able to re-enter its primitive habitat and become a free-living insect (Wygodzinsky and Usinger, 1964).

Besides this, the many new bridges and overcrossings built along interstate highways have produced a series of small ecological niches across the country where no rain can reach the earth, and where *Reduvius personatus,* as well as many other arthropods, can thrive. It has also been found in a wood rat nest in Arizona (Ryckman and Ryckman, 1967*a*).

Tan Assassin Bug, *Reduvius senilis* Van Duzee

This species occurs in the hot Sonoran deserts of the southwestern United States and of Mexico. Ryckman and Ryckman (1967*a*) never found *R. senilis* anywhere but in wood rat nests. The nests provided both protection and an abundant food supply. One or more species of Triatominae were usually also present in the nests, and the latter were readily attacked and fed upon by the assassin bugs. In one encounter described by the Ryckmans, a fourth-instar nymph of *Reduvius senilis* weighing 2.2 mg attacked a fifth-instar nymph of *Triatoma protracta* weighing 141.3 mg, the prey being 64.2 times heavier than the predator. After feeding on the *Triatoma,* the predator weighed 7.3 mg, an increase of 232%. Before feeding on other insects, *R. senilis* injects a powerful proteolytic enzyme that dissolves (lyses) all the internal tissues and leaves only the exoskeleton undisturbed. [See also, Wood (1954*a, b*).] If a starved *Reduvius* is released in a colony of *Triatoma* filled with mammalian blood, the predator will frequently attack and kill more *Triatoma* than it can suck dry. The proteolytic enzyme injected into the *Triatoma* lyses the red blood cells, and there is a release and dispersal of red hemoglobin pigment throughout the bodies of the victims, and the appearance of "red bugs" in the colony.

Four females laid a total of 338 eggs, or 84.5 per female. There were 5 instars, and great variability in the length of each stadium, particularly the fifth, which was by far the longest. As a result of this great variability, some nymphs hatching from eggs during a given week in late summer did not reach the adult stage until the first, second, or third summer thereafter (Ryckman and Ryckman, 1967*a*).

Reduvius sonoraensis Usinger

This species has been collected in California and Arizona, as well as in Sonora and Baja California, along the coast of the Gulf of California. The adults may be micropterous (small-winged) or macropterous (large-winged). As with the preceding species, the period required to reach the adult stage varied greatly, ranging from 157 to 564 days (mean, 415.8). As in the case of *R. senilis, R. sonoraensis* can kill and devour large numbers of *Triatoma* (Ryckman and Ryckman, 1967*a*).

Reduvius vanduzeei Wygodzinsky and Usinger

This species inhabits the cooler coastal region of southern California and the higher interior mountains of California, Arizona, and Utah. It is brownish yellow and 13 mm long.

Subfamily Piratinae (Corsairs)

The Piratinae are ground-dwellers, living under rocks and logs and feeding on insects. The adults are attracted to lights, and sometimes get into houses. The 2 genera of interest in the present context are *Melanolestes* and *Rasahus.* They ordinarily feed on insects, but if handled by a human, they can inflict a painful bite.

Black Corsair, *Melanolestes picipes* (Herrich-Schaeffer)

This species is widely distributed in the United States. There are 2 color forms. One is entirely black, and in the other there is a gradation of color, from those having only a slight tinge of red along the connexivum (abdominal margin at junction of dorsal and ventral sclerites) to those with the abdomen entirely red. There are macropterous and micropterous forms of both sexes. The nymphs invariably have red abdomens with black markings. This is the only corsair having only 4 nymphal instars (Readio, 1927); the other species have 5. There is 1 generation per year, and the winter is spent as an adult or nymph under stones or in even more protected places, such as hollow stumps. The adults are attracted to lights during warm spring evenings.

Envenomation. A *Melanolestes* bite and subsequent symptoms were described by McClure (1935). When he first picked up the bug, "it squealed loudly by rasping its proboscis, the labium, back and forth in the ventral groove by an up-and-down movement of its head." The bug slipped from his fingers, and when he attempted to pick it up again, it drove its beak into the tip of the fourth finger of his right hand. The pain caused by the bite was "excruciating," and a small, red spot appeared at the site of the bite, and finally a blister appeared which contained no blood but only a bit of brownish flesh. The pain continued unabated during the night and following day, by which time the end of the finger was numb and had become swollen to the second joint. By the third day, the entire finger was swollen, and the swelling extended into the muscles along the side of the hand. By noon of the third day, most of the fever was gone, and the swelling began to subside. By the fifth day, the finger had returned to normal size, but the spot at the site of the bite remained, and the last joint was numb and remained so for a week longer. The brown spot fell away with a scab, but a small dent remained in the end of the finger for nearly a month. Four months later, a small bit of scar tissue was all that remained.

Twospotted Corsair, *Rasahus thoracicus* Stål

This is probably the most common species of the subfamily Piratinae throughout its range in California, Arizona, and Mexico. It is 20 mm long, and its body is relatively slender, mostly yellowish tan with black wing membranes, and with a characteristic, circular, yellowish-tan spot in the center of the membrane of each forewing which provides a very good identifying character (plate VIII, 5). In common with *Melanolestes picipes, R. thoracicus* makes a loud, squeaking noise when captured, and may bite, but only in self-defense. Its bite is very painful. It is found most abundantly on rocky hillsides covered with weeds or brush, and may be collected under rocks. It is predaceous on a wide variety of insects. It flies to lights at night, and may then enter homes, or it can be carried in on plants.

Rasahus biguttatus (Say)

This is also a "twospotted corsair," being similar in appearance to *R. thoracicus,* including the 2 pale, circular spots on the wings. It occurs in much of the United States and in Cuba and South America. Readio (1927) studied its life history in Kansas, finding that it spent the winter under rocks as a nymph in its fifth, or less often in its fourth, instar. Adults were taken in Kansas as early as May 29 and as late as July 15. Both nymphs and adults were very active, and Readio believed that on a warm day it would be impossible to pick one up with the fingers without being bitten. As with *R. thoracicus,* its bites are painful, but are inflicted only in self-defense. It is likewise attracted to lights at night, and occasionally enters homes. It has been found between the mattresses of a bed infested with bed bugs, but they were not regarded as its normal prey (Walsh and Riley, 1869).

Subfamily Harpactorinae

This is the largest reduviid subfamily. Its members may be found on leaves, twigs, and tree trunks, in grass, and on flowers. They are predators on phytophagous insects, and the subfamily includes the important predators of the genus *Zelus* (nonvenomous). The most important venomous species is *Arilus cristatus*, which is also a predator on plant pests.

Wheel Bug, *Arilus cristatus* (L.)

This is a large, gray bug, with a high, semicircular ridge or projection on the crest of its thorax that bears 8 to 12 uniformly arranged teeth or coglike protuberances, resembling very short spokes or cogs of a wheel (figure 301), hence the

Fig. 301. Wheel bug, *Arilus cristatus*. (Courtesy Chemagro Corporation.)

common name. The knobs of the head, the cogwheel crest on the prothorax, and the eyes are black, and the legs and antennae are tinged with chestnut reddish (Uhler, 1884; Readio, 1927). The females are often more than 30 mm long, and the males are slightly smaller. When captured, the females sometimes emit a forked, orange-colored, glandular organ from the hind end of the body, having a pungent scent that is probably protective.

The wheel bug is common in the eastern and southern United States, and ranges westward into New Mexico. It lays its bottleshaped, brown eggs in the early fall in masses containing an average of 128 (Girault, 1906), gluing them to twigs, the bark of trees, fence rails, or under the eaves of buildings (Readio, 1927). A "gummy cement" protects the eggs throughout the winter, so that they can hatch when warm weather returns (Uhler, 1884). The eggs and 5 nymphal instars were described by Readio. The nymphs have many red, reddish-yellow, and black markings, approaching somewhat the somber color of the adult in the fifth instar. There is 1 generation per year.

In the South, these bugs can be collected in large numbers as they prey on various insects that are attracted to mercury vapor lamps, such as those used in parking lots (R. E. Wagner and D. A. Reierson, personal communication).

Envenomation. The bite of a wheel bug has been variously described, as follows:

> A sharp pain and subsequent inflammation and swelling, the result being somewhat like that of a bee sting [H. Garman, quoted by Readio (1927).]
>
> Much more painful than the sting of a wasp or hornet. One of these insects having stung the writer, the pain lasted for several hours, and was only alleviated by the application of ammonia. Several days afterward, the flesh immediately surrounding the puncture was so much poisoned that it sloughed off, leaving a small hole in the thumb injured. [T. Glover, quoted by Readio (1927).]
>
> The wound was at once very painful and remained so for 10 days, at first appearing red, and then the portion adjoining the wound became hardened and quite white. [Barber (1919).]
>
> The finger became reddened and felt hot to the touch. In the course of a few days, growths resembling papillomas developed at the sites of the punctures, the largest of these projecting as a small, hornlike structure. Both of these growths persisted for months, the largest slowly disappearing between 6 and 9 months after the infliction of the bite. The injured finger remained warmer than the other fingers during this period and, according to the patient's statement, still felt warmer than the other fingers a year later. The development of pronounced cutaneous growths after a bite appeared indicative of the action of some toxin as a stimulant irritant. [Hall (1924).]

FLEAS

Of the approximately 1,600 described species of fleas in the world, 95% occur on mammals and 5% on birds. Most species attacking man in the United States have been placed in the family Pulicidae (order Siphonaptera). Because of their abundance, world-wide distribution, irritating bites, and ability to transmit disease, fleas are among the principal medically important groups of arthropods.

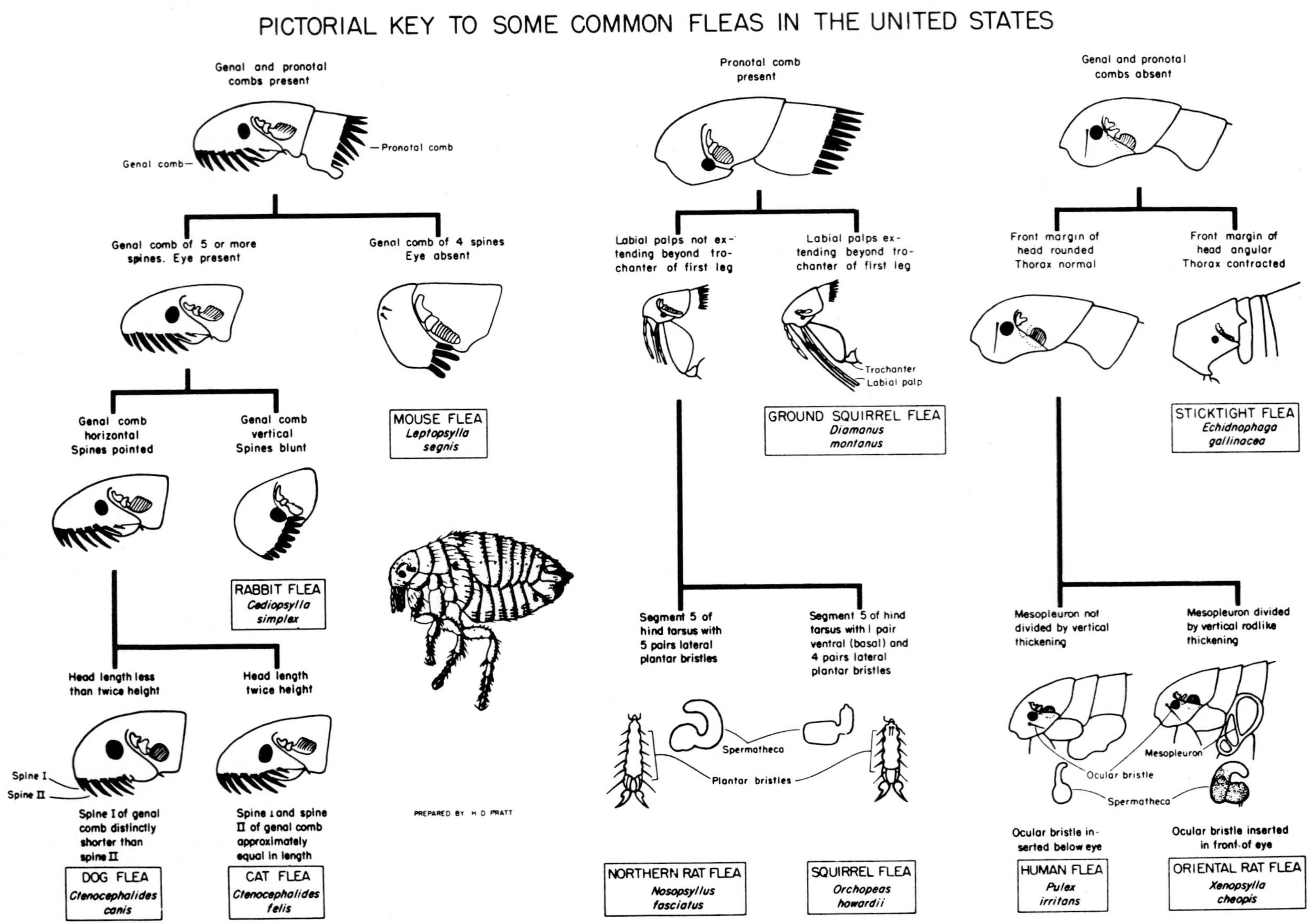

Fig. 302. Pictorial key to help identify common fleas. (From Pratt and Wiseman, 1962.)

General Morphology of Adults. Fleas are small, black to brownish-black insects, with complete metamorphosis. They are wingless, having lost their wings in the course of their evolutionary development. Adult fleas are 1 to 4 mm long, depending on the species, and are laterally compressed. They have stout, spiny legs, adapted for leaping, and short, 3-jointed, clubbed antennae that fit into depressions along the sides of the head. Fleas have piercing–sucking mouthparts (figure 46, *C,* chapter 4), the principal elements of which are the grooved labrum and a pair of sharp, swordshaped mandibles. The concave inner sides of the mandibles, together with the labrum, form the sucking channel (Ewing and Fox, 1943).

In some species, a conspicuous transverse row of spines is located just above the mouthparts, and is called the *genal* comb or ctenidium. In others, the pronotum bears a transverse row of heavy spines called the *pronotal* comb. When flea specimens are "cleared" in lactic acid for microscopic examination, the combs show up prominently, and can be used to identify species with the aid of low magnification. Figure 302 shows that the cat, dog, mouse, and rabbit fleas have both genal and pronotal combs. Certain species that infest rodents have the pronotal but not the genal comb, and the human, oriental rat, and sticktight fleas lack both combs. It has long been assumed that the combs or ctenidia assist fleas in locomotion through fur, hair, and feathers, but this is apparently not the case; locomotion is facilitated by the strong setae on the legs, projecting almost at right angles. The combs serve to make dislodgement or capture by the host more difficult. They are admirably adapted to resist a backward pull, and so must have a considerable survival value (Humphries, 1966).

The Flea Life Cycle. A female flea will lay 4 to 8 eggs after each blood meal, and can usually lay several hundred eggs during her adult life. The smooth, oval or rounded, light-colored eggs (figure 303, *C*), about 0.5 mm long, are deposited on, but not firmly attached to, the body, the bedding, or the nest of the host. Although they are a little sticky, those laid on the host's body may fall or be brushed off. This accounts for their being found in crevices in the floor, under the edges of carpeting, in sofas, or in cat or dog boxes, kennels, etc., where they usually hatch in 10 days or less.

The small, hairy, wormlike larvae (figure 303, *B)* are whitish, with a distinct, brownish head, and do not have eyes or legs. They have 3 thoracic and 10 abdominal segments, with a single row of bristles around each segment. The larvae are about 1.5 mm long when they emerge, finally becoming about 5 mm long. They move forward by using their backward-projecting bristles and a pair of hooked, chitinous processes located at the end of the abdomen by which they can obtain purchase on a surface. When disturbed, they may "flip" in circles to escape. The larvae of some species require dried blood for food, but others do not. Those that do not need blood can feed on the many kinds of organic debris that are present in the crevices in which the eggs commonly hatch. They also feed on their own cast skins. They require high humidity.

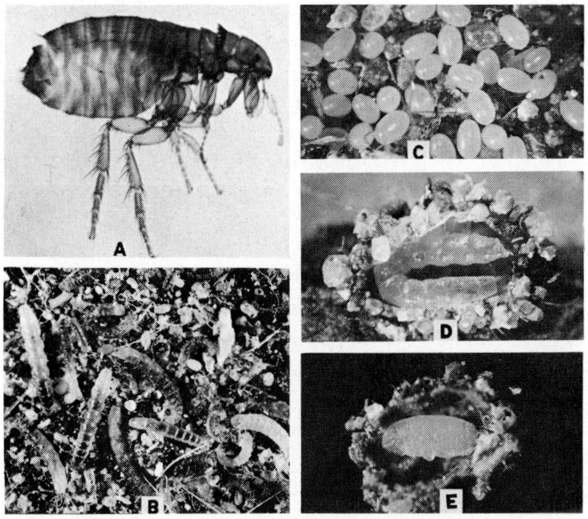

Fig. 303. Cat flea, *Ctenocephalides felis. A,* adult; *B,* larvae; *C,* eggs; *D,* larva in cocoon; *E,* pupa in cocoon.

Within a week to several months, the larvae grow through 3 instars, then spin a pupal cocoon (figure 303, *D*) covered with grains of sand, dust, or organic debris, in which they are quite effectively camouflaged in their natural surroundings. The pupae are initially white (figure 303, *E*), but they change to brownish before the adults emerge. In the pupa, the appendages are not closely pressed to the body, and it has the general shape and characteristics of the adult. Under favorable conditions, the adult emerges in a week or two, but under adverse conditions, the pupal period may be prolonged to as much as a year. The adult may remain in the cocoon for a long time until vibrations indicating the presence of a possible host stimulate it to emerge and become active.

The potentially long pupal stage, besides the fact that adult fleas can live without food for remarkably long periods, accounts for the fact that people may enter a house after it has been unoccupied by humans or pets for months, yet be rapidly and severely attacked by fleas. Depending on the species and weather conditions, 2 or 3 weeks to as many months, and rarely as long as 2 years, are required for many species of fleas to complete a life cycle.

Reaction to Flea Bites

Some people react severely to flea bites, while others hardly notice them. The famous entomologist Karl von Frisch, while on one of his trips to the Naples Zoological Station, shared a room with a friend because that particular room gave them "a marvelous view over the blue Gulf to the lovely contours of Capri." When they found that it was flea-infested, before going to bed they walked barefoot back and forth in the room, and then took off their "nightshirts" and removed the fleas. Von Frisch averaged 4 or 5 per night, and his friend, 30 or 40. However, the friend was annoyed only by their crawling, whereas von Frisch got a big blister from every bite. He pointed out that it would take more than 1.5 million fleas to produce a full-sized drop of the tiny amount of saliva injected in 1 bite, but that this would be enough "to set all the people in Hamburg scratching" (von Frisch, 1960). (See comment in the next paragraph.)

Fleas most often bite people about the legs and ankles, and there are usually 2 or 3 bites in a row. The bites are felt immediately, but tend to become increasingly irritating, and are frequently sore for as much as a week. Flea bites may vary in their effects, from a transient wheal to prolonged symptoms that last for years, depending on the sensitivity of the victim. Children less than 10 years old were found to be more sensitive to the bites than older persons (Hudson *et al.*, 1960*b*). The insects inject a hemorrhagic saliva that can cause severe itching, and repeated bites may produce a generalized rash. A small, red spot usually appears where the flea's mandibles have penetrated the skin. The spot is surrounded by a red halo, but there is never much swelling. Some relief from the itching can be obtained by applying cooling preparations, such as carbolated vaseline, menthol, camphor, or calamine solution, or by placing a piece of ice on the affected area. For people who suffer severe allergic reactions, antigens are available that are sufficiently effective to cause the immunized person to be subsequently unaware of flea bites.

Disease Transmission

The most serious indictment against fleas is their ability to transmit disease organisms. This ability is enhanced by their promiscuous feeding habits; they move from one host species to another. For example, the cat flea, *Ctenocephalides felis,* readily attacks humans, dogs, rats, opossums, raccoons, and foxes. The human flea, *Pulex irritans,* can be found on dogs, rats, pigs, mice, skunks, badgers, deer, foxes, coyotes, prairie dogs, ground squirrels, and burrowing owls. The species normally found on rats and ground squirrels also bite man.

Plague

Probably the most dreadful calamity in all human history was the "Black Death," a series of plagues of the Middle Ages, particularly in the fourteenth century. Although not known at the time, the highly contagious disease was caused by a bacillus, *Yersinia pestis* (=*Pasteurella*), carried primarily by rats and transmitted by the fleas they harbored. The disease caused swellings of the lymph glands and subcutaneous hemorrhages that turned black. The plague first struck in Asia, and then spread to Russia, Persia, Turkey, North Africa, and Europe. The *Encyclopedia Britannica* estimates that one-third of the world's people perished. The most dangerous forms of plague are bubonic, septicemic, and pneumonic. Two types are named for the parts of the body infected. Bubonic plague, confined initially to lymphatic glands (buboes), is not so dangerous as the septicemic, and especially the pneumonic, phases, which are rapidly spread from person to person by airborne infection (Dubos and Hirsch, 1965). Bubonic plague is no longer a major scourge of mankind, but only because we now know how it is transmitted, and therefore how to prevent it. War and dislocation can still increase its potential, but not as in past centuries. The World Health Organization reported 1,318 cases of plague in 1968 and 864 in 1969, the majority from Vietnam. Treatment of the disease is quite effective with modern antibiotics. However, it can sometimes be dangerous, even with early diagnosis and treatment.

The oriental rat flea, *Xenopsylla cheopis,* is the most important vector of the plague bacillus, *Yersinia pestis,* from rat to man, but at least 29 other species of fleas can transmit the disease, including the northern rat flea, *Nosopsyllus fasciatus;* the mouse flea, *Leptopsylla segnis;* the dog flea, *Ctenocephalides canis;* the cat flea, *C. felis;*

and the human flea, *Pulex irritans.* Plague is transmitted while the infected flea is feeding, by regurgitation of the bacillus from the flea's alimentary tract through the proboscis into the new host. *Yersinia pestis* is rapidly eliminated from the proventriculus of the oriental rat flea when the mean monthly ambient temperature exceeds 28° C (82° F), and plague epidemics decline with the advent of hot weather (Cavanaugh, 1971).

"Sylvatic" or "endemic" is the designation used for the plague in wild rodents. Under certain ecological conditions in vacated squirrel burrows, fleas may continue to harbor the plague bacilli for many months, providing a reservoir for the infection. Fleas are particularly suitable for this purpose because of their remarkable longevity, for they can survive more than 6 months, even without food. Sylvatic plague occurs in many parts of the world, remaining somewhat localized in each area. A group of rodents maintains the infection, and if domiciliary rats or mice come in contact with such a focus of infection, the disease may be carried to human habitations and human cases may result. Likewise, if humans enter sylvatic plague territory, they may become infected.

Ground squirrels and chipmunks have been shown to be reservoirs for the plague bacillus in California. In the case of sylvatic plague, a person is more likely to become infected by contact with sick or dead rodents than by flea transmission. Since 1925, all the known cases of plague in the United States have been associated with wild rodents and the fleas that infested them. The United States Public Health Service reported 36 cases of human plague during the 6-year period 1965–1970. Twelve of these cases were reported in 1970. Eight cases have been reported in California since 1940. Areas in California where rodents are periodically killed by plague extend from the northeastern part of the state through the Sierra Nevada to the Tehachapis, and also in the coastal areas between San Francisco and Ventura County. People are warned not to handle sick or dead animals. Sometimes, a sudden widespread increase (epizootic) of a disease will occur among such rodents, and the chance for human infection will increase greatly. During epizootics, control has been attempted by placing bait boxes containing 10% DDT and peanut butter at 500-ft (150-m) intervals in the infested area. The rodents that enter a box to feed pick up not only enough insecticide to kill the fleas on them, but also the fleas living in their nests or burrows. (Anonymous, 1970*a*).

Murine Typhus

The oriental rat flea is also the principal vector of *murine* or *endemic* typhus that may be transmitted from rats to man. This disease is caused by the microorganism *Rickettsia mooseri,* and is widespread in rats and other rodents. The principal mode of transmission is believed to be by rubbing or scratching infected flea feces into wounds made by their bites. Murine typhus should not be confused with the more serious *epidemic* typhus, which has caused the deaths of millions of persons, but is transmitted by lice. The fatality rate for murine typhus, on the other hand, is only about 2%. In the United States, the disease is often confused with tickborne Rocky Mountain spotted fever, both diseases being characterized by rash and fever. However, Rocky Mountain spotted fever usually occurs during the spring and early summer, while murine typhus occurs in late summer and fall. Murine typhus cases in the United States dropped sharply from 5,193 cases in 1945, through a combination of rat control and insecticide dust for fleas. Also, new broad-spectrum antibiotics have been successful in the treatment of clinical cases of the disease (Pratt and Wiseman, 1962).

In California, murine typhus has remained endemic only in the 5 southern counties since 1919, when the first cases were recognized. In the 30 years from 1920 to 1949, there were 487 cases and 21 deaths from the disease. In the 20 years from 1950 through 1969, there were 79 cases and no deaths. Since 1960, there have been about 1.5 cases per year (CPHS, 1971).

Cats as Transmitters. About 75% of the 28 to 52 cases of murine typhus per year that occurred in the United States since 1945 were reported from Texas, particularly in 3 southern counties in which the disease was epidemic, and in some instances under circumstances in which the usual method of spread from rat to rat flea to man did not seem to occur. Prompted by this fact, the possible role of the domestic cat in the transmission of murine typhus to man was investigated. *Rickettsia mooseri* had been found in cat fleas, *Ctenocephalides felis,* in nature (Irons *et al.,* 1944), and it was assumed that the cat fleas obtained the rickettsiae from their normal source, the rat. There is much circumstantial evidence that most of the murine typhus in southern Texas is transmitted via flea-infested cats or from flea-infested grass. (In southern Texas, grass is often so heavily infested with cat fleas that they are locally known as "grass fleas.") The greatest incidence of murine typhus in southern Texas has been noted to be in women and children, who spent more time at home where most of the illnesses were contracted (Older, 1970). The investigation in southern Texas confirmed the conclusion arrived at by Irons *et al.* (1944) that at least 5 cases of murine typhus in Austin in 1942, including 1 laboratory infection, were probably acquired through the agency of cat fleas harbored by kittens carried away from a feed store. In endemic areas, feed stores were said to be particularly serious foci of typhus.

Likewise in southern California, where 34 separate outbreaks of murine typhus were recorded from 1952 to 1969, 21 were in families that owned cats or had frequent contact with them. Again, as in southern Texas, most cases occurred among housewives and children. Seven outbreaks were intimately associated with either dogs, cats, opossums, or skunks, and 2 with rats. With greatly improved rat and flea control, there has been a shift in the incidence of murine typhus in man from central Los Angeles to foothill and orchard areas in eastern Los Angeles and Orange counties. Opossums in such areas are heavily infested with fleas, and are suspected of being responsible for some of the sporadic cases of typhus in man (Adams *et al.,* 1970).

Tapeworms

Fleas, particularly dog and cat fleas, can serve as intermediate hosts for the dog tapeworm, *Dipylidium caninum* (L.), and the rodent tapeworm, *Hymenolepis diminuta* (Rudolphi), which occasionally attack humans, particularly very young children.

Cat Flea, *Ctenocephalides felis* (Bouché), and Dog Flea, *C. canis* (Curtis) (Pulicidae)

These 2 fleas, particularly the former, are the most common species found in and around homes. They are most abundant in summer, and seem to be especially numerous when homes are reoccupied after the residents have been gone for a few weeks. In California, these fleas are much more abundant in the relatively humid coastal areas than in the more arid ones farther inland, and are particularly numerous during years of greatest rainfall and humidity. The cat and dog fleas are so similar in appearance and biology that for practical purposes they can be described under the same heading. A few minor differences are noted in the following text.

Description. The female (figure 303, *A*) is 2.5 mm long, and the male is slightly smaller. The head of the female cat flea is twice as long as high when seen from the side, while that of the female dog flea is less than twice as long as high. In both species, the genal comb consists of 8 pairs of spines, and the pronotal comb also consists of 8 pairs (figure 302). With the aid of low magnification, it can be determined that the first 2 anterior spines of the genal comb of the cat flea are about equal in length, while the first spine is distinctly shorter in the dog flea (Ewing and Fox, 1943; Pratt and Wiseman, 1962).

The larvae of both species (figure 303, *B*) are nearly twice as long as the adults. They feed on particles of dry blood, excrement, and various organic substances collected in corners and crevices in the infested premises. When infestations are very heavy, the accumulations of grayish larvae and white eggs give the sleeping quarters of cats and dogs a "salt-and-pepper" appearance that easily identifies the infestation.

How Infestations Are Spread. Particularly in warm, humid areas, the exclusion of cats and dogs, or their proper management, is necessary to prevent flea infestations. Residential premises are sometimes so heavily infested that cats will leave to seek relief, and will carry the infestation to an uninfested home. If an infested stray cat has kittens under or near an uninfested home, the appearance of fleas within the house is very likely to follow. Fleas often also become established in the lawns of heavily infested residences. Fleas can jump vertically for 5 or 6 inches (about 15 cm), and can attach themselves to skin or clothing. Some pest control operators avoid being bitten by spraying insecticide onto their clothing.

In addition to cats, dogs, and humans, many animals, including the opossum, are attacked by cat fleas and dog fleas. Indigenous to the southeastern United States, opossums are now found throughout the country. They were introduced into California when some of the animals kept as pets escaped in San Jose in 1910, and they are now abundant throughout the state. When people live near wooded areas or stream courses, or where trees and shrubs are abundant, they may have opossums in the attics, wall voids, crawl spaces, or basements of their homes. These animals can then be sources of infestation.

Human Flea, *Pulex irritans* L. (Pulicidae)

The human flea is found all over the world. Besides man, it infests cats, dogs, and many other domestic animals, particularly the pig. It breeds in profusion in pigsties, and people working in them can readily pick up large numbers of fleas and start infestations in their homes. The human flea is usually the most important species in farm areas. While the bites of the cat flea tend to be concentrated on the lower parts of the legs, those of the human flea may be generally distributed over the body (Keh and Barnes, 1961).

Description. The human flea lacks the genal and pronotal combs characteristic of the cat flea and dog flea, and may be distinguished from the oriental rat flea by having its ocular bristle inserted beneath the eye instead of in front of it, as it is in the latter species (figure 302).

Oriental Rat Flea, *Xenopsylla cheopis* (Rothschild) (Pulicidae)

Along with its favored host, the rat, this species (figure 302) is cosmopolitan. It is one of the most abundant fleas in the southern states and southern California, and is the most important vector of bubonic plague and murine typhus. Fortunately, it is not so commonly found in human habitations as are some other fleas, but it has been reported to bite people in buildings. It also infests cottontail rabbits and ground squirrels. High relative humidity (70% or greater) and temperatures of 65 to 80° F (18 to 27° C) are favorable conditions for hatching of rat flea eggs. Although a temperature of 40° F (4° C) was reported to be fatal to the oriental rat flea, the temperature would seldom be that low in the warm nests of rats. The larval period may be 12 to 84 days, the pupal period in the cocoon, 7 to 182 days, and adults may live for 100 days (Hubbard, 1947).

Rabbit Flea, *Cediopsylla simplex* (Baker) (Pulicidae)

This species can be distinguished from the cat and dog fleas by the almost vertical position of its genal comb and the rounded ends of the spines, compared with the horizontal position and sharp points found in the latter (figure 302). This is an eastern species, and is known to bite hunters and hikers. Similar species occur in the western region.

Sticktight Flea, *Echidnophaga gallinacea* (Westwood) (Pulicidae)

This is one of the smallest fleas. The females are 1 to 1.5 mm long, and the males are less than 1 mm. Sticktight fleas are most abundant in the southern states. They infest wild birds, some of which, such as Brewer's blackbird, house sparrow, and bobwhite quail, may be expected to be effective dispersal agents (Stewart, 1932). Nevertheless, sticktight fleas are said to be rare in the Northwest, where these birds are plentiful. Although sticktight fleas were formerly sometimes considered to be severe pests in California, they are now rarely seen, apparently having been decimated by the more effective insecticides used in the treatment of poultry in recent decades (R. N. Hawthorne, personal communication). The adult flea attaches itself firmly to the head and neck of domestic fowls, often causing ulcers. The female oviposits in the ulcers, and the larvae drop to the ground and feed on organic matter. Sticktight fleas (figure 302) can become abundant in poultry yards and adjacent buildings. Besides fowls, they will attack man, cats, dogs, rats, rabbits, ground squirrels, horses, and many other mammals. They are potential transmitters of plague and murine typhus, but the habit of the females of remaining fastened to a single host most of their lives greatly decreases their importance in this respect.

The western hen flea, *Ceratophyllus niger* Fox (Ceratophyllidae), is larger than the sticktight flea, and attaches itself only briefly while feeding. It breeds primarily in fowl droppings. Besides fowls, it readily attacks humans, cats, and dogs.

Chigoe, *Tunga penetrans* (L.) (Tungidae)

The chigoe flea, also known as "jigger," "chigger," "chigu," or "sand flea," is said to have inspired the sailor's oath, "I'll be jiggered!" This flea should not be confused with the true "chigger," which is a species of mite (*Trombicula alfreddugesi*). In fact, the chigoe flea is not definitely known to have become established in the United States. It is a tiny, burrowing flea, found in the tropical and subtropical regions of North and South America, the West Indies, and Africa. It is reddish brown, and about 1 mm long, although the gravid female may become as large as a "small pea." The fertilized female slashes the skin with her mouthparts, then inserts her head and body until only the last 2 abdominal segments are exposed. There she feeds, and continues to oviposit. The eggs hatch in 3 or 4 days, and the entire life cycle requires about 17 days (Hubbard, 1947). The chigoe usually attacks humans between the toes or under the toenails, becomes engorged with blood, causes great pain, and may produce inflammation and localized ulcers. Tetanus and gangrene may result from secondary infections. Autoamputation of toes has been recorded. People should avoid walking in their bare feet where these fleas are known to occur.

Ground Squirrel Flea, *Diamanus montanus* (Baker) (Ceratophyllidae)

This flea is dark brown, medium-sized, and has very long labial palpi (figure 302). It is found on ground squirrels from Nebraska and Texas to the Pacific Coast. When 132 California ground squirrels (*Citellus beecheyi*) were combed, 4,371 *D. montanus* and 226 specimens of another flea, *Hoplopsyllus anomalus* (Baker), were obtained (McCoy, 1909). Both species are vectors of plague. In laboratory tests, *H. anomalus* proved to be about half as efficient as the northern rat flea (described next) as a plague vector (Hubbard, 1947).

Northern Rat Flea, *Nosopsyllus fasciatus* (Bosc) (Ceratophyllidae)

This species is found on domestic rats and house mice throughout North America and Europe, but is most common in temperate regions where plague is not a severe problem. It may be important in transmitting plague organisms among rats, and has occasionally been taken from wild rodents. In experimental tests, *N. fasciatus* transmitted endemic typhus from rat to rat (Dyer *et al.*, 1932). This flea and *Xenopsylla cheopis* are known to be intermediate hosts for *Hymenolepis diminuta*, the common tapeworm of rats and mice, which is rarely found in man. *Nosopsyllus fasciatus* and *Diamanus montanus* both lack the genal comb. The labial palps of the latter extend beyond the trochanter of the first leg, while those of *N. fasciatus* do not (figure 320). The pronotal comb has 18 to 20 spines.

Squirrel Flea, *Orchopeas howardii* (Baker) (Ceratophyllidae)

The squirrel flea is found throughout the United States wherever gray squirrels (*Sciurus griseus*) occur. It is frequently taken from the red squirrel and flying squirrel when the ranges of these hosts overlap that of the gray squirrel. If a gray squirrel builds its nest in an attic and is later killed or excluded, the fleas that breed in the nest material may escape and invade other parts of the house. Attic nests of these squirrels should be removed and insecticides should be applied to their sites (Pratt and Wiseman, 1962). The female's head is well rounded, while the male's is flattened on top. The genal row of 3 bristles is at a level with the top of the eye, and the middle bristle is about half as long as the outer one (Hubbard, 1947). (See also, figure 302.)

Mouse Flea, *Leptopsylla segnis* (Schönherr) (Leptopsyllidae)

The mouse flea can be distinguished from other common fleas having both genal and pronotal combs by its genal comb having only 4 spines (figure 302). It is most commonly found on domestic rats, and less often on the house mouse, *Mus musculus*. This flea is common in Europe, and is most abundant in the United States along the east and west coasts near the original ports of entry. It is scarce in inland areas and during the summer, for it thrives best in cool weather. Although it can be infected with plague in the laboratory and has been found to be naturally infected with murine typhus in China, it is considered to be a poor vector of those diseases (Pratt and Wiseman, 1962).

Flea Bites on the Beach

Fleas may occur on beaches, particularly if there are near-by residences. People using beaches may be bitten by these fleas or by stable flies (*Stomoxys calcitrans*), which can often be found in piles of seaweed. Small amphipod crustaceans (*Orchestia*), known as "sand fleas," also found under beached seaweed, are often blamed for bites, but they are in fact completely harmless.

Control of Fleas

Successful flea control must include not only the treatment of infested animals, but also thorough treatment of the entire infested premises. Pets can, of course, be sent to the veterinarian for treatment. In that case, they should not be allowed to return until the fleas have been controlled in the infested source area.

Even if both pets and premises are free of fleas, the pets can become reinfested if they are allowed to roam about in infested areas and associate with infested animals in the neighborhood. Other animals from outside sources that enter the treated premises can also cause reinfestation.

Treatment of Pets. Dusts are safer than sprays to use for treating pets, because no solvent is present to carry the toxicant into the skin. The safest insecticides are the botanicals (pyrethrum or rotenone). Some pyrethrum dusts may merely paralyze the fleas, allowing them to recover later. To prevent recovery and reinfestation, paralyzed fleas should be combed out of the pet's fur into a piece of newspaper and burned. Several brands of 1% and even 2% pyrethrins, plus 10% piperonyl butoxide (a synergist), and substantial quantities of base oil, incorporated into silica aerogels, are readily available. These preparations have extremely high insecticidal potency, and good results have been reported when they were thoroughly applied.

Another relatively safe insecticide is malathion; 3% malathion dust was used on a female cat and 3 successive litters of kittens with no apparent harm to the animals. Each treatment controlled the fleas for 7 to 10 days (Pratt and Wiseman, 1962). Insecticide dusts should be applied with a shaker, or be rubbed into the fur by hand, keeping the dust out of the animal's eyes, nostrils, and mouth. Treatment should be particularly thorough around the ears and beneath the forelegs. As much as 1 oz (30 cc) of dust is required for a large dog. Particularly when the dust contains pyrethrum, the fleas move about actively for a while, causing the animal some discomfort.

The following insecticides have been suggested for flea control on pets: (1) carbaryl dip or wash 0.5%, or dust 2 to 5%; (2) coumaphos dip 0.2 to 0.5%, spray 1%, or dust 0.5% (the coumaphos dust also contains 1% trichlorfon); (3) lindane dust 1% (neither lindane nor coumaphos should be used on cats or on dogs under 2 months old); (4) malathion dip 0.25%, spray 0.5%, or dust 4 to 5%; (5) pyrethrum spray 0.2% plus 2% synergist, or dust 1%; and (6) rotenone dust 1% (Anonymous, 1970*a*). Ronnel as a 0.25% solution can also be used to dip and sponge dogs and cats. Mallis (1969) warned against the use of rotenone on canaries, for it might cause them to contract pneumonia. He suggested a powder containing pyrethrum only. Pigs are also known to be particularly susceptible to rotenone (Anonymous, 1970*a*).

Dips or washes should be used under the supervision of a veterinarian. He can also treat dogs and cats by oral administration of 25 mg of ronnel per lb (0.45 kg) of body weight of the animal every other day, or trichlorfon tablets at 34 mg per lb of body weight twice weekly.

Flea Collars. Registration has been obtained for dichlorvos-impregnated flea collars for dogs (except whippets and greyhounds), and cats (except Persian), and the collars have been effective (CDC, 1972). Some animals suffer an adverse reaction, initially indicated by a reddening of the skin under the collar. The skin under the collar should be examined frequently, particularly when the collar is being used for the first time. For dogs, the collar should be loosely fastened, to allow for good separation between collar and skin. For cats, it must be fitted tightly, to avoid possible strangulation when the animal is climbing. To prevent deterioration of the insecticide, the collar should not be allowed to become wet. If the skin under the collar does become irritated, it sometimes takes weeks or even months to heal. In most cases, simply removing the collar and topically applying a medicated ointment to the affected area will be all that is needed. Occasionally, a person may be so hypersensitive to dichlorvos that the handling of the flea collar will cause dermatitis in the contacted areas of skin (Anonymous, 1970*h*).

Control of fleas with no risk of dermatitis and with no other clinical effects appeared to be possible when disks of 20% dichlorvos in plastic containers with ventilation holes were hung on dog collars. These plastic containers resulted in complete control of fleas within 120 hours on all dogs tested. Ticks were not killed, but dogs initially without them remained free for the 92-day period of the test (Quick, 1971).

Treatment of Premises. Fleas are most concentrated in and around kennels or other places where animals sleep, and in rugs or under porches where they rest. Infested animal bedding should be burned, or else laundered in hot, soapy water. It should not be left unlaundered for long periods while the animals are away, such as during vacations.

Sometimes accumulations of lint and dust contain flea eggs, larvae, and pupae that can be removed with a vacuum cleaner. The cleaned area can then be sprayed with 0.5% lindane or, if the fleas happen to be resistant to organochlorine insecticides, 2% malathion or ronnel or 0.5% diazinon may be used. The relatively new insecticide Gardona®, of very low mammalian toxicity, is said to be effective for flea control. Baseboards, moldings, bed frames, floor areas, and rugs should be lightly sprayed. If oil-base spray contacts asphalt tile, it should be wiped up immediately. The author prefers oil-base sprays, for a very fine mist can be generated and there is less chance of staining, but open flames should be avoided. A fine, light mist with oil-base is particularly desirable for treating upholstered furniture.

When an area is not easily accessible or where it is not practicable to remove rubbish and clutter, a 0.2% spray of dichlorvos may be desirable because of its fuming action. Pest control operators sometimes add a small amount of dichlorvos to the residual spray. Insecticide dusts are suggested for the treatment of inaccessible places where spraying may be difficult or impossible. Silica aerogel dust (Dri-die® 67) is useful for the treatment of attics and wall voids because it can be so easily and uniformly blown throughout any enclosed area. Because it is inorganic, silica aerogel remains effective indefinitely. The homeowner should make the attic, wall voids, and underarea (crawl space) rodentproof before treatment. Also, stray cats, dogs, and other animals should be kept out of the subfloor area, for they may infest it with fleas.

The author has rapidly and effectively treated a home by allowing a fog from a Hi-Fog machine (figure 21, chapter 3) to drift lightly over the entire area inside the house—over carpeting, under beds, sofas, and other furniture, and into closets, using 2% malathion or 2% ronnel. Using a hose-attachment sprayer, the entire yard area was sprayed with 1% diazinon emulsion, with special attention to the lawn. Fleas were found to be particularly abundant around garbage cans. Most modern insecticides are remarkably effective against fleas, and failure to control them can usually be attributed to faulty application.

TICKS

The ticks constitute the superfamily Ixodoidea of the order Acarina, which includes the mites, but their larger size and leathery cuticle distinguish the ticks from the mites. The Acarina can be distinguished from insects by the head, thorax, and abdomen being fused together, by the absence of antennae, and by the nymphs and adults having 4 pairs of legs. However, the larvae have 3 pairs of legs, as do full-grown insects. There are 2 families of ticks: the "soft" or "softbacked" ticks (Argasidae), with about 20 species in the United States, and the "hard" ticks (Ixodidae), with about 55 species. The latter are of greater economic importance.

Description and Biology. Instead of a true head, ticks have a *capitulum* (see figure 305, for hard ticks). This consists of a basal portion, the *basis capituli,* to which the hypostome, the chelicerae, and the 4-segmented palps are attached. The *hypostome* is not only a piercing organ, but also possesses rows of backward-projecting spines that anchor the tick to the skin of its host. The insertion of the hypostome is facilitated by the *chelicerae,* that serve as cutting agents. At the extremity of each chelicera is a pair of *digits,* used to lacerate the skin of the host (figure 305).

Both male and female ticks are bloodsuckers. After engorgement, fertilization takes place. The male then dies, and the female drops to the ground and seeks a sheltered place for oviposition. After the period of several days required for the eggs to develop, she lays her eggs and also dies.

For their first blood meal, the 6-legged tick larvae or "seed ticks" must depend on chance meetings with suitable hosts, so as might be expected, they are able to live for long periods without food. In nature, the tendency of ticks to congregate along animal paths and roads, drawn by the scent left by passing animals, increases their chances of finding hosts. After engorging, they molt and become 8-legged nymphs.

The "soft ticks" may have several nymphal instars, and may have to endure a period of prolonged starvation each time they await a new host, but the "hard ticks" have only 1 nymphal instar. In either case, the mature nymphs may have to endure months of starvation before finding a suitable host upon which to feed, molt, and become adult. The adult leaves the host to lay eggs on the ground. Upon emerging from the eggs, the larvae commonly climb up on grasses or shrubs in order to be more readily contacted by passing animals.

Economic Importance. Few tick species are ever found in the home, but they may infest dogs and are important pests of livestock. Tick bites can cause great discomfort to humans. They remain attached for long periods of feeding, and therefore have the opportunity to inject large amounts of their venom, causing symptoms ranging from mild irritation to paralysis. If not soon removed, some species (wood ticks) can cause a form of paralysis known as "tick paralysis."

SOFT TICKS (ARGASIDAE)

The "soft" or "softbacked" ticks are 4 to 12 mm long, have a leathery, wrinkled, granulated integument, and no dorsal shield. The capitulum and mouthparts are ventral, whereas they project anteriorly in the "hard" ticks. The pedipalpi are leglike, and the 4 segments are equal in length, whereas in the hard ticks the fourth segment is very small. The spiracles are situated on the sides of the body above the third and fourth pairs of legs. There is no marked morphological difference between the sexes. Most adult soft ticks are long-lived, up to several years, the females laying eggs periodically—20 to 50 after each blood meal. The frequency of feeding depends on the number of hosts that pass by.

Among the soft ticks, only species of the genera *Ornithodoros* and *Argas* are important as vectors of disease. *Argas* spp. are important poultry parasites, but seldom attack man. *Ornithodoros* can be distinguished by the rounded sides of the body, in contrast to the distinct marginal border of *Argas,* which is always visible, even when the tick is fully engorged.

Species of *Ornithodoros*

Ornithodoros spp. are argasids that are perhaps best known through the dreaded African relapsing fever tick *O. moubata* (Murray), that hides in the dust and thatch of native huts and engorges rapidly on the blood of the inhabitants, transmitting an endemic relapsing fever caused by many different strains of a spirochete, *Borrelia recurrentis.* In the United States, tick-transmitted endemic relapsing fever was first reported in central Texas (Weller and Graham, 1930). Since then, it has been reported in widely scattered areas in the western states. Man is infected by the bite or coxal fluid of infected argasid ticks, principally *Ornithodoros turicata* and *O. hermsi,* in the United States (CPHS, 1971). Aside from the fact that *Ornithodoros* includes species that are vectors of relapsing fever, the genus is important because it includes several species that inflict painful bites.

Ornithodoros hermsi Wheeler, Herms, and Meyer

This species is the most serious vector of relapsing fever in the mountain areas of California and some other western states. *Ornithodoros hermsi* is found at elevations of 5,000 ft (1,500 m) and higher, while *O. turicata* Dugès and *O. parkeri* Cooley are found at lower elevations, but are not serious problems.

The female of *O. hermsi* is 5 to 6 mm long and 3 to 4 mm wide. The male is slightly smaller, but is similar in appearance. Unengorged ticks are light sandy in color. Freshly engorged specimens are dull, deep garnet, with a grayish-blue tint a few days after feeding. Absence of food greatly prolongs the life cycle. The larvae may live as long as 95 days without food, the nymphs even longer, and the adults more than 7 months. With occasional feeding, they have been kept alive in a pillbox for over 4 years (Herms, 1939). The duration of the life cycle varies from 202 to 314 days at 30° C (86° F) to 364 to 602 days at 21° C (70° F) (Pratt and Littig, 1962).

Ornithodoros turicata and *O. talaje* (Guérin-Meneville) transmit relapsing fever in southern areas between Florida and California and northward to Kansas, Colorado, and Utah, and *O. parkeri* transmits the disease in the northwestern area.

Relapsing Fever. Relapsing fever is caused by host-specific spirochetes (slender, spirally undulating bacteria). The disease is characterized by chills, fever, sweating, generalized pains, and usually nausea, vomiting, and headache. Ticks not only remain infective for life, but pass the spirochetes to their offspring through the egg. Not only ticks, but also their rodent hosts, serve as reservoirs for the spirochetes. *Ornithodoros hermsi* transmits the spirochete *Borrelia hermsi* and *O. parkeri* transmits *B. parkeri* without cross-infections between the 2 species (Davis, 1939, 1948; Pratt and Littig, 1962).

Mountain cabins in areas where tickborne relapsing fever occurs should be rodentproofed to exclude small rodents. Avoid soiling the fingers with the blood of chipmunks or squirrels. Infection can take place by contamination of a wound in the skin, although it is known that *Ornithodoros hermsi, O. turicata,* and *O. parkeri* definitely infect by their bites (Hunter *et al.,* 1960).

Pajaroello, *Ornithodoros coriaceus* Koch

A considerable number of soft ticks produce local or systemic disturbances by their bites alone —notably, some species of *Ornithodoros,* including *O. turicata,* just mentioned. In the more mountainous coastal counties of California, the pajaroello, *O. coriaceus,* is a much-feared tick of this type, attacking deer, cattle, and man. It has also been reported in South America and Mexico. It is especially abundant in deer beds.

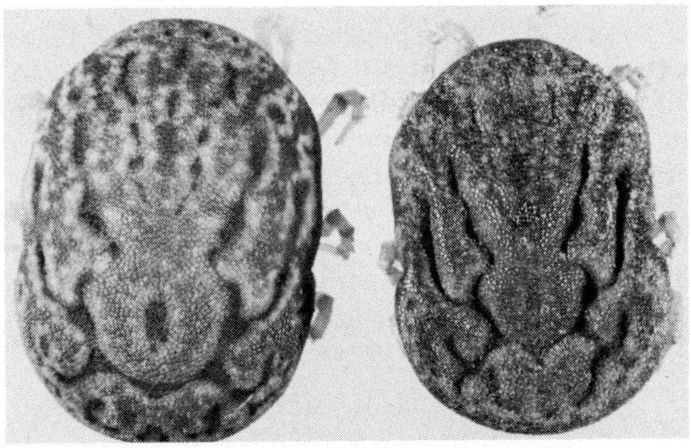

Fig. 304. The pajaroello, *Ornithodoros coriaceus.*

Description. The pajaroello is a large tick, leathery, rough, and granulated (figure 304), and the whole surface, above and below, is a dirty yellowish-earthy color, and has rusty-red spots irregularly distributed (plate VIII, 6). The capitulum and palps are light yellow, and the legs are gray-brown. The unengorged tick is 10 to 12 mm long, rounded at the posterior end, and pointed anteriorly. The sides are parallel, with only a slight median constriction.

Life Cycle. The female deposits large, spherical eggs, sometimes more than 1,000 per season, that hatch in 3 or 4 weeks. The very active larvae attach readily to a host. The ticks become sexually differentiated after the fourth molt, requiring about 4 months, or sometimes not until after the fifth molt. They usually molt once for each feeding (engorgement), but sometimes there are 2 molts between feedings.

Effect of a Bite. Herms (1939) described the symptoms of a victim bitten twice by a pajaroello, 14 days apart. The first time, the tick was partly engorged, and the second time, it was fully engorged when dislodged. The second bite, on the left leg, was the more severe one. The engorged tick was about 20 mm long and 12 mm wide. A bright-red spot was visible at the point of the tick's attachment, surrounded by an irregular, purple ring about 20 mm in diameter. After 3 hours, the lower leg was greatly swollen, and a clear lymph exuded freely from the lesion.

The Genus *Argas*

Species of this genus are important parasites of poultry. Each tick requires a considerable quantity of blood to become replete, and therefore large numbers of them can extract enough blood to cause the death of a bird. Humans have occasionally become infested, generally in or near buildings in which infested fowls were located. In some parts of the world, species of *Argas* have been implicated as carriers of "avian spirochetosis," a disease attacking chickens, geese, turkeys, guinea fowls, and other birds. The spirochetes are transmitted from one generation of ticks to the next through the egg, thence to the host by the tick's bite or by fecal contamination.

Fowl Tick, *Argas persicus* (Oken)

The fowl tick is actually a complex of species, including *A. persicus*. It is cosmopolitan, and is one of the most important poultry parasites. It is also known as "chicken tick," "adobe tick," "tampan," and "blue bug," the latter name referring to the appearance of the female when fully engorged. It will bite humans, and may transmit spirochetes to them (James and Harwood, 1969).

HARD TICKS (IXODIDAE)

Description and Biology. The female hard tick has a hard plate or shield on the anterior region of the back (figure 305, *B*), the remainder of which is smooth, and the body can be greatly extended by feeding. The shield covers the entire back of the male. The mouthparts are visible from above, projecting in front of the body, whereas in the soft ticks they are under the body, and are not visible from above. The pedipalpi (figure 305, *B*, palpus) are rigid, and are varied in form. The spiracles are situated behind the fourth pair of coxae. The coxae and tarsi are generally armed with spurs, which are absent in soft ticks. Pulvilli are always present in hard ticks, but are absent or rudimentary in soft ticks. The females feed only once, and lay one large batch of eggs—often several thousand.

Hard ticks are scarce in forests, being more abundant in shrubby areas, especially along trails or paths, to which they are attracted by the scents

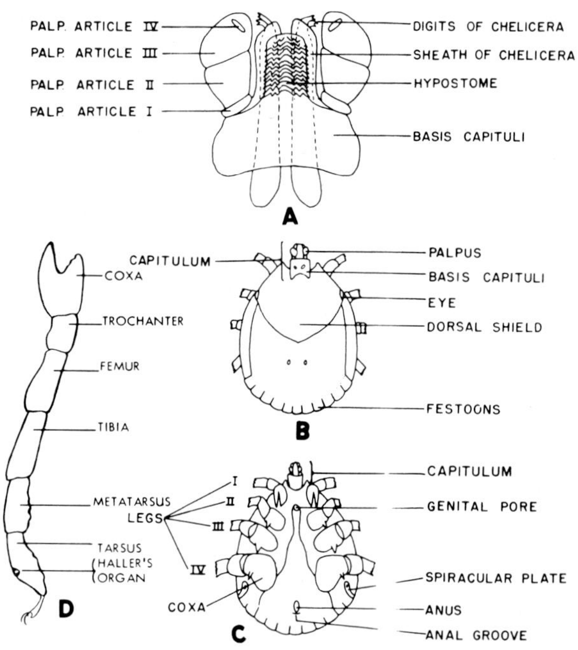

Fig. 305. External morphological features of hard ticks (Ixodidae). *A*, ventral aspect of capitulum; *B*, dorsal aspect of female; *C*, ventral aspect of male; *D*, leg. (From Pratt and Littig, 1962.)

left by animals. Larvae hatching from eggs that have been deposited in some sheltered place will climb to the tips of grass blades or twigs near the trail used by their potential hosts, often clinging together in clusters of 100 or more. Any vibration of their resting place caused by a passing host animal stimulates the larvae to wave their legs about. If the animal brushes against them, they cling to its fur. "One-host ticks" [e.g., the winter tick, *Dermacentor albipictus* (Packard)] develop to maturity on the host. "Three-host ticks" (e.g., *D. andersoni, D. variabilis,* and *Amblyomma americanum*) drop from the host to molt. They may then crawl to a perch near a trail, responding to vibrations as did the larvae, and again cling to the fur of a passing animal. They engorge, and again drop off the host to molt, most species overwintering as unfed adults. Hard ticks cannot stand as much desiccation as soft ticks, and the latter generally survive in much drier situations.

Wood Ticks (*Dermacentor* species)

These ticks are usually ornate, having pale markings on the dorsum, and possess eyes and 11 festoons. The palpi are short, and are either broad or moderate, and the basis capituli is rectangular dorsally. The spiracles are suboval or comma-shaped. As bloodsucking parasites and as disease vectors, species of *Dermacentor,* the wood ticks, constitute the most important group of ticks in the United States. Wood ticks are not only carriers of disease, but the toxins injected with their bites cause itching and a sickness known as pyrexia (a febrile condition). Invertebrates are commonly paralyzed by arthropod venoms, but apparently the only parallel phenomenon in vertebrates is the paralysis caused by the bites of wood ticks. Salivary secretions of the ticks cause a neuromuscular block, presumably at the presynaptic terminal fibers of somatic motor axones, which results in the failure of release of acetylcholine. Recovery is rapid and complete if the ticks are removed before respiratory paralysis ensues (Beard, 1963). (See the section headed "Tick Paralysis," farther on.)

The life history of wood ticks can be illustrated in a general way by that of the American dog tick, *Dermacentor variabilis,* which was worked out in great detail by Smith *et al.* (1946).

American Dog Tick, *Dermacentor variabilis* (Say)

This tick occurs on the Pacific Coast and east of the Rocky Mountains. Like its counterpart in the Rocky Mountain area, *D. andersoni,* it is not only a vector of the causal organisms of Rocky Mountain spotted fever and tularemia, but is also a common cause of tick paralysis. It is commonly found on dogs and, in the eastern United States, it is the species most likely to be found on man. *Dermacentor variabilis* is seldom found in homes unless carried in while it is feeding on a dog and then drops off to seek a hiding place. It is not a domiciliary species as in the case of the brown dog tick, *Rhipicephalus sanguineus,* an account of which see farther on.

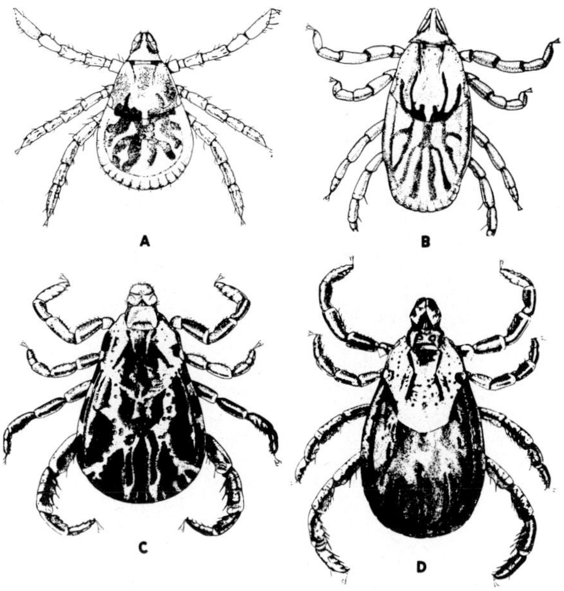

Fig. 306. American dog tick, *Dermacentor variabilis*. *A*, dorsal aspect of larva; *B*, nymph; *C*, adult male; *D*, adult unengorged female. (From Smith *et al.*, 1946.)

Description of Adults. The adult male (figure 306, *C*) is about 4.5 mm long and 2.5 mm wide. The dorsal shield is dark brown, marked with a variable pattern of white. The unfed female (figure 306, *D*) is slightly larger than the male. Her shield, which is restricted to the anterior portion of the dorsum, is much more extensively marked with white than in the male. The male does not become larger as he feeds, but the female enlarges (figure 307) to about 13 mm long by 10 mm wide. The size of the shield does not change, but it becomes inconspicuous in relation to the greatly expanded remainder of the body (Smith *et al.,* 1946).

Life Cycle. Over a 14- to 32-day period, the female lays masses of 4,000 to 6,500 ellipsoidal, yellowish-brown eggs, and then dies. The eggs normally hatch in 36 to 57 days. The larvae (figure 306, *A*) lack spiracles, have only 3 pairs of legs, have red markings near the eyes, and lack white markings on the shield. Unfed larvae are yellow, and about 0.6 mm long; engorged specimens are slate gray to black, and are about 1.2 mm long. The unfed larvae crawl about, seeking hosts, and can live for more than a year (maximum, 540 days) without food. In a simulated meadow, larvae became engorged on mice in an average of 4.4 days, then dropped from their hosts to seek protected places in which to molt. The nymphs lacked a genital opening and the white markings on the

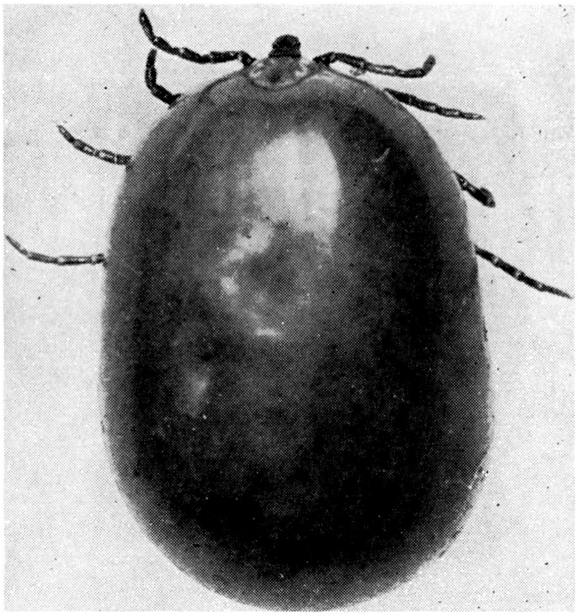

Fig. 307. Engorged female of the American dog tick, *Dermacentor variabilis.* (From Smith *et al.*, 1946.)

shield that characterized the adults. The unfed nymphs (figure 306, *B*) were light yellowish brown, with red markings near the eyes, and were about 1.5 mm long. The engorged nymphs were slate gray, and about 4 mm long. As with the larvae, they crawled about seeking hosts. The engorging period ranged from 3 to 11 days; the greatest number dropped on the sixth day, and found protected places in which to molt. They molted after 3 weeks to several months. Nymphs also could live for more than a year without food; the maximum period was 584 days (Smith *et al.*, 1946).

Unengorged adults may live for more than 2 years if they do not find dogs or other large animals to which they can attach themselves. The engorgement of females requires 5 to 13 days, and mating takes place on the host. In the absence of suitable hosts, the life history of the American dog tick may be prolonged to 2 or more years, but under favorable conditions, it may not take more than 3 months.

Hosts. In the field in Massachusetts, meadow mice were the most important hosts of the larvae and nymphs, while dogs were the most important for adults. Dogs can pick up hundreds of ticks in a day, and horses and cattle also suffer some annoyance (Smith *et al.*, 1946).

Rocky Mountain Wood Tick, *Dermacentor andersoni* Stiles

This tick (plate VIII, 6) is found in the Rocky Mountain states, as well as in Nevada, eastern Oregon and Washington, and in parts of California east of the Sierra Nevada and Cascade mountains. It is superficially almost identical to *D. variabilis.* Cattle, horses, and dogs can become seriously infested. Only the adults feed on such large animals and on man; the larvae and nymphs feed on small wild animals, especially rodents. In areas where *D. andersoni* occurs, it is the principal vector of Rocky Mountain spotted fever, transmitting it from small mammals (principally rodents) to man.

Shoshone Indian braves exposed their squaws to the "evil spirits" believed to be associated with gophers in the foothills of the Rockies, and the braves avoided areas shown by this pragmatic test to be inhabited by the evil spirits. When white men brought large, domesticated animals into the foothills, the tick population multiplied closer to fixed settlements, and Rocky Mountain spotted fever became a much greater problem to white men than it originally had been to the Indians. What originally was thought to be a regional nuisance has now been shown to be a world-wide problem, for RMSF has been recognized in North and South America, and similar infections have occurred in Europe, South Africa, India, and Australia (Aikawa, 1966).

Biology. Nymphs and adult ticks spend the winter among grass and leaves on the ground, without feeding, and not until the first warm days of spring do they climb up on the brush to become attached to passing hosts. They feed only from about the middle of March to the middle of July, and this is the only period when people contract spotted fever.

Lone Star Tick, *Amblyomma americanum* (L.)

The lone star tick, so called because of the silvery spot on its dorsal surface, ranges from west-central Texas to north Missouri and eastward to the Atlantic Coast. Its abundance in some areas, particularly in the Ozark region and eastern Oklahoma, regions characterized by the brushy-type vegetation so conducive to high tick populations, is believed to be a threat to their economic development. In the eastern and southern states, this tick has been a pest since early in the 18th century (Bequaert, 1946), and currently attacks man more frequently than does any other species. Its severe bites are accompanied by extensive inflammation, suppuration, and subsequent development of lesions. Residents and tourists in the Ozark region encounter the pest practically every day during the "tick season" in untreated recreation areas. Hundreds of workers are stricken by any one of several debilitating tickborne diseases, particularly tularemia. Up to 57% of all new-born fawns in certain areas of the Ozark region are lost because of lone star tick infestation. Severe weight losses are suffered by infested cattle. Control with pesticides is made difficult by certain characteristics of the biology of the lone star tick (Hair and Howell, 1970).

Biology. In Oklahoma, engorged females seldom appear on hosts before mid-March, and maximum numbers of fed females drop from domestic livestock and deer during late April, May, and early June. After a lapse of 5 to 16 days, the dropped females oviposit under available ground litter, frequently depositing over 5,000 eggs per egg mass. The eggs generally hatch in about 30 days. The larvae climb to the tops of plants. Heavy rains and strong winds tend to dislodge them, and over 90% apparently perish under such circumstances. Lone star ticks overwinter in the fed larval stage, unfed nymphal stage, and 2 forms of adults: unfed adults, or fed nymphs that become adults during the winter after exposure to adequate warmth.

The ticks normally contact their hosts when the latter brush up against them on the tips of tall grasses and shrubs. However, if the host remains in the area for a considerable time without contacting them, nymph and adult ticks may become stimulated by heat and carbon dioxide from the host's body, and will then fall to the ground, find the host, and climb onto it. Larvae depend almost entirely on the host to make contact with the larval mass as it hangs from the tips of vegetation; they seldom seek a host (Hair and Howell, 1970).

Tick Paralysis

The bite of the wood tick is not felt immediately, even though the tick remains attached, but the bite causes more or less inflammation later. If the tick is not removed with care (see "Removal of Ticks," farther on), its mouthparts usually remain in the skin and cause an ulcer that can become infected. Another possible consequence of the bite, if a gravid female tick attaches at the back of the neck or base of the skull and feeds there for a week or so, is a disease known as tick paralysis. A paralysis develops, and if the tick is not removed, death by respiratory failure may result. Human beings, particularly children, have been affected, but more frequently domestic range animals are the victims. In North America, tick paralysis has been most common in the Pacific Northwest, where it had been known since the first decade of this century, and by 1954, 332 human cases had already been recorded. *Dermacentor andersoni* was almost always the tick that was implicated. The mortality rate was 11.7%, despite the fact that there is usually a latent period of about 5 days before *D. andersoni* can produce enough toxin to cause paralysis, and paralysis can be avoided by removal of the ticks during that period. For some 12 to 24 hours before the onset of actual paralysis, there may be vague complaints, irritability, tiredness, and pain or paresthesia (sensation of prickling, tingling, or creeping on the skin), in the lower extremities. There then occurs a loss of coordination, followed in a few hours by flaccid paralysis in the lower extremities. In another 12 to 24 hours, the paralysis ascends to the arms if the ticks are not removed. If the ticks *are* removed, recovery is rapid (Rose, 1954).

Not all female ticks can cause paralysis, and the reason for this is not known. Tick strains in some areas are more dangerous than in others. Using hamsters as experimental animals, Hughes and Philip (1958) found that ticks (*D. andersoni*) from British Columbia caused a much higher rate of paralysis than ticks from the Bitter Root Valley of Montana. This may account for the statistics presented by Rose (1954), showing about 72% of the cases of tick paralysis in the Pacific Northwest to be from British Columbia.

In California, *D. andersoni* occurs in the northeastern part of the state and southward along the eastern slope of the Sierra Nevada to Mono County. Instances of tick paralysis in deer and cattle, but not humans, are known from California. All clinically observed human cases have been acquired in Oregon or elsewhere (R. F. Peters, correspondence). However, the potential for tick paralysis, as well as the tick fevers (described following) would appear to exist. Vacationers in tick-infested areas should examine themselves daily to make certain that a tick has not embedded itself anywhere on the body. In the eastern and southern states, tick paralysis was not reported until 1938 (Costa, 1952). The species of ticks implicated have been *Dermacentor variabilis, Amblyomma americanum,* and *A. maculatum* Koch.

Colorado Tick Fever

The "Colorado type" of tick fever is caused by a virus, and is known to occur in western Canada and in Washington, Oregon, Idaho, Montana, California, Nevada, Utah, Wyoming, Colorado, and South Dakota. It is described as "an acute, febrile, dengue-like disease, usually without rash; a brief remission is usual, followed by a second bout of fever, each lasting 2 or 3 days. Characteristically mild, but may be severe in children, occasionally with encephalitis or tendency to bleed; deaths are rare." The reservoirs for the virus of Colorado tick fever are many species of ground squirrels, chipmunks, mice, wood rats, and other rodents. Ticks also serve as long-term reservoirs (CPHS, 1971).

The distribution of Colorado tick fever in California has coincided with that of its usual vector, *Dermacentor andersoni.* In 1957–1970, there were 115 laboratory-confirmed cases of Colorado tick fever in California, the majority acquired in that state but some in endemic areas from other states (CPHS, 1971). Since 2 tick fevers—Colorado tick fever and Rocky Mountain spotted fever—occur in California and other western states, the 2 diseases should be differentiated, and each should be given its full name in scientific papers and news releases.

Rocky Mountain Spotted Fever

Rocky Mountain spotted fever (RMSF) is endemic throughout the United States outside of New England, particularly in mountainous regions, as well as in Canada and some parts of Mexico and South America. RMSF, first known in the Bitter Root Valley of Montana as early as 1872, is no longer a regional curiosity, and Aikawa (1966) suggested that it be called "rickettsial spotted fever," honoring the name of H. T. Ricketts, the most active and effective early investigator of the disease. In the West, the Rocky Mountain wood tick, *Dermacentor andersoni,* is the principal vector, both male and female ticks being infective; and in the East, the American dog tick, *D. variabilis,* is the most important vector (Larson, 1955). The rabbit tick, *Haemaphysalis leporispalustris* (Packard), is also a vector in the West, and the Pacific Coast tick, *D. occidentalis* Marx, is suspected. The lone star tick, *Amblyomma americanum,* is considered to be a probable vector in parts of the eastern United States, and in Texas, Oklahoma, and Arkansas (Pratt and Littig, 1962). It appears that any tick that infests the approximately 40 hosts of rickettsia is a potential vector. Birds are believed to be important factors in dissemination of ticks that harbor rickettsial agents, including *Rickettsia rickettsii,* the RMSF pathogen (Clifford *et al.,* 1969). Rickettsias are a group of microorganisms about the size of small bacteria, and on the biologic scale they lie between the viruses and the bacteria. Besides the various spotted fevers, they are the causal organisms for the typhus fever group of rickettsial diseases, scrub typhus, Q fever, and probably trench fever.

Symptoms. RMSF is a severe, acute, infectious disease of the small peripheral blood vessels. It is said to cause greater damage to tissues on the surface of the body and to those of the brain than does any other rickettsial disease. Mortality to humans is commonly 20% or more (Rivers and Horsfall, 1959; Aikawa, 1966). A rash appears about the second to fifth day after a bite, on the wrists, ankles, and less often on the back, later spreading to all parts of the body. This rash is the most characteristic symptom of RMSF. The rash is sometimes preceded by a mottled appearance of the skin. The spots may become necrotic on the extremities. At the outset, the patient complains of frontal and occipital headache, intense aching in the lumbar region, and marked malaise. The incubation period may be 3 to 12 days in the milder infections and 2 to 5 days in the more severe ones. Before the initial chill, only minimal evening fever is experienced, but after the chill, the temperature continues to rise, reaching 102 to 104° F (39 to 40° C) on the second day, and then rises gradually to a maximum of 104 to 105° F (40 to 40.6° C) in the second week. In the virulent cases in the Bitter Root Valley of Montana, the temperature may reach 106 to 107° F (41 to 41.6° C), and may remain this high until the patient dies. The pulse rate may reach 150 and the respiratory rate 60 per minute in the terminal phase of the disease (Aikawa, 1966).

Transmission. Ticks transmit *Rickettsia rickettsii* from any small mammal to man during frost-free periods of the year. The organism enters the gut of a tick with the blood it sucks from its host and then spreads to all parts of the body, including the salivary glands. Fortunately, the disease is not transmitted unless the tick remains attached to its host for 2 hours or more. The disease is passed from one generation of ticks to the next via the egg, without decrease in virulence (Price, 1953). Despite transovarial transmission of *R. rickettsii,* the tick apparently suffers no ill effect, and the vertebrate vectors also appear to be in symbiotic relationship with the tick and suffer no evident clinical disease (Aikawa, 1966).

Vaccination. A high degree of protection to people who are frequently exposed to tick bites is afforded by vaccination—initially 3 injections at 5- to 7-day intervals and each year thereafter a booster shot. As a result, the number of reported cases of RMSF is steadily decreasing (Pratt and Littig, 1962; Aikawa, 1966).

Q Fever

Another rickettsial disease sometimes transmitted to man by ticks is Q fever (query fever), also known as coxiellosis or nine-mile fever, the causal organism for which is *Coxiella burnetii.* It has been reported from all continents, and in the United States, outbreaks have occurred most commonly among stockyard workers. Transmission to man is usually by inhalation of dust. Milk can be a source of infection. Tick species that have been found infected are *Dermacentor andersoni, Amblyomma americanum, Otobius megnini,* and many others (Horsfall, 1962; Pratt and Littig, 1962). Vaccination of persons liable to be exposed to infection has been effective (Hunter *et al.,* 1960).

Tularemia

This is a disease caused by the bacillus *Francisella tularensis,* that occurs in the continental United States, Alaska, Canada, northern Europe, Japan, and many other parts of the world. In the United States, it was first reported in Utah in 1919 and was called "deer fly fever," but was later reported among various rodents in Tulare County, California.

Rabbits are important reservoirs for this bacillus, and it is mechanically transmitted from rabbit to rabbit by the deer fly *Chrysops discalis* (Francis and Mayne, 1921). However, ticks may be more important than deer flies in the transmission of this disease (James and Harwood, 1969). In fact, at least 54 species of arthropods can harbor the tularemia bacillus (Steinhaus, 1946). Besides existing in rabbits, tularemia is also known to be carried by meadow mice, ground squirrels, beavers, coyotes, sheep, and various game birds. When transmitted to humans by deer flies, exposed body surfaces are bitten, and the onset of pain and fever is sudden. Lymph glands become inflamed and swollen. The fever lasts for 3 to 6 weeks, and the patient generally convalesces slowly (Francis, 1919).

Tickborne (Endemic) Relapsing Fever

Louseborne relapsing fever and tickborne relapsing fever are both caused by spirochetes of the genus *Borrelia,* and they cannot be distinguished with certainty on the basis of clinical manifestations. (See comment in the earlier section on lice.) Most authorities believe that the spirochetes transmitted by lice and ticks are variants of the species *Borrelia recurrentis,* but many "species" have been identified by geographic location and by their argasid tick vectors, and were listed by Dubos and Hirsch (1965). It is reasonable to suppose that the isolation resulting from the burrow- or den-infecting habits of argasid ticks may have led to the development of different strains of a single pathogen. Tickborne relapsing fever tends to be endemic, and is completely so in the United States.

In California, relapsing fever is most commonly contracted from the bites of argasid ticks, generally *Ornithodoros hermsi,* and usually at elevations ranging from 5,000 to 8,000 ft (1,500 to 2,400 m). Ticks in rodent-infested mountain cottages are common sources of the spirochetes (Herms and Wheeler, 1936). Infected ticks (*O. hermsi, O. turicata, O. parkeri,* and *O. talaje*) have been found in California, Arizona, New Mexico, Utah, Idaho, Montana, Colorado, Kansas, Oklahoma, Texas, and Florida. In Texas alone, 100 cases of tickborne relapsing fever were diagnosed from June, 1942, to May, 1949 (Dubos and Hirsch, 1965), and in California, there were 65 recorded cases from 1950 through 1970 (CPHS, 1971). In both states, many additional cases probably escaped detection.

Species of *Ixodes*

Members of the genus *Ixodes* can be readily distinguished from all other hard ticks (*Ixodidae*) by the groove surrounding the anus in front, and by the absence of "festoons," which are illustrated in figure 305, *B.* In California, the Pacific tick, *Ixodes pacificus* Cooley and Kohls, is common on deer and cattle and bites humans freely, commonly causing marked systemic disturbances. This species is related to the European castor bean tick, *Ixodes ricinus* (L.), and in fact was formerly considered to be a variety of that species. The castor bean tick is so designated because of its color, shape, and the enormous size of the engorged female, up to 14 mm in length. Particularly serious consequences result from removing the tick when the capitulum is allowed to remain embedded in the skin. This may lead to irritation and infection.

The sensory physiology and behavior of *I. ricinus* were investigated in great detail by Lees (1948). The data are worth recording because they may hold for ticks in general. While crawling, the unfed female waves its first pair of legs like antennae. It uses these legs for crawling when it is not hungry. On the tarsi of its legs is Haller's organ, which contains sensilla stimulated by odor and humidity. As already stated, ixodid ticks climb to the tops of grass blades and twigs and wave their forelegs about when stimulated to do so by an approaching animal, and will cling to the passing host. The tips of the pedipalps of *I. ricinus* have sensory pads used for selection of a suitable feeding site on the host. *Ixodes ricinus* spends most of its life in microclimates of very high humidity, so desiccation forces the tick to periodically move down to the moist base of the herbage upon which it has climbed so that it can absorb moisture through its cuticle. One is reminded of similar periodic trips of subterranean termites, via their shelter tubes, from dry structural timbers to moist subterranean galleries where they can replenish their body moisture by cuticular absorption.

Brown Dog Tick, *Rhipicephalus sanguineus* (Latreille)

The brown dog tick differs from members of the preceding genus in that it rarely attacks man, yet is the species the homeowner is most likely to be obliged to deal with, for it is a common pest of dogs.

Description of Adults. The male is about 3 mm long, flat, uniformly reddish brown, and has tiny pits scattered over the back. The female (figure 308) before feeding resembles the male in size, shape, and color. When engorged with blood, she can reach about 12 mm in length. The engorged part of the body then changes in color to gray-blue or olive.

Life Cycle. The engorged female falls off the dog, and seeks a sheltered place in which to lay a mass of 1,000 to 3,000 tiny, dark-brown eggs. These may be deposited in any available place, but because she tends to crawl upward, they are likely to be deposited in cracks or crevices in the roofs of kennels or the ceilings of porches. The female then dies, and her eggs hatch in 19 to 60 days. Cool and/or dry weather tends to prolong the hatching period. The minute, 6-legged, light-brown larvae or "seed ticks" move to the lower parts of walls, and attach themselves to a dog at the first opportunity, but can survive as long as

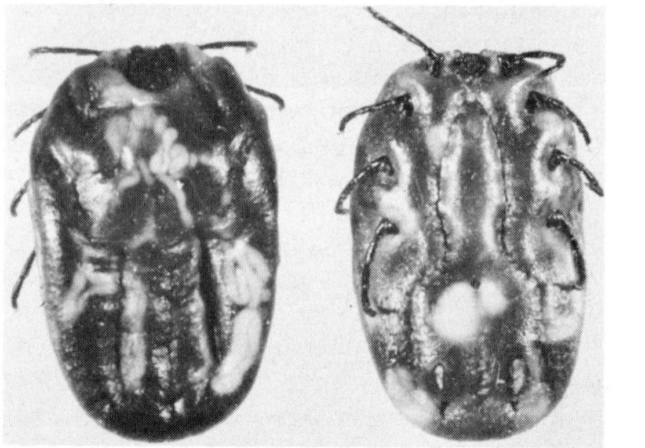

Fig. 308. Brown dog tick, *Rhipicephalus sanguineus*.

8 months without food or water. After finding a host, they engorge themselves in 3 to 6 days, changing from flattened to globular in shape, about 2 mm in diameter, and become bluish in color. They then drop, hide in cracks to molt, and in 6 to 23 days become 8-legged, reddish-brown nymphs.

After a few days of inactivity, the nymphs are ready to attach themselves to a dog, but like the larvae, they can live for months without food or water. Once attached, they engorge in 4 to 9 days and become oval, about 3 mm wide, and dark gray. They then drop off the dog, hide, and molt in 12 to 29 days. They are then adults, for among the hard ticks there is only a single nymphal instar. The adults will attach themselves to a dog at the first opportunity, but they can live as long as 18 months before attachment. They then engorge themselves with blood in 6 to 50 days, mate, and the females drop off to oviposit and repeat the life cycle (Bishopp, 1939). Under favorable conditions, the life cycle may be completed in 63 days at 85° F (29° C). However, under the usual environmental conditions and delays in attachment to the host, there are seldom more than 2 generations per year in the North and 4 in the South. The brown dog tick prefers warm, dry situations, and seldom develops in large numbers outdoors in the United States.

Obnoxious Features. Besides infesting dogs, the brown dog tick can be a serious nuisance in the home, where it may be seen emerging from numerous hiding places. Even if ticks are eliminated from a dog and its kennel, as well as from the home, it can pick up ticks from other infested residences in the neighborhood, from a boarding kennel, or when taken along on a vacation. A dog can pick up ticks that have dropped from another dog, but cannot become infested by direct contact, because a feeding tick drops off and molts before attaching itself to a new host. Ticks may attach themselves anywhere on a dog, but are most often found on the ears and neck and between the toes. Larvae and nymphs are often found in the long hair on the back. By sucking large quantities of blood, ticks can reduce a dog's vitality and cause it to be irritable. They transmit a canine disease known as piroplasmosis in various parts of the world. The causal organism is a protozoan, *Babesia canis,* that infects red blood cells, but the disease is rare in the United States.

Control of Ticks

One obvious means of tick control is to avoid becoming infested. Some degree of protection can be obtained by keeping clothing buttoned and tucking trouser legs into the tops of socks or boots. One should also avoid sitting on the ground or on logs in brushy areas. Periodic inspection of the clothing and body and removal of ticks before they become attached should be done in order to avoid exposure to tick paralysis and various other tickborne diseases. If ticks do become attached, it is important to know the best way to remove them.

Removal of Ticks Attached to the Body. It is desirable to remove attached ticks without leaving the capitulum embedded in the skin. Pratt and Littig (1962) recommended as follows:

> If ticks become attached, the simplest method of removing them is by a slow, steady pull that will not break off the mouthparts and leave them in the wound. There is no certain way to make a tick detach its mouthparts. A drop of chloroform, carbon tetrachloride, ether, benzene, vaseline, or fingernail polish rubbed over it will help remove the tick. Several minutes to a half-hour later, when the tick has withdrawn its mouthparts, it can be removed with less damage to the skin. Touching the tick with the lighted end of a cigarette (or a hot needle) sometimes causes it to release attachment. An antiseptic should always be applied to tick bites just as to open wounds. If the hands have touched the tick during removal, wash them thoroughly with soap and water, since the tick secretions may be infective.

For the removal of *Ixodes* ticks, Busvine (1966) suggested that, after dabbing them with chloroform or ether, the capitulum should be pressed inward to loosen the "teeth," and then the tick should be gently pulled away. Some ticks, such as *Amblyomma* and *Ixodes,* have longer mouthparts than others, such as *Dermacentor*. Their mouthparts penetrate more deeply, even through socks, and the ticks are harder to remove.

Control Outdoors

All types of repellents may be rendered relatively ineffectual by the tendency of ticks to crawl underneath clothing and attach themselves to untreated portions of the body. Therefore, treatment of clothing is recommended. Good protection is afforded by M-1960, Indalone, diethyl toluamide, dimethyl carbate, dimethyl phthalate, or benzyl benzoate (Pratt and Littig, 1962). In areas having frequent visitors, such as city parks, it should be borne in mind that ticks do not travel far after having a blood meal and dropping off their hosts. Therefore, the greatest numbers are found along walks and the edges of lawns. Adequate economical control might be obtained by applying pesticides in swaths 10 to 20 ft (3 to 6 m) wide along paths, roadsides, and the edges of lawns and gardens, where most of the ticks occur and where most exposure of humans is likely (Pratt, 1962; Pratt and Littig, 1962).

Control of ticks in areas having heavy plant growth can be obtained with dust, suspension, or emulsion formulations of various recommended residual pesticides, applied by hand or power equipment, or in certain situations by aircraft. Sprays are generally applied to grassy and brushy areas adjacent to human and domestic animal habitations, along roadsides and footpaths, vegetation surrounding camping and picnic sites, or wherever ticks may be present and annoying. Contamination of feed or foodstuffs, fishbearing water, or blossoms visited by bees should be avoided. A spray consisting of Gardona 75 wettable powder at a concentration of 0.5% of actual toxicant has been recommended for tick control when an insecticide of low mammalian toxicity and without long-lasting residues is desired, such as in recreation areas. The spraying can begin in the early spring when ticks first appear, and can be repeated if necessary. In recreation areas, the first spraying should be done shortly before the early-summer influx of visitors (Hair and Howell, 1970).

Park Maintenance. In parks and other recreation areas, grass should be mowed periodically, for grass over 6 in. (15 cm) high offers considerable protection for ticks against the sun and dryness. Also, mowing will dislodge and disperse masses of "seed ticks," after which most of them succumb to desiccation. Tall grass also has the disadvantage of providing cover for host animals. An excessive "overstory" of trees and shrubs should be avoided in heavily infested areas, so that from 50 to 80% of the ground in a park is exposed when the sun is in any one position.

Control in Buildings

The brown dog tick is the species the home-owner is most likely to have to deal with, although the American dog tick and other species also enter the home on pets. The present recommendations pertain to the brown dog tick, which accounts for the emphasis on organophosphorus pesticides. It is necessary to use pesticides of long residual efficacy, because tick eggs may be protected from the spray in cracks or behind baseboards and moldings. Even if they are hit by the pesticide, they are not always killed, so the residue left by it must be able to kill the larvae hatching from the eggs after they start crawling about. Resistance to chlorinated hydrocarbons is now widespread in the United States, and they are no longer generally recommended. Oil solutions or water emulsions of 1% propoxur, 0.5% diazinon, 0.5% dioxathion, 0.5% chlorpyrifos, 1.5% fenthion, 1 or 2% malathion, 2% ronnel, or a dust containing 5% carbaryl have been suggested (CDC, 1973). Except for malathion and ronnel, when these pesticides are used indoors they should be applied only as "spot treatments" to the usual harborage sites.

Sometimes ticks are present in subfloor crawl spaces, under porches, and in other areas accessible to the dog but difficult to treat with a spray. Dust formulations containing the pesticides just mentioned can then be used most effectively.

When treating for the brown dog tick, the objective should be to apply the spray or dust to areas frequented by the dog, paying special attention to its sleeping quarters, as well as to baseboards, doorways, window casings, floor and wall crevices, and beneath the edges of carpets. If the dog is permitted on overstuffed chairs, the cushions should be removed and all crevices and seams treated. In warmer climates, if the dog spends much time outside, the yard also should be treated, along with the doghouse, subfloor crawl spaces, and other areas accessible to it. Retreatment may be necessary 1 or 2 weeks later for severe infestations. When the premises are treated, the dog must also be treated, or reinfestation will occur. The owner may do this himself, or have it done by a veterinarian. Recommended treatments are washes of 1% coumaphos or 0.5% malathion, or dusts of 5% carbaryl, 0.5% coumaphos, 1% trichlorfon, or 3 to 5% malathion. Better penetration of the dog's fur may be obtained with liquid washes. The coumaphos wash or trichlorfon dust should not be used on dogs under 2 months of age. Veterinarians may use animal dips containing 0.1% dichlorvos, 0.2% naled, 1% carbaryl, 0.15% dioxathion, or 1% ronnel. Veterinarians have also reported success with the oral use of ronnel (CDC, 1973).

MITES

Mites (order Acarina) are very small arthropods, with head and thorax fused into a cephalothorax. They have sucking mouthparts, no antennae, and those of interest as household pests have 4 pairs of legs as adults. Although most of these species have only 3 pairs of legs in the first (larval) stage after hatching from the egg, they gain a fourth pair in the second (nymphal) stage. The life cycle generally consists of the egg, larval stage, one or more nymphal instars or stages, and an adult stage. The life cycle usually requires only 2 or 3 weeks, and results in rapid increase and huge populations of mites under favorable conditions. A thorough discussion on the morphology and development of the free-living mites, on their role as parasites of animals and plants, and as vectors of disease, may be found in *Mites, or the Acari* by T. E. Hughes (1959).

Chiggers (Trombiculidae)

Chiggers or "red bugs," called "harvest mites" in Europe, are the larvae of mites belonging to the suborder Trombidiformes, which are world-wide in distribution. There are over 200 families of mites, but the family to which chiggers belong (Trombiculidae) contains about 10% of all mite species (Sasa, 1961). Some species attack humans and cause a dermatitis (trombidiosis). The red welts and severe itching do not appear until several hours or even a day after exposure; therefore, it is difficult to know exactly when or where the infestation occurred. Several chiggers transmit a rickettsial disease called "scrub typhus" or "tsutsugamushi disease" in the Orient and various areas of the Pacific.

Description. The members of the suborder Trombidiformes are characterized by the respiratory system, when present, opening in the region of the gnathosoma, the portion of the body bearing the mouth and its appendages. Chiggers are very small, 150 to 300 microns (0.15 to 0.3 mm) long when unengorged, and are red to pale yellow or white, depending on the species. Like all mite larvae, they have 6 legs. They are parasitic, but later stages are free-living, 8-legged mites. Only the larvae are harmful and only they are correctly referred to as "chiggers." The adults are bright red, hairy, or granular (Michener, 1946; Wharton and Fuller, 1952; Baker *et al.*, 1956). The various stages of the trombiculid mites in general are adequately represented by figure 309, which shows an unengorged and engorged larva, a nymph, and an adult of *Trombicula batatas* (L.).

Common Chigger, *Trombicula alfreddugesi* (Oudemans)

In the Western Hemisphere, this is the most common and widespread species, ranging from Canada to South America and the West Indies. *Trombicula alfreddugesi* parasitizes many species of mammals, birds, reptiles, and amphibians, as well as man. On humans, chiggers tend to congregate in areas constricted by clothing, such as ankles, crotch, waistline, and armpits. It is unfortunate that when chiggers attach to humans they are not noticed for some time, for they are easily removed. According to Baker *et al.* (1956):

> Itching is usually noted 3 to 6 hours after the chiggers have attached, and may persist for as long as 2 weeks. Part of the irritation is thought to be an allergic response to the salivary secretions of the mite. A papule forms at the site of attachment which may develop into a vesicle. Scratching usually removes the offending mite but, if repeated often enough, may result in an infection.

In some regions, this mite is a pest of chickens and turkeys, affecting the younger birds most seriously. When heavily parasitized, the birds become droopy, refuse to feed, and may eventually die from starvation and exhaustion (Baker *et al.*, 1956). A much more important chigger pest of chickens and turkeys, however, is *Neoschongastia americana* (Hirst), which ranges across the southern United States from California to Georgia, but does not attack man (Kunz *et al.*, 1969).

Description. Chigger larvae are 0.15 to 0.25 mm long before engorgement, and are red to reddish-orange, rarely white. Their mouthparts include 2 pairs of grasping palps provided with forked claws. The nymphs are much more hirsute than the larvae. The body is constricted behind the second pair of legs, giving them and the adults the characteristic shape of trombiculid mites shown in figure 309. The adults are much larger than the nymphs, and are even more hirsute. They are 0.9 to 1.1 mm long, and brilliant red (Jenkins, 1949; Baker *et al.*, 1956).

Life Cycle. The spherical eggs, approximately 0.1 to 0.2 mm in diameter, are usually laid in the soil. The larva crawls about on the surface of

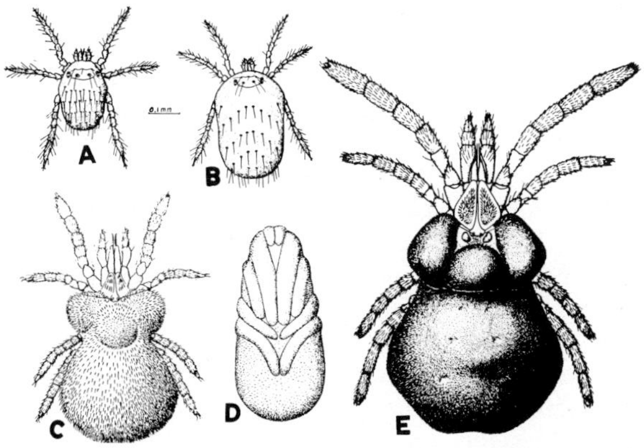

Fig. 309. A chigger mite, *Trombicula batatas*. *A*, unengorged larva; *B*, engorged larva; *C*, nymph; *D*, ventral view of preadult; *E*, adult. (From Michener, 1946.)

the soil until it finds a suitable vertebrate host. It attaches to the host by means of its chelicerae and sucks blood, but as a rule does not burrow under the skin. Engorgement usually takes about 3 days. The larva then drops, enters the soil, and changes, via the nymphochrysalis, to the nymphal stage. The nymphs probably feed on the eggs and young instars of small arthropods. The adult emerges from a dorsal split in the imagochrysalis and nymphal cuticle (Baker *et al.*, 1956).

The life cycle may require 2 to 12 months or longer, depending on the temperature. There may be 1 to 3 generations per year in temperate climatic zones, but reproduction may be continuous throughout the year in warmer regions, with as many as 6 generations. Females kept at suitable temperatures and supplied with water and food were observed to live more than a year and to produce larvae throughout that period. The time when chiggers are active varies from 2 months in Minnesota and Massachusetts to the entire year in southern Florida. Chiggers are most abundant during rainy spells in the area from Kansas to Texas, and may disappear during hot, dry weather (Jenkins, 1948).

Some Related Species

Trombicula splendens Ewing is a related species in the eastern United States. It prefers moister habitats, such as swamps and rotten logs or stumps. It is one of the most common causes of trombidiosis in the southeastern states.

Trombicula lipovskyana (Wolfenbarger) may be found in similar places in Tennessee, Kansas, Oklahoma, and Arkansas.

Trombicula belkini Gould is widely distributed in California, and has also been collected in Utah. Reptiles seem to be its favored hosts, but it also infests rodents and ground birds. It sometimes annoys humans and their pets (CEIR, 1960). This species is closely related to *T. alfreddugesi,* but the larvae lack nude, whiplike setae on the tarsus of leg II (Baker *et al.*, 1956; Gould, 1956).

Trombicula batatas (L.) (figure 309) is common in Central and South America, the state of Puebla, Mexico, and has been reported from the southeastern United States (Michener, 1946; Jenkins, 1948). It has been collected on humans and many domestic and wild animals. One 12-year-old boy had 138 attached larvae (Michener, 1946). It has been reported to attack humans in the San Joaquin Valley of California (Doetschman and Furman, 1949).

Gould (1956) published an extensive monographic study of the larval trombiculid mites of California.

Favored Habitats. Chiggers are most abundant in areas that support thickets or scrub-type vegetation and where the ground is undisturbed, supporting many rabbits, other rodents, and various small host animals. They are generally eliminated automatically by habitat destruction in areas that are heavily populated or intensively farmed. In new urban subdivisions, however, chiggers may persist in lawns for several years. To determine the exact area of chigger infestation, a piece of black cardboard can be placed edgewise on the ground where an infestation is suspected. If chiggers are present, the tiny yellow or pink larvae will crawl rapidly over the cardboard and accumulate on the upper edge. Chiggers can also be easily detected on black, polished shoes (USDA, 1963). Jenkins (1948) suggested the possibility that chiggers might be of value in decreasing mosquito populations. The adults were often abundant in depressions in the ground which had become temporary pools containing *Aedes* and *Psorophora* larvae in the spring. Mosquito eggs laid in such depressions probably were serving as food for *Trombicula* adults.

Repellents. In areas where chiggers are known to be a problem, the avoidance of their favored habitats is, of course, a way of minimizing infestation. Protective clothing and repellents are recommended as already described for protection against mosquitoes and ticks. If infested, a thorough soapy bath as soon as possible is a highly effective treatment. Repeat the lathering and rinsing several times. Most of the chiggers, attached or unattached, will be killed.

Among the best repellents for chiggers are those containing diethyl toluamide (OFF), ethyl hexanediol (6-12), and dimethyl phthalate, applied to the skin and clothing around the ankles, waist, and armpits. To apply dusting sulfur to skin and clothing is an old but effective method of preventing chiggers. Repellents should be applied particularly to the legs, ankles, cuffs, waist, and sleeves. Some relief from itching can be obtained by applying a solution of 5% benzocaine, 2% methyl salicylate, 0.5% salicylic acid, 73% ethyl alcohol, and 19.5% water. This can be prepared by a druggist. It may be applied to each welt with a piece of cotton. Each treatment gives relief for an hour or more (USDA, 1963).

Control. Good control of chiggers in the field can be obtained for 1 or 2 months with toxaphene at 2 lb (0.91 kg) or lindane at 0.25 lb (0.11 kg) of actual toxicant per acre, preferably as emulsions. The amount of water used as a carrier of such quantities depends, of course, on the type of spray equipment available. A given quantity of insecticide can be used with either a large or small quantity of water, as long as the toxicant is thoroughly and uniformly distributed. The following quantities (stated as emulsifiable concentrates) of 4 insecticides that are effective against chiggers as well as insects have been recommended (Anonymous, 1970*d*).

Insecticide and formulation	*For 1,000 sq ft (93 sq m)*	*For 1 acre (0.405 ha)*
Chlordane 45%	10 tsp (50 cc)	3 pt (1,440 cc)
Toxaphene 60%	7 tsp (35 cc)	2 pt (960 cc)
Diazinon 25%	0.5 pt (240 cc)	2.50 gal (9.50 l)
Malathion 57%	0.5 pt (240 cc)	2.50 gal (9.50 l)

A convenient way to treat 1,000 sq ft (93 sq m) of lawn would be to mix any one of the formulations shown in the table with 3 gal (11 l) of water, but if weeds or tall grass were present, the same quantities of insecticide could be more effectively applied in 6 gal (22 l) of water. To spray an acre (0.405 ha), at least 25 gal (95 l) of water are required. Malathion treatments may need to be repeated because malathion is nonpersistent. There are also dust formulations of these insecticides that can be used effectively for chigger control. (Consult appropriate authorities about pesticides currently authorized.)

Straw Itch Mite, *Pyemotes ventricosus* (Newport) (Pyemotidae)

This extremely small mite, almost invisible to the unaided eye, is primarily a parasite of certain insects, including 3 moths, 10 beetles, 4 wasps and bees, a bug, a fly, and a termite. Some of these host insects infest straw, wheat, stored food products, straw mattresses, and wood, and are therefore found in the home. The straw itch mite has also been called "grain itch," "hay itch," and "straw mattress" mite. Humans can become infested, with resulting dermatitis, by coming in contact with materials such as straw, hay, grasses, grains, and even beans, peas, cottonseed, tobacco, and broomcorn that have been infested with insect larvae upon which the mites feed. These mites also attack horses, cattle, and possibly other mammals (Goldberger and Schamberg, 1909; Baker *et al.*, 1956; A. M. Hughes, 1961; Fine and Scott, 1963, 1965; Scott and Fine, 1963, 1964, 1967; Butler, 1972).

Description. The female is an almost microscopically small, elongate mite (figure 310), 0.22 mm long and white to yellow in color. When gravid, she becomes greatly distended behind the fourth pair of legs, and attains a length of up to 2 mm. Her abdomen shows traces of lateral segmentation, and she has clublike hair between the first and second pairs of legs. The male is only 0.16 mm long, but is wider than the female.

Life Cycle. This mite has a strange and unusual biology. The males wander continuously over the distended body of the pregnant female, feeding on it parasitically. The large eggs hatch, and 200 to 300 mites develop to adulthood within the female's enlarged abdomen. They are extruded at the rate of about 50 per day. Only some 3% are males, but they emerge first and remain clustered around the genital opening. With the aid of their hind legs, they drag the females through the opening, even though they can emerge unassisted, and copulation takes place immediately. The females then search for hosts. Only 6 to 10 days are required from the time of fertilization to the hatching of eggs. The mites are active during the warmer months of the year at 80° F (27° C) or above (Baker *et al.*, 1956; Scott and Fine, 1963).

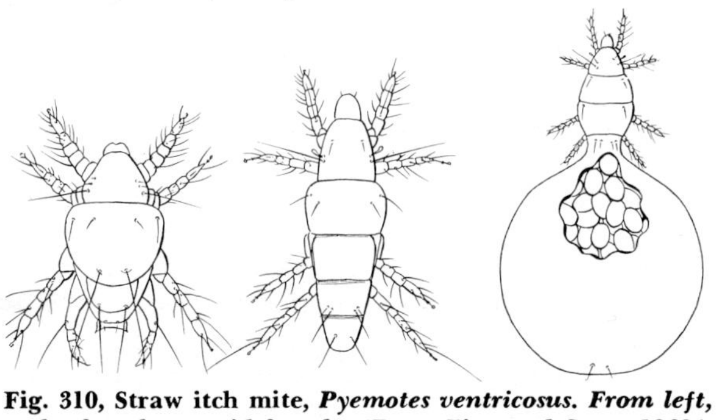

Fig. 310, Straw itch mite, *Pyemotes ventricosus*. *From left*, male; female; gravid female. (From Fine and Scott, 1963.)

Distribution of Bites. The bites of straw itch mites are characteristically distributed almost exclusively on clothed portions of the body, although they occur rarely on other areas, with the exceptions of the palms, soles, and mucous membranes. There is no tendency for the mites to be grouped, although this sometimes occurs fortuitously. A person may feel a prickling sensation at the time of the bite, but otherwise no immediate reaction seems to occur. The period between the time of the bite and the delayed reaction has been variously reported as 10 to 16, 16, 27, and 17 to 28 hours (Fine and Scott, 1965).

Straw Itch Mite Dermatitis. A considerable number of epidemics of dermatitis have been traced to infestation by *Pyemotes ventricosus*. Since many such outbreaks have not been recorded or correctly diagnosed, it is likely that this ailment is more common than is generally realized. Straw itch mite dermatitis is usually associated with sleeping on straw mattresses, harvesting grain, or otherwise handling or coming in contact with grain, straw, hay or other substances such as those just mentioned. The possibility of infestation is particularly strong if there are large numbers of the mites' host insects present, such as the Angoumois grain moth (*Sitotroga cerealella*) and the wheat jointworm (*Harmolita tritici*). The host insect need not necessarily be a species associated with hay or grain. For example, cases of straw itch mite dermatitis have been associated with severe infestations of furniture beetles (*Anobium punctatum*) in the floor joists of houses. The recurrence of such cases during the same season for 3 successive years led investigators to conclude that the mites migrated in search of new hosts as the adult beetles emerged and left the wood. The mites apparently were not able to penetrate the thick exoskeletons of the beetles when the latter were in the pupal and adult stages, and therefore they left and sought new hosts. In one house, the mites were controlled by treating the floors with 2% deodorized malathion emulsion (Fine and Scott, 1963, 1965; Scott and Fine, 1963).

Treatment and Prevention. The treatment of symptoms is not the solution to the problem. Either a person must avoid infested areas, or the mites and their host insects must be eliminated.

Tropical Rat Mite, *Ornithonyssus bacoti* (Hirst) (Macronyssidae)

The tropical rat mite commonly occurs on rats throughout the world, particularly in tropical and subtropical regions, but also in some temperate areas. It is an ectoparasite of rats, and attacks people living in rat-infested buildings. Its bite may cause irritation and sometimes painful dermatitis. It is an important pest of laboratory animals, particularly rats, mice, and hamsters, sometimes deteriorating their health or even causing death by exsanguination (Baker *et al.*, 1956).

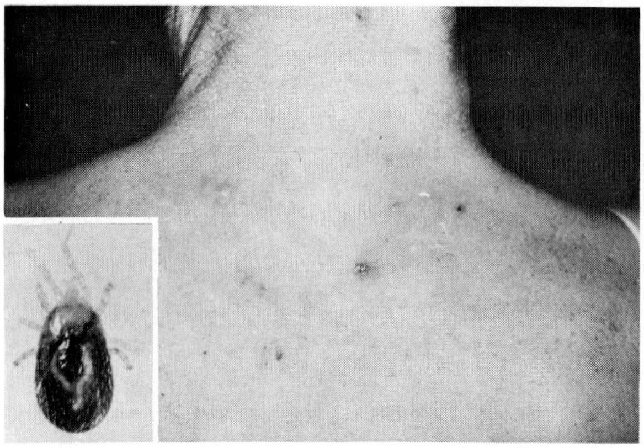

Fig. 311. Back and neck of a woman, showing maculae caused by bites of the tropical rat mite, *Ornithonyssus bacoti*. *Inset:* Engorged mite. (From Ebeling, 1960.)

When rats occur in a house, their fecal pellets may be found in the attic, and can often be seen from the crawl hole. When rats are killed, the mites leave their bodies and may travel great distances, particularly along the heating pipes in the walls, for when they are not engorged with blood they are very active. When searching for mite infestations, a flashlight should be used and warm areas such as those near hot-water and steam pipes should be examined with particular care.

Description. The tropical rat mite is gray to pale yellowish gray, changing to red or black when engorged with blood (figure 311, *inset*). The females vary in length from about 1.15 mm when unfed to 1.41 mm when engorged. The males are about two-thirds as long as the females. A useful taxonomic character is the single dorsal plate, which is relatively narrow and does not cover the entire dorsal surface, even in specimens that have not fed. The dorsal plate bears pairs of long setae, more numerous on the anterior half, and in most specimens with only 6 or 7 pairs on the posterior half (Skaliy and Hayes, 1949; Baker *et al.*, 1956).

Life Cycle. After the adult female engorges, her first eggs may be laid within 2 days at ordinary temperatures (68 to 72° F; 20 to 22° C). They are deposited on debris in rat nests and burrows, but apparently not on the rats themselves. They hatch in about 36 hours. The larvae do not feed, and within a day they molt to enter the first nymphal stage, the *protonymph*. The protonymphs attach to a host and obtain a blood meal before dropping off and molting to become *deutonymphs*. In this stage they do not feed, but in 24 to 36 hours molt to become adult males or females. They mate and engorge within 3 days. Unfertilized females reproduce parthenogenetically. Four or 5 blood meals are required for the completion of the entire life cycle. The life of an adult female was found to average 61.9 days; the number of eggs laid, 98.8; and the life cycle from egg to egg, 10 to 12 days (Shelmire and Dove, 1931; Bertram *et al.*, 1946; Baker *et al.*, 1956).

Tropical Rat Mite Dermatitis. When the mites are abundant, they may be found anywhere in the house, and both nymphs and adults may attack people. Their bites produce irritation, and sometimes a painful dermatitis will continue for 2 or 3 days, leaving red spots on the infested areas (figure 311). Scratching may result in secondary infections.

Within some households, certain individuals are affected while others are not. Sometimes, much time and money will be spent on ineffective medication, and it is usually difficult for the infested person to obtain a correct diagnosis. This acariasis cannot be distinguished from flea bites, and is sometimes misidentified as scabies.

Control. The complete control of rats would, of course, eventually result in the elimination of tropical rat mites from infested premises. However, rat control often proves to be difficult, and "ratproofing" an attic may also be difficult and very expensive. It should also be borne in mind that trapping or otherwise killing rats may increase the attacks on the inhabitants of the house for a time because of the suddenly increased number of mites that leave the bodies of the dead rats. Unfed protonymphs have been observed to survive for as long a period as 43 days without food (Sudd, 1952).

Acaricides that depend on toxic action lose their toxicity too rapidly, particularly in the high summer temperatures of an attic. Reinfestation may then occur. HCN gas fumigation has been used successfully, but it is expensive and leaves no residue.

The successful results of a fluorinated silica aerogel dust, blown into attics for the prevention of drywood termites (*Incisitermes minor*) suggested a similar use of this material for the control of the tropical rat mite (Ebeling, 1960). In 5 infested houses and 1 2-story apartment house, in each of which 1 or more inhabitants had been attacked by rat mites for prolonged periods, the silica aerogel Dri-die 67 was blown into the attic at the rate of 1 lb to 1,000 sq ft (0.45 kg to 93 sq m) of attic area. For 4 of the houses, the dust was also blown into the crawl space under the house at the same rate (for floor space) as that for the attic area. An electric duster with a 1-gal (4-l) hopper was used to apply the dust. In the attic, the dust was applied entirely from the crawl hole, and under the house, from 1 or 2 crawl holes or a larger number of foundation vents. Since in all cases the mites were already distributed throughout the dwelling, some dust was applied with a small bellows hand duster under mattresses, on the spring supports of beds, along the edges and in the 4 corners of bed frames, into the junctures of seats and back or arm rests and under the pillows of sofas and lounges, under furniture and other out-of-the-way places, and in a few spots along the floor boards and ceiling moldings.

The decision to make use of Dri-die 67 dust was made in an effort to bring about the immediate cessation of mite attacks. Principal reliance was placed on the dusting of the attic for long-term control.

Rat mites may live as long as 63 days with no food (Scott, 1949), so those not coming into contact with the dust may continue to infest the inhabitants of a house. In all the buildings treated, severe infestations had been experienced up to the date of treatment, but ceased immediately afterward, and were never resumed except in 1 house where the housewife received a few more bites after treatment. In this instance, she applied more dust behind the electric outlet plates and various other areas that had not been covered the first time. Control was soon obtained, and no reinfestation occurred. Conventional liquid acaricides could have been applied in the living spaces of the treated houses, for people usually try to avoid leaving an unsightly residue. However, dusting is appropriate in attics, wall voids, and other inconspicuous areas. The dusting should be done *before* rat control is attempted, so that mites leaving the bodies of dead rats will contact the dust and will not be able to reach the living space and infest its occupants.

House Mouse Mite, *Liponyssoides sanguineus* (Hirst) (= *Allodermanyssus*) (Macronyssidae)

This mite occurs in northern Africa, Asia, Europe, and the United States. Although the house mouse (*Mus musculus*) is its preferred host, it will also feed on rats and other rodents. The house mouse mite attacks man, and causes a dermatitis in about the same way as does the tropical rat mite.

Importantly, there is also considerable circumstantial evidence that it can transmit rickettsial pox, caused by *Rickettsia akari* (Baker *et al.*, 1956).

Description. Unengorged mites are 0.65 to 0.75 mm long, and engorged females may reach a length of 1 mm or more. Their color may range from red to blackish, depending on how recently blood meals have been taken; they cause the mites to appear black. This species has 2 dorsal plates on the adult female, but even on unengorged specimens, the plates do not cover the entire dorsal surface. The anterior plate on the dorsum is 10 times larger than the posterior plate, and bears several pairs of setae, whereas the posterior plate bears only 1 pair. The chelicerae are long and whiplike (Baker *et al.*, 1956).

Life Cycle. As with most macronyssid mites, there is an egg, larva, protonymph, deutonymph, and adult stage. Unlike the tropical rat mite, both protonymph and deutonymph require blood meals.

The life cycle occupies 17 to 23 days, and unfed females have been observed to live as long as 51 days. The adult mite leaves its host after feeding, and may be found crawling about in mouse nests, runways, or on the walls and ceilings of infested buildings (Baker *et al.*, 1956).

Control. Control measures are the same as for the tropical rat mite.

Northern Fowl Mite, *Ornithonyssus sylviarum* (Canestrini and Fanzago) (Macronyssidae)

This species (figure 312) is an ectoparasite of domestic fowls and many wild birds, but in the absence of bird hosts it will sometimes attack humans, causing an itch. It is similar to the tropical rat mite in appearance and life cycle. It can become a household pest when birds build nests under eaves of a house or in the attic. For control, these nests should be removed. Otherwise, treatment is the same as that recommended for control of the tropical rat mite.

Chicken Mite, *Dermanyssus gallinae* (De Geer) (Dermanyssidae)

This cosmopolitan species is a pest of poultry and wild birds. Poultry roosts or bird nests can be sources of home infestations, and the human occupants can also be infested. The bite of this mite causes painful skin irritation. Unfed adult mites are about 0.75 mm long and nearly white (figure 312). After a blood meal, they may become 1 mm long and bright red. The female oviposits in crevices or under debris in chickenhouses or bird nests. Under favorable conditions, the entire life cycle may require only 7 days. The adults can survive without blood meals for 4 or 5 months (Baker *et al.*, 1956). Chicken mites have been controlled by spraying the chickenhouse with 1% malathion or by dusting infested litter with 2% malathion dust at the rate of 1 lb to 20 sq ft (0.45 kg to 1.85 sq m) (Furman *et al.*, 1955).

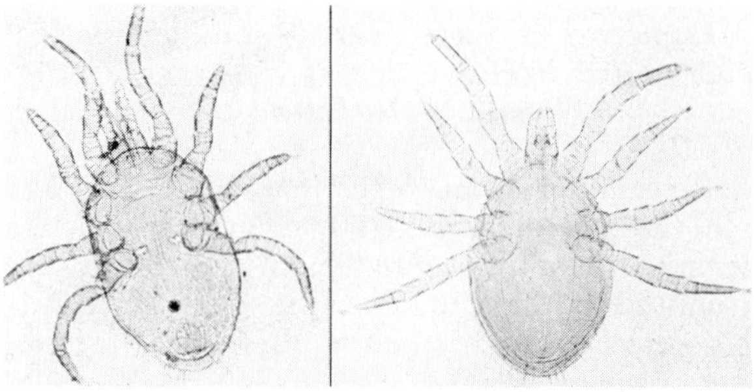

Fig. 312. Northern fowl mite, *Ornithonyssus sylviarum* (*left*), and chicken mite, *Dermanyssus gallinae*.

Human Itch Mite, *Sarcoptes scabiei* var. *hominis* (Hering) (Sarcoptidae)

Different varieties of *Sarcoptes scabiei* (De Geer) are believed to be specific for different mammals, including man and a large variety of domestic and wild animals, but are transferable from one host to another. The variety specific to man is generally referred to as the "itch" or "scab" mite, and acariasis caused by it is sometimes called "scabies." People are most likely to become infested when living in continually crowded quarters, such as slums or jails, or during periods of major calamities that result in prolonged overcrowding.

The human itch mite has a legitimate claim to fame in the history of biology. The original description of the life cycle and habits of this mite and the proof that it was the cause of scabies were accomplished by the Italian pharmacist Diacinto Cestoni and the relatively obscure young physician Giovan Cosino Bonomo. This was in the seventeenth century, when endo- and ectoparasites were usually considered to be produced by spontaneous generation. The observations made by Cestoni and Bonomo became generally known through a letter written by Bonomo to Francesco Redi (1626–1697), an experimental entomologist best known as a debunker of the "spontaneous generation" myth. The letter was reproduced in facsimile by Lane (1928). It has been described as "the birth certificate of parasitology" (Sadun, 1969).

Buxton (1921*a*) made a detailed study of the external anatomy of the equine itch mite, *Sarcoptes scabiei* var. *equi* (Gerlach, 1857). A subsequent study of the anatomy of the human itch mite, *Sarcoptes scabiei* De Geer, 1778, var. *hominis* (Hering, 1880), revealed certain minute differences in scales and spines, but these differences were not constant and measurements overlapped. Buxton (1921*b*) concluded that it was convenient to regard the 2 forms as varieties, but that this was more justifiable on physiological than on morphological grounds. For practical purposes, Buxton's (1921*a*) drawings and descriptions of the variety *equi* serve for the variety *hominis*. Heilesen (1946) made a detailed study of the anatomy of all stages of *hominis*, as well as an investigation of the biology of the species *scabiei*.

Description. Itch mites (figure 313) are broadly oval, somewhat hemispherical, and so small that even the adults are barely visible to the unaided eye. Adult females are 0.33 to 0.45 mm long, and the males, 0.20 to 0.24 mm. The mites are a translucent, dirty-white color, with the more highly chitinized portions brownish. The integument is finely striate over most of its surface. In living specimens, the body is seen to be divided into 2 regions by a fold in the integument; the posterior portion bears the last 2 pairs of the 4 pairs of very short legs. The last 2 pairs of legs do not extend as far as the margins of the body. The anterior 2 pairs of legs on females and all but the third pair on males are provided with delicate, stalked, terminal suction pads. In the females the posterior 2 pairs of legs, and in the males the third pair, terminate in bristles. The adults lack eyes and any special respiratory organs. Characteristic spines and bristles on the dorsal surface aid in identifying the species. The spines are directed backward, and may serve to anchor the mite in position when it is digging burrows in the skin (Munro, 1919; Buxton, 1941*b*; Hand, 1946; Heilesen, 1946).

Life Cycle. Both sexes and all stages of the itch mite tend to burrow into the skin immediately when placed on it, but the nymphs and males make only small, temporary holes, and move about frequently. The largest and longest burrows are made by the egg-laying female. The female always burrows in folds of the skin, preferring the deeper furrows and cracks. She can be induced to enter when a fine scratch has been made with a needle in the surface of the skin. She may also place herself in the acute angle between a sloping hair and the surface of the skin to gain support for initiating the burrowing (Heilesen, 1946). The winding burrow may reach a length of 5 to 15 mm. It is excavated in the deeper part of the horny epidermal layer, rarely as far as the granular layers (Buxton, 1941*b;* Heilesen, 1946).

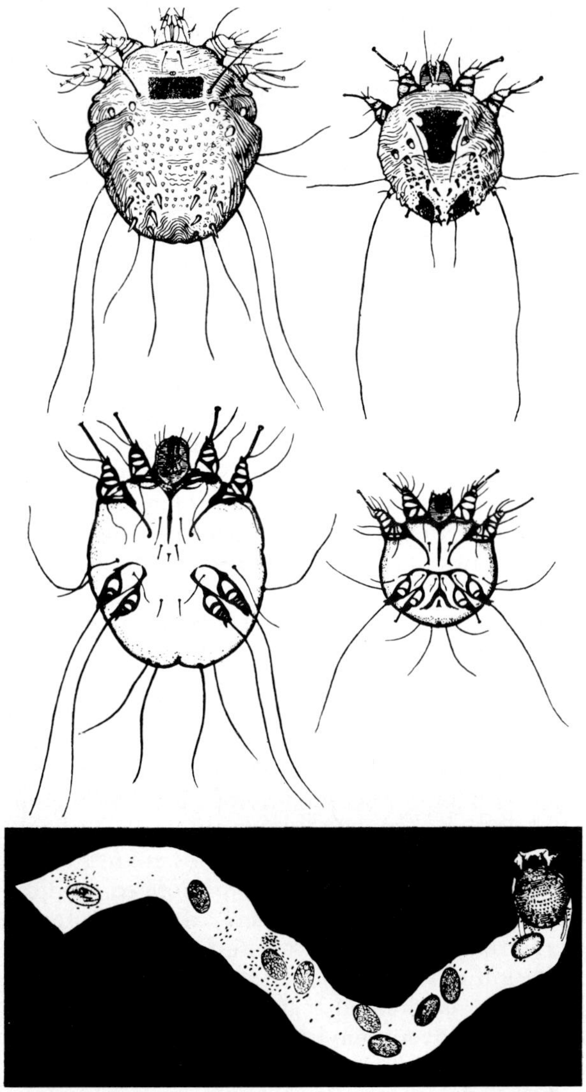

Fig. 313. ***Top,*** **dorsal and ventral views of the human itch mite,** ***Sarcoptes scabiei*** **var.** ***hominis. Left,*** **female;** ***right,*** **male;** ***bottom,*** **egg burrow with 9 eggs and a female mite at the end of the burrow. (Arranged from Munro, 1919.)**

Munro (1919) believed that, if undisturbed, the female would lay all her eggs in 1 burrow (figure 313). A second mating does not take place; she dies in the burrow. It appears from the observations of various authors that the adult life of the mite is from 2 to as many as 6 weeks. It is difficult to determine the number of eggs laid by a female in her lifetime, but it is usually estimated to be between 40 and 50. Munro also observed that the period between the beginning of burrow formation and the finding of the first larva varied from 71 to 78 hours. He also found that 9 eggs removed from burrows and kept at 29° to 30° C (about 85° F) hatched in 68 to 80 (average 74) hours. He gave the following numbers of days for the duration of the various life stages of the female: egg, 2.5 to 3.5; larva, 1.5 to 3; first nymph, 1.5 to 2.5; and second nymph, 2 to 4. He concluded that the life cycle required from 9 to 15 days. The numbers of days for the various life stages of the female, as determined by Heilesen (1946), were as follows: egg, 3 to 4; larva, 3; first nymph, 3 to 4; second nymph, 3 to 4; and from copulation to oviposition, 2; a total of 14 to 17 days. The developmental period for the male was only 9 to 11 days. The male has only 1 nymphal stage, whereas 2 are recognized in the female. In the second nymphal stage, the female can be fertilized by the male, even though the orifice (tocoptome) by which the eggs are laid has not yet been formed (Warburton, 1920).

Means of Transmission. The ease of transmission of body lice via infested clothing and bedding has led many people to assume that itch mites could be transmitted in the same way. An important difference, however, is that body lice live on their host's clothes and contact his body only to feed, whereas itch mites spend most of their lives beneath the host's skin. In experiments with 63 male volunteers, in none of 31 men were itch mites transferred via blankets previously used by infested men, and in only 2 cases out of 32 were mites transmitted when uninfested men used underclothing immediately after it had been used by infested men (Mellanby, 1941). Merely putting clothing away for 2 or 3 days at ordinary room temperature should be sufficient to rid it of mites. Two persons in a bed gave the greatest opportunity for the spread of itch mites. However, transmission is also possible through "dancing, flirtation, and ordinary intimate contact between members of a family" (Heilesen, 1946).

Body Regions Infested. Munro (1919) observed the burrowing procedures of egg-bearing females. The suckers of a female's anterior pair of legs were fixed onto the skin, and she propped her body up with the bristles on her posterior pair, assuming an almost perpendicular position. With her chelate mouthparts she commenced to cut the skin and bore in, becoming completely concealed in as little as 2.5 minutes. (This was a much shorter period than the one recorded by Heilesen, who found that 6 adult females burrowed into his skin in 15 to 40 minutes.) She ceased burrowing at a low temperature or when the body of her host was cold, and recommenced with a slight rise in temperature or warming of the body. Munro was able to activate burrowing of female mites in his wrist by passing from a cold room to a warm one, and was able to regulate the rate of burrowing by alternately warming and cooling an infested wrist over a radiator or other source of heat. He observed that under normal conditions, the burrowing period corresponded more or less to the time spent in bed. He stated:

> The parts of the body selected by the ovigerous female are the interdigital spaces; the wrists and the ulnar margins of the wrists; the elbows and the anterior folds of the axillae; the penis, scrotum, and buttocks; the back of the knee; and the ankles and toes. In young children, the egg burrows may occur on any part of the body, and in women, the undersides of the breasts are very commonly selected.

Localization of Infestation. In an examination of 886 soldiers, 9,978 adult female mites were found and removed, an average of slightly over 11 per patient. About 52% had fewer than 6 mites and only 3.9% had more than 50. One patient had 511. The percentages of mites found in the various areas of the body were: hands and wrists, 63.1; elbows (extensor aspect), 10.9; feet and ankles, 9.2; penis and scrotum, 8.4; buttocks, 4.0; axillae, 2.4; and in the remaining regions of the body, a total of 2 (Johnson and Mellanby, 1942). In another investigation, among 119 women the percentages of mites in various body areas were as follows: hands and wrists (excluding palms), 74.3%; palms of hands, 7.5% (none were found on the palms of men); elbows, 5.8; feet and ankles, 8.8; buttocks, 1.1; and in all other areas, 2.5. Among 18 children, itch mites were found to be more uniformly spread over many parts of the body, as indicated by Munro (1919), with many mites on their ankles and feet (Bartley and Mellanby, 1944).

Active stages of mites confined in cells on parts of the body other than the foregoing will burrow into these parts, but if the cells are removed, they will leave them for the nearest sites usually selected. When Munro confined female mites on his forearm, they always burrowed into his skin enough for concealment, but left these burrows and were recovered on the wrist in from 20 minutes to 2.5 hours.

Symptoms of Infestation. Whereas in animals large numbers of mites give rise to "sarcoptic mange," in humans relatively small numbers of mites can cause unpleasant symptoms, and the disease is known as "scabies." At a certain stage, the irritation may become so severe that the patient becomes frantic and suffers from lack of sleep. If the infestation is long continued, or if a later infestation occurs, an allergic reaction develops, with intense itching and a redness, or rash of follicular papules over much of the body. The rash may develop on areas such as around the armpits, the wrists, the waist, inside the thighs, and backs of the calves, but these areas do not necessarily coincide with those of mite infestation. The rash may occur over much of the body, even though only a few mites may be present in restricted locations between the fingers (Pratt, 1963). In an investigation of 55 volunteers who had not been infested with itch mites before, Mellanby (1944) observed that during the first month of itch mite infestation, there were few or no symptoms and no erythema. Infested persons might even be unaware that mites were burrowing into their skin.

Symptoms began to be evident in about a month, and in about 6 weeks the irritation was sufficiently severe to cause some loss of sleep. Then the itching grew progressively worse, and after 100 days it was practically continuous and almost unbearable. However, when secondary infections and impetigo developed, the mite population decreased, and was sometimes completely eliminated. (If secondary infections are neglected, they may themselves require prolonged medical treatment.) When volunteers already infested were treated and then reinfested, intense local irritation was felt within 24 hours, and a patch of erythema surrounded each mite. This apparently caused an adverse environment for the mites, for in another 2 days most of them had disappeared, some scratched out by the patient and others leaving the burrow. Relatively few mites reached maturity when compared with the original infestation.

Medical Treatment. It is important to diagnose scabies correctly, for neither the irritation nor the liability to skin diseases can cease until the mites have been eliminated. Look for the burrow of a female in such places as between the knuckles and in folds of the wrist and elbow, and then gently prick the burrow open. Toward the end of the burrow, the mite can usually be distinguished as a dull-white spot. Remove it with a needle. A bath before treatment is desirable for hygienic reasons. Thorough treatment is essential, and is best done by a physician or a reliable nurse or orderly. Treatment consists of application of ointments or liquid preparations. Ramsay (1969) prescribed either 25% benzyl benzoate emulsions, Kwell® (1% lindane) cream or lotion, or Eurax® cream or lotion. The latter contains 10% of crotamiton (*N*-ethyl-*o*-crotonotoluide). Ramsay recommended that these preparations be applied in the evening after the patient had taken a warm bath, and that the application be left in place until the next evening, when the treatment would be repeated. All areas of the skin below the neck should be treated, including body folds, palms, and soles. A cleansing bath should be taken 48 hours after the second application. Some tingling of the skin is to be expected after treatment, and it may last as long as 10 to 14 days. Calamine lotion or emulsion may be applied to alleviate this condition. Instructions on the package in which the medication is sold should always be read and followed carefully. Secondary infections may require the skills of a medical doctor or dermatologist. If the treatment is satisfactory and reinfestation occurs, an effort should be made to find untreated persons with whom the patient may have had contact.

Canine Mange Mite, *Sarcoptes scabiei* var. *canis* Gerlach, on Dogs and Humans

Sarcoptes scabiei on domestic animals has generally been referred to as a sarcoptic mange mite, and the symptom as "mange" or "scabies." The mite causes a self-limiting but very uncomfortable eruption if the secondary host is human. Man is particularly vulnerable to infestation by the variety of mange mite that infests the dog, probably because of his closer association with that animal than with others.

Description. This broadly oval mite is very small—about the size of the human itch mite. As in the human itch mite, in both sexes the anterior 2 pairs of legs and, in the male, also the fourth pair, have delicate, terminal, stalked suction pads. In the females, the posterior 2 pairs of legs, and in the males the third pair, end in bristles. Characteristic spines and bristles on the dorsal surface aid in identifying the species.

Life Cycle. The female lays her eggs in burrows she makes in the skin. They hatch in 3 to 5 days, and a complete life cycle requires 8 to 17 days (Smith and Claypoole, 1967).

Symptoms on Dogs. The eruption begins with small, white or erythematous (reddish, inflamed) papules, initially appearing in the groin and "armpit" areas and on the periphery of the ears. The papules may become so numerous that they appear to be contiguous, especially on the relatively hairless portions of the body. Crusts form, composed of dried exudates of serum and blood. The skin becomes lichenous, scaly, and corrugated in appearance. The eruption spreads, but is usually most severe on the ears, head, abdomen, and in groin and "armpit" areas. Hair is lost, or can be easily pulled out in patches. The remaining hair becomes dull and lusterless, and the animal develops a characteristic "mousey" odor. Intense generalized itching accompanies the eruption, and if untreated, the animal may become emaciated and even die of exhaustion from the itching and extensive chronic inflammatory reaction, often with a secondary infection. Canine scabies occurs most commonly in undernourished puppies, particularly if they are suffering from internal parasites (Smith and Claypoole, 1967).

Symptoms on Humans. A rash develops on some persons after only brief contact with an infested dog, but the most severe cases develop in those with prolonged close contact. The eruption is most often in the form of pimples, but may be characterized by blisters and inflammation. There is frequently a sloughing away of the skin. Lesions are most common on the forearms, lower region of the chest, and on the abdomen and thighs, but may be generalized. The distribution of symptoms usually differs from that of human scabies, in that the finger webs and the genitalia are usually not affected, but both these areas may be involved if the infestation is severe. The face is not affected except in children. In view of the facts that (1) the sarcoptic mite can live for 4 to 5 weeks, (2) in no cases have symptoms lasted longer than this, and (3) burrows have not been found in human skin, it appears that *Sarcoptes scabiei* var. *canis* does not reproduce on human skin, but this could be definitely proved only by prolonged observation on heavily infested human volunteers—an unlikely development (Smith and Claypoole, 1967).

Control of Mange on Dogs. In mange or scabies control on dogs, advantage has been taken of the fact that male mites do not do any significant boring into the skin, and may be controlled by continuous exposure to toxic vapors. In one experiment, a dichlorvos resin strip (No-Pest Strip®) was placed in the bedding inside the doghouse of a beagle, 3 × 1.5 × 2 ft (90 × 45 × 60 cm) in area. Within 2 weeks, the dog's skin, which had been red on the lower half of the body and on the legs from a severe case of sarcoptic mange, lost its redness, and new hair began to appear. The strip was moved to one side of the doghouse, and the dog's condition continued to improve; he recovered completely in 6 weeks. Whereas mites were taken from all skin scrapings before treatment, none were taken 6 weeks later. The dog had chewed the edges of the dichlorvos strip, but no ill effect was noted (Whitney, 1969).

The specialist who conducted this experiment had used dichlorvos resin strips for several years to protect dogs and cats from flies and mosquitoes, and had observed no sarcoptic mange or fleas on 60 beagles and 36 cats that had been kept in kennels and a cat room where the strips had been used.

Treatment for Scabies or Mange. Both dogs and humans have been effectively treated with "gamma benzene hexachloride cream." Humans were cured with 1 application. The patient should avoid further close contact with the pet until it is cured (Smith and Claypoole, 1967). The old NBIN lotion for a combination treatment for lice and scabies (Eddy, 1946) is still available under the name of Topocide, but may be procured by prescription only.

(C) PESTIFEROUS ARTHROPODS THAT DO NOT ENVENOMATE

There are several groups of arthropods that do not envenomate their victims, but become pests by merely being present in homes or other locations inhabited by man and his domestic animals. Some species, such as certain midges, become so pestiferous that they may cause recreational areas to be abandoned, or may deface lakeshore property. Others, such as the eye gnats, not only annoy the residents of an infested area, but interfere with the harvesting of crops. Such species are discussed in the present section, along with the larvae of certain insects that invade body tissues or cavities of humans, causing a condition known as *myiasis,* which is discussed near the end of the section.

Phantom Midges (Subfamily Chaoborinae)

Some authors consider these midges to be members of a distinct family of Diptera, and give them the common name, "chaoborid gnats," from the family name, Chaoboridae. In this book, they are considered to a be a subfamily of the Culicidae, along with the dixine midges (Dixinae) and the mosquitoes (Culicinae). (The mosquitoes are by far the most important culicid subfamily, and are practically the only species with mouthparts that can pierce the skin and suck blood.)

Description and Biology. The adult phantom midges are somewhat similar to mosquitoes, but lack the long proboscis and do not bite. Their mouthparts are no longer than their heads. The wing scales so characteristic of Culicinae are present only in the hind margins of the wings of the Chaoborinae; there are few or no scales along the veins. The larvae of some phantom midges resemble certain mosquito larvae so closely that the two might be confused by the casual observer. However, the larvae of phantom midges usually have their antennae inserted near the midline of the head; if inserted elsewhere, the antennae are provided with prehensile hairs with which the larvae can seize aquatic prey. Mosquito larvae always have their antennae inserted laterally on the head.

The term "phantom" is descriptive of the larvae, which are almost transparent and are difficult to see, even in clear water. Their antennae extend downward, and have 3 to 5 apical bristles that aid in securing food and in locomotion. The mouth has a large, eversible pharyngeal sac that functions as a crop for the selection of food and the rejection of extraneous materials. The pupae somewhat resemble those of mosquitoes. They have a large cephalothoracic area, bladderlike respiratory tubes, and 2 wide, flat, anal paddles. The dark, spindleshaped eggs float singly or in small rafts on the surface of the water, but those of some species soon sink to the bottom.

Clear Lake Gnat, *Chaoborus astictopus* Dyar and Shannon (= *lacustris* Freeborn)

This species first came to attention in the Clear Lake area in northern California. The gnats appear early in May and may be found until about mid-October. In the days before effective control, they were so extremely abundant that they got into the eyes, mouths, noses, and ears of the lakeshore residents, and often became so thick on door and window screens that scarcely any light could penetrate (Herms, 1937). At one time, this gnat was a serious problem only near Clear Lake, but as the water resources of the state continue to be developed, the problem is becoming widespread—the insects are becoming as numerous in many dams, reservoirs, and smaller ponds in other parts of California as they originally were in Clear Lake. The species is very adaptable to varied water conditions (Cook, 1967). *Chaoborus punctipennis* (Say), a species of similar appearance and habits, is a pest in Florida and along the Atlantic seacoast.

Description. The adults of *C. astictopus* are brown, with a vestiture of straw-colored hairs. They are 4 to 5 mm long. The males have a longer and more slender abdomen than the females and, as in the case of the mosquitoes, have more plumose antennae; they also have a pair of claspers at the tip of the abdomen (figure 314, *A*).

Biology. The females lay their dark, spindle-shaped eggs (figure 314, *D*) on the surface of the water, where they form geometrical designs similar to those of anopheline mosquitoes. Unlike mosquito eggs, however, they sink to the bottom when the water is slightly agitated. They hatch in 20 to 24 hours. When full grown, the "phantom larvae" (figure 314, *E*) are 8 to 10 mm long. They do not have to come to the surface for air as mosquito larvae do, for they breathe by means of gills, and possibly cutaneously as well. There are 4 larval instars (Herms, 1937). The first-, second-, and some of the third-instar larvae are free-swimming, and most of the third and all of the fourth and pupal instars stay in the bottom mud (Snell and Hazeltine, 1963). In laboratory experiments at 70 to 80° F (21 to 27° C) and with abundant plankton available, the larvae could mature in 11 to 25 days, although sometimes the fourth instar was greatly prolonged. Mature larvae fed readily upon small mosquito larvae that were offered to them. The pupal stage was 2 to 3 days. The adults freed themselves from the pupal skins in a few seconds, and were able to fly almost immediately. The emerging adults flew to shore, where they mated in shoreline trees (Deonier, 1943).

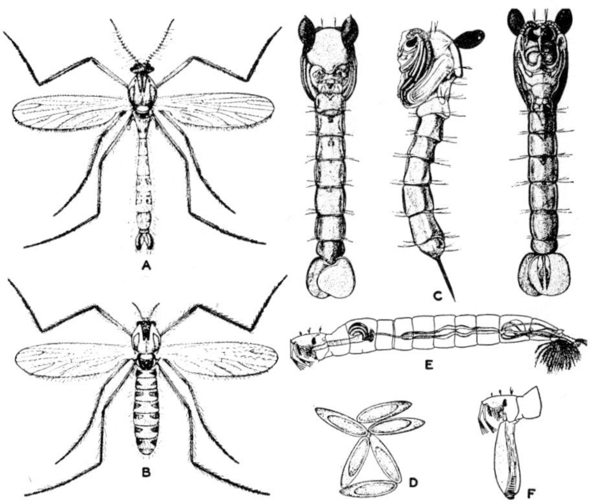

Fig. 314. Clear Lake gnat, *Chaoborus astictopus*. *A*. male; *B*, female; *C*, pupa; *D*, eggs; *E*, larva; *F*, eversible pharyngeal sac extended from the mouth. (From Herms, 1937.)

Snell and Hazeltine (1963) treated the water of a pond with methyl parathion at 4 ppb (parts per billion), which had a residual effect lasting only 9 days and killed only the first- and second-instar larvae and most of the third-instar larvae. By this treatment, an overlapping of instars and stages was eliminated sufficiently to enable them to measure the developmental periods by periodic samplings of the water. Under the natural ambient conditions of the pond, the period required for each stage and instar was found to be as follows: egg, 1 day; first-instar larva, 5–6 days; second-instar larva, 5–6 days; third-instar larva, 10–11 days; fourth-instar larva, 11–19 days; pupa, 2–3 days; adult preoviposition period, 2 days; resulting in 36 to 48 days for a summer generation.

Control. The chlorinated hydrocarbon TDE, in applications of up to 20 ppb in a single late-season treatment of the entire lake, resulted in effective phantom midge control for a 2-year period in Clear Lake in 1949. The lake was treated again in 1954 and 1957. The control was less effective with each treatment, that of 1957 giving only 95% control. Great concentrations of insecticide were found in fish, and large numbers of birds that nested around Clear Lake were killed. The decision was made to change to a rapidly biodegradable organophosphorus insecticide. In 1962, 3 treatments with an acetone solution of methyl parathion, the first with 2.2 ppb and the second and third with 3.3 ppb, resulted in control of the gnats with no ill effects on wildlife (Hazeltine, 1962; 1963). Currently, methyl parathion is being applied in 3 treatments, approximately 3 weeks apart, in July and August, standardizing concentration at 3 ppb. This has broken the cycle of the Clear Lake gnat without measurable harm to the lake's over-all biology.

Midges (Chironomidae)

This is a family of minute to moderately large species, the adults of which superficially resemble mosquitoes. The smaller species are believed by some people to be "young" mosquitoes. However, midges differ from mosquitoes in that the wings are not scaled and the mouthparts are short and not adapted for biting. Adult midges are slender, seldom more than 10 mm long, but usually less than 5 mm; the thorax is relatively large, partially concealing the small, spherical head; and the antennae are slender, 5- to 14-segmented, more or less hairy in the females but plumose in the males. The legs are slender and long, especially the forelegs, and the tarsi are often very long. The wings are long and narrow, bare or sparsely haired in the females and plumose in the males. The abdomen is long and narrow, especially in the males, that of the females being shorter and more robust, and the mouthparts form a short proboscis.

Grodhaus (1963) described in concise tabular form the adults of 13 chironomid species considered to be important as nuisance pests in California, giving color, wing length, larval habits associated with mass emergences, notes on geographic occurrences in California, and the characteristics of the 3 subfamilies. A key to chironomid adults taken in or near rice fields, and keys to the species of those larvae and pupae that were reared to adults, were prepared by Darby (1962). Keys to the adults and larvae of California lagoon-inhabiting species were prepared by Grodhaus (1967) but care should be taken when identifying adults with these keys not to include species from sources other than lagoons.

Similarly to mosquitoes, midges oviposit on water. The midge larvae are scavengers, and are aquatic or rarely terrestrial. Terrestrial species live in and on decayed vegetation, dung, fungi, mosses, under bark, and in the soil. Aquatic larvae are found in quiet or flowing water, and may become abundant in lakes, ponds, reservoirs, and tanks. Most are bottom-feeders, and are sometimes found at depths of nearly 1,000 ft (300 m). For most species, the food is thought to be particulate matter consisting of algae, other plant cells, and miscellaneous detritus. They possess a closed (apneustic) respiratory system and, unlike mosquito larvae, need not come to the surface for air. They absorb oxygen dissolved in the water by means of small "gills." Some larvae are free-living, and others make mud tubes or cases on stones, leaves, and pieces of wood.

In the subfamily Chironominae, which includes the important genus *Chironomus,* the larvae are usually red, with some variation in intensity. They typically inhabit tubes constructed of small particles and located in the bottom sediment. The pupae also inhabit these tubes, leaving just before adult emergence (Grodhaus, 1963).

Chironomid Midges as Pests. Despite not being able to pierce the skin or bite, midges can nevertheless be serious nuisances in urban as well as rural areas. In urban areas, runoff water from the streets of residential and commercial areas finds its way into a network of drainage ditches and flood channels. The drainage water is high in nutrients and creates a very favorable environment for midge production throughout the year except, in California, during the winter floods. Immediately following the winter floods, species of *Chironomus* and *Tanypus* can become so numerous that thousands will rest on the outside walls, under the eaves, and in doorways of houses within a quarter mile (400 m) of the breeding sources, and will enter the home through the slightest crack or opening. This is especially the case at night, for they are attracted to artificial light (Thompson *et al.,* 1970). They fly into people's eyes, ears, and mouths, and are sometimes inhaled. When in large numbers, they soon contaminate food served in illuminated areas. Where outbreaks are frequent, spiders that feed on midges soon become nuisances also. Real estate values tend to be reduced where midges are abundant. Midges also can interfere with the processing of paper, plastics, and food products, and with automotive refinishing operations (Grodhaus, 1963; Anderson *et al.,* 1965).

Chironomids often occur in large numbers in reservoirs, and may be passed into the drinking water. They often emerge in extremely large numbers from waste-water stabilization lagoons or oxidation ponds (Usinger and Kellen, 1955; Grodhaus, 1967; Brumbaugh *et al.,* 1969). These are shallow impoundments receiving domestic sewage or other liquid waste. Bacterial action and photosynthesis alter the organic material, and much of the waste is converted into plankton. The soft layer of organic sediment that accumulates on the bottom provides food and substrate for chironomid larvae. As might be expected, chironomid midges are produced in enormous numbers in rice fields. Darby (1962) recorded 30 species from rice fields of the southern Sacramento Valley, California, 18 of which were considered to be rare.

Chironomus plumosus L.

These midges, often called "lake flies," are common over much of the United States. They can create a serious problem in lakefront communi-

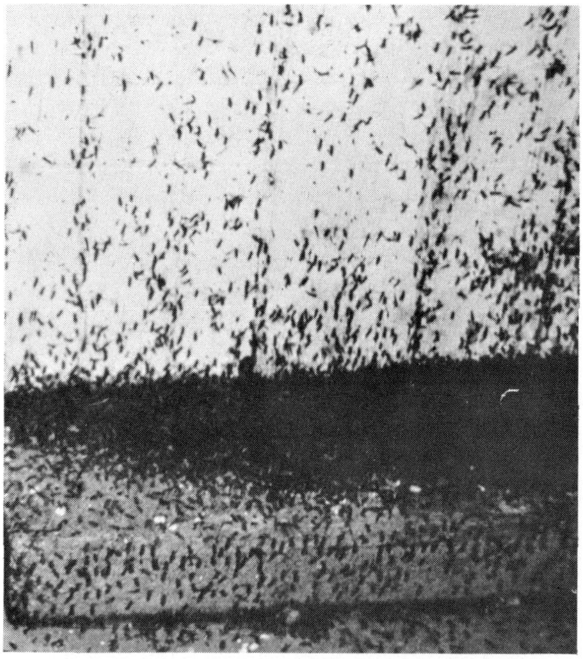

Fig. 315. Chironomid midges, *Chironomus plumosus*, massed against a building. (Courtesy E. C. Bay.)

ties. The problem in one such community in California was graphically described by Bay (1967). *Chironomus plumosus* is a pest not only because of the presence of bothersome swarms of adults, but also because the adults may lay their eggs in large, dark, sticky deposits glued to the sides of buildings (figure 315) and to surfaces such as windows, walls, neon signs, and moored boats. These exposed egg masses, each containing up to 2,000 eggs, never hatch, for hatching is possible only when they are deposited in water. The egg masses either become hard and dry or, if they become wet, dissolve and smear into unsightly brown stains. If there are no lights on shore during the hour following sunset, few egg masses will be deposited on terrestrial structures, but if the females are attracted by shore lights, they will remain and deposit their eggs there, dying afterward. The dry egg masses can be removed by wiping with a dry cloth. If the eggs have been smeared by rain, they should be wet with a 10 to 20% aqueous solution of Clorox or a similar commercial bleach, then washed away after a few minutes (Bay, 1967).

Description. The adults (figure 316) are about 12 mm long—the largest American chironomids. They are predominantly light brown, often tinged with green or yellow, and with the basal two-thirds of each abdominal segment dark brown. The thoracic tergites have median anterior and lateral posterior dark areas. The pale legs have narrow, dark bands at the distal end of each tarsal segment. When it first emerges, the adult has a reddish cast, but this disappears and darkens after a few hours, although some of the reddish color may be visible in the thorax for more than a day (Hilsenhoff, 1966).

Life Cycle. More than 90% of the oviposition occurs within 1 or 2 hours after sundown. The egg mass, 3 to 5 mm long, is ordinarily placed on a

Fig. 316. Female *Chironomus plumosus* completing oviposition. (From Bay, 1967.)

lake or other large body of water, where it swells to several times its original size and becomes transparent, revealing the individual eggs (figure 317). The egg mass slowly sinks to the bottom (Hilsenhoff, 1966; Bay, 1967). The time required for the eggs to hatch depends on the temperature at the bottom, and can vary from 3 days at 24° C (75° F) to 14 at 9° C (48° F) (Hilsenhoff, 1966).

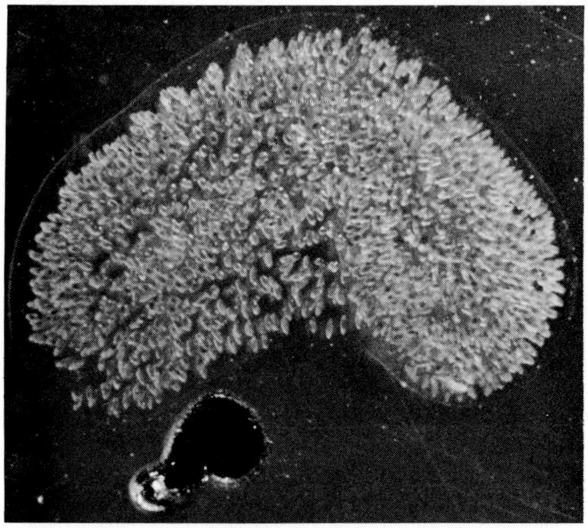

Fig. 317. Expanded egg mass of *Chironomus plumosus*, with unexpanded egg mass below. (Courtesy E. C. Bay.)

The first-instar larvae are colorless, and not until late in the second instar does a pale-pink color appear. In the third instar, the color deepens to light red, becoming very deep red in the fourth instar. There is also considerable green pigment just before pupation, especially in the thoracic region. Full-grown larvae range from 15 to 30 mm in length. The larvae feed on decayed algae and diatoms. At least the last 3 instars, they construct U-shaped tubes in the mud. Fourth-instar larvae were observed to crawl from their tubes to feed while remaining anchored to the tubes by their caudal prolegs. Whereas the tubes of the second- and third-instar larvae are at or just beneath the mud surface, the fourth-instar larvae may construct their U-shaped tubes to extend as far as 8 cm into the mud.

Pupation takes place in the larval burrow. In Lake Winnebago, Wisconsin, pupation may require 6 to 10 days in early May, when mud temperatures at the bottom are about 10° C (50° F), and as little as 1 day in July and August, when the temperatures normally range from 23 to 25° C (73 to 77° F). As with the larvae, the pupa is also red, with some green pigment, and as it ages it becomes darker until it is almost black. Female pupae average 17.7 mm in length, compared with 16.8 mm for the males. After pupation is complete, the pupa leaves the larval burrow and swims to the surface. The adult then emerges in 15 to 30 seconds. The empty pupal cases float for at least 24 hours, and often accumulate in windrows on the surface (Hilsenhoff, 1966).

Chironomus attenuatus Walker [= *Tendipes decorus* Johannsen and *T. attenuatus* (Walker)]

This species is widely distributed in the United States, and is often a troublesome pest. It normally breeds in lakes, ponds, rivers, and streams, but in semiarid regions, seasonal breeding situations can be developed. In California, under favorable rainfall and water conditions, it may develop 2 or more consecutive broods in the organic ooze of its aquatic habitats, with larval populations of 5,000 to 50,000 per square meter. The swarms of adults may interfere not only with normal outdoor activity, such as clothes drying and painting, but may enter houses and annoy the occupants, or enter lighted commercial buildings and cause damage, such as becoming embedded in plastic sheeting or bringing the enameling of cars to a halt (Grant, 1960). Along with other chironomids, particularly *Tanypus grodhausi*, it can be a hazard to motorists when occurring in dense swarms near breeding sites (Bay *et al.*, 1965).

Chironomus attenuatus was among the 3 most important species of related midges in Los Angeles County's water-spreading basins surveyed by Anderson *et al.* (1964), the others being *C. californicus* and *Pentaneura monilis*. It was the most important chironomid pest in the waste-water stabilization lagoons at Stockton, California (Brumbaugh *et al.*, 1969), and was one of the few related species found in the organic material of sewage oxidation ponds (Usinger and Kellen, 1955). It was noted to occur in large numbers in California rice fields (Darby, 1962).

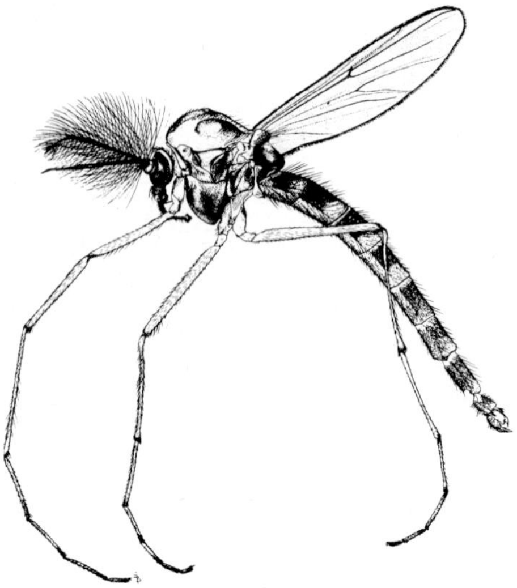

Fig. 318. A widely distributed chironomid midge, *Chironomus attenuatus* (male).

Description. The adults (figure 318) are 5 to 6.5 mm long, and have a ground color of light pruinose green to light brown. The antennae, mouthparts, and thoracic markings are brownish ochraceous to dark brown. The abdomen is usually pale green, with tergites 2 to 7 each having a central transverse brown band, indicated in the figure, occupying from a fourth to most of its length. In occasional specimens, the abdomen is entirely green. The males may be identified by their plumose antennae; otherwise, the 2 sexes are similar in general appearance. The usual absence of a "tarsal beard" and the heavy, dark-brown wing veins distinguish this exceptionally variable species from others in the genus.

Biology. From the time a swarm of adults is noticed at sunset, it grows in size until it may form a column 30 cm in diameter, about 1.5 m in height, and with a base about that distance above the ground. Mating takes place in the swarm. When ovipositing, the female extrudes a brown egg mass and deposits it on the hind femora. It is held there for 2 or 3 minutes, then thrown down on a water surface. The egg mass then gradually expands. There is an average of about 700 eggs in the cluster (Ping, 1917).

The larva, at first pale gray, becomes red, with a brownish head, and is about 5 mm long. In every respect it resembles the mature larva, which attains a length of 11 to 12 mm and has ventral blood gills about 2 mm long. The larva has been reported to feed on aquatic vegetation (Ping, 1917), but it has also been shown that the introduction of raw organic matter, when it overbalances the established agents of decomposition, tends to support high populations of this and other midges (Whitsel *et al.,* 1963).

The mature pupa, about 7 mm long, swims tadpolelike under the surface of the water. After pupation is complete, the pupa leaves the larval burrow and swims to the surface. The adult usually frees itself from the pupal skin in less than 15 seconds, and frequently within 4 or 5 seconds. Adults have lived for as long as 6 days under very humid laboratory conditions, but might live longer in the grasses, rushes, and sedges on the shores of bodies of water. There are usually 5 generations per year (Ping, 1917; Hilsenhoff, 1966).

Chironomus californicus Johannsen (= *Dicrotendipes*)

Large numbers of these midges (figure 319) become nuisances when they rest on vegetation and screen doors in residential properties near reservoirs and other sources. The adults are grayish in color and about 4.5 mm long, apparently too large to pass through ordinary fly screen. According to E. C. Bay (correspondence), peak emergences usually occur between 2 and 3 weeks after newly wetted environments are formed, then decline to about 10 or 15% of their original numbers after the initial outbreak. The drying and rewetting of reservoirs, or large parts thereof, may result in additional outbreaks. Practically all mature (nonteneral) pupae within a pond transform and emerge in an evening, predominantly in a 30- to 60-minute period shortly after sundown. This is also about the time of maximum egg deposition (Frommer and Rauch, 1971).

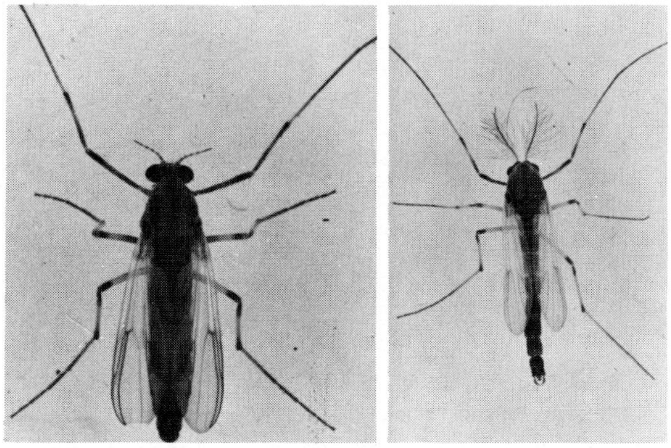

Fig. 319. *Chironomus californicus. Left,* female; *right,* male. (From Ebeling, 1964*a*.)

Control of Chironomid Midges

The midges have been controlled for relatively brief periods by fogging with malathion in fuel oil at 0.5 lb (0.23 kg) of toxicant per acre (0.405 ha), or by applying fenthion in a granular formulation at 0.2 lb (0.09 kg) per acre (Hagmann, 1963; Patterson and Wilson, 1966). Tests in a shallow sewage oxidation pond showed chlorpyrifos, parathion, and methyl parathion to be effective at 0.2 to 0.25 lb (0.09 to 0.11 kg) per acre against *Goeldichironomus holoprassinus* (Rempel), *Tanypus grodhausi* (Sublette), and *Chironomus* sp. 51 (Mulla and Khasawinah, 1969). Chironomid larvae have been controlled in waste-water stabilization lagoons with a specially designed, quick-breaking emulsion of fenthion and methyl parathion (Brumbaugh *et al.,* 1969). Chlorpyrifos, applied as granules or emulsifiable concentrate at 0.2 lb per acre, gave excellent control of chironomid midge larvae in lakes. The concentrate caused slight kills of certain species of fish, but the granular formulation did not (Mulla *et al.,* 1971). Granular formulations have the advantage of being selective, for they quickly sink to the bottom where the larvae live, and are therefore less likely to kill nontarget species (Jamnback, 1969). A fog of base oil containing 5% malathion and applied with a portable fog generator is effective in eliminating adult midges from shrubbery, lawns, and the walls of residences, but only for a few hours.

Chironomid midges are rarely nuisances in a well-balanced aquatic community. Pollution of water with materials that encourage the growth of the algae upon which the midge larvae feed results in excessive midge populations. One of the disadvantages of insecticides is that they kill the larvae without destroying the algae. The food supply thus accumulates, and favors an outbreak of midges when the insecticide is no longer effective and the body of water is recolonized (Jamnback, 1969). The midges have been controlled in small bodies of water by stocking them with carp and goldfish at the rate of 150 to 500 lb per acre (68 to 225 kg per 0.405 ha) of water surface (Bay and Anderson, 1965).

Moth Flies and Allies (Psychodidae)

The psychodids are small, long-legged, hairy, mothlike Diptera (figure 319-A). They are rarely more than 5 mm long. These flies are found in moist places, sometimes in such numbers that it is difficult to avoid inhaling them. This is especially true of some species that breed in sewage filter plants, where they are called "filter flies" or "sewage flies." Swarms of these flies, in the genera *Psychoda* and *Telmatoscopus,* may also emerge from sink and bathtub drains. The species of *Psychoda* have the terminal segments of their antennae reduced in size, whereas those of *Telmatoscopus* do not (figure 319-A). Although the moth flies have been considered only as annoyances in the United States, in South Africa they have been implicated in many cases of inhalant allergy.

Most species of moth flies are in the subfamily Psychodinae. Their wings are usually held rooflike over the body. Their mouthparts are short and are not adapted for bloodsucking. They may be found in shady places, along streams with decaying wastes, or around sap exuding from trees. Large numbers may be seen on dense foliage in swamps, crawling about on the undersurfaces of leaves. Except under artificial conditions, they are of little importance as pests. The eggs of this subfamily are pale or brown, oval, and have inconspicuous reticulations. The larvae are commonly aquatic. They are elongate and cylindrical, but are flattened ventrally and have 8 ventral suckers. The tergal plates and the posterior siphon and spiracle are conspicuous features.

Species in the subfamily Phlebotominae, the "sand flies," are less hairy than the Psychodinae, their wings are not held rooflike over the body, and the larvae are never truly aquatic. The females have rather long, piercing mouthparts, adapted for bloodsucking. The males suck moisture from any available source, such as perspiration from humans. These gnats are active only at night and when there is little or no wind. During the day, they may be found in sheltered places both outdoors and in buildings. There is only 1 genus, *Phlebotomus,* species of which are carriers of diseases such as pappataci fever, kala azar, leishmaniasis, and Oroya fever in many tropical areas of the world, but this does not occur in the United States.

Moth Flies in Sewage Filter Plants

Moth flies may develop in the muck or gelatinous material that accumulates in sewage disposal beds, septic tanks, compost, or dirty garbage containers, or they may emerge from the drains of sinks or bathtubs, from treeholes, rain barrels, or from very moist organic solids and bird nests that have accumulations of moist excreta. They are probably most commonly associated with filter sewage beds where the larvae and pupae thrive in the gelatinous film covering the filter stones or in other places where decomposing organic ma-

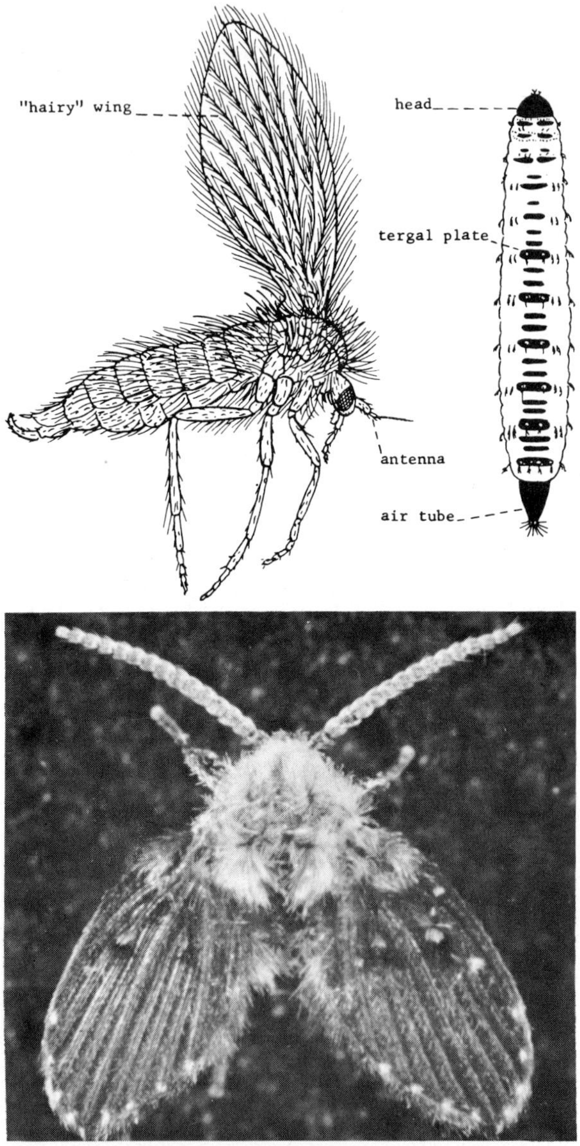

Fig. 319-A. Moth flies (Psychodidae). ***Top,*** **adult fly and larva;** ***bottom, Telmatoscopus albipunctatus.*** **(Top drawing from Scott, 1961*c*.)**

terials are found. The larvae are beneficial in that they feed on algae, fungi, bacteria, and sludge in sewage-disposal beds, breaking down the gelatinous film into small fecal pellets that are easily washed away. Moth flies are pests when the adult flies become so abundant that they get into the eyes, ears, and noses of the workmen in the area. They may also find their way into near-by homes, or may even originate in drain pipes within the homes.

Psychodids go through their life cycles in 1 to 3 weeks, and live about 2 weeks after emerging. They are weak fliers, so that in the home they are generally seen crawling on walls or other surfaces. When they do fly, their flight is in short, jerky lines and for only a few feet at a time. They are attracted to lights, and are so small that they can penetrate ordinary fly screens.

Pacific Drain Fly, *Psychoda phalenoides* (L.) (= *pacifica* Kincaid)

This is the psychodid most commonly referred to as a pest along the Pacific Coast, ranging from Alaska to southern California. The adult is 2 to 2.3 mm long, with brownish-gray wings and with body and wings densely covered with long hairs. As with most other psychodids, the wings are usually borne rooflike over the body, and each segment of the 13-segmented antennae has a bulbous swelling with a whorl of long hairs. Larvae and pupae of the Pacific drain fly may occur in the gelatinous lining of the portion of a drain pipe beneath a sink that is above the permanently water-filled trap. In one experiment, when the gelatinous material was removed by forcing a paper swab with an outer coating of wax paper down the drain, no more flies emerged from the sink (Mallis and Pence, 1941).

Psychoda alternata Say

This species is 2 mm long, and has 15-segmented antennae. It ranges from Florida to Massachusetts and westward to Washington and California. The female lays a gelatinous mass of 20 to 100 eggs in any of the suitable breeding media just described. The eggs hatch in 32 to 48 hours, larval development requires 9 to 15 days, and the pupal period lasts 20 to 40 hours. The life cycle takes 21 to 27 days under room conditions. The adults are weak fliers, and can often be found resting on foliage. They seek shade in the daytime, but are attracted to light at night. They feed on nectar or polluted water. They may enter houses in large numbers (Quate, 1955; Scott, 1961*c*).

Other Species of *Psychoda*

Also common in sewage filter beds are 2 species that are pale yellowish and, as with all species of *Psychoda,* the terminal segments of the antennae are reduced in size. *Psychoda satchelli* Quate has 14-segmented antennae. It ranges from Georgia to Quebec and westward to Alaska and California. *Psychoda cinerea* Banks has 16-segmented antennae. Unlike other species with 16-segmented antennae, the 3 terminal segments are separate and are equal in size. It ranges from Puerto Rico to Ontario and westward to Washington and California (Quate, 1955).

Telmatoscopus albipunctatus (Williston)

This is one of the moth flies that infests sewage filter beds and drain pipes (figure 319-A, *bottom*). It is widely distributed over the United States. The specimen shown in the figure was one of many that emerged from the drain pipe of an infrequently used sink, with a leaking faucet, in a glasshouse on our Los Angeles campus. Its color is generally brown or blackish, but with the white hairs on its thorax and antennae rather conspicuous. These features, along with the white spots at the tips of the wing veins, clearly seen in the figure, make this a well-marked species that is easily distinguished from other North American species of the genus (Quate, 1955).

The larva is 8 to 10 mm long. It has 26 tergal plates: 2 on segments 1 to 4 and 3 on segments 5 to 10 (Quate, 1955). Its brown or blackish color distinguishes it from the pale-yellowish color of the *Psychoda* species (Scott, 1961*c*). Besides being found in sinks and drains, the larvae have been found in mud, in water in treeholes and rain barrels, in shallow pools partly filled with dead leaves and debris, and in the rinds of kukui tree fruit (*Aleurites moluccana* Willdenow; candlenut; Euphorbiaceae) (Williams, 1943).

Allergic Reactions to Moth Flies

Thirteen cases of bronchial asthma were reported among workers at a sewage plant in the Transvaal, South Africa, all the result of allergic response to the moth fly *Psychoda alternata.* The inhalant allergen was demonstrated to be the dust resulting from the disintegration of the flies' bodies. One individual who was investigated with particular thoroughness was found to have no family history of allergy, and was not found to be sensitive to any of the more common causes of allergy, such as feathers, dusts, and pollens (Ordman, 1946).

Control of Moth Flies in Sewage Filter Beds

In sewage filters, flooding and treatment of the slime layer result in good control of moth flies. However, the efficiency of the biofilter depends on the activity of this slime layer, and it must not be destroyed. Treating the slime with insecticides disrupts its biological equilibrium by killing large numbers of key organisms, and the toxicant may poison aquatic life downstream from the filter bed. Scott (1961*c*) suggested that weeds and grass be mowed short throughout the sewage-plant area, and that walls and other structural elements of the filters be sprayed with 2.5% malathion suspension at 1 gal per 1,000 sq ft (4 l per 93 sq m), using a flat-spray nozzle, to give a residual deposit. He suggested that the spray treatment extend out from the edge of the filter for 20 ft (6 m), but that the slime layer not be sprayed. If necessary, the edges and walls of sludge-drying beds could be sprayed, and steps to improve the quality of the product coming from digestors should be taken. The treatment should be repeated as needed, typically at 2-month intervals. Scott recommended DDT for the treatment of sludge beds, but DDT, aldrin, and dieldrin are no longer registered for such uses as the control of filter fly larvae in sewage systems or the control of mosquito or tabanid larvae in aquatic outdoor areas (Anonymous, 1970*a*). The appropriate state authorities should be consulted for information on effective, currently registered insecticides. Scott also noted that the removal of vegetation reduced favored resting sites for filter fly adults, and that residual insecticides killed adult flies so rapidly that large populations could not build up. Some larvae might remain in the filter beds, and occasionally small "blooms" of adults might occur, but they usually persisted in such small numbers that they were no longer pests.

Frit Flies and Eye Gnats (Chloropidae)

The frit flies are very small insects, rarely exceeding 3 mm in length. They are very common, and may be collected almost anywhere on many plants and particularly on grasses, on which the larvae feed. Many species are pests on cereal plants all over the world, the wheat stem maggot, *Meromyza americana* Fitch, the frit fly, *Oscinella frit* (L.), and the American frit fly, *O. soror* (Macquart), being the principal species in the United States (Schread, 1964). Two species, *O. neocoxendix* Sabrosky and *O. coxendix* (Fitch), cause "silver top" of bluegrass (*Poa pratensis*) (Starks and Thurston, 1962).

The eye gnats (*Hippelates* spp.) are strikingly different from other chloropids in habits and in the manner in which they are pests. In California, they occur south of Madera County at lower elevations; the species are *H. collusor* (Townsend), *dorsalis* Loew, *impressus* Becker, *pusio* Loew, *robertsoni* Sabrosky, and *hermsi* Sabrosky. All except *H. hermsi* are pests (Ecke, 1963). Herms (1928) pointed out that *H. collusor* was probably responsible for the epidemics of bacterial conjunctivitis (pinkeye) in the Coachella Valley in southern California, and a few years later the same species was incriminated as the vector of the disease in Florida (Bengston, 1933). Eye gnats later created problems in other cultivated areas, such as the Imperial and San Joaquin valleys. They are present in many desert areas of California, such as the Mojave Desert (Womeldorf and Mortenson, 1963), and could create problems if and when such areas were intensively cultivated and irrigated.

Some gnat-producing areas are adjacent to residential tracts, but many are 1 to 5 mi (1.6 to 8 km) from them. By means of gnats tagged with radioactive phosphorus, it was shown that *H. collusor* could disperse both upwind and downwind, with the greatest distances traveled in 2 experiments being 4.1 and 4.3 mi (6.6 and 6.9 km), respectively, in the direction of the wind (Mulla and March, 1959).

Hippelates impressus, first collected in California in 1963, has been reported from near sea level to 6,000 ft (1,800 m) elevation in large areas of southeastern California, and occurs in border states east to Texas, as well as in Mexico and the Virgin Islands. It is highly pestiferous in foothill and mountain habitats. Among eye gnats in California, it is exceeded in importance only by *H. collusor* and *H. dorsalis.* Nothing is known about the biology and breeding habits of *H. impressus,* but heavy summer rains, which may occur in the mountains despite the seasonal aridity of the general desert region, seem to cause outbreaks of this eye gnat (Mulla, 1971).

Description of Stages. Adult eye gnats are 1.5 to 2.5 mm long. Most species range from shiny black to dull gray, with yellow, orange, or dark-brown and orange legs (plate VIII, 3; figure 320). *Hippelates impressus* is a small, reddish-orange species. Most *Hippelates* eye gnats have a large, black, curved spur on the hind tibia. They can also be distinguished from other small flies by their small mouthparts and short antennae with rounded third segments and bare aristae.

The eggs are 0.45 to 0.52 mm long, at first pearly white or translucent, but becoming snowy white as they mature. They are fluted lengthwise, distinctly curved, and bluntly tapered at either end. The larvae have the typical maggotlike form, and are 2.5 to 3.5 mm long when full grown. Hall (1932) found that the females would oviposit upon various kinds of excrement and decaying meats, fruits, and other vegetable matter. He did not determine the number of eggs a single female could deposit, but found an average of 24 fully formed eggs in gravid females, as well as developing eggs in the "ovarian spaces." Under moist conditions in the insectary, the average incubation period was 3.7 days. Later investigations showed that *H. collusor* oviposited in soil that contained minimum levels of moisture adequate for egg hatching and subsequent larval development (Legner and Bay, 1965; Legner, 1968). Recently cultivated soil or emergent vegetation stimulated oviposition. The eggs were generally "ejected with propulsion" to depths averaging less than 5 mm below the soil surface.

Hall (1932) observed that the larvae burrowed into a food medium or surrounding sand as soon as they emerged from the eggs. If the medium was sufficiently moist, the larvae came to the surface, and as the moisture decreased, they retreated to their tunnels in the soil. Their resistance to desiccation increased with increasing age. The larval period ranged from 5 to 46 days, depending on food medium, moisture, and temperature. On human excrement, it averaged 11.4 days; on canine excrement, 8.7 days; and on decaying oranges, 17

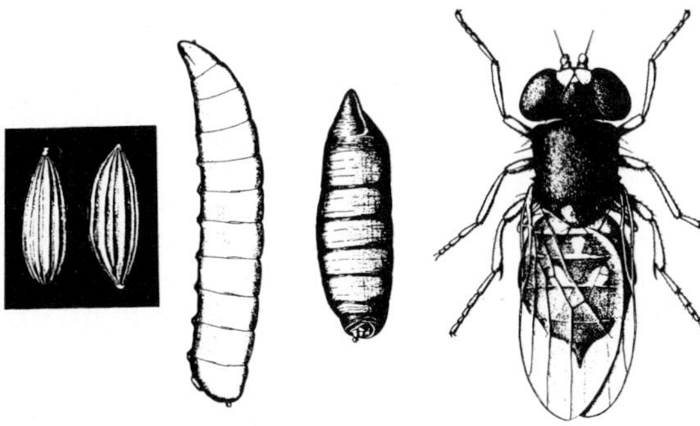

Fig. 320. Eye gnat, *Hippelates collusor*. *Left to right,* eggs; larva; pupa; adult. (From Hall, 1932.)

days. Hall also determined that *H. collusor* (figure 320), under insectary conditions, could complete its development from egg to adult in 11 days. The longest period was slightly more than 3 months and the average period was 18.5 days. Herms and Burgess (1930) determined that in the laboratory at temperatures ranging from 80 to 105° F (27 to 41° C), the life cycle required an average of about 3 weeks. Much more information was obtained by workers in later years on the biology of eye gnats under field conditions.

After emerging from eggs in the soil, the larvae of *H. collusor* move to greater depths and feed on the living roots of plants and on dead organic matter that has been made available through water saturation (Legner *et al.,* 1966; Legner and Olton, 1969). Pupae are usually found in the larval tunnels, either in the food medium or in surrounding sand. After the adult emerges from the puparium, it digs through several centimeters of soil to reach the surface, where it inflates its wings (Legner *et al.,* 1971).

Habitat. *Hippelates* larvae feed on decaying organic matter in soil. It is necessary that the soil be friable, tilled, and with adequate moisture in order to support heavy populations of gnats (Mulla, 1963*c*). Most of the eggs are laid within a few hours after the land has been plowed (Bigham, 1941; Mulla, 1966). Eye gnats probably were not a problem until agricultural operations were begun. Some species of eye gnats breed in limited numbers in alfalfa fields, golf courses, lawns, ditch banks, river basins and banks, and lake shores, but tilled farm lands produce by far the greatest number. However, this is the case only when organic matter is worked into the soil.

Injury Caused by Eye Gnats. Eye gnats are attracted to wounds, scabs, pus, blood, sebaceous material, the natural orifices of the body, and especially to the eyes. They do not bite; in fact, they have sponging mouthparts similar to those of the house fly. They feed by everting the soft, pseudotracheate labellum over moist surfaces and then sucking in liquids. The labellum has spines that produce many minute scratches on the eyeball while the insect is feeding, assisting the entry of pathogenic organisms, and resulting in the malady known as "pinkeye." Even if they were not vectors of pinkeye, the gnats would be very annoying.

Repellents. Chemical repellents for eye gnats were investigated by Mulla (1963*b*), using an olfactometer and skin tests. In the skin tests, 2.5 ml of each repellent solution were rubbed evenly on the face, ears, back of the neck, and on the arms of each of 3 or 4 different people on different occasions. Tryile Mix (dimethyl phthalate 64%, ethyl hexanediol 17%, Indalone 19%), ethyl hexanediol, and dimethyl carbate performed best in both the olfactometer and the skin tests.

Control. Control has so far been based principally on certain cultural measures. In experiments made in a date garden in the Coachella Valley in southern California, Mulla (1963*c*) found that when weeds were controlled by the use of herbicides, gnat control was very good. Herbicides were found to be superior to frequent tillage in controlling weeds and suppressing gnat breeding. Petroleum oils applied to weeds and cover crops up to 9 days before disking the ground resulted in excellent control of *H. collusor,* as well as *H. dorsalis* and other eye gnats of less importance in experiments made in the Coachella Valley. This was probably because the oil caused vegetation to be unfit for food for the larvae after it had been disked into the soil. Certain components of the oils might also have acted as repellents against ovipositing eye gnats or as ovicides or larvicides, even after they had been disked in (Mulla *et al.,* 1966). However, oil treatments applied immediately after disking also resulted in good control. This was probably because the gnats tended to increase oviposition activity after disturbance or disking of their natural breeding habitats. Most of this oviposition took place within 24 hours, and oil applied immediately after disking probably repelled the ovipositing insects. Peak emergence of eye gnats and the duration of the emergence period were influenced by weather, but in the Coachella Valley, most of the emergence took place within 2 to 4 weeks after disking (Mulla, 1966). (Where vegetation is controlled by a herbicide, as in "clean culture" of citrus orchards, disking does not affect eye gnat breeding, for no eye gnats are present.)

The effect of disturbance of the soil on the eye gnat population was again demonstrated in South Carolina, when plots of fairly high grass 100 × 100 ft (30 × 30 m) in area were either plowed under or left undisturbed. Emergence traps placed in the plots at intervals between the eighteenth and thirty-third day after plowing showed striking differences in the catches of flies in the plowed and unplowed plots. Of the 4 species involved—*H. pusio;* a variant of *H. pusio; H. bishoppi* Sabrosky; and *H. dissidens* (Tucker)—the last-named, a relatively unimportant species, was the only one not significantly affected by plowing (Gaydon and Adkins, 1969).

Natural Enemies. Legner *et al.* (1971) considered the natural enemies of *Hippelates collusor* to be of sufficient importance to justify consideration of "integrated control" (see that heading in chapter 2). The whole community of predators and parasites, as well as nonpredatory species, should be taken into consideration, for the latter can be regarded as potential competitors of eye gnats for food and space. Reductions of field populations of *H. collusor,* as well as of the biting midge *Leptoconops kerteszi,* were obtained with granular and spray applications of urea to the soil. The possibility of reducing the breeding potential of pestiferous soil arthropods by this method without creating problems of resistance and destruction of natural enemies was indicated by Legner *et al.* (1970). More natural enemies have since been introduced, but they are helpful only under conditions in which the soil is not disturbed by cultivation (Legner and Bay, 1970).

Other Chloropids

These insects are sometimes so numerous in buildings in Europe that they cover ceilings and windows and make some rooms uninhabitable. The principal pest is *Thaumatomyia notata* Meigen (= *Chloropisca*). Therefore, it is of interest that a closely related species, *T. annulata* (Walker), was found by Sabrosky (1940) in great numbers around window casings of a hotel in Michigan, appearing year after year. These gnats were said to be easily controlled with pyrethrum sprays or dusts. C. W. Sabrosky (correspondence) has since received reports of this species in houses, sometimes in large numbers, in areas ranging from Idaho to New Hampshire. One correspondent from Montreal, Wisconsin, stated that the flies were nuisances in the homes throughout his immediate neighborhood.

Stable Fly, *Stomoxys calcitrans* (L.) (Muscidae)

The stable fly (figure 160, *B,* chapter 6) is often mistaken for the house fly. Distinguishing features and biology are discussed in that chapter. The house fly and the stable fly may be found together in houses, but the latter is primarily an outdoor, day-biting fly that is most likely to be found indoors in the fall and during rainy weather. Unlike the house fly, it has a stiff proboscis (figure 46, *G,* chapter 4), with which it can inflict a painful bite even through socks or stockings, pierce the skin, and draw blood. The small labella at the tip of the proboscis are everted, exposing on their inner surfaces a set of filelike teeth with which the fly rasps a way through the skin. A rapid twisting of the labella aids in the cutting, and the slender anterior part of the labium is soon completely inserted. If the fly is left undisturbed, it may feed for as long as 15 minutes and consume about 25 mg of blood—about 3 times its own weight. Thus, the stable fly takes much bigger meals than the house fly, does not have to seek food continuously, and spends more time resting than the house fly. Both sexes bite and suck blood, and attack domestic animals and man.

The stable fly is often found on piles of seaweed on beaches and along the vegetation-strewn shores of lakes, particularly where wet plant debris has accumulated in compost mounds, and can be a great annoyance in recreation areas. In Florida, the application of 1% methoxychlor emulsion to piles of decaying aquatic vegetation has controlled stable flies for periods of 30 days or longer. Methoxychlor is chemically related to DDT, but has a much lower mammalian toxicity (LD_{50} 5,000–7,000 mg/kg), and is much more readily biodegradable and less persistent in the environment. The stable fly differs from others discussed in this section of chapter 9 in that it inflicts a painful bite but apparently does not envenomate. It should be able to transmit disease germs, at least mechanically, and has been occasionally incriminated as a carrier of certain ailments to animals, but apparently not to man.

March Flies (Bibionidae)

March flies are small to medium in size, robust, mostly black, dark gray, or reddish brown, and hairy. They have large heads, well-developed mouthparts, and large eyes that are holoptic (contiguous) in the males. The antennae are short, and 8- to 16-segmented. The anterior femora are usually thickened and sometimes greatly swollen. The cylindrical, legless larvae have large heads, and feed on decaying vegetable matter, dung, and especially on the roots of grasses. They pupate in the soil. According to Curran (1965), the common name is derived from the frequent occurrence of *Bibio albipennis* Say (= *hirtus* Loew), the most widespread North American species, in large numbers during the month of March. This species is common in California, as well as *B. vestitus* Walker. The adults of both species feed in great numbers on nectar in the blossoms of fruit trees. Both are small, black and reddish in color, and hairy, with a conspicuous black spot near the middle of the costal margin of the wing (Essig, 1926).

The Genus *Dilophus*

Inquiries are sometimes received concerning "small, black flies," identified as *Dilophus occipitalis* Coquillett, that are attracted to windows in large numbers. These insects are about 3 mm long, and both sexes are black throughout except for the reddish thorax of the female. She has a conspicuous black spot along the foremargin of each wing, and has a smaller head than the male. This species is common in southern California, and may account for some of the infestations of small flies and "gnats" that are called to our attention every spring (Ebeling, 1964*a*).

Another bibionid that gives rise to occasional calls from homeowners in southern California is *D. orbatus* Say (figure 321). It is black, robust,

Fig. 321. A March fly, *Dilophus orbatus*. (From Ebeling, 1964*a*.)

very hairy, and about 6 mm long. In one infestation in June, 1964, adults emerged from a lawn in large numbers in the morning and were abundant by 11 o'clock, but had nearly disappeared by 1:30 in the afternoon. They managed to get into the house and garage. An infestation in another residential property was observed in October, at 3 p.m. The flies were clinging in large clumps to blades of grass in the lawn. In both instances, the lawn had been excessively fertilized with manure and bloodmeal. All infestations of bibionids have been recorded from affluent neighborhoods where lawns are likely to be heavily fertilized (Ebeling, 1964*a*). In Florida, sod farm operators report that the larvae of *Dilophus orbatus* feed on grass roots. Flights of adults are annoying to motorists, but less so than flights of the much larger "lovebug," *Plecia nearctica* (L. A. Hetrick, correspondence). In England, *Dilophus* species are believed to be important pollinators of fruit trees (Collyer and Hammond, 1951).

Lovebug, *Plecia nearctica* Hardy

A native bibionid that has captured public imagination in recent years is the "lovebug," as it is known in Florida, or the "honeymoon fly," so named in Louisiana. Other common names are telephonebug, doubleheadedbug, unitedbug, and March fly. These flies range throughout the Gulf states and into Mexico and Central America. They neither bite nor sting, but their mere presence is a nuisance when they occur in large numbers, and they are plagues to motorists and filling-station attendants. The adults in flight splatter on windshields, lights, grills, and radiators of motor vehicles, and their dried remains are hard to remove. They may cause overheating of motors with extensive damage when large numbers are drawn into the cooling systems of liquid-cooled engines. In the northern half of peninsular Florida, there has been a "population explosion" of the lovebug since 1965, but the reason for this is not known. A flight in September, 1969, was estimated to extend over approximately 25% of the land area of Florida, and to reach altitudes of 1,000 to 1,500 ft (300 to 450 m) (Hetrick, 1969, 1970*a*, *b*).

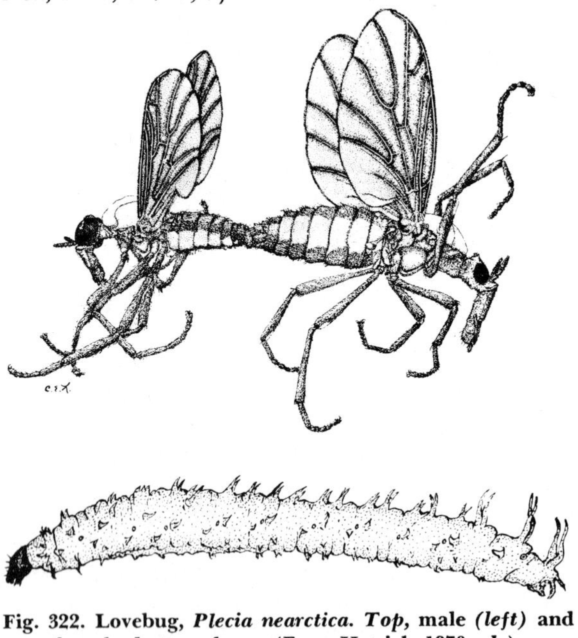

Fig. 322. Lovebug, *Plecia nearctica*. *Top*, male *(left)* and female; *bottom*, larva. (From Hetrick, 1970*a*, *b*.)

Description. The adults are black, with black wings, and with the dorsal portion of the thorax red. The female is 15 mm long and the male is somewhat smaller (figure 322). The larva is slate-gray, with a darker head capsule. It is 11 to 12 mm long.

Biology. The adults of the first generation fly in May and of the second in September, each for a period of about 4 weeks. The pair mates in flight, and copulation continues until the male dies, flight and crawling being controlled by the larger female. The males live only 2 or 3 days, but the females may live for a week or longer, and may mate with more than 1 male. The mating pairs rest at night, usually on low-growing vegetation. The female lays over 300 eggs in decaying vegetation. The larvae feed on the accumulated decomposed material, often skeletonizing dead leaves. They require adequate moisture and favorable soil temperatures. The pupal stage lasts 7 to 9 days (Hetrick, 1970*a*).

Control. The flies may be avoided by traveling at night. There is less splattering of them on a vehicle when driving speeds are reduced. Window screen placed between the grill and radiator of an automobile will prevent clogging of cooling fins and overheating of engines. Several minutes of soaking with water will make the fly remains much easier to remove. In May or September, when the adults are in flight, exterior painting of buildings should be avoided.

Scatopsids (Scatopsidae)

These insects were once included in the family Bibionidae. They are small, black or brownish flies, with the appendages and thorax often partly yellowish. The femora of the legs are robust, as with the bibionids, but the legs are somewhat shorter. The antennae have 7 to 12 segments, and are usually slightly longer than the head. There are 3 ocelli (simple eyes), and the proboscis is short and stout (Curran, 1965). The scatopsids were recognized as a separate family by Melander (1916), who stated that the family included "minute black flies, at present bearing the stigma of human disapproval owing to their habit of living in houses but breeding in excrement. Several species have been reared from privies and exposed human feces, and the same species frequently swarm in windows."

Scatopse fuscipes Meigen

A species of scatopsid was found in large numbers on a screen door of a house in Los Angeles, and many had gained entry into the building. They were identified as *Scatopse fuscipes*, probably the most common scatopsid in this country. The larvae live and feed in all kinds of decaying plant materials, such as cannery waste, winery waste, rotting vegetables, and livestock manure.

The best control is sanitation, but if an unidentifiable infestation occurs, dichlorvos may be applied as a spray, and will be quite effective as an insecticide (E. F. Cook, correspondence).

Myiasis

This is the invasion of living tissues or cavities of man or animals by fly larvae (maggots). Similar conditions caused by beetle or moth larvae are called *canthariasis* and *scoleciasis*, respectively. Insects of other orders could be involved, but fly maggots account for over 90% of invasions of human beings by insects (Scott, 1964*b*). Myiasis may be termed gastric, intestinal, urogenital, rectal, nasal, auricular, ophthalmic, or dermal (cutaneous), depending on the body area invaded. Only accidental or facultative myiasis will be discussed here.

Obligate myiasis, in which the larvae of bot flies or warble flies must live parasitically in the body of an animal, is a veterinary problem, and is not treated in the present volume.

Fly larvae are most likely to enter the human intestinal tract by being swallowed as eggs or young maggots on cold foods, such as meat, cheese, and fruit. The eggs or larvae may be destroyed or digested, in which cases no adverse effects are noted, but they may also survive and cause stomach pains, nausea, vomiting, diarrhea with discharge of blood, and such symptoms as aching of joints and severe nervousness. Myiasis is referred to as "furuncular" if boil-like lesions develop, and "traumatic" if maggots get started in wounds and work their way into healthy tissue. Among cases coming to the attention of the U. S. Public Health Service over an 11-year period, 35% were furuncular. Second in number of cases was enteric myiasis,with 30%.

James (1947) tabulated 112 species of myiasis-producing Diptera in 20 families. The families Calliphoridae and Sarcophagidae contained 34 and 26 species, respectively, or 53.5% of the total. They were particularly important in traumatic myiasis.[24] The third in number of species were the Muscidae, with 15, including the house fly, *Musca domestica*. The muscids were particularly involved in enteric myiasis.

[24] James (1947) reported a case of traumatic myiasis involving the sarcophagid *Lucilia caesar* so serious that it resulted in the amputation of a hand. Merritt (1969) described and illustrated an incredibly horrible infestation by larvae of the calliphorid *Phaenicia sericata* in large, atrophic ulcers in the legs of an elderly woman who had developed stasis of the lower extremities. It was necessary to amputate both legs. Such a severe infestation could take place only under extremely unusual circumstances, but it indicates that traumatic myiasis remains a potential threat to man.

Follicle Mite, *Demodex folliculorum* (Simon) (Demodicidae)

Demodicids are mites that live in the sebaceous pores and hair follicles, usually on the nose, forehead, cheek, and chin. However, they may inhabit follicles, with or without hair, anywhere on the body. When "blackheads" are squeezed out of the skin and examined under the microscope, demodicids are sometimes found, but there is apparently no evidence that the mites cause the blackheads. Demodicids have a curious body form, the long, striate abdomen giving them a somewhat wormlike appearance (figure 323). Their legs are mere stumps. They have been found on as many as 97% and 100% of adults examined in different surveys. Children and adolescents are seldom infested. Among adults, those with oily skin or who use cosmetics such as lubricating creams extravagantly and fail to cleanse the skin properly are the ones particularly subject to infestation (Ayres and Anderson, 1932; Ayres, 1963; Ayres and Mihan, 1967; Morgan and Coston, 1964; Coston, 1967). *Demodex folliculorum* infests hair follicles and *D. brevis* Akbulatova infests sebaceous glands (Desch and Nutting, 1972). Both may be present on the same individual, but *D. folliculorum* is the species that has received the most attention.

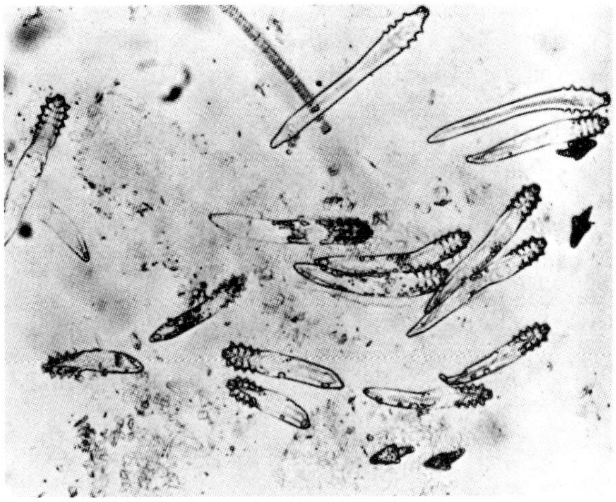

Fig. 323. Follicle mite, *Demodex folliculorum*. Fifteen mature and 4 immature forms, greatly magnified. (From Ayres and Mihan, 1967.)

Description and Habits. The mite is less than 0.4 mm long, and usually escapes notice. Its developmental period is about 14.5 days (Coston, 1967). Inflammation and secondary infection often result when a large number of mites congregate in a single follicle. A *Demodex* type of acne rosacea or "rosacea-like demodicidosis" affecting the medial portion of the face is believed to result from excessive numbers of follicle mites; it disappears rapidly following the topical application of a suitable antiparasitic medication (Ayres, 1963; Ayres and Mihan, 1967; Morgan and Coston, 1964).

Bacteria have been located on the bodies of *D. folliculorum*, suggesting the potential of this mite as a mechanical transmitter of disease germs (English *et al.*, 1970).

The follicle mite can cause an inflammatory condition of the eyelids called demodex blepharitis. Many mites can be found clinging to or migrating about on extracted eyelashes. As many as 25 mites have been seen clinging near the root end of a single eyelash, "packed together like sardines" (Coston, 1967). The infested eyelashes may be soggy or waterlogged, and sometimes can be pulled out with virtually no resistance (English, 1971).

On Domestic Animals. Demodicid mites infest the sebaceous glands and hair follicles of most mammals, e.g., *Demodex canis* Leydig (dog), *D. cati* Mégnin (cat), *D. phylloides* (Sokor) (pig), *D. bovis* Stiles (cattle), *D. equi* Railliet (horse), and *D. caprae* Railliet (goat) (Coston, 1967). With a "maceration flotation technique," the paired eyelids from a total of 269 cattle, horses, sheep, goats, and deer were examined in the southwestern United States. The mites were found in 11.4% of the cattle, 59.7% of the horses, 11.1% of the goats, but none in either sheep or deer (Baker and Fisher, 1969).

Nearly all dogs are infested (Greve and Gaafar, 1964). In the "red mange" or "demodectic mange" of dogs, *Staphylococcus* bacteria are generally associated with the mites and their presence complicates the disease.

The disease is most common among short-haired dogs, and symptoms are first noticed among those that are below the age of 1 year. If lesions are formed, they are usually confined to the head, particularly around the eyes, in the temporal regions, and at the sides of the muzzle, with loss of hair in the affected parts. If the disease is neglected, it may spread from these areas and from other foci that may develop later, and eventually involve most of the body surface (Unsworth, 1946).

Treatment. Most infested persons are women. The greater use of soap and water on the face by men, particularly by those who shave with a razor, results in the face not usually being involved, even when the forehead and scalp show evidence of infestation. After thoroughly washing the face with soap and water, a prescribed ointment or cream can be rubbed in daily. Danish ointment, a compound polysulfide preparation, has given satisfactory results, and seldom causes irritation when kept away from the eyelids and neck. If Danish ointment cannot be tolerated, a 0.5% selenium sulfide cream can be applied sparingly at night and washed off in the morning (Frazier, 1969). An ointment of 10% sulfur and 5% balsam of Peru is said to be an effective remedy (Busvine, 1966). The improvement can be maintained by washing daily with soap and water.

Demodex blepharitis should be treated early, so that the itching and scratching will not result in abscesses of the eyelids. The services of a physican should be obtaned. Anything that cleanses the mouth of the follicle reduces the number of mites. Vigorously rubbing with a cotton applicator moistened with warm water over the bases of emerging cilia, so as to break apart the collar of insulating debris over the follicle mouth, will reduce the number of mites. The fastest and most direct method of treatment is for an ophthalmologist to apply ether to the lid margins, then after 5 or 10 minutes to apply ether or alcohol again, destroying the emerging mites. This treatment is repeated weekly for 3 weeks, along with a twice-a-day cleaning with cotton applicators, followed by appropriate ointments massaged into the lash margin (Coston, 1967).

(D) NONVENOMOUS ARTHROPODS THAT ARE SOURCES OF ALLERGENIC EMANATIONS AND INHALANT ALLERGIES

The enormous numbers of insects and the wide distribution of insects and insect parts in the atmosphere give them, in common with pollens, great potential as excitants of allergic symptoms, particularly those of inhalant or respiratory allergy, that is, asthmalike reactions. The role of insects and insect parts in causing allergies in humans appears to be generally no more than a chance relationship, although some insect parts, such as the urticating (stinging) hairs of certain caterpillars, are apparently specific adaptations for envenomation. Except for the early recognition of such obvious allergenic specializations, scientists have generally been late in recognizing the importance of the arthropods discussed in the present section as sources of allergens. Good examples of this delayed recognition are the house dust mites, the practically universal distribution and importance of which had not become apparent until quite recently. The mites, however, are conspicuous exceptions to the general rule that usually only true insects cause inhalant allergy.

The dispersion of insects by air currents is usually involuntary, particularly for small and feeble-flying species. Insect parts, such as shed skins, scales, and wing particles, may also contaminate the air. Insect fragments, secretions, and excretions, rather than complete and flying insects, are the principal sources of inhalant allergies. More than 1 species is often involved in allergic reactions. Only the more commonly implicated species that are responsible for allergies because of their parts or emanations rather than their bites or stings are discussed here. However, apparently no arthropod species can be definitely stated to be nonallergenic. For example, the house fly was long considered to be incapable of causing allergy, but upon the insistence of an allergy patient, the wing of a house fly was rubbed into a scratch on her skin. Within 15 minutes there was a marked reaction, and several hours later her arm was greatly swollen. An allergic reaction was also obtained by injecting 0.1 ml of fly-wing extract in buffered saline solution into the patient (Jamieson, 1938). The first cases of asthma attacks caused by gnats (*Simulium jenningsi*) and boxelder bugs (*Leptocoris trivittatus*) have been reported (Taub, 1970). Cockroach allergy has been studied extensively for years (see near the beginning of chapter 6). However, house flies, moth flies, *Simulium* gnats, boxelder bugs, cockroaches, dermestid beetles, and many other allergenic insects are discussed in other parts of this volume in connection with pest problems with which they are more commonly associated. Further information on allergic reactions to such insects may be found in those chapters or sections.

Mayflies

Mayflies (order Ephemerida) breed in streams, ponds, and lakes. The young forms (naiads) (figure 324) have 7 pairs of tracheal gills, usually located on the first 7 abdominal segments, by means of which they obtain oxygen from the air dissolved in the water. There is usually 1 generation per year. The adults (figure 324) are attracted to lights, and may enter homes and settle in large numbers on walls or other interior surfaces. This in itself causes mayflies to be pests in communities close to lakes and rivers. In addition, they are among the principal arthropod sources of allergenic emanations and inhalant allergies.

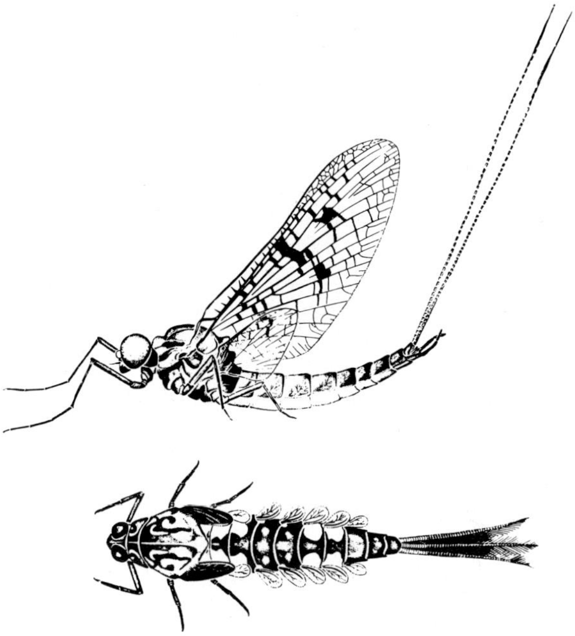

Fig. 324. A mayfly, *Ameletus amador* Mayo. *Above*, adult; *below*, nymph or naiad. (From Chandler, 1956.)

One of the earliest published reports on inhaled insect allergens concerned mayfly swarms (Wilson, 1913). By placing a drop of a mixture of mayfly parts in 2 ml of sterile saline solution to 1 eye of a susceptible person, a marked conjunctival redness resembling that seen in allergy patients was produced, whereas the other eye remained normal. Mayflies are still regarded as important sources of inhalant allergy. These insects do not have allergenic emanations, such as hairs or scales on their wings, as do caddisflies, moths, and butterflies. However, as discussed in the synopsis of orders at the end of chapter 4, the insect molts once more after acquiring wings. After flying a short distance, it alights and sheds its skin, known as a *pellicle*. This is easily fragmented by a breeze, and its particles may be carried long distances and are often inhaled. Figley (1940) proved repeatedly by conjunctival and inhalation tests that epithelial particles from the pellicles were allergenic. He pointed out that mayflies lived only 36 to 72 hours, dried up very rapidly, and that their dried, fragmented bodies added to the total amount of epithelial scales in the air.

Caddisflies

In common with the mayflies, the caddisflies (order Trichoptera) breed in streams, ponds, and lakes, and the adults frequently occur in large numbers in such areas (see the synopsis in chapter 4). The adults (figure 325) are small, dull-colored, mothlike insects, having their wings usually clothed with silky hair; they may be nearly naked in some species, or are rarely clothed with scales. The wings are held rooflike when at rest. Caddisflies have captured the popular imagination because of their aquatic larvae, which build protective cases out of small bits of sticks, leaves, and pebbles which they drag along as they crawl under water with only the head, thorax, and legs protruding (figure 325, *D*). Some species are herbivorous, feeding upon diatoms, algae, and other forms of aquatic plant life. Others are carnivorous, feeding on midge larvae and other small aquatic animals. Some are both herbivorous and carnivorous. The larvae are important sources of food for fresh-water fish, and are also used by anglers as bait.

A book has been published that deals with the biology, ecology, and systematics of caddisflies, with special reference to the larvae, by N. E. Hickin (1967).

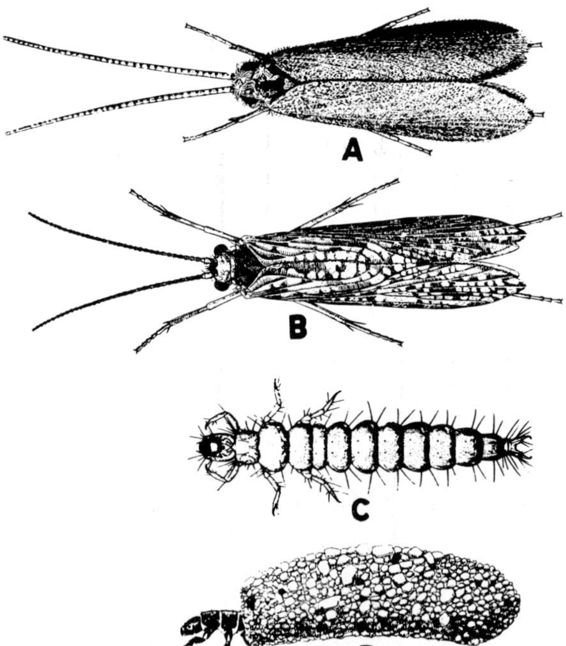

Fig. 325. Caddisflies. *A, Dolophilus shawnee* Ross; *B, Rhyacophila fenestra* Ross; *C,* larva of *Rhyacophila fenestra; D, Ochrotrichia unio* (Ross), larva and case. (From Ross, 1944.)

The adults are strongly attracted to light, and enormous numbers collect about street lights, homes, and hotel or theatre marquees. In discussing the situation in a community along the St. Lawrence River, Peterson (1952) stated that the annoyances caused by caddisflies, mainly *Hydropsyche bifida* Banks, included "complete restriction of all exterior decorating during the summer months; the filth and stench of the masses of dead insects; the necessity, when one drives through the community during the evening flight, of observing the caution normally reserved for dense fog; and the allergic reaction suffered by hypersensitive individuals, locally termed 'sand fever'."

A resident of a river community in Ontario, Canada, suffered asthmatic attacks during the period of caddisfly flights from about the end of June to the middle of September. The patient did not suffer the attacks after she left the infested area. A buffered saline extract of the insects produced positive skin and eye tests. The asthmatic attacks were believed to be caused by hairs and "scaly epithelium" from the wings of the caddisflies. The allergic reaction of the patient ceased about 3 weeks after a series of gradually increasing hypodermic injections of the extract were administered at intervals of 3 to 5 days for a period of nearly 3.5 months. A state of hyposensitiveness was shown by diminished skin and eye reaction, even when the patient was exposed to swarms of the insects (Parlato, 1929). The foregoing test was made in 1929. By 1957, treatment with hyposensitizing injections of caddisfly extract was common in areas infested with these insects. Osgood (1957) reported that of 623 patients suffering from symptoms of respiratory allergy in areas near the Niagara River in New York, 34.5% reacted positively to caddisfly extract, but in the immediate area of heaviest infestation, an even greater percentage (60%) reacted positively.

Moths and Butterflies

Moths and butterflies (order Lepidoptera) are densely covered with scales and hairs, and would be expected to be capable of producing allergenic emanations while in flight or when body or wing fragments are carried about in air currents. Parlato (1932*a*) exposed glass slides in many locations in Buffalo, New York, and surrounding areas over a 3-year period. He found that on some days the amount of hair and scales was similar to the amount of pollen, and that these emanations were found at various places where people usually assembled. He made tests on 6 patients known to be allergic to caddisfly emanations. Four gave positive skin and eye reactions to the extracts of moths and butterflies, and the other 2 had only a moderate skin reaction. A little "dust" was placed on the conjunctiva of 1 eye of each of the 4 most positive reactors, and these conjunctivae also showed immediate and distinctly positive reactions. On the other hand, among the controls, 10 patients known to be sensitive to the pollens of grains and ragweeds (*Ambrosia* spp.), and 15 sensitive to other airborne allergens, such as cat hair, feathers, etc., plus 5 individuals with no history of allergy, all reacted negatively to the moth and butterfly extract. However, an office employee with no history of allergy reacted positively. Passive transfer tests made with her blood serum were also positive, producing wheals on a sensitized site of a normal person and on a hypersensitive individual.

In a later investigation, Parlato (1932*b*) determined that atopic reagins[25] of the serum of persons who were hypersensitive to the emanations of moths and butterflies were readily transferable to the skin of a nonallergic patient. He found that the atopic reagin of a butterfly was identical to that of a moth and similar to that of a caddisfly, and that immunologically the 2 orders (Lepidoptera and Trichoptera) could be handled as 1 group.

Aphids

In 2 studies made in England of insect populations in the air at various elevations, up to 2,000 ft (600 m) in one study and 300 ft (90 m) in the other, aphids (suborder Homoptera) constituted the dominant family of insects at all elevations (Freeman, 1938). In the United States, the astronomical numbers that could be found in the air were noted by Gaillard (1950). He also pointed out that an enormous number of minute particles of aphid exuviae were airborne over wide areas and reached many allergic people. He found that 75 out of 121 allergy patients, whose symptoms were unexplained by tests for known seasonal inhalant allergens, gave positive reactions to aphid extract. Gaillard showed that skin-sensitizing antibodies were present in the blood of skin-sensitive persons, and that the antibody was not antigenically related to ragweed. He believed that the aphid was an important insect inhalant that had theretofore been unnoticed.

[25] Atopic reagins are dermatitis-producing reagins. A reagin is an antibody in the blood of an individual with atopic allergy that is able to sensitize the skin of a normal individual.

Caterpillars with Urticating Hairs

The larvae of Lepidoptera are often called caterpillars, a word derived from the Latin *catta pilosa* or "hairy cat," and many species are indeed hairy, but many others are hairless. Thus the term "caterpillar" is etymologically quite appropriate for the species to be discussed under this heading. Localized epidemics of envenomation resulting from the urticating hairs of certain caterpillars have occasionally occurred, particularly in the eastern United States. The hairs, which may be hollow, contain a poison that is injected into the skin when their sharp points penetrate. People are affected by pain and rashes when the caterpillars or their shed skins fall on them or when the insects are handled. Kemper (1958) classified the urticating hairs of various genera as to physical types and effects on human volunteers. The hairs can produce a reaction long after the death of the caterpillar, and in fact the dry hairs may then be better able to penetrate the skin. There are many families of Lepidoptera that contain species with nettling (urticating) hairs.

There are several species described or noted here that appear to be the most important ones.

Puss Moth, *Megalopyge opercularis* (J. E. Smith) (Megalopygidae)

This moth is widely distributed in the southeastern and south central states, feeding on the foliage of deciduous trees and shrubs, and has caused some of the most serious epidemics of caterpillar dermatitis (Scott, 1964*c*). No one appears to know why the insects can periodically increase in such large numbers, since it is known that they can be heavily parasitized. More stings were reported in South Carolina in 1960 than during the preceding 25 years combined (Anonymous, 1961*b*).

There have occurred a number of epidemics of dermatitis in San Antonio, Texas, so widespread that the public schools were closed until surrounding trees could be sprayed (Foot, 1922).

The mature caterpillar (figures 326, *E*, and 327) is about 25 mm long, and tapers posteriorly from its widest point in the thoracic region. It may be pale yellow, gray, reddish brown, or a mixture of colors. It is densely covered with hair that lies rather flat and extends downward to the surface upon which the insect rests. Among these hairs are smooth, yellowish spines with black tips. The spines are hollow, and a venom passes through them and into the skin of a person contacted

(Anonymous, 1961*b*). Almost immediately after the caterpillar contacts skin, an intense burning is felt. It becomes worse, and is soon accompanied by itching. Small, white papules form, and soon turn red. The inflamed area spreads, and a general swelling of the portion of the body that contacted the hairs may occur; for example, stings on the wrist may cause the entire arm to swell to almost double its normal size and to become numb. The stings are especially severe in young children, who may develop considerable fever and nervous symptoms. The pain is similar to a severe nettle sting. The itching may persist for 1 to 12 hours or even more, and local lesions often remain for several days (Bishopp, 1923). Another stinging caterpillar in the same family of moths is that of the flannel moth, *Norape ovina* Sepp. The larva (figure 326, *C*) feeds chiefly on redbud (*Cercis* sp.; Leguminosae).

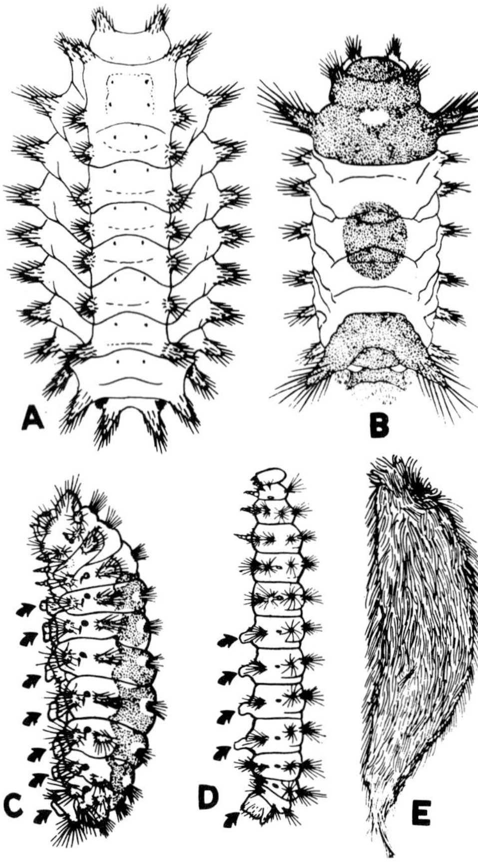

Fig. 326. Some important species of stinging caterpillars in the United States. *A*, a slug caterpillar, *Parasa chloris; B*, saddleback caterpillar, *Sibine stimulea; C*, a flannel moth caterpillar, *Norape ovina; D*, io moth caterpillar, *Automeris io; E*, puss caterpillar, *Megalopyge opercularis*. (Scott, 1966*b*.)

Saddleback Caterpillar, *Sibine stimulea* (Clemens) (Limacodidae)

The larvae of this family of moths are called "slug caterpillars" because they are short, fleshy, and sluglike. The saddleback caterpillar is green, with a brown, saddlelike mark on its back (figure 326, *B*). Its hairs can cause severe irritation to the skin. *Parasa chloris* (Herrich-Schaeffer) (figure 326, *A*) is another urticating slug caterpillar.

Io Moth, *Automeris io* (F.) (Saturniidae)

The larva of the io moth is a spring-green caterpillar (figure 326, *D*), having urticating hairs. A narrow, reddish stripe, edged below with white, extends along each side of the body. It feeds on various deciduous trees and on blackberry and currant, and may be found from July to September.

Tussock Moths (Lymantriidae)

The common name of this family of moths is derived from the long tufts or tussocks of hairs on the caterpillars' bodies. The caterpillars are also conspicuous for their bright, variegated colors. The best known of the lymantriids are the gypsy moth, *Porthetria dispar* (L.), and the browntail moth, *Nygmia phaeorrhoea* (Donovan), both Old World species that are now serious pests of

Fig. 327. Larva of puss moth, *Megalopyge opercularis*, on an orange leaf. (Courtesy Helen L. Greene.)

forest and shade trees in the eastern United States. The browntail moth is well known for its urticating hairs. The adult moth has a white thorax, white wings, and a wingspread of 25 to 40 mm. The abdomen is mostly brown, and the common name is derived from the conspicuous tuft of chestnut-brown hairs at its tip. The full-grown caterpillar is about 40 mm long, dark brown, with a row of white tufts on each side of the body. There is a bright-red tubercle on the back of the eleventh and of the twelfth body segment. The nettling hairs of the caterpillar are responsible for the "caterpillar dermatitis" or "browntail rash" that is common in infested areas. Pruritus occurs 20 to 30 minutes after contact, followed by inflammation and wheals. The eruption can be generalized if the larval hairs become lodged in clothing, and the patient will be exposed to an attack every time the clothing is worn. Attacks from this species usually occur in May and June, the months of larval development (Sutton and Sutton, 1939).

The term "tussock moth" also commonly refers to the species of *Hemerocampa*. The species usually implicated in irritating skin rashes are the whitemarked tussock moth, *H. leucostigma* (J. E. Smith), in the eastern states, and the Douglas fir tussock moth, *H. pseudotsugata* McDunnough, in the western states. The females of these moths are wingless. They lay their eggs on or near the cocoons from which they emerge, and cover them with froth and the hairs from a tuft at the tip of the abdomen. Hibernation is in the egg stage. The caterpillars are serious defoliators, attacking mainly forest and shade trees. They move to lower levels by means of silken threads in search of food, and may often be seen hanging on them. Their long hair and light weight allow them to be carried long distances by the wind, particularly in their early instars, thereby spreading the infestation.

The male of *H. leucostigma* has a wingspread of about 30 mm, and its forewings are gray, with darker wavy crossbands. The full-grown caterpillar is 30 to 35 mm long, and is light brown with yellow and black stripes. It has a bright-red head and a bright-red spot on the sixth and seventh segments of the abdomen. There are 4 short, white, dorsal tufts of hairs on the first 4 abdominal segments, a long pencil-like tuft of black hairs on each side of the head, and another near the posterior end. The caterpillars skeletonize the leaves of deciduous forest, shade, and sometimes fruit trees.

The male of *H. pseudotsugata* has a wingspread of about 25 mm, and is dull brownish gray. The full-grown caterpillar is 20 to 25 mm long, and is gray to light brown. It has a shiny black head, many red spots, and broken, narrow, orange stripes along the sides. There are short, brushlike tufts of light-brown or cream-colored hairs on the first 4 and the last abdominal segments, and 2 anterior, black, pencil-like tufts of black hairs and a longer posterior one. There is 1 generation per year.

Hemerocampa pseudotsugata is principally a pest in the Pacific Northwest, defoliating Douglas firs and true firs. In August, 1973, there was a severe infestation of the pest in about 800,000 acres (324,000 ha) of forest in Oregon and portions of Washington and Idaho, and the infestation has been spreading. The federal ban on DDT leaves authorities with no effective control measure. An infestation is usually controlled by a virus within a 3-year period, but this has not yet happened in the present case. There have been numerous instances of skin rash, particularly among loggers (Hager, 1973).

Blister Beetles (Meloidae)

Blister beetles are narrow, elongate insects, with soft and flexible elytra and with the pronotum narrower than either the head or the elytra. Their bodies contain a crystalline anhydride of cantharidic acid called cantharidin. The beetles exude the vesicating venom when disturbed. They fill their tracheae with air, close their spiracles, and build up body-fluid pressure until breaks occur in the exoskeleton, usually at joints. Body fluids containing cantharidin then spill through the breaks (figure 328). The beetles proceed to relieve the pressure, and the wounds heal. If they are handled during or immediately after this process, the cantharidin contacts the skin and produces fluid-filled blisters that may eventually burst. The released fluid often produces secondary blisters, and sometimes secondary infection, especially impetigo. Persons with such blisters are advised to consult a physician. If the blisters do not break, the fluid is slowly resorbed in 3 or 4 days (Swarts and Wanamaker, 1946; Lehmann *et al.*, 1955; Scott, 1962, 1966*b*).

Some Common Blister Beetles

Epicauta cinerea (Förster), the clematis blister beetle, is the species most commonly seen in the southwestern United States. In common with other species of the genus, it is known to produce symptoms of dermatosis. It is 10 to 17 mm long, black, and uniformly clothed with gray pubescence.

Scott (1962) recorded the case of a woman in Georgia who brushed an *Epicauta* off her neck, then crushed it, picked it up in a piece of paper,

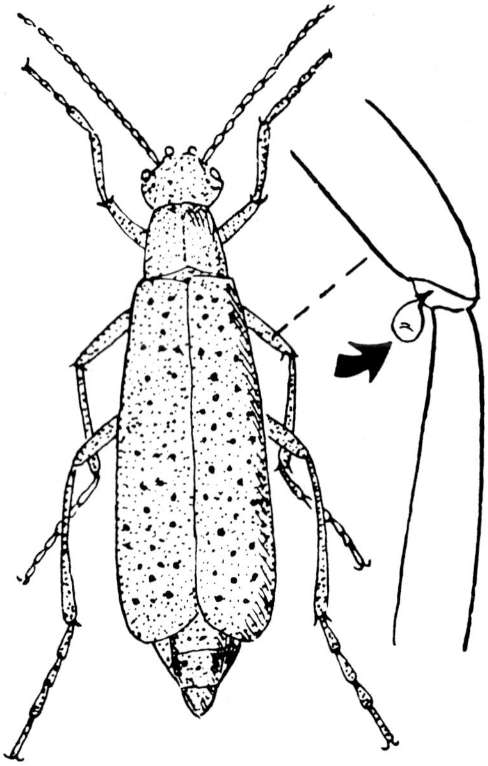

Fig. 328. Spotted blister beetle, *Epicauta maculata*, exuding a drop of vesicating venom from a break in a weak spot of its exoskeleton. (From Scott, 1966*b*.)

and brought it to his laboratory for identification. Within 12 hours, 3 large blisters had developed on her neck, and the affected area was still painful after 48 hours. An entomologist who handled the paper in which the beetle had been wrapped developed a blister on one finger.

Epicauta maculata (Say), the spotted blister beetle, is light brown over-all, including the legs, and is dotted with round, black spots (figure 328). It is medium-sized, varying from 9.5 to 16 mm in length (Gilbertson and Horsfall, 1940).

Epicauta vittata (F.), the striped blister beetle, is said to be the most common species in the eastern United States, and extends westward as far as Montana. It is 12 to 14 mm long, blackish, with each elytron bordered yellow, and has a median yellow stripe.

Epicauta pennsylvanica (De Geer), the black blister beetle, is common not only in the eastern states, but ranges from Mexico through much of the country. It is 7 to 13 mm long, and uniformly black (Essig, 1926; Lehmann *et al.*, 1955).

Biology and Habits. Blister beetles have a comparatively complex life history. They undergo a *hypermetamorphosis,* with a variety of larval forms (figure 329). The eggs, laid in cavities prepared in the ground, hatch into long-legged forms called *triungulins.* The triungulins of most spe-

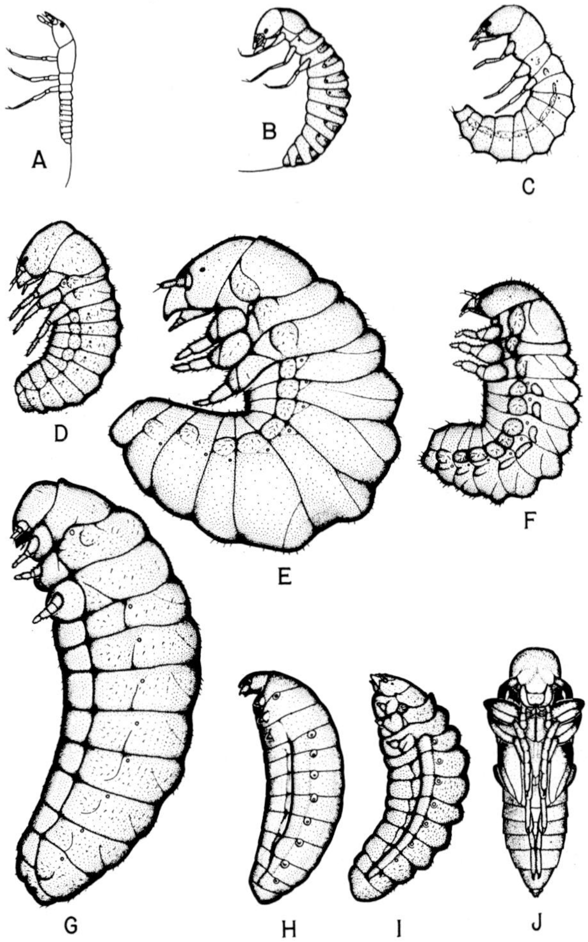

Fig. 329. Larval and pupal stages of the black blister beetle, *Epicauta pennsylvanica*. *A*, unfed first instar (× 13); *B*, fully fed first instar (× 13); *C*, *D*, *E*, second, third, and fourth instars (× 13); *F*, newly molted fifth instar (× 7); *G*, gorged fifth instar (× 7); *H*, sixth instar (× 4); *I*, seventh instar (× 4); *J*, pupa (× 4). (From Horsfall, 1943.)

cies search for pods of grasshopper eggs, but others climb upon flowers and attach themselves to solitary bees, which then carry these very active and agile larvae to their nests, where the triungulins feed on the bees' eggs. The next larval instar is similar to the triungulin, but has shorter legs. The third, fourth, and fifth instars become increasingly scarabaeiform (grublike). The sixth instar, called the *coarctate larva* or *pseudopupa,* is the hibernating instar. It has a darker and thicker exoskeleton and lacks functional appendages. In the spring, it molts and becomes the seventh instar, which is small, white, and legless, but active. This instar transforms into the true pupa. The adult emerges in about 2 weeks, and normally there is 1 generation per year (Gilbertson and Horsfall, 1940; Horsfall, 1943).

Some blister beetles are important pests, feeding on potatoes, tomatoes, and other crops. Outbreaks of the beetles tend to occur in years of severe grasshopper infestations (Gilbertson and Horsfall, 1940). *Epicauta cinerea* often occurs on *Cineraria,* and *E. pennsylvanica* usually prefers goldenrod (*Solidago* spp.), although it is found on other plants, and is sometimes injurious to various truck crops. Workers in potato fields where large numbers of blister beetles are feeding may get blisters on exposed surfaces of their bodies. Barefooted children sometimes crush the beetles and get a large amount of cantharidin on their feet. They should be advised to avoid walking, and to apply wet dressings to their feet for 24 to 48 hours, followed by one of the antibiotic dressings (Lehmann *et al.,* 1955).

Millipedes

Millipedes (class Diplopoda) live in dark, moist places, and feed principally on decaying vegetable matter (see chapter 10). They may climb the sides of houses in large numbers after nightfall, and many may find their way into a home. The species of some genera exude a vesicating venom that causes blisters on persons handling them. However, they are mostly nuisances (Scott, 1966*b*).

House Dust Mites, *Dermatophagoides pteronyssinus* (Trouessart) and *D. farinae* Hughes (Pyroglyphidae)

Among the many constituents of house dust that cause allergic reactions of the respiratory tract (rhinitis and asthma), the most potent allergens were found by European investigators to be contained in certain pyroglyphid mites, principally *Dermatophagoides pteronyssinus* (Voorhorst *et al.,* 1964). On the other hand, all mites found in dust samples in Ohio were classified by D. G. Larson as the closely related species *D. farinae,* although F. T. M. Spieksma, of Leiden, the Netherlands, identified *D. pteronyssinus* in some of the Ohio samples (Mitchell *et al.,* 1969). Samšiňák *et al.* (1972) found *D. farinae* to be the dominant species in Czechoslovakia. They believed that *D. pteronyssinus* favored the coastal climate, whereas *D. farinae* favored the drier central European climate. The taxonomic positions of the species of *Dermatophagoides* were clarified by Fain (1967). Although *Dermatophagoides* was not experimentally demonstrated to be a cause of common house dust allergen until 1964 (Voorhorst *et al.,* 1964), a review of the world literature in 1971 already disclosed 103 references (van Bronswijk and Sinha, 1971).

Dermatophagoides pteronyssinus and *D. farinae* are cosmopolitan, and are sometimes found in enormous numbers (Voorhorst *et al,* 1964; Miyamoto *et al.,* 1969*b*). The fecal pellets of either species are comparable to the mites, or extracts from them, as sources of allergen. Fortunately, most people are not sensitive to these mites—fewer than 5% of children under 17 years old and even fewer adults (Spieksma, 1967; Larson *et al.,* 1969; Mitchell *et al.,* 1969; Miyamoto *et al.,* 1969*b;* Halmai and Alexander, 1971).

In one survey of 64 samples of house dust obtained from Connecticut, New York, New Jersey, Georgia, Washington, California, Nebraska, Wisconsin, and Illinois, 39 contained house dust mites. Twenty-one had *D. farinae* only, 5 had *D. pteronyssinus* only, and 11 had *D. farinae* and *D. pteronyssinus. Dermatophagoides farinae* was not only found in more samples, but also in greater numbers: up to 69 per gram compared with 9 per gram for *D. pteronyssinus.* Besides being found in house dust, *Dermatophagoides* species are also found on birds, in bird nests, on various animals, and humans, as well as in stored food products. *Dermatophagoides evansi* Fain and *D. chelidonis* (Hull) were found most commonly in bird nests (Wharton, 1970). Mites were found in 10 out of 26 house dust samples randomly collected from floors and mattresses with vacuum cleaners in Canadian homes of persons with or without allergies. Four of the 10 samples had *D. farinae,* 3 had *D. pteronyssinus,* and 1 had both species. Of the 57 mites collected, 48% were *D. farinae,* 17% were *D. pteronyssinus,* and most of the others were stored-product species (Sinha *et al.,* 1970). Based on published and unpublished reports of house dust mites in the United States, *Dermatophagoides* species, principally *farinae* and *pteronyssinus,* had been isolated from house dust samples in 22 states by 1971. *Dermatophagoides evansi, D. chelidonis, D. bakeri* Fain, and

D. scheremetewskyi Bogdanov were found in very small numbers. House dust samples ranged from 32 to 100% positive for species of the genus, and they constituted from 43 to 100% of the mite fauna. In California, these mites were found in house dust samples from 24 widely scattered areas (Furumizo and Mulla, 1971).

In Hawaii, house dust was collected with vacuum cleaners over an 8-month period in 40 houses. *Dermatophagoides* was found in all the samples obtained, *D. pteronyssinus* being much more abundant than *D. farinae*. The maximum number of *Dermatophagoides* in a 5-gram dust sample was 4,390, a much higher density than any reported from house dust in other parts of the world. Dust from carpets (wool, cotton, or synthetic fibers) always had more mites than dust from floors without carpeting or with other types of coverings. Old carpets not cared for had more mites than newer ones or those often cleaned and washed (Sharp and Haramoto, 1970).

Dermatophagoides pteronyssinus was found in all of 250 homes visited throughout New Zealand, except for one in which the occupant had been treating the mattress at frequent intervals with sulfur dust. Except for this one house, the number of mites varied from 2 to 480 per 0.1 gram of dust in the 300 dust samples (Cornere, 1972).

Description. The mites are extremely small. The adult female mites of *D. farinae* (figure 330) are about 0.5 mm long; tritonymphs and males, about 0.4; protonymphs (figure 331), 0.3; larvae,

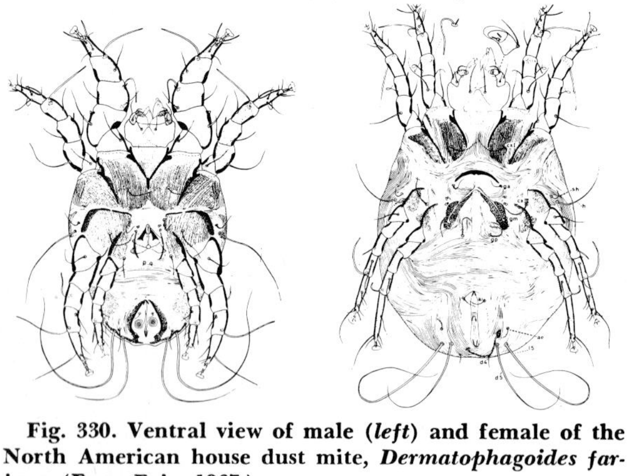

Fig. 330. Ventral view of male (*left*) and female of the North American house dust mite, *Dermatophagoides farinae*. (From Fain, 1967.)

about 0.2; and eggs, close to 0.3 mm long. The adult mites are light cream-colored, except for a few areas of sclerotized integument (propodosomal dorsal plate and parts of the appendages), which are light beige (G. W. Wharton, correspondence).

Life Cycle. Pyroglyphids have 5 distinct life stages: egg, larva, protonymph, tritonymph, and adult. For *D. farinae* and *D. pteronyssinus*, the developmental period from egg to adult ranges from 23 to 30 days. Males and females of both

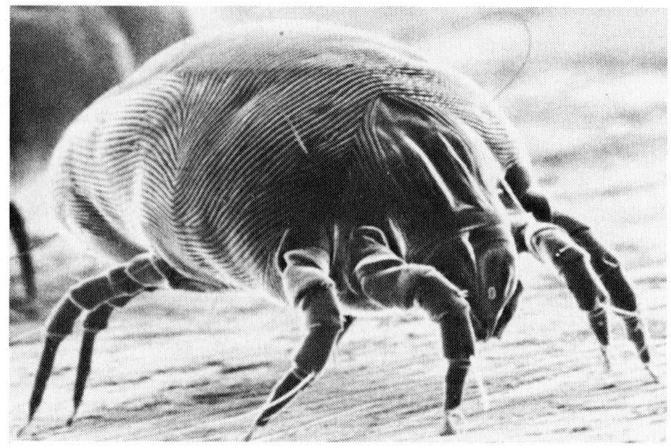

Fig. 331. Protonymph of the house dust mite *Dermatophagoides farinae*. (From Wharton, 1970.)

species have a premating period of 1 to 3 days. The male remains fertile for life, but mating results in egg-laying only during the first half of the adult life of the female.

At a temperature of 25° C, the first oviposition period for *D. farinae* lasts about 30 days, with 0.8 to 1.4 eggs laid per day. The first oviposition period for *D. pteronyssinus* is 20 days, with 1.2 to 2.5 eggs laid per day. Both species have a second and even a third oviposition period, but fewer eggs are laid (van Bronswijk and Sinha, 1971).

The life cycle of *D. farinae* is very similar to that of *D. pteronyssinus*, but the period from egg to adult averages 4 days longer. In both species, no eggs are laid unless copulation has occurred (Larson *et al.*, 1969).

Habitat. Both the mite species are found principally in cotton-stuffed mattresses and furniture (Cunnington and Gregory, 1968; Mitchell *et al.*, 1969).

In a hospital in England, it was shown by means of a cyclone dust extractor that the mites could become airborne during bedmaking, including turning the mattress (Cunnington and Gregory, 1968). In other investigations, it was shown that mites could be inhaled (Carter *et al.*, 1944; Taboada, 1954; Spieksma, 1967).

In central Ohio, Mitchell *et al.* (1969) made a survey of dust samples and, using a dissecting microscope, they removed all mites and all recognizable cuticles or exoskeletons of mites from 0.1-g samples of fine dust. They never found living mites in dry dust. Mite cuticles or dead mites were found in 21 out of 30 samples of dust from mattresses and padded furniture, and 17 samples had more than 15 mites per 0.1 g. Floor sweepings and dust from commercial buildings contained either no mites or negligible numbers. Articles which were used by humans both day and night, such as day beds, contained the most mites. Mites were rare or absent in articles of furniture that had not been used by humans or were not stuffed with vegetable fibers.

Mitchell *et al.* (1969) believed that the mites probably fed primarily on dander (skin scales) from humans and pets. However, they have also been reared on human whiskers in a special microchamber (Shamiyeh *et al.,* 1971). Additional information on possible food sources was provided by van Bronswijk (1973*a*) in an investigation of 470 *Dermatophagoides pteronyssinus* from mattresses and 147 from bedroom floors. Their alimentary canals contained pollen, spores of microorganisms, fungal mycelia, bacteria, and fibers of plant origin, probably from cotton bedsheets. This indicated that the mites were not so strictly monophagous on human skin scales as was suggested by Spieksma (1967).

Laboratory tests revealed that optimal development of *D. pteronyssinus* took place at 25° C (77° F). After 8 weeks at 30° C (86° F), the number of mites was 40%, and at 20° C (68° F), it was only 15% of the number at 25° C. The mites can develop within a range of 50 to 90% relative humidity, the optimum being 70 to 80%. In the Netherlands, only in summer is relative humidity in houses high enough to allow rapid increase in the mite population. Although the average temperature on floors in summer is usually below 20° C (68° F), in occupied beds and upholstered chairs optimal temperatures are soon reached, and relative humidity is increased by 5 to 8% (Koekkoek and van Bronswijk, 1972).

In the Netherlands, *D. pteronyssinus* is present in greatest numbers in old houses, and that is where most of the people live who are hypersensitive to house dust allergen (Voorhorst, 1969). In Europe, "house dust atopic asthma" patients find relief in the dry conditions at altitudes of about 1,000 m (3,300 ft) or higher. The cold air, soil conditions, and a type of construction providing good protection against the penetration of water result in dry conditions in houses. The dryness prevents the development of large numbers of allergen-producing *D. pteronyssinus* (Voorhorst, 1970; Spieksma *et al.,* 1971).

Dust Mites and Inhalant Allergy

Although the allergen is formed by living mites, dead mites or their fragments or feces are allergenic as well as live ones. Maunsell *et al.* (1968) observed that asthmatic patients who reacted negatively to house dust were negative to the mites. Mitchell *et al.* (1969) found that 93% of the patients who showed positive skin reactions to house dust extract also showed positive reactions to the extract of mites, and vice versa. Out of 100 patients who showed reactions to dust mites, 61 reacted positively to 1 or more insects and/or spiders. Bernecker (1970) observed that a certain cross-antigenicity existed, not only between mites occurring in house dust, but also among mites in general, including species such as *Tetranychus urticae, Panonychus ulmi,* and *Dermanyssus gallinae,* which are not ordinarily found in house dust.

Mites constitute the major antigenic substance in house dust in Japan. More than 35 species of mites were found in house dusts collected from 9 different parts of Japan, and 1 g of dust contained from several hundred to over 2,000 mites. The major allergen was possessed by *D. pteronyssinus* (27.2% of the total number of mites) and *D. farinae* (4% of the total), the 2 species having a close antigenic relationship. Cross-antigenicity was found from skin testing, and *in vitro* neutralization of skin-sensitizing antibody, among the 6 species of mites investigated. Molds were found to be of little importance as allergens (Miyamoto *et al.,* 1969*b*). The molecular weight of the allergen produced by *Dermatophagoides farinae* is greater than 10,000 but less than 69,000 (most likely around 40,000 to 50,000), and it is probably a protein and polysaccharide conjugate. Work on the chemistry of the allergen is continuing. *Dermatophagoides farinae* develops readily on dog meal, and is available in larger quantities for scientific investigations than is *D. pteronyssinus,* which must be reared more slowly on human dandruff (Miyamoto *et al.,* 1969*a*).

Reactions to Various Tests. In England, tests with extracts of *D. pteronyssinus* were made on skin, bronchi, nasal mucosa, leucocytes, and on normal human tissue passively sensitized by the patient's serum, on 21 patients with asthma and house dust allergy. Fourteen patients gave positive reactions to every test. In 18 patients, attacks were readily induced by inhalation of the mite extract. In 18, nasal reactions were provoked, and were not followed by asthmatic symptoms. A serum test was positive in all patients, and a leucocyte test was positive in 19. In 6 patients, delayed asthmatic responses to bronchial tests were observed. An extremely small amount of allergen was capable of causing a reaction. In view of the wide distribution of *D. pteronyssinus,* it was concluded that it was an important factor in allergic asthma (McAllen *et al.,* 1970).

Desensitization tests were made with 34 persons suffering from bronchial asthma and showing positive skin reactions to house dust and *D. farinae.* Eighteen patients were given mite extracts and 16 were given house dust vaccine in weekly subdermal injections of 0.1 to 1.0 ml, followed by semimonthly maintenance doses for 6 to 9 months. Hyposensitization with *D. farinae* extracts was found to be safe and effective, but patients did not benefit from house dust vaccine in the doses administered (Maunsell *et al.,* 1971). The absence of positive results with house dust vaccine was not in accord with the results obtained by another investigator, who desensitized 11 patients with house dust extract and 13 others with *Dermatophagoides pteronyssinus* extract. Skin reactions to both extracts decreased to about the same extent over the period of 1.5 to 2 years of the tests (Voorhorst, 1970).

In a hospital in England, patients with asthma and rhinitis who were clinically sensitive to house dust mites were treated with a pyridine-extracted, alum-precipitated extract of house dust to which *D. farinae* had been added before extraction. Among 52 patients, 43 (83%) showed improvement. Symptoms were worse or unchanged in 9 patients (17%) (Munro Ashman *et al.,* 1971). In a "double-blind" controlled trial, 11 asthmatics allergic to house dust who were given an aqueous extract of *D. pteronyssinus* improved and 5 remained well for a year, whereas 11 control patients, given an extract of human skin scales, showed little change (Smith, 1971).

Other Mites Causing Allergy

The family Acaridae (= Tyroglyphidae) contains species best known as pests of stored food products (see chapter 7, under "Acarid Mites"), but some species also cause dermatitis in people handling infested commodities, as well as causing inhalant allergy or bronchial asthma. *Acarus siro* L. causes a rash known as "vanillism" in vanilla-pod handlers. *Tyrophagus castellani* (Hirst) causes "copra itch" among workers handling copra and a dermatitis in people who handle cheese. In the family Glycyphagidae, *Glycyphagus domesticus* (De Geer) causes "grocer's itch" in workers handling dried fruits, skins, and other heavily infested products. Allergen from *Glycyphagus destructor* (Schrank) was found in chaff (Voorhorst, 1970). Allergen obtained from *G. destructor* differs materially from that of the 2 house dust mites, *Dermatophagoides pteronyssinus* and *D. farinae,* whereas the allergens of the latter 2 species are identical (Voorhorst and Spieksma, 1969).

In the family Pyroglyphidae, *Dermatophagoides scheremetewskyi* attacks bats, rodents, and sparrows in the eastern United States. Several cases of human infestation resulting in severe and persistent dermatitis have been reported (Traver, 1951; Baker *et al.,* 1956). However, species of this genus are best known as the ones causing inhalant allergy.

Control of Mites Causing Inhalant Allergy

The 4 sources of dust mite allergen in houses are floors, mattresses, furniture, and clothes (van Bronswijk and Sinha, 1971). The number of mites on floors can be reduced by keeping the house dry. The optimal relative humidity for pyroglyphid mites is from 75 to 95%. Under conditions suitable for all mite species, *Dermatophagoides* can survive only in small numbers in cracks and crevices, but at lower humidities not suitable for *Acarus* and *Glycyphagus, Dermatophagoides* may greatly increase in numbers. None of these mites can survive under very dry conditions (van Bronswijk *et al.,* 1971). Motile stages of *D. farinae* and *D. microceras* were eradicated by 6 hours of exposure to –15° C (5° F), whereas *D. chelidonis, D. pteronyssinus,* and *Euroglyphus maynei* (Correman) required –28° C (–18° F) for the same period of exposure. One-day exposure of mattresses to the outdoor air has been suggested for the control of pyroglyphid mites in areas where sufficiently low temperatures are reached (van Bronswijk and Koekkoek, 1972).

Small cracks and crevices in the floor and under plinths should receive greater attention when vacuum-cleaning. Under conditions obtaining in the Netherlands, it was observed that the habit of *Dermatophagoides* to hide in cracks and crevices tended to favor these mites, relative to *Acarus, Glycyphagus,* and *Tyrophagus,* when a floor was vacuum-cleaned, thus eliminating the latter 3 genera as competitors and also eliminating large numbers of the predatory mite *Cheyletus.* Thus, it was sometimes possible to find the highest numbers of *Dermatophagoides* in regularly vacuum-cleaned rooms (van Bronswijk *et al.,* 1971).

After mattresses were vacuum-cleaned, there was an 8-fold reduction in the number of mites that became airborne during bedmaking. Dead mites and mite excreta possessed as much allergen as live mites, and vacuum-cleaning was necessary after the mites were killed. Ultraviolet radiation for 2 hours almost completely removed the allergen (van Bronswijk and Sinha, 1971). Mattresses appear to be the reservoirs from which bedroom floors are reinfested by dust mites for a brief period during the year (van Bronswijk, 1973*a*). The mites apparently live only in the surface layer of the mattress (Hughes and Maunsell, 1970; van Bronswijk, 1973*a*). It is reasonable to expect that control of dust mites in bedrooms inhabited by persons who are allergic to dust mites may be accomplished by preventing penetration of moisture and accumulation of food in the upper few centimeters of the mattress, as with the aid of a plastic sheet. Experiments on this approach to dust mite control are in progress (van Bronswijk, 1973*a*). Also, the application of fungicides to eliminate microorganisms that are essential to the mites appears to be a promising approach (van Bronswijk and Koekkoek, 1971; van Bronswijk, 1973*b*).

Frankland (1972) recommended that environmental mite control should always be attempted in the home of an allergic person before hyposensitization. Feather and kapok pillows and eiderdowns should be removed. Blankets should be dry-cleaned, and the linen washed whenever possible. At least once a week, the house should be vacuumed and the bedroom more often, paying special attention to cracks and crevices. Helson (1971) suggested ventilation, vacuum-cleaning, dusting of 1% lindane dust into cracks and crevices and, if an acaricide were used, the removal of all dead mites, cast skins, and fecal pellets. On the other hand, van Bronswijk *et al.* (1971) reported little effect from a 24-hour exposure of *Dermatophagoides* to residues of 0.1% and 1.0% of lindane and DDT, in acetone, that had been applied to the ceilings of cages containing approximately 20 mites. Other mites (*Acarus* and *Glycyphagus*) were very susceptible to the residues.

Chapter 10 · *Pests in Excessively Damp Locations*

Modern homes in the United States provide particularly favorable refuges for certain pests attracted to excessively damp locations. In some areas, the problem is aggravated to some extent by hollow walls, as in wood-frame and some other types of construction. Wall voids provide seclusion and sometimes considerable moisture for the growth of fungi upon which certain insects can feed. They are accessible to insects via the space around utility pipes and electrical conduits that enter through the framing, and also through imperfections in construction. Similar entry routes also result in accessibility between wall voids and the attic. Insects may enter the attic through screened or louvered vents, under shingles, or through chinks, cracks, and crevices in the roof and eaves. The typical construction also provides for enclosed spaces under cabinets, pantries, closets, bookshelves, and built-in appliances. These favored areas for cryptobiotic insects may be accessible because of imperfections in design and construction or because of faulty construction materials.

Buildings with sliding glass doors, particularly on slab-on-ground foundations, sometimes allow the entry of certain pests from moist, shady, or secluded places, such as sowbugs, pillbugs, springtails, millipedes, centipedes, earwigs, or oriental cockroaches. Enclosed patios, often with considerable decorative vegetation, have resulted in the introduction of plant pests and soil- or humus-inhabiting pests into the home. The soil adjacent to the foundation is often shaded, damp, and contains abundant humus and fertilizer. Greater emphasis on the beautification of the yard and garden often involves extensive cultivation of trees, shrubs, flowers, and lawns. Wooden and brick planters are becoming increasingly popular. The insect and other fauna associated with such areas may enter the house directly around doors and windows or via subfloor areas which are protected from the sun and are often cool and damp.

Some pests, such as cockroaches and earwigs, that are favored by both moisture and seclusion, and others, such as silverfish, carpet beetles, cluster flies, face flies, and some species of ants, that are favored primarily by seclusion, have been treated in other chapters of this book under what appear to be more appropriate headings. The insects and other arthropods considered in the present chapter are those that are attracted to any damp area, usually feeding on fungi, and are found in such enclosed places as wall voids only when those spaces become excessively damp for long periods. Among them are fungus beetles, fungus gnats, fungus moths, psocids, springtails, and certain mites. These arthropods all have a protective film of lipid covering their entire cuticles and protecting them from a lethal rate of water loss, even in dry environments. With the exception of the springtails, they are not compelled to live in damp areas, but seek them because they find food there, such as leaf mold or fungal mycelia. (See under "Springtails" in this chapter.) There are other arthropods, such as sowbugs, pillbugs, amphipods, millipedes, and centipedes, that must live in damp areas because they lack the protective lipid film possessed by insects and mites. These are not the pests that deliberately infest damp locations in the home, but must live in the even damper conditions found in many areas surrounding or beneath the house, such as under piles of leaves, grass, peat moss, wood chips, in damp leaf mold, or in cracks in damp earth. They enter homes only as accidental intruders, and cannot survive there.

SOURCES OF MOISTURE IN BUILDINGS

Although leaky roofs, poor flashings, inadequate sidewall construction, and plumbing leaks are common sources of excessive moisture in homes and other buildings, condensation within buildings has become a more important factor in much of the United States. Condensation problems are obviously different in cold and temperate regions from those in warm, humid ones. Cold-weather condensation is most likely to occur in North America north of 37° latitude (excluding the Pacific Coast strip) where the average January temperature is below 35° F (1.7° C). In this region, temperature differential results in vapor pressure inside a house being several times higher than outside, so that moisture tends to move outward through cracks, crevices, and building materials, and is most likely to condense on some cold surface, such as the outside face of sheathing, the back and inside faces of siding, on the undersides of roofs, and on chimneys. The moisture generally condenses in the form of ice or frost. The frost melts during warm periods, and in the more severe cases, water in the attic may drip to the ceiling below and cause damage to the finish. Cold-weather condensation can be prevented by placing moisture barriers (treated papers, foils, films, or coatings) on the warm side of insulated walls and ceilings (Anderson, 1972).

In warm, humid regions, such as the Gulf Coast of the United States, air-conditioning is the cause of most condensation problems. Although it might appear that warm-weather condensation should be prevented by placing a moisture barrier on the warm side (outside) of an insulated wall, research has shown that a moisture barrier installed under wooden siding may lead to serious accumulations of moisture within the siding. Progress in the prevention of condensation in insulated walls in warm, humid regions will depend on further research, but it is already evident that such walls will be relatively free of condensation moisture if they are reasonably permeable to water vapor. Serious condensation of moisture can occur in wooden floors in air-conditioned homes in warm, humid areas if inside temperatures are not maintained below 75° F (24° C). Such condensation problems can be prevented by decreasing the humidity of crawl spaces through proper ventilation, soil drainage, soil covers, or mechanical dehumidification (Amburgey, 1972).

A major cause of condensation has been the increase in the daily use of water by occupants of buildings. Each day, a typical family of 4 converts about 25 lb (11 kg) or approximately 3 gal (11 l) of water into water vapor in the air of an average house (Anderson, 1972). High water tables, garden irrigation, etc., can introduce as much as 100 lb (45 kg) of water per day per 1,000 sq ft (93 sq m) into building subareas. If this water is not removed by proper ventilation, it is conducive to infestation by insects and other pests as well as infection by decay fungi (NPCA, 1955).

Rooms are smaller and more airtight than they were a few decades ago, resulting in less air to absorb and disperse the moisture released in household operations. Relative humidity is generally higher than it was in houses built many years ago. The latter generally had large rooms with high ceilings, windows that were not weatherstripped, and their construction details allowed greater air loss and air infiltration. Through central heating, all rooms within a house are likely to be warm, resulting in large-scale condensation on and in the cold walls. Water condensing on windowpanes runs down into the sash, sills, framing, and often onto the walls. More houses are being built on concrete slabs, for which there is no subfloor ventilation, and even for conventional construction, ventilation is often inadequate in subfloor areas and attics, which also helps to retain the moisture within the house. With joist-type construction, an effort should be made to ensure adequate ventilation of the crawl space, and a vapor barrier over the soil may be desirable. The moisture problem can also be minimized by keeping windows closed during the day, using a dehumidifier, and raising the inside temperature (Anderson, 1972).

Excessive moisture can result from humid weather, leaky roofs, wet basements, manufacturing and processing operations, evaporative air coolers, and inside venting of plumbing or of machines such as automatic clothes driers (they should be vented to the outside). Clothes driers should be particularly emphasized in this connection because the vent of the machine not only blows hot, humid air into the building, but also contributes a large amount of organic lint that coats the walls and furniture, and upon which mildews can flourish. Insects such as fungus beetles, psocids, and springtails can subsist on the mildew spores (Scott, 1966*c*).

The construction of concrete or masonry floors, walls, and fireplaces requires large volumes of water. In the construction of a small home with a basement, the concrete floor would have more than 2,000 lb (908 kg) of water when it is poured, the concrete basement walls would have over 4,000 lb (1,816 kg) and, if the house is plastered, over 2,500 lb (1,135 kg) of water would be used for that purpose. The rough frame of a house may be damp when constructed, and rain may add more water before the roof and sidewalls are completed. Because of rapid completion of the building, this water may remain in the house or apartment structure for months after occupancy.

Green lumber is occasionally used in construction, and the building is then quickly enclosed, giving the lumber little chance to dry. This results in "sweating" and high humidities in wall voids, favoring the growth of molds upon which insects can feed. Pests such as fungus beetles and psocids are often problems associated with new construction. When the building is occupied, the artificial heating required for human comfort eventually dries out the lumber, and infestation by mold-feeding insects then seldom reoccurs (NPCA, 1955; Anderson, 1972).

FUNGUS BEETLES
(Mostly Lathridiidae and Cryptophagidae)

Fungus beetles in the above 2 families are feeders on fungi in damp areas within buildings. They are sometimes colloquially called "plaster beetles" by pest control operators. This name was originally suggested by the occurrence of these beetles in the walls and living spaces of new buildings that had been plastered and remained damp longer than usual because of the water released by the plaster or from other sources. Moisture and warmth favor the growth of molds upon which the beetles feed.

Description. Fungus beetles are very small insects, ranging from 1 to 3 mm in length. They are various shades of brown, varying from light yellowish brown to nearly black. Morphological features of the 2 families are illustrated by the lathridiids *Cartodere constricta* (Gyllenhal) and *Aridius nodifer* (Westwood), and the cryptophagid *Cryptophagus laticollis* Lucas (figure 332). Lathridiids usually have rough surfaces, and the elytra are conspicuously marked by rows of pits and are broader than the exposed parts of the thorax and head. The bodies of cryptophagids are always more or less convex, while those of lathridiids may be convex or flattened, depending on the species. Most of them are unable to fly.

The antennae of lathridiids are 8- to 11-segmented, and have 1- to 3-segmented terminal clubs. The antennae of cryptophagids are always 11-segmented, and usually have loose, 3-segmented clubs, but in a few species these clubs are 2- or 4-segmented. The tarsi of lathridiids are usually 3-segmented, but in the males of some species, the tarsi of the first pair or first 2 pairs of legs may be 2-segmented. The tarsi of cryptophagids are 5-segmented, except for the males of some species, in which they have 4 segments. Thus, the number of tarsal segments is a good criterion for the separation of the 2 families (Hinton, 1945).

Some Common Species

The beetles shown in figure 332 were among many taken from 2 newly built, plastered apartment buildings in the Los Angeles area. *Cartodere constricta* is yellowish brown and 1.75 mm long; *Aridius nodifer* is yellowish brown to pitchy black and 1.9 mm long; *Cryptophagus laticollis* is reddish yellow and 2.1 mm long; and the cucujid *Silvanus bidentatus* (F.) (figure 332, *D*) is cinnamon brown and 3.3 mm long. The latter is a fungus-feeder or scavenger, according to T. J. Spilman (correspondence), which was found abundantly in an apartment that was also infested with the lathridiid *Aridius nodifer*, a few cryptophagid beetles, and a mycetophagid fly, *Litargus balteatus* LeConte.

The 5 species were collected 3 months after the apartment was plastered.

According to Hickin (1964), 48 species of lathridiids and 80 species of cryptophagids were found in Great Britain. In buildings he checked, the most common lathridiids were *Aridius nodifer* (Westwood), *Thes bergrothi* (Reitter), *Corticaria serrata* (Paykull), and *Lathridius minutus* (L.). The most common cryptophagids were *Cryptophagus acutangulus* Gyllenhal, *C. cellaris* (Scopoli), and *C. distinguendus* Sturm.

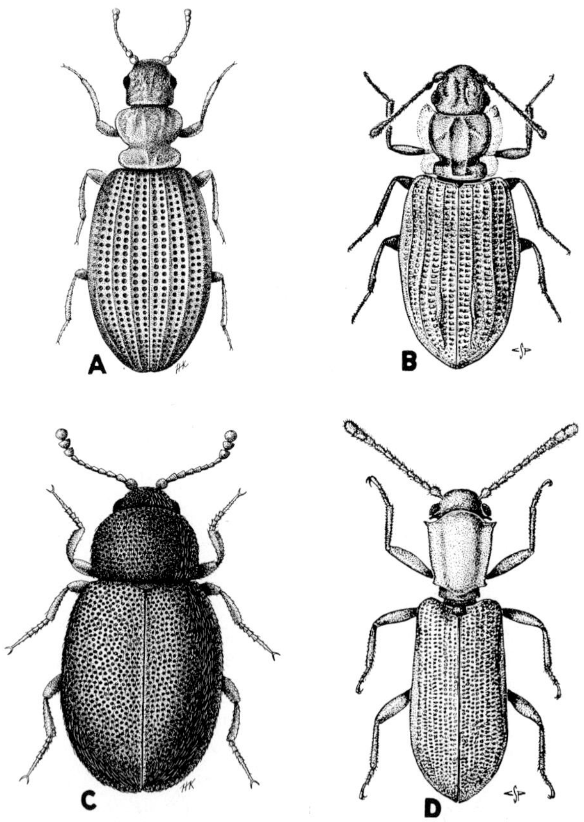

Fig. 332. Fungus beetles. Lathridiidae: *A, Cartodere constricta; B, Aridius nodifer*. Cryptophagidae: *C, Cryptophagus laticollis*. Cucujidae: *D, Silvanus bidentatus*.

Microgramme filum (Aubé) (= *Cartodere*) (figure 333) was considered by Hinton (1941) to be the lathridiid most commonly found in buildings in Europe as a whole. It is also distributed throughout North Africa and North and South America.

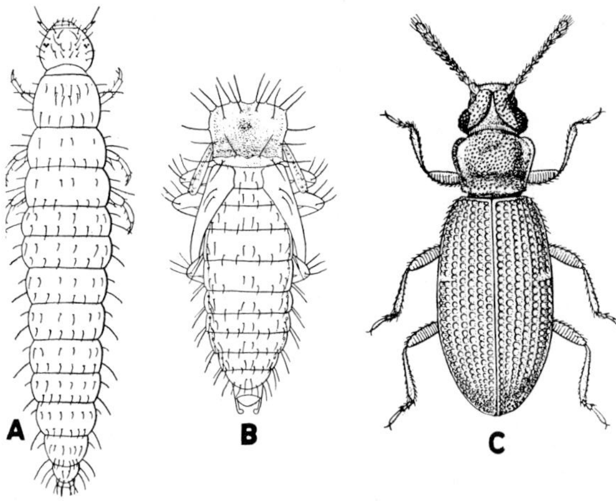

Fig. 333. A fungus beetle, *Microgramme filum*. *A*, larva; *B*, pupa; *C*, adult. (From Hinton, 1941.)

The adult beetles are 1.2 to 1.6 mm in length, brownish, and have 2-segmented antennal clubs, whereas other species of this genus have 3 segments in the clubs. Hindwings are lacking in all species of *Microgramme*. Mature larvae of *M. filum* are 1.7 to 2 mm long, whitish, and have 3-segmented antennae, the second segment twice as long as the first. The pupae are white, and 1 mm long.

Hinton observed that 2 females laid a total of 20 eggs. The shortest life cycle at 75° F (24° C) was 36 days. At 65° F (18° C), a complete life cycle required about 54 days, and at lower temperatures might take as long as 5 months. The life cycles of 4 lathridiid species investigated in the United States varied from 12.8 to 28 days, and averaged 21.2 days at a mean temperature of 74.6° F (23.7° C) (Kerr and McLean, 1956). *Microgramme filum* was found in such environments as damp, moldy warehouses and cellars; damp areas near water taps; around leaky windows; under moldy wallpaper and moldy papier-mâché dishes; in damp stored wheat, corn, and rye; in boxes containing dried beer yeast; on moldy bread; and in many herbaria. *Microgramme arga* (Reitter) (= *Cartodere*) is similar to *M. filum*, but can be distinguished by having 3 instead of 2 segments in the antennal clubs and by having larger eyes. It is widely distributed, occurring in Europe, North Africa, and North America (Hinton, 1941).

Fungus Beetles Feeding on Moldy Stored Food Products

Henoticus californicus Mannerheim, the California fungus beetle, is a uniformly brownish cryptophagid, about 3 mm long, that is probably indigenous to California, but has spread to various parts of the world through commercial channels. The larvae and adults feed on molds on various food products, and the females oviposit on the food (Linsley and Michelbacher, 1943). It is quite likely that this is one of the so-called "plaster beetles."

Alphitophagus bifasciatus (Say) the twobanded fungus beetle, is a reddish-brown tenebrionid, with 2 wide black bands across the elytra. It is 3 mm long, elongate-oval, and convex. Outdoors it normally feeds on fungi growing on moist trees, but indoors it infests fermenting or decayed cereals and cereal products. It is cosmopolitan in distribution, and occurs in coastal California and Oregon (Essig, 1926).

Alphitobius diaperinus (Panzer), the lesser mealworm (Tenebrionidae) (figure 182, chapter 7) is also attracted to damp and moldy cereal products and spoiled foods. (See chapter 7, under "Insects Infesting Broken Grain," for a few more data.)

Ahasverus advena (Waltl), the foreign grain beetle, is a cosmopolitan cucujid, with projections at the anterior corners of the prothorax. It feeds on molds on damp grains, bread, biscuits, and other farinaceous materials, as well as dead insects that may be found in or on them.

Feeding Habits of Fungus Beetles

In nature, fungus beetles are found in vegetable detritus, under bark and stones, and sometimes in ant and termite nests. The lathridiid *Microgramme filum* has been observed feeding on the spores of various species of *Polysaccum*, *Ustilago*, *Arctium*, *Trichothecium*, *Lycoperdon*, *Tilletia*, and on the spores and hyphae of fungi in herbaria, where the beetles are often pests. It was also reared on cultures of *Mucor mucedo* and *Penicillium glaucum* in petri dishes (Hinton, 1941). Four species—*Cartodere constricta*, *Corticaria serrata*, *Adistemia watsoni* (Wallaston), and *Microgramme arga*—were reared successfully on mixed cultures of the molds *Aspergillus*, *Mucor*, *Botrytis*, and *Penicillium*, appearing to prefer *Penicillium* (Kerr and McLean, 1956). As with the psocids, the lathridiids may feed on fungi developing on stored foods and seeds in damp locations. Associated with such stored products in

California are the lathridiids *Enicmus minutus* (L.), *E. suspectus* Fall, and *Holoparamecus caularum* (Aubé), and some widely distributed but undetermined species of cryptophagids (Strong and Okumura, 1958).

Economic Importance

Because both the larvae and adults of fungus beetles feed exclusively on the hyphae and spores of molds, mildews, and other fungi, they can exist only under the damp conditions required for fungal growth. They usually appear 3 or 4 months or as long as a year after the walls of a building are plastered, and disappear after the walls are thoroughly dry and can no longer support the growth of molds. Plastered walls are most likely to support an infestation if they are papered, so wallpaper should not be applied until the walls are completely dry. Walls and other parts of a building can also become damp from sources of moisture other than plaster, and can support infestations of fungus beetles or psocids. The beetles cannot crawl on smooth, vertical surfaces, and are often trapped in ceiling light fixtures, washbasins, bathtubs, or pots and pans. Fungus beetles and psocids can sometimes be found on building materials as they are incorporated into the structure, or on wrappings of plumbing and electrical fixtures, but they can also be brought in later on moldy or mildewed furnishings or other materials.

Mallis (1969) reported the lathridiid *Microgramme arga* from ground cereals in Oregon, on the walls of a house and in a drugstore in Ohio, on heads of wheat in a field in Texas, and in an old brick house in Pennsylvania that had just been remodeled into apartments. The insects were found in June on walls that had been plastered 3 months before the infestation. They were seen around the windows and near a light fixture in the ceiling. The new plaster was apparently not completely dry at the beginning of the infestation, and high humidity caused by heavy rains kept the walls from drying out, thus encouraging the growth of molds. By late summer, the beetles had disappeared.

In Los Angeles, California, lathridiids, cryptophagids, and psocids were found in one plastered apartment building soon after the first apartments were occupied. A few of the beetles could be found in the building as long as a year after its completion. The adults were generally found at lights or windows. In one of the apartments just discussed under the heading, "Some Common Species," 3 months after occupancy, many lathridiids, cucujids, and a few cryptophagids and mycetophagids were collected. The apartment had been sprayed earlier for "plaster beetles," and hundreds had already been killed.

Control of Fungus Beetles

Anything that results in the drying out of the parts of a building that support the growth of molds and mildews upon which fungus beetles subsist, such as a long period of hot, dry weather, can terminate an infestation. (Moldy foods should, of course, be removed.) Artificial heat and ventilation can aid in the drying process. It is of interest in this connection that 98% of adult fungus beetles (*Adistemia watsoni*) died in 7 days at 100° F (38° C), 100% died within 24 hours at 110° F (43° C), and at 130, 140, and 150° F (54, 60 and 66° C), complete mortality was obtained in 10, 5, and 4 minutes, respectively (Kerr and McLean, 1956).

Fungicides such as Cunilate (copper-18-quinolinolate) and G-4 (dichlorophene) (NPCA, 1957), or 2% formaldehyde in petroleum distillate (Busvine, 1966), can be sprayed onto moldy areas, taking care to avoid contamination of foods. Fumigation of a building is rarely warranted for an existing fungus beetle infestation, since it provides no residual protection against reinfestation. Aerosols are convenient, and may kill the beetles contacted, but not those emerging later from concealed locations. Longer-lasting insecticidal residues can be obtained with sprays containing 2% chlordane, 1% lindane, or 0.5% dieldrin, provided the sprays are applied where the beetles can contact the residues, such as in the vicinity of moldy materials and about lights and windows where they congregate. Wall voids and other places difficult to reach with sprays can be dusted, using 5% chlordane, 1% lindane, or 1% dieldrin, adding 3% G-4 (dichlorophene) if a fungicide is desired in addition. Thorough milling to form a fine dust that will drift well into concealed areas should improve the mixture (NPCA, 1957).

PSOCIDS

Psocids (order Psocoptera) are small insects, found throughout the world in damp and secluded places—in grass, dead leaves and litter under trees and shrubs, damp wood, under lichens and moss, or under bark, where they feed on molds and mildews. At least 2 of the species commonly found in buildings are cosmopolitan pests. Psocids have biting mouthparts, but they

Fig. 334. A house-infesting psocid, *Liposcelis* sp.

cannot bite humans or animals. They are harmless except for the contamination of foods, and are nuisances in the same way as many other insects—merely by their presence in a home. These insects are sometimes called "booklice" or "barklice" because of their superficial resemblance to certain species of lice and because they are found, sometimes in very large numbers, on moldy books and papers in damp buildings or their basements and under loose, damp bark. Books are said to be particularly susceptible to mold growth because of the glues and sizes used in binding, upon which molds can develop under damp conditions (Hickin, 1964). Psocids also resemble minute workers of subterranean termites; in fact, psocids and termites, as well as cockroaches, are believed by some investigators to have had a common ancestry.

Psocids are soft-bodied insects, the immature forms of which resemble the adults in form and structure. They are gray or light brown, winged or wingless, with a large abdomen, a narrow thorax, a large head, and long, many-segmented antennae. The outdoor species have wings, but are weak fliers. Their wings are held rooflike over the body, and are often marked with a pattern of some sort. The pearly-white eggs are relatively large, being about one-third the size of the adult insect. They are laid singly, and adhere to the surface upon which they are deposited.

Species Infesting Buildings

Two well-known cosmopolitan species commonly encountered in buildings or as household pests are the cereal psocid, *Liposcelis divinatorius* (Müller) (Liposcelidae) and the larger pale booklouse, *Trogium pulsatorium* (L.) (Trogiidae), also known as the "deathwatch." The latter receives its specific name from the ticking sounds it produces by striking its abdomen against paper and similar materials. Only the females make these sounds, believed to be mating calls. Another species, *Lepinotus inquilinus* (Heyden) (Trogiidae) is believed to be capable of making the same kind of sound (Pearman, 1928). A species of *Psyllipsocus* (Psyllipsocidae) (figure 334) is sometimes found in buildings in southern California. The cereal psocid and the larger pale booklouse are both wingless indoor species. The latter has small, scalelike wing pads, well-developed compound eyes, and is about 2 mm long. The cereal psocid has no wing pads, only a few scattered simple eyes, and is about 1.5 mm long. Both species can move about very swiftly. Both have soft bodies, and are cream-colored to grayish or light brown. Other psocids reported as household pests are *Lachesilla pedicularia* (L.) (Lachesillidae); *Psyllipsocus ramburii* (Selys) and *Psocathropos lachlani* Ribaga (Psyllipsocidae); *Lepinotus patruelis* Pearman and *L. inquilinus* (Heyden) (Trogiidae); and *Liposcelis bostrychophilus* Badonnel (=*granicola* Broadhead and Hobby) (Liposcelidae) (Pearman, 1928; Broadhead and Hobby, 1944; Finlayson, 1949; Gurney, 1950).

Psocids can be serious pests of stored foods under excessively humid conditions. A key was developed to help identify 5 species commonly infesting stored foods in the United States: *Trogium pulsatorium,* the larger pale booklouse, which has 3 distinct thoracic segments (the other 4 species have only 2); *Liposcelis bostrychophilus,* the banded psocid, which has 7-faceted eyes and a brown head and body; *L. paetus* Pearman, the warehouse psocid (figure 335), which has 2- to 4-faceted eyes, a brown head, and yellow body; *L. entomophilus* (Enderlein), the grain psocid, which has 2 to 5 large pronotal bristles; and *L. terricolis* (Badonnel), the "booklouse" (figure 335), which has only 1 large pronotal bristle. [*Liposcelis bostrychophilus* and *L. paetus* do not have large pronotal bristles (Scott, 1963*b*).]

Biology. The cereal psocid, *Liposcelis divinatorius,* reproduces parthenogenetically (males are not known). In one investigation, the life cycle averaged 24.4 days from June to August, with an average of 57 eggs deposited. During cold weather, the adults died, and the eggs hatched in the spring. In winter, only 21 eggs per female were produced on an average, the preoviposition period was 45 days, and the life cycle was 110 days (Ghani and Sweetman, 1951). There may be 6 to 8 generations per year (Candura, 1932). In another investigation, the period from egg to adult required about 1 month at a temperature of 80° F (27° C) and relative humidity of 65%. There were 3 molts. After a preoviposition period of 2 to 3 weeks, eggs were laid at the rate of about 1 every 12 hours until perhaps 75% of the total were laid. Then, a long period followed when only an occasional egg was deposited. An adult life of over 3 months was observed (Finlayson. 1932). In England, *Trogium pulsatorium* was found to have only 1 generation a year and to overwinter in the nymphal stage, whereas *Liposcelis divinatorius* bred continuously, and was found in all stages, even during the winter months (Pearman, 1928).

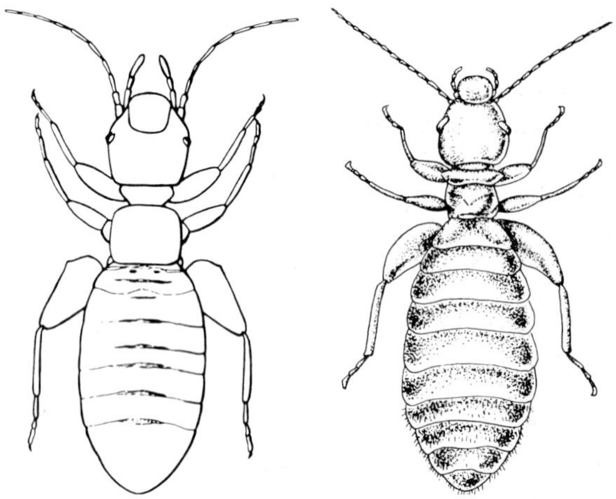

Fig. 335. Two psocids commonly found on stored foods under excessively humid conditions. *Left,* warehouse psocid, *Liposcelis paetus; right,* booklouse, *Liposcelis terricolis.* (From Scott, 1963*b*.)

Influence of Humidity

At humidities near the "critical equilibrium," a decrease in relative humidity of only 5% can result in the decline of a psocid infestation. For example, specimens of *Liposcelis rufus* Broadhead and Hobby lost water at RH's of 53% or less. At 25° C (77° F) and 33% RH, they lost 50% of their water in 11 days, but most of them recovered the lost water within 6 to 7 hours after being transferred to an RH of about 58%. The flat and contracted abdomens of the dehydrated insects became inflated, returning to their original sizes and shapes, when adequately moist air became available.

Desiccated *L. divinatorius* recovered even more rapidly. Insects that had become flattened and lethargic from desiccation became turgid and active within 2 or 3 hours when moisture was restored, and began laying eggs (Finlayson, 1932). With food available, females of *L. rufus* died within 2 to 3 weeks at all RH's below 58%. Other species were even less resistant to desiccation. Females of *L. knullei* Broadhead and Hobby lost water twice as fast as *L. rufus,* and died within 1 week at all humidities below their critical level. *Liposcelis bostrychophilus* survived only 10 days, and egg-laying ceased, below the critical equilibrium humidity (Knülle and Spadafora, 1969).

Economic Importance

Psocids, like fungus beetles, may be pests primarily for a few months after the completion of a building, before the plastered walls have had time to dry out and while the dampness can support a growth of molds, particularly in wall voids. Psocids may escape from the voids and infest large sections of the living space. They are often most abundant on the top floors of multistoried buildings because they are the last to be plastered and have the greatest variation in temperature—both factors that result in greater dampness (Hartnack, 1943). In a house, psocids occur not only in spaces between walls, under floors, and behind door and window trims, but also in cupboards and closets, behind electrical and other fixtures, in upholstered furniture containing tow or Spanish moss, in rugs, books, paper, groceries, flour, cartons of merchandise, straw matting, and contents of trunks taken from storage, feeding on microscopic molds occurring on such materials if they are damp (Back, 1946). Psocids are sometimes found only in 1 area of a house, such as a room used for drying clothes. The regular use of the room for that purpose may result in a continuous infestation. As with silverfish and fungus beetles, psocids may fall into washtubs, washbasins, pots, or pans and be trapped, for they cannot climb on smooth, vertical surfaces.

After having moved from 2 apartment buildings in succession because of psocid infestations, tenants in a beach community in southern California moved to a third building, but first they had their overstuffed furniture, clothing, blankets, linens, and other personal belongings fumigated to avoid the possibility of a third occurrence of what had become a traumatic experience.

Besides being household pests, psocids resemble fungus beetles in occasionally infesting granaries, warehouses, herbaria, insect collections, libraries, stored foods, and the nests of birds and insects (Finlayson, 1932; Linsley, 1942, 1944; Broadhead and Hobby, 1944). The 2 household pests, *Liposcelis divinatorius* and *Trogium pulsatorium*, have been reported as predators of eggs of the Angoumois grain moth, *Sitotroga cerealella* (Finlayson, 1932), and *Liposcelis bostrychophilus* was observed to feed on the eggs of the Indian meal moth, *Plodia interpunctella* (Lovitt and Soderstrom, 1968). In one test, *L. bostrychophilus* consumed 71 and 72% of the eggs when other food was absent or present, respectively.

Control of Psocids

In general, the foregoing recommendations for control of fungus beetles pertain also to psocids. With either pest, merely waiting for a building to dry out may solve the problem. This solution is more likely to be accepted by homeowners than by rent-paying occupants of an infested house or apartment, especially if they have heard psocids referred to as "booklice."

Mallis (1969) cited the results secured by a pest control operator who thoroughly sprayed, with 2% chlordane in base oil, all the exposed woodwork of a 6-month-old house that had been built with green lumber. He obtained excellent control of psocids. In view of the widely recognized efficacy of dichlorvos, Mallis suggested that a 0.5% solution of this insecticide should also be tried in psocid control.

SPRINGTAILS

Springtails (order Collembola) are very small insects, usually less than 2 mm long, and they do not undergo metamorphosis. They are among the most primitive insects, having been identified as fossils from the Devonian period, over 400 million years ago, and thus antedating any other known insect fossils (Jeannel, 1960). According to Scott (1966*c*), of about 500 species reported from North America, 19 have been noted to infest buildings, of which 10 are shown in figure 336. There are species adapted to every climatic condition—high in the Himalayas, in Antarctica, in the intertidal zones of ocean beaches, on the surface of standing water, and in great numbers on snow in winter. They can occur in enormous numbers, for example, up to 50,000 per cu ft (28 cu dm) of forest litter. Most species of springtails lack tracheae, and respire through their cuticles. As might be expected, the cuticles of such species are very permeable to water. Therefore, these insects must spend most of their time in very damp locations. If they find themselves in a dry situation they crawl about actively until a damp environment is found (Joose and Groen, 1970).

Springtails are found outdoors in moist situations, usually feeding on algae, fungi, and decaying vegetable matter. They are among the most troublesome swimming pool pests. In southern California, *Hypogastrura armata* (Nicolet) (figure 336, *E*) appears to be the one most frequently involved. This is a common species of world-wide distribution. It is dark gray to black dorsally, light gray ventrally, and with a reddish hue on the head, antennae, and legs.

If their environment becomes dry, then in the course of their active crawling in search of moisture, springtails may invade the home, entering

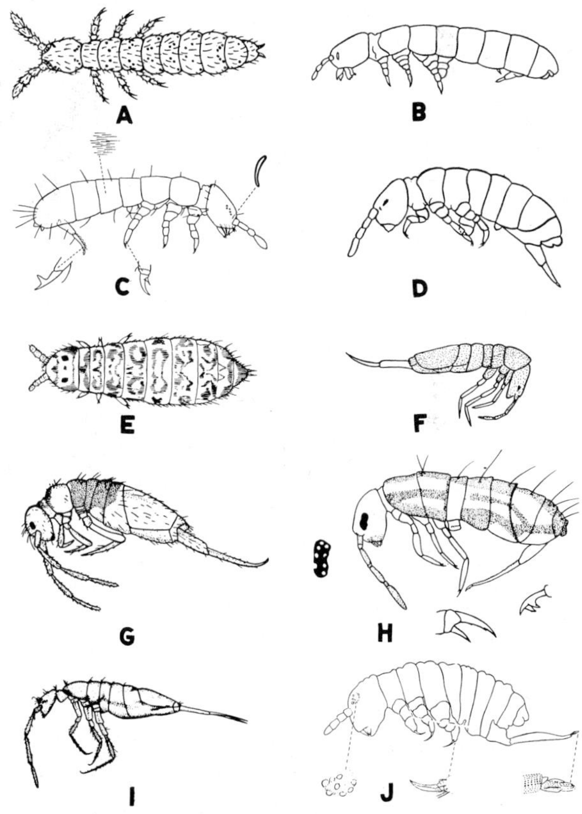

Fig. 336. Building-infesting springtails. *A, Onychiurus armatus* Tullberg; *B, Isotomodes tenuis* Folsom; *C, Folsomia* sp.; *D, Proistoma* sp.; *E, Hypogastrura armata* (Nicolet); *F, Lepidocyrtus* sp.; *G, Entomobrya atrocinta* Schött; *H, Orchesella* sp.; *I, Seira platani* (Nicolet); *J, Podura aquatica* L. (From Scott, 1966*c*.)

through window screens, open doors, vent pipes, or with merchandise or ornamental plants. They may be attracted to light, entering through windows or under doors. After a hot day, they may swarm over the side of a building in tremendous numbers, increasing the chance of indoor infestation. After entering a house, they crawl about, and are often trapped in sinks, washbasins, and bathtubs. They are most commonly found where there are sources of moisture, as in the kitchen and bathroom, where they hide in very small cracks and crevices. They may also occur in damp wall voids. In some homes, potted plants serve as sources, the springtails coming from the damp soil (Scott *et al.*, 1962; Scott, 1966*c*).

Fig. 337. Two springtails commonly found in damp locations in and around buildings. *Top, Onychiurus armatus*; *bottom, Entomobrya atrocinta.*

Ideal conditions for springtails result from high humidity in conjunction with excessive organic debris. In addition to whatever nutritive material that may be present in the organic matter, mildew spores can form, contributing further sustenance. Springtail infestations can be suspected whenever mildew odors are detectable. Infestations tend to increase during hot, humid weather, and decrease during cold weather when the heating system dries the air and the building structure. Even during the drier periods, however, springtails may be found in great abundance around the insulations of steam and water pipes (Scott *et al.*, 1962; Scott, 1966*c*).

A pictorial key has been prepared to help identify the common springtails of the United States (Scott *et al.*, 1962).

Springtails as Pests

These insects are pests principally by virtue of their presence in the home. They have never been incriminated in the transmission of any human disease, but *Entomobrya nivalis* L., a cosmopolitan species, has been reported to cause an itching type of dermatitis in man, and *Orchesella albosa* Guthrie, found in North America and Europe, has been recorded as infesting the head and pubic areas, but without causing dermatitis (Scott *et al.*, 1962). Another cosmopolitan species, *Entomobrya atrocinta* Schött (figures 336 and 337), is a pest of dried milk powder (Scott, 1963*b*). Springtails crawling or hopping on the skin may cause itching, and when crushed on the skin, they may cause a mild, localized, allergic response. (Scott *et al.*, 1962; Scott, 1966*c*).

Control of Springtails

The best control for springtails is to decrease humidity within the building, if possible, and to improve sanitation by not depositing food particles in cracks, crevices, and around floor edges. Large deposits of dust and lint should not be permitted to accumulate anywhere in the building. Moldy bedding, mattresses, couches, and chairs have been found to be good habitats for the pests. Spraying or dusting of infested surfaces, including potted plants, has been found to be effective, using common household insecticides. The plants should be watered only after the soil in the pots appears to be dry. (See also, chapter 11.)

MISCELLANEOUS MITES AND INSECTS

Grocer's Itch Mite, *Glycyphagus domesticus* (De Geer) (Acaridae)

This is a small, white mite found in pantries, feeding on cereals, sugar, cheese, and other foods, and in furniture stuffed with green Algerian fiber, where it feeds on the fungi growing on the fiber. The furniture may become white with myriads of these mites, and they may spread to other parts of the house. According to A. M. Hughes (1961), Algerian fiber and the fiber from which rush furniture is made are highly hygroscopic, taking up and retaining moisture from the air. In a damp environment, molds upon which the mites feed develop on the furniture. Infested fiber should be removed and the furniture restuffed with some other material. However, anything reducing humidity within a room, such as improved ventilation, will decrease the chances of infestation. Infestation does not occur in furniture constructed of either Algerian fiber or rushes if they have been treated with copper sulfate.

The mold mite, *Tyrophagus putrescentiae* (Schrank), is becoming increasingly common as a household pest, causing allergic reactions.

Silverfish

The silverfish are described and illustrated in chapter 8, under "Pests of Fabrics and Paper." They seek warmth and moisture, and are sometimes found in large numbers in damp wall voids of newly constructed buildings. Wall voids can be drilled and treated with insecticide dust or fog.

Fungus Gnats (Mycetophilidae)

Description and Biology. Fungus gnats are small, slender, delicate flies that somewhat resemble mosquitoes, and many species are also about the same size as mosquitoes. They are attracted to light, and do not bite. The thorax is relatively large and arched, and the head is small. The antennae are long and filiform, 12- to 17-segmented, but usually with 10 segments, the 2 basal segments being quite large. The coxae are long, the coxae and femora are thickened, and the tibiae have prominent distal spurs and usually a series of short or long bristles. Adults may be black, yellow, brown, red, or a combination of these colors, depending on the species. They are fairly common, and may be seen moving rapidly over dark and damp places where there is an abundance of decaying vegetation or fungi, on which they lay their eggs singly. The eggs hatch within a few days.

The larvae are usually slender, 12-segmented, footless, with 8 pairs of spiracles, smooth and whitish, but with brown or black sclerotized heads. They infest fungi, damp soil, litter, decaying vegetation, or decayed wood. Under optimum conditions, the larvae can develop through all 5 instars in 6 to 8 days. Some species form silken cocoons on or in the ground in which to pupate, and the adults emerge in about 3 days. Certain species are pests in mushroom cellars (Essig, 1926; Lauret, 1962; Curran, 1965).

Fungus Gnats in Homes

Fungus gnats breed in humus, leaves, and litter under shrubbery around the house, and find their way inside. Within the home they breed in pots, boxes, flower vases, and any container in which house plants are grown. All household sprays, aerosols, or suspended vapor-releasing strips are effective in controlling them indoors.

Some Common Species

The 2 most common and widely distributed species in the United States are *Mycetophila fungorum* (De Geer) (=*punctata* Meigen), a brownish gnat 4 to 6 mm long, and *M. ichneumonea* Say (= *mutica* Loew), a very small, reddish-yellow or fuscous gnat, 2.7 mm long. In addition to these, in California are found *M. contigua* Walker (=*fallax* Loew), 3 to 3.7 mm long and dark brown to black; *Mycomyia mendax* Johannsen, 6 mm long, yellow and black; and *Leia striata* (Williston), 5 mm long, with the head and thorax yellow and the abdomen black.

Darkwinged Fungus Gnats (Sciaridae)

Sciarids (figure 338) are closely related to the mycetophilids, but their eyes meet above the bases of the antennae, whereas in the mycetophilids the antennae are inserted on a level with the middles

Fig. 338. A sciarid fly occasionally found in damp locations in buildings.

of the eyes. Also, the coxae of the sciarids are not so elongate as those of the mycetophilids. The adults are small, usually black, and occur in moist, shady places. The larvae feed on fungi, and those of some species often travel over the ground in snakelike "armies." Some species occasionally become pests in mushroom cellars, while others infest the roots and tubers of plants and cause some injury.

Humpbacked Flies (Phoridae)

The phorids are very small flies that are easily recognized by the small head and prominent pronotum, giving them a humpbacked appearance. Only the veins toward the foremargins of the wings are thickened; the others are weak and are not connected by cross veins. The hind femora are laterally flattened. The dirty-white, slightly flattened larvae are up to 4 mm long. The adults are abundant on and near decaying vegetation. They are attracted to lights in structures and are often a nuisance in houses.

Megaselia rufipes (Meigen) is a cosmopolitan species that is commonly encountered throughout the United States. It is black or dull brown, with yellowish or brown legs.

Megaselia scalaris (Loew) is a tropical and subtropical species and *Dohrniphora cornuta* (Bigot) is a cosmopolitan species; both are common in California.

Fungus Moth, *Aglossa caprealis* (Hübner) (Pyralidae)

Certain species of moths develop on the dry-rot fungus (*Poria incrassata*) (figure 102, chapter 5), on damp straw, or in leaf mold in sufficient numbers to become household pests. *Aglossa caprealis* can subsist entirely on the dry-rot fungus. It can be readily distinguished from a tineid moth, *Nemapogon* sp., which has similar habits, but is smaller and more slender. However, *A. caprealis* could easily be confused with cutworm adults or other moths that accidentally gain entry into a home.

Description. The adult moth is dark brown, with uneven, lighter markings across its broad wings. When at rest, the male measures 13 mm and the female 16 mm in length. The female is 9 or 10 mm across at the tips of the folded wings (figure 339). The mature larva is naked, shiny dark brown to black, and is about 20 mm long. It is ringed with interrupted annulations that give it a wrinkled appearance (Pence and Hogue, 1957).

Fig. 339. A pyralid fungus moth, *Aglossa caprealis*. Adult and larva.

Habits. This moth occurs in the dark, damp locations where dry-rot fungus develops. In one instance in which the insects originated in mycelial growth of fungus under a sink, where leaky tile grouting permitted moisture to penetrate the wood beneath, the larvae were found in much of the house, but particularly in the kitchen and bathroom, and migrated extensively before pupation. In another instance, this species was found infesting a garage adjoining a home. A pile of damp fabrics stored in the garage had developed a growth of fungus, providing food for the larvae. A laboratory experiment demonstrated that these larvae could thrive on a diet consisting only of the mycelium of *Poria incrassata*. They fed directly on fragments of dry mycelia placed on a damp substrate. The fragments were bound together in dense webbing (Pence and Hogue, 1957).

Meal Moth, *Pyralis farinalis* (L.) (Pyralidae)

This moth is found breeding in damp locations, such as in straw, vegetable refuse, or moldy leaves, and it sometimes invades houses. Both this species and *Aglossa caprealis* were reported from grain under damp conditions (Strong and Okumura, 1958). G. T. Okumura (correspondence) believes that all species of the family Pyralidae, even those found on trees, are to some extent mycetophilic and scavenging, feeding on fungi or on dead and moldy material.

Description. The meal moth has a wingspread of about 25 mm. The base and apex of the forewing are reddish brown, with the middle portion pale, but bordered on each side by a wavy, white line. A row of black dots occurs along the posterior margin. The dirty-gray larva, with a dark head and prothoracic shield, is about 25 mm long when full grown (Linsley and Michelbacher, 1943).

Habits. In one instance, where the moths infested all the rooms of a house and persisted despite attempted control measures, a mass of webbed and matted straw was found on moist soil in the crawl space under the building. The straw was teeming with adults and immature stages.

Fig. 340. A tineid fungus moth, *Nemapogon* sp.

The moths had found their way into the rooms by entering forced-air heating vents and pipe openings.

The infestation stopped after the straw was removed (Berns, 1958).

Nemapogon sp. (= *Tinea*) (Tineidae)

The forewings of this species are brown, with many black spots (figure 340). The wing markings distinguish it from the various species of clothes moths, for which it is at times mistaken because of its obvious family resemblance. It has been found infesting homes in which the substructure was infected with dry-rot fungus.

In common with *Aglossa caprealis,* the moth can reach full development on a diet consisting solely of dry-rot fungus (Pence, 1956*a*).

Control of Fungus Moths

Measures taken to control the dry-rot fungus (chapter 5) will also eliminate infestations of fungus moths. Spraying the infested areas will usually eliminate larvae or adults that are moving about in the living space of a house.

SOWBUGS AND PILLBUGS

Sowbugs and pillbugs (class Crustacea, order Isopoda) are said to be indigenous to Europe, and to have been introduced into North America. Hatchett (1947) believed that this accounted for their abundance only around human settlements on this continent.

The commonly used English name, "woodlouse," is more appropriately applied in Europe, where these isopods occur commonly in wooded areas far from human habitations, than it is in North America.

In the United States, sowbugs and pillbugs may be found in and around homes wherever there is a combination of excessive moisture and an abundance of decaying organic matter. The majority of isopods are aquatic forms, and obtain their oxygen through gills, but the sowbugs and pillbugs breathe by means of tubelike invaginations or pseudotracheae, enabling them to live on land. However, the pseudotracheae open to the exterior by a single pore which lacks the spiracular closing device possessed by other arthropods. Some respiration also takes place through the integument of the body (Cloudsley-Thompson, 1968).

These land crustaceans are not so well adapted as insects for controlling water loss, for they not only lack a closing device for the respiratory system but also lack the epicuticular wax layer that protects insects from excessive transpiration, even in dry environments (Edney, 1957; Barnes, 1963; Cloudsley-Thompson, 1968). Therefore, these isopods must remain during the day beneath objects on damp ground, or even bury themselves well beneath the ground, if necessary, to avoid desiccation. They may also huddle together in masses to reduce the evaporation rate, and travel about during the night, taking advantage of the lower temperature and higher humidity. Moist food and water to drink also help to replace the water lost by integumentary evaporation.

Sowbugs and pillbugs normally live outdoors, and sometimes injure tender young plants or their roots. Their scavenging habits are indicated by an observation that pillbugs in large numbers almost completely devoured a dead rat and ate the flesh off a peach pit (Pierce, 1907). They sometimes become accidental intruders into houses, where they do no damage and cannot survive. Slab floors and sliding doors have increased the likelihood of occasional intrusions of these isopods into the home.

Sowbugs, *Porcellio laevis* Koch and *P. scaber* (Latreille) (Porcellionidae)

These cosmopolitan sowbugs are the most common species throughout the country. They are often pests in residential properties where there are damp locations in which they can hide during the day, such as under trash, rocks, boards, flower pots, piles of grass clippings, flower-bed mulches, or other decaying vegetation.

Description. Sowbugs (figure 341) are up to 13 mm long, dorsoventrally flattened, oval, and have a broadly convex, hard exoskeleton composed usually of 10 freely articulating tergites that tend to project laterally. The segments of the thorax and abdomen are usually about the same width, and are not clearly demarcated dorsally. The adults are dark gray to slate color above, with a wide, dark, longitudinal median band becoming lighter in color laterally, and they are light gray beneath. They have well-developed, sessile, compound eyes, well-developed antennae, and 7 pairs of legs, a pair for each thoracic segment (Barnes, 1963). *Porcellio scaber* differs from *P. laevis* in having transverse rows of small tubercles covering the head and body dorsally (Cloudsley-Thompson, 1968).

Fig. 341. Isopods that may invade the home. *Top*, sowbugs, *Porcellio dilatatus; bottom*, pillbugs, *Armadillidium vulgare*.

Biology. The eggs hatch in a brood pouch (*marsupium*). During the last weeks of a gravidity of about 50 days, there may be eggs in the marsupium as well as young sowbugs ready to emerge. In investigations by Hatchett (1947) in Michigan, 24 young per brood were born on an average, but as many as 88 young per brood have actually been recorded (Verhoeff, 1919, 1920). There may be 1, 2, or 3 broods per year, but usually 2 (Hatchett, 1947). Adult sowbugs may live about 2 years.

Pillbug, *Armadillidium vulgare* (Latreille) (Armadillidiidae)

Pillbugs look somewhat like sowbugs. In the pillbug, the last abdominal appendages, the *uropods,* are rounded posteriorly, whereas in the sowbugs they project from the hind end of the body like a pair of small, pointed tails. Thus, the uropods can be used to distinguish between pillbugs and sowbugs, as can be seen in figure 341. When pillbugs are disturbed, they roll up into a tight ball, cover the front part of the head with their uropods, and are then, of course, even easier to distinguish from sowbugs than when they are not rolled up.

Biology. The biology and habits of pillbugs are similar to those of sowbugs. Pierce (1907) noted that pillbugs were white and had 6 pairs of legs when they left the marsupium. Within 24 hours they molted, and still had 6 pairs of legs. The next molt occurred between the fourteenth and eighteenth days, and the isopods then had 7 pairs of legs. There was no regularity in the time of molting after the first molt; the time depended on the food supply. Pierce found that females about 7 mm long could reproduce. The largest pillbug he noticed was 15 mm long, and he believed it to be several years old. From 29 to 79 young per brood were recorded in Texas (Pierce, 1907); 5 to 62 (average 28) in Michigan (Hatchett, 1947); and 48 to 156 in France (Vandel, 1939). There could be from 1 to 3 generations per year, depending on the mean temperature of the region.

Control of Sowbugs and Pillbugs

Much can be accomplished toward the control of these crustaceans by eliminating the moist environments that make their terrestrial life possible. Piles of leaves and grass clippings provide both moist hiding places and food, and should be removed. Boxes, boards, flower pots, and trash should be removed if they are lying on damp soil, or they should be stored off the ground. Proper ventilation of the crawl spaces under buildings will eliminate a common source of excessive moisture. An undue amount of watering should be avoided, particularly where there is a dense growth, such as of ivy.

Residual pesticide treatments may be applied to and near foundation walls, to damp areas surrounding or near the building, and to the crawl spaces underneath it. Most of the common garden insecticides, such as carbaryl, chlordane, diazinon, malathion, and propoxur (Baygon), applied as sprays or dusts, are effective. Treatment of peat moss, wood chips, and redwood bark used as mulches in the garden is particularly important. Subsequent sprinkling will carry the insecticide down into the soil where the crustaceans hide. Known points of entry into the home for sowbugs and pillbugs should be sealed off. These pests can not only be killed with household insecticides, but can also be swept or vacuumed away. Most of them will be dead by the time they are discovered for, as already stated, they cannot survive for more than a day or two away from moist areas.

MILLIPEDES

Millipedes (class Diplopoda) have high moisture needs, in common with the sowbugs and pillbugs. They feed on both living and decomposing vegetation, and occasionally on dead snails, earthworms, and insects. They may attack growing crops, but seldom in numbers sufficient to need control measures. During periods of drought, they are sometimes forced to attack crops to obtain water. The principal complaint against millipedes is their tendency to invade homes, sometimes in incredible numbers.

Description. Millipedes are long, cylindrical, many-jointed, wormlike arthropods. The mature forms vary from 10 or 15 to 100 mm or more in length. Most of them are blackish or shades of brown, but some species are red, orange, or have mottled patterns. They have 2 body regions. The head has a pair of short, 7-segmented antennae, at least 2 pairs of mouthparts, and usually eyes. In the body, the first segment behind the head bears no legs, while the next 4 segments (the next 3 in a few species) each have 1 pair of legs. All other segments except the last (in some species except the last 2 or 3) have 2 pairs of legs. The segments with 2 pairs of legs are *diplosegments,* derived from the fusion of originally separate segments. "Spiracles" leading into tracheae open above the coxae. However, the tracheal system has no closing mechanism such as the spiracles of

Fig. 342. ***A,*** **millipede,** ***Julus hesperus; B,*** **millipede,** ***Tylobolus claremontus; C,*** **centipede,** ***Scolopendra polymorpha.***

insects (Miley, 1930). The integument is hard, rounded above, and flattened below. Like the integument of isopods and chilopods, it possesses no protective lipid barrier. The absence of surface lipid, along with the inability to close the tracheal openings, makes diplopods particularly susceptible to a lethal rate of water loss in dry environments.

The two pairs of legs on most segments, particularly well shown in figure 342, *A,* distinguish millipedes from centipedes. Most millipedes are rounded above, whereas centipedes are flattened. Millipedes crawl very slowly, whereas centipedes crawl quite rapidly. However, the gait of millipedes exerts a surprising pushing force, enabling the animals to force their way through humus, leaves, and loose soil (Barnes, 1963).

Biology. Millipedes lay from 20 to 300 eggs in nests in the soil. The eggs of most species hatch in several weeks, and the newly hatched young usually have only the first 3 pairs of legs and not more than 7 segments. Millipedes have a simple metamorphosis, going through a number of molts during each of which additional segments and legs are added. In many species, there are 7 larval instars. Many species of millipedes reach sexual maturity in 2 years, but some require 4 or 5 and will then live several more years. Adults generally overwinter in the soil. In some orders, the millipedes build molting chambers in the soil, similar to the egg nests, for after each molt the animals are helpless against predators. Millipedes usually eat their cast skins to restore lost supplies of calcium, and if they do not do so, further development is abnormal (Cloudsley-Thompson, 1968).

Protective Devices. Millipedes do not possess venom-bearing claws (toxicognaths) as do centipedes, but they do have certain protective devices. The species of one order of millipedes (Oniscomorpha) can roll up into a ball like the pillbugs, protecting their soft underparts.

In the Colobgnatha, Polydesmoidea, and Juliformia there are repugnatorial glands, either 1 per segment on all but the more anterior segments or on more or less alternate segments. They open on the sides of the tergites, and secrete a mixture of hydrocyanic acid, iodine, and quinone, a brown or white liquid with an iodoform odor. It is believed to be toxic to other small animals, and is reported to be caustic to the human skin and to cause vesicular dermatitis. Most species exude the liquid slowly, but some can discharge it as a spray. Some of the tropical species can produce a severe dermatitis and possibly blindness (Halstead and Ryckman, 1949; Barnes, 1963).

Millipede Migrations

Sometimes millipedes, occasionally accompanied by centipedes and sowbugs, will migrate in great numbers. This is believed to be the result of a heavy buildup of the millipede population because of very favorable environmental conditions, followed by drought (Cloudsley-Thompson, 1949, 1968). Migrations have at times involved such enormous numbers of millipedes as to make it necessary to sprinkle sand on slippery railroad tracks to provide traction for the driving wheels of locomotives. Brooks (1919) described a migration of the millipede *"Fontaria virginiensis* Drury" (*Pleuroloma* sp. inc.) in West Virginia in such great numbers that cattle refused to graze on invaded pastures, wells became filled with drowned corpses of the arthropods, and workers in fields became nauseated and dizzy from the odor of millipedes they crushed while hoeing in a cornfield. In another migration of an estimated 65 million millipedes, a farmer shoveled a half bushel or more of these pests from his porch and surrounding area and carried them away every morning for a period of 2 weeks. In another migration of *"Fontaria brunnea"*[26] in strawberry farms in West Virginia, the millipedes fed on overripe fruit and became so abundant that the picking of the fruit had to be discontinued.

Back (1939) described migrations of millipedes in the eastern United States in which they swarmed over basements and first-floor rooms of houses in great numbers. Sometimes they crawled up walls and dropped from ceilings. These migrations occurred most often in the fall. He considered the migrations to be caused by an urge to seek hibernating quarters, although sometimes heavy rains raised the water level in the soil and forced the millipedes to seek shelter elsewhere. Back found that homes in wooded areas with virgin soil, still filled with quantities of decaying vegetation, were particularly susceptible to infestation. As many as 700 millipedes entered a room in a single evening.

Many millipedes, as well as sowbugs and pillbugs, drop into swimming pools, and are usually more common problems in this respect than are insects. Also, sliding doors in houses with concrete slab foundations allow the entry of many millipedes, centipedes, sowbugs, and pillbugs.

Some Species Common Around Structures

The 2 millipede species found by Bennett and Kerr (1973) to be the greatest nuisances around buildings were the greenhouse millipede, *Oxidus gracilis* (Koch) (= *Orthomorpha),* and *Orthomorpha coarctata* (Saussure), particularly the latter, although it is found only in the southern half of Florida. *Oxidus gracilis* is widespread throughout the southern and western United States in the

[26] *Fontaria brunnea* Bollman is actually a species or subspecies in the genus *Pleuroloma,* native to the Minnesota–Michigan–Iowa region only, casting doubt on the correctness of the identification. All species of *Pleuroloma* seem to have a tendency for great aggregations and "swarming" that is not observed in closely allied genera (R. L. Hoffman, correspondence).

open, and in greenhouses elsewhere. The millipedes shown in figure 343 were found in large numbers congregated under flower pots in commercial orchid glasshouses in the Los Angeles area.

Description. *Oxidus gracilis* is one of the "flat-backed" millipedes (order Polydesmoidea), in which the somites are dorsoventrally flattened. The 20 postcephalic somites bear 30 and 31 pairs of legs in the males and females, respectively. The adult males are about 19 mm long, and the females, about 21 mm. These millipedes vary in color from the creamy white of recently molted specimens to deep chestnut brown or black in the oldest individuals (Causey, 1943).

Fig. 343. The greenhouse millipede, *Oxidus gracilis*.

Orthomorpha coarctata is almost indistinguishable from *Oxidus gracilis*. Two other species common in California are *Julus hesperus* Chamberlin and *Tylobolus claremontus* Chamberlin (figure 342, *A* and *B*).

Life Cycle. As with most tropical animals, *Oxidus gracilis* has no annual breeding season. Oviposition was found to occur during any month of the year if the temperature was maintained above 22° C (71° F). Eggs were deposited in small, rough cavities, 7 to 15 mm below the soil surface, in clutches that varied from 14 to approximately 300. The larvae of each instar differed in the numbers of somites and legs, in size, and in density of pigmentation. Sexual maturity was reached in the eighth instar. Five millipedes reared from eggs collected in September passed through the 7 larval instars in 148 to 177 days at "heated-room" temperatures (Causey, 1943).

Orthomorpha coarctata laid eggs in the soil in holes that were 20 to 40 mm deep and 5 to 7 mm in diameter. There were 25 to 300 eggs per clutch. The period from egg to adult ranged from 119 to 187 days in a rearing room. There were apparently 2 generations per year (Bennett and Kerr, 1973).

Habits. In the limited areas where *Orthomorpha* occurs, it is a greater problem than *Oxidus*. It apparently breeds in the thick thatches of St. Augustine grass (*Stenotaphrum secundatum*) in southern Florida that contain large amounts of decaying organic matter on which it feeds. The greatest migration activity was observed from about 8 to 11 in the forenoon, but very little occurred in the afternoon hours until sunset. Migration activity about one-third of that noted in midmorning continued throughout the night, from 10 p.m. to 7 a.m. The millipedes were commonly observed on patios, outside walls, and foundations of buildings. Their numbers increased during the summer and fall months, and reached pest levels from late September to early December (Bennett and Kerr, 1973).

Oxidus gracilis appeared to breed only in wild and overgrown areas where there was much decaying leaf litter. From these places, it occasionally dispersed to near-by habitations. It fed on decaying organic matter, but could not be induced to feed on tender, living plants (Causey, 1943; Bennett and Kerr, 1973).

Control of Millipedes

Decaying vegetation and other favorable harborages should be removed, and insecticides should be applied in the same manner as that recommended for the control of sowbugs and pillbugs. Control appears to be effective with practically any of the currently popular insecticides at the concentrations most commonly used for the control of garden insects, applied as liquid sprays, dusts, or granules. Granules (for example, 5% carbaryl or 10% diazinon) are convenient to use, and will filter down through turf to the soil surface better than dusts. Homer (1967) recommended a spray of 1% carbaryl or 2% chlordane, applied to the soil around the foundation wall, in addition to a 5 or 10% dust of carbaryl or chlordane in a strip 10 to 15 ft (3 to 4.5 m) wide around the house. Millipedes were not controlled indoors with either space sprays or baits, but sprays or dusts of carbaryl, chlordane, or diazinon were usually effective. Bennett and Kerr (1973) found 3 carbamates (carbaryl, methomyl, and propoxur) to be the most effective among the many pesticides they tested. They all gave a 100% kill in 4 to 5 hours.

CENTIPEDES

The centipedes (class Chilopoda) are the fourth and last of the obligatory dwellers in very moist environments to be considered in this chapter. Like the millipedes, they are long, narrow, and many-segmented, but are nearly always flattened. Their antennae are much longer and have many more segments than those of the millipedes, and they also differ in never having more than 1 pair of legs per abdominal segment (figure 342, *C*). As with the sowbugs, pillbugs, and millipedes, centipedes are nocturnal, and do not have the impervious, cuticular wax layer that is characteristic of insects (Cloudsley-Thompson, 1954). They always live in damp, dark, secluded places, such as under boards, stones, piles of leaves or litter, under logs, or under bark and in crevices in damp soil. Most species are carnivorous, but others will sometimes feed on plant tissues, and may occasionally be injurious to crops. They can be accidental intruders into homes, just as with the millipedes. Since they are mostly carnivorous, they are able to obtain their moisture requirements to some extent by feeding on the live insects they can capture. Also, they crawl more rapidly than millipedes and sowbugs, and are more likely to find sources of moisture. They are thus able to survive in buildings for much longer periods. Because they have venom-bearing claws (toxicognaths), centipedes are more fully treated in chapter 9.

Chapter 11 · *Pests of House Plants*

GROWING PLANTS within the home or an adjacent hobby greenhouse is now so common that the pests that attack these plants may provide a substantial part of the interest in urban entomology. This increasingly important subject was given special attention by Johnson and Smith (1970), and much of the following discussion is based on their recommendations.

The house-plant pest problem can be minimized if cut flowers and new plants brought into the home or greenhouse are examined to determine if pests are present so that they can be immediately eliminated. Isolation of new plants for a few weeks, to make sure they are free of pests, is further insurance against their spread. Infestations by such soil pests as springtails, psocids, and earthworms can be prevented by using sterilized soil for potting. Plants in planters or window boxes are most easily treated for pests if they are kept in pots and placed at a proper depth in sand, soil, or peat. They can then be readily removed for treatment, or replaced if necessary. Treatment for pests is most effectively done at the first signs of infestation or injury.

Spraying with insecticides can sometimes be avoided by washing plants with soapy water and a soft brush or cloth, using 2 teaspoonfuls of a mild detergent to 1 gallon (4 l) of water. Aphids, mealybugs, and scale insects can be removed from broad-leaved plants this way. Some pests can be removed from potted plants by washing them with a spray of lukewarm water from the sprinkler that is commonly an attachment on kitchen sinks.

If only a few lightly infested plants are involved, the pests can be removed by hand. Aphids, mealybugs, and scales can be controlled by removing or by simply wetting the insects with a cotton swab dipped in rubbing alcohol. Sprays may be applied with a hand atomizer or compressed-air sprayer (figures 9 and 14, chapter 3), but it is sometimes more convenient to dip the plants into the spray mixture. Turn the plant upside down and immerse it for a few seconds. A slit cardboard disk to fit around the stem can be used to prevent loose soil from spilling out when the pot is inverted. Aerosol sprays in pressurized cans with pushbutton sprayer tops are available in stores selling garden supplies. Always read the label on the container to make sure the spray mixture was really prepared for use on plants, and be sure to follow carefully the printed directions for application.

When choosing pesticide formulations (see chapter 2, under "Insecticides"), remember that sprays or dips made from emulsifiable concentrates wet and cover foliage and pests better than those made with wettable powders, but are more likely to discolor or burn the leaves or flowers of tender plants. Diluted emulsifiable concentrates should be shaken or stirred occasionally, and preparations made from wettable powders even more frequently, to maintain a uniform mixture of insecticide in the water. If the spray beads up on the leaves and does not readily wet the plant, one-half teaspoonful of a mild household detergent per gallon (4 l) of spray or dip will improve wetting and reduce visible residues. A guide for mixing sprays or dips is given in table 7. Precautions that should be observed in the use of pesticides are given near the end of chapter 2.

TABLE 7

Guide for Mixing Sprays or Dips*

Insecticides, and forms in which they may be purchased	Amount to mix with 1 gallon of water†
CHLORDANE	
45% emulsifiable concentrate..	2 teaspoonfuls
40% wettable powder.........	4 level teaspoonfuls
DICOFOL	
18.5% emulsifiable concentrate	1 teaspoonful‡
LINDANE	
20% emulsifiable concentrate..	1 teaspoonful
25% wettable powder.........	1 level tablespoonful‡
MALATHION	
57% emulsifiable concentrate..	2 teaspoonfuls
ROTENONE	
5% emulsifiable concentrate...	4 teaspoonfuls
Derris or cubé root wettable powder containing 4 or 5% of rotenone................	4 level tablespoonfuls

*From Johnson and Smith (1970).

†For practical purposes: 1 quart equals 1 liter; 1 gallon equals 4 liters.

‡One teaspoonful equals 5 cc; 1 tablespoonful equals 3 teaspoonfuls or 15 cc.

Earthworms (Chaetopoda)

Earthworms may be seen on the surface during the day if the soil is wet, otherwise they emerge only at night. Their castings or droppings are good indications of infestation. They sometimes eat parts of plants growing close to the soil surface.

Their continuous tunneling may injure root systems, or loosen young plants in the soil. By removing organic material from the carefully prepared soil mix used for potting plants, earthworms deteriorate its physical and chemical composition. This is the opposite of their usual role in nature, in which they aerate the soil and increase its organic content. They can be controlled by drenching the soil with a dilute emulsion of two-thirds teaspoonful of a 45% chlordane emulsifiable concentrate in 1 gallon of water.

Slugs and Snails (Gastropoda)

Many homes have a screened or glassed-in patio or an adjacent lathhouse or greenhouse to which pests that require damp locations, such as slugs,

Fig. 344. Gray garden slug, *Deroceras reticulatum.*

snails, sowbugs, pillbugs, millipedes, and springtails, are attracted. Among the slugs, the gray garden slug, *Deroceras reticulatum* (Müller) (figure 344) is the species most commonly encountered in California. This and other species, as well as the European brown snail, *Helix aspersa* Müller, are discussed more fully in chapter 12. Granules with 3 to 7% of metaldehyde, scattered about or placed in little piles in infested areas, after watering or after a rain, give a convenient and effective control of slugs and snails; they will kill on contact or when eaten. Scattering the granules appears to be the better procedure when the gastropods are widely distributed, as in a garden or flower bed, but placing them in small piles may be advantageous where these pests are concentrated in restricted areas, as in a greenhouse.

Sowbugs and Pillbugs (Isopoda)

These are small, oval crustaceans, 7 to 13 mm long, and usually gray in color. The two groups named are similar in appearance (figure 341, chapter 10). Sowbugs scurry for cover when disturbed, and pillbugs roll up into a ball. Both usually hide in loose soil or under vegetable debris, boards, sacks, or pots on damp ground during the day, and are active at night. Although they normally feed on decaying vegetable matter, they sometimes eat roots and tender plant parts, especially those of bedding plants and seedlings. Their hiding places should be eliminated as far as possible. A number of common pesticides can then be sprayed under boardwalks and benches, along foundations, and in other infested areas. (See chapter 10, under "Control of Sowbugs and Pillbugs.")

Mites (Acarina)

Mites are very tiny arthropods, usually no larger than the head of a pin. The adults have 4 pairs of legs instead of the 3 pairs possessed by insects. Some species are ectoparasites of man. Others are among the most important of the plant pests, and commonly infest house plants.

Spider Mites

Tetranychus urticae Koch, the twospotted spider mite (figure 345), is the "red spider" most commonly encountered on house plants. All spider mites are pests of cultivated plants, particularly under dry and warm conditions. They

Fig. 345. Twospotted spider mite, *Tetranychus urticae.*

are oval in shape, greenish, yellowish, or reddish, and are so small that they are barely visible without the aid of a magnifying glass.

Spider mites are found mainly on the undersurfaces of leaves, but in heavy infestations a small percentage may be seen on other parts of the plant. When the mites are very numerous, a silky webbing may be found covering infested areas and extending from leaf to leaf to cover the entire plant. Mites can be seen crawling over this webbing. Spider mites suck out plant juices. The removal of chlorophyll results at first in whitish or yellowish speckled areas on the upper surfaces of leaves, and eventually in a more uniformly bronzed or yellowed discoloration, possible defoliation, and stunting or even death of the plant. For plants that are sufficiently sturdy, a forceful spray of water may be enough to break up the webbing and dislodge the mites. This should be followed within 2 days by a dip or spray with malathion or dicofol (Kelthane®), with special attention to the undersides of the leaves. Tender plants may be treated with the dip or spray without the preliminary water spray. Several applications may be required, at weekly intervals.

False Spider Mites

Brevipalpus obovatus Donnedieu, the privet mite, is the principal species involved as a pest of house plants. False spider mites are flat, oval, and dark red, and are too small to be seen easily with the unaided eye. With the aid of a hand lens, active mites and bright-red eggs can be readily seen, mostly on the undersides of leaves, along veins or over the entire leaves. Infested plants may become stunted and defoliated. Control can be obtained effectively by spraying or dipping with a dicofol or diazinon (Spectracide®) mixture, adding one-half teaspoonful of a mild household detergent to improve wetting. When spraying, be sure to wet the undersides of the leaves.

Cyclamen Mite

Steneotarsonemus pallidus (Banks), the cyclamen mite, may be seen with a magnifying glass in protected places on young, tender leaves, young stem ends, buds, and flowers. The adults are tiny, oval, amber or tan-colored, glistening, and semitransparent. The even smaller immature forms are milky white, and the eggs are oval and pearly white. The mites can spread from plant to plant where the leaves of different plants come in contact, or they can be transferred on hands or clothing. Their attacks cause the leaves of infested plants to become twisted, curled, and brittle, while the buds become deformed and often fail to open. If they do open, the flowers may be deformed and streaked. The injured leaves, buds, and flowers are often blackened. African violets (*Saintpaulia ionantha*) develop small, twisted, excessively hairy leaves that will soon die under a heavy infestation. Injured plant parts should be removed, and the infested plants, pots and all, should be immersed in water at 110° F (43° C) for 15 minutes. Two or 3 applications of dicofol spray at 10-day intervals will also give successful control.

Springtails (Collembola)

Springtails (figures 336 and 337, chapter 10) can be troublesome in the soil in flower pots and in home greenhouses, as well as in houses where no plants are grown but where there are moisture problems. They are tiny, slender, whitish or darker insects, and get their common name from their ability to jump (see chapter 10). They may become abundant in moist places where there is much organic material, for they normally feed on algae, fungi, and decaying vegetable matter. However, they may also chew on seedlings or the tender parts of plants, particularly near the ground level. They may be controlled by spraying the soil surface, pots, saucers, shelves, and affected parts of plants with chlordane or malathion.

Psocids (Psocoptera)

Psocids (figures 334 and 335, chapter 10), also known as barklice or booklice, are soft, yellowish-white to gray insects, oval and somewhat flattened, ranging in length from 1.2 to 5 mm, and may be winged or wingless, depending on the species (see chapter 10). They can sometimes be seen in large numbers on soil or on flower pots and benches, especially in undisturbed locations in the home greenhouse. They may also be found in or on old books and papers in slightly damp places, and on damp and moldy cereals. They normally feed on fungi or dead organic matter, and apparently do not injure living plants. Nevertheless, they can become nuisances because of the large numbers that sometimes develop. For control, spray the infested areas with chlordane or malathion.

Thrips (Thysanoptera)

House plants are sometimes infested by several species of thrips. They are small, slender insects, some species being scarcely visible to the unaided eye. The young nymphs are whitish to yellow or orange, and are not very active. Those of the

Fig. 346. Greenhouse thrips, *Heliothrips haemorrhoidalis.* Adults, nymphs, and "egg blisters" (*arrows*), with exit holes. (From Ebeling, 1959.)

greenhouse thrips, *Heliothrips haemorrhoidalis* (Bouché) (figure 346), carry droplets of black liquid excrement on the tips of their abdomens. Adult thrips may be tan, brown, blackish brown, or black, and tend to fly, leap, or run when disturbed. Both nymphs and adults have conical, rasping–sucking mouthparts, and cause a typical injury characterized by irregular or streaked silvered areas, speckled with little black dots of excrement. They may destroy buds or blossoms, and will whiten, curl, or deform leaves. For control, the plants may be dipped into or sprayed with malathion emulsions or suspensions.

Aphids (Aphididae)

Aphids usually infest the undersides of leaves, or are clustered about tender leaves and the stems of flower buds. A few species feed on roots. They suck out the plant juices, and "sooty-mold" fungus sometimes grows on the sticky, sugary honeydew they excrete. The plant may be stunted, and its leaves will be curled and distorted. These pests may be easily removed by hand rubbing, washing off the plants, or by using an alcohol swab.

Infested plants may also be dipped into malathion or lindane spray mixes, or treated with aerosols designed for direct plant application.

Mealybugs (Pseudococcidae)

These soft-bodied insects, with a covering of white, powdery wax, are about 4.5 mm long when full grown (figure 347). Like aphids, they are usually found at rest along veins on the undersides of leaves, but sometimes they can be seen crawling about slowly. Their eggs are laid in loose, cottony masses of wax. These are among the honeydew-excreting insects, and are therefore responsible for some of the discoloration caused by the "sooty-mold" fungus. Mealybugs suck plant juices, and the constant drain on the sap supply for which they can be responsible results in slow growth of the plant as well as wilting between waterings.

The plants may be sprayed or dipped in a malathion mixture with one-half teaspoonful of mild detergent added per gallon (4 l), for the waxy mealybugs are difficult to wet. These insects can also be hand rubbed, washed off, or killed with an alcohol swab. Soil mealybugs can be treated by dipping the infested plant and pot in dilute malathion emulsion.

Fig. 347. Longtailed mealybug, *Pseudococcus longispinus.*

Scale Insects (Coccidae and Diaspididae)

There are two types of scale insects, unarmored and armored. The females of the unarmored or soft scales (Coccidae) are oval to nearly circular and flat to nearly globular, and are sometimes partially covered with wax. Common examples are the brown soft scale, *Coccus hesperidum* L. (figure 348), and the hemispherical scale, *Saissetia*

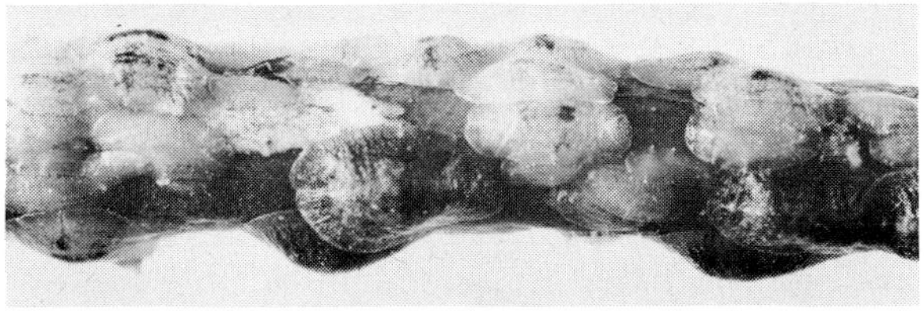

Fig. 348. Twig heavily infested with various stages of the brown soft scale, *Coccus hesperidum*.

coffeae (Walker). Many species become stationary when fully developed. The males usually pupate under a small, thin, transparent, flat scale, and emerge as tiny winged adults. The females of the armored scales (Diaspididae) are flat, elongate to circular, and have a thick, protective shell or armor above and a very thin layer beneath the body. Common examples are the oleander scale, *Aspidiotus nerii* Bouché, and the greedy scale, *Hemiberlesia rapax* (Comstock) (figure 349). Once the motile first-instar nymphs settle down on a plant, the females never move again. The males transform to winged adults as with the unarmored

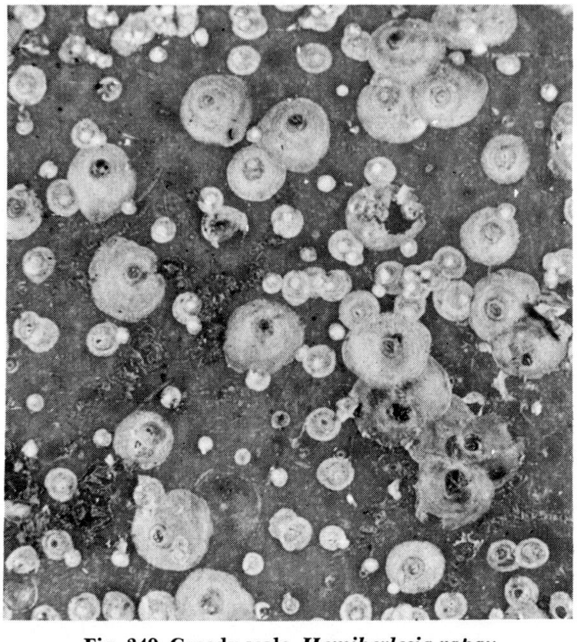

Fig. 349. Greedy scale, *Hemiberlesia rapax*.

scales. Scale insects suck sap from the plant, resulting in retarded growth and even defoliation. The unarmored scales, but not armored scales, excrete honeydew, which results in the growth of "sooty-mold" fungus and attracts ants.

Infested plants may be sprayed or dipped in malathion, adding one-half teaspoonful of a mild detergent per gallon (4 l) to increase the wetting capability of the spray. Scales are more difficult to control than most pests of ornamental plants, and several applications at intervals of 3 or 4 weeks may be required to kill them all. The scales may remain attached to the plant after they die, but their dry bodies can be readily distinguished from the juicy ones of the living insects. If only a few scales are found, they may be rubbed off or dabbed with alcohol.

Whiteflies (Aleyrodidae)

Whiteflies are very small insects, the adults having 2 pairs of wings which, along with the body, are covered with a fine, whitish, powdery wax. The nymphs and pupae are soft, oval, flat,

Fig. 350. Greenhouse whitefly, *Trialeurodes vaporariorum*.

and resemble the unarmored scales. The principal whitefly pest of plants in the home, the greenhouse whitefly, *Trialeurodes vaporariorum* (Westwood) (figure 350), is about 1.5 mm long, with a yellow body and immaculate white wings. The nymphs are oval, thin, flat, semitransparent, and pale green, and bear a fringe of wax rods or filaments. Both nymphs and adults suck sap from foliage, weakening the plant, and the nymphs excrete honeydew, like the unarmored scales they resemble. Whiteflies can be controlled by dipping them in or spraying them with emulsions or suspensions of malathion, rotenone, or lindane. A special effort should be made to wet the undersides of leaves, where most of the whiteflies will be located. Good control has been obtained with repeated applications of pyrethrin-containing aerosols (F. S. Morishita, correspondence).

Cutworms and Other Caterpillars

These moth larvae vary in size from the barely visible, newly hatched young to some that may be 40 to 50 mm or more long. They may be of various colors, depending on the species, and some are covered with dense hair. They are chewing insects that feed on leaves, buds, or flowers, or may even cut off young plants near the soil level. Dark pellets of excrement may be found on the foliage or on the soil beneath infested plants. Hand picking may be practical to control the insects, or the plants may be sprayed or dipped into malathion or pyrethrin emulsions or suspensions. Cutworms (Noctuidae) that hide in the soil, where they usually remain during the day, can be controlled by drenching the soil with malathion.

Fungus Gnats (Mycetophilidae)

Fungus gnats are small, slender, delicate, gray or dark-gray mosquitolike insects, most species being 3 to 6 mm long, with a few as long as 12 mm or more. They are attracted to light, tend to gather at windows, and are nuisances in the home. The larvae are slender and whitish, with dark heads, and live in damp soil. Large quantities of decaying vegetable matter in damp soil are conducive to heavy infestations. The larvae often feed on the roots and crowns of plants, causing them to be stunted, discolored, and possibly defoliated (see chapter 10). They can be controlled by drenching the soil surface with a dilute chlordane emulsion prepared with the emulsifiable concentrate. Gnats flying about in a greenhouse can be killed with common household aerosols or dichlorvos resin strips.

Ants (Formicidae)

Some species of ants are attracted to the sugary excretions (honeydew) of aphids, mealybugs, whiteflies, and unarmored scales. Some can dig up and carry away newly planted seeds or small seedlings, or injure plant roots by their burrowing activities. Sprays applied to control the honeydew-excreting insects should also control ants. If not, the infested soil may be soaked with chlordane, as recommended in this chapter for the control of earthworms.

Chapter 12 · *Miscellaneous Pests*

THIS CHAPTER is devoted to pests, or presumed pests, that do not appear to fit appropriately into the specific categories mentioned elsewhere, although a few are also mentioned briefly in other chapters. Some are included here only because they evoke fear or curiosity, particularly among city dwellers, even though they are harmless. They are therefore responsible for more calls to entomologists and pest control operators than are justified by their importance as pests or nuisances. Others can be serious pests, usually under certain climatic or environmental conditions that result in a rapid increase of a species to epidemic proportions. Residential or other buildings may be entered merely because they are in the path of a mass migration of some obnoxious species. In other cases, the pests are seeking a protected harborage. Although certain invaders, such as crickets, may consume or damage food or fabrics, none enter a building for these specific purposes.

SLUGS AND SNAILS

In areas where climatic conditions are suitable, slugs and snails (Mollusca: Gastropoda) can be pests not only in the garden, but also by crawling onto walks, porches, walls, and into doorways or basements. In such places they are not only unsightly, but also leave slime trails that remain visible for many days. When accidentally crushed underfoot, they are not only messy, but are sources of almost traumatic revulsion in some people.

Slugs and snails hide during the day in damp places, such as under boards, rock piles, low-lying shrubs, damp refuse and litter, and dense vegetation of all kinds. Beds of ivy are particularly conducive to heavy infestations of these gastropods. They may be seen during daylight mostly in the early morning and particularly after a rain, or if the lawn and adjacent vegetation have been sprinkled the preceding day. At that time, the pests are generally on their way back to their daytime hiding places, avoiding dry or dusty areas or sharp objects if they can. Snails may sometimes be seen at midday, usually firmly attached to walls, fences, tree trunks, or their food plants. When dry conditions prevail, they seal the opening of the shell with a mucous secretion that becomes a parchmentlike sheet called the *operculum;* this finally becomes hard and leathery. They can then exist in a dormant condition for as long as 4 years. With the return of favorable conditions, the operculum is rasped away by the snail's mandible. Snails can survive lower humidities and lower temperatures than slugs (White and Davis, 1942).

Snails and slugs are garden pests, particularly of tender, young seedling plants and of berries and other fruits. They may cause only surface scars, or may make deep holes in the fruit, besides covering it with the mucus they deposit when crawling.

Morphological Differences Between Slugs and Snails

Slugs are less conspicuous than snails, and are not so well known to most people. Whereas snails are characterized by the coiling of the "visceral hump" into a spiral shell, in slugs the shell is vestigial, and is concealed by the "mantle" situated on the dorsum. Slugs and snails belong to the same order (Stylommatophora), and are similar in structure and biology. They glide along by means of a long, flat, muscular organ known as a "foot." Mucus, which is constantly secreted by glands in the foot, facilitates movement. Slugs and snails are hermaphrodites, and mutual fertilization takes place; both individuals may lay eggs.

Slugs (Limacidae)

There are at least 20 species and subspecies of slugs in California, but only the limacids are common, and in fact they are often serious pests of home garden plants as well as commercial field crops. Except for the indigenous *Deroceras laeve,* the injurious limacids are all species that have been introduced into North America from the Old World, some at least a century ago. A survey of slugs attacking orange fruit on dooryard and orchard trees in southern California revealed that 3 species (figure 351) appeared to predominate (Ebeling, 1959). (See also, figure 344, chapter 11.) The 3 species were often found together in great abundance beneath orange trees, where their damage to fruit and foliage was visible, usually on the "skirts" of the trees near the points at which they contacted the ground. The orange fruits and foliage were probably only parts of the slugs' diet, for they can also feed on decaying vegetable matter; on fungi or molds; on dead, injured or slow-moving insects; and on sowbugs or earthworms that can be found in the decaying litter in such places as beneath orange trees. Slugs feed on an extremely wide variety of vegetable and animal matter. A check list of California land slugs was prepared by Lange (1944).

Some Important Species

The species found in greatest abundance in the author's survey was the gray garden slug, *Deroceras reticulatum* (Müller) [= *agreste* (L.)] (figure 351, *A* and *B*), native to Europe, and now one of

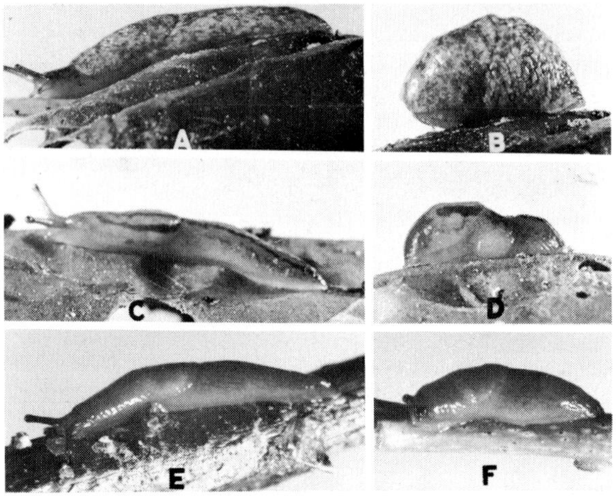

Fig. 351. Common slugs. *A* and *B*, gray garden slug, *Deroceras reticulatum; C* and *D*, banded slug, *Lehmannia poirieri; E* and *F*, greenhouse slug, *Milax gagates.* Note the "extended" and "contracted" positions shown for each species.

the worst pests of garden and field crops, especially in humid coastal areas. It is 35 to 50 mm long, ranges in color from buff to brown or gray, and exudes a milky slime when touched.

The longnecked slug, *Deroceras laeve* (Müller), is an almost cosmopolitan species that invades gardens with *D. reticulatum,* particularly in the San Francisco Bay region. It is about 32 mm long, and is usually various shades of amber, without spots or markings, but is sometimes blackish, with the head and tentacles smoky. In live individuals, the mantle appears to have a circular orange-brown area. A characteristic feature is its long neck which, when extended, is almost as long as the mantle.

The gray field slug, *Deroceras gracile* (Rafinesque) or *Agriolimax campestris* (Binney), considered to be a form of *D. laeve* by Pilsbry (1948), occurs in all parts of the United States. It is a small species, seldom more than 25 mm long, and is buff to almost black. The slime is clear, not milky as in *D. reticulatum.* According to A. G. Smith (correspondence), there have been many introductions of the European *D. laeve* into the United States, particularly into central California, and it is these that are called "longnecked slugs." Smith has found the indigenous form of *D. laeve,* as recognized by Pilsbry, in many places all through the Sierra Nevada and also in his garden in Berkeley. He has never found the "longnecked" race in his garden, although it is prevalent in Golden Gate Park in San Francisco and in nurseries and cultivated areas in Alameda and Contra Costa counties.

The banded slug, *Lehmannia poirieri* (Mabille) [= *Limax marginatus* (Müller)] (figure 351, *C* and *D*), another European species, may attain a length of from 50 to 75 mm, and has a general light-brown color, with parallel, longitudinal, gray, dark-brown, or black bands on the mantle. There are 3 such bands in immature slugs, but the middle band tends to disappear with approaching maturity. These slugs are basically fungus-feeders, but they are also among the principal citrus pests, and will feed on garden plants and underground tubers.

The spotted garden slug, *Limax maximus* L., is the largest of the European "gray" slugs introduced into western America, exceeded in size only by the indigenous *Agriolimax* species (Hanna, 1966). It prefers human habitations and, although like other limacids it is a fungus-feeder by preference, it often causes injury to garden plants. It is yellowish gray or brown, with large, irregular black spots, and is usually about 100 mm long, but may reach a length of 200 mm. The mature specimens have a dark-spotted mantle, but in immature slugs these spots may be united into bands.

The tawny garden slug, *Limax flavus* L., is widely distributed in the eastern and southern parts of the United States, and was recorded in California as early as 1903. It is a large species, 75 to 115 mm long, and is exceeded in size only by *L. maximus* among the introduced European species (Hanna, 1966). It is yellowish green, with yellow spots, and exudes a yellowish slime when disturbed. Its mantle consists of a series of concentric grooves, giving it a reticulate appearance, and the ocular tentacles are bluish. This slug is gregarious, and may be found under garbage cans, in basements, wells, drains, and under doorsteps. It may be found in nurseries, and also does some damage to garden plants.

The greenhouse slug, *Milax gagates* (Draparnaud) (figure 351, *E* and *F*), was introduced into North America as early as 1872. It is very destructive to nursery, garden, and field crops in California, and its burrowing habits make it a pest of root crops. It may attain a length of 40 to 70 mm, is dark gray or black, has an oval or diamond-shaped groove on the center of the mantle, and has a prominent, sharp, dorsal keel extending back the entire distance behind the mantle. The keel is especially conspicuous when the body is contracted (Lange, 1944).

Giant West Coast Land Slugs (Arionidae)

The giant slugs are generally too rare to be even mild nuisances, and only occasionally do some species damage garden plants. However, because of their large size, they are often the subjects of inquiry when found. They do not thrive in cultivated areas, and are most likely to be seen by the urban population in damp, unmolested areas, such as in the state parks of the redwood belt of northern California (e.g., Muir Woods). The only natural enemies of the giant slugs are apparently ducks and garter snakes (*Thamnophis*). The thick, viscous coat of slime on their bodies is very bitter, and they are shunned by most potential predators (A. G. Smith, personal communication).

The largest and most widely distributed of the giant slugs is the giant yellow slug, giant redwood slug, or banana slug, *Ariolimax columbianus* (Gould). It occurs in parts of Alaska, west of the Cascade Mountains in British Columbia, Washington, and Oregon, in California south to the Salinas Valley on the coast, and at least as far south as Tuolumne County on the western slopes of the Sierra Nevada. The adult of this very large slug weighs from 35 to 72 grams, and its foot varies from 18.5 to 26 cm in length when fully extended. The ground color is described as "deep, olivaceous green, tan-green, ochraceous, ochraceous-yellow, slate-green, or intermediate shades," with the mantle of the same color. According to Lange (1944), this species occasionally invades gardens to feed on vegetables and other plants. There is also a form that is spotted or blotched with black. The ground color of *A. columbianus,* including the mantle, is very variable, but never attains the light-yellow color of the other species of the genus, which are less variable. The color of individual slugs will change with age and alterations in food, moisture, and light. There is also a smaller and more slender subspecies, *A. c. stramineus* Hemphill, distributed south and west of the Salinas Valley, from the Monterey peninsula to at least Ventura County and in Santa Cruz and Santa Rosa islands. It weighs 20 to 27 grams, and its foot is 13 to 16 cm long. Its ground color is vitreous lemon-yellow or light straw-color (Mead, 1943; Pilsbry, 1948).

Ariolimax californicus Cooper occurs in San Mateo County and northwestern Santa Clara County in coastal central California. It is a smaller species than *A. columbianus,* the adults weighing 25 to 42 grams and having a foot length of 17.5 to 20 cm. The ground color, including the mantle, is vitreous butter-yellow, rarely ochraceous yellow. This slug occasionally invades vegetable fields and gardens, but usually does not do extensive damage (Lange, 1944). *Ariolimax c. brachyphallus* Mead occurs on the San Francisco peninsula. It is the smallest slug of the genus, weighing 17 to 23 grams, and has a foot length of 12.5 to 15 cm. Its ground color is ochraceous or ochraceous brown, and its mantle is usually ochraceous yellow and rarely the same color as the remainder of the body.

Ariolimax dolichophallus Mead is limited to coastal central California, from northern Santa Cruz and western Santa Clara counties to the Salinas Valley. It is a slender species, with adults weighing 20 to 39 grams and a foot length of 15 to 18 cm. Its ground color, including the mantle, is opaque butter-yellow. It obtains its name because of its extremely attenuate apical phallus (copulatory organ) being very long—occasionally it may be longer than the slug itself (Mead, 1943; Pilsbry, 1948).

European Brown Snail, *Helix aspersa* Müller (Helicidae)

The European brown snail has been widely disseminated throughout the world by being accidentally or intentionally introduced on plants. In California, it is found in most residential and cultivated areas, from San Diego to Sonoma counties. It was intentionally introduced from France between 1850 and 1860 and "planted" among the vineyards along Guadalupe Creek in Santa Clara County, and later in San Francisco and Los Angeles (Basinger, 1931). Although this species has some value as an article of food in certain European countries, other kinds are more popular. It is rarely eaten in California, except by an occasional European immigrant. It is one of the most destructive and prolific dooryard and garden pests in the state, which would be free of destructive snails if it had not been deliberately introduced.

Description. The shell of this snail (figure 352) is grayish yellow and brown. The brown is usually in 5 interrupted bands, the second and third ones being ordinarily confluent, giving the appearance of a single band. All the bands may be united in some individuals. When fully developed, the shells have 5 or 4.5 whorls, and are 2.5 cm or more in diameter. The shell's form is that of a dextral or right-hand spiral, the sinistral or left-hand aspect being very rare. The body is light to dark gray, and is about 6 cm long when fully extended (Basinger, 1931).

Fig. 352. European brown snail, *Helix aspersa*.

Life Cycle. Oviposition usually occurs within 3 to 6 days after fertilization. The snail selects a spot where the soil is damp, and prepares a nest for the eggs in the ground at a depth of 2.5 to 4 cm (figure 353). The eggs are white, somewhat translucent, spherical, and about 3 mm in diameter. The egg mass contains an average of 86 eggs. After ovipositing, the snail closes the opening of the nest with a little mound of earth and excrement to conceal its location. The young snails have shells of somewhat more than 1 whorl when they hatch. They usually remain in the nest for 2 to 4 days, then work their way to the surface. Two years are required to reach maturity.

Control of Slugs and Snails

After removing as many as possible of the favored hiding places of slugs and snails, sprays or poison baits may be applied, the latter usually being preferred. Baits are applied in places frequented by these gastropods, such as along the edges of dense vegetation bordering lawns. This should be done when plenty of moisture is present, such as after a sprinkling or following a rain. Pelletized baits should not be used where dogs are present, for dogs will eat them.

For many years, a bait containing 5% metaldehyde $(C_2H_4O)_n$ or 3.25% metaldehyde and 5% of an arsenate has been used for the control of slugs and snails. The bait containing only metaldehyde is particularly effective against slugs and during the warmer periods of the year. The arsenate increases its effectiveness under adverse conditions or against snails, providing greater toxicity, while the metaldehyde continues to act as an attractant. The bait should be applied every 7 to 10 days until control is accomplished. Also available is a 20% spray concentrate which is diluted with water to produce a 0.25% spray.

The carbamate insecticide Zectran® (table 1, chapter 2) is also an effective molluscicide, and is available as 2% granular bait or a 2E (2 lb per gal or about 1 kg per 4 l) emulsifiable concentrate. The bait kills the pests in 3 to 5 days and lasts for about 2 weeks. It should be reapplied in about 3 weeks. The emulsifiable concentrate is diluted to make a spray of 0.1% toxicant. It is said to be more effective than metaldehyde spray, even though it kills more slowly.

The most recently developed molluscicide is the carbamate insecticide Mesurol®, 3,5-dimethyl-4-(methylthio)phenol methylcarbamate, also sometimes expressed as 4-(methylthio)3,5-xylyl methylcarbamate. (See table 1, chapter 2.) Mesurol, in

Fig. 353. Snail eggs in a nest in damp soil. (From Basinger, 1931.)

both wettable powder and pellet formulations, was found to be superior to metaldehyde against the gray garden slug, particularly under damp or cold conditions (Wouters, 1970). It is available as a 2% granular bait, and is said to be the most rapid-acting of the molluscicides, giving good control in 1 or 2 days. Only a single treatment is usually required (NPCA, 1973).

AMPHIPODS

These are small crustaceans, nearly all of which are laterally compressed, giving them a shrimp-like appearance. Among the Amphipoda is a species that may invade the home following rain or prolonged watering of the vegetation in which it seeks harborage.

Talitroides sylvaticus (Haswell) (Talitridae)

These small amphipods (figure 354) are recorded from Australia, several Pacific islands, including Hawaii, and from California and Louisiana (Mallis, 1969). They are in the same family as the "beach fleas" that live beneath piles of seaweed and other drift material near high-tide mark on the beach.

Fig. 354. An amphipod, *Talitroides sylvaticus*, that sometimes invades houses.

Live specimens of *T. sylvaticus* are brownish black, and approximately 8 mm long when mature, but turn red when they die (plate VIII, 7), and this is the condition in which they are usually seen by the homeowner, for they do not live long after entering a house. They may be found in coastal southern California in and on the ground in damp places, particularly under ivy, where they sometimes occur in enormous numbers, jumping about like fleas.

Although terrestrial amphipods require damp locations, excess moisture forces them to seek drier locations, and they may then become accidental intruders into the home. The lack of a waterproof wax layer which forces them to seek damp locations is also a hazard to them if their environment becomes too wet, for they are not only susceptible to desiccation but also to a lethal rate of water uptake by osmosis. The author observed after an unseasonable early rainfall of 0.3 in. (8 mm) in September, 1963, that *T. sylvaticus* migrated away from an ivy patch, and large numbers found their way into a near-by house. These were dead the following day, apparently, as observed by Mallis (1942, 1969), finding their new environment too dry. Again, following early rain in October, 1966, a local resident brought in large numbers of the amphipods, found in his home, for identification. During the following winter they became so troublesome that he sprayed his ivy with a 0.5% diazinon emulsion and found it to be an effective control measure.

PSEUDOSCORPIONS

Pseudoscorpions or "false scorpions" (order Chelonethida) (figure 35, chapter 4) are arachnids that sometimes invade the home and should be recognized as harmless. Outdoors they may be found in leaf mold, or in the nests of birds, mammals, and insects, under stones, and beneath bark, where their disklike webs, spun with their chelicerae, are often abundant. The web forms a sort of cocoon in which the pseudoscorpion molts and hibernates. Pseudoscorpions prey on other arthropods, such as small insects and mites. Species of the genus *Chelifer* are sometimes found crawling about in the house.

HARVESTMEN

In common with the pseudoscorpions, harvestmen (order Phalangida), also known as "daddy-longlegs," should be recognized as harmless arachnids (figure 38, chapter 4). They are often misidentified as spiders of some sort. Although principally nocturnal, they are often seen in large numbers in cloudy weather, and may invade the home.

They eat insects, both living and dead, particularly aphids, but also eat vegetable matter, and can be injurious to tender plants. Insects are seized by the leglike pedipalps, passed to the chelicerae, macerated, and the liquids are consumed, along with some coarse particles. *Phalangium* and *Liobunum* are familiar genera.

SOLPUGIDS

The solpugids (order Solpugida), sometimes known as "sun spiders" or *arañas del sol,* are very swift and agile arachnids, mostly nocturnal (figure 39, chapter 4). Their huge chelicerae may impart a fearsome appearance, but they do not have venom glands, and are harmless. They feed on other small arthropods. Solpugids are found in warm, dry, desert areas. They do not spin webs. The female lays her eggs in a burrow in the ground and stays there until they are hatched. During the day, solpugids commonly remain under stones, boards, logs, and other objects that may cover their burrows. They seek their prey at night, catching it on the run, and occasionally find their way into homes. The principal genera in the United States Southwest are *Ammotrecha, Eremobates,* and *Hemepotrecha.*

MITES

Mites are discussed at considerable length in chapters 7 and 9, entitled "Pests of Stored Food Products" and "Pests Attacking Man and His Pets." Some of the species discussed in those chapters may occasionally be seen crawling about on walls, cabinets, furniture, and other interior surfaces. There is only 1 important mite species (or complex of forms and species) that normally feeds on plants outdoors and then invades the house—the clover mite. It sometimes enters homes in such numbers as to become an important pest.

Clover Mite, *Bryobia praetiosa* Koch (Tetranychidae)

This is a cosmopolitan species that infests many trees, shrubs, flowers, grasses, and agricultural crops; it is found on at least 200 species of host plants. It is probably the mite of most frequent concern to homeowners. Aside from being pests on surrounding trees and other vegetation, clover mites can invade the home, usually around windows, under doors, or through cracks in walls or masonry, in such large numbers as to give infested surfaces such as walls, windows, and floors a reddish appearance. If crushed, they will leave reddish spots or smears. The problem is accentuated if lawn or other grass is allowed to grow close to the foundation and is heavily fertilized, resulting in lush growth (Spear, 1954).

Biology. The adult clover mite (figure 355) averages about 0.75 mm in length, and may be rusty brown, olive green, or dark red. The best distinguishing feature is the pair of front legs that are longer than the body and about twice the length of the other legs. The young stages are bright red. Ordinarily, the smooth, spherical, red eggs are laid on trees and other vegetation throughout the summer and fall. Those laid in late fall hatch the following spring, and may sometimes be seen in such numbers on the limbs of trees, particularly at crotches, as to give the appearance of brick dust. They are also sometimes seen on the foundations or outer walls of buildings.

Clover mites reproduce parthenogenetically, that is, without fertilization. Males have not been found in the United States, and only rarely in a few other parts of the world. A female will lay about 70 eggs. This species has a larval stage, 2 nymphal stages (protonymph and deutonymph), and an adult stage. About 1 month is ordinarily required to complete a generation outdoors, the optimum temperature for development being 69° F (21° C). Above 75° F (24° C), the eggs become dormant, and they are also inactive below 40° F (4° C). In Illinois and Pennsylvania, the form of clover mite that infests dwellings becomes dormant during May and remains so until September (Snetsinger, 1967). According to Spear (1954), in the eastern states the mites may be found in infested houses from November until May or June. They can invade a house in enormous numbers; 250,000 were estimated to be present at one time on the floor of a bedroom.

Other Species and Forms

Occasionally, what are presumed to be clover mites invading the home are actually the brown wheat mite, *Petrobia latens* (Müller) (Tetranychidae), and the winter grain mite, *Penthaleus major* (Dugès) (Enpodidae), or these species in mixed populations with the clover mite. Of greater interest, however, is the fact that there are 15 to 20 different morphological forms of the clover mite, with different behavioral and seasonal occurrence patterns. Snetsinger (1967) conceived a hypothetical clover mite control job in which there might be 7 different forms (table 8) on the premises, of which only one,[27] the "dwelling and tree-trunk" form, would be a potential pest. This form lays its eggs under bark and in cracks and crevices in building walls. The seasons of major mite activity in the eastern states are spring and fall, with some activity during warm spells in winter. On a sunny winter afternoon, when the air temperature is 32° F (zero C), the

[27] According to R. N. Hawthorne (correspondence), in California both *Bryobia rubrioculus* and *B. praetiosa,* the "tree-trunk form," invade the home.

temperature on the south side near the foundation of a house may be 60° F (16° C) or more because of radiated heat, and the mites may be actively feeding.

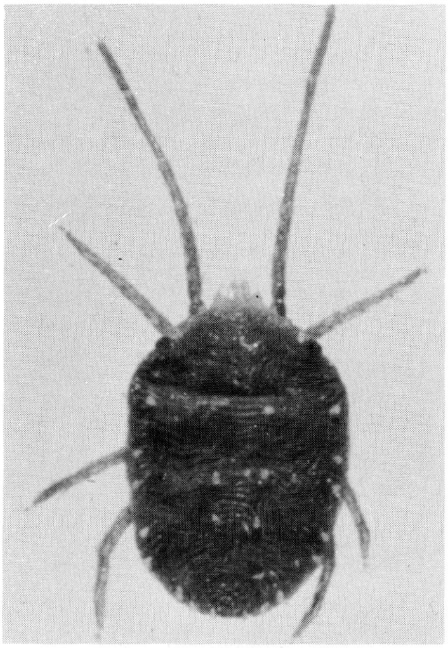

Fig. 355. Clover mite, *Bryobia praetiosa*.

The period when the mites are most likely to enter dwellings varies with the climate—in April or May in Pennsylvania, but earlier farther south and later farther north. In California, pest control operators receive most of their calls for mite control in the spring months, with a secondary peak of activity in the fall. Infestations are usually most severe on recently occupied and landscaped property. After populations of natural enemies[28] have had opportunities to become established, the mites become less of a problem. Also, the grass near the foundation is likely to be replaced by ornamental plantings, and the vigor of the remainder of the lawn decreases, both tending to reduce the mite population (Snetsinger, 1967).

The Clover Mite as an Ectoparasite

A new role for *Bryobia praetiosa* as a pest was revealed in a report from Argentina that, since 1964, this species has occasionally been ectoparasitic on man, always in October (a spring month in that area). Attacks have been recorded in schools, laboratories, and houses, particularly against children. Usually, skin irritation as well as sores resulting from the rubbing of affected parts was observed. In some persons, the attacks were followed by listlessness and loss of interest (Maury and de Alzuet, 1970).

Control of Clover Mites

The dwelling and tree-trunk form of the mite feeds on grass near the foundations of dwellings and near the bases of trees. An acaricidal barrier on a strip of bare soil 18 in. (45 cm) wide around the outside of a building can reduce the number of mites reaching it by about 90%. Acaricides may also be applied under high pressure to the foundation of a dwelling, so as to penetrate into cracks and crevices in the wall where the mites seek protection against cold and desiccation and where they lay their eggs. It may also be advantageous to spray for a distance of about 15 ft (5 m) from the foundation. A grass-free strip around buildings has been found to be about as effective as acaricide applications, but a combination of the 2 control measures will definitely give the most satisfactory results.

Acaricides such as chlorobenzilate and ovex can be safely used around windows and doorsills or similar points of entry. Treatments within the house give only temporary relief if no control measures are practiced outside (English and Snetsinger, 1957; Snetsinger, 1967). Inside residences, mites can be removed with a vacuum cleaner so as not to crush them and to avoid making the red smears that result if a cloth or brush is used.

TABLE 8

Seven Forms of Clover Mites That Might Be Present on Dooryard Grasses, Shrubs, and Trees*

Form
Apple Tree Bryobia: *Bryobia rubrioculus* (Scheuten, 1857); *B. pyri* (Boisduval, 1866); *B. arborea* Morgan and Anderson, 1957.
Elm Tree Bryobia: *Bryobia ulmophila* Reck, 1947.
Unmowed Field Bryobia: *Bryobia gloriosa* Koch, 1836.
Dwelling and Tree-Trunk Bryobia: *Bryobia cristata* (Dugès, 1834); *B. praetiosa* Koch, 1836; *B. borealis* Oudemans, 1930.
Honeysuckle Bryobia: *Bryobia lonicerae* Reck, 1956.
Gooseberry Bryobia: *Bryobia ribis* Thomas, 1894.
Ivy Bryobia: *Bryobia kissophila* Eyndhoven, 1955.

*Adapted from Snetsinger (1967). The common names given were proposed by Snetsinger. These are followed by some scientific names used by other authors, with their names stated.

[28] There are at least 62 species of predators that attack the clover mite. Eight are spiders, 15 are mites, 5 each are thrips, anthocorid bugs, and mirid bugs, 15 are coccinellid beetles, and the rest are in other families and orders of arthropods (Snetsinger, 1960).

EARWIGS

The common name, "earwig," is derived from a European superstition of ancient origin that these insects enter the ears of sleeping persons and bore into the brain. However, despite the formidable appearance of their caudal forceps (figure 356), they are practically harmless to man. There have been reports of some species occasionally nipping people with their forceps and drawing blood. In the United States, apparently the only record of injury by earwigs was reported by Bishopp (1961), one of whose toes was bitten by a nymph of *Labidura riparia* inside the shoe, causing profuse bleeding. The insect appeared to be distended with ingested blood.

Earwigs (order Dermaptera) feed on live or decaying vegetation as well as live or dead insects. When a female was offered living insect prey, she immediately grasped it in her forceps, and while still holding it, she turned her body in a half circle and ate the insect (Klostermeyer, 1942). Earwigs at times cause damage to cultivated plants, but can also become household pests by invading a home in large numbers and being severe nuisances.

Earwigs are medium-sized insects, ranging in the United States from 5 to 25 mm in over-all length. They are elongate and flattened, have a tough, shiny exoskeleton of various shades of brown, or are sometimes black. The mouthparts are of the biting type. The metamorphosis is gradual, usually with 4 or 5 nymphal instars. When wings are present, the first pair form very short, usually truncate, veinless, hard wing covers, making them resemble superficially the staphylinid or rove beetles. The second pair of wings are fan-shaped, with a peculiar radial venation. When not in use, they are folded in a complicated fashion into a compact mass that is almost completely covered by the abbreviated wing covers. The body terminates in a pair of strong, movable forceps or pincers, those of the males being larger and more caliperlike. They are used for defense, capturing prey, probing the area inside a narrow crevice, and to a lesser extent for folding and unfolding the wings (Morgan, 1923; Essig, 1942). Morgan (1924), in an investigation of *Labidura bidens* Olivier, could find no evidence that the forceps were used in mating, as had sometimes been reported.

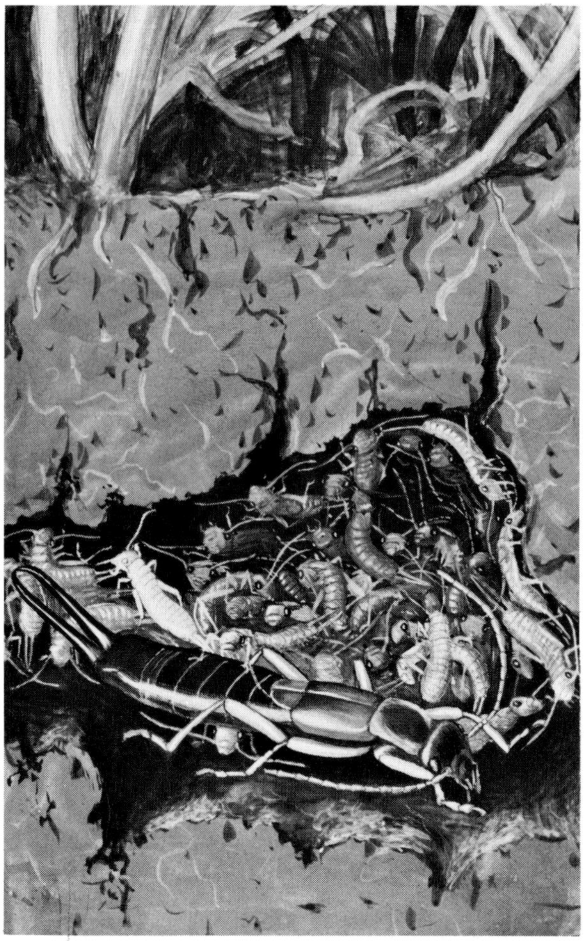

Fig. 357. European earwig, *Forficula auricularia*, and newly hatched nymphs in subterranean nest. An example of maternal care for the young among insects. (From Fulton, 1924*a*.)

Fig. 356. European earwig, *Forficula auricularia*. Female (*left*); short-forceps-type male; long-forceps-type male.

Earwigs are nocturnal in their habits, hiding during the day in moist, shady places under stones or logs, in cracks in the soil, or in almost any secluded place. Their ways are similar to those of cockroaches, including their gregarious nature, their habit of cleaning their appendages with their mouthparts (Bharadwaj, 1966), and the fact that they are repelled by dust (Crumb *et al.*, 1941). Neither eggs nor nymphs are able to withstand prolonged desiccation.

Maternal Care of Eggs and Young

The female lays smooth, oval, pearly-white, yellowish or cream-colored eggs in a chamber in the ground. She exhibits an instinct that is rare among insects—maternal care of the eggs and young (figure 357). Referring to the European earwig, *Forficula auricularia,* Crumb *et al.* (1941) stated:

> The eggs were never known to hatch in the absence of the female, probably because it is necessary for them to be kept in a moist situation, and yet protected from mold. The care of the eggs includes "licking," turning as done by a hen, and frequent shifting of their position, now in a layer on the side of the cell and now, perhaps, in a heap at the bottom. In spite of her solicitude for them, the female will eat her eggs readily if conditions become too unfavorable or even if she is disturbed too much.

For the first few days after the eggs hatch, the female keeps the nest closed. It is opened later, but she guards against intruders, using her forceps as weapons. She brings in food for the young, which remain inside or close to the nest until after their first molt (Fulton, 1924*a, b;* Crumb *et al.,* 1941).

The female of the striped earwig, *Labidura riparia,* also remains with her eggs in a specially constructed chamber, grooming the eggs at intervals until they hatch 8 to 10 days later. She does not eat during the incubation period. The newly hatched nymphs remain in the chamber with the female for another 2 to 4 days. She will eat the nymphs if the nest is disturbed (Caussanel, 1970).

European Earwig, *Forficula auricularia* L. (Forficulidae)

The European earwig (figures 356 and 357) is a cosmopolitan species that was first reported in the United States in Newport, Rhode Island, in 1901 and in Seattle, Washington, in 1907. It is now found in many areas throughout the country and in southern Canada. These earwigs require high humidity for continuous development. In one experiment at an average relative humidity of 48%, they were all dead in 12 days (Crumb *et al.,* 1941). Probably in semiarid areas such as southern California, the insects can survive only because they can find secluded, moist places in which to spend most of their time. They thrive best in cool, moist climates. On the other hand, prolonged wet periods in the spring when the young earwigs are hatching cause high mortality among them.

Description. Adult females, including forceps, are about 16 mm long. Some adult males are about the same size, but others are considerably larger. As with the Dermaptera in general, the males have larger and more caliperlike forceps (figure 356). With regard to the size of the forceps, the males are dimorphic, the forceps being about 3.5 mm long on some and 7 mm long on others (figure 358, *F* and *H*). Those with longer forceps

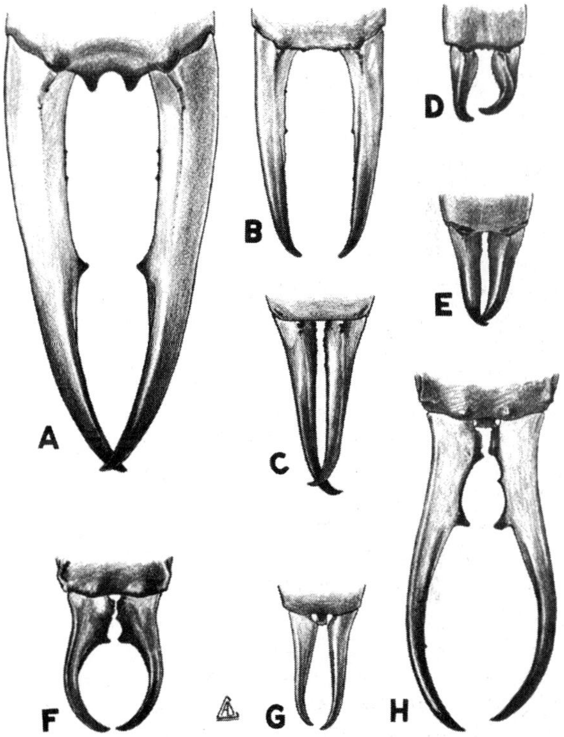

Fig. 358. Forceps of 3 species of earwigs. *A, Labidura riparia* (male, large type); *B, Labidura riparia* (male, small type); *C, Labidura riparia* (female); *D, Euborellia annulipes* (male); *E, Euborellia annulipes* (female); *F, Forficula auricularia* (male, short type); *G, Forficula auricularia* (female); *H, F. auricularia* (male, long type).

usually have much larger bodies. The general color is dark reddish brown, with the head decidedly reddish and the wing covers, legs, and antennae paler. On each side of the dorsal aspect of the third and fourth abdominal segments (second and third visible segments) are the openings of glands that discharge small drops of liquid if the insect is picked up. This liquid is probably responsible for the peculiar odor of earwigs (Fulton, 1924*a*).

Habits and Life Cycle. Pairs of European earwigs dig into the soil in the fall, and usually hibernate in cells 3 or 4 cm beneath the surface. In observations made in mid-January, Fulton (1924*a*) found eggs in some cells and, when no eggs were present, the body of the female was full of them. The eggs are about 1 mm long, pale yellow or cream-colored, opaque, and broadly elliptical. As they develop they become larger, whiter, and more translucent. The males leave the cells in January, and can be seen in numbers until the last of April or May, when some re-enter the soil with the females a second time. The females begin to leave the soil in April, most of them appearing on the surface by the middle of May, having spent about 7 months in the soil without food. Some of the females again enter the soil to deposit a second clutch of eggs. Thus, there may be 1 or 2 generations a year.

Under the climatic conditions at Puyallup, Washington, adults entered the soil for hibernation from the latter part of September to the last of October, the period of maximum entrance apparently depending on the advent of cool weather. Many males, an occasional female, and rarely a nymph remained above the surface throughout the winter. The average incubation period for winter egg masses was 72.8 days, and for the second or spring clutch it was about 20 days. There appeared to be about equal numbers of new males and females. In the field, the period required for the development of the 4 nymphal instars was about 68 days, while in the laboratory between 60 and 70° F (16 and 21° C), it was about 31 days (Crumb *et al.*, 1941).

Few people have seen earwigs in flight, but Crumb's group observed as many as 20 flights in one day, mostly in bright, warm sunshine. An elevated object seemed to be necessary for takeoff, and the insects were able to rise and fly at least 30 ft (9 m).

Economic Importance. European earwigs are omnivorous, but prefer plants such as mosses, lichens, and algae as food. Some animal food seems to be required, and this includes insects, spiders, mites, and protozoa, alive or dead (Crumb *et al.*, 1941). They occasionally do some damage to flowers, fruits, vegetables, ornamental shrubs, and trees, and have been known to feed on honey in beehives. Damage is small when compared with the numbers of earwigs that are sometimes present. In one investigation, digestive-tract analyses showed the diet to consist largely of lichens and pollen (Dimick and Mote, 1934).

The insects are most abundant in the gardens or homes of urban areas, and around buildings in rural communities. They tend to avoid cultivated or dry land, and sometimes invade houses in enormous numbers. Fulton (1924*a*) quoted thus from a correspondent:

> Literally thousands and thousands of those bugs inhabited my premises last summer. They made it almost impossible to live in my home; they inhabited the sleeping porch till we had to leave it. They work mostly at night, but in the daytime might be found in kitchen drawers, and often burrow an inch into a loaf of bread. They crawl over ceilings at night and drop on the bed, or inhabit themselves in a person's clothing during the night, and while their bite has never proved serious, it is entirely uncomfortable.

With regard to infestations in Oregon, Fulton stated that it was almost impossible to keep European earwigs out by the use of screens, and that they had become so annoying in some districts that property values had depreciated considerably. There were successive importations from Europe of a tachinid parasite, *Digonichaeta setipennis* Fall, from 1924 to 1929 in Oregon and California, and as high as 38.4% parasitism was recorded (Ritcher, 1966). The European earwig is believed to be less abundant than before the parasites were introduced, but nevertheless it is often a severe nuisance as a house and garden pest. This is indicated by a survey of the opinions of California structural pest control operators as to the relative importance of the 6 most important pests (termites excluded) that they have been called upon to control. These pests, in descending order of importance, were reported to be: in northern California, German cockroaches, ants, earwigs, mice, fleas, and silverfish; in central California, German cockroaches, earwigs, mice, silverfish, brownbanded cockroaches, and black widow spiders; and in southern California, German cockroaches, ants, fleas, oriental cockroaches, earwigs, and silverfish (PCOC, 1973). Earwigs were not listed as to species, but the European earwig was the principal pest species in most areas.

Striped Earwig, *Labidura riparia* (Pallas) (Labiduridae)

The striped earwig, also known as the shore or riparian earwig, appears to be distributed in tropical and subtropical areas throughout the world. It was first reported in the United States, in Texas, in 1876 (Hebard, 1942). It is now widely distributed in the southern states and in parts of Arizona and California. It apparently became established in the Yuma (Arizona) area in 1952, and spread into the major irrigated farming areas of southern Arizona and southeastern California (Schlinger *et al.*, 1959; Nutting, 1960). It has become widely distributed throughout southern California, and has extended its range northward to scattered areas in the central valley.

Description. This fully winged species is usually larger than the European earwig, ranging from 20 to 25 mm in length—usually slightly less than 25 mm. The color varies from pale brown, with distinct black markings, to chestnut or reddish brown, with the dark markings somewhat merged into the general color. The abdomen is usually banded, and the forceps vary from yellowish to reddish brown and are darker near the tips. The antennae and legs are yellowish. The last segment of the male's abdomen usually bears 2 short spines that are larger and more strongly developed on large specimens. The male's forceps are usually elongate and conspicuously toothed (figure 358, *A*), but some are shorter and weakly toothed (figure 358, *B*). The female's forceps (figure 358, *C*) are close together where they are attached to the abdomen, and bear short, basal teeth along their inner margins, but these are sometimes hardly noticeable (A. B. Gurney, correspondence).

Life Cycle. Striped earwigs live in subterranean burrows beneath rubbish, mulches, and turf thatches. One female was observed to seal the entrance to a burrow she had made, after depositing 51 eggs within the burrow, tending them, washing them, and moving them about. (The eggs become covered with fungi or desiccated if a female is removed after ovipositing.) The eggs hatched in 7 days, and the first-instar nymphs fed on the egg cases. The nymphs remained in the burrow, and were groomed and manipulated until their first molt, which occurred in another 7 days. The female then released them from the burrow and lost her maternal interests, devouring any nymphs she could capture. Depending on the amount of food (maggots) and water provided, the nymphs matured in from 49 to 60 days (average 56) at temperatures varying only a few degrees from 78° F (26° C) (Clements and Kerr, 1969). *Labidura riparia* was observed to lay 3 or 4 clutches of eggs per year in Europe (Caussanel, 1970).

Habits. Apparently, all North American species of earwigs are predatory, but they are also scavengers. Striped earwigs seem to eat any food they encounter, such as garbage, cat and dog foods, or insect carcasses. They are not known to damage plants as the European earwigs do. Most of the earwigs (96.5%) found around structures in Florida in 1967 were *Labidura riparia,* the remainder being principally the ringlegged earwig, *Euborellia annulipes,* and a few spinetailed earwigs, *Doru aculeatum* (Scudder). However, during part of the summer of 1968, *D. aculeatum* was one of the most common earwigs, when large numbers developed and flew to lights (Clements and Kerr, 1969). *Labidura riparia* was trapped in flight (toward lights) only on nights after rainfall had occurred the previous day or night. The initial flight coincided with the maturation of the first generation of earwigs (Gross and Spink, 1971).

Economic Importance. In the Gulf states and California, striped earwigs became important household pests during the 1960's. They appeared to be attracted to lights and/or to other insects that were also attracted to lights. Their disagreeable odor, intensified if they are crushed, resulted in this species being particularly detested. Infestations in a residence could usually be detected, because of the characteristic odor, before the insects were seen. Another noteworthy peculiarity of these insects was their inherent resistance to chlorinated hydrocarbons. As a result, the application of 10% granular heptachlor to urban yard and field test plots, which eliminated important predators (imported fire ant and thief ant), greatly increased the striped earwig population (Gross and Spink, 1969).

Ringlegged Earwig, *Euborellia annulipes* (Lucas) (Carcinophoridae)

This cosmopolitan species was first recorded in the United States in 1884, and is now widely distributed throughout the nation (Bharadwaj, 1966). It has been recorded from every county in southern California as well as in most of the northern counties. It appears to be a native American species (R. N. Hawthorne, correspondence).

Description. The ringlegged earwig is 12 to 15 mm long, wingless, and dark brown to black, with a yellowish-brown undersurface. The legs are yellowish, with femora and tibiae bearing fuscous rings. The black, 16-segmented antennae have the third, and usually the fourth, segment from the apex white. The female has 8 abdominal tergites, and the male has 10. Typical forceps of male and female are shown in figure 358, *D* and *E,* respectively. The nymphs resemble the adults. In the successive nymphal instars, the number of antennal segments is 8 in the first, 11 in the second, 13 in the third, 14 in the fourth, and 16 in the fifth (Bharadwaj, 1966).

Life Cycle. Females can produce as many as 6 clutches of eggs from a single mating (Klostermeyer, 1942), but usually produce 1 to 4 (Bharadwaj, 1966). The eggs average 0.845 mm long, and are at first creamy white and ovate, but later become gray or brownish and kidneyshaped. The average number of eggs laid per female has been recorded as 47 (Klostermeyer, 1942), and as 52.7 (Bharadwaj, 1966). Klostermeyer found the preoviposition period to be an average of 10.8 days; egg stage, 14.3 days; and nymphal stage (5 instars), 80.7 days. Some females lived as long as 7 months after reaching maturity. Females outnumbered males by a ratio of about 4 to 1.

Economic Importance. The ringlegged earwig has been recorded as a pest of Irish and sweet potatoes in storage, in flour mills, breweries, meatpacking plants, and slaughterhouses, damaging the roots of vegetables grown in greenhouses, and as a pest in gardens and nurseries. It was reported as one of the principal pests in corn-processing plants (Gould, 1948). It acts as a carrier of certain endoparasites of fowls in Hawaii. It has become a nuisance by invading houses, particularly in the Gulf states. On the other hand, it is of some value as a predator on certain insect pests, including leafhoppers, beetles, moths, and various stored-grain pests (Klostermeyer, 1942; Bharadwaj, 1966).

African Earwig, *Euborellia cincticollis* (Gerstaecker)

This African species occurs in a few localities in Texas, Arizona, and Nevada. In California, where it was first reported in 1946, it was found in many localities in the southeastern desert. It has since spread as far as the northern part of the central valley. In California, these earwigs have been reported to cause damage to Persian melons where they contact the ground. In Arizona, they have been swept from alfalfa, collected in houses and motels, and seen at lights (Schlinger *et al.*, 1959; Nutting, 1960). Open fields with some type of low-growing plants are favored areas, provided moisture and drainage conditions are favorable. The furrows of seed alfalfa fields are ideal habitats (Knabke and Grigarick, 1971).

Description. These earwigs are black, with a reddish tinge. Their legs are yellow or amber, usually with faint, smoky markings on the femora and tibiae. Mature specimens average 14 mm in length. Some adults are wingless, or have short wings and are easily confused with *Euburellia annulipes,* which is always wingless (in California) and is similar in general appearance. However, normally in both sexes of *E. cincticollis,* antennal segments 15 and 16 are white, the remainder being dark brown, whereas in *E. annulipes,* usually segments 12 and 13 are white. An occasional specimen of either species will have antennae with all segments dark. Specimens of *E. cincticollis* have only faint, smoky areas on the femora and tibiae, whereas the legs of *E. annulipes* are prominently black-spotted (Ting, 1951). A distribution map shows *E. cincticollis* to be limited to interior localities of California, whereas *E. annulipes* occurs both inland and along the coast (Knabke and Grigarick, 1971).

Euborellia cincticollis, as well as *E. annulipes,* can be distinguished from the European earwig, *Forficula auricularia,* in that the latter lacks the white antennal segments, and the forceps also differ greatly in size and shape, being larger and more "forcepslike" in the male (Ting, 1951). Also, in *F. auricularia* the second tarsal segment is lobed, whereas in the genus *Euborellia* it is cylindrical (Knabke and Grigarick, 1971).

Life Cycle. The African earwig normally remains almost entirely hidden in the soil at depths of about 2 to 30 cm, or under vegetative growth. Eggs are laid from late April to July and August. Field-collected females produced and brooded several clutches of eggs in the laboratory at 2- to 4-week intervals, averaging a total of 83.7 eggs per female. The average periods from egg to adult were 128 days at 22 to 27° C, 78 days at 27 to 29°, and 82 days at 32 to 33°. Five nymphal instars were most common at 22 to 27° and 6 to 7 instars were most common at 32 to 33°. One or 2 generations per year were estimated to be produced in the field. Both nymphs and adults were noted to overwinter. Field-collected adults survived as long as 240 days in the laboratory, and were believed to be able to live a year or more in the field. In one investigation, flights of *E. cincticollis* occurred between July 14 and September 26, but major flights did not occur until August 12. Flights took place mainly from sunset to about 10 p.m. (Knabke and Grigarick, 1971).

Control of Earwigs

European earwigs are strongly attracted by fish oil. This characteristic has been utilized in the preparation of an effective poison bait. One lb

(0.45 kg) of sodium fluosilicate is mixed with 12 lb (5.5 kg) of dry bran. A quart (about 1 l) of fish oil is then mixed with the poisoned bran. No water is added. The bait should be applied in the evening beside fences, hedges, woodpiles, around the foundations of buildings, in flower beds, under and in the crotches of trees, and on the lawn. The lawn should not be sprinkled for 2 days after application of the bait. In Oregon, where the bait was developed, most feed and seed stores sell a ready-mixed bait for European earwig control. The bait is poisonous, and should be kept from children, but poultry will not eat enough to be harmed (Anonymous, 1968*b*).

Malathion has been suggested as an effective contact insecticide, but it does not have a long residual action. It should be applied in the evening in the same areas as suggested for application of the poison bait, but of course it should be more uniformly distributed. The earwigs will contact the residue when they come out for their nightly foraging. Malathion may be used as 5% dust or as a spray prepared by mixing 1 cup (240 cc) of 25% wettable powder or 2 tablespoonfuls (30 ml) of 57% emulsifiable concentrate per gal (4 l) of water (Anonymous, 1968*b*). The earwigs are particularly numerous around heavy growths of ivy. R. N. Hawthorne (personal communication) suggests that not only the ground, but also fences and trees (up to the lower crotches) be sprayed in March and April, after turning over all rocks and boards to expose hidden insects. (He has observed in recent years that the earwigs were ascending pear trees in spring to feed on the opening buds.) Organochlorine insecticides are effective, and if the homeowner uses chlordane to spray his lawn for lawn moths, the spray can be extended to other areas of the yard for good control of earwigs and ants.

Although the European and ringlegged earwigs are susceptible to organochlorine insecticides, the striped earwig is not. It can be controlled with organophosphorus or carbamate insecticides, such as malathion, diazinon, and carbaryl. Propoxur was found to be a particularly toxic insecticide for *Labidura riparia*. The LC_{50} for propoxur was 0.04% compared with 0.13% for diazinon. However, treatments applied inside the house, as for cockroaches, did not control striped earwigs. They could be controlled by applying insecticide to the soil in areas 1 to 2 yd (1 to 2 m) wide around the perimeters of structures. Propoxur bait performed well in tests, but would probably have to be applied once a month (Clements and Kerr, 1969).

CRICKETS

The closely related crickets (Gryllidae) and katydids (Tettigoniidae) (order Orthoptera) share with the cicadas the honor of being the musicians of the insect world. The music of crickets and katydids is produced by the males with their stridulatory organs. These consist of a file and a scraper on each of the tegmina (forewings). The tegmina are moved back and forth so that the files are rubbed against the scrapers. This causes a vibration of areas called "tympana," producing the chirping or ticking sound. Species of crickets and katydids, as well as cicadas, which belong to a different order (Homoptera), can be recognized by the differences in the sounds of their songs, and these differences have been responsible for the initial recognition by entomologists of many theretofore undescribed species. The females, which do not sing, are attracted to the "calling" sounds of the males, which are sensed or received by organs of hearing located at the bases of the tibiae of the forelegs. Although a cricket's song may seem pleasant outdoors, it is generally disliked within the house. The female cricket moves to locate the stridulating male, for which reason many more females than males are likely to be found. The males, after finding suitable locations, usually remain in them all their lives (Alexander, 1957).

Crickets and katydids both have long, filiform antennae. The tegmina of crickets lie flat on the back, but are bent down abruptly on the sides, and the tarsi are 3-segmented. The tegmina of katydids are held rooflike over the back, and the tarsi are 4-segmented.

Of the 2 groups, the occasional household pests are the crickets. Two species may be involved—the field cricket, *"Gryllus assimilis"* (quotation marks are explained in the following text), and the house cricket, *Acheta domesticus*. In the western states, the field cricket is the principal pest. When their natural food supply of grasses dries up, crickets may migrate, and homes and other buildings may be invaded by thousands of these insects.

Multitudes can be attracted to lights and will sometimes cover streets and walks several inches deep (see under "Economic Importance"). They usually enter a building during the first cold days in the fall, through cracks and apertures such as around poorly fitted doors or loose basement windows. Some are carried in with firewood. Excessive rainfall may cause household or other building invasion.

Field Crickets, "*Gryllus assimilis* (F.)" (= *Acheta*)

Field crickets are indigenous to most of North, Central, and South America, and can be found almost everywhere except at high altitudes, in the far north, and in extremely moist situations. R. D. Alexander (correspondence) believes that there are about 25 species of *Gryllus* in the United States. Five of them are chiefly east of the Mississippi River, and 6 are chiefly in the Great Plains, Rocky Mountains, and southward. California has at least 10 species, including *assimilis* (F.) (only in the southern areas of the state), *armatus* Scudder, *lineaticeps* Stål, *pennsylvanicus* (Burmeister), *veletis* (Alexander and Bigelow), and 6 undescribed species. In the literature, however, field crickets have usually been included in a single, admittedly variable, species, "*Gryllus assimilis* (F.)," so for the sake of convenience, the name is retained here to represent the entire complex.

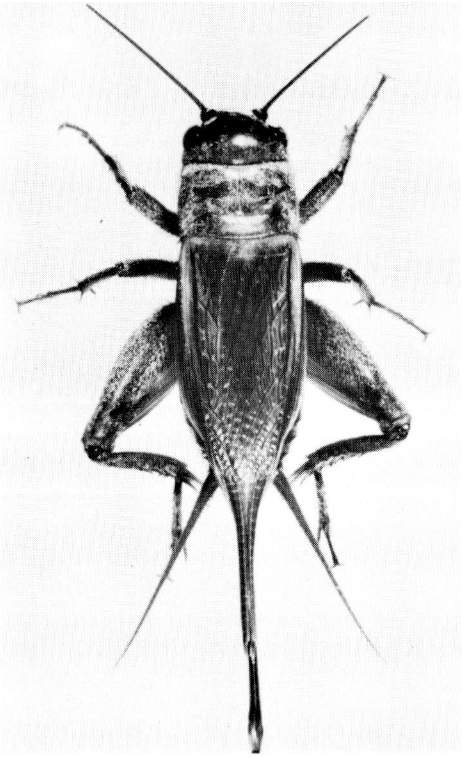

Fig. 359. Field cricket, "*Gryllus assimilis*."

Description. In the United States, field crickets range in length from 13 to 28 mm, and from solid black to pale straw color. They have long, slender antennae, much longer than the body. The females have 3 conspicuous appendages extending from the tip of the abdomen, the middle one being the ovipositor (figure 359). The males may be distinguished by their having only 2 caudal appendages. Alexander (1957) recognized that, among field crickets, forms that were morphologically similar in different localities could be found in 3 types of habitats: "(1) large, light-colored forms in sandy areas (beaches, sand dunes, and deserts); (2) small, black forms in deciduous forests; and (3) forms intermediate between these two extremes in grassy and weedy areas, such as prairies, fields, pastures, roadsides, and lawns."

Life Cycle. Severin (1926) gave an account of the life cycle of "*Gryllus assimilis*" based on 12 years of investigation of the insect in South Dakota. He believed that there were 2 biological races in that state, one of which, probably less than 5% of the cricket population, overwinters as nymphs of usually the fifth and sixth instars, under leaves, trash, and other debris. They become adults during the latter part of May, reproduce during June, and die in July. The other race overwinters in the ground in the egg stage. They reach maturity during late July and early August. Many live out their life span, reproduce, and die by mid-September, but some live until they are destroyed by a heavy freeze. There is very little overlapping of living adults of the 2 races.

In the race that hibernates in the egg stage, the females begin to oviposit within 2 weeks after becoming adults, and continue until they die. They prefer to oviposit in soil that is firm, but soft enough to permit the ovipositor to enter readily. They prefer damp soil, such as that along ditches or at the bases of cracks in baked soil, or they will uncover damp soil by digging pits with the first 2 pairs of legs (Metcalf *et al.*, 1962). The eggs are deposited 7 to 25 mm beneath the surface. They are laid singly, but 50 or more may be laid in an area about 5 cm square. A female normally lays 150 to 400 eggs. When first laid, they are light honey-yellow, and 2.8 to 3.2 mm long. After passing through the winter, they are cream-colored and slightly larger (Severin, 1926). According to Severin, field crickets passed through 8 or 9 instars. There was little difference in the period required by males and females for nymphal development, and it ranged from 82 to 93 days. In the Gulf states and the California deserts, "*Gryllus assimilis*" overwinters in nymphal instars, and is active throughout the year. There may be as many as 3 generations per year.

Economic Importance. The field cricket can destroy field crops such as alfalfa, alfalfa seed, cereals, and vegetable crops such as tomatoes, cu-

cumbers, peas, beans, and strawberries. Great outbreaks of field crickets occasionally occur in several states simultaneously in the Midwest and South (Howell and Hensley, 1953; Hutchins and Langston, 1953). They sometimes invade towns and cities in great swarms. At such times, the sides of buildings may be covered with the insects, and it is necessary to sweep them from store fronts and sidewalks. They can become so abundant on streets and sidewalks under street lights that the slipperiness resulting from crushed crickets causes driving and walking to be hazardous. During such invasions of populated areas, the crickets will also invade houses and other buildings. They have been reported to damage articles made of nylon, wool, plastic fabrics, and leather (Howell and Hensley, 1953), and "to eat holes in paper and rubber and in cotton, linen, woolen, or fur garments, either outdoors or indoors, especially when soiled with perspiration or foods" (Metcalf *et al.*, 1962). The odor of dead crickets may be unbearable if they are not removed. During these invasions, cats may reject food and eat only crickets. The cats become emaciated and are subject to fits (R. N. Hawthorne, correspondence). Fortunately, field crickets cannot adapt themselves like house crickets to conditions within houses, and will die off by early winter (NPCA, 1968).

Folsom and Woke (1939) believed that during summers of extended drought, field crickets developed in large numbers when soils cracked extensively and provided shelter for the insects against weather and predators. Howell and Hensley (1953) observed that great swarms of crickets that invaded cities and towns were usually correlated with severe drought, followed by rainfall sufficient to produce food adequate for cricket development. However, in the desert valleys of southeastern California, where summer thunderstorms occasionally occur on days of especially high temperature, great swarms of crickets will invade towns on the same day as a thunderstorm takes place. This happens nearly every summer. Five mm of rain can trigger such an outbreak.

House Cricket, *Acheta domesticus* (L.)

Unlike field crickets, house crickets can maintain themselves indefinitely indoors. They deposit their eggs singly in dark places and crevices, such as behind baseboards, and can go through their entire life cycles within the building. They are normally found in warm, dark locations, hiding during the day but coming out at night, when their presence is indicated by their chirping.

Description. The adult averages about 20 mm long, and is light yellowish brown or straw-colored, with 3 dark bands on the head (figure 360). As with the field cricket, it has slender antennae that are longer than the body.

Fig. 360. House cricket, *Acheta domesticus*.

Life Cycle. Under laboratory conditions, the female lays an average of 728 eggs at 28° C (82° F), and there are 7 to 8 nymphal instars. The nymphal stage (over-all) requires, on the average, 55.8 days for males and 52.9 days for females. The preoviposition period averages 10 days, oviposition period 35 days, and postoviposition period 19 days (Ghouri and McFarlane, 1958). In the field, overwintering eggs generally hatch in late spring, and the adults appear in late summer, there being only 1 generation per year.

Economic Importance. During a 15-year period of investigation before 1935, the house cricket was more often encountered outdoors than indoors (Back, 1936). Wherever it had become a serious household pest in 35 cities and towns, the infested houses were located near an active city trash dump. In these dumps, the crickets were not seen during the day unless they were exposed by removing objects under which they were hiding. With the approach of darkness, they began to migrate, crawling or flying, particularly after city lights were turned on. They then appeared in large numbers on tree trunks, telephone or electric light poles, and on the sides of houses, sometimes entering second- and third-story windows or skylights on the roof. During such infestations, one could scarcely move an article of furniture or other object in a house without uncovering crickets. Infestations tended to continue until the trash dumps or fills were covered over with earth.

In infestations such as those just described, all types of clothing are damaged. Large areas are eaten out of the fabric, rather than the small holes typical of clothes moth infestation. Heavy damage to wool, cotton, silk, synthetic fabrics, and even carpeting has often been reported (Back, 1936). Clothes stained with perspiration are particularly attractive to crickets.

In a controlled experiment with cotton and synthetic fabrics, house crickets caused severe damage to some synthetic fabrics and some damage to all of them. There was greater damage to clean fabrics than to the same fabrics when stained with an animal fat collected from kitchen food-preparation drippings and strained through an 80-mesh screen. Damage to the unstained acetate, viscose, and triacetate fabrics was extensive. There was slight damage to the polyester, nylon, and acrylic fabrics. Crickets preferred the same 3 fabrics as American cockroaches with which they were compared. There apparently was no damage to cotton fabrics (Finley *et al.*, 1968).

Control of Field and House Crickets

A poison bran bait, with cane molasses as the attractant and sodium fluosilicate as the toxicant, gave good results against field crickets. Sodium fluoride was slightly less effective, and arsenicals were ineffective. A sodium fluosilicate bait was also recommended for control of the house cricket in city dumps (Back, 1936). Some authorities believe that to prevent crickets from entering houses, the bait should be placed under boxes, boards, and other objects under which crickets tend to hide during daylight hours. However, D. A. Reierson (personal communication) was able to stop the movement of field crickets toward a house by broadcasting a pound (0.45 kg) of a proprietary 2% propoxur bait in an adjacent field from which the crickets were coming. The bait was scattered mostly next to a block-wall fence that the insects had to cross to reach the house. Many dead crickets were found in the treated areas for several days after the bait was applied. This house had been infested for several weeks, and cricket feeding scars were evident on carpets, draperies, and upholstered furniture. Poison bran bait with bloodmeal was found to be the most effective bait for field crickets in tests made in the Imperial Valley of California (R. N. Hawthorne, correspondence).

Mallis (1969) quoted D. E. Howell as stating that in 1954, a field cricket infestation in Stillwater, Oklahoma, was controlled with a 0.5% lindane spray. Four hours later, more than 75 cu m of crickets were swept up in the downtown areas of the city. In later tests, malathion, diazinon, and other organophosphorus insecticides were found to be even more effective than lindane.

Removal of ivy and shrubs from buildings is said to aid in cricket control. Openings in buildings, particularly at the thresholds of doorways, should be sealed. Residual sprays applied to baseboards, in closets, under stairways, and around fireplaces are said to be useful, keeping in mind that field crickets are usually restricted to basements and ground-floor levels, whereas house crickets may be found in any part of the house. Any insecticides and concentrations used for cockroaches are effective in cricket control. Insecticide dusts, such as Dri-die 67®, Drione®, 5% chlordane, or sodium fluoride are useful in enclosed spaces, such as attics, crawl spaces, wall voids, or behind baseboards. Kepone® pellets may be applied under drawers, sinks, and in other hiding places, as in cockroach control. Following a residual spray, a space spray or aerosol may be useful to drive many of the crickets out of hiding places and cause them to contact fresh residual deposits. In the southern states and the desert valleys of the Southwest, it is necessary to treat outdoors as well as in buildings. Some infestations of house crickets have been controlled by covering the city dump with earth.

Mole Crickets

There are other crickets, besides the field and house crickets, that may occasionally invade homes and other buildings and become minor pests. Among them are the mole crickets. They belong in the family Gryllotalpidae, which was once considered to be a subfamily of the Gryllidae in which the crickets already discussed belong. As the common name implies, they resemble miniature moles because of their large, shovel-like fore tibiae, adapted for digging, and in their burrowing habits. Mole crickets are usually brown or black, and 25 to 50 mm long. The tegmina are short, and the second pair of wings protrude like tails. They are weak fliers, and seldom attempt to fly. They usually have no stridulatory organs, and do not jump. When crawling, the forelegs are held forward and off the ground.

Habits. Although mole crickets spend most of their time in permanent burrows several inches deep in the soil, they may come to the surface when the soil has become wet, and particularly at night when temperatures are above 70° F (21° C). They will then make temporary burrows in the upper stratum (2.5 or 5 cm) of the soil, and can cause serious damage to lawns and golf courses by their burrowing. They are among the major pests of turf in Florida and some other southeastern states. They also crawl about on the surface and feed on ripening strawberries and the fruits of other crops if they touch the ground. They occasionally enter basements of buildings (Wisecup and Hayslip, 1943; Short, 1973).

Neocurtilia cultriger (Uhler) (= *Gryllotalpa*)

This is the only western mole cricket. It occurs in California, Arizona, New Mexico, and Texas. It is about 25 mm long, cinnamon brown, with dark marks at the bases of the legs, and has wide tegmina that cover the 4 basal abdominal segments.

Southern Mole Cricket, *Scapteriscus acletus* Rehn and Hebard; and Changa, *S. vicinus* Scudder

In the southern states, these 2 species of mole crickets are often found in sandy soils. They are brown, and about 40 mm long. They pass the winter in the soil as nymphs or adults, burrowing deeper during cold weather. In spring and early summer, the females oviposit in cells they construct in the soil from 3 to 10 in. (7.5 to 25 cm) below the surface, laying about 35 eggs per cell. These hatch in 10 to 40 days, depending on the temperature. There is only 1 generation per year.

Control of Mole Crickets

For many years, a sodium fluosilicate–wheat bran bait was used for the control of mole crickets in Florida (Wisecup and Hayslip, 1943). More recently, chlorpyrifos and propoxur baits have been recommended, and were found to be equally effective when applied at 1 lb/acre (1.135 kg/hectare). With a given quantity of toxicant per unit area, a 0.5% bait gives better results than a more concentrated one because of more thorough distribution with larger quantities of bait. Chlordane spray and granules, carbofuran granules, and propoxur spray were also effective. Better controls with sprays and granules in the spring and better controls with baits in the summer and fall were observed. Good control could be obtained only when night temperatures were above 60° F (16° C) and when the soil was moist (Short, 1973).

Camel Crickets

Camel crickets (Gryllacrididae), also called stone or cave crickets, are wingless, have very long antennae, and long hind legs with large femora adapted for jumping. The large head is bent downward between the forelegs and the back is arched, giving these insects a humpbacked appearance. Camel crickets are nocturnal, and live in caves, wells, cracks or holes in the ground, in rotten logs or stumps, or under damp leaves or stones in cool, dark situations. Most species are found in woodland areas, but others occur in grassland or even in desert regions. These dig holes in the ground or inhabit the burrows of larger animals or other cavities, from which they emerge at night. Camel crickets have no "song" to attract attention, are somberly colored, spend the day in close concealment, and are immobile in the presence of strong light. It is not surprising that they are not well known to the layman or even to the professional entomologist. Despite the common name, "cave cricket," only 1 species of *Ceuthophilus* found in Carlsbad Caverns is an obligatory cavernicole or troglobiont; other species are merely troglophiles, and may be found in any humid and concealed situation, sometimes invading damp cellars (Hubbell, 1936). They overwinter as adults or nymphs, oviposit in the spring, and the eggs hatch during April.

Crampton (1923) observed a species of *Ceuthophilus* crawling about in the open privies in a mountain camp in Massachusetts and the same species crawling over food on pantry shelves. He was surprised to find no reference to these insects as vectors of pathogenic organisms.

Ceuthophilus californianus Scudder

Most camel crickets are in this genus. Hubbell (1936) wrote a massive monographic revision of the genus, which contains many species, and is well represented in the western United States. *Ceuthophilus californianus* (figure 361, *top*) is uniformly pale brown, except for the darker chestnut-brown dorsal bands, and is 15 to 25 mm long. It is most commonly seen at the entrances of gopher holes on overcast spring days, and may

Fig. 361. Two California camel crickets. *Top, Ceuthophilus californianus; bottom, Pristoceuthophilus pacificus* (adult male).

also be found in soil or under stones or logs. This species occurs only in California, inhabiting the arid grasslands of the central valley and coastal strip. It is replaced by *C. hesperus* Hubbell from Los Angeles southward (Hubbell, 1936).

Ceuthophilus pallidus Thomas

This species occurs over a very large area in the Great Plains, from southern Canada to Texas, and in the dry grasslands of the southwestern United States and northern Mexico (Hubbell, 1936). It is often found indoors in summer, and was observed eating holes in lace curtains in southern New Mexico.

Ceuthophilus maculatus (Harris)

This is another related species that has shown a preference for textiles. Townsend (1893) reported that it was found feeding on clothes hung out on a line after washing. It occurs in Colorado at high altitudes, from 7,000 to 10,000 ft (2,100 to 3,000 m).

Pristoceuthophilus pacificus (Thomas)

This genus occurs along the Pacific Coast, from British Columbia to Mexico. Most species have dorsal specializations on the males' abdomens, such as spines, tubercles, and peculiar glandular structures. Some species also have strangely formed hind tibiae, usually found only in the males (Caudell, 1916). The species illustrated here is *P. pacificus* (figure 361, *bottom*), and was taken in the author's laboratory in Los Angeles.

Greenhouse Stone Cricket, *Tachycines asynamorus* Adelung

This is probably the best-known camel cricket in the United States. It is reported mainly as a pest in greenhouses in Europe and in the United States as far west as Colorado. A. B. Gurney (correspondence) believes that in Washington, D.C., this insect migrates from sewer outlets into homes. People who find it in their basements generally describe it as a "large, spiderlike creature, without wings, that jumps very high when disturbed."

Jerusalem Cricket, *Stenopelmatus fuscus* Haldeman

The Jerusalem cricket (family Stenopelmatidae) is worthy of consideration because of the great fascination and sometimes fear that it evokes on the part of the layman who may occasionally see one in his yard or garden, or even in his home. It occurs west of the Rocky Mountains and in British Columbia and Mexico. In the latter country, it is called *niña de la tierra* (child of the earth) because its large, round, naked head with 2 beadlike black eyes gives it a fancied resemblance to a miniature child. The southwestern Indians once feared it, and called it "child of the desert." In California, it is commonly called "potato bug" because it occasionally feeds on potatoes and other root crops while still in the soil (Davis, 1927). When a large planting of potatoes near El Monte, California, failed to sprout, every hill that was uncovered and examined was found to contain several Jerusalem crickets (Baker, 1971). However, such extensive damage by these insects is rare.

At Los Angeles, we receive many telephone calls concerning this strange-looking creature, despite its relative scarcity. It is one of the few insects that the layman adequately describes. Common questions are, "Is it dangerous?" and "What shall I feed it?" The Jerusalem cricket is harmless, although when cornered it may rise on its hind legs, facing the annoyer, and jump at the attacker if teased sufficiently. Its powerful mandibles can inflict a minor, nonvenomous wound when it is handled carelessly. It may also make a sound like 2 pieces of sandpaper rubbed together by scraping the hind legs against the rough plates on the sides of the abdomen. These insects are becoming more prevalent in California, and are entering houses in increasing numbers, if one may judge by the inquiries received from homeowners. They have even been found in third-floor apartments.

Description. *Stenopelmatus fuscus* (figure 362) is the largest member of the genus, being 30 to 50 mm long; an occasional specimen may be larger. The head, thorax, and legs are shining, pale amber-yellow to brownish. There is not the slightest vestige of wings. Dorsally, the abdomen is shining amber-brown, with wide, almost black, bands across the anterior margins of the segments, and ventrally it is pale amber. The long hind tibiae have 2 or 3 outer and 4 or 5 inner spines. The male has a more massive head and thorax and a smaller abdomen than the female.

Biology and Habits. Jerusalem crickets are nocturnal and are not often seen, but in winter, spring, and early summer they may be found

Fig. 362. Jerusalem cricket, *Stenopelmatus fuscus*.

under rocks in open, grassy pasture land, or may occasionally be seen crawling about in the evening. Baker (1971) described the interesting trail left by one of these insects in soft dirt: "A snake-like, continuous track made by the dragged abdomen, with a series of cleatlike tracks on either side, made by the legs. By following these tracks on a warm evening, one may sometimes find the insect searching for food." In California, they estivate in the fall, and can then be obtained only by digging in manure heaps and damp places. In captivity, they will eat bread, grass roots, vegetables, and fruit, but prefer any kind of meat or small insects. After copulation, the females may kill and eat the males.

The Jerusalem cricket burrows in the soil, but unlike the mole cricket, which burrows with specialized fore tibiae, the *Stenopelmatus* does its digging mainly with its large and heavily sclerotized head. The closed mandibles are used as a hoe, scraping the earth back and underneath. In southern California, they have been found in moist, light, loamy soil or in deep cracks in adobe soil. The females make nest burrows beneath rocks or boards that extend downward for 6 to 10 in. (15 to 25 cm), then turn sharply to one side and end in an enlarged chamber, often lined with a paperlike substance, in which the eggs are laid. The eggs are whitish, oval, and about 3 mm in diameter (Davis, 1927; Baker, 1971).

BUGS

The true bugs (Hemiptera-Heteroptera) have the forewings thickened in the basal half, while the apical half is thin and membranous. They are sucking insects, and are common pests of vegetation. The species described here are those that sometimes leave vegetation and migrate into near-by houses in large numbers, if one may judge by the inquiries received from homeowners.

Boxelder Bug, *Leptocoris trivittatus* (Say); and Western Boxelder Bug, *L. rubrolineatus* Barber (Rhopalidae)

Boxelder bugs are widely distributed throughout the United States, particularly where pistillate or seedbearing boxelder trees (*Acer negundo*) are found. *Leptocoris trivittatus* ranges throughout most of the United States and as far west as eastern Nevada. The western boxelder bug, *L. rubrolineatus,* which is similar in appearance and habits, is the species occurring in California and Oregon. Boxelder bugs feed on seeds, foliage, and tender twigs. They do not attack the male tree. They also occur to a lesser extent on maple and ash, and will occasionally feed on the young fruits of crops such as apples, pears, cherries, peaches, plums, and grapes, their punctures causing the fruit to be deformed. The adults spend the winter in dry, sheltered places—hollow tree trunks, barns or other outbuildings, and even in homes. They may swarm in large numbers on warm, sunny days, and residences not already chosen by the bugs for hibernation can be invaded during these flight periods. The mere presence of large numbers of boxelder bugs in the home is, of course, a nuisance, and in addition they can discolor curtains, furnishings, and clothing by spotting them with their excrement. When crushed or handled roughly, the bugs produce a strong, disagreeable odor. They sometimes bite people, and some difficulty has been experienced in healing the bites (Knowlton, 1947; Brannon, 1961).

Description. The adults of *Leptocoris trivittatus* average about 12 mm long, are rather flat and narrow, gray-brown to black, and have red margins on the basal halves of the wings. The adults of *L. rubrolineatus* are a little shorter and narrower than those of *L. trivittatus*. They average about 10 mm long, and the basal halves of the wings have not only red margins, but also 3 longitudinal red stripes (plate VIII, 8). The nymphs are bright red at first, but become marked with black when about half grown.

Biology. In the spring, the overwintering females lay their minute, red eggs in cracks and crevices of the bark of the host plant. The eggs hatch in about 2 weeks, close to the time when the new leaves are about to appear. There are 5 nymphal instars.

In the warmer regions of the United States, there are 2 generations per year. In California, all stages mass-migrate periodically, covering the ground, shrubs, and fences. What motivates the bugs to migrate is not known.

Control of Boxelder Bugs

An obvious way to avoid infestations by this pest in residential properties is to get rid of near-by female boxelder trees. Some city ordinances proclaim the female trees to be nuisances, and they can be removed at city expense. If this species is to be planted as an ornamental or shade tree, male trees should be purchased from the nursery. They are propagated by cuttings from staminate trees.

Chemical control can best be obtained by spraying the nymphs on the host trees before the adults have had a chance to migrate. Power spray equipment is usually required. Many residual insecticides have been successfully used. Malathion is suggested for those interested in an insecticide of minimal mammalian toxicity and with a brief residual life. One spray on the ground, tree trunks, and fences will usually control the pests, despite the brief residual life of malathion (R. N. Hawthorne, correspondence). In the home, the insects can be collected with a tank-type vacuum cleaner and then destroyed. Good control has also been obtained by applying aerosols containing TDE plus synergized pyrethrins. Before the advent of organochlorine insecticides, ordinary pyrethrin fly sprays were recommended for indoor treatment of boxelder bugs (Craighead, 1950). When referring to conditions in the eastern United States, Vasvary (1965) suggested, for indoor infestations, 2% chlordane in deodorized kerosene, sprayed or painted into cracks and on surfaces where boxelder bugs were hiding, paying particular attention to attics, especially close to the eaves, and to the cellars near foundation walls.

Weed Bug, *Arhyssus crassus* Harris (Rhopalidae)

The weed bug is a house pest of considerable importance throughout California, particularly in new subdivisions and in areas adjacent to uncultivated fields, ditch banks, or levees. As much as a liter of these bugs has been collected in some houses. The adults (figure 363) are uniformly brown, and about 9 mm long. A favored host is the yellow star thistle, *Centaurea solstitialis*, but the insects occur on many other plants. They migrate into houses in large numbers when their native habitats are disturbed or when the weeds they normally infest begin to dry in summer. They are attracted to light-colored buildings more than to those of darker colors, and tend to crawl through any openings that they find under windows or doors. They stain draperies and rugs with their excrement, which is practically impossible to remove by the usual cleaning operations. The bugs themselves have an objectionable odor.

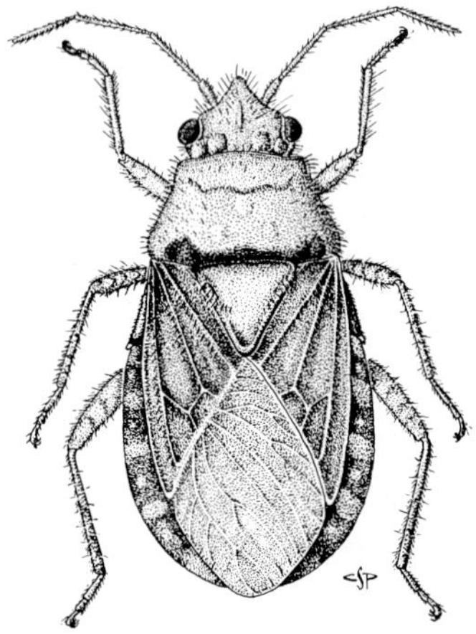

Fig. 363. Weed bug, *Arhyssus crassus*.

Control of Weed Bugs

There is no control measure that is effective after the bugs have moved into populated areas and become household pests. Even when they are sprayed before entering a house, they are still able to go inside, and then die. Their control, in the form of disking or spraying, needs to be applied in the originally infested grasslands. Carbaryl (Sevin®) is an effective insecticide (R. N. Hawthorne, correspondence).

Grass Bug, *Nysius raphanus* Howard (Lygaeidae)

This bug feeds principally on grass, in contrast to *Arhyssus crassus,* which feeds mainly on weeds. As a house pest, this species is most troublesome in new subdivisions. It is a very small bug, about 3 mm long, with a dark body but with whitish, translucent wings that give the insect a generally grayish appearance (figure 364). It is rather in-

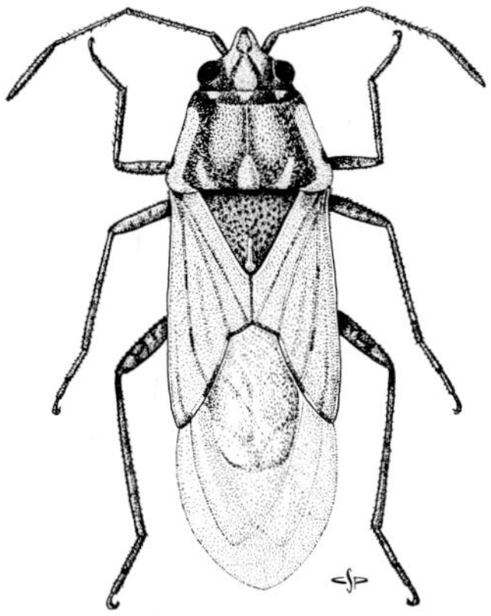

Fig. 364. Grass bug, *Nysius raphanus*.

conspicuous until it becomes very numerous. In rural areas, *Nysius raphanus* usually occurs in pasture lands, but in cities and towns, it is found in vacant lots and along fences. When the wild grass becomes dry, or when it is mowed or burned, the adults and nymphs migrate, covering fences and sides of houses, finally invading the buildings in great numbers. Once an infestation gets under way, it takes about 5 years for the bugs to disappear (R. N. Hawthorne, correspondence).

BEETLES

Certain beetles (order Coleoptera) may invade residences and other buildings and become troublesome seasonal pests in some areas.

Ground Beetles (Carabidae)

The ground beetles are predaceous on many injurious insects and consequently are normally beneficial to man, but they are nevertheless considered to be pests when they enter buildings.

Tule Beetle, *Agonum maculicolle* Dejean

In the central valleys of California, the tule beetle is the most troublesome carabid. It breeds in the tules or marshlands along rivers, and appears in great numbers at dusk following fall rains. It may enter houses, and is particularly annoying because of its offensive odor, which is also a characteristic of various other carabids. Tule beetles usually invade a house from the direction of their breeding grounds. According to C. S. Papp, this or a closely similar species invades houses in southern California. The adults are almost 10 mm long. The pronotum and elytra are dark brown, with light-brown or tan margins (plate VIII, 8). Insecticides do not seem to be able to kill tule beetles quickly enough to keep them from crawling into a building before they die. A successful control measure has been to install an electric light above a container filled with water, with detergent added, about 50 ft (15 m) from the house. The beetles, attracted to the light, fall into the water and drown.

Other Ground Beetles

Stenolophus (*Agonoderus*) *lineola* F. is a common beetle throughout the United States. It sometimes becomes a nuisance by invading houses after being attracted to porch lights (Papp, 1960).

Nomius pygmaeus (Dejean), the stink beetle, occasionally invades houses in the western states. It has a particularly offensive odor that is retained for weeks on household articles upon which the beetle has crawled (Hinton, 1945). Its presence in towns has been associated with near-by forest fires, the insects apparently having been driven out of the forests by smoke (Spencer, 1942).

The control measure just suggested for the tule beetle should also be effective against other carabids that have similar habits.

Darkling Ground Beetles (Tenebrionidae)

In this family are some small black or brown species that may invade the home, particularly when they migrate into green areas from the surrounding hills after the native vegetation dries up following the last spring rains in California. In the southern part of the state, *Metoponium abnorme* (LeConte), *Coniontis subpubescens* LeConte (figure 365), *C. parviceps* Casey, and species of *Blapstinus* are examples of tenebrionids that feed on the buds and foliage of plants such as young avocado trees, grapevines, and seedling tomatoes, and may also find their way into houses in large numbers. (Figure is on next page.)

Elm Leaf Beetle, *Pyrrhalta luteola* (Müller) [= *Galerucella xanthomelaena* (Schrank)] (Chrysomelidae)

This important elm-tree pest was introduced from Europe about 1834, and is widely distributed in the United States. In California, it was first found in Fresno in 1924. The adults are 5 to 6 mm long and yellowish to dullish green, with an indistinct black stripe along each side. The orange-yellow, spindleshaped eggs are laid on end in groups of 5 to 25 on the undersides of leaves.

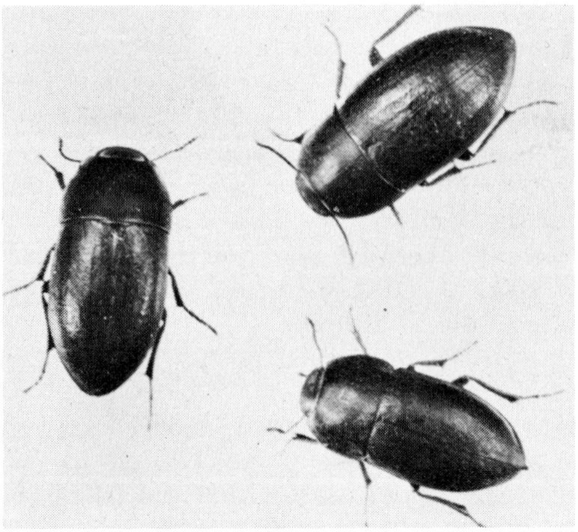

Fig. 365. Darkling ground beetle, *Coniontis subpubescens*.

The yellow to black larvae, about 12 mm long, skeletonize the leaves. The pupae, about 6 mm long, are bright orange-yellow, and have black bristles. The adults may enter dwellings in late summer or fall to hibernate behind curtains, between books, beneath carpets, and in other protected places. They again become active in the spring, and crawl throughout the house. There are 2 to 5 generations per year, depending on climatic conditions. These insects may be killed within the home with the usual household insecticides, but control is difficult because the infestation is being continually renewed. For outdoor control, infested trees can be sprayed with carbaryl 50% wettable powder, using a 2 lb per 100 gal (0.91 kg per 378 l) of water (Koehler *et al.*, 1970; Cress, 1971).

Weevils (Curculionidae)

The weevils comprise a very large family of beetles, characterized by the prolongation of the head into a distinct snout with biting mouthparts at its tip. All weevils are phytophagous, both as larvae and as adults.

Black Vine Weevil, *Brachyrhinus sulcatus* (F.); and Cribrate Weevil, *B. cribricollis* (Gyllenhal)

These weevils belong in a subfamily (Otiorhynchinae) in which the snout is short and wide. They do not fly. The black vine weevil (figure 366-67, *A*) is about 8.5 mm long and brownish black. There are small patches of golden scales scattered about on the elytra. The pronotum is densely covered with tubercles, and the elytra are coarsely punctured and striate. The adults are nocturnal, and feed on many plants. The white, legless larvae live in the soil and feed on plant roots, and are 10 mm long when mature. The cribrate weevil is 6 mm long, and varies from black to brown (figure 366–67, *B*). This weevil was first noticed in Los Angeles County, California, in 1928, and has since spread over much of the state. Weevils of this genus were accidentally introduced into North America, and have subsequently been widely distributed on nursery stock. They are often found on hedges and other dooryard plants, sometimes nearly defoliating them before the pests are noticed. When conditions become unfavorable, the weevils tend to seek shelter by migrating into residences. (Note: Since publication of the 1975 edition, *"Brachyrhimus"* has been changed to *"Otiorhynchus."*)

Egyptian Alfalfa Weevil, *Hypera brunneipennis* (Boheman)

This important alfalfa and clover pest is another imported species. It was discovered near Yuma, Arizona, in 1939, and was confined to the desert areas of Imperial County, California, for many years, but extended its range to San Diego County in 1951 and now occurs over much of the state.

The Egyptian alfalfa weevil (figure 366-67, *C*) resembles the more widely distributed alfalfa weevil, *H. postica* (Gyllenhal), in appearance, life history, and habits. The adult is about 5 mm long, grayish brown to nearly black, with short, grayish hairs. It belongs in a subfamily (Curculioninae) in which the beak is long and slender. In *H. brunneipennis,* the beak projects downward

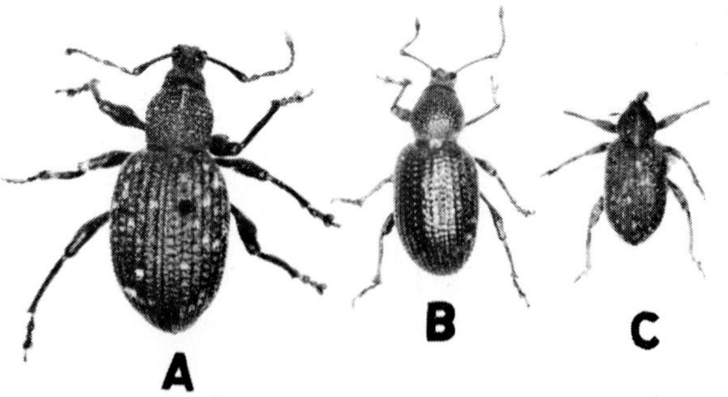

Fig. 366-67. Some crop pests that can invade the home. *A*, black vine weevil, *Brachyrhinus sulcatus; B*, cribrate weevil, *Brachyrhinus cribricollis; C*, Egyptian alfalfa weevil, *Hypera brunneipennis*.

from the front of the head and is about half the length of the thorax. The adults hibernate about the crowns of alfalfa (or clover) plants or under leaves and rubbish, then feed a few days and mate. The females lay as many as 40 oval, yellowish eggs in cavities they make with their beaks in alfalfa stems, and may lay from 400 to 600 eggs each during the spring. The green larvae, with a prominent dorsal stripe, feed on alfalfa and become mature about the time the first crop is cut. They then spin a netlike, nearly spherical, cocoon in the ground and pupate, emerging as adults in about 10 days. The adults also feed on alfalfa plants, and may occur in enormous numbers.

The weevils tend to migrate from alfalfa fields to marginal areas, tree rows, fence rows, buildings, and sheltered locations, and sometimes invade residences. Entomologists of the California Department of Agriculture became aware of over 30 invasions of houses by alfalfa weevils in a single month (November, 1971) from specimens sent to their taxonomists by agricultural commissioners from all areas of the state. Probably many other such invasions were not reported. The beetles cause no spotting problem such as caused by weed bugs, and leave no disagreeable odor as do the weed bugs and tule beetles, but nevertheless they are pests merely because of their unwanted presence in the home.

Strawberry Root Weevils, *Brachyrhinus ovatus* (L.) and *Sciaphilus muricatus* (F.)

In some parts of the country, the strawberry root weevil, *Brachyrhinus ovatus,* can become a pest in the spring and fall. It is 6 or 7 mm long and blackish brown. The larvae feed on the small roots of wild and cultivated strawberries, brambles, and evergreens such as pine and yew. Root weevils sometimes invade a home in enormous numbers. The adults will then be found singly or in groups in such places as bathtubs and cupboards, or even crawling on the ceiling. Another curculionid, *Sciaphilus muricatus,* has the same habits. It is about a fourth longer than *B. ovatus,* and is more slender. The body is tannish gray, the head tending to be slightly lighter.

Destruction of wild strawberries in the vicinity of houses helps to control both these species. Also, thorough spraying of foundations with wettable powders of chlordane or lindane, or applying them as dusts, has proved to be useful in control. The surrounding area within 10 ft (3 m) of the foundation should also be treated. The treatments may need to be repeated in about 3 weeks and possibly again later. If the beetles invade the home, gathering them up with a vacuum cleaner seems to be the best procedure (Cress, 1971).

INSECTS FEEDING IN BEE COMBS IN BUILDINGS

Certain insects are attracted to weakened, abandoned, or fumigated honey bee colonies in inaccessible parts of buildings, feeding on wax, honey, or the waste products in the comb. They can find their way into occupied areas of the buildings and become serious pests. Some of these insects (ants, carpet beetles, and cockroaches) are discussed elsewhere in this book. The greater wax moth and a soldier fly are discussed in this section.

Greater Wax Moth, *Galleria mellonella* (L.) (Pyralidae)

As just outlined, combs of honey bee colonies in wall voids, attics, and chimneys are often infested by the greater wax moth. The moths and

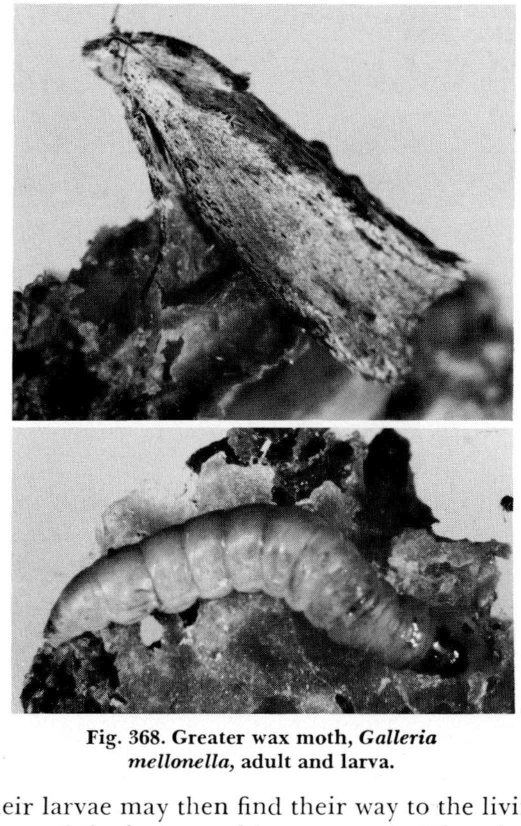

Fig. 368. Greater wax moth, *Galleria mellonella,* adult and larva.

their larvae may then find their way to the living space of the house and become pests. The adults are gray or pale brown, marked with black. "Bumps" on the forewings, which can be seen along the upper margins in figure 368, are composed of tufts of wing scales that are much longer and broader than those covering the rest of the wing areas. The females are about 20 mm long. The males are somewhat smaller, and do not have palpi. The larvae are white when first hatched, but in later instars they may vary from white to yellow, with brown or black heads and prothoracic shields, or their backs may be largely black. The larvae destroy combs by burrowing through the cells. They feed on pollen and waste materials, and construct silken tunnels as they make their way through a comb. If they become pests in the living space of the infested house, the comb should be removed, if possible, or it should at least be sprayed or dusted with any available household insecticide.

A Soldier Fly, *Hermetia illucens* (L.) (Stratiomyidae)

This widespread soldier fly occasionally becomes a house pest after honey bee colonies have been exterminated in wall voids or other hard-to-reach places. The adult (figure 369) is about 15 mm

Fig. 369. A stratiomyid fly, *Hermetia illucens*, adult and larva.

long and black, except for a large, translucent area on the second tergum and white areas on the tibiae. The larvae feed on waste materials in bee combs, and can become very numerous if the wax moth does not reach the combs first. The larvae sometimes find their way into the living space of the home, and the conspicuous pupal cases (shown in the figure) have been found in carpeting. According to James (1960), the adults may be attracted to light, but do not seem to molest food. They are sluggish until induced to fly. Outdoors, they visit excrement, but usually not flowers. James recorded this species as breeding in decaying or putrefying vegetable or other organic matter, including human excrement, but stated that it might at times be a predator in such media. He also had records of its occurrences in decomposing wet rice, soybeans, fruits, vegetables, ketchup, animal cadavers, and in waste materials found in beehives. This species has also been recorded as causing intestinal myiasis in man.

SWIMMING POOL PESTS

Arthropods can be problems in swimming pools, even those that are well cared for, because of their tendency to fall into the water and be trapped there. If the lawn and surrounding areas of shrubbery have an abundance of leaf litter, humus, or other organic material, and are kept damp by watering or rains, infestations of sowbugs, pillbugs, millipedes, or springtails often develop. (Controls for such arthropods are discussed in chapter 10.) Earthworms can also be problems. These and other pests may fall into the pool while crawling about. Many moths, beetles, and leafhoppers may accidentally fall in after being attracted to lights in and around the pool. Bees and wasps are often trapped in pools when they seek water. If swimming pools are not properly cared for, mosquitoes and midges can breed in the water. In areas near lakes, ponds, or streams, aquatic bugs and beetles often fly about in search of other water habitats, and are attracted to swimming pool lights. Among the bugs are the backswimmers (Notonectidae), which can use their sharp beaks to bite with an effect similar to that of a bee sting. Among the water scavenger beetles (Hydrophilidae) is a small, black, shining species, *Tropisternus californicus* (LeConte), the adults of which sometimes bite bathers in swimming pools in southern California (Essig, 1926). Mallis (1973) received some larvae of an unidentified species of *Tropisternus* in Pennsylvania, along with some blue plastic pool-liner material showing tiny holes. The owner of the pool said the holes were made by the larvae, which have long, sharp jaws.

Control of Swimming Pool Pests

Dip nets can be used to skim insects off the surface of the water. Pool skimmers usually do an adequate job of pest removal if they are used frequently. Pools left unused or that are not maintained for extended periods should be drained. Lights near pools should be used as sparingly as possible, particularly on warm nights. Lighting installed 20 to 30 ft (6 to 9 m) from a pool helps to keep attracted insects away from it. Also, yellow lights are less attractive to insects than white ones. Shrubbery and grass should be kept well trimmed, to reduce harborage and breeding areas for insects and other arthropods. Although insecticides should not be applied in the pool, a light malathion spray can be used advantageously on surrounding grass and other vegetation. During pool parties on hot summer nights, a 0.1% pyrethrin aerosol or mist applied around lights may provide temporary relief (Rachesky, 1973).

INSECTS IN MAUSOLEUMS AND CEMETERIES

Pest problems in mausoleums are apparently universal. The author and his co-workers have been asked to suggest means of controlling them on many occasions, and have often collected mausoleum-infesting insects and tested control measures. In any part of a mausoleum, small flies and midges, including mosquitoes, breed in the water or the slime remaining after the water evaporates from the numerous flower vases. A phorid fly, *Megaselia scalaris,* has been the species collected in largest numbers. Phorids are known to feed principally on decayed vegetable matter, such as might be expected where leaves or petals have remained in water for long periods. Insect infestations from this source have noticeably decreased when artificial flowers have been substituted.

In the areas containing the crypts, some kinds of mycetophilids feed on the fungi at imperfect seals on the outsides of the concrete shutters while the crypts are damp inside. (See the detailed discussion of fungus gnats in chapter 10.) Certain common species of beetles may sometimes be seen in the hallways.

The insect species that were collected during the foregoing investigations were as follows:

DIPTERA

Phoridae
- *Megaselia scalaris* (Loew)
- *Dohrniphora cornuta* (Bigot)

Chloropidae
- *Oscinella frit* (L.)

Tethinidae
- *Pelomyia coronata* (Loew)

Mycetophilidae
- *Mycetophila fungorum* (De Geer) (= *punctata* Meigen)

Milichiidae
- *Desmometopa* sp.

COLEOPTERA

Cleridae
- *Necrobia rufipes* (De Geer) (plate IV, 3; figure 189)
- *Necrobia ruficollis* (F.) (figure 189)

Tenebrionidae
- *Alphitobius diaperinus* (Panzer) (figure 182)
- *Alphitobius laevigatus* (F.)

In cemeteries, mosquitoes, particularly such species as *Culex pipiens, C. pipiens quinquefasciatus,* and *Culiseta incidens,* can breed outdoors in large numbers in the flower receptacles when water is left standing in them too long. Cemeteries have occasionally been sources of the yellowfever mosquito, *Aedes aegypti,* in southern states. Such species as *Aedes sierrensis* and *Ae. triseriatus* can breed in treeholes of shade trees if they are not sealed or filled with sand. The sources of these mosquitoes are often difficult to find (Aarons, 1948; Shanafelt, 1969). In the Orange County (California) Mosquito-Abatement District, in cemeteries where all containers are turned upside down when flowers are removed, mosquito production is negligible. In others, an average of 21% of the flower containers have contained mosquito larvae. Containers located in areas shaded from afternoon sunshine were the ones that most often harbored the larvae (Shanafelt, 1969).

Control of Mausoleum Pests

The number of Diptera in mausoleum halls can be reduced by not allowing water to remain in flower vases too long (some midges develop from egg to adult in 10 days), and by making sure that no damp organic matter remains in the bottoms of the vases when they are refilled. Flies and beetles can be controlled by the use of insecticides applied with foggers or mist blowers, as described in chapter 3. Figure 370 shows a vapor generator by means of which lindane vapor was blown from the roof of a mausoleum down through the vent pipes in order to treat the crypts.

Fig. 370. Applying lindane vapor through a vent pipe of a mausoleum with a vapor generator.

Chapter 13 · *Vertebrate Pests: Rodents, Bats, and Birds*

A CHAPTER CONCERNING vertebrate pests in a book titled *Urban Entomology* is technically incongruous, but traditionally a discussion of vertebrate pests has been included in books dealing principally with the biology and control of insects and certain other arthropods. Professional exterminators generally consider the control of both invertebrate and vertebrate pests as a normal part of their activity.

With the possible exception of some of the largest predatory mammals and raptorial birds, which are adversely affected by man's encroachment or by being hunted and killed as real or presumed menaces to his livestock and range animals, wildlife is generally on the increase in the United States. Concomitantly, the boundaries of cities are continuously expanding and encroaching on hilly and wooded areas where wild animals and birds abound. In suburban residential areas within the city limits of Los Angeles, for example, deer browse on ornamental shrubs; rabbits feed on garden plants; raccoons, opossums, rats, squirrels, and many species of birds feed on fruits and nuts; and cedar waxwings, *Bombycilla cedrorum* (Vieillot), strip the berries off *Pyracantha* bushes. Coyotes and foxes are sometimes seen, skunks and snakes invade the crawl spaces under houses, and wood rat (*Neotoma*) lodges in the chaparral are sources of the much-feared conenose bugs. In the present chapter, however, the discussions of vertebrate pests will be confined to rats, mice, bats, and the most commonly incriminated "pest birds" (pigeons, starlings, sparrows, and swallows).

RATS AND MICE

Rats and mice (Rodentia: Muridae) have adapted themselves outstandingly well to an intimate relationship with man in his own environment. In doing so, they have become among the most successful of all wild animals from the standpoint of survival. Only by continuous effort and at great expense has man been able to attain a tolerable coexistence with these rodents, but the loss to the human race in food and commodities destroyed and in fatal diseases transmitted has been staggering.

Rats have always been more important than mice as enemies of man, and will therefore be considered first.

Of the 2 domestic species, the Norway rat is the more important one in the temperate parts of the world or in areas that are predominantly temperate, such as the United States. The roof rat is more important in tropical and subtropical regions. In these areas, as well as in some that may be considered to be temperate, the roof rat tends to become relatively more prominent with increasing urbanization. In California, the Norway rat lives throughout the lowlands, both on the coast and in the interior, in both urban and rural environments. The roof rat occurs in approximately the same localities as the Norway rat, but thrives better in urban areas and is found at higher elevations, up to about 5,000 ft (1,500 m). It is a much better climber in buildings, trees, or vines than the Norway rat.

Norway Rat, *Rattus norvegicus* (Berkenhout)

Description. Of the 2 species of rats in the United States, this one is the larger. It is also known as the brown, house, wharf, or sewer rat (figure 371). The combined head and body length is 7.5 to 10 in. (19 to 25 cm), and the tail is 6 to 8.5 in. (15 to 22 cm) long. The adults weigh 0.75 to 1.25 lb (0.34 to 0.57 kg). The body fur is coarse and brown, with scattered black hairs, and the underside is gray to yellowish white. The muzzle is blunt, the tail is seminaked and bicolored (darker above and lighter below), and the ears are small and closely set. Droppings of the Norway rat are up to 0.75 in. (20 mm) long, and capsule-shaped (Storer, 1968).

Biology. Norway rats normally live 9 to 12 months (much longer in captivity), become sexually mature in 3 to 5 months, and may have 4 to 7 litters, each with 8 to 12 young, indicating their high reproductive potential. They may live seasonally in areas removed from man. Their harborages and nests may be in burrows, with tunnels 2 or 3 in. (5 or 7.5 cm) in diameter, that lead under buildings, piles of lumber, or stones, or sometimes into stream banks; or they may build their nests under buildings, in garbage or rubbish dumps, or in marshes. In buildings, the nests are made of bits of rags, burlap, paper, straw, or similar materials. They are crudely and loosely constructed in a form somewhat resembling a bird nest, and are located in hidden areas, so as to afford harborage and shelter for the adults and protection for their young. A burrow commonly has secondary "bolt holes," hidden under grass, boards, or other coverings, that the rats can use for escape. Norway rats often range 100 to 150 ft (30 to 45 m) from their harborages. However, they can nest and spend their entire lives inside a building and, if they live outdoors, they may enter buildings for food at night.

Roof Rat, *Rattus rattus* L.

Description. This species (figure 372) is smaller than the Norway rat. There are 3 color variants in the United States. These do not appear to be valid "subspecies," as they are sometimes called, for they can interbreed (Caslick, 1956; Marshall, 1962). The color variants and the subspecific names by which they have been known are the roof, gray-bellied, or Alexandrine rat, *Rattus rattus alexandrinus* (Geoffroy St. Hilaire); the black or ship rat, *R. r. rattus* (L.); and the white-bellied rat, *R. r. frugivorus* (Rafinesque). The white-bellied rat is not widely distributed in California, but has been found in Bakersfield. The roof rat is grayish white to yellowish white below. The black rat is almost solid black, with some gray below, and in California it was formerly found mainly near salt water, but has since been seen in inland towns such as Bakersfield, as well as in Downieville and Sonora in the Sierra foothills (J. E. Brooks, correspondence). For convenience, all color variants of *Rattus rattus* will be referred to as "roof rats." The combined head and body length is 6 to 8.5 in. (15 to 22 cm), and the tail is 7 to 10 in. (18 to 25 cm) long. Thus, the tail is generally longer than the head and body, whereas the tail of the Norway rat is generally shorter than the head and body. The adult roof rat weighs

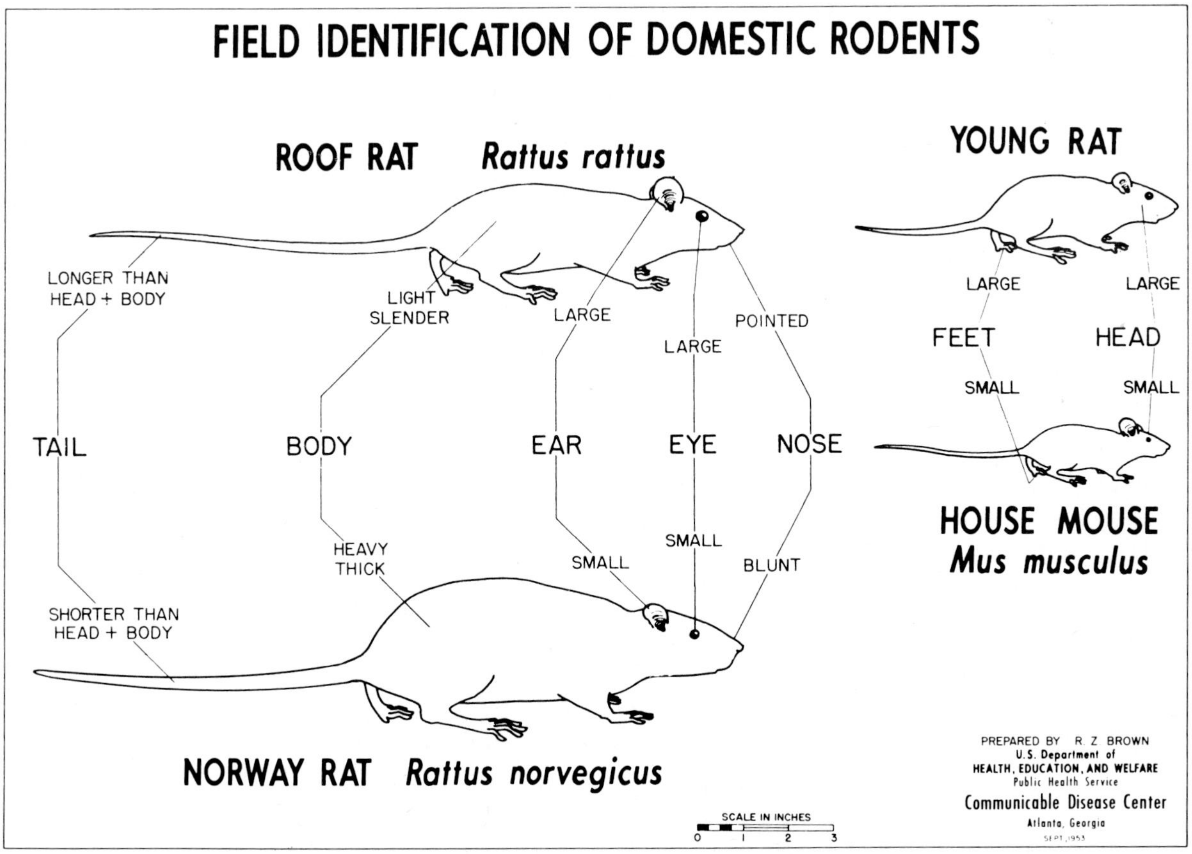

Fig. 371. Comparison of sizes and anatomical features of 3 domestic rodents. (The inch scale converts to 2.5, 5, and 7.5 cm.) (Courtesy USPHS, Communicable Disease Center.)

Fig. 372. Roof rat, *Rattus rattus.*

0.5 to 0.63 lb (0.23 to 0.28 kg). The muzzle is pointed, the tail is seminaked and uniform in color, and the ears are large and prominent. The droppings are up to 0.5 in. (12 mm) long, and spindleshaped.

Biology. Roof rats live 9 to 12 months, become sexually mature in 3 to 5 months, and may have 6 litters per year with 6 to 8 young per litter. Like the Norway rat, the roof rat is omnivorous, and has similar food and water requirements. On the average, the roof rat will consume a higher proportion of vegetable matter (fruits, nuts, vegetables, and grains) and less meat and garbage than the Norway rat. It can live in similar locations, but the roof rat is a better climber, and is more likely to build its nests in walls, attics, vines, or trees (seldom in burrows). In tropical and subtropical areas, palm trees are important harborages, particularly when the old fronds are not removed annually. The roof rat is able to live completely away from man more successfully than the Norway rat, but also benefits more from intense urbanization. The Norway rat is in its ideal environment in rural areas where its tendency to burrow is advantageous. On the other hand, the climbing ability and greater range of roof rats suit them well for city life. They can move along telephone or utility wires, can climb from trees to rooftops, and can climb vertical pipes up to 10 cm in diameter. Both are excellent swimmers, and can swim long distances under water.

Physical Characters Separating Norway Rats and Roof Rats

To summarize the physical differences between the 2 rat species as discussed to this point: The mature Norway rat is about 25% longer in head and body length than the roof rat, but weighs about twice as much. Its tail is shorter than its head and body, whereas the roof rat's tail is longer than its head and body. The muzzle of the Norway rat is blunt, and that of the roof rat is almost pointed. The Norway rat has small ears, densely covered with short hairs, whereas the roof rat has large and almost naked ears (figure 371). The Norway rat usually has 12 mammary glands, and the roof rat usually has 10.

Food and Water

Rats are omnivorous, and will eat any human food. They will feed on fruits, vegetables, nuts, poultry, eggs, or ground-nesting birds and their eggs, to mention some examples. Rats do not need other sources of moisture if fruits, vegetables, and other foods with a high water content are available. They can obtain water from such sources as heavy condensation, leaking faucets, toilet bowls, and sewer pipes. Rats can chew through plastic water pipes and galvanized drains in the plumbing to obtain water.

The Senses of Rats

Rats and mice have poor vision, but keen senses of smell, taste, hearing and touch. Their tactile sensory receptors consist of nerve endings located at the bases of the whiskers (vibrissae) and the long guard hairs. They tend to move along vertical surfaces where the tactile hairs can guide them. This characteristic aids the exterminator in locating rat runs. The rat's auditory range encompasses man's, and extends into the ultrasonic range. Laboratory rats have been found to be able to emit ultrasonic sounds (Anderson, 1954), and blinded rats used echolocation in learning to run a maze (Riley and Rosenzweig, 1957), thus demonstrating an ability that also plays such an important role in the orientation of bats.

Rats as Pests in Buildings

In food storehouses, rats eat, destroy, or contaminate with excrement, urine, and hairs an enormous amount of grain, flour, cereals, sugar, vegetables, fruits, nuts, meats, animal feed, pet foods, and many other types of stored food products. The same kind of damage can occur in the home pantry. In addition, rats can gnaw into upholstery, textiles, paper, and books to provide materials for their nests. Rodent incisors grow continuously, and rats must gnaw hard substances to keep them at a proper length. They sometimes cause damage by gnawing through lead and plastic pipes and garbage cans. By gnawing through electric wiring or a gas pipe, they can cause a fire or asphyxiation. Fires can also result if they carry matches to their nests and gnaw off the flammable ends, as they have been known to do. Their nests are usually highly flammable (Hamilton, 1947*b*).

Human habitations most susceptible to infestation are dilapidated houses or apartment buildings with many points of entry for rats and easy access between wall voids and the living space. Infestations often result from failure to ratproof the outsides of buildings where the necessity is obvious, and from inadequate internal maintenance. In many communities, building codes specify the manner in which ratproofing of new and old structures should be done, but concerted efforts among owners, contractors, and workmen are necessary for such codes to be effective. City ordinances and state laws (e.g., California Health and Safety Code, 1965, Sections 1800 to 1813) make the owner of any premises responsible for controlling rats and mice; failure to do so is a misdemeanor that may result in court action.

Inadequate provisions for disposal of garbage, or in other ways allowing scraps of food to be accessible to rats, also favor rat infestation. The same may be said with regard to mouse and cockroach infestations. Good tenants are penalized by those with careless and substandard housekeeping habits.

Hallways, attics, wall voids, spaces between ceilings and upper floors, trash chutes, raceways, and dropped ceilings may contribute to infestations of rats, mice, cockroaches, or other household pests moving freely between apartments.

One source of economic loss from rats is the decrease in rental value of infested houses and apartments. The tenant may eventually become accustomed to the sound of rats running about in attics and wall voids, but not to their bites. Out of 65 people treated for rat bites in one hospital, 7 developed ratbite fever. As will be explained later, this is only one of many diseases inflicted or spread by rats. Most people are bitten while asleep, and the rat runs away when the victim is awakened by the bite. However, once a rat bites a person, it is more likely to bite others. Most of the victims are infants.

The Problems of New Residential Areas

The foregoing discussion is not meant to imply that rats cannot become established in neighborhoods where modern construction, good housekeeping, and good sanitation predominate. Experience has shown that rats, particularly the roof rat, can find much food and harborage in the usual dooryard trees and vines in such neighborhoods. Fruit and nut trees are not the only sources of food from vegetation, for the roof rat has adapted itself to many plants and animals common in residential areas. The following were offered to an adult male rat, and were readily accepted and eaten: "snails and earwigs; common weeds, such as mallow, wild radish, wild mustard, dandelion, and clover; the fruits of some ornamental shrubs, such as *Raphiolepis,* and flowers of pomegranate and nasturtium (*Tropaeolum*); the buds from *Clematis* vines, the berries from mature English Ivy, *Pyracantha* berries, dog and cat food, cauliflower leaves, and some succulents, such as *Echeveria* and *Sedum*" (Jamieson, 1965).

During recent years, the roof rat has been gradually displacing the Norway rat in California as the principal rat problem. In large cities, it characteristically inhabits hotels and apartment houses (Storer and Davis, 1953). The usual rat control procedures tend to be more effective against Norway rats than against roof rats and mice, and this appears to have resulted in an increase in the proportion of the latter 2 species in some communities (Dykstra, 1950).

Jamieson (1965) found, in a section of San Jose, California, that rat-infested premises invariably had 1 or more of the following in common: "A woodpile, an outbuilding, a thickly growing plant accumulation (juniper, ivy or other vines, etc.), or a walnut tree." Unharvested or improperly stored walnuts were a common year-round food supply for rats. He noted that rats gained entrance to homes "through loosely covered crawlhole openings, broken foundation ventilation screens, openings inside garages, and openings in roofline areas of houses with medium or heavy shake roofs. Rats gained access to roofs by using power or telephone lines or by climbing adjacent trees, high shrubs, or vines." There had been a great increase in complaints about rat infestation during the 5-year period preceding the survey, despite the fact that garbage service in the area was satisfactory and garbage cans were adequately maintained. Garbage was not found to be a food source for roof rats.

According to Brooks (1966), roof rats are common not only in older residential areas, but also in tracts with structures ranging in age from 10 to 18 years. These tracts are often laid out in former orchards, and many fruit or nut trees are allowed to remain. Also, the tract sites are typically landscaped with numerous shrubs and trees, pools and patios, all favoring rat infestation. In some neighborhoods, 25 to 50% of the homes may be frequented by roof rats.

Rats in Sewer Systems

In the United States, sewers that are infested with rats are most commonly infested with Norway rats. These usually live along waterways and in underground burrows, while the roof rats prefer to inhabit trees, shrubs, attics, and similar habitats. However, roof rats were, except in 1 location, the only species found in sewers in San Diego, and were also the only species found in sewers in Ontario, California (Mackie, 1964; Rohe, 1966).

Roof rats are abundant in the sewer systems of Iraq and Iran. In some cities, the rats from the sewers began invading the houses to such an extent that the occupants expressed regret that their houses had been connected with the municipal sewer system (Wichmand, 1969). The keyhole-shaped toilets in the floors, the usual facilities in that part of the world, have no water traps, and provide easy access to or from the sewer system for rats in homes and other buildings.

In city sewer systems, as in buildings, roof rats are found most abundantly in parts of a city where extensive landscaping and abundant fruiting shrubs and trees provide shelter and food. It is suspected that rats enter a sewer system primarily through sewer vents on the roofs of houses, although they have been found to enter through broken laterals near the ground, through unconnected laterals of new construction, and through toilets in vacated premises where there is no water in the traps. Once the rat "runs" become well established, as through sewer vent pipes, new rats entering the area can find them and follow them into the sewers, guided by their keen sense of smell. If rats breed and nest in a sewer system and cease returning to the surface, their runs disappear; other rats are not led to the sewers, and reinfestations are less likely to occur. Where cockroach remnants (primarily wings) but no live insects were found in the manholes of a sewer system, baiting of the manholes invariably demonstrated the presence of rats. On the other hand, rat activity was never discerned where live cockroaches or abundant spider webs (indicating the presence of insects) were observed (Mackie, 1964).

Rats as Pests on Farms

The same type of damage attributed to rats in urban areas occurs, of course, on the farm. In addition, there are opportunities for depredations against crops and livestock. Rats will eat or destroy any type of vegetable in the garden, and can climb trees to reach fruits. In a single night, they can kill large numbers of poultry or destroy their eggs. They can also kill the newborn young of other farm animals, such as pigs and lambs, and can also destroy much wildlife, particularly ground-nesting birds (Hamilton, 1947*b*). In southern California and Florida, avocado orchards are often contiguous to city suburbs, and avocado trees are frequently used as ornamentals within the city limits. Rats can cause great damage to avocado fruit. This is particularly evident when orchards are not cultivated, as is usually the case. The ground is then deeply covered with leaves, and older orchards often become veritable forests, with a continuous canopy of interlacing branches. Broken branches, prunings, and piles of props and boxes left lying on the ground often serve as nesting places for rats. The advent of the anticoagulant rat poisons was a boon to avocado growers (Ebeling, 1959).

Rats in Disease Transmission

Plague. The great outbreaks of bubonic plague in Europe, resulting in the deaths of millions of people, particularly in the fourteenth century, were caused by a highly fatal disease of the circulatory and respiratory systems, the causal organism for which was the bacterium *Yersinia pestis* (= *Pasteurella*). Rats, particularly *Rattus rattus,* are susceptible to the disease, and so are mice, squirrels, horses, dogs, and man. Plague is generally transmitted to man through the bites of fleas from infected rats. When the incidence of plague is high in field rodents, rats also have a greater opportunity to become infected and, because they are domestic rodents, the chance of human infection is increased. Sylvatic plague is a much less virulent form of plague than bubonic, and is the form with which we are concerned in the United States. Sylvatic plague is now endemic in field rodents in the western United States, and is more likely to be transmitted to man by handling field rodents, dead or dying from the disease, than by flea bites. Human mortality from plague has been all but eliminated by the use of antibiotics and sulfa drugs.

Salmonellosis. Food poisoning by bacteria of the genus *Salmonella* can result from consumption of food contaminated with rat excreta. A large number of people can become infected from the same contaminated food source. It is now considered to be the disease most commonly transmitted by rats and other rodents (Scott, 1959). Amoebic dysentery is another disease that can be contracted by eating food contaminated with rat excreta.

Murine Typhus. This is a disease of rodents, especially rats, caused by *Rickettsia mooseri.* The disease is transmitted among rodents and to man when the feces of infected fleas are scratched into the skin. Human cases of murine typhus occur occasionally when the domestic rat population is maintaining the disease at a moderately high level.

Leptospirosis (Weil's Disease). This is an acute systemic spirochetal disease of rats caused by *Leptospira icterohaemorrhagicae,* which is transmitted among rats and to man by contact with infected rat urine. Domestic animals, such as dogs and horses, are also susceptible. The bacteria seem to enter the skin through small scratches, and are most likely to be contacted by humans handling rats or working in wet environments, such as soldiers, sewer workers, miners, exterminators, fish handlers, workers in scientific laboratories where rats are experimental animals, or children playing in contaminated water.

Rat Bite. As previously stated, rat bite is common in areas where rat populations are high. Vastly increased urbanization has resulted in a comparable increase in the rat-bite problem. The incidence of rat bite in the United States is conservatively estimated to be about 14,000 annually. Among the dangers from rat bite are secondary infection and tetanus. The danger of infection can be minimized by proper disinfection and treatment of the bite. Rabies is not a likely aftermath. Ratbite fever, also known as haverhill fever, is a bacterial disease of rats caused by *Streptobacillus moniliformis.* It is transmitted from rat to rat and to man through rat bites. Without treatment, the disease can produce up to about 10% mortality. Fatality from ratbite fever occurred in 11% of the victims of 87 rat bites in Baltimore (Richter, 1945). The disease develops about 3 to 10 days after the rat bite. It involves "a primary lesion, swelling of regional lymph nodes, sharp fever peaks (often alternating with afebrile periods), rash, and arthritic symptoms" (Scott, 1966*a*).

Listeriosis. This is a bacterial disease of rodents, ruminants, fowls, and man caused by *Listeria monocytogenes.* It is believed that food and air contaminated with powdered rodent feces can transmit the disease from infected rodents to man. However, the epidemiology of the disease is still relatively unknown.

Trichinosis. A microscopic young trichina worm lives in muscles of rats, swine, man, or sometimes of cats or black bears. When a susceptible animal eats the flesh of an infested one, the young trichinae mature, breed, and produce many more worms in the new host. They produce poisonous substances that can lead to illness and death. If pigs eat infested rats and a human eats their imperfectly cooked flesh, he may receive many of the worms (Storer, 1968).

Tropical Rat Mite. Rats may carry the tropical rat mite, *Ornithonyssus bacoti.* It is the most common ectoparasite of roof rats in southern California and the San Joaquin Valley, and is also common throughout the southern states. Rat mite dermatitis and its prevention are discussed in chapter 9.

Rat Detection

Rats may first be detected by the sounds they make when running about in the attic or crawling around inside the walls. If they are so abundant that they can be heard, their fecal droppings can usually be observed in the attic, often by proceeding no farther than the ceiling crawl hole. In fact, by then many housewives will have already noticed the droppings on pantry shelves or in food packages. Droppings that are old and desiccated usually indicate an earlier infestation that has been eradicated. A few soft pellets, possibly wet and glistening, indicate the presence of live rats. If droppings are not seen, a thin layer of flour or talc sprinkled in infested areas may reveal new droppings or tracks after the rats have run over the dust. Dri-die 67, blown into attics for drywood termite prevention, covers old droppings and tracks and reveals new signs of rat infestation. It also kills ectoparasites that leave the bodies of dead rats. Rat tracks are easily identified by the fact that the front paws are 4-toed and the rear paws are 5-toed. Telltale dirty and greasy marks (figure 373) made by their bodies may be evident

Fig. 373. Greasy marks (*arrows*) indicating rat runways. (Courtesy USPHS, Communicable Disease Center.)

along rat runways, such as beams and pipes. These runways may lead to holes that can be closed or to suitable places for placing poison baits or traps.

The presence and condition of gnawed materials, such as foods and wood, may indicate not only the presence of rats, but also how long the rats were there. Rat nests, usually made up of rags and bits of paper, indicate that rats are (or were) present, and that a colony has become established. Fresh droppings near the nest show that it is in use. Finally, the sighting of a live rat not only proves that rats are present, but ordinarily would indicate that they are present in large numbers.

Portable, battery-powered black (ultraviolet) lights are available that are particularly useful for inspecting goods coming into a food-processing plant or warehouse. Black light causes some materials, such as rodent urine, to fluoresce. The black-light device should be held within 6 in. (15 cm) of a surface in average room lighting or at a greater distance if the area is darker. Under black light, most rodent hairs fluoresce blue-white and their urine fluoresces blue- to yellow-white, the fresher stains having the bluer hue. Rodents urinate while moving, so the stains occur in typical droplet patterns or even in a continuous trail where rodent traffic is heavy. Buildings that are likely to be rodent-infested have certain contaminants with fluorescence that may be confused with that of rat stains. A secondary test with Urease Brom Thymol test paper can be quickly made. Contaminated goods should be refused, because even though no live rodents are seen, a health inspector detecting stains may cite a food-processing plant or warehouse for possession of rodent-contaminated goods (NPCA, 1971).

Sanitation

It is reasonable to suppose that decreasing the food available to rats would increase competition among them and serve as a check on increase in numbers. Garbage not disposed of in electric garbage disposals should be drained, wrapped, and stored in rustproof containers with tight lids. Governmental publications generally recommend that garbage be picked up twice a week, but in many cities it is picked up only once a week, and that practice appears to be satisfactory. The refuse should be buried in sanitary landfills or burned in efficient incinerators.

Many open dump sites are inhabited by domestic rats and mice. When the hauling of refuse to the dump site is discontinued, these rodents search for food in near-by areas. Dump closing should include a rodent-poisoning program (Walsh and Lusk, 1971).

Stack all foodstuffs 12 to 18 in. (30 to 45 cm) off the floor, and remove all debris and clutter. Foods may be stored in tin cans or in glass jars with screw-on lids. Keep both spilled human and animal foods cleaned up. Keep weeds cut, and dispose of piles of trash and junk. Rats are less likely to use lumber and and other materials for harborage if they are neatly piled well off the ground. Yard or general environmental sanitation is as important as house sanitation. Fruits and nuts should not be left lying on the ground. Nuts, in particular, provide a preferred and persistent form of sustenance for rats.

Reducing the number of rats in an area and controlling their usual ectoparasites should precede environmental sanitation, otherwise displaced rats and their ectoparasites will increase rather than decrease the public health hazard, for the rodents will be driven to other areas in the neighborhood (Bjornson and Wright, 1960). By applying insecticidal dusts containing malathion or carbaryl along runways, in burrows, or across openings to bait boxes, rats can be made to dust themselves and thereby kill the fleas and mites that infest them.

Rats that persistently infest homes and other buildings must have water as well as food. Water may be obtained from sinks, toilets, under refrigerators, and leaks in plumbing, or from areas of heavy condensation. The Norway rat has a greater need for water than the roof rat, and this may account for its preference for living along the banks of rivers and streams or in drains and sewers. However, the roof rat needs no other water supply if fruits, vegetables and other moist foods are available (Williams, 1948). Norway rats are excellent swimmers, and were observed to compete with fish for horse meat thrown into the water in a fish hatchery. They also ate fingerling fish. The rats were observed swimming at temperatures considerably below freezing (Cottam, 1948).

Ratproofing

In ratproofing, structural details of a building are changed to prevent rats from gaining entry. If only Norway rats are involved, the ratproofing of all ground-floor openings will suffice, but if roof rats are present, upper floors must also be ratproofed. Materials used for excluding rats should be those through which they cannot gnaw (cast iron and tile). Emphasis should be placed on the most likely points of entry, for closing every possible entrance may not be economically feasible. When constructing a building, it is wise to consider ratproofing and fireproofing at the same time. When ratproofing, it should be borne in mind that young rats can crawl through openings as small as 0.5 in. (12 mm), and that rats can climb vines, shrubs, or trees adjacent to buildings or climb along wires to gain access to upper stories. They can climb vertical pipes up to 4 in. (10 cm) in diameter, can jump 36 in. (90 cm) vertically or 48 in. (120 cm) horizontally, and can drop 50 ft (15 m) without being killed. They can burrow to a depth of 48 in. in soil, and can emerge from floor drains, water traps, or toilets (Brooks, 1969). They can gnaw through soft metals such as lead and aluminum and, of course, through wood. Ratproofing will not necessarily prevent the animals from making their own entry holes.

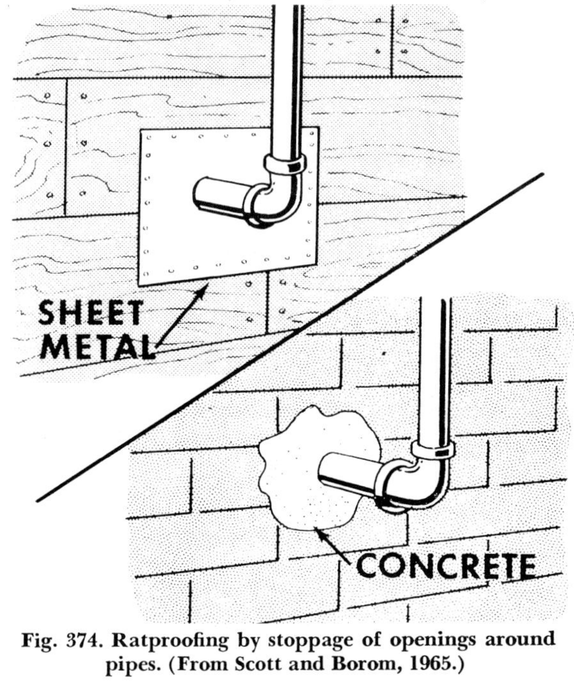

Fig. 374. Ratproofing by stoppage of openings around pipes. (From Scott and Borom, 1965.)

The time and expense required for ratproofing are usually more justifiable in buildings used for food storage or food handling than in those that may principally provide harborages for rats. Likewise, in a building in which the food-storage or food-handling portion can be effectively isolated from the remainder, the expense of doing so should be balanced against that of interior proofing or eradication. Open buildings, such as barns or warehouses, in which food is available to rats, may not be amenable to ratproofing. However, even in such buildings, storerooms or other enclosed areas can be ratproofed. The infestation can be further prevented or limited by using permanent feeding stations containing anticoagulant bait and by environmental sanitation.

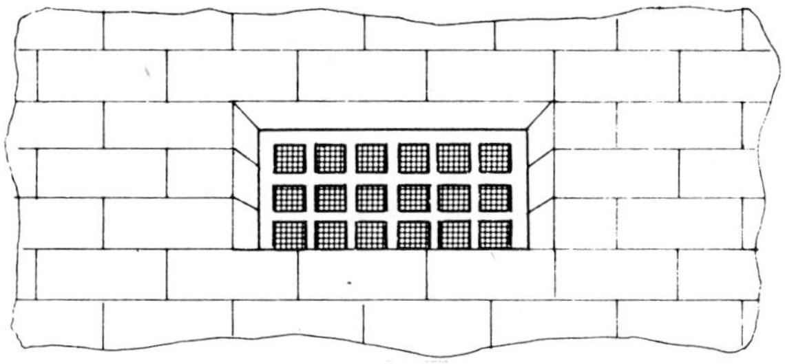

Fig 375. Ratproofing with galvanized 17-gauge hardware cloth placed against a metal grill. (From Scott and Borom, 1965.)

Rat exclusion begins at the foundation wall, particularly when Norway rats are involved. If the lower floor is constructed of material impervious to rats, the foundation walls need treatment only for harborage elimination, repair of cracks or breaks, and the closing of openings around wires and pipes with concrete or metal collars (figure 374). Vents (figure 375) should be checked, and should consist of screens of galvanized 17-gauge hardware cloth, if it can be placed against a metal grill. Otherwise, 18-gauge expanded metal is used, cemented neatly in place (Scott and Borom, 1965).

If the lower floor can be penetrated by rats, the installation of a "curtain wall" (figure 376) is desirable. However, this is an expensive undertaking, and would ordinarily be considered only for the protection of food-handling establishments where rats make particularly intensive efforts to enter. A concrete floor is even more effective. Basement floors should be built of concrete or masonry. Wooden floors offer no protection against rats. Openings in basement walls around pipes or similar installations should be sealed, as explained for exterior walls. Basement windows should be protected by 17-gauge hardware cloth

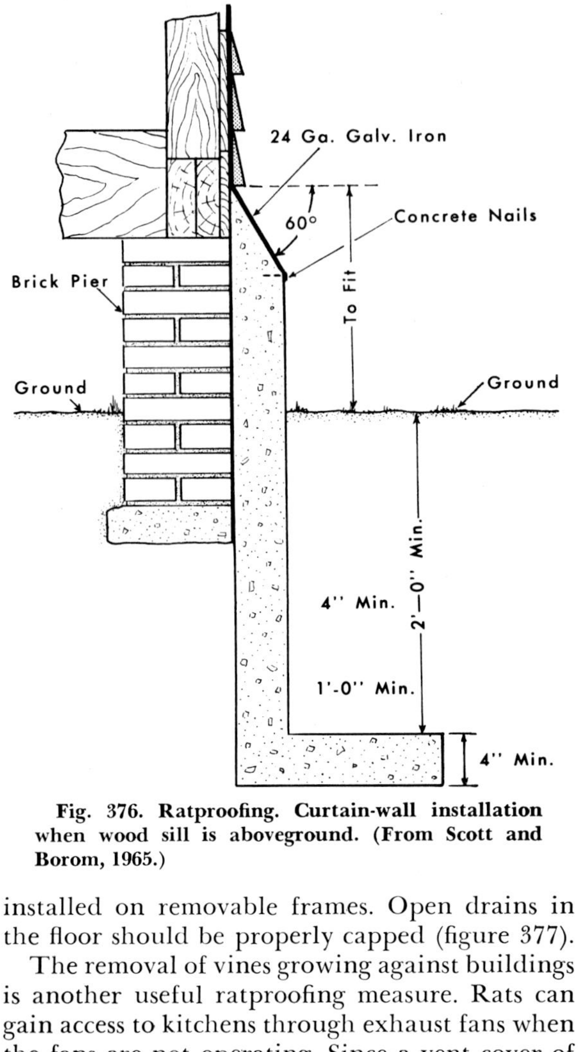

Fig. 376. Ratproofing. Curtain-wall installation when wood sill is aboveground. (From Scott and Borom, 1965.)

installed on removable frames. Open drains in the floor should be properly capped (figure 377).

The removal of vines growing against buildings is another useful ratproofing measure. Rats can gain access to kitchens through exhaust fans when the fans are not operating. Since a vent cover of hardware cloth to protect the fan can become occluded with grease and dirt and thereby interfere with its operation, the opening can be protected by strips of metal about 1.5 in. (4 cm) wide, each of which is hinged and overlaps the one below. The strips open when the fan operates and fall back into place when it stops (Scott and Borom, 1965).

Rat runs have been traced from the ground to as high as the top of an 11-story building. If roofs are accessible to rats, ventilators that project above the surface should be screened with hard-

Fig. 377. Ratproofing. Properly capped drain. (From Scott and Borom, 1965.)

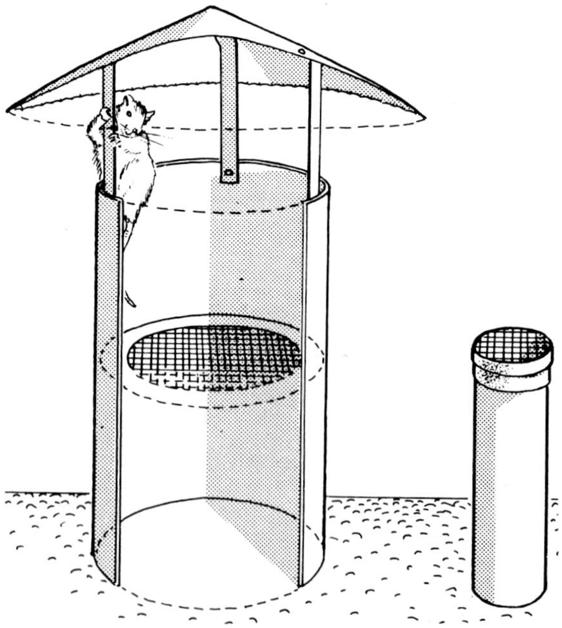

Fig. 378. Ratproofing by stoppage of vent pipes on roof with hardware cloth. (From Scott and Borom, 1965.)

ware cloth (figure 378). Circular guards, such as those used to keep rats from climbing ships' hawsers, can be installed on wires leading to buildings. Wires and pipes against building walls can be made ratproof, as shown in figure 379. Cone-shaped circular guards or barrel guards keep rats from climbing vertical pipes (figure 380), and perpendicular rat guards provide smooth surfaces that they cannot climb or jump over. The guards should be 18 in. (45 cm) wide and at least 36 in. (90 cm) above any surface which rats can use for a jumping-off place.

Periodic inspection and maintenance are desirable after a building has been ratproofed, for rats can dig new tunnels or gnaw new openings. Breaks and cracks can appear in a building after a repair job. Even if ratproofing is complete, the rodents can enter if doors or windows are left open, or they may be brought in with merchandise. Also, after ratproofing, any rodents left

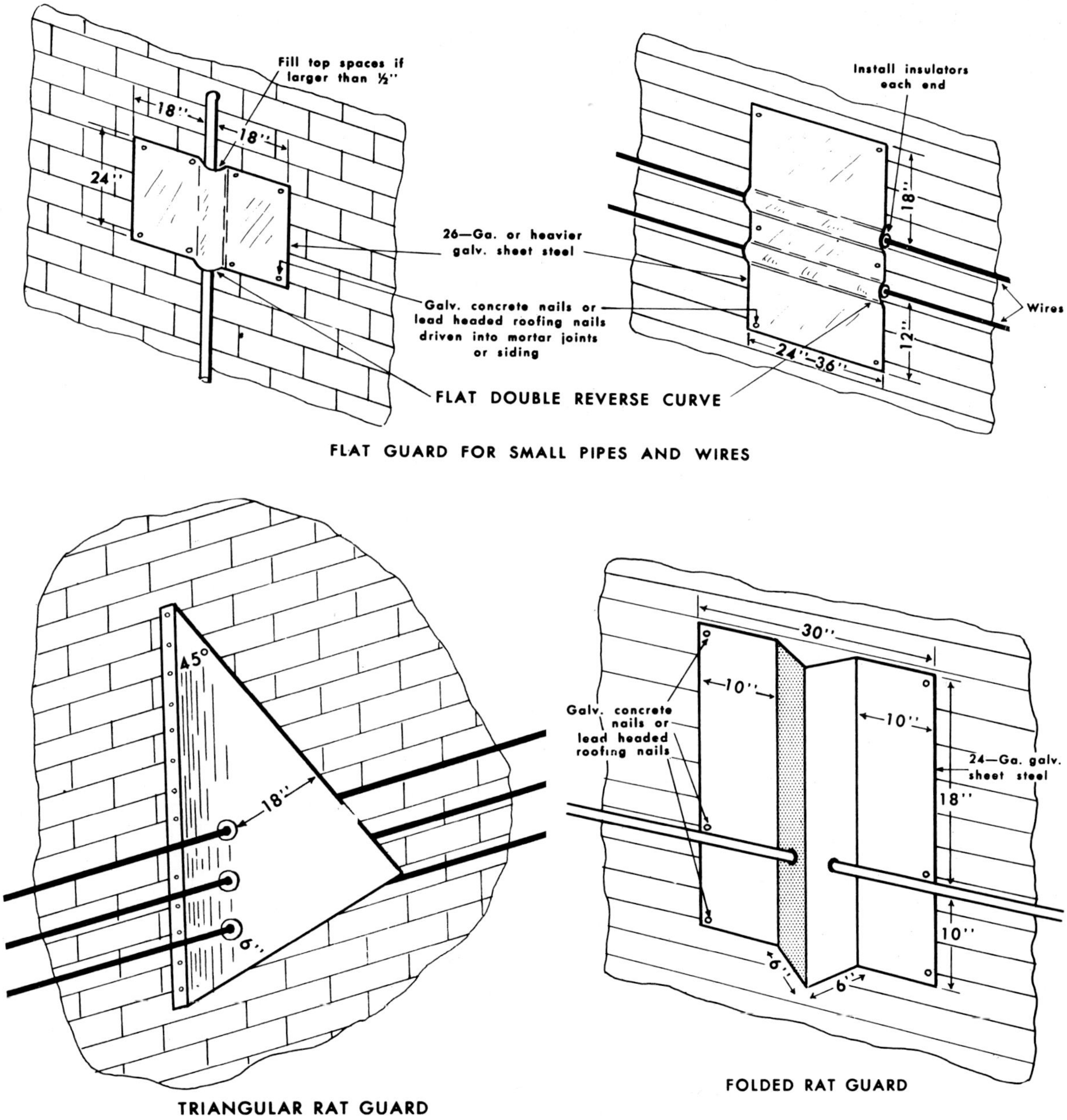

Fig. 379. Rat guards for small pipes and wires. (From Scott and Borom, 1965.)

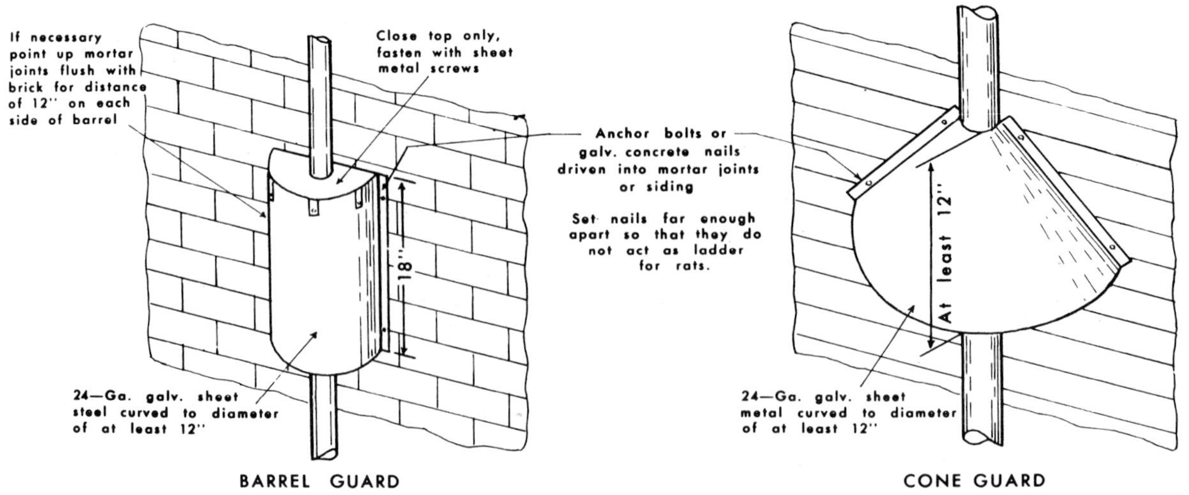

Fig. 380. Ratproofing with curved guards for pipes. (From Scott and Borom, 1965.)

within the building should be exterminated, or they may gnaw their way out and create new "leaks." The elimination of the animals from areas around a building is a further aid to good results.

Poisoning: Materials and Methods

In conjunction with environmental sanitation and ratproofing, suppression by such measures as poisoning with baits, the use of tracking dusts, fumigation, trapping, or a combination of these methods, should be undertaken. Otherwise, the rats with their ectoparasites are merely caused to infest surrounding properties and to become even worse public health hazards.

Rat poisons include "slow-acting," "chronic," or "multiple-dose" compounds, such as warfarin and other anticoagulants; and "fast-acting," "acute," "direct," or "single-dose" compounds, such as norbormide, ANTU, red squill, sodium fluoroacetate (1080), fluoroacetamide (1081), zinc phosphide, and strychnine. These poisons vary greatly in their toxicity and hazard to man and his pets and other domestic animals. The relative hazards and safe procedures for their use should be well known by anyone attempting to use them. Some of the poisons must be used only by experienced pest control operators under proper circumstances. The most common and effective way to use a rat poison is in the form of a food or liquid bait. Liquid baits are most effective in hot, dry areas and where water is scarce. In food baits, the poison is mixed with grain, fish, meat, fruit, or vegetables. If the food alone is offered for a few nights before the poison is added, the effectiveness of the poison is increased, particularly in the case of acute poisons such as zinc phosphide. This procedure is known as "prebaiting."

Rats shun new objects and new foods, suffering from what is known as *neophobia* (dread of or aversion to novelty). Studies of a confined rat colony showed that a period of 36 hours was required for the least suspicious of 8 rats to try a new food container, even though it contained an old, familiar food. In the meantime, the rats fed on a small cache of hoarded corn kernels in their nest box. The investigators believed that if there had been an adequate supply of other food available, it would have taken 7 to 14 days before all rats, including the most suspicious, began to feed freely from a new bait station. On the other hand, mice began feeding from a new bait station in as little as 8 minutes. Mice are not affected by neophobia; they tend to investigate new objects (Knote, 1973).

When treating for rats, strange, loud noises, such as loud voices and clanging and banging of equipment, should be avoided, so as not to increase the rats' suspicions. In the controlled experiment just mentioned, after hearing a loud noise it was 4 hours before the first rat left its nest box and 8.5 hours before all 8 rats exposed themselves.

Anticoagulants. These poisons cause blood to lose its clotting ability, causing the rat to bleed to death from internal hemorrhages. Rats will eat the baits without becoming "poison-shy." They are the safest poisons for general use, and are particularly recommended for use by the nonprofessional. To be effective, most anticoagulant baits must be eaten by the rats over a period of several days, since the concentration of poison is usually so low that it is not detected by them. An anticoagulant is effective only if the rodent eats a little every day for several days. Warfarin may not have a lethal effect for 7 to 14 days. One ordinary dose will kill neither rodent nor man. A human or pet ingesting multiple doses can be treated effectively with vitamin K and, if indicated, whole-blood transfusion. The anticoagulants are therefore considered to be remarkably safe (Brooks and Scott, 1961). A dog that has accidentally consumed a toxic quantity of anticoagulant bait will bleed at the nose or mouth or have a bloody stool. When such symptoms appear, there is still ample time for the administration of the antidote of whole blood from another dog or of vitamin K (Uhlarik, 1970).

The anticoagulants currently available are: (1) warfarin, 3-(alpha-acetonylbenzyl)-4-hydroxycoumarin; (2) coumachlor, 3-(alpha-acetonyl-4-chlorobenzyl)-4-hydroxycoumarin; (3) fumarin, 3-(alpha-acetonylfurfuryl-4-hydroxycoumarin; (4) pindone (Pival), 2-pivalyl-1,3 indandione; (5) PMP, 2-isovaleryl-1,3-indandione; (6) diphacinone, 2-diphenyl-acetyl-1,3-indandione; and (7) chlorophacinone, 2-[(*p*-chlorphenyl) phenylacetyl]-1,3-indandione. Warfarin, coumachlor, pindone, PMP, and diphacinone were found to be more effective against Norway rats than roof rats (Hayes and Gaines, 1959). In one investigation, warfarin was 3 to 4 times more effective against Norway rats than against "black" rats. House mice were even more resistant to warfarin than black rats (Dykstra, 1950). In a special mixture of warfarin called Prolin®, an antibacterial compound is added that is intended to lessen production of vitamin K in the animal's intestine (Storer, 1968).

Unlike other anticoagulants, chlorophacinone is soluble in oil but practically insoluble in water. It often kills rodents in one feeding, although they do not die for several days. A bait (RoZol®) that is effective against Norway and roof rats as well as mice is prepared by impregnating grain with a mineral oil solution of 0.005% chlorophacinone. It is advantageous for a rodenticide to be oil-soluble when a bait is exposed to humid conditions. Also effective against both rats and mice is a bait (Promar®) containing 0.005% diphacinone. It is available in meat, fish, and apple flavors, and is dyed so as not to be easily mistaken for foodstuffs. Although said to be an "all-weather" bait, it does not hold up so well, when exposed to heavy rainfall, as do the paraffin wax baits. However, it is better accepted by rodents. Control is usually obtained in 3 or 4 days.

Poisoned bait, usually placed in "bait boxes," is put in locations that rats are known to frequent. Bait boxes are constructed with entrance holes of a size that no animal larger than a rat can enter. Bait stations should be used when baits are applied indoors in areas where the public and domestic animals are present, and they should have warning labels. Anticoagulant baits may be placed in open trays indoors in locations not accessible to children and pets. Outdoors, baits may be placed in such places as burrows and tunnels, or in covered bait stations.

In the preparation of baits, anticoagulants are mixed with foods favored by rats. Rolled oats, yellow cornmeal, bacon, salmon, and mineral or salmon oils are commonly included in baits, but any food eaten by humans may be used. The anticoagulant is mixed with the foods in the recommended proportions, which are different for multiple-dose (table 9) and single-dose (table 10) rodenticides (CDC, 1969*a*). When the acceptance of a bait declines, the type of bait base should be changed. In tests made in England, oatmeal was preferred to other milled cereals, but up to 50% cornmeal could be substituted for the oatmeal without decreasing its attractiveness. The addition of up to 8% peanut oil increased the attractiveness of this mixture to both rats and mice. (See page 590 for tables.)

Dry, water-soluble concentrates of anticoagulants are available for making solutions of Pivalyn®, warfarin, or Fumasol-G® (0.005 to 0.006%), or PMP (0.012 to 0.015%). Baby-chick drinking fountains are convenient for dispensing the solutions, especially where no other water is available to the rats (Storer, 1968).

Most rat poisoning in the United States for more than a decade has been done with anticoagulants, but no evidence of rat resistance was reported until 1971, when resistance of Norway rats was confirmed in a rural area of North Carolina. The resistant rats remained susceptible to zinc phosphide, which was subsequently used (Jackson *et al.*, 1971). Resistance of Norway rats has since been noted in several other areas of the United States. Rats resistant to warfarin are cross-resistant to other anticoagulants. The resistance is inherited, and is believed to be controlled by a single gene.

In Scotland, England, Wales, Denmark, and Holland, resistance of Norway rats to anticoagulants as a group has been known for over a decade,

TABLE 9
Multiple-Dose Rodenticides Used Against Mice, Roof Rats, and Norway Rats*

Rodenticide	Concentration (per cent)		
	Mice	Roof rats	Norway rats
Diphacinone†	0.012–0.025	0.005–0.01	0.005–0.01
Fumarin or pindone (Pival)†	0.025–0.05	0.025	0.025
Warfarin†	0.025–0.05	0.025	0.005–0.025

*From CDC (1969*a*).
†May be used in dry or liquid bait.
Dilution factors:
0.05 % (500 ppm) = 1 part of 0.5% concentrate to 9 parts of bait.
0.025% (250 ppm) = 1 part of 0.5% concentrate to 19 parts of bait.
0.01 % (100 ppm) = 1 part of 0.5% concentrate to 49 parts of bait.
0.005% (50 ppm) = 1 part of 0.5% concentrate to 99 parts of bait.

as well as resistance of house mice to warfarin in England (CDC, 1969*a;* Jackson, 1969). Resistance of the roof rat to warfarin was suspected in Liverpool in 1970, and was later confirmed there and in several other cities (Greaves *et al.,* 1973). A new anticoagulant, coumatetralyl, under the trade name of Racumin, was developed by Bayer in Germany, has been found to be lethal to rats resistant to warfarin and other anticoagulants, and has higher levels of bait acceptance than the latter (Jackson, 1969).

Mold will develop in warfarin and fumarin baits left under humid conditions for the 15 to 30 days recommended for domestic rodent control, but no mold will develop in pivalyn baits unless they are exposed for longer than 7 weeks. Warfarin and fumarin baits should contain a mold inhibitor if they are to be used in humid areas or locations.

The confused flour beetle, *Tribolium confusum,* often feeds on anticoagulant rodenticide baits and is not killed by them. The beetles develop freely in warfarin baits, are retarded in fumarin baits, and occur only rarely in pivalyn baits. Malathion in the baits at 0.01% concentration prevents flour beetle infestations (Brooks and Scott, 1961).

Whole grains, rolled grains, and seeds have been criticized as baits on the basis that they do not hold enough poison, but Cole (1969*a*), a rodent control officer in New Zealand, defended the use of such baits, pointing out that rats prefer to crush their own grains. He used 15 kinds of grains and seeds, including pumpkin seeds, in a single bait. His stock material was 3.33 oz (100 cc) of technical warfarin in 10 lb (4.5 kg) of cornstarch and 50 lb (23 kg) of other "finer ingredients." This was mixed with 350 lb (159 kg) of bait. His bait-dispensing trays were 12 in. (30 cm) square, which permitted a rat to stand in the food. Rats, like all rodents, use their front feet to assist in feeding. Their feet are always damp. They continually lick their feet, and this adds to the quantity of poison consumed. When they feed on

TABLE 10
Single-Dose Rodenticides Used Against Mice, Roof Rats, and Norway Rats*

Rodenticide	Concentration (per cent)		
	Mice	Roof rats	Norway rats
ANTU	†	†	1.0–3.0
Arsenic trioxide	†	1.0–1.5	1.0–1.5
Fluoroacetamide (1081)	2.0	2.0	2.0
Norbormide	†	†	1.0
Red squill	†	†	5.0–10.0‡
Strychnine	0.3–0.5	†	†
Sodium fluoroacetate (1080)	0.2–0.3	0.2–0.3	0.2–0.3
Zinc phosphide	0.5–1.0	0.5–1.0	0.5–1.0

*From CDC (1969*a*).
†Not used against these species.
‡Depending upon the LD_{50}, 500 mg/kg red squill are used at the 10% level.

whole or rolled grains, the finer materials, which contain the greater concentrations of poison, adhere to their feet and are conveyed to the mouth every time they pick anything up. The only additive Cole found attractive was ground nutmeg, which he used at the rate of 1 tsp (5 cc) to 40 lb (18 kg) of bait.

The Cole bait was used by Pierce (1970) in South Carolina with better success than he had theretofore had with rat baits. On the other hand, Brooks (1973) found ground meal to be more acceptable than whole or cracked grain. He believed that particle size was a more important factor than the kind of seed used in affecting the acceptance of a bait to rats, particularly with hard, dense seeds. A number of ground seeds were each compared with ground yellow corn in the quantity consumed by Norway rats trapped in refuse dumps and in a poultry shed. The rats preferred the following seeds to corn, in descending order of preference: sweet rice (a variety of short-grain rice grown on a limited acreage in California), proso millet, peanuts, barley, and sunflower seeds. Rolled oats gave variable results. This is believed to be the result of high lipase activity, which can render the oats unpalatable unless the enzyme is inactivated by steaming or otherwise heating. Sweet rice and hulled millet were preferred to yellow cornmeal in ratios of 10 or 20 to 1 by both groups of wild rats. In later field trials of these 2 grains with chemosterilants and rodenticides, as well as in many laboratory experiments, they were shown to be superior baits.

Norbormide. This compound, with the trade name Shoxin, is used specifically for Norway rats, killing them in 15 minutes to 1 hour by a shock-like impairment of blood circulation. It is less effective against the roof rat and is nontoxic to mice. It has very little toxicity to other mammals or to birds, and can therefore be safely used around pets, livestock, and poultry. Proprietary products were generally found to be ineffective in field tests in Georgia. However, when experimentally prepared at 1% concentration with yellow cornmeal, norbormide provided good control in 7 out of 8 rural premises (CDC, 1969*a*). Baits are usually placed in small quantities throughout the area where rats travel, instead of in bait boxes.

Red Squill. This is another rodenticide that is relatively safe to man and his domestic animals. Red squill causes the death of rats by respiratory failure, and since it is a moderately slow-acting poison, it permits them to leave the premises before they die. It is a red powder, made from the fleshy inner bulb scales of the red variety of the squill or "sea onion," *Urginea maritima,* which grows wild in the hills bordering the Mediterranean Sea. The objectionable bitter taste and emetic effect of red squill cause most animals to avoid or regurgitate it, but not rats and mice. Rodents cannot vomit baits once they are eaten. Red squill should be used for a rapid reduction of Norway rats over a relatively brief period. It is not recommended for roof rats or mice because their acceptance of red squill is relatively poor. It should not be used against a given rat population more often than once in 6 months, for surviving rats become bait-shy. It should be stored in tightly sealed containers or it will become hard and caked. The powder is very irritating to the nose and skin, and a dust mask and rubber gloves are recommended for anyone preparing the bait (Brooks, 1969).

One part of the fortified red squill powder, thoroughly mixed with 9 parts of meat or fish, is recommended for use against Norway rats where there is risk of human exposure (CDC, 1969*a*). The Los Angeles County Health Department reports that a single treatment with 10% red squill biscuits sufficed, and rats were killed or driven out of attics or basements. The rodents did not die indoors, but left the house. Ratproofing of attics and foundations should follow poisoning in about 5 days.

In an intensive rat-eradication campaign in Memphis, Tennessee, red squill baits were prepared by mixing 1 part of red squill and 1 part of ground yellow cornmeal with 8 parts of jack mackerel. Quantities of 60 lb (27 kg) or more were prepared in each batch, using a large mechanical food mixer. The bait was processed into 0.5-in. (12-mm) balls, and wrapped in 4-in. (10-cm) paper squares cut from the white pages of discarded telephone directories to form "torpedoes." Baits were placed in nonvented plastic freezer bags and stored in a mechanical freezer. The bait was transported each day in styrofoam ice chests. The unthawed baits were placed in and around burrows and other appropriate places because they were attractive to rats while frozen (Moseley, 1970). Scilliroside is the active ingredient of red squill. A new method of concentrating the toxic ingredient has been developed. Laboratory and field experiments with a 0.015% concentration of the stabilized scilliroside indicated that the compound had an exceptional potential for use in the control of Norway rats (Maddock and Schoof, 1970).

ANTU (Alphanaphthyl Thiourea). This is a fine, grayish powder that is stable, nonvolatile, and insoluble in water. It is highly toxic to Norway rats, dogs, and pigs, but much less toxic to roof rats, mice, and cats. It kills rats by acute lung dropsy and an accumulation of fluid in the chest cavity. ANTU may be mixed with baits at 2% to 3% concentrations or blown into burrows and holes as a contact poison. Mixed with equal parts of flour or pyrophyllite, it may be applied to rat runways and, when used in this way, it is reported to be particularly successful in eliminating the last survivors in ratproofed buildings (Anonymous, 1946). ANTU in baits should not be used more than once a year because rats obtaining a sublethal quantity will not eat more. Also, they develop a tolerance to the toxicant.

Arsenic (Arsenic Trioxide) (As_2O_3). The usual commercial grade of white arsenic is a rather slow-acting poison. It acts more rapidly if a finely ground powder is used. It is practically tasteless and odorless, and is highly toxic to rats, humans, domestic animals, and birds. It is effective in a bait against both Norway and roof rats, but not mice. It can be poisonous when absorbed through cuts or breaks in the skin. It should be exposed only in suitable bait boxes, and gloves should be worn when mixing or handling the bait. It should be kept out of reach of children and pets. Rats soon become wary of arsenical baits, so it is desirable to use relatively high concentrations in order to ensure that the rodents will receive lethal quantities. USPHS (1948) suggested 1 oz (30 cc) of arsenic and 0.5 oz of a "binder," such as ground bacon, molasses, peanut oil, ground salmon, or ground sardines, to 1 qt (about 1 l) of grain.

Zinc Phosphide (Zn_3P_2). This is a crystalline, grayish-black powder, insoluble in water, with a garliclike odor and strong taste that repel most animals, but it is apparently attractive to rats. It is moderately fast-acting, releasing phosphine gas when it contacts weak stomach acids. It is more effective against roof rats than Norway rats, and is also effective against mice. The LD_{50} of zinc phosphide to Norway rats is 35 to 48 mg/kg, indicating a much greater toxicity than red squill or arsenic to these species. With strong doses, death occurs in as brief a period as 17 minutes. Rats succumb from heart failure accompanied by kidney damage occurring after longer periods. With low doses, death may take several days; recovery from sublethal doses is complete.

Zinc phosphide is normally used at 0.5 to 1% concentrations in baits. Diced fruit or vegetables may be coated with the powder, or it can be mixed with foods such as hamburger, hamburger plus 10% cornmeal, seeds, grains, or cereals. With dry baits, 3% of a fat or oil can be included as a binder (Schoof, 1970). Storer (1948, 1968) recommended that it be used at a 1% concentration in any dry cereal, and that 12 to 24 oz (360 to 720 cc) of corn (maize) oil, mineral oil, or glycerin be added to each 100 lb (45 kg) of bait to make the poison adhere to the cereal. The mixing and handling of zinc phosphide baits should be done outdoors or in a well-ventilated room, and the bait should be handled or distributed with leather or rubber gloves. This poison can be absorbed through cuts, but not directly through unbroken skin. Zinc phosphide is an effective single-dose rodenticide, but it must be handled and used with care. Ingestion of 4 to 5 grams has resulted in death to humans in 2 instances. A figure of 40 mg/kg as the LD_{50} for rats would be equivalent to 2.8 g for a 154-lb (70-kg) man, or 280 g (0.62 lb) of a 1% bait. Some persons have survived ingestion of 25 and 50 grams of zinc phosphide. Chance of survival is improved if vomiting occurs within an hour after ingestion.

Because zinc phosphide is quickly decomposed in a rodent's stomach, secondary poisoning of valuable predators is prevented. Therefore, by using it for rodent control, the adverse ecological effects of rodenticides such as sodium monofluoroacetate or thallium preparations, which can kill predators feeding on poisoned rats, are avoided (Utagawa, 1971).

Sodium Fluoroacetate (1080). This is a white powder, with the formula $NaC_2H_2FO_2$, but is always colored black for the purpose of identification. It is a stable compound that is highly soluble in water. When exposed to air, it takes up water and becomes sticky. It is a highly toxic stomach poison for Norway rats, roof rats, and mice, usually killing them in 30 to 90 minutes. It has no objectionable taste or odor, and is therefore readily taken by these rodents. The animals are usually immobilized by the first symptoms of illness, so they are frequently found in the open near the bait stations (Simmons and Nicholson, 1947). They may be removed and either burned or buried deeply before pets are allowed in the area.

In this respect, the poison has the same advantage as those that tend to drive the rodents away from the premises entirely, such as red squill, and thereby minimize the likelihood of odors developing in inaccessible places.

This poison is highly toxic to humans and domestic animals that chance to eat the bait. It should be used only by trained and experienced personnel, and only in special situations, such as in a public health emergency. There are no known antidotes for the poison. Anyone using the baits should wear gloves, keep his hands away from his mouth, and wash them thoroughly before smoking or eating. The baits should be exposed only in adequately protected and locked bait stations, except in areas of limited public access, such as sewers or in fenced, locked, and patrolled refuse-disposal sites. The concentration of the poison in a bait should be 0.2 to 0.3% (CDC, 1969*a*). The concentration is so low that it is much easier to dissolve the poison in water than to mix it with solid baits.

Fluoroacetamide (1081). This is a white crystalline material, formulated as a purplish-black powder. It has properties similar to those of sodium fluoroacetate, being soluble in water, odorless, almost tasteless, effective against all 3 species of domestic rodents, and highly poisonous to man and his domestic animals. It does not act so rapidly as sodium fluoroacetate, and is used at 2% concentration in solid or water baits. The same precautions should be exercised in the use of either material (CDC, 1969*a*).

Paraffin Blocks. Poison bait will be eaten by rats if it is mixed with paraffin. Paraffin baits appeal to homeowners because of the ease with which they can be used. Also, the poison cannot be leached away by rain. Jamieson (1965) obtained good control with a bait consisting of small proprietary paraffin blocks containing an attractant and seeds and impregnated with 0.005% diphacinone (Diphacin). The blocks were nailed to fences, boards, or trees, or were wired to shrubbery out of reach of dogs. It was necessary to conceal the bait blocks, for those placed in the open or short distances from areas of rat activity were not visited by rats.

In addition, conventional anticoagulant bait stations can be attached to fences, shrubs, and trees in the same way.

Paraffin blocks have been popular for rat control in municipal sewer systems. Rohe (1966) used them successfully in the city sewers of Ontario, California. The poison bait was prepared according to this formula:

Poultry rolled oats	23.5 lb (10.7 kg)
Fluoroacetamide (1081)	0.25 lb (0.11 kg)
Powdered sugar	1.25 lb (0.57 kg)
Salad oil	50 fluid oz (1.5 l)

One part of melted paraffin was mixed with 2 parts of bait, embedding a corrosion-resistant wire. The paraffin blocks were placed in clear plastic bags, permitting the wire attachment to protrude from the top. A poison label on each block was visible from the outside.

Storer (1968) gave the following instructions for making baits with paraffin blocks:

> For paraffined baits, the anticoagulant is first mixed in a 1:19 ratio with a suitable grain bait, predetermined as acceptable by the rodents. Blocks of paraffin are melted in a double boiler (over a water bath), and then the dry mixed bait is slowly stirred into the melted paraffin to make a thick but flowable mixture. (Excessive heat may affect the poison.) The warm mix is poured into cardboard containers, such as topless milk cartons or large paper cups, to cool and solidify. Thereafter, the prepared bait is exposed where rodent signs are present, and renewed as necessary. Paraffin ratio, by weight, is 1:1 or 1.5:1, according to the kind of grain used.

Brooks (1964), working with Norway rats in Albany, California, concluded that poisoning rats once a year in sewer systems would not keep them in check unless a reduction of 90% or better was achieved. Two treatments per year were recommended.

In Kansas City, Missouri, officials believed that combining sanitary sewer systems and storm-drain systems made it easier for rats to gain access to the surface from catch basins and breaks in private sewer lines, uncovered basement drains, and uncapped sewers. In a section of the city where only sanitary sewer lines existed, there were very few rat problems. The city-wide use of electric home garbage-disposal units resulted in an increase in the rat populations in the sewer systems.

The city contracted with a private company to bait 3,484 manholes 8 times within the contract year: 3 times in the first 30 days, then once every 60 days for the remainder of the year. The baiting was done by the pest control firm submitting the lowest bid, at a cost to the city of $26,100 ($7.49 per manhole). The success of this pilot project induced the officials to contract for the baiting of another 3,338 manholes. The same firm again submitted the lowest bid, this time amounting to $7.25 per manhole. The bait used, after many had been tested, was a 4-oz (113-g) paraffin cake impregnated with cracked corn and diphacinone anticoagulant. The cakes were flat, and had carinated and wavy edges, so as to offer the maximum number of edges on which the rats could gnaw. A hole was drilled in the center of each cake, and a 20-gauge wire was run through as many of them (usually 4) as were considered necessary for a particular manhole. The wire was attached to a masonry nail driven into mortar between bricks in the sewer wall, allowing the cakes to rest on the bottom of the manhole. In some manholes, 5 lb (2.3 kg) of bait were eaten in 15 days. The contract provided for extra treatment for manholes in which all the bait was eaten. Some of these had to be baited 15 to 20 times, with 1 to 3 lb (0.45 to 1.35 kg) of bait used for each retreatment (Anonymous, 1964*a*).

Fumigation

Fumigation may be employed for the rapid elimination of rats and mice and their ectoparasites in inaccessible areas in buildings or in burrows. Fumigants are dangerous, and a thorough knowledge of the materials used and procedures employed, as well as practical experience, is required for safe and effective rodent fumigation. The fumigation of buildings should be done only by licensed pest control operators, but is not commonly practiced nowadays because safer and less expensive control methods are available that are highly effective if correctly applied. One disadvantage of fumigation is that the rodents may die in inaccessible places and cause offensive odors, as contrasted with poisons that tend to drive them away from the treated premises before they die and, of course, trapping, which makes possible the disposal of all the animals that are killed. The following discussion will be confined to fumigation in outdoor burrows. A thorough coverage of methods of fumigation of homes, warehouses, and other buildings by licensed pest control operators is presented in the *Handbook of Pest Control* (Mallis, 1969).

Calcium Cyanide [$Ca(CN)_2$]. Upon exposure of this compound to atmospheric moisture, it generates hydrocyanic acid gas. Granular calcium cyanide can be used, placing 1 oz (30 cc) of the granules 6 to 8 in. (15 to 20 cm) inside the rodent burrow with a long-handled spoon and then closing the entrance with soil (Storer, 1948). Cyanide dust may also be used, applied with a hand or foot-pump duster. The dust may be blown into the burrow entrances, after which they should be closed with soil. Other holes leading into the burrow system should be closed before gassing is begun, to prevent the escape of the gas. If entrance holes, or others that have been closed, are reopened by rats within a day or two, the burrow should be refumigated. Care should be taken to avoid pumping gas into a burrow that may have tunnels leading into an inhabited building, for unlike other fumigants, such as methyl bromide and chloropicrin, cyanide gas is lighter than air and could escape into the building.

Methyl Bromide (CH_3Br). This gas is extensively used for the fumigation of buildings for drywood termites, and this treatment would of course also kill rats, mice, and other pests. It is available as a liquid in 1-lb (0.45-kg) pressurized canisters. A leaflet with directions for its safe and effective use is provided with the canister, or can be obtained from the dealer. For the fumigation of rat burrows, a valve and hose are available to be used with the canister. The hose is placed in the entrance of the burrow, soil is packed around it, and the valve is opened for 1 or 2 seconds. The results should be checked in a day or two, as with cyanide fumigation. Methyl bromide should not be used near the roots of trees and shrubs, for it is a penetrating, phytotoxic gas.

Chloropicrin (CCl_3NO_2). This is the compound commonly referred to as "tear gas." It may be mixed with motor oil and poured into rat burrows, after which the burrows are plugged with soil. The gas adheres tenaciously to soil particles, and remains active in the burrow for a month or more, not only killing the rats but repelling others that might otherwise take over the uninhabited burrow. An important safety factor in the use of chloropicrin is that very small concentrations are detectable by humans.

Engine Exhaust. The exhaust from automobiles has been used to force carbon monoxide (CO) through a hose and into rat burrows under pressure. The gas reaches all parts of the tunnel system and remains effective for some time. In burrows under cement-floored farm buildings, where cyanide would be dangerous to livestock, engine exhaust can be safely applied (Storer, 1968).

Trapping

When it is desirable to avoid the use of rodenticides for fear of contaminating food products or of harming children or pets, or to avoid leaving dead rodents in inaccessible places and thereby causing odor problems, or when the last survivors of a rat-poisoning program refuse bait, it may be necessary to resort to trapping. The traps should not be placed above foods unless they are covered with a tarp. Uncovered foods under the traps may become contaminated with excrement and urine.

The type of trap generally used for rats and mice is the wood-base snap trap. It is not necessary to wash, boil, or smoke traps or to wear gloves, for rats are not repelled by the human scent (Storer, 1968). It is important to secure rat traps to some adjacent "anchor" with strong wire to prevent dying rats from dragging them into inaccessible locations. A bait, consisting of almost any food such as bacon, beef, fish, nut meats, prunes, apples, carrots, or pet food, is securely fastened to the trigger of the trap, and it is placed in a rat runway. It may be necessary to try a number of baits to find the one that is most attractive at a particular time or place. Cheese, the traditional rat or mouse food in the mind of the layman, and the one he generally uses, is the least favored by experienced pest control operators. The type of bait and the locations of the traps should be changed from time to time if the number of rats or mice caught dwindles.

If the rat runway is along a wall, a box placed near the wall will form a narrow passageway. The trap can then be set at the entrance of the passageway at a right angle to the wall, and with the bait pan close to the wall. Traps may also be nailed to upright supports in a vertical position where "rub marks" are seen on vertical posts or "swing marks" on horizontal wooden members occur at their points of intersection with overlying beams (figure 373). Roof rats, in particular, are wary of traps set on the floor, and are more easily trapped in elevated locations (Anonymous, 1964*b*). (Roof rats are also more wary of baits set on the floor than are Norway rats.) When rodents are very numerous, it is better to use many traps for a brief time than to use fewer for a longer one.

Professional trappers sometimes modify their equipment by expanding the trigger devices and using the traps without bait. However, even with this method, a smear of peanut butter, cornmeal, or other common bait material will increase the possibility of success. The traps should be examined daily to remove any captured rodents, and all sprung traps should be reset. If it is necessary to handle dead rats or mice, it should be done with gloves or long forceps or tongs, so at to avoid bites from fleas or mites. The rodents should be released from the traps directly into plastic bags, which are then secured at the tops to prevent ectoparasites from escaping.

Elimination of Dead-Rat Odors

All the methods of rat suppression other than trapping are likely to kill the animals in inaccessible places where they will start to putrefy in 3 or 4 days and cause bad odors. (Mice are so small that they ordinarily do not cause enough odor to fall within the range of human detection.) Breck (1965) described a method for rat-odor elimination. If he could not himself locate the site of the dead rat by the smell, he released several bluebottle flies in the room, and these flew to the spot where the dead animal lay behind the wall. Probably most of the blow flies (Calliphoridae) and flesh flies (Sarcophagidae) would be effective. The spot was marked, the flies were killed with a pyrethrum aerosol, and a small hole was drilled through the wall 6 in. (15 cm) above the spot located by the flies. A small electric sprayer, with a type of nozzle that would spray in many directions from 1 position, was used to inject a disinfectant liquid through the hole and into the wall void. Breck added 0.25 cup (60 cc) of Hyamine 1622 to 1 pt (0.5 l) of warm water to make a disinfectant solution, or Zephiran chloride at 1 oz per gal (30 cc per 4 l), or Bactine diluted 50:50 with water. He used 1 pt of the solution. He usually repeated the treatment at the next interstud void on each side of the suspected area.

Storer (1948) recommended a deodorant solution of 2 oz (60 cc) zinc chloride, 1.25 oz (37 cc) salt, and 18 oz (540 cc) water, to be sprayed or poured on the site of the odor, or the following proprietary deodorants that could be used in the same way: Edison (Thos. A. Edison, Inc.); Metazene (Motomco); Neutroleum (Fritsche Bros.); Nilodor (The Nilodor Co., New York); and Polycide No. 2 (Brooks Chemicals, Cleveland, Ohio).

MICE

There are 3 subspecies of mice: *Mus musculus musculus* L., *M. m. domesticus* (Rutty), and *M. m. brevirostris* Waterhouse. These have overlapping ranges, and are so similar in appearance that they are usually all referred to as the "house mouse," under the name *Mus musculus* L. (figure 381; see also, figure 371). Field mice in California are most likely to be *M. m. brevirostris*, the common subspecies of Spain and Latin America, which probably arrived with the early Spanish-speaking settlers of the region (Fertig and Edmonds, 1970). The house mouse is indigenous to the arid grasslands of southwest Asia, where prehistoric man first harvested the early forms of wheat and barley. Long droughts are characteristic of that region. Small rodents of the area had to adapt themselves to survival for long periods without drinking water, just as in the cases of those in the southwestern United States, such as the kangaroo rat (*Dipodomys*). Thus, mice are admirably suited to be ubiquitous commensals of man, capable of being transported for long periods in closed containers, such as boxes, trunks, or barrels. The man-made environments they now

Fig. 381. House mouse, *Mus musculus*. (Courtesy J. E. Brooks.)

occupy are often low in humidity and contain only dry food, much as in their original environment. It is not likely that mice could be controlled merely by preventing their access to free water (Fertig and Edmonds, 1970).

Mice consume much less food than rats, but do much gnawing and nibbling, making numerous small holes and causing slight damage in many places.[29] They may damage fabrics when gathering materials for nests for their young. When numerous, they are responsible for a disagreeable, pungent odor and fecal contamination.

Description. Mice are similar in appearance to young roof rats, having large ears, pointed muzzles, a tail at least as long as the head and body combined, and short, broad feet. Mice living indoors are usually dark gray above and light gray on the belly. The tail is also lighter below than above. Those living outdoors are likely to have a sandy or yellowish-brown appearance because of their bleached hair tips. The combined length of the head and body of an adult is 2.5 to 3.5 in. (6 to 9 cm), the tail length is 3 to 4 in. (7.5 to 10 cm), and they usually weigh 0.5 to 0.75 oz (14 to 21 g). Fecal droppings are 0.125 to 0.25 in. (3 to 6 mm) long, and rodshaped.

Biology and Habits. Mice become sexually mature when they are 2 or 3 months old. The period of gestation is 19 to 21 days. There may be an average of 5 or 6 young per litter in some areas and 6 to 8 in others, but there can be as many as 12 to 15. In one investigation, females were found to have an average of 7.6 pregnancies in a lifetime and 5.6 embryos per pregnancy (Laurie, 1946). Some captive females produced 100 young per year (Storer, 1968). Sometimes there may be only 1 brood in a nest, but in other cases several females share the same nest with their congregated broods. In 1 nest there may be some newly born young (blind, naked, pink, and only 1 to 1.5 g in weight), while others will be about ready to leave the nest. Shredded cloth, sacking, paper, or other soft materials are used to make the nests. A female usually does not bear young after she is 15 months old, but she may live much longer. Males have been known to live as long as 2.5 to 3 years (Storer, 1931; Laurie, 1946; Eaton and Cabell, 1949).

Mice tend to eat small quantities of food at a time, differing from rats in this respect. They prefer to move about, nibbling a bit here and there. Unlike rats, they do not need to search for water. In the laboratory, they can live in good health on a diet of wheat, which normally contains only about 10% water. If water is provided, they will take a small amount, generally less than 1 ml per day (Southern and Laurie, 1946). Mice prefer to feed on dry cereals if drinking water is also available to them. The attractiveness of a poison bait for mice is increased if it is mixed with water.

Mice have keen senses, except that of sight, and can climb, jump, and swim well, although they rarely enter water. They can live outdoors as well as in buildings, although those that are outdoors tend to move into buildings at the approach of cold weather or heavy rains. Both wild and domestic mice make a whistling sound, particularly at night, that has been compared with the chirping of a canary bird.

Mouse Plagues

When mice are not abundant or not confined near food, they do little damage and are merely nuisances. However, they can rapidly increase in numbers, and then their collective damage can be intolerable. Damage consists of destruction or contamination of foodstuffs, clothing, furniture,

[29] The tendency of mice to nibble and chew on practically everything in their paths led to the composition of the Christmas carol, "Silent Night, Holy Night." In the little town of Oberndorf, Austria, shortly before Christmas in 1818, Father Joseph Mohr found that mice had chewed up the bellows of his church organ. He asked his friend, the schoolmaster, Franz Gruber, to compose a simple melody that could be played on the guitar. The night before Christmas eve, Father Mohr wrote the verse and Franz Gruber composed the music, both men working all night at the task. On the following night, at high Mass, Father Mohr played this most famous of Christmas carols on his guitar while he, Franz Gruber, and 2 friends sang the immortal words for the first time in public.

paper goods, and books in homes, and of stored materials in warehouses, granaries, and store buildings. As in the case of rats, mice usually contaminate more food than they consume, causing it to be condemned by food inspectors.

Occasionally, mild weather, an abundant food supply, and reduction in natural enemies may cause mice to multiply to the proportions of a plague and then to migrate in large numbers. In one such plague in Kern County, California, their numbers were computed to be 82,280 per acre (0.405 ha), and possibly to total several million (Hall, 1927). During the great Australian mouse plagues of 1916 and 1917, a farmer reported picking up 28,000 dead mice on his veranda the morning after putting out poison bait. During this plague, they fed on grain and such unusual materials as seaweed, lead (graphite) pencils, and lead bullets out of cartridges (Hinton, 1931). In 1970, a plague occurred in northwestern Victoria, Australia, that was so severe that motorists reported the highway to be dangerously slippery at speeds above 20 mi (32 km) per hour. In one hospital, the legs of the beds had to be set in metal cans to protect patients. Mice were said to be traveling across fields "like a moving carpet" and feeding on newly harvested grain. They ate enough grain in a silo to cause the level to fall 6 in. (15 cm) overnight, and the next morning a farmer killed 2,000 mice there (Anonymous, 1970*c*).

Diseases Carried by Mice

Mice are not considered to be as important as rats as carriers of some diseases. Nevertheless, as the most highly domesticated of all rodents, their true role as disease carriers should be known. Cameron (1949) discussed these diseases according to the following 8 methods of transmission:

(1) *By Biting Man.* Mice do not bite man as frequently as do rats, and are therefore less often carriers of ratbite fever. However, the bacterium (*Spirillum minus*) occurs in mice. Cats, dogs, ferrets, and weasels can contract the disease from rodents (presumably mice), and convey it to man by their bites.

(2) *By Infecting Human Food with Their Droppings.* Mice have a greater opportunity than any other animal to infect human food with their droppings. The food-poisoning organism (*Salmonella*) and 2 small tapeworms (*Hymenolepis nana* and *H. diminuta*) can be transmitted to man via mouse and rat droppings.

(3) *By Infecting Human Food with Their Urine.* This is less likely than infection by way of mouse excrement. The spirochetelike *Leptospira,* which causes Weil's disease in man, is much less frequently found in mice than in rats, possibly because mice do not drink much water and in any case are not likely to drink foul water from sewers, as is common for rats.

(4) *By Being Eaten.* Cats obtain their most important tapeworm and, in Europe, a roundworm which lives in their lungs, by eating mice. Mice in turn obtain it by feeding on the cats' droppings. Neither worm occurs in man.

(5) *Indirectly via the Cat or Dog.* A fungal disease of the skin, called favus, is caused by a mold that enters hair follicles and surrounding skin and causes small crusts to form that eventually turn into a cup-shaped yellow scab. The scab becomes brown, falls off, and leaves bald patches. The disease can be transmitted to man directly by the mouse or indirectly by the cat. It occurs in all parts of the world.

(6) *Indirectly via Bloodsucking Insects.* [The relation of rats to plague has already been discussed in this volume.] Mice harbor fleas, which can transmit plague, but mice do not seem to be often infected with the disease organisms. The number of infected fleas they can harbor on their small bodies is much smaller than in the case of rats. At least 1 outbreak of endemic or murine typhus has been traced to mice. The oriental rat flea (*Xenopsylla cheopis*), the principal vector, will live on mice as well as rats. Because of modern heating systems, this flea is no longer entirely tropical in distribution, and has extended its range as far north as Canada. Under suitable conditions, the disease could become epidemic.

(7) *Indirectly via Bloodsucking Mites.* Rickettsial pox, a disease that resembles chicken pox, is caused by *Rickettsia akari.* It can be transmitted from mouse to man via the house mouse mite, *Liponyssoides sanguineus* (Hirst) (= *Allodermanyssus*).

(8) *Indirectly by Dying in a Water Supply and Contaminating It with Disease Organisms.* There is not believed to be any evidence that this has happened, but the possibility must be considered.

Control of Mice

The same principles and techniques described for rat prevention pertain also to mice, particularly those applying to sanitation. Mouseproofing is more difficult than ratproofing because the holes through which mice can enter are so much smaller, more numerous, and are more difficult to locate. Their outside harborages are more difficult to find and eliminate. Mice can enter a building through openings around pipes, through torn screens, and through remarkably narrow apertures of various kinds. They can also be carried into a building in household goods, crates, boxes, or overstuffed furniture.

Poisoning. Baits with the poisons discussed for rat control, with the exception of red squill, can also be used against mice. Because mice tend to nibble rather than eat large quantities at a time, a higher concentration of the poison may be required, as shown in tables 9 and 10, than is used for rats. Since mice do not forage as widely as rats, it is desirable to use many small baits, well distributed, rather than a few large ones. Frequent renewal of baits is desirable, for mice are not attracted to old baits (CDC, 1969*a;* Anonymous, 1970*a*).

Tracking Powders. Tracking powders have been popular for mouse control because they can be effectively applied in closed attics, wall voids, and other enclosed or "dead" spaces. DDT micronized powder with 50% of toxicant is lethal to mice when applied to mouse runways and harborage areas. DDT is safer than baits. The mouse picks up the DDT on its body as it runs through it, then licks it off. Disease-bearing ectoparasites are controlled by the insecticide on its body. Faster control of mice is achieved with DDT than with anticoagulants. It is not effective against rats. Now that DDT is no longer labeled for use as a tracking powder, sodium fluosilicate may be considered. It is said to be effective for both rats and mice (Storer, 1968). The anticoagulant chlorophacinone (RoZol) has been registered by the Environmental Protection Agency (EPA) and in most states as a mouse-tracking powder, and is said to be a reliable replacement for DDT. Besides being applied as just described, tracking powders can be applied in bands about 4 in. (10 cm) wide around some foods, such as packages of grain (Boydstun, 1959). A tracking powder should be applied only at floor level, and its use should be avoided where food is prepared or stored.

Trapping. Traps used for catching mice are of the same type, usually wood-base snap traps, as used for rats, but are smaller. Mice are easier to catch than rats, for they are not so wary of the traps. They do not range about so widely as rats, and the traps must be placed close together, 3 or 4 ft (90 or 120 cm) apart, to ensure that the mice will find them. The locations of traps and the types of bait to be used are similar to those recommended for trapping rats.

Glue Boards. Mice can also be trapped with "glue boards." The glue board is generally 12 in. (30 cm) square. One side is covered with a sticky, resinous material. The boards are tacked down to the floor across mouse runways. When the mice attempt to cross them, their feet become stuck in the glue. Rats are better able to pull themselves free from the glue than are mice, although it is said that even rats are unlikely to escape if they get more than 2 of their feet stuck in the adhesive. Glue boards of smaller dimensions can be used effectively for trapping mice. They can be rendered ineffective by very cold weather or by excessive accumulations of dirt and grease (Ives, 1948).

BATS

Bats are nocturnal mammals of the order Chiroptera that inhabit dark, secluded places. They have long been associated with evil, witchcraft, and sorcery, at least in western culture, and have been feared and despised by many people. On the other hand, the Chinese welcome them in their homes as symbols of luck and good times. Among the many superstitions about bats is that they tend to become entangled in women's hair. All bats are considered to be "vampires" by some people, although the true vampire or blood-drinking bats (Desmodontidae) are restricted to tropical and semitropical regions of the Western Hemisphere.

There are over 2,000 species of bats in the world. They have world-wide distribution, and are among the most numerous of land vertebrates. Only 3 of the 16 living families occur in the United States, but bats are common all over the country, and are most abundant in the Southwestern states.

They are usually found feeding over a pond or stream, along a forest edge, beside a cliff or ravine, or among buildings (Barbour and Davis, 1969). They may become pests when they choose a building for roosting or hibernation. A small number of them are infected with rabies virus, and can transmit the disease if they bite people. Bats should be handled only with tongs or when wearing heavy gloves. Some disease organisms, but not rabies, can be transmitted by dusty bat manure.

Biology and Habits. Bats are unique in being the only mammals capable of true sustained flight. Their front legs, and particularly the greatly elongated fingers, are modified to support a large, thin, membranous extension of dorsal and ventral body integument, and are therefore able to function as wings. The same membrane encloses the hind legs and tail. Depending on the species, the wingspan may be from 6 in. (15 cm) to over 5 ft (150 cm). An interesting anatomical modification is the reverse manner of attachment of the hind legs to the hip, so that the knee points backward when it is bent. The bottom of the foot then faces forward to facilitate the bat's alighting upside down and hanging by its toes (Barbour and Davis, 1969; Vaughan, 1970).

Most bats are insectivorous, seeking night-flying insects. Unlike the majority of nocturnal animals, they have small eyes. However, they can find their prey with remarkable accuracy by using their extremely acute sense of hearing. While in flight, bats emit cries that are too high-pitched for the human ear to detect. The high-pitched sound waves they emit reflect from solid objects. They can detect prey, natural enemies, and obstacles in their path by perceiving the reflected sound in a manner similar to radar. Thus, orientation, obstacle avoidance and, in insectivorous species, food location, are attained by means of ultrasonic echolocation. Bats have been observed to turn back at a fog front "as if at a solid wall," suggesting an adverse effect on echolocation. Theoretically, fog containing droplets of a certain range of sizes should have a strong influence on the propagation of echolocation signals from a bat (Pye, 1971). The system of orientation utilized depends on a highly refined skill. Bats that are tired, in poor health, or incompletely awakened from hibernation may fly in a relatively clumsy fashion (Griffin, 1958).

Roeder (1967) made a study of some common noctuid moths with ultrasonic receptors in their tympanic organs that enable them to take evasive action when pursued by bats.[30] The moths have an advantage in that they need only detect the ultrasonic cry of the bat, but the bat must detect the echo of its cry after it has "rebounded" from the moth. The moths turn and loop to make their pulses as unpredictable as possible, and dive toward the ground where the bat cannot follow. Nevertheless, many moths are captured.

Bats catch insects in a wingtip, transfer them to a cup formed by the interfemoral membrane, and then eat them. They must alight to eat insects that are too large to be handled in flight. At least 1 species commonly searches for insects on the ground. Bats drink while in flight, scooping up the water with the lower jaw (Barbour and Davis, 1969).

Most bats produce only 1 offspring at a time. The offspring are born hairless and with their eyes closed. They are usually carried by the mother until they are too heavy, and afterward are left at the roost until they learn to fly. Some bats live as long as 20 years. They may live together in colonies of from several to more than a million individuals, but many are solitary. An estimated 9 million bats inhabit Carlsbad Caverns, New Mexico, and in fact it was an immense flight of bats, appearing from a distance like a large cloud of dust, that led to the discovery of these caverns (Gray, 1961). Most species of hibernating bats seek shelters where water is available and humidity is high. They tend to form clusters, possibly to help stabilize the temperature in the optimum range and to minimize evaporative water loss (Davis, 1970).

Bats have remarkable homing instincts. Some individuals that were removed from their roosts returned from distances of at least 810 miles (1304 km). In temperate climates they are generally nonmigratory, hibernating in winter or becoming torpid during colder periods, but even in California a few species seek warm roosts or migrate south in winter. On the other hand, the hoary bats spend the winter in coastal and southern California and the summer anywhere from northern California to British Columbia. The roosting places of nonmigrating species during winter are not necessarily the same as the ones used during the remainder of the year. Also, bats tend to choose different roosts for temporary nesting after feeding. When at rest in their roosts, bats hang suspended by their hind feet. An accumulation of fecal droppings beneath their roosting sites will indicate their presence. Fresh droppings indicate that they are currently present, whereas old ones show that the colony was normally present at some other season. The droppings are easily crushed into fine fragments, in contrast to those of mice, which are firm and do not disintegrate readily. Dark stains on louvers or other openings where bats enter attics, church belfries, or similar roosting places, and urine and fecal splatterings on the walls below, are also signs of their presence. They also make their presence known by high-pitched squeaking, rustling, or by their odor (Gray, 1961; Murray, 1969; Griffin, 1970).

According to Murray (1969), there are 24 species of bats in California (later reduced to 23

[30] An amazing example of adaptation is provided by one of the moth ear mites, *Dacrocheles phalaenodectes* (Treat). It infests one or the other of the metathoracic tympanic organs (ears) of various species of noctuid moths, but very rarely both. The first mite to find a moth ear apparently leaves a "trail" that is followed by others. Eventually, up to nearly 100 mites may infest an ear and eventually destroy its sensory function. If both ears were infested and destroyed, the moths would be unable to avoid bats and the mites would be destroyed, together with their hosts. There are other species of mites that do not destroy the function of the moths' ears; these may be found in both ears (Treat, 1969).

because of revised synonymy), and 17 of these are reasonably widespread and abundant. He developed an illustrated key for 8 species of greatest interest as pests.

Freetail Bats (Mollosidae)

In these bats, the tail extends beyond the tail membrane.

Tadarida brasiliensis (Geoffroy St. Hilaire) is the most common pest bat in California (figure 382). It is about 3.75 in. (9.5 cm) in total length,

Fig. 382. Freetail bat, *Tadarida brasiliensis*. (From Barbour and Davis, 1969.)

and is dull brownish gray; the ears extend forward rooflike over the eyes; and the lips are wrinkled. It is a rather small bat, with long, narrow wings. The ears almost meet, but are not joined at the midline. On the West Coast and in the southern states it inhabits buildings, but in Texas, Oklahoma, New Mexico, and Arizona it is primarily a cave bat. It is colonial at all seasons. In the summer, several million freetail bats occupy Carlsbad Caverns, and even larger nursery colonies occur in caves in Arizona, Texas, and Mexico. In California and some of the southwestern states, maternity colonies of 1,000 or more are found in some buildings (Barbour and Davis, 1969). These bats frequently inhabit old downtown buildings, particularly those with tile roofs (Murray, 1969).

Eumops perotis (Schinz), the related mastiff bat, is the largest of the common California species, being over 7 in. (18 cm) long. It is brownish in color, and the ears are joined over the forehead and project forward over the eyes. While it prefers rugged, rocky canyons and cliffs, it has become adapted to roost in buildings, and may be found in such cities as Los Angeles and Tucson, usually in colonies of fewer than 100. This bat is so large and its wings are so narrow that regularly used roosts are in locations that allow a drop of about 10 ft (3 m) or more for launching into flight. The massive urine stains and unusually large droppings of *Eumops perotis* are easily recognized signs of its presence (Barbour and Davis, 1969).

Tree Bats (Vespertilionidae)

These bats have furred tail membranes. They are solitary, and sleep during the day while hanging in trees.

Lasiurus cinereus (Palisot de Beauvois), the hoary bat, is about 5.25 in. (13.5 cm) long, and is light brown, heavily frosted with white (figure 383). The ears are relatively short and rounded, and are edged with black. It roosts in the foliage of trees in summer, almost never enters homes, and is seldom seen by man.

Lasiurus borealis (Muller), the red bat, is about 4 in. (10 cm) long, and is brick-red to orange-red.

Fig. 383. Hoary bat, *Lasiurus cinereus*. (From Barbour and Davis, 1969.)

It usually hangs from the petiole of a leaf, but occasionally from a twig or branch. It is well concealed, and when hanging by one foot it closely resembles a dead leaf (Barbour and Davis, 1969).

Bigeared Bats (Vespertilionidae)

In these species, the tail membrane is unfurred. The ears are extremely large, and extend beyond the nose if laid forward.

Antrozous pallidus (LeConte), the pallid bigeared bat, is 4.75 to 5 in. (12 to 13 cm) long, and is dull yellowish buff. Its ears are not joined across the forehead (figure 384, *A* and *B*). It is among the commonest species at lower elevations throughout the southwestern states. The summer roosting places are usually rock crevices and buildings. It forms small colonies, generally of 12 to 100 individuals. A large, dark race, *Antrozous pallidus pacificus* Merriam, occurs on the West Coast. (See group picture on the next page.)

Plecotus townsendi (Cooper), the western bigeared bat, is about 3.75 in. (9.5 cm) long, grayish brown, and the ears are joined acoss the forehead. There are 2 prominent lumps on its nose. In the western states, it is probably the most characteristic bat inhabiting caves and abandoned mine tunnels, occurring as scattered individuals or small colonies (Barbour and Davis, 1969).

Brown Bats (Vespertilionidae)

These bats have unfurred tail membranes, relatively small ears, and a general brownish color.

Myotis californicus (Audubon and Bachman), one of the little brown bats, is 3 to 3.75 in. (7.5 to 9.5 cm) long, and light to blackish brown (figure 384, *C*). It may be either solitary or colonial.

Eptesicus fuscus (Palisot de Beauvois), a big brown bat, is about 4.5 in. (11.5 cm) long, and is glossy, rich brown, with blackish wing and tail membranes (figure 384, *D*). It is gregarious in habit, and in summer forms colonies in attics and barns, between expansion joints beneath bridges, and in similar shelters. Its close association with man results in its probably being familiar to more people in the United States than any other species. In California, however, it is not so well known as the freetail bats.

Economic Importance

Bats are beneficial by feeding on insects or producing "bat guano" for fertilizer. Except for 1 rare nectar-feeding species, all the bats in California eat insects. A foraging bat can capture about 500 insects per hour (Murray, 1969). In some parts of the world, there are chiropterogamous trees, producing large flowers that attract nectar- and pollen-feeding bats. These trees open their flowers only at night, when bats are flying. Vampire bats (not found in the United States) harbor and transmit several viruses, including the one that causes paralytic rabies,[31] which is occasionally responsible for serious losses to livestock (Gray, 1961).

However, a bat need not be a bloodsucker to transmit disease. In the southwestern United States, 1% of resident bats and 2 to 3% of migratory bats were found to be infected with rabies (Constantine, 1967*b*). This disease will be discussed farther on in greater detail.

All bats attempt to bite if threatened. If a disoriented, rabies-infected bat alights on a person, it may bite to avoid falling from its "landing site," or as an aid in climbing, a habit of even normal, healthy bats. In addition, as soon as a bat has landed on him, a person tends to brush it off, causing the bat to bite whether or not it has rabies (Constantine, 1967*b*). When foraging, a bat may dive toward a person in such a way as to suggest an attack to someone unfamiliar with the habits of these animals, but there are very few valid accounts of deliberate attacks upon humans (Murray, 1969).

Bats may be infested with ectoparasites (mites, ticks, bat bed bugs, and fleas). These sometimes attack humans when the bats inhabit buildings, and the argasid ticks can transmit spirochetes (*Borrelia*) that occur in bats and cause relapsing fevers.

Mexican freetail bats can be carriers of St. Louis encephalitis virus in nature throughout the year, and their close relationship with mosquitoes suggests that the bats could be directly involved in the sequence of events that leads to epidemics of the virus (Allen *et al.*, 1970).

Dusty bat manure is a suitable medium for the airborne spores of a systemic fungus disease of man, caused by *Histoplasma capsulatum*. In its

[31] The vampire bat *Desmodus rotundus* (E. Geoffroy) kills almost a million cattle every year. This species has an extremely low tolerance for anticoagulant drugs. Also, bats have a habit of licking themselves and their neighbors in the roost. If a mixture of petroleum jelly and anticoagulant (diphenadione) is spread on the backs of captured bats, each animal in turn will spread the mixture to 2 or 3 dozen others, killing them all. If the anticoagulant is injected into the stomachs of cattle, for several days the blood of the animals contains enough of it to kill the bats that suck their blood. The cost of control is said to be about 0.3% of the value of the annual damage caused by bats (Anonymous, 1972*b*; Kent, 1972).

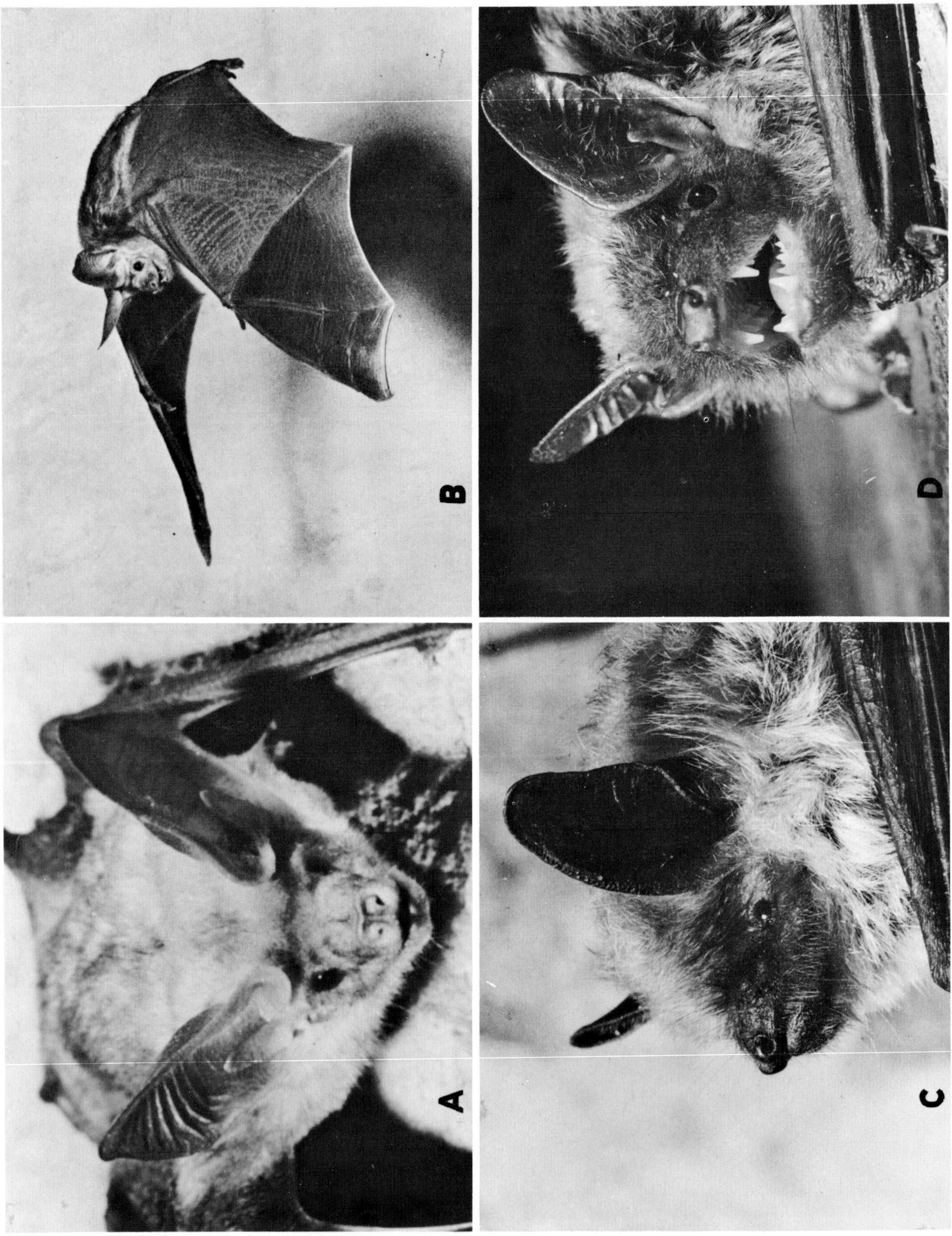
A
B
C
D

mild form, the disease resembles a case of flu, but in a severe case, abscesses and lesions may appear in the lungs, and the symptoms more closely resemble those of tuberculosis. The role of bats in the natural dissemination of the fungus and the creation of new endemic foci remains speculative (Hasenclever, 1972). Nevertheless, workers shoveling or sweeping bat guano should wear dust-proof respirators, change dusty clothes at the end of the work day, and bathe thoroughly (Murray, 1969). Bats possibly also serve as reservoirs of other fungi that cause barber's itch, athlete's foot, and ringworm. They are probably capable of transmitting leptospirosis, an acute systemic spirochetal disease of man, particularly to persons handling bats or living in bat-infested buildings. These animals may also be among the reservoirs of Chagas' disease, transmitted to man by cone-nose bugs, but are probably of minor importance in this respect (Scott, 1961*a*). Aside from rabies, histoplasmosis is probably the only other disease in which bats are implicated that could seriously be called a hazard.

In the United States, bats probably assume their major importance as nuisances in homes and other buildings. Large numbers of them occasionally establish their roosts in attics, wall voids, hollow floors, unused furnaces, belfries, chimneys, or other suitable locations. These are species that would be cave-dwellers in nature. Their excrement and urine cause offensive odors, and their squeaking and scrambling noises may become intolerable.

If various types of insects and mites that infest bats invade the living space of a home and attack its occupants, control measures are required, but these should be confined to control of the ectoparasites. In fairness to bats, it should be stated that many people have lived comfortably for years with these interesting and beneficial animals using their homes, outbuildings, and trees as roosting places.

Control of Bats

The only permanent control is effected through "batproofing." This is difficult, for some species can enter apertures as narrow as 1 centimeter. It may take so long to close off apertures that the job cannot be finished while the bats are out of the building at night. Street (1956) suggested that all but 1 or 2 of the main passageways used by bats should be closed at any convenient time. The entire bat population tends to leave a building within a few minutes after dark. Following this flight, the remaining passageways can be closed. Roosts outside the building should also be eliminated.

It is helpful to be able to identify the species of a bat and to know something about its habits. For instance, crevice-dwellers, such as freetail bats (figure 382), might be eliminated merely by closing up usable crevices between double ceiling beams (Murray, 1969). Screens, such as hardware cloth, can be used to exclude bats, but should not be coarser than 3 meshes to 1 in. (2.5 cm). Wood or sheet metal may also be used. Oakum, a loose fiber obtained by untwisting old hemp ropes, can be used for caulking the narrower passageways. (Steel wool will serve the same purpose.) Batproofing should not be done in summer, for young bats trapped inside a building will produce offensive odors when they die (Scott, 1961*a*). The best time for batproofing is when the animals have left for the season. Fortunately, glass-fiber insulating material, so commonly used in attics, is repellent to bats, probably because of irritation to their skin (Street, 1956; Cope, 1959).

Fumigation is successful, but it can be done only by experienced and licensed pest control operators, and is expensive. It should be resorted to only if batproofing appears to be impractical. As just noted, unless the dead bats are removed, their odor will become offensive. Also, a fumigant leaves no residue to prevent reinfestation. Poison baits are ordinarily not effective, because most species of bats will feed only on live food. Repellents, such as paradichlorobenzene and naphthalene, are effective in confined spaces, but the bats will return after the material has volatilized.

DDT, used as a tracking powder, was formerly the only chemical that was effective, practical, and safe for bat control, but DDT is now banned for structural pest control. The anticoagulant rodenticide chlorophacinone (RoZol), registered by the Environmental Protection Agency and most states as a tracking powder for rats and mice, has also been used experimentally by pest control operators as a tracking powder for bats. It is reported to be successful, but to date (early 1975) it has not yet received EPA registration.

Fig. 384. Some common California bats. *A*, pallid bigeared bat, *Antrozous pallidus; B, Antrozous pallidus* in flight; *C*, little brown bat, *Myotis californicus; D*, big brown bat, *Eptesicus fuscus.* (Picture of bat in flight, courtesy R. T. Orr; others from Barbour and Davis, 1969.)

RABIES

Rabies is a virus that affects all warmblooded animals. It is one of the oldest diseases mentioned in history, and was associated with the bite of a rabid dog by Aristotle. It is now distributed in all continents except Australia. There is said to be no cure once the symptoms become apparent, and death usually occurs in 3 to 5 days after symptoms begin. The symptoms in man are at first mental depression, restlessness and fever, and then uncontrollable excitement, with excessive salivation and muscle spasms that cause excruciating pain and are easily stimulated by drafts or attempts to drink water. Therefore, the victim refuses water even when extremely thirsty.

The greatest danger is from bites of diseased animals that result in deep wounds, particularly if they are near the head, as the disease attacks the central nervous system. Practically all cases of rabies occur when virus-laden saliva enters a wound (Tierkel, 1963). The incidence of rabies in rodents is almost nil. The incubation period for the disease varies from as brief a period as 10 days to as long as 2 years. The average period is about 2 months. Approximately 20% of human cases of rabies can occur within the first 30 days after the person is bitten by an infected animal, 80% within the first 60 days, and only 1% after 4 months (Webster, 1942). Early in the incubation period, a series of rabies vaccine "shots" can be administered before symptoms of the disease appear. It is important to be able to locate and identify any dog that bites you, so that it can be determined if the animal is rabid. If bitten by a small animal, such as a skunk or bat, it is important to capture it and bring it in for tests. Any bite by a skunk or bat should be considered as potentially dangerous. Wounds should be cleaned, bleeding promoted, and antiseptics applied. This should be done within a half hour after the bite, but even then it cannot be entirely depended on for prevention (Merchant and Barner, 1964). Rabies is not highly contagious and, regardless of treatment, appears in only about 15% of lightly exposed to 70% of heavily exposed victims.

Whereas the incidence of reported rabies in domestic animals (principally cattle, dogs, and cats) in the United States during the period 1953–1965 decreased from 7,344 to 1,326, the incidence of reported rabies in wild animals (principally skunks, foxes, and bats) increased from 1,476 to 3,257. The increase was particularly great in skunks and bats. All areas of the country were affected (NPCA, 1967). There is little overlap between areas where skunk rabies is dominant and where fox rabies is dominant. Unlike bats, skunks with rabies sicken and die, but they withstand the disease better than many animals, and may be worse problems, for they are more likely to bite dogs and other pets (Getze, 1972).

In a dog, the first sign of rabies is a change in disposition. He may appear to be troubled or distracted, and become either less friendly with other dogs and people or more gentle and friendly, but soon he becomes restless and quiet. The disease may then assume 1 of 2 general forms. In "furious rabies," the dog strays from home, snaps and bites at whatever he encounters, swallows cloth, stones, or sticks, and froths at the mouth. His lower jaw drops and his bark becomes hoarse or disappears, the hind legs grow weak, and prostration and death soon follow. In "dumb rabies," the dog is not especially irritable, rarely bites, and is lethargic. A period of tremors and paralysis, especially paralysis of the lower jaw, is followed by complete prostration. After lying quietly for a period that can vary from a few hours to 2 days, the animal dies (Webster, 1942).

If wild animals which are normally nocturnal, such as skunks, raccoons, and bats, are seen in the daytime, or if they appear sick or abnormally friendly or aggressive, such behavior is unnatural and is a good indicator of rabies.[32] They should be avoided. Bats should be suspected of being rabid if they appear to be ill or are on the ground or within easy reach on trees, shrubbery, or buildings, or if they are seen in erratic flights, especially with frequent landings. Infectious bites usually occur when people attempt to handle such sick bats. They are so agile that they can bite almost any way they are held, and their teeth are so sharp that the bite may not be felt. If a person is bitten, he should try to capture the biting animal so that it may be tested for rabies infection, and then immediately seek medical attention. As already stated, heavy gloves or tongs should be used when handling bats. Over a 5-year period, about 39 bats per year were found to be rabid in California. Many species were implicated, but most frequently found to be rabid were the freetail and hoary bats (Murray, 1969). The highest rates of infection of bats in the southwestern United States were found in autumn (Constantine, 1967*b*).

[32] Abnormal daytime behavior should be distinguished from the normal behavior of red and hoary bats that roost in concealed locations in trees during the day.

It is not unusual for bats to enter inhabited rooms at night. If they appear to be sick or paralyzed, rabies should be suspected and local health authorities should be contacted. However, inactive bats may merely be torpid. Rabies-infected bats are found in a relatively scattered pattern. The way in which the disease circulates among bats is not well understood. In California, evidence has not yet been produced that bat colonies suffer substantial dieoffs, and the finding of 1 or even of several rabid bats in a community does not justify a local eradication effort (Murray, 1969).

The death of 2 men from rabies after entering Frio Cave near Uvalde, Texas, each insisting that he had not been bitten by a bat or any other animal, led to investigations by the federal Communicable Disease Center of possible transmission of rabies virus by air as well as by bite. Frio Cave is inhabited principally by millions of Mexican freetail bats, *Tadarida brasiliensis mexicana*. Experiments in the cave with various species of wild and domestic animals gave evidence that caves might be potential sources of rabies virus infection for man and for certain predators, and there is evidence that animals susceptible to infection by rabies virus by the air route can transmit the disease to more resistant species by biting them (Constantine 1962, 1967*a*; Tierkel, 1963).

There is also circumstantial evidence that the rabies virus emanates from bats in droplets (urine aerosol), thus permitting transmission by contact. Also, a form of airborne transmission is believed to be possible by the residue remaining after the droplet has dried, referred to as "aerosol transmission" (Langmuir, 1961). The upper respiratory tract can filter out larger droplets, but smaller particles can pass directly into deeper parts of the respiratory system. The principal factors favoring aerosol transmission of the rabies virus in bat-infested caves are a high bat population and minimal air circulation in the caves (Constantine, 1967*a*). In California, colonies of even a few thousand bats are seldom encountered, and the conditions for potential nonbite transmission of rabies are rare, if they occur at all (Murray, 1969).

BIRDS

Although birds are sometimes beneficial and are also admired for their beauty and song, some species can be pests if they (1) eat or destroy agricultural products; (2) successfully compete with and reduce the number of desirable birds; (3) roost or nest around human habitations, where they deface buildings and walkways with their droppings and irritate people with their noise; (4) contaminate foods [about 4% of all United States wheat is contaminated by bird droppings (Scott, 1961*b*)]; (5) cause costly and dangerous accidents by colliding with aircraft; or (6) are carriers of diseases or ectoparasites. The species most often considered to be pests in urban areas are pigeons, starlings, and sparrows. All three were introduced into North America by early European settlers: the pigeons as domestic food birds, and the starlings and sparrows apparently for sentimental reasons. A fourth species, the cliff swallow, is a pest when it attaches its mud nests beneath the eaves of buildings. The monk parakeet, which is indigenous to portions of South America, is at present (1975) observed in only small numbers in the United States, but appears to have great potential as a severe agricultural pest and a general nuisance.

Pigeon, *Columba livia* Gmelin (Columbidae)

The pigeon is by far the most important of the bird pests to human habitations. It is now established over most of the world, particularly in the larger cities. It may spread over the countryside, but has not established itself as a truly wild bird, for it is dependent upon man to a great extent. Those inhabiting open country during the summer tend to seek urban areas or farm buildings in winter. The feral pigeon, as well as the starling and house sparrow, are not among the birds protected by the Federal Migratory Bird Treaty Act. They are, however, protected by some states or by some communities within states. Also, the acceptability of various methods of controlling pigeons varies from state to state. For example, poisoning of pigeons, starlings, or sparrows is illegal in Connecticut, Maryland, New York, North Dakota, Pennsylvania, Rhode Island, and West Virginia. Food-handling industries would be well advised to take bird-protection laws into consideration when choosing a location for a plant, so as to avoid having to deal with both Food and Drug inspections and an aggressively enforced, communitywide bird sanctuary law (Spear, 1968).

Description. The pigeon is easily recognized because of its large size, stout body, and rather short legs, short bill, and smooth, compact plumage (figure 385). There are many domesticated varieties, differing greatly in color patterns, that have been derived from the rock pigeon of the coasts of Europe. Their call is the familiar "coo-a-roo." Pigeons are remarkably long-lived, commonly living for 15 years and occasionally for 30 years or more (Scott, 1961*b*; Reilly, 1968).

Zenaidura macroura (L.), the mourning dove, is the only other bird with which the pigeon might be confused. Like the pigeon, it is commonly found in urban environments, but is not

Fig. 385. Pigeon, *Columba livia*. (Photo by Torrey Jackson.)

considered to be a pest. In fact, it is an important game bird, protected by state and federal legislation and by the provisions of the Federal Migratory Bird Treaty with Canada and Mexico. The mourning dove is a smaller bird, buff-colored and, unlike the pigeon, it has long central tail feathers that give the tail a distinctly pointed appearance, even when in flight. The pigeon's tail is fan-shaped.

Biology and Habits of Pigeons. The nests consist of crude and flimsy platforms made of sticks, twigs, and grasses, built on the ledges of buildings or cliffs, or in caves. There are 1 or 2 white eggs (normally 2), requiring 17 to 19 days for incubation. The young birds (squabs) are fed predigested food ("pigeon milk"), which gradually becomes thicker in consistency until the birds are weaned and ready to leave the nest. The young pigeons are 35 to 37 days old when they make their first flight. Several broods are reared each year. More eggs are laid before the young of the preceding brood are weaned.

Pigeons eat seeds and grain, other vegetable matter, some fruit, some animal food such as insects and spiders, some garbage, and enough grit to ensure proper digestion. About a pound (roughly, half a kilogram) of food is eaten per week by adults. Pigeons are good scavengers. The lack of adequate sanitation, as well as deliberate feeding of the birds by humans, and many "cliff-type" nesting sites, particularly on the older buildings, are the principal factors that attract these birds to cities.

Pigeons as Public Nuisances

Many people have a high level of tolerance for "pest" birds, particularly pigeons. They are sometimes kept as pets, and then are usually marked by leg bands. They also afford pleasure to people who feed them, especially in city parks, and to those who breed them or use them for racing. They are considered to be among the most intelligent of birds, and can be easily trained, as for theater acts. They are used as laboratory animals in disease investigation and as sources of pigeon-pox virus used for immunizing laying hens against chicken-pox. Their droppings are used as manure. More than 1,000,000 lb (454,000 kg) of squab meat are marketed each year as a food delicacy (Scott, 1961*b*).

Probably the greatest indictment against pigeons, as well as other pest birds, is the defacement of buildings, statues, and automobiles caused by their droppings. Bird excrement may also kill lawns and shrubbery when it becomes too abundant, and will cause objectionable odors. Bird nests can obstruct drain pipes, mar window-sills, interfere with awnings, and create hazardous footing on fire-escapes. Bird droppings, feathers, and nesting materials often contaminate grain, and large quantities of it may be consumed by pigeons that live around grain-handling establishments (Scott, 1961*b*).

Pigeons as Carriers of Disease

Pigeons may spread diseases to man, generally through their droppings or by respiratory droplets. However, their importance in this respect can be overemphasized, and Scott (1961*b*) recommended that, in the absence of human disease outbreaks demonstrably associated with pigeons, control should be directed against excessive numbers rather than for eradication. The following medical information is taken from his discussion of the problem:

> A psittacosis-like virus disease known as pigeon ornithosis is transmitted to humans by pigeon droppings or respiratory droplets, and can also be transmitted from man to man. Symptoms are usually mild in humans. Also, only a few pigeons are obviously ill, even though as many as 30 to 75% of the birds may be infected.
>
> Although pigeons are known to be carriers of encephalitis viruses, which are normally transmitted from birds to man by mosquitoes [see chapter 9, under

"Mosquitoborne Diseases"], the exact role of pigeons in the epidemiology of this disease complex is as yet unknown. Nevertheless, it has been suggested that surveillance for encephalitis in pigeons might give warning of incipient epidemics in man and in horses.

Newcastle disease is a highly contagious and fatal virus neuropneumonia of birds that is extremely rare in man. Chickens and turkeys rather than pigeons are considered to be probably the sources of most human infections, and transmission is believed to be effected by respiratory droplets or ectoparasites.

Aspergillosis, a fungus disease of the human ear, sinuses, lungs, or skin, caused by *Aspergillus fumigatus,* is so ubiquitous that pigeons, although commonly infected, are not considered to be significant sources. The same may be said about "thrush," a fungus infection of the mouth, respiratory system, vagina, skin, or fingernails of humans. It is caused by the fungus *Candida albicans,* which can cause high mortality among pigeons, particularly the young birds.

Dusty pigeon manure may harbor airborne spores of *Histoplasma capsulatum,* the causal agent of histoplasmosis, a systemic fungus disease of man. This disease involves the lymph nodes of the tracheae and bronchi, usually without symptoms, but is sometimes accompanied by fever, anemia, a reduction of leucocytes in the blood, and localized lesions of the skin, mouth, and throat (Case, 1969). One investigation revealed that children living in buildings where pigeons roosted or nested had an infection rate 3 times that of children living in pigeon-free buildings. Another systemic fungus infection of man, caused by *Cryptococcus neoformans,* has likewise been associated with pigeon manure.

Pigeons frequently harbor the protozoan *Toxoplasma gondii,* the cause of toxoplasmosis in man and other mammals. Toxoplasmosis is often subclinical in manifestation, or resembles a mild influenza, but can involve damage to the central nervous system and eyes, particularly those of infants.

Transmission probably takes place from pigeon to man, and although the mode of transmission is unknown, it is believed to be by respiratory droplets, fecal contamination, contact with infected tissues, or by ectoparasites.

Pigeons are known to be susceptible to 2 bacterial diseases, pseudotuberculosis (*Pasteurella pseudotuberculosis*) and avian tuberculosis (*Mycobacterium tuberculosis* var. *avium*), both of which are rare in humans. However, the common bacterial food poisoning of man called salmonellosis (*Salmonella typhimurium*) is found in about 2% of pigeon feces.

The protozoan *Eimeria labbeana,* the cause of pigeon coccidiosis, may be found in 10 to 90% of pigeons, and results in the death of many of them. It is apparently not transmissible to man. Pigeon-pox, a widespread viral disease believed to be mechanically transmitted from pigeon to pigeon by mosquitoes, is also not transmissible to man.

Flies often breed in pigeon excrement, pigeon nests, and dead pigeons. They can be carriers of several kinds of disease organisms.

Ectoparasites of Pigeons

These birds harbor many ectoparasites. Only those species that can infest man and cause discomfort or transmit disease will be discussed.

Cimex columbarius (Jenys), the pigeon nest bug, related to the bed bug, may bite but does not transmit human diseases.

Triatoma rubrofasciata (De Geer), a conenose that infests pigeons, bites man, and can transmit Chagas' disease.

Pseudolynchia canariensis (Macquart), the pigeon fly, may bite humans, but is not known to transmit any disease.

Ceratophyllus columbae (Stephens) and *C. gallinae* (Schrank), two fleas that infest pigeons, will also bite humans, but are not transmitters of disease (Scott, 1961*b*).

Argas reflexus (F.), the pigeon tick, has the same peculiarities—it may bite, but will not transmit disease.

Dermanyssus gallinae (De Geer), the chicken mite, infests pigeons, may cause a severe dermatitis in man, and has been found to be infected with eastern, St. Louis, and western encephalitis viruses.

Ornithonyssus sylviarum (Canestrini and Fanzago), the northern fowl mite, another ectoparasite of pigeons, sometimes bites humans, and has been found to carry Newcastle disease as well as western and St. Louis encephalitis.

Insect Pests from Pigeon Nests

The nests of pigeons and other birds are sometimes constructed within buildings, and are not only sources of ectoparasites of man, such as chicken mites and northern fowl mites (just discussed), but also of fabric pests (clothes moths and carpet beetles) and pantry pests (drugstore beetles, spider beetles, cryptophagid beetles, lathridiid beetles, mealworms, cadelle beetles, and dermestids), all of which are discussed in other chapters.

Starling, *Sturnus vulgaris* L. (Sturnidae)

Whereas in Eurasia starlings migrate from one part of their range to another, in America they congregate in large roosting flocks wherever food is plentiful. On very cold days, they may seek places around structures, taking advantage of heat coming from rooftoops and near chimneys. They are nuisances not only in cities, but also cause serious economic losses by consuming grain around grain elevators and on feedlots, and by damaging various agricultural crops in rural areas.

Fig. 386. Starling, *Sturnus vulgaris*, with spangled winter feathers. (Photo by M. V. Critch.)

Description. The starling is 7.5 to 8.5 in. (19 to 21.5 cm) long, with black plumage, including a very short, slightly notched, squaretipped tail. In summer, there is a dorsal purplish gloss which becomes more brownish in winter. Also in winter, the adults become spangled with white and buff (figure 386) (the white spangles suggesting the common name), and the bill changes from yellow to blackish. The summer starling has been described as "a blackbird that has lost its tail" (Cooke, 1928). The voice is cacophonous, but the bird is a good mimic and can imitate many other species. Its more pleasing notes are usually produced during such imitations (Reilly, 1968).

Biology and Habits. Starlings nest in any tree cavity, nesting box, or deserted or pre-empted woodpecker hole of suitable size. The nest is packed with grasses, twigs, straw, and miscellaneous debris. No sanitation is practiced, and when nests are used in successive seasons, they become foul-smelling. The female lays 2 to 8 eggs, usually 4 to 6. These are white, or pale bluish or greenish white, and are incubated for 11 to 13 days by both sexes. The first flight takes place at 19 to 22 days. The food is almost 60% animal matter, mainly insects and other small invertebrates, but vegetable matter, largely berries and other fruits, is also consumed, as well as some seeds (Reilly, 1968).

Starlings as Carriers of Disease

At least some of the diseases carried by pigeons have also been found in starlings and other birds. There has been a high incidence of positive tests for histoplasmosis in people living near starling and blackbird roosts, particularly among school children. Symptoms of the disease are described in the section on pigeonborne diseases. Pest control operators have been warned that eliminating birds from their roosts is not enough to control the disease; the roost area must also be decontaminated. Some authorities recommend that only operators displaying some immunity to histoplasmosis should work in roosts, and that respirators or masks should be used. Application of either 3% creosol or 3% formalin can be made for the decontamination of roost areas (Case, 1969).

House Sparrow, *Passer domesticus*(L.) (Ploceidae)

The house (English) sparrow, first introduced into North America in the 1850's, has since spread across most of the continent and is also present in Hawaii. Early in the twentieth century, the house sparrow was one of the most abundant birds in the United States. It followed the highways, where an abundant supply of half-digested grain was provided in the droppings of horses, and along railroads where grain cars, particularly in autumn, scattered their cargo along the entire route. The house sparrow population decreased as automobiles became common and the number of horses declined. However, house sparrows are still locally common in cities and about farm buildings, and are scattered throughout open country, particularly in the southern states.

Description. The house sparrow is 5.5 to 6.25 in. (14 to 16 cm) long. The crown, center of the nape, rump, and tail of the male are light to medium gray. The back and wings are buffy brown, and the wings have 2 white bars. The throat and adjoining area of the upper breast are black. A small spot behind the eye, as well as the cheeks and the remainder of the underparts, is very pale gray. The female is lighter buffy brown above, has a white throat, and a dark streak through the eye region. The sexes may be distinguished by the black throat and gray crown of the male (figure 387) and the white throat and dark eye stripe of the female. In both sexes, the bill is black in summer and dusky brown in winter, the eyes are brown, and the legs and feet are buffy brown. The voice consists of "twittering, chirping, and loud cheeps" (Reilly, 1968).

Biology and Habits. The courtship of sparrows is worthy of note because of the violent, acrobatic posturings of the male before the female, an activity that deteriorates into vicious fights between competing males and sometimes between females as well. House sparrows rarely build their nests in woods, thickets, or fields as other sparrows do, but swarm to cities for this purpose during the breeding season. The nest may be located near buildings, in hollow trees, in birdhouses, on ledges, or in roof-drain downspouts, crevices, or abandoned woodpecker holes. When the birds become so abundant that such places are no longer available, they may build their rough, bulky, covered nests among the branches of shade trees. In any location, their nests are unsightly collections of grass, straw, and debris, sometimes close enough together to be considered as colonial nests. Nest-building may begin as early as February, and both sexes participate.

The female lays 3 to 7 eggs, commonly 5. These are white, pale blue, or pale green, with a few gray or brown spots. She incubates them for 11 to 12 days, and there are several broods per year. The young birds fly at about 15 days. They are fed mainly insects, but the adults feed on grain, other seeds, and various vegetable materials, as well as some insects and other animal matter (Pearson, 1928; Reilly, 1968).

Fig. 387. House sparrow, *Passer domesticus* (male). (Courtesy J. H. Gerard, National Audubon Society.)

House Sparrows as Pests

House sparrows sometimes enter a house in large numbers via the chimney. A resident of Redlands, California, estimated that there were 400 sparrows in the living and dining room of her home after they had "come down the chimney like crazy." They perched on furniture, fouling the upholstery, and hung upside down from curtain rods and clung to the curtains. Police were called, and sprayed Mace into the chimney to prevent the entrance of more sparrows.

Sparrows tend to destroy the nests of small native birds and drive them out of towns, substituting their noisy chatter for the more pleasant songs of the displaced birds (Pearson, 1928). Around power lines and in electrical substations, their nests have sometimes created serious fire hazards (Truman, 1961*b*). They cause serious economic loss where large populations feed around poultry establishments. They also are suspected to be carriers of certain poultry diseases.

House Sparrows as Carriers of Disease

Encephalitis occurs in man and other mammals only when there is a sufficient rate of infection in birds and when mosquitoes are present to transmit the virus from a bird to the next host. The mortality rate from eastern encephalitis has reached as high as 66% among children under 16 years of age. Because of (1) the high rate of antibodies to encephalitis virus sometimes found in house sparrows, indicating that the birds have had the virus for some time, and (2) the close association of these birds with man, there is reason to believe that sparrows may be among the most important bird reservoirs for this disease (Case, 1969).

Cliff Swallow, *Petrochelidon pyrrhonota* (Vieillot) (Hirundinidae)

Cliff (eave) swallows (figure 388) are migrant, beneficial birds that are protected by law. Their food is composed entirely of insects caught in flight.

During March through June, their nesting period, they can become pests by building their mud nests beneath the eaves of buildings. The mud and fecal matter are dropped down onto the walls beneath the nests and onto walks and patios in the vicinity of the nesting sites. When the nests

Fig. 388. Cliff swallow, *Petrochelidon pyrrhonota*, and mud nest. (Photo by M. V. Critch.)

are removed from their positions on structures, they often leave stained areas, and the paint will come away with them.

Description. The mature swallow is 5 to 6 in. (12.5 to 15 cm) long. Its wings, back, and square-tipped tail are sooty black, the rump and forehead buffy yellow, the cap metallic blue-black, the throat and face brownish red, and the underside light smoke-gray. The immature birds are much duller-colored, and the pattern is not so sharply defined. A buffy-gray shade replaces much of the brownish red (Reilly, 1968).

Biology and Habits. Cliff swallows spend their winters in South America, and annually migrate northward to the United States. Their nesting sites are used year after year, and many of the birds will return to use the same site at which they nested the year before. If old nests are still present, the birds will re-use them. It is only a matter of chance if a bird uses the same nest it formerly occupied. If old nests are not used, the birds set about gathering quantities of mud that are plastered against a vertical wall immediately beneath an overhang, usually an eave.

In California, in areas where there are many bridges and culverts over streams, gullies, and dirt-lined irrigation ditches carrying water and providing an abundant supply of mud during the spring months, as in the rice-growing areas of the Sacramento Valley, almost every swallow colony will build its nests under a bridge or culvert. In southern California, where the winter rainy season is generally over before the swallows arrive, such structures are abundant but are rarely used as nesting sites. The gullies beneath the bridges are generally dry when the birds are looking for mud for nest construction. On the other hand, mud is available near dwellings, where lawns and gardens are watered frequently, and most of the swallow nests are under the eaves of buildings.

Mayhew (1958) noted that most of the nests were found attached to single-family homes, but that approximately 200 birds were making use of buildings on the Riverside campus of the University of California for nesting.

The nests are gourdshaped, about 6 in. (15 cm) in diameter, with a necklike round entrance about 2 in. (5 cm) in diameter (figure 388). During nest construction, the birds do not inhabit the nests at night, and begin to do so only when they are nearly complete. The eggs are laid during April and May, and the birds are very active tending them and feeding the growing young. By the end of June, the young have left and the flock deserts the nests.

Monk Parakeet, *Mycopsitta monachus* Boddaert (Psittacidae)

The monk parakeet is indigenous to eastern Bolivia, Paraguay, Uruguay, and Argentina, south to Rio Negro in open woods, cultivated lands, and palm groves (Meyer de Schauensee, 1970). It is 11 in. (28 cm) long, and is of a general greenish color, except for the gray forecrown and breast, and has a long and pointed green or bluish-green tail. This parakeet is sometimes purchased as a pet in the United States, but is usually soon released because of its habit of biting and its unpleasant squawking. The released birds have been able to thrive in the wild, and an estimated 4,000 monk parakeets are now nesting, breeding, and becoming severe pests along the Atlantic Coast from Florida to Maine and westward to Michigan and Illinois. They are reported to be clipping antenna wires, feeding on fruit of garden trees, and stripping ornamental shrubs for their nest material.

Judging from its history as a severe pest in agricultural areas of South America, the monk parakeet can also become an important agricultural pest in the United States (Anonymous, 1973*c*).

Collisions of Birds and Aircraft

Of the birds discussed in this section, pigeons, starlings, and sparrows are among the many species that can cause accidents to aircraft, although gulls and waterfowl are responsible for most of them (Anonymous, 1967*b*). Hundreds of collisions of birds with aircraft occur every year. Bird problems are most likely to occur around airports near bodies of water during spring and fall migration periods, when bird populations, flocking tendencies, and mass movements are at their peaks. In some cases, garbage dumps in the vicinity or in the flight line of an airport constitute potential hazards from gulls and other scavenging birds. Over one 4-year period, crashes caused by birds flying into planes resulted in the death of almost 100 people. In all cases, an engine or control mechanism was destroyed, or the distraction of the pilot resulted in his loss of control of the plane. Even if the pilot can see the birds approaching, he cannot veer a big jet or even a light plane out of the path of their rapid flight (Anonymous, 1967*b*).

Bodies of water, or trees, shrubs, and brush on airport grounds may contribute to a bird problem because they can provide places in which birds can feed on insects, seeds, or berries, raise their young, or roost. Grasses and weeds, especially between runways, often create problems. If the grass is more than 4 or 5 in. (10 or 12.5 cm) tall, it can provide cover for ground-nesting birds and for mice and rabbits that attract hawks and owls. If it is shorter, it provides areas in which birds such as gulls, starlings, robins, and blackbirds can find seeds, worms, and insects. After rains, worms seek dry places, and may become stranded on runways and attract birds. To prevent bird hazards, airports must be "biological deserts" through the elimination of vegetation and small bodies of water (ponds and ditches bordering the airport) and low spots that can catch rain or waste water. Dead fish tossed up by breakers or hauled to land by fishing boats can lure sea birds which, once they enter an area, can plague near-by airports.

Noise stimuli, such as distress signals and various explosive devices, have been used to disperse birds at airfields. Trapping has also been used to eliminate pigeons and starlings. At one airport, about 2,000 starlings settled in a hangar, causing a mess, fouling aircraft engines, and becoming hazards to planes using the field. A contact avicide reduced the number of starlings to about 75 in a period of 3 weeks (Anonymous, 1967*b*). Avitrol® bait has been used to disperse gulls from airports, garbage dumps, and other congregating areas. The gulls picking up the bait, usually bread or pellets, fly erratically and emit distress cries, frightening away the remainder of the feeding flock (M. W. Cummings, correspondence). Starlings are frightened away in a similar manner. Avitrol is discussed in greater detail under "Poisoning."

Control of Birds

Because of usually adverse public reactions to the killing of birds, various other procedures have been attempted to reduce bird problems. Scott (1964*a*) recommended sanitation and stoppage as the most satisfactory means of controlling pigeons, particularly in view of the public's generally high regard for these birds. He pointed out that periodic community cleanup campaigns could be helpful. Concentration of such activity in winter takes advantage of the low ebb in pigeon populations during cold weather, caused by disease, shortage of food, exposure, and predation. Another advantage is that the dieback of weeds and other plant life reveals accumulations of food and nesting materials, which can be removed. Effective refuse collection and disposal are also important in any campaign against pigeons. If feedlots, stockyards, and other commercial animal-handling facilities are major pigeon-breeding grounds and cannot be transferred outside the city, efforts should be made to modify procedures so as to limit the amount of food available to pigeons. Potential harborage structures should be "pigeon-stopped." In addition to structural modifications, flashings of wood, masonry, plastic, or metal installed at 45 to 50° angles on roosting and nesting surfaces have been found to be effective.

The planning of buildings to eliminate roosting places such as ledges and overhanging cornices would, of course, be a big step toward bird control. Simple structural changes can sometimes be made to eliminate roosting ledges in existing buildings (figure 389), or screens may be used to prevent birds from alighting on favored resting places. Once they have moved elsewhere, they will feel no particular urge to return (Kalmbach, 1954).

Repelling Pigeons, Starlings, and Sparrows

Various adhesives, pastes, and jellies have been used with limited success in attempts to keep pest birds away from buildings. These may consist of

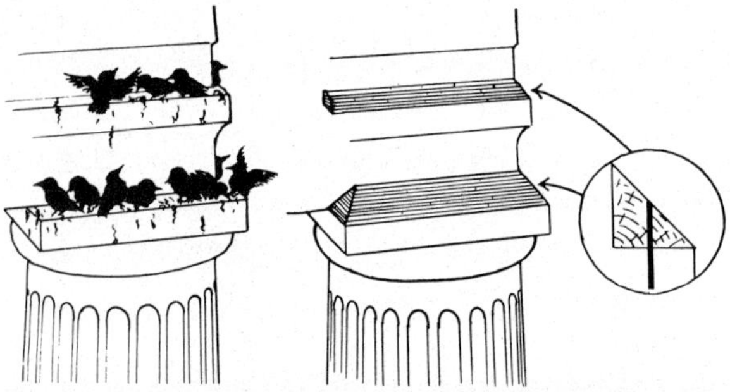

Fig. 389. Modification of a building to eliminate bird roosts. (From Kalmbach, 1954.)

thick, tacky materials or may be jellylike. Some kinds can be sprayed, but these are usually applied to trees. Surfaces of buildings where birds roost or nest should be cleaned and, if they are porous, should be treated with some type of liquid sealer before adhesive materials are applied. Nondrying adhesives, in layers as deep as 0.5 in. (12 mm), may be applied to the sealed surfaces. Strips of glue applied no more than 3 or 4 in. (7.5 or 10 cm) apart, so that birds cannot land between them without becoming entangled, are often effective repellents. However, pigeons sometimes build their nests over the adhesives, and continue to use the place as a nesting area. Glue or jelly applications may deface the building. They can also be hazards to window cleaners and other workmen, who may track the glue or jelly over floors and carpets and be caused to slip and fall (Oderkirk, 1958; Truman, 1961*b*).

Mist blowers can be used to blow a fine mist of weak ammonia solution into trees or around buildings where birds are roosting. Several applications may be needed to discourage the birds, and the treatment will probably have to be repeated a number of times during the season (Truman, 1961*b*).

Repelling Cliff Swallows

As already stated, swallows will nest on the same building in successive years, and deface the walls and areas beneath their nests with mud during nest construction and with their droppings thereafter. Washing the nests down with a stream of water has been effective in some cases, but must be done daily for extended periods because the birds are persistent, and will continue to apply mud to the original nesting site. Two other methods have been tested and shown to be effective in the prevention of swallow nests. Their nests are constructed beneath the eaves of buildings, adhering to the vertical wall face and a portion of the eave. By stringing a wire along the junction of the wall and the roof overhang, and draping a 12-in. (30-cm) curtain of polyethylene or aluminum foil over it, the swallows are prevented from attaching their nests to the wall because the new surface is too smooth. Although this method has been effective, it is somewhat unsightly. The second method is a bit more complex, but does not require the installation of an unsightly material under the eaves, and will keep the swallows away for several years. Two fine nichrome wires are installed under the eaves parallel to the wall and 1 and 3 in. (2.5 and 7.5 cm) from it. The wires are then connected to a cattle-fence charger. When swallows attempt to build their nests, they receive strong but nonlethal shocks from the wires, and will move rapidly to some other nesting site (R. E. Wagner, personal communication).

Trapping

Birds may be trapped alive in such places as city parks and on the roofs of buildings. Kalmbach (1954) recommended trapping pigeons at a number of feeding areas, using different types of traps on a rotational basis and thereby reducing the tendency of the birds to become trap-wise. After a few successful trappings, the feeding area should be left alone until the remaining birds have become reaccustomed to the location.

Traps should be constructed so as to permit the continuous entrance of birds for a long period, for it is usually not practical to visit traps more often than every day or two. One such trap is shown in figure 390. A coneshaped configuration of wire guides the birds through an aperture into a large cage that they can enter but from which they cannot escape. Grain may be placed in the

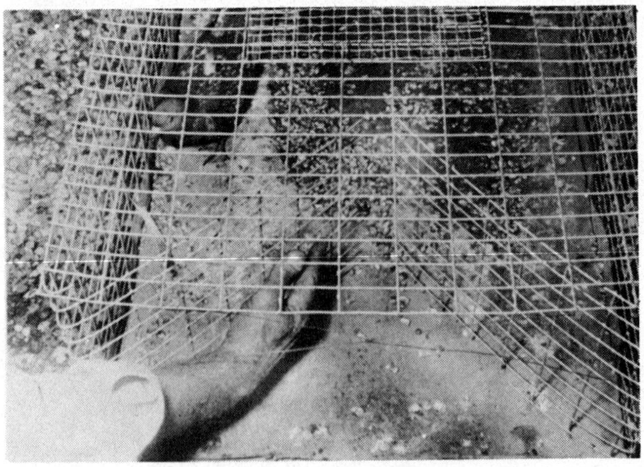

Fig. 390. Wire trap for pigeons or other pest birds. (From Hockenyos, 1962.)

cage as bait, but sometimes a container of water is even more effective (Hockenyos, 1962). Another type of trap utilizes a principle employed at pigeon lofts. The entrance to a screen cage consists of vertical rods that are hinged at the top (figure 391). A bird trying to get into the cage to reach

Fig. 391. Bird trap utilizing hinged rods, as seen from inside. (From Kalmbach, 1954.)

the food or water can easily swing the rods inward, but after it has passed through, the rods return to their original positions. The figure shows the door mechanism as seen from inside the cage. Pest birds that are trapped may be killed, as by gassing, but songbirds should be released, and banded pigeons should be released or returned to their owners.

Poisoning

Poisoning is considered to be the most practical means of getting rid of pigeons and sparrows. Generally, poison baits are used, with strychnine sulfate or strychnine alkaloid as the poison. However, the use of poisons by anyone other than a licensed professional pest control operator should be discouraged. Any interested person should contact a pest control operator or, in California, he may also contact his county agricultural commissioner.

In many cities, exposure of poisoned bait in places accessible to pigeons is illegal. Even when poisoned in secluded and restricted places, pigeons may die where they can be eaten by cats or dogs, and can thereby also poison these animals (Kalmbach, 1954). In cities, birds will usually feed on any grain, while those around farms or feed mills are best attracted by grains that were not present before the baiting was begun. When in doubt, a trial with nonpoisoned wheat, oats, corn, barley, rice, or sorghum, as well as a commercial birdseed mixture, will indicate the preferred grain for a given species at a given place and time of year.

An instance was recorded in Columbus, Ohio, in which an ultra-modern drive-in restaurant was invaded by 800 to 1,000 sparrows that roosted on exposed metal girders or supports of the structure. Droppings were reported to be 1 in. (2.5 cm) thick in some places, including the sidewalk. The control program included the use of a strychnine alkaloid formulation recommended by the United States Fish and Wildlife Service. After a few days of "prebaiting" without poison, the poison bait was placed on the roof and at ground levels at one end of the parking lot. About 75% of the birds were killed (Oderkirk, 1958). In another sparrow control campaign, after 5 days of prebaiting in a large section of the city of Logansport, Indiana, the above poison bait was scattered along all ledges that could be reached from the platform of a fork-lift truck. The baiting was finished by 11:30 a.m., sparrows started dropping at 1:30 p.m. (about the time when they normally flew into the city in large numbers), and by 4:00 p.m., "they were falling in such numbers that unfavorable public opinion developed." The birds were picked up and disposed of. Some were still falling a month later. In the Logansport campaign, no success had resulted from a wide variety of scare devices and noisemakers. Starlings were apparently unaffected by the Logansport poison campaign, for no dead starlings were observed. Starlings normally roost higher than sparrows, and no baits were placed on the upper ledges frequented by them. Thus, it was not determined if starlings could have been poisoned (Fitzwater, 1958).

Starlings nest and hunt for food in the country, and fly into the city only to roost; therefore, they usually cannot be poisoned in the city. They can sometimes be poisoned in outlying public garbage dumps, particularly when the snow cover is heavy (Truman, 1961*b*).

DRC-1347 has been proposed as a toxicant specifically for starlings. The oral LD_{50} (mg/kg) of DRC-1347 for starlings is only 4.2, compared with 13 for pigeons, 320 for house sparrows, and 420 for sparrow hawks. It has been used successfully both in poison baits and as a contact perch toxicant. Because of its selective toxicity, it is believed that this material could be used effectively against sparrows without killing predatory birds and mammals. Applied to the feet of sparrows in laboratory tests, it was even more toxic than endrin and fenthion, which have been used as perch toxicants but are not selective in their toxic action. However, DRC-1347 does not appear to be registered for use in controlling starlings, and thus cannot be used except for qualified research programs. As with many other promising compounds, the current antipesticide sentiment is not conducive to such research.

Tribromomethanol. Milwaukee has had a severe problem with nuisance birds because it is a grain-shipping port and a brewing center. Spillage of grain from boxcars attracts pigeons and other pest birds from a wide area. In order to get rid of such birds without antagonizing bird-lovers or violating state conservation laws, an anesthetic drug, tribromomethanol, was used. One lb (0.45 kg) of the drug was dissolved in 3 pt (1.42 l) of denatured alcohol, and this solution was mixed with 25 lb (11 kg) of grain. A crystalline coating of the drug was left on the grain after the alcohol evaporated. This formulation, used as a bird bait, was effective against house sparrows and starlings, but the concentration had to be doubled for pigeons. The anesthetized birds were killed humanely, and songbirds and banded pigeons were released. Tribromomethanol may also be used in wafer baits by adding 10 oz (300 ml) of a 33.3% solution of the drug to 1 gal (4 l) of water (Penn, 1965).

Thallium Sulfate. In Milwaukee, songbirds are not normally present during the winter, and have not been noted even in summer in railroad marshaling yards. When there was no hazard to songbirds, a poison bait, thallium sulfate in corn (maize) or barley, was used in place of tribromomethanol because of the high cost of the latter. Prebaiting was unnecessary in the railroad yards, where pigeons had become accustomed to feeding on the same grains as those used in the bait. Also, devices such as oversized dove decoys tended to allay the natural suspicion of the pigeons toward new sources of food (Penn, 1965).

Avitrol® 200 is a relatively new compound, registered for use in baits for bird control. Its active ingredient is 4-aminopyridine, an odorless, off-white solid. Avitrol is registered only for certain birds, and for use only by professional pest control operators. It is used at a concentration of 0.5% in wheat or mixed grains for sparrows, blackbirds, and cowbirds, on or in the areas of structures or nesting and roosting sites, or at 1% in whole corn for crows, or in pelletized feed for starlings in feedlots. Some species of birds that ingest this toxicant react in such a manner as to scare off other birds in the area. It is also toxic to other animals and to humans when ingested or inhaled, but does not pass readily through the skin. Nevertheless, gloves should be worn when handling Avitrol 200 or baits containing it. After applying the toxicant, the hands should be thoroughly washed with soap and water before eating or smoking.

Starlings may utter distress calls, flap about, fly in an unusual manner, or otherwise show the effects of Avitrol. These symptoms will continue for 30 minutes or longer. Birds knocked down either die or recover in 1 to 15 hours. The distress symptoms produced by the drug cause the entire flock to leave the area. Pigeons show the same effects, but make little or no noise, and do not cause others in the flock to depart. House sparrows pay little attention to other members of the flock, and nearly all the birds must be affected to achieve control. When treating gulls, it should be borne in mind that even 1 distressed bird will cause a large number to leave the vicinity. The uneaten treated bait should then be removed (NPCA, 1972).

Bird Perches with Toxicant. Perches for controlling pest birds are available. The perch consists of a metal container with a liquid contact toxicant and a wick surface upon which birds can perch and pick up the poison with their feet. These perches are registered for the control of starlings, pigeons, and house sparrows, but should be used only by persons experienced in bird management and trained (or under the supervision of someone trained) in the use of the appliances. The perches should not be used under circumstances in which nontarget birds would be endangered.

Ammonia Gas. A highly successful control project was conducted with anhydrous ammonia in 2 cattle sheds, each measuring $40 \times 72 \times 14$ ft ($12 \times 22 \times 4.3$ m), and it would doubtless be equally successful in other types of closed, uninhabited buildings (Devore *et al.*, 1967). The buildings were infested principally by starlings and sparrows, and had to be treated between the times when the birds went to roost at night and when they began foraging early in the morning. Polyethylene sheets were secured over all entrances. The seal did not have to be tight, but offered some resistance to gas and prevented the escape of birds. About 50 lb (23 kg) of the gas were used in treating each building. The gas was released into the sheds until they were half filled with the dense, white fog that forms when anhydrous ammonia is exposed to the atmosphere. The man operating the release valve wore a gas mask, and others on the job wore respirators. The white fog diffused rapidly through the sheds, and could be seen escaping near the roofs. They were left closed for only 7 minutes, and the polyethylene covers were then removed to permit ventilation. The dead birds were removed 30 minutes later, and the cattle were allowed to re-enter.

A total of 618 starlings, 290 sparrows, 24 mice, and 2 pigeons were killed in the 2 buildings, and no birds survived the treatment. A month after treatment, the bird population had again grown to 75 or 80, mostly sparrows, but such numbers were not considered to be problems. The anhydrous ammonia used in the sheds cost only about 3 or 4 dollars.

Spear (1968) called attention to the fact that anhydrous ammonia in liquid or gaseous state was corrosive to copper, copper alloys, aluminum alloys, and galvanized surfaces, and was explosive between 16 and 25% by volume in air. In view of the hazards related to inhalation and explosion connected with the use of ammonia and other gases and the need to comply with laws pertaining to fumigation, he suggested that the work be done by experienced fumigators.

Chemosterilization

A type of "birth control" formulation for pigeons has been effectively used in pigeon control. It consists of corn kernels coated with 1% by weight of 20,25-diazocholestenol dihydrochloride (azachosterol dihydrochloride) or Ornitrol. It inhibits egg production, and causes temporary sterility in pigeons and perhaps other birds, although the corn kernel is too large to be eaten by most songbirds, and it has no known effect on mammals. Ornitrol's effect on female pigeons can last up to 6 months, whereas it causes only short-term sterility in males. In experiments made from 1967 through 1969 in feeding areas during the mating season, the pigeon population was reduced from 2,500 to 300 in Maine (Anonymous, 1970*j*).

Birds are relatively long-lived, and although chemosterilization may reduce a bird population over a period of a few years, the existing flocks will, of course, continue to do the damage the treatment was designed to eliminate. According to M. W. Cummings (correspondence), Ornitrol treatments must usually be repeated every 1 or 2 months, and the treatment becomes very expensive if any sizable number of birds is involved.

Treating a Nest with Insecticide Dust

When pigeons are being reared and the objective is merely to rid them of ectoparasites, such as mites, lice, and the pigeon fly (*Pseudolynchia canariensis*), as well as other arthropods inhabiting their nests, the nests may be dusted with 4% malathion or 5% carbaryl (Sevin) (Furman *et al.*, 1969). For lice, the dust must be distributed on the wing feathers.

Chapter 14 • *Delusory Parasitosis and "Cable Mite" Dermatitis*

THE TERM "delusion of parasitosis" has been used to refer to an emotional disorder in which the patient has an unwarranted belief that live organisms, such as mites or insects, are present on or in his body (Wilson and Miller, 1946; Wilson, 1952; Obermayer, 1955). The term was later shortened to "delusory parasitosis" by W. G. Waldron (1962), public health entomologist for the Los Angeles County Health Department, who had investigated hundreds of cases of the disorder and made a thorough and systematic study of it. The term "entomophobia" has long been used in referring to the delusion, but it more appropriately applies simply to an unfounded fear of insects, as the Greek roots imply. The visitor who is repulsed and even frightened when viewing for the first time a suddenly agitated mass of American cockroaches in a rearing cage can be said to be suffering from entomophobia in the strict sense of the word. Generally, greater understanding of and familiarity with insects decrease this fear, just as they decrease what is usually a deep-seated fear of harmless snakes and eventually enable people to handle them.

About 5 or 6 victims of delusory parasitosis telephone or call in person for our advice and help each year. Over a 25-year period, certain patterns of behavior have become apparent that are similar to those described by Waldron (1962, 1963), Schrut and Waldron (1963), and other investigators to whom they refer in their literature. Descriptions by telephone of the victim's troubles usually identify cases of delusory parasitosis because of the stereotyped pattern and the absurdity of the symptoms. The supposed "bugs" or "creatures" are believed to change colors; appear and disappear while they are being watched; enter the skin and reappear; invade the hair, nose, and ears; and to persist despite repeated treatments of the supposedly infested house by competent pest control operators and despite frequent bathing and applications of ointments. If the person, in most cases a woman, brings in samples of the supposed pest, it invariably consists of small bits of lint, cloth, dandruff, scabs, sand particles, dirt, or miscellaneous debris. Occasionally, by chance, innocuous insects or insect parts are included with the debris. There is usually a long history of ineffectual visits to medical doctors and/or dermatologists before an entomologist is seen. The skin has often been injured by frequent and sometimes desperate scratching and cleansing. The presumed pests offered for identification are frequently wrapped in tissue paper or stuck onto tape, although occasionally they are contained in a small vial. The patient is confident that microscopic examination of the collected samples will reveal the creatures that seem so real. One patient returned to the author's office 3 times during 6 years, each time accompanied by a dubious and apologetic husband, but each time confident that the cause of her troubles would be found. Another has called repeatedly concerning "termites" that were believed to be "eating out" her insides and occasionally emerging from the joints of her arms and legs. Presumed specimens sent for identification were, of course, scabs and other debris.

Waldron (1962) presented 7 typical delusions of persons suffering from delusory parasitosis. These were:

(1) The "bugs" are black or white when first noted, and then later may change color. [We have frequently noted this same reaction among afflicted persons.]

(2) The "bugs" jump. [One person demonstrating this to the author was using a steel knife blade to prod the inanimate objects she thought were insects along a nylon slip that was stretched on a plastic sink top. The static electricity generated by this activity made the objects "jump." Unfortunately, the simple explanations that were made in an effort to describe the phenomenon sufficed only for a short period of time. The same type of observation was made at the home of a highly successful and well-known playwright who was a victim of delusory parasitosis. He observed that "organisms" on a piece of paper jumped about when he brought his finger close to them. The explanation, of course, was that there was a release of electrostatic charge from his body, which had been gained by walking over a large expanse of carpeting on a dry, windy day. The same individual "observed" that bits of debris developed "stringers" that grew out from the main part of the particle and enabled it to move about. A common characteristic of the victim is that he is desperate for an explanation of each new observation.]

(3) The "bugs" may infest the patient's hair, and can be shaken or combed onto a sheet, towel, or newspaper.

(4) "Bites" or papulelike irritations which develop on the skin usually itch, causing the persons to scratch themselves, even to the point of severe tissue damage.

(5) The "bugs" may come out of such common household items as toothpaste, vaseline, or cosmetics. [One patient claimed that tiny bugs came out from between her teeth after she brushed them.]

(6) The supposed infestation in a home may become so severe as to literally force the person to move to another location. Unfortunately, the "bugs" usually reappear in the new dwelling.

(7) The patient may be so positive of the supposed infestation and give such a lucid description that other members of the family will stoutly support the contention, even though they are not afflicted.

In our experience, the victims of delusory parasitosis are often well educated, responsible, hardworking, and appear to be rational and sensible about matters not pertaining to this one particular obsession. Possibly, these people are very susceptible to the destructive results of prolonged stress.

In the few cases in which our conversations have casually led to the revelation of personal problems, causes for stress and worry have been evident—usually marital troubles. For anyone wishing to pursue this aspect of delusory parasitosis, some interesting female case descriptions were given by Schrut and Waldron (1963), who characterized these cases as having as their basis "an unconscious sexual guilt with an attempt to ward off feared, and at the same time hoped for, invasion of a sexual nature."

In addition to this cause, they also pointed out, however, that sexual problems were by no means the sole sources of difficulty:

> Infantile or early childhood guilt or other effects relating to rage, envy, and aggression, as well as other forbidden impulses, may individually or in combination underlie the symptoms in these cases.
>
> The fixation upon arthropods has some specific and individual meaning in the life history of the person involved, and causes the patient to see a physician, an entomologist, or both. Generally, such problems can be traced to severe infantile or childhood emotional deprivation with lack of basic emotional need fulfillment which is too often glibly explained away solely by present-day problems.

Emotional problems were prominent among the 51 case histories, from the literature and from their own observations, that were discussed by Wilson and Miller (1946). For example, for one patient the delusion was precipitated by the shock of an accident in which his sister lost her vision as the result of his negligence. Among the case histories were several in which auditory hallucinations were prominent features, the patients claiming that they could hear the insects "knocking" or "clicking." Another common characteristic revealed in the investigations was the patients' excessive preoccupation with cleanliness. One patient, a widow aged 51 and "a person of higher than average intelligence," had skin that was dry and rough from the use of disinfectants. "She spends her entire time from early morning to late at night cleaning, scrubbing, and sterilizing her home. All available money is spent on cleaner's bills for clothing, bedding, curtains, and similar items."

Another of these odd patients "boiled the entire family wardrobe each night and all the bedclothes each morning. She insisted that the family bathe in gasoline daily and rub with sulfur and lard."

The entomologist's obligation ends, of course, after he has made his diagnosis from the entomological viewpoint. However, he should make certain that the possibility of insect or mite infestation has been thoroughly investigated. People with delusory parasitosis generally have had insect or mite problems, have reacted violently to them, and have "eradicated" the pests several times over.

The adviser can usually determine from the nature of the conversation whether the caller with itching sensations really has a pest problem, such as fleas, lice, bed bugs, tropical rat mites, straw itch mites, human itch mites, or the like.

The problem of the person really having pests is, of course, generally much easier to solve than the problem of the person who has no pests but believes he has. An example of how obscure a real pest problem can be is illustrated by a report by J. R. Traver (1951) on an infestation of mites, *Dermatophagoides scheremetewskyi* Bogdanov, on her own person, resulting in itching red papules on scalp, eyes, ears, nostrils, shoulders, under the arms, beneath the breasts, on the chest, both upper and lower back, and occasionally around the umbilicus. Other members of her family were likewise infested. The ailment was initially diagnosed by a physician as "psychoneurotic." Fortunately, the victim was a zoologist, had access to a microscope, and found the causative agent. Many attempted treatments failed to eradicate the infestation. However, attacks by the implicated mite species are extremely rare.

"CABLE MITE" DERMATITIS

Probably anyone with extensive experience in household pest control has had his attention called to real or apparent cases of dermatitis or paresthesia (prickling, tingling, or creeping sensation) among persons, usually women, who work with old, dusty records, with elaborate electronic equipment, or in other industrial situations, but for which no insects, mites, or other organisms appear to be responsible. The presumed cause for these symptoms is sometimes referred to as "cable mites." Sometimes by the time the case is called to the attention of an entomologist or pest control operator, the premises have already been treated repeatedly with insecticides and acaricides, but with no results. Likewise, various ointments and frequent bathing may have offered no relief. It is therefore encouraging to find a report of a successful solution to a situation of this type.

Scott and Clinton (1967) investigated reports of paresthesia by 10 female employees in the physics laboratory of a national agency. No possible causal organisms and no visible symptoms on the victims could be found. The only surfaces in the laboratory that were not routinely cleaned by custodians were "scanner mirrors" because they were so fragile and extremely expensive. The mirrors were covered with a thick layer of yellowish, fluffy, fibrous material, found to be mainly rock wool. Scott and Clinton hypothesized as follows:

> When the power cable for the new scanning machines was installed, particulate matter from the rock-wool bats in the ceilings of some cubicles was dislodged and drifted into the ventilating registers around the edges of the cubicle. From these, it showered slowly over the cubicles and their occupants. The scanning projectors caused a slight vibration of the scanning mirrors when they were turned on. This vibration allowed for dislodgment of some of the accumulations of dust into the surroundings.

Following the verdict that the "cable mite" dermatitis of the 10 employees resulted from the "rock-wool aerosol," work areas of the laboratory were thoroughly scrubbed and floors and mirrors were vacuumed weekly. Even though the hypothesis was incompletely tested, personnel were told that the problem had been solved and that recurrences of the dermatitis were unlikely, thus overcoming the effects of what was believed to be a psychological enhancement, resulting from conversations among employees, of a condition originally occasioned by physical agents. Two of the victims continued to complain occasionally of irritations of gradually decreasing intensity. A check of the situation 6 months later revealed no further complaints from employees.

Cable mite dermatitis is not strictly a delusory parasitosis, except insofar as the victims may believe they are being infested by insects or mites, even though no organism can be found. Unlike the victims of delusory parasitosis, people suffering from cable mite dermatitis are being affected by some real—not imaginary—substance in their environment; if the substance can be found and eliminated, they will fully recover from their dermatitis symptoms.

Credits for Photographs

When the legend for any photograph lacks a credit line, it signifies that the photo was made by one of the entomologists associated with the author on the University of California campus at Los Angeles. The black-and-white photos furnished by Roy J. Pence, Donald A. Reierson, and Robert E. Wagner are listed here by their respective figure numbers and the color photos by their respective plate and frame numbers.

Black-and-White Photos

Roy J. Pence

Figures 36, 49, 50, 58, 59, 61, 63, 64, 65, 70, 71, 73, 76, 77, 85, 86, 106, 113, 118, 121, 126, 135, 138, 142, 147, 158, 166, 171, 172, 173, 176, 181, 182, 184, 195, 196, 198, 200, 201, 202, 203, 204, 216, 222, 229, 249, 251, 261, 265, 271, 303, 304, 308, 311, 319, 321, 338, 339, 340, 343, 346, 347, 348, 349, 350, 354, 355, 362, 365.

Donald A. Reierson

Figures 29, 37, 38, 39, 43, 54, 55, 95, 100, 105, 109, 111, 114, 123, 124, 127, 136, 143, 148, 151, 152, 153, 179, 191, 194, 205, 206, 210, 212, 218, 226, 227, 228, 231, 232, 234, 235, 236, 237, 238, 239, 240, 241, 242, 243, 246, 247, 248, 250, 252, 255, 256, 259, 296, 298, 299, 319-A, 334, 337, 341, 342, 344, 345, 351, 352, 356, 359, 360, 361, 366-67, 368, 369, 372.

Robert E. Wagner

Figures 20, 33, 66, 67, 72, 74, 75, 78, 82, 83, 84, 88, 101, 102, 110, 119, 145, 146, 154, 230, 233, 245, 370.

Color Photos

Roy J. Pence

Plates I, 1, 2, 3, 6, 8; II, 1 (right), 3, 4; IV, 1, 4, 5 (right), 7 (left and center); V, 1 (right), 2, 4; VI, 1, 3, 5, 8; VII, 2; VIII, 4 (right), 6 (left).

Donald A. Reierson

Plates II, 1 (left), 2, 5, 6, 8; III (entire plate); IV, 2, 3, 5 (left); 6, 7 (right, bottom), 8; V, 1 (left), 3 (bottom), 5, 6, 7, 8; VI, 2, 4, 6, 7; VII, 1, 3 (left), 4, 5, 6, 7; VIII, 2, 3, 4 (left), 5, 6 (right), 7, 8.

Robert E. Wagner

Plates I, 4, 5, 7; II, 7; V, 3 (top); VII, 3 (right).

Literature Cited

Aarons, T.
1948. Cemetery mosquito control by aerosol. Proc. California Mosquito Control Assoc. 16:84–85.

Aboul-Nasr, A. E., and M. A. S. Erakey
1958. Behaviour and sensory physiology of the bedbug, *Cimex lectularius* L., to some environmental factors. Chemoreception. (Hemiptera-Heteroptera: Cimicidae.) Bull. Soc. Entomol. Égypte 52:353–62.

Ackerman, A. B.
1968. Crabs—the resurgence of *Phthirus pubis*. New England J. Med. 278:950–51.

Adams, J. A.
1933. Biological notes upon the firebrat, *Thermobia domestica* Packard. J. New York Entomol. Soc. 41:557–62.

Adams, W. H., R. W. Emmons, and J. E. Brooks
1970. The changing ecology of murine (endemic) typhus in southern California. Amer. J. Trop. Med. Hyg. 19:311–18.

Aikawa, J. K.
1966. Rocky Mountain spotted fever. C. C. Thomas, Springfield, Illinois. 140 pp.

Akesson, N. B.
1970. Low rates of finely atomized sprays by air and ground rig. *In:* Research on mosquito control (R. F. Smith, chm.), 15 pp. Rept. of Univ. of California Mosquito Control Res. Comm.
1971. Safety precautions when applying pesticide chemicals. Agricultural Engineering, Univ. of California, Davis. (Mimeographed.)

Aldrich, J. M.
1916. The dipterous genus *Symphoromyia* in North America. Proc. U. S. National Museum 49:113–42.

Alexander, R. D.
1957. The taxonomy of the field crickets of the eastern United States (Orthoptera: Gryllidae: *Acheta*). Annals Entomol. Soc. Amer. 50:584–602.

Allen, R., S. K. Taylor, and S. E. Sulkin
1970. Studies of arthropod-borne virus infections in Chiroptera. VIII. Evidence of natural St. Louis encephalitis virus infection in bats. Amer. J. Trop. Med. Hyg. 19:851–59.

Allen, T. C., and J. W. Dolan
1970. Wood-cement composite as a deterrent to termite infestation. J. Econ. Entomol. 63:1669–70.

Allen, T. C., G. R. Esenther, and E. P. Lichtenstein
1964*a*. Toxicity of dieldrin–concrete mixtures to termites. J. Econ. Entomol. 57:26–29.

Allen, T. C., G. R. Esenther, and R. D. Shenefelt
1961. Concrete–insecticide mixtures toxic to termites. J. Econ. Entomol. 54:1055–56.

Allen, T. C., R. V. Smythe, and H. C. Coppel
1964*b*. Response of twenty-one termite species to aqueous extracts of wood invaded by the fungus *Lenzites trabea* Pers. ex Fr. J. Econ. Entomol. 57:1009–11.

Amburgey, T. L.
1972. Preventing wood decay. Pest Control 40(1):19–20, 38, 40, 42.

Amonkar, S. V., and A. Banerji
1971. Isolation and characterization of the larvicidal principle of garlic. Science 174:1343–44.

Amonkar, S. V., and E. L. Reeves
1970. Mosquito control with the active principle of garlic, *Allium sativum*. J. Econ. Entomol. 63: 1172–75.

Amos, T. G.
1968. Some laboratory observations on the rates of development, mortality, and oviposition of *Dermestes frischii* (Kug.) (Coleoptera: Dermestidae). J. Stored Prod. Res. 4:103–117.

Anderson, C. H., and W. Schulte
1971. Teflon® as a surface for deposition of aerosol droplets. Mosquito News 31:499–504.

Anderson, J. R.
1964. The behavior and ecology of various flies associated with poultry ranches in northern California. Proc. California Mosquito Control Assoc. 32: 30–34.
1965. A preliminary study of integrated fly control on northern California poultry ranches. Proc. California Mosquito Control Assoc. 33:42–44.
1966*a*. Recent developments in the control of some arthropods of public health and veterinary importance: Muscoid flies. Entomol. Soc. Amer. Bull. 12:342–48.
1966*b*. Biological interrelationships between feces and flies. *In:* Management of farm animal wastes. Proceedings of the National Symposium, May 6–7. pp. 20–23.

Anderson, J. R., and J. H. Poorbaugh
1964. Biological control: Possibility for house flies. California Agriculture 18(9):2–5.
1968. The face fly, *Musca autumnalis*. A new livestock fly is now moving toward California. California Agriculture 22(3):4–6.

Anderson, J. W.
1954. The production of ultrasonic sounds by laboratory rats and other mammals. Science 119:808–809.

Anderson, L. D., and E. L. Atkins, Jr.
1962. Effects of mosquito abatement pesticide treatments on honey bees. Proc. California Mosquito Control Assoc. 30:78–80.

Anderson, L. D., E. C. Bay, and A. A. Ingram
1964. Studies of chironomid midge control in water-spreading basins near Montebello, California. California Vector Views 11:13–20.

Anderson, L. D., E. C. Bay, and M. S. Mulla
1965. Aquatic midge investigations in southern California. Proc. California Mosquito Control Assoc. 33: 31–33.

Anderson, L. O.
1972. Condensation problems: Their prevention and solution. USDA Forest Service Research Paper No. FPL-132. 36 pp.

Andrew, B. J.
1930. Method and rate of protozoan refaunation in the termite *Termopsis angusticollis* Hagen. Univ. California Publ. Zool. 33:449–70.

Andrewes, C.
1969. The lives of wasps and bees. American Elsevier Publishing Co., New York. 204 pp.

Anonymous
1946. Instructions for using ANTU (alphanaphthyl thiourea) as a rat poison. Pests 14(7):22, 24.
1949. The bed-bug: Its habits and life history, and how to deal with it. British Museum (Nat. Hist.), Econ. Ser. 5. 18 pp.
1953. Twenty years of steady progress with the NACA. Agricultural Chemicals 8(9):40–44.
1961*a*. Organization for mosquito control (H. F. Gray, ed.). Amer. Mosquito Control Assoc. Bull. 4:1–54.
1961*b*. Puss caterpillars: A new problem for PCO's? Pest Control 29(5):15–18.
1962. Purdue staffers tell how to control bees and wasps. Pest Control 30(12):28.
1963. Ventilation and decay control: A natural for termitemen. Pest Control 31(10):68–74.
1964*a*. Kansas City hires PCO's to crack down on sewer rats. Pest Control 32(8):13–15, 34, 36.
1964*b*. A second look at rodent trapping. Pest Control 32(8):16–18, 22, 60, 62.
1964*c*. Biologists put the bite on mosquitoes. Medical World News, May 8. pp. 168–69.
1967*a*. Carpenter ants annoy many operators; PC survey uncovers treatment hints. Pest Control 35(1): 42, 44.
1967*b*. Preventive bird control can reduce costly, dangerous bird–airplane collisions. Pest Control 35(9): 12–16, 18.
1967*c*. Oldhouse borer, Purdue's Formosan termite study, and dry-rot reporting are discussed in Philadelphia. Pest Control 35(12): 28–30.
1967*d*. Leading termites astray. New Scientist 33:398.
1967*e*. Annual report of the Government Pest Infestation Laboratory, Lyngby, Denmark. *See specifically* p. 14.
1968*a*. Phoenix first: Modern cockroaches enjoy high standard of living. PCO News 28(8):6.
1968*b*. Control of the European earwig. Oregon State Univ., Coop. Ext. Serv., Ext. Circ. 655. 5 pp.
1969*a*. An effective method of controlling roaches in sewers. Public Works (July).
1969*b*. A national program of research for bees and other pollinating insects and insects affecting man. Prepared by a joint task force of the USDA, state universities, and land grant colleges. Research Progress Development and Evaluation Staff, Room 318-E, Administration Building, USDA, Washington, D.C. 20250.
1970*a*. Public health pesticides. Pest Control 38(3):16–54.
1970*b*. Encephalitis: California prepares again. Pest Control 38(4):13–14, 16.
1970*c*. "Moving carpet" of mice is observed in Australia. Pest Control 38(5):46.
1970*d*. Chiggers. Purdue Univ. Entomol. Ext. Serv. Publ. E-34. 2 pp.
1970*e*. The cluster fly and the face fly in homes. Purdue Univ. Coop. Ext. Serv. Publ. E-35. 2 pp.
1970*f*. Bees, wasps, hornets, and yellowjackets. Purdue Univ. Coop. Ext. Serv. Publ. E-44. 2 pp.
1970*g*. Bites of spiders and other arthropods. Medical Tribune, January 22. pp. 878–80.
1970*h*. Flea collar risks and benefits. J. Amer. Vet. Med. Assoc. 156:6.
1970*i*. Commentary on a talk by Dr. R. L. Hall. Arizona Agr. Chem. Assoc. Fieldman News 1(18):1–3.
1970*j*. Birth control: Pill for pigeons. Chemical Engineering News 48(8):14–15.
1971*a*. Minnows take on mosquitoes. Agricultural Research 19(7):14.
1971*b*. Rodents help control fleas. Agricultural Research 20(4):14.
1971*c*. Special news report: Encephalitis in Texas. Pest Control 39(8):15.
1971*d*. Judicial officer allows commercial use of lindane vaporizers. Food Chem. News 13(33):34–35.
1972*a*. Pyrethrum's potent cousin. Agricultural Research 20(8)5.
1972*b*. The vampire bat: A new idea works. Pest Control 40(8)20, 22.
1972*c*. Pesticides in the modern world. Symposium prepared by members of the Cooperative Programme of Agro-Allied Industries with FAO and other United Nations organizations. 59 pp.
1973*a*. Cockroaches are linked to the disease, parasitic toxoplasmosis. Pest Control 41(2):22.
1973*b*. Late news (concerning imported fire ants). Pest Control 41(8):50.
1973*c*. Little superbird invading United States called a menace to agriculture. Los Angeles *Times* (December 14).

Arbesman, C. E.
1965. Clinical anaphylaxis and serum sickness. *In:* Immunological diseases (M. Sampter and H. L. Alexander, eds.), pp. 654–62. Little, Brown and Co., Boston.

Arbesman, C. E., C. Langlois, P. Bronoon, and S. Shickman
1966. The allergic response to stinging insects. VII. Fractionation of whole body and venom sac extracts of yellowjacket. Journal of Allergy 38:1–8.

Areán, V. M., and I. Fox.
1955. Dermal alterations in severe reaction to the bite of the sand fly, *Culicoides furens*. Amer. J. Clinical Pathology 25:1359–66.

Arnett, R. H.
1968. The beetles of the United States. American Entomological Institute, 5950 Warren Road, Ann Arbor, Michigan. 1112 pp.

Asel, N. E.
1969. Spider bites (*Loxosceles reclusa*). *In:* Current therapy (H. F. Conn, ed.), p. 880. W. B. Saunders Co., Philadelphia.

Atkins, J. A., C. W. Wingo, and W. A. Sodeman
1957. Probable cause of necrotic spider bite in Midwest. Science 126:73.

Atkins, J. A., C. W. Wingo, W. A. Sodeman, and J. E. Flynn
1958. Necrotic arachnidism. Amer. J. Trop. Med. Hyg. 7:165–84.

Austen, E. E., A. W. McKenny Hughes, and H. Stringer
1935. Clothes moths and house moths. British Museum (Nat. Hist.), Econ. Ser. 19.

Ayres, S., Jr.
1963. Rosacea-like demodicidosis. California Medicine 98:328–30.

Ayres, S., Jr., and N. P. Anderson
1932. *Demodex folliculorum:* Its role in the etiology of acne rosacea. Archives of Dermatology 25:89–98.

Ayres, S., Jr., and R. Mihan
1967. Rosacea-like demodicidosis involving the eyelids. Archives of Dermatology 95:63–66.

Babu, T. H., and K. Sláma
1972. Systemic activity of a juvenile hormone analog. Science 175:78–79.

Back, E. A.
1923. Clothes moths and their control. USDA Farmers' Bull. No. 1353. 25 pp.
1936. *Gryllus domesticus* L. and city dumps. J. Econ. Entomol. 29:198–202.
1938. Carpet beetles. USDA Leaflet No. 50.
1939. Centipedes and millipedes in the house. USDA Leaflet No. 192.
1946. Psocids in dwellings. Pests 14(11):26–27.

Back, E. A., and R. T. Cotton
1926. Biology of the saw-toothed grain beetle, *Oryzaephilus surinamensis* Linné. J. Agr. Res. 38:435–52.
1938. The black carpet beetle, *Attagenus piceus* (Olivier). J. Econ. Entomol. 31:280–86.

Bacon, O. G., A. E. Michelbacher, and W. H. Hart
1948. Catalina cherry moth attacks walnuts. Diamond Walnut News 30(5):10.

Bacot, A.
1917. A contribution to the bionomics of *Pediculus humanus (vestimenti)* and *Pediculus capitis.* Parasitology 9:228–58.

Baerg, W. J.
1922. Regarding the habits of tarantulas and the effects of their poison. Scientific Monthly 14:482.
1936. The black widow. Arkansas Agr. Expt. Sta. Bull. 325. 34 pp.

Bailey, D. L., G. C. LaBrecque, and T. L. Whitfield.
1971. House fly control in a poultry house with insecticides in slow-release plastic formulations. J. Econ. Entomol. 64:138–40.

Bailey, J. B., and J. E. Swift
1968. Pesticide information and safety manual. Univ. California Agr. Ext. Serv. 147 pp.

Bailey, S. F., and D. C. Baerg
1967. The flight habits of *Anopheles freeborni* Aitken. Proc. California Mosquito Control Assoc. 35:55–69.

Bailey, S. F., R. M. Bohart, and L. J. Booher
1954. Mosquito control on the farm. California Agr. Expt. Sta. Ext. Serv. Circ. 439. 27 pp.

Bailey, S. F., D. A. Eliason, and B. L. Hoffmann
1965. Flight and dispersal of the mosquito *Culex tarsalis* Coquillett in the Sacramento Valley of California. Hilgardia 37:73–113.

Baker, D. W., and W. F. Fisher
1969. The incidence of demodectic mites in the eyelids of various mammalian hosts. J. Econ. Entomol. 62:942.

Baker, E. W., T. M. Evans, D. J. Gould, W. B. Hull, and H. K. Keegan
1956. A manual of parasitic mites of medical or economic importance. National Pest Control Association, Inc., Elizabeth, New Jersey.

Baker, N. W.
1971. Jerusalem crickets. Pacific Discovery 24(2):12–13.

Balazuc, J.
1950. Anaphylactic phenomenon caused by *Triatoma* bite. Ann. Inst. Med. Reg. Univ. Nac. Tucumán 3:35–37.

Balch, R.E.
1937. Notes on the wharf borer (*Nacerda melanura* L.). Canadian Entomologist 69:1–5.

Balduf, W. V.
1954. Observations on the white-faced wasp *Dolichovespula maculata* (Linn.). Annals Entomol. Soc. Amer. 47:445–58.

Banks, W. A., B. M. Glancey, C. E. Stringer, D. P. Jouvenaz, C. S. Lofgren, and D. E. Weidhaas
1973. Imported fire ants: Eradication trials with mirex bait. J. Econ. Entomol. 66:785–89.

Banks, W. A., C. E. Stringer, Jr., W. E. Barthel, and C. S. Lofgren
1966. Control of imported fire ants with Nonachlor. J. Econ. Entomol. 59:465–67.

Bano, A., and S. K. Majumder
1965. Pathological changes induced by tricalcium phosphate in insects. J. Invert. Pathol. 7:384–87.
1968. Tricalcium phosphate as an insecticide. *In:* Pesticides (S. K. Majumder, ed.), pp. 177–85. Acad. Pest Control Sci., Mysore, India.

Barber, G. W.
1919. On the bite of *Arilus cristatus.* J. Econ. Entomol. 12:466.

Barbour, R. W., and W. H. Davis
1969. Bats of America. Kentucky University Press, Lexington. 286 pp.

Barkai, H.
1964. A preliminary survey of bedbug resistance to organophosphorus compounds in Israel. WHO Vector Control 58:1–5.

Barnes, R. D.
1963. Invertebrate zoology. W. B. Saunders Co., Philadelphia and London. 632 pp.

Barnhart, C. S., Jr.
1963. How to control cockroaches with dry ice fumigation. Pest Control 31(2):30.

Barr, S. E.
1971. Allergy to Hymenoptera stings: Review of the world literature, 1953–1970. Annals of Allergy 29:49–66.

Barr, W. F., and P. D. Hurd, Jr.
1947. Notes on the *Dasymutilla* of the Palo Verde Valley, California, with the description of a new species. Pan-Pacific Entomologist 23:85–90.

Bartley, W. C., and K. Mellanby
1944. The parasitology of human scabies (women and children). Parasitology 35:207–208.

Basinger, A. J.
1931. The European brown snail in California. California Agr. Expt. Sta. Bull. 515. 24 pp.

Batra, S. W. T.
1970. Behavior of the alkali bee, *Nomia melanderi,* within the nest. (Hymenoptera: Halictidae.) Ann. Entomol. Soc. Amer. 63:400–406.

Batth, S. S., J. Singh, and D. C. Villeneuve
1973. Dichlorvos vaporizers: Method for evaluation under simulated household use. J. Econ. Entomol. 66:146–50.

Baumhover, A. H., A. J. Graham, B. A. Bitter, D. E. Hopkins, W. D. New, F. H. Dudley, and R. C. Bushland
1955. Screw-worm control through the release of sterilized flies. J. Econ. Entomol. 48:462–66.

Bay, E. C.
1965. Mosquito control by fish: A present-day appraisal. WHO Chronicle 21(10):415–23.
1967. What to do about lakeside property defacement by chironomid egg masses. Pest Control 35(4): 10–13.
1970. Mosquito fish. *In:* Research on mosquito control (R. F. Smith, chm.), 15 pp. Rept. of Univ. of California Mosquito Control Res. Comm.

Bay, E. C., and L. D. Anderson
1965. Chironomid control by carp and goldfish. Mosquito News 25:310–16.

Bay, E. C., L. D. Anderson, and J. Sugerman
1965. The abatement of a chironomid nuisance on the highways at Lancaster, California. California Vector Views 12:29–32.

Beadle, L. D.
1957. Legal aspects of compulsory elimination of mosquito breeding areas. Mosquito News 17:277–80.

Beal, R. H.
1969. Treating pine lumber with insecticide and preservative prevents tubing by subterranean termites. J. Econ. Entomol. 62:757–58.
1971. Concrete impregnated with chlordane and dieldrin not tubed over by subterranean termites. J. Econ. Entomol. 64:1289–91.

Beal, R. H., and F. L. Carter
1968. Initial soil penetration by insecticide emulsions used for subterranean termite control. J. Econ. Entomol. 61:380–83.

Beal, R. H., and V. K. Smith
1971. Relative susceptibilities of *Coptotermes formosanus, Reticulitermes flavipes,* and *R. virginicus* to soil insecticides. J. Econ. Entomol. 64:472–75.

Beament, J. W. L.
1955. Wax secretion in the cockroach. J. Exptl. Biol. 32:514–38.

Beard, R. L.
1963. Insect toxins and venoms. Annual Rev. Entomol. 8:1–18.

Bearden, R. H., and C. D. Steelman
1971. Control of the dark rice field mosquito with FLIT® MLO. J. Econ. Entomol. 64:469–72.

Becker, G.
1942. Oekologische und physiologische Untersuchungen über die holzzerstörenden Larven von *Anobium punctatum* De Geer. Zeitschr. Morphol. Oekol. Tiere 39:98–152.
1943. Beobachtungen und experimentelle Untersuchungen zur Kenntnis des Mulmbockkäfers, *Ergates faber* (L.). 2. Die Larvenentwicklung. Z. angew. Entomol. 30:263–96.
1948. Ueber Kastenbildung und Umwelteinfluss bei Termiten. Biologisches Zentralblatt 67:407–44.
1950. Zerstörung des Holzes durch Tiere. *In:* Handbuch der Holzkonservierung (Mahlke, Troschel, Liese, eds.), pp. 111–65. J. Springer, Berlin.
1951. Ein kaum bekannter Käfer im Bauholz breitet sich aus. Holz-Zentralblatt 77:1895–96.
1963. Der Einfluss des Eiweiss-Gehalts von Holz auf das Hausbocklarven-Wachstum. Z. angew. Entomol. 51:368–90.
1964. Termiten-anlochende Wirkung einiger bei Basidiomyceten-Angriff in Holz enstehender Verbindungen. Holzforschung 18:168–72.
1965. Versuche über den Einfluss von Braunfäulepilzen auf Wahl und Ausnutzung der Holznahrung durch Termiten. Material und Organismen 1:95–156.
1968. Einfluss von Ascomyceten und Fungi imperfecti auf Larven von *Hylotrupes bajulus* (L.). Material und Organismen 3(3):229–40.
1969*a*. Rearing of termites and testing methods used in the laboratory. *In:* Biology of termites (K. Krishna and F. M. Weesner, eds.), pp. 351–85. Academic Press, New York and London.
1969*b*. Versuche an Holzspanplatten mit Hausbockkäfer-Larven und Termiten. Material und Organismen Beiheft 2:28–41.
1970*a*. Vergleichende Untersuchungen zur Oekologie einiger *Reticulitermes*-Arten (Insecta: Isoptera). Z. angew. Entomol. 65:183–216.
1970*b*. *Reticulitermes* (Insecta: Isoptera) in Mittel- und West-Europa. Z. angew. Entomol. 65:268–78.
1971. Physiological influences on wood-destroying insects of wood compounds and substances produced by microorganisms. Wood Science and Technology 5:236–46.

Becker, G., and W. Kerner-Gang
1964. Schädigung und Förderung von Termiten durch Schimmelpilze. Z angew. Entomol. 53:429–48.

Becker, G., and R. Mannesmann
1968. Untersuchungen über das Verhalten von Termiten gegenüber einigen spurbildenen Stoffen. Z. angew. Entomol. 62:399–436.

Becker, W. B.
1954. The old-house borer in Massachusetts. J. Econ. Entomol. 47:362–63.

Belkin, J. N.
1972. Fundamentals of entomology. (In two parts.) Bio-Rand Foundation, Baltimore, Maryland. 220 pp.

Belkin, J. N., N. Ehmann, and G. Heid
1951. Preliminary field observations on the behavior of the adults of *Anopheles franciscanus* McCracken in southern California. Mosquito News 11:23–31.

Bellevoye, M. A.
1889. Observations on *Monomorium pharaonis* Latr. [*sic*]. Insect Life 2:230–33.

Belyea, R. M.
1952. Death and deterioration of balsam fir weakened by spruce budworm defoliation in Ontario. Canadian Entomologist 84:325–35.

Bender, E.
1941. Investigations of the biology and morphology of the Microlepidoptera that inhabit wine-cellars. Rev. Appl. Entomol. A-35:71–72.

Bengston, I. A.
1933. Seasonal acute conjunctivitis in the southern states. Public Health Reports (USPHS) 48:917–26.

Bennett, D. R., and S. H. Kerr
1973. Millipedes in and around structures in Florida. Florida Entomologist 56:43–48.

Benson, R. L.
1936. Diagnosis and treatment of sensitization to mosquitoes. Journal of Allergy 8:47–59.
1939. Diagnosis of hypersensitiveness to bee and to mosquito, with report on successful specific treatment. Archives of Internal Medicine 64:1306–27.

Benson, R. L., and H. Semenov
1930. Allergy in its relation to bee sting. Journal of Allergy 1:105–16.

Bequaert, J. C.
1929. A new *Pseudomasaris* from California, with some considerations on the masarid wasps (Hymenoptera). Psyche 36:61–88.
1940. An introductory study of *Polistes* in the United States and Canada, with descriptions of some new North and South American forms. J. New York Entomol. Soc. 48:1–31.
1946. The ticks, or Ixodoidea, of the northeastern United States and eastern Canada. Entomologica Americana 25(2):73–120.

Berger, B. G.
1945. Food preferences of the firebrat. J. Econ. Entomol. 38:577–82.

Berkoff, C. E.
1969. The chemistry and biochemistry of insect hormones. Quarterly Reviews 23:372–91.

Bernecker, C.
1970. The antigenicity of house dust and mites. Acta Allergologica 25:392–403.

Berns, R. E.
1958. Straw meal moths. Pest Control 26(11):16.

Bernton, H. S., and H. Brown
1964. Insect allergy: Preliminary studies of the cockroach. Journal of Allergy 35:506–13.
1970*a*. Cockroach allergy: Age of onset of skin reactivity. Annals of Allergy 28:420–22.
1970*b*. Insect allergy: The allergenicity of the excrement of the cockroach *Blattella germanica*. Annals of Allergy 28:543–47.

Bernton, H. S., T. F. McMahon, and H. Brown
1972. Cockroach asthma. British Journal of Diseases of the Chest 66:61–66.

Beroza, M.
1970. Current usage and some recent developments with insect attractants and repellents in the USDA. *In:* Chemicals controlling insect behavior (M. Beroza, ed.), pp. 145–63. Academic Press, New York and London.

Beroza, M., B. A. Bierl, E. F. Knipling, and J. G. R. Tardif
1971*a*. The activity of the gypsy moth sex attractant disparlure vs. that of the live female moth. J. Econ. Entomol. 64:1527–29.

Beroza, M., B. A. Bierl, J. G. R. Tardif, D. A. Cook, and E. C. Paszek
1971*b*. Activity and persistence of synthetic and natural sex attractants of the gypsy moth in laboratory and field trials. J. Econ. Entomol. 64:1499–1508.

Beroza, M., and E. F. Knipling
1972. Gypsy moth control with the sex attractant pheromone. Science 177:19–27.

Berthold, R., Jr., and B. R. Wilson
1967. Resting behavior of the German cockroach, *Blattella germanica*. Annals Entomol. Soc. Amer. 60: 347–51.

Bertram, D. S., K. Unsworth, and R. M. Gordon
1946. The biology and maintenance of *Liponyssus bacoti* Hirst, 1913, and an investigation into its role as a vector of *Litomosoides carinii* to cotton rats and white rats, together with some observations on the infection in white rats. Annals Trop. Med. and Parasitol. 40:228–54.

Bess, H. A.
1971. Control of the drywood termite, *Cryptotermes brevis*, in Hawaii. Hawaii Agr. Expt. Sta. Tech. Bull. 87:5–31.

Bess, H. A., and J. W. Hylin
1970. Persistence of termiticides in Hawaiian soils. J. Econ. Entomol. 63:633–38.

Bess, H. A., and A. K. Ota
1960. Fumigation of buildings to control the drywood termite, *Cryptotermes brevis*. J. Econ. Entomol. 53:503–10.

Bess, H. A., A. K. Ota, and C. Kawanishi
1966. Persistence of soil insecticides for control of subterranean termites. J. Econ. Entomol. 59:911–15.

Bharadwaj, R. K.
1966. Observations on the bionomics of *Euborellia annulipes*. Annals Entomol. Soc. Amer. 59:441–50.

Bidlingmayer, W. L.
1959. Mosquito penetration tests with louver screening. Florida Entomologist 42:63–67.

Bierl, B. A., M. Beroza, and C. W. Collier
1970. Potent sex attractant of the gypsy moth: Its isolation, identification, and synthesis. Science 170: 87–89.
1972. Isolation, identification, and synthesis of the gypsy moth sex attractant. J. Econ. Entomol. 65:659–64.

Bigham, J. T.
1941. *Hippelates* (eye gnat) investigations in the southeastern states. J. Econ. Entomol. 34:439–44.

Bishopp, F. C.
1923. The puss caterpillar and the effect of its sting on man. USDA Circular No. 288. 14 pp.
1939. The brown dog tick, with suggestions for its control. USDA, Bureau of Entomology and Plant Quarantine. 4 pp.
1961. Injury to man by earwigs (Dermaptera). Proc. Entomol. Soc. Washington 63:114.

Bjornson, B. F., and C. V. Wright
1960. Control of domestic rats and mice. U. S. Government Printing Office, Public Health Service Publ. No. 563. 26 pp.

Blackmore, J. S., and F. J. Winn
1956. A winter isolation of western equine encephalitis virus from hibernating *Culex tarsalis* Coquillett. Proc. Soc. Exptl. Biol. and Med. 91:146–48.

Blum, M. S.
1970. The chemical basis of insect sociality. *In:* Chemicals controlling insect behavior (M. Beroza, ed.), pp. 61–94. Academic Press, New York and London.

Boch, R., D. A. Shearer, and B. C. Stone
1962. Identification of isoamyl acetate as an active component of the sting pheromone of the honey bee. Nature 195:1018–20.

Boeckh, J., K. E. Kaissling, and D. Schneider
1965. Insect olfactory receptors. Cold Spring Harbor Symposium on Quantitative Biology 30:263–80.

Bohart, G. E.
1958. Alfalfa pollinators, with special reference to species other than honey bees. Tenth Intern. Congr. Entomol. 4:929–37.

Bohart, R. M.
1941. Yellowjackets versus campers. Pan-Pacific Entomologist 17:58.
1949. Notes on North American *Polistes,* with descriptions of new species and subspecies. Pan-Pacific Entomologist 25:97–103.

Bohart, R. M., and R. C. Bechtel
1957. The social wasps of California. Bull. California Insect Surv. 4:73–101.

Bohart, R. M., and D. S. Horning, Jr.
1971. California bembicine sand wasps. Bull. California Insect Surv. 13:1–49.

Bolton, M.
1970. The brown spider bite. J. Kansas Med. Soc. 71: 197–201.

Borror, D. T., and D. M. DeLong
1971. An introduction to the study of insects. (Third edition.) Holt, Rinehart, and Winston, New York. 812 pp.

Bowers, W. S.
1968. Juvenile hormone: Activity of natural and synthetic synergists. Science 161:895–97.
1969. Juvenile hormone: Activity of aromatic terpenoid ethers. Science 164:323–25.
1971. Insect hormones and their derivatives as insecticides. WHO Bulletin No. 44:381–89.

Bowers, W. S., and C. C. Blickenstaff
1966. Hormonal termination of diapause in the alfalfa weevil. Science 154:1673–74.

Bowers, W. S., and W. G. Bodenstein
1971. Sex pheromone mimics of the American cockroach. Nature 232:259–61.

Bowers, W. S., H. M. Fales, M. J. Thompson, and E. C. Uebel
1966. Juvenile hormone: Identification of an active compound from balsam fir. Science 154:1020–21.

Bowers, W. S., and M. J. Thompson
1963. Juvenile hormone activity: Effects of isoprenoid and straight-chain alcohols on insects. Science 142:1469–70.

Bowers, W. S., M. J. Thompson, and E. C. Uebel
1965. Juvenile and gonadotropic hormone activity of 10,11-epoxyfarnesenic acid methyl ester. Life Sciences 4:2323–31.

Boyce, J. S.
1961. Forest pathology. McGraw-Hill, New York. 572 pp.

Boyd, M. F.
1949. Malariology. (Two volumes.) W. B. Saunders Co., Philadelphia. 1643 pp.

Boydstun, R. G.
1959. Three PCO's shoptalk rodent control. Pest Control. 27(8):24–26.

Bradley, G. H.
1966. A review of malaria control and eradication in the United States. Mosquito News 26:462–70.

Brannon, D. H.
1961. Boxelder bug. Cooperative Economic Insect Report (CEIR) 11(49):1100.

Breck, P. A.
1965. Forty rat trix. Pest Control 33(8):14–15, 54, 56, 58, 60–61.

Brehm, W. L.
1968. Care and maintenance of the one-gallon stainless-steel sprayer. Pest Control 36(5):16–20.

Brett, C. H.
1962. Damage by and control of silverfish and firebrats. Pest Control 30(10):75–78.

Brimblecombe, A. R.
1938. Hide beetle damage to Oregon pine. J. Australian Inst. Agr. Sci. 4:49–50.

Bristow, J. W.
1966. Importance of mosquito control in the Los Angeles metropolitan area. Mosquito News 26:340–42.

Broadhead, E., and B. M. Hobby
1944. Studies on a species of *Liposcelis* occurring in stored products in Britain. Entomol. Monthly Magazine—Part I, 80:45–59; Part II, 80:163–73.

Brooks, A. J., and H. G. Scott
1960. Inexpensive insecticide duster. The Sanitarian 22: 347–48.
1961. Preliminary studies of insect and mold infestations of anticoagulant rodenticide baits. Florida Entomologist 44:15–23.

Brooks, F. E.
1919. A migrating army of millipedes. J. Econ. Entomol. 12:462–64.

Brooks, G. D., and H. F. Schoof
1964a. Effectiveness of various dosages of dichlorvos–resin against *Culex pipiens quinquefasciatus*. Mosquito News 24:141–43.
1964b. Field evaluation of five types of dichlorvos dispensers against *Culex pipiens quinquefasciatus* in catch basins. Mosquito News 24:144–48.

Brooks, J. E.
1964. Population responses of sewer rats following poisoning. California Vector Views 11:41–46.
1966. Roof rats in residential areas: The ecology of invasion. California Vector Views 13:69–74.
1969. Domestic rodents. *In:* Community pest and related vector control (J. E. Brooks and R.D. Peck, eds.), pp. 161–73. Pest Control Operators of California, Los Angeles.
1973. Preferences of wild Norway rats for grains, seeds, and legumes. Pest Control 41(8): 13, 16, 18, 36, 38–39.

Brooks, M. A., and A. G. Richards
1955. Intracellular symbiosis in cockroaches. I. Production of aposymbiotic cockroaches. Biological Bulletin 109:22–39.

Brooks, R. B., Jr., and J. A. Vick
1972. Toxicological studies of bee venom in mice, dogs, and monkeys. American Bee Journal 112:250–51.

Brown, A. W. A.
1953. Control of insect vectors of disease by WHO. Entomol. Soc. Ontario, Annual Rept. 84:23–27.
1958. Insecticide resistance in arthropods. WHO, Monograph Series No. 38. 240 pp. Geneva, Switzerland.
1966. Attraction of mosquitoes to hosts. J. Amer. Med. Assoc. 196:249–52.
1971. Pest resistance to pesticides. *In:* Pesticides in the environment (R. White-Stevens, ed.), 1:457 ff. Decker, New York.

Brown, A. W. A., and P. E. Morrison
1955. Studies on portable jet aerosol generators for control of adult mosquitoes. Mosquito News 15:1–12.

Brown, A. W. A., and R. Pal
1971. Insecticide resistance in arthropods. WHO, Monograph Series No. 38. 491 pp. Geneva, Switzerland.

Brown, H., and H. S. Bernton
1971. Problems in the diagnosis of allergy to the Hymenoptera. Annals of Allergy 29:438–41.

Brumbaugh, L. R., J. L. Mallars, and A. V. Vieira
1969. Chironomid midge control with quick-breaking emulsions in waste-water stabilization lagoons at Stockton, California. California Vector Views 16: 1–9.

Bry, R. E., R. E. Boatright, J. H. Lang, and R. S. Cail
1973a. Protecting woolen fabric against insect damage with resmethrin. Soap, Cosm., and Chem. Spec. 49(3):40, 42, 44.

Bry, R. E., M. C. Bowman, J. H. Lang, and F. G. Crumley
1971. Evaluation of Ciba C-9491 as a mothproofing agent. Part I. J. Econ. Entomol. 64:177–83.

Bry, R. E., M. C. Bowman, F. G. Crumley, J. H. Lang, and R. S. Cail
1972. Evaluation of Ciba C-9491 as a mothproofing agent. Part II. J. Econ. Entomol. 65:584–90.

Bry, R. E., J. H. Lang, and R. E. Boatright
1973b. Toxicity of resmethrin to carpet beetles and clothes moths. Pest Control 41(11):32, 47.

Bry, R. E., J. H. Lang, and N. M. Dennis
1969. Gardona as a mothproofer of woolen fabric. J. Econ. Entomol. 62:868–72.

Bry, R. E., L. L. McDonald, and J. H. Lang
1968. Allethrin as a protectant of woolen fabric against black carpet beetle larval damage. J. Econ. Entomol. 61:1194–97.

Bryant, E. B.
1951. Redescription of *Chiracanthium mildei* L. Koch, a recent spider immigrant from Europe. Psyche 58:100–23.

Büchner, P.
1928. Holznahrung und Symbiose. J. Springer, Berlin.
1930. Tier und Pflanze in Symbiose. J. Springer, Berlin.

Buffkin, D. C., F. E. Russell, and A. Deshmukh
1971. Preliminary studies on the toxicity of black widow spider eggs. Toxicon 9:393–402.

Bugnion, E.
1922. The growth of the antennae and cerci of the cockroach *Periplaneta americana*. Bull. Soc. Entomol. Égypte 1921:56–66.

Burden, G. S., W. A. Banks, and E. E. Madden
1972. Chlorpyrifos (Dursban) in field tests against German cockroaches. Pest Control 40(3):13–14.

Buren, W. F.
1972. Revisionary studies on the taxonomy of the imported fire ants. J. Georgia Entomol. Soc. 7:1–26.

Burgdorfer, W., and M. G. R. Varma
1967. Trans-stadial and transovarial development of disease agents in arthropods. Annual Rev. Entomol. 12:347–76.

Burgoyne, W. E., and N. B. Akesson
1969. A low-volume spray system for twin-engine aircraft used in mosquito control. California Vector Views 16:17–18.

Burgoyne, W. E., N. B. Akesson, T. D. Mulhern, and K. Phillips
1967. The present status of low-volume (LV) air sprays for California mosquito control. Mosquito News 27:398–407.

Burke, H. E., R. D. Hartman, and T. E. Snyder
1922. The lead cable borer or "short circuit beetle" in California. USDA Bulletin No. 1107. 35 pp.

Burnett, H. S., W. E. Parmelee, R. D. Lee, and E. D. Wagner
1957. Observations on the life cycle of *Thelazia californiensis* Price, 1930. Journal of Parasitology 43:433.

Burnett, R.
1971. DDT residues: Distribution of concentrations in *Emerita analoga* (Stimpson) along coastal California. Science 174:606–608.

Burns Brown, W., and H. K. Heseltine
1964. Fumigation tents for small-scale treatments. Intern. Pest Control 6(4):11–15.

Burton, G. J.
1963. Bedbugs in relation to transmission of human diseases. Public Health Reports (USPHS) 78:513–24.

Burton, V. E., J. R. Anderson, and W. Stanger
1965. Fly control costs in northern California poultry ranches. J. Econ. Entomol. 58:306–309.

Busck, A.
1917. The pink bollworm: *Pectinophora gossypiella*. J. Agr. Res. 9:343–70.

Bushland, R. C., and D. E. Hopkins
1951. Experiments with screw-worm flies sterilized by x-rays. J. Econ. Entomol. 44:725–31.
1953. Sterilization of screw-worm flies with x-rays and gamma-rays. J. Econ. Entomol. 46:648–56.

Busvine, J. R.
1944. Destruction of lice in clothing by hot and cold air. Bull. Entomol. Res. 35:115–25.
1958. Insecticide resistance in bed-bugs. WHO Bulletin No. 19:1041–52.
1966. Insects and hygiene. (Second edition.) Methuen, London. 467 pp.

Butler, L.
1972. Parasitization of the black carpet beetle by the straw itch mite. J. Econ. Entomol. 65:702–705.

Buxton, G. M., and T. J. Freeman
1968. Positive separation of *Blattella vaga* and *Blattella germanica* (Orthoptera: Blattidae). Pan-Pacific Entomologist 44:168–69.

Buxton, P. A.
1921*a*. The external anatomy of the *Sarcoptes* of the horse. Parasitology 13:114–45.
1921*b*. On the *Sarcoptes* of man. Parasitology 13:146–51.
1941*a*. Studies on population of head-lice (*Pediculus humanus capitis:* Anoplura). IV. The composition of populations. Parasitology 33:224–42.
1941*b*. The parasitology of scabies. British Medical Journal (ii):397–401.
1947. The louse. (Second edition.) Arnold, London. 164 pp.

BVC (Bureau of Vector Control)
1951. Mosquito abatement in California. California State Department of Public Health, Bureau of Vector Control, Bulletin No. VC-1. 47 pp.

Calderon, M., S. Navarro, and E. Donahaye
1969. *Ectomyelois ceratoniae* (Zell.) (Lepidoptera: Phycitidae), a major pest of stored almonds in Israel. J. Stored Prod. Res. 5:427–28.

Cameron, E. A.
1967. Notes on *Sirex juvencus californicus* (Hymenoptera: Siricidae), with a description of the male and a key to the California species of *Sirex*. Canadian Entomologist 99:18–24.
1968. *Sirex juvencus californicus* in smog-killed trees in southern California (Hymenoptera: Siricidae). Pan-Pacific Entomologist 44:168.

Cameron, E. A., C. P. Schwalbe, M. Beroza, and E. F. Knipling
1974. Disruption of gypsy moth mating with microencapsulated disparlure. Science 183:972–73.

Cameron, T. M. W.
1949. Diseases carried by house mice. Pest Control 17(9): 9–11.

Campbell, W. G., and S. A. Bryant
1940. A chemical study of the bearing of decay by *Phellinus cryptarum* Karst. and other fungi on the destruction of wood by the deathwatch beetle (*Xestobium rufovillosum* De G.). Biochemical Journal 34:1404–14.

Candura, G. S.
1932. Contributo alla conoscenza biologica del *Troctes divinatorius* (Müller). Bollettino di Zoologia 3: 177–84.

Cantwell, G. E.
1974. The African (Brazilian) bee problem. American Bee J. 114:368–72.

Cantwell, G. E., G. J. Tompkins, and P. N. Watson
1973. Control of the German cockroach with carbon dioxide. Pest Control 41(3):40, 42, 48.

Caro, M. R., V. J. Derbes, and R. Jung
1957. Skin responses to the sting of the imported fire ant. Arch. Dermatol. Syph. 75:475–88.

Carpenter, S. J., and W. J. LaCasse
1952. Mosquitoes of North America north of Mexico. University of California Press, Berkeley. 360 pp., 127 pl.

Carter, F. L., C. A. Stringer, and R. H. Beal
1970. Penetration and persistence of soil insecticides used for termite control. Pest Control 38(10): 18–20, 22, 24, 62.

Carter, H. F., G. Wedd, and V. St. E. d'Abrera
1944. The occurrence of mites (Acarina) in human sputum, and their possible significance. Indian Medical Gazette 79:163–68.

Case, D. S.
1969. Experts say bird control programs developing faster than materials. Pest Control 37(2):47–51.

Casida, J. E.
1969. Toxicology of insecticides commonly used in turf maintenance. *In:* Proceedings of Scott's Turfgrass Research Conference. Vol. 1 (Entomology). (H. T. Streu and R. T. Bangs, eds.), pp. 1–13. O. M. Scott and Sons Co., Marysville, Ohio.

Caslick, T. W.
1956. Color phases of the roof rat, *Rattus rattus*. Journal of Mammalogy 37:255–57.

Caton, J. R., and R. D. Sjogren
1969. A modified box trap for the capture of *Gambusia affinis*. California Vector Views. 16:69–70.

Caudell, A. N.
1916. The genera of the tettigoniid insects of the subfamily Rhaphidophorinae found in America north of Mexico. Proc. U. S. National Museum 49:655–90.

Causey, N. B.
1943. Studies on the life history and the ecology of the hothouse millipede, *Orthomorpha gracilis* (C. L. Koch, 1847). American Midland Naturalist 29: 670–82.

Caussanel, C.
1970. Principales exigences écophysiologiques du forficule des sables, *Labidura riparia* [Derm. Labiduridae]. Ann. Soc. Entomol. France 6:589–612.

Cavanaugh, D. C.
1971. Specific effect of temperature upon transmission of the plague bacillus by the oriental rat flea, *Xenopsylla cheopis*. Amer. J. Trop. Med. Hyg. 20:264–73.

Cavil, G. W. K., and P. L. Robertson
1965. Ant venoms, attractants, and repellents. Science 149:1337–45.

CDC (Center for Disease Control; originally, Communicable Disease Center)
1966. Pictorial keys: Arthropods, reptiles, birds, and mammals of public health significance. U. S. Public Health Service (CDC). 191 pp.
1969*a*. Public health pesticides. Pest Control 37(3):15–49.
1969*b*. Encephalitis—United States, 1968. Morbidity and Mortality Weekly Rept. 18:369–70.
1972. Public health pesticides. Pest Control 40(4):36–65.
1973. Public health pesticides. Pest Control 41(4):17–50.

CEIR (Cooperative Economic Insect Reports, USDA)
1960. A report of the survey and detection operations of the Plant Pest Control Division, USDA.
1973. Special insects of regional importance. CEIR 23(16):226–27.

Chadwick, L. E., and H. Rahn
1954. Temperature dependence of rattling frequency in the rattlesnake, *Crotalus viridis viridis*. Science 119:442–43.

Chamberlain, J. C., C. W. Getzendaner, H. H. Hessig, and V. D. Young
1955. Studies of airplane spray deposit patterns at low flight levels. USDA Tech. Bull. No. 1110:1–45.

Chamberlin, W. J.
1918. Bark beetles infesting Douglas fir. Oregon Agr. Expt. Sta. Bull. 147. 40 pp.
1925. The coniferous trees of the United States, with the scolytid beetles said to attack them. Pan-Pacific Entomologist 2:23–35.
1939. The bark and timber beetles of North America north of Mexico. Oregon State Univ. Coop. Assoc., Corvallis. 513 pp.
1949. Insects affecting forest products and other materials. Oregon State Univ. Coop. Assoc., Corvallis. 159 pp.

Champ, B. R., and C. N. Smithers
1965. Insects and mites associated with stored products in Queensland. I. Psocoptera. Queensland J. Agr. Anim. Sci. 22:259–62.

Chandler, H. P.
1956. Megaloptera. *In:* Aquatic insects of California (R. L. Usinger, ed.), pp. 229–33. University of California Press, Berkeley.

Chandler, L.
1958. Seven species of carpenter bees are found in the United States. Pest Control 26(9):36, 38, 40, 47.

Cheema, P. S.
1956. Studies on the bionomics of the case-bearing clothes moth, *Tinea pellionella* (L.). Bull. Entomol. Res. 47:167–82.

Chemsak, J. A., and E. G. Linsley
1971. Some aspects of adult assembly and sexual behavior of *Rosalia funebris* Motschulsky under artificial conditions. Pan-Pacific Entomologist 47: 149–54.

Child, H. J.
1934. The internal anatomy of termites and the histology of the digestive tract. *In:* Termites and termite control (C. A. Kofoid, ed.; reprinted and partially revised, 1965), pp. 58–88. University of California Press, Berkeley.

Chillcott, J. G.
1960. A revision of the Nearctic species of Fanniinae (Diptera: Muscidae). Canadian Entomologist, Supplement 14 (accompanies Vol. 92). pp. 1–295.

Chittenden, R. H., and W. J. Gies
1898. The influence of borax and boric acid upon nutrition, with special reference to proteid metabolism. Amer. J. Physiol. 1:1–39.

Christian, M. B.
1940*a*. *Lyctus* powder-post beetles and their control. Pests 8(6):8–10.
1940*b*. Biology of the powder-post beetles, *Lyctus planicollis* LeConte and *Lyctus parallelopipedus* (Melsh.). Part I. Louisiana Conservation Review 9(4):56–59.
1941. Biology of the powder-post beetles, *Lyctus planicollis* LeConte and *Lyctus parallelopipedus* (Melsh.). Part II. Louisiana Conservation Review 10(1):40–42.

Chrystal, R. N.
1928. The *Sirex* woodwasps and their importance in forestry. Bull. Entomol. Res. 19:219–47.

Chrystal, R. N., and J. G. Myers
1928. Natural enemies of *Sirex cyaneus* F. in England and their life histories. Bull. Entomol. Res. 19: 67–77.

Clark, P. H., and M. M. Cole
1967. Resistance of body lice to carbaryl. J. Econ. Entomol. 60:398–400.

Clements, R. H., and S. H. Kerr
1969. Earwigs. Pest Control 37(10):26, 28, 30.

Cleveland, L. R., S. R. Hall, E. P. Sanders, and J. Collier
1934. The wood-feeding roach *Cryptocercus*, its protozoa, and the symbiosis between protozoa and roach. Mem. Amer. Acad. Arts Sci. (N.S.) 17: 185–342.

Clifford, C. M., D. E. Sonenshine, E. L. Atwood, C. S. Robbins, and L. E. Hughes
1969. Tests on ticks from wild birds collected in the eastern United States for rickettsiae and viruses. Amer. J. Trop. Med. Hyg. 18:1057–61.

Cloudsley-Thompson, J. L.
1949. The significance of migration in myriapods. Annual Magazine of Natural History (Series 12) 2: 947–62.
1954. The ecological significance of diurnal rhythms in terrestrial arthropods. Science Progress 42:46–52.
1968. Spiders, scorpions, centipedes, and mites. Pergamon Press, Oxford and New York. 278 pp.

Coaton, W. G. H.
1948. *Cryptotermes brevis,* a new wood-borer problem in South Africa. South Africa Dept. Agr. Bull. 290. 18 pp.

Cockran, D. M., and C. J. Goin
1970. The new field book of reptiles and amphibians. G. P. Putnam's Sons, New York. 359 pp.

Cohen, A. C., and B. C. Myres
1970. A function of the horns (supraocular scales) in the sidewinder rattlesnake, *Crotalus cerastes,* with comments on other horned snakes. Copeia 1970: 574–75.

Colburn, R. B., and G. S. Langford
1970. Field evaluation of some mosquito adulticides, with observations on toxicity to honey bees and house flies. Mosquito News 30:518–22.

Cole, D. A.
1969*a*. Thirty years of rodent baiting in New Zealand. Pest Control 37(11):10, 12, 14.

Cole, F. R.
1969*b*. The flies of western North America. University of California Press, Berkeley and Los Angeles. 693 pp.

Cole, J. H.
1962. *Hofmannophila pseudospretella* (Stnt.) (Lepidoptera: Oecophoridae): Its status as a pest of woolen textiles, its laboratory culture, and susceptibility to mothproofers. Bull. Entomol. Res. 53:83–89.

Cole, M. M., P. H. Clark, and D. E. Weidhaas
1969*a*. Failure of laboratory colonies of body lice to develop resistance to malathion. J. Econ. Entomol. 62:568–70.

Cole, M. M., J. M. Hirst, J. G. McWilliams, and I. H. Gilbert
1969*b*. Sleeve tests of insecticidal powders for control of body lice. J. Econ. Entomol. 62:198–200.

Collins, R. P.
1968. Carbonyl compounds produced by the bed bug, *Cimex lectularius.* Annals Entomol. Soc. Amer. 61:1338–39.

Collins, W. J.
1973. German cockroach resistance. I. Resistance to diazinon includes cross-resistance to DDT, pyrethrins, and propoxur in a laboratory colony. J. Econ. Entomol. 66:44–47.

Collyer, C. H., and C. O. Hammond
1951. Flies of the British Isles. Frederick Warme and Co., Ltd., London and New York.

Comstock, J. H.
1949. An introduction to entomology. Comstock Publishing Co., Inc., Ithaca, New York. 1064 pp.

Connin, R. V., O. K. Jantz, and W. S. Bowers
1967. Termination of diapause in the cereal leaf beetle by hormones. J. Econ. Entomol. 60:1752–53.

Constantine, D. G.
1962. Rabies transmission by nonbite route. Public Health Reports (USPHS) 77:287–89.
1967*a*. Rabies transmission by air in bat caves. U. S. Public Health Service (CDC) Publ. No. 1617. 51 pp.
1967*b*. Bat rabies in the southwestern United States. Public Health Reports (USPHS) 82:867–88.

Cook, S. F., Jr.
1967. The increasing chaoborid midge problem in California. California Vector Views 14:39–44.

Cooke, M. T.
1928. The spread of the European starling in North America. USDA Circular No. 40. 9 pp.

Coombs, C. W., and G. E. Woodroffe
1963. An experimental demonstration of ecological succession in an insect population breeding in stored wheat. Journal of Animal Ecology 32:271–79.
1965. Some factors affecting the longevity and oviposition of *Ptinus tectus* Boieldieu (Coleoptera: Ptinidae) which have relevance to succession among grain beetles. J. Stored Prod. Res. 1:111–27.
1970. Some factors affecting larval development in *Ptinus tectus* Boieldieu (Coleoptera: Ptinidae) relevant to succession among grain beetles. J. Stored Prod. Res. 6:199–216.

Cope, J. B.
1959. Build bats out. Pest Control 27(8):28–29.

Coppock, R. H.
1970. Research on mosquito control. Rept. of Univ. of California Mosquito Control Res. Comm. (R. F. Smith, chm.), 15 pp.

Corbet, A. S., and W. H. T. Tams
1943. Keys for the identification of the Lepidoptera infesting stored food products. Proc. Zool. Soc. London (B) 113:55–148. 286 figs., 5 pls.

Corbett, C. H., and H. T. Pagden
1941. A review of some recent entomological investigations and observations. Malayan Agricultural Journal 29:347–75.

Cornere, B. M.
1972. House dust mites: A national survey. New Zealand Med. J. 76:270–74.

Cornwell, P. B.
1968. The cockroach. Hutchinson, London. 391 pp.

Costa, J. A.
1952. Tick paralysis on the Atlantic seaboard. Amer. J. Dis. Child. 83:336–47.

Coston, T. O.
1967. *Demodex folliculorum* blepharitis. Trans. Amer. Ophthal. Soc. 65:361–92.

Cottam, C.
1948. Aquatic habits of the Norway rat. Journal of Mammalogy 29:299.

Cotton, R. T.
1963. Pests of stored grains and grain products. Burgess Publishing Co., Minneapolis. 318 pp.

Cotton, R. T., and N. E. Good
1937. Annotated list of the insects and mites associated with stored grain and cereal products, and their arthropod parasites and predators. USDA Misc. Publ. No. 258. 81 pp.

Cotton, R. T., and B. A. St. George
1929. The mealworms. USDA Tech. Bull. No. 95. 37 pp.

Cowles, R. B., and R. L. Phelan
1958. Olfaction in rattlesnakes. Copeia 1958:77–83.

CPHS (California Public Health Service)
1971. A manual for the control of communicable disease in California. California State Department of Public Health. 560 pp.

Craighead, F. C.
1950. Insect enemies of eastern forests. USDA Misc. Publ. No. 657. 679 pp.

Crampton, G. C.
1923. The cave cricket, *Ceuthophilus,* as a possible vector of pathogenic organisms. J. Econ. Entomol. 16:460.

Crawford, C. S., and J. L. Cloudsley-Thompson
1971. Water-relations and desiccation-avoiding behavior of the vinegarroon, *Mastigoproctus giganteus.* Entomol. Expt. Appl. 14:99–106.

Creighton, J. T.
1943. Household pests. Florida Agr. Ext. Serv. Bull. 122.
1954. Household pests. Florida Dept. Agr. Bull. (N.S.) 156. 72 pp.

Creighton, W. S.
1950. The ants of North America. Harvard: Bull. Mus. Compar. Zool. Vol. 104. 585 pp., 55 pls.
1953. New data on the habits of ants of the genus *Veromessor.* Amer. Mus. Novitates 1612:1–18.

Cress, D. C.
1971. Casual home-invading pests. Univ. Michigan Ext. Bull. E-652.

Crow, J. F.
1957. Genetics of insect resistance to chemicals. Annual Rev. Entomol. 2:227–46.

Cruickshank, P. A., and R. M. Palmere
1971. Terpenoid amines as insect juvenile hormones. Nature 233:488–89.

Crumb, S. E., P. M. Eide, and A. E. Bonn
1941. The European earwig. USDA Tech. Bull. No. 766.

Cunliffe, F.
1952. Biology of the cockroach parasite, *Pimeliaphilus podapolipophagus* Trägårdh, with a discussion of the genera *Pimeliaphilus* and *Hirstiella* (Acarina: Pterygosomidae). Proc. Entomol. Soc. Washington 54:153–69.

Cunnington, A. M.
1965. Physical limits for complete development of the grain mite, *Acarus siro* L. (Acarina: Acaridae), in relation to its world distribution. Journal of Applied Ecology 2:295–306.

Cunnington, A. M., and P. H. Gregory
1968. Mites in bedroom air. Nature 217:1271–72.

Curran, C. H.
1946*a*. Insects in the house: Spider beetles. Natural History 55:46.
1946*b*. Uninvited guests in the house: The house centipede (*Scutigera forceps*). Natural History 55:240.
1949. Clothes moths. Natural History 58:325–31.
1965. The families and genera of North American Diptera. Henry Tripp, Woodhaven, New York.

Curry, J. P.
1971. Development of populations of *Acarus siro* (Acarina: Acaridae) on various foodstuffs. Annals Entomol. Soc. Amer. 64:531–32.

Cushing, C. D.
1970. The case of the mysterious suit sleeves. Pest Control 38(7):36–37.

Cutler, B., and L. E. Cutler
1971. Experimental envenomization of hairless mice by *Loxosceles arizonica* Gertsch and Mulaik and *Loxosceles devia* Gertsch. and Mulaik (Araneae: Scytodidae). J. Med. Entomol. 8:48.

Cutler, J. R., and G. Hosie
1966. Bird nests as sources of infestations of *Ptinus tectus* Boieldieu (Coleoptera: Ptinidae), and the distribution of this insect in stacks of bagged flour. J. Stored Prod. Res. 2:27–34.

Cymorek, S.
1965. Experimente mit *Lyctus.* Holz und Organismen, International Symposium 1:391–413.
1967. Ueber den Einfluss der Holzdichte auf die Entwicklung von Holzinsekten und Versuche darüber mit *Lyctus brunneus* (Steph.) in Pressholz. Material und Organismen 2:195–205.

Damback, C. A., and E. Good
1943. Life history and habits of the cicada killer in Ohio. Ohio Journal of Science 43:32–41.

Darby, R. E.
1962. Midges associated with California rice fields, with special reference to their ecology (Diptera: Chironomidae). Hilgardia 32:1–206.

Darwin, C.
1859. On the origin of species by means of natural selection. Harvard University Press (1964). 502 pp.

David, W. A. L., and P. Bracey
1946. Factors influencing the interaction of insecticidal mists on flying insects. Bull. Entomol. Res. 37: 177–90.

Davidson, A.
1893. The nests and parasites of *Xylocopa orpifex* Smith. Entomological News 4:151–53.
1905. An enemy of the trap-door spider. Entomological News 16:233–34.

Davis, A. C.
1927. Studies on the anatomy and histology of *Stenopelmatus fuscus* Hald. Univ. California Publ. Entomol. 4:159–208.

Davis, G. E.
1939. *Ornithodoros parkeri:* Distributions and host data; spontaneous infection with relapsing fever spirochetes. Public Health Reports (USPHS) 54: 1345–49.
1948. The spirochetes. Annual Rev, Microbiol. 2:305–34.

Davis, H. G., G. W. Eddy, T. P. McGovern, and M. Beroza
1967. 2,4-Hexadienyl butyrate and related compounds highly attractive to yellowjackets (*Vespula* spp.). J. Med. Entomol. 4:275–80.

Davis, W. H.
1970. Hibernation: Ecology and physiological ecology. *In:* Biology of bats (W. A. Wimsatt, ed.), Vol. I:265–300. Academic Press, New York and London. 406 pp.

Deacon, J. E., C. Hubbs, and B. J. Zahuranec
1964. Some effects of introduced fishes on the native fish fauna of southern Nevada. Copeia 1964:384–88.

Dechene, R.
1970. Studies of some behavioural patterns of *Iridomyrmex humilis* Mayr (Formicidae: Dolichoderinae). Wasmann J. Biol. 28:175–84.

DeLong, D. M.
1962. Beer cases and soft-drink cartons as insect distributors. Pest Control 30(7):14–18.

Deonier, C. C.
1943. Biology of the immature stages of the Clear Lake gnat (*Chaoborus astictopus*) (Diptera: Culicidae). Annals Entomol. Soc. Amer. 36:383–88.

Desch, C., and W. B. Nutting
1972. *Demodex folliculorum* (Simon) and *D. brevis* Akbulatova of man: Redescription and re-evaluation. Journal of Parasitology 58:169–77.

Devore, A. L., D. W. Maxson, J. L. Albright, and R. W. Taylor
1967. How about anhydrous ammonia for bird control? Pest Control 35(2):24, 26.

Dicke, R. J., and J. J. Paul
1951. Space spray combinations of chlorinated insecticides. J. Econ. Entomol. 44:896–98.

Dillaha, C. J., C. T. Jansen, W. M. Honeycutt, and C. R. Hayden
1964. North American loxoscelism. J. Amer. Med. Assoc. 188:33–36.

Dimick, R. E., and D. C. Mote
1934. The present status of the European earwig. California Dept. Agr. Monthly Bull. 23:298–300.

Ditmars, R. L.
1931. Snakes of the world. Macmillan, New York. 207 pp.

Doane, R. W., and O. J. Gilliland
1929. Three California ambrosia beetles. J. Econ. Entomol. 22:915–21.

Doane, R. W., E. C. Van Dyke, W. J. Chamberlin, and H. E. Burke
1936. Forest insects. McGraw-Hill, New York. 463 pp.

Dodge, H. R.
1954. Fly identification is easy with CDC pictorial keys. Pest Control 22(4):10–12, 14, 16.

Doetschman, W. H., and D. P. Furman
1949. A tropical chigger, *Eutrombicula batatas* (L.), attacking man in California. Amer. J. Trop. Med. 29:605–608.

Dominik, J.
1966. Results of investigations on the species composition of insects destroying wood in buildings of different ages. Rev. Appl. Entomol. A-56:1041 (1968). (Published in Polish, 1966; the title translated and reprinted in English, 1968.)

Donohoe, H. C.
1946. Notes on dried-fruit insects. Pests I, 14(9):32, 34; II, 14(10):14–16.

Doucette, C. F., and F. F. Smith
1926. Control experiments on the Surinam cockroach, *Pycnoscelus surinamensis* (L.). J. Econ. Entomol. 19:650–56.

Dow, R.
1942. The relation of the prey of *Sphecius speciosus* to the size and sex of the adult wasp (Hymenoptera: Sphecidae). Annals Entomol. Soc. Amer. 35:310–17.

Dubos, R. J., and J. G. Hirsch
1965. Bacterial and mycotic infections of man. Lippincott, Philadelphia. 1025 pp.

Duncan, C. D.
1939. A contribution to the biology of North American vespine wasps. Stanford Univ. Publ. Biol. Sci. 8(1):1–272.

Durr, H. R. R.
1956. The morphology and bionomics of the European house borer, *Hylotrupes bajulus* (Coleoptera: Cerambycidae). South Africa Dept. Agr., Div. Entomol., Entomol. Memoirs. Vol. 4, Pt. I.

Dyar, H. G., and R. C. Shannon
1927. The North American two-winged flies of the family Simuliidae. Proc. U. S. National Museum 69(2636):1–54.

Dyer, R. E., E. T. Ceder, W. G. Workman, A. Rumreich, and L. F. Badger
1932. Transmission of endemic typhus by rubbing either crushed, infected fleas or infected flea feces into wounds. Public Health Reports (USPHS) 47: 131–33.

Dykstra, W. W.
1950. New techniques and developments in rodent control. Pest Control 18(12):9–12, 24, 39.

Eads, R. B., F. J. von Zuben, S. E. Bennett, and O. L. Walker
1954. Studies on cockroaches in a municipal sewerage system. Amer. J. Trop. Med. Hyg. 3:1092–98.

Eaton, C. B.
1959. Observations on the survival of *Arhopalus productus* (LeConte) larvae in Douglas-fir timber. Pan-Pacific Entomologist 35:114–16.

Eaton, C. B., and R. L. Lyon
1955. *Arhopalus productus* (Lec.), a borer in new buildings. U. S. Forest Service Tech. Paper 11.

Eaton, O., and C. A. Cabell
1949. Raising laboratory mice and rats. USDA Leaflet No. 253.

Ebeling, W.
1945. Properties of petroleum oils in relation to toxicity to potato tuber moth larvae. J. Econ. Entomol. 38:26–34.
1959. Subtropical fruit pests. Univ. California Div. Agr. Sci., Berkeley. 436 pp.
1960. Control of the tropical rat mite. J. Econ. Entomol. 53:475–76.
1961. Physiochemical mechanisms for the removal of insect wax by means of finely divided powders. Hilgardia 30:531–64.
1964*a*. Midges and gnats as accidental invaders of homes. PCO News 24(11):28–30.
1964*b*. Tests with Vapona insecticide resin strip. PCO News 24(4):23.
1968. Termites: Identification, biology, and control of termites attacking buildings. California Agr. Expt. Sta. Ext. Serv. Manual 38. 74 pp.
1969. Insectproofing during building construction. California Agriculture 23(5):4–7.
1971. Sorptive dust for pest control. Annual Rev. Entomol. 16:123–58.

Ebeling, W., and R. J. Pence
1958. Laboratory evaluation of insecticide-treated soils against the western subterranean termite. J. Econ. Entomol. 51:207–211.
1965. Prevention and control of the western subterranean termite. Univ. California Agr. Expt. Sta. Circ. 469. (Revised 1965.) 16 pp.

Ebeling, W., and D. A. Reierson
1970. Effect of population density on exploratory activity and mortality rate of German cockroaches in choice boxes. J. Econ. Entomol. 63:350–55.
1973*a*. Should flushing agents be added to blatticides? Pest Control 41(6):24, 46, 48, 50–51.
1973*b*. Leadcable borers infesting houses. National Pest Control Operators News 33(8):16–17.

Ebeling, W., D. A. Reierson, and R. E. Wagner
1967. Influence of repellency on the efficacy of blatticides. II. Laboratory experiments with German cockroaches. J. Econ. Entomol. 60:1375–90.
1968. Influence of repellency on the efficacy of blatticides. III. Field experiments with German cockroaches, with notes on three other species. J. Econ. Entomol. 61:751–61.

Ebeling, W., and R. E. Wagner
1959*a*. Rapid desiccation of drywood termites with inert sorptive dusts and other substances. J. Econ. Entomol. 52:190–207.
1959*b*. Control of drywood termites. California Agriculture 13(1):7–9.
1962. Interim report on five-year tests with pretreatment termiticides. Pest Control 30(2):16, 18, 20.
1963. Methods and equipment for treating houses with Dri-die 67 during construction. PCO News 23(6): 14–17.
1964. Built-in pest control. Pest Control 32(2):20–32.

Ebeling, W., R. E. Wagner, and D. A. Reierson
1966. Influence of repellency on the efficacy of blatticides. I. Learned modification of behavior of the German cockroach. J. Econ. Entomol. 59:1374–88.

Eberhard, M. J. W.
1969. The social biology of polistine wasps. Univ. Michigan Mus. Zool. Misc. Publ. 140:1–101.

Ecke, D. H. (ed.)
1963. A field guide to the domestic flies of California. California Mosquito Control Assoc., Visalia, Field Guide Series Publ. No. 3. 31 pp.

Eckert, J. E.
1940. Studies on the poison system of the honeybee. Annals Entomol. Soc. Amer. 33:258–68.

Eckert, J. E., and A. Mallis
1941. Ants and their control in California. Univ. California Agr. Expt. Sta. Circ. 342. 39 pp.

Eckert, J. E., and F. R. Shaw
1960. Beekeeping. Macmillan, New York. 530 pp. (*See specifically:* pp. 84–85.)

Eddleman, D., and D. Simon
1969. Note on the brown cockroach, *Periplaneta brunnea,* in Texas. Annals Entomol. Soc. Amer. 62: 678–79.

Eddy, G. W.
1946. A combination treatment for lice and scabies. J. Invert. Dermatol. 7:85–91.

Eddy, G. W., and R. C. Bushland
1946. Control of human lice. USDA Bureau of Entomology and Plant Quarantine Publ. No. E-685. 5 pp.

Edmunds, L.R.
1953. Some notes on the Evaniidae as household pests and as a factor in the control of roaches. Ohio J. Sci. 53:121–22.
1957. Observations on the biology and life history of the brown cockroach, *Periplaneta brunnea* Burmeister. Proc. Entomol. Soc. Washington 59: 283–86.

Edney, E. B.
1957. The water relations of terrestrial arthropods. Cambridge University Press, Cambridge, England. 109 pp.

Ehmann, N. R.
1961. A new method for control of powder-post beetles. Pest Control 29(6):42, 46–46*b*.

Eichler, W., and W. Kleinsorge
1973. Pharaoameisenbekämpfung in Krankenhäusern (Schluss). Angewandte Parasitologie 14:23–43.

Eisener, T., F. McKittrick, and R. Payne
1959. Defense sprays of roaches. Pest Control 27(6):9–12, 44–45.

Elgart, M. L., and R. S. Higdon
1973. Pediculosis pubis of the scalp. Archives of Dermatology 107:916–17.

Ellis, R. V., and H. G. Ahrens
1932. Hypersensitiveness to air-borne bee allergen. Journal of Allergy 3:247–52.

Engelmann, F.
1968. Endocrine control of reproduction of insects. Annual Rev. Entomol. 13:1–26.

English, F. P.
1971. *Demodex folliculorum* and oedema of the eyelash. British Journal of Ophthalmology 55:742–46.

English, F. P., T. Iwamoto, R. W. Darrell, and A. G. DeVoe
1970. The vector potential of *Demodex folliculorum*. Archives of Ophthalmology 84:83–85.

English, L. L., and R. Snetsinger
1957. The habits and control of the clover mite in dwellings. J. Econ. Entomol. 50:135–41.

Ennik, F.
1971. Laboratory observations on the biology of *Loxosceles unicolor* Keyserling (Araneae: Loxoscelidae). Santa Barbara Mus. Nat. Hist., Contrib. in Sci. No. 3. 16 pp.

Esenther, G. R.
1969. Termites in Wisconsin. Annals Entomol. Soc. Amer. 62:1274–84.

Esenther, G. R., T. C. Allen, J. E. Casida, and R. D. Shenefelt.
1961. Termite attractant from fungus-infected wood. Science 134:50.

Esenther, G. R., and R. H. Beal
1974. Attractant-mirex bait suppresses activity of *Reticulitermes* spp. J. Econ. Entomol. 67:85–88.

Esenther, G. R., and H. C. Coppel
1964. Current research on termite attractants. Pest Control 32(2):34–46.

Esenther, G. R., and D. E. Gray
1968. Subterranean termite studies in southern Ontario. Canadian Entomologist 100:827–34.

Essig, E. O.
1926. Insects and mites of western North America. Macmillan, New York. 1050 pp.
1932. A small insect which stings severely. Science 75: 242–43.
1942. College entomology. Macmillan, New York. 900 pp.
1943. The California-laurel borer, *Rosalia funebris* Mots. Pan-Pacific Entomologist 19:91–92.

Evans, H. E.
1963. Wasp farm. Natural History Press, Garden City, New York. 178 pp.

Every, R. W., and J. A. Rudinsky
1962. The golden buprestid, a wood-boring beetle. Oregon State Univ. Ext. Circ. 713.

Ewer, D. W., and R. F. Ewer
1942. The biology and behaviour of *Ptinus tectus* Boiel. (Coleoptera: Ptinidae), a pest of stored products. III. The effect of temperature and humidity on oviposition, feeding, and duration of life cycle. J. Exptl. Biol. 18:290–305.

Ewing, H. E., and L. Fox
1943. The fleas of North America. USDA Misc. Publ. No. 500. 143 pp.

Fahey, E. J., and J. V. Osmun
1971. Movement and degradation of chlorinated hydrocarbon insecticides in soil beneath concrete floors of buildings protected against subterranean termite attack. Proc. Twentieth Intern. Congr. Entomol. (Moscow, 1968) 2:226–27.

Fain, A.
1967. Le genre *Dermatophagoides* Bogdanov 1864. Son importance dans les allergies respiratoires et cutanées chez l'homme (Psoroptidae: Sarcoptiformes). Acarologia 9:179–225.

Fales, J. H., D. F. Bodenstein, and W. S. Bowers
1970. Seven juvenile hormone analogues as synergists for pyrethrins against house flies. J. Econ. Entomol. 63:1379–80.

Farrar, M. D.
1953. The granulated-type insecticide for soil treatment. J. Econ. Entomol. 46:377–79.

Farris, S. H., and A. Funk
1965. Repositories of symbiotic fungus of the ambrosia beetle *Platypus wilsoni* Swaine (Coleoptera: Platypodidae). Canadian Entomologist 97:527–32.

Feingold, B. F., E. Benjamini, and D. Michaeli
1968. The allergic responses to insect bites. Annual Rev. Entomol. 13:137–58.

Ferguson, F. F.
1954. Biological factors in the transmission of the American arthropod-borne virus encephalitides. U. S. Public Health Service (CDC) Monograph No. 23. 37 pp.

Fertig, D. S., and V. W. Edmonds
1970. Water requirements of mice: A re-evaluation. Pest Control 38(6):18, 22, 24, 28, 30–31.

Figley, K. D.
1940. Mayfly (Ephemerida) hypersensitivity. Journal of Allergy 11:376–87.

Fine, R. M., and H. G. Scott
1963. Straw itch mite dermatitis caused by *Pyemotes ventricosus*. J. Med. Assoc. Georgia 52:160–66.
1965. Straw itch mite dermatitis caused by *Pyemotes ventricosus:* Comparative aspects. Southern Med. J. 58:416–20.

Finlayson, L. H.
1949. The life history and anatomy of *Lepinotus patruelis* (Psocoptera: Atropidae). Proc. Zool. Soc. London 119:301–23.

Finlayson, L. R.
1932. Some notes on the biology and life history of psocids. Entomol. Soc. Ontario Annual Rept. 63: 56–58.

Finley, E. L., F. G. McDermott, and H. R. Gross
1968. Degradation of fabric by American cockroach, house cricket, and striped earwig. J. Econ. Entomol. 61:1552–57.

Fischer, R. L.
1951. Observations on the nesting habits of megachilid bees. J. Kansas Entomol. Soc. 24:46–50.

Fisher, R. C.
1937. Studies of the biology of the death-watch beetle, *Xestobium rufovillosum* De G. I. A summary of past work and a brief account of the developmental stages. Annals Appl. Biol. 24:600–613.

Fitzpatrick, F. K.
1948. Susceptibility to typhus of rats on deficient diet. Amer. J. Public Health 38:676–81.

Fitzwater, W. D.
1958. Test pattern for bird control: Logansport, Indiana. Pest Control 26(7):9–10, 12, 14–16.

Flaschenträger, V., El Sayed Amin, and H. J. Jarczyk
1957. Ein Lockstoffanalysator für Insekten. Mikrochimica Acta 1957:385–89.

Fletcher, D. J. C.
1971. The glandular source and social functions of trail pheromones of two species of ants (*Leptogenys*). Journal of Entomology A-46:27–37.

Flock, R. A.
1941. The field roach *Blattella vaga*. J. Econ. Entomol. 34:121.
1951. Damage to household goods by the fan palm caterpillar. J. Econ. Entomol. 44:260–61.

Fluno, J. A.
1961. Wasps as enemies of man. Entomol. Soc. Amer. Bull. 7:117–19.

Folsom, J. W., and P. A. Woke
1939. The field cricket in relation to the cotton plant in Louisiana. USDA Tech. Bull. No. 462, 28 pp.

Foot, N. C.
1922. Pathology of the dermatitis caused by *Megalopyge opercularis,* a Texas caterpillar. J. Exptl. Med. 35:737–53.

Foote, R. H., and D. R. Cook
1959. Mosquitoes of medical importance. USDA Agr. Handbook No. 152. 158 pp.

Forbes, S. A.
1920. The common house ant. *In:* Twenty-third report of the State Entomologist on the noxious and beneficial insects of Illinois, p. 157.

Foulk, J. D.
1967. Ecology and control of *Leptoconops kerteszi* biting gnats in southern California. Proc. California Mosquito Control Assoc. 35:88.

Fox, I.
1955. A catalogue of the bloodsucking midges of the Americas (*Culicoides, Leptoconops,* and *Lasiohelea*), with keys to the subgenera and Nearctic species, a geographical index, and biblography. Univ. Puerto Rico J. Agr. 39:214–85.

Fox, I., and I. García-Moll
1961. Ants attacking fleas in Puerto Rico. J. Econ. Entomol. 54:1065–66.

Fox, W. J.
1894. Second report on some Hymenoptera from Lower California, Mexico. Proc. California Acad. Sci. (Second series.) 4:92–121.

Francis, E.
1919. Deer fly fever or Pahvant Valley plague. Public Health Reports (USPHS) 34:2061–62.

Francis, E., and B. Mayne
1921. Experimental transmission of tularemia by flies of the species *Chrysops discalis.* Public Health Reports (USPHS) 36:1738–46.

Francke-Grosmann, H.
1939. Ueber das Zusammenleben von Holzwespen (*Siricidae*) mit Pilzen. Z. angew. Entomol. 25:647–80.
1957. Ueber das Schicksal der Siriciden Pilze während der Metamorphose. Deutsche Akademie Landwirtschaft, Berlin. 7 pp.

Frankland, A. W.
1972. House dust mites and allergy. Archives of Disease in Childhood 47:327–29.

Frazier, C. A.
1969. Insect allergy. Warren H. Green, Inc., St. Louis, Missouri. 493 pp.

Free, J. B., and C. G. Butler
1959. Bumblebees. Collins, London. 208 pp.

Freeborn, S. B.
1926. The mosquitoes of California. Univ. California Publ. Entomol. 3:333–60.

Freeborn, S. B., and R. M. Bohart
1951. The mosquitoes of California. Bull. California Insect Surv. 1:25–78.

Freeborn, S. B., and B. Brookman
1943. Identification guide to the mosquitoes of the Pacific Coast states. U. S. Public Health Service (CDC) mimeograph. 23 pp.

Freeman, J. A.
1938. Composition of the aerial insect fauna up to 300 feet. Nature 142:153–54.
1964. Stored-products entomology in Britain. Annals Appl. Biol. 53:200–215.

Friis, M.
1961. Mysterious horntails. Pest Control 29(2):66.

Frings, H., and M. Frings
1949. The loci of contact chemoreceptors in insects. American Midland Naturalist 41:602–658.

Frommer, S. I., and P. A. Rauch
1971. Pupal duration, adult emergence, and oviposition periods for the midge *Dicrotendipes californicus* (Johannsen) (Diptera: Chironomidae). California Vector Views 18:33–41.

Fulton, R. B.
1924*a*. The European earwig. Oregon Agr. Expt. Sta. Bull. 207. 29 pp.
1924*b*. Some habits of earwigs. Annals Entomol. Soc. Amer. 17:357–67.
1933. Notes on *Habrocytus cerealellae,* parasite of the Angoumois grain moth. Annals Entomol. Soc. Amer. 26:536–52.
1953. The cheese skipper. North Carolina Agr. Expt. Sta. Tech. Bull. 103. 35 pp.

Funk, A.
1970. Fungal symbionts of the ambrosia beetle *Gnathotrichus sulcatus.* Canadian J. Bot. 48:1445–48.

Furman, D. P., and E. P. Catts
1970. Manual of medical entomology. National Press Books, Palo Alto, California. 163 pp.

Furman, D. P., E. C. Loomis, L. A. Riehl, and A. S. Rosenwald
1969. Control of external parasites of chickens and pigeons. Univ. California Agr. Ext. Serv. 12 pp.

Furman, D. P., and W. C. Reeves
1957. Toxic bite of a spider, *Chiracanthium inclusum* Hentz. California Medicine 87:114.

Furman, D. P., L. E. Vincent, and W. S. Coates
1955. External parasites of poultry. California Agriculture 9(1):13.

Furniss, R. L., and R. W. Every
1958. Carpenter ant control. Oregon Agr. Expt. Sta. Ext. Circ. 627. 5 pp.

Furumizo, R. T., and M. S. Mulla
1971. Distribution of *Dermatophagoides* mites in house dust samples. Entomol. Soc. Amer., Proc. North Central Branch 26:67–68.

Gahan, C. J., and F. Laing
1932. Furniture beetles. British Museum (Nat. Hist.) Econ. Ser. No. 2.

Gahan, J. B., and H. G. Wilson
1970. Seven insecticides as residual sprays in buildings naturally infested with *Anopheles quadrimaculatus.* Mosquito News 30:410–16.

Gahan, J. B., H. G. Wilson, and W. C. McDuffie
1954. Dry sugar baits for the control of houseflies. J. Agr. and Food Chem. 2:425–28.

Gaillard, G. E.
1950. The aphid: An insect allergen. Journal of Allergy 21:386–99.

Galbraith, M. N., and D. H. S. Horn
1966. An insect-moulting hormone of insects. Chemical Communications 1966:905–906.

Gary, N. E.
1971. Possible approaches to controlling the African bee. American Bee Journal 111:426–29.

Gaud, M., and M. T. Morgan
1947. Epidemiological study of relapsing fever in North Africa (1943–1945). WHO Bulletin No. 1:69–92.

Gauss, R.
1972. Amerikanische *Polistes*-Art mit ihrem Bislang noch unbekannten Parasiten nach Deutschland verschleppt. Z. angew. Entomol. 70:337–39.

Gay, F. J., and A. H. Wetherly
1959. The termiteproofing of concrete. Construction Review (Australia) 32(9):26–28.
1970. The population of a large mound of *Nasutitermes exitiosus* (Hill) (Isoptera: Termitidae). J. Australian Entomol. Soc. 9:27–30.

Gaydon, D. M., and T. R. Adkins, Jr.
1969. Effect of cultivation on emergence of eye gnats (*Hippelates* spp.) in South Carolina. J. Econ. Entomol. 62:312–14.

Gebhardt, L. P., G. J. Stanton, D. W. Hill, and G. C. Collett
1964. Natural overwintering hosts of the virus of western equine encephalitis. New England J. Med. 271:172–77.

Gentner, L. G., H. E. Morrison, and H. B. Rasmussen
1948. Aerosol generator applications of DDT for codling moth control. J. Econ. Entomol. 41:67–69.

Georghiou, G. P.
1967. Differential susceptibility and resistance to insecticides of coexisting populations of *Musca domestica, Fannia canicularis, F. femoralis,* and *Ophyra leucostoma.* J. Econ. Entomol. 60:1338–44.
1969. Genetics of resistance to insecticides in houseflies and mosquitoes. Experimental Parasitology 26: 224–55.
1971*a*. Isolation, characterization, and resynthesis of insecticide resistance factors in the housefly, *Musca domestica.* Proc. Second Intern. Congr. Pest. Chem. 2:77–94.
1971*b*. Resistance of insects and mites to insecticides and acaricides, and the future of pesticide chemicals. *In:* Agricultural chemicals: Harmony or discord for food, people, environment (J. E. Swift, ed.), pp. 112–24. Univ. California Div. Agr. Sci.
1972. The evolution of resistance to pesticides. Annual Rev. Ecol. and Systematics 3:133–68.

Georghiou, G. P., M. K. Hawley, E. C. Loomis, and M. F. Coombs
1972*a*. Recent developments in insecticide resistance in house flies in California. California Agriculture 26(9):4–6.

Georghiou, G. P., R. D. Sjogren, and J. Barkley
1972*b*. Effects of organophosphorus, carbamate, and organochlorine insecticides on *Leptoconops kerteszi* Kieffer (Diptera: Ceratopogonidae). Mosquito News 32:205–208.

Gerberg, E. J.
1957*a*. A revision of the New World species of powder-post beetles belonging to the family Lyctidae. USDA Tech. Bull. No. 1157. 55 pp.
1957*b*. Lyctid powder-post beetle second only to termite in wood damage. Pest Control 25(5):37–40.
1966. Field and laboratory repellency tests with 2,2,4-trimethyl-1-3-pentanediol (TMPD). J. Econ. Entomol. 59:872–75.

Gertsch, W. J.
1949. American spiders. Van Nostrand, New York. 285 pp.
1958. The spider genus *Loxosceles* in North America, Central America, and the West Indies. Amer. Mus. Novitates 1907:1–46.

Gertsch, W. J., and S. Mulaik
1940. Spiders of Texas. Amer. Mus. Nat. Hist. Bull. 77: 307–40.

Getze, G.
1972. United States bat population called immense rabies reservoir. Los Angeles *Times,* June 5.

Ghani, M. A., and H. L. Sweetman
1951. Ecological studies of the book louse, *Liposcelis divinatorius* (Müll.). Ecology 32:230–44.

Ghouri, A. S. F., and J. E. McFarlane
1958. Observations in the development of crickets. Canadian Entomologist 90:158–65.

Gibo, D. L.
1972. An introduced population of social wasps, *Polistes apachus,* that has persisted for ten years (Hymenoptera: Vespidae). Bull. Southern California Acad. Sci. 71:53.

Gilbert, I. H., H. K. Gouck, and C. E. Schreck
1970*a*. Comparison of four cinchoninates, an oxetanone, and two standard skin repellents against *Aedes aegypti* and *Stomoxys calcitrans.* Florida Entomologist 53:89–92.

Gilbert, I. H., H. K. Gouck, and C. N. Smith
1957. New insect repellent. Soap and Chem. Spec. 33(5): 115–17, 129, 131, 133; 33(6):95, 97, 99, 109.
1966. Attractiveness of men and women to *Aedes aegypti* and relative protection time obtained with deet. Florida Entomologist 49:53–66.

Gilbert, I. H., J. E. Scanlon, and D. L. Bailey
1970*b*. Repellents against mosquitoes in Thailand. J. Econ. Entomol. 63:1207–1209.

Gilbert, I. H., N. Smith, and H. K. Gouck
1970*c*. Skin repellents against *Aedes aegypti.* J. Econ. Entomol. 63:1203–1204.

Gilbert, O.
1951. Observations on the feeding of some British false scorpions. Proc. Zool. Soc. London 121: 547–55.

Gilbertson, G. I., and W. R. Horsfall
1940. Blister beetles and their control. South Dakota Agr. Expt. Sta. Bull. 340. 23 pp.

Gillett, J. D.
1971. Mosquitos. Weidenfeld and Nicolson, London. 274 pp.

Gillies, P. A., G. V. Chambers, and D. J. Womeldorf
1971. Evaluation of aerial applications of FLIT® MLO for control of organophosphorus-resistant *Aedes nigromaculis* and *Culex tarsalis.* Mosquito News 31:528–40.

Gillies, P. A., and D. J. Womeldorf
1968. Field tests with FLIT® MLO on larvae of the pasture mosquito *Aedes nigromaculis.* Proc. California Mosquito Control Assoc. 36:92.

Girault, A. A.
1906. Standards of the number of eggs laid by insects. IV. Entomological News 17:6.

Gjullin, C. M., W. W. Yates, and H. H. Stage
1950. Studies on *Aedes vexans* (Meig.) and *Aedes sticticus* (Meig.), flood-water mosquitoes, in the lower Columbia River Valley. Annals Entomol. Soc. Amer. 43:262–75.

Gladney, W. J., and C. C. Dawkins
1972. Insecticidal tests against the brown recluse spider. J. Econ. Entomol. 65:1491–93.

Glasgow, R. D., and D. L. Collins
1946. The thermal aerosol fog generator for large scale application of DDT and other insecticides. J. Econ. Entomol. 39:227–35.

Glick, L.
1969. Be alert for brown recluse spiders. U. S. Army Materiel Command Safety Digest (March), pp. 1–4.

Glick, P. A., and L. W. Noble
1961. Airborne movement of the pink bollworm and other arthropods. USDA Agr. Res. Serv. Tech. Bull. No. 1255. 20 pp.

Gluud, A.
1968. Zur Feinstruktur der Insektencuticula. Ein Beitrag zur Frage des Eigengiftschutzes der Wanzencuticula. Zool. Jahrb. Abt. Anat. Ontog. Tiere 85: 191–227.

Goforth, R. B., Jr.
1964. Injector voids bees. Pest Control 32(11):50.

Gojmerac, W. L.
1969. Cluster flies in the home. Univ. Wisconsin Agr. Ext. Spec. Circ. 143. 4 pp.

Goldberger, J., and J. F. Schamberg
1909. Epidemic of an urticaroid dermatitis due to a small mite (*Pediculoides ventricosus*) in the straw of mattresses. Public Health Reports (USPHS) 24: 973–75.

Goldfield, M., O. Sussman, and R. P. Kandle
1966. A progress report on arbovirus studies in New Jersey. Proc. New Jersey Exterm. Assoc. 53:47–51.

Good, N. E.
1936. The flour beetles of the genus *Tribolium*. USDA Tech. Bull. No. 498. 57 pp.

Goodwin, D.
1973. Detection of hepatitis antigen in cockroaches. Thesis for M. S. degree, Dept. of Microbiology, California State Univ., Long Beach. 44 pp.

Gordon, R. M., and W. Crewe
1948. The mechanisms by which mosquitoes and tsetse-flies obtain their blood-meal. Annals Trop. Med. and Parasitol. 42:334–56.

Gorham, J. R.
1968. The geographic distribution of the brown recluse spider, *Loxosceles reclusa* (Araneae: Scytodidae), and related species in the United States. Cooperative Economic Insect Report (CEIR) 18:171–75.

Gould, D. J.
1956. The larval trombiculid mites of California (Acarina: Trombiculidae). Univ. California Publ. Entomol. 11:1–115. 26 pls.

Gould, G. E.
1948. Insect problems in corn-processing plants. J. Econ. Entomol. 41:774–78.

Gould, G. E., and H. O. Deay
1937. Notes on the bionomics of roaches inhabiting houses. Proc. Indiana Acad. Sci. 47:281–84.
1940. The biology of six species of roaches which inhabit buildings. Purdue Univ. Agr. Expt. Sta. Bull. 451. 31 pp.

Graham, K.
1963. Concepts of forest entomology. Reinhold, New York. 388 pp.
1967. Fungal–insect mutualism in trees and timber. Annual Rev. Entomol. 12:105–26.

Grant, C. D.
1960. Studies of *Leptoconops torrens* and chironomid midges in California. Proc. and Papers California Mosquito Control Assoc. 28:95–98.

Grassé, P.-P., and C. Noirot
1959. L'évolution de la symbiose chez les Isoptères. Experientia 15:365–72.

Gray, H. E.
1934. Some stored product pests in Canada, with special reference to the hairy spider beetle, *Ptinus villiger* Reit. Entomol. Soc. Ontario, Annual Rept. 65: 59–66.
1942. Spider beetles. Pests 10(7):10–13.
1947. The common dermestid beetles. Pests 15(8): 8, 10, 12, 14.

Gray, P.
1961. The encyclopedia of the biological sciences. Reinhold, New York. 1119 pp.

Greaves, J. H., B. D. Rennison, and R. Redfern
1973. Warfarin resistance in the ship rat in Liverpool. Intern. Pest Control 15(2):17.

Green, H. B.
1962. On the biology of the imported fire ant. J. Econ. Entomol. 55:1003–1004.

Greenberg, B.
1965. Flies and disease. Scientific American 213(1):92–99.
1971. Flies and disease. I. Ecology, classification, and biotic associations. Princeton University Press, Princeton, New Jersey. 856 pp.

Greenfield, H. R., and M. A. Wood
1959. Weed control aids mosquito abatement. California Vector Views 6:83–85.

Greve, J. H., and S. M. Gaafar
1964. Effects of hypothyroidism on canine demodicosis. Amer. J. Vet. Res. 25:520–22.

Griffin, D. R.
1958. Listening in the dark. Yale University Press, New Haven, Connecticut. 413 pp.
1970. Migrations and homing in bats. *In:* Biology of bats (W. A. Wimsatt, ed.), Vol. I, 233–63. Academic Press, New York and London. 406 pp.

Griffiths, J. T., and O. E. Tauber
1942. Fecundity, longevity, and parthenogenesis of the American cockroach, *Periplaneta americana* (L.). Physiological Zoology 15:196–209.

Griswold, G. H.
1944. Studies on the biology of the webbing clothes moth, *Tineola bisselliella* (Hum.). Cornell Agr. Expt. Sta. Memoir 262. 59 pp.

Grob, D.
1963. Cholinesterase and anticholinesterase agents. *In:* Handbuch der experimentellen Pharmakologie (G. B. Koelle, ed.), Vol. 15, Chap. 21. J. Springer, Berlin.

Grodhaus, G.
1963. Chironomid midges as a nuisance. II. The nature of the nuisance and remarks on its control. California Vector Views 10:27–37.
1967. Identification of chironomid midges commonly associated with waste-stabilization lagoons in California. California Vector Views 14:1–12.

Gross, H. R., Jr., and W. T. Spink
1969. Responses of striped earwigs following applications of heptachlor and mirex, and predator–prey relationships between imported fire ants and striped earwigs. J. Econ. Entomol. 62:686–89.
1971. Flight habits of the striped earwig, *Labidura riparia*. Annals Entomol. Soc. Amer. 64:746–48.

Grosser, M.
1970. The control of insects. Zoecon Annual Report, pp. 12–23.

Grothaus, R. H., W. N. Sullivan, M. S. Schecter, and E. L. Cox
1972. Resmethrin odor and performance improved. Soap, Cosm., and Chem. Spec. (September), pp. 54, 56, 58.

Grout, R. A. (ed.)
1949. The hive and the honey bee. Dadant and Sons, Hamilton, Illinois. 652 pp.

Grundemann, A. W.
1947. Studies on the biology of *Triatoma sanguisuga* (LeConte) in Kansas (Reduviidae: Hemiptera). J. Kansas Entomol. Soc. 20:77–85.

Gudgal, E. F., and F. H. Grauer
1954. Acute and chronic reactions to black fly bites (*Simulium* fly). Archives of Dermatology 70:609–15.

Guldin, R. A.
1967. Psyching the German roach. Pest Control 35(10): 26, 28, 30.

Gunn, D. L.
1935. The temperature and humidity relations of the cockroach. III. A comparison of temperature preference and rates of desiccation and respiration of *Periplaneta americana, Blatta orientalis,* and *Blattella germanica*. J. Exptl. Biol. 12:185–90.

Gunn, J. W., H. Smith, R. B. Loibl, Jr., and C. W. Pencille
1947. Report on the semiannual convention of pest control operators in California. Pests 15(9):8, 10, 11.

Gunther, F. A.
1971. Insecticides—past, present, and future: A university viewpoint. Manitoba Entomologist 5:19–27.

Gunther, F. A., and R. C. Blinn
1955. Analysis of insecticides and acaricides. Interscience Publishers, New York and London. 696 pp.

Gunther, F. A., W. E. Westlake, and P. S. Jaglan
1968. Reported solubilities of 738 pesticide chemicals in water. Residue Reviews 20:1–148.

Gupta, A. R., V. T. Das, J. R. Trout, W. R. Gusciora, D. S. Adam, and G. J. Bordach
1973. Effectiveness of spray–dust–bait combination. Pest Control 41(9):20, 22, 24, 26, 58, 60–62.

Gurney, A. B.
1950. Corrodentia. *In:* Pest control technology, pp. 129–63. National Pest Control Association, Inc., Elizabeth, New Jersey.
1953. Distribution, general bionomics, and recognition characters of two cockroaches recently established in the United States. Proc. U. S. National Museum 103:39–56.

Guthrie, D. M., and A. R. Tindall
1968. The biology of the cockroach. St. Martin's Press, New York. 408 pp.

Haber, V. R.
1919. Cockroach pests in Minnesota, with special reference to the German cockroach. Univ. Minnesota Agr. Expt. Sta. Bull. 186. 16 pp.
1926. The tracheal system of the German cockroach, *Blattella germanica* (Linn.). Bull. Brooklyn Entomol. Soc. 21:61–92.

Habermann, E.
1972. Bee and wasp venoms. Science 177:314–22.

Hach, V., and E. C. McDonald
1971. Dimethyl amide of thujic acid: A potent repellent of *Aedes aegypti*. Science 174:144–45.

Hackman, R. H.
1974. Chemistry of the insect cuticle. *In:* The physiology of Insecta (M. Rockstein, ed.), Vol. VI. Academic Press, New York and London.

Haeger, J. S.
1955. The non-bloodfeeding habits of *Aedes taeniorhynchus* (Diptera: Culicidae) on Sanibel Island, Florida. Mosquito News 15:21–26.

Hafez, M., and A. M. Afifi
1956. Biological studies of the furniture cockroach *Supella supellectilium* Serv. in Egypt (Orthoptera: Blattidae). Bull. Entomol. Soc. Egypt 40:365–96.

Hagen, K. S.
1953. Influence of adult nutrition upon the reproduction of three fruit fly species. Senate of California, third special report on the control of the oriental fruit fly in the Hawaiian Islands, pp. 72–76.

Hager, P.
1973. Oregon renews bid for DDT to save forests. Los Angeles *Times,* August 31.

Hagmann, L. E.
1963. The midge problem. Soap and Chem. Spec. 39(2): 89, 91, 104.

Haines, T. W., and E. C. Palmer
1955. Studies of distribution and habitat of cockroaches in southwestern Georgia, 1952–53. Amer. J. Trop. Med. Hyg. 4:1131–34.

Hair, J. A., and D. E. Howell
1970. Lone star ticks: Their biology and control in Ozark recreation areas. Oklahoma Agr. Expt. Sta. Bull. B-679:1–47.

Hall, D. G., Jr.
1932. Some studies on the breeding media, development, and stages of the eye gnat *Hippelates pusio* Loew (Diptera: Chloropidae). Amer. J. Hyg. 16:854–64.

Hall, E. R.
1927. An outbreak of mice in Kern County, California. Univ. California Publ. Zool. 30:189–203.

Hall, M. C.
1924. Lesions due to the bite of the wheel bug, *Arilus cristatus* (Hemiptera: Reduviidae). Archives of Internal Medicine 33:513–15.

Hall, W. J., and E. E. Wehr
1949. Diseases and parasites of poultry. USDA Farmers' Bull. No. 1652 (revised). 97 pp.

Halmai, Z., and F. A. R. Alexander
1971. Studies on the house dust allergen (preliminary report). Allergie und Immunol. 17:69–71.

Halstead, B. W., and R. Ryckman
1949. Injurious effects from contacts with millipedes. Med. Arts Sci. 3:16–18.

Halstead, D. G. H.
1968. Some observations on the biology of *Lophocateres pusillus* (Klug) (Coleoptera: Trogositidae). J. Stored Prod. Res. 4:197–202.
1969. A new species of *Tribolium* from North America, previously confused with *Tribolium madens* (Charp.) (Coleoptera: Tenebrionidae). J. Stored Prod. Res. 4:295–304.

Hamilton, H. W.
1947*a*. Economic poison laws. Soap and Sanit. Chem. 23(11):114–15.

Hamilton, W. J., Jr.
1947*b*. Rats and their control. New York Agr. Coll. Ext. B. 353 (revised). 34 pp.

Hamlin, J. C., W. D. Reed, and M. E. Phillips
1931. Biology of the Indian meal moth on dried fruits in California. USDA Tech. Bull. No. 242. 26 pp.

Hammon, W. M., and W. C. Reeves
1945. Recent advances in the epidemiology of the arthropod-borne virus encephalitides, including certain exotic types. Amer. J. Public Health 35:994–1004.
1952. California encephalitis virus: A newly described agent. I. Evidence of natural infection in man and other animals. California Medicine 77:303–309.

Hammon, W. M., W. C. Reeves, B. Brookman, and E. M. Izumi
1951. Isolation of the viruses of western equine and St. Louis encephalitis from *Culex tarsalis* mosquitoes. Science 94:328–30.

Hand, E. A.
1946. Diagnosis of infestation with "*Sarcoptes scabiei*" var. "*hominis*." Discussion of life cycle of organism. U. S. Naval Med. Bull. 46:834–44.

Hanna, G. D.
1966. Introduced mollusks of western North America. California Acad. Sci. Occas. Paper No. 48. 108 pp.

Hansens, E. J.
1963. Area control of *Fannia canicularis*. J. Econ. Entomol. 56:541.

Happ, J.
1966. Wax moth lumps wallpaper. Pest Control 34(10): 108, 110.

Harris, K. L.
1946. An annotated bibliography of methods for the examination of foods for filth. J. Assoc. Offic. Agr. Chemists 29:420–28.
1955. Additional bibliography of methods for the examination of foods for filth (1946–1954). J. Assoc. Offic. Agr. Chemists 38:1016–19.

Harris, K. L., J. F. Nicholson, L. K. Randolph, and J. L. Trawick
1952. An investigation of insect and rodent contamination of wheat and wheat flour. J. Assoc. Offic. Agr. Chemists 35:115–58.

Harris, K. L., J. F. Nicholson, and M. G. Yakowitz
1969. Principles of isolation and detection of food and drug contaminants. *In:* Grain sanitation (S. K. Majumder and J. S. Venugopal, eds.), pp. 48–80. Acad. Pest Control Sci., Mysore, India.

Harris, K. L., J. L. Trawick, J. F. Nicholson, and W. Weiss
1953. An investigation of rodent and insect contamination of corn and cornmeal. J. Assoc. Offic. Agr. Chemists 36:1037–69.

Harris, W. V.
1961. Termites—Their recognition and control. Longmans, London. 187 pp.

Harrow, K. M., H. V. Brewerton, and J. J. W. McGrath
1970. Insecticides in the glue line to protect plywood from *Anobium punctatum* (De Geer). New Zealand J. Sci. 13:27–35.

Hartnack, H.
1939. 202 common household pests in North America. Hartnack Publishing Co., Chicago. 319 pp.
1943. Unbidden house guests. Hartnack Publishing Co., Tacoma, Washington. 542 pp.

Harvey, P. A.
1934. Economic significance of the common dry-wood termite. *In:* Termites and termite control (C. A. Kofoid, ed.; reprinted and partially revised, 1965), pp. 234–65. University of California Press.

Hase, A.
1933. Ueber heftige blasige Hautreaktionen nach Culicoidesstichen. Zeitschrift für Parasitenkunde 6:119–28.

Hasenclever, H. F.
1972. Histoplasmosis in bats. Health Laboratory Science 9:125–32.

Hatch, M. H.
1946. *Hadrobregmus gibbicollis* infesting woodwork. J. Econ. Entomol. 39:274.

Hatchett, S. P.
1947. Biology of the Isopoda of Michigan. Ecological Monographs 17:47–79.

Haviland, E. E.
1947. Biology and control of the Allegheny mound ant. J. Econ. Entomol. 40:413–19.

Hawthorne, R. N.
1969. Along came a spider. PCO News 29(8–9):6–7.

Hayes, W. J., Jr.
1963. Clinical handbook on economic poisons. U. S. Public Health Service (CDC) Publ. No. 476. 144 pp. (Revised 1963.)

Hayes, W. J., and T. B. Gaines
1959. Laboratory studies of five anticoagulant rodenticides. Public Health Reports (USPHS) 74(2): 105–13.

Hazeltine, W. E.
1962. Safety of wildlife is important to planners of gnat-control programs. Agricultural Chemicals 17(12): 12–14, 76.
1963. The development of a new concept for control of the Clear Lake gnat. J. Econ. Entomol. 56: 621–26.

Heal, R. E. (ed.)
1951. Approved reference procedures for subterranean termite control. National Pest Control Association, Inc., Elizabeth, New Jersey. 250 pp.

Hebard, M.
1917. The Blattidae of North America north of the Mexican boundary. Mem. Amer. Entomol. Soc. 2:1–284.
1935. Studies in the Orthoptera of Arizona. I. New genera, species, and geographical races. Trans. Amer. Entomol. Soc. 61:111–53.
1942. The Dermaptera and orthopterous families Blattidae, Mantidae, and Phasmidae of Texas. Trans. Amer. Entomol. Soc. 68:239–319.

Hecht, O.
1929. Ueber Insektenstiche. Dermatologische Wochenschrift 88:793–810.
1930. Die Hautreaktion auf Insektenstiche als allergische Erscheinung. Zoologischer Anzeiger 87:94–109, 143–57, 231–46.

Hegner, R. W., and J. G. Engemann
1968. Invertebrate zoology. (Second edition.) Macmillan, New York. 619 pp.

Heilesen, B.
1946. Studies on *Acarus scabiei* and scabies. Acta Dermato-Venereologica 26 (Supplement 14), 7–370.

Helson, G. A. H.
1971. House dust mites. New Zealand J. Agr. 122(5):45.

Henderson, J. R., N. Karabatsos, A. T. C. Bourke, R. C. Wallis, and R. M. Taylor
1962. A survey for arthropod-borne viruses in south-central Florida. Amer. J. Trop. Med. Hyg. 11: 800–810.

Hensill, G. S., and S. J. Leland
1952. Lindane converters for insect control. Soap and Sanit. Chem. 28(11):119–21, 141, 143.

Heran, H., and L. Wanke
1952. Beobachtungen über die Entfernungsmeldung der Sammelbienen. Z. vergleich. Physiol. 34:383–93.

Herms, W. B.
1928. The Coachella Valley (California) *Hippelates* fly project. J. Econ Entomol. 21:690–93.
1937. The Clear Lake gnat. Univ. California Agr. Expt. Sta. Bull. 607. 22 pp.
1939. Medical entomology. Macmillan, New York.

Herms, W. B., S. F. Bailey, and B. McIvor
1935. The black widow spider. Univ. California Agr. Expt. Sta. Bull. 591. 30 pp.

Herms, W. B., and R. W. Burgess
1930. A description of the immature stages of *Hippelates pusio* Loew and a brief account of its life history. J. Econ. Entomol. 23:600–603.

Herms, W. B., and C. M. Wheeler
1936. *Ornithodoros hermsi* Wheeler as a vector of relapsing fever in California. Journal of Parasitology 22:276–82.

Hetrick, L. A.
1969. "Love-bugs," plague or blessing? Sunshine State Agr. Res. Rept. 14(5):6–7.
1970*a*. Biology of the "love-bug," *Plecia nearctica* (Diptera: Bibionidae). Florida Entomologist 53:23–26.
1970*b*. The "love-bug," *Plecia nearctica* Hardy (Diptera: Bibionidae). Florida Dept. Agr. and Consumer Serv., Entomol. Circ. No. 102. 2 pp.

Heuermann, R. F., and O. L. Kurtz
1955. Identification of stored product insects by the micromorphology of the exoskeleton. I. Elytral patterns. J. Assoc. Offic. Agr. Chemists 38:766–81.

Hickin, N. E.
1942. The food and water requirements of *Ptinus tectus* Boieldieu (Coleoptera: Ptinidae). Proc. Roy. Entomol. Soc. London A-17:99–108.
1949. The common furniture beetle, *Anobium punctatum* (De Geer) (Coleoptera: Anobiidae) in Britain. Entomol. Monthly Magazine 85:213–14.
1963*a*. The woodworm problem. Hutchinson, London and New York. 123 pp. (Second edition, 1972.)
1963*b*. The insect factor in wood decay. Hutchinson, London and New York. 336 pp.
1963*c*. The dry rot problem. Hutchinson, London and New York. 116 pp. (Second edition, 1972.)
1964. Household insect pests. Hutchinson, London and New York. 172 pp. (Second edition, 1974.)
1967. Caddis larvae. Hutchinson, London and New York. 475 pp.

Hickman, C. P.
1967. Biology of the invertebrates. C. V. Mosby Co., St. Louis, Missouri. 673 pp.

Hicks, C. H.
1927. *Pseudomasaris vespoides* (Cresson), a pollen-provisioning wasp. Canadian Entomologist 59: 75–79.

Hildebrand, S. F.
1919. Fishes in relation to mosquito control in ponds. Public Health Reports (USPHS) 34:1113–28.

Hill, G. F.
1925. Notes on *Mastotermes darwiniensis* Froggatt (Isoptera). Proc. Roy. Soc. Victoria (N.S.) 37: 119–24.

Hilsenhoff, W. L.
1966. The biology of *Chironomus plumosus* (Diptera: Chironomidae) in Lake Winnebago, Wisconsin. Annals Entomol. Soc. Amer. 59:465–73.

Hilton, D. F. J.
1970. A method for providing personal protection against simuliids. Mosquito News 30:474.

Hinds, W. E., and W. F. Turner
1911. Life history of the rice weevil (*Calandra oryzae* L.) in Alabama. J. Econ. Entomol. 4:230–36.

Hinman, E. H., and A. D. Hess
1949. Biological activities of the public health program of the Tennessee Valley Authority. J. Tennessee Acad. Sci. 24:195–205.

Hinton, H. E.
1940. The Ptinidae of economic importance. Bull. Entomol. Res. 31:331–81.
1941. The Lathridiidae of economic importance. Bull. Entomol. Res. 32:191–247.
1945. A monograph of the beetles associated with stored products. British Museum (Nat. Hist.). Vol. I.

Hinton, H. E., and A. S. Corbet
1963. Common insect pests of stored food products. British Museum (Nat. Hist.), Econ. Ser. No. 15. (Fourth edition.)

Hinton, M. A. C.
1931. Rats and mice as enemies of mankind. British Museum (Nat. Hist.), Econ. Ser. No. 18.

Hite, J. M., W. J. Gladney, J. L. Lancaster, Jr., and W. H. Whitcomb
1966. Biology of the brown recluse spider. Univ. Arkansas Agr. Expt. Sta. Bull. 711. 26 pp.

Hixon, H.
1943. The tropical bed bug established in Florida. Florida Entomologist 26:47.

Hockenyos, G. L.
1962. Pigeons, starlings, and English sparrows. Proc. Vertebrate Pest Control Conference, February 6–7, pp. 271–307. National Pest Control Association, Inc., Elizabeth, New Jersey.

Hocking, B.
1960. An insect-proof doorway. Bull. Entomol. Res. 51: 135–44.
1971. Bloodsucking behavior of terrestrial arthropods. Annual Rev. Entomol. 16:1–26.

Hodel, C.
1949. Drywood termite control in southern California. Pests 17(3):30, 32.

Hoffman, W. A.
1933. *Rhyzopertha dominica* as a library pest. J. Econ. Entomol. 26:293–94.

Hölldobler, B.
1970. *Steatoda fulva* (Theridiidae), a spider that feeds on harvester ants. Psyche 77:202–208.
1971. Homing in the harvester ant, *Pogonomyrmex badius*. Science 171:1149–51.

Hölldobler, B., and E. O. Wilson
1970. Recruitment trails in the harvester ant, *Pogonomyrmex badius*. Psyche 77:385–99.

Holm, S. N.
1966. The utilization and management of bumble bees for red clover and alfalfa seed production. Annual Rev. Entomol. 11:155–82.

Homer, C.
1967. Bug of the month: Millipedes. PCO News 27(12): 31.
1969. Bug of the month: Scorpions. PCO News 29(5):31.

Honigberg, B. M.
1970. Protozoa associated with termites, and their role in digestion. *In:* Biology of termites (H. Krishna and F. M. Weesner, eds.), Vol. II, pp. 1–36. Academic Press, New York and London.

Hopkins, A. D.
1909. Contributions toward a monograph of the scolytid beetles. I. The genus *Dendroctonus*. USDA Tech. Ser. No. 17. Part 1, 164 pp., 8 pls.

Hopkins, L.
1955. Biology and control of *Polistes exclamans* Viereck in Arizona citrus groves. J. Econ. Entomol. 48: 161–63.

Horn, D. H. S., E. J. Middleton, J. A. Wunderlich, and F. Hampshire
1966. Identity of the moulting hormones of insects and crustaceans. Chemical Communications 1966: 339–41.

Horner, N. V., and K. W. Stewart
1967. Life history of the brown spider, *Loxosceles reclusa* Gertsch and Mulaik. Texas J. Sci. 19:333–47.

Horsfall, W. R.
1942. Biology and control of mosquitoes in the rice area. Univ. Arkansas Agr. Expt. Sta. Bull. 427.
1943. Biology and control of common blister beetles in Arkansas. Univ. Arkansas Agr. Expt. Sta. Bull. 436. 55 pp.
1955. Mosquitoes: Their bionomics and relation to disease. Ronald Press, New York. 723 pp. (Republished 1972 by Hafner Publishing Co., N.Y.)
1962. Medical entomology. Ronald Press, New York.

Houghton, C. W.
1939. The porter beetle, *Hylotrupes bajulus* (L.). Pests 7(10):19.

Howard, L. O.
1895. A migration of cockroaches. Insect Life 7:349.
1900. The insects to which the name "kissing-bug" became applied during the summer of 1899. USDA Div. Entomol. Bull. No. 22:24–30.
1901. Preventive and remedial work against mosquitoes. Bur. Entomol. Bull. No. 88. U. S. Government Printing Office, Washington, D.C.

Howard, L. S., H. G. Dyar, and F. Knab
1912. The mosquitoes of North and Central America and the West Indies. Carnegie Institute, Washington, D.C. Vol. II, i–x, 713 pls.

Howe, R. W.
1943. Life history data for *Ptinus tectus* Boiel. (Coleoptera: Ptinidae) at 70% relative humidity at 21° C and 25° C. Proc. Roy. Entomol. Soc. London A-18:63–65.
1950. Studies on beetles of the family Ptinidae. III. A two-year study of the distribution and abundance of *Ptinus tectus* Boiel. in a warehouse. Bull. Entomol. Res. 41:371–94.
1956*a*. Studies on beetles of the family Ptinidae. XV. The biology of *Ptinus hirtellus* Sturm and some notes on *P. latro* F. Entomol. Monthly Magazine 92: 369–72.
1956*b*. The biology of two common storage species of *Oryzaephilus* (Coleoptera: Cucujidae). Annals Appl. Biol. 44:341–55.
1957. A laboratory study of the cigarette beetle, *Lasioderma serricorne* (F.) (Coleoptera: Anobiidae), with a critical review of the literature on its biology. Bull. Entomol. Res. 48:9–56.
1959. Studies on beetles of the family Ptinidae. XVII. Conclusions and additional remarks. Bull. Entomol. Res. 50:267–326.
1965. *Sitophilus granarius* (L.) (Coleoptera: Curculionidae) breeding in acorns. J. Stored Prod. Res. 1:99–100.

Howe, R. W., and H. D. Burges
1952. Studies on beetles of the family Ptinidae. VII. The biology of five ptinid species found in stored products. Bull. Entomol. Res. 43:153–86.
1953. Studies on beetles of the family Ptinidae. IX. A laboratory study of the biology of *Ptinus tectus* Boiel. Bull. Entomol. Res. 44:461–516.

Howell, D. E.
1961. Fly baits. Pest Control 29(4):9–11.

Howell, D. E., and S. D. Hensley
1953. Field cricket control in buildings. Proc. Oklahoma Acad. Sci. 34:105–107.

Howlett, F. M.
1917. Notes on head- and body-lice and upon temperature reactions of lice and mosquitoes. Parasitology 10:186–88.

Howse, P. E.
1970. Termites: A study in social behaviour. Hutchinson, London. 150 pp.

Hoy, J. B., and J. R. Anderson
1966. Snipe flies (*Symphoromyia*) attacking man and other animals in California. Proc. California Mosquito Control Assoc. 34:61–64.

Hoy, J. B., and D. E. Reed
1971. The efficacy of mosquito fish for control of *Culex tarsalis* in California rice fields. Mosquito News 31:567–72.

Hrdý, I., and J. Křeček
1972. Development of superfluous soldiers induced by juvenile hormone analogues in the termite *Reticulitermes lucifugus santonensis*. Insectes Sociaux 19:105–109.

Hubbard, C. A.
1947. Fleas of western North America. Iowa State College Press, Ames. 533 pp.

Hubbell, T. H.
1936. A monographic revision of the genus *Ceuthophilus*. Univ. Florida Publ. Biol. Sci. Ser., Vol. 2, No. 1. 551 pp, 38 pls.

Hubbs, C. L.
1919. The stickleback: A fish eminently fitted by nature as a mosquito destroyer. California Fish and Game 5:21–24.

Hudson, A., L. Bowman, and C. W. M. Orr
1960*a*. Effects of absence of saliva on blood feeding by mosquitoes. Science 131:1730–31.

Hudson, B. W., B. F. Feingold, and L. Kartman
1960*b*. Allergy to flea bites. II. Investigations of the bite sensitivity to humans. Experimental Parasitology 9:264–70.

Huff, C. G.
1931. A proposed classification of disease transmission by arthropods. Science 74:456–57.

Hughes, A. M.
1961. The mites of stored food. Ministry of Agriculture, Fisheries, and Food (Great Britain) Tech. Bull. No. 9. 287 pp.

Hughes, L. E., and C. B. Philip
1958. Experimental tick paralysis in laboratory animals and native Montana rodents. Proc. Soc. Exptl. Biol. and Med. 99:316–19.

Hughes, M., and K. Maunsell
1970. House dust and asthma. British Medical Journal 1970(3):466.

Hughes, T. E.
1959. Mites, or the Acari. Athlone Press, London.

Hull, G., Jr., and R. H. Davidson
1958. The biology of the brown-banded cockroach and its relative susceptibility to five organic insecticides. J. Econ. Entomol. 51:608–610.

Hummel, H. E., L. K. Gaston, H. H. Shorey, R. S. Kaae, K. J. Byrne, and R. M. Silverstein
1973. Clarification of the chemical status of the pink bollworm sex pheromone. Science 181:873–75.

Humphrey, C. J.
1923. Decay of lumber and building timbers due to *Poria incrassata* (B. & C.) Burt. Mycologia 15: 257–77.

Humphries, D. A.
1966. The function of combs on fleas. Entomol. Monthly Magazine 102:232–36.

Hunt, R. W.
1949. The common drywood termite as a pest. J. Econ. Entomol. 42:959–62.
1959. Wood preservatives as deterrents to drywood termites in the Southwest. J. Econ. Entomol. 52: 1211–12.

Hunter, G. W. III, W. W. Frye, and J. C. Swartzwelder
1960. A manual of tropical medicine. W. B. Saunders Co., Philadelphia. 892 pp.

Hurd, P. D., Jr.
1948. Systematics of the California species of the genus *Pepsis* Fabricius (Hymenoptera: Pompilidae). Univ. California Publ. Entomol. 8:123–50.
1951. The California velvet ants of the genus *Dasymutilla* Ashmead (Hymenoptera: Mutillidae). Bull. California Insect Surv. 1:89–114.
1952. The Scoliidae of California (Hymenoptera: Aculeata). Bull. California Insect Surv. 1:141–47.
1955. The carpenter bees of California (Hymenoptera: Apoidea). Bull. California Insect Surv. 4:35–72.
1958. Observations on the nesting habits of some New World carpenter bees, with remarks on their importance in the problem of species formation (Hymenoptera: Apoidea). Annals Entomol. Soc. Amer. 51:365–75.

Hurd, P. D., Jr., and C. D. Michener
1955. The megachiline bees of California. Bull. California Insect Surv. 3:1–247.

Hurd, P. D., Jr., and J. S. Moure
1963. A classification of the large carpenter bees (Xylocopini) (Hymenoptera: Apoidea). Univ. California Publ. Entomol. 29:1–365.

Hutchins, R. T., and J. M. Langston
1953. An unusual occurrence of the field cricket. J. Econ. Entomol. 46:169.

IAC (Insect Allergy Committee, American Academy of Allergy)
1965. Insect-sting allergy. Questionnaire study of 2,606 cases. J. Amer. Med. Assoc. 193:115–20.

Ikeshoji, T., and M. S. Mulla
1970. Overcrowding factors of mosquito larvae. J. Econ. Entomol. 63:90–96.

Illingworth, J. F.
1915. Use of cockroaches in medicine. Proc. Hawaiian Entomol. Soc. 3:112–13.
1918. The leather beetle (*Dermestes vulpinus* Fab.), a troublesome pest of dried fish in Hawaii. Proc. Hawaiian Entomol. Soc. 3:375–78.
1927. A report on insects and other animal organisms collected in the pineapple-growing section at Mauna Loa, Molokai, June, 1926. Proc. Hawaiian Entomol. Soc. 6:390–97.
1929. Pests of pineapple in Hawaii. Proc. Hawaiian Entomol. Soc. 7:254–66.
1942. An outbreak of cockroaches, *Nauphaeta cinerea* (Olivier), in Hawaii. Proc. Entomol. Soc. Hawaii 11:169–70.

Imms, A. D.
1957. A general textbook of entomology. (Ninth edition, revised by G. W. Richards and R. G. Davies.) Methuen, London. 866 pp.

Irons, J. V., S. W. Bohls, D. C. Thurman, and T. McGregor
1944. Probable role of the cat flea, *Ctenocephalides felis*, in transmission of murine typhus. Amer. J. Trop. Med. 24:359–62.

Isherwood, H. R.
1957. Pressure-treated wood tops shields in home mechanical termite control. Pest Control 25(11):32, 34.

Ishii, S., and Y. Kuwahara
1968. Aggregation of German cockroach nymphs. Experientia 24:88–89.

Ives, H. B.
1948. House mice and their control. Pests 16(10):11–12, 14, 16, 18, 54.

Izutsu, M., S. Ueda, and S. Ishii
1970. Aggregation effects on the growth of the German cockroach, *Blattella germanica* (L.). Appl. Entomol. Zool. 5:159–71.

Jackson, W. B.
1969. Anticoagulant resistance in Europe. Pest Control 37(3):51–55.

Jackson, W. B., P. J. Spear, and C. G. Wright
1971. Resistance of Norway rats to anticoagulant rodenticides confirmed in the United States. Pest Control 39(9):13–14.

Jacobson, M.
1972. Insect sex pheromones. Academic Press, New York. 392 pp.

Jacobson, M., N. Green, D. Warthen, C. Harding, and H. H. Toba
1970. Sex pheromones of the Lepidoptera. Recent progress and structure–activity relationships. *In:* Chemicals controlling insect behavior (M. Beroza, ed.), pp. 3–20. Academic Press, New York.

Jakob, W. L., and H. F. Schoof
1971. Studies with juvenile hormone-type compounds against mosquito larvae. Mosquito News 31:540–43.
1972. Mosquito larvicide studies with MON 585, a juvenile hormone mimic. Mosquito News 32:6–10.

James, M. T.
1947. The flies that cause myiasis in man. USDA Misc. Publ. No. 631. 175 pp.
1955. The blowflies of California (Diptera: Calliphoridae). Bull. California Insect Surv. 4:1–30.
1960. The soldier flies or Stratiomyidae of California. Bull. California Insect Surv. 6:79–122.

James, M. T., and R. F. Harwood
1969. Herms's medical entomology. Macmillan, New York. 484 pp.

Jamieson, D.
1965. A history of roof rat problems in residential areas of Santa Clara County, California. California Vector Views 12:25–28.

Jamieson, H. C.
1938. The house fly as a cause of nasal allergy. Journal of Allergy 9:273–74.

Jamnback, H.
1969. Bloodsucking flies and other outdoor nuisance arthropods of New York state. New York State Mus. Sci. Serv. Memoir 19. Univ. State of New York, State Educ. Dept., Albany. 90 pp.
1973. Recent developments in control of blackflies. Annual Rev. Entomol. 18:281–304.

Jamnback, H., and T. Wathews
1963. Studies of populations of adult and immature *Culicoides sanguisuga* (Diptera: Ceratopogonidae). Annals Entomol. Soc. Amer. 56:728–32.

Jeanne, R. L.
1972. Social biology of the Neotropical wasp *Mischocyttarus drewseni*. Bull. Harvard Mus. Comp. Zool. 144:63–150.

Jeannel, R.
1960. Introduction to entomology. Hutchinson, London. 344 pp.

Jenkins, D. W.
1948. Trombiculid mites affecting man. I. Bionomics with reference to epidemiology in the United States. Amer. J. Hyg. 48:22–35.
1949. Trombiculid mites affecting man. IV. Revision of *Eutrombicula* in the American hemisphere. Annals Entomol. Soc. Amer. 42:289–318.
1950. Bionomics of *Culex tarsalis* in relation to western equine encephalomyelitis. Amer. J. Trop. Med. 30:909–16.

Jiménez-Porras, J. M.
1970. Biochemistry of snake venoms. Clinical Toxicology 3:389–431.

Johnson, C. G.
1942. The ecology of the bed-bug, *Cimex lectularius* L., in Britain. Journal of Hygiene 41:345–461.

Johnson, C. G., and K. Mellanby
1942. The parasitology of human scabies. Parasitology 34:285–90.

Johnson, D. R.
1969. Malaria eradication: What it has achieved. Mosquito News 29:523–32.

Johnson, G. V., and F. F. Smith
1970. Insects and related pests of house plants: How to control them. USDA Home and Garden Bull. No. 67. 16 pp.

Johnson, J. P.
1945. The incinerator beetle, *Dermestes cadaverinus* Fabr. Connecticut Agr. Expt. Sta. Bull. 488:411–15.

Johnson, K. P., M. L. Lepow, and R. T. Johnson
1968. California encephalitis. I. Clinical and epidemiological studies. Neurology 18:250–54.

Johnson, M. S., and A. J. Hill
1948. Partial resistance of a strain of bedbugs to DDT residual. Medical News Letter 12(1): 26–28.

Johnston, H. R.
1960. Soil treatment for subterranean termites. USDA Southern Forest Expt. Sta. Occas. Paper No. 152: 1–6. (Revised.)

Johnston, H. R., V. K. Smith, and R. A. Beal
1971. Chemicals for subterranean termite control: Results of long-term tests. J. Econ. Entomol. 64: 745–48.

Jones, C. M.
1969. Biology of the face fly: Migration of larvae. J. Econ. Entomol. 62:255–56.

Joose, E. N. G., and J. B. Groen
1970. Relationship between saturation deficit and the survival and locomotory activity of surface-dwelling Collembola. Entomol. Expt. Appl. 13: 229–35.

Joseph, B.
1973. Preview of improvements incorporated into Cessna aircraft for '74. National Pest Control Operators News 33(11):23.

Kalmbach, E. R.
1954. Pigeon, sparrow, and starling control. Pest Control 22(5):9–10, 12, 31; 22(6):32, 36, 38.

Kandle, R. P.
1964. Continual arbovirus research activities in New Jersey. Proc. New Jersey Mosquito Exterm. Assoc. 51:15–18.

Kapoor, I. P., R. L. Metcalf, R. F. Nystrom, and G. K. Sangha
1970. Comparative metabolism of methoxychlor, methiochlor, and DDT in mouse, insects, and in a model ecosystem. J. Agr. and Food Chem. 18: 1145–52.

Karlson, P., and A. Butenandt
1959. Pheromones (ectohormones) in insects. Annual Rev. Entomol. 4:39–58.

Kaston, B. J.
1937. Notes on a new variety of black widow spider from southern Florida. Florida Entomologist 21: 60–61.
1968. Remarks on black widow spiders, with an account of some anomalies. Entomological News 79: 113–23.
1970. Comparative biology of American black widow spiders. Trans. San Diego Soc. Nat. Hist. 16(3): 33–82.
1973. How to know the spiders. (Second edition.) Wm. C. Brown Co., Dubuque, Iowa. 289 pp.

Keele, C. A.
1963. Venoms and the causes of pain. New Scientist 17:396–99.

Keen, F. P.
1938. Insect enemies of western forests. USDA Misc. Publ. No. 273. 209 pp.

Keh, B.
1956. Cone-nosed bugs of California. California Vector Views 3:47–49.

Keh, B., and A. M. Barnes
1961. Fleas as household pests in California. California Vector Views 8:55–58.

Keh, B., N. T. Brownfield, and M. E. Person
1968. Experimental use of bait with mirex lethal to both adult and immature *Vespula pensylvanica* (Hymenoptera: Vespidae). California Vector Views 15:115–18.

Keh, B., and J. H. Poorbaugh
1971. Understanding and treating infestations of lice on humans. California Vector Views 18:23–31.

Keiding, J.
1963. Investigations of housefly behaviour in relation to new methods of chemical fly control. 1. Investigations on farms. *In:* Annual Report for 1961–62, Government Pest Infestation Laboratory, Springforbi, Denmark, pp. 55–60.
1964. Investigations of housefly behaviour in relation to new methods of chemical fly control. 2. Distribution of flies in stables by day and by night. *In:* Annual Report for 1963, Government Pest Infestation Laboratory, Springforbi, Denmark, pp. 48–50.

Keilin, D., and G. H. F. Nuttall
1919. Hermaphroditism and other abnormalities in *Pediculus humanus.* Parasitology 11:279–328.

Kelsey, J. M., D. Spiller, and R. W. Doane
1945. Biology of *Anobium punctatum.* New Zealand J. Sci. Tech. Progress Report 27(B) 1:59–68.

Kemper, H.
1936. Die Bettwanze und ihre Bekämpfung. Kleintierkunde und Pelztierkunde 9(3):1–107.
1958. Experimentelle Untersuchungen über die Wirkung von Raupenhaaren auf die menschliche Haut. Proc. 10th Intern. Congr. Entomol. (Montreal) 3:719–23.

Kenaga, E. E., and W. E. Allison
1969. Commercial and experimental organic insecticides. Entomol. Soc. Amer. Bull. 15:85–148.

Kent, F. B.
1972. Latins fighting rabies borne by vampire bats. Los Angeles *Times,* May 20. Part I, p. 19.

Kent, N. L., and R. A. McCance
1941. The absorption and excretion of "minor" elements by man. 1. Silver, gold, lithium, boron, and vanadium. Biochemical Journal 35:837–44.

Kern, R. A.
1921. Dust sensitization in bronchial asthma. Med. Clin. N. Amer. 5:751–58.
1962. Environment in relation to allergic disease. Archives of Environmental Health 4:28–49.

Kerr, T. W., Jr., and D. L. McLean
1956. Biology and control of certain Lathridiidae. J. Econ. Entomol. 49:269–70.

Kilpatrick, J. W., and K. D. Quarterman
1952. Field studies on the resting habits of flies in relation to chemical control. II. In rural areas. Amer. J. Trop Med. Hyg. 1:1026–31.

Kimball, J. H.
1965. Vector control in Orange County [California]. Pest Control 33(7):28–32; 33(8):28–40; 33(9):38–46.

Kimball, J. H., and N. D. Perruzzi
1970. FLIT® MLO preferred for mosquito control in residential areas. Orange County [California] Mosquito Abatement District, Santa Ana. 5 pp. (Mimeographed.)

Kimball, J. H., and A. H. Thompson
1967. Thiokol Swamp Spryte equipped with automatic compressed-air spraying units for applying granule and liquid insecticides. Mosquito News 27: 420–22.

King, E. G., Jr., and W. T. Spink
1969. Foraging galleries of the Formosan subterranean termite. *Coptotermes formosanus,* in Louisiana. Annals Entomol. Soc. Amer. 62:536–42.

Kingsbury, J. M.
1964. Poisonous plants of the United States and Canada. (Third edition.) Prentice-Hall, Englewood Cliffs, New Jersey.

Kiritani, K.
1965. Biological studies on the *Sitophilus* complex (Coleoptera: Curculionidae) in Japan. J. Stored Prod. Res. 1:169–76.

Kissling, R. E., R. W. Chamberlain, W. D. Sudia, and D. D. Stamm
1957. Western equine encephalitis in wild birds. Amer. J. Hyg. 66:48–55.

Klauber, L. M.
1956. Rattlesnakes: Their habits, life histories, and influence on mankind. Vol. I, 707 pp. University of California Press, Berkeley and Los Angeles. (Revised edition, 1973, titled: The way of the serpent.)

Kliewer, J. W., T. Miura, and H. C. Chapman
1969. Seasonal occurrence and physiology of *Culex tarsalis* in foothills of Fresno County, California. Annals Entomol. Soc. Amer. 62:13–18.

Klostermeyer, E. C.
1942. The life history and habits of the ringlegged earwig, *Euborellia annulipes* (Lucas) (Order Dermaptera). J. Kansas Entomol. Soc. 15:13–18.

Knabke, J. J., and A. A. Grigarick
1971. Biology of the African earwig, *Euborellia cincticollis* (Gerstaecker), in California, and comparative notes on *Euborellia annulipes* (Lucas). Hilgardia 41:157–94.

Knipling, E. F.
1960*a*. Use of insects for their own destruction. J. Econ. Entomol. 53:415–20.
1960*b*. The eradication of the screw-worm fly. Scientific American 203(4):54–61.
1962. Potentialities and progress in the development of chemosterilants for insect control. J. Econ. Entomol. 55:782–86.
1968. The role of chemicals in the general insect control picture. Entomol. Soc. Amer. Bull. 14:102–107.

Knopf, J. A. E., and G. B. Pitman
1972. Aggregation pheromone for manipulation of the Douglas-fir beetle. J. Econ. Entomol. 65:723–26.

Knote, C.
1973. Rodent fear and the practical difference between rats and mice. Pest Control 41(10):23, 67–72.

Knowlton, G. F.
1947. Boxelder bug (*Leptocoris trivittatus*) "bites" man. Bull. Brooklyn Entomol. Soc. 42:33.

Knudsen, P.
1967. Distribution and abundance of *Hylotrupes bajulus* (L.) (Coleoptera: Cerambycidae) and *Anobium punctatum* (De Geer) (Coleoptera: Anobiidae) along the Sognefjord in west Norway. Rev. Appl. Entomol. A-56:188–89.

Knülle, W.
1963*a*. Die Beziehungen zwischen dem Wassergehalt von Vorratsgutern, der Luftfeuchte, und dem Milbenbefall. Z. angew. Entomol. 51:300–312.
1963*b*. Untersuchungen über den Einfluss von Raumfeuchte, Temperatur, und Lagerhöhe auf die Vermilbung von Trockenpflaumen. Z. angew. Entomol. 52:275–85.

Knülle, W., and R. R. Spadafora
1969. Water vapor sorption and humidity relationships in *Liposcelis* (Insecta: Psocoptera). J. Stored Prod. Res. 5:49–55.

Koehler, C. S., P. D. Smith, C. S. Davis, and L. R. Brown
1970. Elm leaf beetle. Univ. California Agr. Ext. Serv. OSA No. 173.

Koekkoek, H. M. M., and J. E. M. H. van Bronswijk
1972. Temperature requirements of a house dust mite, *Dermatophagoides pteronyssinus,* compared with the climate in different habitats of houses. Entomol. Exp. Appl. 15:438–42.

Kolbezen, J. J., R. L. Metcalf, and T. R. Fukuto
1954. Insecticidal activity of carbamate cholinesterase inhibitors. J. Agr. and Food Chem. 2:864–70.

König, W.
1936. Biologische Studien über *Ptinus tectus* Boiel. Z. Wiss. Zool. 148:556–99.

Korting, A.
1962. Zur Entwicklung und Schadtätigkeit des Hausbockkäfers, *Hylotrupes bajulus* (L.), in Dachstuhlen verschiedenen Alters. Anzeiger für Schädlingskunde 39(10):150–53.

Kovoor, J.
1964. Modifications chimiques provoquées par un termite (*Microcerotermes edentatus* Was.) dans du bois de peuplier sain ou partiellement dégradé par des champignons. Bull. Biol. France Belg. 98: 491–510.

Kreyenberg, J.
1928. Experimental investigation on the biology of *Dermestes lardarius* and *D. vulpinus.* Z. angew. Entomol. 14:140–78.

Krishna, K.
1966. Formosan subterranean termite and key to eight termite genera. PCO News 26(12):18–20.
1969. Introduction. *In:* Biology of termites (K. Krishna and F. M. Weesner, eds.), Vol. I, pp. 1–17. Academic Press, New York and London.

Krishnakumaran, A., and H. A. Schneiderman
1965. Prothoracotrophic activity of compounds that mimic juvenile hormone. J. Insect Physiol. 11: 1517–32.

Kunz, S. E., M. A. Price, and O. H. Graham
1969. Biology and economic importance of the chigger *Neoschongastia americana* on turkeys. J. Econ. Entomol. 62:872–75.

Kurtz, O. L., N. A. Carson, and H. C. Van Dame
1952. Identification of cereal insects in stored grain by their mandible characteristics. J. Assoc. Offic. Agr. Chemists 35:817–26.

Kurtz, O. L., and K. L. Harris
1962. Microanalytical entomology for food sanitation control. Association of Official Agricultural Chemists, Washington, D.C. 576 pp., 800 illus.

Labarthe, J.
1964. Textiles: Origins and usage. Macmillan, N.Y.

LaFollette, J. R.
1952. Wasps in citrus groves in Arizona. Sunkist Pest Control Circ. 181:2.

LaHue, D. W.
1970. Evaluation of malathion, diazinon, a silica aerogel, and a diatomaceous earth as protectants on wheat against lesser grain borer attack in small bins. USDA Agr. Res. Serv., Marketing Res. Rept. No. 860. 11 pp.

LaMer, V. K., S. Hochberg, K. Hodges, I. Wilson, J. A. Fales, and R. Latta
1947. The influence of the particle size of homogeneous insecticidal aerosols on the mortality of mosquitoes in confined atmospheres. J. Colloid. Sci. 2: 539–49.

Lane, J. E.
1928. Bonomo's letter to Redi: Important document in history of scabies. Arch. Dermatol. Syph. 18:1–25.

Lange, W. H., Jr.
1944. Land slugs in California. Bull. Southern California Acad. Sci. 43:33–40.

Langlois, C., S. Shulman, and C. E. Arbesman
1963. Immunologic studies of caddis fly. Journal of Allergy 34:385–94.

Langmuir, A. D.
1961. Epidemiology of airborne infection. Bacteriological Reviews 25:173–81.

LaPlante, A. A., Jr.
1949. Contact toxicity of several formulations of DDT to first-instar oriental fruit moth larvae. J. Econ. Entomol. 42:786–95.

Larson, C.
1955. Rocky Mountain spotted fever. Amer. J. Nursing 55:716–19.

Larson, D. G., W. F. Mitchell, and G. W. Wharton
1969. Preliminary studies on *Dermatophagoides farinae* Hughes, 1961 (Acari) and house dust allergy. J. Med. Entomol. 6:295–99.

Latta, R., L. D. Anderson, E. E. Rogers, V. K. LaMer, S. Hochberg, H. Lauterbach, and I. Johnson
1947. The effect of particle size and velocity of movement of DDT aerosols in a wind tunnel on the mortality of mosquitoes. J. Washington Acad. Sci. 37:397–407.

Laudani, H.
1961. Biology and habits of dermestids. Pest Control 29(10):58–61.

Lauret, T. H. (ed.)
1962. A field guide to common mosquito-like gnats of California. California Mosquito Control Assoc., Visalia. 19 pp.

Laurie, E. M. O.
1946. The reproduction of the house-mouse (*Mus musculus*) living in different environments. Proc. Soc. Biol. Sci. (Series B) 133(872):248–81.

Lavigne, R. J.
1966. Individual mound treatments for control of the western harvester ant, *Pogonomyrmex occidentalis*, in Wyoming. J. Econ. Entomol. 59:525–32.
1969. Bionomics and nest structure of *Pogonomyrmex occidentalis* (Hymenoptera: Formicidae). Annals Entomol. Soc. Amer. 62:1166–75.

Law, J. H., C. Yuan, and C. M. Williams
1966. Synthesis of a material with high juvenile hormone activity. Proc. Nat. Acad. Sci. (U. S.) 55: 576–78.

Lawson, P. B.
1933. Notes on the life history of the hourglass spider. Annals Entomol. Soc. Amer. 26:568–74.

Lebrun, D.
1972. Effets de l'implantation de glandes mandibulaires sur la différenciation imaginale de *Kalotermes flavicollis* (Fabr.). Compt. Rend. Acad. Sci. Paris (Sér. D) 274:2077–79.

Ledoux, A.
1945. Etude expérimentale du grégairisme et de l'interattraction sociale chez les blattides. Ann. Sci. Natur. Zool. 7:75–104.

Lees, A. D.
1948. The sensory physiology of the sheep tick *Ixodes ricinus* (L.). J. Exptl. Biol. 25:145–207.

Leeson, H. S.
1941. The effect of temperature upon the hatching of the eggs of *Pediculus humanus corporis* De Geer (Anoplura). Parasitology 33:243–49.

Legner, E. F.
1968. Parasite activity related to ovipositional responses in *Hippelates collusor*. J. Econ. Entomol. 61: 1160–63.

Legner, E. F., and E. C. Bay
1965. Predatory and parasitic agents attacking the *Hippelates pusio* complex in Puerto Rico. Univ. Puerto Rico J. Agr. 49:377–85.
1970. Dynamics of *Hippelates* eye gnat breeding in the Southwest. California Agriculture 24(5):4–6.

Legner, E. F., and H. W. Brydon
1966. Suppression of dung-inhabiting fly populations by pupal parasites. Annals Entomol. Soc. Amer. 59:638–51.

Legner, E. F., L. Moore, and R. A. Medved
1971. Observations on predation of *Hippelates collusor* and distribution in southern California of associated fauna. J. Econ. Entomol. 64:461–68.

Legner, E. F., and G. S. Olton
1968. The biological method and integrated control of house and stable flies in California. California Agriculture 22(6):2–4.
1969. Migrations of *Hippelates collusor* larvae from moisture and trophic stimuli and their encounter by *Trybliographa* parasites. Annals Entomol. Soc. Amer. 62:136–41.
1970. World-wide survey and comparison of adult predator and scavenger insect populations associated with domestic animal manure where livestock is artificially congregated. Hilgardia 40:225–66.
1971. Distribution and relative abundance of dipterous pupae and their parasitoids in accumulations of domestic animal manure in the southwestern United States. Hilgardia 40:505–35.

Legner, E. F., G. S. Olton, and F. M. Eskafi
1966. Influence of physical factors on the developmental stages of *Hippelates collusor* in relation to the activities of its natural parasites. Annals Entomol. Soc. Amer. 59:851–61.

Legner, E. F., and J. H. Poorbaugh, Jr.
1972. Biological control of vector and noxious synanthropic flies: A review. California Vector Views 19:81–100.

Legner, E. F., R. D. Sjogren, and J. T. Wiles
1970. Effects of low biuret urea on natural populations of *Hippelates collusor* (Townsend) and *Leptoconops kerteszi* Kieffer. Mosquito News 30:633–40.

Lehmann, C. F., J. L. Pipkin, and A. C. Ressmann
1955. Blister beetle dermatosis. Arch. Dermatol. Syph. 71:36–38.

Lemmon, A. B.
1951. Respirators for pesticides. California State Bur. Chem. Announc. PC-3. (Processed.)

Leopold, A. C., and R. Ardrey
1972. Toxic substances in plants and the food habits of early man. Science 176:512–13.

Levi, H. W.
1957. The spider genera *Crustulina* and *Steatoda* in North America, Central America, and the West Indies (Araneae: Theridiidae). Bull. Harvard Mus. Comp. Zool. 117:367–424.
1958. Number of species of black-widow spiders (Theridiidae: *Latrodectus*). Science 127:1055.
1959. The spider genus *Latrodectus*. Trans. Amer. Micr. Soc. 78:7–43.

Levi, H. W., L. R. Levi, and H. S. Zim
1968. Spiders and their kin. Golden Press, New York. 160 pp.

Levi, H. W., and A. Spielman
1964. The biology and control of the South American brown spider, *Loxosceles laeta* (Nicolet), in a North American focus. Amer. J. Trop. Med. Hyg. 13:132–36.

Levine, M. I.
1971. Insect stings. J. Amer. Med. Assoc. 217:964.

Levinson, H. Z.
1966. Studies on the juvenile hormone (neotenin) activity of various hormonomimetic materials. Rivista di Parassitologia 27:47–63.

Lewallen, L. L.
1954. Biological and toxicological studies of the little house fly. J. Econ. Entomol. 47:1137–41.

Liener, I. E. (ed.)
1969. Toxic constituents of plant foodstuffs. Academic Press, New York. 500 pp.

Light, S. F.
1933. Termites of western Mexico. Univ. California Publ. Entomol. 6:79–164.
1934*a*. The termite fauna of Mexico and its economic significance. *In:* Termites and termite control (C. A. Kofoid, ed.; reprinted and partially revised, 1965), pp. 334–39. University of California Press, Berkeley.
1934*b*. Dry-wood termites, their classification and distribution. *In:* Termites and termite control (C. A. Kofoid, ed.; reprinted and partially revised, 1965), pp. 206–209. University of California Press. Berkeley.
1934*c*. The damp-wood termites of western United States, genus *Zootermopsis* (formerly *Termopsis*). *In:* Termites and termite control (C. A. Kofoid, ed.; reprinted and partially revised, 1965), pp. 273–310. University of California Press, Berkeley.

Light, S. F., M. Randall, and F. G. White
1930. Termites and termite damage, with preliminary recommendations for prevention and control. Univ. California Agr. Expt. Sta. Circ. 318:1–64.

Light, S. F., and F. M. Weesner
1947. Methods for culturing termites. Science 106: 131–32.

Lindauer, M.
1960. Communication among social bees. Harvard University Press, Cambridge, Massachusetts. 143 pp.
1967. Recent advances in bee communication and orientation. Annual Rev. Entomol. 12:439–70.

Lindgren, D. L., and L. E. Vincent
1953. Nitidulid beetles infesting California dates. Hilgardia 22:97–118.

Lindgren, D. L., L. E. Vincent, and R. G. Strong
1958. Studies on hydrogen phosphide as a fumigant. J. Econ. Entomol. 51:900–903.

Lindsay, E.
1940. The biology of the silverfish, *Ctenolepisma longicaudata* Esch., with particular reference to its feeding habits. Proc. Roy. Soc. Victoria (N.S.) 52: 35–78. (Part 1.)

Linley, J. R., and J. B. Davies
1971. Sandflies and tourism in Florida and the Bahamas and Caribbean area. J. Econ. Entomol. 64:264–78.

Linsley, E. G.
1942. Insect food caches as reservoirs and original sources of some stored products pests. J. Econ. Entomol. 35:434–39.
1943*a*. The dried fruit moth breeding in nests of the mountain carpenter bee in California. J. Econ. Entomol. 36:122–23.
1943*b*. The recognition and control of deathwatch, powderpost, and false powderpost beetles. Pests 11(3):11–14.
1943*c*. Delayed emergence of *Buprestis aurulenta* from structural timbers. J. Econ. Entomol. 36:348–49.

1944. Natural sources, habitats, and reservoirs of insects associated with stored food products. Hilgardia 16:187–224.
1958. The ecology of solitary bees. Hilgardia 27:543–99.
1962. The Cerambycidae of North America. II. Taxonomy and classification of the Parandrinae, Prioninae, Spondylinae, and Aseminae. Univ. California Publ. Entomol. 19. 102 pp.
1964. The Cerambycidae of North America. V. Taxonomy and classification of the subfamily Cerambycinae, tribes Callichromini through Ancylocerini. Univ. California Publ. Entomol. 22. 197 pp.

Linsley, E. G., and A. E. Michelbacher
1943. Insects affecting stored food products. California Agr. Expt. Sta. Bull. 676. 44 pp.

Lisella, F. S.
1973. What roach insecticides are safe for use by allergic persons? J. Amer. Med. Assoc. 224:410.

Locke, M.
1965. Permeability of insect cuticle to water and lipids. Science 147:295–98.
1974. The structure and formation of the integument in insects. *In:* The physiology of Insecta (M. Rockstein, ed.), Vol. VI. Academic Press, New York and London.

Lofgren, C. S., C. E. Stringer, and F. J. Bartlett
1962. Imported fire ant toxic bait studies: GC-1283, a promising toxicant. J. Econ. Entomol. 55:405–407.

Longshore, W. A., Jr., I. M. Stevens, A. C. Hollister, Jr., A. Gittelsohn, and E. H. Lennette
1956. Epidemiologic observations on acute infectious encephalitis in California, with special reference to the 1952 outbreak. Amer. J. Hyg. 63:69–86.

Loomis, E. C. (ed.)
1967. A field guide to common mosquitoes of California. California Mosquito Control Assoc., Visalia. (Third edition.) 26 pp.

Loomis, E. C., J. R. Anderson, and A. S. Deal
1967. Identification of common flies associated with livestock and poultry. Univ. California Agr. Ext. Serv. AXT-236. 11 pp.

Loomis, E. C., R. M. Bohart, and J. N. Belkin
1956. Additions to the taxonomy and distribution of California mosquitoes. California Vector Views 3:37–45.

Loschiavo, S. R., and L. B. Smith
1970. Distribution of the merchant grain beetle, *Oryzaephilus mercator* (Silvanidae: Coleoptera), in Canada. Canadian Entomologist 102:1041–47.

Lovitt, A. E., and E. L. Soderstrom
1968. Predation on Indian meal moth eggs by *Liposcelis bostrychophilus.* J. Econ. Entomol. 61:1444–45.

Luke, J. E., and R. Snetsinger
1972. Laboratory evaluation of six insecticides for control of the firebrat. J. Econ. Entomol. 65:917–18.

Lund, A. E., and N. T. Engelhardt
1962. Subterranean termites and *Absidia coerulea* Bainer (Mucorales). J. Insect Pathol. 4:131–32.

Lüscher, M.
1956*a*. Zur Frage der Uebertragung sozialer Wirkstoffe bei Termiten. Naturwissenschaften 42:186–87.
1956*b*. Die Entstehung von ersatz Geschlechtstieren bei der Termite *Kalotermes flavicollis* Fabr. Insectes Sociaux 3:119–28.
1961. Social control of polymorphism in termites. *In:* Insect polymorphism (J. S. Kennedy, ed.), pp. 57–67. Royal Entomological Society, London.

Lüscher, M., and B. Müller
1960. Ein spurbildendes Sekret bei Termiten. Naturwissenschaften 47:503.

Macchiavello, A.
1937. La *Loxosceles laeta,* causa del aracnoidismo cutáneo, o mancha gangrenosa, de Chile. Rev. Chil. Hist. Nat. 41:11–19.

MacConnell, J. G., M. S. Blum, and H. M. Fales
1970. Alkaloid from fire ant venom: Identification and synthesis. Science 168:840–41.
1971. The chemistry of fire ant venom. Tetrahedron 27: 1125–39; Entomology Abstracts 2(11):33.

MacGillivray, N. A.
1964. The importance of water use studies in mosquito control. California Vector Views 11:61–64.

Macht, D. I., and D. B. Kehoe
1943. The toxicity of snake venoms after administration by stomach. Federal Proceedings 2:31.

Mackie, D. B.
1932. Annual report of the Division of Entomology and Pest Control. California Dept. Agr. Monthly Bull. 21:474–88.

Mackie, D. B., and W. B. Carter
1937. Methyl bromide as a fumigant: A preliminary report. California Dept. Agr. Bull. 26:153–62.

Mackie, R. A.
1964. Control of the roof rat, *Rattus rattus,* in the sewers of San Diego. California Vector Views, 11: 7–10.
1969. Biology and control of cockroaches in the San Diego city sewer system. California Vector Views 16:57–66.

Maddock, D. R., C. M. Elmore, Jr., and H. F. Schoof
1963. Preliminary tests with DDVP vapor for control of *Culex pipiens quinquefasciatus* in catch basins. Mosquito News 23:69–74.

Maddock, D. R., and H. F. Schoof.
1970. New red squill derivative: Laboratory and field studies on stabilized scilliroside against Norway rats. Pest Control 38(8):32–35, 45–46.

Mahaney, J. H.
1972. Termites on USNS *Sunnyvale.* Pest Control 40(7): 17–18, 36, 38.

Maier, P. P., W. C. Baker, M. D. Bogue, J. W. Kirkpatrick, and K. D. Quarterman
1952. Field studies on the resting habits of flies in relation to chemical control. I. In urban areas. Amer. J. Trop. Med. Hyg. 1:1020–25.

Mail, G. A.
1940. Infestation of a high school by *Oeciacus vicarius* Horv. J. Econ. Entomol. 33:949.

Majumder, S. K., and A. Bano
1964. Toxicity of calcium phosphate to some pests of stored grain. Nature 202:1359–60.
1966. Reversion of the toxicity of tricalcium phosphate to insects by trehalose. Nature 210:1052–53.

Majumder, S. K., and K. Krishnamurthy
1960. Bleaching earths as insecticides. J. Sci. Ind. Res. 19*c*:29–30.

Majumder, S. K., K. S. Narasimhan, and V. Subrahmanyan
1959. Insecticidal effects of activated charcoal and clays. Nature 184:1165–66.

Mallars, J. L., and J. R. Fowler (eds.)
1970. Mosquito-eating fishes of California. California Mosquito Control Assoc. Entomol. Comm. 27 pp.

Mallis, A.
1938. The California fire ant and its control. Pan-Pacific Entomologist 14:87–91.
1941*a*. A list of the ants of California, with notes on their habits and distribution. Bull. Southern California Acad. Sci. 40:61–100.
1941*b*. Preliminary experiments on the silverfish *Ctenolepisma urbani* Slabaugh. J. Econ. Entomol. 34: 787–91.
1942. An amphipod household pest in California. J. Econ. Entomol. 35:595.
1969. Handbook of pest control. (Fifth edition.) MacNair Dorland, New York. 1158 pp.
1971. Do cockroaches eat bed bugs? Pennsylvania Pest Control Quarterly 14(1):5.
1973. Unusual insect problems in Pennsylvania. Pennsylvania Pest Control Quarterly 16(2):3, 6.

Mallis, A., B. T. Burton, and A. C. Miller
1962. The attraction of salts and other nutrients to the larvae of fabric insects. J. Econ. Entomol. 55:351–55.

Mallis, A., and A. C. Miller
1964. Prolonged resistance in the house fly and bed bug. J. Econ. Entomol. 57:608–609.

Mallis, A., A. C. Miller, and R. C. Hill
1958. Feeding of four species of fabric pests on natural and synthetic textiles. J. Econ. Entomol. 51:248–49.

Mallis, A., and R. J. Pence
1941. The Pacific drain fly in homes. J. Econ. Entomol. 34:586–87.

Malloch, J. R.
1914. American black flies or buffalo gnats. USDA Bur. Entomol. Tech. Ser. No. 26:1–83.

Markin, G. P.
1968. Nest relationship of the Argentine ant *Iridomyrmex humilis* (Hymenoptera: Formicidae). J. Kansas Entomol. Soc. 41:511–16.
1970*a*. Foraging behavior of the Argentine ant in a California citrus grove. J. Econ. Entomol. 63:740–44.
1970*b*. The seasonal life cycle of the Argentine ant, *Iridomyrmex humilis* (Hymenoptera: Formicidae), in southern California.. Annals Entomol. Soc. Amer. 63:1238–42.

Markin, G. P., and J. H. Dillier
1971. The seasonal life cycle of the imported fire ant, *Solenopsis saevissima richteri,* on the Gulf coast of Mississippi. Annals Entomol. Soc. Amer. 64: 562–65.

Markin, G. P., and S. O. Hill
1971. Microencapsulated oil bait for control of the imported fire ant. J. Econ. Entomol. 64:193–96.

Marlatt, C. L.
1908. Cockroaches. USDA Agr. Div. Entomol. Circ. No. 51.
1916. The bedbug. USDA Farmers' Bull. No. 754. 12 pp.

Marshall, J. F.
1938. The British mosquitoes. British Museum (Natural History), London. 341 pp.

Marshall, J. T., Jr.
1962. Geographic distribution of color phases of Micronesian rodents. *In:* Pacific island rat ecology (T. I. Storer, ed.), pp. 39–44. Bishop Museum Bulletin 225, Honolulu.

Marshall, T. K.
1957. Wasp and bee stings. The Practitioner 178:712–22.

Martin, H., and H. Shaw
1947. Developments in methods and materials for the control of pests and diseases in Germany. Final Report, British Intelligence Obj. Sub. Commission (London) 1095:1–93.

Mason, H. C., T. J. Henneberry, F. F. Smith, and W. L. McGovern
1968. Suppression of *Drosophila melanogaster* in tomato field plots by the release of flies sterilized by apholate. J. Econ. Entomol. 61:166–70.

Masterson, R. A., H. W. Stegmiller, M. A. Parsons, C. C. Croft, and C. B. Spencer
1971. California encephalitis: An endemic puzzle in Ohio. Health Laboratory Science 8:89–96.

Matheson, R.
1944. Handbook of the mosquitoes of North America. Comstock Publishing Co., Inc., Ithaca, New York. 314 pp.
1951. Entomology for introductory courses. (Second edition.) Comstock Publishing Co., Inc., Ithaca, New York. 629 pp.

Mathur, R. N.
1954. Immature stages of Indian Coleoptera. Curculionidae. Indian Forest Records (Entomol.) 8:227–31.

Matsumura, F., D. M. Jewett, and H. C. Coppel
1972. Interspecific response of termite to synthetic trail-following substances. J. Econ. Entomol. 65:600–602.

Matsumura, F., Akira Tai, and H. C. Coppel
1969. Termite trail-following substance, isolation, and purification from *Reticulitermes virginicus* and fungus-infected wood. J. Econ. Entomol. 62:599–603.

Maunder, J. W.
1971*a*. Resistance to organochlorine insecticides in head lice, and trials using alternative compounds. Medical Officer 125:27–29.
1971*b*. Use of malathion in the treatment of lousy children. Community Medicine 126:145–47.

Maunsell, K., D. G. Wraith, and A. M. Cunnington
1968. Mites and house dust allergy in bronchial asthma. Lancet No. 7555:1267–70.

Maunsell, K., D. G. Wraith, and A. M. Hughes
1971. Hyposensitisation in mite asthma. Lancet No. 7706:967–68.

Maury, R., and A. B. de Alzuet
1970. Una plaga de la agricultura: *Bryobia praetiosa* Koch 1936, atacando al hombre. Neotrópica 16: 90–91.

Mautz, D.
1971. Der Kommunikationseffekt der Schwänzeltänze bei *Apis mellifica carnica* (Pollm.). Z. vergl. Physiol. 72:197–220.

Mayer, J.
1972. Decision-making in the biological field. BioScience 22:141–43, 152.

Mayhew, W. W.
1958. The biology of the cliff swallow in California. Condor 60:7–37.

McAllen, M. L., E. S. K. Assem, and K. Maunsell
1970. House dust mite asthma: Results of challenge tests on five criteria with *Dermatophagoides pteronyssinus*. British Medical Journal 2:501–504.

McClure, H. E.
1935. The bite of *Melanolestes picipes*. Entomological News 46:138–39.

McCoy, G. W.
1909. Siphonaptera observed in the plague campaign in California, with a note upon host transference. U. S. Publ. Hlth. and Marine Hosp. Serv., Publ. Hlth. Repts. 24:1013–30.

McCrone, J. D.
1969. Spider venoms: Biochemical aspects. American Zoologist 9:153–56.

McCrone, J. D., and H. W. Levi
1964. North American widow spiders of the *Latrodectus curacaviensis* group (Araneae: Theridiidae). Psyche 71:12–27.

McDonald, J. L., and T. H. Dickens
1970. Field evaluations of Dursban insecticide briquettes when used as mosquito larvicide material. Mosquito News 30:536–66.

McDowell, J. W., H. G. Scott, C. J. Stojanovich, and H. B. Weinburgh
1964. Tularemia. U. S. Public Health Service (CDC), Training Guide. 73 pp.

McGregor, S. E.
1970. The African bee and American beekeeping. American Bee Journal 110:460–61.

McHaffey, D. G.
1972. Photoperiod and temperature influences on diapause in eggs of the floodwater mosquito *Aedes nigromaculis* (Ludlow) (Diptera: Culicidae). Mosquito News 32:51–61.

McIntyre, T., and R. A. St. George
1961. The old-house borer. USDA Leaflet No. 501. (Slightly revised in 1970.)

McKittrick, F. A.
1964. Evolutionary studies of cockroaches. Cornell Agr. Expt. Sta. Memoir No. 389. 197 pp.
1965. A contribution to the understanding of cockroach–termite affinities. Annals Entomol. Soc. Amer. 58:18–22.

McLean, D. M., and M. T. Wilson
1972. Multiplication of California encephalitis virus in Yukon mosquitoes. Mosquito News 32:27–30.

McNew, G. L.
1970. The Boyce Thompson Institute program in forest entomology that led to the discovery of pheromones in bark beetles. Boyce Thompson Inst. Contrib. 24:251–62.

Mead, A. R.
1943. Revision of the giant West Coast land slugs of the genus *Ariolimax* Moerch (Pulmonata: Arionidae). American Midland Naturalist 30:675–717.

Mease, J. A.
1943. Deer fly desensitization. J. Amer. Med. Assoc. 122: 227.

Medler, J. T.
1958. Principles and methods for the utilization of bumblebees in cross-pollination of crops. Tenth Intern. Congr. Entomol. 4:973–81.

Melander, A. L.
1916. The dipterous family Scatopsidae. Washington Agr. Expt. Sta. Bull. 130:3–21.

Melichar, B., and J. Willomitzer
1967. Bewertung der physikalischen Insektizide. Sci. Pharmaceut. 2:589–97. Proc. Twenty-fifth Congr. Pharmaceut. Sci. (Prague), 1965.

Mellanby, K.
1941. The transmission of scabies. British Medical Journal 1941(II):405–406.
1944. The development of symptoms, parasitic infection, and immunity in human scabies. Parasitology 35:197–206.
1946. Man's reaction to mosquito bites. Nature 158: 554.

Merchant, I. A., and R. D. Barner
1964. Infectious diseases of domestic animals. Iowa State University Press, Ames. 478 pp.

Merritt, R. W.
1969. A severe case of human cutaneous myiasis caused by *Phaenicia sericata* (Diptera: Calliphoridae). California Vector Views 16:24–26.

Mertins, J. W., H. C. Coppel, and F. Matsumura
1971. Sternal gland in *Coptotermes formosanus* (Isoptera: Rhinotermitidae). Annals Entomol. Soc. Amer. 64:478–80.

Messenger, K.
1963. Low-volume aerial spraying will be a boon to applicators. Agricultural Chemicals 18(12):63–66.

Metcalf, C. L., W. P. Flint, and R. L. Metcalf
1962. Destructive and useful insects. (Fourth edition.) McGraw-Hill, New York, Toronto, and London. 1087 pp.

Metcalf, R. L., and R. B. March
1952. Behavior of octamethyl pyrophosphoramide in citrus plants. J. Econ. Entomol. 45:988–97.

Meyer de Schauensee, R.
1970. A guide to the birds of South America. Livingston Publishing Co., Wynnewood, Pennsylvania. 470 pp.

Michener, C. D.
1942. The history and behavior of a colony of harvester ants. Scientific Monthly 55:248–58.
1946. Observations on the habits and life history of a chigger mite, *Eutrombicula batatas* (Acarina: Trombiculinae). Annals Entomol. Soc. Amer. 39: 101–118.
1969. Comparative social behavior of bees. Annual Rev. Entomol. 14:299–342.
1973. The Brazilian honey bee. BioScience 23:523–27.

Michener, C. D., and M. H. Michener
1951. American social insects. Van Nostrand, New York, Toronto, and London. 267 pp.

Mickel, C. E.
1928. Biological and taxonomic investigations on the mutillid wasps. Bull. U. S. National Museum 143.

Micks, D. W., and J. A. Berlin
1970. Continued susceptibility of *Culex pipiens quinquefasciatus* to petroleum hydrocarbons. J. Econ. Entomol. 63:1936–37.

Micks, D. W., G. V. Chambers, J. Jennings, and K. Barnes
1968. Mosquito control agents derived from petroleum hydrocarbons. II. Laboratory evaluation of a new petroleum derivative, FLIT® MLO. J. Econ. Entomol. 61:647–50.

Middlekauff, W. W.
1950. The horse flies and deer flies of California (Diptera: Tabanidae). Bull. California Insect Surv. 1:1–21.
1960. The siricid wood wasps of California (Hymenoptera: Symphyta). Bull. California Insect Surv. 6: 59–77.
1962. Notes and description of the previously unknown male of *Sirex longicauda* Middlekauff. Pan-Pacific Entomologist 38:31–32.

Mielke, U.
1970. Aktive Wanderungen bei *Blattella germanica*. Angewandte Parasitologie 11:27–28.

Miley, H. H.
1930. Internal anatomy of *Euryurus erythropygus* (Brandt) (Diplopoda). Ohio Journal of Science 30:229–49.

Miller, C. D. F.
1961. Taxonomy and distribution of Nearctic *Vespula*. Canadian Entomologist, Supplement to Vol. 93.

Miller, D. G., Jr.
1956. Massive anaphylaxis from insect stings. *In:* Venoms (E. E. Buckley and N. Porges, eds.), pp. 117–21. Amer. Assoc. Adv. Sci. Publ. 44. 467 pp.

Miller, E. M.
1967. *Prorhinotermes simplex* may be taken for *Coptotermes formosanus* in Florida. Pest Control 35(7):50.
1969. Caste differentiation in the lower termites. *In:* Biology of termites (K. Krishna and F. M. Weesner, eds.), Vol. I, pp. 283–310. Academic Press, New York and London.

Miller, R. W., C. H. Gordon, N. O. Morgan, M. C. Bowman, and M. Beroza
1970. Coumaphos as a feed additive for the control of house fly larvae in cow manure. J. Econ. Entomol. 63:853–55.

Mills, H. B.
1943. An outbreak of the snipe fly *Symphoromyia hirta*. J. Econ. Entomol. 36:806.

Milner, K. C., W. L. Jellison, and B. Smith
1957. The role of lice in transmission of *Salmonella*. J. Infect. Dis. 101:181–92.

Minnick, D. R., S. H. Kerr, and R. C. Wilkinson
1972. Control of *Cryptotermes brevis*. J. Econ. Entomol. 65:1577–79.

Minton, S. A., Jr.
1970. Identification of poisonous snakes. Clinical Toxicology 3:347–62.
1972. Poisonous spiders of Indiana, and a report of a bite by *Chiracanthium mildei*. J. Indiana State Med. Assoc. 65:425–26.

Minton, S. A., Jr., and M. R. Minton
1969. Venomous reptiles. Charles Scribner's Sons, New York. 274 pp.

Mitchell, W. F., G. W. Wharton, D. G. Larson, and R. Modic
1969. House dust, mites, and insects. Annals of Allergy 27:93–99.

Miura, T., and R. M. Takahashi
1973. Insect-developmental inhibitors. 3. Effects on nontarget aquatic organisms. J. Econ. Entomol. 66: 917–22.

Miyamoto, T., S. Oshima, and T. Ishizaki
1969*a*. Antigenic relation between house dust and a dust mite, *Dermatophagoides farinae* Hughes, 1961, by a fractionization method. Journal of Allergy 44:282–91.

Miyamoto, T., S. Oshima, K. Mizuno, M. Sasa, and T. Ishizaki
1969*b*. Cross-antigenicity among six species of dust mites and house dust antigens. Journal of Allergy 44: 228–38.

Monro, H. A. U.
1969. Manual of fumigation for insect control. (Second edition, revised.) FAO Agr. Studies No. 79, United Nations (Rome). 381 pp.

Montgomery, T. H.
1908. Further studies on the activities of araneads. American Naturalist 42:697–709.

Montgomery, V. H.
1936. Spider beetles. Exterminators' Log 4:20.

Moore, B. P.
1967. Chemical communication in insects. Science Journal 3(9):44–49.
1969. Biochemical studies in termites. *In:* Biology of termites (K. Krishna and F. M. Weesner, eds.), Vol. I, pp. 407–32. Academic Press, New York and London.

Moore, R. C.
1971. Chemical control of German cockroaches in urban apartments. Connecticut Agr. Expt. Sta. Bull. 717. 11 pp.
1972. Boric acid–silica dusts for control of German cockroaches. J. Econ. Entomol. 65:458–61.

Moore, W.
1918. An interesting reaction to louse bites. J. Amer. Med. Assoc. 71:1481–82.

Morgan, F. D.
1968. Bionomics of Siricidae. Annual Rev. Entomol. 13:239–56.

Morgan, P. N.
1969. Preliminary studies on venom from the brown recluse spider; *Loxosceles reclusa*. Toxicon 6: 161–65.

Morgan, R. J., and T. O. Coston
1964. Demodex blepharitis. Southern Med. J. 57:694–99.

Morgan, W. P.
1923. Notes on the function of the forceps of earwigs. Proc. Indiana Acad. Sci. 33:303–306.
1924. Further observations on the function of the earwig forceps. Proc. Indiana Acad. Sci. 34:347–48.

Morrill, A. W., Jr., R. G. Hahl, and K. A. Lotze
1955. Studies on insecticidal fog generation for military use. Mosquito News 15:85–90.

Mosconi-Bernardini, P., and M. L. Vecchi
1966. Osservazioni istologiche e fluoromicroscopiche sulla ghiandola sternale di *Reticulitermes lucifugus* (Rhinotermitidae). Symp. Genet. Biol. Ital. 13:169–77.

Moseley, I. K.
1970. Government rodent control: A profile of a Memphis program in progress. Pest Control 38(5): 24–32.

Mount, G. A.
1970. Optimum droplet size for adult mosquito control with space sprays or aerosols of insecticides. Mosquito News 30:70–75.

Mount, G. A., and C. S. Lofgren
1967. Ultralow-volume and conventional aerial sprays for control of adult salt-marsh mosquitoes, *Aedes solicitans* (Walker) and *Aedes taeniorhynchus* (Wiedemann), in Florida. Mosquito News 27: 473–77.

Mount, G. A., and N. W. Pierce
1971. Eleven promising insecticides evaluated as ultra-low-volume cold aerosols against *Aedes taeniorhynchus* (Wiedemann). Mosquito News 31:77–80.

Mount, G. A., J. G. Williams, and C. T. Adams
1969. Control of adult mosquitoes in Alaska with malathion. Mosquito News 29:84–86.

Mrak, E. M.
1969. Report of the Secretary's Commission on Pesticides and Their Relationship to Environmental Health (E. M. Mrak, chm.). U. S. Dept. of Health, Education, and Welfare, Washington, D.C. 677 pp.

Mueller, H. L.
1959*a*. Serial intracutaneous testing for bee and wasp sensitivity. Journal of Allergy 30:123–29.
1959*b*. Further experiences with severe allergic reactions to insect stings. New England J. Med. 261:374–77.
1965. Insect allergy. *In:* Immunologic diseases (M. Samler and H. L. Alexander, eds.). Little, Brown and Co., Boston. 682 pp.

Muesebeck, C. F. W., K. V. Krombein, and H. K. Townes
1951. Hymenoptera of America north of Mexico. USDA Agr. Monograph No. 2. 1420 pp.

Mulhern, T. D.
1966. Public health aspects of pest mosquito control. Mosquito News 26:497–500.

Mulla, M. S.
1962. Mosquito control with granular larvicides. Proc. California Mosquito Control Assoc. 30:70–74.
1963*a*. Some physiochemical factors influencing the performance of new granular mosquito larvicides. Proc. California Mosquito Control Assoc. 31: 94–98.
1963*b*. Chemical repellents for *Hippelates* eye gnats: A method and procedure for evaluation. J. Econ. Entomol. 56:753–57.
1963*c*. An ecological basis for the suppression of *Hippelates* eye gnats. J. Econ. Entomol. 56:768–70.
1966. Oviposition and emergence period of the eye gnat *Hippelates collusor*. J. Econ. Entomol. 59:93–96.
1967. Biocidal and biostatic activity of aliphatic amines against southern house mosquito larvae and pupae. J. Econ. Entomol. 60:515–22.
1970. Integrated control of mosquitoes: Chemical measures against preimaginal stages. Entomol. Soc. Amer. Misc. Publ. 7:78–98.
1971. *Hippelates impressus,* an eye gnat new to California (Diptera: Chloropidae). California Vector Views 18(1):3–4.

Mulla, M. S., and H. Axelrod
1960. Efficiency of granulated insecticides influenced by solvents used for impregnation. J. Econ. Entomol. 53:938–49.

Mulla, M. S., and M. F. B. Chaudhury
1968. Ovicidal activity of aliphatic amines and petroleum oils against two species of mosquitoes. J. Econ. Entomol. 61:510–15.

Mulla, M. S., and H. A. Darwazeh
1971. Field evaluation of aliphatic amines: Petroleum oil formulations against preimaginal mosquitoes. Proc. California Mosquito Control Assoc. 39:120–36.

Mulla, M. S., H. A. Darwazeh, A. F. Geib, and W. E. Westlake
1969. Control of pasture *Aedes* mosquitoes by dripping larvicides into flowing water, with notes on residues in a pasture habitat. J. Econ. Entomol. 62: 365–70.

Mulla, M. S., H. A. Darwazeh, and P. A. Gillies
1970*a*. Evaluation of aliphatic amines against larvae and pupae of mosquitoes. J. Econ. Entomol. 63:1472–75.

Mulla, M. S., H. A. Darwazeh, and D. R. Peters
1970*b*. Mosquito control in sewage oxidation ponds with drip and pour-in larvicides. Mosquito News 30: 456–60.

Mulla, M. S., M. J. Garber, H. Axelrod, and F. G. Andrews
1966. Control of *Hippelates* eye gnats with herbicidal oils. J. Econ. Entomol. 59:552–56.

Mulla, M. S., L. W. Isaak, and H. Axelrod
1963. Field studies on the effects of insecticides on some aquatic wildlife species. J. Econ. Entomol. 56:184–88.

Mulla, M. S., and A. M. Khasawinah
1969. Laboratory and field evaluation of larvicides against chironomid midges. J. Econ. Entomol. 62: 37–41.

Mulla, M. S., and R. B. March
1959. Flight range, dispersal patterns, and population density of the eye gnat, *Hippelates collusor*. Annals Entomol. Soc. Amer. 52:641–46.

Mulla, M. S., R. L. Norland, D. M. Fanara, H. A. Darwazeh, and D. W. McKean
1971. Control of chironomid midges in recreational lakes. J. Econ. Entomol. 64:300–307.

Muller, G. H.
1970. Flea collar dermatitis in animals. J. Amer. Vet. Med. Assoc. 157:1616–26.

Müller, H.
1873. Probosces capable of sucking the nectar of *Anagraecum sesquipedale*. Nature 8:223.

Muller, H. J.
1927. Artificial transmutation of the gene. Science 66: 84–87.

Munro, J. W.
1919. Report of scabies investigation. J. Roy. Army Med. Corps 33:1–41.
1966. Pests of stored products. Hutchinson, London.

Munro Ashman, D., A. W. Frankland, H. Morrow Brown, J. G. Feinberg, and M. J. S. Langman
1971. Treatment of house dust allergy with pyridine-extracted, alum-precipitated, mite-fortified house dust (allpyral). Annals of Allergy 29:578–81.

Murray, D.
1968. Field tests of FLIT® on organophosphorus-resistant *Aedes nigromaculis* in the Delta Mosquito Abatement District. Proc. California Mosquito Control Assoc. 36–93.

Murray, K. F.
1969. Bats in California. *In:* Community pest and related vector control (J. E. Brooks and T. D. Peck, eds.), pp. 181–265. Pest Control Operators of California, Los Angeles.

Muthu, M., P. R. P. Pillai, and B. V. Hiranniah
1970. Relative toxicity of ethylene dibromide, methyl bromide, and phosphine to the adults of *Caloglyphus krameri* (Berlese) and *Tyrophagus putrescentiae* (Schrank) (Acarina: Acaridae). J. Stored Prod. Res. 6:93–96.

Myers, G. S.
1965. *Gambusia,* the fish destroyer. Tropical Fish Hobbyist (January), pp. 31–32, 53–54.

NACA (National Agricultural Chemicals Association)
1971. Pesticide industry profile study by NACA. Compiled by Ernst and Ernst, Trade Association Dept., Washington, D.C.

Nauk, E. G., and F. Zumpt
1940. Versuche zur Uebertragung des murinen Fleckfiebers durch die Bettwanze. Zentralblatt für Bacteriologie, I Abteilung (Originale) 146:97–103.

Navy (United States)
1965. Poisonous snakes of the world. U. S. Navy Bureau of Medicine and Surgery. 212 pp.

Nelson, H. D.
1967. Malathion-treated drying trays to protect raisins from insects: Commercial tests in 1966. USDA Agr. Res. Serv. Publ. 51–9. 7 pp.
1970. Fumigation of natural raisins with phosphine. USDA Agr. Res. Serv., Marketing Res. Rept. 886.

Nelson, H. D., G. H. Spitler, and A. P. Yerington
1967. Use of malathion-treated drying trays to protect raisins from insects during drying and storage. USDA Agr. Res. Serv., Marketing Res. Rept. No. 789. 14 pp.

Nelson, M. J.
1971. The winter biology of *Culex tarsalis* (Diptera: Culicidae) in Imperial Valley, California. Contrib. Amer. Entomol. Inst. 7(6):1–56.

Neumann, W., and E. Habermann
1956. Paper electrophoresis separation of pharmacologically and biochemically active compounds of bee and snake venom. *In:* Venoms (E. E. Buckley and N. Porges, eds.), pp. 171–74. Amer. Assoc. Adv. Sci. Publ. 44. 467 pp.

Nichols, N., and T. W. Green
1963. Allergic reactions to "kissing bug" bites. California Medicine 98:267–68.

Nickle, W. R.
1967. *Heterotylenchus autumnalis* sp. n (Nematoda: Sphaerulariidae), a parasite of the face fly *Musca autumnalis* De Geer. Journal of Parasitology 53: 398–401.

Nielsen, E. T., and A. T. Nielsen
1953. Field observations on the habits of *Aedes taeniorhynchus*. Ecology 34:141–56.

Nieschulz, O.
1935. Ueber die Temperaturabhängigkeit der Activität und die Vorzugstemperatur von *Musca domestica* und *Fannia canicularis*. Zoologischer Anzeiger 110:225–33.

Nigam, L. N.
1933. The life history of a common cockroach, *Periplaneta americana* (Linnaeus). Indian J. Agr. Sci. 3:530–43.

Nininger, H. H.
1916. Studies in the life histories of two carpenter bees of California, with notes on certain parasites. Pomona J. Entomol. Zool. 8:158–65.

Noble Nesbitt, J.
1969. Water balance in the firebrat, *Thermobia domestica* (Packard). Exchange of water with the atmosphere. J. Exptl. Biol. 50:745–69.

Noirot, C.
1969. Glands and secretions. *In:* Biology of termites (K. Krishna and F. M. Weesner, eds.), Vol. I, pp. 89–123. Academic Press, New York and London.

Noirot, C., and C. Noirot-Timothée
1965. Le gland sternal dans l'évolution des termites. Insectes Sociaux 12:265–72.

Norment, B. R., and T. L. Pate
1968. Residual activity of diazinon and lindane for control of *Loxosceles reclusa*. J. Econ. Entomol. 61:574–75.

Norris, D. M., J. M. Baker, and H. M. Chu
1969. Symbiontic interrelationships between microbes and ambrosia beetles. III. Ergosterol as the source of sterol to the insect. Annals Entomol. Soc. Amer. 62:413–14.

NPCA (National Pest Control Association)
1955. Condensation and other moisture problems in structures. NPCA Tech. Release 11-55. 9 pp.
1957. Scavenger and fungus beetles. NPCA Tech. Release 9-57. 3 pp.
1961. Lyctid powder-post beetles: Their biology and control. NPCA Tech. Release 19-61. 17 pp.
1964*a*. The horntails. NPCA Tech. Release 14-64. 4 pp.
1964*b*. Flatheaded or metallic wood borers, family Buprestidae. NPCA Tech. Release 15-64. 4 pp.
1965. The old-house borer. NPCA Tech. Release 11-65. 10 pp.
1966. Cockroaches and their control. NPCA Tech. Release 9-66. 26 pp.
1967. Rabies. NPCA Tech. Release 2-67. 9 pp.
1968. Crickets. NPCA Tech. Release 2-68. 5 pp.
1970. Good practice for thermal fogging of insecticides indoors. NPCA Tech. Release 21-70. 4 pp.
1971. The use of black light to detect rodent contamination. NPCA Tech. Release 18-71. 4 pp.
1972. Avitrol. NPCA Tech. Release 5-72. 10 pp.
1973. Slugs and snails. NPCA Tech. Release 15-73. 4 pp.

Nuttall, G. H. F.
1917*a*. The part played by *Pediculus humanus* in the causation of disease. Parasitology 10:43–79.
1917*b*. The biology of *Pediculus humanus*. Parasitology 10:180–85.
1918. The biology of *Phthirus pubis*. Parasitology 10: 383–405.

Nutting, W. L.
1956. Reciprocal protozoan transfaunations between the roach *Cryptocercus* and the termite *Zootermopsis*. Biological Bulletin 110:83–90.
1960. Notes on the occurrence of four adventive earwigs in Arizona. Pan-Pacific Entomologist 36: 202–204.
1965. Observations on the nesting site and biology of the Arizona damp-wood termite *Zootermopsis laticeps* (Banks) (Hodotermitidae)). Psyche 72: 113–25.

Obermayer, M. E.
1955. Psychocutaneous medicine. Charles C. Thomas, Springfield, Illinois. 478 pp.

O'Brien, R. D., and R. W. Fisher.
1958. The relation between ionization and toxicity to insects for some neuropharmacological compounds. J. Econ. Entomol. 51:169–75.

Oderkirk, G.
1958. Bird nuisance not on wing. Pest Control 26(3): 41–42.

Okumura, G. T.
1955. Illustrated key to lepidopterous larvae found in stored almonds and walnuts. California Dept. Agr. Publ. E-52. 3 pp. (Processed.)
1967. A report on canthariasis and allergy caused by *Trogoderma* (Coleoptera: Dermestidae). California Vector Views 14:19–22.

Older, J. J.
1970. The epidemiology of murine typhus in Texas, 1969. J. Amer. Med. Assoc. 214:2011–17.

Olsen, R. F.
1964. The epizootiology of Chagas' disease in the southeastern United States, with particular emphasis on the *Trypanosoma cruzi–Triatoma sanguisuga–Didelphis marsupialis* complex in Alabama. Ph.D. thesis, Auburn (Alabama) University. 230 pp.

Oppenoorth, F. J.
1965. Biochemical genetics of insecticide resistance. Annual Rev. Entomol. 10:185–206.

Ordman, D.
1946. Bronchial asthma caused by the trickling sewage filter fly (*Psychoda*): Inhalant insect allergy. Nature 157:441.

Osgood, H.
1957. Allergy to caddis fly (Trichoptera). II. Clinical aspects. Journal of Allergy 28:292–300.

Osmun, J. V.
1954. Carpenter bee, *Xylocopa virginica* (L.). Pest Control Technology, Entomol. Section: 300-A-B.
1958. Factors affecting subslab fumigant dispersion in soil. Pest Control 26(2):23–24, 56.

Osmun, J. V., and W. L. Butts
1966. Pest control. Annual Rev. Entomol. 11:515–48.

Ouye, M. T., and B. A. Butt
1962. A natural sex lure extracted from female pink bollworms. J. Econ. Entomol. 55:419–21.

Palmer, E. D.
1946. Intestinal canthariasis due to *Tenebrio molitor*. Journal of Parasitology 32:54–55.

Papp, C. S.
1959. Concerning North American *Necrobia*. Bull. Southern California Acad. Sci. 58:162–65.
1960. Distributional records on some North American beetles, with remarks on their ecology in southern California. Entomological News 71:69–77.

Papp, C. S., and G. T. Okumura
1959. A preliminary study of the Ptinidae of California. California Dept. Agr. Bull. 48:228–48.

Parker, B. M., and F. L. Campbell
1940. Relative susceptibility of the oötheca and adult female of the German cockroach to liquid household insecticides. J. Econ. Entomol. 33:610–14.

Parkin, E. A.
1933. The larvae of some wood-boring Anobiidae (Coleoptera). Bull. Entomol. Res. 24:33–68.
1934. Observations on the biology of the lyctus powder-post beetles, with special reference to oviposition and the egg. Annals Appl. Biol. 21:495–518.
1940. The digestive enzymes of some wood-boring beetle larvae. J. Exptl. Biol. 17:364–77.
1942. Symbiosis and siricid woodwasps. Annals Appl. Biol. 29:268–74.

Parlato, S. J.
1929. A case of coryza and asthma due to sand flies (caddis flies). Journal of Allergy 1:35–42.
1932*a*. Emanations of flies as exciting causes of allergic coryza and asthma. III. Hypersensitiveness to moths and butterflies (Lepidoptera). Journal of Allergy 3:125–38.
1932*b*. Emanations of flies as exciting causes of allergic coryza and asthma. V. A study of the atopic reagins of the caddis fly, butterfly, and moth. Journal of Allergy 3:459–68.

Parrish, H. M.
1959. Deaths from bites and stings of venomous animals and insects in the United States. Archives of Internal Medicine 104:198–207.
1963. Analysis of 460 fatalities from venomous animals in the United States. Amer. J. Med. Sci. 245:129–41.

Partida, G. J., and R. G. Strong
1970. Distribution and relative abundance of *Trogoderma* spp. in relation to climatic zones in California. J. Econ. Entomol. 63:1553–60.

Patterson, R. A.
1960. Physiological action of scorpion venom. Amer. J. Trop. Med. Hyg. 9:410–14.

Patterson, R. S., and F. L. Wilson
1966. Fogging and granule applications are teamed to control chironomid midges on Florida lakefronts. Pest Control 34(6):26, 29–30, 32.

Patton, W. S.
1931. Insects, ticks, mites, and venomous animals. II. A. R. Grubb, London. 786 pp.

Pausch, R. D.
1971. Local house fly control with baited chemosterilants. I. Preliminary laboratory studies. J. Econ. Entomol. 64:1462–65.

Payne, N. M.
1936. Injury to lumber by *Hadrobregmus carinatus* Say.. J. Econ. Entomol. 29:1027.
1946. Life history and habits of the flat grain beetle (*Laemophloeus minutus* Oliv.), J. New York Entomol. Soc. 54:9–12.

PCOC (Pest Control Operators of California, Inc.)
1973. General pest control survey results. PCOC News Letter (December). 2 pp.

Peacock, A. D., and A. T. Baxter
1949. Studies in Pharaoh's ant, *Monomorium pharaonis* (L.). I. The rearing of artificial colonies. Entomol. Monthly Magazine 85:256–60.
1950. Studies in Pharaoh's ant, *Monomorium pharaonis* (L.). Entomol. Monthly Magazine 86:129–35, 171–78, 294–98.

Pearman, J. V.
1928. On sound production in the Psocoptera and on a presumed stridulatory organ. Entomol. Monthly Magazine 64:179–86.

Pearson, T. G.
1928. Portraits and habits of our birds: English sparrow. Nat. Assoc. Audubon Soc. Educ. Leaflet No. 90, Vol. II.

Peckham, G. W., and E. G. Peckham
1905. Wasps, social and solitary. Houghton, Mifflin, Boston. 311 pp.

Pelsue, F. W., G. C. McFarland, and H. T. Magy
1970. Buffalo gnat (Simuliidae) control in the Southeast Mosquito Abatement District. Proc. and Papers California Mosquito Control Assoc. 38:102–104.

Pence, R. J.
1955. Bees and the PCO. PCO News 15(10):8–10.
1956*a*. Increased occurrence of the fungus moth, *Tinea* sp. PCO News 16(4):5.
1956*b*. The tolerance of the drywood termite, *Kalotermes minor* Hagen, to desiccation. J. Econ. Entomol. 49:553–54.
1957. The prolonged maintenance of the western subterranean termite in the laboratory with moisture gradient tubes. J. Econ. Entomol. 50:238–40.
1958. Fabric pests will attack synthetic fibers. Pest Control 26(9):27–28, 30.
1966. Analyzing fur damage with a microscope. Univ. California Agr. Expt. Sta. Ext. Circ. 541. 18 pp.
1967. A simple rule for classifying most pesticides. PCO News 27(8):24.
1969. Excised honey bee abdomens and the biosonic analyzer system aid pharmacological and toxicological investigation. California Agriculture 23(12):6–8.

Pence, R. J., R. D. Chambers, and M. S. Viray
1963. "Psychogenetic stress" and autointoxication in the honey bee. Nature 200:930–32

Pence, R. J., and C. S. Davis
1969. Carpet beetles and clothes moths. Univ. California Agr. Ext. Serv. AXT-301. 7 pp.

Pence, R. J., and C. L. Hogue
1957. A new record of a fungus moth as a household pest in California. PCO News 17(12):6.

Pence, R. J., and P. Lomax
1971. The honey bee abdomen preparation as a pharmacological test system. Z. angew. Entomol. 69: 337–42.

Pence, R. J., and J. Morganroth
1962. Field effects of methyl bromide on carpet beetle eggs. Pest Control 30(7):20–24.

Pence, R. J., and M. S. Viray
1963. Modified base oil for the control of carpet beetles. Pest Control 31(10):32, 34–36.
1969. Atrophy and developmental abnormality in insects exposed to dodecyl alcohol. J. Econ. Entomol. 62:622–29.

Pencille, C.
1947. Structural fumigation to eliminate dry-wood termites (*Kalotermes*) with liquid hydrocyanic acid gas. Pest Control and Sanitation 2(8):10–12.

Penn, L. A.
1965. Pigeon control in Milwaukee. Pest Control 33(8): 16–18, 61.

Pennak, R. W.
1964. Collegiate dictionary of zoology. Ronald Press, New York. 581 pp.

Perlman, E.
1955. Near-fatal allergic reactions to bee and wasp stings: A review and report of seven cases. J. Mt. Sinai Hospital 22:336–47.

Perlman, F.
1962. Treatment for severe reactions to the bites and stings of arthropods. Medical Times 90:813–20.

Perry, A. S.
1964. The physiology of insecticide resistance by insects. *In:* The physiology of Insecta (M. Rockstein, ed.), pp. 285–378. Academic Press, New York and London.

Peters, H.
1965. The house fly as a disease carrier. Pest Control 33(5):16–20.

Peters, R. F.
1959. Public health vector control: A career opportunity for entomologists. California Vector Views 6:49–51.
1963. Urbanization's impact on the poultry industry. California Vector Views 10:69–72.

Peters, R. H.
1957. A decade of mosquito source reduction in a local mosquito abatement program. Mosquito News 17: 280–82.

Peterson, D. J.
1952. Observations on the biology and control of pest Trichoptera at Fort Erie, Ontario. Canadian Entomologist 84:103–107.

Pfeiffer, C. C.
1951. Is boric acid harmless? Modern Hospital 74(3):106, 110, 140, 142.

Phelps, R. J., and G. R. DeFoliart
1964. Nematode parasitism of Simuliidae. Univ. Wisconsin Res. Bull. 245. 78 pp.

Philip, C. B.
1928. Methods of collecting and rearing the immature stages of Tabanidae (Diptera). Journal of Parasitology 14:243–53.

Pickens, A. L.
1934*a*. The biology and economic significance of the western subterranean termite, *Reticulitermes hesperus*. *In:* Termites and termite control (C. A. Kofoid, ed.; reprinted and partially revised, 1965), pp. 157–83. University of California Press.
1934*b*. The barren-lands subterranean termite, *Reticulitermes tibialis*. *In:* Termites and termite control (C. A. Kofoid, ed.; reprinted and partially revised, 1965), pp. 184–86. University of California Press.

Pierce, T. D.
1970. Dry bait rat control in a feed warehouse. Pest Control 38(8):11–12, 14, 16.

Pierce, W. D.
1907. Notes on the economic importance of sowbugs. USDA Bur. Entomol. Bull. No. 64:15–22.
1942. Utilization of the black widow parasite, and further data on spiders and parasites. Bull. Southern California Acad. Sci. 41:14–28.

Pilsbry, H. A.
1948. Land Mollusca of North America north of Mexico. Philadelphia Acad. Nat. Sci. Monograph (Part 2):521–1113. (*See specifically:* pp. 706–22.)

Pimentel, D.
1955. Relationship of ants to fly control in Puerto Rico. J. Econ. Entomol. 48:28–30.

Ping, C.
1917. Observations of *Chironomus decorus* Johannsen. Canadian Entomologist 49:418–26.

Pippin, W. F.
1970. The biology and vector capability of *Triatoma sanguisuga texana* Usinger and *Triatoma gerstaeckeri* (Stål) compared with *Rhodnius prolixus* (Stål). J. Med. Entomol. 7:30–45.

Pitman, G. B., and J. P. Vité
1970. Field response of *Dendroctonus pseudotsugae* (Coleoptera: Scolytidae) to synthetic frontalin. Annals Entomol. Soc. Amer. 63:661–64.

Plank, H. K.
1948. Biology of the bamboo powder-post beetle in Puerto Rico. Puerto Rico Fed. Expt. Sta. (Mayaguez) Bull. 44.

Plapp, F. W., Jr.
1970. On the molecular biology of insecticide resistance. *In:* Biochemical toxicology of insecticides (R. D. O'Brien and I. Yamamoto, eds.), pp. 179–92. Academic Press, New York.

Poorbaugh, J. H., and C. R. Smith
1968. The face fly (*Musca autumnalis* De Geer) found in California. California Vector Views 15:54.

Pope, P.
1953. Studies on the life histories of some Queensland Blattidae (Orthoptera). I. The domestic species. II. Some native species. Proc. Roy. Soc. Queensland 63:23–59.

Potter, C.
1935. The biology and distribution of *Rhyzopertha dominica* (Fab.). Trans. Roy. Entomol. Soc. London A-83:449–82.

Pratt, H. D.
1955. Cockroach identification with CDC pictorial key. Pest Control 23(5):9–12.
1962. Ticks: Biology, habits, and control. Pest Control 30(10):34–38, 92.
1963. Mites of public health importance and their control. U. S. Public Health Service (CDC) Publ. No. 772. Insect Control Series: Part IX. 28 pp.

Pratt, H. D., R. C. Barnes, and K. S. Littig
1963. Mosquitoes of public health importance and their control. U. S. Public Health Service (CDC) Publ. No. 772, Insect Control Series: Part VI. 64 pp.

Pratt, H. D., and K. S. Littig
1961. Lice of public health importance and their control. U. S. Public Health Service (CDC) Publ. No. 772, Insect Control Series: Part VIII. 16 pp.
1962. Ticks of public health importance and their control. U. S. Public Health Service (CDC) Publ. No. 772, Insect Control Series: Part X. 42 pp.

Pratt, H. D., and H. G. Scott
1962. A key to some beetles commonly found in stored foods. Proc. Entomol. Soc. Washington 64:43–50.

Pratt, H. D., and J. S. Wiseman
1962. Fleas of public health importance and their control. U. S. Public Health Service (CDC) Publ. No. 772, Insect Control Series: Part VII. 36 pp.

Price, R. D.
1956. The multiplication of *Pasteurella tularensis* in human body lice. Amer. J. Hyg. 63:186–97.

Price, W. H.
1953. A quantitative analysis of the factors involved in the variations in virulence of rickettsiae. Science 118:49–52.

Pricer, J. L.
1908. The life history of the carpenter ant. Biological Bulletin 14:177–218.

Pruzansky, J., A. R. Feinberg, G. Schick, and S. M. Feinberg
1958. Antigenic relationships in insect extracts. Proc. Soc. Exptl. Biol. and Med. 97:312–14.

Pryor, M. G. M.
1940. On the hardening of the cuticle of insects. Proc. Roy. Soc. London B-128:393–407.

Pumphrey, R. J., and A. F. Rawdon-Smith
1936*a*. Hearing in insects: The nature of the response of certain receptors to auditory stimuli. Proc. Roy. Soc. London B-121:18–27.
1936*b*. Sensitivity of insects to sound. Nature 137:990.

Pye, J. D.
1971. Bats and fog. Nature 229:572–74.

Quate, L. W.
1955. A revision of the Psychodidae (Diptera) in America north of Mexico. Univ. California Publ. Entomol. 10:103–273.

Quick, H. L.
1971. Clinical evaluation of a free-hanging 20% dichlorvos (DDVP) disk for control of flea infestation of dogs. Vet. Med. Small Animal Clinician 66:773–74.

Quraishi, M. S., and Z. T. Poonawalla
1969. Radioautographic study of the diffusion of topically applied DDT-C^{14} into the house fly and its distribution in internal organs. J. Econ. Entomol. 62:988–94.

Rachesky, S.
1969. The cluster fly. Pest Control 37(3):55.
1970*a*. March of the ants. Pest Control 38(2):36.
1970*b*. Basic fly control. Pest Control 38(9):42–43.
1973. Swimming pool pests. Pest Control 41(8):39–40.

Rafatjah, H.
1971. The problem of resurgent bed-bug infestation in malaria eradication programmes. J. Trop. Med. Hyg. 74:53–56.

Ramsay, C.
1969. How I treat scabies. Postgraduate Medicine 45: 258–59.

Rasmussen, S.
1967. Fumigation of houses with methyl bromide against the house longhorn beetle, *Hylotrupes bajulus*. Material und Organismen 2:65–78.

Rau, P.
1924. Biology of the roach *Blatta orientalis*. Trans. St. Louis Acad. Sci. 25:57–79.
1933. The jungle bees and wasps of Barro Colorado Island. Published by the author, Kirkwood, Missouri. 324 pp.
1940. The life history of the wood-roach, *Parcoblatta pennsylvanica* (De Geer) (Orthoptera: Blattidae). Entomological News 51:4–9, 33–35.
1941. Cockroaches: The forerunners of termites (Orthoptera: Blattidae: Isoptera). Entomological News 52:256–59.

Rau, P., and N. Rau
1918. Wasp studies afield. Princeton University Press, Princeton, New Jersey. 372 pp.

Readio, P. A.
1927. Studies on the biology of the Reduviidae of America north of Mexico. Univ. Kansas Sci. Bull. 17: 5–291.

Redd, J. C.
1957. Why termite control is prescription work. Pest Control 25(9):37–38, 46.

Reeves, E. L.
1970. Pathogens of mosquitoes. California Mosquito Control Assoc., Proc. Thirty-Eighth Annual Conf., pp. 20–22.

Reeves, E. L., and C. Garcia
1969. Mucilaginous seeds of the Cruciferae family as potential biological control agents for mosquito larvae. Mosquito News 29:601–607.

Reeves, W. C.
1965. Ecology of mosquitoes in relation to arboviruses. Annual Rev. Entomol. 10:25–46.
1967. Factors that influence the probability of epidemics of western equine, St. Louis, and California encephalitis in California. California Vector Views 14:13–18.

Reeves, W. C., B. Brookman, and W. McD. Hammon
1948. Studies on the flight range of certain *Culex* mosquitoes, using a fluorescent dye marker, with notes on *Culiseta* and *Anopheles*. Mosquito News 8: 61–69.

Reeves, W. C., and W. McD. Hammon
1962. Epidemiology of the arthropod-borne viral encephalitides in Kern County, California, in 1943–52. Univ. California Publ. Public Health 4:1–257.

Reeves, W. M.
1971. The impact of mosquito-borne diseases on organized mosquito control districts. Mosquito News 31:319–25.

Rehn, J. A. G.
1945. Man's uninvited fellow traveler: The cockroach. Scientific Monthly 61:265–76.

Rehn, J. A. G., and M. Hebard
1909. An orthopterological reconnaissance of the southwestern United States. II. New Mexico and western Texas. Proc. Philadelphia Acad. Nat. Sci. 61:111–75.

Reierson, D. A.
1966. Feeding preferences of drywood termites and the termiticidal effect of boric acid. PCO News 26(11): 14–15.
1973. Field tests to control German cockroaches with ULV aerosol generators. Pest Control 41(1):26–32.

Reierson, D. A., and W. Ebeling
1970. The brown cockroach, *Periplaneta brunnea* Burmeister, found in California. PCO News 30(7): 6–9.

Reilly, E. M., Jr.
1968. The Audubon illustrated handbook of American birds. McGraw-Hill, New York. 524 pp.

Rhoades, W. C., and D. R. Davis
1967. Effects of meteorological factors on the biology and control of the imported fire ant. J. Econ. Entomol. 60:554–58.

Richards, A. G.
1947. The organization of arthropod cuticle: A modified interpretation. Science 105:170–71.

Richter, C. P.
1945. Incidence of rat bites and rat-bite fever in Baltimore. J. Amer. Med. Assoc. 128:324–26.

Ricketts, H. T., and R. M. Wilder
1910. Further investigation regarding the etiology of tabardillo, Mexican typhus fever. J. Amer. Med. Assoc. 55:309–311.

Riddiford, L. M.
1967. *Trans*-2-hexanal: Mating stimulant for polyphemus moths. Science 158:139–41.

Riley, C. V.
1889. The little red ant (*Monomorium pharaonis* L.). Insect Life 2:106–108.

Riley, D. A., and M. R. Rosenzweig
1957. Echolocation in rats. J. Compar. Physiol. Psych. 50:323–28.

Ripper, W. E., R. M. Greenslade, and G. S. Hartley
1951. Selective insecticides and biological control. J. Econ. Entomol. 44:448–59.

Ritcher, P. O.
1966. Biological control of insects and weeds in Oregon. Oregon Agr. Expt. Sta. Tech. Bull. 90. 39 pp.

Ritter, F. J., and C. M. A. Coenen-Saraber
1969. Food attractants and a pheromone as trail-following substances for the Saintonge termite. Multiplicity of the trail-following substances in *Lenzites trabea*-infected wood. Entomol. Expt. et Appl. 12:611–22.

Rivers, T. M., and F. L. Horsfall, Jr. (eds.)
1959. Viral and rickettsial infections of man. (Third edition.) Lippincott, Philadelphia. 967 pp.

Roark, R. C.
1942. Definition of aerosol. J. Econ. Entomol. 35:105–106.

Robbins, W. E., J. N. Kaplanis, W. J. Thompson, T. J. Shortino, and S. C. Joyner
1970. Ecdysones and synthetic analogs: Molting hormone activity and inhibitive effects on insect growth, metamorphosis, and reproduction. Steroids 16:105–25.

Rockwell, E. M.
1952. Some investigational studies concerning reactions to insect bites. Annals of Allergy 10:404–410.

Rodriguez, J. G., P. Singh, and B. Taylor
1970. Manure mites and their role in fly control. J. Med. Entomol. 7:335–41.

Rodriguez, J. L., and L. A. Riehl
1959*a*. Results with cockerels for house fly control in poultry droppings. J. Econ. Entomol. 52:542–43.
1959*b*. Malathion dust for the control of lice on domestic game birds. J. Econ. Entomol. 52:774–75.
1962. Control of flies in manure of chickens and rabbits by cockerels in southern California. J. Econ. Entomol. 55:473–77.

Roeder, K. D.
1967. Prey and predator. Entomol. Soc. Amer. Bull. 13:6–9.

Roeser, G.
1940. Zur Kenntnis der Lebensweise der Gewachshausschabe *Pycnoscelus surinamensis* L. Gartenbauwissenschaft 15:184–225.

Rogers, P. A., and J. Cornelius
1970. Tentative guidelines for the safe handling and disposal of used pesticide containers in California. California State Department of Public Health, Berkeley. 39 pp.

Rohe, D. L.
1966. Field evaluation of the rodenticides fluoroacetamide and norbormide against roof rats in sewers. California Vector Views 13:79–82.

Röller, H., and K. H. Dahm
1968. The chemistry and biology of juvenile hormones. Rec. Prog. Hormone Res. 24:651–80.

Röller, H., K. H. Dahm, C. C. Sweeley, and B. M. Trost
1967. The structure of the juvenile hormone. Angewandte Chemie (international edition in English 6:179–80; German version 79:190).

Romanuk, M., K. Sláma, and F. Sorm
1967. Constitution of a compound with a pronounced juvenile hormone activity. Proc. Nat. Acad. Sci. (U. S.) 57:349–52.

Root, A. I., and E. R. Root
1935. The ABC and XYZ of bee culture. A. I. Root Co., Medina, Ohio. 815 pp.

Rose, I.
1954. A review of tick paralysis. J. Canadian Med. Assoc. 70:175–76.

Ross, H. H.
1929. The life history of the German cockroach. Trans. Illinois State Acad. Sci. 21:84–92.
1940. The Rocky Mountain "black fly," *Symphoromyia atripes*. Annals Entomol. Soc. Amer. 33:254–57.
1944. The caddis flies, or Trichoptera, of Illinois. Illinois Nat. Hist. Surv. 23:1–326.
1965. Textbook of entomology. (Third edition.) John Wiley and Sons, New York. 539 pp.

Roth, I. M.
1951. Loci of sensory end-organs used by mosquitoes, *Aedes aegypti* (L.) and *Anopheles quadrimaculatus* Say, in receiving host stimuli. Annals Entomol. Soc. Amer. 44:59–74.

Roth, L. M., and E. R. Willis
1950. The oviposition of *Dermestes ater* De Geer, with notes on bionomics under laboratory conditions. American Midland Naturalist 44:427–47.
1952. A study of cockroach behavior. American Midland Naturalist 47:66–129.
1955. Water relations of cockroach oöthecae. J. Econ. Entomol. 48:33–36.
1956. Parthenogenesis in cockroaches. Annals Entomol. Soc. Amer. 49:195–204.
1957. The medical and veterinary importance of cockroaches. Smithsonian Misc. Coll. 134(10):1–147.
1960. The biotic associations of cockroaches. Smithsonian Misc. Coll. 141:1–470.

Rothschild, M.
1965. Fleas. Scientific American 213(6):44–53.

Roy, D. N., and A. W. A. Brown
1954. Entomology: Medical and veterinary. Excelsior Press, Calcutta.

Roy, D. N., and S. M. Ghosh
1945. Studies on the population of head-lice, *Pediculus humanus* var. *capitis* De G. Parasitology 36:69–71.

Ruckelshaus, W. D.
1972*a*. The environmental record, 1971. National Pest Control Operators News 32(2):6–7.
1972*b*. Shell Chemical Co., *et al.* Federal Register 37(239): 26463–26465.

Rueger, M. E., and T. A. Olson
1969. Cockroaches (Blattaria) as vectors of food poisoning and food infection organisms. J. Med. Entomol. 6:185–89.

Runner, G. A.
1916. Effect of röntgen rays on the tobacco, or cigarette beetle, and the results of experiments with a new form of the röntgen tube. J. Agr. Res. 6:383–88.

Russell, F. E.
1967*a*. First aid for snake venom poisoning. Toxicon 4:285–89.
1967*b*. Comparative pharmacology of some animal toxins. Federal Proceedings 26:1206–24.
1969. Bites of spiders and other arthropods. *In:* Current therapy (H. F. Conn, ed.), pp. 878–79. W. B. Saunders Co., Philadelphia.
1970. Bite by the spider *Phidippus formosus:* Case history. Toxicon 8:193–94.

Russell, F. E., C. B. Alexander, and F. W. Buess
1968. Venom of the scorpion *Vejovis spinigerus*. Science 159:90–91.

Russell, F. E., and H. W. Puffer
1970. Pharmacology of snake venoms. Clinical Toxicology 3:433–44.

Russell, F. E., W. G. Waldron, and M. B. Madon
1969. Bites by the brown spiders *Loxosceles unicolor* and *Loxosceles arizonica* in California and Arizona. Toxicon 7:109–17.

Russell, F. E., J. Wainschel, and W. J. Gertsch
1972. Bites of spiders and other arthropods. *In:* Current therapy (H. F. Conn, ed.), pp. 841–43. W. B. Saunders Co., Philadelphia.

Rutledge, J. H.
1967. Potential uses of phosphine by pest control operators. Pest Control 35(7):11–14, 58, 60.

Ryckman, R. E.
1962. Biosystematics and hosts of the *Triatoma protracta* complex in North America. Univ. California Publ. Entomol. 27:93–240.
1969. Vapona for the control of museum pests. J. Med. Entomol. 6:98.

Ryckman, R. E., and A. E. Ryckman
1967*a*. Epizootiology of *Trypanosoma cruzi* in southwestern North America. XI. Biology of the genus *Reduvius* in North America and the possible relationship of *Reduvius* to the epizootiology of *Trypanosoma cruzi*. J. Med. Entomol. 4:326–34.

Ryckman, R. E., and J. V. Ryckman
1967*b*. Epizootiology of *Trypanosoma cruzi* in southwestern North America. XII. Does Gause's rule apply to the ectoparasitic Triatominae? J. Med. Entomol. 4:379–86.

Sabrosky, C. W.
1940. Chloropids swarming in houses. J. Econ. Entomol. 33:946–47.

Sacher, R. M.
1971. A mosquito larvicide with favorable environmental properties. Mosquito News 31:513–16.

Sadun, E. H.
1969. Back to antiquity. Military Medicine 134:731–36.

Sailer, R. I.
1952. The bedbug: An odd bedfellow that's still with us. Pest Control 20(10):22, 24, 70, 72.

Samšinák, K., F. Dusbabek, and E. Vobrázková
1972. Note on the house dust mites in Czechoslovakia. Folia Parasitologica 19:383–84.

Sands, W. A.
1969. The association of termites and fungi. *In:* Biology of termites (K. Krishna and F. M. Weesner, eds.), Vol. I, pp. 495–524. Academic Press, New York and London.

Sasa, M.
1961. Biology of chiggers. Annual Rev. Entomol. 6: 221–44.

Sauer, R. J.
1970. Poisonous spiders of Michigan: Identification, bite symptoms, and treatment. Michigan State Univ. Ext. Bull. C-674. 3 pp.

Schaefer, C. H., and D. J. Ramke
1971. An operational evaluation of FLIT® MLO for the control of mosquito larvae in California, with supplemental comments on problems involved with the aerial application of oils. California Vector Views 18:13–16.

Schaefer, C. H., and W. H. Wilder
1972. Insect-developmental inhibitors. 1. A practical evaluation as mosquito control agents. J. Econ. Entomol. 65:1066–71.
1973. Insect-developmental inhibitors. 2. Effects on target mosquito species. J. Econ. Entomol. 66:913–16.

Scharrer, B.
1951. The cockroach. Scientific American 185(6):58–62.

Schenone, H., A. Rojas, H. Reyes, F. Villaroel, and G. Suarez
1970. Prevalence of *Loxosceles laeta* in houses in central Chile. Amer. J. Trop. Med. Hyg. 19:564–67.

Schildknecht, H.
1964. Defensive substances of the arthropods, their isolation and identification. Angewandte Chemie (international edition in English) 3:73–82.

Schliessmann, D. J.
1966. Progress report of the *Aedes* eradication program in the United States for 1965. Mosquito News 26:486–89.

Schlinger, E. I., R. van den Bosch, and E. J. Dietrick
1959. Biological notes on the predaceous earwig *Labidura riparia* (Pallas), a recent immigrant to California (Dermaptera: Labiduridae). J. Econ Entomol. 52:247–49.

Schmauss, L. F.
1929. Case of arachnoidism (spider bite). J. Amer. Med. Assoc. 92:1265–66.

Schmialek, P.
1961. Die Identifizierung zweier im Tenebriokot und in Hefe vorkommender Substanzen mit Juvenilhormonwirkung. Zeitschrift für Naturforschung 16*b*:461–64.
1963. Ueber Verbindungen mit Juvenilhormonwirkung. Zeitschrift für Naturforschung 18*b*:516–19.

Schneider, D.
1966. Chemical sense communication in insects. Symp. Soc. Exptl. Biol. 20:273–97.

Schneider, F.
1940. Schadinsekten und ihre Bekämpfung in ostindischen Gambirkulturen. Schweitzer Z. Forst. Nos. 2 and 3.

Schoof, H. F.
1959. How far do flies fly? Pest Control 27(4):16, 18, 20, 22, 66.
1970. Zinc phosphide as a rodenticide. Pest Control 38(5):38, 42–44.

Schoof, H. F., and R. E. Siverly
1954. The occurrence and movement of *Periplaneta americana* within an urban sewerage system. Amer. J. Trop. Med. Hyg. 3:367–71.

Schread, J. C.
1964. Insect pests of Connecticut lawns. Connecticut Agr. Expt. Sta. Circ. 212 (revised). 10 pp.

Schrut, A. H., and W. G. Waldron
1963. Psychiatric and entomological aspects of delusory parasitosis. J. Amer. Med. Assoc. 186:429–30.

Scott, H.
1932. Carpenter bees eating lead cable covers. Entomol. Monthly Magazine 68:9.

Scott, H. G.
1959. Domestic rodents: Their public health importance and control. U. S. Public Health Service (CDC). 4 pp.
1961*a*. Bats: Public health importance, identification, and control. Pest Control 29(8):23–26.
1961*b*. Pigeons: Public health importance and control. Pest Control 29(9):9–20, 60–61.
1961*c*. Filter fly control at sewage plants. The Sanitarian 24(1):14–17.
1962. Blister beetle dermatitis produced by *Epicauta cinerea* (Coleoptera: Meloidae). J. Econ. Entomol. 55:145–46.
1962–63. How to control insects in stored foods. Pest Control, Part I, 30(12):18–28; Part II, 31(1):24–34.
1963*a*. The control of insects of public health importance. U. S. Public Health Service (CDC). 3 pp.
1963*b*. Household and stored-food insects of public health importance and their control. U. S. Public Health Service (CDC) Publ. No. 772, Insect Control Series: Part XII. 55 pp.
1964*a*. Pigeon-borne disease control through sanitation and pigeon stoppage. Pest Control 32(9):14–15, 19, 38–44.
1964*b*. Human myiasis in North America. Florida Entomologist 47:255–61.
1964*c*. Stinging caterpillars. Pest Control 32(9):24–25.
1966*a*. Rat bite: Epidemiology and control. U. S. Public Health Service (CDC). 5 pp.
1966*b*. Envenomization. U. S. Public Health Service (CDC). 22 pp.
1966*c*. Insect pests. I. Springtails. Modern Maintenance Management 18(9):19–21.

Scott, H. G., and M. R. Borom
1965. Rodent-borne disease control through rodent stoppage. U. S. Public Health Service (CDC). 34 pp.

Scott, H. G., and J. M. Clinton
1967. An investigation of "cable mite" dermatitis. Annals of Allergy 25:409–14.

Scott, H. G., and R. M. Fine
1963. Straw itch mite dermatitis caused by *Pyemotes ventricosus*. Dermatología Trópica 2:226–31.
1964. *Pyemotes ventricosus* (straw itch mite) ontogeny. Florida Entomologist 47:93–95.
1967. Straw itch mite dermatitis. Pest Control 35(7): 19–23.

Scott, H. G., and K. S. Littig
1959. Insecticidal equipment for the control of insects of public health importance. U. S. Public Health Service (CDC) Publ. No. 772, Insect Control Series: Part III. 33 pp.
1962. Flies of public health importance and their control. U. S. Public Health Service (CDC) Publ. No. 772, Insect Control Series: Part V. 40 pp.

Scott, H. G., and H. D. Pratt
1957. Scorpions, spiders, and other arthropods of minor public health importance. U. S. Public Health Service (CDC) Training Guide. 29 pp.

Scott, H. G., J. S. Wiseman, and C. J. Stojanovich
1962. Collembola infesting man. Annals Entomol. Soc. Amer. 55:428–30.

Scott, J. A.
1949. Longevity of tropical rat mites kept without food. Journal of Parasitology 35:434.

Scott, K. G.
1960. Southern California TO's view of FHA preconstruction termite preventives. Pest Control 28(1): 35–36, 38.

Sein, F., Jr.
1923. *Cucarachas*. Puerto Rico Insect Expt. Sta. Circ. 64. 12 pp.

Sell, W.
1931. Staubausscheidung an einfachen Körpern und in Luftfiltern. Forsch. Verein Deutschen Ingenieuren V.I.D., Verlag No. 347:1.

Severin, H. C.
1926. The common black field cricket, *Gryllus assimilis* (Fabr.), and its control. J. Econ. Entomol. 19: 218–27.

Sferra, P. R.
1962. Granular insecticides: Their history and literature. Agricultural Chemicals 16(4):21, 24, 102.

Shafer, G. D.
1949. The ways of a mud-dauber. Stanford University Press, Palo Alto, California. 78 pp.

Shamiyeh, N. B., S. E. Bennett, R. P. Hornsby, and N. L. Woodiel
1971. Isolation of mites from house dust. J. Econ- Entomol. 64:53–55.

Shanafelt, J. G., Jr.
1969. A new approach to mosquito control in cemeteries. Proc. California Mosquito Control Assoc. 37:52–55.

Shannon, F. A.
1965. Scorpions, snakes, and Gila monsters. Arizona Medicine 22:968–74.

Sharp, J. L., and F. H. Haramoto
1970. *Dermatophagoides pteronyssinus* (Trouessart) and other Acarina in house dust in Hawaii. Proc. Hawaiian Entomol. Soc. 20:583–89.

Shelmire, B., and W. E. Dove
1931. The tropical rat mite, *Liponyssus bacoti* (Hirst). J. Amer. Med. Assoc. 96:579–84.

Shields, T. L., and E. N. Walsh
1956. "Kissing bug" bite. Archives of Dermatology 74: 14–21.

Shorey, H. H.
1970. Sex pheromones of Lepidoptera. *In:* Control of insect behavior by natural products (D. L. Wood, R. M. Silverstein, and M. Nakajima, eds.), pp. 249–84. Academic Press, New York and London.
1973. Behavioral responses to insect pheromones. Annual Rev. Entomol. 18:349–80.

Shorey, H. H., R. S. Kaae, and L. K. Gaston
1974. Sex pheromones of Lepidoptera. Development of a method for pheromonal control of *Pectinophora gossypiella* in cotton. J. Econ. Entomol. 67:347–50.

Short, D.
1973. Field evaluation of insecticides for controlling mole crickets in turf. Down to Earth 29(1):26–29.

Shulman, S.
1967. Allergic responses to insects. Annual Rev. Entomol. 12:323–46.

Shuyler, H. R.
1956. Are German and oriental roaches changing their habits? Pest Control 24(9):9–10.

Silveira, S. M., G. W. Force, and D. J. Womeldorf
1964. Field test of dichlorvos resin strips for controlling mosquito larvae in catch basins in Turlock, California. California Vector Views 11:57–59.

Silverborg, S. B.
1953. Fungi associated with the decay of wooden buildings in New York state. Phytopathology 43:20–22.

Silverstein, R. M.
1970. Methodology for isolation and identification of insect pheromones: Examples from Coleoptera. *In:* Control of insect behavior by natural products (D. L. Wood, R. M. Silverstein, and M. Nakajima, eds.), pp. 285–99. Academic Press, New York and London.

Simmons, P.
1927. The cheese skipper as a pest in cured meats. USDA Dept. Bull. No. 1453. 55 pp.

Simmons, P., and G. W. Ellington
1925. The ham beetle, *Necrobia rufipes* De Geer. J. Agr. Res. 30:845–63.

Simmons, P., C. K. Fisher, and J. G. Tyler
1948. Notes on the Apache wasp in California. Annals Entomol. Soc. Amer. 41:450–54.

Simmons, P., W. D. Reed, and E. A. McGregor
1931. Fig insects in California. USDA Circular No. 157. 71 pp.

Simmons, S. W., and H. P. Nicholson
1947. The use of rodenticides 1080 and ANTU. Pests 15(5):8, 10.

Simmons, S. W., and W. M. Upholt
1951. Disease control with insecticides. A review of the literature. WHO Bulletin No. 3:535–56.

Sinha, R. N.
1971. Fungus as food for some stored-product insects. J. Econ. Entomol. 64:3–6.

Sinha, R. N., B. Berck, and H. A. H. Wallace
1967. Effect of phosphine on mites, insects, and microorganisms. J. Econ. Entomol. 60:125–32.

Sinha, R. N., J. E. M. H. van Bronswijk, and H. A. H. Wallace
1970. House dust allergy, mites, and their fungal associations. J. Canadian Med. Assoc. 103:300–301.

Sjogren, R. D.
1971. An effective repellent for *Leptoconops kerteszi* Kieffer. Mosquito News 31:115–16.

Sjogren, R. D., and J. D. Foulk
1967. Colonization studies of *Leptoconops kerteszi,* biting gnats. Mosquito News 27:394–97.

Sjogren, R. D., and M. S. Mulla
1968. Drip application of three organophosphorus insecticides for mosquito control. Mosquito News 28:172–77.

Sjogren, R. D., and R. E. Ryckman
1966. Epizootiology of *Trypanosoma cruzi* in southwestern North America. VIII. Nocturnal flights of *Triatoma protracta* (Uhler) as indicated by collections at black-light traps (Hemiptera: Reduviidae: Triatominae). J. Med. Entomol. 3:81–92.

Skaliy, P., and W. J. Hayes, Jr.
1949. The biology of *Liponyssus bacoti* (Hirst, 1913) (Acarina: Liponyssidae). Amer. J. Trop. Med. 29: 759–72.

Sláma, K.
1971. Insect juvenile hormone analogues. Annual Rev. Biochem. 40:1079–1102.

Sláma, K., and C. M. Williams
1965. Juvenile hormone activity for the bug *Pyrrhocoris apterus.* Proc. Nat. Acad. Sci. (U. S.) 54:411–14.
1966. "Paper factor" as an inhibitor of the embryonic development of the European bug *Pyrrhocoris apterus.* Nature 210:329–30.

Slifer, E. H., J. J. Prestage, and H. W. Beams
1959. The chemoreceptors and other sense organs on the antennal flagellum of the grasshopper (Orthoptera: Acrididae), Journal of Morphology 105: 145–91.

Slominski, J. W., and W. L. Gojmerac
1972. The effect of surfaces on the activity of insecticides. Univ. Wisconsin Coll. Agr. and Life Sci. Res. Rept. 2376. 8 pp.

Smallman, B. N.
1948. Residual insecticides for the control of spider beetles in cereal warehouses. J. Econ. Entomol. 41:869–74.

Smart, J.
1948. Insects of medical importance. (Second edition.) British Museum (Natural History), London. 295 pp.

Smith, A. P.
1971. Hyposensitization with *Dermatophagoides pteronyssinus* antigen: Trial in asthma induced by house dust. British Medical Journal 1971(4): 204–206.

Smith, C. A.
1914. The development of *Anopheles punctipennis* (Say). Psyche 21:1–19.

Smith, C. N., M. N. Cole, and H. K. Gouck
1946. Biology and control of the American dog tick. USDA Tech. Bull. No. 905. 74 pp.

Smith, C. N., G. C. LaBrecque, and A. B. Borkovec
1964. Insect chemosterilants. Annual Rev. Entomol. 9:269–84.

Smith, C. W., and D. W. Micks
1968. A comparative study of the venom and other components of three species of *Loxosceles*. Amer. J. Trop. Med. Hyg. 17:651–56.

Smith, E. B., and T. F. Claypoole
1967. Canine scabies in dogs and in humans. J. Amer. Med. Assoc. 199:59–64.

Smith, E. L.
1970. Biology and structure of some California bristletails and silverfish. Pan-Pacific Entomologist 46: 212–25.

Smith, H. S.
1907. Contribution toward a monograph of the Ceratinidae of North and Middle America. Trans. Amer. Entomol. Soc. 33:115–24.

Smith, J. D., and E. B. Smith
1971. Multiple fire ant stings: A complication of alcoholism. Archives of Dermatology 103:438–41.

Smith, L. M., and H. Lowe
1948. The black gnats of California. Hilgardia 18:157–83.

Smith, M. E., and H. F. Chao
1956. Outdoor roaches. Pest Control 24(9):50.

Smith, M. R.
1965. House-infesting ants of the eastern United States: Their recognition, biology, and economic importance. USDA ERS Tech. Bull. No. 1326. 105 pp.

Smith, R. F.
1944. A note on the hibernation of *Mischocyttarus flavitarsis* in a nest of *Dolichovespula arenaria*. Pan-Pacific Entomologist 20:80.
1972. The impact of the green revolution on plant protection in tropical and subtropical areas. Entomol. Soc. Amer. Bull. 18:7–14.

Smith, R. F., and W. W. Allen
1954. Insect control and the balance of nature. Scientific American 190(6):38–42.

Smith, T. A.
1969. The maturation of fly larvae following removal from the larval medium. California Vector Views 16:73–78.

Smith, V. K., Jr.
1968. Lindane controls *Lyctus* powder-post beetles during air-drying of rough ash lumber. J. Econ. Entomol. 61:323–24.

Smith, V. K., Jr., R. H. Beal, and H. R. Johnston
1972. Twenty-seven years of termite control tests. Pest Control 40(6):28, 42, 44.

Smith, V. K., Jr., and H. R. Johnston
1962. Eastern subterranean termite. USDA Forest Pest Leaflet No. 68. 7 pp.

Smith, W. W., and V. G. Perry
1967. Intersexes in *Culicoides* spp. caused by mermithid parasitism in Florida. J. Econ. Entomol. 60:1025–27.

Smythe, R. V., T. C. Allen, and H. C. Coppel
1965. Response of the eastern subterranean termite to an attractive extract from *Lenzites trabea*-invaded wood. J. Econ. Entomol. 58:420–23.

Smythe, R. V., and F. L. Carter
1970. Feeding responses to sound wood by *Coptotermes formosanus, Reticulitermes flavipes,* and *R. virginicus* (Isoptera: Rhinotermitidae). Annals Entomol. Soc. Amer. 63:841–47.

Smythe, R. V., F. L. Carter, and C. C. Baxter
1971. Influence of wood decay on feeding and survival of the eastern subterranean termite, *Reticulitermes flavipes*. Annals Entomol. Soc. Amer. 64: 59–62.

Smythe, R. V., and H. C. Coppel
1966*a*. Some termites may secrete trail-blazing attractants to lead others to food sources. Pest Control 34(10): 73–78.
1966*b*. A preliminary study of the sternal gland of *Reticulitermes flavipes* (Isoptera: Rhinotermitidae). Annals Entomol. Soc. Amer. 59:1008–10.

Smythe, R. V., H. C. Coppel, and T. C. Allen
1967*a*. The responses of *Reticulitermes* spp. and *Zootermopsis angusticollis* (Isoptera) to extracts from woods decayed by various fungi. Annals Entomol. Soc. Amer. 60:8–9.

Smythe, R. V., H. C. Coppel, S. H. Lipton, and F. M. Strong
1967*b*. Chemical studies of attractants associated with *Reticulitermes flavipes* and *R. virginicus*. J. Econ. Entomol. 60:228–33.

Snell, J. B., and W. E. Hazeltine
1963. Use of insecticide to determine the life history of an aquatic gnat, *Chaoborus astictopus*. Annals Entomol. Soc. Amer. 56:816–18.

Snelling, R. R.
1954. Wasps of the genus *Polistes* in California and Arizona (Hymenoptera: Vespidae). J. Kansas Entomol. Soc. 27:151–55.
1970. Ant warfare offensive and defensive maneuvers. Los Angeles County Museum Quarterly (spring issue), pp. 10–17.

Snetsinger, R.
1960. Some predators of the clover mite. Entomological News 71:149–56.
1967. Some answers to the clover mite problem. Pest Control 35(7):15–18, 56.

Snodgrass, R. E.
1935. Principles of insect morphology. McGraw-Hill, New York. 667 pp.
1944. The feeding apparatus of biting and sucking insects affecting man and animals. Smithsonian Misc. Coll. (Publ. No. 3773) 102(7):113 pp.
1949. The anatomy of the honey bee. *In:* The hive and the honey bee (R. A. Grout, ed.), pp. 473–518. Dadant and Sons, Hamilton, Illinois.
1956. Anatomy of the honey bee. Comstock Publishing Co., Inc. Ithaca, New York. 334 pp.
1967. Insects: Their ways and means of living. Dover Publications, New York. 362 pp.

Snyder, T. E.
1926. Preventing damage by *Lyctus* powder-post beetles. USDA Farmers' Bull. No. 1477. 12 pp.
1935. Our enemy, the termite. (Revised 1948.) Comstock Publishing Co., Inc., Ithaca, New York. 257 pp.
1955*a*. Introduced wood-borers. Pest Control 23(1):28.
1955*b*. Termite attacks on plastics and fabrics. Pest Control 23(3):48, 56.

Solomon, M. E.
1946. Tyroglyphid mites in stored products: Ecological studies. Annals Appl. Biol. 33:82–97.
1962. Ecology of the flour mite (*Acarus siro* L.). Annals Appl. Biol. 50:178–84.

Southern, H. N., and E. M. O. Laurie
1946. The house-mouse (*Mus musculus*) in corn ricks. Journal of Animal Ecology 15:135–49.

Sparrow, H.
1955. Foyer de fièvre récurrente à poux en Éthiope. Compt. Rend. Acad. Sci. Paris 241:1636–39.

Spear, P. J.
1952. Insecticide vaporizers. Soap and Sanit. Chem. 28(8):147–48, 173.
1954. Some problems in the control of clover mites. Chem. Spec. Mfr. Assoc., Proc. Annual Meeting 41:123–24.
1968. Current bird control measures. Cereal Science Today 13(7):270–72, 284.

Spencer, G. J.
1930. The status of the barn swallow bug, *Oeciacus vicarius*. Canadian Entomologist 62:20–21.
1942. Insects and other arthropods in buildings in British Columbia. Proc. Entomol. Soc. British Columbia 39:23–29.

Spieksma, F. Th. M.
1967. The house dust mite *Dermatophagoides pteronyssinus* (Trouessart, 1897), producer of the house dust allergen (Acari: Psoroptidae). Thesis, Univ. of Leiden, the Netherlands. 65 pp.

Spieksma, F. Th. M., and M. I. A. Spieksma-Boezeman
1967. The mite fauna of house dust, with particular reference to the house dust mite *Dermatophagoides pteronyssinus* (Trouessart, 1897) (Psoroptidae: Sarcoptiformes). Acarologia 9:226–41.

Spieksma, F. Th. M., P. Zuidema, and M. J. Leupen
1971. High altitude and house dust mites. British Medical Journal 1971(1):82–84.

Spielman, A., and H. W. Levi
1970. Probable envenomation by *Chiracanthium mildei*, a spider found in houses. Amer. J. Trop. Med. Hyg. 19:729–32.

Spiller, D.
1949. Effect of humidity on hatching of eggs of the common house borer, *Anobium punctatum*. Rev. Appl. Entomol. A-38:263.
1964. Numbers of eggs laid by *Anobium punctatum* (De Geer). Bull. Entomol. Res. 55:305.

Spink, W. T., H. R. Gross, and L. D. Kirst
1966. *Xyletinus peltatus* in structural timbers. Pest Control 34(2):12–13, 15, 44.

Spitler, G. H.
1969. Laboratory evaluation of malathion as a storage protectant for figs. J. Econ. Entomol. 62:754–56.

Spitler, G. H., and J. D. Clark
1970. Laboratory evaluation of malathion as a protectant for prunes during storage. J. Econ. Entomol. 63:1668–69.

Spitler, G. H., and P. L. Hartsell
1967. Laboratory evaluation of malathion as a protectant for almonds during storage. J. Econ. Entomol. 60:1456–58.
1969. Laboratory evaluation of malathion as a protectant for stored walnuts. J. Econ. Entomol. 62:305–307.
1970. Laboratory evaluation of malathion as a protectant for natural raisins. J. Econ. Entomol. 63:1502–1505.

Spitz, W. J.
1958. How we get complete coverage of subslab soil with termite chemicals. Pest Control 26(2):38, 40, 43, 51.

Springhetti, A.
1970. Influence of the king and queen on the differentiation of soldiers in *Kalotermes flavicollis* (Fabr.). Monitore Zool. Ital. 4:99–105.

Stahnke, H. L.
1948. Scorpions. Arizona State College (Tempe) Bull., Gen. Ser. 71. 16 pp.
1956*a*. A new species of scorpion of the Buthidae: *Centruroides pantheriensis*. Entomological News 67: 15–19.
1956*b*. Scorpions. Arizona State College (Tempe), Poison Animals Research Laboratory. 40 pp.
1965. Stress and toxicity of venoms. Science 150:1456–57.
1966. The treatment of venomous bites and stings. Arizona State University (Tempe). 117 pp.
1971. Some observations of the genus *Centruroides* Marx (Buthidae: Scorpionida) and *C. sculpturatus* Ewing. Entomological News 82:281–307.

Stanger, W.
1967. How to remove bees from buildings. Univ. California Agr. Ext. Serv., One Sheet Answer No. 161.

Stark, R. W., P. R. Miller, F. W. Cobb, Jr., D. L. Wood, and J. R. Parmeter, Jr.
1968. Photochemical oxidant injury and bark beetle (Coleoptera: Scolytidae) infestation of ponderosa pine. I. Incidence of bark beetle infestation in injured trees. Hilgardia 39:121–26.

Starks, K. J., and R. Thurston
1962. Silver top of bluegrass. J. Econ. Entomol. 55: 865–67.

Stedman, E., and G. Barger
1925. Physostigmine (eserine). J. Chem. Soc. (London) 127:247–58.

Steelman, C. D., and A. J. Poché
1970. Insecticide–fertilizer mixtures for controlling the dark ricefield mosquito. J. Econ. Entomol. 63: 1986–87.

Steinberg, M., M. M. Cole, E. S. Evans, J. T. Whitlaw, and J. J. Yoon
1971. Toxicological and entomological field evaluation of the effects of Mobam® powder against body lice (Anoplura: Pediculidae). J. Med. Entomol. 8:68–72.

Steinberg, M., M. M. Cole, T. A. Miller, and R. A. Godke
1972. Toxicological and entomological field evaluation of Mobam® and Abate® powders used as body-louse toxicants (Anoplura: Pediculidae). J. Med. Entomol. 9:73–77.

Steinbrecht, R. A., and B. Müller
1971. On the stimulus-conducting structures in insect olfactory receptors. Zeitschrift für Zellforschung 117:570–75.

Steiner, L. F., W. C. Mitchell, E. J. Harris, T. T. Kozuma, and M. S. Fujimoto
1965. Oriental fruit fly eradication by male annihilation. J. Econ. Entomol. 58:961–64.

Steiner, L. F., G. G. Rohwer, E. L. Ayers, and L. D. Christensen
1961. The role of attractants in the recent Mediterranean fruit fly eradication program in Florida. J. Econ. Entomol. 54:30–35.

Steinhaus, E. A.
1946. Insect microbiology. Comstock Publishing Co., Inc., Ithaca, New York. 776 pp.

Steinweden, J. B.
1945. Methyl bromide fumigation. California Dept. Agr. Bull. 34:4–14.

Stenburg, R. L.
1947. The techniques of application and the control of roaches and bedbugs with DDT. Pests 15(8):16–24.

Stephen, W. P.
1957. Bumble bees of western America (Hymenoptera: Apoidea). Oregon Agr. Expt. Sta. Tech. Bull. 40. 163 pp.

Stephen, W. P., G. E. Bohart, and P. F. Torchio
1969. The biology and external morphology of bees. Oregon Agr. Expt. Sta., Corvallis. 140 pp.

Sterling, H. R.
1970. Insecticide evaluation and population study with the brown recluse spider. M. S. thesis, Dept. of Entomol., Oklahoma State Univ.

Steve, P. C.
1960. Biology and control of the little house fly, *Fannia canicularis,* in Massachusetts. J. Econ. Entomol. 53:999–1004.

Stevenson, D. A., and K. P. Mathews
1967. Occupational asthma following inhalation of moth particles. Journal of Allergy 39:274–83.

Stewart, D.
1957. Sulfuryl fluoride: A new fumigant for control of the drywood termite *Kalotermes minor* Hagen. J. Econ. Entomol. 50:7–11.
1966. Balanced fumigation for better termite control. Down to Earth 22(2):8–10.

Stewart, M. A.
1932. Dispersal of the sticktight fleas of hens (*Echidnophaga gallinacea* Westw.). J. Econ. Entomol. 25:164–67.

St. George, R. A.
1970. Protecting log cabins, rustic work, and unseasoned wood from injurious insects in the eastern United States. USDA Farmers' Bull. No. 2104. 18 pp.

St. George, R. A., H. R. Johnston, and R. J. Kowal
1963. Subterranean termites, their prevention and control in buildings. USDA Home and Garden Bull. No. 64. (Revised.) 30 pp.

St. George, R. A., H. R. Johnston, and T. McIntyre
1957. Old-house borer. Pest Control 25(2):29–31.

Stickney, F. S., D. F. Barnes, and P. Simmons
1950. Date palm insects in the United States. USDA Circular No. 846:1–57.

Stillwell, M. A.
1966. Woodwasps (Siricidae) in conifers and the associated fungus, *Stereum chailletii,* in eastern Canada. Forest Science 12:121–28.
1967. The pigeon tremex, *Tremex columba* (Hymenoptera: Siricidae), in New Brunswick. Canadian Entomologist 99:685–89.

Stombler, V., C. D. Pelekassis, and E. S. Doyle
1957. *Drosophila* control on harvested tomatoes for processing in California. J. Econ. Entomol. 50: 476–80.

Stone, A.
1930. The bionomics of some Tabanidae (Diptera). Annals Entomol. Soc. Amer. 23:261–304.

Stone, A., K. L. Knight, and H. Starcke
1959. A synoptic catalog of the mosquitoes of the world. Published by the Entomological Society of America. 358 pp.

Stone, A., C. W. Sabrosky, W. W. Wirth, R. H. Foote, and J. R. Coulson
1965. A catalog of the Diptera of America north of Mexico. USDA Agr. Handbook No. 276. 1696 pp.

Storer, T. I.
1931. Community nests of the house mouse. Journal of Mammalogy 12:317.
1948. Control of rats and mice. Univ. California Agr. Ext. Serv. Circ. 142.
1968. How to control rats and mice. Univ. California Agr. Expt. Sta. Ext. Serv. Leaflet 127. (Replaces former Circular No. 410.) 28 pp.

Storer, T. I., and D. E. Davis
1953. Studies on rat reproduction in San Francisco. Journal of Mammalogy 34:365–73.

Storer, T. I., and R. L. Usinger
1957. General zoology. McGraw-Hill, New York. 664 pp.

Storm van Leeuwen, W.
1922. Oorzaken en behandeling van asthma bronchiale, hooikoorts, en verwante toestanden in verband met "overgevoeligheid" tegen proteinen en stoffen van bekende chemische samenstelling. [Causes and treatment of bronchial asthma, hay fever, and related conditions in connection with "oversensitiveness" to proteins and substances of known chemical structure.] J. B. Wolters, Groningen, The Hague, Netherlands. 209 pp.

Street, T. M.
1956. Bats and their control in California. California Vector Views 3:19, 21–22.

Strickland, W. B.
1969. The occurrence of *Aedes sierrensis* larvae in stump cavities of the California coastal redwood, *Sequoia sempervirens.* California Vector Views 16: 33–34.

Strong, R. G.
1970. Distribution and relative abundance of stored-product insects in California: A method of obtaining sample populations. J. Econ. Entomol. 63:591–96.

Strong, R. G., and G. T. Okumura
1958. Insects and mites associated with stored foods and seeds in California. California Dept. Agr. Bull. 47:233–49.

Stuart, A. M.
1961. Mechanism of trail-laying in two species of termites. Nature 189:419.
1963. Origin of the trail in the termites *Nasutitermes corniger* (Motschulsky) and *Zootermopsis nevadensis* (Hagen) (Isoptera). Physiological Zoology 36:69–84.
1964. The structure and function of the sternal gland of *Zootermopsis nevadensis* (Isoptera). Proc. Zool. Soc. London 143:43–52.
1967. Alarm, defense, and construction behavior relationships in termites (Isoptera). Science 156:1123.
1969. Social behavior and communication. *In:* Biology of termites (K. Krishna and F. M. Weesner, eds.), Vol. I, pp. 193–232. Academic Press, New York.

Stuart, A. M. and P. Satir
1968. Morphological and functional aspects of an insect epidermal gland. J. Cell. Biol. 36:527–49.

Subcommittee on Insect Pests
1969. Insect-pest management and control. National Academy of Sciences, Washington, Publ. 1695.

Sudd, J. H.
1952. Laboratory studies of adult female *Bdellonyssus bacoti* (Hirst, 1916). Annals Trop. Med. and Parasitol. 46:158–64.

Sudia, W. D., and V. F. Newhouse
1971. Venezuelan equine encephalitis in Texas, 1971: Informational report. Mosquito News 31:350–51.

Summers, F. M., and D. W. Price
1970. Review of the mite family Cheyletidae. Univ. California Publ. Entomol. 61:1–153.

Sutherland, K. H.
1965. Los Angeles County Health Department history and functions. Los Angeles County Health Department, Division of Public Health Education.

Sutton, R. L., and R. L. Sutton, Jr.
1939. Diseases of the skin. C. V. Mosby Co., St. Louis, Missouri. 1549 pp.

Swarts, W. B., and J. F. Wanamaker
1946. Skin blisters caused by vesicant beetles. J. Amer. Med. Assoc. 131:594–95.

Sweetman, H. L.
1938. Physical ecology of the firebrat, *Thermobia domestica* (Packard). Ecological Monographs 8:285–311.
1939. Responses of the silverfish, *Lepisma saccharina* L., to its physical environment. J. Econ. Entomol. 32:698–700.
1958. Principles of biological control. Wm. C. Brown Co., Dubuque, Iowa. 560 pp. (*See:* Chapter 13.)
1965. Recognition of structural pests and their damage. Wm. C. Brown Co., Dubuque, Iowa. 371 pp.

Sweetman, H. L., and W. M. Kulash
1944. The distribution of *Ctenolepisma urbana* Slabaugh and certain other Lepismatidae. J. Econ. Entomol. 37:444.

Swingle, M. C.
1938. Low temperatures as possible means of controlling cigarette beetles in stored tobacco. USDA Circular No. 462. 8 pp.

Taboada, M. de F.
1954. Pulmonary acariasis in Spain: An illustrative case report. British Medical Journal 1954(4859):437–38.

Tai, A., F. Matsumura, and H. C. Coppel
1971. Synthetic analogues of the termite trail-following pheromone, structure and biological activity. J. Insect Physiol. 17:181–88.

Tarshis, I. B.
1961. Laboratory and field studies with sorptive dusts for the control of arthropods affecting man and animals. Experimental Parasitology 11:10–33.

Taub, S. J.
1970. Severe asthma attacks caused by gnats (*Simulium jenningsi*) and boxelder bugs (*Leptocoris trivittatus*). Eye, Ear, Nose, and Throat Monthly 49:475.

Taylor, R. T., and H. F. Schoof
1968. Evaluation of thermal and nonthermal fogs against four species of mosquitoes. Mosquito News 28:8–11.

Telford, A. D.
1957. The pasture *Aedes* of central and northern California. The egg stage: gross embryology and resistance to desiccation. Annals Entomol. Soc. Amer. 50:537–43.
1958. The pasture *Aedes* of central and northern California: Seasonal history. Annals Entomol. Soc. Amer. 51:360–65.
1963. A consideration of diapause in *Aedes nigromaculis* and other aedine mosquitoes (Diptera: Culicidae). Annals Entomol. Soc. Amer. 56:409–18.

Terriere, L. C.
1968. Insecticide–cytoplasmic interactions in insects and vertebrates. Annual Rev. Entomol. 13:75–98.

Thomas, C. M., and R. J. Dicke
1971. Response of the grain mite, *Acarus siro* (Acarina: Acaridae), to fungi associated with stored-food commodities. Annals Entomol. Soc. Amer. 64:63–68.

Thomas, L. A.
1963. Distribution of the virus of western equine encephalomyelitis in the mosquito vector *Culex tarsalis*. Amer. J. Hyg. 78:150–65.

Thompson, A. H.
1967. A low-cost airboat for use in mosquito control. Mosquito News 27:417–19.

Thompson, A. H., C. L. Barnes, and D. A. Mathews
1970. Injection of Dursban spray emulsion at half-mile intervals controls mosquitoes and chironomid larvae in large drainage channels. Presented at Thirty-Eighth Annual Conference, California Mosquito Control Association, Sacramento. (January.) (Mimeographed.)

Thompson, F. M.
1932. Pine oils as agents for protecting rustic furniture and log cabins from various wood-borers. J. Econ. Entomol. 25:347–51.

Thompson, G. A.
1968. A note on fog application. Mosquito News 28:643.

Thompson, W. H., D. O. Trainer, V. Allen, and J. B. Hale
1963. The exposure of wildlife workers in Wisconsin to the zoonotic diseases. Trans. N. Amer. Wildlife Nat. Res. Conf. 28:215.

Thorp, R. W.
1969. The identity of *Bombus vandykei* (Hymenoptera: Apidae). Pan-Pacific Entomologist 45:87–96.

Tierkel, E. S.
1963. Rabies. *In:* Diseases transmitted from animal to man (fifth edition, T. G. Hull, ed.), 907 pp. Charles C. Thomas, Springfield, Illinois.

Tilden, J. W.
1969. Common pest ants. *In:* Community pest and related vector control (J. E. Brooks and T. D. Peck, eds.), pp. 72–81. Pest Control Operators of California, Los Angeles.

Tillyard, R. J.
1926. Insects of Australia and New Zealand. Angus and Robertson, Sydney. 560 pp.
1936. Are termites descended from true cockroaches? Nature 137:655.
1937. Kansas Permian insects. XX. The cockroaches, order Blattaria. Amer. J. Sci. 34:169–202.

Times (L. A.)
1972. Flies "drinking" selves to death. Los Angeles *Times,* January 16, Section A, page 1.

Ting, P. C.
1951. An African earwig in California (Dermaptera: Labiduridae). California Dept. Agr. Bull. 40:4–9.

Tompkins, G. J., and G. E. Cantwell
1973. The use of dry ice to control German cockroaches in hospital food service carts. Pest Control 41(11): 24, 26.

Tooke, F. G. C.
1949. Beetles injurious to timbers in South Africa. South Africa Dept. Agr. Sci. Bull. 293:28–39, 72–88.

Torchio, P. F.
1965. Observations on the biology of *Colletes ciliatoides* (Hymenoptera: Apoidea: Colletidae). J. Kansas Entomol. Soc. 38:182–87.

Torsney, P. J.
1968. Allergic reactions to insect stings. New York State J. Med. 68:2765–68.

Townsend, C. H. T.
1893. Note on *Ceuthophilus* eating curtains and other fabrics. Insect Life 6:58.

Trager, W.
1932. A cellulose from the symbiotic intestinal flagellates of termites and the roach *Cryptocercus punctulatus.* Biochemical Journal 26:1762–71.

Traver, J. R.
1951. Unusual scalp dermatitis in humans caused by the mite *Dermatophagoides.* Proc. Entomol. Soc. Washington 53:1–25.

Travis, B. V., and S. M. Schuchman
1968. Tests (1967) with black fly larvicides. J. Econ. Entomol. 61:843–45.

Travis, B. V., A. L. Smith, and A. H. Madden
1951. Effectiveness of insect repellents against black flies. J. Econ. Entomol. 44:813–19.

Treat, A. E.
1969. Behavioral aspects of the association of mites with noctuid moths. Proc. Second Intern. Congr. Acarol. (1967), pp. 275–86.

Triplehorn, C. A.
1965. Insects found in Ohio grain elevators and feed mills. J. Econ. Entomol. 58:578–79.

Truman, L. C.
1961*a.* Cockroaches. Pest Control 29(6):21–28.
1961*b.* Birds and other vertebrates. Pest Control 29(9): 29–35.

Truman, L. C., and W. L. Butts
1967. Scientific guide to pest control operations. Published by Pest Control Magazine, Cleveland, Ohio.

Tsai, Y.-H., and K. M. Cahill
1970. Parasites of the German cockroach, *Blattella germanica* (L.), in New York city. Journal of Parasitology 56:375–77.

Tsuji, H., and T. Mizuno
1971. Laboratory use of first-instar nymphs of the American cockroach, *Periplaneta americana.* Japanese J. Sanit. Zool. 22:1–7.

Tsuji, H., and S. Ono
1970*a.* Laboratory evaluation of several bait factors against the German cockroach, *Blattella germanica* (L.). Japanese J. Sanit. Zool. 20:240–47.
1970*b.* Glycerol and related compounds as feeding stimulants for cockroaches. Japanese J. Sanit. Zool. 21:149–55.

Tsukamoto, M.
1969. Biochemical genetics of insecticide resistance in the housefly. Residue Reviews 25:289–314.

UCAE (University of California Agricultural Extension)
1971–72. Fly control on poultry ranches. UCAE AXT-72.

Uhlarik, J.
1970. What makes an anticoagulant safe? Pest Control 38(11):18.

Uhler, P. R.
1884. Order IV: Hemiptera. *In:* Standard Natural History 2:204–96.

Unsworth, K.
1946. Studies on the clinical and parasitological aspects of canine demodectic mange. J. Comp. Pathol. Therap. 56:114–27.

USDA (United States Department of Agriculture)
1954. How to spray the aircraft way. USDA Farmers' Bull. No. 2062. 31 pp.
1962. Granular insecticides: Research developments and uses. Agr. Res. Serv. Special Rept. No. 22-78.
1963. Chiggers: How to fight them. USDA Leaflet 403.
1966. Protecting woolens against clothes moths and carpet beetles. USDA Home and Garden Bull. No. 113. 8 pp. (Revised in 1970.)
1967. Safe use of agricultural and household pesticides. USDA Handbook No. 321. 65 pp.
1968. Stored-grain pests. USDA Farmers' Bull. No. 1260 (revised; originally 1922). 46 pp.

Usinger, R. L.
1944. The Triatominae of North and Central America and the West Indies and their public health significance. Public Health Reports (USPHS) Bull. No. 288. 83 pp.
1966. Monograph of Cimicidae. Thomas Say Foundation, Vol. VII. Published by the Entomological Society of America. 585 pp.

Usinger, R. L., and V. R. Kellen
1955. The role of insects in sewage disposal beds. Hilgardia 23:263–321.

USPHS (United States Public Health Service)
1948. Rodent control manual. Typhus Control Unit, USPHS.

Utagawa, T.
1971. Effects of environmental pollution on birds and animals. Kogai to Taisaku 7(1):35–39. *Reviewed in:* Health Aspects of Pesticides 5(1):5 (1972).

Uvarov, B. P.
1928. Locusts and grasshoppers. Imperial Bur. Entomol., London. 352 pp.

Vanat, S. V., and I. M. Vanat
1971. Contribution to the toxic–allergic reaction induced by DDT. (Translated.) Klin. Med. (Moscow) 49:126–27. *From:* Health Aspects of Pesticides 5:123.

van Bronswijk, J. E. M. H.
1973*a*. *Dermatophagoides pteronyssinus* (Trouessart, 1897) in mattress and floor dust in a temperate climate (Acari: Pyroglyphidae). J. Med. Entomol. 10:63–70.
1973*b*. Hausstaub-Oekosystem und Hausstaub-Allergene. Acta Allergologica 27:219–28.

van Bronswijk, J. E. M. H., and H. M. M. Koekkoek
1971. Nipagin (*p*-methyl hydroxy benzoate) as a pesticide against a house dust mite: *Dermatophagoides pteronyssinus*. J. Med. Entomol. 8:748.
1972. Effects of low temperatures on the survival of house dust mites of the family Pyroglyphidae (Acari: Sarcoptiformes). Netherlands J. Zool. 22: 207–12.

van Bronswijk, J. E. M. H., J. M. C. P. Schoonen, M. A. F. Berlie, and F. S. Lukoschus
1971. On the abundance of *Dermatophagoides pteronyssinus* (Trouessart, 1897) (Pyroglyphidae: Acarina) in house dust. Res. Popul. Ecol. 13:67–79.

van Bronswijk, J. E. M. H., and R. N. Sinha
1971. Pyroglyphid mites (Acari) and house dust allergy. Journal of Allergy 47:31–52.

Vandel, A.
1939. Sur le mode de répartition des sexes chez l'isopode terrestre, *Armadillidium vulgare* (Latr.). Compt. Rend. Acad. Sci. Paris 208:1050–52.

Varma, M. G. R.
1962. Transmission of relapsing fever spirochaetes by ticks. Symp. Zool. Soc. London 6:61–82.

Vasvary, L. M.
1965. Boxelder bug control. Rutgers Coll. Agr. Ext. Leaflet No. 371.

Vaughan, T. A.
1970. The skeletal system. *In:* Biology of bats (W. A. Wimsatt, ed.), Vol. I, pp. 97–138. Academic Press, New York and London.

Venkatachalam, P. S., and B. Belavady
1962. Loss of hemoglobin iron due to excessive biting by bedbugs. Trans. Roy. Soc. Trop. Med. Hyg. 56:218–21.

Verhoeff, K. W.
1919. Ueber die Larven, das Marsupium, und die Bruten der Oniscoidea (27 Isopoden-Aufsatz). Archiv für Naturgeschichte 83A(12):1–54.
1920. Zur Kenntniss der Larven, des Brutsackes, und der Bruten der Oniscoidea (28 Isopoden-Aufsatz). Zoologischer Anzeiger 51:169–89.

Verrall, A. F.
1968. Keep wood dry or use preservatives to control decay fungi in structures. Pest Control 36(4):54, 69–70.

Vianna, N., E. Whitney, T. Bast, R. Deibel, J. Doll, and J. Culver
1971. California encephalitis in New York state. Amer. J. Epidem. 94:50–55.

Vick, J. A., and R. B. Brooks, Jr.
1972. Pharmacological studies of the major fractions of bee venom. American Bee Journal 112:288–89.

Vinson, J. W., and C. M. Williams
1967. Lethal effects of synthetic juvenile hormone on the human body louse. Proc. Nat. Acad. Sci. (U. S.) 58:294–97.

Vité, J. P.
1970. Pest management systems using synthetic pheromones. Boyce Thompson Inst. Contrib. 24:343–50.

von Frisch, K.
1948. Gelöste und ungelöste Rätsel der Bienensprache. Naturwissenschaften 35:12–23, 38–43.
1950. Die Sonne als Kompass im Leben der Bienen. Experientia 6:210–21.
1960. Ten little housemates. [Translated from the fourth edition (1955) of *Zehn Kleine Hausgenossen.*] Macmillan, New York, 146 pp.
1967. The dance language and orientation of bees. [Translated from *Tanzsprache und Orientierung der Bienen* (J. Springer, Berlin, 1965) by L. E. Chadwick.] Belknap-Harvard University Press, Cambridge, Massachusetts. 566 pp.

von Frisch, K., and M. Lindauer
1955. Ueber die Fluggeschwindigkeit der Bienen und über ihre Richtungsweisung bei Seitenwind. Naturwissenschaften 42:377–85.

von Frisch, K., M. Lindauer, and F. Schmiedler
1960. Wie erkennt die Biene den Sonnenstand bei geschlossener Wolkendecke? Naturwissenschäftlicher Rundschau 10:1–7.

von Geldern, C. E.
1927. Systemic effects following the sting of a species of *Epyris.* Science 65:302–303.

von Rümker, R., H. R. Guest, and W. M. Upholt
1970. The search for safer, more selective, and less persistent pesticides. BioScience 20:1004–1007.

Voorhorst, R.
1969. Further research on house-dust atrophy in relation to the house-dust mite. Entomology Abstracts 2(4):80.
1970. Quantitative aspects of the problems of house-dust atrophy and house-dust mites. Acta Allergologica 25:237–54.

Voorhorst, R., and F. Th. M. Spieksma
1969. Recent progress in the house dust mite problem. Acta Allergologica 24:115–23.

Voorhorst, R., M. I. A. Spieksma-Boezeman, and F. Th. M. Spieksma
1964. Is a mite (*Dermatophagoides* sp.) the producer of the house dust allergen? Allergie Asthma 10:329.

Vrat, V.
1949. Reproductive behavior and development of eggs of the three-spined stickleback (*Gasterosteus aculeatus*) of California. Copeia 1949:252–60.

Wade, W. H.
1961. Biology of the navel orangeworm, *Paramyelois transitella* (Walker), on almonds and walnuts in northern California. Hilgardia 31:129–71.

Wagner, R. E.
1961. Control of the yellowjacket, *Vespula pensylvanica,* in public parks. J. Econ. Entomol. 54:628–30.

Wagner, R. E., and W. Ebeling
1959. Lethality of inert dust materials to *Kalotermes minor* Hagen and their role as preventives in structural pest control. J. Econ. Entomol. 52: 208–12.

Wagner, R. E., W. Ebeling, and W. R. Clark
1964. An electric barrier for confining cockroaches in large rearing or field collecting cans. J. Econ. Entomol. 57:1007–1009.

Wagner, R. E., W. Ebeling, and D. A. Reierson
1966. Control of cockroaches in sewers. Public Works 97(1):82–84.

Wagner, R. E., and D. A. Reierson
1969. Yellowjacket control by baiting. I. Influence of toxicants and attractants on bait acceptance. J. Econ. Entomol. 62:1192–97.
1971. Recognizing and controlling yellowjackets around buildings. National Pest Control Operators News 31(11):6–7, 30–32.

Waldron, W. G.
1962. The role of the entomologist in delusory parasitosis (entomophobia). Entomol. Soc. Amer. Bull. 8:81–83.
1963. The problem of delusory parasitosis (entomophobia) in arthropod control work. Proc. and Papers, Thirty-First Annual Conference, California Mosquito Control Assoc., pp. 75–76.
1965. Observations on spider bites in southern California. California Vector Views 12:66–67.
1969*a*. *Loxosceles laeta* (Nicolet), an introduced species in Los Angeles County. Entomol. Soc. Amer. Bull. 15(4):377–79.
1969*b*. Rattlesnakes. *In:* Community pest and related vector control (J. E. Brooks and T. D. Peck, eds.), pp. 186–89. Pest Control Operators of California, Los Angeles.

Waldron, W. G., and F. Hall
1972. Mode of entry into Los Angeles County, California, of the brown cockroach, *Periplaneta brunnea* Burmeister. California Vector Views 19:1–2.

Walker, P. T.
1971. The use of granular pesticides from the point of view of residues. Residue Reviews 40:65–131.

Wall, W. J., Jr.
1953. Damage to carpet materials by silverfish. J. Econ. Entomol. 46:1121–22.

Walsh, B. D., and C. V. Riley
1869. Parasites of the human animal. American Entomologist 1:84–88.

Walsh, J. D., and J. P. Jones, Jr.
1962. Public health significance of the cone-nosed bug, *Triatoma protracta* (Uhler), in the Sierra Nevada foothills of California. California Vector Views 9:33–37.

Walsh, J. D., and E. E. Lusk
1971. Rodent control prior to the closing of dumps. California Vector Views 18:77–80.

Walter, V.
1968. A roach by any other name. Pest Control 36(2):32.

Wang, J. S., and T. S. Brook
1970. Toxicological and biological studies of the odorous house ant, *Tapinoma sessile.* J. Econ. Entomol. 63:1971–73.

Warburton, C.
1920. Sarcoptic scabies in man and animals. Parasitology 12:265–300.

Ware, G. W.
1966. Power-mower flies. J. Econ. Entomol. 59:477–78.

Wasti, S. S., and F. R. Shaw
1971. Encapsulation: A step toward safer insecticides for use in animals. BioScience 21:118–19.

Watanabe, T., and J. E. Casida
1963. Response of *Reticulitermes flavipes* to fractions from fungus-infected wood and synthetic chemicals. J. Econ. Entomol. 56:300–307.

Waterhouse, D. F.
1958. Wool digestion and moth-proofing. Adv. Pest Control Res. 2:207–62. Interscience Publishers, New York.

Watt, M. J.
1965. Notes on pests of stored grain, *Liposcelis bostrychophilus* and *Sitophilus* spp. Agr. Gaz. New South Wales 76:693–96.

Webb, J. L., and R. H. Hutchison
1916. A preliminary note on the bionomics of *Pollenia rudis* (Fabr.) in America. Proc. Entomol. Soc. Washington 18:197–99.

Webster, L. T.
1942. Rabies. Macmillan, New York. 168 pp.

Weesner, F. M.
1956. The biology of colony foundation in *Reticulitermes hesperus* Banks. Univ. California Publ. Zool. 61:253–306.
1965. The termites of the United States. National Pest Control Association, Inc., Elizabeth, New Jersey. 70 pp.

Weir, R. J., and R. S. Fisher
1972. Toxicologic studies on borax and boric acid. Toxicol. and Appl. Pharmacol. 23:351–64.

Weis-Fogh, T.
1964. Diffusion in insect wing muscle, the most active tissue known. J. Exptl. Biol. 41:229–56.

Welch, H. E.
1960. Notes on the identities of mermithid parasites of North American mosquitoes, and a redescription of *Agamomermis culicis* Stiles, 1903. Helminth. Soc. Washington 27:203–206.

Weller, B., and G. M. Graham
1930. Relapsing fever in central Texas. J. Amer. Med. Assoc. 95:1834–35.

Wene, G. P.
1969. Control of the *Polistes* wasp. PCO News 29(4):6.

Went, F. W., J. Wheeler, and G. C. Wheeler
1972. Feeding and digestion in some ants (*Veromessor* and *Manica*). BioScience 22:82–88.

West, Irma
1959. Health hazards of the agricultural aircraft industry in the application of toxic pesticides. California Vector Views 6:13–19.

West, L. S.
1951. The housefly: Its natural history, importance, and control. Comstock Publishing Co., Inc., Ithaca, New York. 584 pp.

Wharton, G. W.
1970. Mites and commercial extracts of house dust. Science 167:1382–83.

Wharton, G. W., and H. S. Fuller
1952. A manual of the chiggers. Mem. Entomol. Soc. Washington No. 4. 185 pp.

Wheeler, C. M., and M. Thebault
1971. Efficacy of synthetic substances with hormonal activity against IV-instar larvae of *Culex pipiens quinquefasciatus* in an outdoor insectary. Mosquito News 31:170–77.

Wheeler, G. C., and J. Wheeler
1973. Ants of Deep Canyon. Univ. California, Riverside.

Wheeler, W. M.
1906. The habits of the tent-building ant, *Crematogaster lineolata* (Say). Amer. Mus. Nat. Hist. Bull. 22:1–18.
1930. The ant *Prenolepis imparis* Say. Annals Entomol. Soc. Amer. 23:1–26.

White, R. E.
1962. The Anobiidae of Ohio. Ohio Biological Survey (N.S.) 1(4):1–58.

White, W. H., and A. C. Davis
1942. Land slugs and snails and their control. USDA Farmers' Bull. No. 1895. 8 pp.

Whitney, L. F.
1969. Vapona Bars: A possible treatment for sarcoptic mange? Vet. Med. Small Animal Clinician 64:993. *Reviewed in:* Health Aspects of Pesticides 4:78 (1971).

Whitsel, R. H., C. A. Vickery, Jr., C. J. Rogers, and C. D. Grant
1963. Studies on the biology and control of chironomid midges in the San Francisco Bay region. Proc. California Mosquito Control Assoc. 31:83–94.

Whitsel, R. H., and R. F. Schoeppner
1966. Summary of a study of the biology and control of the valley black gnat, *Leptoconops torrens* Townsend (Diptera: Ceratopogonidae). California Vector Views 13:17–25.

Whittaker, R. H., and P. P. Feeny
1971. Allelochemics: Chemical interactions between species. Science 171:757–70.

WHO (World Health Organization)
1957. WHO, Expert Committee on Insecticides: Resistance of insects to insecticides. Seventh Rept., WHO Tech. Rept. Series 125, Geneva, Switzerland.
1967. WHO, Expert Committee on Insecticides: Safe use of pesticides in public health. Sixteenth Rept., WHO Tech. Rept. Series 356. 63 pp. Geneva, Switzerland.
1970. WHO, Expert Committee on Insecticides: Insecticide resistance and vector control. Seventeenth Rept., WHO Tech. Rept. Series 443. 279 pp. Geneva, Switzerland.
1971. WHO, Expert Committee on Insecticides: Application and dispersal of pesticides. Eighteenth Rept., WHO Tech. Rept. Series 465. 6 pp. Geneva, Switzerland.

Wichmand, H.
1969. Planning of rat control in some towns in Iran and Iraq. *In:* Annual Report for 1968, Government Pest Infestation Laboratory, Lyngby, Denmark, p. 79.

Wigglesworth, V. B.
1934. The physiology of ecdysis in *Rhodnius prolixus* (Hemiptera). II. Factors controlling moulting and "metamorphosis." Quarterly J. Microscopic Sci. 77:191–222.
1936. The function of the *corpus allatum* in the growth and reproduction of *Rhodnius prolixus* (Hemiptera). Quarterly J. Microscopic Sci. 79:91–121.
1939. The principles of insect physiology. E. P. Dutton and Co., New York. (*See specifically:* p. 344.)
1940. The determination of characters at "metamorphosis" in *Rhodnius prolixus* (Hemiptera). J. Exptl. Biol. 17:201–22.
1948. The insect cuticle. Biological Reviews 23:408–51.
1964. The life of insects. Weidenfeld and Nicolson, London. 360 pp.

Wilcocks, C.
1959. Summary of recent abstracts. VII. Typhus group of fevers. Trop. Disease Bull. 56:1001–1006.

Wiley, H. W.
1907. The excretion of boric acid from the human body. J. Biol. Chem. 3:11–19.

Wilkinson, R. N., W. W. Barnes, A. R. Gillogly, and C. D. Minnemeyer
1971. Field evaluation of slow-release mosquito larvicides. J. Econ. Entomol. 64:1–3.

Wille, J.
1920. Biologie und Bekämpfung der Deutschen Schabe (*Phyllodromia germanica*). Monogr. angew. Entomol. No. 5, Z. angew. Entomol. 7:1–140.

Williams, C. B.
1964. Patterns in the balance of nature and related problems in quantitative biology. Academic Press, London. 324 pp.

Williams, C. L.
1948. The ship rat. Pest Control and Sanitation 3(4): 8–13.

Williams, C. M.
1956. The juvenile hormone of insects. Nature 178: 212–13.
1967. Third-generation pesticides. Scientific American 217(1):13–17.

Williams, F. X.
1943. Biological studies on Hawaiian water-loving insects. III. Diptera or flies—*C,* Tipulidae and Psychodidae. Proc. Hawaiian Entomol. Soc. 11:313–38.

Williams, L. H.
1973. Anobiid beetles should be controlled. Pest Control 41(6):18–22, 38–44.

Willis, E. R., and N. Lewis
1957. The longevity of starved cockroaches. J. Econ. Entomol. 50:438–40.

Willis, E. R., G. R. Riser, and L. M. Roth
1958. Observations on reproduction and development in cockroaches. Annals Entomol. Soc. Amer. 51: 53–69.

Wilson, E. O.
1971. The insect societies. Belknap-Harvard University Press, Cambridge, Massachusetts. 548 pp.

Wilson, H.
1913. Preliminary report of a case of sensitization to the mayfly (Ephemerida). J. Amer. Med. Assoc. 61:1648.

Wilson, J. W.
1952. Delusion of parasitosis (acarophobia). Archives of Dermatology (Chicago) 66:577–85.

Wilson, J. W., and H. E. Miller
1946. Delusion of parasitosis (acarophobia). Archives of Dermatology (Chicago) 54:39–56.

Wilson, N. L., J. H. Dillier, and G. P. Markin
1971. Foraging territories of imported fire ants. Annals Entomol. Soc. Amer. 64:660–65.

Wilson, N. L., and A. D. Oliver
1969. Food habits of the imported fire ant in pasture and pine forest areas in southeastern Louisiana. J. Econ. Entomol. 62:1268–71.

Wingo, C. W.
1960. Poisonous spiders. Missouri Agr. Expt. Sta. Bull. 738.
1964. The status of *Loxosceles reclusa,* the brown recluse spider, as a public health problem. Entomol. Soc. Amer., Proc. North Central Branch 19:115–18.

Winkler, W. P.
1970. The successful control of malaria in the 173rd Airborne Brigade. Military Medicine 135:107–11.

Wirth, W. W., and A. Stone
1956. Aquatic Diptera. *In:* Aquatic insects of California (R. L. Usinger, ed.), pp. 372–482. University of California Press, Berkeley and Los Angeles.

Wisecup, C. B., and N. C. Hayslip
1943. Control of mole crickets by use of poisoned baits. USDA Leaflet No. 237. 6 pp. (Revised in 1953, 8 pp.)

Wolcott, G. N.
1955. Organic termite repellents tested against *Cryptotermes brevis* Walker. Univ. Puerto Rico J. Agr. 39:115–49.

Womeldorf, D. J., and E. W. Mortenson
1963. The occurrence of *Hippelates* gnats in the Mojave Desert, Kern County, California. California Vector Views 10:25.

Wood, D. L.
1964. *Arhopalus productus* (LeConte), a pest of increasing importance in new home construction. PCO News 24(3):22–23.
1970. Pheromones of bark beetles. *In:* Control of insect behavior by natural products (D. L. Wood, R. M. Silverstein, and M. Nakajima, eds.), pp. 301–16. Academic Press, New York and London.

Wood, D. L., R. M. Silverstein, and M. Nakajima (eds.)
1970. Control of insect behavior by natural products. Academic Press, New York and London. 346 pp.

Wood, F. D., and S. F. Wood
1941. Present knowledge of the distribution of *Trypanosoma cruzi* in reservoir animals and vectors. Amer. J. Trop. Med. 21:225.

Wood, S. F.
1942. Reactions of man to the feeding of reduviid bugs. Journal of Parasitology 28:43–49.
1950*a*. The distribution of California insect vectors harboring *Trypanosoma cruzi* Chagas. Bull. Southern California Acad. Sci. 49:98–100.
1950*b*. Allergic sensitivity to the saliva of the western conenose bug. Bull. Southern California Acad. Sci. 49:71–74.
1954*a*. Additional bug (Hemiptera: Reduviidae) annoyance and *Trypanosoma cruzi* in southwestern National Monuments. Bull. Southern California Acad. Sci. 53:52–56.
1954*b*. Experimental destruction of the conenose bug. *Triatoma,* by the assassin bugs, *Reduvius personatus* and *R. senilis* (Hemiptera): Reduviidae). Bull. Southern California Acad. Sci. 53:174–76.

Wood, S. F., and F. D. Wood
1964. Nocturnal aggregation and invasion of homes in southern California by insect vectors of Chagas' disease. J. Econ. Entomol. 57:775–76.

Woodroffe, G. E.
1951. A life-history of the brown house moth, *Hofmannophila pseudospretella* (Staint.) (Lepidoptera: Oecophoridae). Bull. Entomol. Res. 41:529–53.
1953. An ecological study of the insects and mites in the nests of certain birds in Britain. Bull. Entomol. Res. 44:739–72.

Woodruff, R. E.
1967. An oriental wood-borer, *Heterobostrychus aequalis* (Waterhouse), recently established in Florida (Coleoptera: Bostrichidae). Florida Dept. Agr., Div. Plant Ind. and Entomol. Circ. No. 58. 2 pp.

Wouters, L.
1970. Slug control in Oost-Flevoland, Holland. Pflanzenschutz Nachrichten 23:165–69.

Wright, C. G.
1959. Beetles found in yellow pine floor joists of buildings in North Carolina. J. Econ. Entomol. 52:452.
1960. Biology of the southern lyctus beetle, *Lyctus planicollis.* Annals Entomol. Soc. Amer. 53:285–92.
1965*a*. A survey of cockroach species found in some North Carolina apartment projects. Pest Control 33(6): 14–15.
1965*b*. Identification and occurrence of cockroaches in dwellings and business establishments in North Carolina. J. Econ. Entomol. 58:1032–33.
1968. Comparative life histories of chlordane-resistant and nonresistant German cockroaches. J. Econ. Entomol. 61:1317–21.
1971. Efficacy of dichlorvos ministrips for German cockroach control in enclosed kitchen cabinets. J. Econ. Entomol. 64:278–80.

Wright, C. G., and R. C. Hillmann
1973. German cockroaches: Efficacy of chlorpyrifos spray and dust and boric acid powder. J. Econ. Entomol. 66:1075–76.

Wright, C. G., and H. C. McDaniel
1969. Abundance and habitat of five species of cockroaches on a permanent military base. J. Econ. Entomol. 62:277–78.

Wright, W. H.
1944. The bedbug: Its habits and life history, and methods of control. Public Health Reports (USPHS) Suppl. No. 175. 9 pp.

Wygodzinsky, P., and R. L. Usinger
1964. The genus *Reduvius* Fabricius in western North America (Reduviidae: Hemiptera: Insecta). Amer. Mus. Novitates 2175:1–15.

Yaeger, R. G.
1961. The present status of Chagas' disease in the United States. Bull. Tulane Univ. Med. Faculty (New Orleans) 21:9–13.

Yamaguti, S.
1952. Illustrated keys to the adult culicine mosquitoes in America north of Mexico. Published by the author (Okayama University, Japan). 48 pp., 149 pls.

Yamamoto, I.
1970. Mode of action of pyrethroids, nicotinoids, and rotenoids. Annual Rev. Entomol. 15:257–68.

Yeomans, A. H.
1948. Field-model aerosol machines. USDA Bureau of Entomology and Plant Quarantine Publ. No. ET-258. 11 pp. (Processed.)

Yeomans, A. H., E. E. Rogers, and W. H. Ball
1949. Deposition of aerosol particles. J. Econ. Entomol. 42:591–96.

Yerington, A. P.
1969. Insecticidal control of vinegar flies and dried-fruit beetles in cull fruit dumps. California Vector Views 16:79–82.
1971. Evaluation of pyrethrins. Thermal aerosols for control of vinegar flies and dried-fruit beetles in wine cellars. J. Econ. Entomol. 64:986.

Young, E. D.
1973. A great step. National Pest Control Operators News 33(3):6, 12–13.

Zavortink, T. J.
1969. A new species of treehole-breeding *Aedes* (*Ochlerotatus*) from southern California. Amer. Entomol. Inst. Contrib. 5(1):1–8.

Zeigler, T. W.
1955. Silverfish control in dwellings. Pest Control 23(6): 9–12.

Zinsser, H.
1935. Rats, lice, and history. Little, Brown and Co., Boston. 301 pp.

Zwemer, R. J., and N. I. Handelman
1971. Insect sting hypersensitivity and its management. J. Kentucky Med. Assoc. 69:505–509.

Index

N.B.: The page numbers in boldface type indicate the more important or significant text references and/or illustrations.

(Continued)